CW01660039

BEDFORD • FOWLER

Statics

ENGINEERING MECHANICS

INSTRUCTOR'S SOLUTIONS MANUAL

BEDFORD • FOWLER

Statics

ENGINEERING MECHANICS

INSTRUCTOR'S SOLUTIONS MANUAL

EUGENE L. DAVIS

Addison-Wesley Publishing Company

Reading, Massachusetts
Menlo Park, California ◆ New York ◆ Don Mills, Ontario
Wokingham, England ◆ Amsterdam ◆ Bonn ◆ Sydney ◆ Singapore ◆ Tokyo ◆ Madrid
San Juan ◆ Milan ◆ Paris

ISBN 0-201-58194-9
1 2 3 4 5 6 7 8 9 10—BAH—98979695

Foreword

The objective in each problem solution is to provide a framework for an expanded discussion of principles and insights. The goal is to provide the lecturer with an opportunity to point out shortcuts, alternative solutions, and engineering insights against the background of a detailed formal solution. To meet this goal, the solutions are presented as follows:

(1) *The emphasis is on procedures.* Formulas (equations) are avoided unless asked for in the problem statement. Procedures are taken from the corresponding chapter and section of the text, and if a required peripheral procedure is not in the text, the one used is consistent with the level of the text. If a choice existed between procedures, the one with the broadest usage is applied.

(2) *Each problem solution is a logically complete progression.* Each step is taken to the lowest detail level consistent with the level of the text, with a clear progression between steps. This guideline is compromised in the later members of some related solution groups, but the guideline was followed insofar as feasible.

(3) *Each problem solution is self contained*, with a minimum of dependence on other solutions. This is compromised for some later members of related solution groups, but the guideline is followed insofar as feasible.

Dedication

To my grandchildren.

Contents

Problem 1.1 The value of π is 3.141592654... What is the value of π to four significant digits?
Solution: Start with the first non-zero digit on the left, record the next four digits. Round the last digit to the nearest integer. The result: $\boxed{\pi = 3.142}$ to four significant digits.

Problem 1.2 What is the value of e (the base of the natural logarithms) to five significant digits?
Solution: The value of e is 2.718281828... . Record the first five digits, and round the last digit to the nearest integer. The result: $\boxed{e = 2.7183}$ to five significant digits.

Problem 1.3 Determine the value of the expression $\frac{1}{(2-\pi)}$ to three significant digits.
Solution: The value of the expression is $\frac{1}{(2-\pi)} = -0.87596919....$ Start with the first non-zero digit on the left, and record the next three digits left to right. Round the last digit to the nearest integer. The result: $\boxed{\frac{1}{(2-\pi)} = -0.876}$ to three significant digits.

Problem 1.4 If $x = 3$, what is the value of the expression $1 - e^{-x}$ to three significant digits?
Solution: The value of the expression is $1 - e^{-3} = 9.502129....$ Record the first three digits left to right, and round the last one. The result: $\boxed{1 - e^{-3} = 0.950}$ to three significant digits.

Problem 1.5: Suppose that you have just purchased a Ferrari Dino 246GT coupe and you want to know whether you can use your set of SAE (U.S. Customary Units) wrenches to work on it. You have wrenches with widths w = 1/4 in., 1/2 in., 3/4 in. and 1 in., and the car has nuts with dimensions n = 5 mm, 10 mm, 15 mm, 20 mm, and 25 mm. Defining a wrench to fit if w is no more than 2 percent larger than n, which of your wrenches can you use?
Solution: Convert the metric size n to inches, and compute the percentage difference between the metric sized nut and the SAE wrench. The results:

$$5\ mm\left(\frac{1\ inch}{25.4\ mm}\right) = 0.19685...\ in.,\quad \left(\frac{0.19685 - 0.25}{0.196}\right)100 = -27.0\%$$

$$10\ mm\left(\frac{1\ inch}{25.4\ mm}\right) = 0.3937...\ in.,\quad \left(\frac{0.3937 - 0.5}{0.3937}\right)100 = -27\%$$

$$15\ mm\left(\frac{1\ inch}{25.4\ mm}\right) = 0.5905...\ in.,\quad \left(\frac{0.59 - 0.50}{0.59}\right)100 = +15.3\%$$

$$20\ mm\left(\frac{1\ inch}{25.4\ mm}\right) = 0.7874...\ in.,\quad \left(\frac{0.787 - 0.75}{0.787}\right)100 = +4.7\%$$

$$25\ mm\left(\frac{1\ inch}{25.4\ mm}\right) = 0.9843...\ in.,\quad \left(\frac{0.9843 - 1.00}{0.9843}\right)100 = -1.6\%$$

A negative percentage implies that the metric nut is smaller than the SAE wrench; a positive percentage means that the nut is larger then the wrench. Thus within the definition of the 2 % fit, the 1 in. wrench will fit the 25 mm nut. The others cannot be used.

Problem 1.16 In SI units, the universal gravitational constant G has the value, $G = 6.67(10^{-11})\left(\frac{N\text{-}m^2}{kg^2}\right)$. Determine the value of G in U.S. Customary units.

Solution: Convert units using Table 1.2. The result:

$$6.67(10^{-11})\left(\frac{N\text{-}m^2}{kg^2}\right)\left(\frac{1\ lb}{4.448\ N}\right)\left(\frac{1\ ft}{0.3048\ m}\right)^2\left(\frac{14.59\ kg}{1\ slug}\right)^2 = 3.43590...(10^{-8})\left(\frac{lb\text{-}ft^2}{slug^2}\right)$$

$$\boxed{= 3.44(10^{-8})\left(\frac{lb\text{-}ft^2}{slug^2}\right)}$$

Problem 1.17 If the earth is modeled as a homogenous sphere, the velocity of a satellite in a circular orbit is $v = \sqrt{\frac{gR_E^2}{r}}$ where R_E is the radius of the earth and r is the radius of the orbit. (a) If g is in m/s^2 and R_E and r are in meters, what are the units of v? (b) If $R_E = 6370$ km and $r = 6670$ km what is the value of v to three significant digits? (c) For the orbit described in Part (b), what is the value of v in miles/s to three significant digits?

Solution: For (a), substitute the units into the expression and reduce:

(a) $$\sqrt{\frac{g\left(\frac{m}{s^2}\right)(R_E\ m)^2}{(r\ m)}} = \sqrt{\frac{gR_E^2}{r}\left(\frac{m^3}{m\text{-}s^2}\right)} = v\left(\frac{m}{s}\right)$$ Hence, the units are m/s

For (b)., substitute the numerical values into the expression, using $g = 9.81\ m/s^2$

$$v = \sqrt{\frac{(9.81\ \frac{m}{s^2})((6370\ km)(10^3\ \frac{m}{km}))^2}{(6670\ km)(10^3\ \frac{m}{km})}} = \sqrt{59.679...(10^6)}\left(\frac{m}{s}\right) = 7.7252...(10^3)\left(\frac{m}{s}\right)$$

(b) $\boxed{v = 7730\left(\frac{m}{s}\right)}$

For (c), convert units using Table 1.2. The result:

(c) $$v = 7730\left(\frac{m}{s}\right)\left(\frac{1\ ft}{0.3048\ m}\right)\left(\frac{1\ mile}{5280\ ft}\right) = 4.803...\left(\frac{mile}{s}\right) \boxed{= 4.80\left(\frac{mile}{s}\right)}$$

Problem 1.18 In the equation $T = \left(\frac{1}{2}\right)I\omega^2$ the term I is in kg-m^2 and ω is in s^{-1} .(a) What are the SI units of T? (b) The value of T is 100 when I is in kg-m^2 and ω is in s^{-1}. What is the value of T when it is expressed in U. S. Customary units?

Solution: For (a), substitute the units into the expression for T.:

$$(a)\ \mathbf{T} = \left(\frac{1}{2}\right)(I\ kg\text{-}m^2)(\omega\ s^{-1})^2 \boxed{= \left(\frac{1}{2}\right)I\omega^2\left(\frac{kg\text{-}m^2}{s^2}\right)}$$

For (b), convert units using Table 1.2. The result:

$$(b)\ 100\left(\frac{kg\text{-}m^2}{s^2}\right)\left(\frac{1\ slug}{14.59\ kg}\right)\left(\frac{1\ ft}{0.3048\ m}\right)^2 = 73.7759...\left(\frac{slug\text{-}ft^2}{s^2}\right) \boxed{= 73.8\left(\frac{slug\text{-}ft^2}{s^2}\right)}$$

===================================<>===================================

Problem 1.19 The "crawler" developed to transport the Saturn V launch vehicle from the vehicle assembly building to the launch pad is the largest land vehicle ever built, weighing 4.9 (10^6) lb at sea level. (a) What is its mass in slugs? (b) What is its mass in kilograms? (c) A typical car has a mass of about 1000 kg. How many such cars does it take to have the same weight as the crawler at sea level?

Solution: Convert the units, using Table 1.2. The results:

$$(a)\ 4.9(10^6)\ lb\left(\frac{1\ slug}{32.2\ lb}\right) = 1.52173...(10^5)\ slug \quad \boxed{= 1.52(10^5)\ \ slug}$$

$$(b)\ 4.9(10^6)\ lb\left(\frac{1\ slug}{32.2\ lb}\right)\left(\frac{14.59\ kg}{1\ slug}\right) = 2.22...(10^6)\ kg \quad \boxed{= 2.22\ (10^6)\ kg}$$

$$(c)\ \text{Use the result above: No. of cars} = \left(\frac{2.22(10^6)\ kg}{1000\ kg}\right) \quad \boxed{=\ 2220\ cars}$$

===================================<>===================================

Problem 1.20 The acceleration due to gravity is 13.2 ft/s^2 on the surface of Mars and 32.2 ft/s^2 on the surface of the earth. If a woman weighs 125 lb on earth, how much would she weigh on Mars?

Solution: Use Eq (1.6) and note that the mass is the same on earth and Mars. Thus $W_e = g_e m = 125\ lb$, and $W_m = g_m m$ Take the ratio, $\left(\frac{W_e}{W_m}\right) = \left(\frac{g_e}{g_m}\right)$ Solve:

$$W_m = W_e\left(\frac{g_m}{g_e}\right) = 125\left(\frac{13.2\ ft/s^2}{32.2\ ft/s}\right) lb = 51.2422...lb \quad \boxed{=\ 51.2\ lb}$$

===================================<>===================================

Problem 1.21 The acceleration due to gravity is 13.2 ft/s^2 on the surface of Mars and 32.2 ft/s^2 on the surface of earth. A woman weighs 125 lb on earth. To survive and work on the surface of Mars, she must wear life-support equipment and carry tools. What is the maximum allowable weight on earth if the engineers don't want the total weight on Mars to exceed 125 lb?

Solution: Use Eq 1.6 and note that the mass remains the same on earth and Mars. Thus the ratio of the weights is $\left(\frac{W_e}{W_m}\right) = \left(\frac{g_e}{g_m}\right)$, $\quad W_e = W_m\left(\frac{g_e}{g_m}\right) = 125\ lbs\left(\frac{32.2\ ft/s^2}{13.2\ ft/s^2}\right) = 304.924...lbs$

Subtract the woman's weight to determine max allowable weight of equipment on earth:

$(304.9 - 125)\ lb\ = 179.924...\ lb \quad \boxed{=\ 180\ lb}$

===================================<>===================================

Problem 1.22 A person has a mass of 50 kg. (a) The acceleration due to gravity at sea level is $g = 9.81\ m/s^2$. What is the person's weight at sea level? (b) The acceleration due to gravity on the Moon is $g = 1.62\ m/s^2$. What would the person weigh on the Moon?

Solution: Use Eq (1.6). $(a)\ W_e = gm = (9.81\ m/s^2)(50\ kg) = 490.50....N \quad = 491\ N$

$(b)\ W_L = g_L m = (1.62\ m/s^2)(50\ kg) \quad \boxed{=\ 81\ N}$

===================================<>===================================

Problem 1.23 The acceleration due to gravity at sea level is $g = 9.81\ m/s^2$. The radius of the earth is 6370 km. The Universal Gravitation Constant is $G = 6.67(10^{-11})\ N\text{-}m^2/kg^2$. Use this information to obtain the mass of the earth.

Solution: Use Eq (1.3) $a = \dfrac{Gm_E}{R^2}$. Solve for the mass,

$$m_E = \frac{gR^2}{G} = \frac{(9.81\ m/s^2)(6370\ km)^2\left(10^3\ \dfrac{m}{km}\right)^2}{6.67(10^{-11})\left(\dfrac{N\text{-}m^2}{kg^2}\right)} = 5.9679\ldots(10^{24})kg \boxed{= 5.97(10^{24})\ kg}$$

Problem 1.24 A person weighs 180 lb at sea level. The radius of the Earth is 3960 miles. What force is exerted on the person by the gravitational attraction of the Earth if he is in a space station in near-Earth orbit 200 miles above the surface of the Earth?

Solution: Use Eq (1.5).

$$W = mg\left(\frac{R_E}{r}\right)^2 = \left(\frac{W_E}{g}\right)g\left(\frac{R_E}{R_E+H}\right)^2 = W_E\left(\frac{3960}{3960+200}\right)^2 = (180)(0.90616) = 163.1\ lb$$

$\boxed{= 163\ lb}$

Problem 1.25 The acceleration due to gravity on the surface of the Moon is 1.62 m/s^2. The radius of the Moon is $R_M = 1738\ km$. Determine the acceleration due to gravity at a point 1738 km above its surface.

Solution: Use Eq (1.4), rewritten to apply to the Moon... $a = g_M\left(\dfrac{R_M}{r}\right)^2$

$$a = (1.62\ m/s^2)\left(\frac{R_M}{R_M+R_M}\right)^2 = (1.62\ m/s^2)\left(\frac{1}{2}\right)^2 \boxed{= 0.405\ m/s^2}$$

Problem 1.26 If an object is near the surface of the Earth, the variation of its weight with distance from the center of the Earth can often be neglected. The acceleration due to gravity at the surface of the Earth is 9.81 m/s^2 .The radius of the Earth is 6370 km. The weight of an object at sea level is ***mg***, where *m* is its mass. At what height above the Earth does the weight decrease to 0.99 ***mg***?

Solution: Use a variation of Eq (1.5). $W = mg\left(\dfrac{R_E}{R_E+h}\right)^2 = 0.99mg$ Solve for the radial height,

$$h = R_E\left(\frac{1}{\sqrt{0.99}} - 1\right) = (6370)(1.0050378 - 1.0) = 32.09\ldots\ km \boxed{= 32100\ m}$$

Problem 1.27 The centers of two oranges are one meter apart. The mass of each orange is 0.2 kg. What gravitational force do they exert on each other? (The Universal Gravitational Constant is $G = 6.67 \times 10^{-11} \dfrac{N\text{-}m^2}{kg^2}$.)

Solution: Use Eq (1.1) $F = \dfrac{G\, m_1 m_2}{r^2}$. Substitute:

$$F = \frac{(6.67)(10^{-11})(0.2)(0.2)}{1^2} = 2.668(10^{-12})\ N \boxed{= 2.67\ (10^{-12})\ N}$$

Problem 1.28 One inch equals 25.4 mm. The mass of one cubic meter of water is 1000 kg. The acceleration due to gravity at sea level is $g = 9.81\ m/s^2$. The weight of one cubic foot of water is approximately 62.4 lb. Using this information, determine how many newtons equal one pound.

Solution: Use the data in the problem: $1\ m^3 \left(\dfrac{1\ ft}{0.3048\ m}\right)^3 = 35.31\ ft^3$ of water which will have a weight of $\left(\dfrac{62.4\ lb}{1\ ft^3}\right)(35.31\ ft^3) = 2204\ lb$ in customary units, and will have a weight of $(1000\ kg)(9.81\ m/s^2) = 9800\ N$ in SI units. Since $2204\ lb = 9800\ N$, then $\boxed{1\ lb = 4.45\ N}$.

Problem 2.1 The magnitudes $|\vec{F}_A| = 60\ N$ and $|\vec{F}_B| = 80\ N$. The angle $\alpha = 45^o$. Graphically determine the magnitude of the sum of the forces $\vec{F} = \vec{F}_A + \vec{F}_B$ and the angle between $\vec{F}_B$ and $\vec{F}$.

Solution: The graphical construction is shown:

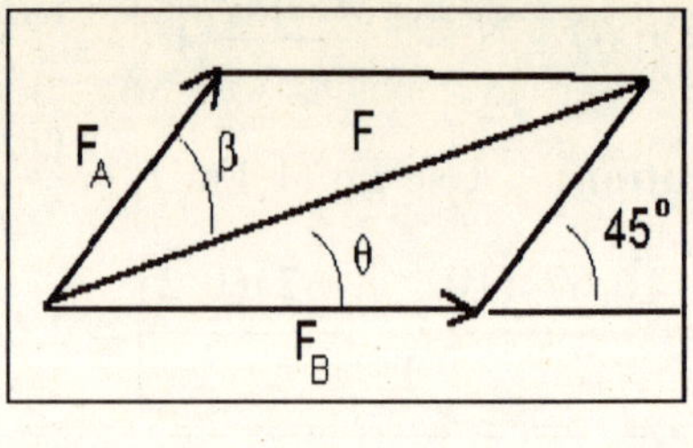

The angle β is graphically determined to be about 26° and the angle θ is about $\boxed{\theta = 19^o}$. The magnitude of the sum $|\vec{F}| = |\vec{F}_A + \vec{F}_B|$ is about $\boxed{|\vec{F}| = 130\ N}$. (These values check with a determination using trigonometry, $\beta = 25.9^o$, $\boxed{\theta = 19.1^o}$, and $\boxed{|\vec{F}| = 129.6\ N}$.)

Problem 2.2 The magnitudes $|\vec{F}_A| = 60\ N$ and $|\vec{F}_B| = 80\ N$. The angle $\alpha = 45^o$. Graphically determine the magnitude of the force $\vec{F} = 2\vec{F}_A - 3\vec{F}_B$ and the angle between $\vec{F}_B$ and $\vec{F}$.

Solution: The graphical construction is shown The angle θ is measured to be about 29° deg. The counterclockwise angle between F_B and the resultant F is $\boxed{180^o - 29^o = 151^o}$ The measured length of the vector $\vec{F}$ is about $\boxed{|\vec{F}| = 177\ N}$ (See Problem 2.4)

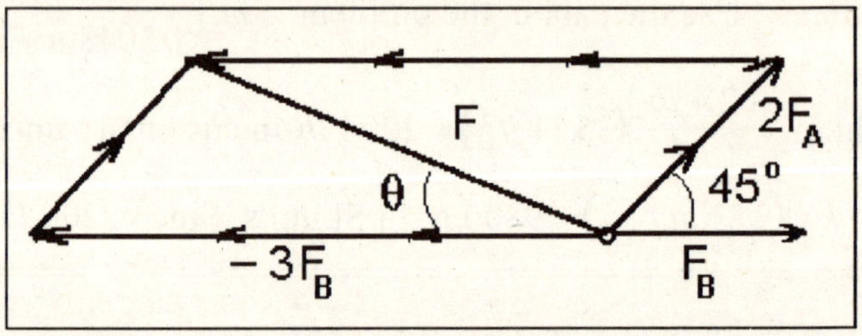

Problem 2.3 The magnitudes $|\vec{F}_A| = 100\ lb$ and $|\vec{F}_B| = 140\ lb$. The angle $\alpha = 40^o$. Use trigonometry to determine the magnitude of the sum of the forces $\vec{F} = \vec{F}_A + \vec{F}_B$. and the angle between $\vec{F}_B$ and $\vec{F}$.

Solution: The construction is shown. Use the cosine law to determine the magnitude $|\vec{F}|$

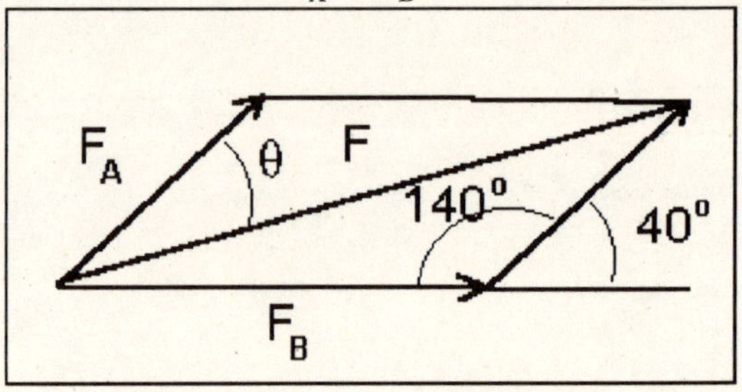

$$|\vec{F}|^2 = |\vec{F}_A|^2 + |\vec{F}_B|^2 - 2|\vec{F}_A||\vec{F}_B|\cos 140^o$$

$$|\vec{F}|^2 = (100)^2 + (140)^2 - 2(100)(140)\cos 140^o$$

$$|\vec{F}| = \sqrt{5.1049...(10^4)} = 225.94... = 225.9\ lb$$

Use the law of sines to determine the angle θ $\dfrac{|\vec{F}|}{\sin 140^o} = \dfrac{|\vec{F}_B|}{\sin\theta}$ From which

$\sin\theta = \dfrac{|\vec{F}_B|}{|\vec{F}|}\sin 140^o = 0.398$, and $\theta = 23.47^o$ The angle between $\vec{F}_B$ and $\vec{F}$ is

Thus $\boxed{\beta = 40 - 23.47 = 16.53^o}$

Problem 2.4 The magnitudes $|\vec{F}_A| = 60\ N$ and $|\vec{F}_B| = 80\ N$. The angle $\alpha = 45°$. Use trigonometry to determine the magnitude of the force $\vec{F} = 3\vec{F}_A - 2\vec{F}_B$, and the angle between $\vec{F}_B$ and $\vec{F}$.

Solution: The construction is shown. Use the cosine law to determine the magnitude of $\vec{F}$.

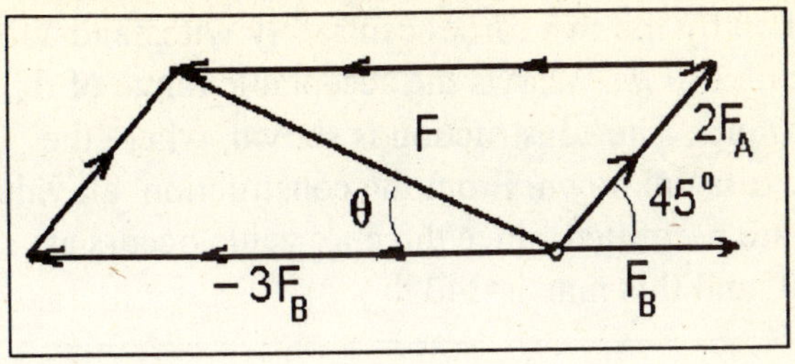

$$|\vec{F}|^2 = |2\vec{F}_A|^2 + |3\vec{F}_B|^2 - 2|2\vec{F}_A||3\vec{F}_B| \cos 45°$$

$$|\vec{F}|^2 = (120)^2 + (240)^2 - 2(120)(240)(0.707)$$

$\boxed{|\vec{F}| = 176.8\ N}$ Use the law of sines

$$\frac{|2\vec{F}_A|}{\sin\theta} = \frac{|\vec{F}|}{\sin 45°}, \sin\theta = \frac{|2\vec{F}_A|}{|\vec{F}|}\sin 45°$$

$\sin\theta = \dfrac{120}{176.8}(0.707) = 0.47986...$ Thus $\theta = 28.7°$. The angle between $\vec{F}_B$ and $\vec{F}$ is $\boxed{180 - 28.7 = 151.3°}$

Problem 2.5 The magnitudes $|\vec{F}_A| = 100\ lb$ and $|\vec{F}_B| = 140\ lb$. If the angle α can have any value, what are the possible maximum and minimum values of the magnitude of the sum of the forces $\vec{F} = \vec{F}_A + \vec{F}_B$ and what are the corresponding values of α?

Solution: A graphical construction shows that the magnitude is a minimum when the two force vectors are opposed, and a maximum when both act in the same direction. The corresponding values are

$\boxed{|\vec{F}|_{max} = |\vec{F}_A + \vec{F}_B| = |100 + 140| = 240\ lb}$, and $\boxed{\alpha = 0°}$.

$|\vec{F}|_{min} = |100 - 140| = 40\ lb$. and $\boxed{\alpha = 180°}$.

Problem 2.6 The magnitude $|\vec{F}_A| = 60\ N$ and the angle $\alpha = 45°$. If the magnitude of the sum of the forces $|\vec{F}_A + \vec{F}_B| = 180\ N$, what is the magnitude of $\vec{F}_B$?

Solution: The construction is shown, where the length of $\vec{F}_B$ is unknown. Use the cosine law:

$$|\vec{F}|^2 = |\vec{F}_A|^2 + |\vec{F}_B|^2 - 2|\vec{F}_A||\vec{F}_B| \cos 135°,\text{ or,}$$

rearranging,

$$|\vec{F}_B|^2 + 2(0.707)|\vec{F}_A||\vec{F}_B| + \left(|\vec{F}_A|^2 - |\vec{F}|^2\right) = 0$$

which is a quadratic equation in $|\vec{F}_B|$, with two roots. Solve and reduce algebraically,

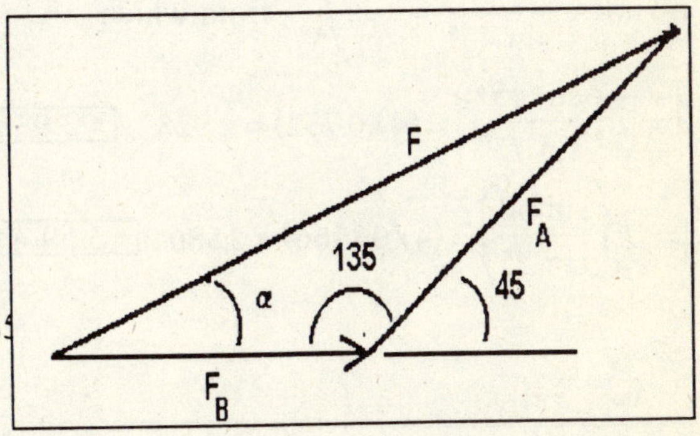

$$|\vec{F}_B| = -\frac{|\vec{F}_A|}{\sqrt{2}} \pm \sqrt{|\vec{F}|^2 - \frac{|\vec{F}_A|^2}{2}} = -42.426...\pm 174.928$$

thus $\boxed{|\vec{F}_B| = 132.5\ N}$

Problem 2.7 The magnitude $|\vec{F}_A| = 100\ lb$ and the magnitude $|\vec{F}_B| = 140\ lb$. Suppose that the support acted on by the two forces can safely withstand a total force of 240 *lb*. What is the acceptable range of α?

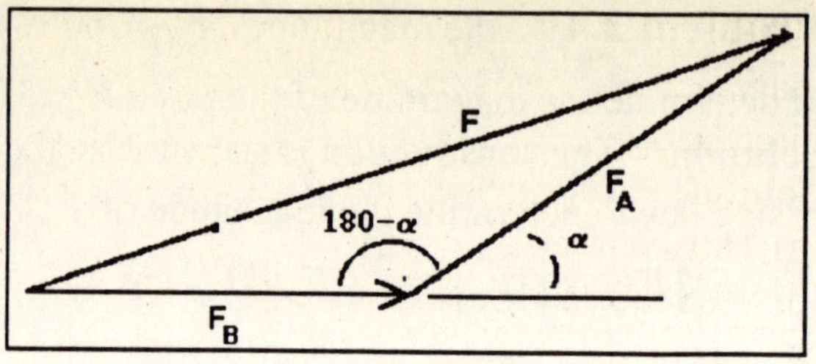

Solution: The construction is shown, where the angle α is unknown. From the construction, all values of α are permitted, since the max value occurs at α = 0, and this max is 240 lb.

Problem 2.8 The force $\vec{F}$ of magnitude 8 kN lies in the plane defined by the intersecting lines L_A and L_B. Suppose that you want to resolve $\vec{F}$ into a vector component $\vec{F}_A$ parallel to L_A and a component $\vec{F}_B$ parallel to L_B. Determine the magnitudes of $\vec{F}_A$ and $\vec{F}_B$ (a) graphically and (b) by using trigonometry.

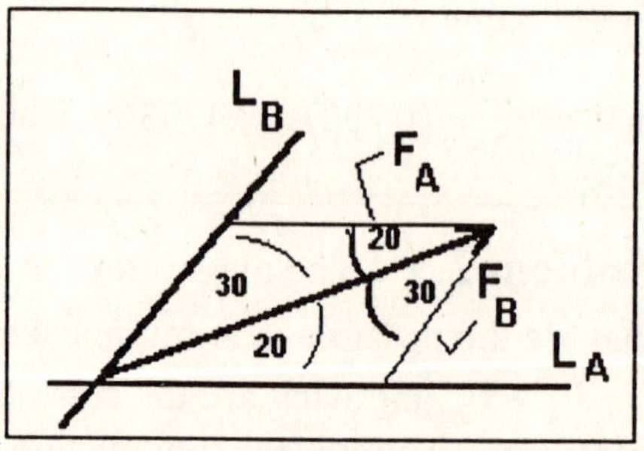

Solution: (a) The graphical construction is shown. A measurement yields $\boxed{|\vec{F}_B| = 3.57\ kN}$ and $\boxed{|\vec{F}_A| = 5.21\ kN}$

(b) The interior angle between $\vec{F}_B$ and L_A is $180° - 20° - 30° = 130°$. Use the law of sines:

$$\frac{\vec{F}}{\sin 130°} = \frac{\vec{F}_B}{\sin 20°},$$ from which $\boxed{|\vec{F}_B| = 3.57\ kN}$, and $$\frac{\vec{F}}{\sin 130°} = \frac{\vec{F}_A}{\sin 30°},$$ from which $\boxed{\vec{F}_A = 5.22\ kN}$

Problem 2.9 The rocket engine exerts an upward force of 4 MN (meganewtons) magnitude on the test stand. If you resolve the force into vector components parallel to the bars AB and CD, what are the magnitudes of the components?

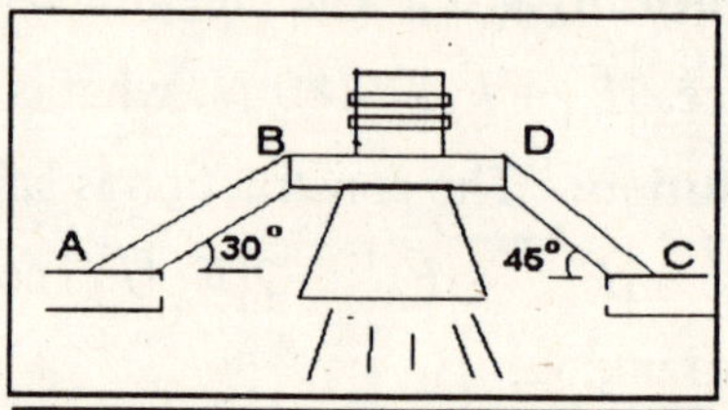

Solution: The vector diagram construction is shown. From the law of sines, $$\frac{|\vec{F}_{BA}|}{\sin 45°} = \frac{|\vec{F}|}{\sin 75°},$$ from which

$$|\vec{F}_{BA}| = |\vec{F}|\left(\frac{\sin 45°}{\sin 75°}\right) = (4)(0.732) = 2.928\ldots \boxed{= 2.93\ MN}$$

$$|\vec{F}_{CD}| = |\vec{F}|\left(\frac{\sin 60}{\sin 75}\right) = (4)(0.8966) = 3.586\ldots \boxed{= 3.59\ MN}$$

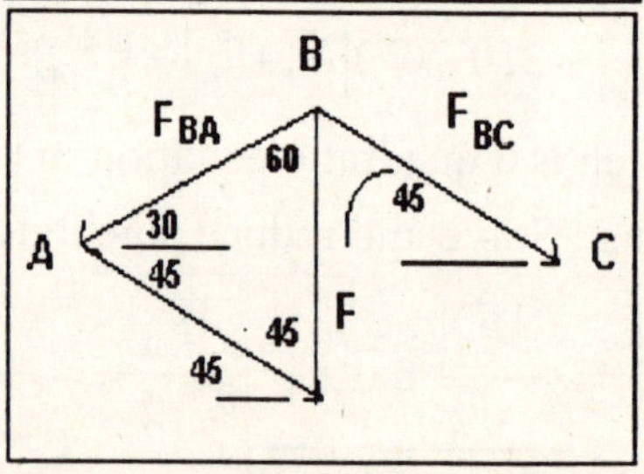

Problem 2.10 Two vectors $\vec{r}_A$ and $\vec{r}_B$ have magnitudes $|\vec{r}_A| = 30\ m$ and $|\vec{r}_B| = 40\ m$. Determine the magnitude of their sum $\vec{r}_A + \vec{r}_B$ if (a) $\vec{r}_A$ and $\vec{r}_B$ have the same direction, and if (b) $\vec{r}_A$ and $\vec{r}_B$ are perpendicular.

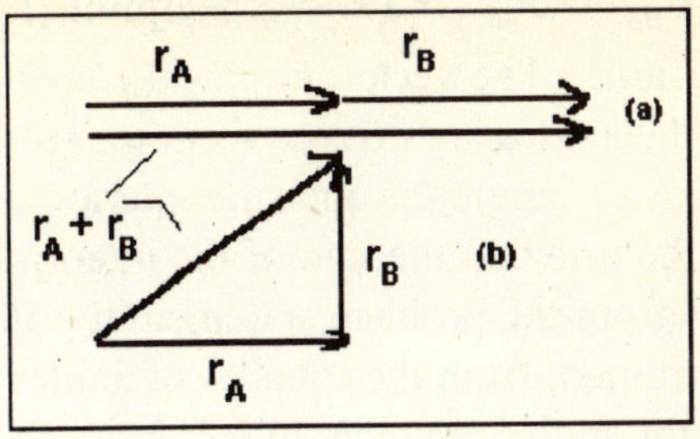

Solution: The vector constructions are shown.
(a) The magnitude of the sum is the sum of the magnitudes for vectors in the same direction:

$\boxed{|\vec{r}_A + \vec{r}_B| = 30 + 40 = 70\ m}$ From the cosine law (which reduces to the Pythagorean Theorem for a right triangle)

$$|\vec{r}_A + \vec{r}_B|^2 = |\vec{r}_A|^2 + |\vec{r}_B|^2 = (30)^2 + (40)^2 = 2500$$

$\boxed{|\vec{r}_A + \vec{r}_B| = 50\ m}$

Problem 2.11 A spherical storage tank is supported by cables. The tank is subjected to three forces: the forces $\vec{F}_A$ and $\vec{F}_B$ exerted by the cables and the weight $\vec{W}$. The weight of the tank is $|\vec{W}| = 600\ lb$. The vector sum of the forces acting on the tank equals zero. Determine the magnitudes of the forces $\vec{F}_A$ and $\vec{F}_B$ (a) graphically and (b) by using trigonometry.

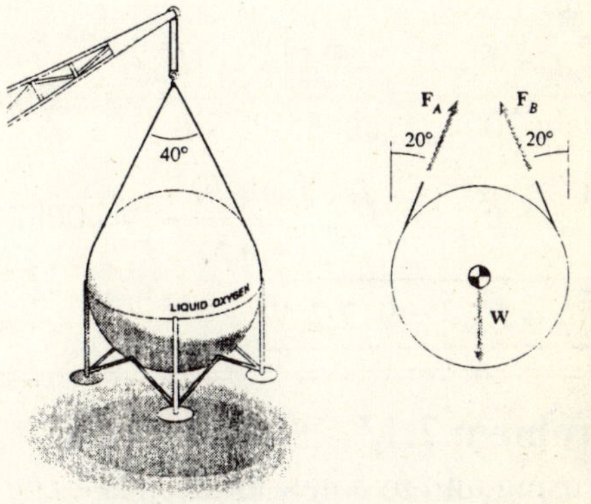

Solution: The vector construction is shown.
(a) The graphical solution is obtained from the construction by the recognition that since the opposite interior angles of the triangle are equal, the sides (magnitudes of the forces exerted by the cables) are equal. A measurement determines the magnitudes. (b) The trigonometric solution is obtained from the law of sines:

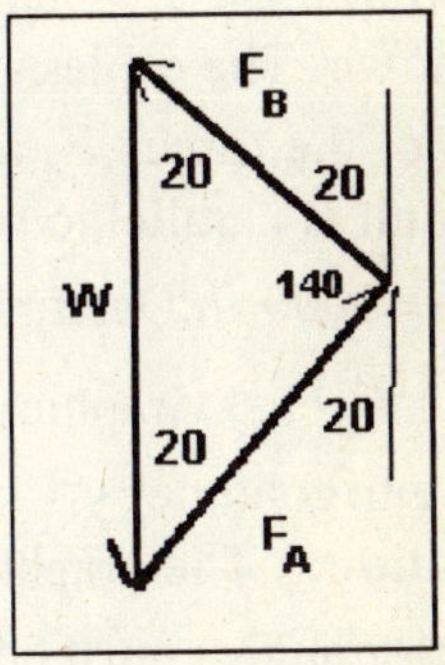

$$\frac{|\vec{W}|}{\sin 140^\circ} = \frac{|\vec{F}_A|}{\sin 20^\circ} = \frac{|\vec{F}_B|}{\sin 20^\circ}$$ Solving:

$$|\vec{F}_A| = |\vec{F}_B| = |\vec{W}|\left(\frac{\sin 20}{\sin 140}\right) = 319.25\ldots \quad \boxed{= 319.3\ lb}$$

Problem 2.12 The rope ABC exerts the forces $\vec{F}_{BA}$ and $\vec{F}_{BC}$ on the block at B. Their magnitudes are $|\vec{F}_{BA}| = |\vec{F}_{BC}| = 800\ N$. Determine $|\vec{F}_{BA} + \vec{F}_{BC}|$ (a) graphically, and (b) by trigonometry.

Solution: The vector graphical construction is shown.
(a) The angles are derived from the rule that for equal legs in a triangle the opposite interior angles are equal, and the rule that the sum of the interior angles is 180 deg. Thus from the problem statement the 70° angle is determined; from the equality of angles and the sum of interior angles the other two 55° angles are derived. The magnitude of the sum of the two vectors is then measured from the graph.

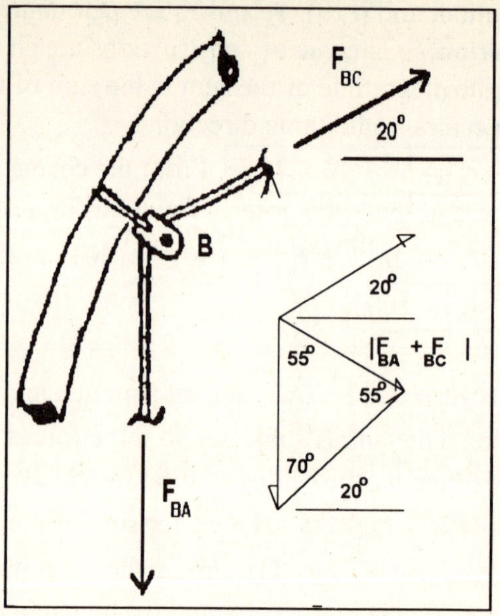

(b) The trigonometric solution follows from the law of sines:

$$\frac{|\vec{F}_{AB} + \vec{F}_{BC}|}{\sin 70^o} = \frac{|\vec{F}_{AB}|}{\sin 55^o} = \frac{|\vec{F}_{BC}|}{\sin 55^o}.\quad \text{Solve:}$$

$$|\vec{F}_{AB} + \vec{F}_{BC}| = |\vec{F}_{AB}|\left(\frac{\sin 70^o}{\sin 55^o}\right) = 800(1.1471...) = 917.72...$$

$$\boxed{|\vec{F}_{AB} + \vec{F}_{BC}| = 917.7\ N}$$

Problem 2.13 Snow cats tow a housing unit to a new location at McMurdo Base, Antarctica. (The top view is shown. The cables are horizontal.) The sum of the forces $\vec{F}_A$ and $\vec{F}_B$ exerted on the unit is parallel to the line L, and $|\vec{F}_A| = 1000\ lb$. Determine $|\vec{F}_B|$ and $|\vec{F}_A + \vec{F}_B|$ (a) Graphically, and (b) by using trigonometry.

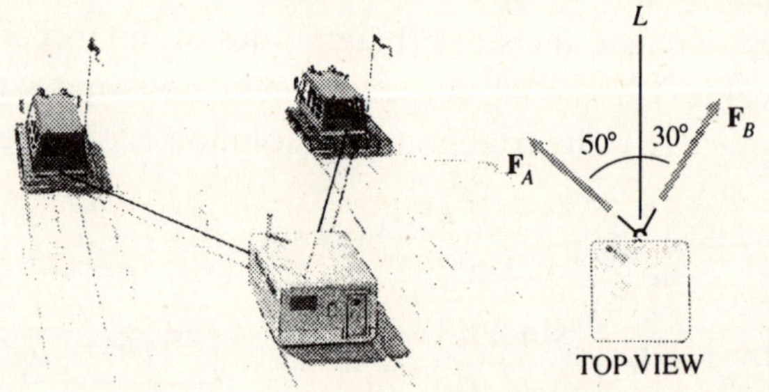

Solution: The graphical construction is shown. The sum of the interior angles must be 180°. (a) The magnitudes of $|\vec{F}_B|$ and $|\vec{F}_A + \vec{F}_B|$ are determined from measurements. (b) The trigonometric solution is obtained from the law of sines:

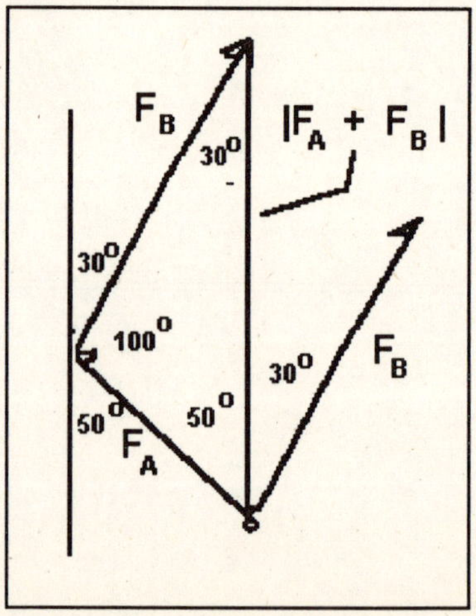

$$\frac{|\vec{F}_A + \vec{F}_B|}{\sin 100} = \frac{|\vec{F}_A|}{\sin 30} = \frac{|\vec{F}_B|}{\sin 50} \text{ from which}$$

$$|\vec{F}_B| = |\vec{F}_A|\left(\frac{\sin 50}{\sin 30}\right) = 1000(1.532) \quad \boxed{= 1532\ lb}$$

$$|\vec{F}_A + \vec{F}_B| = |\vec{F}_A|\left(\frac{\sin 100}{\sin 30}\right) = 1000(1.9696) \quad \boxed{= 1970\ lb}$$

Problem 2.14 A surveyor determines the horizontal distance from A to B is 400 m the horizontal distance from A to C is 600 m. Determine the magnitude of the horizontal vector $\vec{r}_{BC}$ and the angle α (a) graphically, and (b) by using trigonometry.

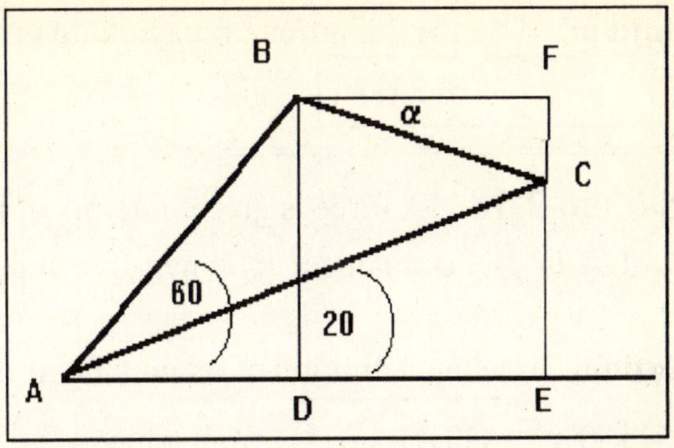

Solution: (a) The graphical solution is obtained by drawing the figure to scale and measuring the unknowns. (b) The trigonometric solution is obtained by breaking the figure into three separate right triangles. The magnitude $|\vec{r}_{BC}|$ is obtained by the cosine law:

$|\vec{r}_{BC}|^2 = (400)^2 + (600)^2 - 2(400)(600)\cos 40^\circ$ or

$\boxed{|\vec{r}_{BC}| = 390.25 = 390.3\ m}$

The three right triangles are shown. The distance BD is $BD = (400)\sin 60^\circ = 346.41\ m$. The distance CE is $CE = 600\sin 20^\circ = 205.2\ m$. The distance FC is $FC = (346.4 - 205.2) = 141.2\ m$.

The angle α is $\sin\alpha = \dfrac{141.2}{390.3} = 0.36177...$, or $\boxed{\alpha = 21.2^\circ}$

Problem 2.15 The vector $\vec{r}$ extends from point A to the midpoint between points B and C. Prove that

$\vec{r} = \frac{1}{2}(\vec{r}_{AB} + \vec{r}_{AC})$.

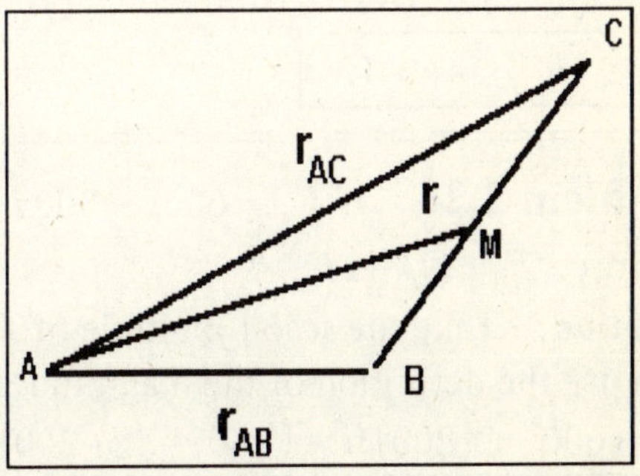

Solution: The proof is straightforward:

$\vec{r} = \vec{r}_{AB} + \vec{r}_{BM}$, and $\vec{r} = \vec{r}_{AC} + \vec{r}_{CM}$. Add the two equations and note that $\vec{r}_{BM} + \vec{r}_{CM} = 0$, since the two vectors are equal and opposite in direction. Thus $2\vec{r} = \vec{r}_{AC} + \vec{r}_{AB}$, or

$\boxed{\vec{r} = \left(\tfrac{1}{2}\right)(\vec{r}_{AC} + \vec{r}_{AB})}$

Problem 2.16 By drawing sketches of the vectors, explain why

$$\vec{U} + (\vec{V} + \vec{W}) = (\vec{U} + \vec{V}) + \vec{W}.$$

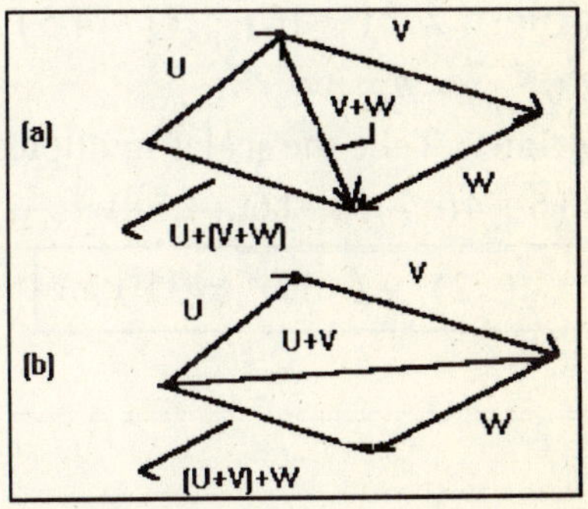

Solution: Additive associativity for vectors is usually given as an axiom in the theory of vector algebra, and of course axioms are not subject to proof. However we can by sketches show that associativity for vector addition is intuitively reasonable: Given the three vectors to be added, (a) shows the addition first of $\vec{V} + \vec{W}$, and then the addition of $\vec{U}$. The result is the vector $\vec{U} + (\vec{V} + \vec{W})$.

(b) shows the addition of $\vec{U} + \vec{V}$, and then the addition of $\vec{W}$, leading to the result $(\vec{U} + \vec{V}) + \vec{W}$.

The final vector in the two sketches is the same vector, illustrating that associativity of vector addition is intuitively reasonable.

Problem 2.17 A force is given in terms of its scalar components by $\vec{F} = 4\vec{i} - 3\vec{j}\ \ (kN)$. What is its magnitude?

Solution: Use the definition of magnitude given in Eq (2.8). The result

$$\left|\vec{F}\right| = \sqrt{4^2 + 3^2} = 5\ (kN)$$

Problem 2.18 A force is given in terms of its scalar components by $\vec{F} = F_x\vec{i} + F_y\vec{j}$. The component $F_x = 120\ lb$. The component F_y is negative and the magnitude of $\vec{F}$ is 150 *lb*. What is the component F_y?

Solution: Use the definition of magnitude in Eq (2.8), and solve algebraically.

$\left|\vec{F}\right| = 150 = \sqrt{(120)^2 + (F_y)^2}$, from which $(F_y)^2 = (150)^2 - (120)^2 = 8100$. Thus $F_y = \pm 90$, or

$$F_y = -90\ lb$$

Problem 2.19 A support is subjected to a force $\vec{F} = F_x\vec{i} + 80\vec{j}\ \ (N)$. If the support will safely support a force of 100 *N*, what is the allowable range of the component F_x?

Solution: Use the definition of magnitude in Eq (2.8) and reduce algebraically.

$100 \ge \sqrt{(F_x)^2 + (80)^2}$, from which $(100)^2 - (80)^2 \ge (F_x)^2$. Thus $|F_x| \le \sqrt{3600}$, or

$$-60 \le (F_x) \le +60\ \ (N)$$

Problem 2.20 If $\vec{F}_A = 600\vec{i} - 800\vec{j}\ \ (kip)$, and $\vec{F}_B = 200\vec{i} - 200\vec{j}\ \ (kip)$, what is the magnitude of the force $\vec{F} = \vec{F}_A - 2\vec{F}_B$?

Solution: Take the scalar multiple of $\vec{F}_B$, add the components of the two forces as in Eq (2.9), and then use the definition of the magnitude.

$\vec{F} = (600 - 2(200))\vec{i} + (-800 - 2(-200))\vec{j} = 200\vec{i} - 400\vec{j}$

$$\left|\vec{F}\right| = \sqrt{(200)^2 + (-400)^2} = 447.2\ kip$$

Problem 2.21 If $\vec{F}_A = \vec{i} - 4.5\vec{j}\ \ (kN)$ and $\vec{F}_B = -2\vec{i} - 2\vec{j}\ \ (kN)$, what is the magnitude of the force $\vec{F} = 6\vec{F}_A + 4\vec{F}_B$?

Solution: Take the scalar multiples and add the components.

$\vec{F} = (6 + 4(-2))\vec{i} + (6(-4.5) + 4(-2))\vec{j} = -2\vec{i} - 35\vec{j}$

$$\left|\vec{F}\right| = \sqrt{(-2)^2 + (-35)^2} = 35.1\ kN$$

Problem 2.22 Two perpendicular vectors $\vec{U}$ and $\vec{V}$ lie in the x,y plane. The vector $\vec{U}=6\vec{i}-8\vec{j}$ and $|\vec{V}|=20$. What are the components of $\vec{V}$?

Solution: The two possible values of $\vec{V}$ are shown in the sketch. The strategy is to (a) determine the unit vector associated with $\vec{U}$, (b) express this vector in terms of an angle, (c) add $\pm 90^o$ to this angle, (d) determine the two unit vectors perpendicular to $\vec{U}$, and (e) calculate the components of the two possible values of $\vec{V}$.

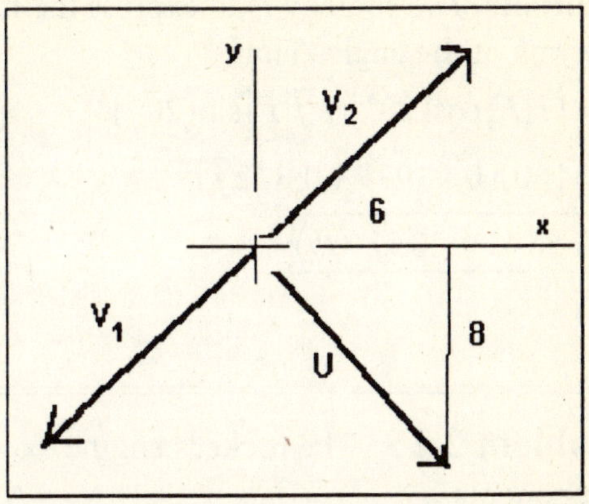

The unit vector parallel to $\vec{U}$ is

$$\vec{e}_U=\frac{6\vec{i}}{\sqrt{6^2+(-8)^2}}-\frac{8\vec{j}}{\sqrt{6^2+(-8)^2}}=0.6\vec{i}-0.8\vec{j}$$

Expressed in terms of an angle,

$$\vec{e}_U=\vec{i}\cos\alpha-\vec{j}\sin\alpha=\vec{i}\cos(53.1^o)-\vec{j}\sin(53.1^o)$$

Add $\pm 90^o$ to find the two unit vectors that are perpendicular to this unit vector:

$$\vec{e}_{p1}=\vec{i}\cos(143.1^o)-\vec{j}\sin(143.1^o)=-0.8\vec{i}-0.6\vec{j}$$

$$\vec{e}_{p2}=\vec{i}\cos(-36.9^o)-\vec{j}\sin(-36.9^o)=0.8\vec{i}+0.6\vec{j}$$

Take the scalar multiple of these unit vectors to find the two vectors perpendicular to $\vec{U}$.

$\vec{V}_1=|\vec{V}|(-0.8\vec{i}-0.6\vec{j})=-16\vec{i}-12\vec{j}$. The components are $\boxed{V_x=-16,\ V_y=-12}$

$\vec{V}_2=|\vec{V}|(0.8\vec{i}+0.6\vec{j})=16\vec{i}+12\vec{j}$ The components are $\boxed{V_x=16,\ V_y=12}$

Problem 2.23 A fish exerts a 200 *N* force on the pole. Express $\vec{F}$ in terms of scalar components.

Solution: Express the vector in terms of the angle, which is (360 - 60)=300°:

$$\vec{F}=\left(\vec{i}|\vec{F}|\cos(300^o)+\vec{j}|\vec{F}|\sin(300^o)\right)$$

$$\boxed{\vec{F}=(200)\left(0.5\vec{i}-0.866\vec{j}\right)=100\vec{i}-173.2\vec{j}\ (N)}$$

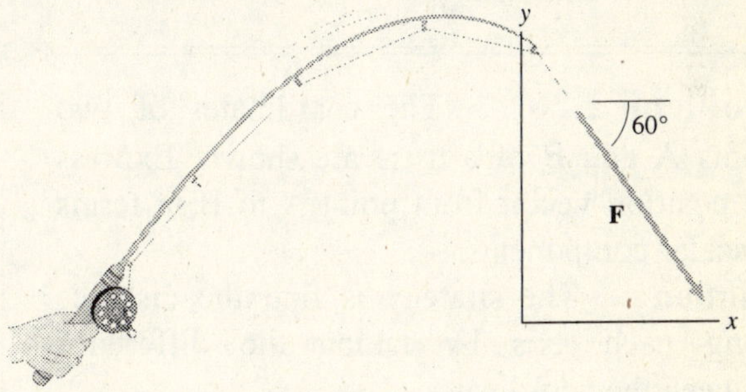

Problem 2.24 A person exerts a 60 *lb* force $\vec{F}$ to push a crate onto a truck. Express $\vec{F}$ in terms of scalar components.

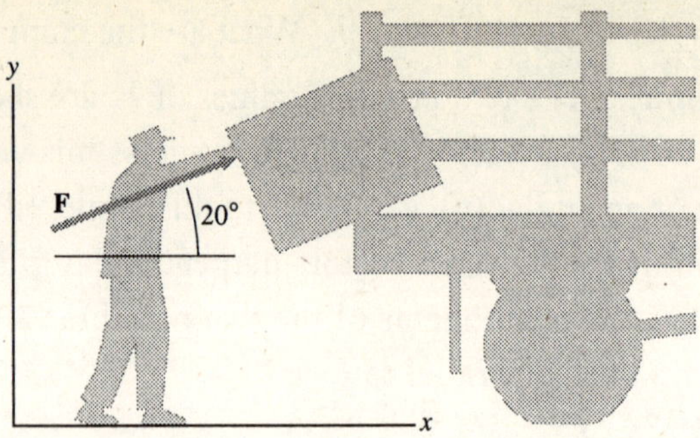

Solution: The strategy is to express the force $\vec{F}$ in terms of the angle. Thus

$$\vec{F} = \left(\vec{i}\left|\vec{F}\right|\cos\left(20^o\right) + \vec{j}\left|\vec{F}\right|\sin\left(20^o\right)\right)$$

$$\vec{F} = (60)\left(0.9397\vec{i} + 0.342\vec{j}\right)$$

$$\boxed{\vec{F} = 56.4\vec{i} + 20.5\vec{j}\ \ (lb)}$$

Problem 2.25 The rocket's engine exerts a 40 *kN* force $\vec{F}$. Express $\vec{F}$ in terms of scalar components.

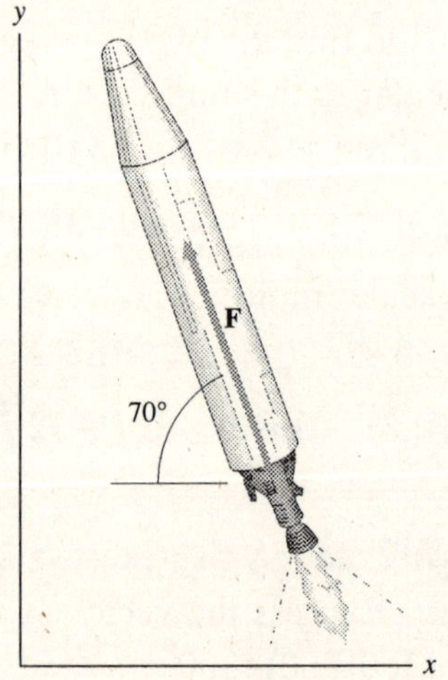

Solution: The strategy is to express the force in terms of the angle, which, measured counter clockwise from the *x* axis is

$$(180 - 70) = 110^o$$

$$\vec{F} = \left(\vec{i}\left|\vec{F}\right|\cos\left(110^o\right) + \vec{j}\left|\vec{F}\right|\sin\left(110^o\right)\right)$$

$$\boxed{\vec{F} = (40)\left(-0.3420\vec{i} + 0.9397\vec{j}\right) = -13.7\vec{i} + 37.6\vec{j}\ (kN)}$$

Problem 2.26 The coordinates of two points A and B of a truss are shown. Express the position vector from point A to B in terms of scalar components.

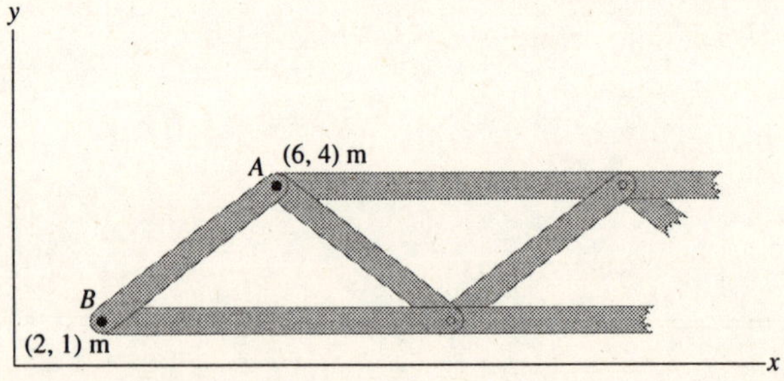

Solution The strategy is find the distance along each axis by taking the difference between the coordinates.

$$\vec{r}_{AB} = (2-6)\vec{i} + (1-4)\vec{j} = -4\vec{i} - 3\vec{j}\ \ (m)$$

Problem 2.27 The position vector from point A to point B is $\vec{r}_{AB} = 12\vec{i} - 16\vec{j}\ (m)$ (a) What is the distance from point A to point b? (b) What is the position vector from point B to point A?

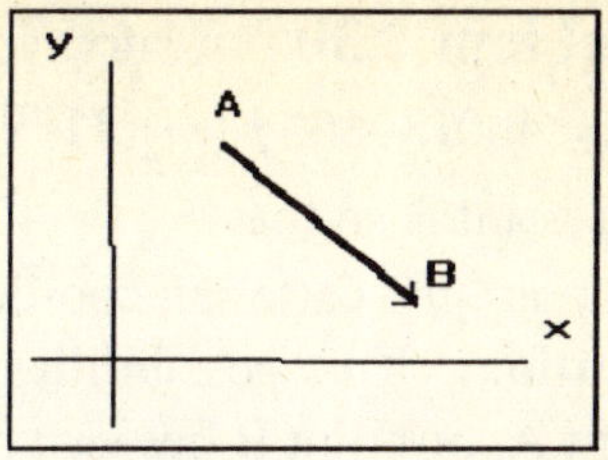

Solution: (a) The distance is the magnitude of the vector:

$$D = |\vec{r}_{AB}| = \sqrt{(12)^2 + (16)^2} = 20\ m$$

(b)The distance from B to A is the negative vector,

$$\vec{r}_{BA} = -\vec{r}_{AB} = -12\vec{i} + 16\vec{j}\ (m)$$

Problem 2.28 (a) Express the position vector from point A to point B in terms of scalar components. (b) Express the position vector from point B to point C. (c) Use the results of Parts (a) and (b) to determine the distance from point A to point C.

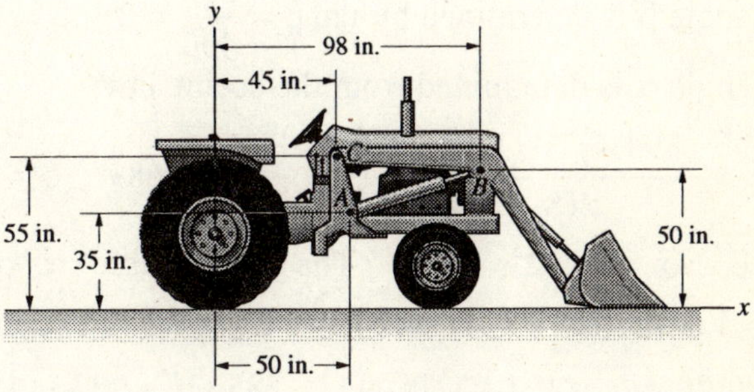

Solution: The coordinates are A(50, 35); B(98, 50); C(45, 55).

(a) The vector from point A to B:

$$\vec{r}_{AB} = (98-50)\vec{i} + (50-35)\vec{j} = 48\vec{i} + 15\vec{j}\ (in.)$$

(b) The vector from point B to C is

$$\vec{r}_{BC} = (45-98)\vec{i} + (55-50)\vec{j} = -53\vec{i} + 5\vec{j}\ (in.)$$

. (c) The distance from A to C is the magnitude of the sum of the vectors, $\vec{r}_{AC} = \vec{r}_{AB} + \vec{r}_{BC} = (48-53)\vec{i} + (15+5)\vec{j} = -5\vec{i} + 20\vec{j}$. The distance $|\vec{r}_{AC}| = \sqrt{(-5)^2 + (20)^2} = 20.62\ in.$

Problem 2.29 Consider the front end loader shown in Problem 2.28. To raise the bucket, the operator increases the length of the hydraulic cylinder AB. The distance between B and C remains constant. If the length of the cylinder AB is 65 *in.*, what is the position vector from A to B?

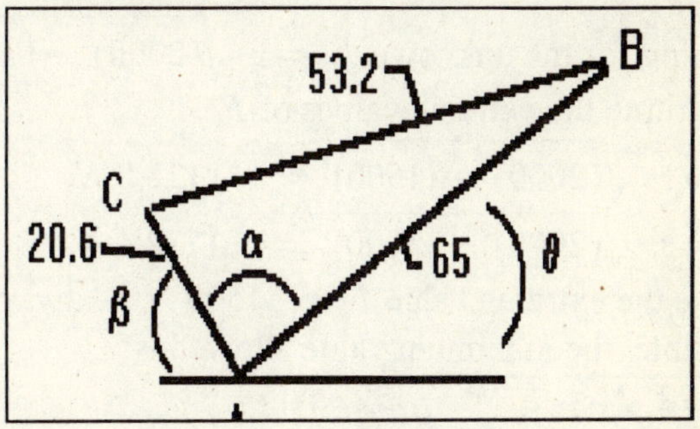

Solution: Assume that the two points A and C are fixed. The strategy is to determine the unknown angle θ from the geometry. From Problem 2.28 $|\vec{r}_{AC}| = 20.6$ and the angle β is $\tan\beta = \left(\frac{-20}{5}\right) = -4$, $\beta = 76°$. Similarly, $|\vec{r}_{CB}| = \sqrt{53^2 + 5^2} = 53.2$. The angle α is found from the cosine law: $\cos\alpha = \dfrac{(20.6)^2 + (65)^2 - (53.2)^2}{2(20.6)(65)} = 0.6776$, $\alpha = 47.3°$ Thus the angle θ is $\theta = 180° - 47.34° - 75.96° = 56.69... = 56.7°$. The vector

$$\vec{r}_{AB} = 65(\vec{i}\cos\theta + \vec{j}\sin\theta) = 35.69...\vec{i} + 54.32...\vec{j} = 35.7\vec{i} + 54.3\vec{j}\ (in.)$$

Problem 2.30 A surveyor measures the location of point A and determines that $\vec{r}_{OA} = 400\vec{i} + 800\vec{j}\ (m)$. He wants to determine the location of a point B so that $|\vec{r}_{AB}| = 400\ m$ and $|\vec{r}_{OA} + \vec{r}_{AB}| = 1200\ m$. What are the cartesian coordinates of point B?

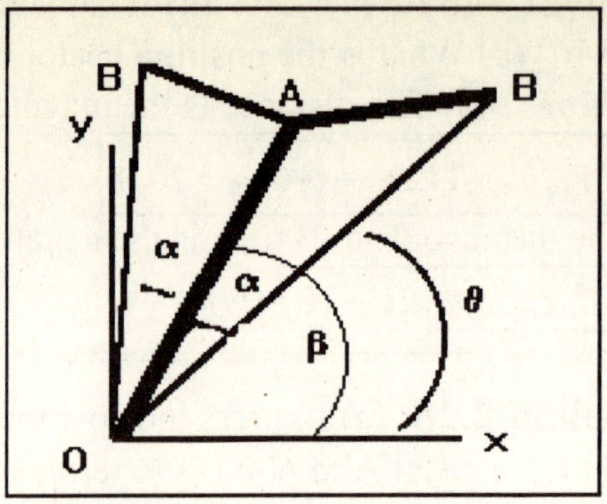

Solution: Two possibilities are: The point B lies west of point A, or point B lies east of point A, as shown. The strategy is to determine the unknown angles α, β, and θ.

The magnitude of OA is $|\vec{r}_{OA}| = \sqrt{(400)^2 + (800)^2} = 894.4$

The angle β is determined by $\tan\beta = \dfrac{800}{400} = 2,\quad \beta = 63.4°$

The angle α is determined from the cosine law:

$$\cos\alpha = \frac{(894.4)^2 + (1200)^2 - (400)^2}{2(894.4)(1200)} = 0.9689.$$ $\alpha = 14.3°$. The angle θ is

$\theta = \beta \pm \alpha = 49.12°,\ 77.74°$. The two possible sets of coordinates of point B are

$$\vec{r}_{OB} = 1200(\vec{i}\cos 77.7 + \vec{j}\sin 77.7) = 254.67\vec{i} + 1172.66\vec{j}\ (m)$$

$$\vec{r}_{OB} = 1200(\vec{i}\cos 49.1 + \vec{j}\sin 49.1) = 785.33\vec{i} + 907.34\vec{j}\ (m)$$

The two possibilities lead to $\boxed{B(254.7\ m,\ 1172.7\ m)}$ $\boxed{B(785.3\ m,\ 907.3\ m)}$

Problem 2.31 You have designed the fixture shown to safely support a force $\vec{F} = F_x\vec{i} + F_y\vec{j}$ of magnitude 2000 N. If it will be subjected to components F_x in the range $-1500\ N \le F_x \le 1000\ N$, what range of values of F_y can it safely be subjected to?

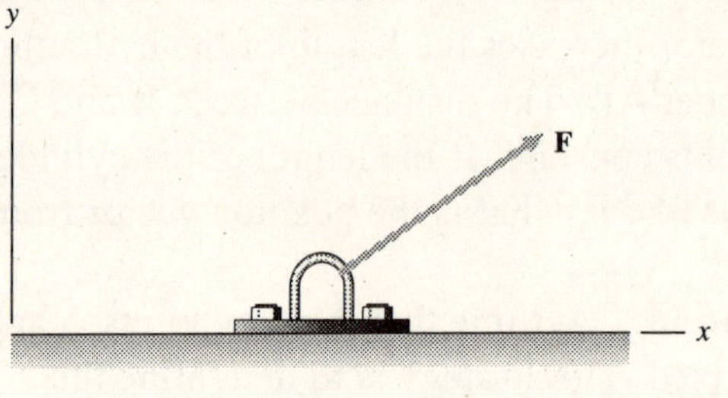

Solution: The magnitude is $2000 = \sqrt{(F_x)^2 + (F_y)^2}$

Solving for the unknown $F_y = \pm\sqrt{(2000)^2 - (F_x)^2}$

Substitute the extreme values of F_x

$$F_y = \pm\sqrt{(2000)^2 - (1500)^2} = \pm\ 1322.8\ N$$

$$F_y = \pm\sqrt{(2000)^2 - (1000)^2} = \pm\ 1732.1\ N$$

Since the extreme value $F_x \ge -1500\ N$ is always possible, the maximum value allowed is:

$\boxed{-1322\ N \le F_y \le +1322\ N}$

Problem 2.32 The coordinates of two points A and B are shown. Determine a unit vector $\vec{e}$ that points from point A to point B.

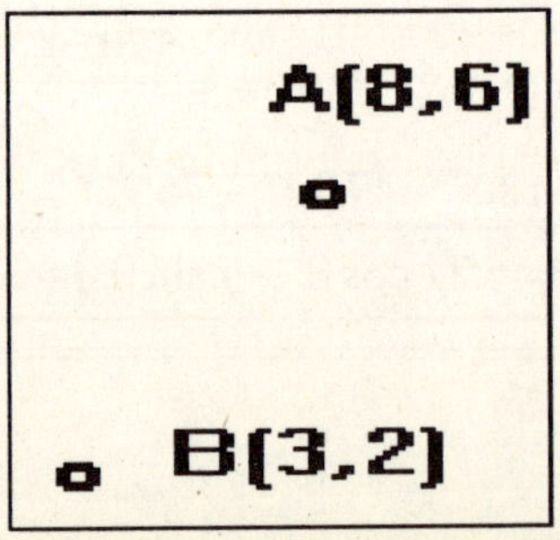

Solution: The vector from A to B is $\vec{r}_{AB} = (3-8)\vec{i} + (2-6)\vec{j}$ $= -5\vec{i} - 4\vec{j}$

The unit vector is $\boxed{\vec{e}_{AB} = -\dfrac{5}{\sqrt{5^2+4^2}}\vec{i} - \dfrac{4}{\sqrt{5^2+4^2}}\vec{j} = -0.781\vec{i} - 0.625\vec{j}}$

Problem 2.33 The x and y coordinates of point A, B, and C of the sailboat are shown. (a) Determine a unit vector that is parallel to the forestay AC and points from A toward C. (b) Determine a unit vector that is parallel to the backstay BC and points from B toward C.

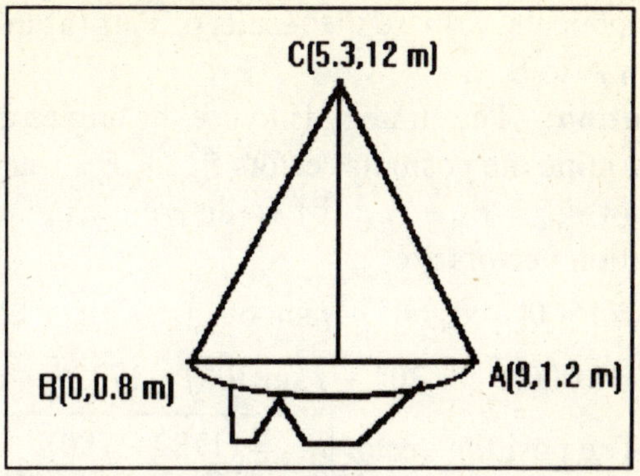

Solution: The coordinates are A(9,1.2), B(0,0.8),C(5.3,12) m. The vectors are:

$$\vec{r}_{AC} = (5.3-9)\vec{i} + (12-1.2)\vec{j} = -3.7\vec{i} + 10.8\vec{j}\ (m)$$

$$\vec{r}_{BC} = (5.3-0)\vec{i} + (12-0.8)\vec{j} = 5.3\vec{i} + 11.2\vec{j}\ (m)$$

The magnitudes are:

$$|\vec{r}_{AC}| = \sqrt{(2.7)^2 + (10.8)^2} = 11.4\ m$$

$$|\vec{r}_{BC}| = \sqrt{(5.3)^2 + (11.2)^2} = 12.4\ m$$

The unit vectors are:

(a) $\vec{e}_{AC} = \dfrac{-3.7}{11.4}\vec{i} + \dfrac{10.8}{11.4}\vec{j} = -0.325\vec{i} + 0.947\vec{j}$

(b) $\vec{e}_{BC} = \dfrac{5.3}{12.4}\vec{i} + \dfrac{11.2}{12.4}\vec{j} = 0.43\vec{i} + 0.90\vec{j}$

Problem 2.34 Consider the force vector $\vec{F} = 3\vec{i} - 4\vec{j}\ (kN)$. Determine a unit vector that has the same direction as $\vec{F}$.

Solution: The magnitude of the force vector is $|\vec{F}| = \sqrt{3^2 + 4^2} = 5$. The unit vector is

$\vec{e} = \dfrac{\vec{F}}{|\vec{F}|} = \dfrac{3}{5}\vec{i} - \dfrac{4}{5}\vec{j} = 0.6\vec{i} - 0.8\vec{j}$ *check:* $|\vec{e}| = \sqrt{0.6^2 + 0.8^2} = 1$

Problem 2.35 The position vector from a point A to point B is $\vec{r}_{AB} = -8\vec{i} + 6\vec{j}\ (ft)$. (a)Determine the unit vector $\vec{e}_{AB}$ that points from point A toward point B. (b)Determine the unit vector $\vec{e}_{BA}$ that points from point B toward point A.

Solution: The magnitude of the position vector is $|\vec{r}| = \sqrt{8^2 + 6^2} = 10\ ft$.

(a) The unit vector $\vec{e}_{AB} = \dfrac{\vec{r}_{AB}}{|\vec{r}_{AB}|} = \dfrac{-8}{10}\vec{i} + \dfrac{6}{10}\vec{j} = -0.8\vec{i} + 0.6\vec{j}$.

(b) By definition, $\vec{e}_{BA} = -\vec{e}_{AB}$, thus $\vec{e}_{BA} = 0.8\vec{i} - 0.6\vec{j}$

Problem 2.36 Two cars A and B are on a circular track of 1000 ft radius. The distance between the two cars measured along the track is 2000 ft. What is the position vector from car A to car B in terms of the coordinate system shown?

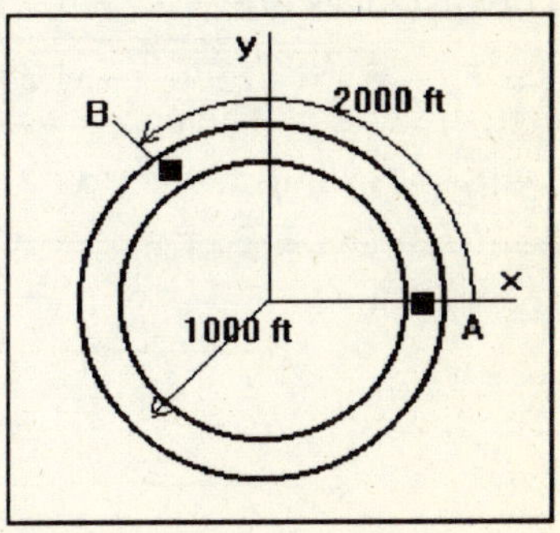

Solution: Determine the vectors $\vec{r}_{OA}$, $\vec{r}_{OB}$, , using track radius and the angular separation around the track, and then find the vector $\vec{r}_{AB} = \vec{r}_{OB} - \vec{r}_{OA}$. The vector $\vec{r}_{OA} = 1000\vec{i} + 0\vec{j}$. The angle from A to B in radians is by definition $\theta = \dfrac{2000}{1000} = 2\ rads$. The vector

$\vec{r}_{OB} = (1000)(\vec{i}\cos 2) + (1000)(\vec{j}\sin 2) = -416.1\vec{i} + 909.2\vec{j}$. The vector from car A to car B is

$\vec{r}_{AB} = -1416.1\vec{i} + 909.2\vec{j}\ ft$

Problem 2.37 A surveyor finds that the length of the line OA is 1500 m and the length of the line OB is 2000 m. (a) Express the position vector from point A to point B in terms of scalar components. (b) Use the results of Part (a) to determine the distance from A to B.

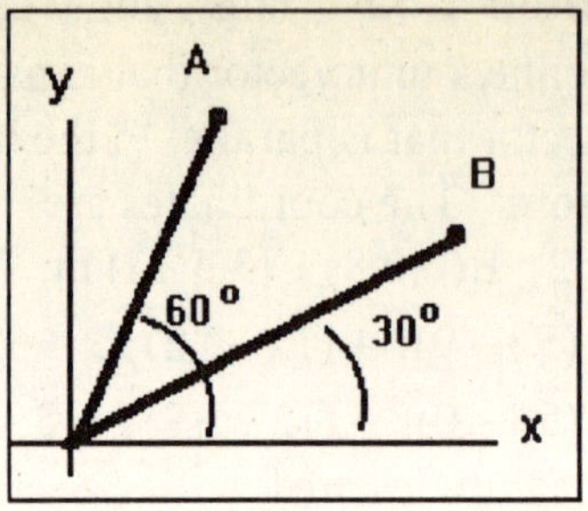

Solution: The strategy is to use the angles and magnitudes to determine the position vectors $\vec{r}_{OA}$, $\vec{r}_{OB}$, and (a) determine the vector $\vec{r}_{AB} = \vec{r}_{OB} - \vec{r}_{OA}$. (b) To determine the magnitude of $\vec{r}_{AB}$. The position vectors are

$$\vec{r}_{OA} = 1500(\vec{i}\cos 60^o + \vec{j}\sin 60^o) = 750\vec{i} + 1299\vec{j}$$

$$\vec{r}_{OB} = 2000(\vec{i}\cos 30^o + \vec{j}\sin 30^o) = 1732\vec{i} + 1000\vec{j}$$

(a) The position vector $\boxed{\vec{r}_{AB} = (1732 - 750)\vec{i} + (1000 - 1299)\vec{j} = 982\vec{i} - 299\vec{j}\ (m)}$

(b) The magnitude is $\boxed{|\vec{r}_{AB}| = \sqrt{982^2 + 299^2} = 1026.5\ m}$

Problem 2.38 The distance from the Sun (S) to Mercury (M) is 57 (10^6) km, the distance from the Sun to Venus (V) is 108 (10^6) km, and the distance from the Sun to Earth (E) is 150 (10^6) km. Assume that the planets lie in the x-y plane. (a) Determine the components of the position vector $\vec{r}_M$ from the Sun to Mercury, the position vector $\vec{r}_V$ from the Sun to Venus, and the position vector $\vec{r}_E$ from the Sun to Earth. (b) Use the results of Part(a) to determine the distance from the Earth to Mercury and the distance from Earth to Venus.

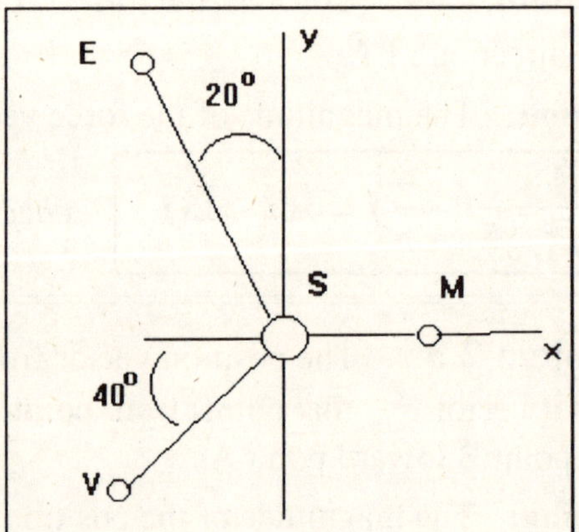

Solution: The strategy is (a) to use the angles and the magnitudes to determine the position vectors of the planets from the Sun, and then (b) to take magnitudes of the differences $\vec{r}_M - \vec{r}_E$, $\vec{r}_V - \vec{r}_E$.

(a) The vectors are

$$\boxed{\vec{r}_M = 57(10^6)\vec{i}\ (km)}$$

$$\boxed{\vec{r}_V = 108(10^6)(\vec{i}\cos 220^o + \vec{j}\sin 220^o) = -82.7(10^6)\vec{i} - 69.4(10^6)\vec{j}\ \ (km)}$$

$$\boxed{\vec{r}_E = 150(10^6)(\vec{i}\cos 110^o + \vec{i}\sin 110^o) = -51.3(10^6)\vec{i} + 140.9(10^6)\vec{j}\ \ (km)}$$

(b) The distances are

$$|\vec{r}_{EM}| = |\vec{r}_M - \vec{r}_E| = \sqrt{(57 - (-51.3))^2 + (-140.9)^2}\ (10^6)\ \boxed{= 177.7(10^6)\ (km)}$$

$$|\vec{r}_{EV}| = |\vec{r}_V - \vec{r}_E| = \sqrt{(-82.7 - (-51.3)^2 + (-69.4 - 140.9)^2}\ (10^6)\ \boxed{= 212.6(10^6)\ (km)}$$

Problem 2.39 The rope exerts forces $\vec{F}_A$ and $\vec{F}_B$ on the pulley. Their magnitudes are $|\vec{F}_A| = |\vec{F}_B| = 80\ lb$. What is the magnitude of the vector sum of the forces?

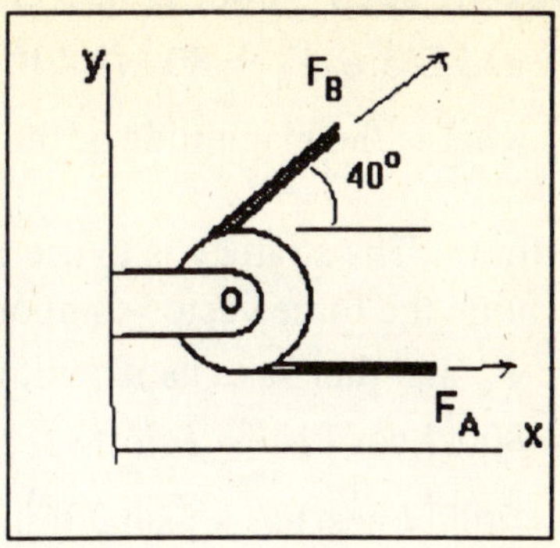

Solution: The strategy is to use the angles and magnitudes to determine the vectors, and then to take the magnitude of their sum. The vectors are

$\vec{F}_A = 80\vec{i} + 0\vec{j}$

$\vec{F}_B = (80)(\vec{i}\cos 40^\circ + \vec{j}\sin 40^\circ) = 61.3\vec{i} + 51.4\vec{j}$

The sum of the vectors is $\vec{F}_R = \vec{F}_A + \vec{F}_B = 141.3\vec{i} + 51.4\vec{j}$

The magnitude is $|\vec{F}_R| = \sqrt{(141.3)^2 + (51.4)^2} = 150.4\ lb$

Problem 2.40 The rope ABC exerts forces $\vec{F}_{BA}$ and $\vec{F}_{BC}$ on the block at B. Their magnitudes are $|\vec{F}_{BA}| = |\vec{F}_{BC}| = 800\ N$. Determine the magnitude of the vector sum of the forces by resolving the forces into components and compare your answer with that of Problem 2.12.

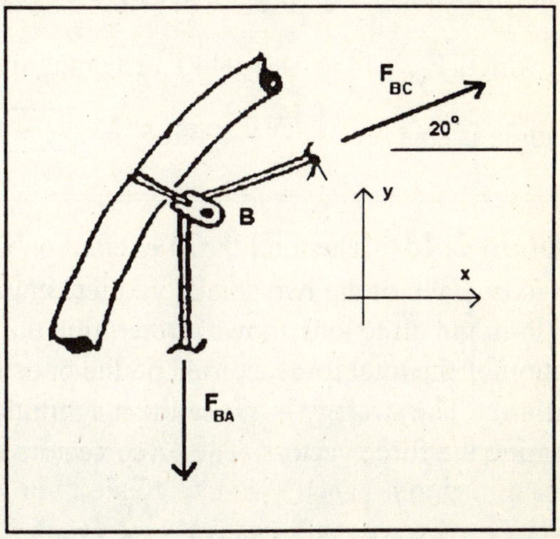

Solution: The strategy is to use the magnitudes and the angles to determine the force vectors, and then to determine the magnitude of their sum. The force vectors are:

$\vec{F}_{BA} = 0\vec{i} - 800\vec{j}$

$\vec{F}_{BC} = 800(\vec{i}\cos 20^\circ + \vec{j}\sin 20^\circ) = 751.75\vec{i} + 273.6\vec{j}$

The sum: $\vec{F}_{BA} + \vec{F}_{BC} = 751.75\vec{i} - 526.4\vec{j}$

The magnitude:

$|\vec{F}_{BA} + \vec{F}_{BC}| = \sqrt{(751.75)^2 + (526.4)^2} = 917.7\ N$

Problem 2.41 The magnitudes of the forces are $|\vec{F}_1| = |\vec{F}_2| = |\vec{F}_3| = 5\ kN$. What is the magnitude of the vector sum of the three forces?

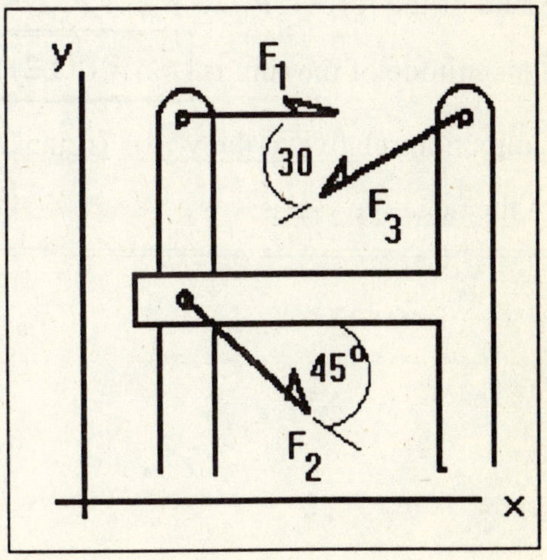

Solution: The strategy is to use the magnitudes and the angles to determine the force vectors, and then to take the magnitude of their sum. The force vectors are:

$\vec{F}_1 = 5\vec{i} + 0\vec{j}\ \ (kN)$

$\vec{F}_2 = 5(\vec{i}\cos(-45^\circ) + \vec{j}\sin(-45^\circ)) = 3.54\vec{i} - 3.54\vec{j}$

$\vec{F}_3 = 5(\vec{i}\cos 210^\circ + \vec{j}\sin 210^\circ) = -4.33\vec{i} - 2.50\vec{j}$

The sum: $\vec{F}_1 + \vec{F}_2 + \vec{F}_3 = 4.21\vec{i} - 6.04\vec{j}$

The magnitude:

$|\vec{F}_1 + \vec{F}_2 + \vec{F}_3| = \sqrt{(4.21)^2 + (6.04)^2} = 7.36\ kN$

Problem 2.42 Four groups engage in a tug of war. The magnitudes of the forces exerted by groups B, C, and D are $|\vec{F}_B| = 800\ lb,\ |\vec{F}_C| = 1000\ lb,\ |\vec{F}_D| = 900\ lb$. If the vector sum of the four forces equals zero, what is the magnitude of the force $\vec{F}_A$ and what is the angle α?

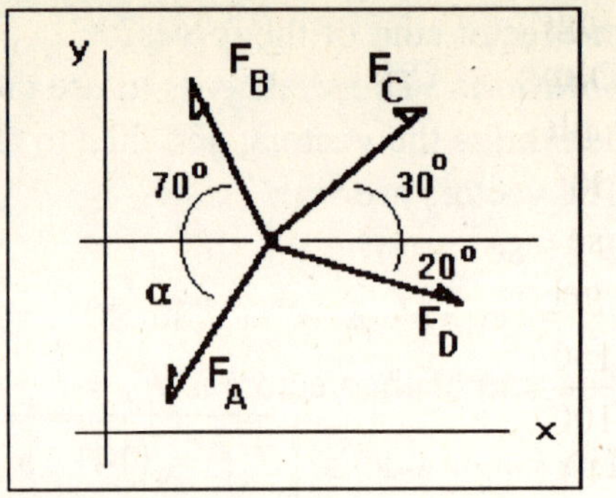

Solution: The strategy is to use the angles and magnitudes to determine the force vector components, to solve for the unknown force $\vec{F}_A$ and then take its magnitude. The force vectors are

$$\vec{F}_B = 800(\vec{i}\cos 110^o + \vec{j}\sin 110^o) = -273.6\vec{i} + 751.75\vec{j}$$

$$\vec{F}_C = 1000(\vec{i}\cos 30^o + \vec{j}\sin 30^o) = 866\vec{i} + 500\vec{j}$$

$$\vec{F}_D = 900(\vec{i}\cos(-20^o) + \vec{j}\sin(-20^o)) = 845.72\vec{i} - 307.8\vec{j}$$

$$\vec{F}_A = |\vec{F}_A|(\vec{i}\cos(180+\alpha) + \vec{j}\sin(180+\alpha)) = |\vec{F}_A|(-\vec{i}\cos\alpha - \vec{j}\sin\alpha)$$

The sum vanishes: $\vec{F}_A + \vec{F}_B + \vec{F}_C + \vec{F}_D = \vec{i}(1438.1 - |\vec{F}_A|\cos\alpha) + \vec{j}(944 - |\vec{F}_A|\sin\alpha) = 0$

From which $\vec{F}_A = 1438.1\vec{i} + 944\vec{j}$. The magnitude is $\boxed{|\vec{F}_A| = \sqrt{(1438)^2 + (944)^2} = 1720\ lb}$

The angle is: $\tan\alpha = \dfrac{944}{1438} = 0.6565$, or $\boxed{\alpha = 33.3^o}$

Problem 2.43 The total thrust exerted on the booster by its main engines is 200,000 lb parallel to the y-axis. Each of the two small "vernier" engines exert a thrust of 5000 lb in the directions shown. Determine the magnitude and direction of the total force exerted on the booster by the engines.

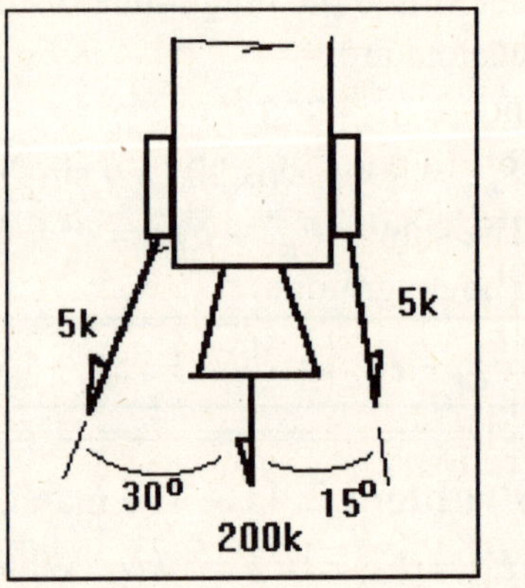

Solution: The strategy is to use the magnitudes and angles to determine the force vectors. The force vectors are:

$$\vec{F}_{ME} = 0\vec{i} - 200\vec{j}\ (kip)$$

$$\vec{F}_{LV} = 5(\vec{i}\cos 240^o + \vec{j}\sin 240^o) = -2.5\vec{i} - 4.33\vec{j}\ (kip)$$

$$\vec{F}_{RV} = 5(\vec{i}\cos 285^o + \vec{j}\sin 285^o) = 1.29\vec{i} - 4.83\vec{j}\ (kip)$$

The sum of the forces: $\vec{F}_{ME} + \vec{F}_{RV} + \vec{F}_{LV} = -1.21\vec{i} - 209.2\vec{j}\ (kip)$

The magnitude of the sum is $\boxed{|\vec{F}_R| = \sqrt{(1.21)^2 + (209.2)^2} = 209.2\ (kip)}$

The direction relative to the y-axis is $\tan\alpha = \dfrac{1.21}{209.2} = 0.005784$, or $\boxed{\alpha = 0.3314^o}$ measured clockwise from the negative y-axis.

Problem 2.44 The magnitudes of the forces acting on the support are $|\vec{F}_1| = |\vec{F}_2| = 100\ lb$. The support will fail if the magnitude of the total force acting on it exceeds 150 lb. Determine the acceptable range of values of the angle α.

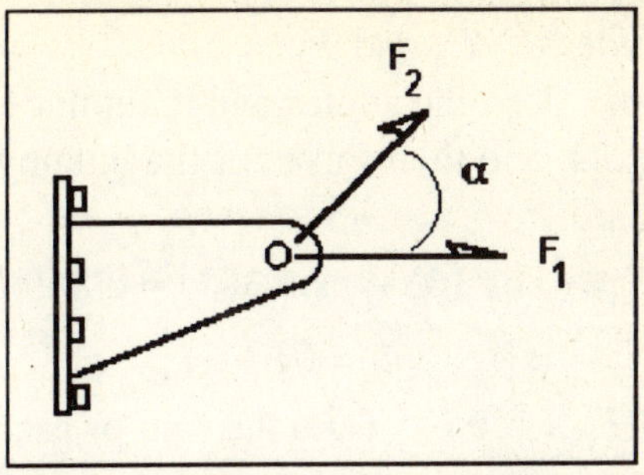

Solution: Use the cosine law to determine the range of angles.

$$150^2 \geq 100^2 + 100^2 - 2(100)(100)\cos(180 - \alpha)$$

Use the trigonometric sum of angles formula, and reduce algebraically:

$$\left(\frac{150}{100}\right)^2 \geq 2(1 + \cos\alpha), \text{ or}$$

$$\left(\frac{1}{8}\right) \geq \cos\alpha$$

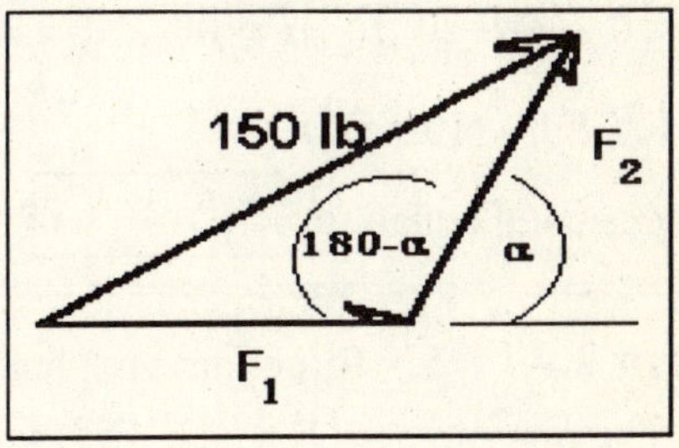

The arc cosine function has two values:

$$82.8^o \leq \alpha \leq 360^o - 82.8^o$$

$$\boxed{82.8^o \leq \alpha \leq 277.2^o}$$

Problem 2.45 Three forces act on the sphere. The magnitude of $\vec{F}_B$ is 60 lb. The vector sum of the three forces equals zero. What are the magnitudes of $\vec{F}_A$ and $\vec{F}_C$?

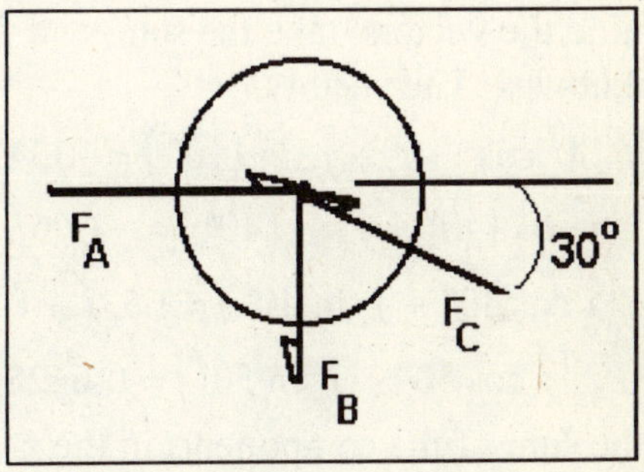

Solution: The sum of three forces with a resultant of zero magnitude forms a closed triangle. In this instance, they form a right triangle. Use the law of sines:

$$\frac{|\vec{F}_B|}{\sin 30} = \frac{|\vec{F}_A|}{\sin 60} = \frac{|\vec{F}_C|}{\sin 90}$$

From which:

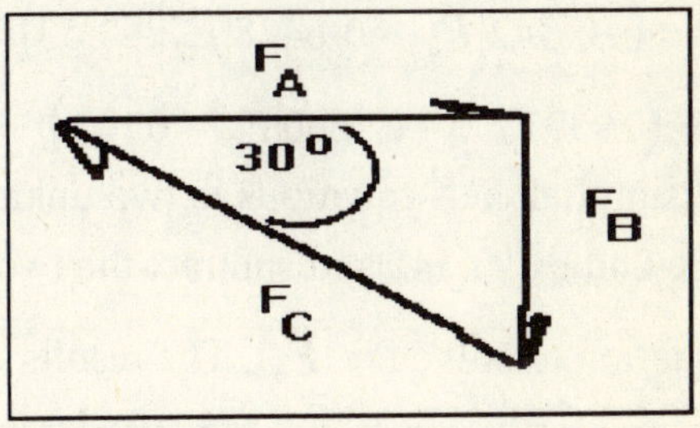

$$\boxed{\vec{F}_A = \frac{60}{0.5}(0.866) = 103.9\ lb}$$

$$\boxed{|\vec{F}_C| = \frac{60}{0.5} = 120\ lb}$$

Problem 2.46 Four forces act on a beam. The vector sum of the forces is zero. The magnitudes $|\vec{F}_B| = 10\ kN$ and $|\vec{F}_C| = 5\ kN$. Determine the magnitudes of $\vec{F}_A$ and $\vec{F}_D$.

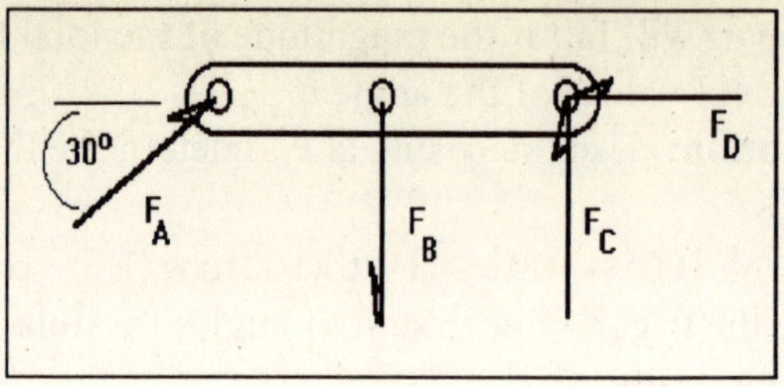

Solution: Use the angles and magnitudes to determine the vectors, and then solve for the unknowns. The vectors are:

$\vec{F}_A = |\vec{F}_A|(\vec{i}\cos 30^o + \vec{j}\sin 30^o) = 0.866|\vec{F}_A|\vec{i} + 0.5|\vec{F}_A|\vec{j}$

$\vec{F}_B = 0\vec{i} - 10\vec{j}$, $\vec{F}_C = 0\vec{i} + 5\vec{j}$,

$\vec{F}_D = -|\vec{F}_D|\vec{i} + 0\vec{j}$. Take the sum of each component in the x- and y- directions:

$\sum \vec{F}_x = (0.866|\vec{F}_A| - |\vec{F}_D|)\vec{i} = 0$

$\Sigma \vec{F}_y = (0.5|\vec{F}_A| - (10-5))\vec{j} = 0.$

From the second equation $\boxed{|\vec{F}_A| = 10\ kN}$. Using this value in the first equation: $\boxed{|\vec{F}_D| = 8.7\ kN}$

Problem 2.47 Six forces act on a beam that forms part of a building. The vector sum of the forces is zero. The magnitudes $|\vec{F}_B| = |\vec{F}_E| = 5\ kip$, $|\vec{F}_C| = 4\ kip$, and $|\vec{F}_D| = 2\ kip$. Determine the magnitude of $\vec{F}_A$ and $\vec{F}_G$.

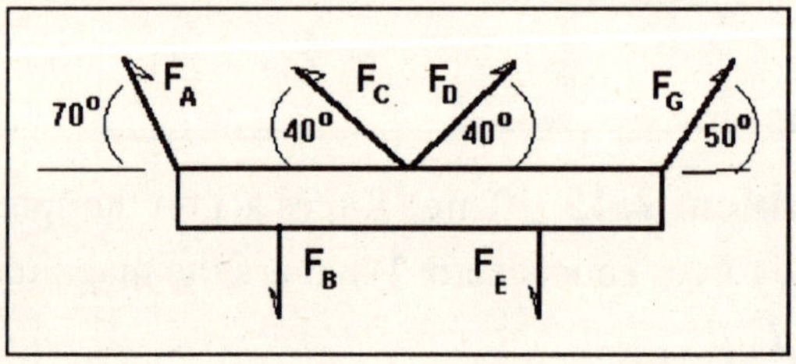

Solution: Use the angles and magnitudes to determine the vectors, take the sum, and then solve for the unknowns. The vectors:

$\vec{F}_A = |\vec{F}_A|(\vec{i}\cos 110^o + \vec{j}\sin 110^o) = -0.342|\vec{F}_A|\vec{i} + 0.9397|\vec{F}_A|\vec{j}$, $\vec{F}_B = 0\vec{i} - 5\vec{j}$,

$\vec{F}_C = 4(\vec{i}\cos 140^o + \vec{j}\sin 140^o) = -3.06\vec{i} + 2.57\vec{j}$,

$\vec{F}_D = 2(\vec{i}\cos 40^o + \vec{j}\sin 40^o) = 1.53\vec{i} + 1.29\vec{j}$, $\vec{F}_E = 0\vec{i} - 5\vec{j}$

$\vec{F}_G = |\vec{F}_G|(\vec{i}\cos 50^o + \vec{j}\sin 50^o) = 0.6428|\vec{F}_G|\vec{i} + 0.7660|\vec{F}_G|$

Take the sum of the components in the x- and y- directions:

$\sum \vec{F}_x = (-0.342|\vec{F}_A| + 0.6428|\vec{F}_G| - 1.53)\vec{i} = 0$

$\Sigma \vec{F}_y = (0.9397|\vec{F}_A| + 0.7660|\vec{F}_G| - 6.143)\vec{j} = 0$

These simultaneous equations in two unknowns may be solved as follows: multiply the 1st by 0.9397, and the 2nd by -0.342 and subtract the two equations. Solve for $|\vec{F}_G|$, then substitute that result into the 1st equation to solve for $|\vec{F}_A|$. The results: $\boxed{|\vec{F}_A| = 3.206... = 3.21\ kip}$,$\boxed{|\vec{F}_G| = 4.086... = 4.09\ kip}$

Problem 2.48 The total weight of a man and a parasail is $|\vec{W}| = 230\ lb$. The drag force $\vec{D}$ is perpendicular to the lift force $\vec{L}$. If the vector sum of the three forces is zero, what are the magnitudes of $\vec{D}$ and $\vec{L}$?

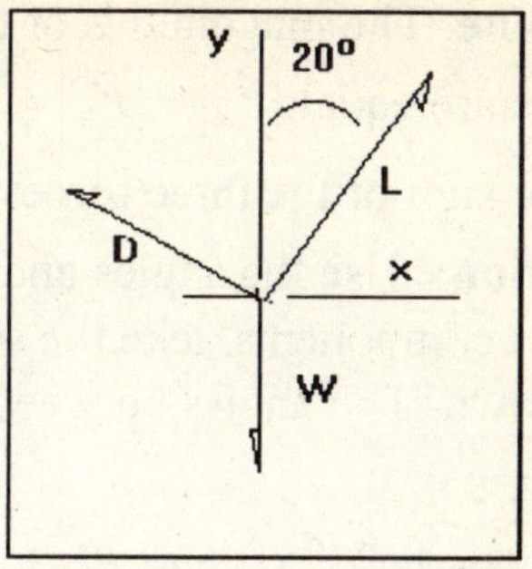

Solution: Three forces in equilibrium form a closed triangle. In this instance it is a right triangle. The law of sines is

$$\frac{|\vec{W}|}{\sin 90^o} = \frac{|\vec{L}|}{\sin 70^o} = \frac{|\vec{D}|}{\sin 20^o}$$

From which :

$$\boxed{|\vec{L}| = |\vec{W}|\sin 70^o = (230)(0.9397) = 216.1\ lb}$$

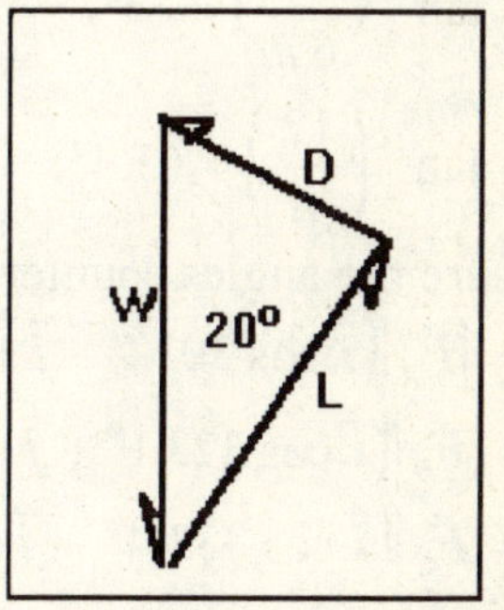

$$\boxed{|\vec{D}| = |\vec{W}|\sin 20^o = (230)(0.3420) = 78.66\ lb}$$

Problem 2.49 Two cables AB and CD extend from the rocket gantry to the ground. Cable AB exerts a force of magnitude 10000 lb on the gantry and the cable CD exerts a force of magnitude 5000 lb. (a) Using the coordinate system shown, express each of the two forces exerted by the cables on the gantry in terms of scalar components. (b) What is the magnitude of the total force exerted on the gantry by the two cables?

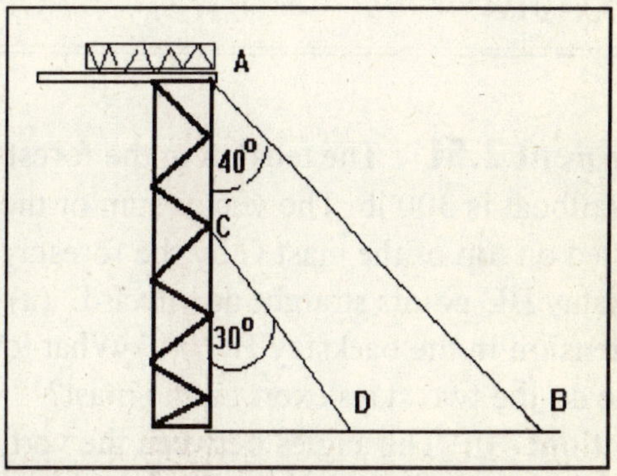

Solution: Use the angles and magnitudes to determine the components of the two forces, and then determine the magnitude of their sum. The forces:

(a) $\boxed{\vec{F}_{AB} = 10(\vec{i}\cos(-50^o) + \vec{j}\sin(-50^o)) = 6.428\vec{i} - 7.660\vec{j}\ kip}$

$\boxed{\vec{F}_{CD} = 5(\vec{i}\cos(-60^o) + \vec{j}\sin(-60^o)) = 2.50\vec{i} - 4.330\vec{j}\ kip}$

The sum: $\vec{F}_{AB} + \vec{F}_{CD} = 8.928\vec{i} - 11.990\vec{j}$, The magnitude is:

(b) $\boxed{|\vec{F}_{AB} + \vec{F}_{CD}| = \sqrt{(8.928)^2 + (11.99)^2} = 14.95\ kip}$

Problem 2.50 The cables A, B, and C help support a pillar that forms part of the supports of a structure. The magnitudes of the forces exerted by the cables are equal: $|\vec{F}_A| = |\vec{F}_B| = |\vec{F}_C|$. The magnitude of the vector sum of the three forces is 200 kN. What is $|\vec{F}_A|$?

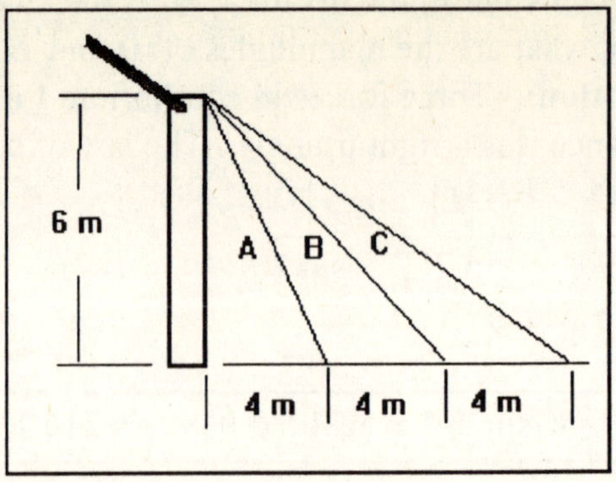

Solution: Use the angles and magnitudes to determine the vector components, take the sum, and solve for the unknown. The angles between each cable and the pillar are:

$$\theta_A = \tan^{-1}\left(\frac{4\,m}{6\,m}\right) = 33.7^\circ, \quad \theta_B = \tan^{-1}\left(\frac{8}{6}\right) = 53.1^\circ$$

$$\theta_C = \tan^{-1}\left(\frac{12}{6}\right) = 63.4^\circ.$$

Measure the angles counterclockwise form the x-axis. The force vectors acting along the cables are :

$$\vec{F}_A = |\vec{F}_A|\left(\vec{i}\cos 303.7 + \vec{j}\sin 303.7\right) = 0.5548|\vec{F}_A|\vec{i} - 0.8319|\vec{F}_A|\vec{j}$$

$$\vec{F}_B = |\vec{F}_B|\left(\vec{i}\cos 323.1^\circ + \vec{j}\sin 323.1^\circ\right) = 0.7997|\vec{F}_B|\vec{i} - 0.6004|\vec{F}_B|\vec{j}$$

$$\vec{F}_C = |\vec{F}_C|\left(\vec{i}\cos 333.4^\circ + \vec{j}\sin 333.4^\circ\right) = 0.8944|\vec{F}_C|\vec{i} - 0.4472|\vec{F}_C|\vec{j}$$

The sum of the forces are, noting that each is equal in magnitude, is

$\sum \vec{F} = \left(2.2489|\vec{F}_A|\vec{i} - 1.8795|\vec{F}_A|\vec{j}\right)$. The magnitude of the sum is given by the problem:

$200 = |\vec{F}_A|\sqrt{(2.2489)^2 + (1.8795)^2} = 2.931|\vec{F}_A|$, from which $\boxed{|\vec{F}_A| = 68.24\ kN}$

Problem 2.51 The tension in the forestay AC of the sailboat is 300 lb. The vector sum of the forces exerted on top of the mast C by the forestay and the backstay BC points straight downward. (a) What is the tension in the backstay BC? (b) What total vertical force do the two stays exert on the mast?

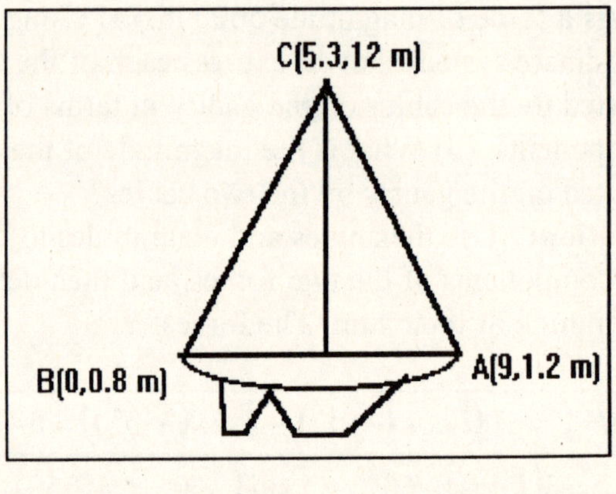

Solution: (a) The angles between the vertical mast and the cables are $\theta_{AC} = \tan^{-1}\left(\frac{9-5.3}{12-1.2}\right) = 18.9^\circ$,

$\theta_{BC} = \tan^{-1}\left(\frac{5.3-0}{12-0.8}\right) = 25.3^\circ$ The force in the mast is vertical, $\vec{F}_M = 0\vec{i} + |\vec{F}_M|\vec{j}$, The forces in the cables are

$$\vec{F}_{CA} = 300\left(\vec{i}\cos 288.9^\circ + \vec{j}\sin 288.9^\circ\right) = 97.17\vec{i} - 283.8\vec{j}$$

$\vec{F}_{CB} = |\vec{F}_{CB}|\left(\vec{i}\cos 244.7^\circ + \vec{j}\sin 244.7^\circ\right) = -0.42736|\vec{F}_{CB}|\vec{i} - 0.9041|\vec{F}_{CB}|\vec{j}$ The horizontal components of the cable tensions vanish, hence the sum of the components along the x-axis vanish,

$0 = \left(97.17 - 0.4274|\vec{F}_{CB}|\right)$, from which $\boxed{|\vec{F}_{CB}| = 227.35\ lb}$ (b) The sum of forces in the vertical direction is the force exerted on the mast: $\sum \vec{F}_y = \left(-283.8 - 0.9041|\vec{F}_{CB}| + |\vec{F}_M|\right) = 0$ from which

$\boxed{|\vec{F}_M| = 489.3\ lb}$

Problem 2.52 The structure shown forms part of a truss designed by an architectural engineer to support the roof of an orchestra shell. The members AB, AC, and AD exert forces $\vec{F}_{AB}$, $\vec{F}_{AC}$, and $\vec{F}_{AD}$ on the joint A. The coordinates of the points are: B(4,2), C(-4,1),D(-2,-3). The magnitude $|\vec{F}_{AB}| = 4\ kN$. If the vector sum of the three forces equals zero, what are the magnitudes of $\vec{F}_{AC}$ and $\vec{F}_{AD}$?

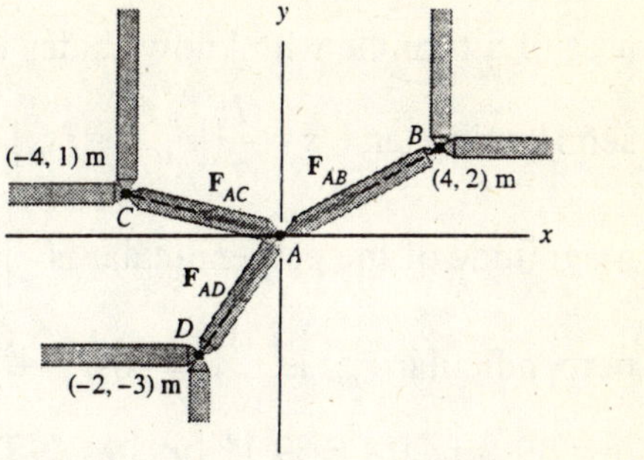

Solution: Determine the unit vectors parallel to each force:

$$\vec{e}_{AD} = \frac{-2}{\sqrt{2^2+3^2}}\vec{i} + \frac{-3}{\sqrt{2^2+3^2}}\vec{j} = -0.5547\vec{i} - 0.8320\vec{j}$$

$$\vec{e}_{AC} = \frac{-4}{\sqrt{4^2+1^2}}\vec{i} + \frac{1}{\sqrt{4^2+1^2}}\vec{j} = -0.9701\vec{i} + 0.2425\vec{j}$$

$$\vec{e}_{AB} = \frac{4}{\sqrt{4^2+2^2}}\vec{i} + \frac{2}{\sqrt{4^2+2^2}}\vec{j} = 0.89443\vec{i} + 0.4472\vec{j}$$

The forces are $\vec{F}_{AD} = |\vec{F}_{AD}|\vec{e}_{AD}$, $\vec{F}_{AC} = |\vec{F}_{AC}|\vec{e}_{AC}$, $\vec{F}_{AB} = |\vec{F}_{AB}|\vec{e}_{AB} = 3.578\vec{i} + 1.789\vec{j}$. Since the vector sum of the forces vanishes, the x- and y- components vanish separately:

$$\sum \vec{F}_x = \left(-0.5547|\vec{F}_{AD}| - 0.9701|\vec{F}_{AC}| + 3.578\right)\vec{i} = 0$$

$$\sum \vec{F}_y = (-0.8320|\vec{F}_{AD}| + 0.2425|\vec{F}_{AC}| + 1.789)\vec{j} = 0$$

These simultaneous equations in two unknowns can be solved by any standard procedure. An HP-28S hand held calculator was used here: The results: $\boxed{|\vec{F}_{AC}| = 2.108\ kN}$, $\boxed{|\vec{F}_{AD}| = 2.764\ kN}$

Problem 2.53 The position vector $\vec{r}$ goes from point A to a point on the straight line between B and C. Its magnitude $|\vec{r}| = 6\ ft$. Express $\vec{r}$ in terms of scalar components.

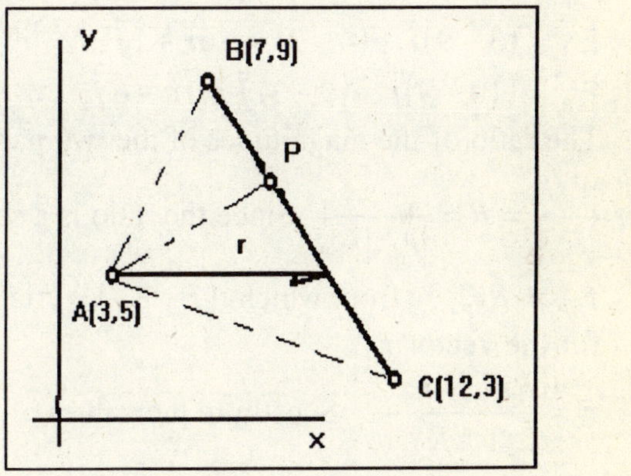

Solution: Determine the perpendicular vector to the line BC from point A, and then use this perpendicular to determine the angular orientation of the vector $\vec{r}$. The vectors are

$$\vec{r}_{AB} = (7-3)\vec{i} + (9-5)\vec{j} = 4\vec{i} + 4\vec{j}, \quad |\vec{r}_{AB}| = 5.6568$$

$$\vec{r}_{AC} = (12-3)\vec{i} + (3-5)\vec{j} = 9\vec{i} - 2\vec{j}, \quad |\vec{r}_{AC}| = 9.2195$$

$$\vec{r}_{BC} = (12-7)\vec{i} + (3-9)\vec{j} = 5\vec{i} - 6\vec{j}, \quad |\vec{r}_{BC}| = 7.8102$$

The unit vector parallel to BC is $\vec{e}_{BC} = \frac{\vec{r}_{BC}}{|\vec{r}_{BC}|} = 0.6402\vec{i} - 0.7682\vec{j} = \vec{i}\cos 50.19^\circ - \vec{j}\sin 50.19^\circ$. Add $\pm 90^\circ$ to the angle to find the two possible perpendicular vectors:

$\vec{e}_{AP1} = \vec{i}\cos 140.19^\circ - \vec{j}\sin 140.19^\circ$, *or* $\vec{e}_{AP2} = \vec{i}\cos 39.8^\circ + \vec{j}\sin 39.8^\circ$ Choose the latter, since it points from A to the line......... ***(Solution to Problem 2.53 continued next page.)***......

====================================<>====================================

...continuation of solution to Problem 2.53...

Given the triangle defined by vertices A, B, C, then the magnitude of the perpendicular corresponds to the altitude when the base is the line BC. The altitude is given by $h=\frac{2(area)}{base}$. From geometry, the area of a triangle with known sides is given by $area=\sqrt{s(s-|\vec{r}_{BC}|)(s-|\vec{r}_{AC}|)(s-|\vec{r}_{AB}|)}$, where *s* is the semi-perimeter, $s=\frac{1}{2}(|\vec{r}_{AC}|+|\vec{r}_{AB}|+|\vec{r}_{BC}|)$. Substituting values, $s=11.343$, and $area=22.0$ and the magnitude of the perpendicular is $|\vec{r}_{AP}|=\frac{2(22)}{7.8102}=5.6333$. The angle between the vector $\vec{r}$ and the perpendicular $\vec{r}_{AP}$ is $\beta=\cos^{-1}\frac{5.6333}{6}=20.1^o$. Thus the angle between the vector $\vec{r}$ and the x-axis is $\alpha=39.8\pm 20.1=59.1^o$ *or* 19.7^o. The first angle is ruled out because it causes the vector $\vec{r}$ to lie above the vector $\vec{r}_{AB}$, which is at a 45° angle relative to the x-axis. Thus:

$$\boxed{\vec{r}=6(\vec{i}\cos 19.7^o+\vec{j}\sin 19.7^o)=5.65\vec{i}+2.02\vec{j}}$$

====================================<>====================================

Problem 2.54 Let $\vec{r}$ be the position vector from point C to the point that is a distance s meters from point A along the straight line between A and B. Express $\vec{r}$ in terms of scalar components. (Your answer will be in terms of s.)

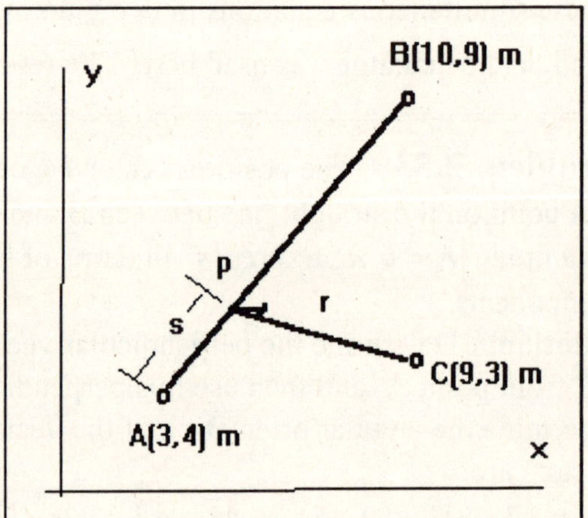

Solution: Determine the ratio of the parts of the line AB and use this value to determine $\vec{r}$. The vectors are:

$\vec{r}_{AB}=(10-3)\vec{i}+(9-4)\vec{j}=7\vec{i}+5\vec{j}$, $|\vec{r}_{AB}|=8.602$

$\vec{r}_{CA}=(3-9)\vec{i}+(4-3)=-6\vec{i}+1\vec{j}$, $|\vec{r}_{CA}|=6.0828$

$\vec{r}_{CB}=(10-9)\vec{i}+(9-3)\vec{j}=1\vec{i}+6\vec{j}$, $|\vec{r}_{CB}|=6.0828$

The ratio of the magnitudes of the two parts of the line is

$\frac{|\vec{r}_{BP}|}{|\vec{r}_{PA}|}=R=\frac{s}{|\vec{r}_{BC}|-s}$ Since the ratio is a scalar, then

$\vec{r}_{BP}=R\vec{r}_{PA}$,, from which $(\vec{r}-\vec{r}_{CA})=R(\vec{r}_{CB}-\vec{r})$. Solve for the vector $\vec{r}$,

$\vec{r}=\frac{R\vec{r}_{CB}+\vec{r}_{CA}}{1+R}$. Substitute the values of the vectors,

note that $R=\frac{s}{8.602-s}$, and reduce algebraically: $\boxed{\vec{r}=(0.8138s-6)\vec{i}+(0.5813s+1)\vec{j}\quad (m)}$.

Check: An alternate solution: Find the angle of the line AB: $\theta=\tan^{-1}\left(\frac{5}{7}\right)=35.54^o$. The components of $\vec{s}$, $\vec{s}=|\vec{s}|(\vec{i}\cos\theta+\vec{j}\sin\theta)=|\vec{s}|(0.8138\vec{i}+0.5812\vec{j})$. The coordinates of point *P* $(3+0.8138|\vec{s}|, 4+0.5812|\vec{s}|)$. Subtract coordinates of point *C* to get $\vec{r}=(0.8135|\vec{s}|-6)\vec{i}+(0.5812|\vec{s}|+1)\vec{j}$. *check.*

====================================<>====================================

Problem 2.55 A vector $\vec{U} = 3\vec{i} - 4\vec{j} - 12\vec{k}$. What is its magnitude?

Solution: Use definition given in Eq (14). The magnitude:

$$\boxed{|\vec{U}| = \sqrt{3^2 + (-4)^2 + (-12)^2} = 13}$$

Problem 2.56 A force vector $\vec{F} = 20\vec{i} + 60\vec{j} - 90\vec{k}$ (N). Determine its magnitude.

Solution: Use definition given in Eq (14). The magnitude:

$$\boxed{|\vec{F}| = \sqrt{(20)^2 + (60)^2 + (-90)^2} = 110\ N}$$

Problem 2.57 A force vector $\vec{F} = F_x\vec{i} - 120\vec{j} - 40\vec{k}$ (lb). Its magnitude $|\vec{F}| = 130\ lb$. What is F_x ?

Solution: The magnitude is: $130 = \sqrt{(F_x)^2 + (-120)^2 + (-40)^2}$. Solve for the unknown:

$(F_x)^2 = (130)^2 - (-120)^2 - (-40)^2 = 900$, or $\boxed{F_x = \pm\ 30\ \ lb}$

Problem 2.58 A vector $\vec{U} = U_x\vec{i} + U_y\vec{j} + U_z\vec{k}$. Its magnitude is $|\vec{U}| = 30$. Its scalar components are related by $U_y = -2U_x$, and $U_z = 4U_y$. Determine the scalar components.

Solution: Substitute the relations between the components, determine the magnitude, and solve for the unknowns. Thus

$\vec{U} = U_x\vec{i} + (-2U_x)\vec{j} + (4(-2U_x))\vec{k} = U_x(1\vec{i} - 2\vec{j} - 8\vec{k})$ where U_x can be factored out since it is a scalar. Take the magnitude, noting that the absolute value of $|U_x|$ must be taken:

$30 = |U_x|\sqrt{1^2 + 2^2 + 8^2} = |U_x|(8.31)$ Solving: $|U_x| = 3.612$, or $U_x = \pm\ 3.61$. The two possible vectors are:

$\vec{U} = +3.61\vec{i} + (-2(3.61))\vec{j} + (4(-2)(3.61))\vec{k}$ $\boxed{= 3.61\vec{i} - 7.22\vec{j} - 28.9\vec{k}}$

$\vec{U} = -3.61\vec{i} + (-2(-3.61))\vec{j} + 4(-2)(-3.61)\vec{k}$ $\boxed{= -3.61\vec{i} + 7.22\vec{j} + 28.9\vec{k}}$

Problem 2.59 A vector $\vec{U} = 100\vec{i} + 200\vec{j} - 600\vec{k}$ and a vector $\vec{V} = -200\vec{i} + 450\vec{j} + 100\vec{k}$. Determine the magnitude of $-2\vec{U} + 3\vec{V}$.

Solution: The resultant is

$-2\vec{U} + 3\vec{V} = (-2(100) + 3(-200))\vec{i} + (-2(200) + 3(450))\vec{j} + (-2(-600) + 3(100))\vec{k}$

$-2\vec{U} + 3\vec{V} = -800\vec{i} + 950\vec{j} + 1500\vec{k}$ The magnitude is:

$$\boxed{|-2\vec{U} + 3\vec{V}| = \sqrt{(-800)^2 + (950)^2 + (1500)^2} = 1947.4}$$

Problem 2.60 Two vectors $\vec{U} = 3\vec{i} - 2\vec{j} + 6\vec{k}$ and $\vec{V} = 4\vec{i} + 12\vec{j} - 3\vec{k}$. (a) Determine the magnitudes of $\vec{U}$ and $\vec{V}$. (b) Determine the magnitude of $3\vec{U} + 2\vec{V}$.

Solution: The magnitudes:

(a) $\boxed{|\vec{U}| = \sqrt{3^2 + 2^2 + 6^2} = 7}$ $\boxed{|\vec{V}| = \sqrt{4^2 + 12^2 + 3^2} = 13}$

The resultant vector $3\vec{U} + 2\vec{V} = (9 + 8)\vec{i} + (-6 + 24)\vec{j} + (18 - 6)\vec{k} = 17\vec{i} + 18\vec{j} + 12\vec{k}$

(b) The magnitude $\boxed{|3\vec{U} + 2\vec{V}| = \sqrt{17^2 + 18^2 + 12^2} = 27.51}$

================================<>================================

Problem 2.61 A vector $\vec{U} = 40\vec{i} - 70\vec{j} - 40\vec{k}$. (a) What is its magnitude? (b) What are the angles $\theta_x, \theta_y, \theta_z$ from Eqs (2.15)?

Solution: The magnitude:

(a) $\boxed{|\vec{U}| = \sqrt{40^2 + 70^2 + 40^2} = 90}$

(b) The direction cosines: $\vec{U} = 90\left(\frac{40}{90}\vec{i} - \frac{70}{90}\vec{j} - \frac{40}{90}\right) = 90(0.4444\vec{i} - 0.7777\vec{j} - 0.4444\vec{k})$

$\boxed{\vec{U} = 90(\vec{i}\cos 63.6^o + \vec{j}\cos 141.1^o + \vec{k}\cos 116.4^o)}$

================================<>================================

Problem 2.62 A force $\vec{F} = 600\vec{i} - 700\vec{j} + 600\vec{k}$ *(lb)*. What are the angles θ_x, θ_y, θ_z between the vector $\vec{F}$ and the coordinate axes?

Solution: The magnitude: $|\vec{F}| = \sqrt{600^2 + 700^2 + 600^2} = 1100$ The unit vector is:

$\vec{e} = \frac{\vec{F}}{|\vec{F}|} = \frac{600}{1100}\vec{i} - \frac{700}{1100}\vec{j} + \frac{600}{1100}\vec{k} = 0.5455\vec{i} - 0.6364\vec{j} + 0.5455\vec{k}$ The angles are

$\boxed{\theta_x = \cos^{-1}(0.5455) = 56.9^o}$, $\boxed{\theta_y = \cos^{-1}(-0.6364) = 129.5^o}$, $\boxed{\theta_z = \cos^{-1}(0.5455) = 56.9^o}$

================================<>================================

Problem 2.63 The cable exerts a 50 *lb* force $\vec{F}$ on the metal hook at O. The angle between $\vec{F}$ and the x-axis is 40° and the angle between $\vec{F}$ and the y-axis is 70°. The z-component of $\vec{F}$ is positive. (a) Express $\vec{F}$ in terms of scalar components. (b) What are the direction cosines of $\vec{F}$?

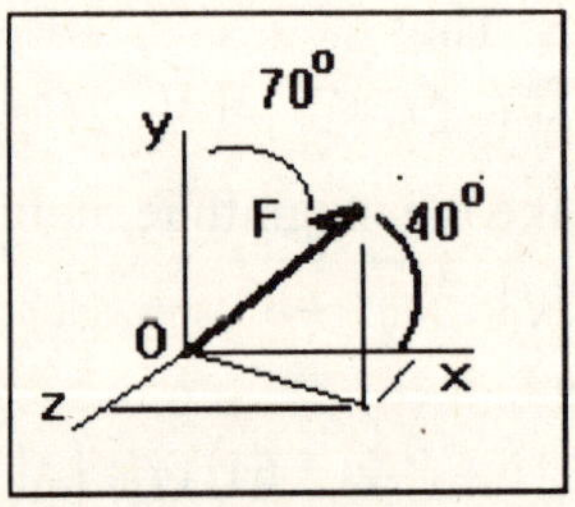

Solution: Use Eqs (2.15) and (2.16).

The force $\vec{F} = 50(\vec{i}\cos 40^o + \vec{j}\cos 70^o + \vec{k}\cos\theta_z)$. Since $1^2 = \cos^2 40^o + \cos^2 70^o + \cos^2\theta_z$,, by definition (see Eq (2.16)) then $\cos\theta_z = \pm\sqrt{1 - 0.5868 - 0.1170} = \pm\, 0.5442$. Thus the components of $\vec{F}$ are (a)

$\vec{F} = 50(0.7660\vec{i} + 0.3420\vec{j} + 0.5442\vec{k})$, $\boxed{= 38.3\vec{i} + 17.1\vec{j} + 27.2\vec{k}\ (lb)}$

(b) the direction cosines are $\boxed{\cos\theta_x = 0.7660, \quad \cos\theta_y = 0.3420, \quad \cos\theta_z = 0.5442}$

================================<>================================

Problem 2.64 A unit vector has direction cosines $\cos\theta_x = -0.5$ and $\cos\theta_y = 0.2$. Its z-component is positive. Express it in terms of scalar components.

Solution: Use Eq (2.15) and (2.16). The third direction cosine is

$\cos\theta_z = \pm\sqrt{1 - (0.5)^2 - (0.2)^2} = +\,0.8426$. Thus the unit vector is $\boxed{\vec{u} = -0.5\vec{i} + 0.2\vec{j} + 0.8426\vec{k}}$

================================<>================================

==================================<>==================================

Problem 2.65 The airplanes engines exert a total thrust force $\vec{T}$ of 200 kN. The angle between $\vec{T}$ and the x-axis is 120 deg and the angle between $\vec{T}$ and the y-axis is 130 deg. The z-component of $\vec{T}$ is positive. (a) What is the angle between $\vec{T}$ and the z-axis? (b) Express $\vec{T}$ in terms of scalar components.

Solution: The x- and y- direction cosines are $l = \cos 120^o = -0.5$, $m = \cos 130^o = -0.6428$ from which the z-direction cosine is $n = \cos\theta_z = \pm\sqrt{1-(0.5)^2-(0.6428)^2} = +0.5804$. . Thus the angle between $\vec{T}$ and the z-axis is (a) $\boxed{\theta_z = \cos^{-1}(0.5804) = 54.5^o}$, and the thrust is $\vec{T} = 200(-0.5\vec{i} - 0.6428\vec{j} + 0.5804\vec{k})$, or: (b) $\boxed{\vec{T} = -100\vec{i} - 128.6\vec{j} + 116.1\vec{k}\ (kN)}$

==================================<>==================================

Problem 2.66 The position vector from a point A to a point B is $3\vec{i} + 4\vec{j} - 4\vec{k}\ (ft)$. The position vector from point A to point C is $-3\vec{i} + 13\vec{j} - 2\vec{k}$. (a) What is the distance from point B to point C? (b) What are the direction cosines of the position vector from point B to point C?

Solution: The vector from point B to point C is $\vec{r}_{BC} = \vec{r}_{AC} - \vec{r}_{AB}$. Thus $\vec{r}_{BC} = (-3-3)\vec{i} + (13-4)\vec{j} + (-2-(-4))\vec{k} = -6\vec{i} + 9\vec{j} + 2\vec{k}$. The distance between points B and C is (a) $\boxed{|\vec{r}_{BC}| = \sqrt{6^2+9^2+2^2} = 11\ (ft)}$ The direction cosines are

(b) $\boxed{\cos\theta_x = \frac{-6}{11} = -0.5454, \quad \cos\theta_y = \frac{9}{11} = 0.8182, \quad \cos\theta_z = \frac{2}{11} = 0.1818}$

==================================<>==================================

Problem 2.67 A vector $\vec{U} = 3\vec{i} - 2\vec{j} + 6\vec{k}$. Determine the components of the unit vector that has the same direction as $\vec{U}$.

Solution: By definition, the unit vector is the vector whose components are the direction cosines of $\vec{U}$. (See discussion following Eq (2.15)).The magnitude is $|\vec{U}| = \sqrt{3^2+2^2+6^2} = 7$. Thus the unit vector is

$$\boxed{\vec{u} = \frac{\vec{U}}{|\vec{U}|} = \frac{3}{7}\vec{i} - \frac{2}{7}\vec{j} + \frac{6}{7}\vec{k}}$$

==================================<>==================================

Problem 2.68 A force vector $\vec{F} = 3\vec{i} - 4\vec{j} - 2\vec{k}\ (N)$. (a) What is the magnitude of $\vec{F}$? (b) Determine the components of the unit vector that has the same direction as $\vec{F}$.

Solution: By definition, the unit vector is the vector whose components are the direction cosines of $\vec{F}$. The magnitude is (a) $\boxed{|\vec{F}| = \sqrt{3^2+4^2+2^2} = 5.385\ (N)}$ The unit vector is (b)

$$\vec{e} = \frac{3}{5.385}\vec{i} - \frac{4}{5.385}\vec{j} - \frac{2}{5.385}\vec{k}\ \boxed{= 0.5571\vec{i} - 0.7428\vec{j} - 0.3714\vec{k}}$$

==================================<>==================================

Problem 2.69 A force vector points in the same direction as the unit vector $\vec{e} = \frac{2}{7}\vec{i} - \frac{6}{7}\vec{j} - \frac{3}{7}\vec{k}$. The magnitude of $\vec{F}$ is 700 lb. Express $\vec{F}$ in terms of scalar components.

Solution: By definition, $\vec{F} = |\vec{F}|\vec{e}$, where $\vec{e}$ is a unit vector in the direction of $\vec{F}$. (See discussion following Eq (2.16).) Thus $\boxed{\vec{F} = 700\left(\frac{2}{7}\vec{i} - \frac{6}{7}\vec{j} - \frac{3}{7}\vec{k}\right) = 200\vec{i} - 600\vec{j} - 300\vec{k}}$

==================================<>==================================

Problem 2.70 A force vector $\vec{F}$ points in the same direction as a position vector $\vec{r} = 4\vec{i} + 4\vec{j} - 7\vec{k}$ (m). The magnitude of $\vec{F}$ is 90 kN. Express $\vec{F}$ in terms of scalar components.

Solution: By definition, $\vec{F} = |\vec{F}|\vec{e}$, where $\vec{e}$ is a unit vector in the direction of $\vec{F}$. Find the unit vector from the position vector. The magnitude is $|\vec{r}| = \sqrt{4^2 + 4^2 + 7^2} = 9$; the unit vector is $\vec{e} = \frac{4}{9}\vec{i} + \frac{4}{9}\vec{j} - \frac{7}{9}\vec{k}$. The components are $\vec{F} = 90\left(\frac{4}{9}\vec{i} + \frac{4}{9}\vec{j} - \frac{7}{9}\vec{k}\right)$ $\boxed{= 40\vec{i} + 40\vec{j} - 70\vec{k}\ (kN)}$

Problem 2.71 Astronauts on the shuttle use radar to determine the magnitudes and direction cosines of the position vectors of two satellites A and B. The vector $\vec{r}_A$ from the shuttle to satellite A has magnitude 2 km and direction cosines $\cos\theta_x = 0.768$, $\cos\theta_y = 0.384$, and $\cos\theta_z = 0.512$. The vector $\vec{r}_B$ from shuttle to satellite B has magnitude 4 km and direction cosines $\cos\theta_x = 0.743$, $\cos\theta_y = 0.557$, $\cos\theta_z = -0.371$. What is the distance between the two satellites?

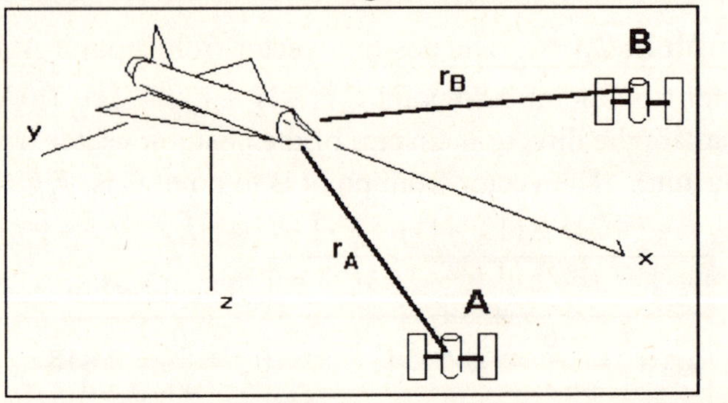

Solution: The two position vectors are:

$$\vec{r}_A = 2(0.768\vec{i} + 0.384\vec{j} + 0.512\vec{k}) = 1.536\vec{i} + 0.768\vec{j} + 1.024\vec{k}\ (km)$$

$$\vec{r}_B = 4(0.743\vec{i} + 0.557\vec{j} - 0.371\vec{k}) = 2.972\vec{i} + 2.228\vec{j} - 1.484\vec{k}\ (km)$$

The distance is the magnitude of the difference:

$$|\vec{r}_A - \vec{r}_B| = \sqrt{(1.536 - 2.927)^2 + (0.768 - 2.228)^2 + (1.024 - (-1.484))^2}\ \boxed{= 3.24\ (km)}$$

Problem 2.72 Archaeologists make measurements of a pre-Colombian structure and obtain the dimensions shown. Determine (a) the magnitude and (b) the direction cosines of the position vector from point A to point B.

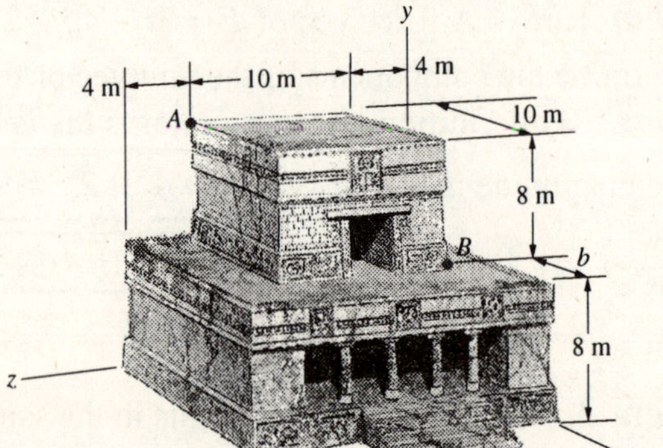

Solution: The coordinates are A(0, 16, 14), and B(10, 8, 4). The vector from A to B is $\vec{r}_{AB} = (10 - 0)\vec{i} + (8 - 16)\vec{j} + (4 - 14)\vec{k}$ $= 10\vec{i} - 8\vec{j} - 10\vec{k}$. The magnitude is

(a) $\boxed{|\vec{r}_{AB}| = \sqrt{10^2 + 8^2 + 10^2} = 16.2\ m}$

The direction cosines are

(b) $\boxed{\cos\theta_x = \frac{10}{16.2} = 0.6155,}$

$\boxed{\cos\theta_y = \frac{-8}{16.2} = -0.4938,}$ $\boxed{\cos\theta_z = \frac{-10}{16.2} = -0.6155}$.

==============================<>==============================

Problem 2.73 Consider the structure described in Problem 2.72. After returning to the United States, an archaeologist discovers that he lost the notes containing the dimension b, but other notes indicate that the distance from point B to point C is 16.4 m. What are the direction cosines of the vector from B to C?

Solution: The coordinates of B and C are B(10,8, 4) and C(10 + b, 0, 18). The vector from B to C is $\vec{r}_{BC} = (10+b-10)\vec{i} + (0-8)\vec{j} + (18-4)\vec{k} = b\vec{i} - 8\vec{j} + 14\vec{k}$. The magnitude of this vector is known:

$16.4 = \sqrt{b^2+8^2+14^2} = \sqrt{b^2+260}$, from which

$b^2 = (16.4)^2 - 260 = 8.96$, or $b = \pm 3 = +3\ m$. The direction cosines are

$$\boxed{\cos\theta_x = \frac{3}{16.4} = 0.1829,} \quad \boxed{\cos\theta_y = \frac{-8}{16.4} = -0.4878,} \quad \boxed{\cos\theta_z = \frac{14}{16.4} = 0.8537}$$

==============================<>==============================

Problem 2.74 The height of Mount Everest was measured by a surveyor using the following procedure. He first measured the distance between two points A and B of equal altitude. Suppose that they are 10000 ft above sea level and are 32000 ft apart. He then used a theodolite to measure the direction cosines of the vectors from point A to the top of the mountain P and from point B to P. Suppose that for $\vec{r}_{AP}$ the direction cosines are ($\cos\theta_x = 0.509$, $\cos\theta_y = 0.509$, $\cos\theta_z = 0.694$), and for $\vec{r}_{BP}$ the direction cosines are ($\cos\theta_x = -0.605$, $\cos\theta_y = 0.471$, $\cos\theta_z = 0.642$). The z-axis of the coordinate system is vertical. What is the height of Mount Everest?

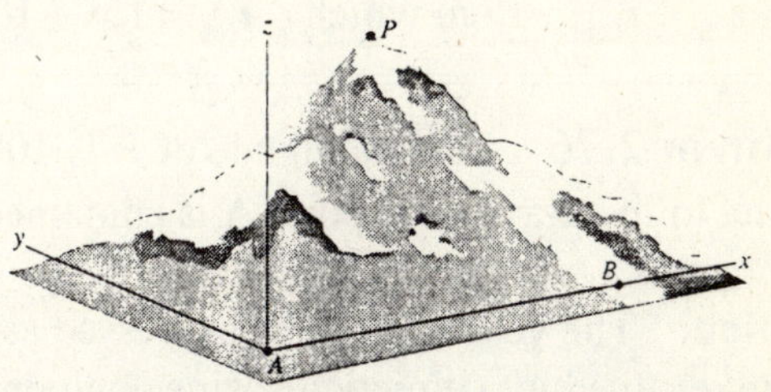

Solution: Construct the two triangles: (a) Triangle ABP, which has one known side, AB, and two known adjacent interior angles θ_A and θ_B (b) Triangle AOP, which is a right triangle with a derived known interior angle θ_{AO}.

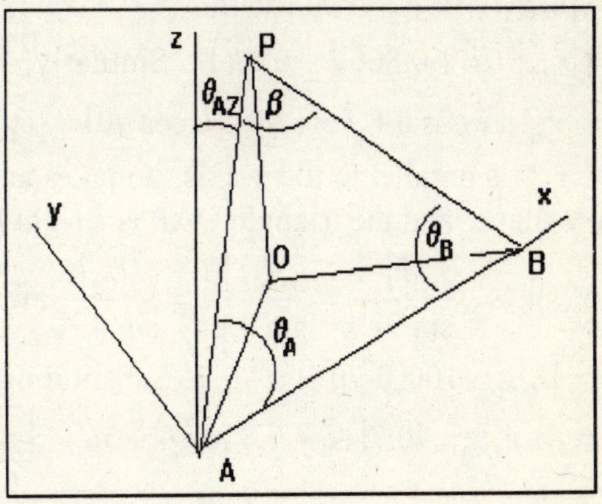

From triangle ABP, determine the length of AP, and from triangle APO and the derived interior angle, determine the height above the base, OP.

The interior angles of the triangle ABP are

$\theta_A = \theta_{AX} = \cos^{-1}(0.509) = 59.4°$,

$\theta_B = 180 - \theta_{BX} = 180 - \cos^{-1}(-0.605)$

$= 180° - 127.2° = 52.77°$ and

$\beta = 180 - 59.4 - 52.77 = 67.83°$. From the law of sines: $\frac{|\vec{r}_{AB}|}{\sin 67.83°} = \frac{|\vec{r}_{BP}|}{\sin 59.4°} = \frac{|\vec{r}_{AP}|}{\sin 52.77°}$. Therefore the length of side AP is $|\vec{r}_{AP}| = |\vec{r}_{AB}|\left(\frac{\sin 52.77}{\sin 67.83}\right) = 32000(0.8598) = 27512.9\ ft$. The interior angle of the triangle APO is $\theta_{AO} = 90 - \theta_{AZ} = 90 - \cos^{-1}(0.694) = 90 - 46.05 = 43.95°$. Therefore the length of the side OP is $h_{OP} = |\vec{r}_{AP}|\sin 43.95° = 27512.9(0.6940) = 19093.9\ ft$. *Check:* The z-component of $|\vec{r}_{AP}|$ is $h_{op} = |\vec{r}_{AP}|\sin\theta_{AZ} = 19093.9\ ft$. *check.* The base is 10000 ft above sea level, hence the height of P above sea level is $\boxed{P = 19093.9 + 10000 = 29094\ ft}$

==============================<>==============================

Problem 2.75 The distance from point O to point A is 20 ft. The straight line AB is parallel to the y-axis and the point B is in the x-z plane. Express the vector $\vec{r}_{OA}$ in terms of scalar components.

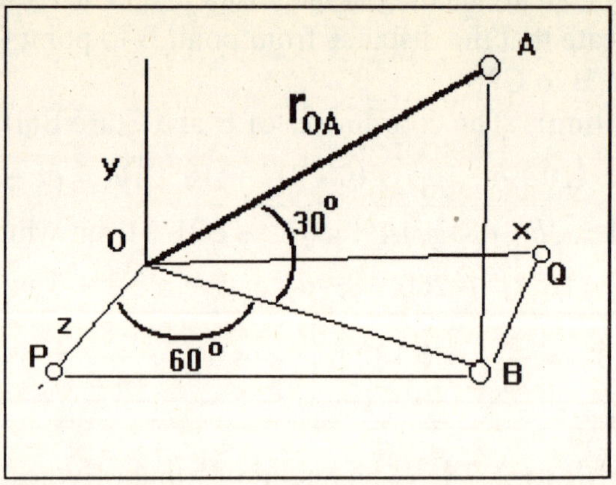

Solution: See Example 2.10. The length BA is, from the right triangle OAB,

$|\vec{r}_{AB}| = |\vec{r}_{OA}|\sin 30^{o} = 20(0.5) = 10\ ft.$

Similarly, the length OB is

$|\vec{r}_{OB}| = |\vec{r}_{OA}|\cos 30^{o} = 20(0.866) = 17.32\ ft$

The vector $\vec{r}_{OB}$ can be resolved into components along the axes by the right triangles OBP and OBQ. and the condition that it lies in the x-z plane

$\vec{r}_{OB} = |\vec{r}_{OB}|\left(\vec{i}\cos 30^{o} + \vec{j}\cos 90^{o} + \vec{k}\cos 60^{o}\right)$

$\vec{r}_{OB} = 15\vec{i} + 0\vec{j} + 8.66\vec{k}$ The vector $\vec{r}_{BA}$ can be resolved into components from the condition that it is parallel to the y-axis.

$\vec{r}_{BA} = |\vec{r}_{BA}|\left(\vec{i}\cos 90^{o} + \vec{j}\cos 0^{o} + \vec{k}\cos 90^{o}\right) = 0\vec{i} + 10\vec{j} + 0\vec{k}$. The vector $\vec{r}_{OA}$ is given by

$\vec{r}_{OA} = \vec{r}_{OB} + \vec{r}_{BA}$, from which $\vec{r}_{OA} = 15\vec{i} + 10\vec{j} + 8.66\vec{k}\ (ft)$

Problem 2.76 The magnitude of $\vec{r}$ is 100 in. The straight line from the head of $\vec{r}$ to point A is parallel to the x-axis, and point A is contained in the y-z plane. Express $\vec{r}$ in terms of scalar components.

Solution: The vector $\vec{r}$ can be expressed as the sum of the two vectors, $\vec{r} = \vec{r}_{OA} + \vec{r}_{AP}$, both of which can be resolved into direction cosine components. The magnitudes can be determined from the law of sines for the triangle OAP.

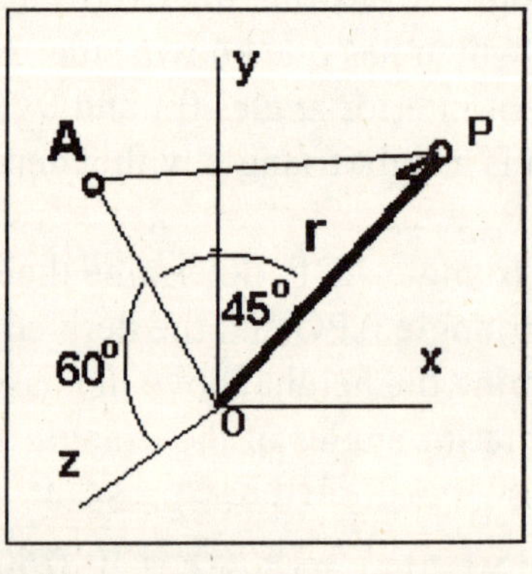

$\vec{r}_{OA} = |\vec{r}_{OA}|\left(\vec{i}\cos 90^{o} + \vec{j}\cos 30^{o} + \vec{k}\cos 60^{o}\right)$

$\vec{r}_{OA} = |\vec{r}_{OA}|\left(0\vec{i} + 0.866\vec{j} + 0.5\vec{k}\right)$. Similarly,

$\vec{r}_{AP} = |\vec{r}_{AP}|\left(\vec{i}\cos 0 + \vec{j}\cos 90 + \vec{k}\cos 90\right) = |\vec{r}_{AP}|\left(1\vec{i} + 0\vec{j} + 0\vec{k}\right)$

Since $\vec{r}_{AP}$ is parallel to the x-axis, it makes an angle of 90° with the y-z plane, and the triangle OAP is a right triangle. From the

law of sines $\dfrac{|\vec{r}|}{\sin 90^{o}} = \dfrac{|\vec{r}_{AP}|}{\sin 45^{o}} = \dfrac{|\vec{r}_{OA}|}{\sin 45^{o}}$, from which

$|\vec{r}_{AP}| = |\vec{r}_{OA}| = 100(0.707) = 70.7$. Substituting these values into the vectors

$\vec{r} = \vec{r}_{OA} + \vec{r}_{AP} = 70.7\left(1\vec{i} + 0.866\vec{j} + 0.5\vec{k}\right)$ $\boxed{= 70.7\vec{i} + 61.2\vec{j} + 35.4\vec{k}\ (in.)}$

Problem 2.77 The straight line from the head of $\vec{F}$ to point A is parallel to the y-axis, and point A is contained in the x-z plane. The x-component of $\vec{F}$ is $F_x = 100\ N$. (a) What is the magnitude of $\vec{F}$?. (b) Determine the angles $\theta_x, \theta_y, \theta_z$ between $\vec{F}$ and the positive coordinate axes.

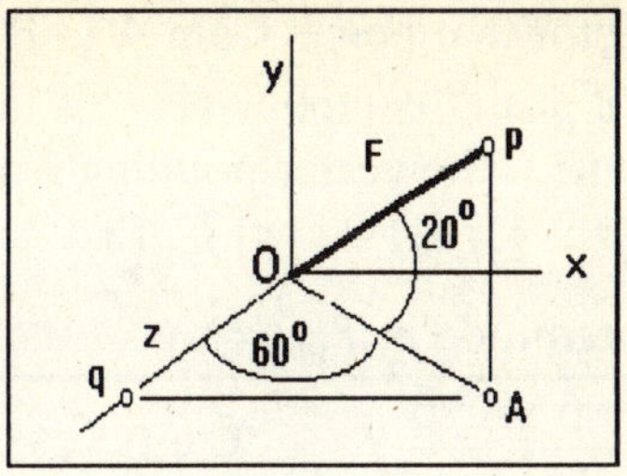

Solution: The triangle OpA is a right triangle, since OA lies in the x-z plane, and Ap is parallel to the y-axis. Thus the magnitudes are given by the sine law:

$$\frac{|\vec{r}_{Ap}|}{\sin 20^o} = \frac{|\vec{F}|}{\sin 90^o} = \frac{|\vec{r}_{OA}|}{\sin 70^o},$$ thus $|\vec{r}_{Ap}| = |\vec{F}|(0.342)$ and

$|\vec{r}_{OA}| = |\vec{F}|(0.9397)$. The components of the two vectors are from the geometry

$\vec{r}_{OA} = |\vec{r}_{OA}|(\vec{i} \cos 30^o + \vec{j} \cos 90^o + \vec{k} \cos 60^o) = |\vec{r}_{OA}|(0.866\vec{i} + 0\vec{j} + 0.5\vec{k})$ and

$\vec{r}_{Ap} = |\vec{r}_{Ap}|(\vec{i} \cos 90^o + \vec{j} \cos 0^o + \vec{k} \cos 90^o) = |\vec{r}_{Ap}|(0\vec{i} + 1\vec{j} + 0\vec{k})$ Noting $\vec{F} = \vec{r}_{OA} + \vec{r}_{Ap}$, then

from above $\vec{F} = |\vec{F}|(0.3420)(0\vec{i} + 1\vec{j} + 0\vec{k}) + |\vec{F}|(0.9397)(0.866\vec{i} + 0\vec{j} + 0.5\vec{k})$

$\vec{F} = |\vec{F}|(0.8138\vec{i} + 0.342\vec{j} + 0.4699\vec{k})$ The x-component is given to be 100 N. Thus,

(a) $\boxed{|\vec{F}| = \frac{100}{0.8138} = 122.9\ N}$ The angles are given by

(b) $\boxed{\theta_x = \cos^{-1}(0.8138) = 35.5^o}$, $\boxed{\theta_y = \cos^{-1}(0.342) = 70^o}$ and $\boxed{\theta_z = \cos^{-1}(0.4699) = 62^o}$

Problem 2.78 The position of a point P on the surface of the earth is specified by the two angles shown. The longitude λ is measured from the point G on the equator directly south of Greenwich, England. The latitude L is measured from the equator. Longitude is given as west (W) longitude or east (E) longitude, indicating whether the angle is measure east or west from point G. Latitude is given as north (N) latitude or south (S) latitude, indicating whether the angle is measured north or south. Suppose that point P is at longitude 30° W and latitude 45° N. Let R_E be the radius of the earth. Using the coordinate system shown, determine the components of the position vector of P relative to the center of the earth. (Your answer will be in terms of R_E.)

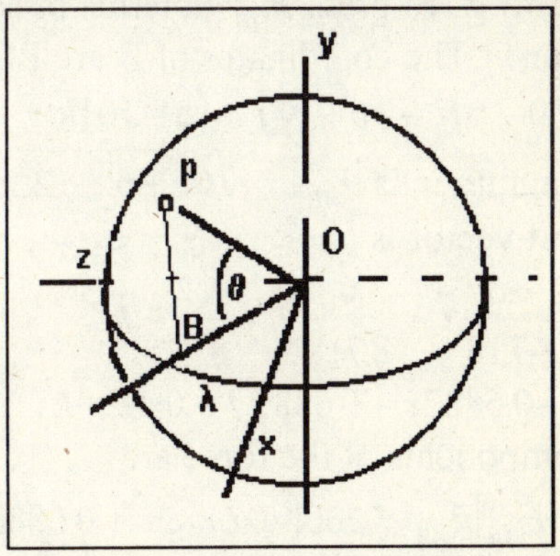

Solution: Drop a vertical line from point P to the equatorial plane. Let the intercept be B (see figure). The vector position of P is the sum of the two vectors: $\vec{P} = \vec{r}_{OB} + \vec{r}_{BP}$. The vector $\vec{r}_{OB} = |\vec{r}_{OB}|(\vec{i} \cos \lambda + 0\vec{j} + \vec{k} \sin \lambda)$. From geometry, the magnitude is $|\vec{r}_{OB}| = R_E \cos \theta$. The vector $\vec{r}_{BP} = |\vec{r}_{BP}|(0\vec{i} + 1\vec{j} + 0\vec{k})$. From geometry, the magnitude is $|\vec{r}_{BP}| = R_E \sin \theta_P$. Substitute and reduce to obtain: $\vec{P} = \vec{r}_{OB} + \vec{r}_{BP} = R_E(\vec{i} \cos \lambda \cos \theta + \vec{j} \sin \theta + \vec{k} \sin \lambda \cos \theta)$.

Substitute from the problem statement: $\lambda = +30^o$, $\theta = 45^o$

$\boxed{\vec{P} = R_E(0.6124\vec{i} + 0.707\vec{j} + 0.3536\vec{k})}$

Problem 2.79 The part of the cable from A to B exerts a 110 lb force $\vec{T}$ at A. (a) Determine the unit vector that points from A to B. (b) Express the force $\vec{T}$ in terms of scalar components.

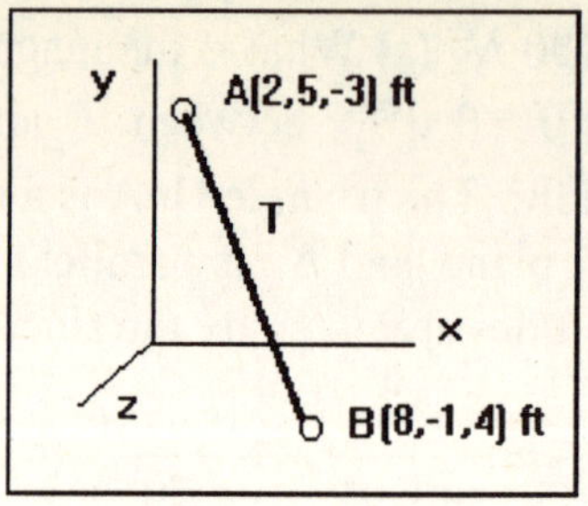

Solution: The vector pointing from A to B is

$$\vec{r}_{AB} = (8-2)\vec{i} + (-1-5)\vec{j} + (4-(-3))\vec{k} = 6\vec{i} - 6\vec{j} + 7\vec{k}$$

The magnitude is $|\vec{r}_{AB}| = \sqrt{6^2 + 6^2 + 7^2} = 11$. The unit vector is

$$\text{(a) } \boxed{\vec{e} = \frac{6}{11}\vec{i} - \frac{6}{11}\vec{j} + \frac{7}{11}\vec{k} = 0.5455\vec{i} - 0.5455\vec{j} + 0.6364\vec{k}}$$

The components of the tension $\vec{T}$ in the cable are

$$\text{(b) } \vec{T} = |\vec{T}|\vec{e} = 110(0.5455\vec{i} - 0.5455\vec{j} + 0.6364\vec{k})$$

$$\boxed{\vec{T} = 60\vec{i} - 60\vec{j} + 70\vec{k}\ (lb)}$$

Problem 2.80 A cable extends from point A to point B. It exerts a 1 kN force $\vec{F}$ at A that is directed along the line from A to B. Express $\vec{F}$ in terms of scalar components.

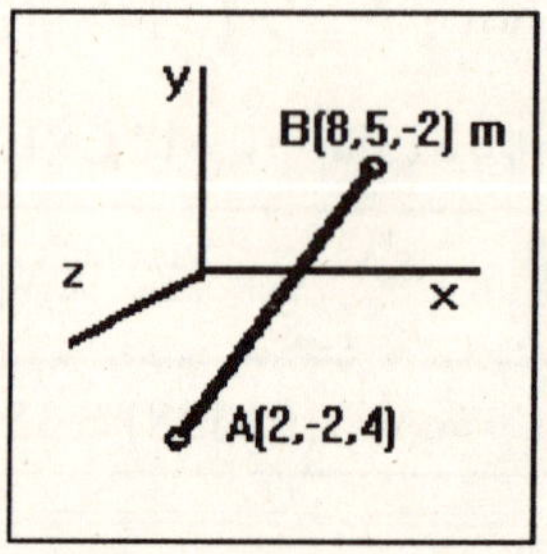

Solution: The position vector from A to B is

$$\vec{r}_{AB} = (8-2)\vec{i} + (5-(-2))\vec{k} + (-2-4)\vec{k} = 6\vec{i} + 7\vec{j} - 6\vec{k}$$

The magnitude is $|\vec{r}_{AB}| = \sqrt{6^2 + 7^2 + 6^2} = 11$ The unit vector is

$$\vec{u}_{AB} = \frac{6}{11}\vec{i} + \frac{7}{11}\vec{j} - \frac{6}{11}\vec{k} = 0.5455\vec{i} + 0.6364\vec{j} - 0.5455\vec{k}$$

The force is $\vec{F} = |\vec{F}|\vec{u}_{AB} = 1(0.5455\vec{i} + 0.6364\vec{j} - 0.5455\vec{k})$

$$\boxed{\vec{F} = 0.5455\vec{i} + 0.6364\vec{j} - 0.5455\vec{k}\ (kN)}$$

Problem 2.81 The cable AB exerts a 200 lb force $\vec{F}_{AB}$ at point A that is directed along the line from A to B. Express $\vec{F}_{AB}$ in terms of scalar components.

Solution: The coordinates of B are B(0,6,8). The position vector from A to B is

$$\vec{r}_{AB} = (0-6)\vec{i} + (6-0)\vec{j} + (8-10)\vec{k} = -6\vec{i} + 6\vec{j} - 2\vec{k}$$

The magnitude is $|\vec{r}_{AB}| = \sqrt{6^2 + 6^2 + 2^2} = 8.718\ ft$

The unit vector is

$$\vec{u}_{AB} = \frac{-6}{8.718}\vec{i} + \frac{6}{8.718}\vec{j} - \frac{2}{8.718}\vec{k}$$

$$\vec{u}_{AB} = -0.6882\vec{i} + 0.6882\vec{j} - 0.2294\vec{k}$$

The components of the force are

$$\vec{F}_{AB} = |\vec{F}_{AB}|\vec{u}_{AB} = 200(-0.6882\vec{i} + 0.6882\vec{j} - 0.2294\vec{k})$$

$$\boxed{= -137.6\vec{i} + 137.6\vec{j} - 45.9\vec{k}}$$

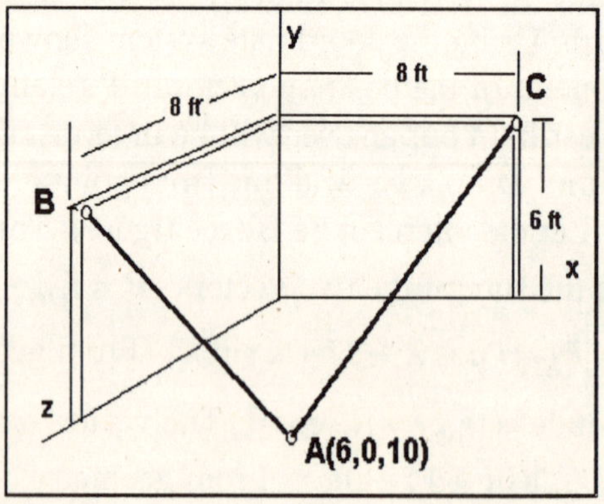

Problem 2.82 Consider the cables and wall described in Problem 2.81. Cable AB exerts a 200 lb force $\vec{F}_{AB}$ at point A that is directed along the line from A to B. The cable AC exerts a 100 lb force $\vec{F}_{AC}$ that is directed along the line A to C. Determine the magnitude of the total force exerted at point A by the two cables.

Solution: Refer to the figure in Problem 2.81. From Problem 2.81 the force $\vec{F}_{AB}$ is

$\vec{F}_{AB} = -137.6\vec{i} + 137.6\vec{j} - 45.9\vec{k}$

The coordinates of C are C(8,6,0). The position vector from A to C is

$\vec{r}_{AC} = (8-6)\vec{i} + (6-0)\vec{j} + (0-10)\vec{k} = 2\vec{i} + 6\vec{j} - 10\vec{k}$ The magnitude is

$|\vec{r}_{AC}| = \sqrt{2^2 + 6^2 + 10^2} = 11.83\ ft$ The unit vector is

$\vec{u}_{AC} = \frac{2}{11.83}\vec{i} + \frac{6}{11.83}\vec{j} - \frac{10}{11.83}\vec{k} = 0.1691\vec{i} + 0.5072\vec{j} - 0.8453\vec{k}$ The force is

$\vec{F}_{AC} = |\vec{F}_{AC}|\vec{u}_{AC} = 100\vec{u}_{AC} = 16.9\vec{i} + 50.7\vec{j} - 84.5\vec{k}$ The resultant of the two forces is

$\vec{F}_R = \vec{F}_{AB} + \vec{F}_{AC} = (-137.6 + 16.9)\vec{i} + (137.6 + 50.7)\vec{j} + (-84.5 - 45.9)\vec{k}$.

$\vec{F}_R = -120.7\vec{i} + 188.3\vec{j} - 130.4\vec{k}$ The magnitude is $|\vec{F}_R| = \sqrt{120.7^2 + 188.3^2 + 130.4^2} = 258.9\ lb$

Problem 2.83 The 70 m tall tower is supported by three cables that exert forces $\vec{F}_{AB}$, $\vec{F}_{AC}$, and $\vec{F}_{AD}$ on it. The magnitude of each force is 2 kN. Express the total force exerted on the tower by the three cables in terms of scalar components.

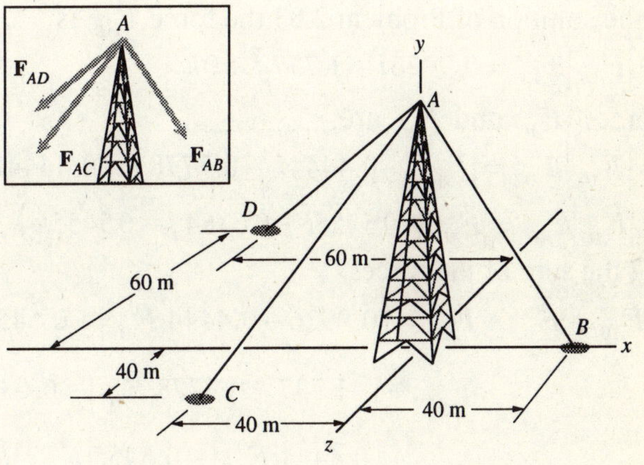

Solution: The coordinates of the points are A(0, 70, 0), B(40, 0, 0), C(-40, 0, 40) D(-60, 0, -60). The position vectors corresponding to the cables are:

$\vec{r}_{AD} = (-60-0)\vec{i} + (0-70)\vec{j} + (-60-0)\vec{k}$

$\vec{r}_{AD} = -60\vec{i} - 70\vec{k} - 60\vec{k}$

$\vec{r}_{AC} = (-40-0)\vec{i} + (0-70)\vec{j} + (40-0)\vec{k}$

$\vec{r}_{AC} = -40\vec{i} - 70\vec{j} + 40\vec{k}$

$\vec{r}_{AB} = (40-0)\vec{i} + (0-70)\vec{j} + (0-0)\vec{k}$

$\vec{r}_{AB} = 40\vec{i} - 70\vec{j} + 0\vec{k}$ The unit vectors corresponding to these position vectors are:

$$\vec{u}_{AD} = \frac{\vec{r}_{AD}}{|\vec{r}_{AD}|} = \frac{-60}{110}\vec{i} - \frac{70}{110}\vec{j} - \frac{60}{110}\vec{k} = -0.5455\vec{i} - 0.6364\vec{j} - 0.5455\vec{k}$$

$$\vec{u}_{AC} = \frac{\vec{r}_{AC}}{|\vec{r}_{AC}|} = -\frac{40}{90}\vec{i} - \frac{70}{90}\vec{j} + \frac{40}{90}\vec{k} = -0.4444\vec{i} - 0.7778\vec{j} + 0.4444\vec{k}$$

................***Solution to Problem 2.83 continued on next page***

...*continuation of solution to Problem 2.83...*

$$\vec{u}_{AB} = \frac{\vec{r}_{AB}}{|\vec{r}_{AB}|} = \frac{40}{80.6}\vec{i} - \frac{70}{80.6}\vec{j} + 0\vec{k} = 0.4963\vec{i} - 0.8685\vec{j} + 0\vec{k}$$

The forces are:

$$\vec{F}_{AB} = |\vec{F}_{AB}|\vec{u}_{AB} = 0.9926\vec{i} - 1.737\vec{j} + 0\vec{k}$$

$$\vec{F}_{AC} = |\vec{F}_{AC}|\vec{u}_{AC} = -0.8888\vec{i} - 1.5556\vec{j} + 0.8888$$

$$\vec{F}_{AD} = |\vec{F}_{AD}|\vec{u}_{AD} = -1.0910\vec{i} - 1.2728\vec{j} - 1.0910\vec{k}$$

The resultant force exerted on the tower by the cables is:

$$\boxed{\vec{F}_R = \vec{F}_{AB} + \vec{F}_{AC} + \vec{F}_{AD} = -0.9872\vec{i} - 4.5654\vec{j} - 0.2022\vec{k}\ kN}$$

Problem 2.84 Consider the tower described in Problem 2.83. The magnitude of the force $\vec{F}_{AB}$ is 2 kN. The x- and z-components of the vector sum of the forces exerted on the tower by the cables are zero. What are the magnitudes of $\vec{F}_{AC}$ and $\vec{F}_{AD}$?

Solution: From the solution of Problem 2.83, the unit vectors are:

$$\vec{u}_{AC} = \frac{\vec{r}_{AC}}{|\vec{r}_{AC}|} = -\frac{40}{90}\vec{i} - \frac{70}{90}\vec{j} + \frac{40}{90}\vec{k} = -0.4444\vec{i} - 0.7778\vec{j} + 0.4444\vec{k}$$

$$\vec{u}_{AD} = \frac{\vec{r}_{AD}}{|\vec{r}_{AD}|} = \frac{-60}{110}\vec{i} - \frac{70}{110}\vec{j} - \frac{60}{110} = -0.5455\vec{i} - 0.6364\vec{j} - 0.5455\vec{k}$$

From the solution of Problem 2.83 the force $\vec{F}_{AB}$ is

$$\vec{F}_{AB} = |\vec{F}_{AB}|\vec{u}_{AB} = 0.9926\vec{i} - 1.737\vec{j} + 0\vec{k}$$

The forces $\vec{F}_{AC}$ and $\vec{F}_{AD}$ are:

$$\vec{F}_{AC} = |\vec{F}_{AC}|\vec{u}_{AC} = |\vec{F}_{AC}|(-0.4444\vec{i} - 0.7778\vec{j} + 0.4444\vec{k})$$

$$\vec{F}_{AD} = |\vec{F}_{AD}|\vec{u}_{AD} = |\vec{F}_{AD}|(-0.5455\vec{i} - 0.6364\vec{j} - 0.5455\vec{k})$$

Taking the sum of the forces:

$$\vec{F}_R = \vec{F}_{AB} + \vec{F}_{AC} + \vec{F}_{AD} = (0.9926 - 0.4444|\vec{F}_{AC}| - 0.5455|\vec{F}_{AD}|)\vec{i}$$
$$+ (-1.737 - 0.7778|\vec{F}_{AC}| - 0.6364|\vec{F}_{AD}|)\vec{j}$$
$$+ (0.4444|\vec{F}_{AC}| - 0.5455|\vec{F}_{AD}|)\vec{k}$$

The sum of the x- and z-components vanishes, hence the set of simultaneous equations:

$$0.4444|\vec{F}_{AC}| + 0.5455|\vec{F}_{AD}| = 0.9926$$

$$0.4444|\vec{F}_{AC}| - 0.5455|\vec{F}_{AD}| = 0$$

These can be solved by means of standard algorithms, or by the use of commercial packages such as TK Solver® or Mathcad®. Here the HP-28S hand held calculator was used to obtain the solution:

$\boxed{|\vec{F}_{AC}| = 1.1168\ kN}$ $\boxed{|\vec{F}_{AD}| = 0.9098\ kN}$

Problem 2.85 Express the position vector from O to the collar at A in terms of scalar components.

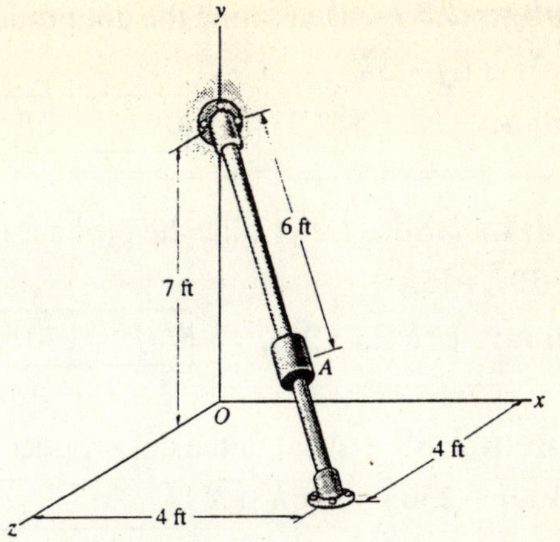

Solution: The vector from O to A can be expressed as the sum of the vectors $\vec{r}_{OT}$ from O to the top of the slider bar, and $\vec{r}_{TA}$ from the top of the slider bar to A. The coordinates of the top and base of the slider bar are: T(0, 7, 0), B(4, 0,4). The position vector of the top of the bar is: $\vec{r}_{OT} = 0\vec{i} + 7\vec{j} + 0\vec{k}$ The position vector from the top of the bar to the base is:

$\vec{r}_{TB} = (4-0)\vec{i} + (0-7)\vec{j} + (4-0)\vec{k}$. or

$\vec{r}_{TB} = 4\vec{i} - 7\vec{j} + 4\vec{k}$ The unit vector pointing from the top of the bar to the base is

$\vec{u}_{TB} = \frac{\vec{r}_{TB}}{|\vec{r}_{TB}|} = \frac{4}{9}\vec{i} - \frac{7}{9}\vec{j} + \frac{4}{9}\vec{k} = 0.4444\vec{i} - 0.7778\vec{j} + 0.4444\vec{k}$. The collar position is

$\vec{r}_{TA} = |\vec{r}_{TA}|\vec{u}_{TB} = 6(0.4444i - 0.7778\vec{j} + 0.4444\vec{k}) = 2.6667\vec{i} - 4.6667\vec{j} + 2.6667$, measured along the bar. The sum of the two vectors is the position vector of A from origin O:

$\vec{r}_{OA} = (2.6667+0)\vec{i} + (-4.6667+7)\vec{j} + (2.6667+0)\vec{k}$ $\boxed{= 2.67\vec{i} + 2.33\vec{j} + 2.67\vec{k}\ ft}$

Problem 2.86 The cable AB exerts a 32 lb force $\vec{T}$ on the collar at A. Express $\vec{T}$ in terms of scalar components.

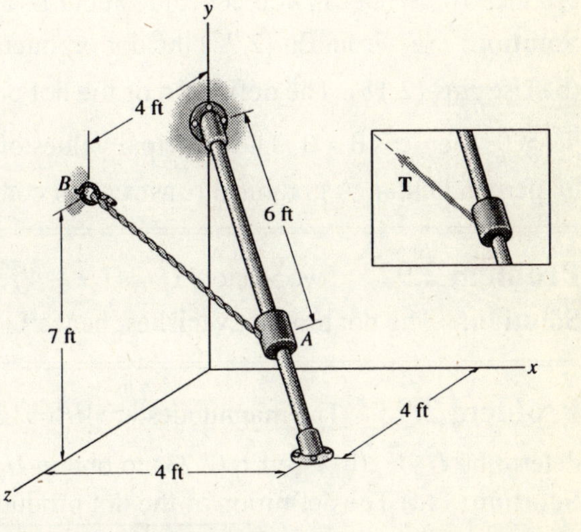

Solution: The coordinates of point B are B(0,7,4). The vector position of B is $\vec{r}_{OB} = 0\vec{i} + 7\vec{j} + 4\vec{k}$.

The vector from point A to point B is given by

$\vec{r}_{AB} = \vec{r}_{OB} - \vec{r}_{OA}$.

From Problem 2.86, $\vec{r}_{OA} = 2.67\vec{i} + 2.33\vec{j} + 2.67\vec{k}$.

Thus $\vec{r}_{AB} = (0-2.67)\vec{i} + (7-2.33)\vec{j} + (4-2.67)\vec{j}$

$\vec{r}_{AB} = -2.67\vec{i} + 4.67\vec{j} + 1.33\vec{k}$ The magnitude is

$|\vec{r}_{AB}| = \sqrt{2.67^2 + 4.67^2 + 1.33^2} = 5.54\ ft$. The unit vector pointing from A to B is

$\vec{u}_{AB} = \frac{\vec{r}_{AB}}{|\vec{r}_{AB}|} = -0.4819\vec{i} + 0.8429\vec{j} + 0.2401\vec{k}$

The force $\vec{T}$ is given by $\boxed{\vec{T}_{AB} = |\vec{T}_{AB}|\vec{u}_{AB} = 32\vec{u}_{AB} = -15.4\vec{i} + 27.0\vec{j} + 7.7\vec{k}\ (lb)}$

Problem 2.87 Determine the dot product $\vec{U}\cdot\vec{V}$ of the vectors $\vec{U} = 2\vec{i} - 4\vec{j} + 3\vec{k}$ and $\vec{V} = -3\vec{i} + 6\vec{j} + 3k$

Solution: Use Eq (2.23). $\boxed{\vec{U}\cdot\vec{V} = (2)(-3)+(-4)(6)+(3)(3) = -21}$

Problem 2.88 Determine the product $\vec{U}\cdot\vec{V}$ of the vectors $\vec{U} = 40\vec{i} + 20\vec{j} + 60\vec{k}$ and $\vec{V} = -30\vec{i} + 15\vec{k}$.

Solution: Use Eq 2.23). $\boxed{\vec{U}\cdot\vec{V} = (40)(-30)+(20)(0)+(15)(60) = -300}$

Problem 2.89 What is the dot product of the position vector $\vec{r} = -10\vec{i} + 25\vec{j}$ (m) and the force $\vec{F} = 300\vec{i} + 250\vec{j} + 300\vec{k}$ (N)?

Solution: Use Eq (2.23). $\boxed{\vec{F}\cdot\vec{r} = (300)(-10)+(250)(25)+(300)(0) = 3250\ N\text{-}m}$

Problem 2.90 What is the dot product of the position vector $\vec{r} = 4\vec{i} - 12\vec{j} - 3\vec{k}$ (ft) and the force vector $\vec{F} = 20\vec{i} + 30\vec{j} - 10\vec{k}$ (lb) ?

Solution: Use Eq (2.23). $\boxed{\vec{r}\cdot\vec{F} = 4(20)+30(-12)-10(-3) = -250\ ft\ lb}$

Problem 2.91 Two vectors $\vec{U} = -6\vec{i} + \vec{j} + 8\vec{k}$ and $\vec{V} = 3\vec{i} + 2\vec{j} + 2\vec{k}$. (a) Determine the dot product. (b) What can you conclude about $\vec{U}$ and $\vec{V}$ from the result of Part (a)?

Solution: (a) From Eq (2.23) the dot product is $\vec{U}\cdot\vec{V} = (-6)(3)+(2)(1)+(8)(2) = 0$

(b) Use Eqs (2.18): The definition of the dot product is $\vec{U}\cdot\vec{V} = |\vec{U}||\vec{V}|\cos\theta$. If $\vec{U}\cdot\vec{V} = 0$ and $|\vec{U}| \neq 0$ and $|\vec{V}| \neq 0$, then $\cos\theta = 0$. The principal values of the arc cosine are $\theta = \pm 90^\circ$, and the vectors are said to be perpendicular. A graphical construction confirms this interpretation.

Problem 2.92 Two vectors $\vec{U} = U_x\vec{i} - 4\vec{j}$ and $\vec{V} = -2\vec{i} + 6\vec{j}$ are perpendicular. What is U_x ?

Solution: The dot product vanishes, hence $\vec{U}\cdot\vec{V} = (U_x)(-2)+(-4)(6) = 0$, from which $\boxed{U_x = -12}$.

Problem 2.93 The magnitudes are $|\vec{U}| = 10$ and $|\vec{V}| = 20$. (a) Use the definition of the dot product to determine $\vec{U}\cdot\vec{V}$. (b) Use Eq (2.23) to obtain $\vec{U}\cdot\vec{V}$.

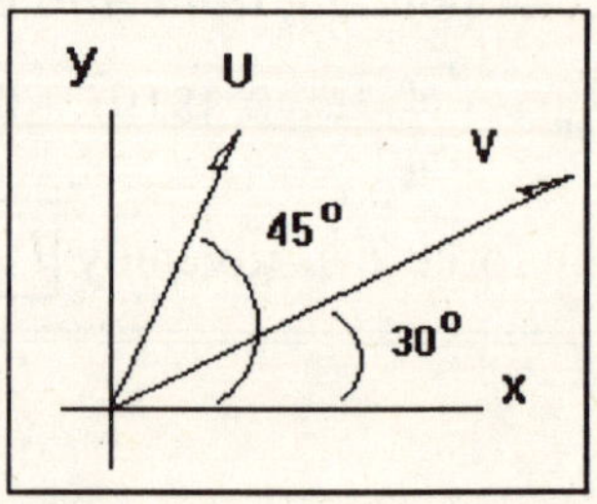

Solution: (a) The definition of the dot product (Eq (2.18)) is

$\vec{U}\cdot\vec{V} = |\vec{U}||\vec{V}|\cos\theta$ Thus $\boxed{\vec{U}\cdot\vec{V} = (10)(20)\cos(45^\circ - 30^\circ) = 193.2}$

(b) The components of $\vec{U}$ and $\vec{V}$ are

$\vec{U} = 10(\vec{i}\cos 45^\circ + \vec{j}\sin 45^\circ) = 7.07\vec{i} + 7.07\vec{j}$

$\vec{V} = 20(\vec{i}\cos 30^\circ + \vec{j}\sin 30^\circ) = 17.32\vec{i} + 10\vec{j}$

From Eq (2.23) $\boxed{\vec{U}\cdot\vec{V} = (7.07)(17.32)+(7.07)(10) = 193.2}$

Problem 2.94 By evaluating the dot product $\vec{U}\cdot\vec{V}$, prove the identity $\cos(\theta_1-\theta_2)=\cos\theta_1\cos\theta_2+\sin\theta_1\sin\theta_2$.

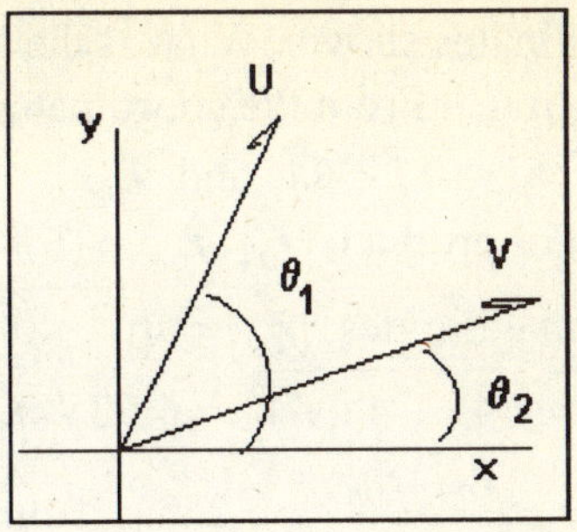

Solution: The strategy is to use the definition Eq (2.18) and the Eq (2.23). From Eq (2.18) and the figure,

$\vec{U}\cdot\vec{V}=|\vec{U}||\vec{V}|\cos(\theta_1-\theta_2)$.

From Eq (2.23) and the figure,

$\vec{U}=|\vec{U}|(\vec{i}\cos\theta_1+\vec{j}\sin\theta_2)$,

$\vec{V}=|\vec{V}|(\vec{i}\cos\theta_2+\vec{j}\sin\theta_2)$,

and the dot product is $\vec{U}\cdot\vec{V}=|\vec{U}||\vec{V}|(\cos\theta_1\cos\theta_2+\sin\theta_1\sin\theta_2)$.

Equating the two results: $\vec{U}\cdot\vec{V}=|\vec{U}||\vec{V}|\cos(\theta_1-\theta_2)=|\vec{U}||\vec{V}|(\cos\theta_1\cos\theta_2+\sin\theta_1\sin\theta_2)$, from which if $|\vec{U}|\neq 0$ and $|\vec{V}|\neq 0$, it follows that $\cos(\theta_1-\theta_2)=\cos\theta_1\cos\theta_2+\sin\theta_1\sin\theta_2$, Q.E.D.

Problem 2.95 Two vectors $\vec{U}=U_x\vec{i}-6\vec{j}+\vec{k}$ and $\vec{V}=-3\vec{i}+V_y\vec{j}+\vec{k}$. Their dot product $\vec{U}\cdot\vec{V}=-35$ and the magnitude of their sum is $|\vec{U}+\vec{V}|=3$. What are the components U_x and V_y ?

Solution: The dot product (using Eq (2.24)) is $\vec{U}\cdot\vec{V}=(U_x)(-3)+(-6)(V_y)+(1)(1)=-35$, or, reducing, (A) $U_x+2V_y=12$. The sum is $\vec{U}+\vec{V}=(U_x-3)\vec{i}+(V_y-6)\vec{j}+2\vec{k}$. The magnitude is $|\vec{U}+\vec{V}|=\sqrt{(U_x-3)^2+(V_y-6)^2+2^2}=3$, or (B) $(U_x-3)^2+(V_y-6)^2=5$.

From (A) above, $V_y=6-0.5U_x$. Substitute into (B) and reduce to obtain the quadratic equation:

$U_x^2-\frac{24}{5}U_x+\frac{16}{5}=0$. The solutions are: $U_x=\frac{12}{5}\pm\sqrt{\left(\frac{12}{5}\right)^2-\frac{16}{5}}$. Substitute into (A) above to obtain, $\boxed{U_x=0.8,\quad V_y=5.6}$ $\boxed{U_x=4,\quad V_x=4}$

Problem 2.96 What is the angle θ between the straight lines AB and AC?

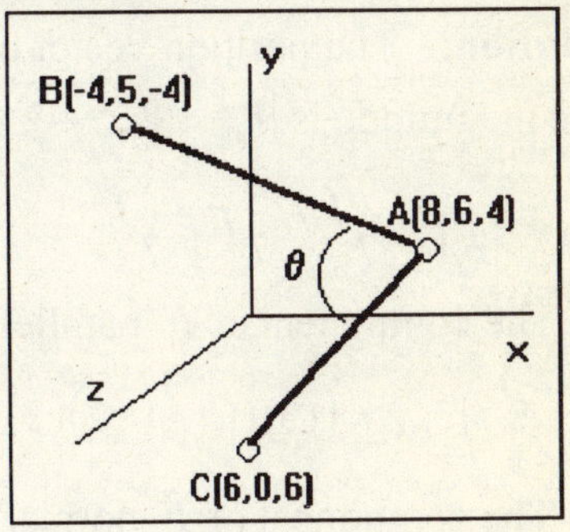

Solution: From the given coordinates, the position vectors are:

$\vec{r}_{OB}=-4\vec{i}+5\vec{j}-4\vec{k}$,

$\vec{r}_{OA}=8\vec{i}+6\vec{j}+4\vec{k}$,

$\vec{r}_{OC}=6\vec{i}+0\vec{j}+6\vec{k}$.

The straight lines correspond to the vectors:

$\vec{r}_{AB}=\vec{r}_{OB}-\vec{r}_{OA}=-12\vec{i}-\vec{j}-8\vec{k}$,

$\vec{r}_{AC}=\vec{r}_{OC}-\vec{r}_{AC}=-2\vec{i}-6\vec{j}+2\vec{k}$

The dot product:

$\vec{r}_{AB}\cdot\vec{r}_{AC}=(-2)(-12)+(-1)(-6)+(+2)(-8)=14$

The magnitudes:

$\left|\vec{r}_{AC}=\sqrt{2^2+6^2+2^2}\right|=6.6333$, $\left|\vec{r}_{AB}=\sqrt{12^2+1^2+8^2}\right|=14.456$. From Eq (2.24), the angle is

$\cos\theta=\frac{\vec{r}_{AC}\cdot\vec{r}_{AB}}{|\vec{r}_{AC}||\vec{r}_{AB}|}=\frac{14}{(14.456)(6.633)}=0.1460$. Take the principal value: $\boxed{\theta=81.6^o}$

Problem 2.97 The ship O measures the positions of the ship A and the airplane B and obtains the coordinates shown. What is the angle θ between the lines of sight OA and OB?

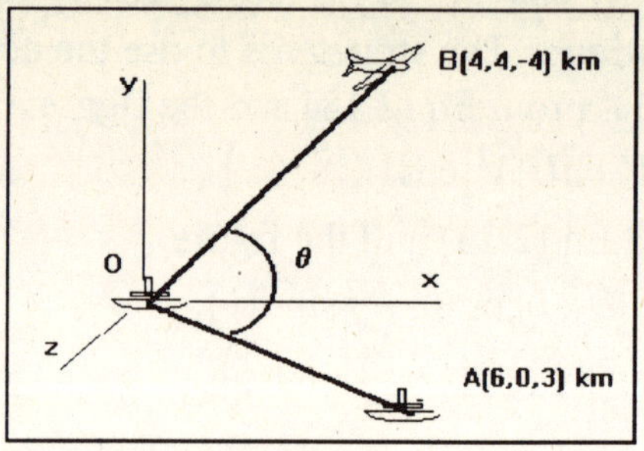

Solution: From the coordinates, the position vectors are:

$\vec{r}_{OA} = 6\vec{i} + 0\vec{j} + 3\vec{k}$ and $\vec{r}_{OB} = 4\vec{i} + 4\vec{j} - 4\vec{k}$

The dot product: $\vec{r}_{OA} \cdot \vec{r}_{OB} = (6)(4) + (0)(4) + (3)(-4) = 12$

The magnitudes: $|\vec{r}_{OA}| = \sqrt{6^2 + 0^2 + 3^2} = 6.71\ km$ and

$|\vec{r}_{OA}| = \sqrt{4^2 + 4^2 + 4^2} = 6.93\ km$. From Eq (2.24)

$cos\,\theta = \dfrac{\vec{r}_{OA} \cdot \vec{r}_{OB}}{|\vec{r}_{OA}||\vec{r}_{OB}|} = 0.2581$, from which $\theta = \pm 75.^o$. From the problem and the construction, only the positive angle makes sense, hence $\boxed{\theta = 75.^o}$

Problem 2.98 The force vector $\vec{F} = F_x\vec{i} + F_y\vec{j} + F_z\vec{k}$. (a) Use Eq (2.26) to determine the component of $\vec{F}$ parallel to the x-axis. (b) Determine the component of $\vec{F}$ parallel to the line L.

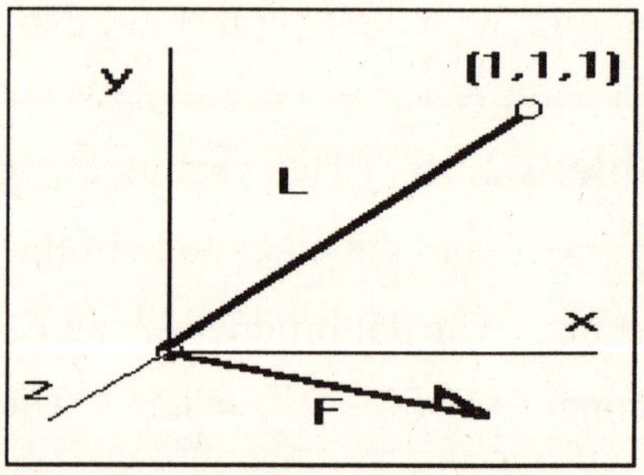

Solution: Eq (2.26) defines the component of a vector $\vec{U}$ parallel to a line L to be $(\vec{U} \cdot \vec{e})\vec{e}$, where $\vec{e}$ is a unit vector parallel to the line L. (a) The unit vector parallel to the x-axis is $\vec{e}_x = 1\vec{i} + 0\vec{j} + 0\vec{k}$, Hence $\vec{F} \cdot \vec{e}_x = F_x$ and $\boxed{(\vec{F} \cdot \vec{e}_x)\vec{e} = F_x\vec{i}}$ (b) The position vector of the terminating line point is: $\vec{r}_P = 1\vec{i} + 1\vec{j} + 1\vec{k}$, The magnitude: $|\vec{r}_P| = \sqrt{3}$, whence the unit vector:

$\vec{e}_x = \dfrac{\vec{r}_P}{|\vec{r}_P|} = \left(\dfrac{1}{\sqrt{3}}\right)(1\vec{i} + 1\vec{j} + 1\vec{k})$. Thus $\boxed{(\vec{F} \cdot \vec{e}_x)\vec{e}_x = \left(\dfrac{F_x + F_y + F_z}{3}\right)(1\vec{i} + 1\vec{j} + 1\vec{k})}$

Problem 2.99 The force $\vec{F} = 21\vec{i} + 14\vec{j}\ (kN)$. Resolve it into vector components parallel and normal to the line OA.

Solution: The position vector of point A is $\vec{r}_A = 6\vec{i} - 2\vec{j} + 3\vec{k}$

The magnitude is $|\vec{r}_A| = \sqrt{6^2 + 2^2 + 3^2} = 7$. The unit vector parallel to OA is

$$\vec{e}_{OA} = \frac{\vec{r}_A}{|\vec{r}_A|} = \frac{6}{7}\vec{i} - \frac{2}{7}\vec{j} + \frac{3}{7}\vec{k}$$

(a) The component of $\vec{F}$ parallel to OA is

$(\vec{F} \cdot \vec{e}_{OA})\vec{e}_{OA} = ((3)(6) + (-2)(2))\left(\dfrac{1}{7}\right)(6\vec{i} - 2\vec{j} + 3k)$ $\boxed{\vec{F}_P = 12\vec{i} - 4\vec{j} + 6\vec{k}\ (kN)}$

(b) The component of $\vec{F}$ normal to OA is (see Eq (2.27))

$\vec{F}_N = \vec{F} - \vec{F}_p = (21 - 12)\vec{i} + (14 - (-4))\vec{j} + (0 - 6)\vec{k}$ $\boxed{= 9\vec{i} + 18\vec{j} - 6\vec{k}\ (kN)}$

Problem 2.100 Cables extend from A to B and from A to C. The cable AC exerts a 1000 lb force $\vec{F}$ at A. (a) What is the angle between the cables AB and AC? (b) Determine the vector component of $\vec{F}$ parallel to the cable AB.

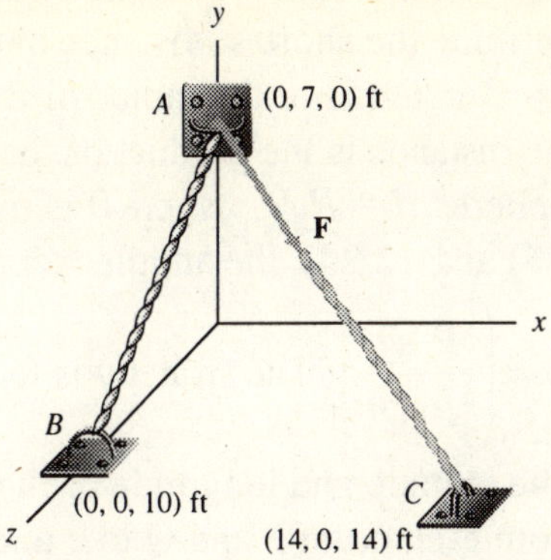

Solution: Use Eq (2.24) to solve (a). From the coordinates of the points, the position vectors are:

$\vec{r}_{AB} = (0-0)\vec{i} + (0-7)\vec{j} + (10-0)\vec{k}$

$\vec{r}_{AB} = 0\vec{i} - 7\vec{j} + 10\vec{k}$

$\vec{r}_{AC} = (14-0)\vec{i} + (0-7)\vec{j} + (14-0)\vec{k}$

$\vec{r}_{AC} = 14\vec{i} - 7\vec{j} + 14\vec{k}$

The magnitudes are:

$|\vec{r}_{AB}| = \sqrt{7^2 + 10^2} = 12.2 \ (ft)$

$|\vec{r}_{AB}| = \sqrt{14^2 + 7^2 + 14^2} = 21$

The dot product: $\vec{r}_{AB} \cdot \vec{r}_{AC} = (14)(0) + (-7)(-7) + (10)(14) = 189$. The angle is

$\cos\theta = \dfrac{189}{(12.2)(21)} = 0.7377$, from which $\theta = \pm 42.5^o$ From the construction: $\boxed{\theta = +42.5^o}$

(b) The unit vector associated with AB is $\vec{e}_{AB} = \dfrac{\vec{r}_{AB}}{|\vec{r}_{AB}|} = 0\vec{i} - 0.5738\vec{j} + 0.8197\vec{k}$. The unit vector associated with AC is $\vec{e}_{AC} = \dfrac{\vec{r}_{AC}}{|\vec{r}_{AC}|} = 0.6667\vec{i} - 0.3333\vec{j} + 0.6667\vec{k}$. Thus the force vector along AC is

$\vec{F}_{AC} = |\vec{F}|\vec{e}_{AC} = 666.7\vec{i} - 333.3\vec{j} + 666.7\vec{k}$ The component of this force parallel to AB is

$\boxed{(\vec{F}_{AC} \cdot \vec{e}_{AB})\vec{e}_{AB} = (737.5)\vec{e}_{AB} = 0\vec{i} - 423.2\vec{j} + 604.5\vec{k} \ (lb)}$

Problem 2.101 Consider the cables AB and AC shown in Problem 2.100. Let $\vec{r}_{AB}$ be the position vector from point A to point B. Determine the vector component of $\vec{r}_{AB}$ parallel to the cable AC.

Solution: From Problem 2.100, $\vec{r}_{AB} = 0\vec{i} - 7\vec{j} + 10\vec{k}$, and $\vec{e}_{AC} = 0.6667\vec{i} - 0.3333\vec{j} + 0.6667\vec{k}$.. Thus $\vec{r}_{AB} \cdot \vec{e}_{AC} = 9$, and $\boxed{(\vec{r}_{AB} \cdot \vec{e}_{AC})\vec{e}_{AC} = 6\vec{i} - 3\vec{j} + 6\vec{k}}$

Problem 2.102 Consider the cables AB and Ac shown in Problem 2.100. Use the dot product to determine the perpendicular distance from point B to the cable AC. That is, determine the length of the shortest straight line from point B to a point on the cable AC.

Solution: From Problems 2.100 and 2.101 $\vec{r}_{AB} = 0\vec{i} - 7\vec{j} + 10\vec{k}$ and the component of $\vec{r}_{AB}$ that is parallel to AC is $\vec{r}_p = (\vec{r}_{AB} \cdot \vec{e}_{AC})\vec{e}_{AC} = 6\vec{i} - 3\vec{j} + 6\vec{k}$ The component perpendicular to $\vec{r}_{AB}$ is, from Eq (2.27), $\vec{r}_N = \vec{r}_{AB} - \vec{r}_p$. Thus $\vec{r}_N = (0-6)\vec{i} + (-7-(-3))\vec{j} + (10-6)\vec{k} = -6\vec{i} - 4\vec{j} + 4\vec{k}$

The distance is the magnitude: $\boxed{|\vec{r}_N| = \sqrt{6^2 + 4^2 + 4^2} = 8.25 \ ft}$ *Check:*

$|\vec{r}_N| = |\vec{r}_{AB}|\sin\theta = 12.2\sin 42.5^o = 8.25$. *check.*

Problem 2.103

The point P is at longitude 30 W and latitude 45 N on the Atlantic Ocean between Nova Scotia and France. (See Problem 2.78.) Point Q is at longitude 60 E and latitude 20 N in the Arabian Sea. Use the dot product to determine the shortest distance along the surface of the Earth from P to Q in terms of the radius of the Earth R_E.

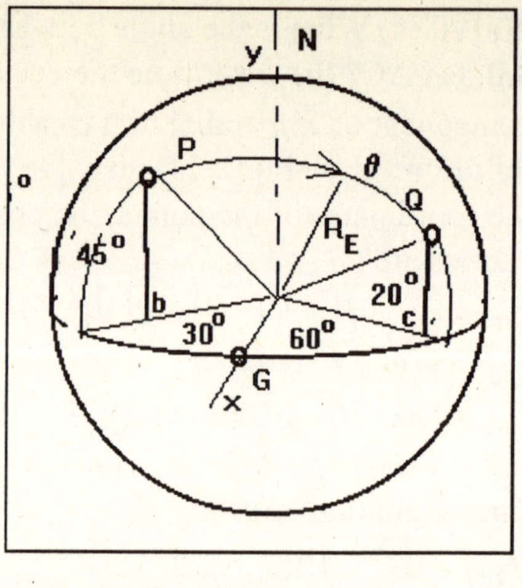

Solution: The distance is the product of the angle and the radius of the sphere, $d = R_E\theta$, where θ is in radian measure. From Eqs (2.18) and (2.24), the angular separation of P and Q is given by $\cos\theta = \left(\dfrac{\vec{P}\cdot\vec{Q}}{|\vec{P}||\vec{Q}|}\right)$. The strategy is to determine the angle θ in terms of the latitude and longitude of the two points.Drop a vertical line from each point P and Q to b and c on the equatorial plane. The vector position of P is the sum of the two vectors: $\vec{P} = \vec{r}_{OB} + \vec{r}_{BP}$. The vector $\vec{r}_{OB} = |\vec{r}_{OB}|\left(\vec{i}\cos\lambda_P + 0\vec{j} + \vec{k}\sin\lambda_P\right)$. From geometry, the magnitude is $|\vec{r}_{OB}| = R_E\cos\theta_P$. The vector $\vec{r}_{BP} = |\vec{r}_{BP}|(0\vec{i} + 1\vec{j} + 0\vec{k})$. From geometry, the magnitude is $|\vec{r}_{BP}| = R_E\sin\theta_P$. Substitute and reduce to obtain:

$$\vec{P} = \vec{r}_{OB} + \vec{r}_{BP} = R_E(\vec{i}\cos\lambda_P\cos\theta_P + \vec{j}\sin\theta_P + \vec{k}\sin\lambda_P\cos\theta_P).$$

A similar argument for the point Q yields

$$\vec{Q} = \vec{r}_{OC} + \vec{r}_{CQ} = R_E\left(\vec{i}\cos\lambda_Q\cos\theta_Q + \vec{j}\sin\theta_Q + \vec{k}\sin\lambda_Q\cos\theta_Q\right)$$

Using the identity $\cos^2\beta + \sin^2\beta = 1$, the magnitudes are $|\vec{P}| = |\vec{Q}| = R_E$

The dot product is $\vec{P}\cdot\vec{Q} = R_E^2\left(\cos\left(\lambda_P - \lambda_Q\right)\cos\theta_P\cos\theta_Q + \sin\theta_P\sin\theta_Q\right)$

$$\text{Substitute : } \cos\theta = \frac{\vec{P}\cdot\vec{Q}}{|\vec{P}||\vec{Q}|} = \cos\left(\lambda_P - \lambda_Q\right)\cos\theta_P\cos\theta_Q + \sin\theta_P\sin\theta_Q$$

Substitute $\lambda_P = +30^\circ$, $\lambda_Q = -60^\circ$, $\theta_P = +45^\circ$, $\theta_Q = +20^\circ$, to obtain $\cos\theta = 0.2418$, or $\theta = 1.326$ *radians*. Thus the distance is $\boxed{d = 1.326R_E}$

Problem 2.104

Determine the cross product $\vec{U}\times\vec{V}$ of the vectors $\vec{U} = 3\vec{i} - 10\vec{j}$ and $\vec{V} = -6\vec{j} + 2\vec{k}$.

Solution: Use Eq (2.34) and expand into 2 by 2 determinants.

$$\vec{U}\times\vec{V} = \begin{vmatrix} \vec{i} & \vec{j} & \vec{k} \\ 3 & -10 & 0 \\ 0 & -6 & 2 \end{vmatrix} = \vec{i}((-10)(2) - (-6)(0)) - \vec{j}((3)(2) - (0)(0)) + \vec{k}((3)(-6) - (0)(-10))$$

$$\boxed{\vec{U}\times\vec{V} = -20\vec{i} - 6\vec{j} - 18\vec{k}}$$

Problem 2.105 Two vectors $\vec{U} = 3\vec{i} + 2\vec{j}$ and $\vec{V} = 2\vec{i} + 4\vec{j}$. (a) What is the cross product $\vec{U} \times \vec{V}$? (b) What is the cross product $\vec{V} \times \vec{U}$?

Solution: Use Eq (2.34) and expand into 2 by 2 determinants..

$$\vec{U} \times \vec{V} = \begin{vmatrix} \vec{i} & \vec{j} & \vec{k} \\ 3 & 2 & 0 \\ 2 & 4 & 0 \end{vmatrix} = \vec{i}((2)(0)-(4)(0)) - \vec{j}((3)(0)-(2)(0)) + \vec{k}((3)(4)-(2)(2)) \boxed{= 8\vec{k}}$$

$$\vec{V} \times \vec{U} = \begin{vmatrix} \vec{i} & \vec{j} & \vec{k} \\ 2 & 4 & 0 \\ 3 & 2 & 0 \end{vmatrix} = \vec{i}((4)(0)-(2)(0)) - \vec{j}((2)(0)-(3)(0)) + \vec{k}((2)(2)-(3)(4)) \boxed{= -8\vec{k}}$$

Problem 2.106 What is the cross product $\vec{r} \times \vec{F}$ of the position vector $\vec{r} = 2\vec{i} + 2\vec{j} + 2\vec{k}$ (m) and the force $\vec{F} = 20\vec{i} - 40\vec{k}$ (N)?

Solution: Use Eq (2.34) and expand into 2 by 2 determinants.

$$\vec{r} \times \vec{F} = \begin{vmatrix} \vec{i} & \vec{j} & \vec{k} \\ 2 & 2 & 2 \\ 20 & 0 & -40 \end{vmatrix} = \vec{i}((2)(-40)-(0)(2)) - \vec{j}((2)(-40)-(20)(2)) + \vec{k}((2)(0)-(2)(20))$$

$$\boxed{\vec{r} \times \vec{F} = -80\vec{i} + 120\vec{j} - 40\vec{k} \ (N\text{-}m)}$$

Problem 2.107 What is the cross product of the position vector $\vec{r} = 4\vec{i} - 12\vec{j} - 3\vec{k}$ (ft) and the force $\vec{F} = 20\vec{i} + 30\vec{j} - 10\vec{k}$ (lb)?

Solution: Use Eq (2.34) and expand into 2 by 2 determinants.

$$\vec{r} \times \vec{F} = \begin{vmatrix} \vec{i} & \vec{j} & \vec{k} \\ 4 & -12 & -3 \\ 20 & 30 & -10 \end{vmatrix} = \vec{i}((-12)(-10)-(30)(-3)) - \vec{j}((4)(-10)-(20)(-3)) + \vec{k}((4)(30)-(20)(-12))$$

$$\boxed{\vec{r} \times \vec{F} = 210\vec{i} - 20\vec{j} + 360\vec{k} \ (ft\text{-}lb)}$$

Problem 2.108 Consider the vectors $\vec{U} = 6\vec{i} - 2\vec{j} - 3\vec{k}$ and $\vec{V} = -12\vec{i} + 4\vec{j} + 6\vec{k}$. (a) Determine the cross product $\vec{U} \times \vec{V}$. (b) What can you conclude about $\vec{U}$ and $\vec{V}$ from Part (a)?

Solution: For (a) Use Eq (2.34) and expand into 2 by 2 determinants.

$$\vec{U} \times \vec{V} = \begin{vmatrix} \vec{i} & \vec{j} & \vec{k} \\ 6 & -2 & -3 \\ -12 & 4 & 6 \end{vmatrix} = \vec{i}((-2)(6)-(4)(-3)) + \vec{j}((6)(6)-(-12)(-2)) + \vec{k}((6)(4)-(-12)(-2))$$

$$\boxed{\vec{U} \times \vec{V} = 0\vec{i} + 0\vec{j} + 0\vec{k}}$$

(b) From the definition of the cross product (see Eq (2.28)) $\vec{U} \times \vec{V} = |\vec{U}||\vec{V}|\sin\theta\ \vec{e}$, where θ is the angle between the two vectors, and $\vec{e}$ is a unit vector perpendicular to both $\vec{U}$ and $\vec{V}$. If $\vec{U} \times \vec{V} = 0$ and if $|\vec{U}| \neq 0$ and $|\vec{V}| \neq 0$ then since by definition $\vec{e} \neq 0$, $\sin\theta$ must be zero: $\sin\theta = 0$, and $\theta = 0^\circ$ or $\theta = 180^\circ$, and *the two vectors are said to be parallel.* (A graphical construction confirms this interpretation.)

Problem 2.109 The vectors $\vec{U} = U_x\vec{i} - 6\vec{j} + U_z\vec{k}$ and $\vec{V} = 2\vec{i} - 3\vec{j} + \vec{k}$ are parallel. Use the cross product to determine U_x and U_z.

Solution: Use Eq (2.34) and expand into 2 by 2 determinants.

$$\vec{U}\times\vec{V} = \begin{vmatrix} \vec{i} & \vec{j} & \vec{k} \\ U_X & -6 & U_Z \\ 2 & -3 & 1 \end{vmatrix} = \vec{i}(-6+3U_Z) - \vec{j}(U_X - 2U_Z) + \vec{k}(-3U_X + 12)$$ Since each component must vanish, $(-6+3U_Z) = 0$, hence $\boxed{U_Z = 2}$. Also $(U_X - 2U_Z) = 0$, hence $\boxed{U_X = 2U_Z = 4}$

Problem 2.110 The magnitudes $|\vec{U}| = 10$ and $|\vec{V}| = 20$. (a) Use the definition of the cross product to determine $\vec{U}\times\vec{V}$. (b) Use the definition of the cross product to determine $\vec{V}\times\vec{U}$. (c) Use Eq (2.34) to determine $\vec{U}\times\vec{V}$. (d) Use Eq (2.34) to determine $\vec{V}\times\vec{U}$.

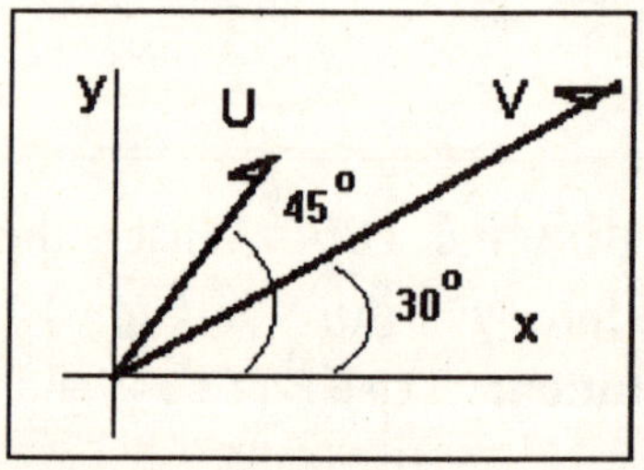

Solution: From Eq (228)) $\vec{U}\times\vec{V} = |\vec{U}||\vec{V}|\sin\theta\,\vec{e}$, From the sketch, the positive z-axis is out of the paper. For $\vec{U}\times\vec{V}$, $\vec{e} = -1\vec{k}$ (points into the paper); for $\vec{V}\times\vec{U}$, $\vec{e} = +1\vec{k}$ (points out of the paper). The angle $\theta = 15°$, hence (a) $\vec{U}\times\vec{V} = (10)(20)(0.2588)(\vec{e}) = 51.8\vec{e}$ $\boxed{= -51.8\vec{k}}$ Similarly, (b) $\vec{V}\times\vec{U} = 51.8\vec{e}$ $\boxed{= 51.8\vec{k}}$ (c) The two vectors are: $\vec{U} = 10(\vec{i}\cos 45° + \vec{j}\sin 45\ ) = 7.07\vec{i} + 0.707\vec{j}$, $\vec{V} = 20(\vec{i}\cos 30° + \vec{j}\sin 30°) = 17.32\vec{i} + 10\vec{j}$

$$\vec{U}\times\vec{V} = \begin{vmatrix} \vec{i} & \vec{j} & \vec{k} \\ 7.07 & 7.07 & 0 \\ 17.32 & 10 & 0 \end{vmatrix} = \vec{i}(0) - \vec{j}(0) + \vec{k}(70.7 - 122.45) = -\vec{k}51.8$$

(d) $$\vec{V}\times\vec{U} = \begin{vmatrix} \vec{i} & \vec{j} & \vec{k} \\ 17.32 & 10 & 0 \\ 7.07 & 7.07 & 0 \end{vmatrix} = \vec{i}(0) - \vec{j}(0) + \vec{k}(122.45 - 70.7) = 51.8\vec{k}$$

Problem 2.111 The force $\vec{F} = 10\vec{i} - 4\vec{j}\ (N)$. Determine the cross product $\vec{r}_{AB}\times\vec{F}$.

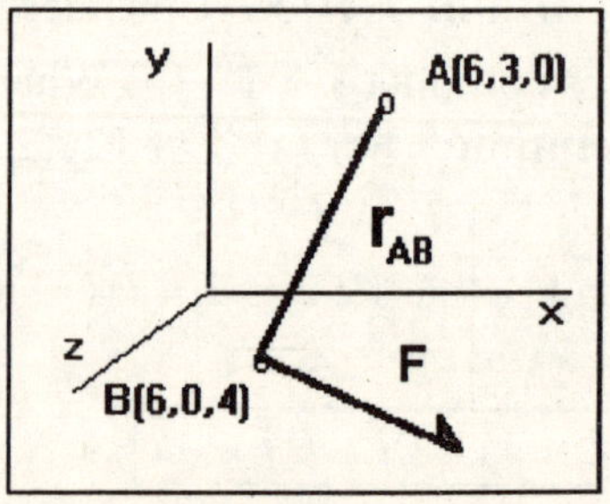

Solution: The position vector is

$\vec{r}_{AB} = (6-6)\vec{i} + (0-3)\vec{j} + (4-0)\vec{k} = 0\vec{i} - 3\vec{j} + 4\vec{k}$

The cross product: $$\vec{r}_{AB}\times\vec{F} = \begin{vmatrix} \vec{i} & \vec{j} & \vec{k} \\ 0 & -3 & 4 \\ 10 & -4 & 0 \end{vmatrix} = \vec{i}(16) - \vec{j}(-40) + \vec{k}(30)$$

$\boxed{= 16\vec{i} + 40\vec{j} + 30\vec{k}\ (N\text{-}m)}$

Problem 2.112

By evaluating the cross product, prove the identity $\sin(\theta_1 - \theta_2) = \sin\theta_1 \ \cos\theta_2 - \cos\theta_1 \ \sin\theta_2$.

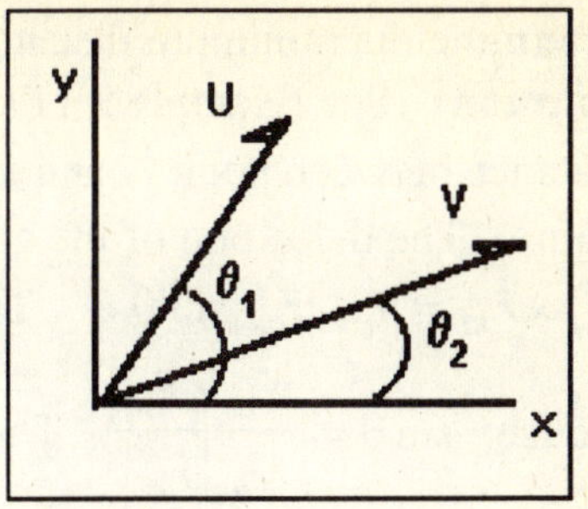

Solution: Assume that both $\vec{U}$ and $\vec{V}$ lie in the x-y plane. The strategy is to use the definition of the cross product (Eq 2.28) and the Eq (2.34), and equate the two. From Eq (2.28) $\vec{U} \times \vec{V} = |\vec{U}||\vec{V}|\sin(\theta_1 - \theta_2)\,\vec{e}$. Since the positive z-axis is out of the paper, and $\vec{e}$ points into the paper, then $\vec{e} = -\vec{k}$. Take the dot product of both sides with $\vec{e}$, and note that $\vec{k} \cdot \vec{k} = 1$. Thus

$\sin(\theta_1 - \theta_2) = -\left(\dfrac{(\vec{U} \times \vec{V}) \cdot \vec{k}}{|\vec{U}||\vec{V}|}\right)$ The vectors are: $\vec{U} = |\vec{U}|(\vec{i}\cos\theta_1 + \vec{j}\sin\theta_2)$, and $\vec{V} = |\vec{V}|(\vec{i}\cos\theta_2 + \vec{j}\sin\theta_2)$. The cross product is

$$\vec{U} \times \vec{V} = \begin{vmatrix} \vec{i} & \vec{j} & \vec{k} \\ |\vec{U}|\cos\theta_1 & |\vec{U}|\sin\theta_1 & 0 \\ |\vec{V}|\cos\theta_2 & |\vec{V}|\sin\theta_2 & 0 \end{vmatrix} = \vec{i}(0) - \vec{j}(0) + \vec{k}(|\vec{U}||\vec{V}|)(\cos\theta_1 \ \sin\theta_2 - \cos\theta_2 \ \sin\theta_1)$$

Substitute into the definition to obtain: $\boxed{\sin(\theta_1 - \theta_2) = \sin\theta_1 \ \cos\theta_2 - \cos\theta_1 \ \sin\theta_2}$ Q.E.D.

Problem 2.113

Use the cross product to determine the components of a unit vector $\vec{e}$ that is perpendicular to the vectors $\vec{U} = 3\vec{i} - 10\vec{j}$ and $\vec{V} = -6\vec{j} + 2\vec{k}$.

Solution: The cross product $\vec{R} = \vec{U} \times \vec{V} = \begin{vmatrix} \vec{i} & \vec{j} & \vec{k} \\ 3 & -10 & 0 \\ 0 & -6 & 2 \end{vmatrix} = \vec{i}(-20) - \vec{j}(6) + \vec{k}(-18)$ The magnitude is:

$|\vec{R}| = \sqrt{20^2 + 6^2 + 18^2} = 27.57$. The unit vector is $\boxed{\vec{e} = \pm\left(\dfrac{\vec{R}}{|\vec{R}|}\right) = \pm(0.7254\vec{i} + 0.2176\vec{j} + 0.6529\vec{k})}$

Problem 2.114

(a) What is the cross product $\vec{r}_{OA} \times \vec{r}_{OB}$? (b) Determine a unit vector that is perpendicular to $\vec{r}_{OA}$ and $\vec{r}_{OB}$.

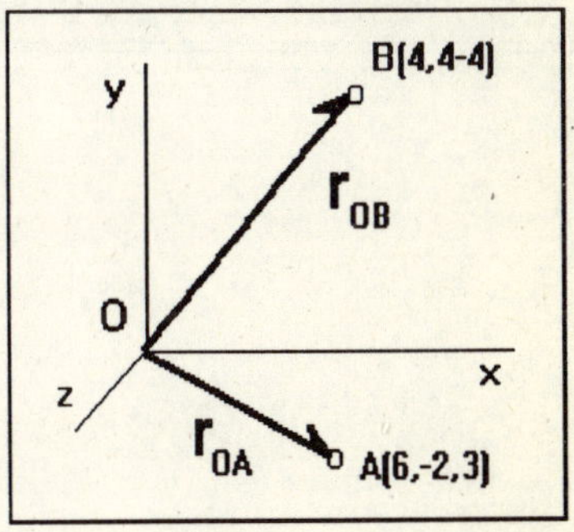

Solution: The two radius vectors are

$\vec{r}_{OB} = 4\vec{i} + 4\vec{j} - 4\vec{k}$, $\vec{r}_{OA} = 6\vec{i} - 2\vec{j} + 3\vec{k}$

(a) The cross product is

$$\vec{r}_{OA} \times \vec{r}_{OB} = \begin{vmatrix} \vec{i} & \vec{j} & \vec{k} \\ 6 & -2 & 3 \\ 4 & 4 & -4 \end{vmatrix} = \vec{i}(8-12) - \vec{j}(-24-12) + \vec{k}(24+8)$$

$= -4\vec{i} + 36\vec{j} + 32\vec{k}\ (m^2)$ The magnitude is

$\boxed{|\vec{r}_{OA} \times \vec{r}_{OB}| = \sqrt{4^2 + 36^2 + 32^2} = 48.33 \ m^2}$ (b) The unit vector is

$\boxed{\vec{e} = \pm\left(\dfrac{\vec{r}_{OA} \times \vec{r}_{OB}}{|\vec{r}_{OA} \times \vec{r}_{OB}|}\right) = \pm(-0.0828\vec{i} + 0.7448\vec{j} + 0.6621\vec{k})}$ (Two vectors.)

Problem 2.115 Consider the points O, A, and B shown in Problem 2.114. Use the cross product to determine the minimum distance from point A to the line OB.

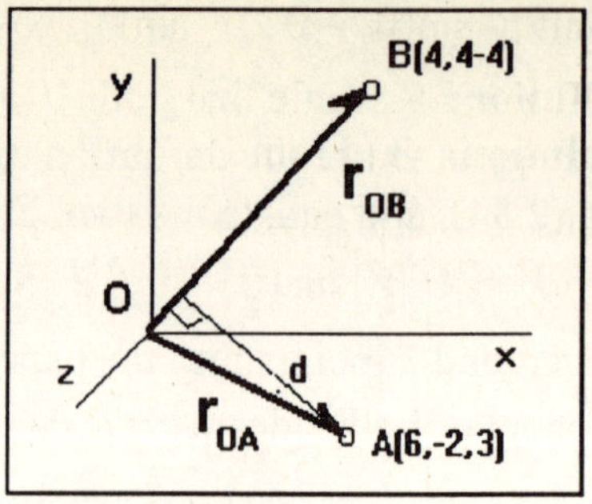

Solution: See Example 2.117. From geometry, the minimum distance between point A and the line OB is $d = |\vec{r}_{OA}|\sin\theta$. (see figure). The definition of the cross product is $\vec{r}_{OA} \times \vec{r}_{OB} = |\vec{r}_{OA}||\vec{r}_{OB}|\sin\theta\,\vec{e}$. Take the magnitude of both sides and reduce: $\sin\theta = \dfrac{|\vec{r}_{OA} \times \vec{r}_{OB}|}{|\vec{r}_{OA}||\vec{r}_{OB}|}$. Thus the minimum distance from point A to the line OB is $d = \dfrac{|\vec{r}_{OA} \times \vec{r}_{OB}|}{|\vec{r}_{OB}|}$ From Problem 2.114,

$\vec{r}_{OB} = 4\vec{i} + 4\vec{j} - 4\vec{k}$, $|\vec{r}_{OB}| = \sqrt{4^2 + 4^2 + 4^2} = 6.928$, and $|\vec{r}_{OA} \times \vec{r}_{OB}| = 48.33$, Substituting,

$$\boxed{d = \frac{48.33}{6.93} = 6.97\ m}$$

Problem 2.116 The cable exerts a 1000 lb force $\vec{F}$ on the hook at B. Determine $\vec{r}_{AB} \times \vec{F}$.

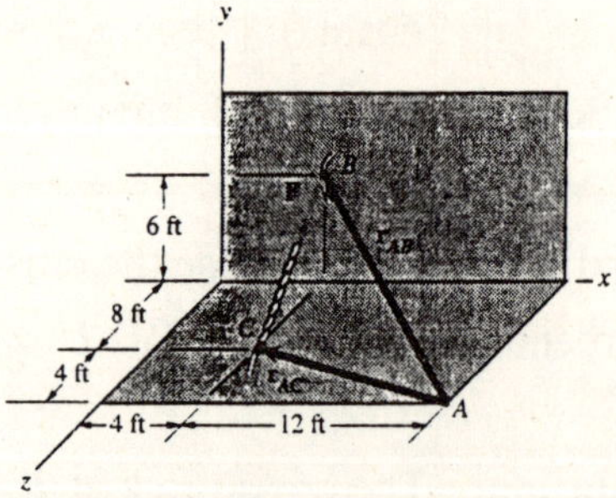

Solution: The coordinates of points A, B, and C are A(16, 0, 12, B(4, 6, 0), C(4, 0, 8). The position vectors are $\vec{r}_{OA} = 16\vec{i} + 0\vec{j} + 12\vec{k}$, $\vec{r}_{OB} = 4\vec{i} + 6\vec{j} + 0\vec{k}$, $\vec{r}_{OC} = 4\vec{i} + 0\vec{j} + 8\vec{k}$.

The force $\vec{F}$ acts along the unit vector $\vec{e}_{BC} = \dfrac{\vec{r}_{BC}}{|\vec{r}_{BC}|} = \dfrac{\vec{r}_{OC} - \vec{r}_{OB}}{|\vec{r}_{OC} - \vec{r}_{OB}|}$

Noting $\vec{r}_{OC} - \vec{r}_{OB} = (4-4)\vec{i} + (0-6)\vec{j} + (8-0)\vec{k} = 0\vec{i} - 6\vec{j} + 8\vec{k}$

$|\vec{r}_{OC} - \vec{r}_{OB}| = \sqrt{6^2 + 8^2} = 10$. Thus $\vec{e}_{BC} = 0\vec{i} - 0.6\vec{j} + 0.8\vec{k}$, and $\vec{F} = |\vec{F}|\vec{e}_{BC} = 0\vec{i} - 600\vec{j} + 800\vec{k}\ (lb)$.

The vector $\vec{r}_{AB} = (4-16)\vec{i} + (6-0)\vec{j} + (0-12)\vec{k} = -12\vec{i} + 6\vec{j} - 12\vec{k}$ Thus the cross product is

$$\vec{r}_{AB} \times \vec{F} = \begin{vmatrix} \vec{i} & \vec{j} & \vec{k} \\ -12 & 6 & -12 \\ 0 & -600 & 800 \end{vmatrix} \boxed{= -2400\vec{i} + 9600\vec{j} + 7200\vec{k}\ (ft\text{-}lb)}$$

==================================<>==================================

Problem 2.117 The cable BC shown in Problem 2.116 exerts a 300 lb force $\vec{F}$ on the hook at B. (a) Determine $\vec{r}_{AB} \times \vec{F}$ and $\vec{r}_{AC} \times \vec{F}$. (b) Use the definition of the cross product to explain the result of Part (a).

Solution: (a) From Problem 2.116, the unit vector $\vec{e}_{BC} = 0\vec{i} - 0.6\vec{j} + 0.8\vec{k}$, and $\vec{r}_{AB} = -12\vec{i} + 6\vec{j} - 12\vec{k}$. Thus $\vec{F} = |\vec{F}|\vec{e}_{BC} = 0\vec{i} - 180\vec{j} + 240\vec{k}$, and the cross product is

$$\vec{r}_{AB} \times \vec{F} = \begin{vmatrix} \vec{i} & \vec{j} & \vec{k} \\ -12 & 6 & -12 \\ 0 & -180 & 240 \end{vmatrix} \boxed{= -720\vec{i} + 2880\vec{j} + 2160\vec{k} \ (ft\text{-}lb)}$$

The vector $\vec{r}_{AC} = (4-16)\vec{i} + 0\vec{j} + (8-12)\vec{k} = -12\vec{i} + 0\vec{j} - 4\vec{k}$. Thus the cross product is

$$\vec{r}_{AC} \times \vec{F} = \begin{vmatrix} \vec{i} & \vec{j} & \vec{k} \\ -12 & 0 & -4 \\ 0 & -180 & 240 \end{vmatrix} \boxed{= -720\vec{i} + 2880\vec{j} + 2160\vec{k} \ (ft\text{-}lb)}$$

(b) The definition of the cross product is $\vec{r} \times \vec{F} = |\vec{r}||\vec{F}|\sin\theta\,\vec{e}$. Since the two cross products above are equal, $|\vec{r}_{AB}||\vec{F}|\sin\theta_1\vec{e} = |\vec{r}_{AC}||\vec{F}|\sin\theta_2\vec{e}$. Note that $\vec{r}_{AC} = \vec{r}_{AB} + \vec{r}_{BC}$ from Problem 2.116, hence $\vec{r}_{AC} \times \vec{F} = \vec{r}_{AB} \times \vec{F} + \vec{r}_{BC} \times \vec{F} = |\vec{r}_{AB}||\vec{F}|\sin\theta_1\vec{e} + |\vec{r}_{BC}||\vec{F}|\sin 0\vec{e} = |\vec{r}_{AB}||\vec{F}|\sin\theta_1\vec{e}$, since $\vec{r}_{BC}$ and $\vec{F}$ are parallel. Thus the two results are equal.

t==================================<>==================================

Problem 2.118 The bar AB is 6 m long and is perpendicular to the bars AC and AD. Use the cross product to determine the coordinates x_B, , y_B , and z_B of point B.

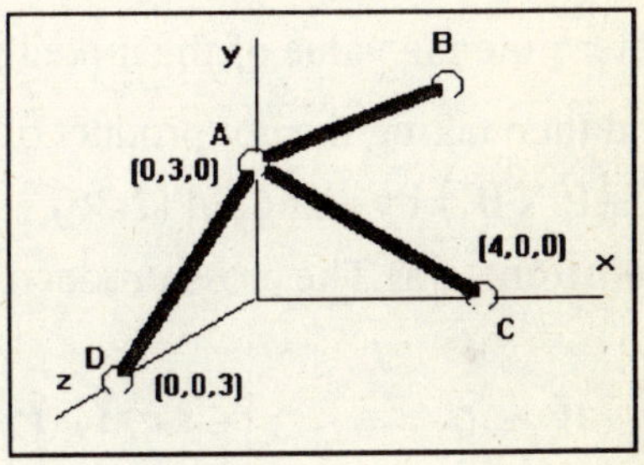

Solution: The strategy is to determine the unit vector perpendicular to both AC and AD, and then determine the coordinates that will agree with the magnitude of AB. The position vectors are:

$\vec{r}_{OA} = 0\vec{i} + 3\vec{j} + 0\vec{k}$, $\vec{r}_{OD} = 0\vec{i} + 0\vec{j} + 3\vec{k}$, and $\vec{r}_{OC} = 4\vec{i} + 0\vec{j} + 0\vec{k}$. The vectors collinear with the bars are:

$\vec{r}_{AD} = (0-0)\vec{i} + (0-3)\vec{j} + (3-0)\vec{k} = 0\vec{i} - 3\vec{j} + 3\vec{k}$,

$\vec{r}_{AC} = (4-0)\vec{i} + (0-3)\vec{j} + (0-0)\vec{k} = 4\vec{i} - 3\vec{j} + 0\vec{k}$. The vector collinear with $\vec{r}_{AB}$ is

$$\vec{R} = \vec{r}_{AD} \times \vec{r}_{AC} = \begin{vmatrix} \vec{i} & \vec{j} & \vec{k} \\ 0 & -3 & 3 \\ 4 & -3 & 0 \end{vmatrix} = 9\vec{i} + 12\vec{j} + 12\vec{k}$$ The magnitude $|\vec{R}| = 19.21\ (m)$. The unit vector is

$$\vec{e}_{AB} = \frac{\vec{R}}{|\vec{R}|} = 0.4685\vec{i} + 0.6247\vec{j} + 0.6247\vec{k}.$$ Thus the vector collinear with AB is

$\vec{r}_{AB} = 6\vec{e}_{AB} = +2.811\vec{i} + 3.75\vec{j} + 3.75\vec{k}$. Using the coordinates of point A: $\boxed{x_B = 2.81 + 0 = 2.81\ (m)}$ $\boxed{y_B = 3.75 + 3 = 6.75\ (m)}$ $\boxed{z_B = 3.75 + 0 = 3.75\ (m)}$

==================================<>==================================

Problem 2.119 Determine the minimum distance from the point P to the plane defined by the three points A, B, C.

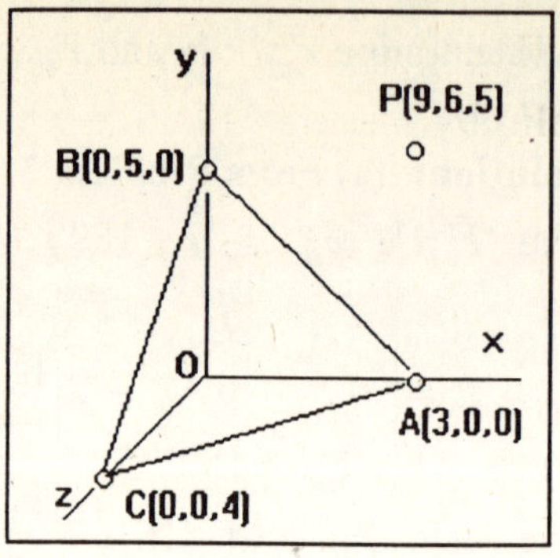

Solution: The strategy is to find the unit vector perpendicular to the plane. The projection of this unit vector on the vector OP: $\vec{r}_{OP} \cdot \vec{e}$ is the distance from the origin to P along the perpendicular to the plane. The projection on $\vec{e}$ of any vector into the plane ($\vec{r}_{OA} \cdot \vec{e}$, $\vec{r}_{OB} \cdot \vec{e}$, or $\vec{r}_{OC} \cdot \vec{e}$) is the distance from the origin to the plane along this same perpendicular. Thus the distance of P from the plane is $d = \vec{r}_{OP} \cdot \vec{e} - \vec{r}_{OA} \cdot \vec{e}$.

The position vectors are : $\vec{r}_{OA} = 3\vec{i}$, $\vec{r}_{OB} = 5\vec{j}$, $\vec{r}_{OC} = 4\vec{k}$ and $\vec{r}_{OP} = 9\vec{i} + 6\vec{j} + 5\vec{k}$. The unit vector perpendicular to the plane is found from the cross product of any two vectors lying in the plane. Noting: $\vec{r}_{BC} = \vec{r}_{OC} - \vec{r}_{OB} = -5\vec{j} + 4\vec{k}$, and $\vec{r}_{BA} = \vec{r}_{OA} - \vec{r}_{OB} = 3\vec{i} - 5\vec{j}$. The cross product: $\vec{r}_{BC} \times \vec{r}_{BA} = \begin{vmatrix} \vec{i} & \vec{j} & \vec{k} \\ 0 & -5 & 4 \\ 3 & -5 & 0 \end{vmatrix} = 20\vec{i} + 12\vec{j} + 15\vec{k}$. The magnitude is $|\vec{r}_{BC} \times \vec{r}_{BA}| = 27.73$, thus the unit vector is $\vec{e} = 0.7212\vec{i} + 0.4327\vec{j} + 0.5409\vec{k}$. The distance of point P from the plane is $\boxed{d = \vec{r}_{OP} \cdot \vec{e} - \vec{r}_{OA} \cdot \vec{e} = 11.792 - 2.164 = 9.63\ m}$ The second term is the distance of the plane from the origin; the vectors $\vec{r}_{OB}$, or $\vec{r}_{OC}$ could have been used instead of $\vec{r}_{OA}$.

Problem 2.120 Consider vectors $\vec{U} = 3\vec{i} - 10\vec{j}$, $\vec{V} = -6\vec{j} + 2\vec{k}$, and $\vec{W} = 2\vec{i} + 6\vec{j} - 4\vec{k}$. (a) determine the value of the mixed triple product $\vec{U} \cdot (\vec{V} \times \vec{W})$ by first evaluating the cross product $\vec{V} \times \vec{W}$ and then taking the dot product of the result with $\vec{U}$. (b) Determine the value of the mixed triple product $\vec{U} \cdot (\vec{V} \times \vec{W})$ by using Eq (2.36).

Solution: (a) The cross product

$$\vec{V} \times \vec{W} = \begin{vmatrix} \vec{i} & \vec{j} & \vec{k} \\ 0 & -6 & 2 \\ 2 & 6 & -4 \end{vmatrix} = (+24 - 12)\vec{i} - (0 - 4)\vec{j} + (0 + 12)\vec{k} = 12\vec{i} + 4\vec{j} + 12\vec{k}$$

Take the dot product: $\boxed{\vec{U} \cdot (\vec{V} \times \vec{W}) = (3)(12) + (4)(-10) + 0 = -4}$

(b)Eq (2.36) expresses the mixed triple product as a 3X3 determinant.

$$\vec{U} \cdot (\vec{V} \times \vec{W}) = \begin{vmatrix} 3 & -10 & 0 \\ 0 & -6 & 2 \\ 2 & 6 & -4 \end{vmatrix} = (3)(24 - 12) - (-10)(-4) + (0) \boxed{= 36 - 40 = -4}$$

Problem 2.121 For the vectors $\vec{U} = 6\vec{i} + 2\vec{j} - 4\vec{k}$, $\vec{V} = 2\vec{i} + 7\vec{j}$, and $\vec{W} = 3\vec{i} + 2\vec{k}$, evaluate the following mixed triple products: (a) $\vec{U} \cdot (\vec{V} \times \vec{W})$; (b) $\vec{W} \cdot (\vec{V} \times \vec{U})$; (c) $\vec{V} \cdot (\vec{W} \times \vec{U})$.

Solution: Use Eq (2.36).

(a) $$\vec{U} \cdot (\vec{V} \times \vec{W}) = \begin{vmatrix} 6 & 2 & -4 \\ 2 & 7 & 0 \\ 3 & 0 & 2 \end{vmatrix} = 6(14) - 2(4) + (-4)(-21) \boxed{= 160}$$

(b) $$\vec{W} \cdot (\vec{V} \times \vec{U}) = \begin{vmatrix} 3 & 0 & 2 \\ 2 & 7 & 0 \\ 6 & 2 & -4 \end{vmatrix} = 3(-28) - (0) + 2(4 - 42) \boxed{= -160}$$

(c) $$\vec{V} \cdot (\vec{W} \times \vec{U}) = \begin{vmatrix} 2 & 7 & 0 \\ 3 & 0 & 2 \\ 6 & 2 & -4 \end{vmatrix} = 2(-4) - 7(-12 - 12) + (0) \boxed{= 160}$$

Problem 2.122 By using Eqs (2.23) and (2.34) show that

$$\vec{U} \cdot (\vec{V} \times W) = \begin{vmatrix} U_X & U_Y & U_Z \\ V_X & V_Y & V_Z \\ W_X & W_Y & W_Z \end{vmatrix}$$

Solution: One strategy is to expand the determinant in terms of its components, take the dot product, and then collapse the expansion. Eq (2.23) is an expansion of the dot product: Eq (2.23): $\vec{U} \cdot \vec{V} = U_X V_X + U_Y V_Y + U_Z V_Z$. Eq (2.34) is the determinant representation of the cross product:

Eq (2.34) $$\vec{U} \times V = \begin{vmatrix} \vec{i} & \vec{j} & \vec{k} \\ U_X & U_Y & U_Z \\ V_X & V_Y & V_Z \end{vmatrix}$$

For notational convenience, write $\vec{P} = (\vec{U} \times \vec{V})$ Expand the determinant about its first row:

$$\vec{P} = \vec{i} \begin{vmatrix} U_Y & U_Z \\ V_Y & V_Z \end{vmatrix} - \vec{j} \begin{vmatrix} U_X & U_Z \\ V_X & V_Z \end{vmatrix} + \vec{k} \begin{vmatrix} U_X & U_Z \\ V_X & V_Z \end{vmatrix}$$

Since the two-by-two determinants are scalars, this can be written in the form: $\vec{P} = \vec{i} P_X + \vec{j} P_Y + \vec{k} P_Z$ where the scalars P_X, P_Y, and P_Z are the two-by-two determinants. Apply Eq (2.23) to the dot product of a vector $\vec{Q}$ with $\vec{P}$. Thus $\vec{Q} \cdot \vec{P} = Q_X P_X + Q_Y P_Y + Q_Z P_Z$ Substitute P_X, P_Y, and P_Z into this dot product

$$\vec{Q} \cdot \vec{P} = Q_X \begin{vmatrix} U_Y & U_Z \\ V_Y & V_Z \end{vmatrix} - Q_Y \begin{vmatrix} U_X & U_Z \\ V_X & V_Z \end{vmatrix} + Q_z \begin{vmatrix} U_X & U_Z \\ V_X & V_Z \end{vmatrix}$$

But this expression can be collapsed into a three-by-three determinant directly, thus:

$$\vec{Q} \cdot (\vec{U} \times \vec{V}) = \begin{vmatrix} Q_X & Q_Y & Q_Z \\ U_X & U_Y & U_Z \\ V_X & V_Y & V_Z \end{vmatrix}$$

This completes the demonstration.

=================================<>=================================

Problem 2.123 (a) By using the definition of the dot product and the cross product, show why $\vec{U}\cdot(\vec{U}\times\vec{V})=0$ for any two vectors $\vec{U}$ and $\vec{V}$. (b) Use Eq (2.36) to show that $\vec{U}\cdot(\vec{U}\times\vec{V})=0$ for any two vectors $\vec{U}$ and $\vec{V}$

Solution: (a) The definition of the dot product is $\vec{U}\cdot\vec{V}=|\vec{U}||\vec{V}|\cos\theta$, where θ is the angle between the vectors $\vec{U}$ and $\vec{V}$. Thus if the angle is any odd multiple of 90°, the dot product vanishes.

The definition of the cross product is $\vec{U}\times\vec{V}=|\vec{U}||\vec{V}|\sin\theta\,\vec{e}$, where as above, θ is the angle between the vectors, and $\vec{e}$ is a unit vector which is perpendicular to both $\vec{U}$ and $\vec{V}$.

Consider the mixed triple product: $\vec{U}\cdot(\vec{U}\times\vec{V})$. The cross product is a vector at right angles to either of its vectors, hence it is normal to $\vec{U}$. The dot product of $\vec{U}$ with any vector that is normal to $\vec{U}$ vanishes by definition of the dot product. Hence . $\vec{U}\cdot(\vec{U}\times\vec{V})=0$, always.

(b) The Eq (2.36) is the determinant representation of the mixed triple product:

$$\vec{U}\cdot(\vec{V}\times W)=\begin{vmatrix} U_X & U_Y & U_Z \\ V_X & V_Y & V_Z \\ W_X & W_Y & W_Z \end{vmatrix}$$

If any two rows (or any two columns) in a determinant are scalar multiples of one another, for example: if $\vec{V}=\alpha\vec{U}$, where α is any non-zero scalar, then the determinant vanishes. This is illustrated as follows for the case $\vec{V}=\alpha\vec{U}$, which is of interest here. Expand the determinant about its first row:

$$\vec{U}\cdot(\vec{V}\times W)=\begin{vmatrix} U_X & U_Y & U_Z \\ \alpha U_X & \alpha U_Y & \alpha U_Z \\ W_X & W_Y & W_Z \end{vmatrix}=U_X(\alpha U_Y W_Z-W_Y\alpha U_Z)-U_Y(\alpha U_X W_Z-W_X\alpha U_Z)+U_Z(\alpha U_X W_Y-W_X\alpha U_Y)$$

rearranging and collecting terms two-by-two,

$$=(\alpha U_X U_Y W_Z-\alpha U_X U_Y W_Z)-(\alpha U_X U_Z W_y-\alpha U_X U_Z W_Y)+(\alpha U_Y U_Z W_X-\alpha U_Z U_Y W_X)=0$$

since by inspection the terms cancel two-by-two.

Since α includes the case $\alpha=1$, then it follows that $\vec{U}\cdot(\vec{U}\times\vec{V})=0$, always.

=================================<>=================================

Problem 2.124 The vectors $\vec{U}=\vec{i}+U_Y\vec{j}+4\vec{k}$, $\vec{V}=2\vec{i}+\vec{j}-2\vec{k}$, and $\vec{W}=-3\vec{i}+\vec{j}-2\vec{k}$ are coplanar (they lie in the same plane) What is the component U_Y ?

Solution: Since the non-zero vectors are coplanar, the cross product of any two will produce a vector perpendicular to the plane, and the dot product with the third will vanish, by definition of the dot product. Thus $\vec{U}\cdot(\vec{V}\times\vec{W})=0$, for example.

$$\vec{U}\cdot(\vec{V}\times\vec{W})=\begin{vmatrix} 1 & U_Y & 4 \\ 2 & 1 & -2 \\ -3 & 1 & -2 \end{vmatrix}=1(-2+2)-(U_Y)(-4-6)+(4)(2+3)=+10U_Y+20=0$$

Thus $\boxed{U_Y=-2}$

=================================<>=================================

Problem 2.125 The magnitude of $\vec{F}$ is 8 kN. Express $\vec{F}$ in terms of scalar components.

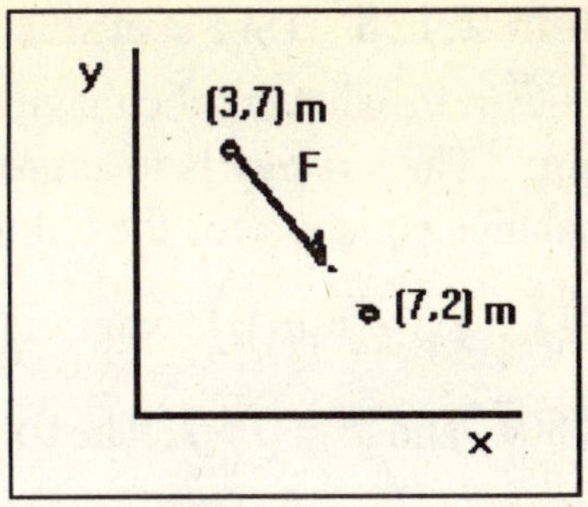

Solution: The unit vector collinear with the force $\vec{F}$ is developed as follows: The collinear vector is

$\vec{r} = (7-3)\vec{i} + (2-7)\vec{j} = 4\vec{i} - 5\vec{j}$

The magnitude: $|\vec{r}| = \sqrt{4^2 + 5^2} = 6.403\ m$. The unit vector is

$\vec{e} = \frac{\vec{r}}{|\vec{r}|} = 0.6247\vec{i} - 0.7809\vec{j}$. The force vector is

$\boxed{\vec{F} = |\vec{F}|\vec{e} = 4.997\vec{i} - 6.247\vec{j} = 5\vec{i} - 6.25\vec{j}\ \ (kN)}$

Problem 2.126 The magnitude of the vertical force $\vec{W}$ is 600 lb and the magnitude of the force $\vec{B}$ is 1500 lb. Given that $\vec{A} + \vec{B} + \vec{W} = 0$, determine the magnitude of the force $\vec{A}$ and the angle α.

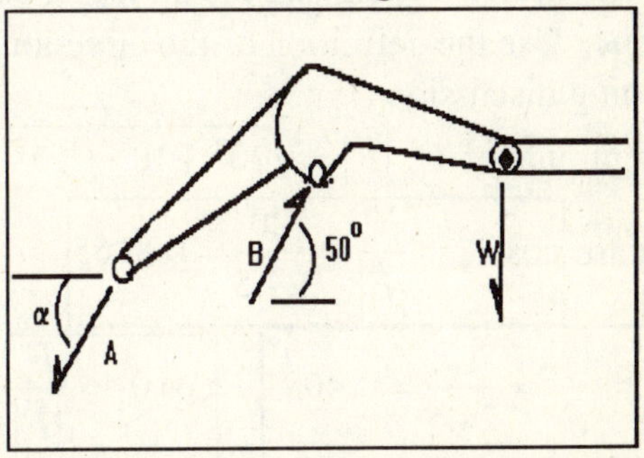

Solution: The strategy is to use the condition of force balance to determine the unknowns. The weight vector is $\vec{W} = -600\vec{j}$ The vector $\vec{B}$ is

$\vec{B} = 1500(\vec{i}\cos 50^o + \vec{j}\sin 50^o) = 964.2\vec{i} + 1149.1\vec{j}$

The vector $\vec{A}$ is

$\vec{A} = |\vec{A}|(\vec{i}\cos(180+\alpha) + \vec{j}\sin(180+\alpha))$

$\vec{A} = |\vec{A}|(-\vec{i}\cos\alpha - \vec{j}\sin\alpha)$. The forces balance, hence $\vec{A} + \vec{B} + \vec{W} = 0$, or $(964.2 - |\vec{A}|\cos\alpha)\vec{i} = 0$, and $(1149.1 - 600 - |\vec{A}|\sin\alpha)\vec{j} = 0$. Thus $|\vec{A}|\cos\alpha = 964.2$, and $|\vec{A}|\sin\alpha = 549.1$. Take the ratio of the two equations to obtain $\tan\alpha = 0.5695$, or $\boxed{\alpha = 29.7^o}$ Substitute this angle to solve : $\boxed{|\vec{A}| = 1110\ lb}$

Problem 2.127 The magnitude of the vertical force vector $\vec{A}$ is 200 lb. If $\vec{A} + \vec{B} + \vec{C} = 0$, what are the magnitudes of the force vectors $\vec{B}$ and $\vec{C}$?

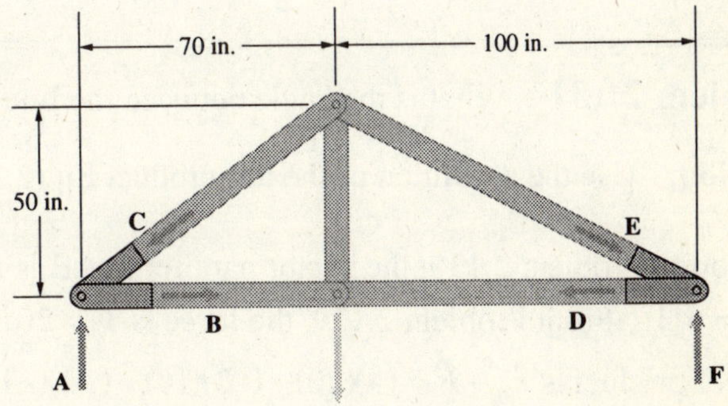

Solution: The strategy is to express the forces in terms of scalar components, and then solve the force balance equations for the unknowns. $\vec{C} = |\vec{C}|(-\vec{i}\cos\alpha - \vec{j}\sin\alpha)$, where $\tan\alpha = \frac{50}{70} = 0.7143$, or $\alpha = 35.5^o$. Thus $\vec{C} = |\vec{C}|(-0.8137\vec{i} - 0.5812\vec{j})$. Similarly, $\vec{B} = +|\vec{B}|\vec{i}$, and $\vec{A} = +200\vec{j}$. The force balance equation is $\vec{A} + \vec{B} + \vec{C} = 0$. Substituting, $(-0.8137|\vec{C}| + |\vec{B}|)\vec{i} = 0$, and $(-0.5812|\vec{C}| + 200)\vec{j} = 0$ Solving, $\boxed{|\vec{C}| = 344.1\ lb}$, $\boxed{|\vec{B}| = 280\ lb}$

Problem 2.128 (See sketch in Problem 2.127) The magnitude of the force $\vec{D}$ is 280 lb. If $\vec{D}+\vec{E}+\vec{F}=0$, what are the magnitudes of the force vectors $\vec{E}$ and $\vec{F}$?

Solution: The strategy is to express the force vectors in terms of scalar components, and then solve the force balance equation for the unknowns. The force vectors are:

$\vec{E}=|\vec{E}|(\vec{i}\cos\beta-\vec{j}\sin\beta)$, where $\tan\beta=\dfrac{50}{100}=0.5$, or $\beta=26.6°$. Thus $\vec{E}=|\vec{E}|(0.8944\vec{i}-0.4472\vec{j})$

$\vec{D}=-280\vec{i}$, and $\vec{F}=|\vec{F}|\vec{j}$. The force balance equation is $\vec{D}+\vec{E}+\vec{F}=0$. Substitute and resolve into two equations: $(0.8944|\vec{E}|-280)\vec{i}=0$, and $(-0.4472|\vec{E}|+|\vec{F}|)\vec{j}=0$. Solve: $\boxed{|\vec{E}|=313.1\ lb}$,

$\boxed{|\vec{F}|=140\ lb}$

Problem 2.129 What are the direction cosines of $\vec{F}$?

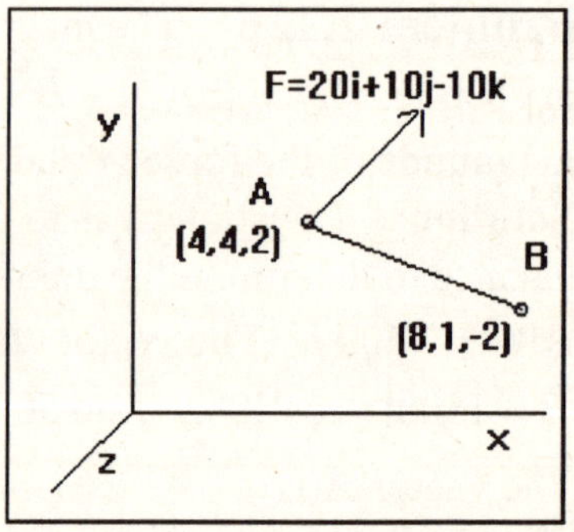

Solution: Use the definition of the direction cosines in Eq (2.15) and the ensuing discussion.

The magnitude of $\vec{F}$: $|\vec{F}|=\sqrt{20^2+10^2+10^2}=24.5$. The direction cosines are $\boxed{\cos\theta_x=\dfrac{F_x}{|\vec{F}|}=\dfrac{20}{24.5}=0.8165}$,

$\boxed{\cos\theta_y=\dfrac{F_y}{|\vec{F}|}=\dfrac{10}{24.5}=0.4081}$, $\boxed{\cos\theta_z=\dfrac{F_z}{|\vec{F}|}=\dfrac{-10}{24.5}=-0.4082}$

Problem 2.130 Determine the scalar components of a unit vector parallel to AB that points from A to B.

Solution: Use the definition of the unit vector in Eq (2.16) and the ensuing discussion.

The position vectors are: $\vec{r}_A=4\vec{i}+4\vec{j}+2\vec{k}$, $\vec{r}_B=8\vec{i}+1\vec{j}-2\vec{k}$. The vector from A to B is $\vec{r}_{AB}=(8-4)\vec{i}+(1-4)\vec{j}+(-2-2)\vec{k}=4\vec{i}-3\vec{j}-4\vec{k}$. The magnitude: $|\vec{r}_{AB}|=\sqrt{4^2+3^2+4^2}=6.4$. The unit vector is $\vec{e}_{AB}=\dfrac{\vec{r}_{AB}}{|\vec{r}_{AB}|}=\dfrac{4}{6.4}\vec{i}-\dfrac{3}{6.4}\vec{j}-\dfrac{4}{6.4}\vec{k}$ $\boxed{=0.6247\vec{i}-0.4688\vec{j}-0.6247\vec{k}}$

Problem 2.131 What is the angle between the line AB and the force $\vec{F}$?

Solution: Use the definition of the dot product Eq (2.18), and Eq (2.24): $\cos\theta=\dfrac{\vec{r}_{AB}\cdot\vec{F}}{|\vec{r}_{AB}||\vec{F}|}$. From the solution to Problem 2.130, the vector parallel to AB is $\vec{r}_{AB}=4\vec{i}-3\vec{j}-4\vec{k}$, with a magnitude $|\vec{r}_{AB}|=6.4$.. From Problem 2.129, the force is $\vec{F}=20\vec{i}+10\vec{j}-10\vec{k}$, with a magnitude of $|\vec{F}|=24.5$. The dot product is $\vec{r}_{AB}\cdot\vec{F}=(4)(20)+(-3)(10)+(-4)(-10)=90$. Substituting,

$\cos\theta=\dfrac{90}{(6.4)(24.5)}=0.574$, $\boxed{\theta=55°}$

Problem 2.132 Determine the vector component of $\vec{F}$ that is parallel to the line AB.

Solution: Use the definition in Eq (2.26): $\vec{U}_P = (\vec{e}\cdot\vec{U})\vec{e}$, where $\vec{e}$ is parallel to a line L. From Problem 2.130 the unit vector parallel to line AB is $\vec{e}_{AB} = 0.6247\vec{i} - 0.4688\vec{j} - 0.6247\vec{k}$. The dot product is $\vec{e}\cdot\vec{F} = (0.6247)(20) + (-0.4688)(10) + (-0.6247)(-10) = 14.053$. The parallel vector is

$$\boxed{(\vec{e}\cdot\vec{F})\vec{e} = (14.053)\vec{e} = 8.78\vec{i} - 6.59\vec{j} - 8.78\vec{k}\ (lb)}$$

Problem 2.133 Determine the vector component of $\vec{F}$ that is normal to the line AB.

Solution: Use the Eq (2.27) and the solution to Problem 2.132.

$$\vec{F}_N = \vec{F} - \vec{F}_P = (20-8.78)\vec{i} + (10+6.59)\vec{j} + (-10+8.78)\vec{k}\ \boxed{= 11.22\vec{i} + 16.59\vec{j} - 1.22\vec{k}\ (lb)}$$

Problem 2.134 Determine the vector $\vec{r}_{BA}\times\vec{F}$ where $\vec{r}_{BA}$ is the position vector from B to A.

Solution: Use the definition in Eq (2.34). Noting $\vec{r}_{BA} = -\vec{r}_{AB}$, from Problem 2.130 $\vec{r}_{BA} = -4\vec{i} + 3\vec{j} + 4\vec{k}$. The cross product is

$$\vec{r}_{BA}\times\vec{F} = \begin{vmatrix} \vec{i} & \vec{j} & \vec{k} \\ -4 & 3 & 4 \\ 20 & 10 & -10 \end{vmatrix} = (-30-40)\vec{i} - (40-80)\vec{j} + (-40-60)\ \boxed{= -70\vec{i} + 40\vec{j} - 100\vec{k}\ (ft\text{-}lb)}$$

Problem 2.135 An engineer calculates that the magnitude of the axial force in one of the beams of a geodesic dome is $|\vec{P}| = 7.65\ kN$. The cartesian coordinates of the endpoints A and B of the straight beam are (-12.4, 22.0, -18.4) m and (-9.2, 24.4, -15.6) m, respectively. Express the force $\vec{P}$ in terms of scalar components.

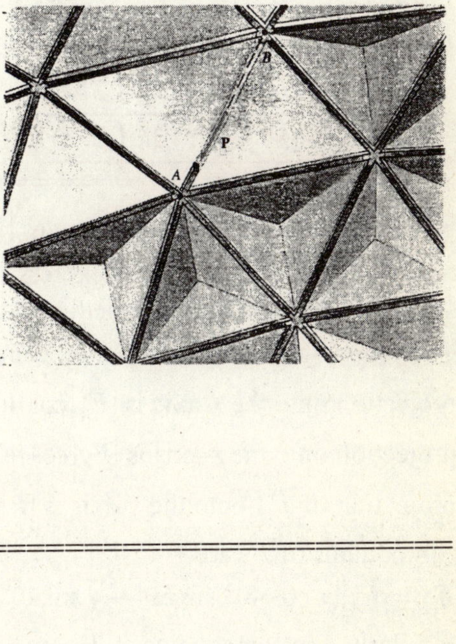

Solution: The strategy is to determine the unit vector pointing from B to A, and then express the force in terms of this unit vector.

$$\vec{r}_{BA} = (-12.4+9.2)\vec{i} + (22.0-24.4)\vec{j} + (-18.4+15.6)\vec{k}$$

$\vec{r}_{BA} = -3.2\vec{i} - 2.4\vec{j} - 2.8\vec{k}$. The magnitude is $|\vec{r}_{BA}| = \sqrt{3.2^2 + 2.4^2 + 2.8^2} = 4.883$. The unit vector is

$\vec{e}_{BA} = \dfrac{\vec{r}_{BA}}{|\vec{r}_{BA}|} = -0.6554\vec{i} - 0.4915\vec{j} - 0.5734\vec{k}$. The force vector is

$$\boxed{\vec{P} = |\vec{P}|\vec{e}_{BA} = 7.65\vec{e}_{BA} = -5.01\vec{i} - 3.76\vec{j} - 4.39\vec{k}\ (kN)}$$

Problem 2.136 The rope exerts a force of magnitude $|\vec{F}| = 200\ lb$ on the top of the pole at B.

(a) Determine the vector $\vec{r}_{AB} \times \vec{F}$ where $\vec{r}_{AB}$ is the position vector from A to B. (b) Determine the vector $\vec{r}_{AC} \times \vec{F}$, where $\vec{r}_{AC}$ is the position vector from A to C.

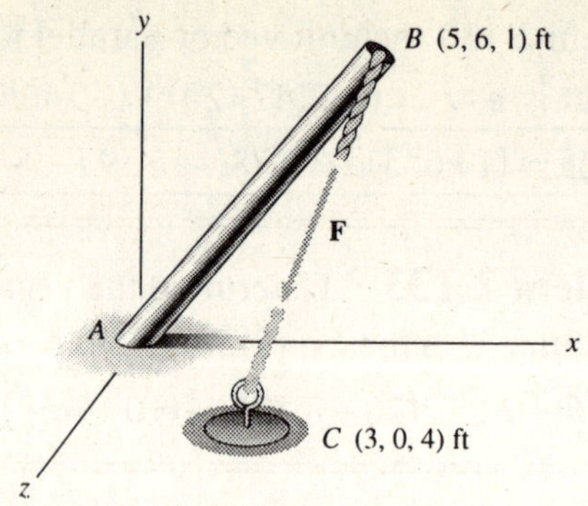

Solution: The strategy is to define the unit vector pointing from B to A, express the force in terms of this unit vector, and take the cross product of the position vectors with this force. The position vectors

$\vec{r}_{AB} = 5\vec{i} + 6\vec{j} + 1\vec{k}$, $\vec{r}_{AC} = 3\vec{i} + 0\vec{j} + 4\vec{k}$,

$\vec{r}_{BC} = (3-5)\vec{i} + (0-6)\vec{j} + (4-1)\vec{k} = -2\vec{i} - 6\vec{j} + 3\vec{k}$.

The magnitude $|\vec{r}_{BC}| = \sqrt{2^2 + 6^2 + 3^2} = 7$. The unit vector is $\vec{e}_{BC} = \dfrac{\vec{r}_{BC}}{|\vec{r}_{BC}|} = -0.2857\vec{i} - 0.8571\vec{j} + 0.4286\vec{k}$. The force vector is $\vec{F} = |\vec{F}|\vec{e}_{BC} = 200\vec{e}_{BC} = -57.14\vec{i} - 171.42\vec{j} + 85.72\vec{k}$. The cross products:

$$\vec{r}_{AB} \times \vec{F} = \begin{vmatrix} \vec{i} & \vec{j} & \vec{k} \\ 5 & 6 & 1 \\ -57.14 & -171.42 & 85.72 \end{vmatrix} = 685.74\vec{i} - 485.74\vec{j} - 514.26\vec{k}$$

$$\boxed{= 685.7\vec{i} - 485.7\vec{j} - 514.3\vec{k}\ (ft\text{-}lb)}$$

$$\vec{r}_{AC} \times \vec{F} = \begin{vmatrix} \vec{i} & \vec{j} & \vec{k} \\ 3 & 0 & 4 \\ -57.14 & -171.42 & 85.72 \end{vmatrix} = 685.68\vec{i} - 485.72\vec{j} - 514.26\vec{k}$$

$$\boxed{= 685.7\vec{i} - 485.7\vec{j} - 514.3\vec{k}\ (ft\text{-}lb)}$$

Problem 2.137 The magnitudes $|\vec{F}_A| = 600\ N$ and $|\vec{F}_B| = 400\ N$. Determine $\vec{F}_A + \vec{F}_B$.

Solution: The strategy is to express the forces in terms of scalar components and add them vectorially.

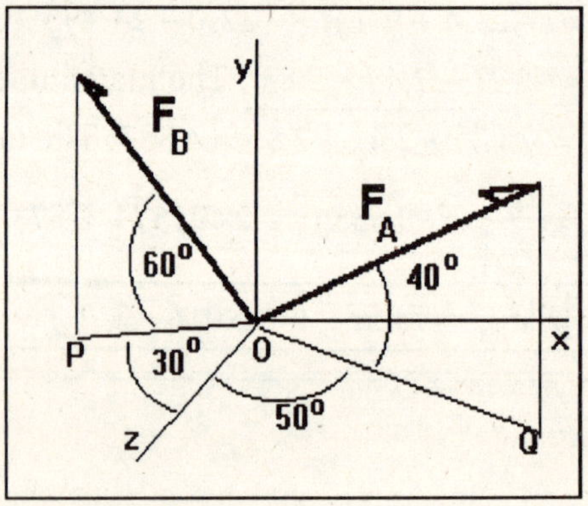

The projection of $\vec{F}_A$ onto the x-z plane is $|\vec{F}_A|\cos 40^o$. The projection onto the x-axis is $|\vec{F}_A|\cos 40^o \sin 50^o$ The projection onto the z-axis is $|\vec{F}_A|\cos 40^o \cos 50^o$ The projection of $\vec{F}_A$ onto the y-axis is $|\vec{F}|\sin 40^o$. Thus the components of $\vec{F}_A$ are

$\vec{F}_A = |\vec{F}_A|(\vec{i}\cos 40^o \sin 50^o + \vec{j}\sin 40^o + \vec{k}\cos 40^o \cos 50^o)$

Similarly, the components of $\vec{F}_B$ are

$\vec{F}_B = |\vec{F}_B|(-\vec{i}\cos 60^o \sin 30^o + \vec{j}\sin 60^o + \vec{k}\cos 60^o \cos 30^o)$ Substituting values:

$\vec{F}_A = 352.1\vec{i} + 385.7\vec{j} + 295.4\vec{k}$

$\vec{F}_B = -100\vec{i} + 346.4\vec{j} + 173.2\vec{k}$ From which $\boxed{\vec{F}_A + \vec{F}_B = 252.1\vec{i} + 731.8\vec{j} + 468.6\vec{k}\ (N)}$

Problem 2.138 (See Problem 2.137) The forces $\vec{F}_A$ and $\vec{F}_B$ have the same magnitude, and $\vec{F}_A \cdot \vec{F}_B = 600\ N^2$. What are $\vec{F}_A$ and $\vec{F}_B$?

Solution: From Problem 2.137, the forces are:

$\vec{F}_A = |\vec{F}_A|(\vec{i}\cos 40^o \sin 50^o + \vec{j}\sin 40^o + \vec{k}\cos 40^o \cos 50^o) = |\vec{F}_A|(0.5868\vec{i} + 0.6428\vec{j} + 0.4924\vec{k})$

$\vec{F}_B = |\vec{F}_B|(-\vec{i}\cos 60^o \sin 30^o + \vec{j}\sin 60^o + \vec{k}\cos 60^o \cos 30^o) = |\vec{F}_B|(-0.25\vec{i} + 0.866\vec{j} + 0.433\vec{k})$

The dot product: $\vec{F}_A \cdot \vec{F}_B = |\vec{F}_A||\vec{F}_B|(0.6233) = 600\ N^2$, from $|\vec{F}_A| = |\vec{F}_B| = \sqrt{\frac{600}{0.6233}} = 31.03\ N$,

and $\boxed{\vec{F}_A = 18.21\vec{i} + 19.95\vec{j} + 15.28\vec{k}\ (N)}$ $\boxed{\vec{F}_B = -7.76\vec{i} + 26.87\vec{j} + 13.44\vec{k}\ (N)}$

Problem 2.139 The magnitude of the force vector $\vec{F}_B$ is 2 kN. Express it in terms of scalar components.

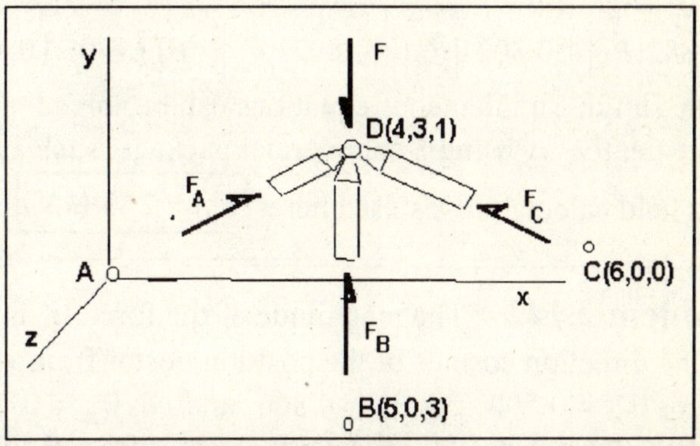

Solution: The strategy is to determine the unit vector collinear with $\vec{F}_B$ and then express the force in terms of this unit vector.

The radius vector collinear with $\vec{F}_B$ is

$\vec{r}_{BD} = (4-5)\vec{i} + (3-0)\vec{j} + (1-3)\vec{k}$

$\vec{r}_{BD} = -1\vec{i} + 3\vec{j} - 2\vec{k}$. The magnitude is

$|\vec{r}_{BD}| = \sqrt{1^2 + 3^2 + 2^2} = 3.74$. The unit vector is

$\vec{e}_{BD} = \frac{\vec{r}_{BD}}{|\vec{r}_{BD}|} = -0.2673\vec{i} + 0.8018\vec{j} - 0.5345\vec{k}$

The force is $\vec{F}_B = |\vec{F}_B|\vec{e}_{BD} = 2\vec{e}_{BD}\ (kN)$ $\vec{F}_B = -0.5345\vec{i} + 1.6036\vec{j} - 1.0693\vec{k}$

$\boxed{= -0.53\vec{i} + 1.60\vec{j} - 1.07\vec{k}\ (kN)}$

Problem 2.140 (See Problem 2.139.) The magnitude of the vertical force vector $\vec{F}$ is 6 kN. Determine the vector components of the force $\vec{F}$ parallel to and normal to the line from B to D.

Solution: The projection of the force $\vec{F}$ onto the line from B to D is $\vec{F}_P = (\vec{F} \cdot \vec{e}_{BD})\vec{e}_{BD}$. The vertical force has the component $\vec{F} = -6\vec{j}\ (kN)$. From Problem 2.139, the unit vector pointing from B to D is $\vec{e}_{BD} = -0.2673\vec{i} + 0.8018\vec{j} - 0.5345\vec{k}$ .The dot product is $\vec{F} \cdot \vec{e}_{BD} = -4.813$. Thus the component parallel to the line BD is $\boxed{\vec{F}_P = -4.813\vec{e}_{BD} = +1.29\vec{i} - 3.86\vec{j} + 2.57\vec{k}\ (kN)}$ The component perpendicular to the line is : $\vec{F}_N = \vec{F} - \vec{F}_P$. Thus $\vec{F}_N = -1.29\vec{i} - 2.14\vec{j} - 2.57\vec{k}\ (kN)$

Problem 2.141 (See Problem 2.139.) The magnitude of the force vector $\vec{F}$ is 6 kN. Given that $\vec{F}+\vec{F}_A+\vec{F}_B+\vec{F}_C=0$, what are the magnitudes of $\vec{F}_A$, $\vec{F}_B$, and $\vec{F}_C$?

Solution: The strategy is to expand the forces into scalar components, and then use the force balance equation to solve for the unknowns. The unit vectors are used to expand the forces into scalar components. The position vectors, magnitudes, and unit vectors are:

$\vec{r}_{AD}=4\vec{i}+3\vec{j}+1\vec{k}$, $|\vec{r}_{AD}|=\sqrt{26}=5.1$, $\vec{e}_{AD}=0.7845\vec{i}+0.5883\vec{j}+0.1961\vec{k}$.

$\vec{r}_{BD}=-1\vec{i}+3\vec{j}-2\vec{k}$, $|\vec{r}_{BD}|=\sqrt{14}=3.74$, $\vec{e}_{BD}=-0.2673\vec{i}+0.8018\vec{j}-0.5345\vec{k}$.

$\vec{r}_{CD}=-2\vec{i}+3\vec{j}+1\vec{k}$, $|\vec{r}_{CD}|=\sqrt{14}=3.74$, $\vec{e}_{CD}=-0.5345\vec{i}+0.8018\vec{j}+0.2673\vec{k}$

The forces are: $\vec{F}_A=|\vec{F}_A|\vec{e}_{AD}$, $\vec{F}_B=|\vec{F}_B|\vec{e}_{BD}$, $\vec{F}_C=|\vec{F}_C|\vec{e}_{CD}$, $\vec{F}=-6\vec{j}\ (kN)$. Substituting into the force balance equation $\vec{F}+\vec{F}_A+\vec{F}_B+\vec{F}_C=0$, $\left(0.7843|\vec{F}_A|-0.2674|\vec{F}_B|-0.5348|\vec{F}_C|\right)\vec{i}=0$

$\left(0.5882|\vec{F}_A|+0.8021|\vec{F}_B|+0.8021|\vec{F}_C|-6\right)\vec{j}=0$ $\left(0.1961|\vec{F}_A|-0.5348|\vec{F}_B|+0.2674|\vec{F}_C|\right)\vec{k}=0$

These simple simultaneous equations can be solved a standard method (e.g., Gauss elimination) or, conveniently, by using a commercial package, such as TK Solver®, Mathcad®, or other. An HP-28S hand held calculator was used here: $\boxed{|\vec{F}_A|=2.83\ (kN)}$, $\boxed{|\vec{F}_B|=2.49\ (kN)}$, $\boxed{|\vec{F}_C|=2.91\ (kN)}$

Problem 2.142 The magnitude of the force $\vec{W}$ is 160 N. The direction cosines of the position vector from A to B are $\cos\theta_X=0.500$, $\cos\theta_Y=0.866$, and $\cos\theta_Z=0$, and the direction cosines for the position vector from B to C are $\cos\theta_X=0.707$, $\cos\theta_Y=0.619$, and $\cos\theta_Z=-0.342$. Point G is the midpoint of the line from B to C. Determine the vector $\vec{r}_{AG}\times\vec{W}$, where $\vec{r}_{AG}$ is the position vector from A to G.

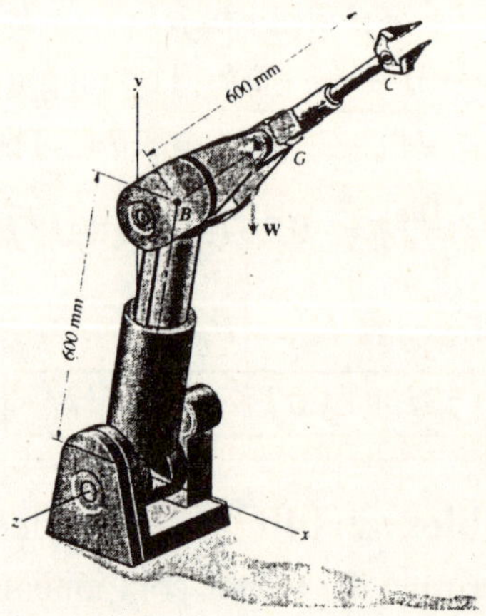

Solution: Express the position vectors in terms of scalar components, calculate $\vec{r}_{AG}$, and take the cross product. The position vectors are: $\vec{r}_{AB}=0.6(.5\vec{i}+0.866\vec{j}+0\vec{k})$

$\vec{r}_{AB}=0.3\vec{i}+0.5196\vec{j}+0\vec{k}$, $\vec{r}_{BG}=0.3(0.707\vec{i}+0.619\vec{j}-0.342\vec{k})$,

$\vec{r}_{BG}=0.2121\vec{i}+0.1857\vec{j}-0.1026\vec{k}$. $\vec{r}_{AG}=\vec{r}_{AB}+\vec{r}_{BG}=0.5121\vec{i}+0.7053\vec{j}-0.1026\vec{k}$. $\vec{W}=-160\vec{j}$

$$\vec{r}_{AG}\times\vec{W}=\begin{vmatrix}\vec{i} & \vec{j} & \vec{k}\\ 0.5121 & 0.7053 & -0.1026\\ 0 & -160 & 0\end{vmatrix}=-16.44\vec{i}+0\vec{j}-81.95\vec{k}\quad \boxed{=-16.4\vec{i}+0\vec{j}-82\vec{k}\ (N\,m)}$$

================================<>================================

Problem 3.1 An object in equilibrium is acted upon by three external forces: $\vec{F}_1 = 20\vec{i} - 30\vec{j}$ N, $\vec{F}_2 = 10\vec{i} - 10\vec{j}$ N and another unknown force $\vec{F}_3$ What is the magnitude of the force vector $\vec{F}_3$?
Solution: Resolve the forces into scalar components, and solve the equilibrium equations. By definition, for forces in equilibrium, the forces vanish. Thus $\sum \vec{F} = \vec{F}_1 + \vec{F}_2 + \vec{F}_3 = 0$. Write for the unknown force: $\vec{F}_3 = f_x\vec{i} + f_y\vec{j}$ Substitute and collect like terms in $\vec{i}$,$\vec{j}$:
$\sum \vec{F}_x = (20 + 10 + f_x)\vec{i} = 0$, $\sum \vec{F}_y = (-30 - 10 + f_y)\vec{j} = 0$. Solve for the unknowns $f_x = -30$ N, $f_y = +40$ N. The magnitude of $\vec{F}_3$ is $|\vec{F}_3| = \sqrt{(f_x^2 + f_y^2)} = \sqrt{((-30)^2 + (40)^2)} = \sqrt{2500}$ $\boxed{= 50\ N}$

================================<>================================

Problem 3.2 An object in equilibrium is acted upon by three external forces:
$\vec{F}_1 = 40\vec{i} + 30\vec{j}$ (kip), and the forces $\vec{F}_2$ and $\vec{F}_3$. The force $\vec{F}_2$ is parallel to the unit vector $\vec{u}_2 = -\left(\frac{2}{\sqrt{5}}\right)\vec{i} + \left(\frac{1}{\sqrt{5}}\right)\vec{j}$, and the force $\vec{F}_3$ is parallel to the unit vector $\vec{u}_3 = -\left(\frac{2}{\sqrt{5}}\right)\vec{i} - \left(\frac{1}{\sqrt{5}}\right)\vec{j}$
Determine the forces $\vec{F}_2$ and $\vec{F}_3$.
Solution: : Express the forces in terms of the unit vectors, apply the equilibrium conditions, and solve for the unknowns. Write $|\vec{F}_2|$ and $|\vec{F}_3|$ in terms of unit vectors,

$\vec{F}_2 = |\vec{F}_2|\left(\frac{-2}{\sqrt{5}}\vec{i} + \frac{1}{\sqrt{5}}\vec{j}\right)$, $\vec{F}_3 = |\vec{F}_3|\left(\frac{-2}{\sqrt{5}}\vec{i} + \frac{-1}{\sqrt{5}}\vec{j}\right)$. Since the object is in equilibrium,
$\sum \vec{F} = \vec{F}_1 + \vec{F}_2 + \vec{F}_3$ Collect and combine like terms,

$$\sum \vec{F}_x = (40 + |\vec{F}_2|\left(\frac{-2}{\sqrt{5}}\right) + |\vec{F}_3|\left(\frac{-2}{\sqrt{5}}\right))\vec{i} = 0$$

$$\sum \vec{F}_y = (30 + |\vec{F}_2|\left(\frac{1}{\sqrt{5}}\right) + |\vec{F}_3|\left(\frac{-1}{\sqrt{5}}\right))\vec{j} = 0$$

These are simultaneous equations in $|\vec{F}_2|$ and $|\vec{F}_3|$. Solve by a variant of Gaussian elimination by multiplying each equation by the coefficient of a like term in the other equation and subtracting: multiply the first by $\frac{\mathbf{1}}{\sqrt{\mathbf{5}}}$ and the second by $\frac{\mathbf{-2}}{\sqrt{\mathbf{5}}}$, subtract the second from the first, and reduce algebraically. The result is $|\vec{F}_3| = 25\sqrt{5}$. Similarly, multiply the first by $\frac{\mathbf{-1}}{\sqrt{\mathbf{5}}}$ and the second by $\frac{\mathbf{-2}}{\sqrt{\mathbf{5}}}$, subtract the second from the first, and reduce algebraically. The result is $|\vec{F}_2| = -5\sqrt{5}$. Substitute into the expressions for $|\vec{F}_1|$ and $|\vec{F}_2|$: $\boxed{\vec{F}_2 = -5\sqrt{5}\left(\frac{-2}{\sqrt{5}}\vec{i} + \frac{1}{\sqrt{5}}\vec{j}\right) = 10\vec{i} - 5\vec{j}\ (kip)}$

$\boxed{\vec{F}_3 = 25\sqrt{5}\left(\frac{-2}{\sqrt{5}}\vec{i} + \frac{-1}{\sqrt{5}}\vec{j}\right) = -50\vec{i} - 25\vec{j}\ (kip)}$ *Check:* An obvious and algebraically simpler approach specifically applicable here is to multiply the second equation by 2 and add the two equations. This is less cumbersome, but the strategy of multiplying each equation by the coefficient of a like term in the other equation lends itself to a general, more widely applicable approach. *check.*

================================<>================================

Problem 3.3 The forces acting on a body in equilibrium are shown. If $|\vec{F}_1| = 75\ N$, what are $\vec{F}_2$ and $\vec{F}_3$.?

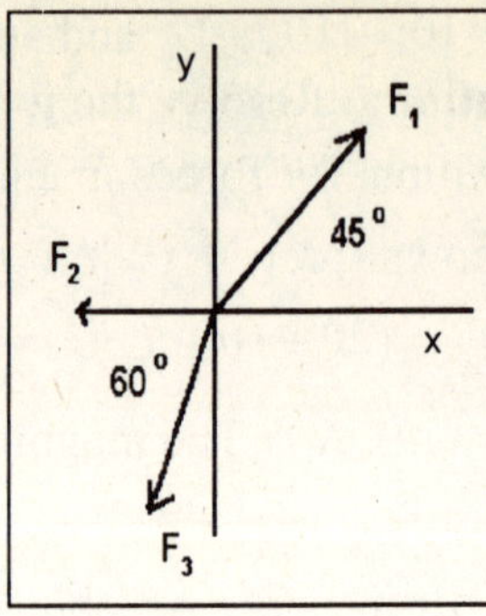

Solution: Resolve the forces into scalar components, and solve the equilibrium equations.. The angle between the force $\vec{F}_1$ and the positive x axis is 45°.Project the force onto the coordinates $\vec{F}_1 = |\vec{F}_1|\vec{e}$,

$\vec{F}_1 = |\vec{F}_1|(\vec{i}\cos 45^\circ + \vec{j}\sin 45^\circ)$ $\vec{F}_1 = 75(0.707\vec{i} + 0.707\vec{j}) = 53\vec{i} + 53\vec{j}$

The angles measured counter-clockwise between the x axis and $\vec{F}_2$ and $\vec{F}_3$ are 180° and 240°, respectively, Thus

$\vec{F}_2 = |\vec{F}_2|(\vec{i}\cos 180^\circ + \vec{j}\sin 180^\circ) = -|\vec{F}_2|\vec{i} + 0\vec{j}$

$\vec{F}_3 = |\vec{F}_3|(\vec{i}\cos 240^o + \vec{j}\sin 240^o) = |\vec{F}_3|(-0.5\vec{i} - 0.866\vec{j})$

The equilibrium conditions $\sum\vec{F} = \vec{F}_1 + \vec{F}_2 + \vec{F}_3 = 53\vec{i} + 53\vec{j} - |\vec{F}_2|\vec{i} + |\vec{F}_3|(-0.5\vec{i} - 0.866\vec{j}) = 0$

Collect and combine like terms in $\vec{i}$, $\vec{j}$, $\sum\vec{F}_x = (53 - |\vec{F}_2| - 0.5|\vec{F}_3|)\vec{i} = 0$

$\sum\vec{F}_y = (53 - 0.866|\vec{F}_3|)\vec{j} = 0$ Solve for the unknowns, $\boxed{|\vec{F}_3| = \frac{53}{0.866} = 61.2\ N}$

$\boxed{|\vec{F}_2| = 53 - 0.5|\vec{F}_3| = 53 - 30.6 = 22.4\ N}$

Problem 3.4 Three forces act on an object. The force $|\vec{F}_1| = 100\ lb$.(a) What is the smallest value of $|\vec{F}_3|$ for which the free body diagram can be in equilibrium?

(b) If $|\vec{F}_3|$ has the value determined in Part (a), what is the corresponding angle α?

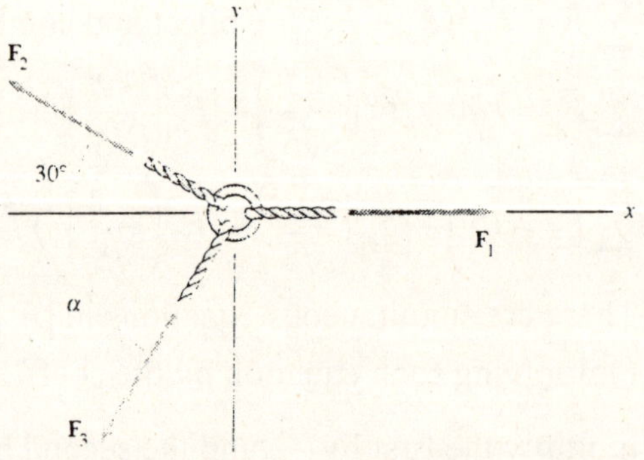

Solution: Apply the cosine law to a closed triangle. Differentiate the cosine law to find the minimum. Apply the sine law to find the value of α.

(a) The closed triangle is shown. The law of cosines in terms of the unknown $|\vec{F}_3|$ is $F_3^2 = F_1^2 + F_2^2 - 2F_1F_2\cos 30^\circ$

where the F_1 , F_2 F_3 are the magnitudes of the corresponding vectors. Since the minimum of F_3^2 is also the minimum of F_3

$\frac{\partial F_3^2}{\partial F_2} = 2F_2 - 2F_1\cos 30^o = 0$ Solve:

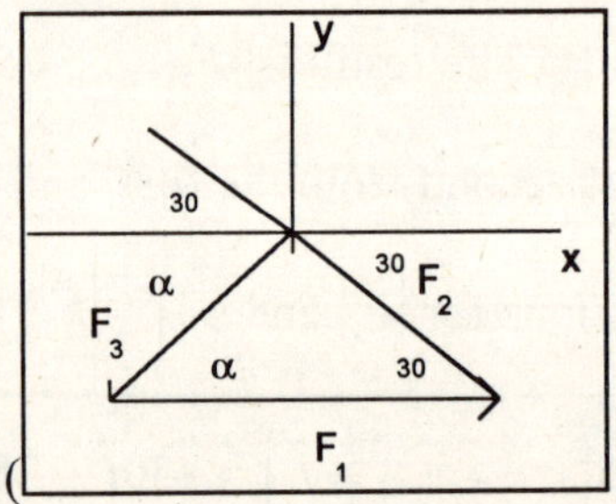

$F_2 = F_1\cos 30 = 100(0.866) = 86.6\ lb$ Substitute into the cosine law, $F_3^2 = F_1^2(1 - \cos^2 30) = F_1^2\sin^2 30^o$

$\boxed{F_3 = \sqrt{F_1^2\sin^2 30^o} = \pm 100(0.5) = 50\ lb}$ since $|\vec{F}_3| \geq 0$

(b) The law of sines $\frac{F_3}{\sin 30} = \frac{F_2}{\sin a}$ from which $\sin\alpha = \frac{F_2}{F_3}\sin 30^o = \frac{86.6}{50}(0.5) = 0.866$, $\boxed{\alpha = 60^o}$

Problem 3.5 The beam is in equilibrium. If B = 260 N and C = 100 N, what are the forces $\mathbf{A_x}$ and $\mathbf{A_y}$?

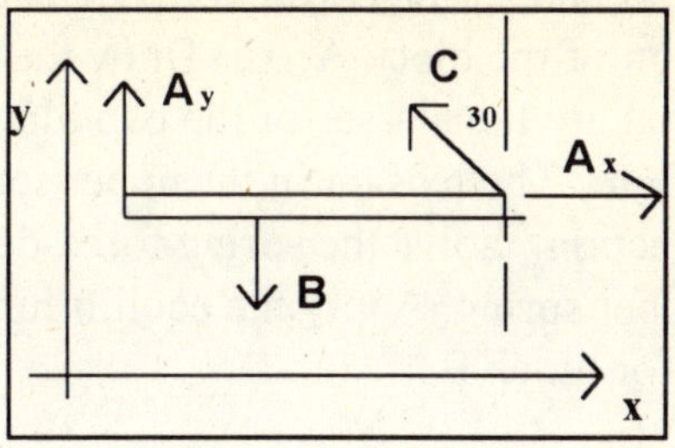

Solution: Resolve the forces into scalar components, and solve the equilibrium equations. The angle between the vector force **B** and the positive x axis is $-90^o = +\ 270^o$. The force **B** is $\vec{B} = (260)(\vec{i}\cos 270^o + \vec{j}\sin 270^o) = 0\,\vec{i} - 260\,\vec{j}$

Similarly, the force C in the reference coordinates is

$\vec{C} = (100)(\vec{i}\cos 120^o + \vec{j}\sin 120^o) = (100)(-0.5\,\vec{i} + 0.866\,\vec{j})$

$\vec{C} = -50\,\vec{i} + 86.6\,\vec{j}$ Similarly, $\vec{A}_x = |\vec{A}_x|(\vec{i}\cos 0^o + \vec{j}\sin 0^o) = \vec{i}\,|\vec{A}_x| + 0\vec{j}$,

$\vec{A}_y = |\vec{A}_y|(\vec{i}\cos 90^o + \vec{j}\sin 90^o) = 0\,\vec{i} + \vec{j}\,|\vec{A}_y|$ .The equilibrium conditions are

$\sum\vec{F} = 0 = \vec{A}_x + \vec{A}_y + \vec{B} + \vec{C} = 0$ Collect and combine like terms in $\vec{i}$, $\vec{j}$: $\sum\vec{F}_x = (|\vec{A}_x| - 50)\,\vec{i} = 0$

$\sum\vec{F}_y = (|\vec{A}_y| - 260 + 86.6)\vec{j} = 0$ Solve for the unknowns: $|\vec{A}_x| = 50\ N$, $|\vec{A}_y| = 173.4\ N$

Problem 3.6 A zoologist estimates that the jaw of a predator, *Martes*, is subjected to a force **P** as large as 800 N. What forces **T** and **M** must be exerted by the temporalis and masseter muscles to support this value of **P** ?

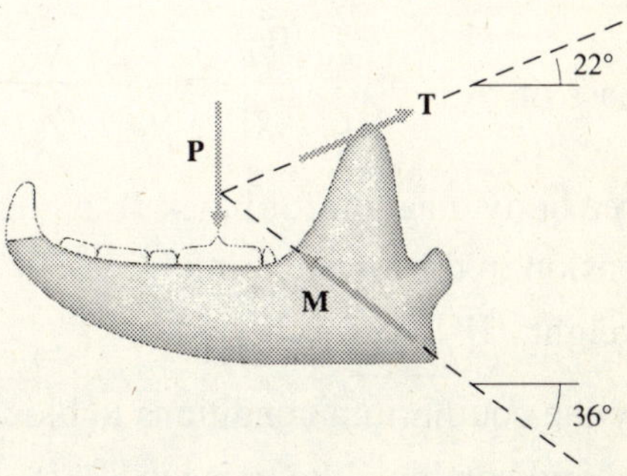

Solution: Resolve the forces into scalar components, and solve the equilibrium equations...Express the forces in terms of the unit vectors:

$\vec{T} = |\vec{T}|(\vec{i}\cos 22^o + \vec{j}\sin 22^o) = |\vec{T}|(0.927\vec{i} + 0.375\vec{j})$

$\vec{P} = 800(\vec{i}\cos 270^o + \vec{j}\sin 270^o) = 0\vec{i} - 800\vec{j}$

$\vec{M} = |\vec{M}|(\vec{i}\cos 144^o + \vec{j}\sin 144^o) = |\vec{M}|(-0.809\vec{i} + 0.588\vec{j})$

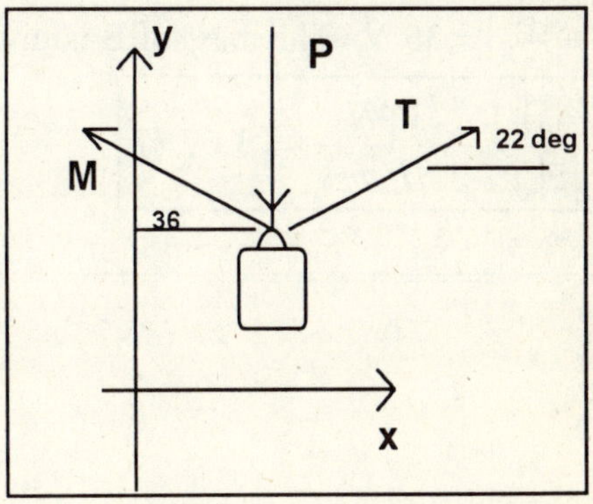

Apply the equilibrium conditions,

$\sum\vec{F} = 0 = \vec{T} + \vec{M} + \vec{P} = 0$ Collect like terms:

$\sum\vec{F}_x = (0.927|\vec{T}| - 0.809|\vec{M}|)\vec{i} = 0$

$\sum\vec{F}_y = (0.375|\vec{T}| + 0.588|\vec{M}| - 800)\vec{j} = 0$

Solve the first equation , $|\vec{T}| = \left(\dfrac{0.809}{0.927}\right)|\vec{M}| = 0.873|\vec{M}|$

Substitute this value into the second equation, reduce algebraically, and solve

:

$|\vec{M}| = 874\ N,\quad |\vec{T}| = 763.3\ N$

Problem 3.7 The two springs are identical, with unstretched lengths 250 mm and spring constants $k = 1{,}200$ N/m. (a) Draw the free body diagram of the block A. (b) Draw the free body diagram of the block B. . (c)What are the masses of the two blocks?

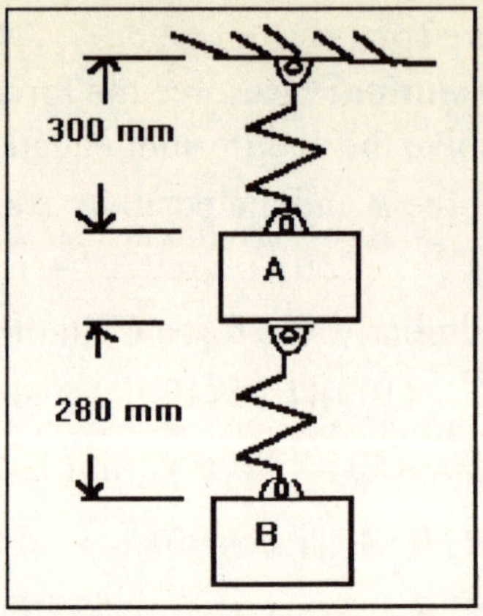

Solution: The tension in the upper spring acts on block **A** in the positive Y direction, Solve the spring force-deflection equation for the tension in the upper spring. Apply the equilibrium conditions to block A. Repeat the steps for block B.

$$\vec{T}_{UA} = 0\vec{i} + \left(1200\ \frac{N}{m}\right)(0.3\ m - 0.25\ m)\vec{j} = 0\vec{i} + 60\vec{j}\ N$$

Similarly, the tension in the lower spring acts on block **A** in the negative Y direction $\vec{T}_{LA} = 0\vec{i} - \left(1200\ \frac{N}{m}\right)(0.28\ m - 0.25\ m)\vec{j} = 0\vec{i} - 36\vec{j}\ N$

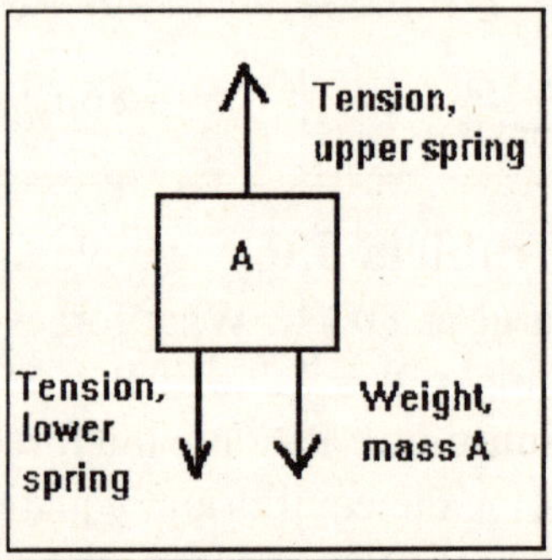

The weight is $\vec{W}_A = 0\vec{i} - |\vec{W}_A|\vec{j}$ The equilibrium conditions are

$$\sum\vec{F} = \sum\vec{F}_x + \sum\vec{F}_y = 0, \sum\vec{F} = \vec{W}_A + \vec{T}_{UA} + \vec{T}_{LA} = 0$$

Collect and combine like terms in $\vec{i}$, $\vec{j}$

$\sum\vec{F}_y = \left(-|\vec{W}_A| + 60 - 36\right)\vec{j} = 0$ Solve $|\vec{W}_A| = (\ 60 - 36) = 24\ N$

The mass of A is $m_A = \frac{|\vec{W}_L|}{|\vec{g}|} = \frac{24\ N}{9.81\ m/s^2}$ $\boxed{= 2.45\ kg}$

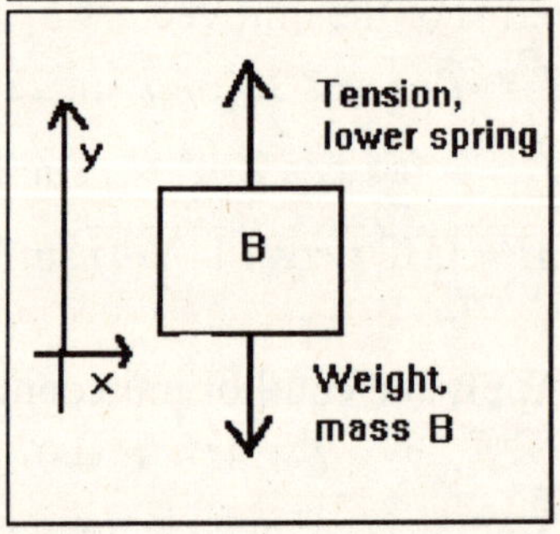

The free body diagram for block **B** is shown.

The tension in the lower spring $\vec{T}_{LB} = 0\vec{i} + 36\vec{j}$

The weight: $\vec{W}_B = 0\vec{i} - |\vec{W}_B|\vec{j}$

Apply the equilibrium conditions to block B. $\sum\vec{F} = \vec{W}_B + \vec{T}_{LB} = 0$

Collect and combine like terms in $\vec{i}$, $\vec{j}$: $\sum\vec{F}_y = \left(-|\vec{W}_B| + 36\right)\vec{j} = 0$

Solve: $|\vec{W}_B| = 36\ N$ The mass of B is given by

$$\boxed{m_B = \frac{|\vec{W}_B|}{|\vec{g}|} = \frac{36\ N}{9.81\ m/s^2} = 3.67\ kg}$$

Problem 3.8 The 200 lb horizontal bar is suspended by the springs A, B, and C. The unstretched lengths of the springs are equal. The spring constants are $k_A = k_C = 400\frac{lb}{ft}$, and $k_B = 300\frac{lb}{ft}$. What are the tensions in the springs?

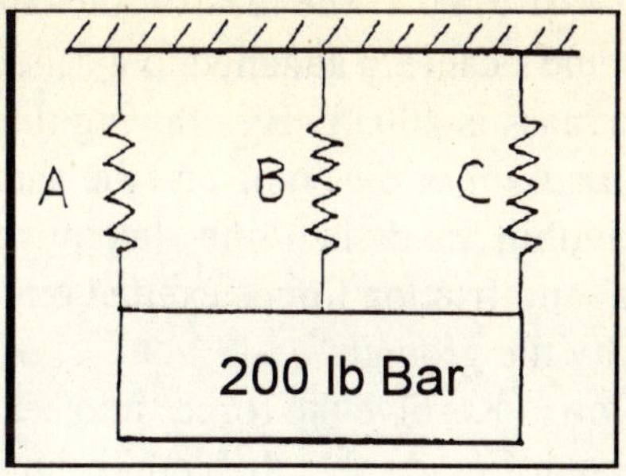

Solution: Apply the linear spring force deflection relation. The extension of each spring is the same. The tensions in springs A and C are equal, thus the ratios of the tensions are known constants. Solve the equilibrium equations for the unknowns. Since the extension of each spring is identical, then the tensions are $|\vec{A}| = k_A \Delta L$, $|\vec{B}| = k_B \Delta L$, and $|\vec{C}| = k_C \Delta L$. Since $k_A = k_C$,then $|\vec{A}| = |\vec{C}|$. The ratio of the tensions is $\frac{|\vec{A}|}{|\vec{B}|} = \frac{k_A \Delta L}{k_B \Delta L} = \frac{k_A}{k_B}$

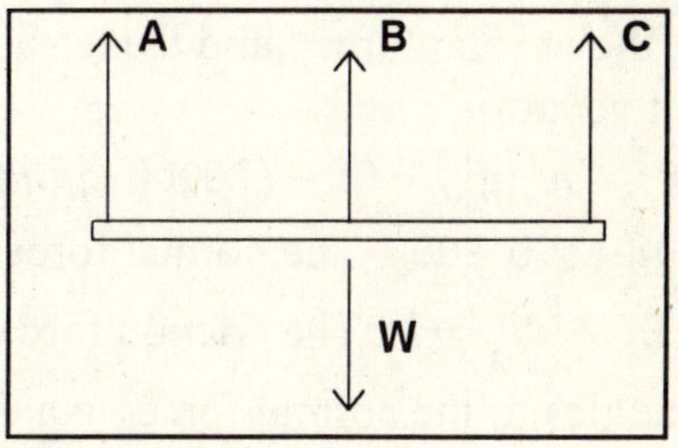

The equilibrium conditions, used in t$\sum \vec{F}_x = 0$, $\sum \vec{F}_y = (|\vec{A}| + |\vec{B}| + |\vec{C}| - |\vec{W}|)\vec{j} = 0$

Substitute $|\vec{A}| = |\vec{C}|$, and $|\vec{A}| = \left(\frac{k_A}{k_B}\right)|\vec{B}|$, to obtain $\sum \vec{F}_y = \left(2\frac{k_A}{k_B} + 1\right)|\vec{B}| - |\vec{W}| = 0$ Solve:

$$|\vec{B}| = \frac{|\vec{W}|}{2\frac{k_A}{k_B} + 1} = \frac{200}{3.67} = 54.55\ lb$$

$$|\vec{A}| = |\vec{C}| = \left(\frac{k_A}{k_B}\right)|\vec{B}| = 72.73\ lb$$

Problem 3.9 If you neglect the weight of the pulleys, what force $\vec{F}$ is needed to support the weight $\vec{W}$?

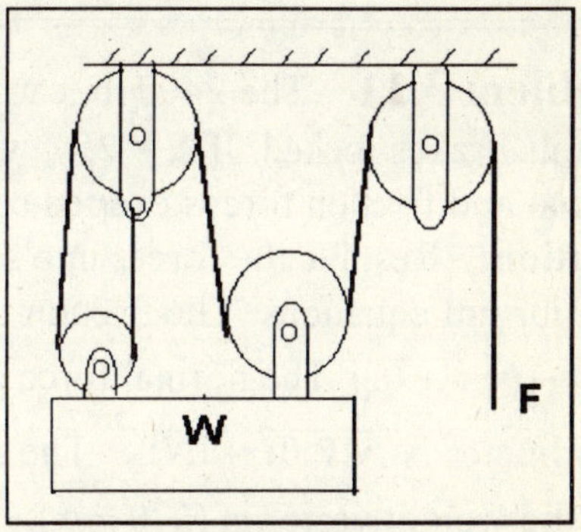

Solution: A frictionless pulley changes the direction of a force, but not its magnitude. Isolate a pulley, solve the equilibrium conditions. Repeat for the weight. From the free body diagram of a pulley, the equilibrium condition is $\sum \vec{F} = (|\vec{F}| + |\vec{F}| - |\vec{F}_D|)\vec{j} = 0$, hence $|\vec{F}_D| = 2|\vec{F}|$, and the force acting upward on the weight is $\vec{F}_W = (|\vec{F}_D| + |\vec{F}_D|)\vec{j}$.The equilibrium condition is

$\sum \vec{F} = (2|\vec{F}| + 2|\vec{F}| - |\vec{W}|)\vec{j} = 0$. Solve:

$$|\vec{F}| = \left(\frac{1}{4}\right)|\vec{W}|$$

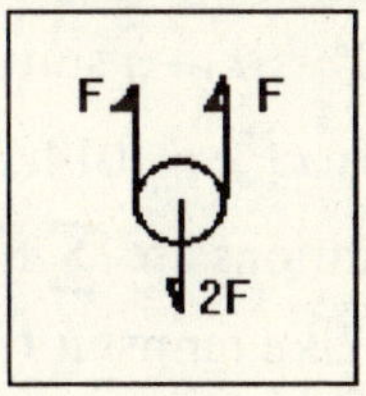

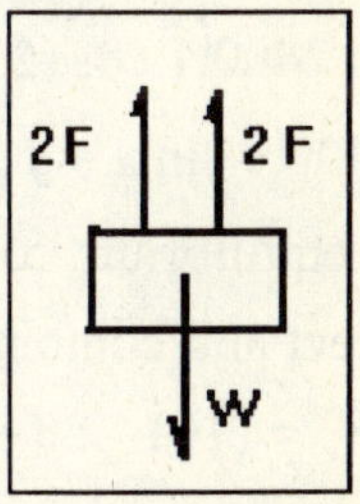

The same result can be had by inspection: each pulley is subjected to twice the tension $\vec{F}$ acting upward. There are two pulleys, and for force balance the weight must equal the upward force, or four times the rope tension.

Problem 3.10 The mass of the crane is 20 Mg (megagrams) and the tension in the cable is 1 kN. The crane's cable is attached to a caisson whose mass is 400 kg. By drawing the free body diagram of the crane and the part of the cable within the dashed line, determine the normal and friction forces exerted on the crane by the ground.

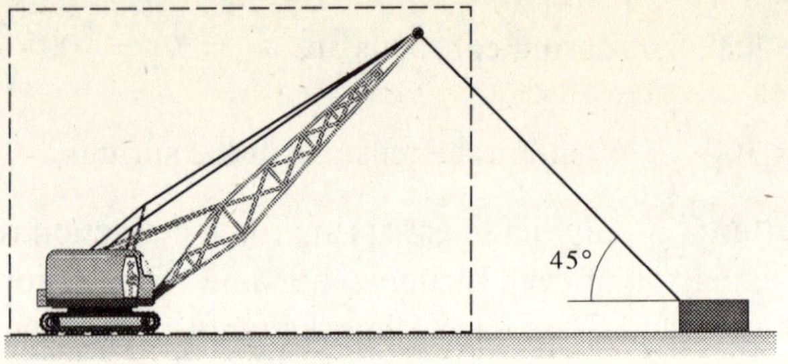

Solution: Resolve the forces into scalar components, and solve the equilibrium equations.

The external forces are the weight, the friction force, the normal force, and the tension in the cable. The weight vector is

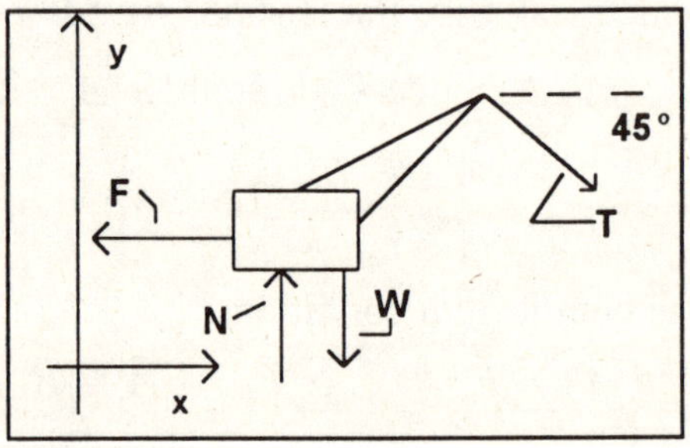

$\vec{W} = 0\vec{i} - m_c|\vec{g}|\vec{j} = 0\vec{i} - (20000\ kg)(9.81\ m/s^2)\vec{j}$

$\vec{W} = 0\vec{i} - 196{,}200\vec{j}$ The normal force vector is

$\vec{N} = \left(0\vec{i} + |\vec{N}_y|\,\vec{j}\right)$. The friction force by definition acts at right angles to the normal force, in a direction that holds the crane in place. $\vec{F}_x = -|\vec{F}_x|\vec{i} + 0\vec{j}$ The angle between the tension vector and the positive x axis is $-45^o = 315^o$, hence the tension vector projection is

$\vec{T} = |\vec{T}|\left(\vec{i}\cos 315^\circ + \vec{j}\sin 315^\circ\right) = (1000)\left(0.707\,\vec{i} - 0.707\,\vec{j}\right) = 707\,\vec{i} - 707\,\vec{j}\ N$ The equilibrium conditions are $\sum \vec{F}_x = \left(-|\vec{F}_x| + 707\right)\vec{i} = 0$, $\sum \vec{F}_y = \left(|\vec{N}_y| - 707 - 196200\right)\vec{j} = 0$ Solve:

$\boxed{|\vec{F}_x| = 707\ N}$ $\boxed{|\vec{N}_y| = 196200 + 707 = 196{,}907\ N}$

Thus the friction force is directed toward the left, and the normal force acts upward.

Problem 3.11 The 2400 lb. car is parked on the sloping road with its brakes locked. If $\alpha = 25°$, what are the magnitudes of the normal and friction forces exerted on the car by the road?

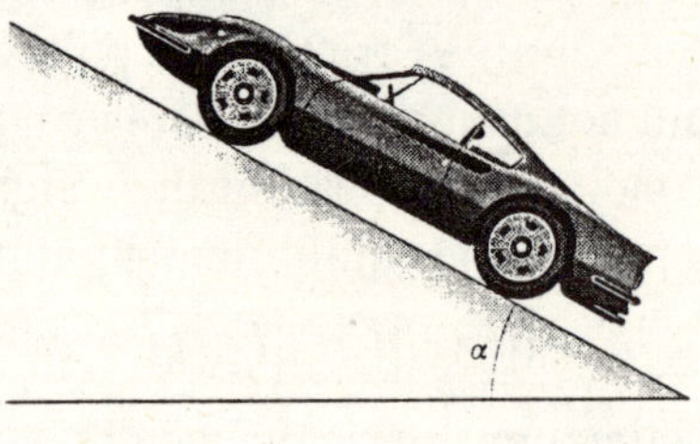

Solution: Resolve the forces into scalar components, and solve the equilibrium equations. The friction force is

$\vec{F} = -|\vec{F}|\,\vec{i} + 0\,\vec{j}$ The normal force projected onto the reference coordinates is $\vec{N} = 0\vec{i} + |\vec{N}|\,\vec{j}$ The angle between the positive x axis and the weight vector is $(270 + \alpha)$. $\alpha = 25°$,

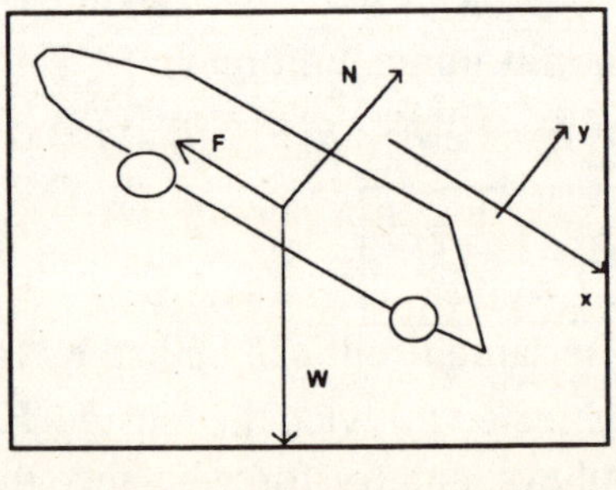

$\vec{W} = 2400\left(\vec{i}\cos(270^o + \alpha) + \vec{j}\sin(270^o + \alpha)\right)$

$= 2400\left(\vec{i}\sin\alpha - \vec{j}\cos\alpha\right) = 1014.28\,\vec{i} - 2175.14\vec{j}$

The equilibrium conditions are $\sum \vec{F} = 0 = \vec{W} + \vec{N} + \vec{F}$

Collect and combine like terms in $\vec{i}$, $\vec{j}$:

$\sum \vec{F}_x = \left(1014.28 - |\vec{F}|\right)\vec{i} = 0$, $\sum \vec{F}_y = \left(-2175.14 + |\vec{N}|\right)\vec{j} = 0$

Solve $\boxed{|\vec{F}| = 1014.3\ lb,\ \ |\vec{N}| = 2175.1\ lb}$

Problem 3.12 Suppose that the 2400 lb car shown in Problem 3.11 will remain in equilibrium only if the friction force exerted on the car is 0.6 times the normal force. What is the largest angle α for which the car will remain in equilibrium?

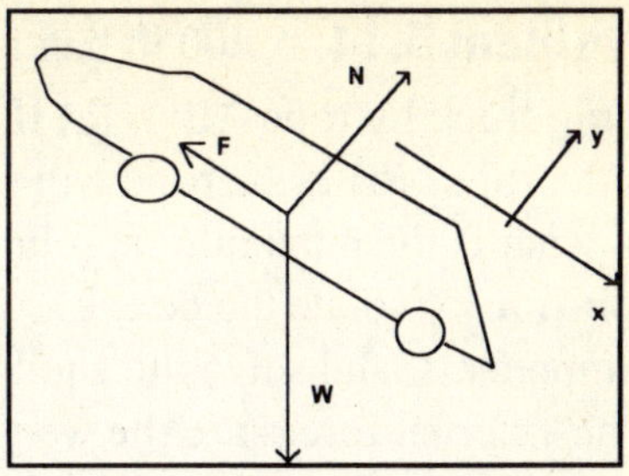

Solution: Resolve the forces into scalar components, and solve the equilibrium equations.
From Problem 3.10, the friction force is $\vec{F} = -|\vec{F}|\,\vec{i} + 0\,\vec{j}$

The normal force projected is $\vec{N} = 0\vec{i} + |\vec{N}|\,\vec{j}$ The angle between the positive x axis and the weight vector is $-(90-\alpha)$, or $(270+\alpha)$. The weight vector is,
$\vec{W} = 2400\left(\vec{i}\cos(270^\circ+\alpha) + \vec{j}\sin(270^\circ+\alpha)\right) = 2400\left(\vec{i}\sin\alpha - \vec{j}\cos\alpha\right)$ .The equilibrium conditions are $\sum\vec{F} = 0 = \vec{W} + \vec{N} + \vec{F}$ Collect like terms: $\sum\vec{F}_x = \left(2400\sin\alpha - |\vec{F}|\right)\vec{i} = 0$ and
$\sum\vec{F}_y = \left(-2400\cos\alpha + |\vec{N}|\right)\vec{j} = 0$ From the conditions given: $|\vec{F}| = 0.6|\vec{N}|$ Solve for the unknown $|\vec{N}|$
$|\vec{N}| = \dfrac{2400\sin\alpha}{0.6} = 2400\cos\alpha$ From which $\dfrac{\sin\alpha}{\cos\alpha} = 0.6$ or $\tan\alpha = 0.6$, $\boxed{\alpha = 31^\circ}$

Problem 3.13 The mass of the crate is 40 kg. The inclined surface is rough. The length of the spring is 180 mm, the unstretched length is 200 mm, and the spring constant is k = 2500 N/m. What is the magnitude of the friction force exerted on the crate by the surface?

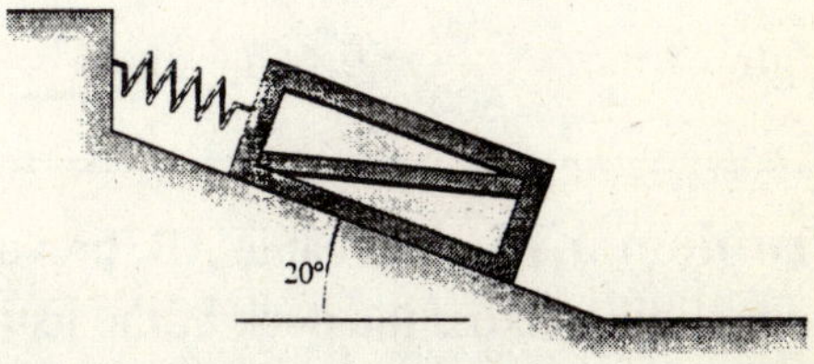

Solution: Isolate the crate., The external forces are the weight, the tension in the spring, the friction force, and the normal force exerted by the surface. The magnitude of the weight is $|\vec{W}| = m|\vec{g}| = (40\ kg)(9.81\ m/s^2) = 392.4\ N$
The angle between the x axis and the weight vector is $-(90^\circ - 20^\circ) = -70^\circ$ (or 290°). The projection of the weight vector is $\vec{W} = |\vec{W}|\left(\vec{i}\cos 290^\circ + \vec{j}\sin 290^\circ\right)$
$= (392.4)\left(0.342\vec{i} - 0.94\vec{j}\right) = 134.2\vec{i} - 368.7\vec{j}$ The projections of the cable tension , the friction force and the normal force are $\vec{T} = |\vec{T}_x|\vec{i} + 0\vec{j}$, $\vec{F} = -|\vec{F}_x|\vec{i} + 0\vec{j}$,$\vec{N} = 0\vec{i} + |\vec{N}_y|\vec{j}$ The spring tension is found from the spring force-deflection relation,
$|\vec{T}_x| = kD = (2500)(0.2\ m - 0.18\ m) = (2500)(0.02) = 50\ N$

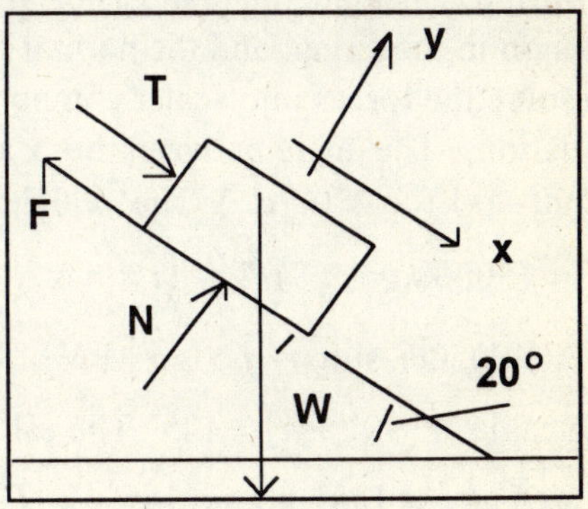

The equilibrium conditions are $\sum\vec{F} = \vec{W} + \vec{N} + \vec{T} + \vec{F}$
Substitute, and collect like terms
$\sum\vec{F}_x = \left(134.2 + 50 - |\vec{F}_x|\right)\vec{i} = 0$, $\sum\vec{F}_y = \left(-368.7 + |\vec{N}_y|\right)\vec{j} = 0$ Solve: $\boxed{|\vec{F}| = 184.2\ N}$

Problem 3.14 A 600 lb box is held in place on the bed of a dump truck by rope AB. (a) If $\alpha = 25^o$, what is the tension in the rope? (b) If the rope will safely support a tension of 400 lb, what is the maximum absolute value of α?

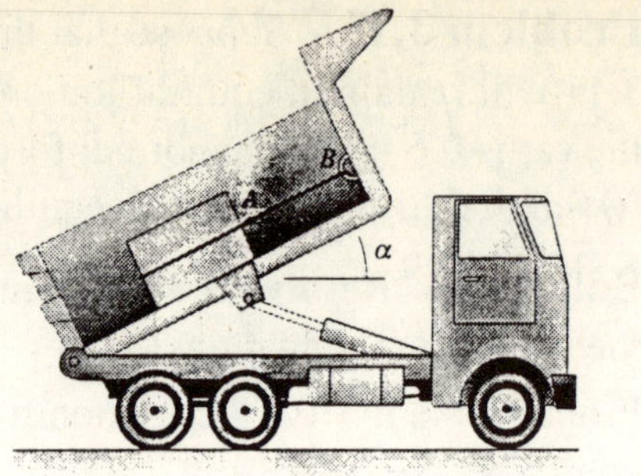

Solution: Isolate the box. Resolve the forces into scalar components, and solve the equilibrium equations.

The external forces are the weight, the tension in the rope, and the normal force exerted by the surface. The angle between the x axis and the weight vector is $-(90-\alpha)$ (or $270+\alpha$). The weight vector is

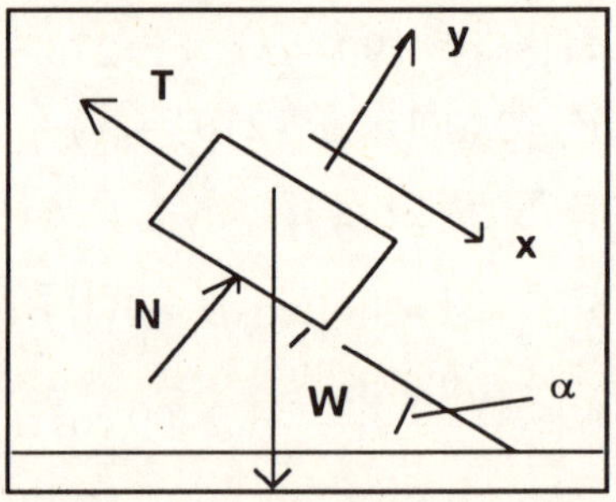

$\vec{W} = |\vec{W}|(\vec{i} \sin\alpha - \vec{j} \cos\alpha) = (600)(\vec{i} \sin\alpha - \vec{j} \cos\alpha)$

The projections of the rope tension and the normal force are

$\vec{T} = -|\vec{T}_x|\vec{i} + 0\vec{j}$ $\vec{N} = 0\vec{i} + |\vec{N}_y|\vec{j}$ The equilibrium conditions are

$\sum \vec{F} = \vec{W} + \vec{N} + \vec{T} = 0$ Substitute, and collect like terms

$\Sigma \vec{F}_x = (600 \sin\alpha - |\vec{T}_x|)\vec{i} = 0$ $\sum \vec{F}_y = (-600 \cos\alpha + |\vec{N}_y|)\vec{j} = 0$

Solve for the unknown tension when $\alpha = 25°$

$\boxed{|\vec{T}_x| = 600 \sin\alpha = 253.6\ lb}$ For a tension of 400 lb, $(600 \sin\alpha - 400) = 0$. Solve for the unknown angle $\sin\alpha = \dfrac{400}{600} = 0.667$ or, $\boxed{\alpha = 41.84^o}$

Problem 3.15 The cable AB prevents the 5 Mg (Megagram) crate from sliding across the deck of the listing ship's hold. If the cable will support a tension of 40 kN, what is the largest value of the list angle α?.

Solution: Isolate the crate. The external forces are the weight, the tension in the string, and the normal force exerted by the surface. Resolve the forces into scalar components, and solve the equilibrium equations. The angle between the x axis and the weight vector is $-(90-\alpha)$ (or $270+\alpha$). The weight vector is

$\vec{W} = (5000\ kg)(9.81\ ^{m}/_{s^2})(\vec{i} \sin\alpha - \vec{j} \cos\alpha)$

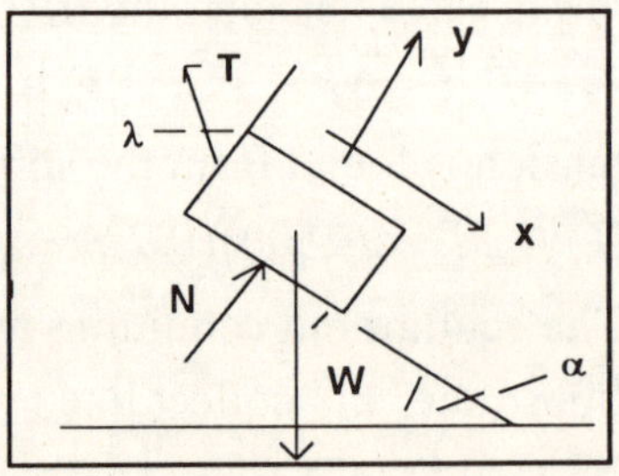

$\vec{W} = 49050(\vec{i} \sin\alpha - \vec{j} \cos\alpha)\ (N)$ .The angle between the x axis and the cable is $(90^o + \lambda) = 145^o$.The cable tension and the normal force are

$\vec{T} = |\vec{T}|(\vec{i} \cos 145^o + \vec{j} \sin 145^o) = |\vec{T}|(-0.819\vec{i} + 0.574\vec{j})$

$\vec{N} = 0\vec{i} + N_y\vec{j}$ The equilibrium conditions are $\sum \vec{F} = \vec{W} + \vec{N} + \vec{T}$,

from which: $\sum \vec{F}_x = (49050 \sin\alpha - 0.819|\vec{T}|)\vec{i} = 0$,

$\sum \vec{F}_y = (-49050 \cos\alpha + 0.574|\vec{T}| + N_y)\vec{j} = 0$. Solve: $|\vec{T}| = (49050)\left(\dfrac{\sin\alpha}{0.819}\right)$ .For a cable tension of 40 kN, $\sin\alpha = \left(\dfrac{40000}{49050}\right)(0.819) = 0.668$, $\boxed{\alpha = 41.9^o}$

Problem 3.16 The weights of the two blocks are $\vec{W}_1 = 200\ lb$ and $\vec{W}_2 = 50\ lb$. What force must the man exert on the rope to hold the blocks in place?

Solution: Isolate block W_2 .and apply equilibrium conditions. Repeat for block W_1.

For W_2 : The weight vector: $\vec{W}_2 = 0\ \vec{i} - 50\ \vec{j}$ The rope tension : $\vec{T}_2 = 0\vec{i} + |\vec{T}_2|\vec{j}$ The equilibrium conditions are

$\sum \vec{F}_x = 0, \quad \sum \vec{F}_y = (-50 + |\vec{T}_2|)\vec{j} = 0$, or $|\vec{T}_2| = 50$

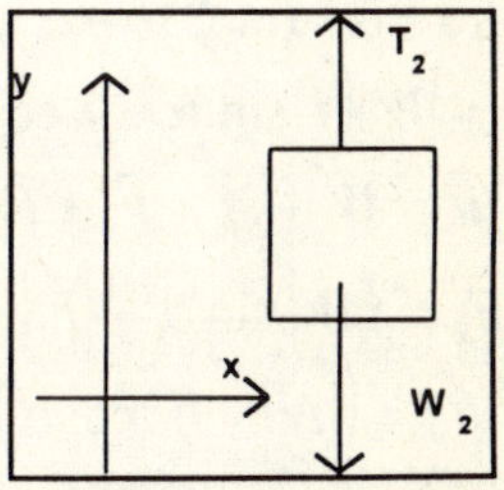

For W_1 : The magnitude of the rope tension $|\vec{T}_2|$ is unchanged by passage over the frictionless lower pulley, hence, $\vec{T}_2 = |\vec{T}_2|\ \vec{i} + 0\ \vec{j} = 50\ \vec{i} + 0\ \vec{j}$ .The rope tension $\vec{T}_1$:

$\vec{T}_1 = -|\vec{T}_1|\ \vec{i} + 0\vec{j}$ The normal force is

$\vec{N} = 0\vec{i} + |\vec{N}|\vec{j}$. The angle between the x axis and the weight vector is $-(90 - \alpha)$ (or $270 + \alpha$). The projection of the weight vector is

$\vec{W} = |\vec{W}|(\vec{i}\ \sin\alpha - \vec{j}\ \cos\alpha) = 200(0.5\vec{i} - 0.866\vec{j}) = 100\ \vec{i} - 172.3\ \vec{j}$

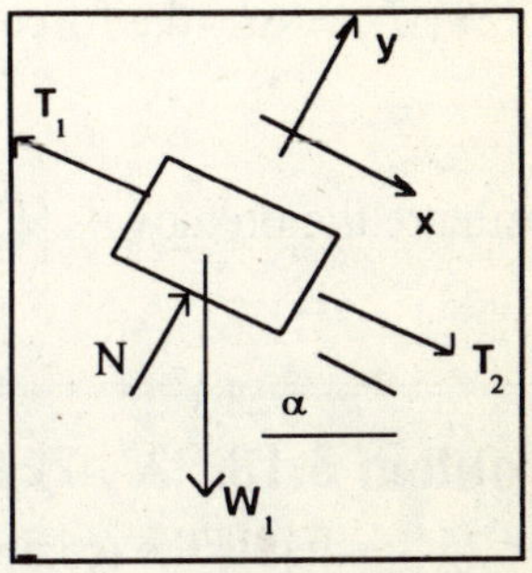

The equilibrium conditions are $\sum \vec{F} = \vec{T}_1 + \vec{T}_2 + \vec{N} + \vec{W}_1 = 0$ Substitute and collect like terms, $\sum \vec{F}_x = (-|\vec{T}_1| + 50 + 100)\ \vec{i} = 0$,

$\sum \vec{F}_y = (|\vec{N}| - 173.2)\vec{j} = 0$ Solve: $|\vec{T}_1| = 150\ lb$ Since the frictionless pulley does not change the magnitude of the rope tension, then the tension at the man's hands is $\boxed{|\vec{T}_1| = 150\ lb}$

Problem 3.17 The two springs have the same unstretched length. Show that the magnitudes of the forces exerted by the springs are

$$|\vec{F}_1| = \frac{|\vec{W}|\sin\alpha}{\left(1+\frac{k_2}{k_1}\right)}, \qquad |\vec{F}_2| = \frac{|\vec{W}|\sin\alpha}{\left(1+\frac{k_1}{k_2}\right)}$$

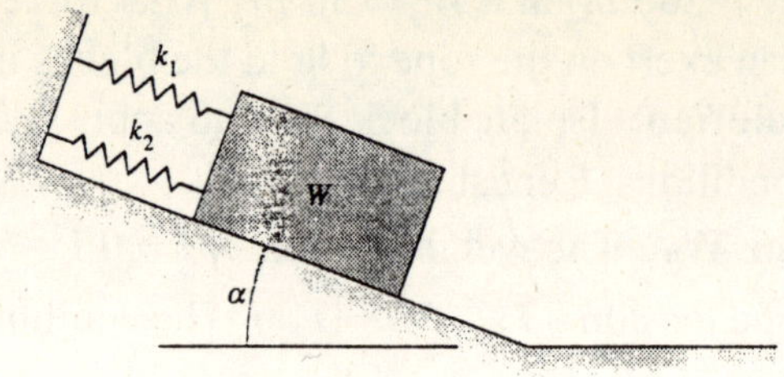

Solution: Isolate the block. Apply the linear spring force-deflection relations to find the ratios of the spring forces. The spring forces are, $\vec{F}_1 = -|\vec{F}_1|\vec{i} + 0\vec{j}$, $\vec{F}_2 = -|\vec{F}_2|\vec{i} + 0\vec{j}$ The normal force is, $\vec{N} = 0\vec{i} + |\vec{N}|\vec{j}$. The angle between the x axis and the weight vector is $-(90-\alpha)$ (or $270+\alpha$). The weight vector is $\vec{W} = |\vec{W}|(\vec{i}\ \sin\alpha - \vec{j}\ \cos\alpha)$ .The equilibrium conditions are $\sum\vec{F} = \vec{W} + \vec{N} + \vec{F}_1 + \vec{F}_2 = 0$. Substitute and collect like terms,

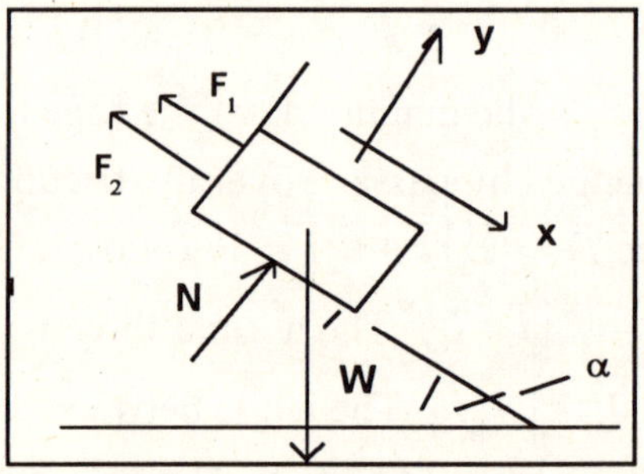

$\Sigma\vec{F}_x = \left(|\vec{W}|\sin\alpha - |\vec{F}_1| - |\vec{F}_2|\right)\vec{i} = 0$,

$\sum\vec{F}_y = \left(|\vec{N}| - W\cos\alpha\right)\vec{j} = 0$. For equal extensions,

$\Delta L_1 = \Delta L_2 = \Delta L$, the forces are $|\vec{F}_1| = k_1\Delta L$, and $|\vec{F}_2| = k_2\Delta L$. The ratio is, $\frac{|\vec{F}_1|}{|\vec{F}_2|} = \left(\frac{k_1}{k_2}\right)$. Substitute to eliminate the unknowns $|\vec{F}_2|$., and $|\vec{F}_1|$. Solve, $\boxed{|\vec{F}_2| = \frac{|\vec{W}|\sin\alpha}{\left(1+\frac{k_1}{k_2}\right)}}$, $\boxed{|\vec{F}_1| = \frac{|\vec{W}|\sin\alpha}{\left(1+\frac{k_2}{k_1}\right)}}$

Problem 3.18 A 10 kg painting is suspended by a wire. If $\alpha = 25^o$ what is the tension in the wire.?

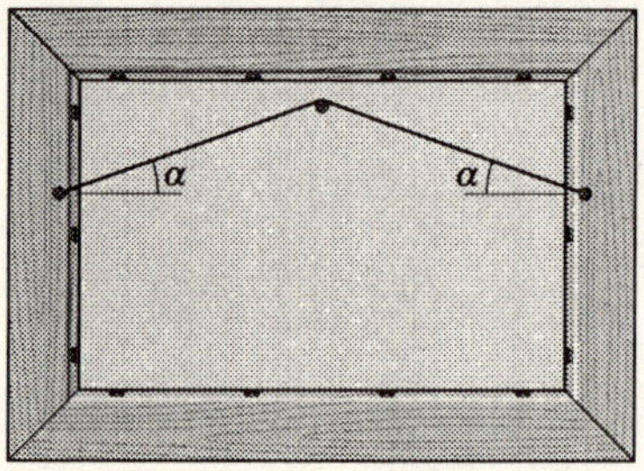

Solution: Isolate support pin fixed to the wall or other support. The angle of the right hand wire with the positive x axis is -α, hence the tension is $\vec{F}_2 == |\vec{F}_2|(\vec{i}\cos\alpha - \vec{j}\sin\alpha)$

The angle of the left hand wire is $(180^o + \alpha)$ hence $\vec{F}_1 = |\vec{F}_1|(-\vec{i}\ \cos\alpha - \vec{j}\sin\alpha)$ The weight is $\vec{W} = 0\vec{i} + |\vec{W}|\vec{j}$

The equilibrium conditions are $\sum\vec{F} = \vec{W} + \vec{F}_1 + \vec{F}_2 = \mathbf{0}$

Substitute the vector forces, and collect like terms,

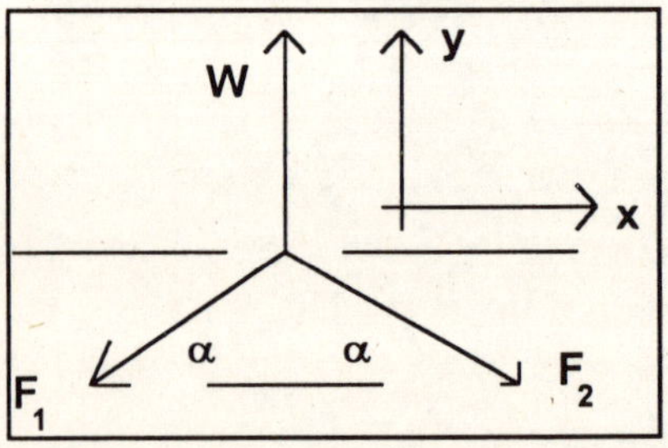

$\sum\vec{F}_x = \left(|\vec{F}_2|\cos\alpha - |\vec{F}_1|\cos\alpha\right)\vec{i} = 0$,

$\sum\vec{F}_y = \left(|\vec{W}| - |\vec{F}_2|\sin\alpha - |\vec{F}_1|\sin\alpha\right)\vec{j} = 0$ Thus $|\vec{F}_1| = |\vec{F}_2|$, and

$|\vec{F}_1| = |\vec{F}_2| = \frac{1}{2}\left(\frac{|\vec{W}|}{\sin\alpha}\right)$ With $\alpha = 25^o$ and

$|\vec{W}| = (10\ kg)(9.81\ m/s^2) = 98.1\ N$ $\boxed{|\vec{F}_1| = |\vec{F}_2| = \left(\frac{1}{2}\right)\left(\frac{98.1}{0.423}\right) = 116.06\ N}$

Problem 3.19 If the wire supporting the painting in Problem 3.18 breaks when the tension exceeds 150 N and you want a 100% safety factor (that is, you want the wire to support twice the actual weight of the painting), what is the smallest value of α you can use?

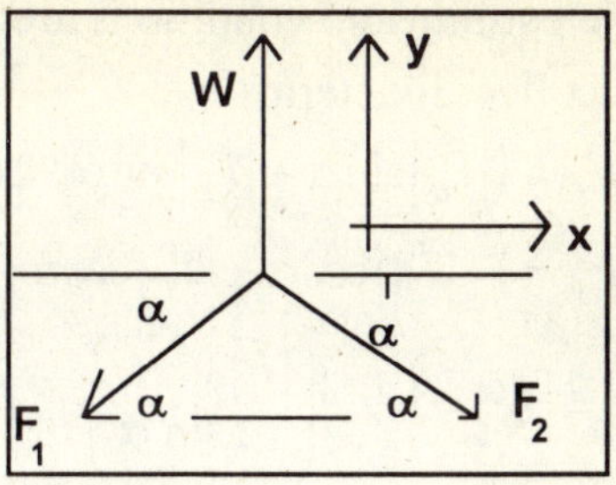

Solution: From Problem 3.18

$$|\vec{F}_1| = |\vec{F}_2| = \tfrac{1}{2}\left(\tfrac{\vec{W}}{\sin\alpha}\right) \text{ and } |\vec{W}| = (10\ kg)(9.81\tfrac{m}{s^2}) = 98.1\ N$$

Thus $\sin\alpha = \left(\frac{1}{2}\right)\left(\frac{98.1}{|\vec{F}|}\right)$ For a tension of $|\vec{F}| = \frac{150}{2} = 75$,

$\sin\alpha = \left(\frac{1}{2}\right)\left(\frac{98.1}{75}\right) = 0.654$ or $\boxed{\alpha = 40.8^o}$

Problem 3.20 A 50 lb. pull is required to draw the bow to the position shown. Determine the tension in the bowstring (a) by drawing a free-body diagram of the string, (b) by drawing a free-body diagram of the bow.

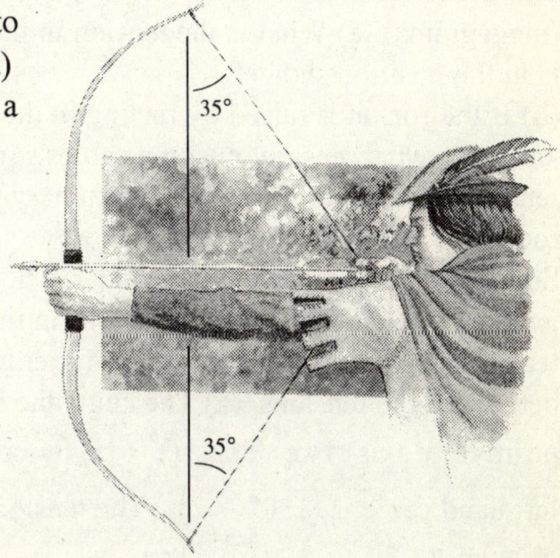

Solution: (a) The upper string has the angle $(90+\alpha)$ with respect to the positive x axis. The tension is

$$\vec{T}_U = |\vec{T}_U|\left(\vec{i}\cos(90+\alpha) + \vec{j}\sin(90+\alpha)\right)$$

$$\vec{T}_U = |\vec{T}_U|\left(-\vec{i}\sin\alpha + \vec{j}\cos\alpha\right)$$

The angle of the lower string with the positive x axis is $(270^o - \alpha)$, and the tension in the lower string is

$$\vec{T}_L = |\vec{T}_L|\left(\vec{i}\cos(270^o-\alpha) + \vec{j}\sin(270^o-\alpha)\right)$$

$\vec{T}_L = |\vec{T}_L|\left(-\vec{i}\sin\alpha - \vec{j}\cos\alpha\right)$ The pull is

$\vec{P} = |\vec{P}|\vec{i} + 0\vec{j} = 50\vec{i} + 0\vec{j}$ The equilibrium conditions are $\sum\vec{F} = \vec{T}_U + \vec{T}_L + \vec{P} = \mathbf{0}$ Substitute and collect like terms

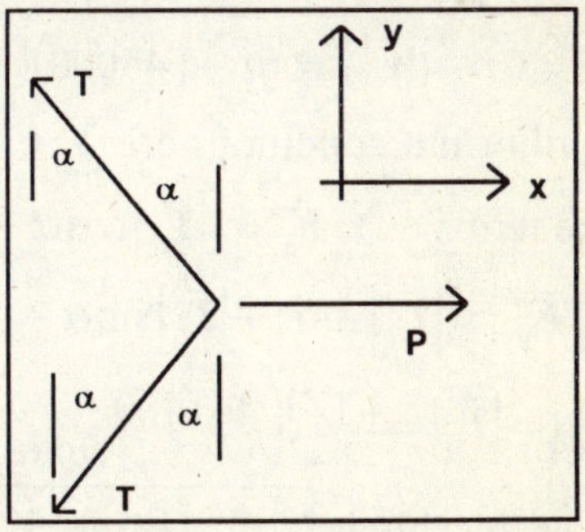

$$\sum\vec{F}_x = \left(-|\vec{T}_U|\sin\alpha - |\vec{T}_L|\sin\alpha + 50\right)\vec{i} = 0$$

$\sum\vec{F}_y = \left(|\vec{T}_U|\cos\alpha - |\vec{T}_L|\cos\alpha\right)\vec{j} = 0$. Solve:

$|\vec{T}| = |\vec{T}_U| = |\vec{T}_L| = \left(\tfrac{1}{2}\right)\left(\tfrac{50}{\sin\alpha}\right)$ For $\alpha = 35^o$,

$$\boxed{|\vec{T}| = \left(\tfrac{1}{2}\right)\left(\tfrac{50}{0.574}\right) = 43.6\ lb}$$

(b) Isolate the bow (See following page.). The forces are the string tensions and the pull force on the center of the bow. The angle between the upper string and the positive x axis is $-(90^o - \alpha) = (270^o + \alpha)$. The angle between the lower string and the positive x axis is $(90^o - \alpha)$. The tensions are: $\vec{T}_U = |\vec{T}_U|\left(\vec{i}\cos(270+\alpha) + \vec{j}\sin(270+\alpha)\right)$. $\vec{T}_U = |\vec{T}_U|\left(\vec{i}\sin\alpha - \vec{j}\cos\alpha\right)$

$\vec{T}_L = |\vec{T}_L|\left(\vec{i}\cos(90-\alpha) + \vec{j}\sin(90-\alpha)\right)$ $\vec{T}_L = |\vec{T}_L|(\vec{i}\sin\alpha + \vec{j}\cos\alpha)$ The pull is $\vec{P} = -|\vec{P}|\vec{i} + 0\vec{j}$

Problem 3.20 continued on next page....

continuation of solution to Problem 3.20...

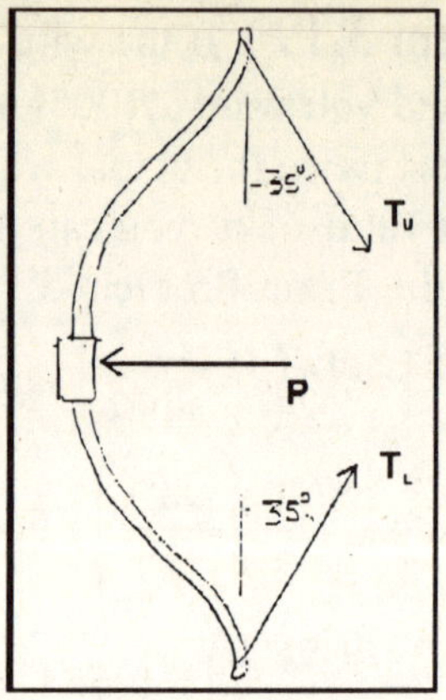

The equilibrium conditions are $\sum \vec{F} = \vec{P} + \vec{T}_U + \vec{T}_L = 0$ Substitute and collect like terms

$\sum \vec{F}_x = \left(|\vec{T}_U|\sin\alpha + |\vec{T}_L|\sin\alpha - |\vec{P}|\right)\vec{i} = 0$

$\sum \vec{F}_y = \left(-|\vec{T}_U|\cos\alpha + |\vec{T}_L|\cos\alpha\right)\vec{j} = 0$ .Solve:

$|\vec{T}_U| = |\vec{T}_L| = \left(\frac{1}{2}\right)\left(\frac{|\vec{P}|}{\sin\alpha}\right)$ for $\alpha = 35^o$ and $|\vec{P}| = 50\ lb$, the tension in the string is $\boxed{|\vec{T}| = 43.6\ lb}$

Problem 3.21 The rocket is suspended by two cables. The mass of the rocket is 45 Mg (megagrams).(a) What is the tension in the cables when the rocket is in the position shown?
(b) If the rocket is raised by reeling in the two cables by equal amounts, what tension must the cables support (based on the tension required to support the stationary rocket) to raise the rocket 2 m above its present position?

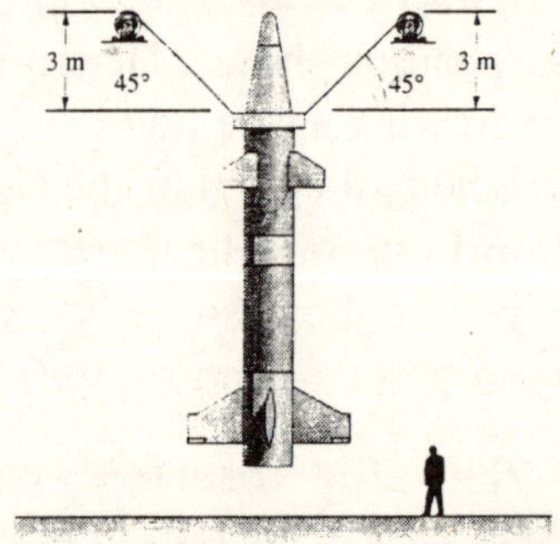

Solution: Isolate the support collar. The forces acting are the weight of the rocket and the tensions in the cables. Solve for tensions in current position. Solve for angles in new position, and determine the tensions. (a) The angle the right hand cable makes with the positive x axis is α. Thus $\vec{T}_R = |\vec{T}_R|(\vec{i}\cos\alpha + \vec{j}\sin\alpha)$ The angle for the left hand cable is $(180^o - \alpha)$. The tension is $\vec{T}_L = |\vec{T}_L|\left(-\vec{i}\cos\alpha + \vec{j}\sin\alpha\right)$ The weight is $\vec{W} = 0\vec{i} - |\vec{W}|\vec{j} = 0\vec{i} - (45000)(9.81)\vec{j} = 0\vec{i} - 441450\vec{j}\ N$ The equilibrium conditions are $\sum \vec{F} = \vec{W} + \vec{T}_L + \vec{T}_R = 0$ .Substitute, collect like terms, $\sum \vec{F}_x = \left(|\vec{T}_R|\cos\alpha - |\vec{T}_L|\cos\alpha\right) = 0$

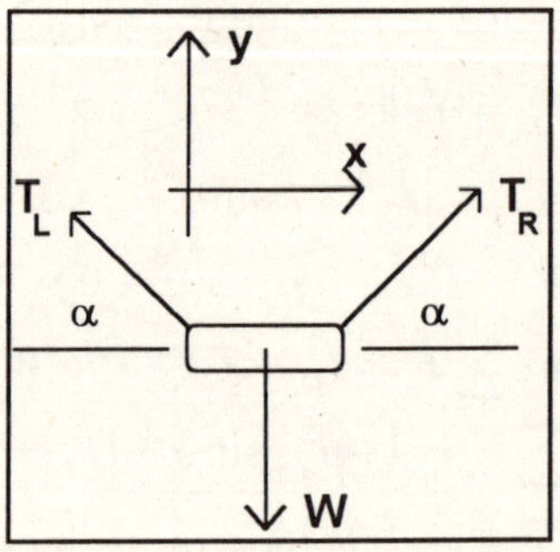

$\sum \vec{F}_y = \left(|\vec{T}_R|\sin\alpha + |\vec{T}_L|\sin\alpha - 441450\right)\vec{j} = 0$ Solve:

$|\vec{T}_R| = |\vec{T}_L| = \left(\frac{1}{2}\right)\left(\frac{441450}{\sin\alpha}\right)$ For $\alpha = 45^o$, the tension in each cable is $\boxed{|\vec{T}_R| = |\vec{T}_L| = 312{,}152\ N = 312\ kN}$

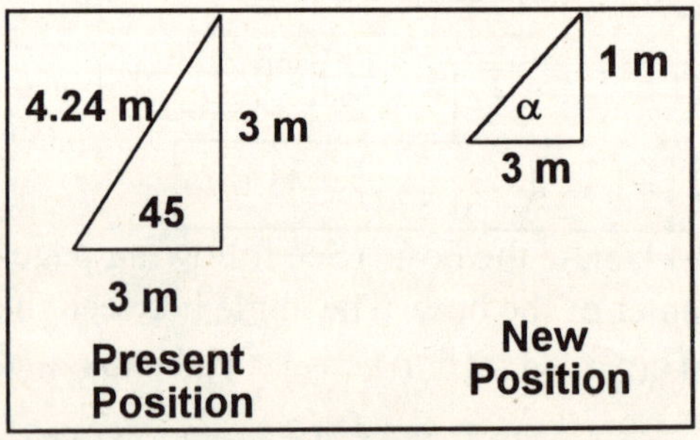

(b) The solution of part (b) can be found from the results of part (a) with new values of the angle α. The horizontal distance between the collar and the cable pulleys remains 3 m before and after. The new vertical distance is 1 m. The new angle is $\alpha = \tan^{-1}\left(\frac{1}{3}\right) = 18.4^o$ Substitute into part (a) $\boxed{|\vec{T}_R| = |\vec{T}_L| = \left(\frac{1}{2}\right)\left(\frac{441450}{\sin\alpha}\right) = 698\ kN}$

Problem 3.22 A construction worker holds a 500 lb. crate in the position shown. What force must she exert on the cable?

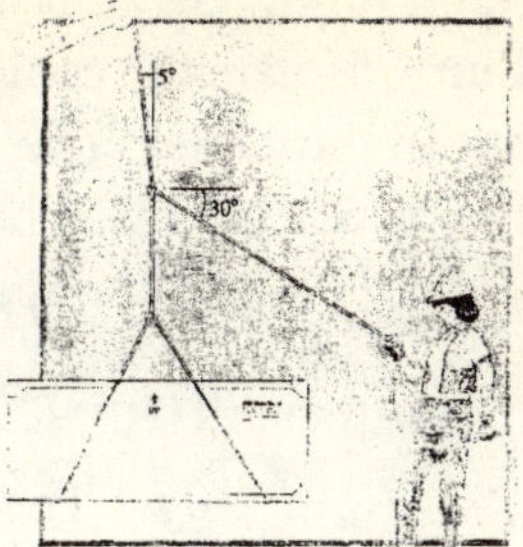

Solution: Isolate the cable juncture. The angle between the upper cable and the positive x axis is 95°. The tension is

$\vec{T} = |\vec{T}|(\vec{i}\cos 95^o + \vec{j}\sin 95^o) = |\vec{T}|(-0.087\vec{i} + 0.996\vec{j})$

The angle between the cable held by the construction worker and the positive x axis is -30°, or 330°. The force is

$\vec{P} = |\vec{P}|(\vec{i}\cos 330^o + \vec{j}\sin 330^o) = |\vec{P}|(0.866\vec{i} - 0.5\vec{j})$

The weight is $\vec{W} = 0\vec{i} - |\vec{W}|\vec{j} = 0\vec{i} - 500\vec{j}\ lb$.

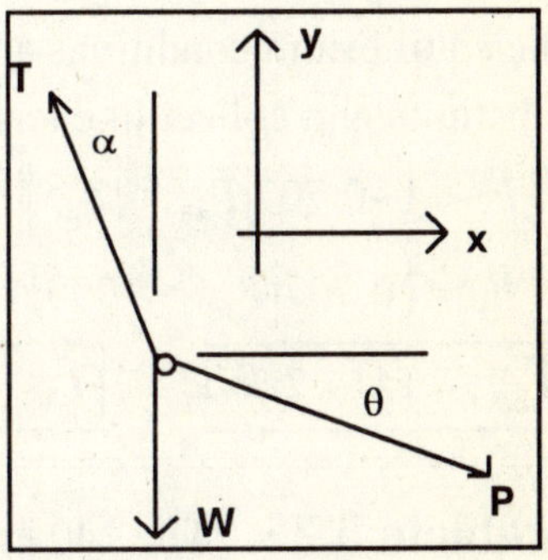

The equilibrium conditions are $\sum \vec{F} = \vec{W} + \vec{T} + \vec{P} = 0$

Substitute and collect like terms: $\sum \vec{F}_x = (-0.087|\vec{T}| + 0.866|\vec{P}|)\vec{i} = 0$

$\sum \vec{F}_y = (0.996|\vec{T}| - 0.5|\vec{P}| - 500)\vec{j} = 0$ Solve: From the first equation $|\vec{T}| = 9.94|\vec{P}|$. Substitute this into the second equation to obtain

$$|\vec{P}| = \left(\frac{|\vec{W}|}{((0.996)(9.94) - 0.5)}\right) = \left(\frac{500}{9.40}\right) \quad \boxed{|\vec{P}| = 53.2\ lb} \quad \textit{Check:}$$

From the law of sines, $|\vec{P}| = \frac{\sin 5^o}{\sin 55^o}|\vec{W}| = \frac{0.0872}{0.8192}(500) = 53.2\ lb.\ check.$

Problem 3.23 A construction worker on the Moon (acceleration due to gravity = 5.32 ft/s²) holds the same crate described in Problem 3.23 in the position shown. What force must she exert on the cable?

Solution: From the solution to Problem 3.23. $|\vec{P}| = \left(\frac{|\vec{W}|}{((0.996)(9.94) - 0.5)}\right) = \left(\frac{|\vec{W}|}{9.40}\right)$ The weight on the moon is the reduced by the ratio of the accelerations due to gravity on the moon to that on the earth,

$|\vec{W}| = \left(\frac{5.32}{32.17}\right)(500) = 82.69\ lb$. Thus the tension in the cable held by the astronaut is

$$\boxed{|\vec{P}| = \left(\frac{82.69}{9.40}\right) = 8.8\ lb}$$

Problem 3.24 Determine the tensions in the cables AB and AC.

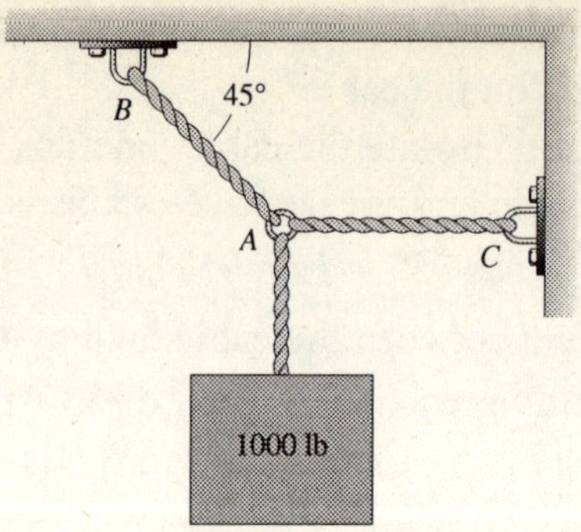

Solution: Isolate the cable juncture. The forces are the tensions in the cables. The angle between cable AB and the positive x axis is 135°. The tension in AB is

$\vec{T}_{AB} = |\vec{T}_{AB}|(\vec{i}\cos 135^{\circ} + \vec{j}\sin 135)$

$\vec{T}_{AB} = |\vec{T}_{AB}|(-0.707\vec{i} + 0.707\vec{j})$ The tension in the cable AC is $\vec{T}_{AC} = |\vec{T}_{AC}|\vec{i} + 0\vec{j}$. The weight:

$\vec{W} = 0\vec{i} - |\vec{W}|\vec{j} = 0\vec{i} - 1000\vec{j}\ lb$

The equilibrium conditions are $\sum\vec{F} = \vec{W} + \vec{T}_{AB} + \vec{T}_{AC} = 0$

Substitute and collect like terms,

$\sum\vec{F}_x = (-0.707|\vec{T}_{AB}| + |\vec{T}_{AC}|)\vec{i} = 0$

$\sum\vec{F}_y = (0.707|\vec{T}_{AB}| - 1000)\vec{j} = 0$ Solve:

$|\vec{T}_{AB}| = 1414.2\ lb$. $|\vec{T}_{AC}| = 1000\ lb$

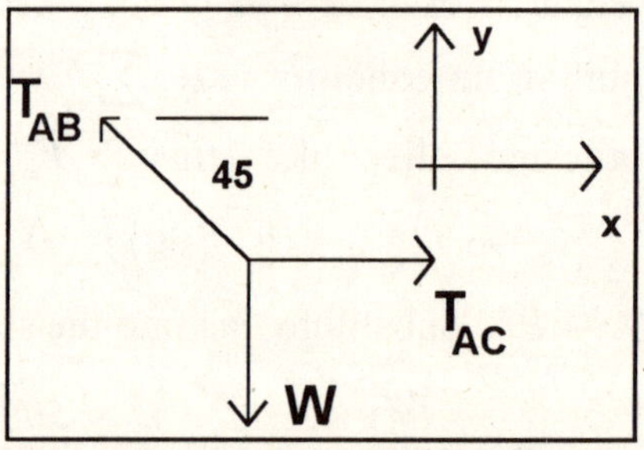

Problem 3.25 The 140 kg traffic light is suspended above the street by two cables. What is the tension in the cables?

Solution: Isolate the traffic light. From symmetry, the angles α formed by the suspension cables are equal. $\tan\alpha = \left(\frac{12\ m}{20\ m}\right) = 0.6$:, $\alpha = 30.964^{o} \cong 31^{o}$

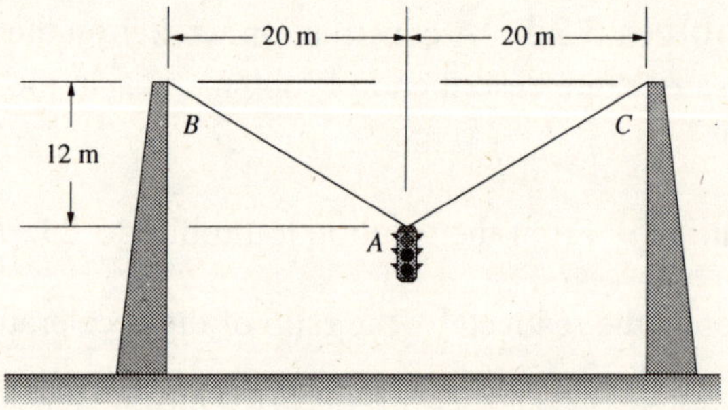

The angle formed by cable **C** and the positive x axis is α .The tension is

$\vec{C} = |\vec{C}|(\vec{i}\cos\alpha + \vec{j}\sin\alpha)$

The angle formed by cable B and the positive x axis is $(180^{o} - \alpha)$. The tension is

$\vec{B} = |\vec{B}|(\vec{i}\cos(180-\alpha) + \vec{j}\sin(180-\alpha))$ The weight is $\vec{W} = 0\vec{i} - \vec{j}|\vec{W}|$. The equilibrium conditions are

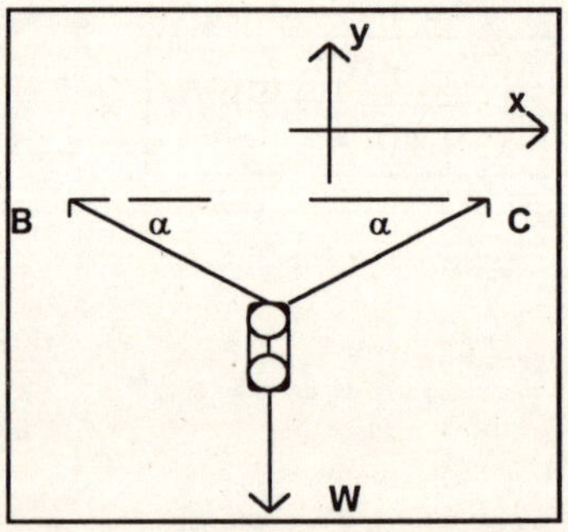

$\sum\vec{F} = \vec{W} + \vec{B} + \vec{C} = 0$ Substitute, and collect like terms:

$\sum\vec{F}_x = (-|\vec{B}|\cos\alpha + |\vec{C}|\cos\alpha)\vec{i} = 0$

$\sum\vec{F}_y = (|\vec{B}|\sin\alpha + |\vec{C}|\sin\alpha - |\vec{W}|)\vec{j} = 0$ From the first equation,

$|\vec{B}| = |\vec{C}|$..Substitute this into the second equation

$|\vec{B}| = |\vec{C}| = \left(\frac{1}{2}\right)\left(\frac{|\vec{W}|}{\sin\alpha}\right)$. For values of

$|\vec{W}| = (140\ kg)\left(9.81\ \frac{m}{s^2}\right) = 1373.4\ N$ and $\alpha = 30.96 \cong 31^{o}$, $|\vec{B}| = |\vec{C}| = \left(\frac{1}{2}\right)\left(\frac{1373.4}{\sin\alpha}\right) = 1334.7\ N$

Problem 3.26 Consider the suspended traffic light in Problem 3.25. To temporarily raise the traffic light during a parade, an engineer wants to connect the 17 m length of cable DE to the midpoints of cables AB and AC as shown in lower sketch. However, for safety considerations, he doesn't want to subject any of the cables to a tension larger than 4 kN. Can he do it?

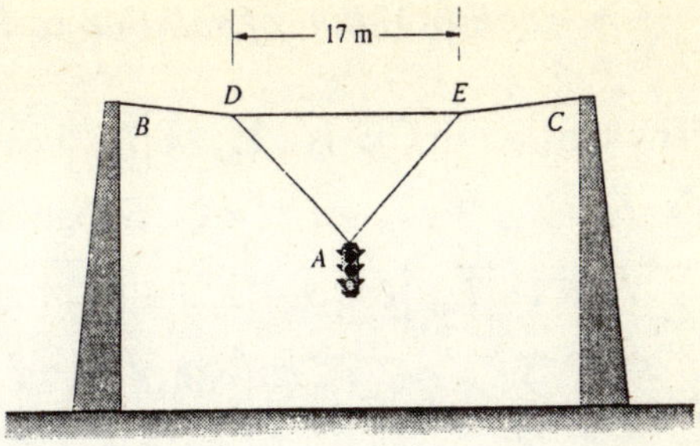

Solution: Determine the length of AC and AB from Problem 3.26: The distance between support poles is 40 m. The vertical drop distance of the light is 12 m. Each triangle is a right triangle so that the length of the cables is $D_{AC} = D_{AB} = \sqrt{(20)^2 + (12)^2} = 23.32\ m$.

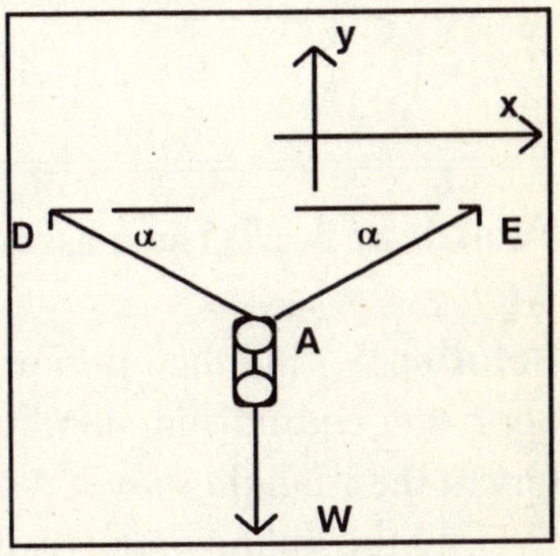

Since the cable DE is attached to the midpoint of the cables AC and AB, AD and AE are each half of this distance, or 11.66 m. The cable DE is given to be 17 m. From this, the angle α is found:

$\cos\alpha = \left(\dfrac{8.5\ m}{11.66\ m}\right) = 0.729$ $\alpha = 43.2^{\circ}$ Isolate the traffic light as shown.

The angle formed by cable AD and the positive x axis is $(180^{\circ} - \alpha)$.

The tension is: $\vec{T}_{AD} = |\vec{T}_{AD}|(\vec{i}\cos(180-\alpha) + \vec{j}\sin(180-\alpha))$. The angle formed by cable AE and the positive x axis is α., hence the tension is $\vec{T}_{AE} = |\vec{T}_{AE}|(\vec{i}\cos\alpha + \vec{j}\sin\alpha)$ The weight is $\vec{W} = 0\vec{i} - \vec{j}|\vec{W}|$

The equilibrium conditions are $\sum\vec{F} = \vec{W} + \vec{T}_{AD} + \vec{T}_{AE} = 0$. Substitute and collect like terms

$\sum\vec{F}_x = (-|\vec{T}_{AD}|\cos\alpha + |\vec{T}_{AE}|\cos\alpha)\vec{i} = 0$, $\sum\vec{F}_y = (|\vec{T}_{AD}|\sin\alpha + |\vec{T}_{AE}|\sin\alpha - |\vec{W}|)\vec{j} = 0$.

Solve, $|\vec{T}_{AD}| = |\vec{T}_{AE}|$.., $|\vec{T}_{AD}| = |\vec{T}_{AE}| = \left(\dfrac{1}{2}\right)\left(\dfrac{|\vec{W}|}{\sin\alpha}\right)$. For

$|\vec{W}| = (140\ kg)\left(9.81\ \dfrac{m}{s^2}\right) = 1373.4\ N$ and

$\alpha = 43.2^{\circ}$, $\boxed{|\vec{T}_{AD}| = |\vec{T}_{AE}| = \left(\dfrac{1}{2}\right)\left(\dfrac{1373.4}{\sin 43.2}\right) = 1003\ N}$

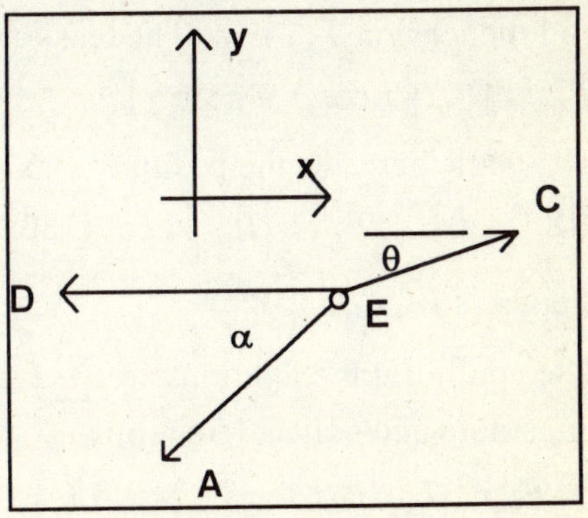

Isolate the cable juncture E as shown. The angle θ is found as follows: The cable EC is 11.66 m. The distance between poles 40 m. The cable DE is 17 m, and cable DE is horizontal. Thus EC projects onto the x-axis

$H_{EC} = \left(\dfrac{1}{2}\right)(40 - 17) = 11.5\ m$ .The ratio is the cosine of the

angle, $\cos\theta = \left(\dfrac{11.5}{11.66}\right) = 0.9863$, or, $\theta = 9.5^{\circ}$ . .From the law of sines: $\dfrac{|\vec{T}_{AD}|}{\sin\theta} = \dfrac{|\vec{T}_{CE}|}{\sin(180-\alpha)}$ from

which $\boxed{|\vec{T}_{CE}| = 4159.6 N}$. Thus the tension in the cable CE exceeds the allowable limit of 4 kN.

Check: Use components: The angle between AE and the positive x axis is $(180^{\circ} + \alpha)$. The tension in AE is $\vec{T}_{AE} = |\vec{T}_{AE}|(\vec{i}\cos(180+\alpha) + \vec{j}\sin(180+\alpha))$ $\vec{T}_{AE} = |\vec{T}_{AE}|(-\vec{i}\cos\alpha - \vec{j}\sin\alpha)$ The tension in ED is: $\vec{T}_{ED} = -|\vec{T}_{ED}|\vec{i} + 0\vec{j}$ .The angle between CE and the positive x axis is θ.

Problem 3.26 continued on next page....

continuation of solution to Problem 3.26...

The tension in CE is : $\vec{T}_{CE} = |\vec{T}_{CE}|(\vec{i}\cos\theta + \vec{j}\sin\theta)$ The equilibrium conditions are $\sum\vec{F} = \vec{T}_{AE} + \vec{T}_{ED} + \vec{T}_{CE} = 0$. Substitute and collect like terms: $\sum\vec{F}_x = (-|\vec{T}_{AE}|\cos\alpha - |\vec{T}_{ED}| + |\vec{T}_{CE}|\cos\theta)\vec{i} = 0$, $\sum\vec{F}_y = (-|\vec{T}_{AE}|\sin\alpha + |\vec{T}_{CE}|\sin\theta)\vec{j} = 0$. From the second equation, $|\vec{T}_{CE}| = \left(\frac{\sin\alpha}{\sin\theta}\right)|\vec{T}_{AE}|$ For $\alpha = 43.2^\circ$, $\theta = 9.5^\circ$, and $|\vec{T}_{AE}| = 1003.2\ N$,

$|\vec{T}_{CE}| = \left(\frac{0.6845}{0.1651}\right)(1003.2) = 4159.6\ N$ *check.*

Problem 3.27 The mass of the suspended crate is 5 kg. What are the tensions in the cables AB and AC?

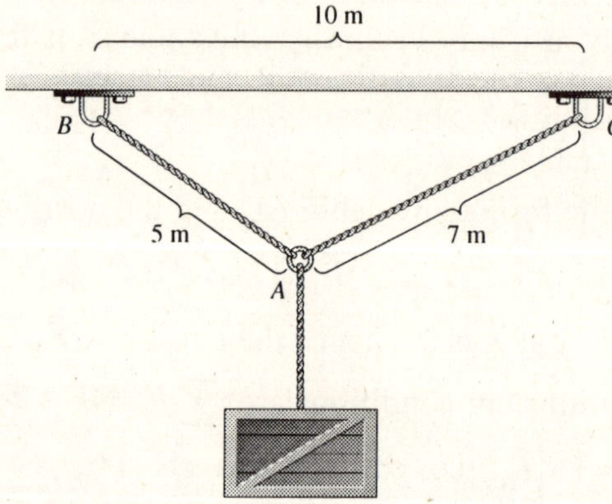

Solution: Find the interior angles in the figure, then apply the equilibrium conditions to the isolated crate. Given the triangle shown, with known sides A, B, and C, find the unknown interior angles α, β, and γ using the . law of cosines

$B^2 = A^2 + C^2 - 2AC\cos\beta$

Solve: $\cos\beta = \frac{A^2 + C^2 - B^2}{2AC}$ Similarly,

$\cos\gamma = \frac{A^2 + B^2 - C^2}{2AB}$. For $A = 10$, $B = 7$, $C = 5$,

$\gamma = 27.66^\circ$ and $\beta = 40.54^\circ$. The third angle is $\alpha = (180 - 27.66 - 40.64) = 111.8^\circ$

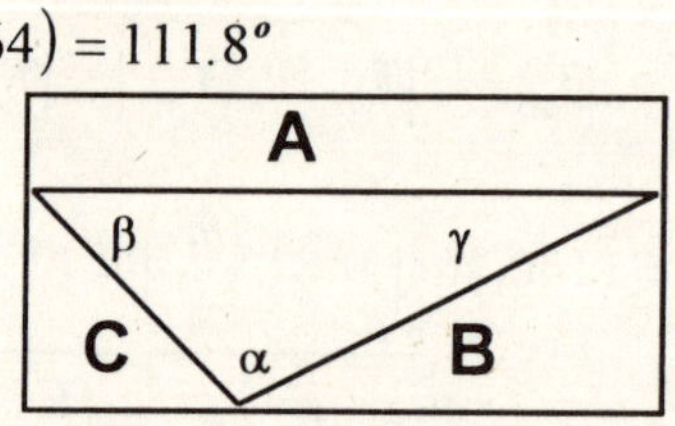

Isolate the cable juncture at A. The angle between the positive x axis and the tension $\vec{T}_{AC}$ is γ. The tension is

$\vec{T}_{AC} = |\vec{T}_{AC}|(\vec{i}\cos\gamma + \vec{j}\sin\gamma)$

The angle between the positive x axis and the tension $\vec{T}_{AB}$ is $(180^\circ - \beta)$., $\vec{T}_{AB} = |\vec{T}_{AB}|(\vec{i}\cos(180^\circ - \beta) + \vec{j}\sin(180^\circ - \beta))$ $\vec{T}_{AB} = |\vec{T}_{AB}|(-\vec{i}\cos\beta + \vec{j}\sin\beta)$ The weight is $\vec{W} = 0\vec{i} - |\vec{W}|\vec{j}$

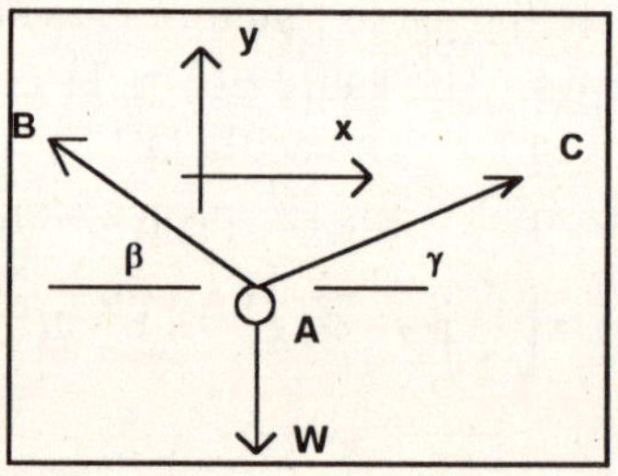

The equilibrium conditions are $\sum\vec{F} = \vec{W} + \vec{T}_{AB} + \vec{T}_{AC} = 0$

Substitute and collect like terms,

$\sum\vec{F}_x = (|\vec{T}_{AC}|\cos\gamma - |\vec{T}_{AB}|\cos\beta)\vec{i} = 0$

$\sum\vec{F}_y = (|\vec{T}_{AC}|\sin\gamma + |\vec{T}_{AB}|\sin\beta - |\vec{W}|) = 0$ Solve:,

$|\vec{T}_{AC}| = \left(\frac{\cos\beta}{\cos\gamma}\right)|\vec{T}_{AB}|$, $|\vec{T}_{AB}| = \left(\frac{|\vec{W}|\cos\gamma}{\sin(\beta+\gamma)}\right)$, $|\vec{T}_{AC}| = \left(\frac{|\vec{W}|\cos\beta}{\sin(\beta+\gamma)}\right)$ For

$|\vec{W}| = (5kg)\left(9.81\frac{m}{s^2}\right) = 49.05\ N$, and $\beta = 40.54^\circ$, $\gamma = 27.66^\circ$ $\boxed{|\vec{T}_{AB}| = 46.79\ N,\ |\vec{T}_{AC}| = 40.15\ N}$

Problem 3.28 What are the tensions in the upper and lower cables? (Your answers will be in terms of W. Neglect the weight of the pulley.)

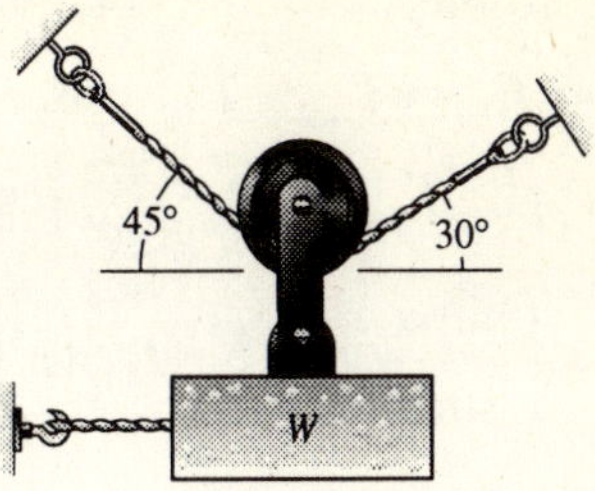

Solution: Isolate the weight. The frictionless pulley changes the direction but not the magnitude of the tension. The angle between the right hand upper cable and the x axis is α, hence

$\vec{T}_{UR} = |\vec{T}_U|(\vec{i}\cos\alpha + \vec{j}\sin\alpha)$

The angle between the positive x and the left hand upper pulley is $(180^o - \beta)$,

hence $\vec{T}_{UL} = |\vec{T}_U|(\vec{i}\cos(180-\beta) + \vec{j}\sin(180-\beta))$

$= |\vec{T}_U|(-\vec{i}\cos\beta + \vec{j}\sin\beta)$ The lower cable exerts a force:

$\vec{T}_L = -|\vec{T}_L|\vec{i} + 0\vec{j}$ The weight: $\vec{W} = 0\vec{i} - |\vec{W}|\vec{j}$

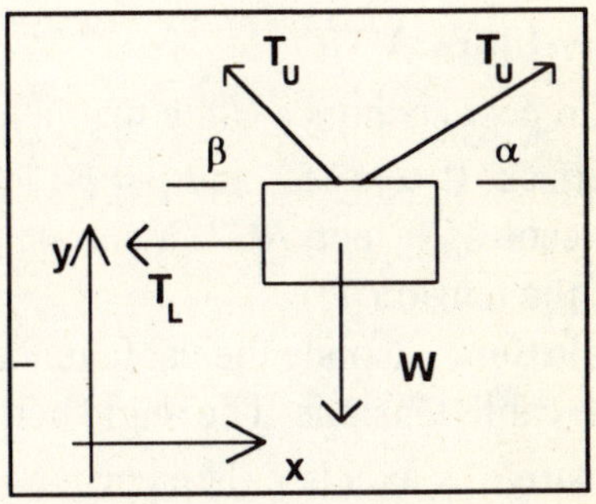

The equilibrium conditions are $\sum \vec{F} = \vec{W} + \vec{T}_{UL} + \vec{T}_{UR} + \vec{T}_L = 0$

Substitute and collect like terms,

$\sum \vec{F}_x = (-|\vec{T}_U|\cos\beta + |\vec{T}_U|\cos\alpha - |\vec{T}_L|)\vec{i} = 0$

$\sum \vec{F}_y = (|\vec{T}_U|\sin\alpha + |\vec{T}_U|\sin\beta - |\vec{W}|)\vec{j} = 0$ Solve:

$|\vec{T}_U| = \left(\dfrac{|\vec{W}|}{(\sin\alpha + \sin\beta)}\right)$, $|\vec{T}_L| = |\vec{T}_U|(\cos\alpha - \cos\beta)$ From which

$|\vec{T}_L| = |\vec{W}|\left(\dfrac{\cos\alpha - \cos\beta}{\sin\alpha + \sin\beta}\right)$. For $\alpha = 30^o$ and $\beta = 45^o$ $\boxed{|\vec{T}_U| = 0.828\ |\vec{W}|,\ |\vec{T}_L| = 0.132\ |\vec{W}|}$

Problem 3.29 An astronaut candidate conducts experiments on an air bearing platform. While he carries out calibrations, the platform is held in place by the horizontal tethers AB, AC, and AD. The forces exerted by the tethers are the only horizontal forces acting on the platform. If the tension in tether AC is 2 N, what are the tensions in the other two tethers?

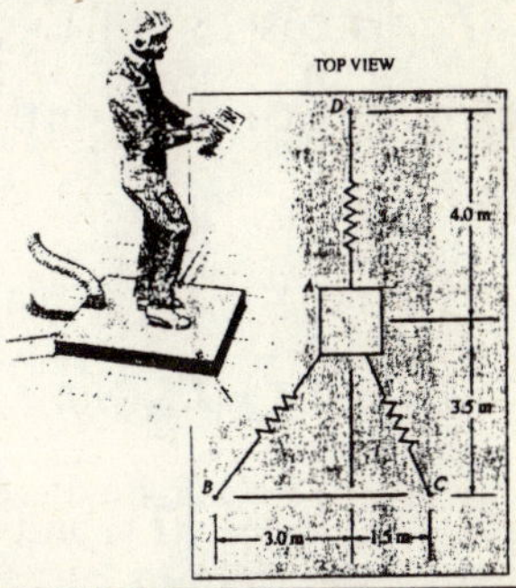

Solution: Isolate the platform. The angles α and β are

$\tan\alpha = \left(\dfrac{1.5}{3.5}\right) = 0.429$, $\alpha = 23.2^o$. Similarly

$\tan\beta = \left(\dfrac{3.0}{3.5}\right) = 0.857$., $\beta = 40.6^o$. The angle between the tether AB and the positive x axis is $(180^o - \beta)$, hence

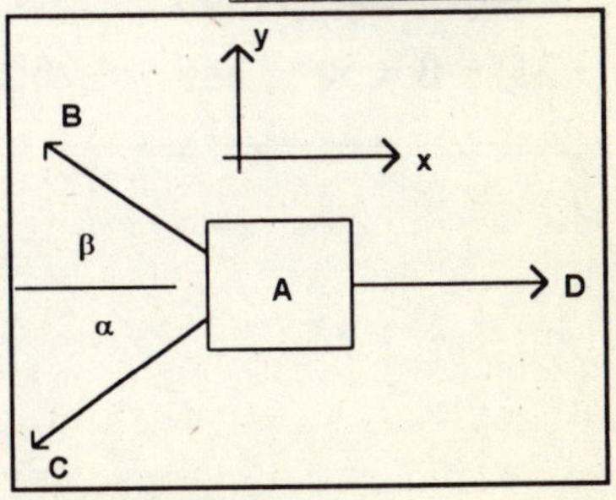

$\vec{T}_{AB} = |\vec{T}_{AB}|(\vec{i}\cos(180^o - \beta) + \vec{j}\sin(180^o - \beta))$

$\vec{T}_{AB} = |\vec{T}_{AB}|(-\vec{i}\cos\beta + \vec{j}\sin\beta)$. The angle between the tether AC and the positive x axis is $(180^o + \alpha)$.

Problem 3.29 continued on next page....

continuation of solution to Problem 3.29...

The tension is $\vec{T}_{AC} = |\vec{T}_{AC}|(\vec{i}\cos(180^o + \alpha) + \vec{j}\sin(180^o + \alpha)) = |\vec{T}_{AC}|(-\vec{i}\cos\alpha - \vec{j}\sin\alpha)$. The tether AD is aligned with the positive x axis, $\vec{T}_{AD} = |\vec{T}_{AD}|\vec{i} + 0\vec{j}$. The equilibrium condition: $\sum \vec{F} = \vec{T}_{AD} + \vec{T}_{AB} + \vec{T}_{AC} = 0$ Substitute and collect like terms, $\sum \vec{F}_x = (-|\vec{T}_{AB}|\cos\beta - |\vec{T}_{AC}|\cos\alpha + |\vec{T}_{AD}|)\vec{i} = 0$, $\sum \vec{F}_y = (|\vec{T}_{AB}|\sin\beta - |\vec{T}_{AC}|\sin\alpha)\vec{j} = 0$ Solve:

$$|\vec{T}_{AB}| = \left(\frac{\sin\alpha}{\sin\beta}\right)|\vec{T}_{AC}|,\ |\vec{T}_{AD}| = \left(\frac{|\vec{T}_{AC}|\sin(\alpha+\beta)}{\sin\beta}\right)$$

For $|\vec{T}_{AC}| = 2\ N$, $\alpha = 23.2^o$ and $\beta = 40.6^o$; $|\vec{T}_{AB}| = 1.21\ N$, $|\vec{T}_{AD}| = 2.76\ N$

Problem 3.30

The post anchors a cable that helps support an oil derrick. If $\alpha = 35^o$ and $\beta = 50^o$,what are the tensions in cables AB and AC? (Your answers will be in terms of the tension T.)

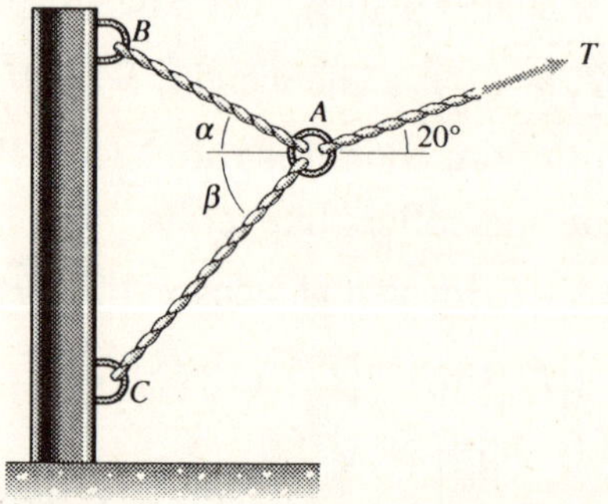

Solution: Isolate the cable juncture. The forces are the cable tensions. The angle between AT and the positive x axis is γ. The tension in AT is $\vec{T}$: $\vec{T} = |\vec{T}|(\vec{i}\cos\gamma + \vec{j}\sin\gamma)$ The angle between AB and the positive x axis is $(180^o - \alpha)$. The tension in AB: $\vec{T}_{AB} = |\vec{T}_{AB}|(\vec{i}\cos(180-\alpha) + \vec{j}\sin(180-\alpha))\vec{T}_{AB} = |\vec{T}_{AB}|(-\vec{i}\cos\alpha + \vec{j}\sin\alpha)$ The angle between AC and the positive x axis is $(180^o + \beta)$. The tension in AC is

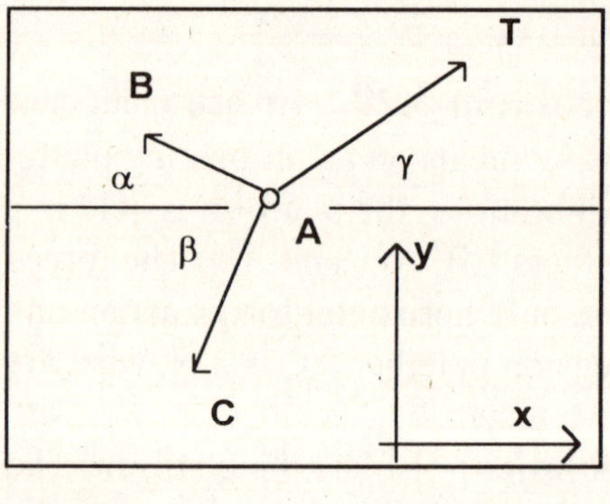

$\vec{T}_{AC} = |\vec{T}_{AC}|(\vec{i}\cos(180+\beta) + \vec{j}\sin(180+\beta))$

$\vec{T}_{AC} = |\vec{T}_{AC}|(-\vec{i}\cos\beta - \vec{j}\sin\beta)$ The equilibrium conditions are $\sum \vec{F} = \vec{T}_{AB} + \vec{T}_{AC} + \vec{T} = 0$. Substitute and collect like terms,

$\sum \vec{F}_x = (|\vec{T}|\cos\gamma - |\vec{T}_{AB}|\cos\alpha - |\vec{T}_{AC}|\cos\beta)\vec{i} = 0$,

$\sum \vec{F}_y = (|\vec{T}|\sin\gamma + |\vec{T}_{AB}|\sin\alpha - |\vec{T}_{AC}|\sin\beta)\vec{j} = 0$ Solve:

$$|\vec{T}_{AC}| = \left(\frac{\sin(\alpha+\gamma)}{\sin(\alpha+\beta)}\right)|\vec{T}| \text{ and } |\vec{T}_{AB}| = \left(\frac{\sin(\beta-\gamma)}{\sin(\alpha+\beta)}\right)|\vec{T}| \text{ For}$$

$\alpha = 35^o$, $\beta = 50^o$, and $\gamma = 20^o$, $|\vec{T}_{AC}| = 0.8223|\vec{T}|$, $|\vec{T}_{AB}| = 0.5019|\vec{T}|$

Problem 3.31 Consider the system shown in Problem 3.30. The angle $\alpha = 35^o$.(a) Determine the value of β that minimizes tension in the cable AC. (b) When β has that value determined in Part (a), what are the tensions in cables AB and AC? (Your answers will be in terms of the tension T.)

Solution: Isolate cable juncture. (a) From Problem 3.30, $\left|\vec{T}_{AC}\right| = \left(\dfrac{\sin(\alpha+\gamma)}{\sin(\alpha+\beta)}\right)\left|\vec{T}\right|$ and

$\left|\vec{T}_{AB}\right| = \left(\dfrac{\sin(\beta-\gamma)}{\sin(\alpha+\beta)}\right)\left|\vec{T}\right|$ The value of the angle β that minimizes the tension is obtainable by inspection.:

When the denominator of the equation for $\left|\vec{T}_{AC}\right|$ is a maximum, the value of $\left|\vec{T}_{AC}\right|$ is minimized. The maximum value of $\sin(\alpha+\beta) = 1$ occurs for $(\alpha + \beta) = 90 + 360\,n \quad (n = 0,1,2...)$, but only the value $n = 0$ has meaning for this problem. Thus the minimizing value is

$\boxed{\beta = 90^o - \alpha = 90^o - 35^o = 55^o}$. (b) The tension for this value of β:

$\boxed{\left|\vec{T}_{AC}\right| = 0.819\left|\vec{T}\right|, \quad \left|\vec{T}_{AB}\right| = 0.574\left|\vec{T}\right|}$

Problem 3.32 The unstretched length of the spring AB is 660 mm and the spring constant is k = 1000 N/m. What is the mass of the object?

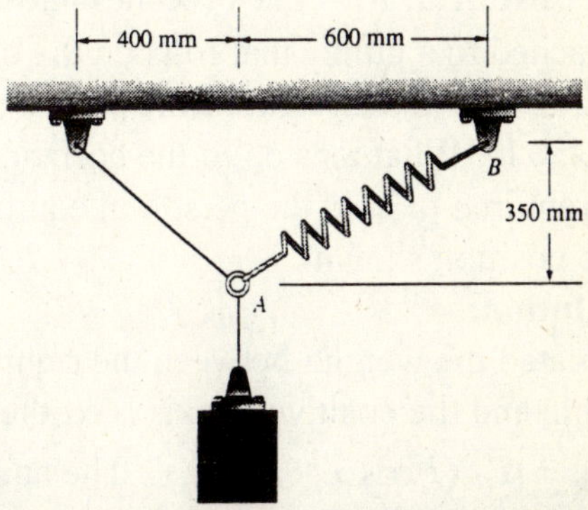

Solution: Use the linear spring force-extension relation to find the magnitude of the tension in spring AB. Isolate juncture A. The forces are the weight and the tensions in the cables. The angles are

$$\tan\alpha = \left(\frac{350}{600}\right) = 0.5833 \quad , \quad \alpha = 30.26^o \ .$$

$$\tan\beta = \left(\frac{350}{400}\right) = 0.875 \quad , \beta = 41.2^o \ .$$

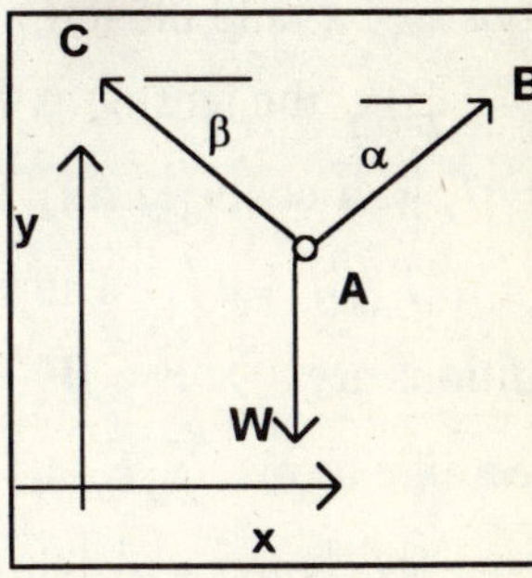

The angle between the x axis and the spring is α. The tension is $\vec{T}_{AB} = \left|\vec{T}_{AB}\right|\left(\vec{i}\cos\alpha + \vec{j}\sin\alpha\right)$ The angle between the x axis and AC is $(180-\beta)$. The tension is $T_{AC} = \left|\vec{T}_{AC}\right|\left(\vec{i}\cos(180-\beta) + \vec{j}\sin(180-\beta)\right)$

$\vec{T}_{AC} = \left|\vec{T}_{AC}\right|\left(-\vec{i}\cos\beta + \vec{j}\sin\beta\right)$ The weight is: $\vec{W} = 0\vec{i} - \left|\vec{W}\right|\vec{j}$

The equilibrium conditions: $\sum\vec{F} = \vec{W} + \vec{T}_{AB} + \vec{T}_{AC} = 0$ Substitute and collect like terms $\sum\vec{F}_x = \left(\left|\vec{T}_{AB}\right|\cos\alpha - \left|\vec{T}_{AC}\right|\cos\beta\right)\vec{i} = 0$,

$\sum\vec{F}_y = \left(\left|\vec{T}_{AC}\right|\sin\alpha + \left|\vec{T}_{AB}\right|\sin\beta - \left|\vec{W}\right|\right)\vec{j} = 0$. Solve: $\left|\vec{T}_{AC}\right| = \left(\dfrac{\cos\alpha}{\cos\beta}\right)\left|\vec{T}_{AB}\right|$ and

$\left|\vec{W}\right| = \left(\dfrac{\sin(\alpha+\beta)}{\cos\beta}\right)\left|\vec{T}_{AB}\right|$ The tension $\left|\vec{T}_{AB}\right|$ is found from the linear spring force-deflection relation.

The spring extension is $\Delta L = \sqrt{(350)^2 + (600)^2} - 660 = 694.62 - 660 = 34.62\ mm$

The tension is $\left|\vec{T}_{AB}\right| = k\,\Delta L = (1000)(0.03462) = 34.6\ N$. For $\alpha = 30.26^o$ and $\beta = 41.2^o$; the weight is

$\left|\vec{W}\right| = \left(\dfrac{0.948}{0.752}\right)(34.6) = 43.62\ N$; The mass is $\boxed{m = \left(\dfrac{\left|\vec{W}\right|}{\left|\vec{g}\right|}\right) = \left(\dfrac{43.62}{9.81}\right) = 4.447\ kg}$

================================<>================================

Problem 3.33 If the system described in Problem 3.32 is on Mars (acceleration due to gravity 4.02 m/s²), what is the mass of the suspended object?

Solution: Isolate juncture A. From Problem 3.32, $|\vec{T}_{AC}| = \left(\frac{\cos\alpha}{\cos\beta}\right)|\vec{T}_{AB}|$ and $|\vec{W}| = \left(\frac{\sin(\alpha+\beta)}{\cos\beta}\right)|\vec{T}_{AB}|$ From Problem 3.32 the tension $|\vec{T}_{AB}|$ is $|\vec{T}_{AB}| = k\,DL = (1000)(0.03462) = 34.6\ N$. For $\alpha = 30.26^\circ$ and $\beta = 41.2^\circ$, the weight is $|\vec{W}| = \left(\frac{0.948}{0.752}\right)(34.6) = 43.62\ N$ If the system above is on Mars, the mass is

$$m = \left(\frac{|\vec{W}|}{|\vec{g}|}\right) = \left(\frac{43.62}{4.02}\right) = 10.85\ kg$$

================================<>================================

Problem 3.34 The breeches buoy is used to transfer the person B between two ships. The person is attached to a pulley that rolls on the overhead cable. The total weight of the person and the buoy is 250 lb. What tension in the horizontal line AB is required to hold the person in equilibrium in the position shown?

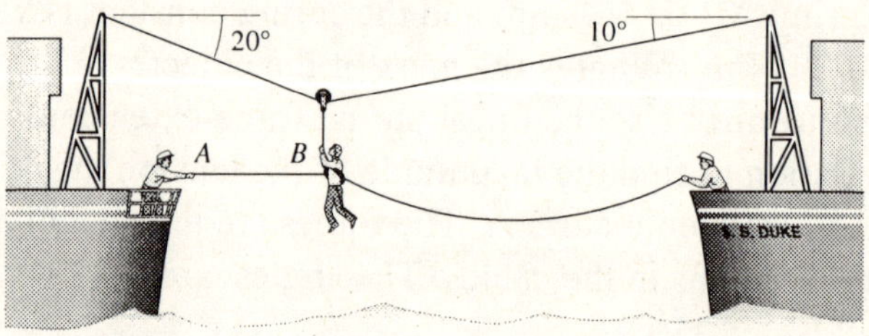

Solution:

Isolated the weight. between the right hand upper cable and the positive x axis is α; the tension is $\vec{T}_{UR} = |\vec{T}_U|(\vec{i}\cos\alpha + \vec{j}\sin\alpha)$. The angle between the positive x and the left hand upper pulley is $(180^\circ - \beta)$; the tension is $\vec{T}_{UL} = |\vec{T}_U|(\vec{i}\cos(180-\beta) + \vec{j}\sin(180-\beta))$ $\vec{T}_U = |\vec{T}_U|(-\vec{i}\cos\beta + \vec{j}\sin\beta)$. The tension in the lower cable is $\vec{T}_{AB} = -|\vec{T}_{AB}|\vec{i} + 0\vec{j}$. The weight is $\vec{W} = 0\vec{i} - |\vec{W}|\vec{j}$. The equilibrium conditions are $\sum\vec{F} = \vec{W} + \vec{T}_{UL} + \vec{T}_{UR} + \vec{T}_{AB} = 0$. Substitute and collect like terms, $\sum\vec{F}_x = (-|\vec{T}_U|\cos\beta + |\vec{T}_U|\cos\alpha - |\vec{T}_{AB}|)\vec{i} = 0$ $\sum\vec{F}_y = (|\vec{T}_U|\sin\alpha + |\vec{T}_U|\sin\beta - |\vec{W}|)\vec{j} = 0$. Solve:

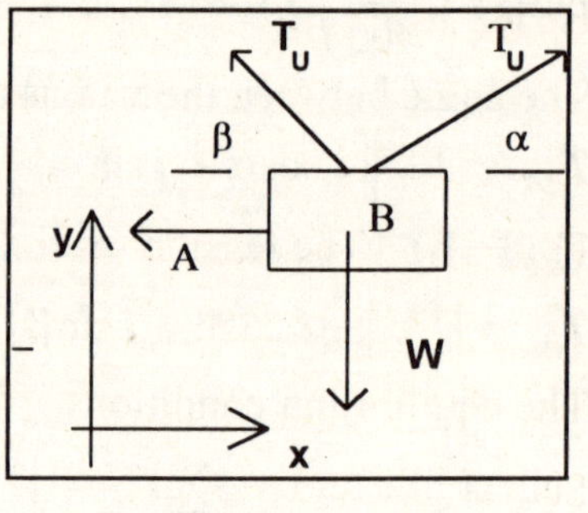

$|\vec{T}_U| = \left(\frac{|\vec{W}|}{(\sin\alpha + \sin\beta)}\right)$, and $|\vec{T}_{AB}| = |\vec{T}_U|(\cos\alpha - \cos\beta)$, from which $|\vec{T}_{AB}| = |\vec{W}|\left(\frac{\cos\alpha - \cos\beta}{\sin\alpha + \sin\beta}\right)$ For $\alpha = 10^\circ$ and , and $|\vec{W}| = 250\ lb$., $|\vec{T}_{AB}| = 21.87\ lb$

================================<>================================

Problem 3.35 The system of cables suspends a 1000 lb bank of lights above a movie set. Determine the tension in cables AB, CD, and CE.

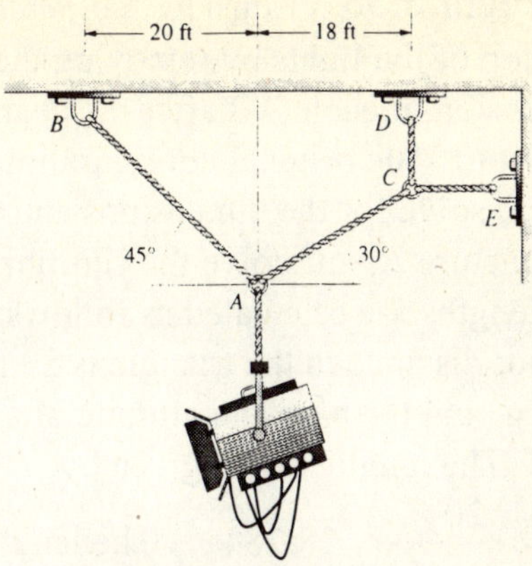

Solution: Isolate juncture A, and solve the equilibrium equations. Repeat for the cable juncture C.

The angle between the cable AC and the positive x axis is α. The tension in AC is $\vec{T}_{AC} = |\vec{T}_{AC}|(\vec{i}\cos\alpha + \vec{j}\sin\alpha)$

The angle between the x axis and AB is $(180^o - \beta)$. The tension is $\vec{T}_{AB} = |\vec{T}_{AB}|(\vec{i}\cos(180-\beta) + \vec{j}\sin(180-\beta))$

$\vec{T}_{AB} = (-\vec{i}\cos\beta + \vec{j}\sin\beta)$ The weight is $\vec{W} = 0\vec{i} - |\vec{W}|\vec{j}$.

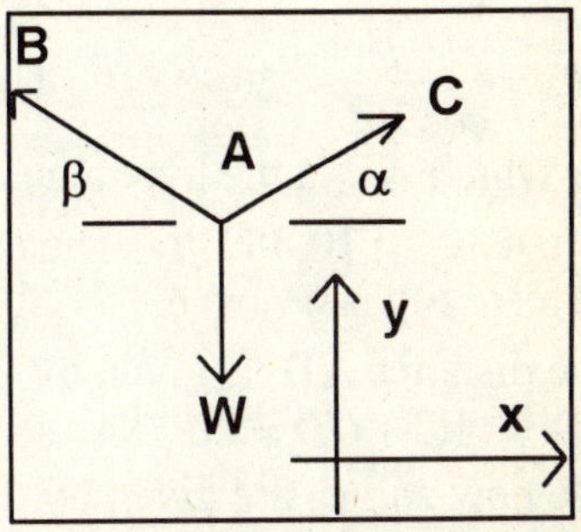

The equilibrium conditions are $\sum\vec{F} = 0 = \vec{W} + \vec{T}_{AB} + \vec{T}_{AC} = 0$

Substitute and collect like terms,

$\sum\vec{F}_x = (|\vec{T}_{AC}|\cos\alpha - |\vec{T}_{AB}|\cos\beta)\vec{i} = 0$

$\sum\vec{F}_y = (|\vec{T}_{AB}|\sin\beta + |\vec{T}_{AC}|\sin\alpha - |\vec{W}|)\vec{j} = 0$ Solve:

$$|\vec{T}_{AB}| = \left(\frac{\cos\alpha}{\cos\beta}\right)|\vec{T}_{AC}| \quad \text{and} \quad |\vec{T}_{AC}| = \left(\frac{|\vec{W}|\cos\beta}{\sin(\alpha+\beta)}\right),$$

$|\vec{W}| = 1000\ lb$, and $\alpha = 30^o$, $\beta = 45^o$

$$|\vec{T}_{AC}| = (1000)\left(\frac{0.7071}{0.9659}\right) = 732.05\ lb$$

$$|\vec{T}_{AB}| = (732)\left(\frac{0.866}{0.7071}\right) = 896.5\ lb$$

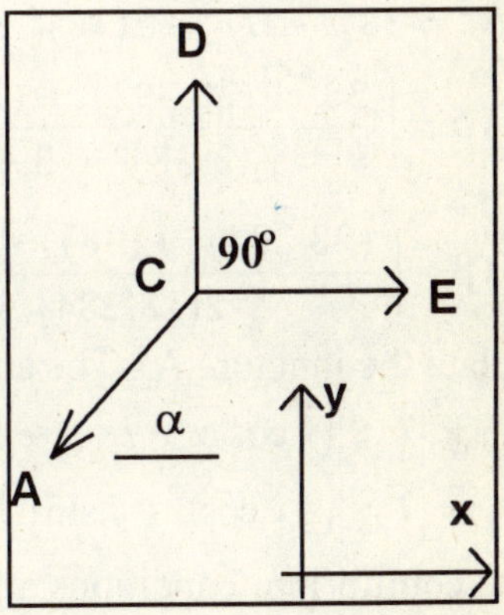

Isolate juncture C. The angle between the positive x axis and the cable CA is $(180^o - \alpha)$. The tension is

$\vec{T}_{CA} = |\vec{T}_{CA}|(\vec{i}\cos(180^o + \alpha) + \vec{j}\sin(180^o + \alpha))$, or

$\vec{T}_{CA} = |\vec{T}_{CA}|(-\vec{i}\cos\alpha - \vec{j}\sin\alpha)$ The tension in the cable CE is

$\vec{T}_{CE} = \vec{i}|\vec{T}_{CE}| + 0\vec{j}$ The tension in the cable CD is $\vec{T}_{CD} = 0\vec{i} + \vec{j}|\vec{T}_{CD}|$

The equilibrium conditions are $\sum\vec{F} = 0 = \vec{T}_{CA} + \vec{T}_{CE} + \vec{T}_{CD} = 0$

Substitute t and collect like terms, $\sum\vec{F}_x = (|\vec{T}_{CE}| - |\vec{T}_{CA}|\cos\alpha)\vec{i} = 0$, $\sum\vec{F}_y = (|\vec{T}_{CD}| - |\vec{T}_{CA}|\sin\alpha)\vec{j} = 0$

Solve: $|\vec{T}_{CE}| = |\vec{T}_{CA}|\cos\alpha$, $|\vec{T}_{CD}| = |\vec{T}_{CA}|\sin\alpha$; for $|\vec{T}_{CA}| = 732\ lb$ and $\alpha = 30^o$,

$$\boxed{|\vec{T}_{AB}| = 896.6\ lb,\ |\vec{T}_{CE}| = 634\ lb,\ |\vec{T}_{CD}| = 366\ lb}$$

Problem 3.36

Problem 3.36 Consider the 1000 lb bank of lights in Problem 3.35. A technician changes the position of the lights by removing the cable CE. What is the tension in cable AB after the change?

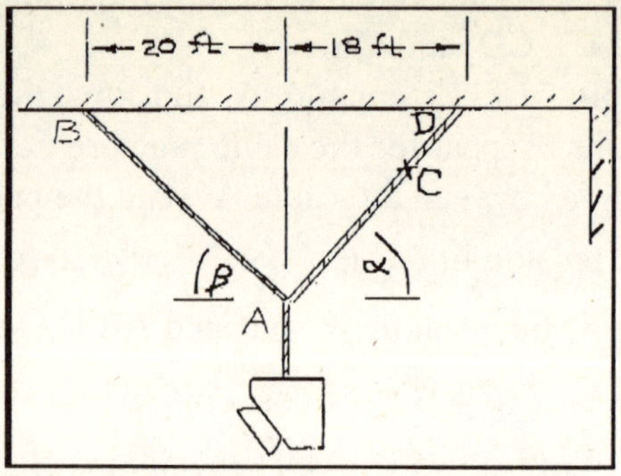

Solution: The original configuration in Problem 3.35 is used to solve for the dimensions and the angles. Isolate the juncture A, and solve the equilibrium conditions. The lengths are calculated as follows: The vertical interior distance in the triangle is 20 ft, since the angle is 45 deg. and the base and altitude of a 45 deg triangle are equal. The length AB is given by

$\overline{AB} = \dfrac{20\ ft}{\cos 45^o} = 28.284\ ft$. The length AC is given by

$\overline{AC} = \dfrac{18\ ft}{\cos 30^o} = 20.785\ ft$. The altitude of the triangle for which AC is the hypotenuse is

$18 \tan 30^o = 10.392\ ft$. The distance CD is given by $20 - 10.392 = 9.608\ ft$.

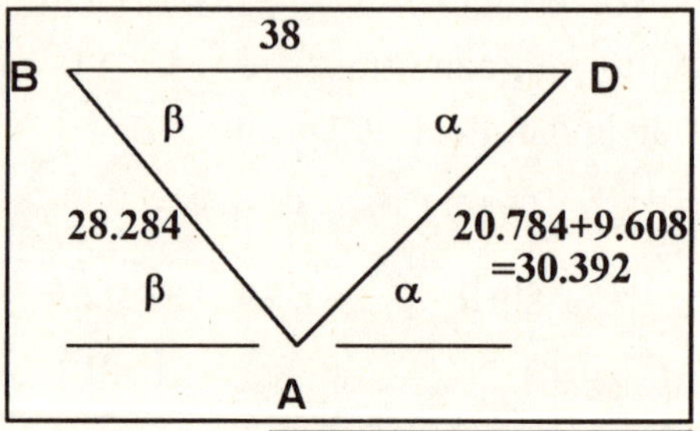

The distance **AD** is given by

$AD = AC + CD = 20.784 + 9.608 = 30.392$

The new angles are given by the cosine law

$AB^2 = 38^2 + AD^2 - 2(38)(AD)\cos\alpha$. Reduce and solve:

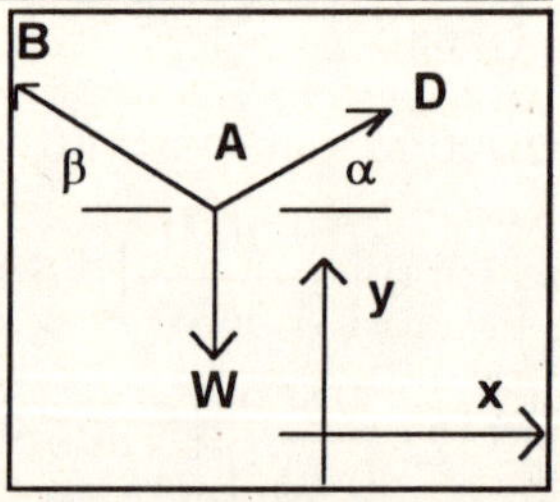

$$\cos\alpha = \left(\frac{38^2 + (30.392)^2 - (28.284)^2}{2(38)(30.392)}\right) = 0.6787\ ,\ \alpha = 47.23^o.$$

$$\cos\beta = \left(\frac{(28.284)^2 + (38)^2 - (30.392)^2}{2(28.284)(38)}\right) = 0.6142\ ,\ \beta = 52.1^o.$$

Isolate the juncture A. The angle between the cable AD and the positive x axis is α. The tension is: $\vec{T}_{AD} = |\vec{T}_{AD}|(\vec{i}\cos\alpha + \vec{j}\sin\alpha)$. The angle between x and the cable AB is $(180^o - \beta)$ The tension is $\vec{T}_{AB} = |\vec{T}_{AB}|(-\vec{i}\cos\beta + \vec{j}\sin\beta)$. The weight is $\vec{W} = 0\vec{i} - |\vec{W}|\vec{j}$

The equilibrium conditions are $\sum\vec{F} = 0 = \vec{W} + \vec{T}_{AB} + \vec{T}_{AD} = 0$ Substitute and collect like terms,

$\sum\vec{F}_x = (|\vec{T}_{AD}|\cos\alpha - |\vec{T}_{AB}|\cos\beta)\vec{i} = 0$, $\sum\vec{F}_y = (|\vec{T}_{AB}|\sin\beta + |\vec{T}_{AD}|\sin\alpha - |\vec{W}|)\vec{j} = 0$. Solve:

$|\vec{T}_{AB}| = \left(\dfrac{\cos\alpha}{\cos\beta}\right)|\vec{T}_{AD}|$, and $|\vec{T}_{AD}| = \left(\dfrac{|\vec{W}|\cos\beta}{\sin(\alpha+\beta)}\right)$. For $|\vec{W}| = 1000\ lb$, and $\alpha = 51.2^o$, $\beta = 47.2^o$

$|\vec{T}_{AD}| = (1000)\left(\dfrac{0.6142}{0.987}\right) = 622.5\ lb$, $\boxed{|\vec{T}_{AB}| = (622.3)\left(\dfrac{0.6787}{0.6142}\right) = 687.9\ lb}$

Problem 3.37 While working on another exhibit, a curator at the Smithsonian Institute pulls the suspended ***Voyager*** aircraft to one side by attaching three horizontal cables as shown. The mass of the airplane is 1250 kg. Determine the tensions in the cable segments AB, BC, and CD.

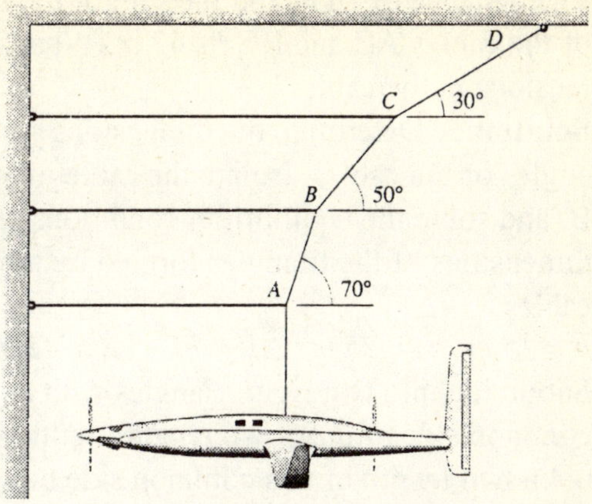

Solution: Isolate each cable juncture, beginning with A and solve the equilibrium equations at each juncture. The angle between the cable AB and the positive x axis is $\alpha = 70^o$; the tension in cable AB is

$\vec{T}_{AB} = |\vec{T}_{AB}|(\vec{i}\cos\alpha + \vec{j}\sin\alpha)$ The weight is

$\vec{W} = 0\vec{i} - |\vec{W}|\vec{j}$ The tension in cable AT is

$\vec{T} = -|\vec{T}|\vec{i} + 0\vec{j}$ The equilibrium conditions are

$\sum \vec{F} = \vec{W} + \vec{T} + \vec{T}_{AB} = 0$. Substitute and collect like terms

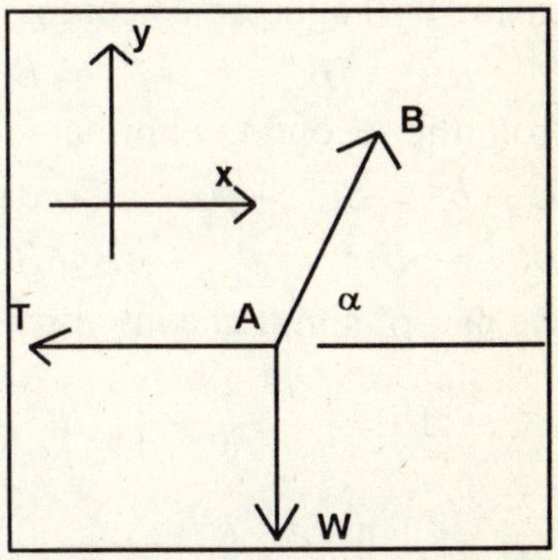

$\sum \vec{F}_x (|\vec{T}_{AB}|\cos\alpha - |\vec{T}|)\vec{i} = 0$,

$\sum \vec{F}_y = (|\vec{T}_{AB}|\sin\alpha - |\vec{W}|)\vec{j} = 0$. Solve: the tension in cable AB is

$|\vec{T}_{AB}| = \left(\frac{|\vec{W}|}{\sin\alpha}\right)$ For $|\vec{W}| = (1250\ kg)\left(9.81\ \frac{m}{s^2}\right) = 12262.5\ N$ and

$\alpha = 70^o$ $\boxed{|\vec{T}_{AB}| = \left(\frac{12262.5}{0.94}\right) = 13049.5\ N}$

Isolate juncture B . The angles are $\alpha = 50^o$, $\beta = 70^o$, and the tension cable BC is $\vec{T}_{BC} = |\vec{T}_{BC}|(\vec{i}\cos\alpha + \vec{j}\sin\alpha)$ The angle between the cable BA and the positive x axis is $(180 + \beta)$; the tension is

$\vec{T}_{BA} = |\vec{T}_{BA}|(\vec{i}\cos(180+\beta) + \vec{j}\sin(180+\beta)) = |\vec{T}_{BA}|(-\vec{i}\cos\beta - \vec{j}\sin\beta)$

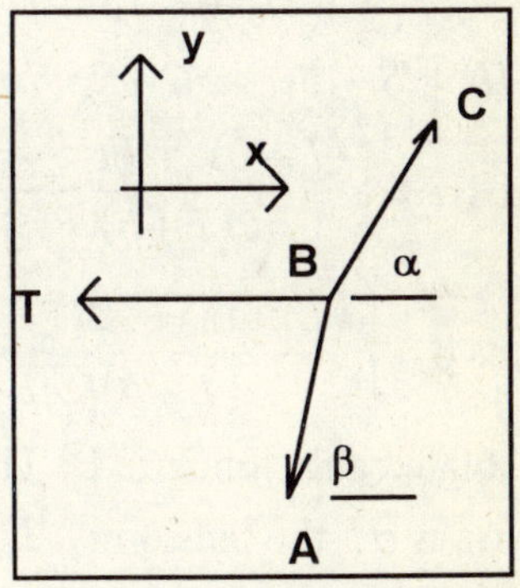

The tension in the left horizontal cable is $\vec{T} = -|\vec{T}|\vec{i} + 0\vec{j}$ The equilibrium conditions are $\sum \vec{F} = \vec{T}_{BA} + \vec{T}_{BC} + \vec{T} = 0$ Substitute and collect like terms $\sum \vec{F}_x = (|\vec{T}_{BC}|\cos\alpha - |\vec{T}_{BA}|\cos\beta - |\vec{T}|)\vec{i} = 0$

$\sum \vec{F}_y = (|\vec{T}_{BC}|\sin\alpha - |\vec{T}_{BA}|\sin\beta)\vec{j} = 0$. Solve: $|\vec{T}_{BC}| = \left(\frac{\sin\beta}{\sin\alpha}\right)|\vec{T}_{BA}|$

For $|\vec{T}_{BA}| = 13049.5\ N$, and $\alpha = 50^o$, $\beta = 70^o$,

$\boxed{|\vec{T}_{BC}| = (13049.5)\left(\frac{0.9397}{0.7660}\right) = 16007.6\ N}$

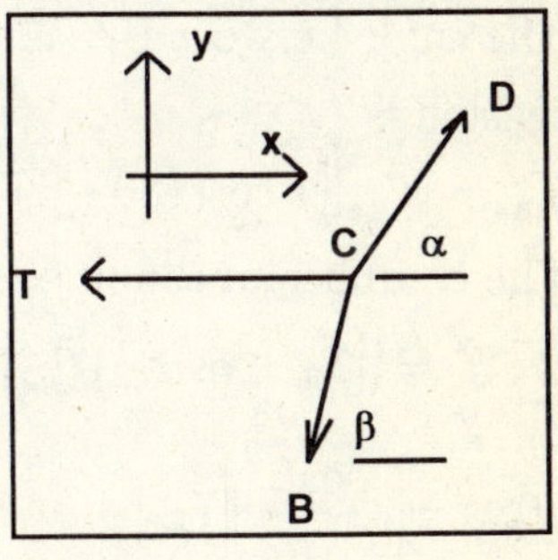

Isolate the cable juncture C. The angles are $\alpha = 30^o$, $\beta = 50^o$. By symmetry with the cable juncture B above, the tension in cable CD is

$|\vec{T}_{CD}| = \left(\frac{\sin\beta}{\sin\alpha}\right)|\vec{T}_{CB}|$. Substitute: $\boxed{|\vec{T}_{CD}| = (16007.6)\left(\frac{0.7660}{0.5}\right) = 24525.0\ N}$

This completes the problem solution.

Problem 3.38 A truck dealer wants to suspend a 4-Mg (megagrams) truck as shown for advertising. The distance b=15 m and the sum of the lengths of the cables AB and BC is 42 m. What are the tensions in the cables?

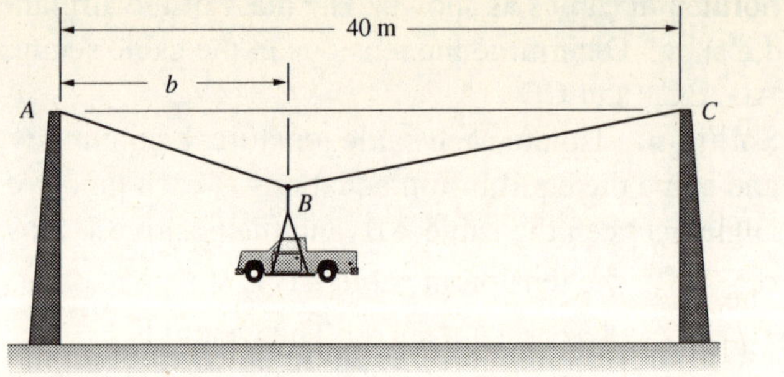

Solution: Determine the dimensions and angles of the cables. Isolate the cable juncture B, and solve the equilibrium conditions. The dimensions of the triangles formed by the cables:

$b = 15\ m,\ \ L = 25\ m,\ \ AB + BC = S = 42\ m$

Subdivide into two right triangles with a common side of unknown length. Let the unknown length of this common side be ***d***, then by the Pythagorean Theorem

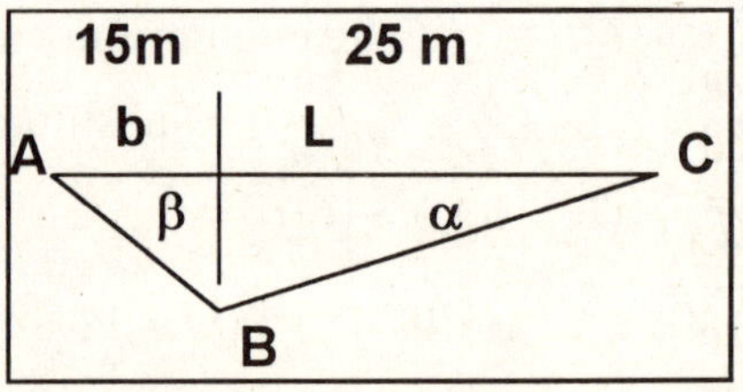

$b^2 + d^2 = AB^2,\ L^2 + d^2 = BC^2$ Subtract the first equation from the second to eliminate the unknown ***d***,

$L^2 - b^2 = BC^2 - AB^2$ Note that

$BC^2 - AB^2 = (BC - AB)(BC + AB)$: substitute and reduce to the pair of simultaneous equations in the unknowns

$BC - AB = \left(\dfrac{L^2 - b^2}{S}\right),\quad BC + AB = S$ Solve:

$$BC = \left(\frac{1}{2}\right)\left(\frac{L^2 - b^2}{S} + S\right) = \left(\frac{1}{2}\right)\left(\frac{25^2 - 15^2}{42} + 42\right) = 25.762\ m \text{ and}$$

$AB = S - BC = 42 - 25.762 = 16.238\ m$ The interior angles are found from the cosine law:

$$\cos\alpha = \left(\frac{(L+b)^2 + BC^2 - AB^2}{2(L+b)(BC)}\right) = 0.9704 \quad \alpha = 13.97^\circ$$

$$\cos\beta = \left(\frac{(L+b)^2 + AB^2 - BC^2}{2(L+b)(AB)}\right) = 0.9238 \quad \beta = 22.52^\circ$$

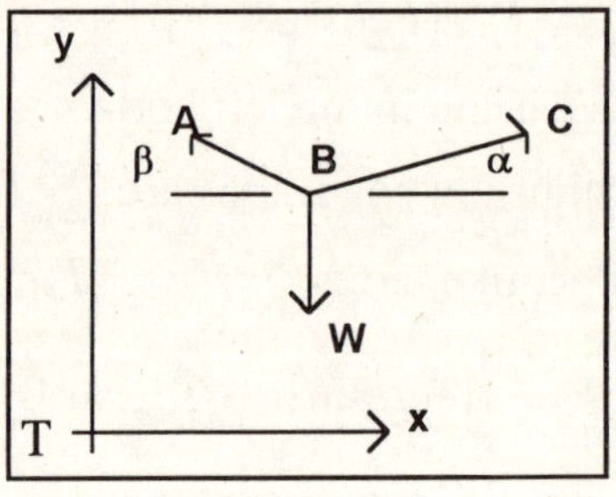

Isolate cable juncture B. The angle between t BC and the positive x axis is α; the tension is $\vec{T}_{BC} = |\vec{T}_{BC}|(\vec{i}\cos\alpha + \vec{j}\sin\alpha)$

The angle between BA and the positive x axis is $(180^\circ - \beta)$; the tension is

$\vec{T}_{BA} = |\vec{T}_{BA}|(\vec{i}\cos(180-\beta) + \vec{j}\sin(180-\beta)) = |\vec{T}_{BA}|(-\vec{i}\cos\beta + \vec{j}\sin\beta)$. The weight is $\vec{W} = 0\vec{i} - |\vec{W}|\vec{j}$.

The equilibrium conditions are $\sum\vec{F} = \vec{W} + \vec{T}_{BA} + \vec{T}_{BC} = 0$ Substitute and collect like terms

$\sum\vec{F}_x = (|\vec{T}_{BC}|\cos\alpha - |\vec{T}_{BA}|\cos\beta)\vec{i} = 0$, $\sum\vec{F}_y = (|\vec{T}_{BC}|\sin\alpha + |\vec{T}_{BA}|\sin\beta - |\vec{W}|)\vec{j} = 0$ Solve:

$|\vec{T}_{BC}| = \left(\dfrac{\cos\beta}{\cos\alpha}\right)|\vec{T}_{BA}|$, and $|\vec{T}_{BA}| = \left(\dfrac{|\vec{W}|\cos\alpha}{\sin(\alpha+\beta)}\right)$. For $|\vec{W}| = (4000)(9.81) = 39240\ N$, and

$\alpha = 13.97^\circ,\ \ \beta = 22.52^\circ$, $\boxed{|\vec{T}_{BA}| = 64033 = 64\ kN,\ \ |\vec{T}_{BC}| = 60953 = 61\ kN}$

Problem 3.39 The distance h=12 in and the tension in cable AD is 200 lb. What are the tensions in the cables AB and AC?

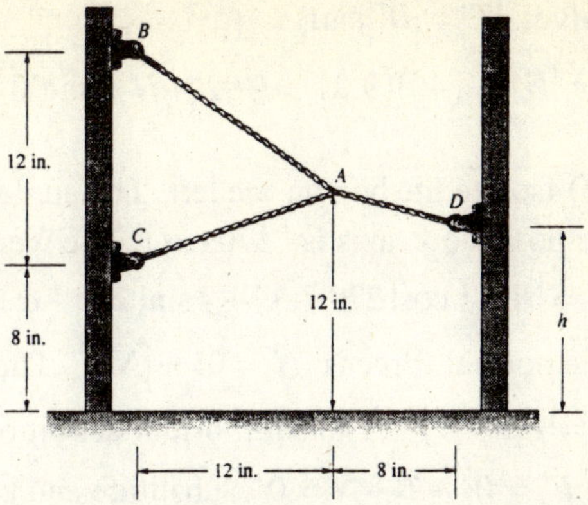

Solution: Isolated the cable juncture. From the sketch, the angles are found from

$\tan\alpha = \left(\dfrac{8}{12}\right) = 0.667 \quad \alpha = 33.7^o$

$\tan\beta = \left(\dfrac{4}{12}\right) = 0.333 \quad \beta = 18.4^o$

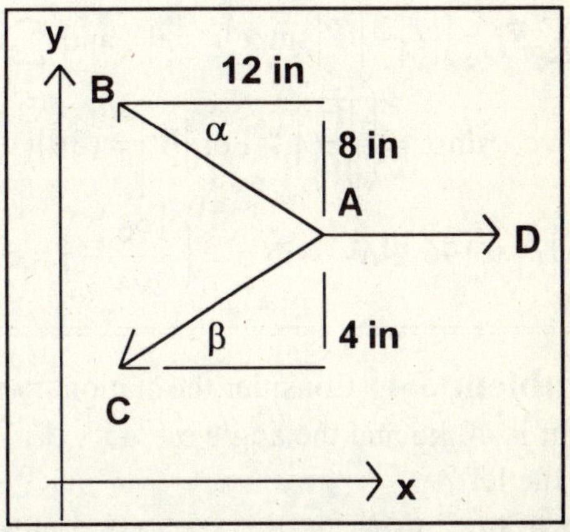

The angle between the cable AB and the positive x axis is $(180^o - \alpha)$, the tension in AB is:

$\vec{T}_{AB} = |\vec{T}_{AB}|(\vec{i}\cos(180-\alpha) + \vec{j}\sin(180-\alpha))$

$\vec{T}_{AB} = |\vec{T}_{AB}|(-\vec{i}\cos\alpha + \vec{j}\sin\alpha)$. The angle between AC and the positive x axis is $(180+\beta)$. The tension is $\vec{T}_{AC} = |\vec{T}_{AC}|(\vec{i}\cos(180+\beta) + \vec{j}\sin(180+\beta))$

$\vec{T}_{AC} = |\vec{T}_{AC}|(-\vec{i}\cos\beta - \vec{j}\sin\beta)$. The tension in the cable AD is $\vec{T}_{AD} = |\vec{T}_{AD}|\vec{i} + 0\vec{j}$. The equilibrium conditions are $\sum\vec{F} = \vec{T}_{AC} + \vec{T}_{AB} + \vec{T}_{AD} = 0$. Substitute and collect like terms,

$\sum\vec{F}_x = (-|\vec{T}_{AB}|\cos\alpha - |\vec{T}_{AC}|\cos\beta + |\vec{T}_{AD}|)\vec{i} = 0$

$\Sigma\vec{F}_y = (|\vec{T}_{AB}|\sin\alpha - |\vec{T}_{AC}|\sin\beta)\vec{j} = 0$. Solve:

$|\vec{T}_{AB}| = \left(\dfrac{\sin\beta}{\sin\alpha}\right)|\vec{T}_{AC}|$, and $|\vec{T}_{AC}| = \left(\dfrac{\sin\alpha}{\sin(\alpha+\beta)}\right)|\vec{T}_{AD}|$. For $|\vec{T}_{AD}| = 200\ lb,\ \alpha = 33.7^o,\ \beta = 18.4^o$

$\boxed{|\vec{T}_{AC}| = 140.6\ lb,\ |\vec{T}_{AB}| = 80.1\ lb}$

Problem 3.40 The mass of the box on the left is 30 kg and the mass of the box on the right is 40 kg. (a) What is the tension in the cable? (b)What is the angle α?

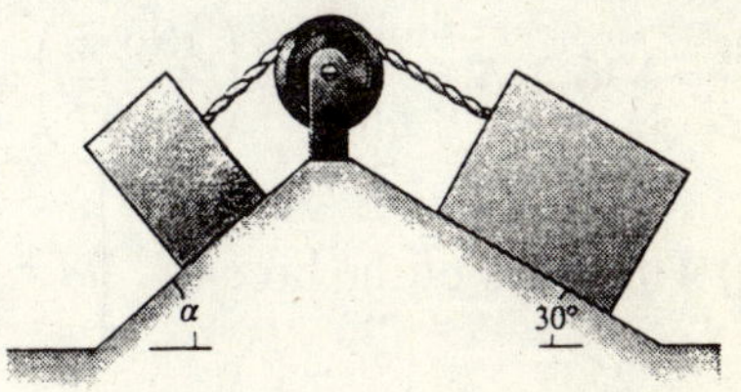

Solution: Isolate the right box, and solve the equilibrium conditions for the cable tension. Repeat for the left box (a): The angle between the weight and the x axis is $-(90-\beta) = (270+\beta)$; the weight is

$\vec{W} = |\vec{W}|(\vec{i}\cos(270+\beta) + \vec{j}\sin(270+\beta))$

$= |\vec{W}|(\vec{i}\sin\beta - \vec{j}\cos\beta)$

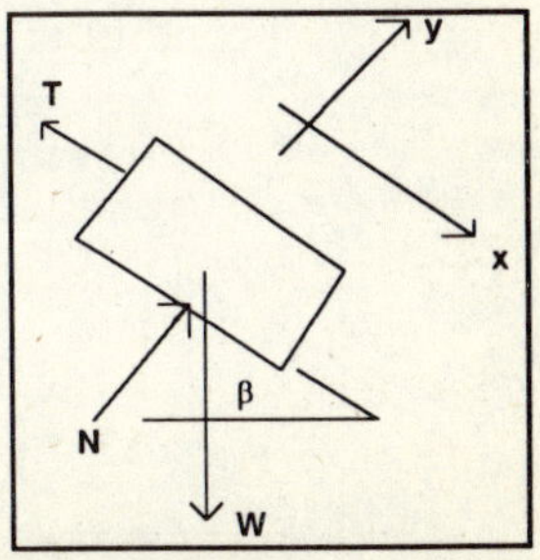

The normal force is $\vec{N} = 0\vec{i} + |\vec{N}|\vec{j}$. The cable tension is $\vec{T} = -|\vec{T}|\vec{i} + 0\vec{j}$ The equilibrium conditions are $\sum\vec{F} = \vec{W} + \vec{T} + \vec{N} = 0$. Substitute and collect like terms,

$\sum\vec{F}_x = (-|\vec{T}| + |\vec{W}|\sin\beta)\vec{i} = 0$, $\sum\vec{F}_y = (|\vec{N}| - |\vec{W}|\cos\beta)\vec{j} = 0$.

Problem 3.40 continued on next page....

continuation of solution to Problem 3.40...

Solve: $|\vec{T}| = |\vec{W}|\sin\beta$

For $|\vec{W}| = (40)(9.81) = 392.4\ N$, and $\beta = 30^o$, the tension in the cable is $\boxed{|\vec{T}| = 196.2\ N}$

(b) Isolate the box on the left. The angle between the weight vector and the positive $\boldsymbol{x}$ axis is $(270-\alpha)$; the weight is

$\vec{W} = |\vec{W}|(\vec{i}\cos(270-\alpha) + \vec{j}\sin(270-\alpha)) = |\vec{W}|(-\vec{i}\sin\alpha - j\cos\alpha)$.

The normal force is $\vec{N} = 0\vec{i} + |\vec{N}|\vec{j}$. The cable tension is

$\vec{T} = |\vec{T}|\vec{i} + 0\vec{j}$. The equilibrium conditions are.

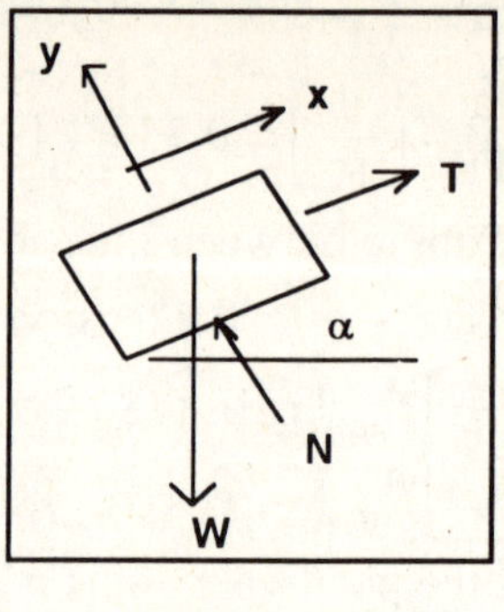

$\sum\vec{F} = \vec{W} + \vec{T} + \vec{N} = 0$. Substitute and collect like terms,

$\sum\vec{F}_x = (|\vec{T}| - |\vec{W}|\sin\alpha)\vec{i} = 0$, and $\sum\vec{F}_y = (|\vec{N}| - |\vec{W}|\cos\alpha)\vec{j} = 0$

Solve: $\sin\alpha = \left(\dfrac{|\vec{T}|}{|\vec{W}|}\right)$. For $|\vec{W}| = (30)(9.81) = 294.3\ N$, and from Part

(a) $|\vec{T}| = 196.2\ N$ $\quad \sin\alpha = \left(\dfrac{196.2}{294.3}\right) = 0.667$ or $\boxed{\alpha = 41.8^o}$

Problem 3.41 Consider the stationary system shown in Problem 3.40. The mass of the box on the right is 40 kg and the angle $\alpha = 45°$. (a) What is the tension in the cable? (b)What is the mass of the box on the left?

Solution: From Problem 3.40 the cable tension for the box on the right is $|\vec{T}| = |\vec{W}|\sin\beta$. For $|\vec{W}| = (40\ kg)(9.81\ m/s^2) = 392.4\ N$, and $\beta = 30^o$, the cable tension is

(a) $\boxed{|\vec{T}| = (392.4)(0.5) = 196.2\ N}$

From Problem 3.40 the weight of the box on the left is $|\vec{W}| = \left(\dfrac{|\vec{T}|}{\sin\alpha}\right)$. For $\alpha = 45^o$ and

$|\vec{T}| = 196.2\ N \quad |\vec{W}| = \left(\dfrac{196.2}{0.707}\right) = 277.5\ N$

(b) The mass of the box is $\boxed{m = \left(\dfrac{|\vec{W}|}{|\vec{g}|}\right) = \left(\dfrac{277.5}{9.81}\right) = 28.3\ kg}$

Problem 3.42 The 50-lb cylinder rests on two surfaces. (a) Draw the free body diagram of the cylinder. (b) If $\alpha = 30°$, what are the magnitudes of the forces exerted on the cylinder by the left and right surfaces?

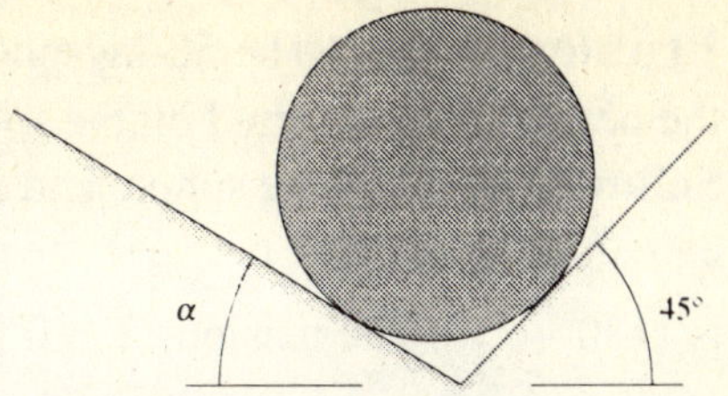

Solution: Isolate the cylinder. (a) The free body diagram of the isolated cylinder is shown. (b) The forces acting are the weight and the normal forces exerted by the surfaces. The angle between the normal force on the right and the x axis is $(90+\beta)$. The normal force is

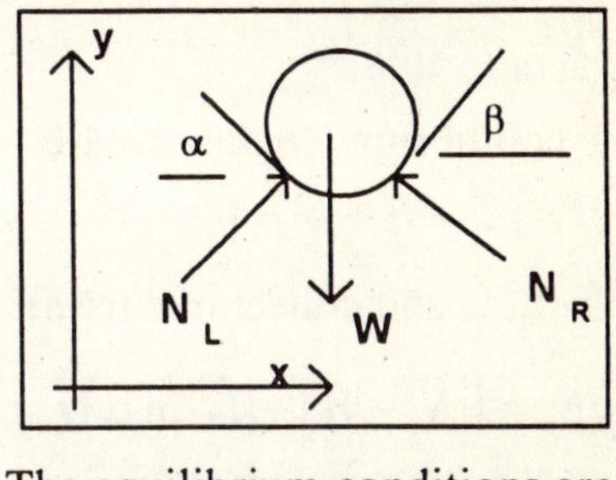

$\vec{N}_R = |\vec{N}_R|(\vec{i}\cos(90+\beta)+\vec{j}\sin(90+\beta))$

$\vec{N}_R = |\vec{N}_R|(-\vec{i}\sin\beta+\vec{j}\cos\beta)$ The angle between the positive x axis and the left hand normal force is $(90-\alpha)$; the normal force is $\vec{N}_L = |\vec{N}_L|(\vec{i}\sin\alpha+\vec{j}\cos\alpha)$. The weight is $\vec{W} = 0\vec{i} - |\vec{W}|\vec{j}$ The equilibrium conditions are $\sum \vec{F} = \vec{W}+\vec{N}_R+\vec{N}_L = 0$ Substitute and collect like terms, $\sum \vec{F}_x = (-|\vec{N}_R|\sin\beta+|\vec{N}_L|\sin\alpha)\vec{i} = 0$,

$\sum \vec{F}_y = (|\vec{N}_R|\cos\beta+|\vec{N}_L|\cos\alpha-|\vec{W}|)\vec{j} = 0$. Solve: $|\vec{N}_R| = \left(\frac{\sin\alpha}{\sin\beta}\right)|N_L|$, and $|\vec{N}_L| = \left(\frac{|\vec{W}|\sin\beta}{\sin(\alpha+\beta)}\right)$

For $|\vec{W}| = 50\ lb$, and $\alpha = 30°$, $\beta = 45°$, the normal forces are $|\vec{N}_L| = 36.6\ lb$, $|\vec{N}_R| = 25.9\ lb$

Problem 3.43 The 50-lb cylinder in Problem 3.42 rests on two surfaces. Obtain an equation for the force exerted on the left hand cylinder in two ways: (a) Using a coordinate system with the y axis vertical; (b) Using a coordinate system with the y axis parallel to the right surface.

Solution: The solution for Part (a) is given in Problem 3.42 (See free body sketch).

$$|\vec{N}_R| = \left(\frac{\sin\alpha}{\sin\beta}\right)|N_L| \quad |\vec{N}_L| = \left(\frac{|\vec{W}|\sin\beta}{\sin(\alpha+\beta)}\right)$$

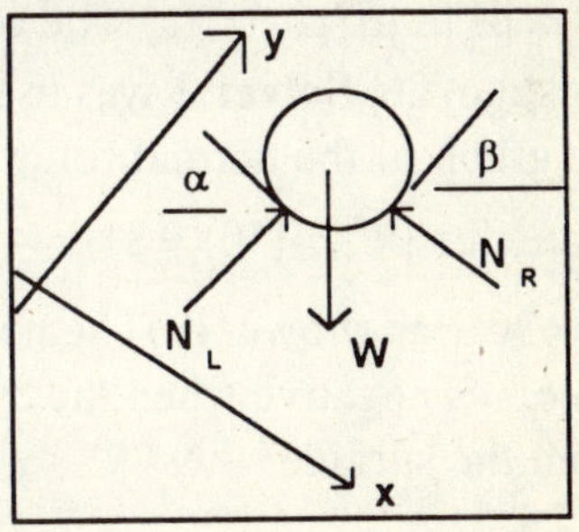

Part(b): The isolated cylinder with the coordinate system is shown. . The angle between the right hand normal force and the positive x axis is $180°$. The normal force: $\vec{N}_R = -|\vec{N}_R|\vec{i}+0\vec{j}$ The angle between the left hand normal force and the positive x is $180-(\alpha+\beta)$. The normal force is $\vec{N}_L = |\vec{N}_L|(-\vec{i}\cos(\alpha+\beta)+\vec{j}\sin(\alpha+\beta))$ The angle between the weight vector and the positive x axis is $-\beta$. The weight vector is $\vec{W} = |\vec{W}|(\vec{i}\cos\beta-\vec{j}\sin\beta)$ The equilibrium conditions are $\sum \vec{F} = \vec{W}+\vec{N}_R+\vec{N}_L = 0$ Substitute and collect like terms, $\sum \vec{F}_x = (-|\vec{N}_R|-|\vec{N}_L|\cos(\alpha+\beta)+|\vec{W}|\cos\beta)\vec{i} = 0$,

$\sum \vec{F}_y = (|\vec{N}_L|\sin(\alpha+\beta)-|\vec{W}|\sin\beta)\vec{j} = 0$. Solve: $|\vec{N}_L| = \left(\frac{|\vec{W}|\sin\beta}{\sin(\alpha+\beta)}\right)$

Problem 3.44 The 50-kg sphere is at rest on the horizontal surface. The force F=500 N. What is the normal force exerted on the sphere by the surface?

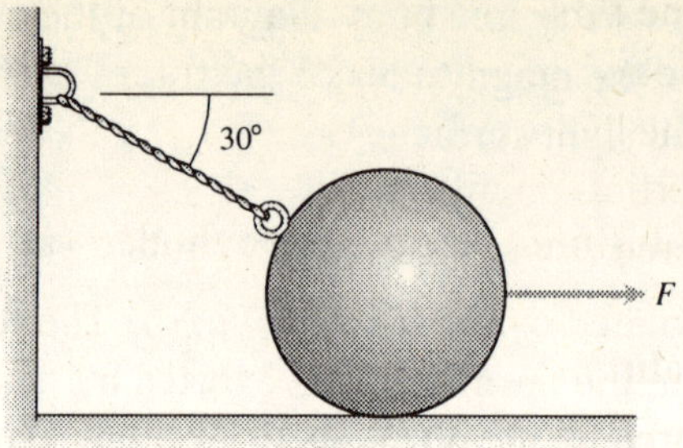

Solution: Isolate the sphere and solve the equilibrium equations. The angle between the cable and the positive x is $(180-\alpha)$. The tension: $\vec{T} = |\vec{T}|(-\vec{i}\cos\alpha + \vec{j}\sin\alpha)$. The other forces are $\vec{F} = |\vec{F}|\vec{i} + 0\vec{j}$, $\vec{N} = 0\vec{i} + |\vec{N}|\vec{j}$, $\vec{W} = 0\vec{i} - |\vec{W}|\vec{j}$

The equilibrium conditions are

$$\sum\vec{F} = \vec{T} + \vec{F} + \vec{N} + \vec{W} = 0$$

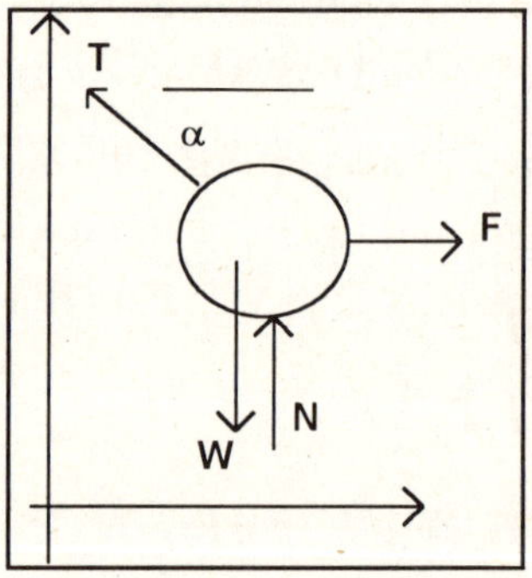

Substitute and collect like terms, $\sum\vec{F}_x = (-|\vec{T}|\cos\alpha + |\vec{F}|)\vec{i} = 0$,

$$\sum\vec{F}_y = (|\vec{N}| - |\vec{W}| + |\vec{T}|\sin\alpha)\vec{j} = 0$$

Solve: $|\vec{T}| = \left(\dfrac{|\vec{F}|}{\cos\alpha}\right)$, and $|\vec{N}| = |\vec{W}| - |\vec{F}|\tan\alpha$

For $|\vec{W}| = (50)(9.81) = 490.5\ N$, $|\vec{F}| = 500\ N$, and $\alpha = 30^o$

$$\boxed{|\vec{N}| = 490.5 - (500)(0.577) = 201.8\ N}$$

Problem 3.45 Consider the stationary sphere shown in Problem 3.44.
(a) Draw a graph of the normal force exerted on the sphere by the surface as a function of the horizontal force $|\vec{F}|$. (b) In the result of Part(a), notice that the normal force decreases to zero and then becomes negative. What does that mean?

Solution: From Problem 3.44, $|\vec{N}| = |\vec{W}| - |\vec{F}|\tan\alpha$ (a) The commercial package **TK Solver 2** was used to produce the graph of the normal force vs. the applied force, for $|\vec{W}| = (50)(9.81) = 490.5\ N$ and $\alpha = 30^o$, as shown. (b)The normal force becomes negative when the cylinder is lifted from the surface.

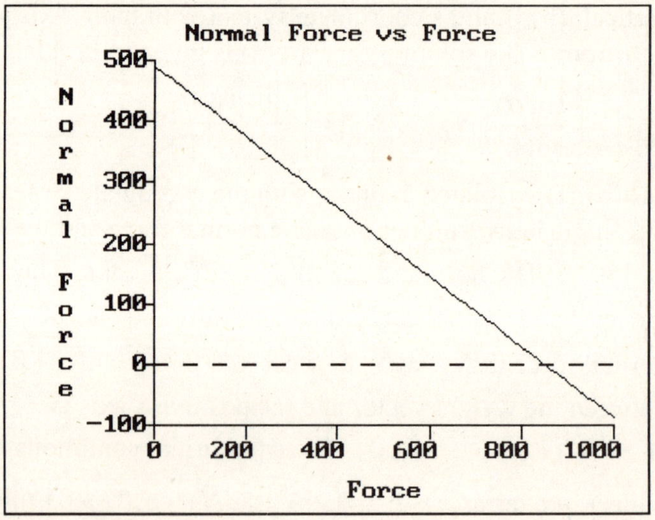

Problem 3.46 The device shown is towed beneath a ship to measure water temperature and salinity. The mass of the device is 130-kg The angle $\alpha = 20^o$. The motion of water past the device causes a horizontal drag D. The hydrostatic pressure distribution in the water exerts a vertical "bouyancy" force $B = 500\ N$. Determine the drag force D and the tension in the cable.

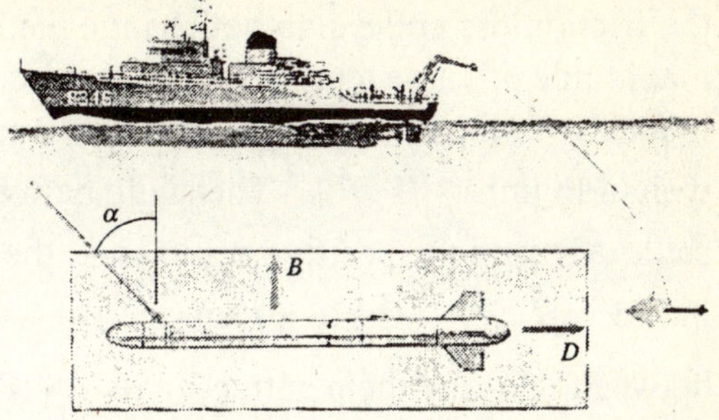

Solution: Isolate the device. The drag, bouyancy and drag forces are $\vec{D} = |\vec{D}|\vec{i} + 0\vec{j}$, $\vec{B} = 0\vec{i} + |\vec{B}|\vec{j}$, $\vec{W} = 0\vec{i} - \vec{j}|\vec{W}|$. The angle between the tow cable and the positive x axis is $(90^o + \alpha)$; the cable tension is

$\vec{T} = |\vec{T}|(\vec{i}\cos(90 + \alpha) + \vec{j}\sin(90 + \alpha))$

$\vec{T} = |\vec{T}|(-\vec{i}\sin\alpha + \vec{j}\cos\alpha)$. The equilibrium conditions are $\sum\vec{F} = \vec{W} + \vec{B} + \vec{T} + \vec{D} = 0$. Substitute and collect terms

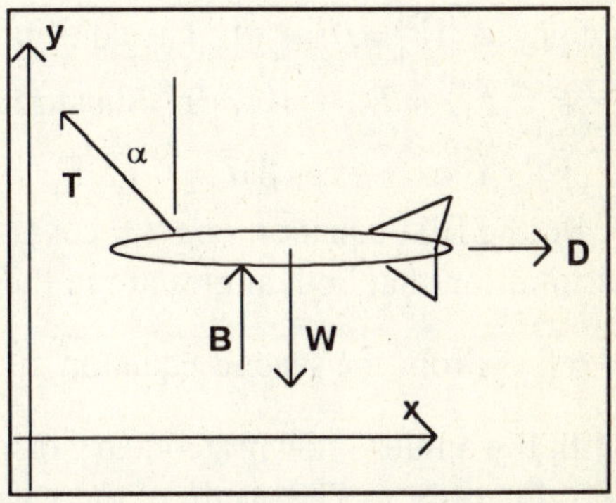

$\sum\vec{F}_x = (|\vec{D}| - |\vec{T}|\sin\alpha)\vec{i} = 0$

$\sum\vec{F}_y = (|\vec{T}|\cos\alpha + |\vec{B}| - |\vec{W}|)\vec{j} = 0$ Solve: $|\vec{D}| = |\vec{T}|\sin\alpha$

$|\vec{T}| = \left(\frac{|\vec{W}| - |\vec{B}|}{\cos\alpha}\right)$. For $|\vec{W}| = (130)(9.81) = 1275.3\ N$,

$|\vec{B}| = 500\ N$, and $\alpha = 20^o$, the tension in the cable and the drag are $\boxed{|\vec{T}| = 825\ N,\ |\vec{D}| = 282.2\ N}$

Problem 3.47 Consider the device described in Problem 3.46. The device weighs 300-lb, the drag force D=50-lb, and the bouyant force B=100-lb. Determine the tension in the cable and the angle.

Solution: From Problem 3.46, the expressions for the tension are: $|\vec{T}| = \left(\frac{|\vec{D}|}{\sin\alpha}\right)$ and

$|\vec{T}| = \left(\frac{|\vec{W}| - |\vec{B}|}{\cos\alpha}\right)$ from which $\cot\alpha = \left(\frac{|\vec{W}| - |\vec{B}|}{|\vec{D}|}\right)$. For $|\vec{W}| = 300\ lb$, $|\vec{B}| = 100\ lb$, and

$|\vec{D}| = 50\ lb$, $\cot\alpha = 4$, $\tan\alpha = \left(\frac{1}{4}\right)$, $\boxed{\alpha = 14^o}$ and $\boxed{|\vec{T}| = 206.15\ lb}$

Problem 3.48 The system is in equilibrium. What are the coordinates of A?

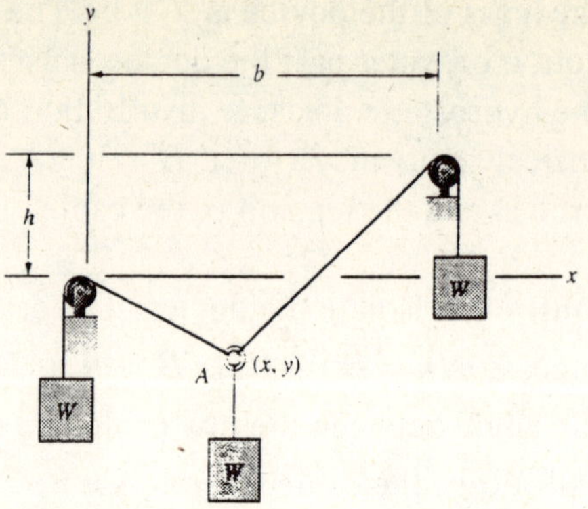

Solution: Determine from geometry the coordinates x,y. Isolate the cable juncture A. Since the frictionless pulleys do not change the magnitude of cable tension, and since each cable is loaded with the same weight, arbitrarily set this weight to unity, $|\vec{W}| = 1$. The angle between the cable AB and the positive x axis is α; the tension in AB is $|\vec{T}_{AB}| = \vec{i}\cos\alpha + \vec{j}\sin\alpha$. The angle between AC and the positive x axis is $(180^o - \beta)$; the tension is $\vec{T}_{AC} = |\vec{T}_{AC}|(-\vec{i}\cos\beta + \vec{j}\sin\beta)$ The weight is $|\vec{W}| = 0\vec{i} - \vec{j}1$ The equilibrium conditions are $\sum \vec{F} = \vec{T}_{AB} + \vec{T}_{AC} + \vec{W} = 0$ Substitute and collect like terms, $\sum \vec{F}_x = (\cos\alpha - \cos\beta)\vec{i} = 0$,$\Sigma \vec{F}_y = (\sin\alpha + \sin\beta - 1)\vec{j} = 0$ From the first equation $\cos\alpha = \cos\beta$. On the realistic assumption that both angles are in the same quadrant, then $\alpha = \beta$. From the second equation $\sin\alpha = \left(\frac{1}{2}\right)$ or $\alpha = 30^o$.

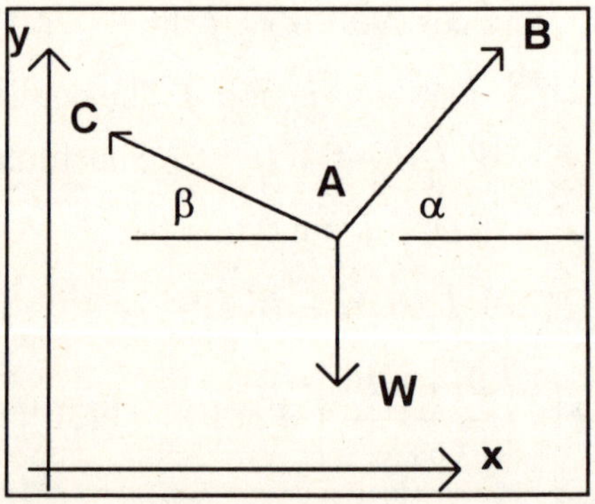

With the angles known, geometry can be used to determine the coordinates x,y. The origin of the x,y coordinate system is at the pulley B, so that the coordinate x of the point A is positive. Define the positive distance ε as shown, so that $\left(\frac{\varepsilon}{x}\right) = \tan\alpha$ Similarly, $\left(\frac{h+\varepsilon}{b-x}\right) = \tan\alpha$. Reduce to obtain $x = b - h\cot\alpha - \varepsilon\cot\alpha$ Substitute into the first equation to obtain $x = \left(\frac{1}{2}\right)(b - h\cot\alpha)$ Multiply this equation by $\tan\alpha$ and use $\varepsilon = x\tan\alpha$ to obtain $\varepsilon = \left(\frac{\tan\alpha}{2}\right)(b - h\cot\alpha)$ The sign of the coordinate y is determined as follows: Since the coordinate x is positive, the condition $(b - h\cot\alpha) > 0$ is required; with this inequality satisfied (as it must be, or the problem is invalid), ε is also positive, as required. But the angle α is in the first quadrant, so that the point A is below the pulley B. Thus $y = -\varepsilon$ and the coordinates of the point A are: $\boxed{x = \left(\frac{1}{2}\right)(b - h\cot\alpha)}$, $\boxed{y = -\frac{1}{2}(b\tan\alpha - h)}$, $\boxed{\alpha = 30^o}$

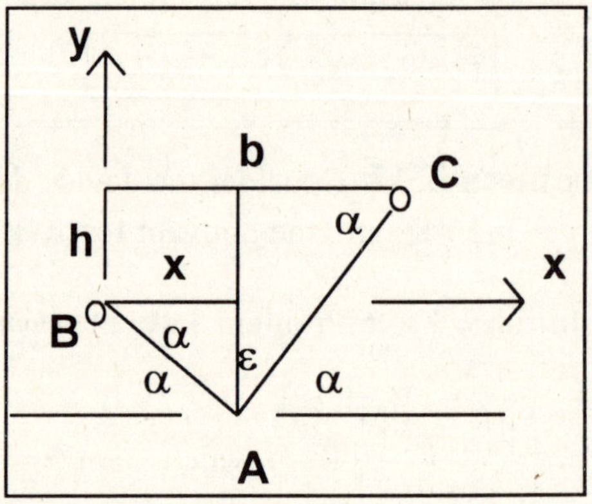

Problem 3.49 The light fixture of weight W is suspended from a circular arch by a large number N of equally spaced cables. The tension T in each cable is the same. Show that $|\vec{T}| = \left(\frac{\pi|\vec{W}|}{2N}\right)$.

Solution: The angle between any cable and the positive x axis is $k\,\delta\theta$, where $k = 0,1,2,3 \ldots K$, where $K = \left(\frac{\pi}{\delta\theta}\right)$ is the number of intervals, one less than the number of cables.
The tension in the kth cable is
$\vec{T}_k = |\vec{T}_k|(\vec{i}\cos k\,\delta\theta + \vec{j}\sin k\,\delta\theta)$ The weight is
$\vec{W} = 0\vec{i} - |\vec{W}|\vec{j}$ The equilibrium conditions are $\Sigma\vec{F} = \vec{W} + \sum_{k=0}^{K}\vec{T}_k = 0$ where N is the number of cables.
Substitute and collect like terms:

$$\Sigma\vec{F}_x = \sum_{k=0}^{K}\left(|\vec{T}_k|\cos k\,\delta\theta\right)\vec{i} = 0$$

$$\Sigma\vec{F}_y = \left(\sum_{k=0}^{K}\left(|\vec{T}_k|\sin k\,\delta\theta\right) - |\vec{W}|\right)\vec{j} = 0$$

Since the tension in each cable is the same, $|\vec{T}_k| = |\vec{T}|$, the tension can be removed from the sum, and the second equation solved for the tension: $|\vec{T}| = \left(\frac{|\vec{W}|}{\sum_{k=0}^{K}\sin k\,\delta\theta}\right)$ The trigonometric sum can be found in handbooks[1]:

$$\sum_{k=0}^{K}\sin k\delta\theta = \left(\frac{\sin\left(\frac{1}{2}(N+1)\delta\theta\right)\sin\left(\frac{1}{2}N\,\delta\theta\right)}{\sin\left(\frac{1}{2}\delta\theta\right)}\right)$$

The angle is divided into K intervals over the arc, $K = \frac{\pi}{\delta\theta}$. Substitute into the sum to obtain $\sum_{k=0}^{K}\sin k\delta\theta = \left(\frac{\cos\left(\frac{1}{2}\delta\theta\right)}{\sin\left(\frac{1}{2}\delta\theta\right)}\right)$ Substitute into the solution for the tension $|\vec{T}| = \left(\frac{|\vec{W}|}{\sum_{k=0}^{K}\sin k\,\delta\theta}\right) = \left(\frac{|\vec{W}|\sin\left(\frac{1}{2}\delta\theta\right)}{\cos\left(\frac{1}{2}\delta\theta\right)}\right) = |\vec{W}|\tan\left(\frac{1}{2}\delta\theta\right)$ If $\delta\theta = \frac{\pi}{K} \cong \frac{\pi}{N} << 1$, $\tan\left(\frac{\delta\theta}{2}\right) \cong \frac{\delta\theta}{2}$,

therefore: $\boxed{|\vec{T}| = \frac{|\vec{W}|\pi}{2N}}$

[1]See, for example, No 420.1, page 92, of Herbert Bristol Dwight: **Tables of Integrals and Other Mathematical Data** (The Macmillan Company, New York, Fourth Edition, 1961).

==================================<>==================================

Problem 3.50 The solution to Problem 3.49 is an asymptotic result whose accuracy increases as N increases. Determine the exact tension $\vec{T}_{exact}$ for N =3,5,9 and 17, and the ratio of the exact solution to the asymptotic solution.

N	3	5	9	17
$\dfrac{\lvert\vec{T}_{exact}\rvert}{\left(\dfrac{\pi\lvert\vec{W}\rvert}{2N}\right)}$	1.91	1.32	1.14	1.07

Solution: From Problem 3.49, the tension is $|\vec{T}| = \left(\dfrac{|\vec{W}|}{\sum_{k=01}^{K} \sin k\,\delta\theta}\right)$ where the denominator is

$$\sum_{k=0}^{K}\sin(k\,\delta\theta) = \left(\frac{\sin\left(\frac{1}{2}(K+1)\delta\theta\right)\sin\left(\frac{1}{2}K\,\delta\theta\right)}{\sin\left(\frac{1}{2}\delta\theta\right)}\right)$$

where N is the number of cables. The angle is divided into K segments over the interval, thus $K = \dfrac{\pi}{\delta\theta}$ and $N = \left(\dfrac{\pi}{\delta\theta}\right) + 1$, since the number of cables is one more than the number of intervals Substitute this into the sum to obtain $\sum_{k=0}^{K}\sin(k\,\delta\theta) = \left(\dfrac{\cos\left(\frac{1}{2}\delta\theta\right)}{\sin\left(\frac{1}{2}\delta\theta\right)}\right)$

Substitute this into the expression for the tension:

$$|\vec{T}| = \left(\frac{|\vec{W}|}{\sum_{k=0}^{N}\sin k\,\delta\theta}\right) = \left(\frac{|\vec{W}|\sin(\frac{1}{2}\delta\theta)}{\cos(\frac{1}{2}\delta\theta)}\right) = |\vec{W}|\tan(\tfrac{1}{2}\delta\theta)$$

Since $\delta\theta = \dfrac{\pi}{N-1}$, the exact solution for the tension in a cable is given by $\boxed{|\vec{T}| = |\vec{W}|\tan\left(\dfrac{\pi}{2(N-1)}\right)}$ where $N-1$ is the number of intervals in an arc of π radians. If the angle increment $\delta\theta$ is sufficiently small, $\tan\left(\dfrac{\delta\theta}{2}\right) \cong \dfrac{\delta\theta}{2}$, and

$\boxed{|\vec{T}| = \dfrac{|\vec{W}|\pi}{2(N-1)} \cong \dfrac{|\vec{W}|\pi}{2N}}$ is the asymptotic solution. The asymptotic solution, the exact solution, and the ratio of the exact solution to the asymptotic solution for the two configurations are given in the table below for 3,5,9, and 17 cables for the two configurations.

N	$\pi/2N$	$\tan\left(\pi/2(N-1)\right)$	**Ratio**
3	**0.5239**	**1**	**1.909**
5	**0.3142**	**0.4142**	**1.318**
7	**0.2244**	**0.2679**	**1.194**
9	**0.1745**	**0.1989**	**1.140**
17	**0.0924**	**0.0985**	**1.066**

==================================<>==================================

Problem 3.51 The system shown provides lateral support for a load resting on the bed of a truck. The spring constant k = 100 lb/ft and the unstretched length of each spring is 2 ft. When the load is subjected to an effective lateral load F the distance from the original position of the load to its equilibrium point is $\delta = 1$ ft. What is F?

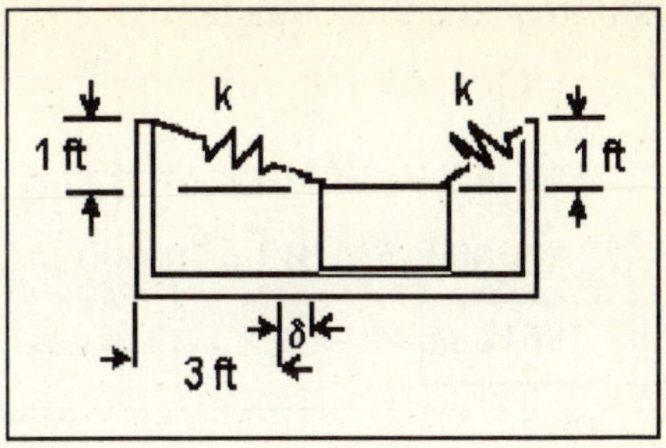

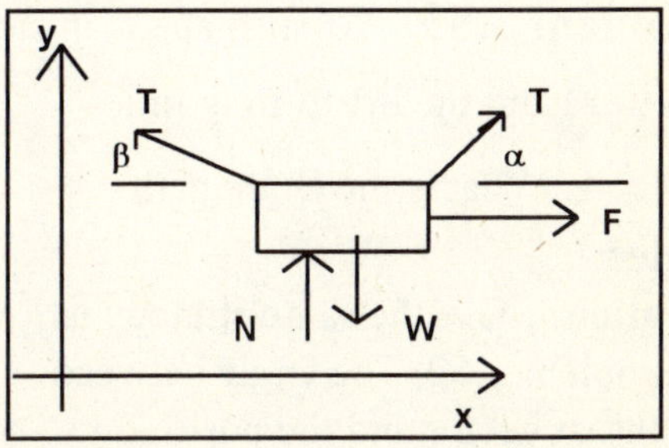

Solution: After the load shift, the extension of the left spring is $\sqrt{(3+\delta)^2+1^2}-2 = 2.123\ ft$. The right spring extension is $\sqrt{(3-\delta)^2+1^2}-2 = 0.236$ The magnitude of the force on the left spring is $|\vec{T}_L| = k\ \Delta L = (100)(2.123) = 212.3\ lb$ The magnitude of the force on right spring is $|\vec{T}_R| = k\ \Delta L = (100)(0.236) = 23.6\ lb$ The angle for the new position of the box for the left spring, $\tan\beta = \dfrac{1}{3+\delta} = 0.25$, $\beta = 14^o$ The angle for the right spring in the new position is $\tan\alpha = \dfrac{1}{3-\delta} = 0.5,\ \alpha = 26.6^o$. The angle between the right spring force and the positive x axis is α; the spring force is $\vec{T}_R = |\vec{T}_R|(\vec{i}\cos\alpha + \vec{j}\sin\alpha)$ The angle between the left spring force and the positive x axis is (180-β); the left spring force is $\vec{T}_L = |\vec{T}_L|(\vec{i}\cos(180-\beta)+\vec{j}\sin(180-\beta)) = |\vec{T}_L|(-\vec{i}\cos\beta+\vec{j}\sin\beta)$ The lateral load $\vec{F} = |\vec{F}|\vec{i}+0\vec{j}$ The weight is $\vec{W} = 0\vec{i}-|\vec{W}|\vec{j}$ The normal force is $\vec{N} = 0\vec{i}+|\vec{N}|\vec{j}$ The equilibrium conditions are $\sum\vec{F} = \vec{T}_L+\vec{T}_R+\vec{W}+\vec{N}+\vec{F} = 0$ Substitute and collect like terms: $\sum\vec{F}_x = (|\vec{T}_R|\cos\alpha - |\vec{T}_L|\cos\beta + |\vec{F}|)\vec{i} = 0$, $\sum\vec{F}_y = (|\vec{T}_R|\sin\alpha + |\vec{T}_L|\sin\beta - |\vec{W}| + |\vec{N}|)\vec{j} = 0$ .Solve for the lateral load, $|\vec{F}| = |\vec{T}_L|\cos\beta - |\vec{T}_R|\sin\alpha$ For $|\vec{T}_L| = 212.3\ lb,\ |\vec{T}_R| = 23.6\ lb,\ \alpha = 26.6^o,\ \beta = 14^o$, the lateral load is $\boxed{|\vec{F}| = 184.9\ lb}$

Problem 3.52 An airplane is in steady level flight $(\gamma = 0)$. Its weight $|\vec{W}| = 30{,}000\ lb$, thrust $|\vec{T}| = 8{,}000\ lb$, and angle of attack α=10°. What are the drag $\boldsymbol{D}$ and lift $\boldsymbol{L}$?

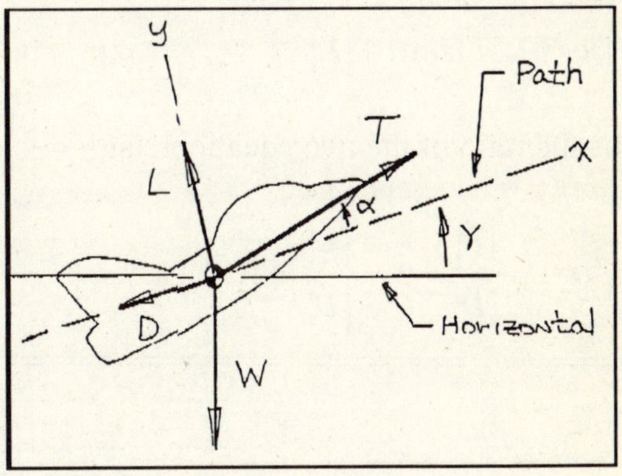

Solution: The angle between the thrust vector and the positive x axis is α; $\vec{T} = |\vec{T}|(\vec{i}\cos\alpha + \vec{j}\sin\alpha)$ The lift vector is $\vec{L} = 0\vec{i} + |\vec{L}|\vec{j}$. The drag: $\vec{D} = -|\vec{D}|\vec{i} + 0\vec{j}$
The angle between the weight vector and the positive x axis is $(270-\gamma)$: $\vec{W} = |\vec{W}|(-\vec{i}\sin\gamma - \vec{j}\cos\gamma)$
The equilibrium conditions are $\sum\vec{F} = \vec{T}+\vec{L}+\vec{D}+\vec{W} = 0$

Problem 3.52 continued on next page....

continuation of solution to Problem 3.52...

Substitute and collect like terms:

$\sum \vec{F}_x = \left(|\vec{T}|\cos\alpha - |\vec{D}| - |\vec{W}|\sin\gamma\right)\vec{i} = 0$

$\sum \vec{F}_y = \left(|\vec{T}|\sin\alpha + |\vec{L}| - |\vec{W}|\cos\gamma\right)\vec{j} = 0$. Solve:

For $|\vec{T}| = 8{,}000\ lb$, $|\vec{W}| = 30{,}000\ lb$, $\alpha = 10^o$, $\gamma = 0$, the lift and drag are $|\vec{D}| = 7878.5\ lb$ $|\vec{L}| = 28611\ lb$

Problem 3.53 An airplane is in steady flight, the angle of attack α=0, the thrust-to-drag ratio $T/D = 2$, and the lift-to-drag ratio $L/D = 4$. What is the flight path angle γ?

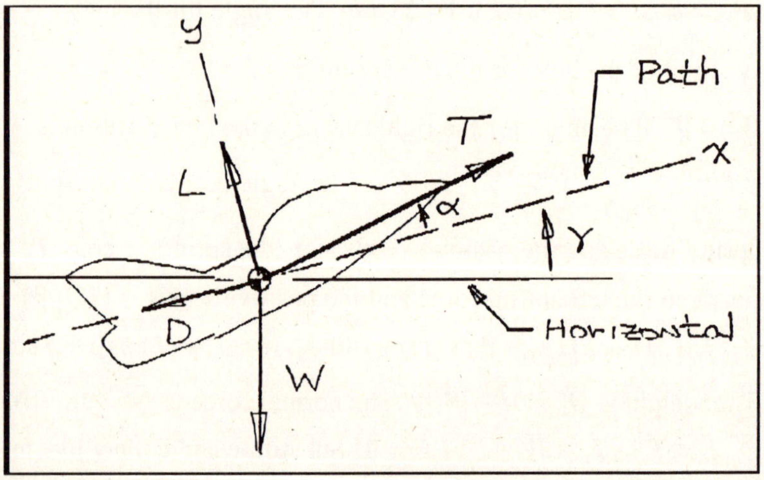

Solution: Use the same strategy as in Problem 3.52. The angle between the thrust vector and the positive x axis is α, $\vec{T} = |\vec{T}|\left(\vec{i}\cos\alpha + \vec{j}\sin\alpha\right)$

The lift vector: $\vec{L} = 0\vec{i} + |\vec{L}|\vec{j}$

The drag: $\vec{D} = -|\vec{D}|\vec{i} + 0\vec{j}$

The angle between the weight vector and the positive x axis is $(270-\gamma)$; $\vec{W} = |\vec{W}|\left(-\vec{i}\sin\gamma - \vec{j}\cos\gamma\right)$. The equilibrium conditions are $\sum \vec{F} = \vec{T} + \vec{L} + \vec{D} + \vec{W} = 0$ Substitute and collect like terms

$\sum \vec{F}_x = \left(|\vec{T}|\cos\alpha - |\vec{D}| - |\vec{W}|\sin\gamma\right)\vec{i} = 0$

$\sum \vec{F}_y = \left(|\vec{T}|\sin\alpha + |\vec{L}| - |\vec{W}|\cos\gamma\right)\vec{j} = 0$

Solve the equations for the terms in γ:

$|\vec{W}|\sin\gamma = |\vec{T}|\cos\alpha - |\vec{D}|$, and

$|\vec{W}|\cos\gamma = |\vec{T}|\sin\alpha + |\vec{L}|$.

Take the ratio of the two equations $\tan\gamma = \left(\dfrac{|\vec{T}|\cos\alpha - |\vec{D}|}{|\vec{T}|\sin\alpha + |\vec{L}|}\right)$ Divide top and bottom on the right by $|\vec{D}|$.

For α=0, $\dfrac{|\vec{T}|}{|\vec{D}|} = 2$, $\dfrac{|\vec{L}|}{|\vec{D}|} = 4$, $\tan\gamma = \left(\dfrac{2-1}{4}\right) = \dfrac{1}{4}$ or $\gamma = 14^o$

Problem 3.54 The airplane glides in steady flight ($T=0$). The lift-to-drag ratio is $L/D=4$. (a) What is the flight path angle γ? (b) If the airplane glides from an altitude of 1000 m to zero altitude, what horizontal distance does it cover?

Solution: Use the same strategy as in Problem 3.52. The angle between the thrust vector and the positive x axis is α: $\vec{T} = |\vec{T}|(\vec{i}\cos\alpha + \vec{j}\sin\alpha)$. The lift vector: $\vec{L} = 0\vec{i} + |\vec{L}|\vec{j}$. The drag: $\vec{D} = -|\vec{D}|\vec{i} + 0\vec{j}$. The angle between the weight vector and the positive x axis is $(270-\gamma)$: $\vec{W} = |\vec{W}|(-\vec{i}\sin\gamma - \vec{j}\cos\gamma)$. The equilibrium conditions are $\sum\vec{F} = \vec{T} + \vec{L} + \vec{D} + \vec{W} = 0$.

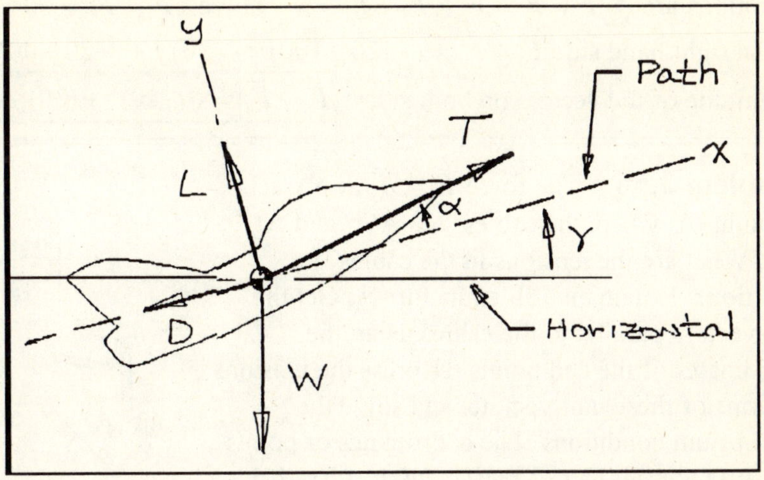

Substitute and collect like terms:

$$\sum\vec{F}_x = (|\vec{T}|\cos\alpha - |\vec{D}| - |\vec{W}|\sin\gamma)\vec{i} = 0$$

$$\sum\vec{F}_y = (|\vec{T}|\sin\alpha + |\vec{L}| - |\vec{W}|\cos\gamma)\vec{j} = 0$$

Solve the equations for the terms in γ, $|\vec{W}|\sin\gamma = |\vec{T}|\cos\alpha - |\vec{D}|$, and $|\vec{W}|\cos\gamma = |\vec{T}|\sin\alpha + |\vec{L}|$

Part (a): Take the ratio of the two equilibrium equations: $\tan\gamma = \left(\dfrac{|\vec{T}|\cos\alpha - |\vec{D}|}{|\vec{T}|\sin\alpha + |\vec{L}|}\right)$. Divide top and bottom on the right by $|\vec{D}|$. For $\alpha=0$,

$$|\vec{T}| = 0,\quad \frac{|\vec{L}|}{|\vec{D}|} = 4,\quad \tan\gamma == \left(\frac{-1}{4}\right) \quad \boxed{\gamma = -14^{\circ}}$$

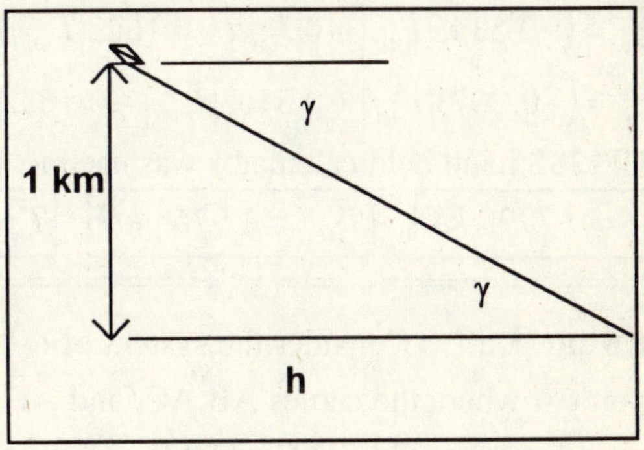

Part (b): The flight path angle is a negative angle measured from the horizontal, hence from the equality of opposite interior angles the angle γ is also the positive elevation angle of the airplane measured at the point of landing.

$$\tan\gamma = \frac{1}{h},\quad \boxed{h = \frac{1}{\tan\gamma} = \frac{1}{\left(\frac{1}{4}\right)} = 4\ km}$$

==================================<>==================================

Problem 3.55 The external forces acting on an object in equilibrium are $\vec{F}_1 = 120\vec{i} - 30\vec{j} + 20\vec{k}$, $\vec{F}_2 = -100\vec{i} + 40\vec{k}$, and $\vec{F}_3$, $\vec{F}_4$. What is the magnitude of $\vec{F}_3 + \vec{F}_4$?

Solution: Solve the equilibrium equations for the unknowns, and take the magnitude. The equilibrium conditions are $\sum \vec{F} = \vec{F}_1 + \vec{F}_2 + \vec{F}_3 + \vec{F}_4 = 0$. Solve: $\vec{F}_3 + \vec{F}_4 = -\vec{F}_1 - \vec{F}_2$. Substitute the known vectors on the right hand side, $\vec{F}_3 + \vec{F}_4 = (-120 + 100)\vec{i} + (30)\vec{j} + (-20 - 40)\vec{k} = -20\vec{i} + 30\vec{j} - 60\vec{k}$. Take the magnitude of the vectors on both sides: $\boxed{|\vec{F}_3 + \vec{F}_4| = \sqrt{(-20)^2 + (30)^2 + (-60)^2} = 70\ lb}$

==================================<>==================================

Problem 3.56 The force $\vec{F} = 5\vec{i}$ (kN) acts on point A where the cables AB, AC, and AD join. What are the tensions in the cables?

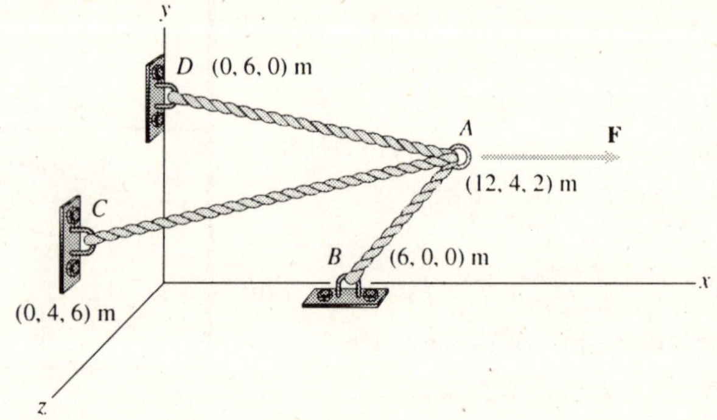

Solution: Isolate the cable juncture A. Get the unit vectors parallel to the cables using the coordinates of the end points. Express the tensions in terms of these unit vectors, and solve the equilibrium conditions. The coordinates of points A,B,C,D are: $A(12,4,2)$, $B(6,0,0)$, $C(0,4,6)$, $D(0,6,0)$. The unit vector $\vec{e}_{AB}$ is, by definition,

$$\vec{e}_{AB} = \frac{\vec{r}_B - \vec{r}_A}{|\vec{r}_B - \vec{r}_A|} = \frac{(6-12)\vec{i} + (0-4)\vec{j} + (0-2)\vec{k}}{\sqrt{(6-12)^2 + (4)^2 + (2)^2}} = \frac{-6}{7.483}\vec{i} - \frac{4}{7.483}\vec{j} - \frac{2}{7.483}\vec{k}$$

$\vec{e}_{AB} = -0.8018\vec{i} - 0.5345\vec{j} - 0.2673\vec{k}$. Similarly, the other unit vectors are $\vec{e}_{AC} = -0.9487\vec{i} + 0\vec{j} + 0.3163\vec{k}$, $\vec{e}_{AD} = -0.9733\vec{i} + 0.1622\vec{j} - 0.1622\vec{k}$. The tensions in the cables are expressed in terms of the unit vectors, $\vec{T}_{AB} = |\vec{T}_{AB}|\vec{e}_{AB}$, $\vec{T}_{AC} = |\vec{T}_{AC}|\vec{e}_{AC}$, $\vec{T}_{AD} = |\vec{T}_{AD}|\vec{e}_{AD}$. The external force acting on the juncture is, $\vec{F} = 5\vec{i} + 0\vec{j} + 0\vec{k}$. The equilibrium conditions are $\sum \vec{F} = 0 = \vec{T}_{AB} + \vec{T}_{AC} + \vec{T}_{AD} + \vec{F} = 0$. Substitute and collect like terms,

$$\sum \vec{F}_x = \left(-0.8018|\vec{T}_{AB}| - 0.9487|\vec{T}_{AC}| - 0.9733|\vec{T}_{AD}| + 5\right)\vec{i} = 0$$

$$\sum \vec{F}_y = \left(-0.5345|\vec{T}_{AB}| + 0|\vec{T}_{AC}| + 0.1622|\vec{T}_{AD}|\right)\vec{j} = 0$$

$$\sum \vec{F}_z = \left(-0.2673|\vec{T}_{AB}| + 0.3163|\vec{T}_{AC}| - 0.1622|\vec{T}_{AD}|\right)\vec{k} = 0$$

An HP-28S hand held calculator was used to solve these simultaneous equations. The results are:

$\boxed{|\vec{T}_{AB}| = 0.7795\ kN}$, $\boxed{|\vec{T}_{AC}| = 1.9765\ kN}$, $\boxed{|\vec{T}_{AD}| = 2.5688\ kN}$

==================================<>==================================

Problem 3.57 Consider the system of cables shown in Problem 3.56. The force $\vec{F} = |\vec{F}|\vec{i}$ acts at the point A where the cables AB, AC, and AD are joined. The tension in cable AB is 10 kN. Determine the tension in the cables AC and AD and the force $|\vec{F}|$.

Solution: From Problem 3.56 the unit vectors are: $\vec{e}_{AB} = -0.8018\vec{i} - 0.5345\vec{j} - 0.2673\vec{k}$, $\vec{e}_{AC} = -0.9487\vec{i} + 0\vec{j} + 0.3163\vec{k}$, $\vec{e}_{AD} = -0.9733\vec{i} + 0.1622\vec{j} - 0.1622\vec{k}$.

The tensions in the cables are expressed in terms of the unit vectors, $\vec{T}_{AB} = |\vec{T}_{AB}|\vec{e}_{AB} = 10\vec{e}_{AB}$, $\vec{T}_{AC} = |\vec{T}_{AC}|\vec{e}_{AC}$, $\vec{T}_{AD} = |\vec{T}_{AD}|\vec{e}_{AD}$. The unknown external force acting on the juncture is $\vec{F} = |\vec{F}|\vec{i} + 0\vec{j} + 0\vec{k}$

Problem 3.57 continued on next page....

continuation of solution to Problem 3.57...

The equilibrium conditions are $\sum \vec{F} = 0 = \vec{T}_{AB} + \vec{T}_{AC} + \vec{T}_{AD} + \vec{F} = 0$. Substitute and collect like terms,

$$\sum \vec{F}_x = \left(-0.8018(10) - 0.9487\left|\vec{T}_{AC}\right| - 0.9733\left|\vec{T}_{AD}\right| + \left|\vec{F}\right|\right)\vec{i} = 0$$

$$\sum \vec{F}_y = \left(-0.5345(10) + 0\left|\vec{T}_{AC}\right| + 0.2622\left|\vec{T}_{AD}\right|\right)\vec{j} = 0$$

$$\sum \vec{F}_z = \left(-0.2673(10) + 0.3163\left|\vec{T}_{AC}\right| - 0.1622\left|\vec{T}_{AD}\right|\right)\vec{k} = 0$$

This set of simultaneous equations may be solved by starting with the second equation and substituting the results into the others. The results $\boxed{\left|\vec{F}\right| = 64.14\ kN}$, $\boxed{\left|\vec{T}_{AC}\right| = 25.35\ kN}$, $\boxed{\left|\vec{T}_{AD}\right| = 32.95\ kN}$

Problem 3.58 To support the tent, the tension in rope AB must be 40 lb. What are the tensions in the cables AC, AD, and AE?

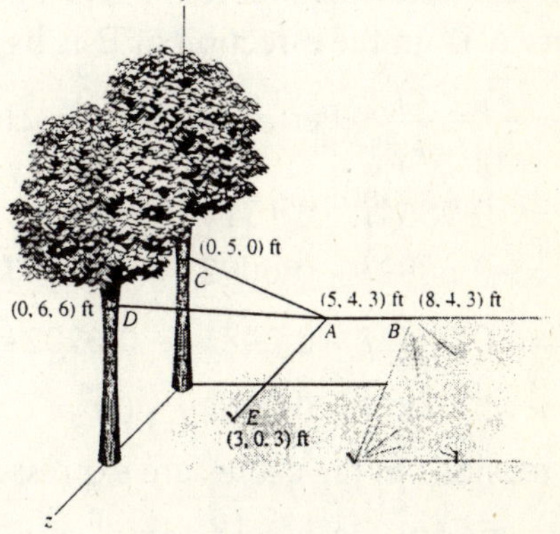

Solution: Get the unit vectors parallel to the cables using the coordinates of the end points. Express the tensions in terms of these unit vectors, and solve the equilibrium conditions. The coordinates of points A,B,C,D,E are: $A(5,4,3)$, $B(8,4,3)$, $C(0,5,0)$, $D(0,6,6)$, $E(3,0,3)$.

The vector locations of these points are,

$\vec{r}_A = 5\vec{i} + 4\vec{j} + 3\vec{k}$, $\vec{r}_B = 8\vec{i} + 4\vec{j} + 3\vec{k}$,

$\vec{r}_C = 0\vec{i} + 5\vec{j} + 0\vec{k}$, $\vec{r}_D = 0\vec{i} + 6\vec{j} + 6\vec{k}$,

$\vec{r}_E = 3\vec{i} + 0\vec{j} + 3\vec{k}$. The unit vector parallel to the tension acting between the points A,B in the direction of B is by definition

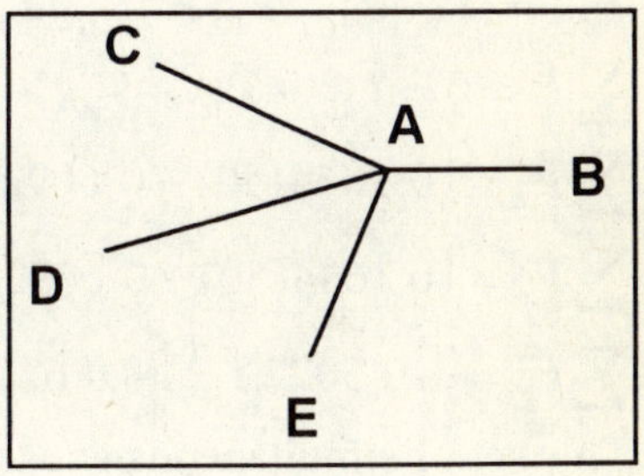

$\vec{e}_{AB} = \dfrac{\vec{r}_B - \vec{r}_A}{\left|\vec{r}_B - \vec{r}_A\right|}$. Perform this operation for each unit vector:

$$\vec{e}_{AB} = 1\vec{i} + 0\vec{j} + 0\vec{k}$$

$$\vec{e}_{AC} = -0.8452\vec{i} + 0.1690\vec{j} - 0.5071\vec{k}$$

$$\vec{e}_{AD} = -0.8111\vec{i} + 0.3244\vec{j} + 0.4867\vec{k}$$

$$\vec{e}_{AE} = -0.4472\vec{i} - 0.8944\vec{j} + 0\vec{k}$$

The tensions in the cables are, $\vec{T}_{AB} = \left|\vec{T}_{AB}\right|\vec{e}_{AB} = 40\vec{e}_{AB}$, $\vec{T}_{AC} = \left|\vec{T}_{AC}\right|\vec{e}_{AC}$, $\vec{T}_{AD} = \left|\vec{T}_{AD}\right|\vec{e}_{AD}$, $\vec{T}_{AE} = \left|\vec{T}_{AE}\right|\vec{e}_{AE}$.

The equilibrium conditions are $\sum \vec{F} = 0 = \vec{T}_{AB} + \vec{T}_{AC} + \vec{T}_{AD} + \vec{T}_{AE} = 0$. Substitute the tensions,

$$\sum \vec{F}_x = \left(40 - 0.8452\left|\vec{T}_{AC}\right| - 0.8111\left|\vec{T}_{AD}\right| - 0.4472\left|\vec{T}_{AE}\right|\right)\vec{i} = 0$$

$$\Sigma \vec{F}_y = \left(+0.1690\left|\vec{T}_{AC}\right| + 0.3244\left|\vec{T}_{AD}\right| - 0.8944\left|\vec{T}_{AE}\right|\right)\vec{j} = 0$$

$$\Sigma \vec{F}_z = \left(-0.5071\left|\vec{T}_{AC}\right| + 0.4867\left|\vec{T}_{AD}\right|\right)\vec{k} = 0$$

This set of simultaneous equations in the unknown forces may be solved using any of several standard algorithms.: The results are: $\boxed{\left|\vec{T}_{AE}\right| = 11.7\ lb}$, $\boxed{\left|\vec{T}_{AC}\right| = 20.6\ lb}$, $\boxed{\left|\vec{T}_{AD}\right| = 21.4\ lb}$

================================<>================================

Problem 3.59 Cable AB is attached to the top of a vertical 3-m post and its tension is 50 kN. What are the tensions in cables AO, AC, and AD?

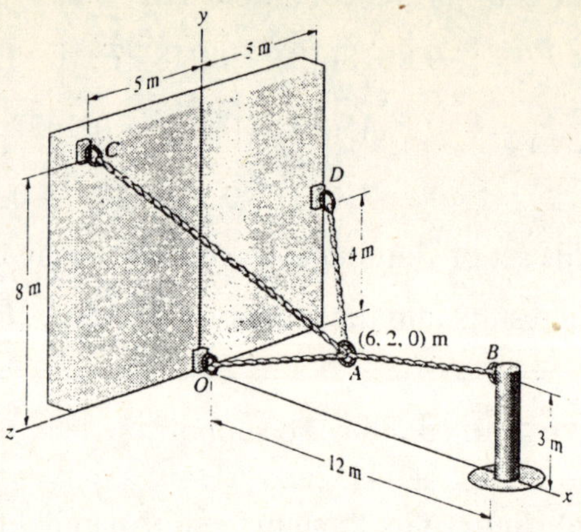

Solution: Get the unit vectors parallel to the cables using the coordinates of the end points. Express the tensions in terms of these unit vectors, and solve the equilibrium conditions. The coordinates of points A,B,C,D,O are found from the problem sketch: The coordinates of the points are . $A(6,2,0)$, $B(12,3,0)$, $C(0,8,5)$, $D(0,4,-5)$, $O(0,0,0)$. The vector locations of these points are:

$\vec{r}_A = 6\vec{i} + 2\vec{j} + 0\vec{k}$, $\vec{r}_B = 12\vec{i} + 3\vec{j} + 0\vec{k}$,

$\vec{r}_C = 0\vec{i} + 8\vec{j} + 5\vec{k}$, $\vec{r}_D = 0\vec{i} + 4\vec{j} - 5\vec{k}$, $\vec{r}_O = 0\vec{i} + 0\vec{j} + 0\vec{k}$.

The unit vector parallel to the tension acting between the points A,B in the direction of B is by definition

$\vec{e}_{AB} = \dfrac{\vec{r}_B - \vec{r}_A}{|\vec{r}_B - \vec{r}_A|}$. Perform this for each of the unit vectors

$\vec{e}_{AB} = +0.9864\vec{i} + 0.1644\vec{j} + 0\vec{k}$

$\vec{e}_{AC} = -0.6092\vec{i} + 0.6092\vec{j} + 0.5077\vec{k}$

$\vec{e}_{AD} = -0.7442\vec{i} + 0.2481\vec{j} - 0.6202\vec{k}$

$\vec{e}_{AO} = -0.9487\vec{i} - 0.3162\vec{j} + 0\vec{k}$

The tensions in the cables are expressed in terms of the unit vectors, $\vec{T}_{AB} = |\vec{T}_{AB}|\vec{e}_{AB} = 50\vec{e}_{AB}$, $\vec{T}_{AC} = |\vec{T}_{AC}|\vec{e}_{AC}$, $\vec{T}_{AD} = |\vec{T}_{AD}|\vec{e}_{AD}$, $\vec{T}_{AO} = |\vec{T}_{AO}|\vec{e}_{AO}$. The equilibrium conditions are

$\sum \vec{F} = 0 = \vec{T}_{AB} + \vec{T}_{AC} + \vec{T}_{AD} + \vec{T}_{AO} = 0$ Substitute and collect like terms,

$\sum \vec{F}_x = \left(0.9864(50) - 0.6092|\vec{T}_{AC}| - 0.7442|\vec{T}_{AD}| - 0.9487|\vec{T}_{AO}|\right)\vec{i} = 0$

$\sum \vec{F}_y = \left(0.1644(50) + 0.6092|\vec{T}_{AC}| + 0.2481|\vec{T}_{AD}| - 0.3162|\vec{T}_{AO}|\right)\vec{j} = 0$

$\sum \vec{F}_z = \left(+0.5077|\vec{T}_{AC}| - 0.6202|\vec{T}_{AD}|\right)\vec{k} = 0$

This set of simultaneous equations in the unknown forces may be solved using any of several standard algorithms. The results are: $\boxed{|\vec{T}_{AO}| = 43.3\ kN}$, $\boxed{|\vec{T}_{AC}| = 6.8\ kN}$, $\boxed{|\vec{T}_{AD}| = 5.5\ kN}$

================================<>================================

================================<>====================================

Problem 3.60 A weather balloon is tethered by the lines AB, AC, and AD. The mass of the balloon and the gas it contains is 80 kg, and the (upward) bouyancy force on the balloon is 1000 N. The coordinate $h = 10\ m$. What are the tensions in the three cables?

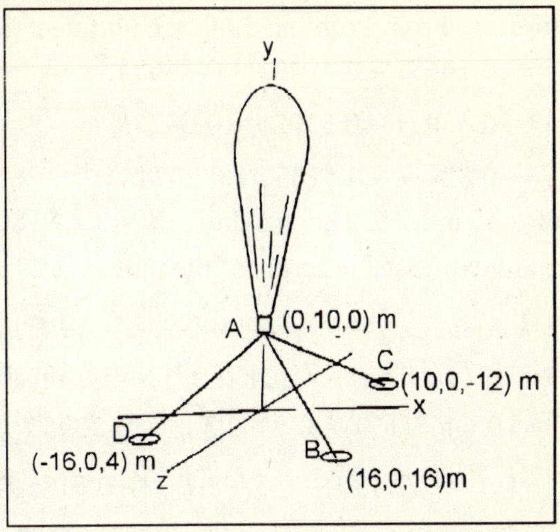

Solution: Get the unit vectors parallel to the cables using the coordinates of the end points. Express the tensions in terms of these unit vectors, and solve the equilibrium conditions.

The coordinates of points A,B,C,D are . $A(0,10,0)$, $B(16,0,16)$, $C(10,0,-12)$, $D(-16,0,4)$. The vector locations of these points are:

$\vec{r}_A = 0\vec{i} + 10\vec{j} + 0\vec{k}$, $\vec{r}_B = 16\vec{i} + 0\vec{j} + 16\vec{k}$,

$\vec{r}_C = 10\vec{i} + 0\vec{j} + -12\vec{k}$, $\vec{r}_D = -16\vec{i} + 0\vec{j} + 4\vec{k}$.

The unit vector parallel to the tension acting between the points A,B in the direction of B is by definition $\vec{e}_{AB} = \dfrac{\vec{r}_B - \vec{r}_A}{|\vec{r}_B - \vec{r}_A|}$. Perform this operation for each unit vector

$\vec{e}_{AB} = +0.6468\vec{i} - 0.4042\vec{j} + 0.6468\vec{k}$

$\vec{e}_{AC} = +0.5392\vec{i} - 0.5392\vec{j} - 0.6470\vec{k}$

$\vec{e}_{AD} = -0.8296\vec{i} - 0.5185\vec{j} + 0.2074\vec{k}$

The tensions in the cables are expressed in terms of the unit vectors, $\vec{T}_{AB} = |\vec{T}_{AB}|\vec{e}_{AB}$, $\vec{T}_{AC} = |\vec{T}_{AC}|\vec{e}_{AC}$, $\vec{T}_{AD} = |\vec{T}_{AD}|\vec{e}_{AD}$. The weight vector is: $\vec{W} = 0\vec{i} - (80)(9.81)\vec{j} + 0\vec{k} = 0\vec{i} - 784.8\vec{j} + 0\vec{k}$ The bouyant force is: $\vec{B} = 0\vec{i} + 1000\vec{j} + 0\vec{k}$. The equilibrium conditions are

$\sum \vec{F} = 0 = \vec{T}_{AB} + \vec{T}_{AC} + \vec{T}_{AD} + \vec{B} + \vec{W} = 0$. Substitute and collect like terms,

$\sum \vec{F}_x = \left(0.6468|\vec{T}_{AB}| + 0.5392|\vec{T}_{AC}| - 0.8296|\vec{T}_{AD}|\right)\vec{i} = 0$

$\sum \vec{F}_y = \left(-0.4042|\vec{T}_{AB}| - 0.5392|\vec{T}_{AC}| - 0.5185|\vec{T}_{AD}| + 215.2\right)\vec{j} = 0$

$\sum \vec{F}_z = \left(0.6468|\vec{T}_{AB}| - 0.6470|\vec{T}_{AC}| + 0.2074|\vec{T}_{AD}|\right)\vec{k} = 0$

This set of simultaneous equations in the unknown forces may be solved using any of several standard algorithms, for example by Gauss' Method, or with a commercial software package. . The commercial program **TK Solver 2** was used here to solve these equations. The results are

$\boxed{|\vec{T}_{AB}| = 98.2\ N}$, $\boxed{|\vec{T}_{AC}| = 155.0\ N}$, $\boxed{|\vec{T}_{AD}| = 177.3\ N}$.

================================<>====================================

Problem 3.61 Consider the tethered balloon of Problem 3.60. The mass of the balloon and the gas it contains is 80 kg. The coordinate h=10 m. The tension in the line AB is measured and determined to be 150 N. What is the (upward) bouyant force acting on the balloon?

Solution: From Problem 3.60, the unit vectors are.

$$\vec{e}_{AB} = +0.6468\vec{i} - 0.4042\vec{j} + 0.6468\vec{k}$$

$$\vec{e}_{AC} = +0.5392\vec{i} - 0.5392\vec{j} - 0.6470\vec{k}$$

$$\vec{e}_{AD} = -0.8296\vec{i} - 0.5185\vec{j} + 0.2074\vec{k}$$

From Problem 3.60, The weight is: $\vec{W} = 0\vec{i} - (80)(9.81)\vec{j} + 0\vec{k} = 0\vec{i} - 784.8\vec{j} + 0\vec{k}$ .The tensions in the cables are expressed in terms of the unit vectors, $\vec{T}_{AB} = |\vec{T}_{AB}|\vec{e}_{AB} = 150\vec{e}_{AB}$, $\vec{T}_{AC} = |\vec{T}_{AC}|\vec{e}_{AC}$, $\vec{T}_{AD} = |\vec{T}_{AD}|\vec{e}_{AD}$. . The bouyant force is $\vec{B} = 0\vec{i} + |\vec{B}|\vec{j} + 0\vec{k}$ .The equilibrium conditions are $\sum \vec{F} = 0 = \vec{T}_{AB} + \vec{T}_{AC} + \vec{T}_{AD} + \vec{B} + \vec{W} = 0$. Substitute and collect like terms,

$$\sum \vec{F}_x = \left(0.6468(150) + 0.5392|\vec{T}_{AC}| - 0.8296|\vec{T}_{AD}|\right)\vec{i} = 0$$

$$\sum \vec{F}_y = \left(-0.4042(150) - 0.5392|\vec{T}_{AC}| - 0.5185|\vec{T}_{AD}| + |\vec{B}| - 784.8\right)\vec{j} = 0$$

$$\sum \vec{F}_z = \left(0.6468(150) - 0.6470|\vec{T}_{AC}| + 0.2074|\vec{T}_{AD}|\right)\vec{k} = 0$$

This set of simultaneous equations in the unknown forces can be solved by a standard procedure, for example by Gauss' Method, or with a commercial software package. The commercial program **TK Solver® 2** was used here to solve these equations. The results are $|\vec{T}_{AC}| = 236.8\ N$, $|\vec{T}_{AD}| = 270.8\ N$, and the upward bouyancy force, $\boxed{|\vec{B}| = 1113.5\ N.}$

Problem 3.62 The small sphere A weighs 20 lb and its coordinates are (4,0,6). It is supported by two flat plates labeled 1 and 2 and the cable AB. The unit vector $\vec{e}_1 = (\frac{4}{9})\vec{i} + (\frac{7}{9})\vec{j} + (\frac{4}{9})\vec{k}$ is perpendicular to the plate 1, and the unit vector $\vec{e}_2 = -(\frac{9}{11})\vec{i} + (\frac{2}{11})\vec{j} + (\frac{6}{11})\vec{k}$ is perpendicular to plate 2. What is the tension in the cable?

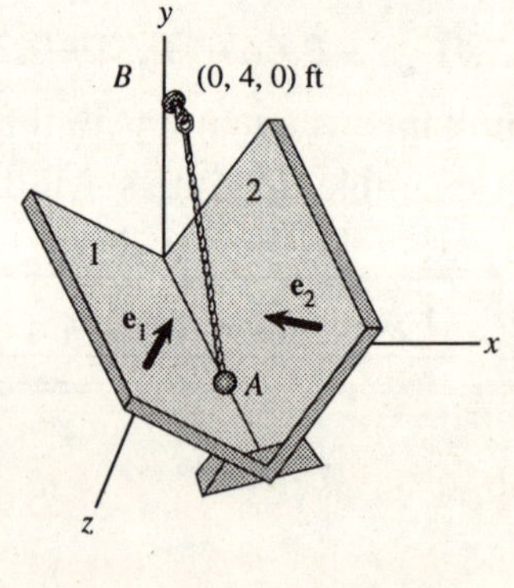

Solution: Get the unit vectors parallel to the cable using the coordinates of the end points. Express the tensions in terms of these unit vectors, and solve the equilibrium conditions. The cable tension, two normal forces, and the weight act on the sphere. The coordinates of points A,B are ,. . $A(4,0,6)$, $B(0,4,6)$. The vector locations of the points A,B are: $\vec{r}_A = 4\vec{i} + 0\vec{j} + 6\vec{k}$, $\vec{r}_B = 0\vec{i} + 4\vec{j} + 0\vec{k}$. The unit vector parallel to the tension acting between the points A,B in the direction of B is by definition $\vec{e}_{AB} = \dfrac{\vec{r}_B - \vec{r}_A}{|\vec{r}_B - \vec{r}_A|}$ The weight is $\vec{W} = 0\vec{i} - |\vec{W}|\vec{j} + 0\vec{k} = 0\vec{i} - 20\vec{j} + 0\vec{k}$ The unit vectors are

Problem 3.62 continued on next page..........

==================================<>==================================

continuation of solution to Problem 3.62...

$\vec{e}_{AB} = -0.4851\,\vec{i} + 0.4851\,\vec{j} - 0.7276\,\vec{k}$

$\vec{e}_1 = 0.4444\,\vec{i} + 0.7778\,\vec{j} + 0.4444\,\vec{k}$

$\vec{e}_2 = -0.8182\,\vec{i} + 0.1818\vec{j} + 0.5455\,\vec{k}$

where the values of the last two were given by the problem statement. The forces are expressed in terms of the unit vectors, $\vec{T}_{AB} = |\vec{T}_{AB}|\,\vec{e}_{AB}$, $\vec{N}_1 = |\vec{N}_1|\,\vec{e}_1$, $\vec{N}_2 = |\vec{N}_2|\,\vec{e}_2$. The equilibrium conditions are

$\sum \vec{F} = 0 = \vec{T}_{AB} + \vec{N}_1 + \vec{N}_2 + \vec{W} = 0$ Substitute the force vectors and collect like terms,

$\sum \vec{F}_x = \left(-0.4851|\vec{T}_{AB}| + 0.4444|\vec{N}_1| - 0.8182|\vec{N}_2|\right)\vec{i} = 0$

$\sum \vec{F}_y = \left(0.4851|\vec{T}_{AB}| + 0.7778|\vec{N}_1| + 0.1818|\vec{N}_2| - 20\right)\vec{j} = 0$

$\sum \vec{F}_z = \left(-0.7276|\vec{T}_{AB}| + 0.4444|\vec{N}_1| + 0.5455|\vec{N}_2|\right)\vec{k} = 0$

The HP-28S hand held calculator was used to solve these equations. was used to solve these equations.

The solution: $\boxed{|\vec{T}_{AB}| = 12.34\ lb}$, $|\vec{N}_1| = 17.51\ lb$, $|\vec{N}_2| = 2.19\ lb$

==================================<>==================================

Problem 3.63 The climber A is being helped up an icy slope by two friends. His mass is 80 kg and the direction cosines of the force exerted on him by the slope are $\cos\theta_x = -0.286$, $\cos\theta_y = +0.429$, $\cos\theta_z = +0.857$. The y axis is vertical. If the climber is in equilibrium in the position shown, what are the tensions in the ropes AB and AC and the magnitude of the force exerted on him by the slope?

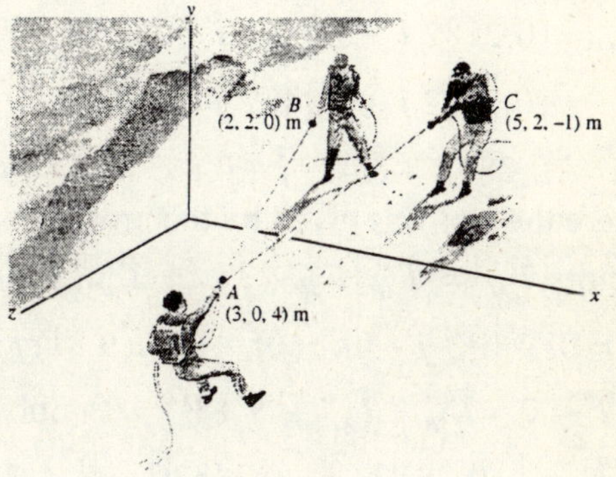

Solution: Get the unit vectors parallel to the ropes using the coordinates of the end points. Express the tensions in terms of these unit vectors, and solve the equilibrium conditions. The rope tensions, the normal force, and the weight act on the climber. The coordinates of points A,B,C are given by the problem,. $A(3,0,4)$, $B(2,2,0)$, $C(5,2,-1)$

The vector locations of the points A, B, C are: $\vec{r}_A = 3\vec{i} + 0\vec{j} + 4\vec{k}$, $\vec{r}_B = 2\vec{i} + 2\vec{j} + 0\vec{k}$, $r_C = 5\vec{i} + 2\vec{j} - 1\vec{k}$. The unit vector parallel to the tension acting between the points A,B in the direction of B is $\vec{e}_{AB} = \dfrac{\vec{r}_B - \vec{r}_A}{|\vec{r}_B - \vec{r}_A|}$ The unit vectors are

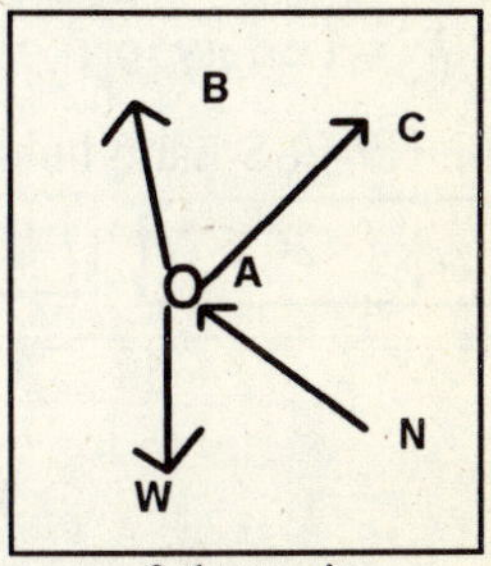

$\vec{e}_{AB} = -0.2182\,\vec{i} + 0.4364\,\vec{j} - 0.8729\,\vec{k}$

$\vec{e}_{AC} = 0.3482\,\vec{i} + 0.3482\,\vec{j} - 0.8704\,\vec{k}$

$\vec{e}_N = -0.286\,\vec{i} + 0.429\vec{j} + 0.857\,\vec{k}$

where the last was given by the problem statement. The forces are expressed in terms of the unit vectors, $\vec{T}_{AB} = |\vec{T}_{AB}|\,\vec{e}_{AB}$, $\vec{T}_{AC} = |\vec{T}_{AC}|\,\vec{e}_{AC}$, $\vec{N} = |\vec{N}|\,\vec{e}_N$. The weight is

$\vec{W} = 0\vec{i} - |\vec{W}|\,\vec{j} + 0\vec{k} = 0\vec{i} - (80)(9.81)\vec{j} + 0\vec{k} = 0\vec{i} - 784.8\vec{j} + 0\vec{k}$. The equilibrium conditions are

$\sum \vec{F} = 0 = \vec{T}_{AB} + \vec{T}_{AC} + \vec{N} + \vec{W} = 0$. Substitute and collect like terms,

Problem 3.63 continued on next page....

================================<>================================

continuation of solution to Problem 3.63...

$$\sum \vec{F}_x = \left(-0.2182\left|\vec{T}_{AB}\right| + 0.3482\left|\vec{T}_{AC}\right| - 0.286\left|\vec{N}\right|\right)\vec{i} = 0$$

$$\sum \vec{F}_y = \left(0.4364\left|\vec{T}_{AB}\right| + 0.3482\left|\vec{T}_{AC}\right| + 0.429\left|\vec{N}\right| - 784.8\right)\vec{j} = 0$$

$$\sum \vec{F}_z = \left(-0.8729\left|\vec{T}_{AB}\right| - 0.8704\left|\vec{T}_{AC}\right| + 0.857\left|\vec{N}\right|\right)\vec{k} = 0$$

The commercial program **TK Solver 2** was used to solve these simultaneous equations. The solution:

$\boxed{\left|\vec{T}_{AB}\right| = 100.7\ N}$, $\boxed{\left|\vec{T}_{AC}\right| = 889.0\ N}$, $\boxed{\left|\vec{N}\right| = 1005.5\ N}$.

================================<>================================

Problem 3.64 Consider the climber being helped by his friends in Problem 3.63. To try to make the tensions in the rope more equal, the friend at B moves to the position (4,2,0) What are the new tensions in the ropes AB and AC and the magnitude of the force exerted on the climber by the slope?
(In Problem 3.65 it was given that the climber's mass is 80 kg and the direction cosines of the force exerted on him by the slope are $\cos\theta_x = -0.286$, $\cos\theta_y = +0.429$, $\cos\theta_z = +0.857$. The y axis is vertical.)

Solution: Get the unit vectors parallel to the ropes using the coordinates of the end points. Express the tensions in terms of these unit vectors, and solve the equilibrium conditions. The coordinates of points A,B,C are $A(3,0,4)$, $B(4,2,0)$, $C(5,2,-1)$. The vector locations of the points A, B, C are:

$\vec{r}_A = 3\vec{i} + 0\vec{j} + 4\vec{k}$, $\vec{r}_B = 4\vec{i} + 2\vec{j} + 0\vec{k}$, $r_C = 5\vec{i} + 2\vec{j} - 1\vec{k}$. The unit vectors are

$$\vec{e}_{AB} = +0.2182\,\vec{i} + 0.4364\,\vec{j} - 0.8729\,\vec{k}$$

$$\vec{e}_{AC} = +0.3482\,\vec{i} + 0.3482\,\vec{j} - 0.8704\,\vec{k}$$

$$\vec{e}_N = -0.286\,\vec{i} + 0.429\vec{j} + 0.857\,\vec{k}$$

where the last was given by the problem statement. The forces are expressed in terms of the unit vectors, $\vec{T}_{AB} = \left|\vec{T}_{AB}\right|\vec{e}_{AB}$, $\vec{T}_{AC} = \left|\vec{T}_{AC}\right|\vec{e}_{AC}$, $\vec{N} = \left|\vec{N}\right|\vec{e}_N$. The weight is

$\vec{W} = 0\vec{i} - \left|\vec{W}\right|\vec{j} + 0\vec{k} = 0\vec{i} - (80)(9.81)\vec{j} + 0\vec{k} = 0\vec{i} - 784.8\vec{j} + 0\vec{k}$ The equilibrium conditions are

$\sum \vec{F} = 0 = \vec{T}_{AB} + \vec{T}_{AC} + \vec{N} + \vec{W} = 0$. Substitute and collect like terms,

$$\sum \vec{F}_x = \left(+0.281\left|\vec{T}_{AB}\right| + 0.3482\left|\vec{T}_{AC}\right| - 0.286\left|\vec{N}\right|\right)\vec{i} = 0$$

$$\sum \vec{F}_y = \left(0.4364\left|\vec{T}_{AB}\right| + 0.3482\left|\vec{T}_{AC}\right| + 0.429\left|\vec{N}\right| - 784.8\right)\vec{j} = 0$$

$$\sum \vec{F}_z = \left(-0.8729\left|\vec{T}_{AB}\right| - 0.8704\left|\vec{T}_{AC}\right| + 0.857\left|\vec{N}\right|\right)\vec{k} = 0$$

The HP-28S hand held calculator was used to solve these simultaneous equations. The solution:

$\boxed{\left|\vec{T}_{AB}\right| = 420.5\ N}$, $\boxed{\left|\vec{T}_{AC}\right| = 532.5\ N}$, $\boxed{\left|\vec{N}\right| = 969.3\ N}$.

================================<>================================

Problem 3.65 The 200 kg slider is held in place on the vertical bar by the cable AB. (a) Determine the tension in the cable. (b) Determine the force exerted on the slider by the bar.

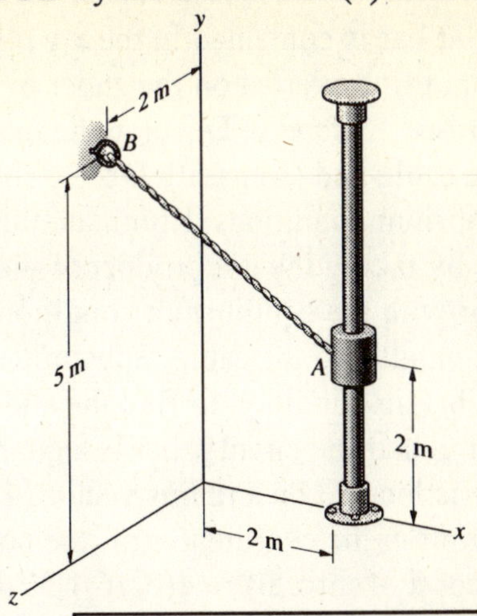

Solution: Strategy: Develop the unit vectors (i) parallel to the cable and (ii) parallel to the slider bar. Apply the equilibrium conditions. Eliminate the slider bar normal force by taking the dot product of the slider bar unit vector with the equilibrium conditions. Solve for the force parallel to the cable. Substitute this force into the equilibrium condition to find the slider bar normal force.

The coordinates of the points A, B are A(2,2,0), B(0,5,2)

The vector positions $\vec{r}_A = 2\vec{i} + 2\vec{j} + 0\vec{k}$, $\vec{r}_B = 0\vec{i} + 5\vec{j} + 2\vec{k}$

The equilibrium conditions are: $\sum \vec{F} = \vec{T} + \vec{N} + \vec{W} = 0$

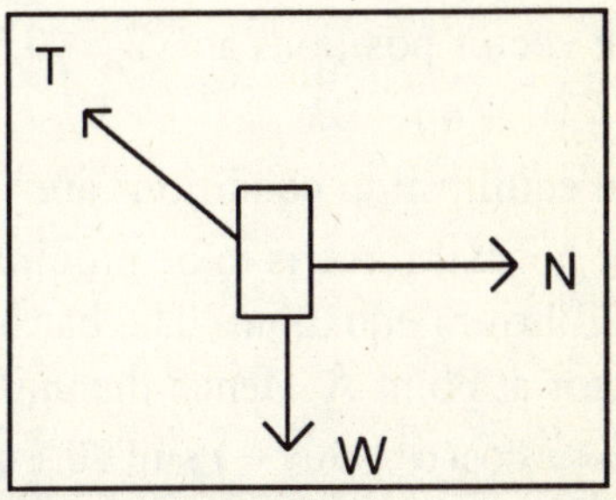

Eliminate the slider bar normal force as follows: The bar is parallel to the y axis, hence the unit vector parallel to the bar is $\vec{e}_B = 0\vec{i} + 1\vec{j} + 0\vec{k}$. The dot product of the unit vector and the normal force vanishes: $\vec{e}_B \cdot \vec{N} = 0$. Take the dot product of $\vec{e}_B$ with the equilibrium conditions: $\sum \vec{e}_B \cdot \vec{F} = \vec{e}_B \cdot \vec{T} + \vec{e}_B \cdot \vec{W} = 0$. The weight is $\vec{e}_B \cdot \vec{W} = 1\vec{j} \cdot (-\vec{j}|\vec{W}|) = -|\vec{W}| = -(200)(9.81) = -1962\ N$

The unit vector parallel to the cable is by definition, $\vec{e}_{AB} = \dfrac{\vec{r}_B - \vec{r}_A}{|\vec{r}_B - \vec{r}_A|}$.

Substitute the vectors and carry out the operation: $\vec{e}_{AB} = -0.4851\vec{i} + 0.7278\vec{j} + 0.4851\vec{k}$.

(a) The tension in the cable is $\vec{T} = |\vec{T}|\vec{e}_{AB}$. Substitute into the modified equilibrium condition

$\sum \vec{e}_B \vec{F} = \left(0.7276|\vec{T}| - 1962\right) = 0$. Solve: $\boxed{|\vec{T}| = 2696.5\ N}$ from which the tension vector is

$\vec{T} = |\vec{T}|\vec{e}_{AB} = -1308\vec{i} + 1962\vec{j} + 1308\vec{k}$.

(b) Substitute into the original equilibrium conditions,

$\sum \vec{F} = 0 = \vec{T} + \vec{N} + \vec{W} = -1308\vec{i} + 1308\vec{k} + \vec{N} = 0$. Solve for the normal force: $\boxed{\vec{N} = 1308\vec{i} - 1308\vec{k}}$.

The magnitude is $|\vec{N}| = 1349.8\ N$ *Note:* For this specific configuration, the problem can be solved without eliminating the slider bar normal force, since it does not appear in the y-component of the equilibrium equation (the slider bar is parallel to the y-axis). However, in the general case, the slider bar will not be parallel to an axis, and the unknown normal force will be projected onto all components of the equilibrium equations (see Problem 3.66) . In this general situtation, it will be necessary to eliminate the slider bar normal force by some procedure equivalent to that used above. *End Note.*

Problem 3.66 The 100 lb slider at A is held in place on the circular bar by the cable AB. The circular bar is contained in the *x,y* plane. (a) Determine the tension in the cable. (b) Determine the normal force exerted on the slider by the bar.

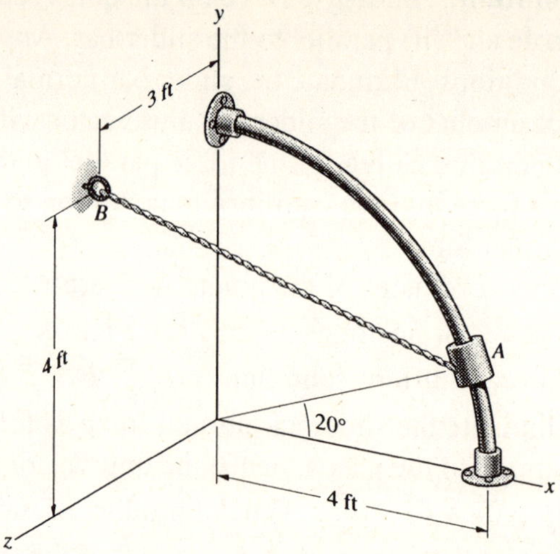

Solution: Strategy: Develop the unit vectors (i) parallel to the cable and (ii) parallel to the slider bar. Apply the equilibrium conditions. Eliminate the slider bar normal force by taking the dot product of the slider bar unit vector with the equilibrium conditions. Solve for the force parallel to the cable. Substitute this force into the equilibrium condition to find the slider bar normal force. Assume that the circular bar is a quarter circle, so that the slider is located on a radius vector (4 ft). With this assumption the coordinates of the points A, B are $A(4\cos\alpha, 4\sin\alpha, 0) = A(3.76, 1.37, 0)$, $B(0,4,3)$
The vector positions are $\vec{r}_A = 3.76\vec{i} + 1.37\vec{j} + 0\vec{k}$, $\vec{r}_B = 0\vec{i} + 4\vec{j} + 3\vec{k}$
The equilibrium conditions are: $\sum \vec{F} = \vec{T} + \vec{N} + \vec{W} = 0$.
The normal force is to be eliminated from the equilibrium equations. The bar is normal to the radius vector at point A. Hence the unit vector parallel to the bar is
$\vec{e}_B = \vec{i}\cos(\alpha + 90) + \vec{j}\sin(90 + \alpha) + 0\vec{k} = -0.3420\vec{i} + 0.9397\vec{j} + 0\vec{k}$
The dot product with the normal force is zero, $\vec{e}_B\vec{N} = 0$. Take the dot product of the unit vector and the equilibrium condition:
$\sum \vec{e}_B\vec{F} = \vec{e}_B\vec{T} + \vec{e}_B\vec{W} = 0$. The weight is
$\vec{e}_B\vec{W} = \vec{e}_B(-\vec{j}|\vec{W}|) = -0.9397|\vec{W}| = -(0.9397)(100) = -94\ lb$. The unit vector parallel to the cable is by definition, $\vec{e}_{AB} = \dfrac{\vec{r}_B - \vec{r}_A}{|\vec{r}_B - \vec{r}_A|}$. Substitute the

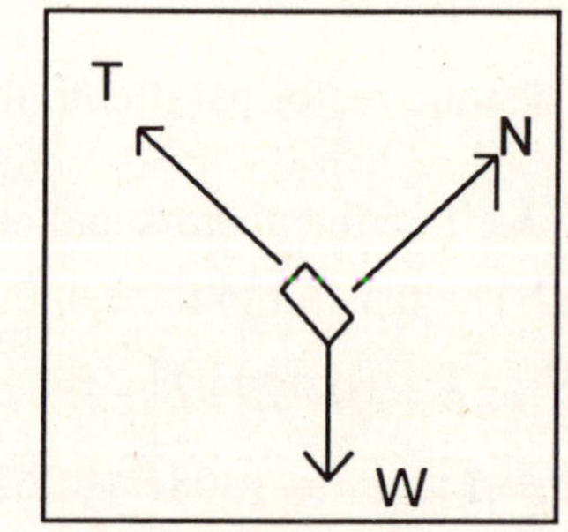

vectors and carry out the operation $\vec{e}_{AB} = -0.6856\vec{i} + 0.4801\vec{j} + 0.5472\vec{k}$. (a) The tension in the cable is $\vec{T} = |\vec{T}|\vec{e}_{AB}$. Substitute into the modified equilibrium condition $\sum \vec{e}_B\vec{F} = (0.6854|\vec{T}| - 94) = 0$.
Solve: $\boxed{|\vec{T}| = 137.1\ lb}$, from which the tension vector is $\vec{T} = |\vec{T}|\vec{e}_{AB} = -94\vec{i} + 65.8\vec{j} + 75\vec{k}$ (b) Substitute $\vec{T}$ into the original equilibrium conditions,
$\sum \vec{F} = 0 = \vec{T} + \vec{N} + \vec{W} = -94\vec{i} + 65.8\vec{j} + 75\vec{k} + \vec{N} - 100\vec{j} = 0$. Solve for the normal force exerted by the bar on the slider $\vec{N} = 94\vec{i} + 34.2\vec{j} - 75\vec{k}$ (lb)

Problem 3.67 (a) Obtain graphs of the tensions in the cables AB and AC for values of d from d=0 to d=1.8 m. (b) Each cable will safely support a tension of 1 kN. Use your graph to estimate the acceptable range of values of d.

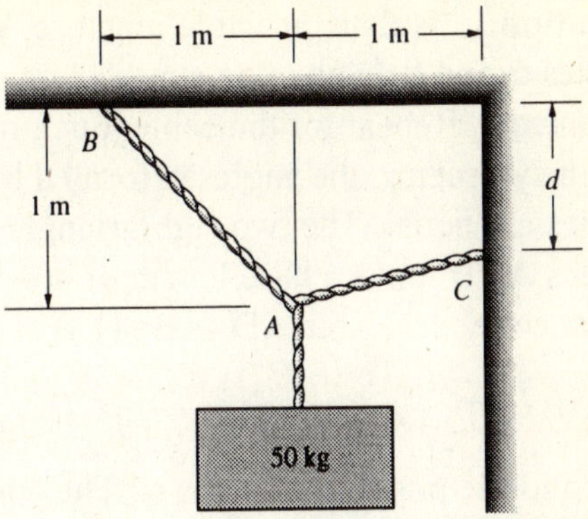

Solution: Isolate the cable juncture A. Find the interior angles α and β. Solve the equilibrium conditions in terms of distance d. Plot the result.

The angle between the positive x axis and the tension $\vec{T}_{AC}$ is α. The tension: $\vec{T}_{AC} = |\vec{T}_{AC}|(\vec{i}\cos\alpha + \vec{j}\sin\alpha)$

The angle between the positive x axis and AB is $(180^o - \beta)$. The tension is

$\vec{T}_{AB} = |\vec{T}_{AB}|(-\vec{i}\cos\beta + \vec{j}\sin\beta)$. The weight is

$\vec{W} = 0\vec{i} - |\vec{W}|\vec{j}$. The equilibrium conditions are

$\sum \vec{F} = \vec{W} + \vec{T}_{AB} + \vec{T}_{AC} = 0$. Substitute the vectors and collect like terms,

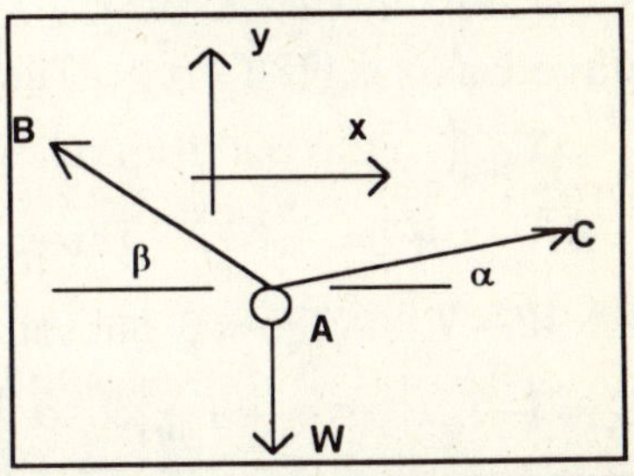

$$\sum \vec{F}_x = (|\vec{T}_{AC}|\cos\alpha - |\vec{T}_{AB}|\cos\beta)\vec{i} = 0$$

$$\sum \vec{F}_y = (|\vec{T}_{AC}|\sin\alpha + |\vec{T}_{AB}|\sin\beta - |\vec{W}|)$$. Solve:

$$|\vec{T}_{AC}| = \left(\frac{\cos\beta}{\cos\alpha}\right)|\vec{T}_{AB}|, \quad |\vec{T}_{AB}| = \left(\frac{|\vec{W}|\cos\alpha}{\sin(\beta+\alpha)}\right),$$ and

$$|\vec{T}_{AC}| = \left(\frac{|\vec{W}|\cos\beta}{\sin(\beta+\alpha)}\right)$$. The weight is

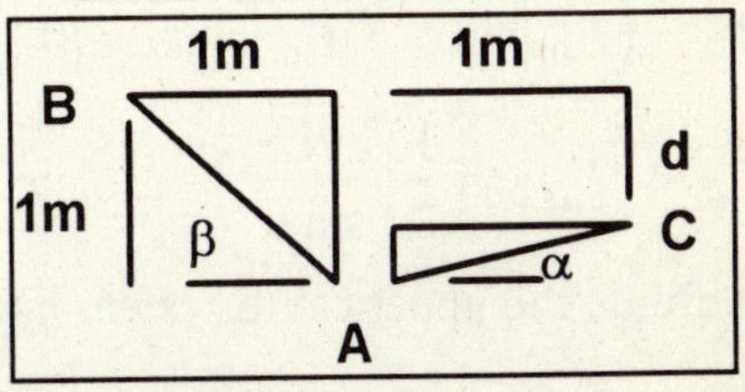

$|\vec{W}| = (50\ kg)\left(9.81\frac{m}{s^2}\right) = 490\ N$. The angles α and β are to be determined. Subdivide the cable interior into two right triangles as shown. From geometry, $\tan\beta = \frac{1\,m}{1\,m} = 1, \quad \beta = 45^o$,

$\tan\alpha = \left(\frac{1-d}{1}\right) = (1-d), \quad \alpha = \tan^{-1}(1-d)$

(Note that the argument $(1-d)$ is dimensionless, since it has been divided by 1 m.) The commercial package **TK Solver 2** was used to produce a graph of the tensions vs. the distance d. From the intersection of the tension line with the 1000 N line, the range of d for safe tension is $\boxed{0 \le d \le 1.31\ m}$.

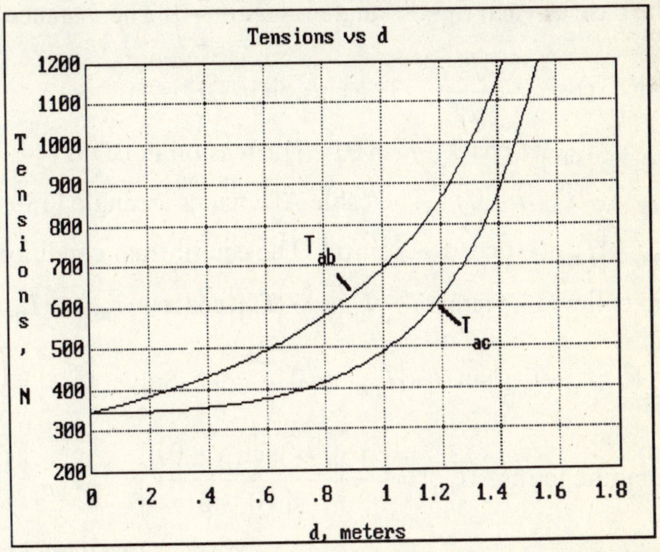

==================================<>==================================

Problem 3.68 The suspended traffic light weighs 100 lb. The cables AB, BC, AD, and DE are each 11 ft long. Determine the smallest permissible length of the cable BD if the tensions in the cables must not exceed 1000 lb.

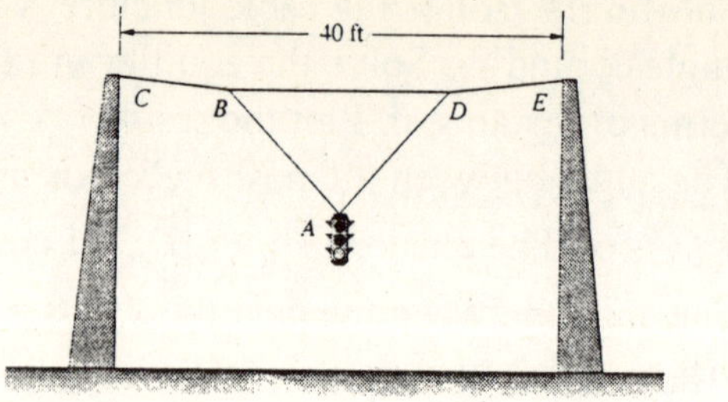

Solution: Isolate the cable juncture A. Find the interior angles α and β. Solve the equilibrium conditions in terms of distance d. Repeat for the cable junctures B and D. From symmetry, the angles α formed by the suspension cables are equal. The two right triangles formed by the cables ABD have a base length of one half BD. Denote this distance by d/2. Note AD=AB=11 ft. The angle α is given by $\cos\alpha = \dfrac{d}{22}$, $\alpha = \cos^{-1}\left(\dfrac{d}{22}\right)$. The angle formed by cable AD and the positive x axis is α. The tension is

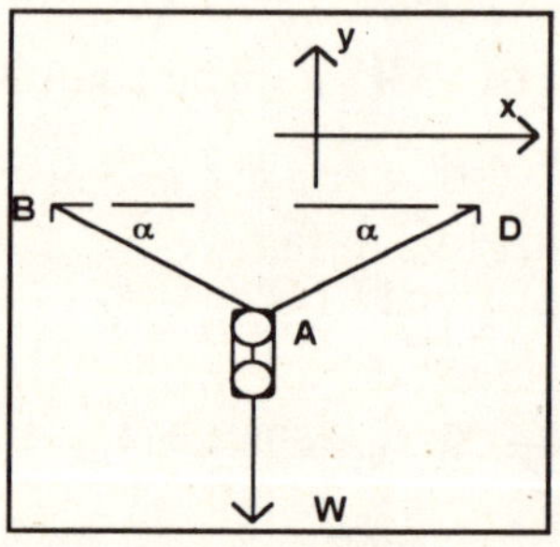

$\vec{T}_{AD} = |\vec{T}_{AD}|(\vec{i}\cos\alpha + \vec{j}\sin\alpha)$. The angle formed by cable AB and the positive x axis is $(180-\alpha)$., The tension is

$\vec{T}_{AB} = |\vec{T}_{AB}|(-\vec{i}\cos\alpha + \vec{j}\sin\alpha)$. The weight is

$\vec{W} = 0\vec{i} - \vec{j}|\vec{W}| = -100\vec{j}$. The equilibrium conditions are

$\Sigma\vec{F} = \vec{W} + \vec{T}_{AD} + \vec{T}_{AB} = 0$ Substitute and collect like terms

$\Sigma\vec{F}_x = \left(-|\vec{T}_{AD}|\cos\alpha + |\vec{T}_{AB}|\cos\alpha\right)\vec{i} = 0$

$\Sigma\vec{F}_y = \left(|\vec{T}_{AD}|\sin\alpha + |\vec{T}_{AB}|\sin\alpha - |\vec{W}|\right)\vec{j} = 0$. Solve:

$|\vec{T}_{AD}| = |\vec{T}_{AB}| = \left(\dfrac{1}{2}\right)\left(\dfrac{|\vec{W}|}{\sin\alpha}\right)$ Isolate the cable juncture D.

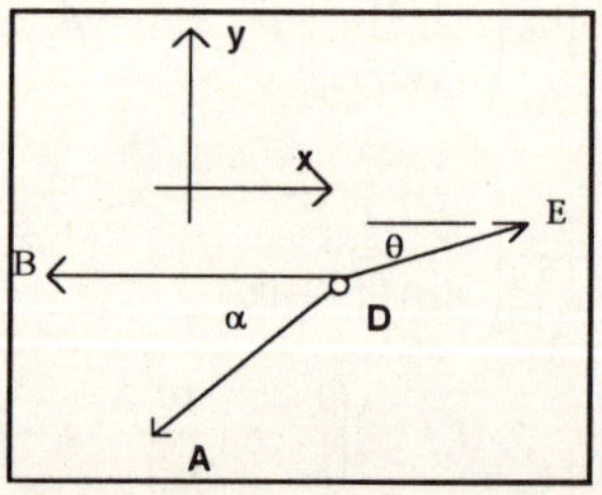

Subdivide the upper cable system into two right triangles and a rectangle

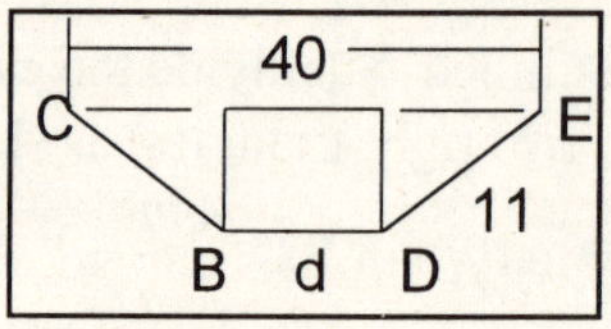

The base of each right triangle is $\left(20 - \dfrac{d}{2}\right)$. The distance DE is 11 ft, hence $\cos\theta = \dfrac{40-d}{22}$. The tension in DE is

$\vec{T}_{DE} = |\vec{T}_{DE}|(\vec{i}\cos\theta + \vec{j}\sin\theta)$. The tension in BD is

$\vec{T}_{BD} = -|\vec{T}_{BD}|\vec{i} + 0\vec{j}$. The cable AD makes an angle $(180+\alpha)$ with the positive x axis:

$\vec{T}_{DA} = |\vec{T}_{DA}|(-\vec{i}\cos\alpha - \vec{j}\sin\alpha)$. The equilibrium conditions are $\sum\vec{F} = \vec{T}_{DE} + \vec{T}_{BD} + \vec{T}_{DA} = 0$. Substitute and collect like terms. $\sum\vec{F}_x = \left(-|\vec{T}_{DA}|\cos\alpha - |\vec{T}_{BD}| + |\vec{T}_{DE}|\cos\theta\right)\vec{i} = 0$

$\sum\vec{F}_y = \left(-|\vec{T}_{DA}|\sin\alpha + |\vec{T}_{DE}|\sin\theta\right)\vec{j} = 0$. Solve: $|\vec{T}_{DE}| = \left(\dfrac{\sin\alpha}{\sin\theta}\right)|\vec{T}_{DA}|$, and $|\vec{T}_{BD}| = |\vec{T}_{DA}|\left(\dfrac{\sin(\alpha-\theta)}{\sin\theta}\right)$

Combine terms, $|\vec{T}_{BD}| = \left(\dfrac{1}{2}\right)\left(\dfrac{|\vec{W}|\sin(\alpha-\theta)}{\sin\theta\ \sin\alpha}\right)$. $|\vec{T}_{DE}| = \left(\dfrac{1}{2}\right)\left(\dfrac{|\vec{W}|}{\sin\theta}\right)$. $|\vec{T}_{AD}| = \left(\dfrac{1}{2}\right)\left(\dfrac{|\vec{W}|}{\sin\alpha}\right)$

Problem 3.68 continued on next page....

continuation of solution to Problem 3.68...

For $|\vec{W}| = 100\ lb$, $\alpha = \cos^{-1}\left(\frac{d}{22}\right)$, and

$\theta = \cos^{-1}\left(\frac{40-d}{22}\right)$ the commercial package **TK Solver 2** was used to produce a graph of the tensions vs. the distance d over the interval $18 < d \le 18.5$.

The shortest length of BD for the maximum tension not to exceed 1000 lb is $\boxed{d = 18.028\ ft}$ From the graph, it is clear that for lengths near 18 ft, a few hundredths of a foot change in length can have tremendous effect on the maximum cable tension.

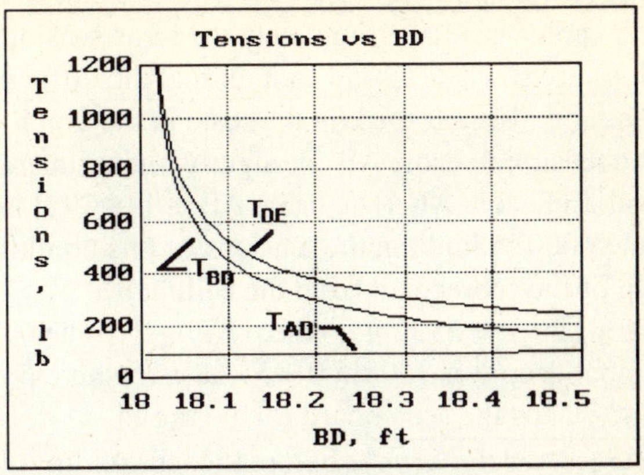

Problem 3.69 The 2000 lb scoreboard **A** is suspended above a sports arena by the cables AB and AC. Each cable is 160 ft long. Suppose you want to move the scoreboard out of the way for a tennis match by shortening cable AB while keeping the length of cable AC constant. (a) Obtain a graph of the tension in cable AB as a function of its length for values of length from 142 ft to 160 ft. (b) Use your graph to estimate how much you can raise the scoreboard relative to its original position if you don't want to subject the cable AB to a tension greater than 6000 lb.

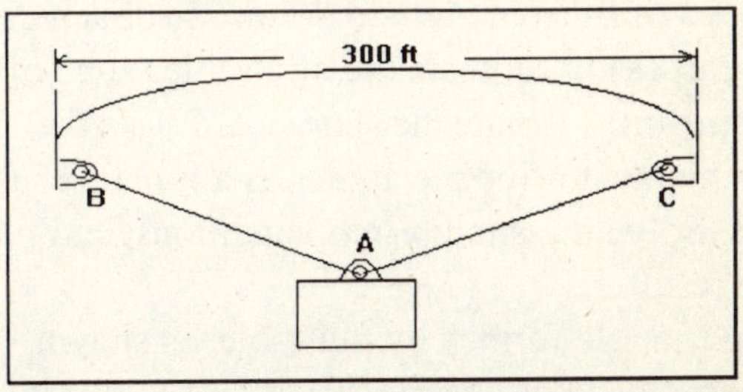

Solution: Isolate the scoreboard. Use the cosine law to find the angles as a function of AB. Solve the equilibrium conditions, and plot the result.

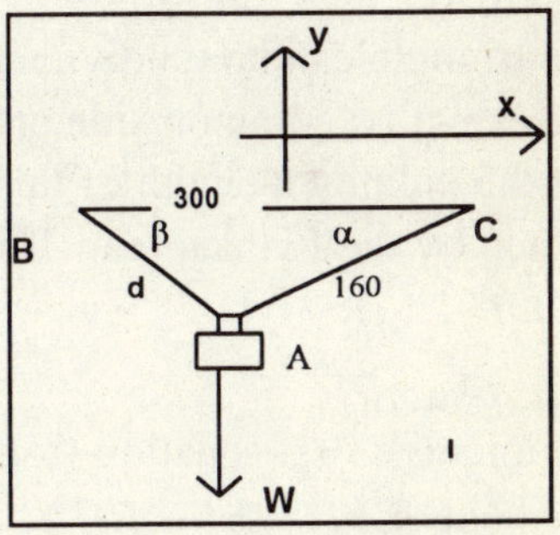

The cosine law is: $\cos\beta = \frac{d^2 + 300^2 - 160^2}{(2)(300)(d)} = \frac{d^2 + 64400}{600d}$

$\cos\alpha = \frac{300^2 + 160^2 - d^2}{2(300)(160)} = \frac{115600 - d^2}{96000}$. The angle formed by cable AB and the positive x axis is $(180° - \beta)$. The tension is $\vec{T}_{AB} = |\vec{T}_{AB}|(-\vec{i}\cos\beta + \vec{j}\sin\beta)$ The angle formed by cable C and the positive x axis is α. The tesnsion is $\vec{T}_{AC} = |\vec{T}_{AC}|(\vec{i}\cos\alpha + \vec{j}\sin\alpha)$. The weight is $\vec{W} = 0\vec{i} - \vec{j}|\vec{W}| = 0\vec{i} - 2000\vec{j}$. The equilibrium conditions are $\sum\vec{F} = \vec{W} + \vec{T}_{AB} + \vec{T}_{AC} = 0$

Substitute : $\sum\vec{F}_x = (-|\vec{T}_{AB}|\cos\alpha + |\vec{T}_{AC}|\cos\beta)\vec{i} = 0$

$\sum\vec{F}_y = (|\vec{T}_{AB}|\sin\alpha + |\vec{T}_{AC}|\sin\beta - |\vec{W}|)\vec{j} = 0$ Solve: $|\vec{T}_{AB}| = |\vec{T}_{AC}|\left(\frac{\cos\beta}{\cos\alpha}\right)$ .. and

$|\vec{T}_{AC}| = \left(\frac{|\vec{W}|\cos\alpha}{\sin(\alpha+\beta)}\right)$ where $\alpha = \cos^{-1}\left(\frac{115600 - d^2}{96000}\right)$, $\beta = \cos^{-1}\left(\frac{64400 + d^2}{600d}\right)$

Problem 3.69 continued on next page....

continuation of solution to Problem 3.69...

The commercial package **TK Solver 2** was used to produce a graph of the tensions vs. the length AB. Both cables have approximately the same tension; the values are so close to one another that the difference cannot be distinguished on the scale in the graph. (The tension in cable AB is slightly higher than the tension in cable AC.)The cable AB is 144.29 ft long at the 6000 lb limit on the tension. At this point the drop of the scoreboard from the ceiling is $144.26 \sin\beta = 25.32\ ft$, where β=10.1°. The original drop was $160 \sin\beta = 55.68\ ft$, where β = 20.4°. Thus the scoreboard can be raised $\boxed{55.68 - 25.32 = 30.36\ ft}$ before the tension limit is exceeded.

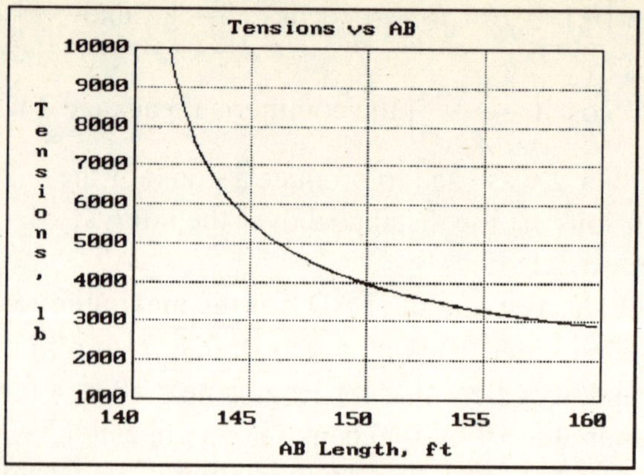

Problem 3.70 Consider the suspended 4-Mg (megagrams) truck in Problem 3.38. The sum of the lengths of the cables AB and BC is 42 m. (a) Obtain graphs of the tensions in cables AB and AC for values of b from zero to 15 m. (b) Each cable will safely support a tension of 60 kN. Use the results of Part (a) to estimate the allowable range of b.

Solution: Isolate the scoreboard. Use the cosine law to find the angles as a function of AB. Solve the equilibrium conditions, and plot the result.

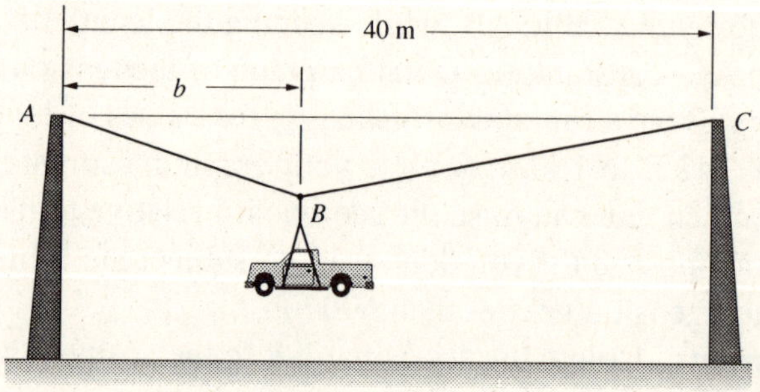

The triangle formed by the cables is shown, where $AC = 40\ m,\ AB + BC = S = 42\ m$

The triangle can be subdivided into two right angles with a common side of unknown length. Let the unknown length of this common side be *d*, then by the Pythagorean Theorem

$$b^2 + d^2 = AB^2$$

$$L^2 + d^2 = BC^2$$

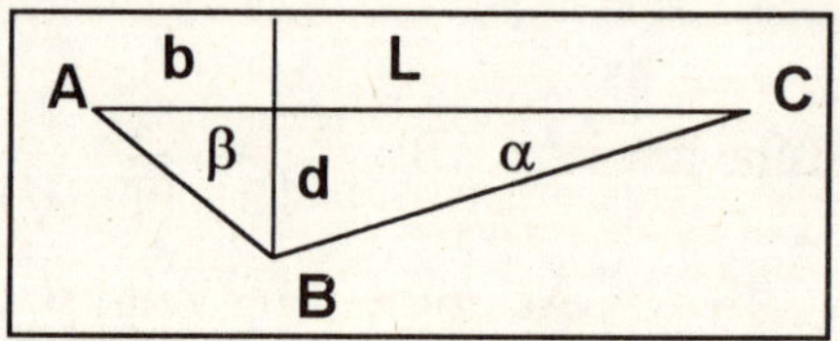

Subtract the first equation from the second to eliminate the unknown *d*, $L^2 - b^2 = BC^2 - AB^2$. Note that

$$L^2 - b^2 = (L - b)(L + b) = 40(40 - 2b) = 1600 - 80b$$

$$BC^2 - AB^2 = (BC - AB)(BC + AB) = 42(BC - AB)$$

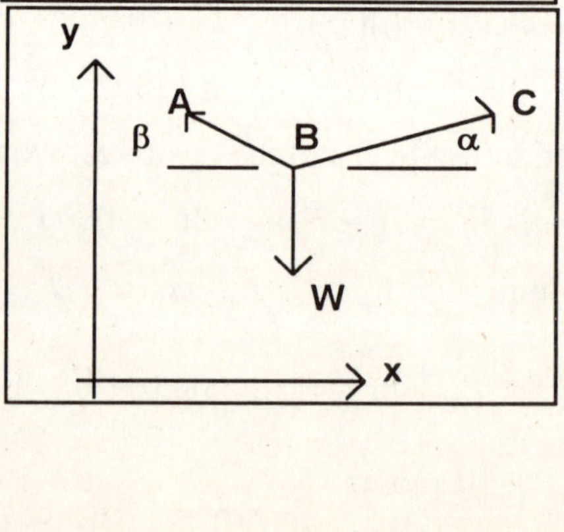

Substitute and reduce to obtain the pair of simultaneous equations in the unknowns

$$BC - AB = \frac{800 - 40b}{21},$$

$BC + AB = 42$. Solve: $BC = \dfrac{841 - 20b}{21}$,

$$AB = \frac{41 + 20b}{21}.$$

Problem 3.70 continued on next page....

continuation of solution to Problem 3.70...

$$\cos\alpha = \left(\frac{(L+b)^2 + BC^2 - AB^2}{2(L+b)(BC)}\right) = \left(\frac{40^2 + (40^2 - 80b)}{2(40)\left(\frac{841-20b}{21}\right)}\right) = \left(\frac{(21)(40-b)}{841-20b}\right)$$

$$\cos\beta = \left(\frac{(L+b)^2 + AB^2 - BC^2}{2(L+b)(AB)}\right) = \left(\frac{40^2 - (40^2 - 80b)}{2(40)\left(\frac{41+20b}{21}\right)}\right) = \left(\frac{21b}{41+20b}\right)$$

The angle between the cable BC and the positive x axis is α, hence the tension $\vec{T}_{BC} = |\vec{T}_{BC}|(\vec{i}\cos\alpha + \vec{j}\sin\alpha)$. The angle between the cable BA and the positive x axis is $(180^\circ - \beta)$, hence the tension is $\vec{T}_{BA} = |\vec{T}_{BA}|(-\vec{i}\cos\beta + \vec{j}\sin\beta)$. The weight is $\vec{W} = 0\vec{i} - |\vec{W}|\vec{j}$.

The equilibrium conditions are $\sum\vec{F} = \vec{W} + \vec{T}_{BA} + \vec{T}_{BC} = 0$. Substitute and collect like terms

$$\sum\vec{F}_x = (|\vec{T}_{BC}|\cos\alpha - |\vec{T}_{BA}|\cos\beta)\vec{i} = 0,$$

$$\sum\vec{F}_y = (|\vec{T}_{BC}|\sin\alpha + |\vec{T}_{BA}|\sin\beta - |\vec{W}|)\vec{j} = 0$$

Solve: $|\vec{T}_{BA}| = \left(\frac{|\vec{W}|\cos\alpha}{\sin(\alpha+\beta)}\right)$, $|\vec{T}_{BC}| = \left(\frac{|\vec{W}|\cos\beta}{\sin(\alpha+\beta)}\right)$

For $|\vec{W}| = (4000)(9.81) = 39240\ N$, and

$$\alpha = \cos^{-1}\left(\frac{21(40-b)}{841-20b}\right),\quad \beta = \cos^{-1}\left(\frac{21b}{41+20b}\right)$$

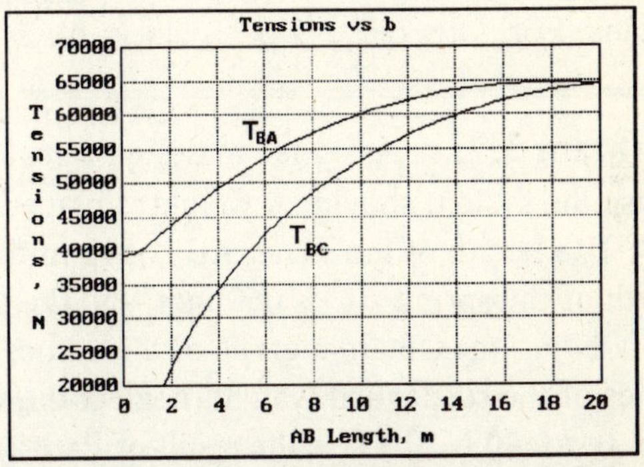

.The commercial package **TK Solver 2** was used to produce a graph of the tension in the two cables as a function of the distance b. The values of b for which the maximum tension is less that 60 kN are $\boxed{0 \le b \le 10\ m}$. By symmetry, the distance b may be measured from either the left or right side.

Problem 3.71 Consider the system shown in Problem 3.39. The tension in cable AD is 200 lb. (a) Obtain graphs of the tension in cables AB and AC for h from 0 to 20 in. (b) For what values of h are the tensions in cables AB and AC positive? What happens when h is not in that range?

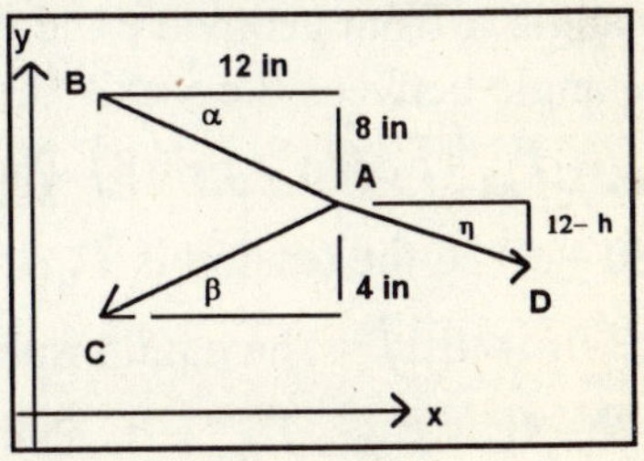

Solution: Isolate the cable juncture A. Determine the angular configuration of the cables in the terms of h. Solve the equilibrium conditions and plot the results. The angles are found from $\tan\alpha = \left(8/12\right) = 0.667$, $\alpha = 33.7^\circ$.

$\tan\beta = \left(4/12\right) = 0.333$, $\beta = 18.4^\circ$. The angle between AB and the positive x axis is $(180^\circ - \alpha)$. The tension is $\vec{T}_{AB} = |\vec{T}_{AB}|(-\vec{i}\cos\alpha + \vec{j}\sin\alpha)$. The angle between AC and the positive x axis is $(180 + \beta)$. The tension is $\vec{T}_{AC} = |\vec{T}_{AC}|(-\vec{i}\cos\beta - \vec{j}\sin\beta)$. The angle between the positive x axis and $\vec{T}_{AD}$ is $(-\eta)$ where $\tan\eta = \left((12-h)/8\right)$. The tension is $\vec{T}_{AD} = |\vec{T}_{AD}|(\vec{i}\cos\eta - \vec{j}\sin\eta)$

Problem 3.71 continued on next page....

continuation of solution to Problem 3.71...

The equilibrium conditions: $\sum \vec{F} = \vec{T}_{AC} + \vec{T}_{AB} + \vec{T}_{AD} = 0$.

Thus

$$\sum \vec{F}_x = \left(-\left|\vec{T}_{AB}\right|\cos\alpha - \left|\vec{T}_{AC}\right|\cos\beta + \left|\vec{T}_{AD}\right|\cos\eta\right)\vec{i} = 0$$

$$\sum \vec{F}_y = \left(\left|\vec{T}_{AB}\right|\sin\alpha - \left|\vec{T}_{AC}\right|\sin\beta - \left|\vec{T}_{AD}\right|\sin\eta\right)\vec{j} = 0 .$$

Solve: $\left|\vec{T}_{AB}\right| = \left(\dfrac{\left|\vec{T}_{AD}\right|\sin(\eta+\beta)}{\sin(\alpha+\beta)}\right)$, and

$\left|\vec{T}_{AC}\right| = \left(\dfrac{\left|\vec{T}_{AD}\right|\sin(\alpha-\eta)}{\sin(\alpha+\beta)}\right)$, For $\tan\eta = \left({(12-h)}/{8}\right)$ and

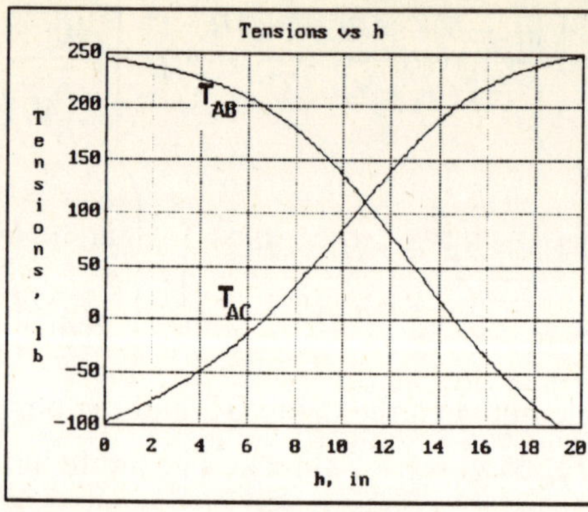

$\left|\vec{T}_{AD}\right| = 200\ lb$,, $\alpha = 33.7°$, $\beta = 18.4°$, the commercial package **TK Solver 2** was used to graph the tensions vs. the distance h. The graph shows that the tensions are positive over the interval $\boxed{6.7 \le h \le 14.7 \text{ (inches)}}$ Outside this interval, one of the cables is slack. For $h \le 6.7$,, $\left|\vec{T}_{AC}\right|$ is slack; for $h \ge 14.7$,, $\left|\vec{T}_{AB}\right|$ is slack.

Problem 3.72 Assigned to design a scale to measure small weights, a group of engineering students devise the system shown. A weight is placed in the pan and the angle α is measured using the protractor at B. The length of the string AB is 500 mm. The unstretched length of the spring AC is 660 mm, and the spring constant is k=20 N/m. (a) Obtain a graph of the values of the angle α for values of the suspended weight (neglect the value of the pan) from 0 to 20 N. (b) Use the result of Part(a) to estimate the angle α corresponding to a weight of 10 N.

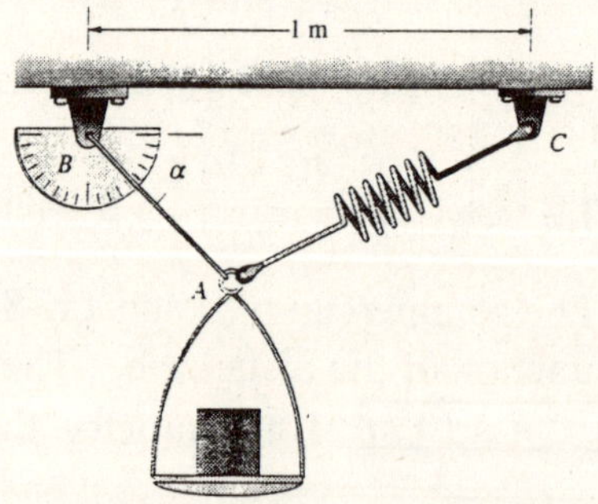

Solution: Solve the equilibrium equations for the magnitude of the tension in the spring. Use the linear spring force-deflection to eliminate the magnitude of the tension in spring AB. Determine the angle α from geometry and the linear spring deflection equation. Solve for the angle α by iteration. The angle between the x axis and the spring is β; the tension is $\vec{T}_{AB} = \left|\vec{T}_{AB}\right|\left(\vec{i}\cos\beta + \vec{j}\sin\beta\right)$ The angle between the x axis and AC is $(180-\alpha)$, ; the tension is $T_{AC} = \left|\vec{T}_{AC}\right|\left(-\vec{i}\cos\alpha + \vec{j}\sin\alpha\right)$. The weight is $\vec{W} = 0\vec{i} - \left|\vec{W}\right|\vec{j}$. The equilibrium conditions are $\sum \vec{F} = \vec{W} + \vec{T}_{AB} + \vec{T}_{AC} = 0$. Substitute the vectors and collect like terms

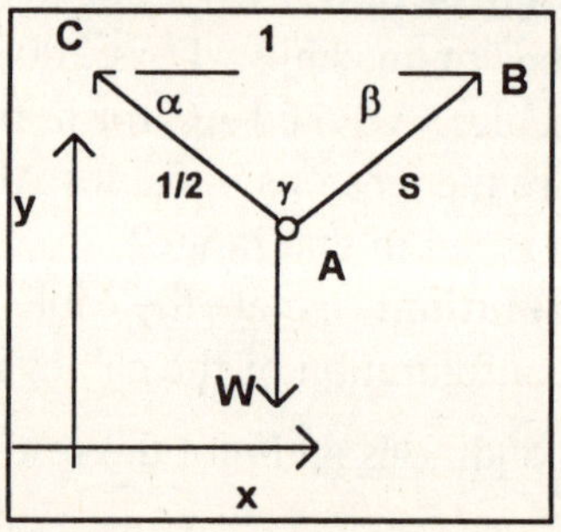

$$\sum \vec{F}_x = \left(\left|\vec{T}_{AB}\right|\cos\beta - \left|\vec{T}_{AC}\right|\cos\alpha\right)\vec{i} = 0$$

$$\Sigma \vec{F}_y = \left(\left|\vec{T}_{AB}\right|\sin\beta + \left|\vec{T}_{AC}\right|\sin\alpha - \left|\vec{W}\right|\right)\vec{j} = 0 \text{ . Solve :}$$

$$\left|\vec{T}_{AC}\right| = \left(\frac{\cos\beta}{\cos\alpha}\right)\left|\vec{T}_{AB}\right| \text{ and } \left|\vec{T}_{AB}\right| = \left(\frac{\left|\vec{W}\right|\cos\alpha}{\sin(\alpha+\beta)}\right)$$

Problem 3.72 continued on next page....

continuation of solution to Problem 3.72...

The tension $\left|\vec{T}_{AB}\right|$ determines the spring deflection, $\left|\vec{T}_{AB}\right| = kDL = k(S-L)$ Thus,

$S = L + \left(\dfrac{\left|\vec{W}\right|\cos\alpha}{k\sin(\alpha+\beta)}\right)$. The angle β is eliminated: $\sin(\alpha+\beta) = \sin(180-\gamma) = \sin\gamma$. From the law of sines $\dfrac{S}{\sin\alpha} = \dfrac{1}{\sin\gamma}$. $S = \dfrac{L}{1-\left(\dfrac{\left|\vec{W}\right|\cos\alpha}{k\sin\alpha}\right)} = \dfrac{L\tan\alpha}{\tan\alpha - \left(\dfrac{\left|\vec{W}\right|}{k}\right)}$. The stretched length of the spring S is also related to the other sides and the angle α by the cosine law: $S^2 = 0.5^2 + 1^2 - 2(0.5)(1)\cos\alpha$, or, $S = \sqrt{1.25-\cos\alpha}$. Substitute this into the expression for S get the expression for the angle α in terms of the weight: $f(\alpha) = L\tan\alpha - \left(\tan\alpha - \dfrac{\left|\vec{W}\right|}{k}\right)\sqrt{1.25-\cos\alpha} = 0$

This transcendental equation must be solved by trial and error. If the **Newton-Raphson** algorithm is selected , with a guess α and a correction to the guess δα, then

$$\delta\alpha = -\left(\frac{f(\alpha)}{\left(\dfrac{\partial f(\alpha)}{\partial\alpha}\right)}\right), \text{ where } \frac{\partial f(\alpha))}{\partial\alpha} = +(L-Q)\sec^2\alpha - \left(\frac{P}{2Q}\right)\sin\alpha$$

and $P = \tan\alpha - \dfrac{\left|\vec{W}\right|}{k}$, $Q = \sqrt{1.25-\cos\alpha}$, $L = 0.66\ m$, . $k = 20\ N/m$, $\left|\vec{W}\right| = weight$

The Newton-Raphson algorithm converges slowly for this problem. A computer program is essential if a solution for more than a few values of α is to be computed. The commercial program **TK Solver 2** was used here to solve for the angle. A graph of the results is shown .At a weight of 10 N, the angle α is $\alpha = 62.7^o$ degrees.

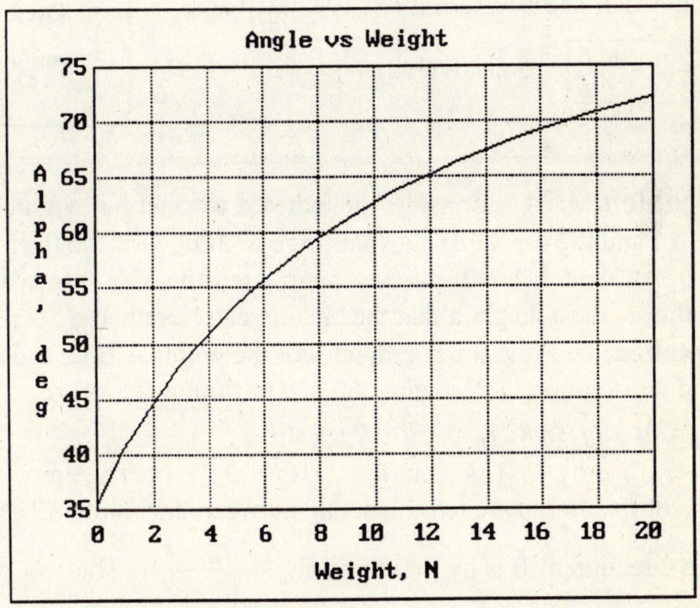

Problem 3.73

The system shown in Problem 3.51 provides lateral support for a load resting on the bed of a truck. When the load is subjected to an effective lateral load F (Figure (b)) the distance from the original position of the load to its equilibrium position is δ. The unstretched length of each spring is 1 ft. Suppose that the load is subjected to an effective lateral load of F = 200 lb. (a) Obtain a graph of the spring constant k for values of δ =0.5 to δ=3 ft. (b) Use the results of (a) to estimate the values of k for δ=1 ft and δ = 2 ft.

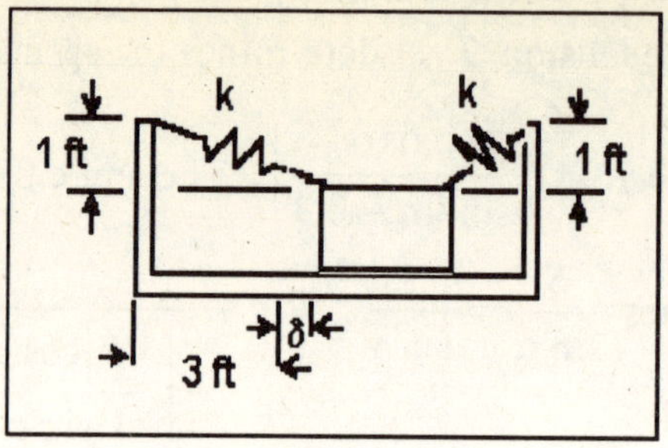

Solution: From the solution of Problem 3.51 the lateral load is given by $|\vec{F}| = |\vec{T}_L|\cos\beta - |\vec{T}_R|\cos\alpha$, where the tension in the left hand spring is $|\vec{T}_L| = k\Delta L_L = k(\sqrt{(3+\delta)^2+1^2}-1)$, and the tension in the right hand spring is $|\vec{T}_L| = k\Delta L_L = k(\sqrt{(3+\delta)^2+1^2}-1)$. The angles are also given in the solution of Problem 3.51: $\tan\alpha = \dfrac{1}{3-\delta}$, and $\tan\beta = \dfrac{1}{3+\delta}$. For $|\vec{F}| = 200\ lb$, the equation to be solved for k is

$$f(k) = k\left(\Delta L_L \cos\beta - \Delta L_R \cos\alpha\right) - 200 = 0$$

The commercial package **TK Solver 2** was used to graph the spring constant vs. the displacement δ, as shown below. The values of k at δ=1 ft and δ=2 ft are $\boxed{k = 104\ lb/ft}$ and $\boxed{k = 54\ lb/ft}$, respectively.

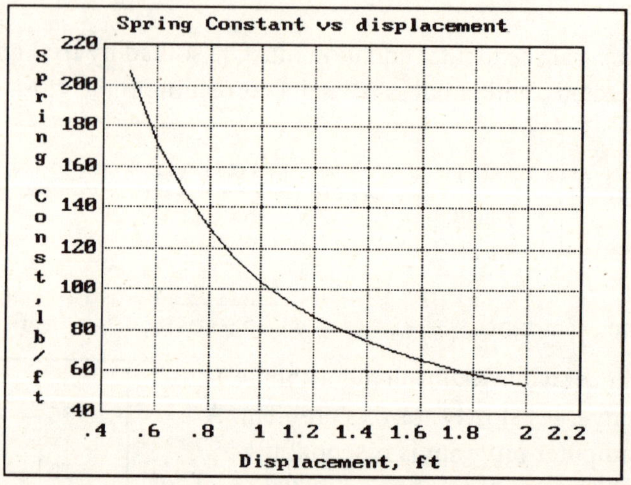

Problem 3.74

Consider the tethered weather balloon in Problem 3.60. The mass of the balloon is 80 kg and the (upward) bouyant force is 1000 N If the lines AB, AC, and AD will safely support a tension of 500 N, what is the smallest height h that the balloon can be tethered?

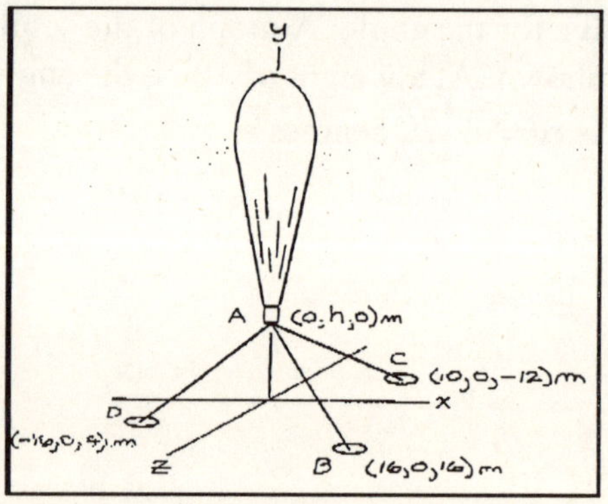

Solution: From the coordinates of the points A,B,C, and D and the solution to Problem 3.60, the position vectors are $\vec{r}_A = 0\vec{i} + h\vec{j} + 0\vec{k}$, $\vec{r}_B = 16\vec{i} + 0\vec{j} + 16\vec{k}$, $\vec{r}_C = 10\vec{i} + 0\vec{j} + -12\vec{k}$, and $\vec{r}_D = -16\vec{i} + 0\vec{j} + 4\vec{k}$. The unit vector parallel to the tension acting between the points A,B in the direction of B is by definition $\vec{e}_{AB} = \dfrac{\vec{r}_B - \vec{r}_A}{|\vec{r}_B - \vec{r}_A|}$. The magnitudes of the position vectors are

Problem 3.74 continued on next page....

continuation of solution to Problem 3.74

$M_{AB} = \sqrt{16^2 + h^2 + 16^2} = \sqrt{512 + h^2}$, $M_{AC} = \sqrt{10^2 + h^2 + 12^2} = \sqrt{244 + h^2}$..., and $M_{AD} = \sqrt{16^2 + h^2 + 4^2} = \sqrt{272 + h^2}$. The unit vectors are $\vec{e}_{AB} = \left(\frac{1}{M_{AB}}\right)(16\vec{i} - h\vec{j} + 16\vec{k})$,

$\vec{e}_{AC} = \left(\frac{1}{M_{AC}}\right)(10\vec{i} - h\vec{j} - 12\vec{k})$, and $\vec{e}_{AD} = \left(\frac{1}{M_{AD}}\right)(-16\vec{i} - h\vec{j} + 4\vec{k})$. The tensions in the cables are expressed in terms of the unit vectors, $\vec{T}_{AB} = |\vec{T}_{AB}|\vec{e}_{AB}$, $\vec{T}_{AC} = |\vec{T}_{AC}|\vec{e}_{AC}$, and $\vec{T}_{AD} = |\vec{T}_{AD}|\vec{e}_{AD}$. The weight is $\vec{W} = 0\vec{i} - (80)(9.81)\vec{j} + 0\vec{k} = 0\vec{i} - 784.8\vec{j} + 0\vec{k}$. The bouyant force is $\vec{B} = 0\vec{i} + 1000\vec{j} + 0\vec{k}$. The equilibrium conditions are $\sum \vec{F} = 0 = \vec{T}_{AB} + \vec{T}_{AC} + \vec{T}_{AD} + \vec{B} + \vec{W} = 0$. Substitute the force vectors and collect like terms,

$$\sum \vec{F}_x = \left(\left(\frac{16}{M_{AB}}\right)|\vec{T}_{AB}| + \left(\frac{10}{M_{AC}}\right)|\vec{T}_{AC}| - \left(\frac{16}{M_{AD}}\right)|\vec{T}_{AD}|\right)\vec{i} = 0$$

$$\sum \vec{F}_y = \left(-\left(\frac{h}{M_{AB}}\right)|\vec{T}_{AB}| - \left(\frac{h}{M_{AC}}\right)|\vec{T}_{AC}| - \left(\frac{h}{M_{AD}}\right)|\vec{T}_{AD}| + 215.2\right)\vec{j} = 0$$

$$\sum \vec{F}_z = \left(\left(\frac{16}{M_{AB}}\right)|\vec{T}_{AB}| - \left(\frac{12}{M_{AC}}\right)|\vec{T}_{AC}| + \left(\frac{4}{M_{AD}}\right)|\vec{T}_{AD}|\right)\vec{k} = 0$$

The commercial program **TK Solver 2** was used to produce a graph of the tensions as a function of the tether height. For all tensions less than 500 N, $\boxed{h \geq 3.1\ m}$

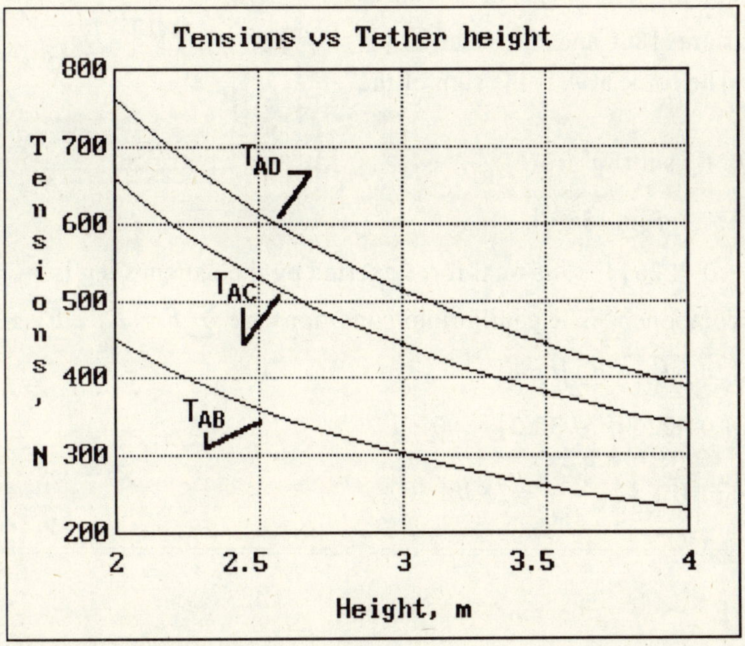

Problem 3.75 The assembly A, including the pulley, weighs 60 lb. What force $|\vec{F}|$ is necessary for the system to be in equilibrium?

Solution: Isolate the bottom pulley. Solve the equilibrium equations. The forces acting on the bottom pulley are the three rope tensions and the weight. Each rope tension is $\vec{F} = 0\vec{i} + |\vec{F}|\vec{j}$. The weight vector is $\vec{W} = 0\vec{i} - |\vec{W}|\vec{j}$. The equilibrium conditions are:

$$\sum \vec{F}_Y = \left(|\vec{F}| + |\vec{F}| + |\vec{F}| - |\vec{W}|\right)\vec{j} = 0. \text{ Solve: } \boxed{|\vec{F}| = \frac{|\vec{W}|}{3} = 20\ lb}$$

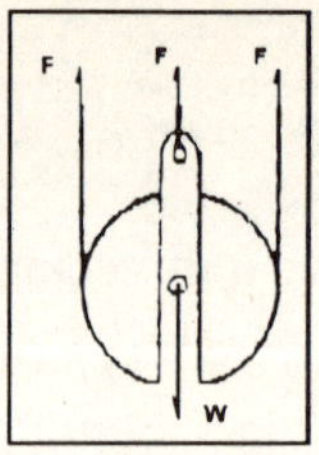

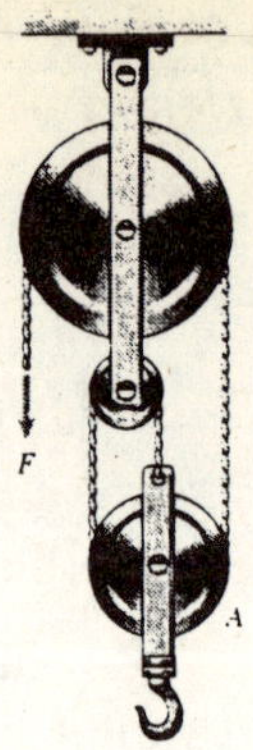

Problem 3.76 The systems shown is called Russell's traction. If the sum of the downward forces exerted at A and B by the patient's leg is 32.2 lb, what is the weight $|\vec{W}|$?

Solution: Isolate the leg. Express the tensions at A and B in scalar components. Solve the equilibrium conditions. The pulleys change the direction but not the magnitude of the force $|\vec{W}|$. The force at B is $\vec{F}_B = |\vec{W}|(\vec{i}\cos 60° + \vec{j}\sin 60°)$. $\vec{F}_B = |\vec{W}|\left(0.5\vec{i} + 0.866\vec{j}\right)$. The angles at A relative to the positive x axis are: $180°$ and $180° - 25° = 155°$. The force at A is the sum of the two forces:

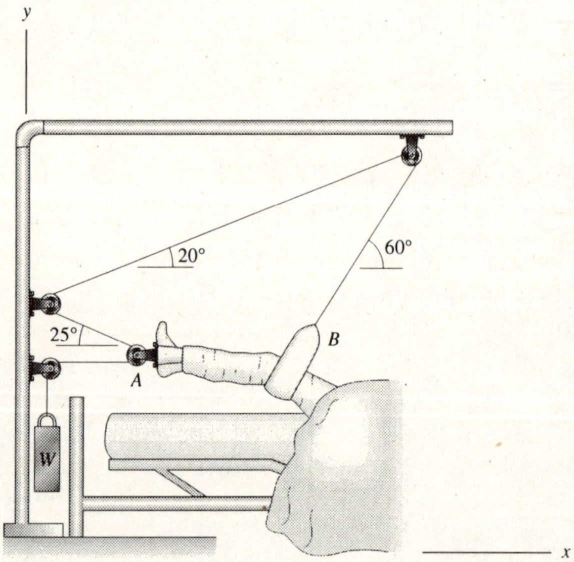

$$\vec{F}_A = |\vec{W}|\left(\vec{i}\cos 180° + \vec{j}\sin 180°\right) + |\vec{W}|\left(\vec{i}\cos 155° + \vec{j}\sin 155°\right)$$

$\vec{F}_A = |\vec{W}|\left(-1.906\vec{i} + 0.4226\vec{j}\right)$. The total force exerted by the patient's leg is $\vec{F}_P = F_H\vec{i} - 32.2\vec{j}$, where F_H is an unknown component. The equilibrium conditions are $\sum \vec{F} = \vec{F}_A + \vec{F}_B + \vec{F}_P = 0$, from which:

$$\sum \vec{F}_X = \left(0.5|\vec{W}| - 1.906|\vec{W}| + F_H\right)\vec{i} = 0$$

$$\sum \vec{F}_Y = \left(0.866|\vec{W}| + 0.4226|\vec{W}| - 32.2\right)\vec{j} = 0.$$

Solve for the weight: $\boxed{|\vec{W}| = \frac{32.2}{1.2886} = 25\ lb}$

Problem 3.77 A heavy rope used as a hawser for a cruise ship weighs 200 lbs. What are the tensions in the rope at A and B?

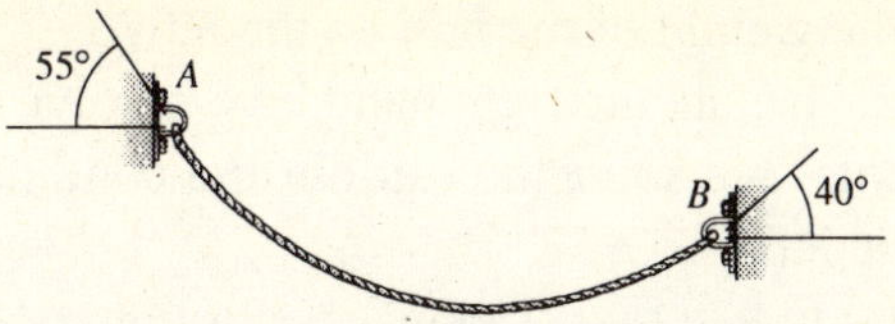

Solution: Resolve the tensions at A and B into scalar components. Solve the equilibrium equations. The tension at B is

$\vec{T}_B = |\vec{T}_B|(\vec{i}\cos 40^o + \vec{j}\sin 40^o)$

$\vec{T}_B = |\vec{T}_B|(0.7660\vec{i} + 0.6428\vec{j})$. The angle at A relative to the positive x axis is $180^o - 55^o = 125^o$. The tension at A:

$\vec{T}_A = |\vec{T}_A|(\vec{i}\cos 125^o + \vec{j}\sin 125^o) = |\vec{T}_A|(-0.5736\vec{i} + 0.8192\vec{j})$. The weight is: $\vec{W} = 0\vec{i} - 200\vec{j}$. The equilibrium conditions are $\sum \vec{F} = \vec{T}_A + \vec{T}_B + \vec{W} = 0$, from which $\Sigma \vec{F}_x = (0.766|\vec{T}_B| - 0.5736|\vec{T}_A|)\vec{i} = 0$

$\sum \vec{F}_y = (0.6428|\vec{T}_B| + 0.8192|\vec{T}_A| - 200)\vec{j} = 0$. Solve: $\boxed{|\vec{T}_B| = 115.1\ lb}$, $\boxed{|\vec{T}_A| = 153.8\ lb}$

Problem 3.78 The 100-lb crate is held in place on the surface by the rope AB. Determine the tension in the string and the magnitude of the normal force exerted on the crate by the surface.

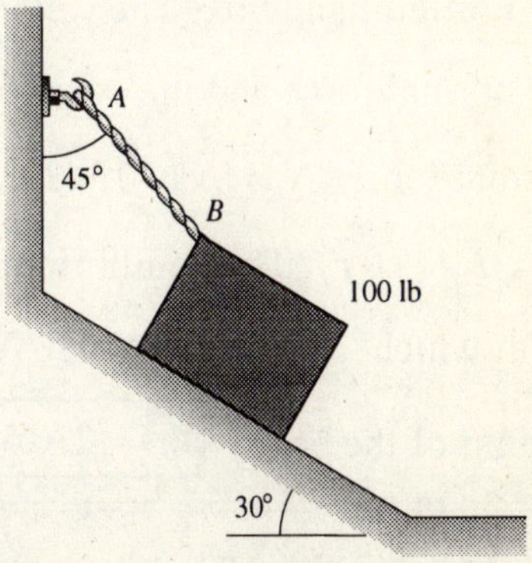

Solution: Isolate the crate, and solve the equilibrium conditions. The weight is $\vec{W} = 0\vec{i} - 100\vec{j}$. The angle between the normal force and the positive x axis is $(90 - 30) = 60^o$. The normal force is

$\vec{N} = |\vec{N}|(\vec{i}\cos 60 + \vec{j}\sin 60) = |\vec{N}|(0.5\vec{i} + 0.866\vec{j})$ The angle between the string tension and the positive x axis is $(180^o - 45^o) = 135^o$, hence the tension is

$\vec{T} = |\vec{T}|(\vec{i}\cos 135^o + \vec{j}\sin 135^o) = |\vec{T}|(-0.7071\vec{i} + 0.7071\vec{j})$. The equilibrium conditions are $\sum \vec{F} = \vec{W} + \vec{N} + \vec{T} = 0$. Substituting, and collecting like terms

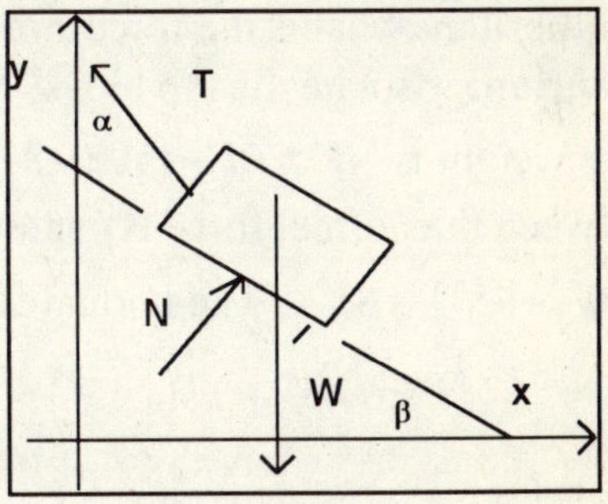

$\sum \vec{F}_x = (0.5|\vec{N}| - 0.7071|\vec{T}|)\vec{i} = 0$

$\sum \vec{F}_y = (0.866|\vec{N}| + 0.7071|\vec{T}| - 100)\vec{j} = 0$

Solve: $\boxed{|\vec{T}| = 51.8\ lb}$, $\boxed{|\vec{N}| = 73.2\ lb}$ *Check:* Use a coordinate system with the x axis parallel to the inclined surface. The equilibrium equation for the x-coordinate is

$\Sigma F_x = |\vec{W}|\sin 30^o - |\vec{T}|\cos 15^o = 0$ from which $|\vec{T}| = \left(\dfrac{\sin 30^o}{\cos 15^o}\right)100 = 51.76 = 51.8\ lb$. The equilibrium equation for the y-coordinate is $\Sigma F_y = |\vec{N}| - |\vec{W}|\cos 30^o + |\vec{T}|\sin 15^o = 0$, from which $|\vec{N}| = 73.2\ lb$. *check.*

Problem 3.79 The cable **AB** is horizontal and the box on the right weighs 100 *lb*. (a) What is the tension in the cable?(b) What is the weight of the box on the left?

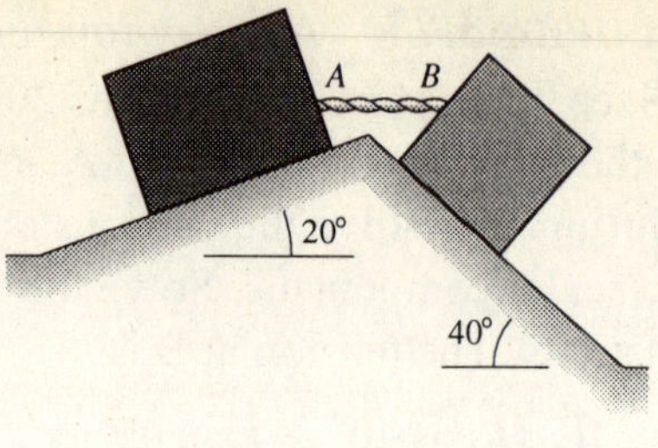

Solution: Isolate the right hand box, resolve the forces into components, and solve the equilibrium conditions. Repeat for the box on the left.

(a) For right hand box. The weight is $\vec{W} = 0\vec{i} - 100\vec{j}$. The angle between the normal force and the positive x axis is $(90^o - 40^o) = 50^o$. The force: $\vec{N} = |\vec{N}|(\vec{i}\cos 50^o + \vec{j}\sin 50^o)$

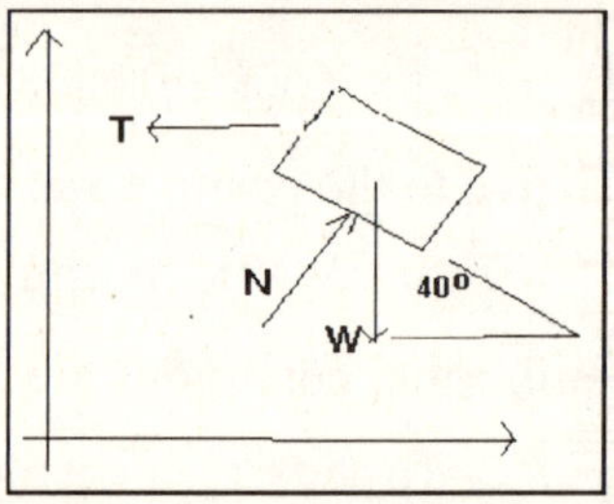

$= |\vec{N}|(0.6428\vec{i} + 0.7660\vec{j})$. The cable tension is $\vec{T} = -|\vec{T}|\vec{i} + 0\vec{j}$. The equilibrium conditions are $\sum\vec{F} = \vec{T} + \vec{N} + \vec{W} = 0$, from which

$$\sum\vec{F}_x = (0.6428|\vec{N}| - |\vec{T}|)\vec{i} = 0$$

$$\sum\vec{F}_y = (0.7660|\vec{N}| - 100)\vec{j} = 0 \quad \text{Solve: } \boxed{|\vec{T}| = 83.9\ lb}$$

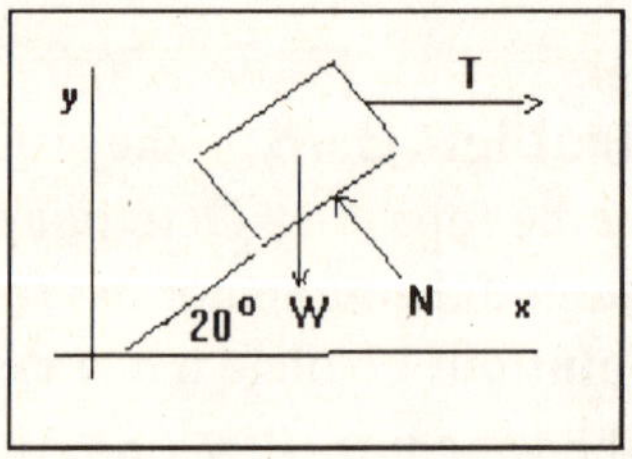

(b) For left hand box: The weight $\vec{W} = 0\vec{i} - |\vec{W}|\vec{j}$. The angle between the normal force and the positive x axis is $(90^o + 20^o) = 110^o$. The normal force: $\vec{N} = |\vec{N}|(-0.3420\vec{i} + 0.9397\vec{j})$. The cable tension is: $\vec{T} = |\vec{T}|\vec{i} + 0\vec{j}$. The equilibrium conditions are: $\sum\vec{F} = \vec{W} + \vec{N} + \vec{T} = 0$, from which: $\sum\vec{F}_x = (-0.3420|\vec{N}| + 83.9)\vec{i} = 0$, and $\sum\vec{F}_y = (0.9397|\vec{N}| - |\vec{W}|)\vec{j} = 0$ Solve for the weight of the box: $\boxed{|\vec{W}| = 230.6\ lb}$

Problem 3.80 The mass of block A is 42 kg and the mass of block B is 50 kg. If the blocks are in equilibrium, what is the force $\vec{F}$?

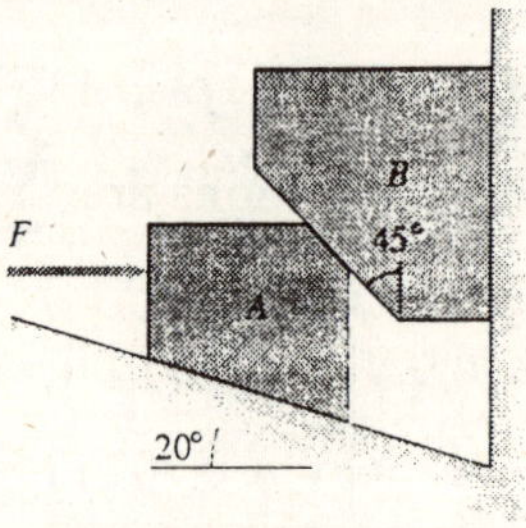

Solution: Isolate the top block. Solve the equilibrium equations. The weight is $\vec{W} = 0\vec{i} - (50)(9.81)\vec{j} = 0\vec{i} - 490.5\vec{j}$. The angle between the normal force $\mathbf{N_1}$ and the positive x axis is $(90^o - 45^o) = 45^0$. The normal force is

$\vec{N}_1 = |\vec{N}_1|(\vec{i}\cos 45^o + \vec{j}\sin 45^o) = |\vec{N}_1|(0.7071\vec{i} + 0.7071\vec{j})$. The force N_2 is $\vec{N}_2 = -|\vec{N}_2|\vec{i} + 0\vec{j}$. The equilibrium conditions are $\sum\vec{F} = \vec{N}_1 + \vec{N}_2 + \vec{W} = 0$, from which

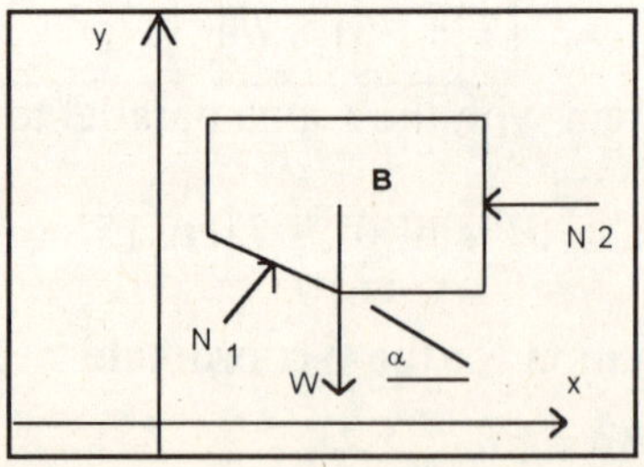

$$\sum\vec{F}_x = (0.7071|\vec{N}_1| - |\vec{N}_2|)\vec{i} = 0$$

$$\sum\vec{F}_y = (0.7071|\vec{N}_1| - 490.5)\vec{j} = 0 \text{. Solve:}$$

$|\vec{N}_1| = 693.7\ N$, $|\vec{N}_2| = 490.5\ N$

Isolate the bottom block. The weight is $\vec{W} = 0\vec{i} - |\vec{W}|\vec{j} = 0\vec{i} - (42)(9.81)\vec{j} = 0\vec{i} - 412.02\vec{j}\ (N)$. The angle between the normal force $\mathbf{N_1}$ and the positive x axis is $(270^o - 45^o) = 225^o$.

Problem 3.80 continued on next page....

continuation of solution to Problem 3.80...

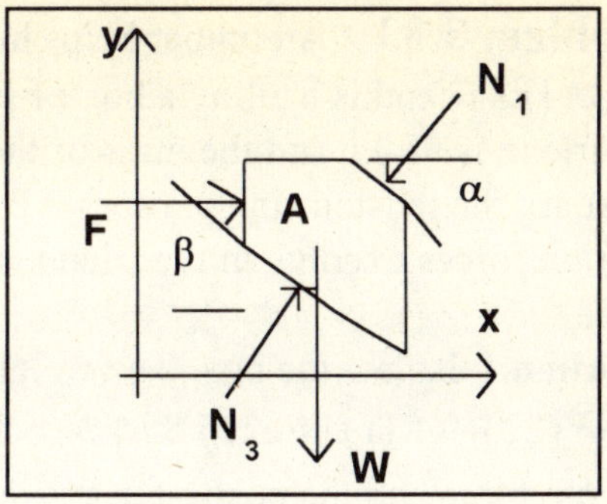

The normal force:

$\vec{N}_1 = |\vec{N}_1|(\vec{i}\cos 225^o + \vec{j}\sin 225^o) = |\vec{N}_1|(-0.7071\vec{i} - 0.7071\vec{j})$

.The angle between the normal force $\vec{N}_3$ and the positive *x*-axis is $(90^o - 20^o) = 70^o$. The normal force is

$\vec{N}_3 = |\vec{N}_3|(\vec{i}\cos 70^o + \vec{j}\sin 70^o) = |\vec{N}_3|(0.3420\vec{i} + 0.9397\vec{j})$.

The force is ... $\vec{F} = |\vec{F}|\vec{i} + 0\vec{j}$. The equilibrium conditions are

$\sum \vec{F} = \vec{W} + \vec{N}_1 + \vec{N}_3 + \vec{F} = 0$, from which:

$\sum \vec{F}_x = (-0.7071|\vec{N}_1| + 0.3420|\vec{N}_3| + |\vec{F}|)\vec{i} = 0$

$\sum \vec{F}_y = (-0.7071|\vec{N}_1| + 0.9397|\vec{N}_3| - 412)\vec{j} = 0$ For $|\vec{N}_1| = 693.7\ N$ from above: $|\vec{F}| = 162\ N$

Problem 3.81 The 680 kg load suspended from the helicopter is in equilibrium. The aerodynamic drag force is horizontal. The y axis is vertical, and the cable OA lies in the x-y plane. Determine the magnitude of the drag force and the tension in cable OA.

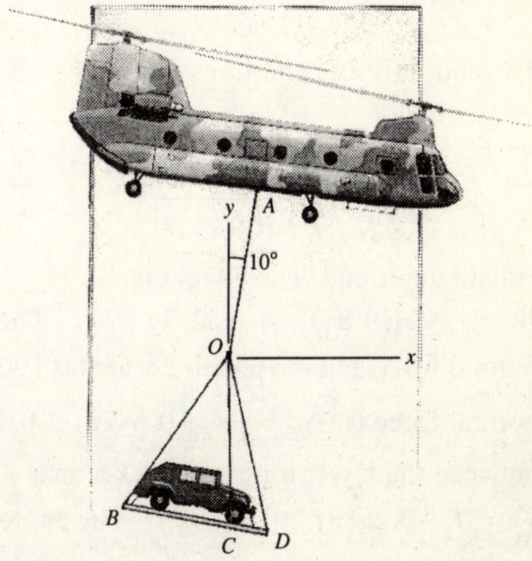

Solution: Isolate the load. Solve the equilibrium equations.

The weight is $\vec{W} = (680)(9.81)(0\vec{i} - \vec{j}) = -6670.8\vec{j}$. The drag force: $\vec{D} = -|\vec{D}|\vec{i}$. The equilibrium conditions:

$\sum \vec{F} = \vec{W} + \vec{D} + \vec{T} = 0$, from which

$\sum \vec{F}_x = (-|\vec{D}| + 0.1736|\vec{T}|)\vec{i} = 0$

$\sum \vec{F}_y = (0.9848|\vec{T}| - 6670.8)\vec{j} = 0$. Solve:

$|\vec{T}| = 6773.7\ N$, $|\vec{D}| = 1176.2\ N$

Problem 3.82 In Problem 3.81, the coordinates of the three cable attachment points B, C, and D are (−3.3, − 4.5, 0), (1.1, − 5.3, 1), and (1.6, − 5.4, − 1), respectively. What are the tensions in the cables?

Solution: Isolate the cable juncture 0. Express the tensions in terms of unit vectors parallel to the cables. The equilibrium conditions to be solved are $\sum \vec{F} = \vec{T}_B + \vec{T}_C + \vec{T}_D + \vec{T} = 0$. From Problem 3.81,

$\vec{T} = 6773.8(0.1736\vec{i} + 0.9848\vec{j}) = 1175.9\vec{i} + 6670.8\vec{j}$ The position vectors are:

$\vec{r}_{OB} = -3.3\vec{i} - 4.5\vec{j} + 0\vec{k}$, $\vec{r}_{OC} = +1.1\vec{i} - 5.3\vec{j} + 1\vec{k}$, $\vec{r}_{OD} = 1.6\vec{i} - 5.4\vec{j} - 1\vec{k}$. The unit vectors are:

$\vec{e}_{OB} = -0.5914\vec{i} - 0.8064\vec{j}$, $\vec{e}_{OC} = 0.1998\vec{i} - 0.9628\vec{j} + 0.1817\vec{k}$, $\vec{e}_{OD} = 0.2797\vec{i} - 0.9440\vec{j} - 0.1748\vec{k}$.

Express the tensions in terms of the unit vectors, and substitute into the equilibrium conditions:

$\sum \vec{F}_x = (-0.5914|\vec{T}_B| + 0.1998|\vec{T}_C| + 0.2797|\vec{T}_D| + 1175.9)\vec{i} = 0$

$\sum \vec{F}_y = (-0.8064|\vec{T}_B| - 0.9628|\vec{T}_c| - 0.9440|\vec{T}_D| + 6670.8)\vec{j} = 0$

$\sum \vec{F}_z = (0|\vec{T}_B| + 0.1817|\vec{T}_C| - 0.1748|\vec{T}_D|)\vec{k} = 0$. Solve: $|\vec{T}_B| = 3598\ N$, $|\vec{T}_C| = 1939\ N$, $|\vec{T}_D| = 2015\ N$.

(An HP-28S hand calculator was used to solve these equations.)

Problem 3.83 A climber helps his friend up an icy slope. His friend is hauling a box of supplies. If the mass of the friend is 90-kg and the mass of the supplies is 22-kg, what are the tensions in the ropes AB and CD? Neglect friction forces exerted on the friend and supplies by the icy slope.

Solution: Isolate the box. The weight vector is $\vec{W}_2 = (22)(9.81)\vec{j} = -215.8\vec{j}\ (N)$. The angle between the normal force and the positive x axis is $(90^o - 60^o) = 30^o$

The normal force is $\vec{N}_B = |\vec{N}_B|(0.866\vec{i} + 0.5\vec{j})$.

The angle between the rope CD and the positive x axis is $(180^o - 75^o) = 105^o$; the tension is:

$\vec{T}_2 = |\vec{T}_2|(\vec{i}\cos 105^o + \vec{j}\sin 105^o) = |\vec{T}_2|(-0.2588\vec{i} + 0.9659\vec{j})$

The equilibrium conditions are $\sum\vec{F} = \vec{W}_B + \vec{T}_2 + \vec{N}_B = 0$

from which

$\sum\vec{F}_x = (0.866|\vec{N}_B| - 0.2588|\vec{T}_2|)\vec{i} = 0, \sum\vec{F}_y = (0.5|\vec{N}_B| + 0.9659|\vec{T}_2| - 215.8)\vec{j} = 0$. Solve:

$|\vec{N}_B| = 57.8\ N$, $\boxed{|\vec{T}_2| = 193.5\ N}$.

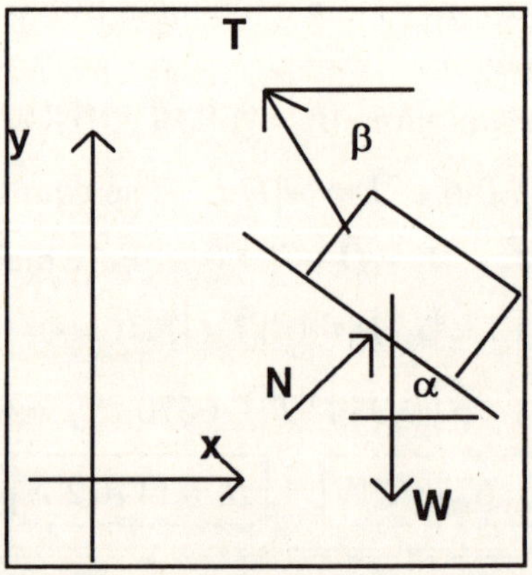

Isolate the friend. The weight is $\vec{W} = -(90)(9.81)\vec{j} = -882.9\vec{j}\ (N)$. The angle between the normal force and the positive x axis is $(90^o - 40^o) = 50^o$ The normal force is :$\vec{N}_F = |\vec{N}_F|(0.6428\vec{i} + 0.7660\vec{j})$ The angle between the lower rope and the x axis is -75° ; the tension is $\vec{T}_2 = |\vec{T}_2|(0.2588\vec{i} - 0.9659\vec{j})$. The angle between the tension in the upper rope and the positive x axis is $(180^o - 20^o) = 160^o$, the tension is

$\vec{T}_1 = |\vec{T}_1|(-0.9397\vec{i} + 0.3420\vec{j})$. The equilibrium conditions are $\sum\vec{F} = \vec{W} + \vec{T}_1 + \vec{T}_2 + \vec{N}_F = 0$. From which :

$\sum\vec{F}_x = (0.6428|\vec{N}_F| + 0.2588|\vec{T}_2| - 0.9397|\vec{T}_1|)\vec{i} = 0$

$\sum\vec{F}_y = (0.7660|\vec{N}_F| - 0.9659|\vec{T}_2| + 0.3420|\vec{T}_1| - 882.9)\vec{j} = 0$

Solve, for $|\vec{T}_2| = 193.5\ N$ .. The result:,

, $|\vec{N}_F| = 1051.6\ N$, $\boxed{|\vec{T}_1| = 772.6\ N}$

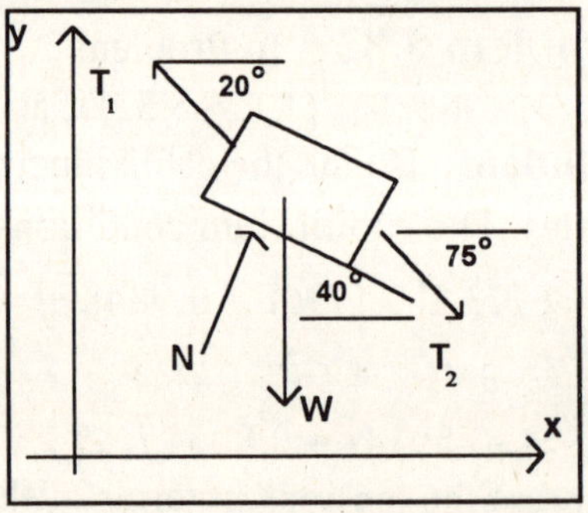

==================================<>==================================

Problem 3.84 The small sphere of mass m is attached to a string of length L and rests on the surface of a sphere of radius R. Determine the tension in the string in terms of m,L,h, and R.

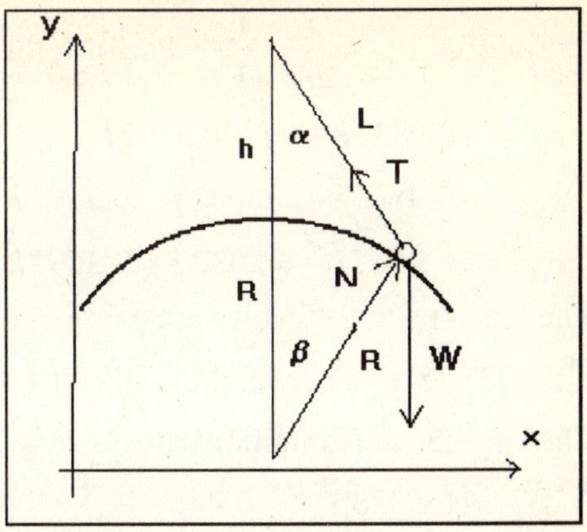

Solution: Isolate the small sphere. Use the law of sines to determine the interior angles that appear in the solution. The weight is $\vec{W} = 0\vec{i} - mg\vec{j}$. The angle between the normal force and the positive x axis is $(90^o - \beta)$; the normal force is $\vec{N} = |\vec{N}|(\vec{i}\sin\beta + \vec{j}\cos\beta)$. The angle between the string tension and the positive x axis is $(90^o + \alpha)$ (use the rule of equality of opposite interior angles from geometry), hence the string tension is $\vec{T} = |\vec{T}|(-\vec{i}\sin\alpha + \vec{j}\cos\alpha)$. The equilibrium conditions are $\sum\vec{F} = \vec{T} + \vec{N} + \vec{W} = 0$, from which: $\sum\vec{F}_x = (|\vec{N}|\sin\beta - |\vec{T}|\sin\alpha)\vec{i} = 0$

$\sum\vec{F}_y = (|\vec{N}|\cos\beta + |\vec{T}|\cos\alpha - mg)\vec{j} = 0$ Solve: $|\vec{T}| = \left(\dfrac{mg\sin\beta}{\sin(\alpha+\beta)}\right)$. From the law of sines for the triangle with sides R, R+h, and L, $\dfrac{(R+h)}{\sin(\alpha+\beta)} = \dfrac{L}{\sin\beta}$. Substitute into the tension: $\boxed{|\vec{T}| = \left(\dfrac{mgL}{(R+h)}\right)}$

==================================<>==================================

Problem 3.85 An engineer doing preliminary design studies for a new radio telescope envisions a triangular receiving platform suspended by cables from three equally spaced 40 m towers. The receiving platform has a mass of 20 Mg (megagrams) and is 10 m below the tops of the towers. What tension would each cable be subjected to?

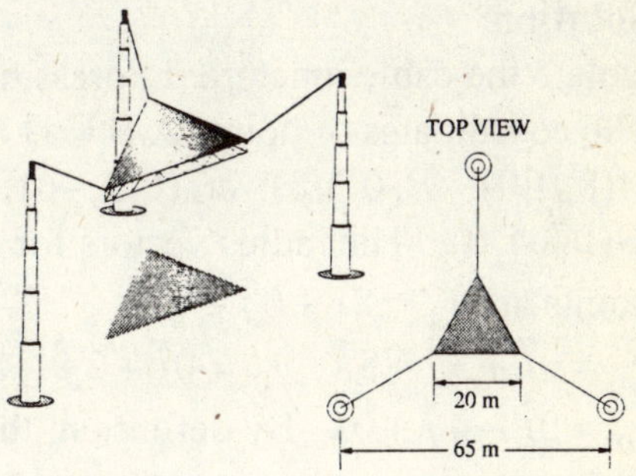

Solution: Isolate the platform. Choose a coordinate system with the origin at the center of the platform, with the Z axis vertical, and the x,y axes as shown. Express the tensions in terms of unit vectors, and solve the equilibrium conditions. The cable connections at the platform are labeled a,b,c, and the cable connections at the towers are labeled A,B,C. The horizontal distance from the origin at the center of the platform to any tower is given by

$L = \dfrac{65}{2\sin(60)} = 37.5\ m$. The coordinates of points A,B,C are . $A(37.5,0,10)$,

$B(37.5\cos(120^o), 37.5\sin(120^o), 10)$, $C(37.5\cos(240^o), 37.5\sin(240^o), 10)$,

The vector locations are: $\vec{r}_A = 37.5\vec{i} + 0\vec{j} + 10\vec{k}$, $\vec{r}_B = -18.764\vec{i} + 32.5\vec{j} + 10\vec{k}$, $\vec{r}_C = -18.764\vec{i} + -32.5\vec{j} + 10\vec{k}$. The distance from the origin to any cable connection on the platform is

$d = \dfrac{20}{2\sin(60^o)} = 11.547\ m$. The coordinates of the cable connections are $a(11.547,0,0)$,

$b(11.547\cos(120^o), 11.547\sin(120^o), 0)$, $c(11.547\cos(240^o), 11.547\sin(240^o), 0)$. The vector locations of these points are, $\vec{r}_a = 11.547\vec{i} + 0\vec{j} + 0\vec{k}$, $\vec{r}_b = -5.774\vec{i} + 10\vec{j} + 0\vec{k}$, $\vec{r}_c = -5.774\vec{i} - 10\vec{j} + 0\vec{k}$. The unit vector parallel to the tension acting between the points A,a in the direction of A is by

Problem 3.85 continued on next page....

continuation of solution to Problem 3.85...

definition $\vec{e}_{aA} = \frac{\vec{r}_A - \vec{r}_a}{|\vec{r}_A - \vec{r}_a|}$. Perform this operation for each of the unit vectors to obtain

$\vec{e}_{aA} = +0.9333\vec{i} + 0\vec{j} + 0.3592\vec{k}$

$\vec{e}_{bB} = -0.4667\vec{i} + 0.8082\vec{j} + 0.3592\vec{k}$

$\vec{e}_{cC} = -0.4667\vec{i} - 0.8082\vec{j} + 0.3592\vec{k}$

The tensions in the cables are expressed in terms of the unit vectors, $\vec{T}_{aA} = |\vec{T}_{aA}|\vec{e}_{aA}$, $\vec{T}_{bB} = |\vec{T}_{bB}|\vec{e}_{bB}$, $\vec{T}_{cC} = |\vec{T}_{cC}|\vec{e}_{cC}$. The weight is $\vec{W} = 0\vec{i} + 0\vec{j} - (20000)(9.81)\vec{k} = 0\vec{i} + 0\vec{j} - 196200\vec{k}$

The equilibrium conditions are $\sum \vec{F} = 0 = \vec{T}_{aA} + \vec{T}_{bB} + \vec{T}_{cC} + \vec{W} = 0$, from which:

$\sum \vec{F}_x = \left(0.9333|\vec{T}_{aA}| - 0.4666|\vec{T}_{bB}| - 0.4666|\vec{T}_{cC}|\right)\vec{i} = 0$

$\sum \vec{F}_y = \left(0|\vec{T}_{aA}| + 0.8082|\vec{T}_{bB}| - 0.8082|\vec{T}_{cC}|\right)\vec{j} = 0$

$\Sigma \vec{F}_z = \left(0.3592|\vec{T}_{aA}| + 0.3592|\vec{T}_{bB}| + 0.3592|\vec{T}_{cC}| - 196200\right)\vec{k} = 0$ The commercial package **TK Solver 2** was used to solve these equations. The results: $\boxed{|\vec{T}_{aA}| = 182.1\ kN}$, $\boxed{|\vec{T}_{bB}| = 182.1\ kN}$, $\boxed{|\vec{T}_{cC}| = 182.1\ kN}$.

Check: For this geometry, where from symmetry all cable tensions may be assumed to be the same, only the *z*-component of the equilibrium equations is required: $\Sigma F_z = 3|\vec{T}|\sin\theta - 196200 = 0$, where $\theta = \tan^{-1}\left(\frac{10}{37.5 - 11.547}\right) = 21.07^\circ$, from which each tension is $|\vec{T}| = 182.1\ kN$. *check.*

Problem 3.86 The bulldozer exerts a force $\vec{F} = 2\vec{i}\ (kip)$ on point A where the cables AB, AC, and AD join. What are the tensions in the cables?

Solution:

Isolate the cable juncture. Express the tensions in terms of unit vectors. Solve the equilibrium equations. The coordinates of points A,B,C,D are: $A(8,0,0)$, $B(0,3,8)$, $C(0,2,-6)$, $D(0,-4,0)$. The radius vectors for these points are $\mathbf{r}_A = 8\vec{i} + 0\vec{j} + 0\vec{k}$, $\mathbf{r}_B = 0\vec{i} + 3\vec{j} + 8\vec{k}$, $\mathbf{r}_C = 0\vec{i} + 2\vec{j} - 6\vec{k}$, $\mathbf{r}_D = 0\vec{i} - 4\vec{j} + 0\vec{k}$. By definition, the unit vector parallel to the tension in cable AB is $\vec{e}_{AB} = \frac{\vec{r}_B - \vec{r}_A}{|\vec{r}_B - \vec{r}_A|}$. Carrying out the operations for each of the cables, the results are:

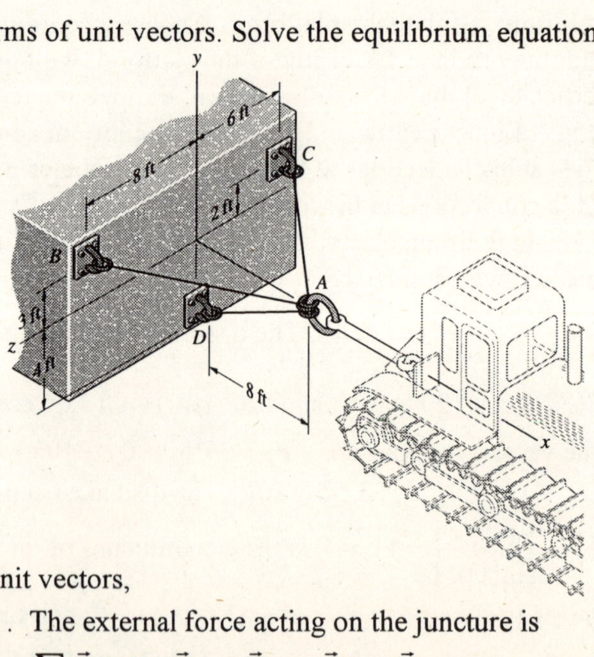

$\vec{e}_{AB} = -0.6835\vec{i} + 0.2563\vec{j} + 0.6835\vec{k}$

$\vec{e}_{AC} = -0.7845\vec{i} + 0.1961\vec{j} - 0.5883\vec{k}$

$\vec{e}_{AD} = -0.8944\vec{i} - 0.4472\vec{j} + 0\vec{k}$. The tensions in the cables are expressed in terms of the unit vectors,

$\vec{T}_{AB} = |\vec{T}_{AB}|\vec{e}_{AB}$, $\vec{T}_{AC} = |\vec{T}_{AC}|\vec{e}_{AC}$, $\vec{T}_{AD} = |\vec{T}_{AD}|\vec{e}_{AD}$. The external force acting on the juncture is $\vec{F} = 2000\vec{i} + 0\vec{j} + 0\vec{k}$. The equilibrium conditions are $\sum \vec{F} = 0 = \vec{T}_{AB} + \vec{T}_{AC} + \vec{T}_{AD} + \vec{F} = 0$

Problem 3.86 continued on next page....

=====================================<>=====================================

continuation of solution to Problem 3.86...

Substitute the vectors into the equilibrium conditions:

$$\sum \vec{F}_x = \left(-0.6835\left|\vec{T}_{AB}\right| - 0.7845\left|\vec{T}_{AC}\right| - 0.8944\left|\vec{T}_{AD}\right| + 2000\right)\vec{i} = 0$$

$$\sum \vec{F}_y = \left(0.2563\left|\vec{T}_{AB}\right| + 0.1961\left|\vec{T}_{AC}\right| - 0.4472\left|\vec{T}_{AD}\right|\right)\vec{j} = 0$$

$$\sum \vec{F}_z = \left(0.6835\left|\vec{T}_{AB}\right| - 0.5883\left|\vec{T}_{AC}\right| + 0\left|\vec{T}_{AD}\right|\right)\vec{k} = 0$$

The commercial program **TK Solver** was used to solve these equations. The results are

$\left|\vec{T}_{AB}\right| = 780.31\ lb$, $\left|\vec{T}_{AC}\right| = 906.49\ lb$, $\left|\vec{T}_{AD}\right| = 844.74\ lb$

=====================================<>=====================================

Problem 3.87 Consider the bulldozer in Problem 3.86. For safety, the site engineer doesn't want to subject any of the cables to a tension larger than 5 kip. Based on this criterion, what is the largest force the bulldozer can apply at A?

Solution: The simplest way to work this problem is to recognize that the tensions are linearly proportional to the applied force. Since the cables are initially unloaded, the solution given for Problem 3.89 can be used in a ratio. For the same configuration, if one loading force $\left|\vec{F}_1\right|$ produces a tension $\left|\vec{T}_1\right|$, the tension for another loading is given by the ratio: $\left(\dfrac{\left|\vec{T}_1\right|}{\left|\vec{F}_1\right|}\right) = \left(\dfrac{\left|\vec{T}\right|}{\left|\vec{F}\right|}\right)$ Applying this to the solution of Problem 3.89 and the limiting load of 5 kip: $\left(\dfrac{\left|\vec{T}_{AC}\right|}{2}\right) = \left(\dfrac{5}{\left|\vec{F}\right|}\right)$ thus, $\left|\vec{F}\right| = \left(\dfrac{10}{0.9065}\right) = 11\ kip$.

The standard procedure for three dimensional problems confirms this answer, as follows: Isolate the cable juncture A. The unit vectors are given in the solution of Problem 3.86. The new external force acting on the juncture is $\vec{F} = \left|\vec{F}\right|\vec{i} + 0\vec{j} + 0\vec{k}$ The equilibrium conditions are $\sum \vec{F} = 0 = \vec{T}_{AB} + \vec{T}_{AC} + \vec{T}_{AD} + \vec{F} = 0$. Substitute the force vectors and collect like terms,

$$\sum \vec{F}_x = \left(-0.6835\left|\vec{T}_{AB}\right| - 0.7845\left|\vec{T}_{AC}\right| - 0.8944\left|\vec{T}_{AD}\right| + \left|\vec{F}\right|\right)\vec{i} = 0$$

$$\sum \vec{F}_y = \left(0.2563\left|\vec{T}_{AB}\right| + 0.1961\left|\vec{T}_{AC}\right| - 0.4472\left|\vec{T}_{AD}\right|\right)\vec{j} = 0$$

$$\sum \vec{F}_z = \left(0.6835\left|\vec{T}_{AB}\right| - 0.5883\left|\vec{T}_{AC}\right| + 0\left|\vec{T}_{AD}\right|\right)\vec{k} = 0$$

This set of simultaneous equations in the unknown forces can be solved using standard procedures. The commercial program **TK Solver 2** was used to graph the tensions as a function of the applied force. At T_{AC} = 5 kip, F = 11 kip.

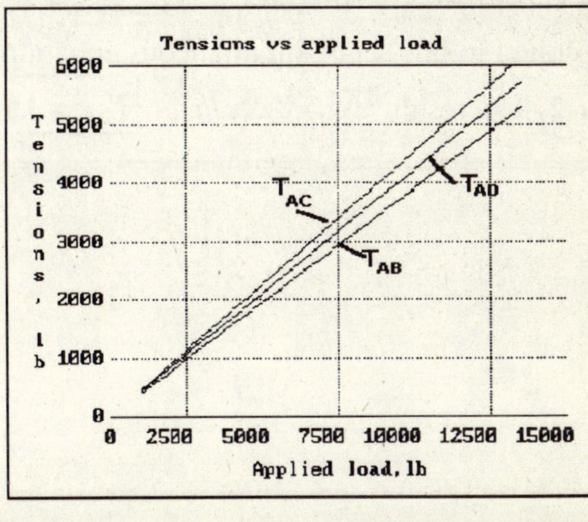

=====================================<>=====================================

Problem 3.88 A metal disk weighs 10 lb. It is held in place at the center of the inclined surface by the strings AB and AC. What are the tensions in the strings?

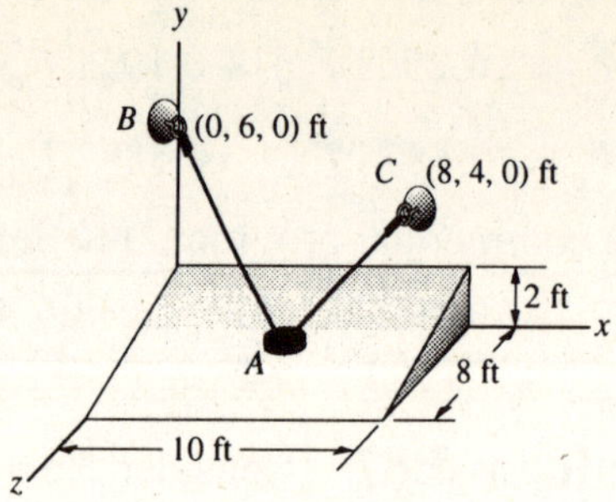

Solution

Isolate the disk, express the tensions in terms of the unit vectors, and solve the equilibrium equations. The coordinates of points A,B,C are: $A(5,1,4)$, $B(0,6,0)$, $C(8,4,0)$, where the coordinates of A are determined from the geometry of the inclined plane. The radius vectors corresponding to these coordinates are $\vec{r}_A = 5\vec{i} + 1\vec{j} + 4\vec{k}$, $\vec{r}_B = 0\vec{i} + 6\vec{j} + 0\vec{k}$, $\vec{r}_C = 8\vec{i} + 4\vec{j} + 0\vec{k}$. The unit vector $\vec{e}_{AB}$ is, by definition, $\vec{e}_{AB} = \dfrac{\vec{r}_B - \vec{r}_A}{|\vec{r}_B - \vec{r}_A|}$ Apply this to find the unit vectors parallel to the cables,

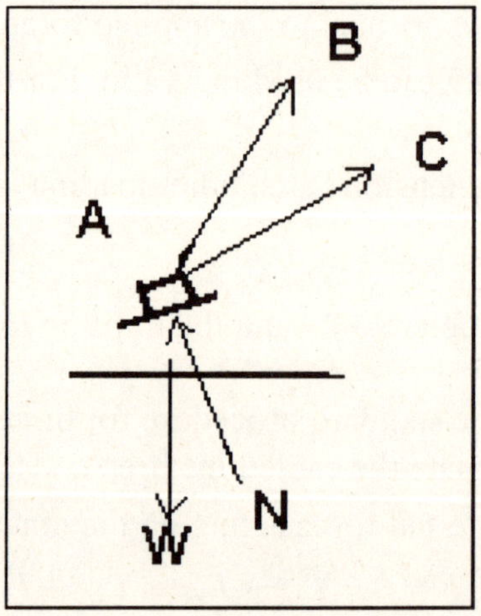

$\vec{e}_{AB} = -0.6155\vec{i} + 0.6155\vec{j} - 0.4924\vec{k}$

$\vec{e}_{AC} = 0.5145\vec{i} + 0.5145\vec{j} - 0.6860\vec{k}$. The weight is $\vec{W} = 0\vec{i} - 10\vec{j} + 0\vec{k}$. The normal force acts normally to the inclined surface, $\vec{N} = |\vec{N}|(0\vec{i} + \vec{j}\cos\alpha + \vec{k}\sin\alpha)$ where $\tan\alpha = \left(\dfrac{2}{8}\right) = 0.25$, $\alpha = 14^{o}$. The tensions in the cables are expressed in terms of the unit vectors, $\vec{T}_{AB} = |\vec{T}_{AB}|\vec{e}_{AB}$, $\vec{T}_{AC} = |\vec{T}_{AC}|\vec{e}_{AC}$. The equilibrium conditions are $\sum \vec{F} = 0 = \vec{T}_{AB} + \vec{T}_{AC} + \vec{W} + \vec{N} = 0$. From which

$$\sum \vec{F}_x = \left(-0.6155|\vec{T}_{AB}| + 0.5145|\vec{T}_{AC}|\right)\vec{i} = 0$$

$$\sum \vec{F}_y = \left(0.6155|\vec{T}_{AB}| + 0.5145|\vec{T}_{AC}| + 0.9703|\vec{N}| - 10\right)\vec{j} = 0$$

$$\sum \vec{F}_z = \left(-0.4924|\vec{T}_{AB}| - 0.6860|\vec{T}_{AC}| + 0.2425|\vec{N}|\right)\vec{k} = 0$$

The solution to this set of simultaneous equations was obtained using the commercial program **TK Solver 2.** The result: $|\vec{N}| = 8.35\ lb$, $\boxed{|\vec{T}_{AB}| = 1.54\ lb}$, $\boxed{|\vec{T}_{AC}| = 1.85\ lb}$

Problem 4.1 Determine the moment of the 50 N force about (a) point A, (b) point B.

Solution: Use 2-dimensional moment strategy: determine normal distance to line of action D; calculate magnitude DF; determine sign.

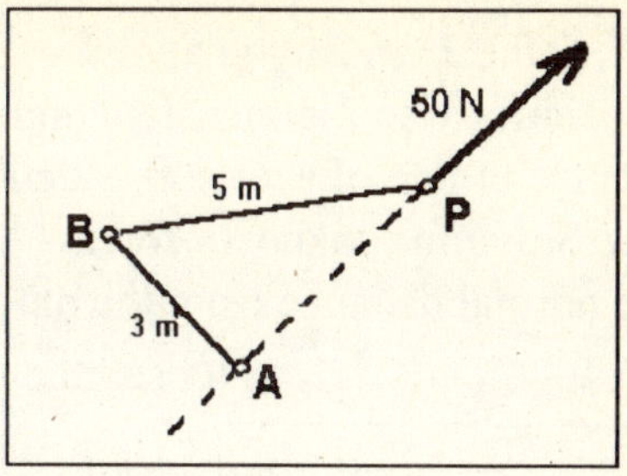

(a) The perpendicular distance from A to line of action is $D = 0$, hence the moment $\boxed{M_A = DF = 0}$.

(b) The perpendicular distance from B to line of action is 3 m (the triangle is a 3-4-5 right triangle), and the action is counter clockwise, hence $\boxed{M_B = +(3)(50) = +150\ N\text{-}m}$.

Problem 4.2 Determine the moment of the 50 lb force about (a) point A, (b) point B, and (c) point C.

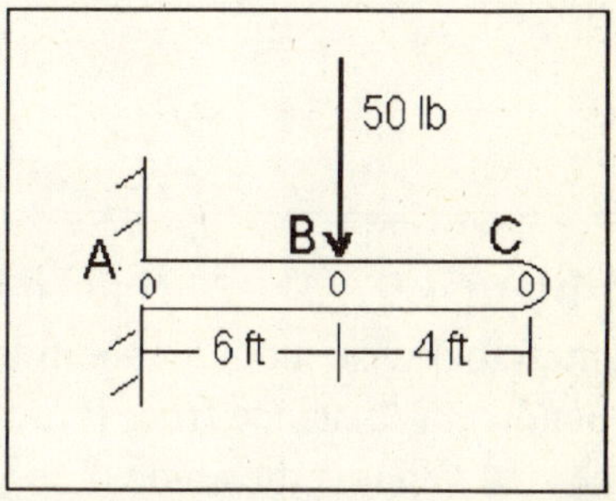

Solution: Use 2-dimensional moment strategy: determine normal distance to line of action D; calculate magnitude DF; determine sign. (a) The normal distance is 6 ft; the action is negative; hence $\boxed{M_A = -(6)(50) = -300\ ft\text{-}lb}$. (b) The normal distance is zero; hence $\boxed{M_B = 0}$. (c) The normal distance is 4 ft; the action is positive, hence $\boxed{M_C = (4)(50) = 200\ ft\text{-}lb}$

Problem 4.3 The wheels of the crane exert 40 kip forces on the horizontal **I**-beam at B and C. Determine the sum of the moments of the forces on the beam due to the action of the crane wheels (a) about A, (b) about D.

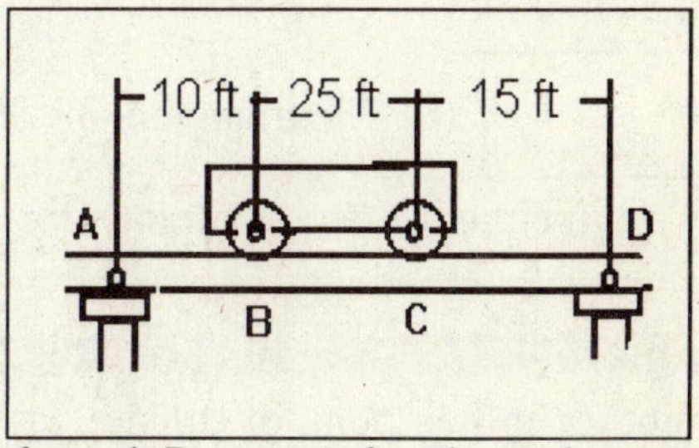

Solution: Use 2-dimensional moment strategy: determine normal distance to line of action D; calculate magnitude DF; determine sign. Add moments.

(a) The normal distances from A to the lines of action are $D_{AB} = 10\ ft$, and $D_{AC} = 35\ ft$. The actions are negative; hence $\boxed{\sum M_A = -(10)(40) - (35)(40) = -1800\ ft\text{-}kip}$. (b) The normal distances from D to the lines of action are $D_{DB} = 40\ ft$, and $D_{DC} = 15\ ft$. The actions are positive; hence $\boxed{\sum M_D = +(40)(40) + (15)(40) = +2200\ ft\text{-}kip}$

Problem 4.4 If you exert a 90 N force on the wrench in the direction shown, what moment do you exert about the center of the nut?

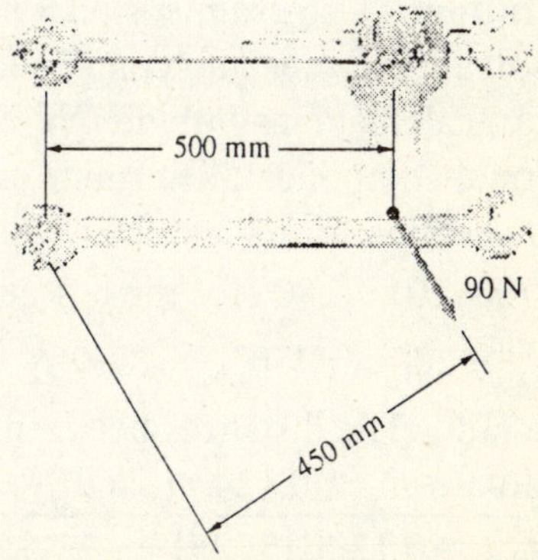

Solution: Use 2-dimensional moment strategy: determine normal distance to line of action D; calculate magnitude DF; determine sign.

The normal distance from nut center to the line of action is $D = 0.45\ m$, and the action is negative, hence $\boxed{M = -(0.45)(90) = -40.5\ N\text{-}m}$

Problem 4.5 If you exert a force $\vec{F}$ on the wrench in the direction shown and a 50 N-m moment is required to loosen the nut, what force $\vec{F}$ must you apply?

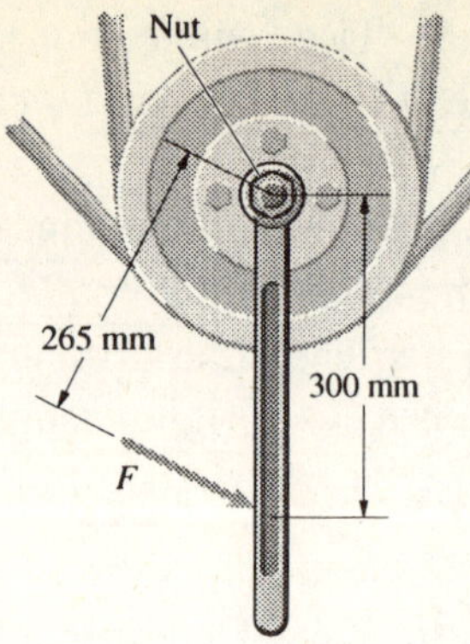

Solution: Use 2-dimensional moment strategy: determine normal distance to line of action *D;* calculate magnitude *DF*; determine sign. Solve for unknown force.

The normal distance from the nut center to the line of action is $D = 0.265\ m$. Thus to loosen the nut, $50 = D|\vec{F}| = 0.265|\vec{F}|$. Solve:

$$\boxed{|\vec{F}| = \frac{50}{0.265} = 188.7\ N}$$ in the direction shown.

Problem 4.6 The 20 N force exerts a 20 N-m counterclockwise moment about P. (a) What is the perpendicular distance from P to the line of action of the force? (b) What is the angle α?

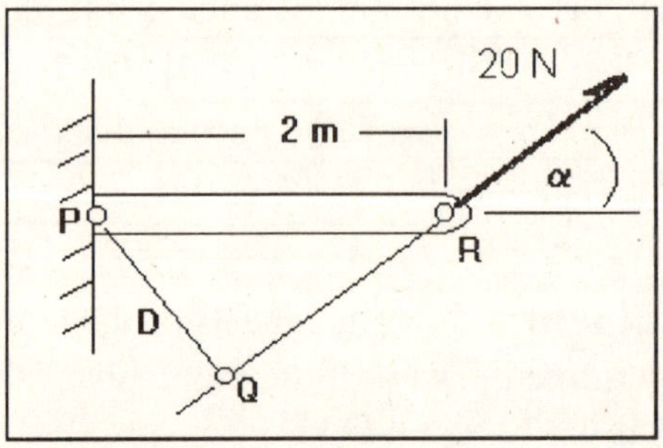

Solution: Use 2-dimensional moment strategy: calculate magnitude *DF*; determine sign. Solve for the unknown normal distance to line of action *D*.

(a) The moment is : $M = -20 = -D|\vec{F}| = -20D\ N\text{-}m$. Solve:

$\boxed{D = \frac{20}{20} = 1\ m}$ (b) The triangle PQR is a right triangle, hence the angle is $\sin\alpha = \frac{1\ m}{2\ m} = 0.5$. The angle may lie in either of two quadrants: for $\sin\alpha > 0$, $\boxed{\alpha = 30^o}$, or $\boxed{\alpha = 150^o}$

Problem 4.7 The gears exert 200 N forces on each other at their point of contact. (a) Determine the moment about A due to the force exerted on the left gear. (b) Determine the moment about B due to the force exerted by the right gear.

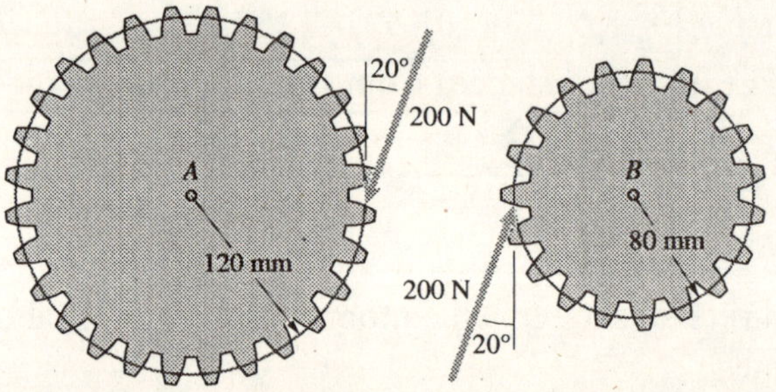

Solution: Use 2-dimensional moment strategy: resolve the forces into components normal to the radii; calculate magnitude *DF,* where *F* is the normal component; determine sign. The angles between the forces and the x-axis are $(270-20) = 250^o$ for the left gear and $(90-20) = 70^o$ for the right gear. The forces are $F_{AY} = 200|\sin 250^o| = 187.9\ N$, and $F_{BY} = 200|\sin 70^o| = 187.9\ N$ These magnitudes are normal to the radii. The distances between the points A and B and their respective action lines are the radii. The radii are $R_A = 0.120\ m$, and $R_B = 0.080\ m$. The actions are negative. Thus

$\boxed{M_A = -0.120(|187.9|) = -22.55\ N\text{-}m}$, and $\boxed{M_B = -0.080(|187.9|) = -15.0\ N\text{-}m}$.

================================<>================================

Problem 4.8 The 5 ft beam AB will fail at A if the moment of the force $\vec{F}$ about A exceeds 10 ft kips. Based on this criterion, how large can the force be?

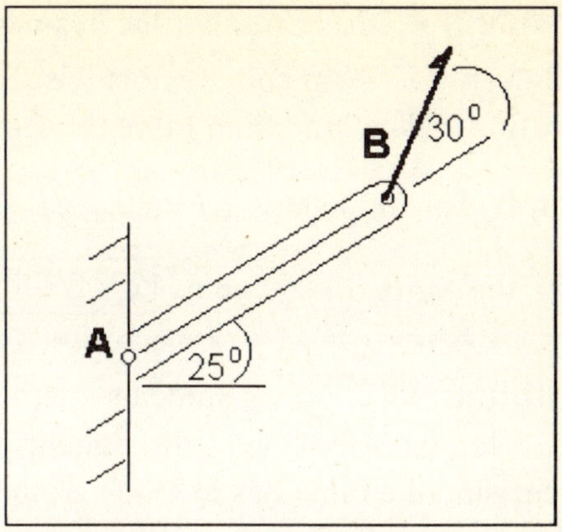

Solution: Use 2-dimensional moment strategy: resolve the force into a component normal to the beam; calculate magnitude DF, where F is the component normal to the beam; determine sign; solve for F. The components of $\vec{F}$ parallel to and normal to the beam are $F = |\vec{F}|(\sin 30^\circ) = 0.5|\vec{F}|$ The normal distance from A to the action line is the length of the beam. The action is positive. The moment about A is $M_A = (5)(0.5|\vec{F}|) = 2.5|\vec{F}|$. For a failure moment of 10 ft-kips, the value of the force is $\boxed{|\vec{F}| = \frac{10}{2.5} = 4 \; kips}$

================================<>================================

Problem 4.9 Determine the moment of the 80 lb force about P.

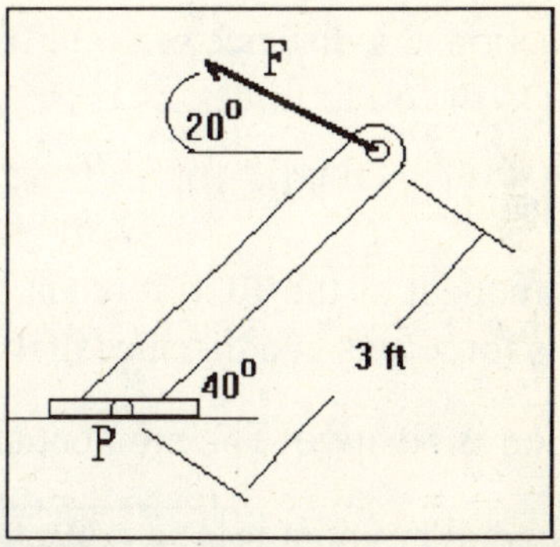

Solution: Use 2-dimensional moment strategy: resolve the force into a component normal to the beam; calculate magnitude DF, where F is the component normal to the beam; determine sign. The angle between the beam and the force is $(180^\circ - 40^\circ - 20^\circ) = 120^\circ$. The component of the force normal to the beam is $F_N = |\vec{F}|\sin 120^\circ = (80)(0.866) = 69.3 \; lb$. The normal distance from P to the action line is the length of the beam, and the action is positive. Thus $\boxed{M = (3)(69.3) = 207.8 \; ft\text{-}lb}$

================================<>================================

Problem 4.10 The 20 N force $\vec{F}$ exerts a 20 N-m counterclockwise moment about P. (a) What is the perpendicular distance from P to the line of action? (b) What is the angle α?

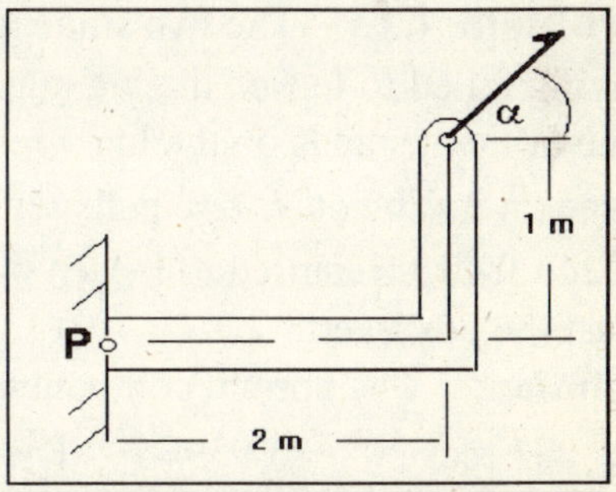

Solution: Use 2-dimensional moment strategy: determine normal distance to line of action D; calculate magnitude DF; determine sign.

(a) The moment is $|\vec{M}| = 20 = 20D$, from which the perpendicular distance is $\boxed{D = \frac{20 \; N\text{-}m}{20 \; N} = 1 \; m}$ (b) The angle between the force and the line from P is $(\alpha - \beta)$, where $\beta = \tan^{-1}\left(\frac{1}{2}\right) = 26.6^\circ$. The component of the force normal to the line from P is $F_N = |\vec{F}|\sin(\alpha - \beta)$., thus $M = 20 = \sqrt{1^2 + 2^2}|\vec{F}|\sin(\alpha - \beta)$ $= \sqrt{5}(20)\sin(\alpha - \beta)$. Solve: $\alpha = \beta + \sin^{-1}(0.4472)$. Thus $\boxed{\alpha = 26.6^\circ + 26.6^\circ = 53.1^\circ}$, $\boxed{\alpha = 26.6^\circ + 153.4^\circ = 180^\circ}$.

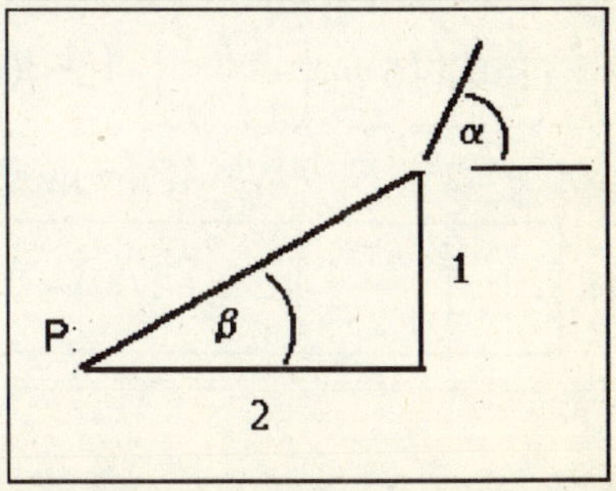

================================<>================================

Problem 4.11 Consider the beam and force shown in Problem 4.10. If $\vec{F}$ is the minimum force that will exert a 20 N-m counterclockwise moment about P, what are $\vec{F}$ and α?

Solution: The minimum force for the moment about P occurs when the force is normal to the line from P. The moment is $M = 20 = \sqrt{1^2 + 2^2}|\vec{F}| = \sqrt{5}|\vec{F}|$, from which $\boxed{|\vec{F}| = \frac{20}{\sqrt{5}} = 8.94\ N}$. From Problem 4.10, the angle α is given by $\boxed{\alpha = \beta + 90^o = 26.6^o + 90^o = 116.6^o}$.

Problem 4.12 Two students attempt to loosen a lug nut with a lug wrench. One of the students exerts the two 60 lb forces, and the other has to reach around his friend and can only exert the 30 lb forces. What torque (moment) do they exert on the nut?

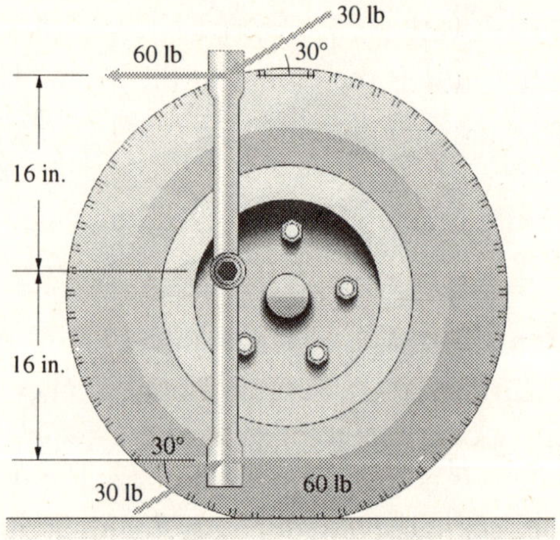

Solution: Determine the normal distance from line of action of the normal force to the lug nut. Calculate moment; determine sign. The two 60 lb forces act in a positive direction at a distance of 16 in from the lug nut. The moment due to the 60 lb forces is

$M_{60} = 2(60\ lb)(16\ in)\left(\frac{1\ ft}{12\ in}\right) = 160\ ft\text{-}lb$. The normal component of the 30 lb force is $F_{30} = 30 \cos 30^o = 26\ lb$. This force acts at a distance of 16 in from the lug nut. The action is positive. The moment due to the 30 lb forces is $M_{30} = 2(26\ lb)(16\ in)\left(\frac{1\ ft}{12\ in}\right) = 69.3\ ft\text{-}lb$.

The total moment is $\boxed{M_T = 69.3 + 160 = 229.3\ ft\text{-}lb}$

Problem 4.13 The two students in Problem 4.12, having failed to loosen the lug nut, try a different tactic. One of them stands on the lug wrench, exerting a 150 lb force on it. The other one pulls with a force $\vec{F}$. If a torque of 245 ft-lb is required to loosen the nut, what force $\vec{F}$ must the boy exert?

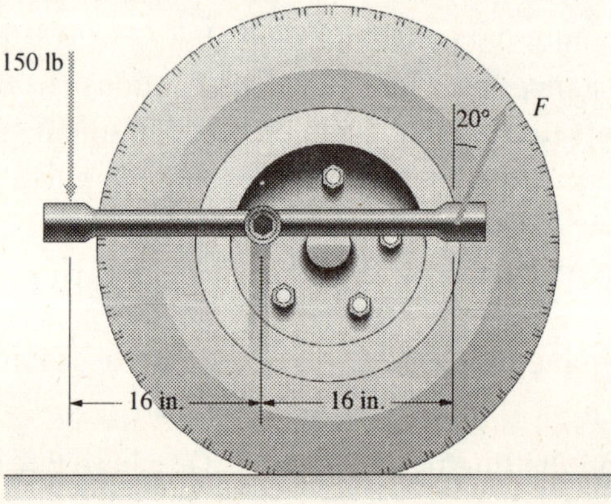

Solution: The normal component of the force is

$F_N = |\vec{F}| \cos 20^o = 0.9397|\vec{F}|$ The total moment is

$M = (150)(16\ in)\left(\frac{1\ ft}{12\ in}\right) + (|\vec{F}|)(0.9397)(16\ in)\left(\frac{1\ ft}{12\ in}\right)$.

$= 200 + 1.253|\vec{F}|\ ft\text{-}lb$. If the moment required is 245 *ft-lb*, then $\boxed{|\vec{F}| = \left(\frac{1}{1.253}\right)(245 - 200) = 35.9\ lb}$

Problem 4.14

Determine the moment exerted by the 30 lb weight (a) about E, (b) about S.

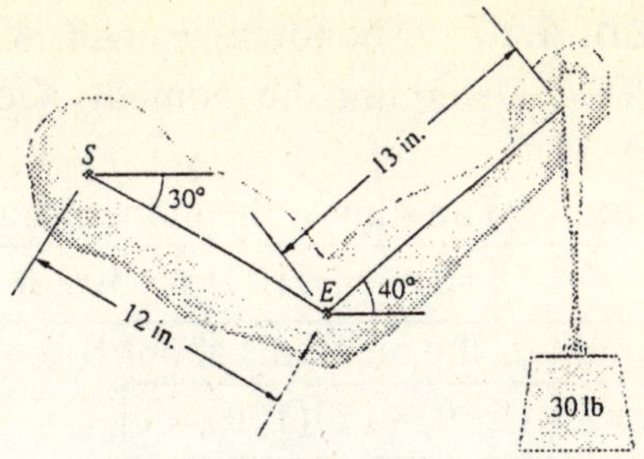

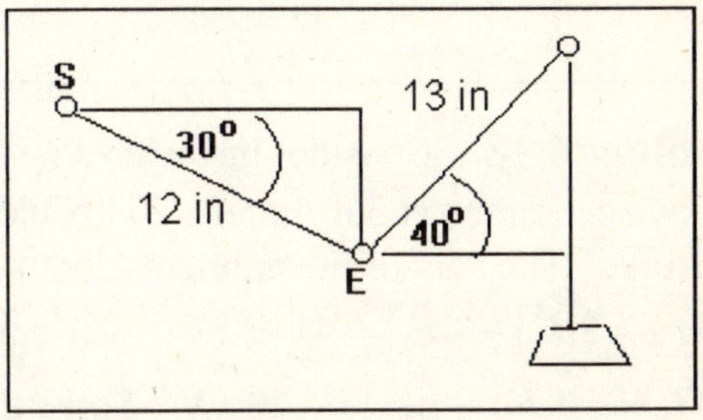

Solution: (a) Use 2-dimensional moment strategy: find normal distance D from line of action to E; calculate magnitude DF, determine sign.

The perpendicular distance D from E to line of action is $D = 13 \cos 40^o = 9.96\ in$. The action is negative. The moment about E is

$$\boxed{M = -(9.96\ in)(30\ lb)\left(\frac{1\ ft}{12\ in}\right) = -24.9\ ft\text{-}lb}$$

(b) The perpendicular distance D from S to the line of action is $D = 12 \cos 30^o + 9.96 = 20.35\ in$. The action is negative.

The moment about S is

$$\boxed{M = -(20.35\ in)(30\ lb)\left(\frac{1\ ft}{12\ in}\right) = -50.88\ ft\text{-}lb}$$

Problem 4.15

Three forces act on a square plate. Determine the sum of the moments of the forces (a) about A, (b) about B, (c) about C.

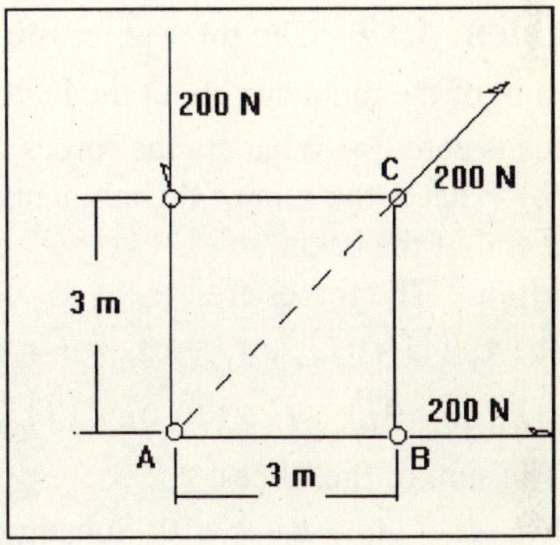

Solution: Determine the perpendicular distance between the points and the lines of action. Determine sign, and calculate moment. (a) The distances from point A to the lines of action is zero, hence the moment about A is $\boxed{M_A = 0}$. (b) The perpendicular distances of the lines of action from B are: 3 m for the force through A, with a positive action, and for the force through C, $D_C = \left(\frac{1}{2}\right)\sqrt{3^2 + 3^2} = 2.12\ m$ with a negative action. The moment about B is $\boxed{M_B = (3)(200) - 2.12(200) = 175.74\ N\text{-}m}$ (c) The distance of the force through A from C is 3 m, with a positive action, and the distance of the force through B from C is 3 m, with a positive action. The moment about C is $\boxed{M_C = 2(3)(200) = 1200\ N\text{-}m}$.

Problem 4.16

Determine the sum of moments of the three forces about (a) point A, (b) point B, (c) point C.

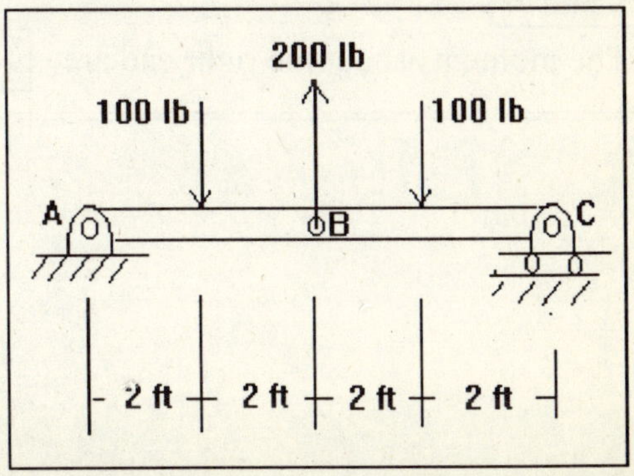

Solution: (a) The sum of the moments about A:

$$\boxed{\sum M_A = -(2)(100) + (4)(200) - (6)(100) = 0}$$

(b) The sum of the moments about B:

$$\boxed{\sum M_B = +(2)(100) - (2)(100) = 0}$$

(c) The sum of the moments about C:

$$\boxed{\sum M_C = +(6)(100) - (4)(200) + (2)(100) = 0}$$

Problem 4.17 The forces $F_1 = 50\ N$ and $F_2 = 100\ N$. Determine the moments about (a) about A, (b) about B.

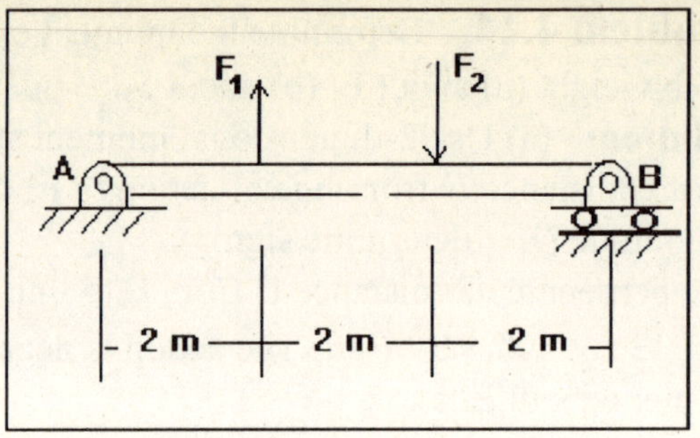

Solution: (a)The sum of the moments about A:

$\boxed{\sum M_A = +(2)(50) - (4)(100) = -300\ N\text{-}m}$

(b) The sum of the moments about B:

$\boxed{\sum M_B = -(4)(50) + (2)(100) = 0}$

Problem 4.18 Consider the beam shown in Problem 4.17. If the two forces exert a 140 kN clockwise moment about A and a 20 kN clockwise moment about B, what are F_1 and F_2?

Solution: The sum of the moments about A:

$\sum M_A = 2|\vec{F}_1| - 4|\vec{F}_2| = -140\ kN$. The sum of the moments about B:

$\sum M_B = -4|\vec{F}_1| + 2|\vec{F}_2| = -20\ kN$. Solve the simultaneous equations in the two unknowns:

$\boxed{|\vec{F}_1| = 30\ kN}$, $\boxed{|\vec{F}_2| = 50\ kN}$

Problem 4.19 The force $|\vec{F}| = 140\ lb$. The vector sum of the forces acting on the beam is zero, and the sum of the moments about the left end of the beams is zero. (a) What are the forces $\vec{A}_X$, $\vec{A}_Y$, and B? (b) What is the sum of the moments about the right end of the beam?

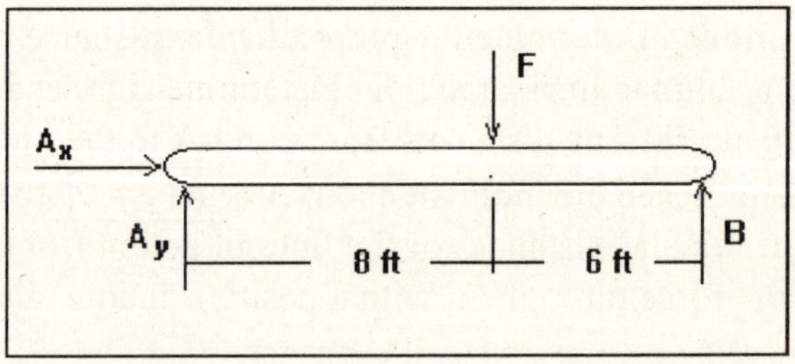

Solution: The forces are:

$\vec{A}_X = |\vec{A}_X|(1\vec{i} + 0\vec{j})$, $\vec{A}_Y = |\vec{A}_Y|(0\vec{i} + 1\vec{j})$

$\vec{B} = |\vec{B}|(0\vec{i} + 1\vec{j})$, $\vec{F} = 140(0\vec{i} - 1\vec{j})$.

(a) The sum of the forces:

$\sum \vec{F} = \vec{A}_X + \vec{A}_Y + \vec{B} + \vec{F} = 0$. Substitute and collect like terms:

$\sum \vec{F}_X = (|\vec{A}_X|)\vec{i} = 0$. $\sum \vec{F}_Y = (\vec{A}_Y + \vec{B} - 140)\vec{j} = 0$. It follows that $\boxed{|\vec{A}_X| = 0}$.

The sum of the moments about the left end is

$\sum M = (+0|\vec{A}_Y| - 140(8) + |\vec{B}|14) = 0$. Solve: $\boxed{|\vec{B}| = 80\ lb}$, from which the sum of forces equation yields $\boxed{|\vec{A}_Y| = 60\ lb}$

(b) The moments about the right end are: $\boxed{\sum M = (+6)((140)) - (14)(60) = 0}$

Problem 4.20 The vector sum of the three forces is zero and the sum of the moments of the three forces about A is zero. (a) What are $\vec{F}_A$ and $\vec{F}_B$? (b) What is the sum of the three moments about B?

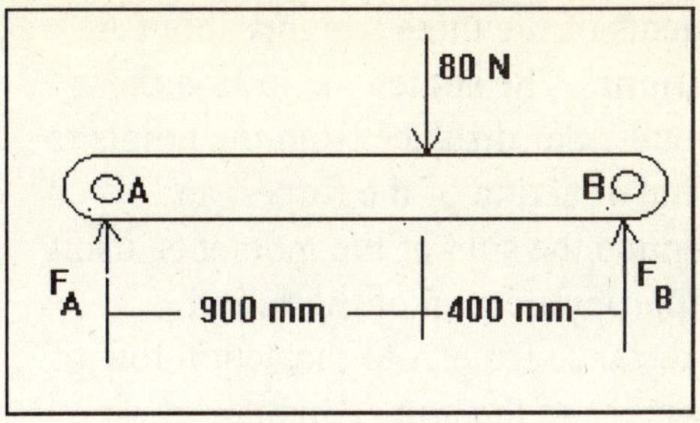

Solution: The forces are:
$\vec{F}_A = |\vec{F}_A|(0\vec{i} + 1\vec{j})$, $\vec{F}_B = |\vec{F}_B|(0\vec{i} + 1\vec{j})$, and
$\vec{F} = 80(0\vec{i} - 1\vec{j})$. The sum of the forces is:
$\sum \vec{F} = \vec{F}_A + \vec{F}_B + \vec{F} = 0$, from which
$\sum \vec{F}_Y = (|\vec{F}_A| + |\vec{F}_B| - 80)\vec{j} = 0$. The sum of the moments:
$\sum M_A = -(0.9)(80) + (1.3)(|\vec{F}_B|) = 0$. (a) Solve these two equations to obtain: $\boxed{|\vec{F}_B| = 55.4\ N}$, and $\boxed{|\vec{F}_A| = 24.6\ N}$ (b) The moments about B:
$\boxed{\sum M_B = (80)(0.4) - (1.3)|\vec{F}_A| = 0}$

Problem 4.21 The force $\vec{F}$ exerts a 240 N-m clockwise moment about A and a 120 N-m counterclockwise moment about B. Determine $\vec{F}$ and the distance from A to its point of application.

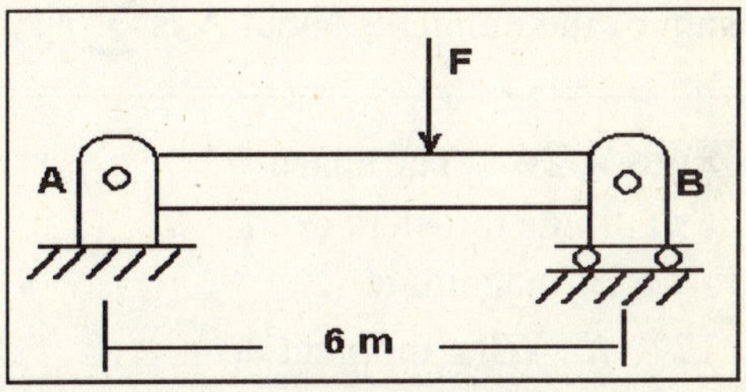

Solution: Let d_A be the distance from A to the line of action of the force. The moment about A is
$M_A = -(d_A)(|\vec{F}|) = -240\ N\text{-}m$
The moment about B is:
$M_B = (6 - d_A)(|\vec{F}|) = 120\ N\text{-}m$. Subtract the first equation from the second to obtain: $6|\vec{F}| = 120 + 240 = 360\ N\,m$, from which $\boxed{|\vec{F}| = 60\ N}$
Substitute this into the first equation to obtain: $\boxed{d_A = \frac{240}{60} = 4\ m}$

Problem 4.22 Determine the sum of the moments of the three forces about (a) point A, (b) point B.

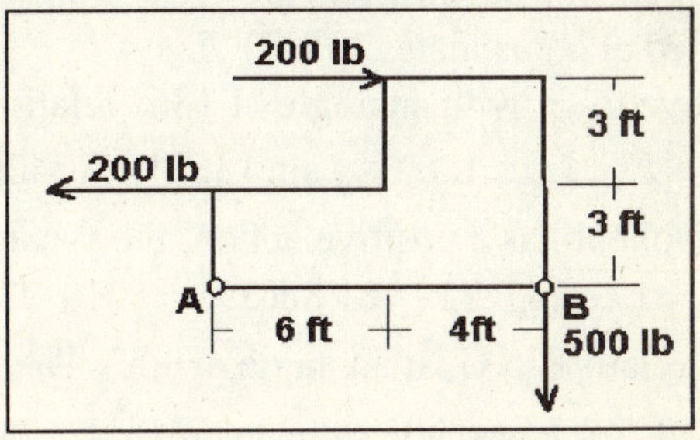

Solution: The strategy is to use the distance from the points to the line of action of the forces to determine the moments. The moments about A:
$\boxed{\sum M_A = +(200)(3) - (200)(6) - (500)(10) = -5600\ ft\text{-}lb}$
The moments about B are:
$\boxed{\sum M_B = (200)(3) - (200)(6) = -600\ ft\text{-}lb}$

Problem 4.23 The 160 N weights of the arms AB and BC of the robotic manipulator act at their midpoint. Determine the sum of the moments of the three weights about A.

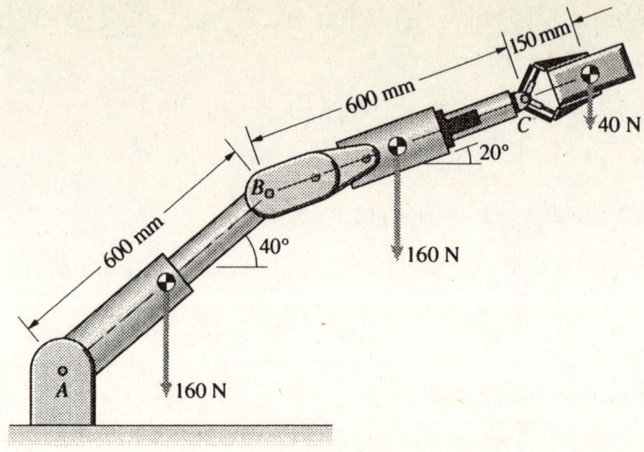

Solution: The strategy is to find the perpendicular distance from the points to the line of action of the forces, and determine the sum of the moments, using the appropriate sign of the action.

The distance from A to the action line of the weight of the arm AB is:

$d_{AB} = (0.300)\cos 40^{\circ} = 0.2298\ m$

The distance from A to the action line of the weight of the arm BC is

$d_{BC} = (0.600)(\cos 40^{\circ}) + (0.300)(\cos 20^{\circ}) = 0.7415\ m$. The distance from A to the line of action of the force is $d_F = (0.600)(\cos 40^{\circ}) + (0.600)(\cos 20^{\circ}) + (0.150)(\cos 20^{\circ}) = 1.1644\ m$

The sum of the moments about A is $\sum M_A = -d_{AB}(160) - d_{BC}(160) - d_F(40)$ $\boxed{= -202\ N\text{-}m}$

Problem 4.24 The space shuttle's attitude thrusters exert two forces of magnitude $|\vec{F}| = 1740\ lb$. What moment do the thrusters exert about G?

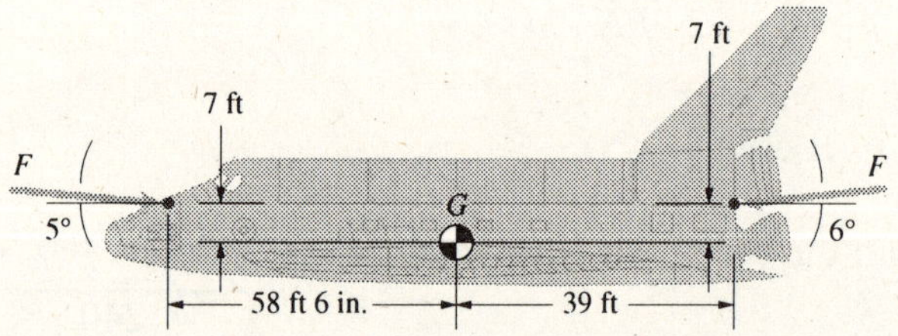

Solution: The strategy is to resolve the components of the thrust about the (x,y) axes, and compute the moment due to each about G. The aft thrusters are located at coordinates (39,7) ft relative to G, with an angle of 186° relative to the positive x-axis. The components are $\vec{F}_A = |\vec{F}_A|(\vec{i}\cos 186^{\circ} + \vec{j}\sin 186^{\circ}) = 1740(-0.9945\vec{i} - 0.1045\vec{j}) = -1730.47\vec{i} - 181.88\vec{j}$. The x-component has a positive action, the y-component has a negative action. The moment is $M_A = 1730.47(7) - 181.88(39) = 5019.97\ ft\ lb$. The forward thrusters are located at coordinates (-58.5, 7) ft relative to G, at an angle of -5°. The components are $\vec{F}_F = 1740(\vec{i}\cos(-5^{\circ}) + \vec{j}\sin(-5^{\circ})) = 1733.38\vec{i} - 151.65\vec{j}$. The action of the x-component is negative; the action of the y-component is positive. The moment $M_F = -1733.38(7) + 151.65(58.5) = -3262.08\ ft\ lb$. The total moment is

$\boxed{\sum M_T = 5019.97 - 3262.08 = 1757.9\ ft\ lb}$

Problem 4.25 The force exerts a 200 *ft-lb* counterclockwise moment about A and a 100 *ft-lb* moment clockwise moment about B. What are $|\vec{F}|$ and θ ?

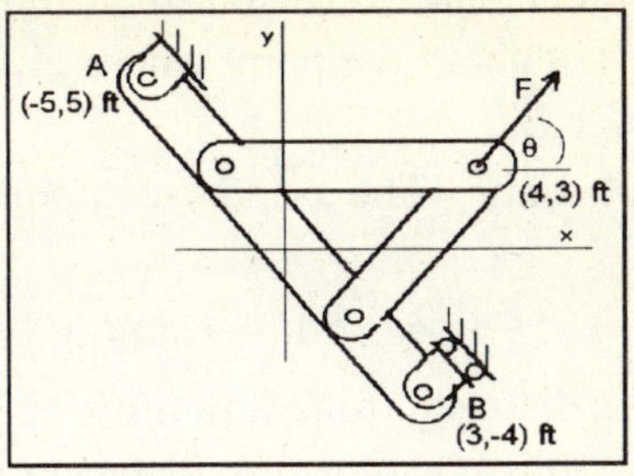

Solution: The strategy is to resolve $\vec{F}$ into x- and y- components, and compute the perpendicular distance to each component from A and B. The components of $\vec{F}$ are: $\vec{F} = \vec{i}F_X + \vec{j}F_Y$. . The vector from A to the point of application is: $\vec{r}_{AF} = (4-(-5))\vec{i} + (3-5)\vec{j} = 9\vec{i} - 2\vec{j}$. The perpendicular distances are $d_{AX} = 9\ ft$, and $d_{AY} = 2\ ft$., and the actions are positive. The moment about A is $M_A = (9)F_Y + (2)F_X = 200\ ft\text{-}lb$. The vector from B to the point of application is $\vec{r}_{BF} = (4-3)\vec{i} + (3-(-4))\vec{j} = 1\vec{i} + 7\vec{j}$; the distances $d_{BX} = 1\ ft$ and $d_{BY} = 7\ ft$, the action of F_Y is positive and the action of F_X is negative.. The moment about B is $M_B = (1)F_Y - (7)F_X = -100\ ft\text{-}lb$. The two simultaneous equations have solution: $F_Y = 18.46\ lb$ and $F_X = 16.92\ lb$. Take the ratio to find the angle: $\boxed{\theta = \tan^{-1}\left(\frac{F_Y}{F_X}\right) = \tan^{-1}\left(\frac{18.46}{16.92}\right) = \tan^{-1}(1.091) = 47.5^o}$

From the Pythagorean theorem $\boxed{|\vec{F}| = \sqrt{F_Y^2 + F_X^2} = \sqrt{18.46^2 + 16.92^2} = 25.04\ lb}$

Problem 4.26 Five forces act on a link in a gear shifting mechanism of a lawn mower. The vector sum of the five forces on the bar is zero. (a) Determine the forces $\vec{A}_X$, $\vec{A}_Y$, and B. (b) Determine the sum of the moments of the forces about the point where the force B acts.

Solution: The strategy is to resolve the forces into x- and y-components, determine the perpendicular distances from B to the line of action, determine the sign of the action, and compute the moments. The angles are measured counterclockwise from the x- axis. The forces are

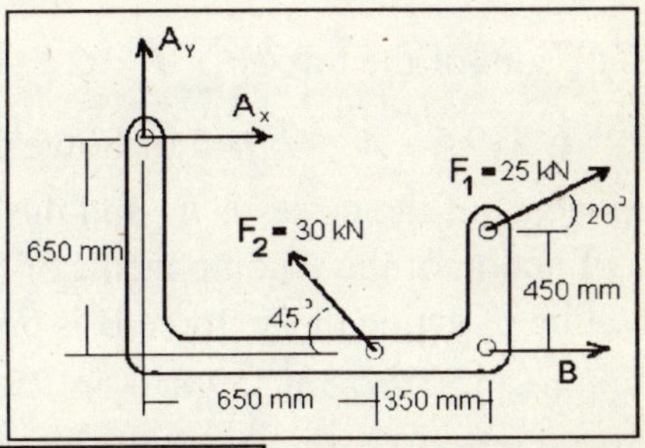

$\vec{F}_1 = 30(\vec{i}\cos 135^o + \vec{j}\sin 135^o) = -21.21\vec{i} + 21.21\vec{j}$

$\vec{F}_2 = 25(\vec{i}\cos 20^o + \vec{j}\sin 20^o) = 23.50\vec{i} + 8.55\vec{j}$. (a) The sum of the forces is $\sum \vec{F} = \vec{A} + \vec{B} + \vec{F}_1 + \vec{F}_2 = 0$. Substituting:

$\sum \vec{F}_X = (A_X + B_X + 23.5 - 21.2)\vec{i} = 0$, and

$\sum \vec{F}_Y = (A_Y + 21.2 + 8.55)\vec{j} = 0$. Solve the second equation: $\boxed{A_Y = -29.76\ kN}$ The distances of the forces from A are: the triangle has equal base and altitude, hence the angle is 45°, so that the line of action of $\vec{F}_1$ passes through A. The distance to the line of action of $\vec{B}$ is 0.65 *m*, with a positive action. The distance to the line of action of the y-component of $\vec{F}_2$ is $(0.650 + 0.350) = 1\ m$, and the action is positive. The distance to the line of action of the x-component of $\vec{F}_2$ is $(0.650 - 0.450) = 0.200\ m$, and the action is positive. The moment about A is $\sum M_A = (8.55)(1) + (23.5)(0.2) + (B_X)(0.65) = 0$. Solve: $\boxed{B_X = -20.38\ kN}$. Substitute into the force equation to obtain $\boxed{A_X = 18.09\ kN}$ (b) The distance from B to the line of action of the y-component of $\vec{F}_1$ is 0.350 *m* , and the action is negative. The distance from B to the line of action of A_X is 0.650 *m* and the action is negative. The distance from B to the line of action of A_Y is 1 *m* and the action is positive. The distance from B to the line of action of the x-component of $\vec{F}_2$ is 0.450 *m* and the action is negative. The sum of the moments about B:

$\boxed{\sum M_B = -(0.350)(21.21) - (0.650)(18.09) + (1)(29.76) - (0.450)(23.5) = 0}$

Problem 4.27 Five forces act on a model truss built by an engineering student as part of a design project. The dimensions $b = 300\ mm$ and $h = 400\ mm$ and $F = 100\ N$. The sum of the moments of the forces about the point where A_X and A_Y act is zero. If the weight of the truss is negligible, what is the force B?

Solution: The x-and y-components of the force $\vec{F}$ are

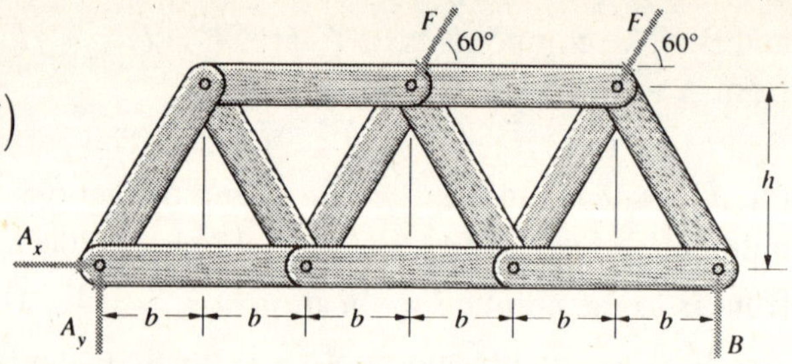

$\vec{F} = -|\vec{F}|(\vec{i}\cos 60^o + \vec{j}\sin 60^o) = -|\vec{F}|(0.5\vec{i} + 0.866\vec{j})$

The distance from A to the x-component is h and the action is positive. The distances to the y-component are $3b$ and $5b$. The distance to B is $6b$. The sum of the moments about A is

$\Sigma M_A = 2|\vec{F}|(0.5)(h) - 3b|\vec{F}|(0.866) - 5b|\vec{F}|(0.866) + 6bB = 0$. Substitute and solve:

$$\boxed{B = \frac{1.6784|\vec{F}|}{1.8} = 93.2\ N}$$

Problem 4.28 Consider the truss shown in Problem 4.27. The dimensions $b = 3\ ft$ and $h = 4\ ft$ and $|\vec{F}| = 300\ lb$. The vector sum of the forces acting on the truss is zero and the sum of the moments of the forces about the point where A_x and A_y act is zero. (a) Determine the forces A_x, A_y and B. (b) Determine the sum of the moments of the forces about the point where the force B acts.

Solution: The forces are resolved into x- and y-components:

$\vec{F} = -300(\vec{i}\cos 60^o + \vec{j}\sin 60^o) = -150\vec{i} - 259.8\vec{j}$.

(a) The sum of the forces: $\sum\vec{F} = 2\vec{F} + \vec{A} + \vec{B} = 0$. The x-and y-components: $\Sigma\vec{F}_x = (A_x - 300)\vec{i} = 0$, $\Sigma\vec{F}_y = (-519.6 + A_y + B)\vec{j} = 0$. Solve the first: $\boxed{A_x = 300\ lb}$. The distance from point A to the x-components of the forces is h, and the action is positive. The distances between the point A and the lines of action of the y-components of the forces are $3b$ and $5b$. The actions are negative. The distance to the line of action of the force B is $6b$. The action is positive. The sum of moments about point A is $\sum M_A = 2(150)h - 3b(259.8) - 5b(259.8) + 6bB = 0$. Substitute and solve: $\boxed{B = 279.7\ lb}$. Substitute this value into the force equation and solve: $\boxed{A_x = 519.6 - 279.7 = 239.9\ lb}$

(b) The distances from B and the line of action of A_Y is $6b$ and the action is negative. The distance between B and the x-component of the forces is h and the action is positive. The distance between B and the y-components of the forces is b and $3b$, and the action is positive. The sum of the moments about B:

$$\boxed{\sum M_B = -6b(239.9) + 2(150)h + b(259.8) + 3b(259.8) = 0}$$

Problem 4.29 The mass $m = 70\ kg$. What is the moment about A due to the force exerted on the beam by the cable at B?

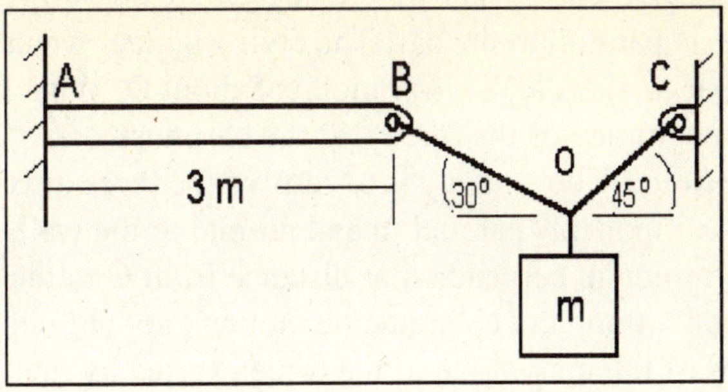

Solution: The strategy is to resolve the force at B into components parallel to and normal to the beam, and solve for the moment using the normal component of the force..

The force at B is to be determined from the equilibrium conditions on the cable juncture O. Angles are measured from the positive x-axis. The forces at the cable juncture are:

$\vec{F}_{OB} = |\vec{F}_{OB}|(\vec{i}\cos 150^o + \vec{j}\sin 150^o) = |\vec{F}_{OB}|(-0.866\vec{i} + 0.5\vec{j})$..

$\vec{F}_{OC} = |\vec{F}_{OC}|(\vec{i}\cos 45^o + \vec{j}\sin 45^o) = |\vec{F}_{OC}|(0.707\vec{i} + 0.707\vec{j})$.

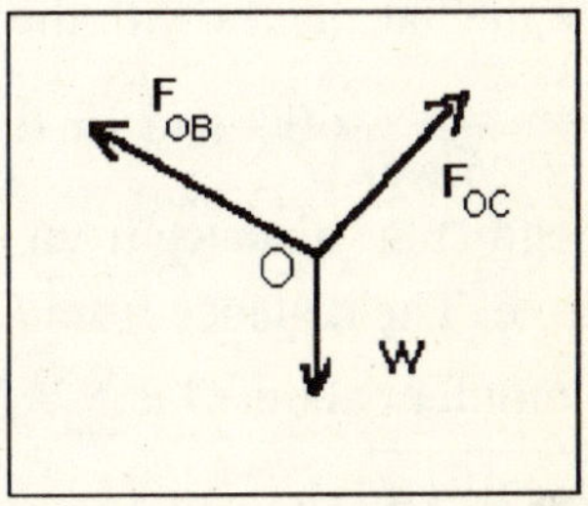

$\vec{W} = (70)(9.81)(0\vec{i} - 1\vec{j}) = -686.7\vec{j}\ (N)$. The equilibrium conditions are: $\sum \vec{F}_x = (-0.866|\vec{F}_{OB}| + 0.7070|\vec{F}_{OC}|)\vec{i} = 0$

$\sum \vec{F}_Y = (0.500|\vec{F}_{OB}| + (.707|\vec{F}_{OC}|) - 686.7)\vec{j} = 0$. Solve:

$|\vec{F}_{OB}| = 502.70\ N$. This is used to resolve the cable tension at B:

$\vec{F}_B = 502.7(\vec{i}\cos 330^o + \vec{j}\sin 330^o) = 435.4\vec{i} - 251.4\vec{j}$. The distance from A to the action line of the y-component at B is $3\ m$, and the action is negative. The x-component at passes through A, so that the action line distance is zero. The moment at A is $\boxed{M_A = -3(251.4) = -754.0\ N\text{-}m}$

Problem 4.30 Consider the system shown in Problem 4.29. The beam will collapse at A if the magnitude of the moment about A due to the force exerted on the beam by the cable at B exceeds 2 kN. What is the largest mass m that can be suspended?

Solution: The strategy is to determine the tension in the cable at B that corresponds to the 2 *kN-m* moment at A, and then determine the mass that will exert this tension. From the solution to Problem 4.29, the y-component of the cable tension at B that corresponds to the 2 *kN* moment is

$-0.5|\vec{F}_{OB}| = \dfrac{-2000\ N\text{-}m}{3\ m} = 666.67\ N$, from which $|\vec{F}_{OB}| = 1333\ N$. The two equilibrium equations for the cable juncture O in the solution for Problem 4.29 are:

$\sum \vec{F}_X = (-0.866|\vec{F}_{OB}| + 0.7070|\vec{F}_{OC}|)\vec{i} = 0$

$\sum \vec{F}_Y = (0.500|\vec{F}_{OB}| + (.707|\vec{F}_{OC}|) - |\vec{W}|)\vec{j} = 0$. Substitute the value of $|\vec{F}_{OB}|$, and rewrite these in the form of simultaneous equations in two unknowns:

$(0.707|\vec{F}_{OC}| + 0|\vec{W}|) = 1154.3$

$(0.707|\vec{F}_{OC}| - |\vec{W}|) = -666.7$. Solve: $|\vec{W}| = 1821\ N$, from which: $\boxed{m = \dfrac{1821}{9.81} = 185.66\ kg}$

=================================<>=================================

Problem 4.31 The bar AB exerts a force at B that helps support the vertical retaining wall. The force is parallel to the bar. The civil engineer wants the bar to exert a 38 *kN-m* moment about O. What is the magnitude of the force that the bar must exert?

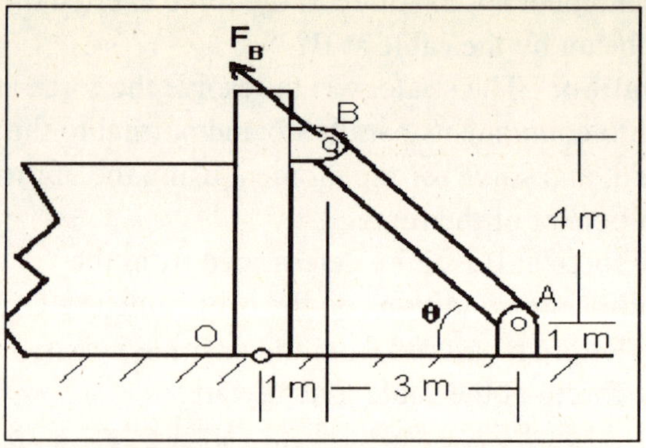

Solution: The strategy is to resolve the force at B into components parallel to and normal to the wall, determine the perpendicular distance from O to the line of action, and compute the moment about O in terms of the magnitude of the force exerted by the bar.

By inspection, the bar forms a 3,4,5 triangle. The angle the bar makes with the horizontal is

$\cos\theta = \frac{3}{5} = 0.600$, and $\sin\theta = \frac{4}{5} = 0.800$. The force at B is $\vec{F}_B = |\vec{F}_B|(-0.600\vec{i} + 0.800\vec{j})$. The perpendicular distance from O to the line of action of the x-component is $(4+1) = 5\ m$, and the action is positive. The distance from O to the line of action of the y-component is $1\ m$, and the action is positive. The moment about O is $\sum M_O = 5(0.600)|\vec{F}_B| + 1(0.800)|\vec{F}_B| = 3.8|\vec{F}_B| = 38\ kN$, from which

$\boxed{|\vec{F}_B| = 10\ kN}$

=================================<>=================================

Problem 4.32 A contestant in a fly fishing contest snags his line in some grass. If the tension in the line is 5 *lb*, what moment does the force exerted on the rod by the line exert about point H where he holds the rod?

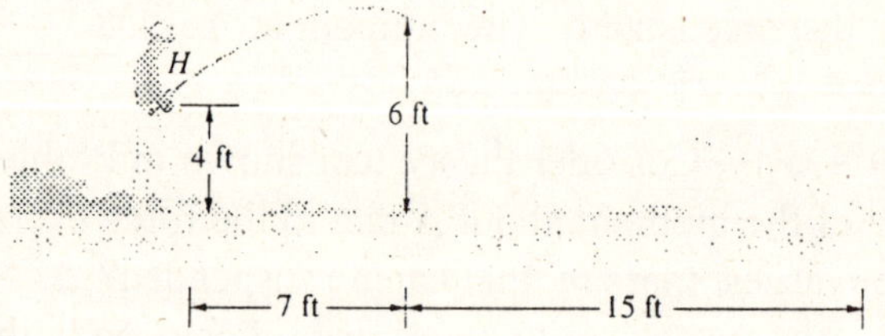

Solution: The strategy is to resolve the line tension into a component normal to the rod; use the length from H to tip as the perpendicular distance; determine the sign of the action, and compute the moment.

The line and rod form two right triangles, as shown in the sketch. The angles are:

$\alpha = \tan^{-1}\left(\frac{2}{7}\right) = 15.95°$

$\beta = \tan^{-1}\left(\frac{6}{15}\right) = 21.8°$. The angle between the perpendicular distance line and the fishing line is $\theta = \alpha + \beta = 37.7°$. The force normal to the distance line is $F = 5(\sin 37.7°) = 3.061\ lb$. The distance is $d = \sqrt{2^2 + 7^2} = 7.28\ ft$, and the action is negative. The moment about H is $\boxed{M_H = -7.28(3.061) = -22.3\ ft\text{-}lb}$ *Check:* The tension can be resolved into x and y components, $F_x = F\cos\beta = 4.642\ lb$, $F_y = -F\sin\beta = -1.857\ lb$. The moment is $M = -2F_x + 7F_y = -22.28 = -22.3\ ft\text{-}lb$. *check.*

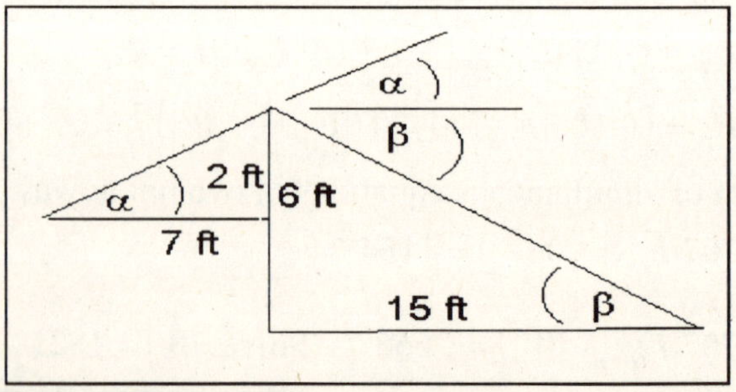

=================================<>=================================

Problem 4.33 The cables AB and BC help support the tower. The tension in cable AB is 5 *kN*. The points A, B, C, and O are contained in the same vertical plane. (a) What is the moment about O due to the force exerted on the tower by the cable AB? (b) If the sum of the moments about O due to the forces exerted on the tower by the two cables is zero, what is the tension in the cable AC?

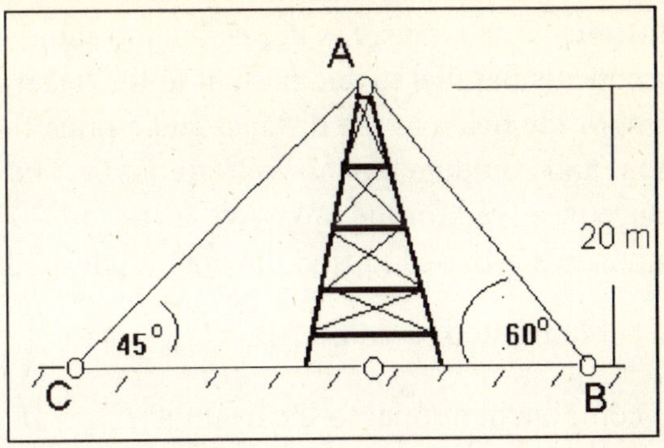

Solution: The strategy is to resolve the cable tensions into components normal to the vertical line through OA; use the height of the tower as the perpendicular distance; determine the sign of the action, and compute the moments.

(a) The component normal to the line OA is $F_{BN} = 5(\cos 60^o) = 2.5\ kN$. The action is negative. The moment about O is

$$\boxed{M_{OA} = -2.5(20) = -50\ kN\text{-}m}$$

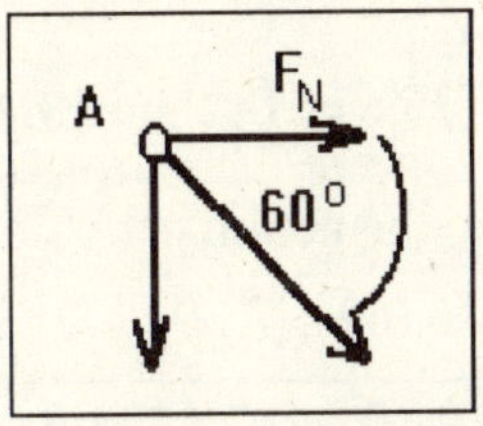

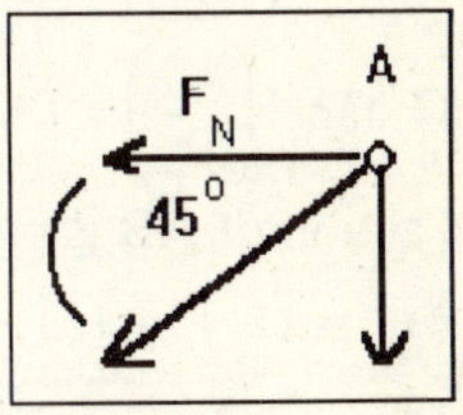

(b) By a similar process, the normal component of the tension in the cable AC is

$F_{CN} = |\vec{F}_C|\cos 45^o = 0.707|\vec{F}_C|$. The action is positive. If the sum of the moments is zero,

$$\sum M_O = (0.707(20)|\vec{F}_C| - 50) = 0, \text{ from which } \boxed{|\vec{F}_C| = \frac{50\ kN\ m}{(0.707)(20\ m)} = 3.54\ kN}$$

Problem 4.34 The tension in each cable is the same. The forces exerted on the beam by the three cables exert a 1.2 *kN-m* counterclockwise moment about O. What is the tension in the cable?

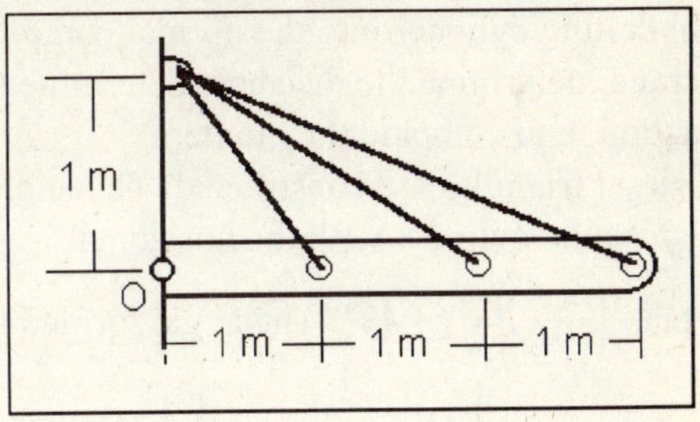

Solution: The strategy is to resolve cable tensions into components normal to the beam; use the distances from O to attachment point; determine action, and compute the moment in terms of cable tensions. From the known moment, solve for the tensions.

Denote the cables as 1, 2, and 3, starting near the root. The angle formed by the first cable with the beam is

$\theta_1 = \tan^{-1}\left(\frac{1}{1}\right) = 45^o$. The component normal to the

beam is: $F_1 = |\vec{T}|\sin 45^o = 0.707|\vec{T}|$. Similarly, $\theta_2 = \tan^{-1}\left(\frac{1}{2}\right) = 26.57^o$,

$F_2 = |\vec{T}|\sin 26.57^o = 0.4472|\vec{T}|$, $\theta_3 = \tan^{-1}\left(\frac{1}{3}\right) = 18.43^o$, and $F_3 = |\vec{T}|\sin 18.43^o = 0.3162|\vec{T}|$. The actions are positive. The sum of the moments about O

$$\Sigma M_O = (1)(0.707|\vec{T}|) + 2(0.4472|\vec{T}|) + 3(0.3162|\vec{T}|) = 1.2\ kN\ m. \text{ Solving: } \boxed{|\vec{T}| = \frac{1.2}{2.55} = 0.4706\ kN}$$

Problem 4.35

Problem 4.35 The tension in cable AB is 300 *lb*. The sum of the moments about O due to the forces exerted on the beam by the two cables is zero. What is the magnitude of the sum of the forces exerted on the beam by the two cables?

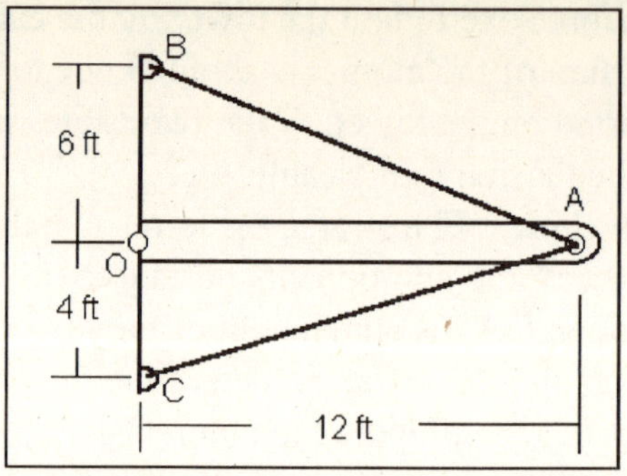

Solution: The strategy is to resolve the cable tensions into a components parallel to and normal to the beam; use the length of the beam as the distance; determine the sign of the action, and compute the moments about O. The moment balance is solved for the unknown force in CA. The parallel components are used to find the magnitude of the force.

The angle formed by cable AB is $\theta_{AB} = \tan^{-1}\left(\frac{6}{12}\right) = 26.6^o$.

The component normal to the beam is $F_{NB} = |\vec{T}_{AB}|\sin 26.6^o = (300)(0.4472) = 134.16\ lb$. The parallel component is $F_{PB} = |\vec{T}_{AB}|\cos 26.6^o = (300)(0.8944) = 268.32\ lb$. Similarly for the cable AC,

$\theta_{CA} = \tan^{-1}\left(\frac{4}{12}\right) = 18.43^o$, $F_{NC} = |\vec{T}_{AC}|\sin 18.43^o = 0.3162|\vec{T}_{AC}|$, $F_{PC} = |\vec{T}_{AC}|\cos 18.43^o = 0.9487|\vec{T}_{AC}|$

The action of AB is positive, the action of AC is negative. The moments about O $\sum M_O = 12(134.16) - 12(0.3162)(|\vec{T}_{AC}|) = 0$. Solve: $|\vec{T}_{AC}| = 424.3\ lb$. The sum of the forces acting on the beam is $\boxed{\sum \vec{F}_P = 268.32 + (0.9487)(424.3) = 670.8\ lb}$

Problem 4.36

Problem 4.36 The hydraulic cylinder BC exerts a 300 *kN* force on the boom of the crane at C. The force is parallel to the cylinder. What is the moment of the force about A?

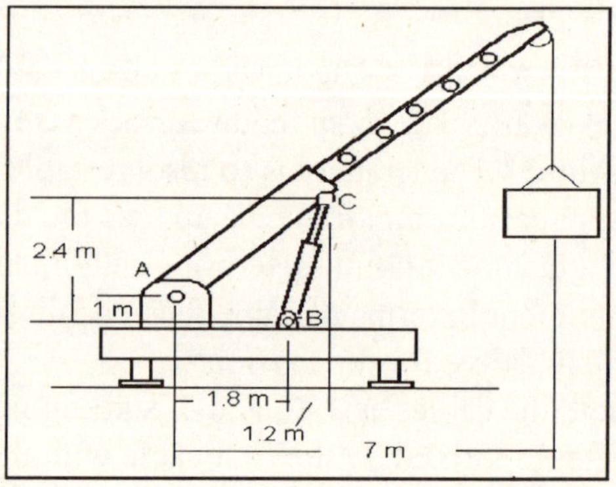

Solution: The strategy is to resolve the force exerted by the hydraulic cylinder into the normal component about the crane; determine the distance; determine the sign of the action, and compute the moment.

Two right triangles are constructed: The angle formed by the hydraulic cylinder with the horizontal is $\beta = \tan^{-1}\left(\frac{2.4}{1.2}\right) = 63.43^o$. The angle formed by the crane with the horizontal is $\alpha = \tan^{-1}\left(\frac{1.4}{3}\right) = 25.02^o$. The angle between the hydraulic cylinder and the crane is $\theta = \beta - \alpha = 38.42^o$. The normal component of the force is:

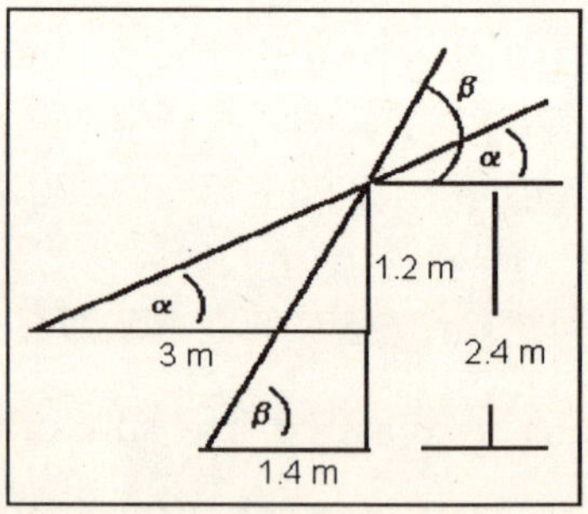

$F_N = (300)(\sin 38.42^o) = 186.42\ kN$. The distance from point A is $d = \sqrt{1.4^2 + 3^2} = 3.31\ m$. The action is positive. The moment about A is $\boxed{M_O = +3.31(186.42) = 617.15\ kN\text{-}m}$ *Check:* The force exerted by the actuator can be resolved into x- and y-components, $F_x = F\cos\beta = 134.16\ kN$, $F_y = F\sin\beta = 268.33\ kN$. The moment about the point A is $M = -1.4F_x + 3.0F_y = 617.15\ kN\,m$. *check.*

================================<>================================

Problem 4.37 The unstretched length of spring is 1 m and the spring constant is $k = 20\ N/m$. If $\alpha = 30^\circ$, what is the moment about A due to the force exerted by the spring on the circular bar at A?

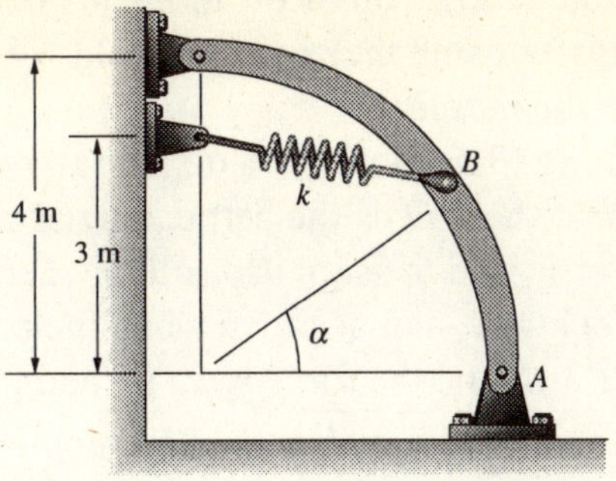

Solution: Assume that the bar is a quarter circle, with a radius of 4 m. The stretched length of the spring is found from the Pythagorean Theorem: The vertical height from the floor to the attachment point on the bar is $h = 4\sin\alpha = 2\ m$, and the distance from the wall is $4\cos\alpha$. The stretched length of the spring is $L = \sqrt{(3-h)^2 + (4\cos\alpha)^2} = 3.6055\ m$.

The spring force is $F = (20)(3.6055 - 1) = 52.1\ N$. The angle that the spring makes with the horizontal is $\beta = \tan^{-1}\left(\dfrac{3-h}{4\cos\alpha}\right) = \tan^{-1}(0.2887) = 16.1^\circ$. The horizontal component of the spring force is

$F_X = F\cos 16.1^\circ = (52.1)(0.9608) = 50.07\ N$. The vertical component of the force is $F_Y = F\sin 16.1^\circ = (52.1)(0.2773) = 14.45\ N$. The distance from A to the spring attachment point to the left of A is $d = 4(1-\cos\alpha) = 0.5259\ m$, hence the action of the vertical component is negative, and the action of the horizontal component is positive. The moment about A is

$\boxed{\Sigma M_A = -0.5359(14.45) + 2(50.07) = 92.4\ N\text{-}m}$.

================================<>================================

Problem 4.38 The exercise equipment shown is used by resting the elbow on the fixed pad and rotating the forearm to stretch the elastic cord AB. The cord behaves like a linear spring. Its unstretched length is 1 ft and its spring constant is 100 *lb/ft*. If $\alpha = 45^\circ$, what moment is exerted about the elbow joint E by the force exerted by the elastic cord at B?

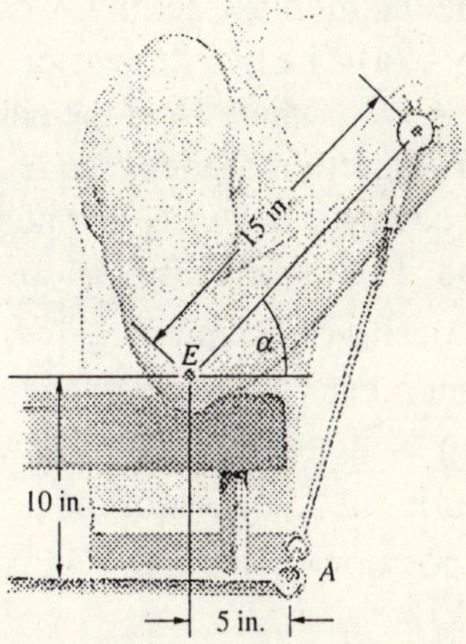

Solution: The strategy is to express the force exerted by the elastic cord into x and y components, and find the moment about E from the two dimensional moment strategy. .The position vectors are :

$\vec{r}_{EB} = 15(\vec{i}\cos\alpha + \vec{j}\sin\alpha) = 10.6066\vec{i} + 10.6066\vec{j}\ (in)$.,

$\vec{r}_{AE} = -5\vec{i} + 10\vec{j}\ (in)$, from which

$\vec{r}_{AB} = \vec{r}_{AE} + \vec{r}_{EB} = 5.6066\vec{i} + 20.6066\vec{j}\ (in)$. The angle of AB with the horizontal is obtained from $\vec{r}_{AB}$, $\beta = \tan^{-1}\left(\dfrac{20.6066}{5.5066}\right) = 74.78^\circ$. The magnitude of $\vec{r}_{AB}$ is the stretched length of the cord $|\vec{r}_{AB}| = \sqrt{5.6066^2 + 20.6066^2} = 21.3557\ in = 1.7796\ ft$. The tension in the cord is $|\vec{T}_{AB}| = k\Delta L = 100(1.7796 - 1) = 77.96\ lb$ .The x- and y-components of the tension are

$F_x = |\vec{T}_{AB}|\cos\beta = 20.47\ lb$, $F_y = |\vec{T}_{AB}|\sin\beta = 75.23\ lb$.. The moment about the elbow joint is $M_E = -10.606F_y + 10.606F_x = -580.8\ in\ lb \boxed{= -48.4\ ft\ lb}$

================================<>================================

Problem 4.39 Determine the moment of the 50 *lb* about the origin O. (a) What is the two dimensional description of the moment? (b) Express the moment as a vector without using the cross product.(c) Use the cross product to determine the moment.

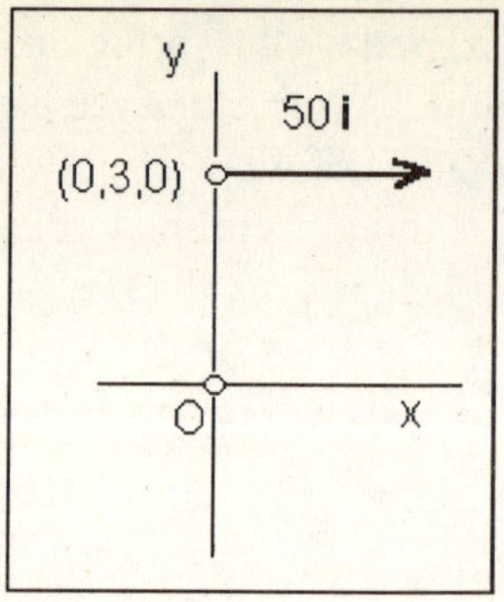

Solution: (a) The moment is defined in two dimensions to be $M = \pm DF$, where D is the perpendicular distance from O to the line of action of F, and the sign is positive when the action is counterclockwise, and negative when the action is clockwise. The distance is 3 *ft*, and the force is 50 *lb*, and the action is negative. Thus, $\boxed{M = -3(50) = -150\ ft\text{-}lb}$. (b) The moment is a vector perpendicular in a right hand sense to the line of action of the force, and the position vector of any point on the line from the origin. The action is clockwise, hence the moment vector is directed into the paper, or $\boxed{\vec{M} = -150\vec{k}\ ft\text{-}lb}$ (c) The cross product is

$$\vec{r} \times \vec{F} = \begin{vmatrix} \vec{i} & \vec{j} & \vec{k} \\ 0 & 3 & 0 \\ 50 & 0 & 0 \end{vmatrix} = (0)\vec{i} - (0)\vec{j} - (3)(50)\vec{k} \boxed{= -150\vec{k}\ ft\text{-}lb}.$$

Problem 4.40 Determine the moment of the 80 *N* force about the origin O. (a) What is the two-dimensional description of the moment? (b) Express the moment as a vector without using the cross product. (c) Use the cross product to determine the moment, letting $\vec{r}$ be the vector from O to A. (d) Use the cross product to determine the moment, letting $\vec{r}$ be the vector from O to B.

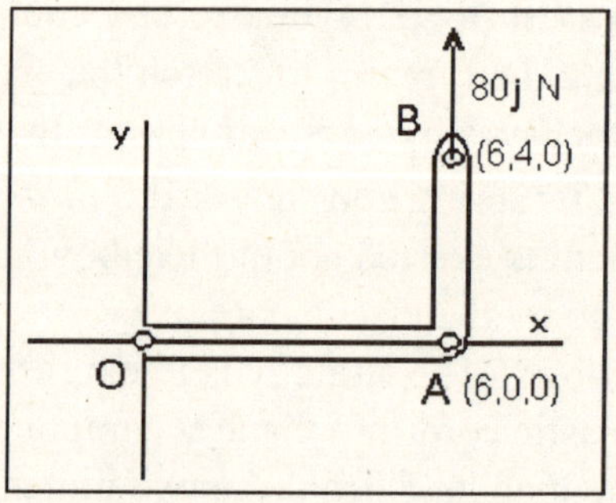

Solution: (a) The two dimensional description of the moment is $M_O = \pm DF$, where D is the perpendicular distance from O to the line of action of the force F, and the sign is positive if the action is counterclockwise, and negative if the action is clockwise. The distance is 6 *m*, and the force is 80 *N*, and the action is positive. The moment: $\boxed{M_O = (6)(80) = 480\ N\text{-}m}$

(b) The moment vector is perpendicular to the position vector and the force in a right handed sense. Since the action is positive, the moment vector is directed out of the paper, in the direction of the positive z-axis. Thus $\vec{M}_O = +480\vec{k}\ (N\text{-}m)$.

(c) The position vector from O to A is $\vec{r}_{OA} = 6\vec{i} + 0\vec{j} + 0\vec{k}$. The cross product is

$$\vec{r}_{OA} \times \vec{F} = \begin{vmatrix} \vec{i} & \vec{j} & \vec{k} \\ 6 & 0 & 0 \\ 0 & 80 & 0 \end{vmatrix} = (0)\vec{i} - (0)\vec{j} + (6)(480)\vec{k} \boxed{= 480\vec{k}\ N\text{-}m}$$

(d)The position vector from point O to B is determined from the coordinates of B: $\vec{r}_{OB} = 6\vec{i} + 4\vec{j} + 0\vec{k}$.

The cross product is $\vec{r}_{OB} \times \vec{F} = \begin{vmatrix} \vec{i} & \vec{j} & \vec{k} \\ 6 & 4 & 0 \\ 0 & 80 & 0 \end{vmatrix} = (0)\vec{i} - (0)\vec{j} + (6)(80)\vec{k} \boxed{= 480\vec{k}\ N\text{-}m}$

Problem 4.41 A bioengineer studying an injury sustained in throwing the javelin estimates that the athlete's shoulder was subjected to an 1800 *in-lb* moment and the magnitude of the force exerted was $|\vec{F}| = 80\ lb$. The vector $\vec{F}$ and the point O are contained in the x-y plane. (a) What was the perpendicular distance from O to the line of action of $\vec{F}$? (b) Express the moment of $\vec{F}$ about O as a vector.

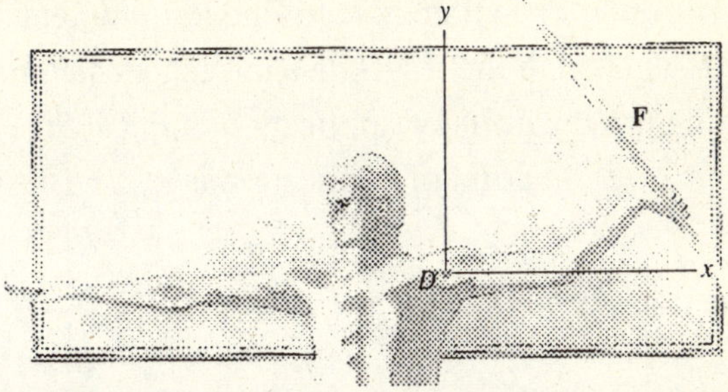

Solution: (a) The magnitudes of the moment and the force are given, hence

$$\boxed{D = \frac{1800\ in\text{-}lb}{80\ lb} = 22.5\ in}.$$

(b) The moment vector is perpendicular to both the position vector to the line of action and the force in a right handed sense. Since the action is positive, the moment vector is directed in the positive z-axis direction. Thus $\boxed{\vec{M}_O = 1800\vec{k}\ (in\text{-}lb)}$

Problem 4.42 Use the cross product to determine the moment of the 100 *kN* force (a) about A, (b) about B.

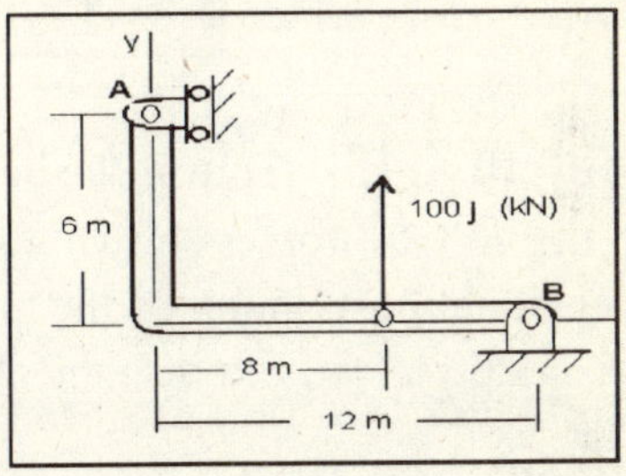

Solution: (a) The coordinates of A are (0,6,0). The coordinates of the point of application of the force are (8,0,0). The position vector from A to the point of application of the force is $\vec{r}_{AF} = (8-0)\vec{i} + (0-6)\vec{j} = 8\vec{i} - 6\vec{j}$. The force is $\vec{F} = 100\vec{j}\ (kN)$.

The cross product is $\vec{r}_{AF} \times \vec{F} = \begin{vmatrix} \vec{i} & \vec{j} & \vec{k} \\ 8 & -6 & 0 \\ 0 & 100 & 0 \end{vmatrix} \boxed{= 800\vec{k}\ (kN\text{-}m)}$

(b) The coordinates of B are (12,0,0). The position vector from B to the point of application of the force is $\vec{r}_{BF} = (8-12)\vec{i} = -4\vec{i}$. The cross product is: $\vec{r}_{BF} \times \vec{F} = \begin{vmatrix} \vec{i} & \vec{j} & \vec{k} \\ -4 & 0 & 0 \\ 0 & 100 & 0 \end{vmatrix} \boxed{= -400\vec{k}\ (kN\text{-}m)}$

Problem 4.43 The line of action of the 100 *lb* force lies in the x-y plane. (a) Use the cross product to determine the moment of the force about O. (b) Use the result of Part (a) to determine the perpendicular distance from O to the line of action.

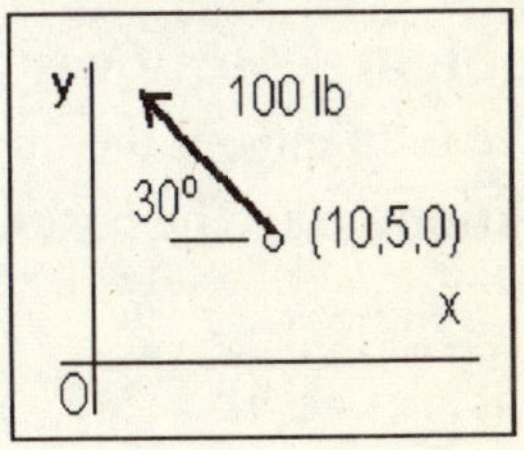

Solution: The resolved force is $\vec{F} = 100(\vec{i}\cos 150^o + \vec{j}\sin 150^o)$ $= -0.86.6\vec{i} + 50\vec{j}$. The position vector to the point of application: $\vec{r} = 10\vec{i} + 5\vec{j}$. (a) The cross product: $\vec{r} \times \vec{F} = \begin{vmatrix} \vec{i} & \vec{j} & \vec{k} \\ 10 & 5 & 0 \\ -86.6 & 50 & 0 \end{vmatrix} = (500 + 433)\vec{k}$. $\boxed{= 933\vec{k}\ ft\text{-}lb}$ (b) The magnitude of the moment is $|\vec{M}| = 933\ ft\text{-}lb$. The magnitude of the force is $|\vec{F}| = 100\ lb$. The distance is

$$\boxed{D = \frac{933\ ft\text{-}lb}{100\ lb} = 9.33\ ft}.$$

Problem 4.44 The line of action of $\vec{F}$ is contained in the x-y plane. The moment of $\vec{F}$ about O is $140\vec{k}$ *(N-m)* and the moment of $\vec{F}$ about A is $280\vec{k}$ *(N-m)*. What are the components of $\vec{F}$?

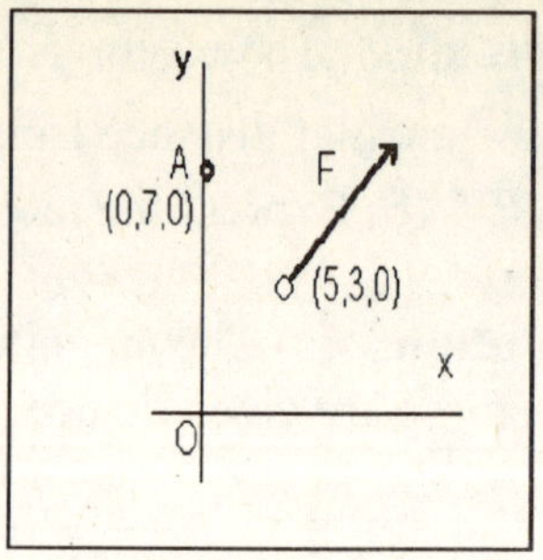

Solution: The strategy is to find the moments in terms of the components of $\vec{F}$ and solve the resulting simultaneous equations. The position vector from O to the point of application is $\vec{r}_{OF} = 5\vec{i} + 3\vec{j}$. The position vector from A to the point of application is $\vec{r}_{AF} = (5-0)\vec{i} + (3-7)\vec{j} = 5\vec{i} - 4\vec{j}$.

The cross products: $\vec{r}_{OF} \times \vec{F} = \begin{vmatrix} \vec{i} & \vec{j} & \vec{k} \\ 5 & 3 & 0 \\ F_X & F_Y & 0 \end{vmatrix} = (5F_Y - 3F_X)\vec{k} = 140\vec{k}$, and

$\vec{r}_{AF} \times \vec{F} = \begin{vmatrix} \vec{i} & \vec{j} & \vec{k} \\ 5 & -4 & 0 \\ F_X & F_Y & 0 \end{vmatrix} = (5F_Y + 4F_X)\vec{k} = 280\vec{k}$. Take the dot product of both sides with $\vec{k}$ to eliminate $\vec{k}$. The simultaneous equations are: $5F_Y - 3F_X = 140$, $5F_Y + 4F_X = 280$. Solving: $F_Y = 40$, $F_X = 20$, from which $\boxed{\vec{F} = 20\vec{i} + 40\vec{j}\ (N)}$

Problem 4.45 To test the bending stiffness of a light composite beam, engineering students subject it to the vertical forces shown. (a) What is the two dimensional description of the moment of the 200 *lb* force about A? (b) Express the moment in Part (a) as a vector.

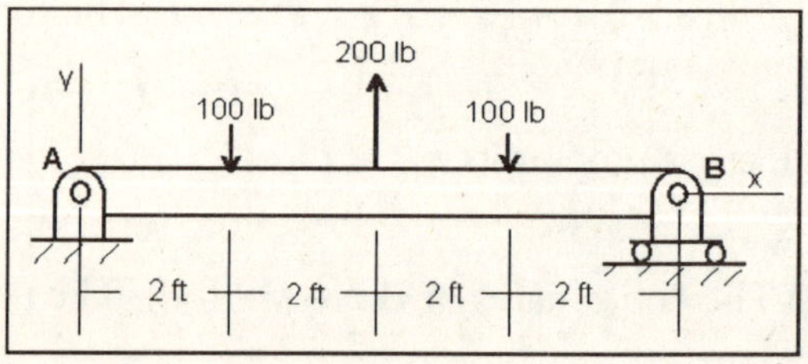

Solution: The two dimensional description of the moment is $M_A = \pm D_{AF}F$, where D_{AF} is the perpendicular distance from A to the line of action of F, and the sign is positive if the action is counterclockwise, and negative if it is clockwise.

(a) The moment of the 200 *lb* force about A is positive, $\boxed{M = 4(200) = 800\ ft\text{-}lb}$

(b) The moment vector is directed along the positive z- axis, hence $\boxed{\vec{M} = 800\vec{k}\ ft\text{-}lb}$

Problem 4.46 Consider the beam and forces in Problem 4.45. Use the cross product to find the sum of the moments of the three forces (a) about A, (b) about B.

Solution: (a) The position vectors are $\vec{r}_{A1} = 2\vec{i}$, $\vec{r}_{A2} = 4\vec{i}$, and $\vec{r}_{A3} = 6\vec{i}$. (a) The sum of the moments about A is $\Sigma\vec{M} = \begin{vmatrix} \vec{i} & \vec{j} & \vec{k} \\ 2 & 0 & 0 \\ 0 & -100 & 0 \end{vmatrix} + \begin{vmatrix} \vec{i} & \vec{j} & \vec{k} \\ 4 & 0 & 0 \\ 0 & +200 & 0 \end{vmatrix} + \begin{vmatrix} \vec{i} & \vec{j} & \vec{k} \\ 6 & 0 & 0 \\ 0 & -100 & 0 \end{vmatrix} \boxed{= -200\vec{k} + 800\vec{k} - 600\vec{k} = 0}$

(b) About B: the position vectors are $\vec{r}_{B1} = -6\vec{i}$, $\vec{r}_{B2} = -4\vec{i}$, and $\vec{r}_{A3} = -2\vec{i}$. The sum of moments about B: $\sum\vec{M}_B = \begin{vmatrix} \vec{i} & \vec{j} & \vec{k} \\ -6 & 0 & 0 \\ 0 & -100 & 0 \end{vmatrix} + \begin{vmatrix} \vec{i} & \vec{j} & \vec{k} \\ -4 & 0 & 0 \\ 0 & +200 & 0 \end{vmatrix} + \begin{vmatrix} \vec{i} & \vec{j} & \vec{k} \\ -2 & 0 & 0 \\ 0 & -100 & 0 \end{vmatrix} \boxed{= 600\vec{k} - 800\vec{k} + 200\vec{k} = 0}$

=================================<>=================================

Problem 4.47 Three forces are applied to the plate. Use the cross product to determine the sum of the moments of the three forces about O.

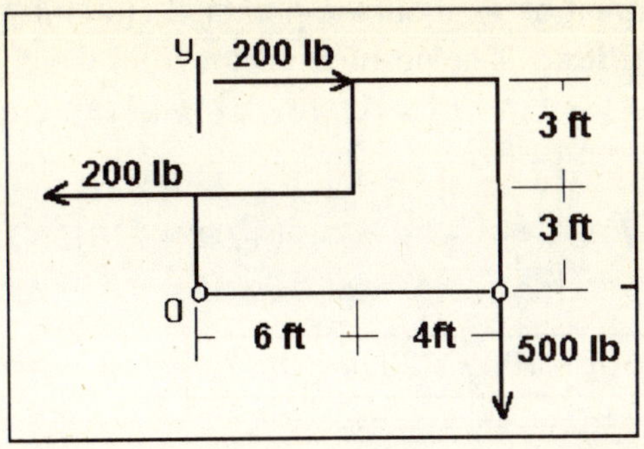

Solution: The position vectors from O to the points of application of the forces are: $\vec{r}_{O1} = 3\vec{j}$, $\vec{F}_1 = -200\vec{i}$; $\vec{r}_{O2} = 10\vec{i}$, $\vec{F}_2 = -500\vec{j}$; $\vec{r}_{O3} = 6\vec{i} + 6\vec{j}$, $\vec{F}_3 = 200\vec{i}$. The sum of the moments about O is

$$\sum \vec{M}_O = \begin{vmatrix} \vec{i} & \vec{j} & \vec{k} \\ 0 & 3 & 0 \\ -200 & 0 & 0 \end{vmatrix} + \begin{vmatrix} \vec{i} & \vec{j} & \vec{k} \\ 10 & 0 & 0 \\ 0 & -500 & 0 \end{vmatrix} + \begin{vmatrix} \vec{i} & \vec{j} & \vec{k} \\ 6 & 6 & 0 \\ 200 & 0 & 0 \end{vmatrix}$$

$$\boxed{= 600\vec{k} - 5000\vec{k} - 1200\vec{k} = -5600\vec{k}\ ft\text{-}lb}$$

=================================<>=================================

Problem 4.48 (a) Determine the magnitude of the moment of the 150 N about A by calculating the perpendicular distance from A to the line of action of the force. (b) Use the cross product to determine the moment of the 150 N about A. (c) Use the result of Part (b) to determine the magnitude of the moment.

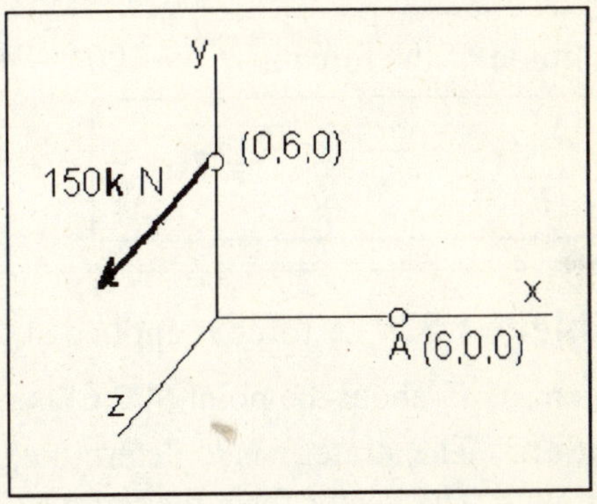

Solution: (a) The perpendicular distance lies in the x-y plane. The distance by the Pythagorean Theorem is $D = \sqrt{6^2 + 6^2} = 8.485\ m$. The action is positive. The moment about A is $\boxed{M_A = 8.485(150) = 1272.8\ N\text{-}m}$

(b) The position vector from A to the point of application of the force is $\vec{r}_{AF} = (0-6)\vec{i} + (6-0)\vec{j}$. The force is $\vec{F} = 150\vec{k}\ N$. The cross product is $\vec{r}_{AF} \times \vec{F} = \begin{vmatrix} \vec{i} & \vec{j} & \vec{k} \\ -6 & 6 & 0 \\ 0 & 0 & 150 \end{vmatrix} = (6)(150)\vec{i} - (-6)(150)\vec{j}$

$\boxed{= 900\vec{i} + 900\vec{j}\ (N\text{-}m)}$ (c) The magnitude of the moment is $\boxed{|\vec{M}| = \sqrt{900^2 + 900^2} = 1272.8\ N\text{-}m}$

=================================<>=================================

Problem 4.49 A force $\vec{F} = 4\vec{i} + 6\vec{j} - 2\vec{k}\ (kN)$ is applied at the point (8,4,4) m. What is the moment of $\vec{F}$ about the point (2,2,2) m.

Solution: The position vector from the point (2,2,2) to the point (8,4,4) is $\vec{r} = 6\vec{i} + 2\vec{j} + 2\vec{k}$. The cross product is: $\vec{r} \times \vec{F} = \begin{vmatrix} \vec{i} & \vec{j} & \vec{k} \\ 6 & 2 & 2 \\ 4 & 6 & -2 \end{vmatrix} = (-4-12)\vec{i} - (-12-8)\vec{j} + (36-8)\vec{k}$ $\boxed{= -16\vec{i} + 20\vec{j} + 28\vec{k}\ (kN\text{-}m)}$

=================================<>=================================

Problem 4.50 A force $\vec{F} = 20\vec{i} - 30\vec{j} + 60\vec{k}$ (lb) is applied at the point (2,3,6) *ft*. What is the moment of $\vec{F}$ about the point (-2,-1,-1) *ft*?

Solution: The position vector from the point (-2,-1,-1) to the point (2,3,6) is $\vec{r} = (2+2)\vec{i} + (3+1)\vec{j} + (6+1)\vec{k} = +4\vec{i} + 4\vec{j} + 7\vec{k}$. The cross product is

$$\vec{r} \times \vec{F} = \begin{vmatrix} \vec{i} & \vec{j} & \vec{k} \\ +4 & +4 & +7 \\ 20 & -30 & 60 \end{vmatrix} = (+240+210)\vec{i} - (+240-140)\vec{j} + (-120-80)\vec{k}$$

$$\boxed{= 450\vec{i} - 100\vec{j} - 200\vec{k} \ (ft\text{-}lb)}$$

Problem 4.51 A force $\vec{F} = 20\vec{i} - 30\vec{j} + 60\vec{k}$ (lb) The moment of $\vec{F}$ about a point P is $\vec{M}_P = 450\vec{i} - 100\vec{j} - 200\vec{k}$ $(ft\text{-}lb)$. What is the perpendicular distance from point P to line of action?

Solution: The magnitude of the moment is $|\vec{M}_P| = \sqrt{450^2 + 100^2 + 200^2} = 502.5$ $(ft\text{-}lb)$. The magnitude of the force is $|\vec{F}| = \sqrt{20^2 + 30^2 + 60^2} = 70$ (lb). The perpendicular distance is

$$\boxed{D = \frac{|\vec{M}_P|}{|\vec{F}|} = \frac{502.5 \ ft\text{-}lb}{70 \ lb} = 7.18 \ ft}$$

Problem 4.52 A force is applied at the point (8,6,13) *m*. Its magnitude is $|\vec{F}| = 90$ N, and the moment of $\vec{F}$ about the point (4,2,6) is zero. What are the components of $\vec{F}$?

Solution: The strategy is to determine the moment in terms of the components of $\vec{F}$ and solve for the unknowns. The vector from the point (4,2,6) to (8,6,13) is $\vec{r} = (8-4)\vec{i} + (6-2)\vec{j} + (13-6)\vec{k} = 4\vec{i} + 4\vec{j} + 7\vec{k}$. The cross product is

$$\vec{r} \times \vec{F} = \begin{vmatrix} \vec{i} & \vec{j} & \vec{k} \\ 4 & 4 & 7 \\ F_X & F_Y & F_Z \end{vmatrix} = (4F_Z - 7F_Y)\vec{i} - (4F_Z - 7F_X)\vec{j} + (4F_Y - 4F_X)\vec{k} = 0$$

If the vector sum of the components is zero, each component must be zero. Hence the three simultaneous equations:

$$0F_X - 7F_Y + 4F_Z = 0$$
$$+7F_X + 0F_Y - 4F_Z = 0$$
$$-4F_X + 4F_Y - 0F_Z = 0$$

These equations are indeterminate; *the only determinate solution is* $F_X = F_Y = F_Z = 0$, which is unacceptable. However, a standard trick for indeterminate equations is not to seek a solution for the unknowns, but instead to seek a solution for the ratios of the unknowns. Divide through by F_Z to obtain the three equations: $0\frac{F_X}{F_Z} - 7\frac{F_Y}{F_Z} = -4$, $+7\frac{F_X}{F_Z} + 0\frac{F_Y}{F_Z} = 4$, $-4\frac{F_X}{F_Z} + 4\frac{F_Y}{F_Z} = 0$. These have the solution: $\frac{F_Y}{F_Z} = \frac{F_X}{F_Z} = \frac{4}{7}$. The magnitude of the force is $90 = \sqrt{F_X^2 + F_Y^2 + F_Z^2}$. Factor out F_Z^2 to obtain

$$90 = (\pm\sqrt{F_Z^2})\sqrt{\left(\frac{4}{7}\right)^2 + \left(\frac{4}{7}\right)^2 + 1},$$

from which $F_Z = \pm\frac{90}{1.2857} = \pm 70$ N, and $F_X = F_Y = \pm 40$ N; hence

$\boxed{\vec{F} = 40\vec{i} + 40\vec{j} + 70\vec{k} \ (N)}$, or $\boxed{\vec{F} = -40\vec{i} - 40\vec{j} - 70\vec{k} \ (N)}$

==================================<>==================================

Problem 4.53 The force $\vec{F} = 30\vec{i} + 20\vec{j} + 10\vec{k}\ (N)$. (a) Determine the moment of $\vec{F}$ about A. (b) What is the product of the magnitude of $\vec{F}$ and the perpendicular distance from point A to the line of action of $\vec{F}$?

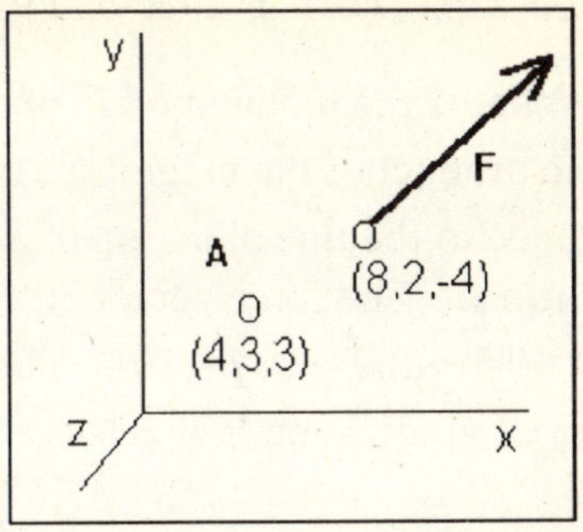

Solution: The position vector from point A to $\vec{F}$ is $\vec{r}_{AF} = (8-4)\vec{i} + (2-3)\vec{j} + (-4-3)\vec{k} = 4\vec{i} - 1\vec{j} - 7\vec{k}$.

(a)The cross product is

$$\vec{r}_{AF} \times \vec{F} = \begin{vmatrix} \vec{i} & \vec{j} & \vec{k} \\ 4 & -1 & -7 \\ 30 & 20 & 10 \end{vmatrix} = (-10+140)\vec{i} - (40+210)\vec{j} + (80+30)\vec{k}$$

$$\boxed{= 130\vec{i} - 250\vec{j} + 110\vec{k}\ (N\text{-}m)}$$

(b) The strategy is to resolve the position vector into components parallel with and normal to the force vector line of action. The magnitude of the normal component is the perpendicular distance. The parallel component of the radius vector is $\vec{r}_P = (\vec{r}_{AF} \cdot \vec{e})\vec{e}$, where $\vec{e}$ is a unit vector parallel to $\vec{F}$. The magnitude of the force vector is $|\vec{F}| = \sqrt{30^2 + 20^2 + 10^2} = 37.42\ N$. The unit vector parallel to $\vec{F}$ is

$\vec{e} = \frac{\vec{F}}{|\vec{F}|} = \frac{30}{37.42}\vec{i} + \frac{20}{37.42}\vec{j} + \frac{10}{37.42}\vec{k} = 0.8017\vec{i} + 0.5345\vec{j} + 0.2672\vec{k}$. The dot product is

$\vec{r}_{AF} \cdot \vec{e} = 0.8019$. The projection on the line of action is $\vec{r}_P = 0.6429\vec{i} + 0.4286\vec{j} + 0.2143\vec{k}$. The perpendicular vector is $\vec{r}_N = \vec{r} - \vec{r}_P = (4 - 0.6429)\vec{i} + (-1 - 0.4286)\vec{j} + (-7 - 0.2143)\vec{k}$,

$\vec{r}_N = 3.3571\vec{i} - 1.4286\vec{j} - 7.2143\vec{k}$. The magnitude is $|\vec{r}_N| = \sqrt{3.3571^2 + 1.4286^2 + 7.2143^2} = 8.08\ m$. The product of $|\vec{F}|$ and the perpendicular is $\boxed{M = (8.08)(37.42) = 302.5\ N\text{-}m}$ *Check:* The moment M is the magnitude of the moment vector found in Part (a), as expected. *check.*

==================================<>==================================

Problem 4.54 The direction cosines of the force $\vec{F}$ are $\cos\theta_X = 0.818$, $\cos\theta_Y = 0.182$, and $\cos\theta_Z = -0.545$. The support of the beam at O will fail if the magnitude of the moment of $\vec{F}$ about O exceeds 100 *kN-m*. Determine the magnitude of the largest force $\vec{F}$ that can be applied to the beam.

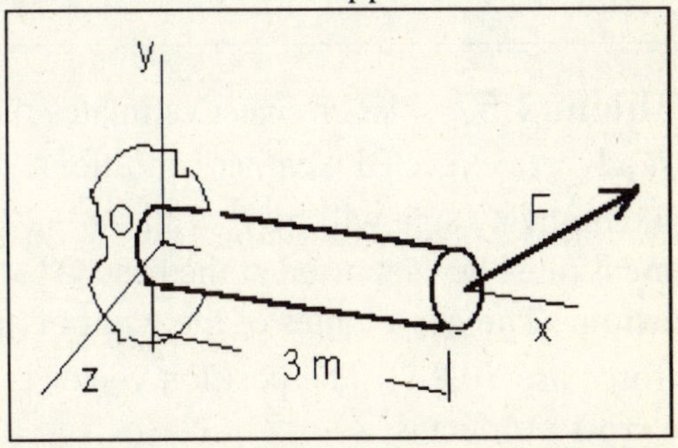

Solution: The strategy is to determine the perpendicular distance from O to the action line of $\vec{F}$, and to calculate the largest magnitude of $\vec{F}$ from $M_O = D|\vec{F}|$. The position vector from O to the point of application of $\vec{F}$ is $\vec{r}_{OF} = 3\vec{i}\ (m)$. Resolve the position vector into components parallel and normal to $\vec{F}$. The component parallel to $\vec{F}$ is $\vec{r}_P = (\vec{r}_{OF} \cdot \vec{e}_F)\vec{e}_F$, where the unit vector $\vec{e}_F$ parallel to $\vec{F}$ is

$\vec{e}_F = \vec{i}\cos\theta_X + \vec{j}\cos\theta_Y + \vec{k}\cos\theta_Z = 0.818\vec{i} + 0.182\vec{j} - 0.545\vec{k}$. The dot product is $\vec{r}_{OF} \cdot \vec{e}_F = 2.454$. The parallel component is $\vec{r}_P = 2.007\vec{i} + 0.4466\vec{j} - 1.3374\vec{k}$. The component normal to $\vec{F}$ is $\vec{r}_N = \vec{r}_{OF} - \vec{r}_P = (3-2)\vec{i} - 0.4466\vec{j} + 1.3374\vec{k}$. The magnitude of the normal component is the perpendicular distance: $D = \sqrt{1^2 + 0.4466^2 + 1.337^2} = 1.7283\ m$. The maximum moment allowed is

$M_O = 1.7283|\vec{F}| = 100\ kN\text{-}m$, from which $\boxed{|\vec{F}| = \frac{100\ kN\text{-}m}{1.7283\ m} = 57.86 \cong 58\ kN}$

==================================<>==================================

Problem 4.55

Problem 4.55 The force $\vec{F}$ exerted on a grip of an exercise machine points in the direction of a unit vector $\vec{e} = \frac{2}{3}\vec{i} - \frac{2}{3}\vec{j} - \frac{1}{3}\vec{k}$ and its magnitude is 30 lb.

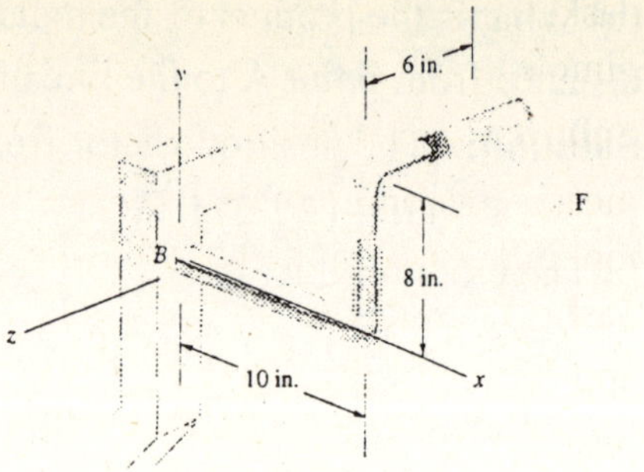

(a) What is the moment of $\vec{F}$ about the origin O? (b) What is the product of the magnitude of $\vec{F}$ and the perpendicular distance to the line of action of $\vec{F}$?

Solution: The force vector is $\vec{F} = 30\vec{e} = 20\vec{i} - 20\vec{j} - 10\vec{k}$. The position vector of the point of application is $\vec{r} = 10\vec{i} + 8\vec{j} - 6\vec{k}\ (in)$. The moment about O is

$$\vec{M}_O = \vec{r} \times \vec{F} = \begin{vmatrix} \vec{i} & \vec{j} & \vec{k} \\ 10 & 8 & -6 \\ 20 & -20 & -10 \end{vmatrix}$$

$$\boxed{= -200\vec{i} - 20\vec{j} - 360\vec{k}\ (in\text{-}lb)}$$

(b) The minimum distance from O to the line of action of $\vec{F}$ is $d = |\vec{r}|\sin\theta$ (See figure. See also Example 2.117 and Problem 2.115). From the definition of the cross product, $\sin\theta = \frac{|\vec{r} \times \vec{F}|}{|\vec{r}||\vec{F}|}$, from which $d = \frac{|\vec{r} \times \vec{F}|}{|\vec{F}|}$, from which $\boxed{|\vec{F}|d = |\vec{r} \times \vec{F}| = 412.31\ in\text{-}lb}$, which is the magnitude of the moment.

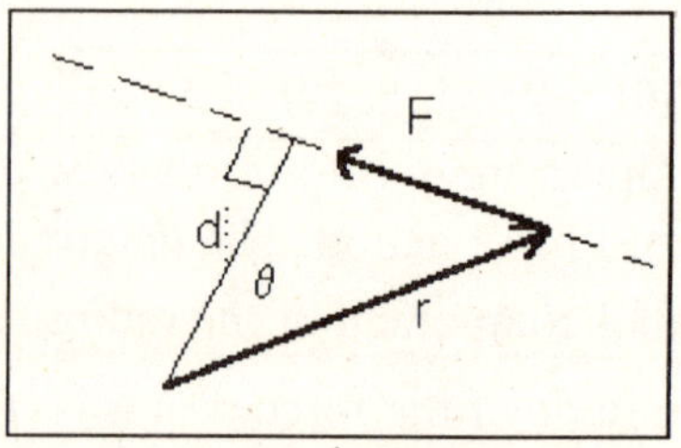

Problem 4.56

Problem 4.56 In Problem 4.55, the force $\vec{F}$ points in the direction of the unit vector $\vec{e} = \frac{2}{3}\vec{i} - \frac{2}{3}\vec{j} - \frac{1}{3}\vec{k}$. The support at O will safely support a moment of magnitude 5000 *in lb*. Based on this criterion, what is the largest safe magnitude of $\vec{F}$?

Solution: This can be solved using the results of Problem 4.55. The distance to the line of action of $\vec{F}$ is $d = \frac{|\vec{r} \times \vec{F}|}{|\vec{F}|} = \frac{412.31}{30} = 13.7437\ in$. The maximum force is $\boxed{|\vec{F}_{MAX}| = \frac{5000\ in\ lb}{13.7437\ in} = 363.80\ lb}$

Problem 4.57

Problem 4.57 An engineer estimates that under the most adverse expected weather conditions, the total force on the highway sign will be $\vec{F} = \pm 1.4\vec{i} - 2.0\vec{j}\ kN$. What moment must be supported at the base O?

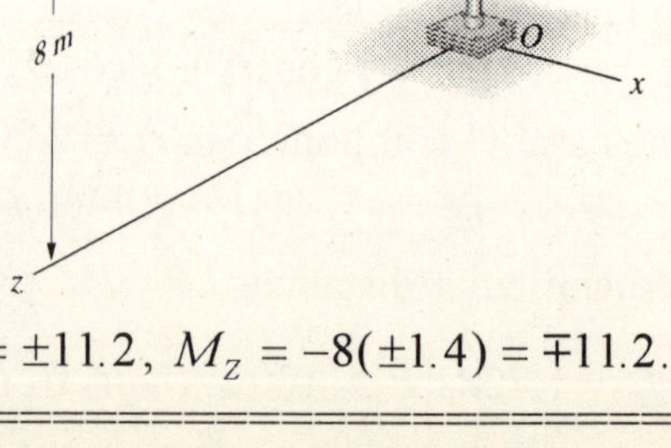

Solution: The coordinates of the point of application of the force are: (0,8,8). The position vector is $\vec{r}_{OF} = 8\vec{j} + 8\vec{k}$. The cross product is

$$\vec{r}_{OF} \times \vec{F} = \begin{vmatrix} \vec{i} & \vec{j} & \vec{k} \\ 0 & 8 & 8 \\ \pm 1.4 & -2 & 0 \end{vmatrix} = 16\vec{i} - (\mp 1.4)(8)\vec{j} + (\mp 1.4)(8)\vec{k}$$

$\boxed{\vec{M}_O = 16\vec{i} \pm 11.2\vec{j} \mp 11.2\vec{k}\ (N\text{-}m)}$ *Check:* Use perpendicular distances to forces: $M_X = 8(2) = 16$, $M_Y = 8(\pm 1.4) = \pm 11.2$, $M_Z = -8(\pm 1.4) = \mp 11.2$.

Problem 4.58 After executing a "slam dunk" shot, a basketball player hangs on the rim of the basket, exerting the forces shown. Determine the moment he exerts about the origin O in terms of the dimensions h and r and the angle α.

Solution: The force acts downward, hence $\vec{F} = -|\vec{F}|\vec{j}$. r is the radius of the basket, and α is the angle measured from the farthest edge of the rim, counterclockwise looking down on the basket. h is the offset of the center of the basket from the origin O. The coordinates of the point of application of the force are $(h+r\cos\alpha, 0, -r\sin\alpha)$. The position vector is $\vec{r}_{OF} = (h+r\cos\alpha)\vec{i} - (r\sin\alpha)\vec{k}$.

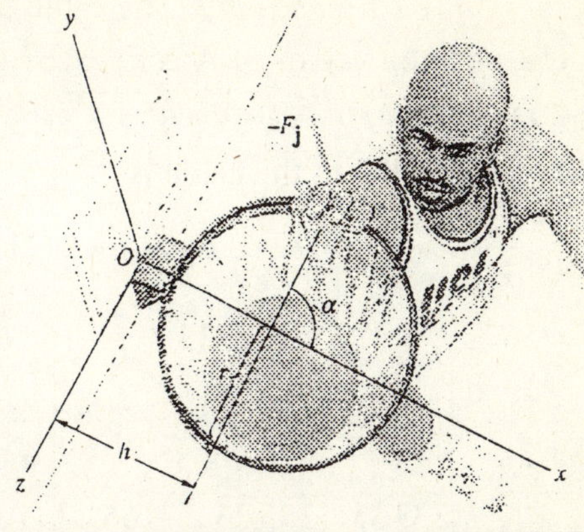

The cross product is $\vec{r}_{OF} \times \vec{F} = \begin{vmatrix} \vec{i} & \vec{j} & \vec{k} \\ h+r\cos\alpha & 0 & -r\sin\alpha \\ 0 & -|\vec{F}| & 0 \end{vmatrix}$

$$\boxed{\vec{M}_O = -|\vec{F}|(r\sin\alpha)\vec{i} - |\vec{F}|(h+r\cos\alpha)\vec{k}}$$

Problem 4.59 The tension in cable AC is 100 *lb.* Determine the moment about the origin O due to the force exerted at A by cable AC. Use Eq (4.2) letting $\vec{r}$ be the vector (a) from O to A, (b) from O to C

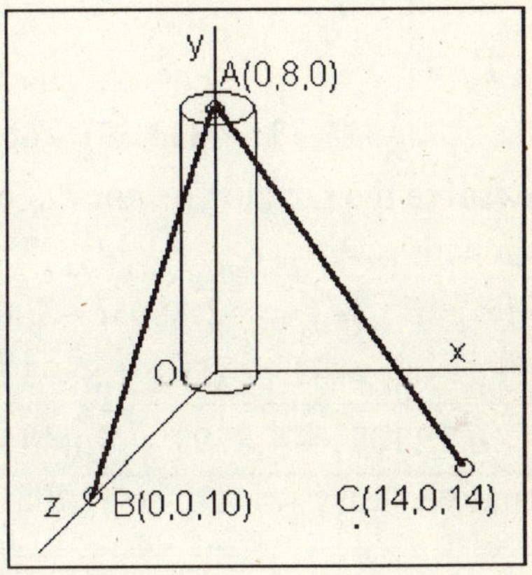

Solution: The position vectors of the point A and C are: $\vec{r}_{OA} = 8\vec{j}$, $\vec{r}_{OC} = 14\vec{i} + 14\vec{k}$. The vector parallel to cable AC is $\vec{r}_{AC} = \vec{r}_{OC} - \vec{r}_{OA} = 14\vec{i} - 8\vec{j} + 14\vec{k}$, with magnitude $|\vec{r}_{AC}| = 21.3542\ ft$. The unit vector parallel to AC is $\vec{e}_{AC} = \frac{\vec{r}_{AC}}{|\vec{r}_{AC}|} = 0.6556\vec{i} - 0.3746\vec{j} + 0.6556\vec{k}$ The tension in cable AC is $\vec{T}_{AC} = 100\vec{e}_{AC} = 65.56\vec{i} - 37.46\vec{j} + 65.56\vec{k}$. (a) The moment using the vector from O to A is

$$\vec{M}_o = \vec{r}_{OA} \times \vec{T}_{AC} = \begin{vmatrix} \vec{i} & \vec{j} & \vec{k} \\ 0 & 8 & 0 \\ 65.56 & -37.46 & 65.56 \end{vmatrix}$$

$$\boxed{= 524.4\vec{i} - 524.4\vec{k}\ (ft\text{-}lb)}$$

(b) The moment using the vector from O to C is

$$\vec{M}_O = \vec{r}_{OC} \times \vec{T}_{AC} = \begin{vmatrix} \vec{i} & \vec{j} & \vec{k} \\ 14 & 0 & 14 \\ 65.56 & -37.46 & 65.56 \end{vmatrix} \boxed{= 524.4\vec{i} - 524.4\vec{k}\ (ft\text{-}lb)}$$

Problem 4.60 The tension in cable OA is 500 N. Determine the moment about B due to the force exerted by the cable (a) on point O, (b) on point A. (c) Determine the perpendicular distance from B to the cable OA.

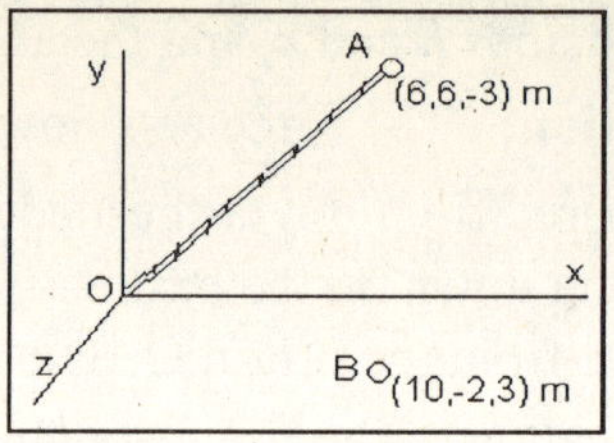

Solution: The radius vector from B to the origin O is $\vec{r}_{BO} = -10\vec{i} + 2\vec{j} - 3\vec{k}$ (m). The cable tension acts on point O toward A. The position vector of A is $\vec{r}_{OA} = 6\vec{i} + 6\vec{j} - 3\vec{k}$ (m). The distance from O to A is $|\vec{r}_{OA}| = \sqrt{6^2 + 6^2 + 3^2} = 9$ m. The unit vector acting toward A parallel to the cable is

$\vec{e}_{OA} = \dfrac{\vec{r}_{OA}}{|\vec{r}_{OA}|} = 0.6667\vec{i} + 0.6667\vec{j} - 0.3333\vec{k}$. The tension acting along the cable from O is

$\vec{T}_O = 500\vec{e}_{OA} = 333.33\vec{i} + 333.33\vec{j} - 166.67\vec{k}$.(a) The moment about B due to tension at O is

$$\vec{r}_{BO} \times \vec{T}_O = \begin{vmatrix} \vec{i} & \vec{j} & \vec{k} \\ -10 & 2 & -3 \\ 333.33 & 333.33 & -166.67 \end{vmatrix} = 666.66\vec{i} - 2666.69\vec{j} - 3999.96\vec{k} = 667\vec{i} - 2667\vec{j} - 4000\vec{k} \text{ (b)}$$

The tension at A acts toward O, and therefore is opposite in sign from the tension acting at O. $\vec{T}_A = -\vec{T}_O$. The position vector from B to A is $\vec{r}_{BA} = (6-10)\vec{i} + (6+2)\vec{j} + (-3-3)\vec{k} = -4\vec{i} + 8\vec{j} - 6\vec{k}$.

$$\vec{r}_{BA} \times \vec{T}_A = \begin{vmatrix} \vec{i} & \vec{j} & \vec{k} \\ -4 & 8 & -6 \\ -333.33 & -333.33 & 166.67 \end{vmatrix} = -666.6\vec{i} + 2666.6\vec{j} + 3999.9\vec{k} \boxed{= -667\vec{i} + 2667\vec{j} + 4000\vec{k}}$$

(c) Resolve the position vector $\vec{r}_{OB}$ into components parallel and normal to the cable. The parallel component is $\vec{r}_P = (\vec{r}_{OB} \cdot \vec{e}_{OA})\vec{e}_{OA}$. The dot product: $\vec{r}_{OB} \cdot \vec{e}_{OA} = -4.3337$ m. The parallel component is $\vec{r}_p = (\vec{r}_{OB} \cdot \vec{e}_{OA})\vec{e}_{OA} = -2.8893\vec{i} - 2.8893\vec{j} + 1.4444\vec{k}$. The perpendicular component is $\vec{r}_N = \vec{r}_{OB} - \vec{r}_P = -7.1107\vec{i} + 4.8893\vec{j} - 4.4444\vec{k}$. The perpendicular distance is $\boxed{|\vec{r}_N| = \sqrt{7.1107^2 + 4.8893^2 + 4.4444^2} = 9.71 \ m}$ (*Note:* See the solution to Problem 4.55(b) for an alternative, "easy way" to solve this Part(c).)

Problem 4.61 Consider the system shown in Problem 4.59. The tension in cable AB is 100 lb and the tension in cable AC is 140 lb. Determine the magnitude of the sum of the moments about O due to the forces exerted at A by the two cables.

Solution: From the solution to Problem 4.60 the unit vector parallel to AC is $\vec{e}_{AC} = 0.6556\vec{i} - 0.3746\vec{j} + 0.6556\vec{k}$. The tension in cable AC is $\vec{T}_{AC} = 140\vec{e}_{AC} = 91.784\vec{i} - 52.444\vec{j} + 91.784\vec{k}$. The position vector parallel to cable AB is $\vec{r}_{AB} = -8\vec{j} + 10\vec{k}$. The magnitude is $|\vec{r}_{AB}| = 12.8062$ ft. The unit vector is $\vec{e}_{AB} = -0.6247\vec{j} + 0.7809\vec{k}$. The tension in cable AB is $\vec{T}_{AB} = 100\vec{e}_{AB} = -62.47\vec{j} + 78.09\vec{k}$ (lb). The sum of the moments is

$$\Sigma\vec{M}_o = (\vec{r}_{OA} \times \vec{T}_{AB}) + (\vec{r}_{OA} \times \vec{T}_{AC}) = (\vec{r}_{OA} \times (\vec{T}_{AB} + \vec{T}_{AC})) = \begin{vmatrix} \vec{i} & \vec{j} & \vec{k} \\ 0 & 8 & 0 \\ 91.78 & -114.91 & 169.87 \end{vmatrix} = 1359\vec{i} - 734.2\vec{k}$$

The magnitude of the sum of moments: $\boxed{|\vec{M}_O| = 1544.65 \ ft\ lb}$

Problem 4.62 Consider the system shown in Problem 4.59. Cables AB and AC have the same tension $\vec{T}$. The post will fail at O if the magnitude of the moment about O due to the forces acting on the post by the cables exceeds 5000 *ft lb*. Based on this criterion, what is the maximum allowable value of $\vec{T}$?

Solution: The strategy is to determine the unit vectors parallel to the cables, determine the sum of the force components, and take the cross product of the position vector from O and this sum. The position vector from O to A is $\vec{r}_{OA} = 8\vec{j}$. The vector from A to C is $\vec{r}_{AC} = (14-0)\vec{i} + (0-8)\vec{j} + (14-0)\vec{k} = +14\vec{i} - 8\vec{j} + 14\vec{k}$. The magnitude is $|\vec{r}_{AC}| = \sqrt{14^2 + 8^2 + 14^2} = 21.35\ ft$. The unit vector parallel to the cable is $\vec{e}_{AC} = 0.6556\vec{i} - 0.3746\vec{j} + 0.6556\vec{k}$. The tension is $\vec{T}_{AC} = |\vec{T}|\vec{e}_{AC}$. The vector from B to A is $\vec{r}_{AB} = 0\vec{i} + (0-8)\vec{j} + (10-0)\vec{k} = 0\vec{i} - 8\vec{j} + 10\vec{k}$. The magnitude is $|\vec{r}_{AB}| = \sqrt{8^2 + 10^2} = 12.81\ ft$. The unit vector is $\vec{e}_{AB} = 0\vec{i} - 0.6247\vec{j} + 0.7809\vec{k}$. The tension is $\sum \vec{T}_A = |\vec{T}|\vec{e}_{AB} + |\vec{T}|\vec{e}_{AC} = |\vec{T}|(\vec{e}_{AB} + \vec{e}_{AC})$.
$\vec{T}_A = |\vec{T}|(0.6556i - 0.9993\vec{j} + 1.4365\vec{k})$. The moment about O is $\vec{M}_O = \vec{r}_{OA} \times \vec{T}_A$

$$\vec{r}_{OA} \times \vec{T}_A = |\vec{T}| \begin{vmatrix} \vec{i} & \vec{j} & \vec{k} \\ 0 & 8 & 0 \\ 0.6556 & -0.9993 & 1.4365 \end{vmatrix} = |\vec{T}|\left(8(1.4365)\vec{i} - 0\vec{j} - 8(0.6556)\vec{k}\right) = |\vec{T}|(11.49\vec{i} - 5.24\vec{k})$$

The magnitude is $|\vec{M}_O| = |\vec{T}|(12.63)\ ft\ lb$. At the limit the tension is $\boxed{|\vec{T}| = \dfrac{5000\ ft\ lb}{12.63\ ft} = 395.9\ lb}$

Problem 4.63 The tower is 70 *m* tall. The tensions in cables AB, AC, and AD are 4 *kN*, 2 *kN*, and 2 *kN* respectively. Determine the sum of the moments about the origin O due to the forces exerted by the cables at A.

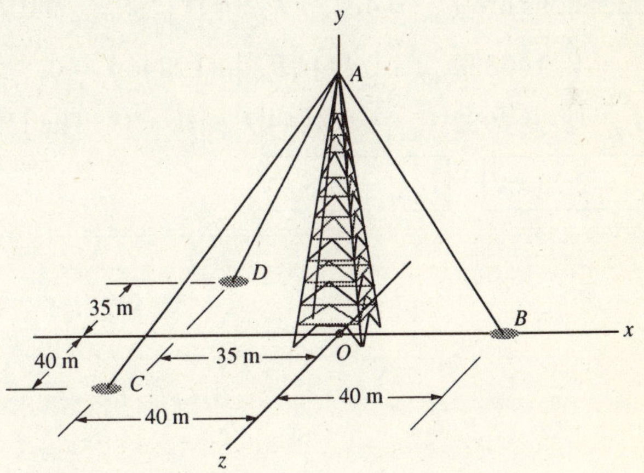

Solution: The coordinates of the points are A(0, 70, 0), B(40, 0, 0), C(-40, 0, 40) D(-35, 0, -35). The position vectors corresponding to the cables are:

$\vec{r}_{AD} = (-35-0)\vec{i} + (0-70)\vec{j} + (-35-0)\vec{k}$

$\vec{r}_{AD} = -35\vec{i} - 70\vec{k} - 35\vec{k}$

$\vec{r}_{AC} = (-40-0)\vec{i} + (0-70)\vec{j} + (40-0)\vec{k}$

$\vec{r}_{AC} = -40\vec{i} - 70\vec{j} + 40\vec{k}$

$\vec{r}_{AB} = (40-0)\vec{i} + (0-70)\vec{j} + (0-0)\vec{k}$

$\vec{r}_{AB} = 40\vec{i} - 70\vec{j} + 0\vec{k}$ The unit vectors corresponding to these position vectors are:

$$\vec{e}_{AD} = \frac{\vec{r}_{AD}}{|\vec{r}_{AD}|} = \frac{-35}{85.73}\vec{i} - \frac{70}{85.73}\vec{j} - \frac{35}{85.73} = -0.4082\vec{i} - 0.8165\vec{j} - 0.4082\vec{k}$$

$$\vec{e}_{AC} = \frac{\vec{r}_{AC}}{|\vec{r}_{AC}|} = -\frac{40}{90}\vec{i} - \frac{70}{90}\vec{j} + \frac{40}{90}\vec{k} = -0.4444\vec{i} - 0.7778\vec{j} + 0.4444\vec{k}$$

$$\vec{e}_{AB} = \frac{\vec{r}_{AB}}{|\vec{r}_{AB}|} = \frac{40}{80.6}\vec{i} - \frac{70}{80.6}\vec{j} + 0\vec{k} = 0.4962\vec{i} - 0.8682\vec{j} + 0\vec{k}$$

Problem 4.63 continued on next page....

================================<>================================

continuation of solution to Problem 4.63...

The forces at point A are

$\vec{T}_{AB} = 4\vec{e}_{AB} = 1.9846\vec{i} - 3.4729\vec{j} + 0\vec{k}$

$\vec{T}_{AC} = 2\vec{e}_{AB} = -0.8889\vec{i} - 1.5556\vec{j} + 0.8889\vec{k}$

$\vec{T}_{AD} = 2\vec{e}_{AD} = -0.8165\vec{i} - 1.6330\vec{j} - 0.8165\vec{k}$. The sum of the forces acting at A are

$\vec{T}_A = 0.2792\vec{i} - 6.6615\vec{j} + 0.07239\vec{k}$ $(kN\text{-}m)$

The position vector of A is $\vec{r}_{OA} = 70\vec{j}$. The moment about O is $\vec{M} = r_{OA} \times \vec{T}_A$

$$\vec{M} = \begin{vmatrix} \vec{i} & \vec{j} & \vec{k} \\ 0 & 70 & 0 \\ 0.2792 & -6.6615 & 0.07239 \end{vmatrix} = (70)(0.07239)\vec{i} - \vec{j}0 - \vec{k}(70)(0.2792) \boxed{= 5.067\vec{i} - 19.54\vec{k}}$$

================================<>================================

Problem 4.64 Consider the 70 *m* tower in Problem 4.63. Suppose that the tension in cable AB is 4 *kN* and you want to adjust the tensions in cables AC and AD so that the sum of the moments about the origin O due to the forces exerted by the cables at point A is zero. Determine the tensions.

Solution: From Varignon's theorem, the moment is zero only if the resultant of the forces normal to the vector $\vec{r}_{OA}$ is zero. From Problem 4.63 the unit vectors are:

$$\vec{e}_{AD} = \frac{\vec{r}_{AD}}{|\vec{r}_{AD}|} = \frac{-35}{85.73}\vec{i} - \frac{70}{85.73}\vec{j} - \frac{35}{85.73} = -0.4082\vec{i} - 0.8165\vec{j} - 0.4082\vec{k}$$

$$\vec{e}_{AC} = \frac{\vec{r}_{AC}}{|\vec{r}_{AC}|} = -\frac{40}{90}\vec{i} - \frac{70}{90}\vec{j} + \frac{40}{90}\vec{k} = -0.4444\vec{i} - 0.7778\vec{j} + 0.4444\vec{k}$$

$$\vec{e}_{AB} = \frac{\vec{r}_{AB}}{|\vec{r}_{AB}|} = \frac{40}{80.6}\vec{i} - \frac{70}{80.6}\vec{j} + 0\vec{k} = 0.4963\vec{i} - 0.8685\vec{j} + 0\vec{k}$$

The tensions are $\vec{T}_{AB} = 4\vec{e}_{AB}$, $\vec{T}_{AC} = |\vec{T}_{AC}|\vec{e}_{AC}$, and $\vec{T}_{AD} = |\vec{T}_{AD}|\vec{e}_{AD}$. The components normal to $\vec{r}_{OA}$ are

$\Sigma\vec{F}_X = (-0.4082|\vec{T}_{AD}| - 0.4444|\vec{T}_{AC}| + 1.9846)\vec{i} = 0$

$\sum\vec{F}_Z = (-0.4082|\vec{T}_{AD}| + 0.4444|\vec{T}_{AC}|)\vec{k} = 0$. The HP-28S calculator was used to solve these equations:

$\boxed{|\vec{T}_{AC}| = 2.23\ kN}$, $\boxed{|\vec{T}_{AD}| = 2.43\ kN}$

================================<>================================

Problem 4.65 The tension in cable AB is 250 *N*. The tension in cable AC is 100 *N*. Determine the sum of the moments about D due to the forces exerted on the wall by the cables.

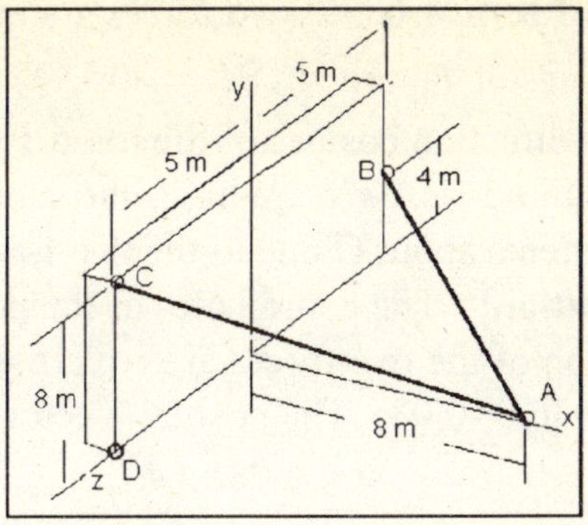

Solution: The coordinates of the points A, B, C are A(8,0,0), B(0,4,-5), C(0,8,5), D(0,0,5). The point A is the intersection of the lines of action of the forces. The position vectors DA is $\vec{r}_{DA} = 8\vec{i} + 0\vec{j} - 5\vec{k}$. The position vectors AB and AC are $\vec{r}_{AB} = -8\vec{i} + 4\vec{j} - 5\vec{k}$, $|\vec{r}_{AB}| = \sqrt{8^2 + 4^2 + 5^2} = 10.247\ m$. $\vec{r}_{AC} = -8\vec{i} + 8\vec{j} + 5\vec{k}$, $|\vec{r}_{AC}| = \sqrt{8^2 + 8^2 + 5^2} = 12.369\ m$. The unit vectors parallel to the cables are: $\vec{e}_{AB} = -0.7807\vec{i} + 0.3904\vec{j} - 0.4879\vec{k}$, $\vec{e}_{AC} = -0.6468\vec{i} + 0.6468\vec{j} + 0.4042\vec{k}$. The tensions are $\vec{T}_{AB} = 150\vec{e}_{AB} = -117.11\vec{i} + 58.56\vec{j} - 73.19\vec{k}$, $T_{AC} = 100\vec{e}_{AC} = -64.68\vec{i} + 64.68\vec{j} + 40.42\vec{k}$ The sum of the forces *exerted by the wall on A* is $\vec{T}_A = -181.79\vec{i} + 123.24\vec{j} - 32.77\vec{k}$. The force exerted on the wall by the cables is $-\vec{T}_A$ The moment about D is $\vec{M}_D = -\vec{r}_{DA} \times \vec{T}_A$,

$$\vec{M}_D = \begin{vmatrix} \vec{i} & \vec{j} & \vec{k} \\ 8 & 0 & -5 \\ 181.79 & -123.24 & +32.77 \end{vmatrix} = (-123.24)(5)\vec{i} - ((8)(+32.77) - (-5)(181.79))\vec{j} + (8)(-123.24)\vec{k}$$

$\boxed{\vec{M}_D = -616.2\vec{i} - 1171.1\vec{j} - 985.9\vec{k}\ (N\text{-}m)}$ (*Note*: An altenate method of solution is to express the moment in terms of the sum: $\vec{M}_D = (\vec{r}_{DC} \times \vec{T}_C) + (\vec{r}_{DB} \times \vec{T}_B)$.)

Problem 4.66 The force $|\vec{F}| = 800\ lb$. The sum of the moments about O due to the force $\vec{F}$ and the forces exerted at A by the cables AB and AC is zero. What are the tensions in the cables?

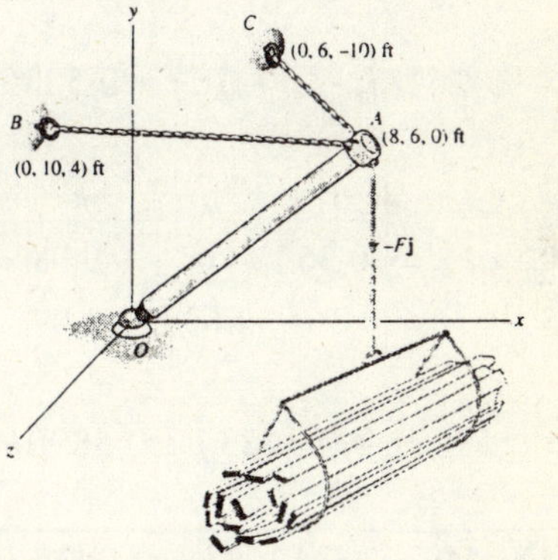

Solution: The coordinates of the points O,A, B, C are O(0,0,0), A(8,6,0), B(0,10,4), C(0,6,-10). The position vectors are: $\vec{r}_{OA} = 8\vec{i} + 6\vec{j} + 0\vec{k}$. $|\vec{r}_{OA}| = 10\ ft$

$\vec{r}_{AB} = (0-8)\vec{i} + (10-6)\vec{j} + (4-0)\vec{k} = -8\vec{i} + 4\vec{j} + 4\vec{k}$

$|\vec{r}_{AB}| = 9.798\ ft$

$\vec{r}_{AC} = (0-8)\vec{i} + (6-6)\vec{j} + (-10-0)\vec{k} = -8\vec{i} + 0\vec{j} - 10\vec{k}$.

$|\vec{r}_{AC}| = 12.806\ ft$.

The line of action of the forces intersect at point A. By Varignon's theorem, the forces at point A are in equilibrium if the moments vanish about O. The unit vectors are $\vec{e}_{OA} = 0.8\vec{i} + 0.6\vec{j} + 0\vec{k}$, $\vec{e}_{AB} = -0.8165\vec{i} + 0.4082\vec{j} + 0.4082\vec{k}$, $\vec{e}_{AC} = -0.6247\vec{i} + 0\vec{j} - 0.7809\vec{k}$. The forces are: $\vec{F}_{OA} = |\vec{F}_{OA}|\vec{e}_{OA}$, $\vec{F}_{AB} = |\vec{F}_{AB}|\vec{e}_{AB}$, $\vec{F}_{AC} = |\vec{F}_{AC}|\vec{e}_{AC}$. The equilibrium conditions are $\sum\vec{F} = \vec{F} + \vec{F}_{OA} + \vec{F}_{AB} + \vec{F}_{AC} = 0$. Combining and collecting terms

$\Sigma\vec{F}_X = (0.8|\vec{F}_{OA}| - 0.8165|\vec{F}_{AB}| - 0.6247|\vec{F}_{AC}|)\vec{i} = 0$

$\Sigma\vec{F}_Y = (0.6|\vec{F}_{OA}| + 0.4082|\vec{F}_{AB}| - 800)\vec{j} = 0$

$\sum\vec{F}_Z = (0.4082|\vec{F}_{AB}| - 0.7809|\vec{F}_{AC}|)\vec{k} = 0$ These equations were solved using the HP-28S hand held calculator. The result: $|\vec{F}_{OA}| = 903.25\ lb$. $\boxed{|\vec{F}_{AB}| = 632.17\ lb}$, $\boxed{|\vec{F}_{AC}| = 330.45\ lb}$

Problem 4.67 Two forces are exerted on the crankshaft by the connecting rods. The direction cosines of $\vec{F}_A$ are $\cos\theta_X = -0.182$, $\cos\theta_Y = 0.818$, and $\cos\theta_Z = 0.545$, and its magnitude is 4 *kN*. The direction cosines of the force $\vec{F}_B$ are $\cos\theta_X = 0.182$, $\cos\theta_Y = 0.818$, and $\cos\theta_Z = -0.545$, and its magnitude is 2 *kN*. . What is the sum of the moments about O due to the two forces?

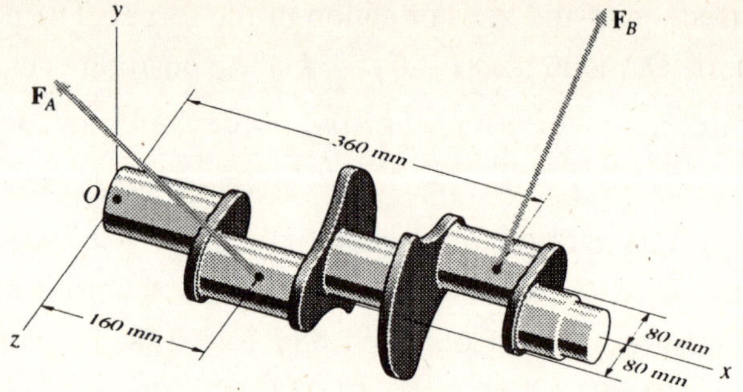

Solution: The coordinates of the points of action of the two forces are A(0.16,0,0.08), B(0.36,0,-0.08). The position vectors are $\vec{r}_{OA} = 0.16\vec{i} + 0\vec{j} + 0.08\vec{k}\ (m)$, $\vec{r}_{OB} = 0.36\vec{i} + 0\vec{j} - 0.08\vec{k}\ (m)$. The unit vectors parallel to the forces are given by the direction cosines:

$\vec{e}_A = -0.182\vec{i} + 0.818\vec{j} + 0.545\vec{k}$,

$\vec{e}_B = 0.182\vec{i} + 0.818\vec{j} - 0.545\vec{k}$

The forces are

$\vec{F}_A = 4\vec{e}_A = -0.728\vec{i} + 3.272\vec{j} + 2.18\vec{k}\ (kN)$

$\vec{F}_B = 2\vec{e}_B = 0.364\vec{i} + 1.636\vec{j} - 1.09\vec{k}\ (kN)$

The sum of the moments is $\sum \vec{M}_O = \vec{r}_{OA} \times \vec{F}_A + \vec{r}_{OB} \times \vec{F}_B$

$$\vec{r}_{OA} \times \vec{F}_A = \begin{vmatrix} \vec{i} & \vec{j} & \vec{k} \\ 0.16 & 0 & 0.08 \\ -0.728 & 3.272 & 2.18 \end{vmatrix} = -(0.08)(3.272)\vec{i} - (0.16(2.18) + 0.08(0.728))\vec{j} + 0.16(3.272)\vec{k}$$

$= -0.2618\vec{i} - 0.4070\vec{j} + 0.5235\vec{k}$.

$$\vec{r}_{OB} \times \vec{F}_B = \begin{vmatrix} \vec{i} & \vec{j} & \vec{k} \\ 0.36 & 0 & -0.08 \\ 0.364 & 1.636 & -1.09 \end{vmatrix} = 0.08(1.636)\vec{i} - (0.36(-1.09) + 0.08(0.364))\vec{j} + 0.36(1.636)\vec{k}$$

$= 0.1309\vec{i} + 0.3633\vec{j} + 0.5890\vec{k}$ Adding the two moments

$$\boxed{\sum \vec{M}_O = -0.1309\vec{i} + 0.0437\vec{j} + 1.113\vec{k}\ (kN\text{-}m)}$$

Problem 4.68 The weight of the arms OA and AB of the robotic manipulator act at their midpoints. The direction cosines of the centerline of arm OA are $\cos\theta_X = 0.500$, $\cos\theta_Y = 0.866$, and $\cos\theta_Z = 0$, and the direction cosines of the centerline of the arm AB are $\cos\theta_X = 0.707$, $\cos\theta_Y = 0.619$, and $\cos\theta_Z = -0.342$. What is the sum of the moments about O due to the two forces?

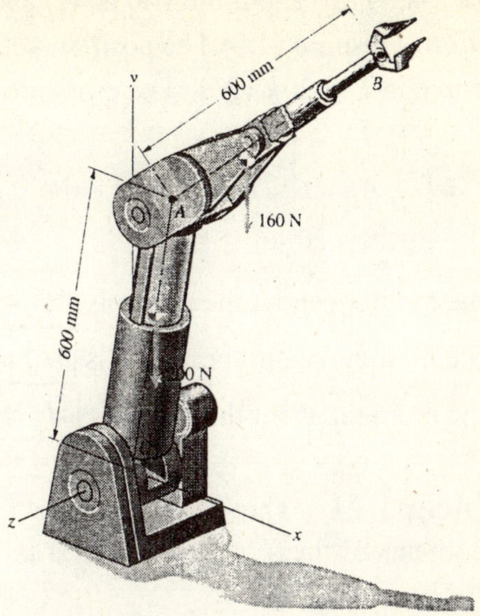

Solution: By definition, the direction cosines are the scalar components of the unit vectors. Thus the unit vectors are $\vec{e}_1 = 0.5\vec{i} + 0.866\vec{j}$, and $\vec{e}_2 = 0.707\vec{i} + 0.619\vec{j} - 0.342\vec{k}$ The position vectors of the midpoints of the arms are

$\vec{r}_1 = 0.3\vec{e}_1 = 0.3(0.5\vec{i} + 0.866\vec{j}) = 0.15\vec{i} + 0.2598\vec{j}$

$\vec{r}_2 = 0.6\vec{e}_1 + 0.3\vec{e}_2 = 0.512\vec{i} + 0.7053\vec{j} - 0.0943\vec{k}$.

The sum of moments is $\vec{M} = \vec{r}_1 \times \vec{W}_1 + \vec{r}_2 \times \vec{W}_2$

$$= \begin{vmatrix} \vec{i} & \vec{j} & \vec{k} \\ 0.15 & 0.2598 & 0 \\ 0 & -200 & 0 \end{vmatrix} + \begin{vmatrix} \vec{i} & \vec{j} & \vec{k} \\ 0.512 & 0.7053 & -0.1026 \\ 0 & -160 & 0 \end{vmatrix} = \boxed{-16.42\vec{i} - 111.92\vec{k}\ (N\text{-}m)}$$

Problem 4.69 Determine the moment of the 40 *lb* force about the z-axis (a) without using Eqs. (4.4)-(4.6), (b) by using Eqs (4.5) and (4.6). (See Example 4.7).

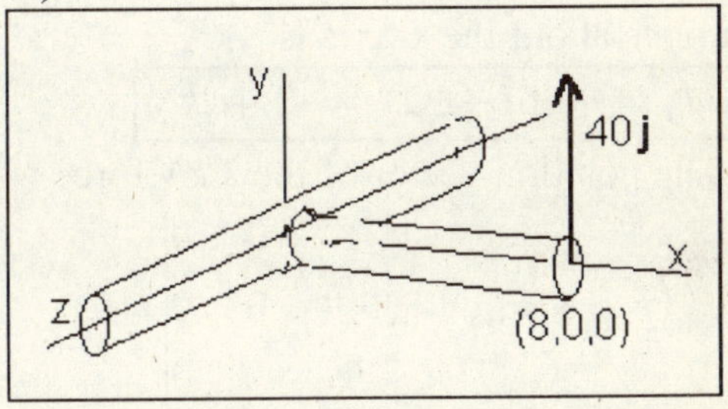

Solution: (a) The moment about the z-axis is the product of the perpendicular distance to the line of action of $\vec{F}$ and the magnitude of $\vec{F}$. The action is negative if the moment is clockwise and positive if the moment is counterclockwise. Thus the moment is $\boxed{M_Z = 8(40) = 320\ ft\ lb}$. It is positive because it points along the positive z-axis by the right hand rule. (b) The position vector of the point of application of the force is $\vec{r} = 8\vec{i} + 0\vec{j} + 0\vec{k}$. The force is $\vec{F} = 40\vec{j}$. The moment is $\vec{M}_Z = \vec{r} \times \vec{F}$. The unit vector along the z-axis is $\vec{e}_Z = 1\vec{k}$. By Eq (4.5),

$\vec{M}_Z = [\vec{e}_Z \cdot (\vec{r} \times \vec{F})]\vec{e}_Z$. From Eq (4.6), $\vec{e}_Z \cdot (\vec{r} \times \vec{F}) = \begin{vmatrix} 0 & 0 & 1 \\ 8 & 0 & 0 \\ 0 & 40 & 0 \end{vmatrix} = 320$. From which, by Eq (4.5)

$\boxed{\vec{M}_Z = 320\vec{e}_Z = 320\vec{k}\ (ft\ lb)}$

Problem 4.70 Determine the moment of the 20 N force about (a) the x-axis, (b) about the y-axis, (c) about the z-axis.

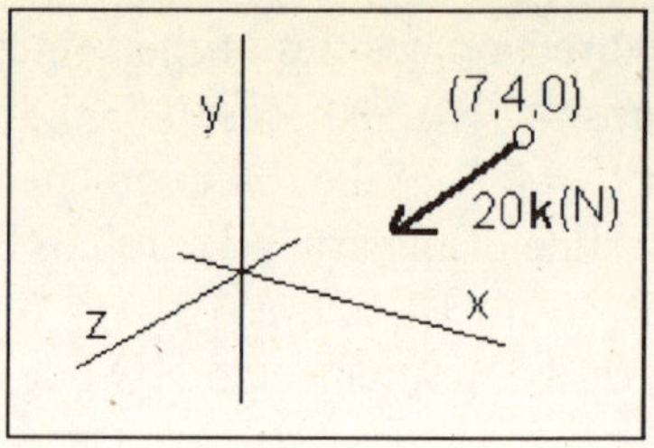

Solution: Use Eq (4.5) The position vector is $\vec{r} = 7\vec{i} + 4\vec{j} + 0\vec{k}$. The force is $\vec{F} = 20\vec{k}\ (N)$. The cross product is

$$\vec{r} \times \vec{F} = \begin{vmatrix} \vec{i} & \vec{j} & \vec{k} \\ 7 & 4 & 0 \\ 0 & 0 & 20 \end{vmatrix} = 80\vec{i} - 140\vec{j} + 0\vec{k}.$$

(a) The moment about the x-axis is $\boxed{\vec{M}_X = [\vec{e}_X \cdot (\vec{r} \times \vec{F})]\vec{e}_X = 80\vec{i}\ (N\text{-}m)}$.

(b) The moment about the y-axis is $\boxed{\vec{M}_Y = [\vec{e}_Y \cdot (\vec{r} \times \vec{F})]\vec{e}_Y = -140\vec{j}\ (N\text{-}m)}$

(c) The moment about the z-axis is $\boxed{\vec{M}_Z = [\vec{e}_Z \cdot (\vec{r} \times \vec{F})]\vec{e}_Z = 0\ (N\text{-}m)}$

Problem 4.71 Three forces parallel to the y-axis act on the rectangular plate. Determine the sum of the moments of the forces about the x-axis (a) without using Eqs (4.4)-(4.6), (b) by using Eqs (4.5) and (4.6).

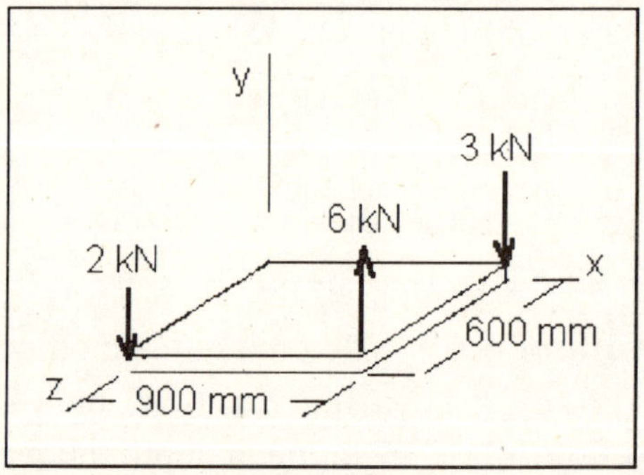

Solution: (a) The perpendicular distance from the x-axis to the 6 kN force is 0.600 m. The action is negative. The moment is $M_{6kN} = -0.6(6)\vec{i} = -3.6\vec{i}\ (kN\text{-}m)$. The perpendicular distance to the 2 kN force is 0.6 m. The action is positive. The moment is $M_{2kN} = 0.6(2)\vec{i} = 1.2\vec{i}\ (kN\text{-}m)$. The 3 kN force action line is on the x-axis, hence the moment is zero. The sum of the moments about the x-axis is

$\boxed{\sum M_X = -3.6\vec{i} + 1.2\vec{i} = -2.4\vec{i}\ kN}$

(b) The position vector of the 3 kN force is $\vec{r}_{3kN} = 0.9\vec{i}$. The magnitude of the moment about the x-axis is $\vec{e}_X \cdot (\vec{r}_{3kN} \times \vec{F}_{3kN}) = \begin{vmatrix} 1 & 0 & 0 \\ 0.9 & 0 & 0 \\ 0 & -3 & 0 \end{vmatrix} = 0.$

The position vector to the line of action of the 2 kN force is $\vec{r}_{2kN} = 0\vec{i} + 0\vec{j} + 0.6\vec{k}$. The magnitude of the moment is $\vec{e}_X \cdot (\vec{r}_{2kN} \times \vec{F}_{2kN}) = \begin{vmatrix} 1 & 0 & 0 \\ 0 & 0 & 0.6 \\ 0 & -2 & 0 \end{vmatrix} = +1.2\ kN\text{-}m$.

The moment about the x-axis is $\vec{M}_X = [\vec{e}_X \cdot (\vec{r}_{2kN} \times \vec{F}_{2kN})]\vec{e}_X = 1.2\vec{i}\ (kN\text{-}m)$

The position vector to the 6 kN force is $\vec{r}_{6kN} = 0.9\vec{i} + 0.6\vec{k}$. The magnitude of the moment is

$$\vec{e}_X \cdot (\vec{r}_{6kN} \times \vec{F}_{6kN}) = \begin{vmatrix} 1 & 0 & 0 \\ 0.9 & 0 & 0.6 \\ 0 & 6 & 0 \end{vmatrix} = -3.6\ kN\text{-}m.$$

The moment about the x-axis is $\vec{M}_{6kN} = [\vec{e}_X \cdot (\vec{r}_{6kN} \times \vec{F}_{6kN})]\vec{e}_X = -3.6\vec{i}\ (kN\text{-}m)$. The sum of the moments is $\boxed{\sum \vec{M}_X = (1.2 - 3.6)\vec{i} = -2.4\vec{i}\ (kN\text{-}m)}$

Problem 4.72 Consider the rectangular plate shown in Problem 4.71. The three forces are parallel to the y-axis. Determine the sum of the moments (a) about the y-axis, (b) about the z-axis.

Solution: (a) The magnitude of the moments about the y-axis is $M = \vec{e}_Y \cdot (\vec{r} \times \vec{F})$. The position vectors of the three forces are given in the solution to Problem 4.71. The magnitude for each force is:

$$\vec{e}_Y \cdot (\vec{r} \times \vec{F}) = \begin{vmatrix} 0 & 1 & 0 \\ 0.9 & 0 & 0 \\ 0 & -3 & 0 \end{vmatrix} = 0,\quad \vec{e}_Y \cdot (\vec{r} \times \vec{F}) = \begin{vmatrix} 0 & 1 & 0 \\ 0.9 & 0 & 0.6 \\ 0 & 6 & 0 \end{vmatrix} = 0,\quad \vec{e}_Y \cdot (\vec{r} \times \vec{F}) = \begin{vmatrix} 0 & 1 & 0 \\ 0 & 0 & 0.6 \\ 0 & -2 & 0 \end{vmatrix} = 0$$

Thus the moment about the y-axis is zero, since the magnitude of each moment is zero.

(b) The magnitude of each moment about the z-axis is

$$\vec{e}_Z \cdot (\vec{r} \times \vec{F}) = \begin{vmatrix} 0 & 0 & 1 \\ 0.9 & 0 & 0 \\ 0 & -3 & 0 \end{vmatrix} = -2.7,\quad \vec{e}_Z \cdot (\vec{r} \times \vec{F}) = \begin{vmatrix} 0 & 0 & 1 \\ 0.9 & 0 & 0.6 \\ 0 & 6 & 0 \end{vmatrix} = 5.4,\quad \vec{e}_Z \cdot (\vec{r} \times \vec{F}) = \begin{vmatrix} 0 & 0 & 1 \\ 0 & 0 & 0.6 \\ 0 & -2 & 0 \end{vmatrix} = 0$$

Thus the moment about the z-axis is $\boxed{\sum \vec{M}_Z = -2.7\vec{e}_Z + 5.4\vec{e}_Z = 2.7\vec{k}\ (kN\text{-}m)}$

Problem 4.73 A person exerts a force $\vec{F} = -2\vec{i} - 4\vec{j} - 12\vec{k}\ lb$ on the gate at a C. What moment do they exert on the hinge line AB of the gate?

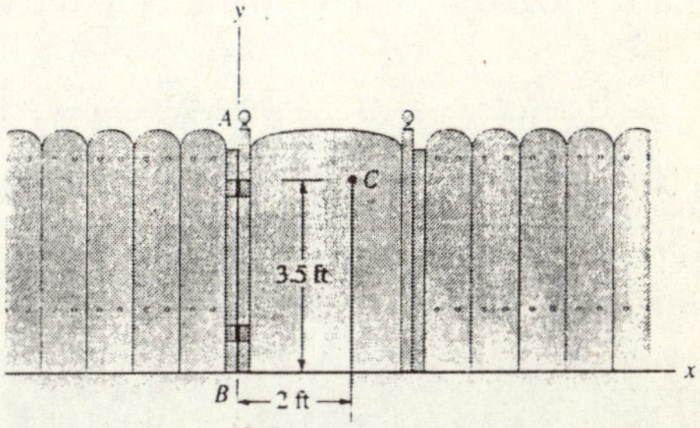

Solution: The position vector of C is $\vec{r}_C = 2\vec{i} + 3.5\vec{j}$. The magnitude of the moment about the hinge is

$$\vec{e}_Y \cdot (\vec{r}_C \times \vec{F}) = \begin{vmatrix} 0 & 1 & 0 \\ 2 & 3.5 & 0 \\ -2 & -4 & -12 \end{vmatrix} = 24\ ft\ lb$$

The moment is $\boxed{\vec{M}_H = 24\vec{e}_Y = 24\vec{j}\ \ (ft\ lb)}$

Problem 4.74 Three forces parallel to the y-axis act on the rectangular plate. The vector sum of the forces is zero and the magnitude of the sum of the moments of the forces about O is 600 *ft lb.* What is the sum of the moments of the three forces about the x-axis?

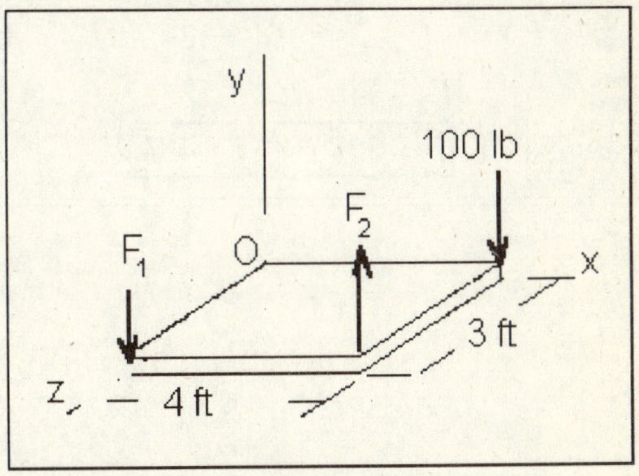

Solution: The position vectors are $\vec{r}_1 = 3\vec{k}$, $\vec{r}_2 = 4\vec{i} + 3\vec{k}$, and $\vec{r} = 4\vec{i}$. The moments about O are

$$\vec{M}_1 = \vec{r}_1 \times \vec{F}_1 = \begin{vmatrix} \vec{i} & \vec{j} & \vec{k} \\ 0 & 0 & 3 \\ 0 & F_1 & 0 \end{vmatrix} = -3F_1\vec{i}$$

$$\vec{M}_2 = \vec{r}_2 \times \vec{F}_2 = \begin{vmatrix} \vec{i} & \vec{j} & \vec{k} \\ 4 & 0 & 3 \\ 0 & F_2 & 0 \end{vmatrix} = -3F_2\vec{i} + 4F_2\vec{k}$$

$$\vec{M} = \vec{r} \times \vec{F} = \begin{vmatrix} \vec{i} & \vec{j} & \vec{k} \\ 4 & 0 & 0 \\ 0 & -100 & 0 \end{vmatrix} = -400\vec{k}$$

Problem 4.74 continued on next page....

continuation of solution to Problem 4.74...

The sum of the moments is $\Sigma\vec{M}_O = (-3F_1 - 3F_2)\vec{i} + (4F_2 - 400)\vec{k}$. The moment about the x-axis is $\vec{e}_x \cdot \vec{M}_o = M_x = -3F_1 - 3F_2$. The sum of the forces $\Sigma\vec{F} = (F_1 + F_2 - 100)\vec{j} = 0$, from which $F_1 = 100 - F_2$. . Substitute into the moment equation $\boxed{M_x = -3(100) + 3F_2 - 3F_2 = -300\ (ft\ lb)}$

Problem 4.75 The force $\vec{F} = 100\vec{i} + 60\vec{j} - 40\vec{k}\ (lb)$.

What is the moment of $\vec{F}$ about the y-axis. Draw a sketch to show the sense of the moment.

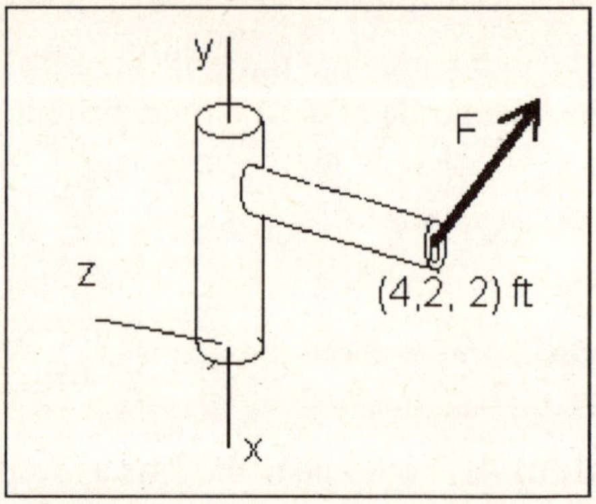

Solution: The radius vector to the point of application of the force is $\vec{r} = 4\vec{i} + 2\vec{j} - 2\vec{k}$. The magnitude of the moment is

$$\vec{e}_Y \cdot (\vec{r} \times \vec{F}) = \begin{vmatrix} 0 & 1 & 0 \\ 4 & 2 & 2 \\ 100 & 60 & -40 \end{vmatrix} = -360\ ft\ lb$$

The moment is $\boxed{\vec{M}_Y = -360\vec{e}_Y = -360\vec{j}}$ The sense of the moment is the direction of the curled fingers of the right hand if the thumb is held parallel to the negative y-axis (pointing thumb down in the sketch).

Problem 4.76 Suppose that the moment of the force $\vec{F}$ shown in Problem 4.75 about the x-axis is $240\vec{i}$ *ft lb,* the moment about the y-axis is $-320\vec{j}$ *(ft lb)*, and the moment about the z-axis is $160\vec{k}$ *(ft lb)*. If $F_Y = 80\ lb$, what are F_X and F_Y ?

Solution: The magnitudes of the moments: $\vec{e}_Y \cdot (\vec{r} \times \vec{F}) = \begin{vmatrix} 0 & 1 & 0 \\ 4 & 2 & -2 \\ F_X & 80 & F_Z \end{vmatrix} = -(4F_Z + 2F_X) = -320$,

$$\vec{e}_Z \cdot (\vec{r} \times \vec{F}) = \begin{vmatrix} 0 & 0 & 1 \\ 4 & 2 & -2 \\ F_X & 80 & F_Z \end{vmatrix} = 320 - 2F_X = 160$$

Solve: $\boxed{F_X = 80\ lb}$, $\boxed{F_Z = 40\ lb}$, from which the force vector is $\boxed{\vec{F} = 80\vec{i} + 80\vec{j} + 40\vec{k}}$

Problem 4.77 Two forces are exerted on the crankshaft by the connecting rods. The direction cosines of $\vec{F}_A$ are $\cos\theta_X = -0.182$, $\cos\theta_Y = 0.818$, and $\cos\theta_Z = 0.545$, and its magnitude is 4 *kN*. The direction cosines of the force $\vec{F}_B$ are $\cos\theta_X = 0.182$, $\cos\theta_Y = 0.818$, and $\cos\theta_Z = -0.545$, and its magnitude is 2 *kN*. . What is the sum of the moments of the two forces about the x-axis due to the two forces?(This is the moment that causes the crankshaft to rotate.)

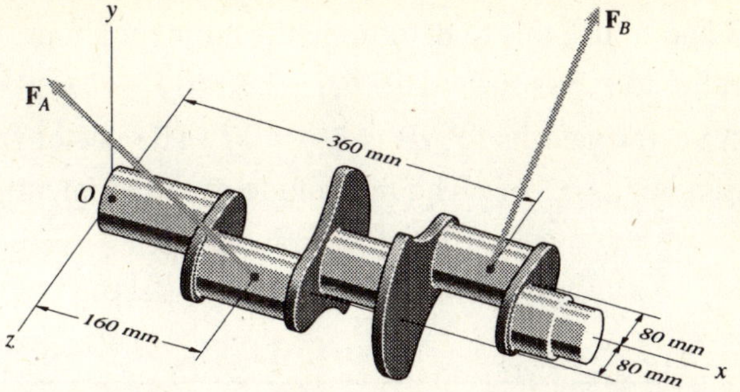

Solution: The coordinates of the points of action of the two forces are A(0.16,0,0.08), B(0.36,0,-0.08). The position vectors are $\vec{r}_{OA} = 0.16\vec{i} + 0\vec{j} + 0.08\vec{k}\ (m)$, $\vec{r}_{OB} = 0.36\vec{i} + 0\vec{j} - 0.08\vec{k}\ (m)$. The unit vectors parallel to the forces are given by the direction cosines: $\vec{e}_{FA} = -0.182\vec{i} + 0.818\vec{j} + 0.545\vec{k}$, $\vec{e}_{FB} = 0.182\vec{i} + 0.818\vec{j} - 0.545\vec{k}$ The forces are $\vec{F}_A = -0.728\vec{i} + 3.272\vec{j} + 2.18\vec{k}\ (kN)\ \vec{F}_B = 0.364\vec{i} + 1.636\vec{j} - 1.09\vec{k}\ (kN)$

The magnitude of the moments:

$$\vec{e}_X \cdot (\vec{r}_A \times \vec{F}_A) = \begin{vmatrix} 1 & 0 & 0 \\ 0.16 & 0 & 0.08 \\ -0.728 & 3.272 & 2.18 \end{vmatrix} = -0.2618, \quad \vec{e}_X \cdot (\vec{r}_B \times \vec{F}_B) = \begin{vmatrix} 1 & 0 & 0 \\ 0.36 & 0 & -0.08 \\ 0.364 & 1.636 & -1.09 \end{vmatrix} = 0.1309$$

The sum of the moments about the x-axis is $\boxed{\vec{M}_X = -0.2618\vec{e}_X + 0.1309\vec{e}_X = -0.1309\vec{i}\ kN\text{-}m}$.

Problem 4.78 Determine the moment of the 20 *N* force about the line AB. Draw a sketch to indicate the sense of the moment.

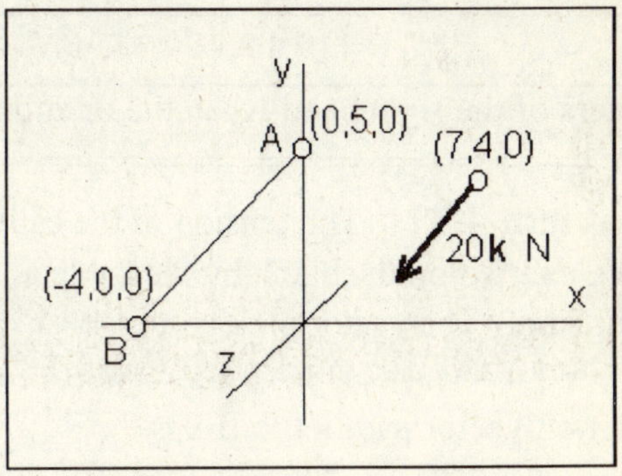

Solution: The strategy is to determine the unit vector parallel to the line AB, and use this to determine the moment about the line AB. The position vector parallel to AB is $\vec{r}_{AB} = (-4-0)\vec{i} + (0-5)\vec{j} + (0-0)\vec{k} = -4\vec{i} - 5\vec{j}$. The magnitude $|\vec{r}_{AB}| = \sqrt{4^2 + 5^2} = 6.403\ m$. The unit vector: $\vec{e} = \frac{\vec{r}_{AB}}{|\vec{r}_{AB}|} = -0.6247\vec{i} - 0.7809\vec{j}$. The vector AF is

$\vec{r}_{AF} = 7\vec{i} - 1\vec{j}$. The moment about the point A is $\vec{M}_A = \vec{r}_{AF} \times \vec{F} = \begin{vmatrix} \vec{i} & \vec{j} & \vec{k} \\ 7 & -1 & 0 \\ 0 & 0 & 20 \end{vmatrix} = -20\vec{i} - 140\vec{j}$

The magnitude of the moment about the line AB $|\vec{M}_{AB}| = \vec{e}_{AB} \cdot \vec{M}_A = 121.8\ N\text{-}m$. The moment about AB is $\boxed{\vec{M}_{AB} = |\vec{M}_{AB}|\vec{e}_{AB} = -76.1\vec{i} - 95.1\vec{j}\ (N\text{-}m)}$ The sense of the moment is in the direction of the curled fingers when the right hand is aligned with AB with the thumb pointing toward B.

Problem 4.79 The force $\vec{F} = -10\vec{i} + 5\vec{j} - 5\vec{k}\ (kip)$. Determine the moment of $\vec{F}$ about the line AB.

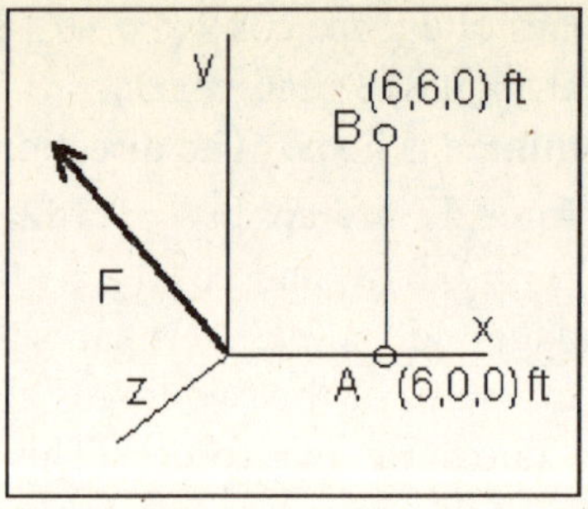

Solution: The strategy is to determine the unit vector parallel to AB and to use this to determine the moment about AB. The vector parallel to AB is $\vec{r}_{AB} = (6-6)\vec{i} + (0-6)\vec{j} + (0-0)\vec{k} = -6\vec{j}$. The unit vector parallel to AB is $\vec{e}_{AB} = 1\vec{j}$. The vector from A to the origin is $\vec{r}_{AO} = -6\vec{i}$. The magnitude of the moment of about AB is

$$\left|\vec{M}_{AB}\right| = \vec{e}_{AB} \cdot (\vec{r}_{AO} \times \vec{F}) = \begin{vmatrix} 0 & 1 & 0 \\ -6 & 0 & 0 \\ -10 & 5 & -5 \end{vmatrix} = -30.$$

The moment about AB is $\boxed{\vec{M}_{AB} = \left|\vec{M}_{AB}\right|\vec{e}_{AB} = -30\vec{e}_{AB} = -30\vec{j}\ (ft\ kip)}$

Problem 4.80 The force $\vec{F} = 10\vec{i} + 12\vec{j} - 6\vec{k}\ (N)$. What is the moment $\vec{F}$ about the line OA?

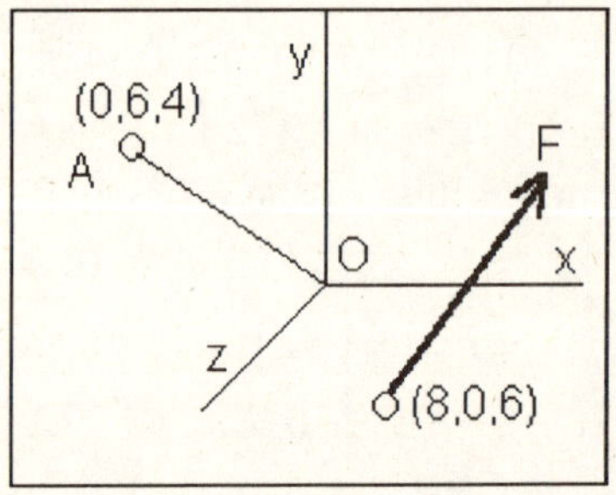

Solution: The strategy is to determine a unit vector parallel to OA and to use this to determine the moment about OA. The vector parallel to OA is $\vec{r}_{OA} = 6\vec{j} + 4\vec{k}$. The magnitude: $\vec{r}_{OA} = \sqrt{6^2 + 4^2} = 7.211\,m$. The unit vector parallel to OA is $\vec{e}_{OA} = 0.8321\vec{j} + 0.5547\vec{k}$. The vector from O to the point of application of $\vec{F}$ is $\vec{r}_{OF} = 8\vec{i} + 6\vec{k}$. The magnitude of the moment about OA is

$$\left|\vec{M}_{O}\right| = \vec{e}_{OA} \cdot (\vec{r}_{OF} \times \vec{F}) = \begin{vmatrix} 0 & 0.8321 & 0.5547 \\ 8 & 0 & 6 \\ 10 & 12 & -6 \end{vmatrix}.$$

$= 89.8614 + 53.251 = 143.1\ N\text{-}m$ The moment about OA is

$\boxed{\vec{M}_{OA} = \left|\vec{M}_{OA}\right|\vec{e}_{OA} = 119.1\vec{j} + 79.4\vec{k}\ (N\text{-}m)}$ The sense of the moment is in the direction of the curled fingers of the right hand when the thumb is parallel to OA, pointing to A.

Problem 4.81 The tension in the cable is 1 *kN*. Determine the moment about the x-axis due to the force exerted on the hatch by the cable at point B. Draw a sketch to indicate the sense of the moment.

Solution: The vector parallel to BA is

$\vec{r}_{BA} = (0.4-1)\vec{i} + 0.3\vec{j} - 0.6\vec{k} = -0.6\vec{i} + 0.3\vec{j} - 0.6\vec{k}$.

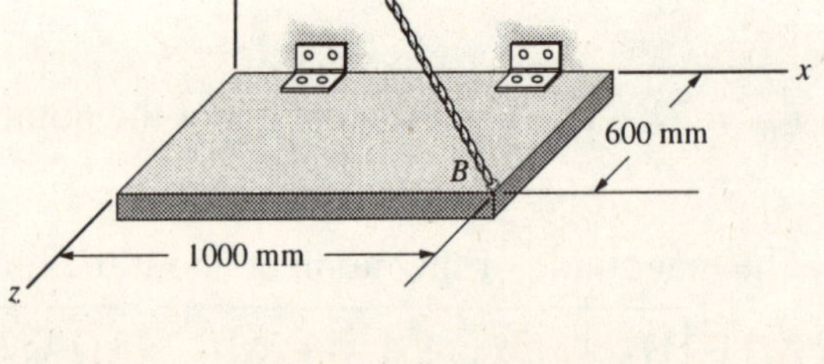

The unit vector parallel to BA is

$\vec{e}_{BA} = -0.6667\vec{i} + 0.3333\vec{j} - 0.6667\vec{k}$. The moment about O is

$$\vec{M}_O = \vec{r}_{OB} \times \vec{T} = \begin{vmatrix} \vec{i} & \vec{j} & \vec{k} \\ 1 & 0 & 0.6 \\ -0.6667 & 0.3333 & -0.66667 \end{vmatrix}$$

$\vec{M}_O = -0.2\vec{i} + 0.2667\vec{j} + 0.3333\vec{k}$. The magnitude is $\left|M_X\right| = \vec{e}_X \cdot \vec{M}_O = -0.2\ kN\text{-}m$. The moment is

$\boxed{\vec{M}_X = -0.2\vec{i}\ \ kN\text{-}m}$ The sense is clockwise when viewed along the x-axis toward the origin.

Problem 4.82 The coordinates of A are (-2.4, 0, -0.6) *m* and the coordinates of B are (-2.2, 0.7, -1.2) *m*. The force exerted at B by the sailboat's main sheet AB is 130 *N*. Determine the moment of the force about the centerline of the mast (the y -axis). Draw a sketch to indicate the sense of the moment.

Solution: The position vectors:

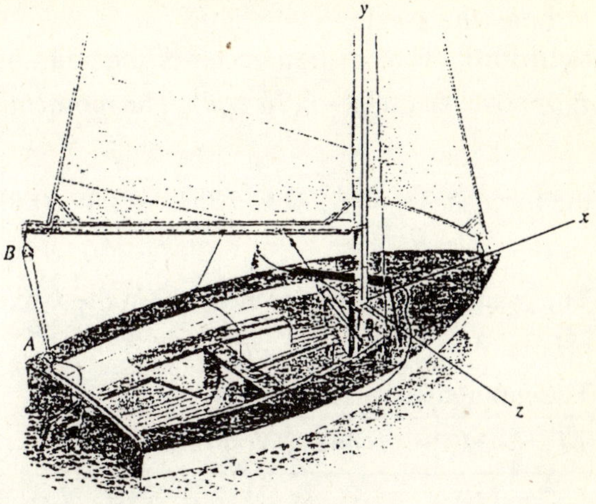

$\vec{r}_{OA} = -2.4\vec{i} - 0.6\vec{k}\ (m)$,

$\vec{r}_{OB} = -2.2\vec{i} + 0.7\vec{j} - 1.2\vec{k}\ (m)$,

$\vec{r}_{BA} = (-2.4+2.2)\vec{i} + (0-0.7)\vec{j} + (-0.6+1.2)\vec{k}\ (m)$

$= -0.2\vec{i} - 0.7\vec{j} + 0.6\vec{k}\ (m)$. The magnitude is

$|\vec{r}_{BA}| = 0.9434\ m$. The unit vector parallel to BA is

$\vec{e}_{BA} = -0.2120\vec{i} - 0.7420\vec{j} + 0.6360\vec{k}$. The tension is

$\vec{T}_{BA} = 130\vec{e}_{BA}$. The moment of $\vec{T}_{AB}$ about the origin is

$$\vec{M}_O = \vec{r}_{OB} \times \vec{T}_{BA} = \begin{vmatrix} \vec{i} & \vec{j} & \vec{k} \\ -2.2 & 0.7 & -1.2 \\ -27.56 & -96.46 & 82.68 \end{vmatrix}$$

$\vec{M}_O = -57.88\vec{i} + 214.97\vec{j} + 231.5\vec{k}$. The magnitude of the moment about the y-axis is $|\vec{M}_Y| = \vec{e}_Y \cdot \vec{M}_O = 214.97\ N\text{-}m$. The moment is $\boxed{\vec{M}_Y = \vec{e}_Y(214.97) = 214.97\vec{j}\ N\text{-}m}$ The sense of the moment is along the direction of the curled fingers of the right hand when the thumb points along the positive y-axis.

Problem 4.83 The tension in cable AB is 200 *lb*. Determine the moments about each of the coordinate axes due to the force exerted on Point B by the cable. Draw sketches to indicate the senses of the moments.

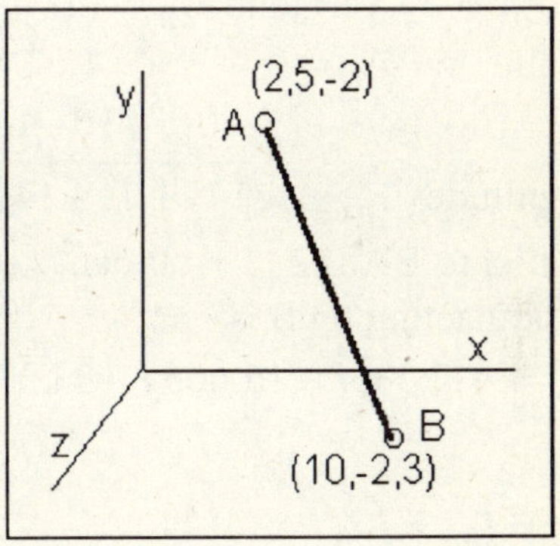

Solution: The position vectors are: $\vec{r}_{OA} = 2\vec{i} + 5\vec{j} - 2\vec{k}\ m$,

$\vec{r}_{OB} = 10\vec{i} - 2\vec{j} + 3\vec{k}\ m$,

$\vec{r}_{BA} = (2-10)\vec{i} + (5-2)\vec{j} + (-2-3)\vec{k} = -8\vec{i} + 7\vec{j} - 5\vec{k}$.

$|\vec{r}_{AB}| = 11.747\ ft$. The unit vector parallel to AB is

$\vec{e}_{BA} = -0.6810\vec{i} + 0.5959\vec{j} - 0.4256\vec{k}$. The tension in the cable is

$\vec{T}_{BA} = 200\vec{e}_{BA} = -136.20\vec{i} + 119.18\vec{j} - 85.12\vec{k}$.

The moment about the origin of the force exerted at B is

$$\vec{M}_O = \vec{r}_{OB} \times \vec{T}_{BA} = \begin{vmatrix} \vec{i} & \vec{j} & \vec{k} \\ 10 & -2 & 3 \\ -136.2 & 119.18 & -85.12 \end{vmatrix} = -187.3\vec{i} + 442.6\vec{j} + 919.4\vec{k}$$

The moments about each of the coordinate axes: (a) about x: $\boxed{\vec{M}_X = -187.3\vec{i}}$. (b) About y: $\boxed{\vec{M}_Y = +442.6\vec{j}}$, (c) About z: $\boxed{\vec{M}_Z = +919.4\vec{k}}$. The sense of the moment about each axis is in the direction of the curled fingers of the right hand (a) about x when the thumb points in the negative x direction, (b) about y when the thumb points in the positive y direction, (c) about z when the thumb points in the positive z-direction.

Problem 4.84 The total force exerted on the blades of the turbine by the steam nozzle is $\vec{F} = 20\vec{i} - 120\vec{j} + 100\vec{k}$ and it effectively acts at the point (100, 80, 300) *mm*. What moment is exerted about the axis of the turbine (the x-axis)?

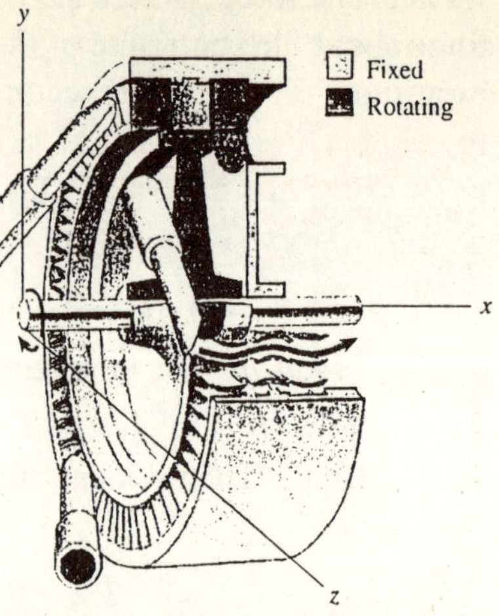

Solution: The position vector of the point of application is $\vec{r}_{OF} = 0.1\vec{i} + 0.08\vec{j} + 0.3\vec{k}$ (m). The moment about the origin of

the force is $\vec{M}_O = \vec{r}_{OF} \times \vec{F} = \begin{vmatrix} \vec{i} & \vec{j} & \vec{k} \\ 0.1 & 0.08 & 0.3 \\ 20 & -120 & 100 \end{vmatrix} = 44\vec{i} - 4\vec{j} - 13.6\vec{k}$

The magnitude of the moment about the x-axis is

$|\vec{M}_X| = \vec{e}_X \cdot \vec{M}_O = 44$ $N\text{-}m$.

The moment about the x-axis is

$$\boxed{\vec{M}_X = |\vec{M}_X|\vec{e}_X = 44\vec{i}\ (N\text{-}m)}$$

Problem 4.85 The tension in cable AB is 50 *N*. Determine the moment about the line OC due to the force exerted by the cable at B. Draw a sketch to indicate the sense of the moment.

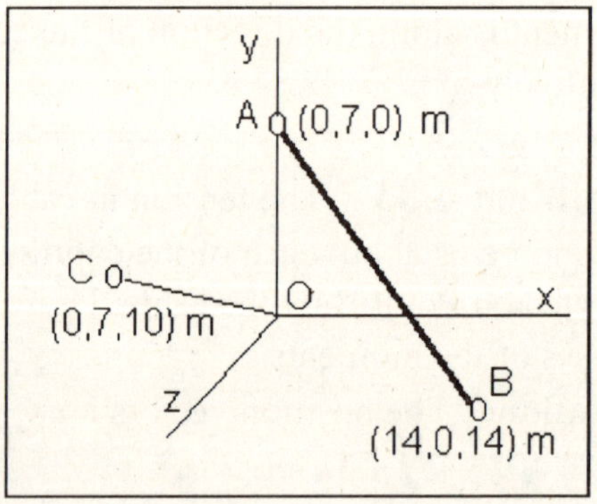

Solution: The vector OC is $\vec{r}_{OC} = 7\vec{j} + 10\vec{k}$, with $|\vec{r}_{OC}| = \sqrt{7^2 + 10^2} = 12.21$ m. The unit vector parallel to OC is $\vec{e}_{OC} = 0.5735\vec{j} + 0.8192\vec{k}$. The position vectors of A and B are $\vec{r}_{OA} = 7\vec{j}$ (m), and $\vec{r}_{OB} = 14\vec{i} + 0\vec{j} + 14\vec{k}$ (m). The vector parallel to BA is

$\vec{r}_{BA} = (0-14)\vec{i} + (7-0)\vec{j} + (0-14)\vec{k} = -14\vec{i} + 7\vec{j} - 14\vec{k}$. The magnitude: $|\vec{r}_{BA}| = \sqrt{14^2 + 7^2 + 14^2} = 21$ m. The unit vector parallel to BA is $\vec{e}_{BA} = -0.6667\vec{i} + 0.3333\vec{j} - 0.6667\vec{k}$. The tension acting on B is

$\vec{T}_{BA} = -33.335\vec{i} + 16.665\vec{j} - 33.335\vec{k}$ (N). The moment about the origin is

$\vec{M}_O = \vec{r}_{OB} \times \vec{T}_{BA} = \begin{vmatrix} \vec{i} & \vec{j} & \vec{k} \\ 14 & 0 & 14 \\ -33.335 & 16.665 & -33.335 \end{vmatrix} = -233.31\vec{i} + 0\vec{j} + 233.31\vec{k}$. The magnitude of the

moment about the line OC is $|\vec{M}_{OC}| = \vec{e}_{OC} \cdot \vec{M}_O = 191.2$ $N\text{-}m$. The moment about the line OC is

$\boxed{\vec{M}_{OC} = 191.1\vec{e}_{OC} = 109.6\vec{j} + 156.5\vec{k}\ (N\text{-}m)}$ The sense of the moment is in the direction of the curled fingers of the right hand when the thumb points from O toward C.

Problem 4.86 The tension in cable AB is 80 *lb*. What is the moment about the line CD due to the force exerted by the cable on the wall at B?

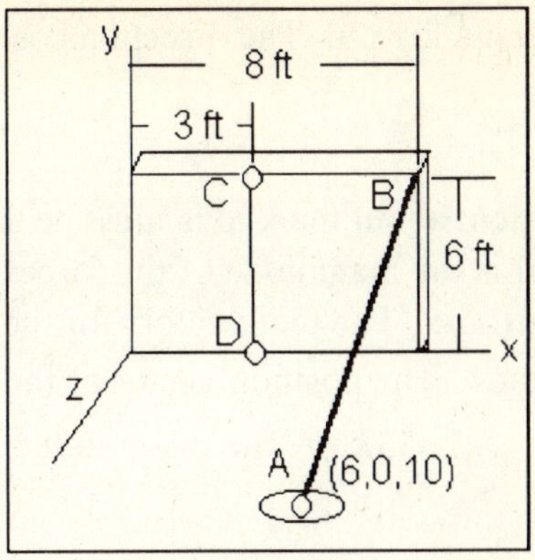

Solution: The strategy is to find the moment about the point C exerted by the force at B, and then to find the component of that moment acting along the line CD. The coordinates of the points B, C, D are B(8, 6, 0), C(3, 6, 0), D(3, 0, 0). The position vectors are: $\vec{r}_{OB} = 8\vec{i} + 6\vec{j}$, $\vec{r}_{OC} = 3\vec{i} + 6\vec{j}$, $\vec{r}_{OD} = 3\vec{i}$. The vector parallel to CD is $\vec{r}_{CD} = \vec{r}_{OD} - \vec{r}_{OC} = -6\vec{j}$. The unit vector parallel to CD is $\vec{e}_{CD} = -1\vec{j}$. The vector from point C to B is $\vec{r}_{CB} = \vec{r}_{OB} - \vec{r}_{OC} = 5\vec{i}$. The position vector of A is $\vec{r}_{OA} = 6\vec{i} + 10\vec{k}$. The vector parallel to BA is $\vec{r}_{BA} = \vec{r}_{OA} - \vec{r}_{OB} = -2\vec{i} - 6\vec{j} + 10\vec{k}$. The magnitude is $|\vec{r}_{BA}| = 11.832\ ft$. The unit vector parallel to BA is $\vec{e}_{BA} = -0.1690\vec{i} - 0.5071\vec{j} + 0.8452\vec{k}$. The tension acting at B is $\vec{T}_{BA} = 80\vec{e}_{BA} = -13.52\vec{i} - 40.57\vec{j} + 67.62\vec{k}$. The magnitude of the moment about CD due to the tension acting at B is $\left|\vec{M}_{CD}\right| = \vec{e}_{CD} \cdot (\vec{r}_{CB} \times \vec{T}_{BA}) = \begin{vmatrix} 0 & -1 & 0 \\ 5 & 0 & 0 \\ -13.52 & -40.57 & 67.62 \end{vmatrix} = 338.1\ (ft\ lb)$. The moment about CD is $\boxed{\vec{M}_{CD} = 338.1\vec{e}_{CD} = -338.1\vec{j}\ \ (ft\ lb)}$ The sense of the moment is along the curled fingers of the right hand when the thumb is parallel to CD, pointing toward D.

Problem 4.87 The universal joint is connected to the drive shaft at A and A'. The coordinates of A are (0, 40, 0) *mm* and the coordinates of A' are (0, -40, 0) *mm.* The forces exerted on the drive shaft by the universal joint are $-30\vec{j} + 400\vec{k}\ (N)$ at A and $30\vec{j} - 400\vec{k}\ (N)$ at A'. What is the magnitude of the torque (moment) exerted by the universal joint on the drive shaft about the shaft axis O-O'?

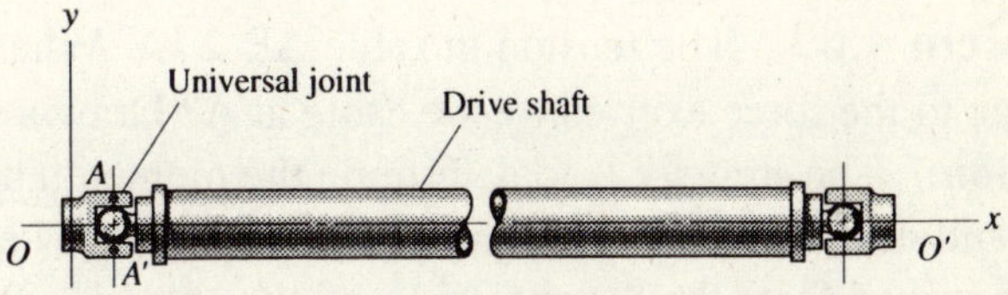

Solution: The position vectors of A and A' are $\vec{r}_{OA} = 40\vec{j}\ (mm)$, and $\vec{r}_{OA'} = -40\vec{j}\ (mm)$. The magnitudes of the moments about the origin are:

$$\left|\vec{M}_{OO}\right| = \vec{e}_X \cdot (\vec{r}_{OA} \times \vec{F}) = \begin{vmatrix} 1 & 0 & 0 \\ 0 & 40 & 0 \\ 0 & -30 & 400 \end{vmatrix} = 16000\ (N\text{-}mm),\ \vec{M}_{OO} = 16000\vec{e}_X = 16000\vec{i}$$

$$\left|\vec{M}_{OO}\right| = \vec{e}_X \cdot (\vec{r}_{OA'} \times \vec{F}) = \begin{vmatrix} 1 & 0 & 0 \\ 0 & -40 & 0 \\ 0 & 30 & -400 \end{vmatrix} = 16000\ (N\text{-}mm)\ \ \vec{M}_{OO} = 16000\vec{e}_X = 16000\vec{i}$$

The sum of the moments is $\sum \vec{M}_{OO} = 32000\vec{i}\ (N\text{-}mm)$ $\boxed{= 32\vec{i}\ \ (N\text{-}m)}$

==================================<>=====================================

Problem 4.88 A motorist applies the two forces shown to loosen a lug nut. The direction cosines of $\vec{F}$ are $\cos\theta_x = \frac{4}{13}$, $\cos\theta_Y = \frac{12}{13}$, and $\cos\theta_Z = \frac{3}{13}$. If the magnitude of the moment about the x-axis must be 32 *ft lb* to loosen the nut, what is the magnitude of the forces she must apply?

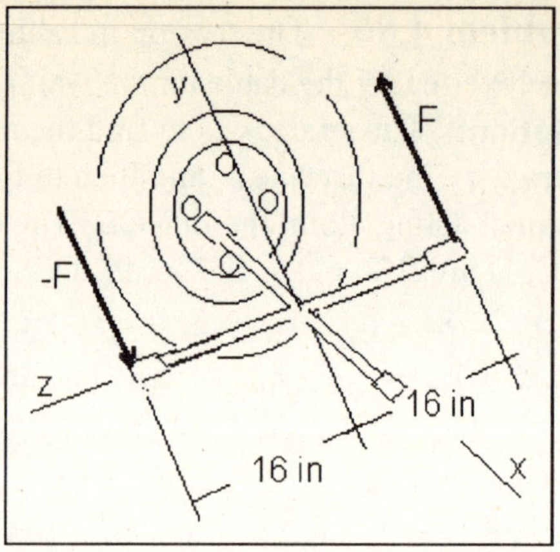

Solution: The unit vectors for the forces are the direction cosines. The position vector of the force $\vec{F}$ is $\vec{r}_{OF} = -1.333\vec{k}$ *ft*. The magnitude of the moment due to $\vec{F}$ is

$$\left|\vec{M}_{OF}\right| = \vec{e}_X \cdot \left(\vec{r}_{OF} \times \vec{F}\right) = \begin{vmatrix} 1 & 0 & 0 \\ 0 & 0 & -1.333 \\ 0.3077F & 0.9231F & 0.2308F \end{vmatrix}$$

$\left|\vec{M}_{OF}\right| = 1.230F$ *ft lb*. The magnitude of the moment due to $-\vec{F}$ is

$$\left|\vec{M}_{-OF}\right| = \vec{e}_X \cdot (\vec{r}_{-OF} \times -\vec{F}) = \begin{vmatrix} 1 & 0 & 0 \\ 0 & 0 & 1.333 \\ -.3077F & -0.9231F & -0.2308F \end{vmatrix} = 1.230F \text{ ft lb}.$$

The total moment about the x-axis is $\sum \vec{M}_X = 1.230F\vec{i} + 1.230F\vec{i} = 2.46F\vec{i}$, from which, for a total magnitude of 32 *ft lb*, the force to be applied is $\boxed{F = \frac{32}{2.46} = 13 \text{ lb}}$

==================================<>=====================================

Problem 4.89 The tension in cable AB 2 *kN*. What is the magnitude of the moment about the shaft CD due to the force exerted by the cable at A? Draw a sketch to indicate the sense of the moment.

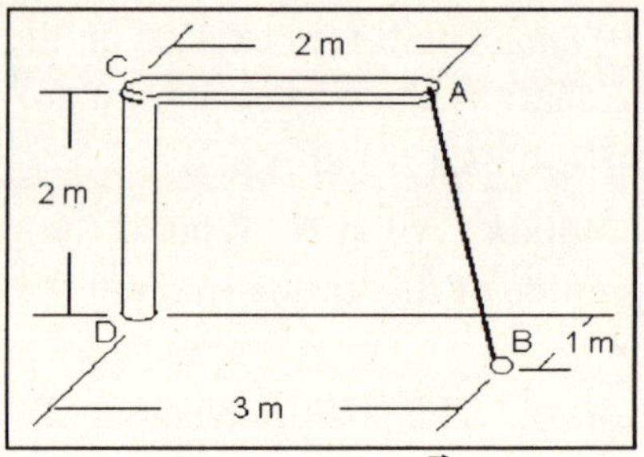

Solution: The strategy is to determine the moment about C due to A, and determine the component parallel to CD. The moment is determined from the distance CA and the components of the tension, which is to be found from the magnitude of the tension and the unit vector parallel to AB. The coordinates of the points A, B, C, and D are: A(2, 2, 0), B(3, 0, 1), C(0, 2, 0), and D(0,0,0). The unit vector parallel to CD is by inspection $\vec{e}_{CD} = -1\vec{j}$. The position vectors parallel to OC, OA, and OB: $\vec{r}_{OC} = 2\vec{j}$, $\vec{r}_{OA} = 2i + 2\vec{j}$, $\vec{r}_{OB} = 3\vec{i} + 1\vec{k}$. The vector parallel to CA is $\vec{r}_{CA} = \vec{r}_{OA} - \vec{r}_{OC} = 2\vec{i}$. The vector parallel to AB is $\vec{r}_{AB} = \vec{r}_{OB} - \vec{r}_{OA} = 1\vec{i} - 2\vec{j} + 1\vec{k}$. The magnitude: $\left|\vec{r}_{AB}\right| = 2.4495$ *m*. The unit vector parallel to AB is $\vec{e}_{AB} = 0.4082\vec{i} - 0.8165\vec{j} + 0.4082\vec{k}$. The tension is $\vec{T}_{AB} = 2\vec{e}_{AB} = 0.8165\vec{i} - 1.633\vec{j} + 0.8165\vec{k}$. The magnitude of the moment about CD is

$$\left|\vec{M}_{CD}\right| = \vec{e}_{CD} \cdot \left(\vec{r}_{CA} \times \vec{T}_{AB}\right) = \begin{vmatrix} 0 & -1 & 0 \\ 2 & 0 & 0 \\ 0.8164 & -1.633 & 0.8165 \end{vmatrix}. \boxed{= 1.633 \text{ kN - m}}$$

The moment about CD is $\vec{M}_{CD} = \vec{e}_{CD}\left|\vec{M}_{CD}\right| = -1.633\vec{j}$ $(kN\text{ - }m)$ The sense is in the direction of the curled fingers of the right hand when the thumb is parallel to DC, pointed toward D.

==================================<>=====================================

═════════════════<>═════════════════

Problem 4.90 The force $\vec{F} = -100\vec{i} + 200\vec{j} + 200\vec{k}$ (N). The moment of $\vec{F}$ about the line from O to A is $470\vec{i} + 315\vec{j} + 160\vec{k}$ $(N\text{-}m)$. What are the direction cosines of the position vector from O to A?

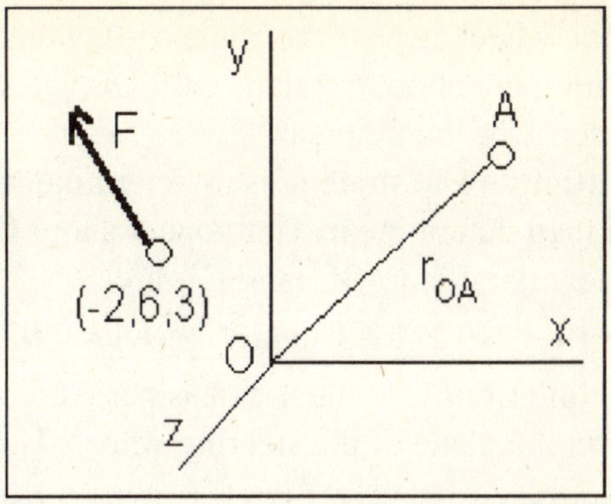

Solution: The strategy is to recognize that with the moment about OA given, the force and its location are not required to solve the problem. The moment about OA is $\vec{M}_{OA} = |\vec{M}_{OA}|\vec{e}_{OA} = |\vec{M}_{OA}|e_X\vec{i} + |\vec{M}_{OA}|e_Y\vec{j} + |\vec{M}_{OA}|e_Z\vec{k}$ $= 470\vec{i} + 315\vec{j} + 160\vec{k}$. But the magnitude is known: $|\vec{M}_{OA}| = \sqrt{470^2 + 315^2 + 160^2} = 587.984$ $N\text{-}m$, . Equating components in $\vec{i}, \vec{j}, \vec{k}$, , and solving:

$e_X = \dfrac{470}{587.984} = 0.7993$, $e_Y = \dfrac{315}{587.984} = 0.5357$, and

$e_Z = \dfrac{160}{587.984} = 0.2721$. By definition, the unit vector components are the direction cosines. Thus to three significant figures: $\boxed{\cos\theta_X = 0.799}$, $\boxed{\cos\theta_Y = 0.536}$, and $\boxed{\cos\theta_Z = 0.272}$.

═════════════════<>═════════════════

Problem 4.91 The direction cosines of the centerline OA are $\cos\theta_x = 0.500$, $\cos\theta_Y = 0.866$, and $\cos\theta_Z = 0$, and the direction cosines of the line AG are $\cos\theta_X = 0.707$, $\cos\theta_Y = 0.619$, and $\cos\theta_Z = -0.342$. What is the moment about OA due to the 250 N weight? Draw a sketch to indicate the sense of the moment about the shaft.

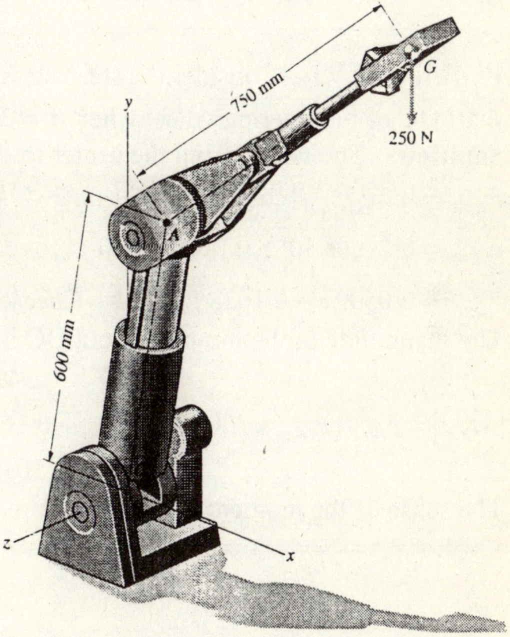

Solution: By definition, the direction cosines are the scalar components of the unit vectors. Thus the unit vectors are $\vec{e}_1 = 0.5\vec{i} + 0.866\vec{j}$, and $\vec{e}_2 = 0.707\vec{i} + 0.619\vec{j} - 0.341\vec{k}$. The force is $\vec{W} = 250\vec{j}$ (N) The position vector of the 250 N weight is $\vec{r}_W = 0.600\vec{e}_1 + 0.750\vec{e}_2$ $= 0.8303\vec{i} + 0.9839\vec{j} - 0.2565\vec{k}$

The moment about OA is $\vec{M}_{OA} = \vec{e}_{OA}(\vec{e}_{OA} \cdot (\vec{r}_W \times \vec{W}))$

$$= \begin{vmatrix} 0.5 & 0.866 & 0 \\ 0.8303 & 0.9839 & -0.2565 \\ 0 & -250 & 0 \end{vmatrix} \vec{e}_1 = -32.06\vec{e}_1$$

$\boxed{= -16\vec{i} - 27.77\vec{j} \ (N\text{-}m)}$

The moment is anti parallel to the unit vector parallel to OA, with the sense of the moment in the direction of the curled fingers when the thumb of the right hand is directed oppositely to the direction of the unit vector.

═════════════════<>═════════════════

==============================<>==============================

Problem 4.92 The radius of the steering wheel is 200 *mm*. The distance from O to C is 1 *m*. The center C of the steering wheel lies in the x-y plane. The driver exerts as force $\vec{F} = 10\vec{i} + 10\vec{j} - 5\vec{k}\ (N)$ on the wheel at A. If the angle α=0, what is the magnitude of the moment about the shaft OC? Draw a sketch to indicate the sense of the moment.

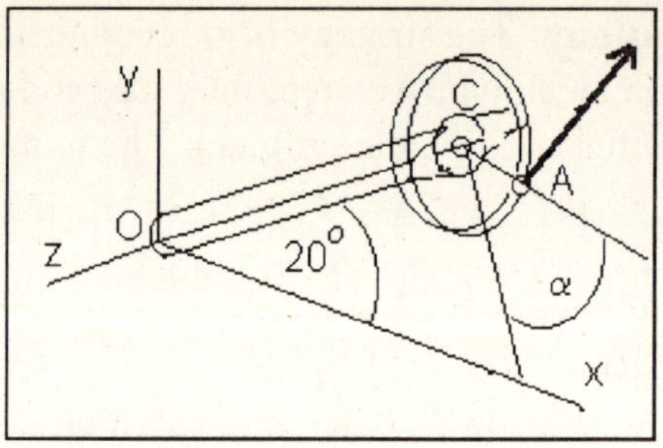

Solution: The strategy is to determine the moment about C, and then determine its component about OC. The radius vectors parallel to OC and CA are:

$\vec{r}_{OC} = 1\left(\vec{i}\cos 20^o + \vec{j}\sin 20^o + 0\vec{k}\right) = 0.9397\vec{i} + 0.3420\vec{j}$.

The line from C to the x-axis is perpendicular to OC since it lies in the plane of the steering wheel. The unit vector from C to the x-axis is $\vec{e}_{CX} = \vec{i}\cos(20-90) + \vec{j}\sin(20-90) = 0.3420\vec{i} - 0.9397\vec{j}$, where the angle is measured positive counterclockwise from the x-axis. The vector parallel to CA is $\vec{r}_{CA} = 0.2\vec{e}_{CX} = +0.0684\vec{i} - 0.1879\vec{j}\ (m)$. The magnitude of the moment about OC

$$\left|\vec{M}_{OC}\right| = \vec{e}_{OC}\cdot(\vec{r}_{CA}\times\vec{F}) = \begin{vmatrix} 0.9397 & 0.3420 & 0 \\ 0.0684 & -0.1879 & 0 \\ 10 & 10 & -5 \end{vmatrix} = 0.9998 \ \boxed{= 1\ N\text{-}m}$$

The sense of the moment is in the direction of the curled fingers of the right hand if the thumb is parallel to OC, pointing from O to C.

==============================<>==============================

Problem 4.93 Consider the steering wheel in Problem 4.92. Determine the moment of $\vec{F}$ about the shaft OC of the steering wheel when $\alpha = 30^o$.

Solution: The vector from the center to the new position A is, from trigonometry:

$\vec{r}_{NEW} = 0.2\vec{e}_{CAOLD}\left(\cos 30^o\right) - 0.2\left(\sin 30^o\right)\vec{k}$. Expanding

$\vec{r}_{NEW} = 0.2(\cos 30^o)(0.0684)\vec{i} - 0.2(\cos 30^o)(0.1879)\vec{j} - 0.2(\sin 30^o)\vec{k}$

$\vec{r}_{NEW} = 0.0592\vec{i} - 0.1628\vec{j} - 0.1\vec{k}$. *Check:* $\left|\vec{r}_{NEW}\right| = \sqrt{0.0592^2 + 0.1628^2 + 0.1^2} = 0.2\ m$.

The magnitude of the moment about OC is

$$\left|\vec{M}_{OC}\right| = \vec{e}_{OC}\cdot\left(\vec{r}_{NEW}\times\vec{F}\right) = \begin{vmatrix} 0.9397 & 0.3420 & 0 \\ 0.0592 & -0.1628 & -0.1 \\ 10 & 10 & -5 \end{vmatrix} = \boxed{1.464\ N\text{-}m}$$

The sense of the moment is the same as in Problem 4.92.

==============================<>==============================

Problem 4.94 The weight causes a tension of 100 *lb* in cable CD. If d=2 *ft*, what is the moment about the z-axis due to the force exerted by the cable CD at point C?

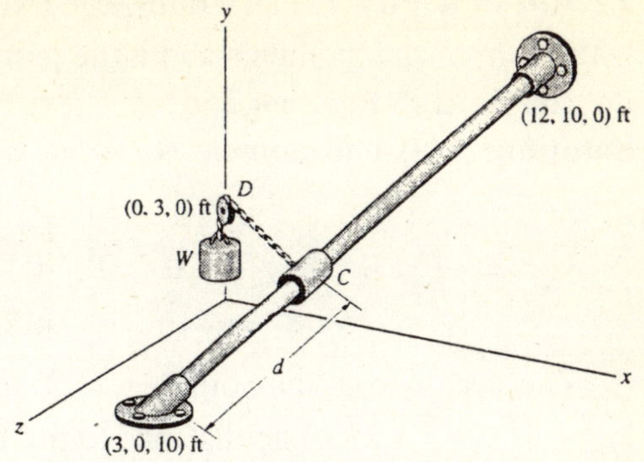

Solution: The strategy is to use the unit vector parallel to the bar to locate point C relative to the origin, and then use this location to find the unit vector parallel to the cable CD. With the tension resolved into components about the origin, the moment about the origin can be resolved into components along the z-axis. Denote the top of the bar by T and the bottom of the bar by B. The position vectors of the ends of the bar are: $\vec{r}_{OB} = 3\vec{i} + 0\vec{j} + 10\vec{k}$, $\vec{r}_{OT} = 12\vec{i} + 10\vec{j} + 0\vec{k}$. The vector from the bottom to the top of the bar is $\vec{r}_{BT} = \vec{r}_{OT} - \vec{r}_{OB} = 9\vec{i} + 10\vec{j} - 10\vec{k}$. The magnitude: $|\vec{r}_{BT}| = \sqrt{9^2 + 10^2 + 10^2} = 16.763\ ft$. The unit vector parallel to the bar, pointing toward the top, is $\vec{e}_{BT} = 0.5369\vec{i} + 0.5965\vec{j} - 0.5965\vec{k}$. The position vector of the point C relative to the bottom of the bar is $\vec{r}_{BC} = 2\vec{e}_{BT} = 1.074\vec{i} + 1.193\vec{j} - 1.193\vec{k}$. The position vector of point C relative to the origin is $\vec{r}_{OC} = \vec{r}_{OB} + \vec{r}_{BC} = 4.074\vec{i} + 1.193\vec{j} + 8.807\vec{k}$. The position vector of point D is $\vec{r}_{OD} = 0\vec{i} + 3\vec{j} + 0\vec{k}$. The vector parallel to CD is $\vec{r}_{CD} = \vec{r}_{OD} - \vec{r}_{OC} = -4.074\vec{i} + 1.807\vec{j} - 8.807\vec{k}$. The magnitude is $|\vec{r}_{CD}| = \sqrt{4.074^2 + 1.807^2 + 8.807^2} = 9.87\ ft$. The unit vector parallel to CD is $\vec{e}_{CD} = -0.4123\vec{i} + 0.1831\vec{j} - 0.8923\vec{k}$. The tension is $\vec{T}_{CD} = 100\vec{e}_{CD} = -41.27\vec{i} + 18.31\vec{j} - 89.23\vec{k}\ lb$. The magnitude of the moment about the z-axis is

$$|\vec{M}_O| = \vec{e}_Z \cdot (\vec{r}_{OC} \times \vec{T}_{CD}) = \begin{vmatrix} 0 & 0 & 1 \\ 4.074 & 1.193 & 8.807 \\ -41.27 & 18.31 & -89.23 \end{vmatrix} \boxed{= 123.83\ ft\ lb}$$

Problem 4.95 The rod AB supports the open hood of the car. The force exerted by the rod on the hood at B is parallel to the rod. If the rod must exert a moment of magnitude 100 *ft lb* about the x-axis to support the hood and the distance d=2 *ft*, what is the magnitude of the force the rod must exert on the hood?

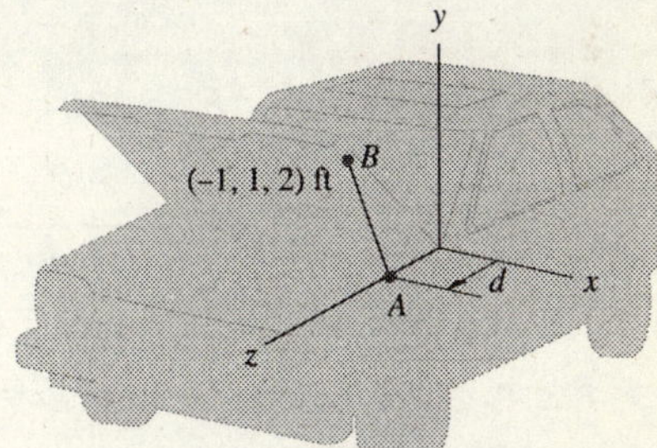

Solution: The coordinates of B are $B(-1, 1, 2)$. The position vectors of A, B are $\vec{r}_{OA} = 2\vec{k}$, $\vec{r}_{OB} = -1\vec{i} + 1\vec{j} + 2\vec{k}$. The vector parallel to AB is $\vec{r}_{AB} = \vec{r}_{OB} - \vec{r}_{OA} = -1\vec{i} + 1\vec{j}$. The unit vector is $\vec{e}_{AB} = -0.7071\vec{i} + 0.7071\vec{j}$. The force is $\vec{F}_{AB} = F\vec{e}_{AB}$. The moment about the origin is

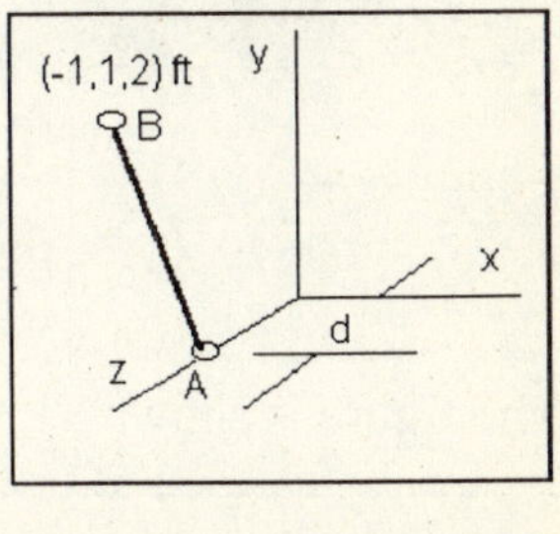

$$|\vec{M}_X| = \vec{e}_X \cdot (\vec{r}_{OA} \times \vec{F}_{AB}) = \begin{vmatrix} 1 & 0 & 0 \\ 0 & 0 & 2 \\ -.7071F & 0.7071F & 0 \end{vmatrix} = -1.414F, \text{ from}$$

which, for $|\vec{M}_X| = 100\ ft\ lb$, $\boxed{F = \dfrac{100}{1.414} = 70.71\ lb}$

==============================<>==============================

Problem 4.96 (a) Determine the moment of the couple. (b) What is the sum of the moments about the point (-4, 6, -2) ft? (c) Represent the couple as in Figure 4.26c.

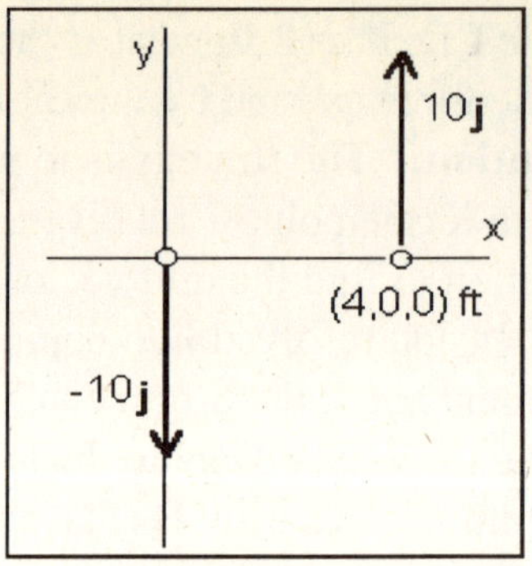

Solution: (a) The moment is

$$\sum \vec{M} = \vec{r} \times \vec{F} + 0 \times \vec{F} = \begin{vmatrix} \vec{i} & \vec{j} & \vec{k} \\ 4 & 0 & 0 \\ 0 & 10 & 0 \end{vmatrix} + \begin{vmatrix} \vec{i} & \vec{j} & \vec{k} \\ 0 & 0 & 0 \\ 0 & -10 & 0 \end{vmatrix} \boxed{= 40\vec{k}}$$

(b) The vector from the point (-4, 6, 2) to the force at the origin is $\vec{r}_{PO} = 4\vec{i} - 6\vec{j} + 2\vec{k}$. The distance to the force at (4, 0, 0) is $\vec{r}_{P4} = \vec{r}_4 - \vec{r}_{OP} = (4+4)\vec{i} - 6\vec{j} + 2\vec{k} = 8\vec{i} - 6\vec{j} + 2\vec{k}$. The moment about the point is $\sum \vec{M}_P = \vec{r}_{PO} \times \vec{F}_O + \vec{r}_{P4} \times \vec{F}_4 = \begin{vmatrix} \vec{i} & \vec{j} & \vec{k} \\ 4 & -6 & 2 \\ 0 & -10 & 0 \end{vmatrix} + \begin{vmatrix} \vec{i} & \vec{j} & \vec{k} \\ 8 & -6 & 2 \\ 0 & 10 & 0 \end{vmatrix}$

$= 20\vec{i} - 40\vec{k} - 20\vec{i} + 80\vec{k} \boxed{= 40\vec{k}}$

(c) The figure is shown .

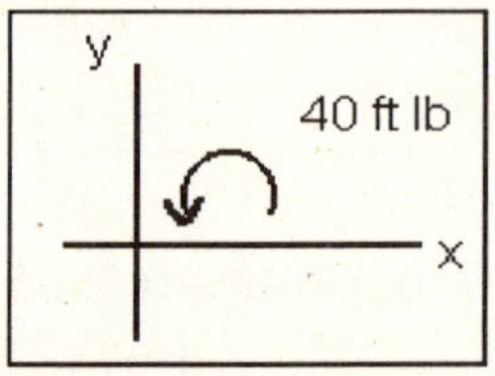

==============================<>==============================

Problem 4.97 The forces are contained in the x-y plane. (a) Determine the moment of the couple. (b) What is the sum of the moments of the two forces about the point (10. -40, 20) ft? (c) Represent the couple as shown in Figure 4.26(c).

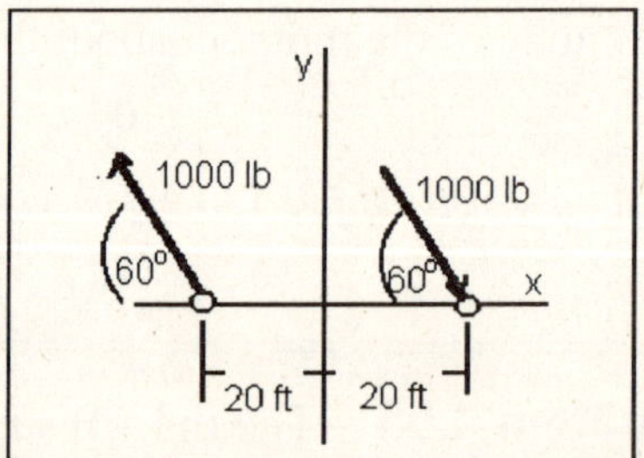

Solution: (a) The forces are

$\vec{F}_1 = 1000\left(\vec{i}\cos(-60^o) + \vec{j}\sin(-60^o)\right) = 500\vec{i} - 866\vec{j}$

$\vec{F}_2 = 1000\left(\vec{i}\cos 120^o + \vec{j}\sin 120^o\right) = -500i + 866\vec{j}$. The position vectors are $\vec{r}_1 = 20\vec{i}$, $\vec{r}_2 = -20\vec{i}$. The sum of the moments about the origin

is $\sum \vec{M}_O = \vec{r}_1 \times \vec{F}_1 + \vec{r}_2 \times \vec{F}_2 = \begin{vmatrix} \vec{i} & \vec{j} & \vec{k} \\ 20 & 0 & 0 \\ 500 & -866 & 0 \end{vmatrix} + \begin{vmatrix} \vec{i} & \vec{j} & \vec{k} \\ -20 & 0 & 0 \\ -500 & 866 & 0 \end{vmatrix}$

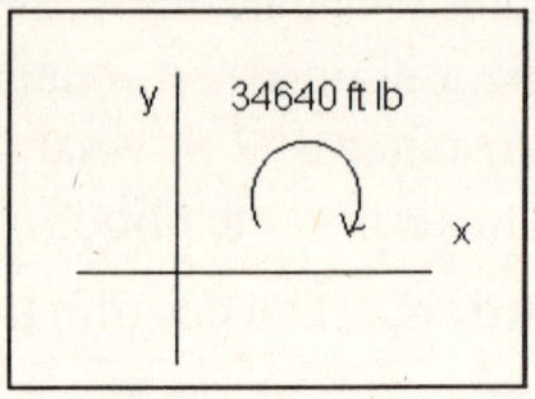

$= -17320\vec{k} - 17320\vec{k} \boxed{= -34640\vec{k}\ ft\ lb}$

(b) The distance from the point (10,-40, 20) ft and the force 1 is $\vec{r}_{P1} = \vec{r}_1 - \vec{r}_P = (20-10)\vec{i} + 40\vec{j} - 20\vec{k} = 10\vec{i} + 40\vec{j} - 20\vec{k}$. The distance to force 2 is $\vec{r}_{P2} = \vec{r}_2 - \vec{r}_p = (-20-10)\vec{i} + 40\vec{j} - 20\vec{k} = -30\vec{i} + 40\vec{j} - 20\vec{k}$. The moment about the point is $\vec{M}_P = \vec{r}_{p1} \times \vec{F}_1 + \vec{r}_{P2} \times \vec{F}_2$. Noting that $\vec{F}_2 = -\vec{F}_1$, this reduces to

$$\vec{M}_P = (\vec{r}_{P1} - \vec{r}_{P2}) \times \vec{F}_1 = \begin{vmatrix} \vec{i} & \vec{j} & \vec{k} \\ 40 & 0 & 0 \\ 500 & -866 & 0 \end{vmatrix} \boxed{= -34640\vec{k}\ (ft\ lb)}$$

(c) The figure is shown.

==============================<>==============================

Problem 4.98 The forces are contained in the x-y plane and the moment of the couple is $-90\vec{k}$ *N-m*. (a) What is the distance b? What is the sum of the moments of the two forces about the point (3, 3, -2) *m*?

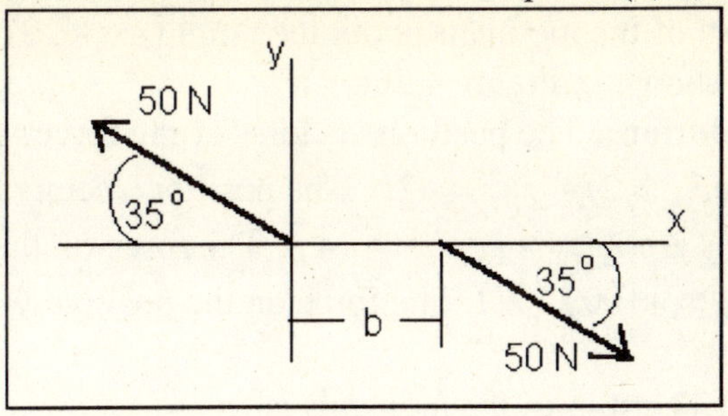

Solution: The force is

$$\vec{F} = 50\left(\vec{i}\cos\left(-35^o\right) + \vec{j}\sin\left(-35^o\right)\right) = 40.96\vec{i} - 28.68\vec{j}$$

The position vector is $\vec{r} = b\vec{i}$. The moment is

$$\vec{M} = \vec{r} \times \vec{F} = \begin{vmatrix} \vec{i} & \vec{j} & \vec{k} \\ b & 0 & 0 \\ 40.96 & -28.68 & 0 \end{vmatrix} = -28.68b\vec{k}.$$ From

which $b = \dfrac{90}{28.68} = 3.138\ m$ $\boxed{= 3.14\ m}$

(b) The distances to the two points of application of the forces are $\vec{r}_O = -3\vec{i} - 3\vec{j} + 2\vec{k}$, and $\vec{r}_1 = (3.14 - 3)\vec{i} - 3\vec{j} + 2\vec{k} = 0.14\vec{i} - 3\vec{j} + 2\vec{k}$. The moment is

$$\vec{M} = (\vec{r}_1 - \vec{r}_O) \times \vec{F} = \begin{vmatrix} \vec{i} & \vec{j} & \vec{k} \\ 3.14 & 0 & 0 \\ 40.96 & -28.68 & 0 \end{vmatrix} \boxed{= -90\vec{k}\ N\text{-}m}$$

Problem 4.99 Point P is contained in the x-y plane. $|\vec{F}| = 100\ N$, and the moment of the couple is $-500\vec{k}\ N\text{-}m$. What are the coordinates of P?

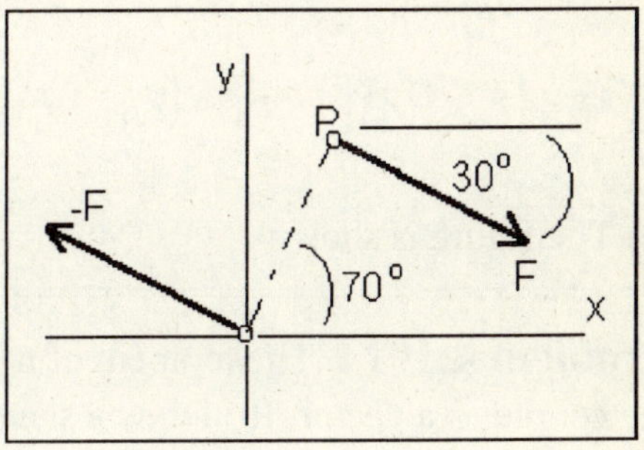

Solution: The force is

$\vec{F} = 100\left(\vec{i}\cos(-30^o) + \vec{j}\sin(-30^o)\right) = 86.6\vec{i} - 50\vec{j}$. Let r be the distance OP. The vector parallel to OP is $\vec{r} = r\left(\vec{i}\cos 70^o + \vec{j}\sin 70^o\right) = r\left(0.3420\vec{i} + 0.9397\vec{j}\right)$. The moment is

$$\vec{M} = \vec{r} \times \vec{F} = \begin{vmatrix} \vec{i} & \vec{j} & \vec{k} \\ 0.3420r & 0.9397r & 0 \\ 86.6 & -50.0 & 0 \end{vmatrix} = -98.48r\vec{k}$$ From

which, $r = \dfrac{500}{98.48} = 5.077\ m$. From above, $\vec{r} = 5.077\left(0.3420\vec{i} + 0.9397\vec{j}\right)$. The coordinates of P are

$\boxed{x = 5.077(0.3420) = 1.74\ m}$, $\boxed{y = 5.077(0.9397) = 4.77\ m}$

Problem 4.100 The forces are contained in the x-y plane. (a) Determine the sum of the moments of the two couples. (b) What is the sum of the moments about the point (-6, -6, 2)? (c) Represent the result as shown in Figure 4.26c.

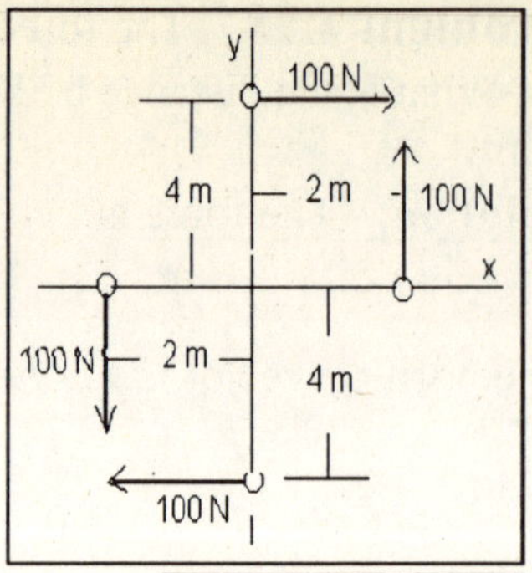

Solution: The position vectors for the forces on the x-axis are $\vec{r}_{X1} = 2\vec{i}$, $\vec{r}_{X2} = -2\vec{i}$. The position vectors for the forces on the y-axis are $\vec{r}_{Y1} = 4\vec{j}$, $\vec{r}_{Y2} = -4\vec{j}$. The force on the positive x-axis is $\vec{F}_X = +100\vec{j}\ (N)$. The force on the positive y-axis is $\vec{F}_Y = +100\vec{i}\ (N)$.

(a) The sum of the moments is

$$\sum \vec{M} = (\vec{r}_{X1} - \vec{r}_{X2}) \times \vec{F}_X + (\vec{r}_{Y1} - \vec{r}_{Y2}) \times \vec{F}_Y = \begin{vmatrix} \vec{i} & \vec{j} & \vec{k} \\ 4 & 0 & 0 \\ 0 & 100 & 0 \end{vmatrix} + \begin{vmatrix} \vec{i} & \vec{j} & \vec{k} \\ 0 & 8 & 0 \\ 100 & 0 & 0 \end{vmatrix}$$

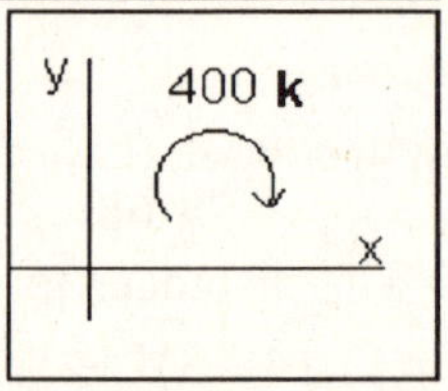

$\boxed{= -400\vec{k}\ (N\text{-}m)}$

(b) The vector from the point P(-6, -6, 2) to the force on the positive x-axis is $\vec{r}_{PX1} = \vec{r}_{X1} - \vec{r}_p = (2+6)\vec{i} + 6\vec{j} - 2\vec{k}$. The vector from the point P to the force on the negative x-axis is $\vec{r}_{pX2} = \vec{r}_{X2} - \vec{r}_p = (-2+6)\vec{i} + 6\vec{j} - 2\vec{k}$. The vector from point P to the force on the positive y-axis is $\vec{r}_{pY1} = \vec{r}_{Y1} - \vec{r}_p = +6\vec{i} + (4+6)\vec{j} - 2\vec{k} = 6\vec{i} + 10\vec{j} - 2\vec{k}$. The vector from P to the force on the negative y axis is $\vec{r}_{PY2} = \vec{r}_{Y2} - \vec{r}_P = 6\vec{i} + (-4+6)\vec{j} - 2\vec{k} = 6\vec{i} + 2\vec{j} - 2\vec{k}$. The sum of the moments:

$$\sum \vec{M} = (\vec{r}_{pX1} - \vec{r}_{pX2}) \times \vec{F}_X + (\vec{r}_{pY1} - \vec{r}_{pY2}) \times \vec{F}_Y = \begin{vmatrix} \vec{i} & \vec{j} & \vec{k} \\ 4 & 0 & 0 \\ 0 & 100 & 0 \end{vmatrix} + \begin{vmatrix} \vec{i} & \vec{j} & \vec{k} \\ 0 & 8 & 0 \\ 100 & 0 & 0 \end{vmatrix} \boxed{= -400\vec{k}\ (N\text{-}m)}$$

(c) The figure is shown.

Problem 4.101 The moment of a couple is 40 *kN-m* counterclockwise. (a) Express the moment of the couple as a vector. (b) Draw a sketch showing two equal and opposite forces that exert the given couple.

Solution: (a) The moment is directed along the positive z-axis, $\boxed{\vec{M} = 40\vec{k}\ (kN\text{-}m)}$

(b) A candidate pair of forces is shown in the sketch: 40 *kN* forces directed oppositely at 1 *m* apart.

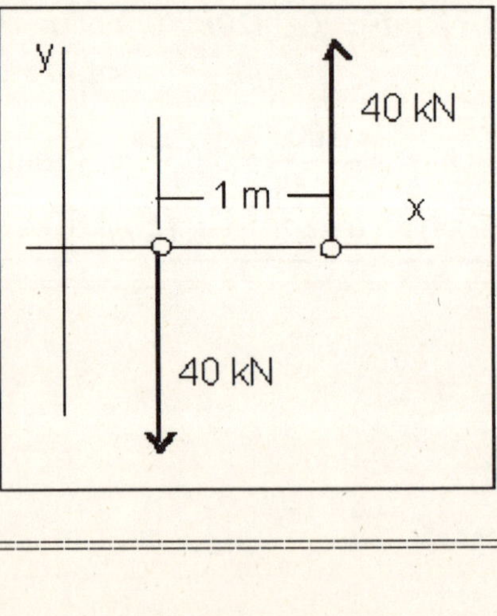

Problem 4.102 The moments of two couples are shown. What is the sum of the moments about the point P?

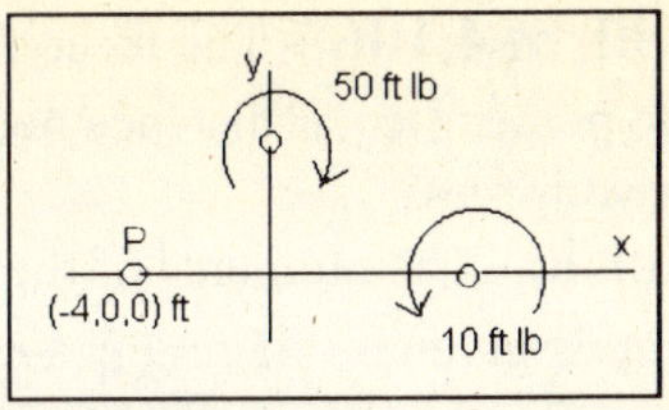

Solution: The moment of a couple is the same anywhere in the plane. Hence the sum about the point P is

$\boxed{\sum \vec{M} = -50\vec{k} + 10\vec{k} = -40\vec{k} \text{ ft lb}}$

Problem 4.103 Determine the sum of the moments exerted on the plate by the two couples.

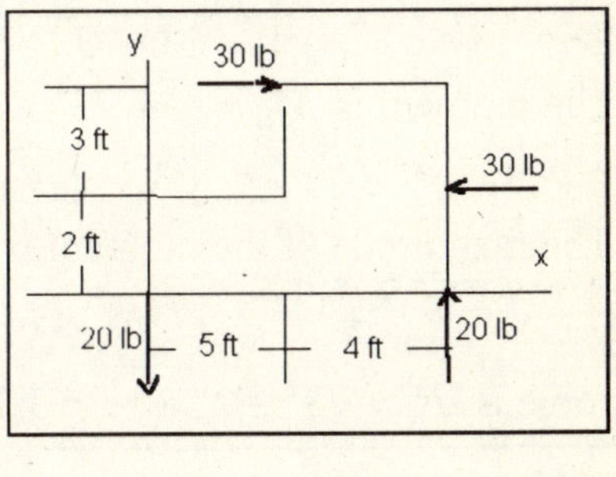

Solution: The moment due to the 30 *lb* couple, which acts in a clockwise direction is $\vec{M}_{30} = -3(30)\vec{k} = -90\vec{k}\ ft\ lb$. The moment due to the 20 *lb* couple, which acts in a counterclockwise direction, is $\vec{M}_{20} = 9(20)\vec{k} = 180\vec{k}\ ft\ lb$. The sum of the moments is $\boxed{\sum \vec{M} = -90\vec{k} + 180\vec{k} = +90\vec{k}\ ft\ lb}$. The sum of the moments is the same anywhere on the plate.

Problem 4.104 What is the sum of the moment exerted on the object?

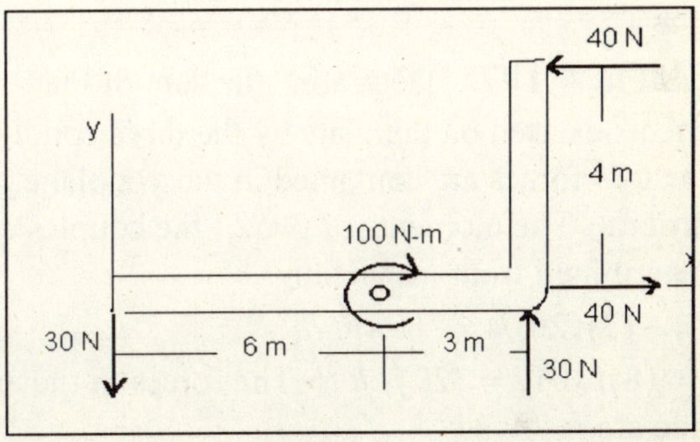

Solution: Three couples act on the object. The moments due to the couples are: $\vec{M}_1 = -100\vec{k}\ N\text{-}m$.

$\vec{M}_2 = 9(30)\vec{k} = 270\vec{k}\ N\text{-}m$,

$\vec{M}_3 = 4(40)\vec{k} = 160\vec{k}\ N\text{-}m$

The sum of the moments:

$\boxed{\sum \vec{M} = -100\vec{k} + 270\vec{k} + 160\vec{k} = 330\vec{k}\ N\text{-}m}$

Problem 4.105 Four forces and a couple act on the beam. The vector sum of the forces is zero and the sum of the moments about the left end of the beam is zero. What are the forces A_X, A_Y, and B?

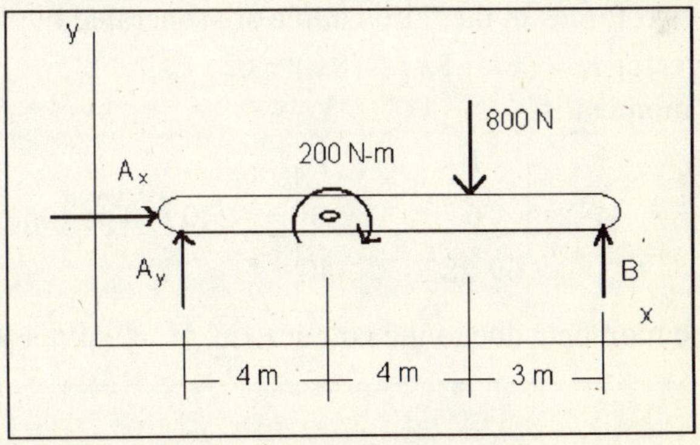

Solution: The sum of the forces about the y-axis is $\sum \vec{F}_X = A_Y + B - 800 = 0$. The sum of the forces about the x-axis is $\sum \vec{F}_X \boxed{= A_X = 0}$. The sum of the moments about the left end of the beam is $\sum \vec{M}_L = 11B - 8(800) - 200 = 0$. From the moments: $\boxed{B = \frac{6600}{11} = 600\ N}$ Substitute into the forces balance equation to obtain:

$\boxed{A_Y = 800 - 600 = 200\ N}$

Problem 4.106 The force $\vec{F} = 3\vec{i} + 2\vec{j} - 1\vec{k}$ (kN). (a) What is the moment of the couple? (b) What is the perpendicular distance between the lines of action of the two forces?

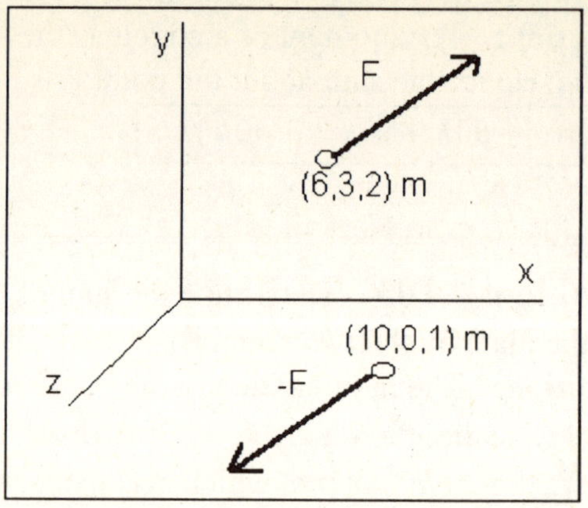

Solution: The moments of the two forces about the origin is given by $\vec{M}_O = (\vec{r}_1 - \vec{r}_2) \times \vec{F}$, where the two vectors are $\vec{r}_1 = 6\vec{i} + 3\vec{j} + 2\vec{k}$, and $\vec{r}_2 = 10\vec{i} + 0j + 1\vec{k}$. The difference: $\vec{r}_1 - \vec{r}_2 = -4\vec{i} + 3\vec{j} + 1\vec{k}$.

(a) The moment is $\vec{M}_O = \begin{vmatrix} \vec{i} & \vec{j} & \vec{k} \\ -4 & 3 & 1 \\ 3 & 2 & -1 \end{vmatrix} \boxed{= -5\vec{i} - 1\vec{j} - 17\vec{k}}$

(b) The magnitude of the moment is $\left|\vec{M}_O\right| = \sqrt{5^2 + 1^2 + 17^2} = 17.75\ kN\text{-}m$. The magnitude of the force is $\left|\vec{F}\right| = \sqrt{3^2 + 2^2 + 1^2} = 3.74\ kN$. The perpendicular distance is

$$\boxed{D = \frac{\left|\vec{M}_O\right|}{\left|\vec{F}\right|} = \frac{17.75\ kN\text{-}m}{3.74\ kN} = 4.74\ m}$$

Problem 4.107 Determine the sum of the moments exerted on the plate by the three couples. (The 80 *lb* forces are contained in the x-z plane.)

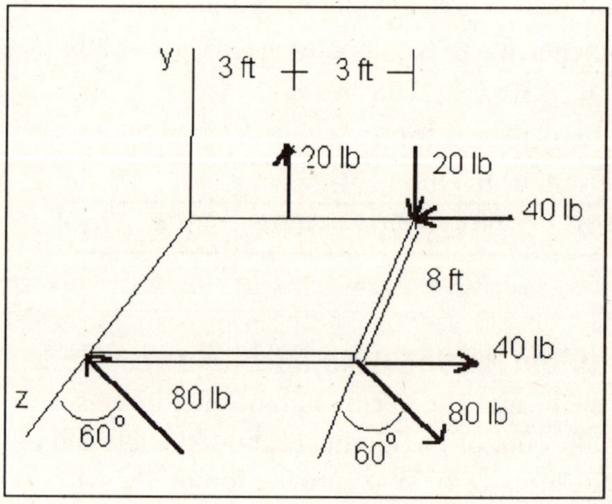

Solution: The moments of two of the couples can be determined from inspection:

$\vec{M}_1 = -(3)(20)\vec{k} = -60\vec{k}\ ft\ lb$.

$\vec{M}_2 = (8)(40)\vec{j} = 320\vec{j}\ ft\ lb$ The forces in the 3rd couple are resolved:

$\vec{F} = (80)(\vec{i}\sin 60^o + \vec{k}\cos 60^o) = 69.282\vec{i} + 40\vec{k}$

The two forces in the third couple are separated by the vector $\vec{r}_3 = (6\vec{i} + 8\vec{k}) - (8\vec{k}) = 6\vec{i}$

The moment is

$\vec{M}_3 = \vec{r}_3 \times \vec{F}_3 = \begin{vmatrix} \vec{i} & \vec{j} & \vec{k} \\ 6 & 0 & 0 \\ 69.282 & 0 & 40 \end{vmatrix} = -240\vec{j}$. The sum of the moments due to the couples: $\sum \vec{M} = -60\vec{k} + 320\vec{j} - 240\vec{j} \boxed{= 80\vec{j} - 60\vec{k}\ ft\ lb}$

Problem 4.108 What is the magnitude of the sum of the moments exerted on the T-shaped structure by the two couples? (The force $\vec{F}$ in the sketch is $\vec{F} = 50\vec{i} + 20\vec{j} - 10\vec{k}\ lb.$)

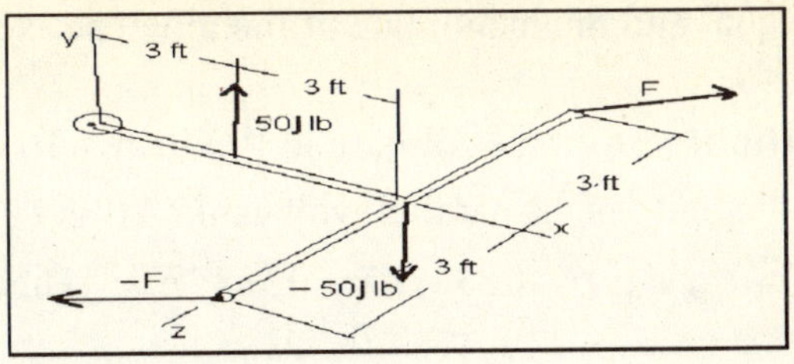

Solution: The moment of the 50 lb couple can be determined by inspection: $\vec{M}_1 = -(50)(3)\vec{k} = -150\vec{k}\ ft\ lb$. The vector separating the other two force is $\vec{r} = 6\vec{k}$. The moment is

$$\vec{M}_2 = \vec{r} \times \vec{F} = \begin{vmatrix} \vec{i} & \vec{j} & \vec{k} \\ 0 & 0 & 6 \\ 50 & 20 & -10 \end{vmatrix} = -120\vec{i} + 300\vec{j}$$

The sum of the moments is $\sum \vec{M} = -120\vec{i} + 300\vec{j} - 150\vec{k}$. The magnitude is $\boxed{|\vec{M}| = \sqrt{120^2 + 300^2 + 150^2} = 356.23\ ft\ lb}$

Problem 4.109 The tension in cables AB and CD is 500 *N*. (a) Show that the two forces exerted by the cables on the rectangular hatch at B and C form a couple. (b) What is the moment exerted on the plate by the cables?

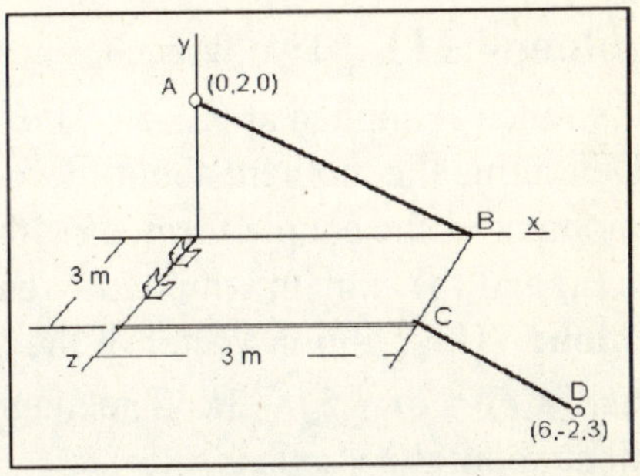

Solution: One condition for a couple is that the sum of a pair of forces vanish; another is for a non-zero moment to be the same anywhere. The first condition is demonstrated by determining the unit vectors parallel to the action lines of the forces. The vector position of point B is $\vec{r}_B = 3\vec{i}\ m$. The vector position of point A is $\vec{r}_A = 2\vec{j}$. The vector parallel to cable AB is $\vec{r}_{BA} = \vec{r}_A - \vec{r}_B = -3\vec{i} + 2\vec{j}$. The magnitude is:

$|\vec{r}_{AB}| = \sqrt{3^2 + 2^2} = 3.606\ m$. The unit vector: $\vec{e}_{AB} = \dfrac{\vec{r}_{AB}}{|\vec{r}_{AB}|} = -0.8321\vec{i} + 0.5547\vec{j}$. The tension is

$\vec{T}_{AB} = |\vec{T}_{AB}|\vec{e}_{AB} = -416.5\vec{i} + 277.35\vec{j}$. The vector position of points C and D are: $\vec{r}_C = 3\vec{i} + 3\vec{k}$, $\vec{r}_D = 6\vec{i} - 2\vec{j} + 3\vec{k}$. The vector parallel to the cable CD is $\vec{r}_{CD} = \vec{r}_D - \vec{r}_C = 3\vec{i} - 2\vec{j}$. The magnitude is $|\vec{r}_{CD}| = 3.606\ m$, and the unit vector parallel to the cable CD is $\vec{e}_{CD} = +0.8321\vec{i} - 0.5547\vec{j}$. The magnitude of the tension in the two cables is the same, and $\vec{e}_{BA} = -\vec{e}_{CD}$, hence the sum of the tensions vanish on the plate. The second condition is demonstrated by determining the moment at any point on the plate. By inspection, the distance between the action lines of the forces is

$$\vec{r}_{CB} = \vec{r}_B - \vec{r}_C = 3\vec{i} - 3\vec{i} - 3\vec{k} = -3\vec{k}.$$ The moment is $$\vec{M} = \vec{r}_{CB} \times \vec{T}_{AB} = \begin{vmatrix} \vec{i} & \vec{j} & \vec{k} \\ 0 & 0 & -3 \\ -416.05 & 277.35 & 0 \end{vmatrix}$$

$\boxed{= 832.05\vec{i} - 1248.15\vec{j}\ (N\text{-}m)}$ The moment about the origin is $\vec{M}_o = (\vec{r}_B - \vec{r}_C) \times \vec{T}_{AB} = \vec{r}_{CB} \times \vec{T}_{AB}$, which is identical with the above expression for the moment. Let $\vec{r}_{PC}$ and $\vec{r}_{PB}$ be the distances to points C and B from an arbitrary point P on the plate. Then $\vec{M}_P = (\vec{r}_{PB} - \vec{r}_{PC}) \times \vec{T}_{AB} = \vec{r}_{CB} \times \vec{T}_{AB}$ which is identical to the above expression. Thus the moment is the same everywhere on the plate, and the forces form a couple.

==============================<>==============================

Problem 4.110 Consider the system shown in Problem 4.109. The tensions in cables AB and CD are equal and the magnitude of the moment exerted on the plate is 6 *kN m*. What is the tension in the cables?

Solution: From the solution to Problem 4.109 the vector from point C to point B is $\vec{r}_{CB} = -3\vec{k}\ (m)$, and the unit vector parallel with cable AB is $\vec{e}_{AB} = -0.8321\vec{i} + 0.5547\vec{j}$. The tension in the cable is $\vec{T}_{AB} = |\vec{T}_{AB}|\vec{e}_{AB} = -0.8321T\vec{i} + 0.5547T\vec{j}$. The moment exerted on the plate is

$$\vec{M} = \vec{r}_{CB} \times \vec{T}_{AB} = \begin{vmatrix} \vec{i} & \vec{j} & \vec{k} \\ 0 & 0 & -3 \\ -0.8321T & 0.5547T & 0 \end{vmatrix} = +1.6641T\vec{i} - 2.4963T\vec{j}$$

From the solution of Problem 4.109 the unit vector parallel to the moment is $\vec{e}_M = 0.5547\vec{i} - 0.832\vec{j}$, from which the moment is $\vec{M} = 6\vec{e}_B = 3.3282\vec{i} - 4.9926\vec{j}$. Solving for the tension components:

$$\boxed{T = \frac{3.3282}{1.6641} = 2\ kN}. \quad Check:\ T = \frac{4.9926}{2.4963} = 2\ kN\ \ check.$$

==============================<>==============================

Problem 4.111 Two wrenches are used to tighten an elbow fitting. The force $\vec{F} = 10\vec{k}\ (lb)$ on the right wrench is applied at (6, -5, -3) *in* and the force $-\vec{F}$ on the left wrench is applied at (4, -5, 3) *in*. (a) Determine the moment about the x-axis due to the force exerted on the right wrench. (b) Determine the moment of the couple formed by the forces exerted on the two wrenches. (c) Based on the results of Parts (a) and (b) explain why two wrenches are used.

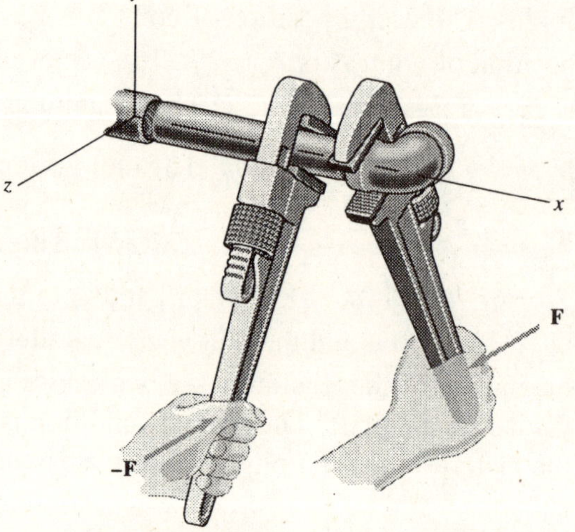

Solution: The position vector of the force on the right wrench is $\vec{r}_R = 6\vec{i} - 5\vec{j} - 3\vec{k}$. The magnitude of the moment about the x-axis is

$$|\vec{M}_R| = \vec{e}_X \cdot (\vec{r}_R \times \vec{F}) = \begin{vmatrix} 1 & 0 & 0 \\ 6 & -5 & -3 \\ 0 & 0 & 10 \end{vmatrix} \boxed{= -50\ in\ lb}$$

(a) The moment about the x-axis is

$$\boxed{\vec{M}_R = |\vec{M}_R|\vec{e}_X = -50\vec{i}\ (in\ lb)}.$$

(b) The moment of the couple is

$$\vec{M}_C = (\vec{r}_R - \vec{r}_L) \times \vec{F}_R = \begin{vmatrix} \vec{i} & \vec{j} & \vec{k} \\ 2 & 0 & -6 \\ 0 & 0 & 10 \end{vmatrix} \boxed{= -20\vec{j}\ in\ lb}$$

(c) The objective is to apply a moment to the elbow relative to connecting pipe, and zero resultant moment to the pipe itself. A resultant moment about the x-axis will affect the joint at the origin. However the use of two wrenches results in a net zero moment about the x-axis-- the moment is absorbed at the juncture of the elbow and the pipe. This is demonstrated by calculating the moment about the x-axis due to the left wrench:

$$|\vec{M}_X| = \vec{e}_X \cdot (\vec{r}_L \times \vec{F}_L) = \begin{vmatrix} 1 & 0 & 0 \\ 4 & -5 & 3 \\ 0 & 0 & -10 \end{vmatrix} = 50\ in\ lb$$

from which $\vec{M}_{XL} = 50\vec{i}\ in\ lb$, which is opposite in direction and equal in magnitude to the moment exerted on the x-axis by the right wrench. The left wrench force is applied 2 *in* nearer the origin than the right wrench force, hence the moment must be absorbed by the space between, where it is wanted.

==============================<>==============================

Problem 4.112 Two systems of forces act on the beam. Are they equivalent?

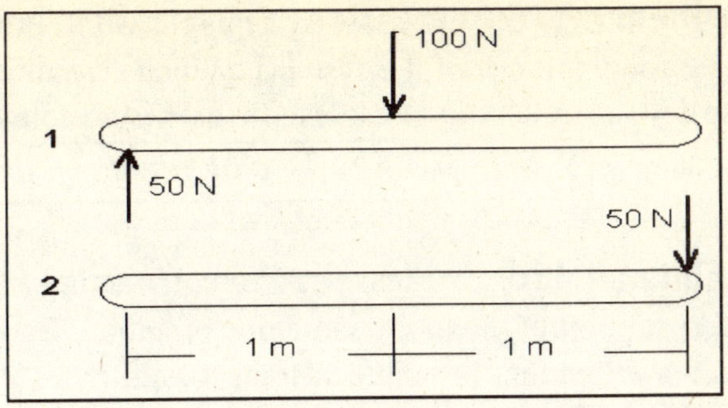

Solution: The strategy is to check the two conditions for equivalence: (a) the sums of the forces must be equal and (b) the sums of the moments about an arbitrary point must be equal. The sums of the forces of the two systems:

$\sum \vec{F}_X = 0$, (both systems) and

$\sum \vec{F}_{Y1} = -100\vec{j} + 50\vec{j} = -50\vec{j}\ (N)$

$\sum \vec{F}_{Y2} = -50\vec{j}\ (N)$. The sums of the forces are equal. The sums of the moments about the left end are:

$\sum \vec{M}_1 = -(1)(100)\vec{k} = -100\vec{k}\ (N\text{-}m)$

$\sum \vec{M}_2 = -(2)(50)\vec{k} = -100\vec{k}\ (N\text{-}m)$ The sums of the moments about the left end are equal. Choose any point P at the same distance $\vec{r} = x\vec{i}$ from the left end on each beam. The sums of the moments about the point P are $\sum \vec{M}_1 = (-50x + 100(x-1))\vec{k} = (50x - 100)\vec{k}\ (N\text{-}m)$

$\Sigma \vec{M}_2 = (-50(2 - x))\vec{k} = (50x - 100)\vec{k}\ (N\text{-}m)$ Thus the sums of the moments about any point on the beam are equal for the two sets of forces; the systems are equivalent. *Yes*

Problem 4.113 Two systems of forces act on the beam. Are they equivalent?

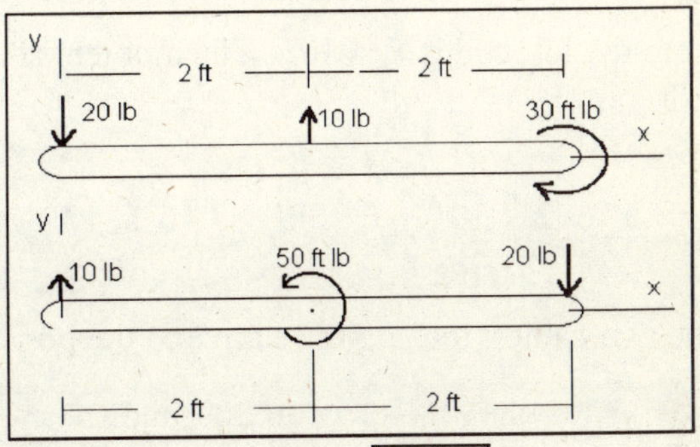

Solution: The sums of the forces are:

$\sum \vec{F}_X = 0$ (both systems)

$\sum \vec{F}_{Y1} = 10\vec{j} - 20\vec{j} = -10\vec{j}\ (lb)$

$\sum \vec{F}_{Y2} = -20\vec{j} + 10\vec{j} = -10\vec{j}\ (lb)$

Thus the sums of the forces are equal. The sums of the moments about the left end are:

$\sum \vec{M}_1 = (-20)(4)\vec{k} + 50\vec{k} = -30\vec{k}\ (ft\ lb)$

$\sum \vec{M}_2 = (+10(2))\vec{k} - 30\vec{k} = -10\vec{k}\ (ft\ lb)$

The sums of the moments are not equal, hence the systems are not equivalent *No*

Problem 4.114 Four systems of forces and moments act on an 8 *ft* beam. Which systems are equivalent?

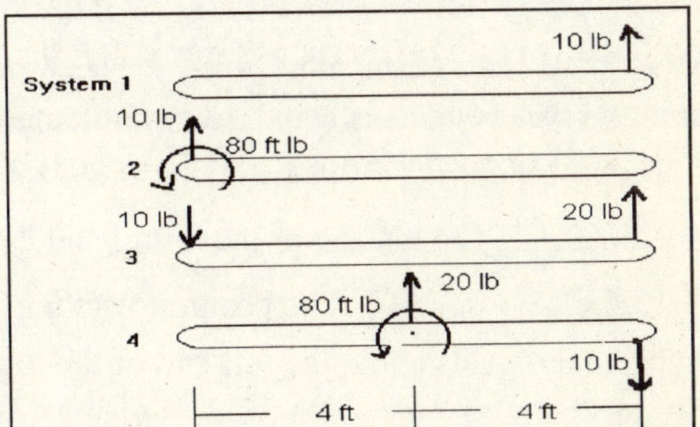

Solution: By inspection, the sums of the forces are equal in all four systems. The sums of the moments about the left end are

$\sum \vec{M}_1 = 80\vec{k}\ (ft\ lb)$

$\sum \vec{M}_2 = +80\vec{k}\ (ft\ lb)$

$\sum \vec{M}_3 = 160\vec{k}\ (ft\ lb)$

$\sum \vec{M}_4 = 80\vec{k} + 20(4)\vec{k} - (10)(8)\vec{k} = 80\vec{k}\ (ft\ lb)$

The first, second and fourth systems are equivalent.

Problem 4.115

The four systems shown in Problem 4.14 can be made equivalent by adding a single couple to one of the systems. Which system is it, and what couple must be added?

Solution: An 80 *ft lb* clockwise (negative) couple must be added to System 3. The new sum of moments is $\boxed{\sum \vec{M}_3 = +(8)(20)\vec{k} - 80\vec{k} = 80\vec{k}\ (ft\ lb)}$.

Problem 4.116

System 1 is a force $\vec{F}$ acting at a point O. System 2 is the force $\vec{F}$ acting at a different point O' along the same line of action. Explain why these systems are equivalent. (This simple result is called the "principle of transmissibility".)

Solution: The sum of forces is obviously equal for both systems. Let P be any point on the beam. The moment about P is the cross product of the distance from P to the line of action of a force times the force, that is, $\vec{M} = \vec{r}_{PL} \times \vec{F}$, where $\vec{r}_{PL}$ is the distance from P to the line of action of $\vec{F}$. Since both systems have the same line of action, and the forces are equal, the systems are equivalent.

Problem 4.117

The vector sum of the forces on the log by the cables is the same in the two cases. Show that the systems of forces exerted on the log in the two cases are equivalent.

Solution: The angle formed by the single cable with the positive x-axis is $\theta = 180^o - \tan^{-1}\left(\frac{12}{16}\right) = 143.13^o$. The single cable tension is $\vec{T}_1 = |\vec{T}|(\vec{i}\cos 143.13^o + \vec{j}\sin 143.13^o)$ $= |\vec{T}|(-0.8\vec{i} + 0.6\vec{j})$. The position vector to the center of the log from the left end is $\vec{r}_c = 10\vec{i}$. The moment about the end of the log is

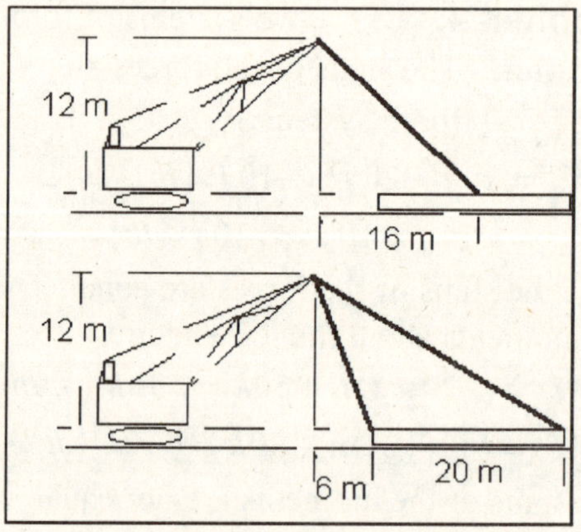

$$\vec{M} = \vec{r} \times \vec{T}_1 = |\vec{T}_1| \begin{vmatrix} \vec{i} & \vec{j} & \vec{k} \\ 10 & 0 & 0 \\ -0.8 & 0.6 & 0 \end{vmatrix} = |\vec{T}|(6)\vec{k}\ (N\text{-}m)$$

For the two cables, the angles relative to the positive x-axis are $\theta_1 = 180^o - \tan^{-1}\left(\frac{12}{6}\right) = 116.56^o$, and

$\theta_2 = 180 - \tan^{-1}\left(\frac{12}{26}\right) = 155.22^o$. The two cable vectors are

$\vec{T}_L = |\vec{T}_L|(\vec{i}\cos 116.56^o + \vec{j}\sin 116.56^o) = |\vec{T}_L|(-0.4472\vec{i} + 0.8945\vec{j})$. ,

$\vec{T}_R = |\vec{T}_R|(\vec{i}\cos 155.22^o + \vec{j}\sin 155.22^o) = |\vec{T}_R|(-0.9079\vec{i} + 0.4191\vec{j})$. Since the vector sum of the forces in the two systems is equal, two simultaneous equations are obtained:

$0.4472|\vec{T}_L| + 0.9079|\vec{T}_R| = 0.8|\vec{T}_1|$, and $0.8945|\vec{T}_L| + 0.4191|\vec{T}_R| = 0.6|\vec{T}_1|$ Solve: $|\vec{T}_L| = 0.3353|\vec{T}_1|$, and $|\vec{T}_R| = 0.7160|\vec{T}_1|$. The tension in the right hand cable is $\vec{T}_R = |\vec{T}_1|(0.7160)(-0.9079\vec{i} + 0.4191\vec{j})$ $= |\vec{T}_1|(-0.6500\vec{i} + 0.3000)$. The position vector of the right end of the log is $\vec{r}_R = 20\vec{i}\ m$ relative to the left end The moments about the left end of the log for the second system are

$$\vec{M}_2 = \vec{r}_R \times \vec{T}_R = |\vec{T}_1| \begin{vmatrix} \vec{i} & \vec{j} & \vec{k} \\ 20 & 0 & 0 \\ -0.6500 & 0.3000 & 0 \end{vmatrix} = |\vec{T}_1|(6)\vec{k}\ (N\text{-}m)$$

This is equal to the moment about the left end of the log for System 1, hence the systems are equivalent.

Problem 4.118

System 1 and System 2 each consist of a couple. If they are equivalent, what is $\vec{F}$?

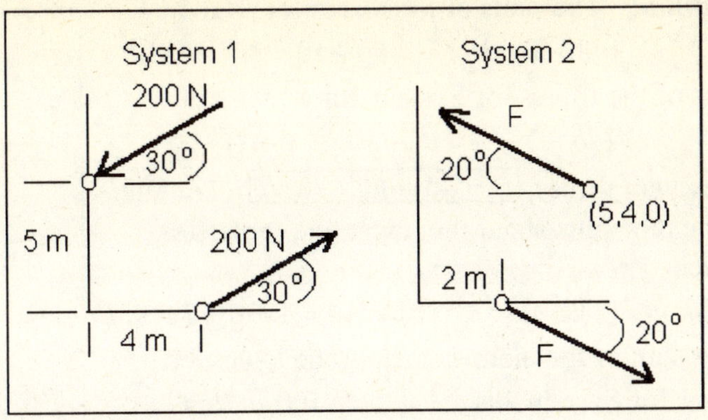

Solution: For couples, the sum of the forces vanish for both systems. For System 1, the two forces are located at $\vec{r}_{11} = 4\vec{i}$, and $\vec{r}_{12} = +5\vec{j}$. The forces are $\vec{F}_1 = 200(\vec{i}\cos 30^o + \vec{j}\sin 30^o) = 173.21\vec{i} + 100\vec{j}$. The moment due to the couple in System 1 is

$$\vec{M}_1 = (\vec{r}_{11} - \vec{r}_{12}) \times \vec{F}_1 = \begin{vmatrix} \vec{i} & \vec{j} & \vec{k} \\ 4 & -5 & 0 \\ 173.21 & 100 & 0 \end{vmatrix}$$

$= 1266.05\vec{k}$ $(N\text{-}m)$ For System 2, the positions of the forces are $\vec{r}_{21} = 2\vec{i}$, and $\vec{r}_{22} = 5\vec{i} + 4\vec{j}$. The forces are $\vec{F}_2 = F(\vec{i}\cos(-20^o) + \vec{j}\sin(-20^o)) = F(0.9397\vec{i} - 0.3420\vec{j})$. The moment of the couple in System 2 is $\vec{M}_2 = (\vec{r}_{21} - \vec{r}_{22}) \times \vec{F}_2 = F\begin{vmatrix} \vec{i} & \vec{j} & \vec{k} \\ -3 & -4 & 0 \\ 0.9397 & -0.3420 & 0 \end{vmatrix} = 4.7848F\vec{k}$, from which, if the systems are to be equivalent, $\boxed{F = \dfrac{1266}{4.7848} = 264.6\ N}$

Problem 4.119

Two equivalent systems of forces act on the L-shaped bar. Determine the forces F_A and F_B and the couple M.

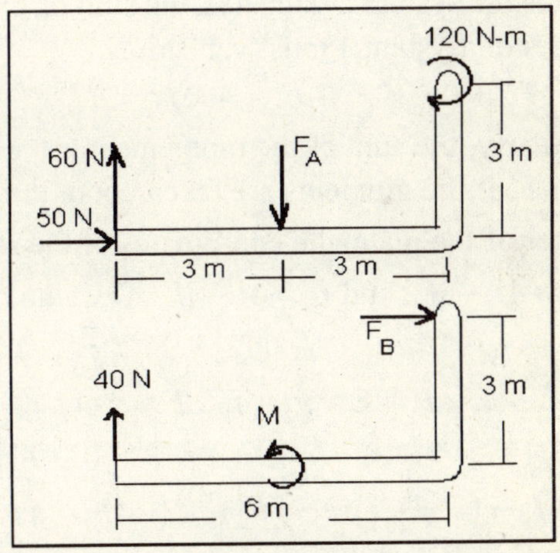

Solution: The sums of the forces for System 1 are $\sum F_X = 50$, and $\sum F_Y = -F_A + 60$. The sums of the forces for System 2 are $\sum F_X = F_B$, and $\sum F_Y = 40$. For equivalent systems: $\boxed{F_B = 50\ N}$, and $\boxed{F_A = 60 - 40 = 20\ N}$ The sum of the moments about the left end for System 1 is $\sum M_1 = -(3)F_A - 120 = -180\ N\text{-}m$. The sum of the moments about the left end for System 2 is $\sum M_2 = -(3)F_B + M = -150 + M\ N\text{-}m$. Equating the sums of the moments, $\boxed{M = 150 - 180 = -30\ N\text{-}m}$

Problem 4.120 Two equivalent systems act on the plate. Determine the force F and the couple M.

Solution: The sums of the forces for System 1 are $\sum F_X = 30\ lb$, $\sum F_Y = 50 - 10 = 40\ lb$. The sums of the forces for System 2 are $\sum F_X = 30\ lb$, $\sum F_Y = F - 30\ lb$. For equivalent forces, $\boxed{F = 30 + 40 = 70\ \ lb}$. The sum of the moments about the lower left corner for System 1 is

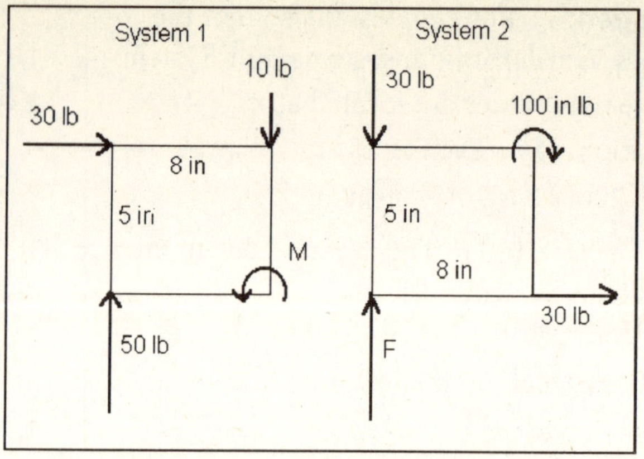

$\sum M_1 = -(5)(30) - (8)(10) + M = -230 + M\ in\ lb$. The sum of the moments about the lower left corner for System 2 is $\sum M_2 = -100\ in\ lb$. Equating the sum of moments, $\boxed{M = 230 - 100 = 130\ in\ lb}$

Problem 4.121 In System 1, four forces act on the rectangular flat plate. The forces are perpendicular to the plate and the 400 *lb* forces acts at its midpoint. In System 2, no forces or couples act on the plate. Systems 1 and 2 are equivalent. Determine the forces F_1, F_2, and F_3.

Solution: The strategy is to develop the force and moment conditions to obtain three simultaneous equations for the three unknown forces. Since the sum of the forces on System 2 vanishes, the sum of the forces on System 1 must vanish:

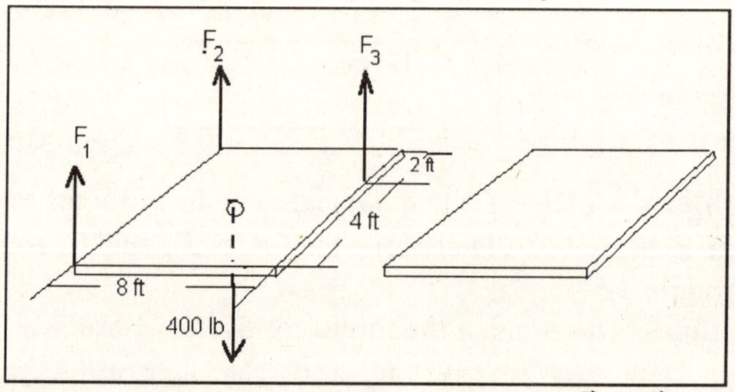

$$\sum \vec{F}_Y = (F_1 + F_2 + F_3 - 400)\vec{j} = 0$$

Similarly, the sum of the moments must vanish. If the moments are taken about the center of the plate, the coordinates of the forces relative to the plate mid point are $\vec{r}_1 = -4\vec{i} + 3\vec{k}$, $\vec{r}_2 = -4\vec{i} - 3\vec{k}$, and $\vec{r}_3 = 4\vec{i} + 1\vec{k}$. The sum of the moments about the midpoint of the plate is

$$\Sigma \vec{M}_1 = \vec{r}_1 \times \vec{F}_1 + \vec{r}_2 \times \vec{F}_2 + \vec{r}_3 \times \vec{F}_3 = \begin{vmatrix} \vec{i} & \vec{j} & \vec{k} \\ -4 & 0 & 3 \\ 0 & F_1 & 0 \end{vmatrix} + \begin{vmatrix} \vec{i} & \vec{j} & \vec{k} \\ -4 & 0 & -3 \\ 0 & F_2 & 0 \end{vmatrix} + \begin{vmatrix} \vec{i} & \vec{j} & \vec{k} \\ 4 & 0 & -1 \\ 0 & F_3 & 0 \end{vmatrix}$$

$\sum \vec{M}_1 = (-3F_1 + 3F_2 + F_3)\vec{i} + (-4F_1 - 4F_2 + 4F_3)\vec{k} = 0$. The three simultaneous equations are:

$F_1 + F_2 + F_3 = 400$

$-3F_1 + 3F_2 + F_3 = 0$

$-4F_1 - 4F_2 + 4F_3 = 0$ (Clearly, this last equation may be divided through by 4; it is left in this form for an audit trail.) Solve: $\boxed{F_1 = 133.33\ lb}$, $\boxed{F_2 = 66.67\ lb}$, $\boxed{F_3 = 200\ lb}$. (The HP-28S hand held calculator was used to obtain the solution.)

=====================================<>=====================================

Problem 4.122 Three forces and a couple are applied to a beam (System 1). (a) If you represent System 1 by a force applied at A and a couple (System 2), what are $\vec{F}$ and $\vec{M}$? (b) If you represent System 1 by the force $\vec{F}$ (System 3) what is the distance D?

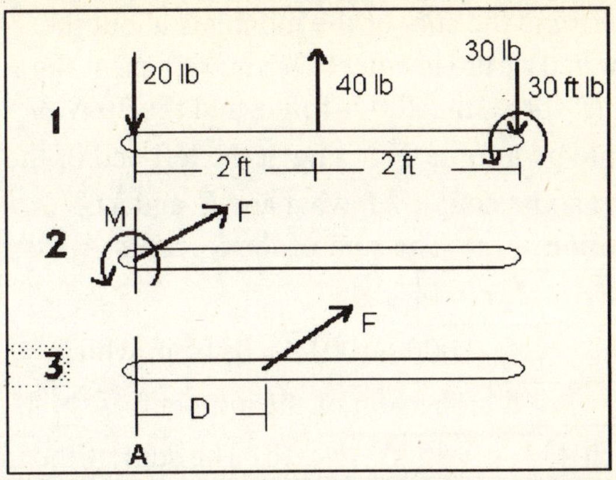

Solution: The sum of the forces in System 1 is $\sum \vec{F}_X = 0\vec{i}$, $\sum \vec{F}_Y = (-20+40-30)\vec{j} = -10\vec{j}\ lb$.

The sum of the moments about the left end for System 1 is

$\sum \vec{M}_1 = (2(40) - 4(30) + 30)\vec{k} = -10\vec{k}\ ft\ lb$.

(a) For System 2, the force at A is $\boxed{\vec{F} = -10\vec{j}\ lb}$

The moment at A is $\boxed{\vec{M}_2 = -10\vec{k}\ ft\ lb}$

(b) For System 3 the force at D is $\vec{F} = -10\vec{j}\ lb$. The distance D is the ratio of the magnitude of the moment to the magnitude of the force, where the magnitudes are those in System 1:

$$\boxed{D = \frac{10}{10} = 1\ ft}$$

=====================================<>=====================================

Problem 4.123 Represent the two forces and couple acting on the beam by a force $\vec{F}$. Determine $\vec{F}$ and determine where its line of action intersects the x axis.

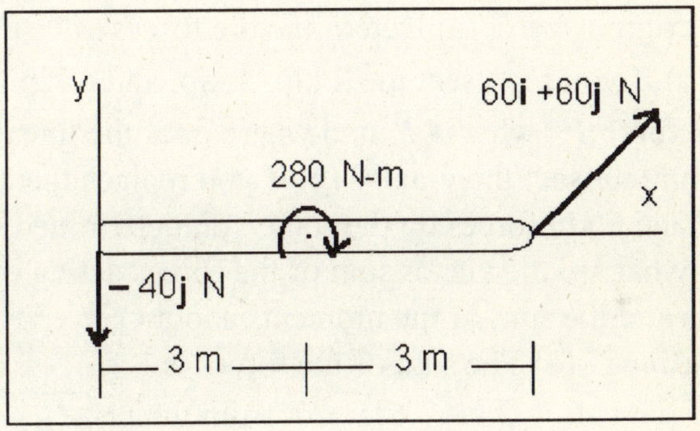

Solution: The sum of the forces is $\sum \vec{F}_X = 60\vec{i}$, and $\sum \vec{F}_Y = (60-40)\vec{j} = 20\vec{j}$. Thus the single force is $\boxed{\vec{F} = 60\vec{i} + 20\vec{j}}$ The sum of the moments about the left end is

$\sum \vec{M} = ((6)(60) - 280)\vec{k} = 80\vec{k}\ N\text{-}m$. From which the intersection on the x-axis of the line of action of the equivalent force is $\boxed{D = \frac{80}{20} = 4\ m}$.

Thus the force intersects the x-axis 4 *m* from the left end of the beam.

=====================================<>=====================================

Problem 4.124 The vector sum of the forces acting on the beam is zero and the sum of the moments about the left end of the beam is zero. (a) Determine the forces A_X, A_Y, and B. (b) If you represent the forces A_X, A_Y, and B by a force $\vec{F}$ acting at the right end of the beam and a couple *M*, what are $\vec{F}$ and *M*?

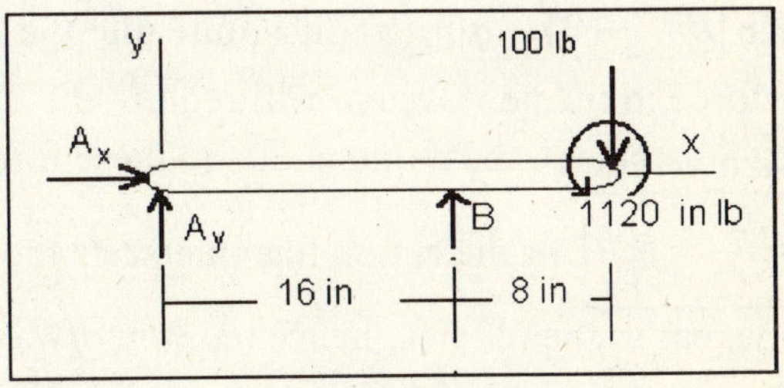

Solution: (a) The sum of the forces $\boxed{\sum \vec{F}_X = A_X = 0}$

$\sum \vec{F}_Y = (A_Y + B - 100) = 0$. The sum of the moments: $\sum \vec{M} = ((16)B - 24(100) + 1120)\vec{k} = (16B - 1280)\vec{k} = 0$, from which $\boxed{B = 80\ \ lb}$ From the force balance equation: $\boxed{A_Y = -80 + 100 = 20\ lb}$. (b) The force at the right end of the beam must balance the 100 lb force, $\boxed{\vec{F} = 100\vec{j}\ lb}$. The couple must balance the existing couple $\boxed{\vec{M} = -1120\vec{k}\ \ in\ lb}$

=====================================<>=====================================

================================<>================================

Problem 4.125 The vector sum of the forces acting on the beam is zero and the sum of the moments about the left end are zero. (a) Determine the forces A_X, A_Y, and the couple M_A. (b) Determine the sum of the moments about the right end of the beam. (c) If you represent the 600 *N* force, the 200 *N* force, and the 30 *N-m* couple by a force $\vec{F}$ acting at the left end of the beam and a couple *M*, what are $\vec{F}$ and *M*?

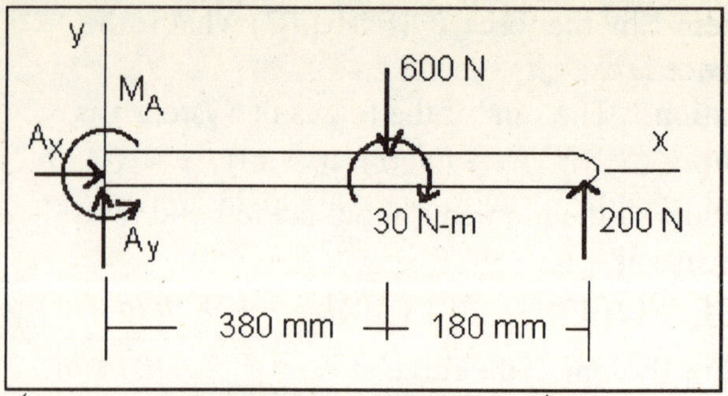

Solution: (a) The sum of the forces is $\boxed{\sum \vec{F}_X = A_X\vec{i} = 0}$ and $\sum \vec{F}_Y = (A_Y - 600 + 200)\vec{j} = 0$, from which $\boxed{A_Y = 400\ N}$. The sum of the moments is $\sum \vec{M}_L = (M_A - 0.38(600) - 30 + 0.560(200))\vec{k} = 0$, from which $\boxed{M_A = 146\ N\text{-}m}$. (b) The sum of the moments about the right end of the beam is $\boxed{\sum \vec{M}_L = (0.18(600) - 30 + 146 - 0.56(400)) = 0}$. (c) The sum of the forces for the new system is $\sum \vec{F}_Y = (A_Y + F)\vec{j} = 0$, from $\boxed{F = -A_Y = -400\ N}$, or $\boxed{\vec{F} = -400\vec{j}\ N}$ The sum of the moments for the new system is $\sum M = (M_A + M) = 0$, from which $\boxed{M = -M_A = -146\ N\text{-}m}$

================================<>================================

Problem 4.126 The vector sum of the forces acting on the truss is zero and the sum of the moments about the origin is zero. (a) Determine the forces A_X, A_Y, and *B*. (b) If you represent the 2 kip, 4 kip, and 6 kip forces by a force $\vec{F}$, what is $\vec{F}$ and where does the line of action intersect the y-axis? (c) If you replace the 2 *kip*, 4 *kip*, and 6 *kip* forces by the force you determined in Part (b), what are the vector sum of the forces acting on the truss and the sum of the moments about O?

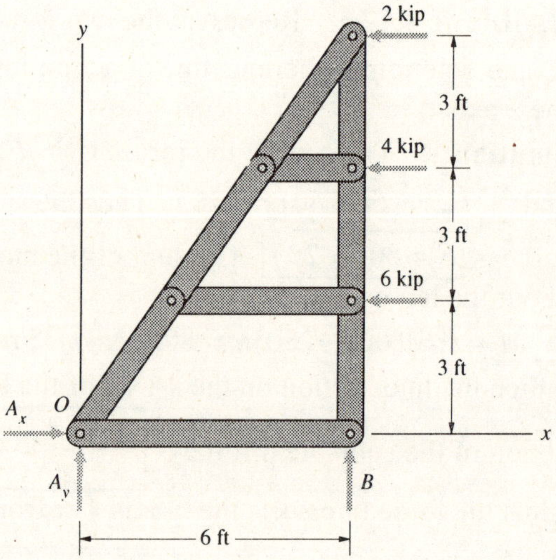

Solution: (a) The sum of the forces is $\sum \vec{F}_X = (A_X - 2 - 4 - 6)\vec{i} = 0$, from which $\boxed{A_X = 12\ kip}$ $\sum \vec{F}_Y = (A_Y + B)\vec{j} = 0$. The sum of the moments about the origin is $\sum M_O = (3)(6) + (6)(4) + (9)(2) + 6(B) = 0$, from which $\boxed{\vec{B} = -10\vec{j}\ kip}$. (b) Substitute into the force balance equation to obtain $\boxed{A_Y = -B = 10\ kip}$. (b) The force in the new system will replace the 2, 4, and 6 kip forces, $\boxed{\vec{F} = (-2 - 4 - 6)\vec{i} = -12\vec{i}\ kip}$. The force must match the moment due to these forces: $FD = 3(6) + (6)(4) + (9)(2) = 60\ kip\ ft$, from which $\boxed{D = \frac{60}{12} = 5\ ft}$, or the action line intersects the y-axis 5 ft above the origin. (c) The new system is equivalent to the old one, hence t*he sum of the forces vanish and the sum of the moments about O are zero.*

================================<>================================

==================================<>==================================

Problem 4.127 The distributed force exerted on part of a building foundation by the soil is represented by five forces. If you represent them by a force $\vec{F}$, what is $\vec{F}$ and where does its line of action intersect the x-axis?

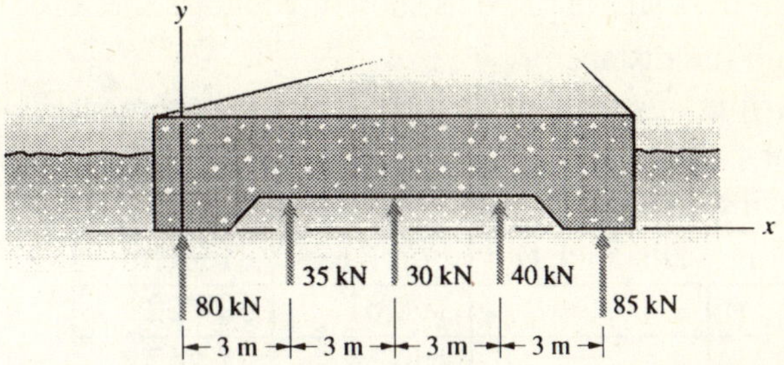

Solution: The equivalent force must equal the sum of the forces exerted by the soil:

$\boxed{\vec{F} = (80+35+30+40+85)\vec{j} = 270\vec{j}\ kN}$

The sum of the moments about any point must be equal for the two systems. The sum of the moments are $\sum M = 3(35)+6(30)+9(40)+12(85) = 1665\ \ kN\text{-}m$. Equating the moments for the two systems $FD = 1665\ kN\text{-}m$. from which $\boxed{D = \frac{1665\ kN\text{-}m}{270\ kN} = 6.167\ m}$

Thus the action line intersects the x axis at a distance $D = 6.167\ m$ to the right of the origin.

==================================<>==================================

Problem 4.128 The thrust forces exerted on an airplane by its four engines are parallel to the x-axis. Their magnitudes are:

Engine No	Thrust
1	160 kN
2	175 kN
3	185 kN
4	160 kN

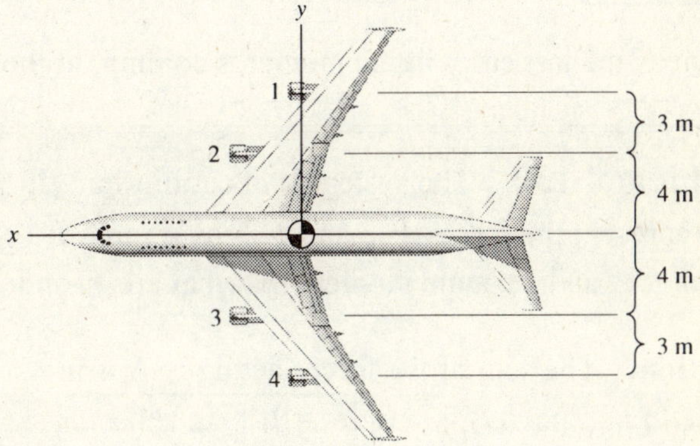

(a) Represent the four thrust forces by a force $\vec{F}$. What is $\vec{F}$ and where does its line of action intersect the y-axis? (b) The pilot wants to adjust the thrust of engine No 1 so that the four thrust forces can be represented by a force acting at the origin. What is the necessary thrust of engine No 1?

Solution: The force $\vec{F}$ must have the same value as the sum of the engine thrusts, $\boxed{\vec{F} = (160+175+185+160)\vec{i} = 680\vec{i}\ \ kN}$ The sum of the moments about the origin is $\sum M = 4(175)+7(160)-4(185)-7(160) = -40\ kN\text{-}m$ The distance from the y-axis is $\boxed{D = \frac{-40}{680} = -0.0588\ m}$ (b) The thrust in engine No 1 must be increased enough to cause an increase in the moment of 40 *kN*. Thus $\Delta T = \frac{40}{7} = 5.71\ kN$, so that the new thrust is $\boxed{T = 160 + \Delta T = 165.71\ \ kN}$

==================================<>==================================

Problem 4.129 The lift forces on an airplanes wing are represented by eight forces. The magnitude of each force is given in terms of its position on the wing by $200[1-\left(\frac{x}{17}\right)^2]^{\frac{1}{2}}\ lb$. If you represent the forces by a single force as shown, what are the force $\vec{F}$ and the distance D ?

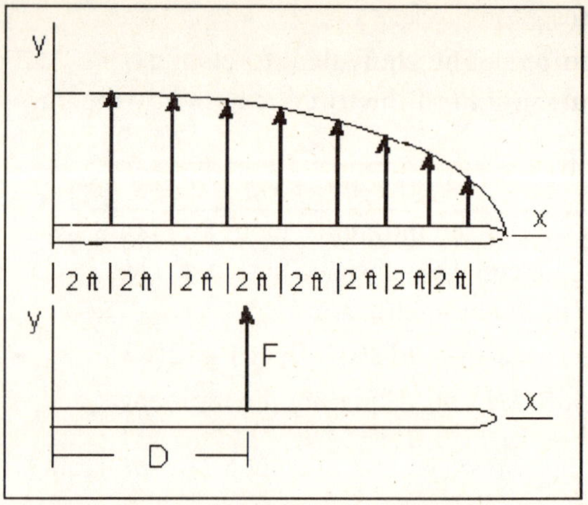

Solution: A table of the forces and the moments about the wing root is:

Station	Distance D	Force F	Moment
1	2	198.61	397.22
2	4	194.38	777.52
3	6	187.13	1122.78
4	8	176.47	1411.76
6	10	161.74	1617.40
7	12	141.67	1700.04
8	14	113.45	1588.37
9	16	67.58	1081.33
Sum		1241.03	9696.3

The equivalent force is the sum of the eight forces, which is the last entry in the forces column, $\boxed{F = 1241\ lb}$. The distance out from the root is the ratio of the sum of the moments about the root, shown as the last entry in the moments column, to the equivalent force: $\boxed{D = \frac{9696.3}{1241} = 7.81\ ft}$

Problem 4.130 The system is in equilibrium. If you represent the forces $\vec{F}_{AB}$ and $\vec{F}_{AC}$ by a force $\vec{F}$ acting at A and a couple moment M, what are $\vec{F}$ and M?

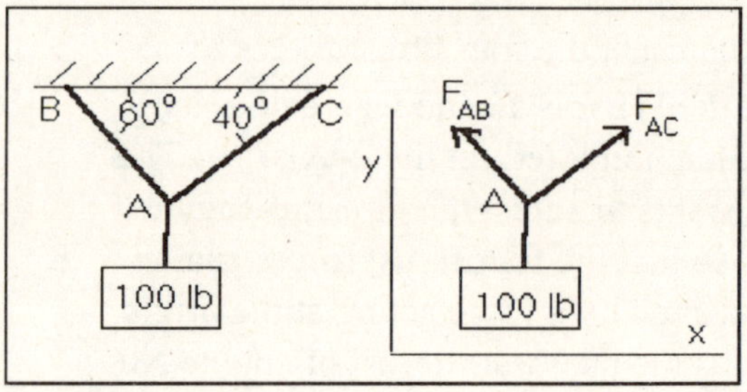

Solution: The sum of the forces acting at A is in opposition to the weight, or $\boxed{\vec{F} = |\vec{W}|\vec{j} = 100\vec{j}\ lb}$ The moment about point A is zero.

Problem 4.131 Three forces act on a beam. (a) represent the system by a force $\vec{F}$ acting at the origin O and a couple M. (b) Represent the system by a single force. Where does the line of action intersect the x-axis?

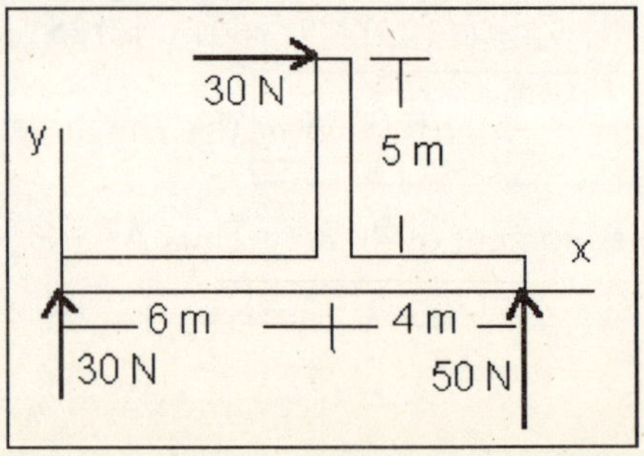

Solution: (a)The sum of the forces is $\sum \vec{F}_X = 30\vec{i}\ N$, and $\sum \vec{F}_Y = (30+50)\vec{j} = 80\vec{j}\ N$. The equivalent at O is $\boxed{\vec{F} = 30\vec{i} + 80\vec{j}\ (N)}$ The sum of the moments about O: $\sum M = (-5(30)+10(50))\ \boxed{= 350\ N\text{-}m}$ (b) The solution of Part (a) is the single force. The intersection is the moment divided by the y-component of force: $\boxed{D = \frac{350}{80} = 4.375\ m}$

Problem 4.132 The tension in cable AB is 400 *N* and the tension in cable CD is 600 *N*. (a) If you represent the forces exerted on the left post by the cables by a force $\vec{F}$ acting at the origin O and a couple moment M, what are $\vec{F}$ and M? (b) If you represent the forces exerted on the left post by the cables by the force $\vec{F}$ alone, where does the line of action intersect the y-axis?

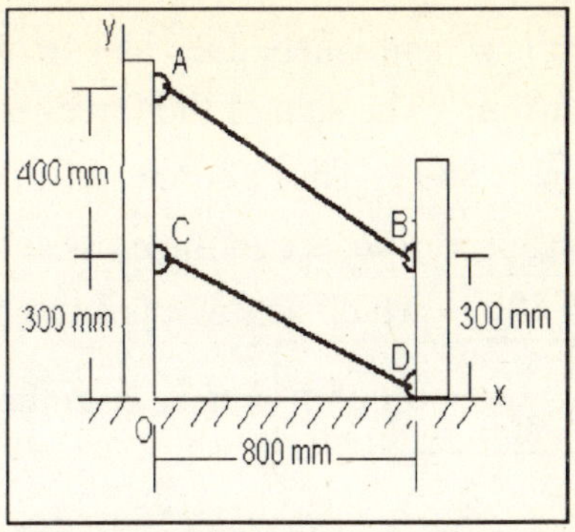

Solution: From the right triangle, the angle between the positive x-axis and the cable AB is $\theta = -\tan^{-1}\left(\frac{400}{800}\right) = -26.6^\circ$.

The tension in AB is $\vec{T}_{AB} = 400(\vec{i}\cos(-26.6^\circ) + \vec{j}\sin(-26.6^\circ))$ $= 357.77\vec{i} - 178.89\vec{j}$ (N). The angle between the positive x-axis and the cable CD is $\alpha = -\tan^{-1}\left(\frac{300}{800}\right) = -20.6^\circ$ The tension in CD is $\vec{T}_{CD} = 600(\vec{i}\cos(-20.6^\circ) + \vec{j}\sin(-20.6^\circ))$ $= 561.8\vec{i} - 210.67\vec{j}$. The equivalent force acting at the origin O is the sum of the forces acting on the left post: $\vec{F} = (357.77 + 561.8)\vec{i} + (-178.89 - 210.67)\vec{j}$ $\boxed{= 919.6\vec{i} - 389.6\vec{j}\ (N)}$ The sum of the moments acting on the left post is the product of the moment arm and the x-component of the tensions: $\sum \vec{M} = -0.7(357.77)\vec{k} - 0.3(561.8)\vec{k}$. $\boxed{= -419\vec{k}\ N\text{-}m}$ *Check:* The position vectors at the point of application are $\vec{r}_{AB} = 0.7\vec{j}$, and $\vec{r}_{CD} = 0.3\vec{j}$. The sum of the moments

is $\sum \vec{M} = (\vec{r}_{AB} \times \vec{T}_{AB}) + (\vec{r}_{CD} \times \vec{T}_{CD}) = \begin{vmatrix} \vec{i} & \vec{j} & \vec{k} \\ 0 & 0.7 & 0 \\ 357.77 & -178.89 & 0 \end{vmatrix} + \begin{vmatrix} \vec{i} & \vec{j} & \vec{k} \\ 0 & 0.3 & 0 \\ 561.8 & -210.67 & 0 \end{vmatrix}$

$= -0.7(357.77)\vec{k} - 0.3(561.8)\vec{k} = -419\vec{k}$ *Check.* (b) The equivalent single force retains the same scalar components, but must act at a point that duplicates the sum of the moments. The distance on the y-axis is the ratio of the sum of the moments to the x-component of the equivalent force. Thus

$\boxed{D = \frac{419}{919.6} = 0.456\ m}$ *Check:* The moment is

$\vec{M} = \vec{r}_F \times \vec{F} = \begin{vmatrix} \vec{i} & \vec{j} & \vec{k} \\ 0 & D & 0 \\ 919.6 & -389.6 & 0 \end{vmatrix} = -919.6D\vec{k} = -419\vec{k}$, from which $\boxed{D = \frac{419}{919.6} = 0.456\ m}$, *Check.*

Problem 4.133 Consider the system shown in Problem 4.132. The tension in cables AB and CD is 400 *N*. If you represent the forces on the right post by a force $\vec{F}$, what is $\vec{F}$ and where does its line of action intersect the y-axis?

Solution: From the solution of Problem 4.132, the tensions are

$\vec{T}_{AB} = -400(\vec{i}\cos(-26.6^\circ) + \vec{j}\sin(-26.6^\circ)) = -357.77\vec{i} + 178.89\vec{j}$, and

$\vec{T}_{CD} = -400(\vec{i}\cos(-20.6^\circ) + \vec{j}\sin(-20.6^\circ)) = -374.42\vec{i} + 140.74\vec{j}$. The equivalent force is equal to the sum of these forces: $\vec{F} = (-357.77 - 374.42)\vec{i} + (178.77 + 140.74)\vec{j}$ $\boxed{= -732.19\vec{i} + 319.5\vec{j}\ (N)}$.

The sum of the moments about O is $\sum \vec{M} = (0.3(357.77) + 0.8(140.74 + 178.89)\vec{k} = 363\vec{k}\ (N\text{-}m)$.

The intersection is $\boxed{D = \frac{363}{732.19} = 0.496\ m}$ on the positive y-axis.

Problem 4.134 If you represent the three forces acting on the beam cross section by a force $\vec{F}$, what is $\vec{F}$ and where does its line of action intersect the y-axis?

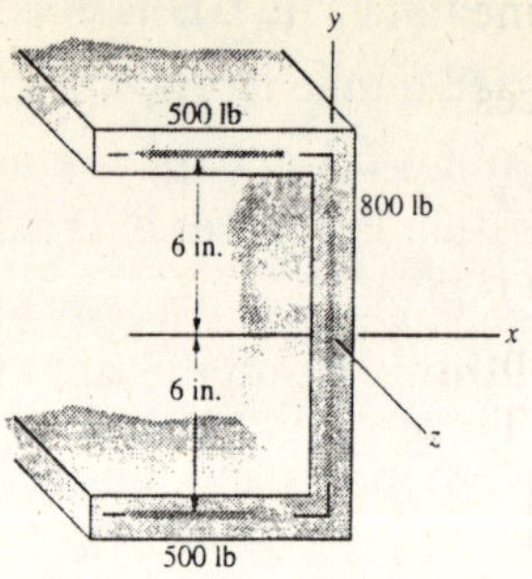

Solution: The sum of the forces is $\sum \vec{F}_X = (500 - 500)\vec{i} = 0$. $\sum \vec{F}_Y = 800\vec{j}$. Thus a force and a couple with moment $\vec{M} = 500\vec{k}$ *ft lb* act on the cross section. The equivalent force is $\boxed{\vec{F} = 800\vec{j}}$ which acts at a positive x-axis location of $\boxed{D = \frac{500}{800} = 0.625\ ft} = 7.5$ *in* to the right of the origin.

Problem 4.135 The two systems of forces and moments acting on the beam are equivalent. Determine the force $\vec{F}$ and moment $\vec{M}$.

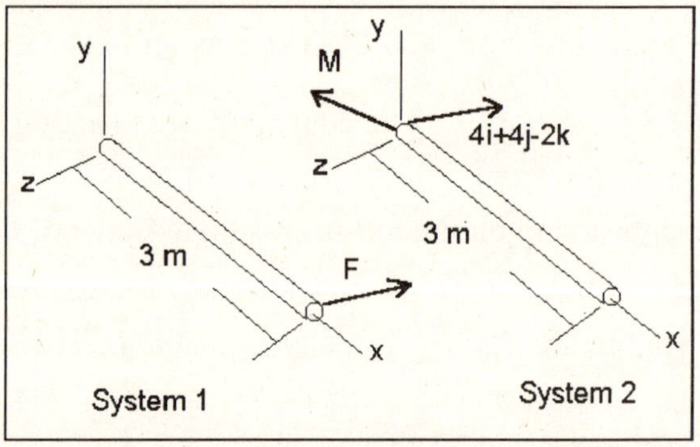

Solution: The sum of the forces on the two systems are equivalent: the force on System 1 is $\boxed{\vec{F}_1 = 4\vec{i} + 4\vec{j} - 2\vec{k}\ (kN)}$. The moments on the two systems are equivalent: the moment about the origin for System 1 is the product of the moment arm and the y- and z- components of the force: $\vec{M} = 3(2)\vec{j} + 3(4)\vec{k} = 6\vec{j} + 12\vec{k}$. Hence the couple moment on System 2 is $\boxed{\vec{M}_2 = 6\vec{j} + 12\vec{k}\ (kN\text{-}m)}$

Problem 4.136 The wall bracket is subjected to the force shown. Determine the moment exerted by the force (a) about the z-axis, (b) about the y-axis. (c) If you represent the force by a force $\vec{F}$ acting at O and a couple $\vec{M}$, what are $\vec{F}$ and $\vec{M}$?

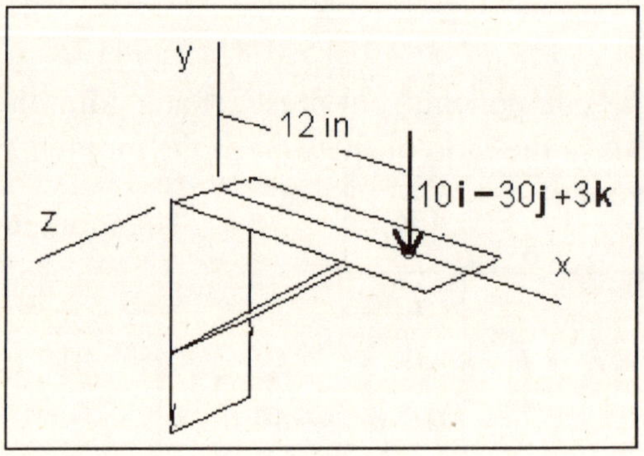

Solution: (a) The moment about the z-axis is is negative, $\boxed{M_Z = -1(30) = -30\ ft\ lb}$, (b) The moment about the y-axis is negative, $\boxed{M_Y = -1(3) = -3\ ft\ lb}$
(c) The equivalent force at O must be equal to the force at $x = 12$ *in*, thus $\boxed{\vec{F}_{EQ} = 10\vec{i} - 30\vec{j} + 3\vec{k}\ (lb)}$
The couple moment must equal the moment exerted by the force at $x = 12$ *in*. This moment is the product of the moment arm and the y- and z-components of the force: $\vec{M} = -1(30)\vec{k} - 1(3)\vec{j}$.
$\boxed{= -3\vec{j} - 30\vec{k}\ (ft\ lb)}$.

Problem 4.137 Michael Jordan of the Chicago Bulls executes a "slam dunk" shot, then hangs momentarily on the rim, exerting the two 100 *lb* forces shown. The dimensions are $h = 14\frac{1}{2}$ *in*, and $r = 9\frac{1}{2}$ *in*, and the angle is $\alpha = 120°$. (a) If you represent the forces he exerts by a force $\vec{F}$ acting at O and a couple $\vec{M}$, what are $\vec{F}$ and $\vec{M}$? (b) The glass blackboard will shatter if $|\vec{M}| > 4000$ *in lb*. Does it break?

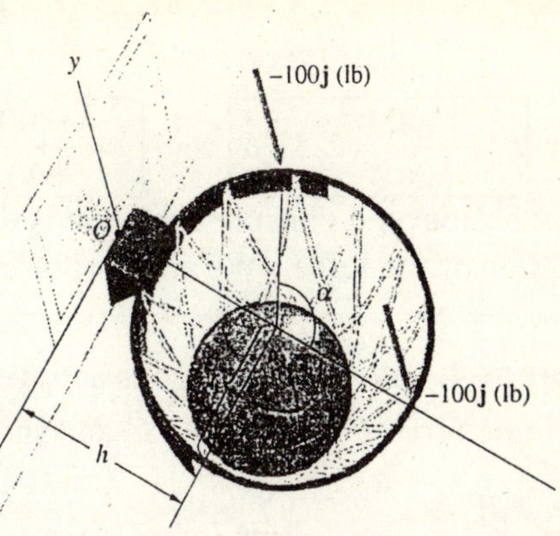

Solution: The equivalent force at the origin must equal the sum of the forces applied: $\boxed{\vec{F}_{EQ} = -200\vec{j}}$

The position vectors of the points of application of the forces are $\vec{r}_1 = (h+r)\vec{i}$, and $\vec{r}_2 = \vec{i}(h + r\cos\alpha) - \vec{k}\, r\sin\alpha$. The moments about the origin are

$$\vec{M} = (\vec{r}_1 \times \vec{F}_1) + (\vec{r}_2 \times \vec{F}_2) = (\vec{r}_1 + \vec{r}_2) \times \vec{F} = \begin{vmatrix} \vec{i} & \vec{j} & \vec{k} \\ 2h + r(1+\cos\alpha) & 0 & -r\sin\alpha \\ 0 & -100 & 0 \end{vmatrix}$$

$= -100(r\sin\alpha)\vec{i} - 100(2h + r(1+\cos\alpha))\vec{k}$. For the values of *h*, *r*, and α given, the moment is $\boxed{\vec{M} = -822.72\vec{i} - 3375\vec{k} \text{ in lb}}$ This is the couple moment required. (b) The magnitude of the moment is $|\vec{M}| = \sqrt{822.72^2 + 3375^2} = 3473.8$ *in lb*. <u>The backboard does not break.</u>

Problem 4.138 The three forces are parallel to the x- axis. (a) If you represent the three forces by a force $\vec{F}$ acting at the origin O and a couple $\vec{M}$, what are $\vec{F}$ and $\vec{M}$? (b) If you represent the three forces by a single force, what is the force and where does its line of action intersect the y-z plane?

STRATEGY: In Part (b) assume that the force acts at a point (0,y,z) of the y-z plane and use the conditions for equivalence to determine the force and the coordinates y and z. See example 4.19.

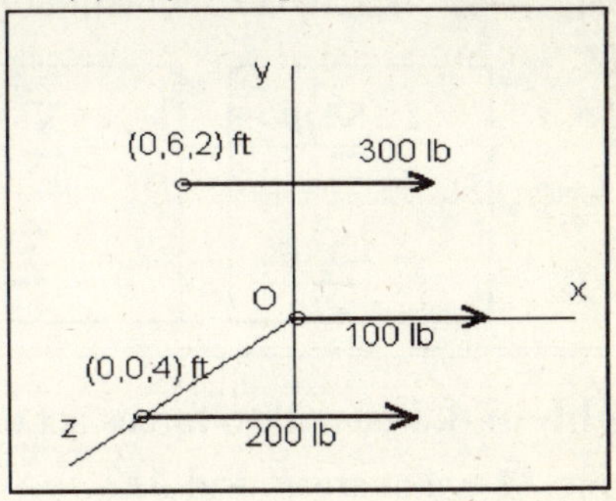

Solution: (a) The equivalent force at the origin is equal to the sum of the forces: $\boxed{\vec{F} = 100\vec{i} + 200\vec{i} + 300\vec{i} = 500\vec{i}}$

The position vectors of the forces are: $\vec{r}_1 = 0$, $\vec{r}_2 = 6\vec{j} + 2\vec{k}$, and $\vec{r}_3 = 4\vec{k}$. The moment about the origin is

$$\sum \vec{M}_O = (\vec{r}_2 \times 300\vec{i}) + (\vec{r}_3 \times 200\vec{i}) = \begin{vmatrix} \vec{i} & \vec{j} & \vec{k} \\ 0 & 6 & 2 \\ 300 & 0 & 0 \end{vmatrix} + \begin{vmatrix} \vec{i} & \vec{j} & \vec{k} \\ 0 & 0 & 4 \\ 200 & 0 & 0 \end{vmatrix}, \quad \boxed{\sum \vec{M}_o = 1400\vec{j} - 1800\vec{k}}$$

(b) The problem has many solutions for the x-value, each corresponding to a point on the action line parallel to the x-axis. By arbitrarily selecting the point $x = 0$, we obtain a practical solution. This is demonstrated as follows. The single force $\boxed{\vec{F} = 600\vec{i}}$ representation must produce the same moment about the origin as the three forces,

Problem 4.138 continued on next page....

continuation of solution to Problem 4.138...

$$\vec{M}_{EQ} = \vec{r} \times \vec{F} = \begin{vmatrix} \vec{i} & \vec{j} & \vec{k} \\ x & y & z \\ 0 & 1400 & \end{vmatrix} \vec{M}_{EQ} = \vec{r} \times \vec{F} = \begin{vmatrix} \vec{i} & \vec{j} & \vec{k} \\ x & y & z \\ 600 & 0 & 0 \end{vmatrix} = 600z\vec{j} - 600y\vec{k} = 1400\vec{j} - 1880\vec{k}$$, from

which $\boxed{z = \dfrac{1400}{600} = 2.33\ ft}$, and $\boxed{y = \dfrac{1800}{600} = 3\ ft}$. Note that x is indeterminate, in that it can have any value whatsoever. (This phenomenon is to be expected for a coordinate when its axis is parallel to the action line of the force.) We arbitrarily select $x = 0$, from which the solution is the point (0, 3, 2.33) *ft*.

Problem 4.139 The positions and weights of three particles are shown. If you represent the weights by a single force $\vec{F}$, determine $\vec{F}$ and show that its line of action intersects the x-z plane at

$$x = \frac{\sum_{i=1}^{3} x_i W_i}{\sum_{i=1}^{3} W_i}, \text{ and } z = \frac{\sum_{i=1}^{3} z_i W_i}{\sum_{i=1}^{3} W_i}.$$

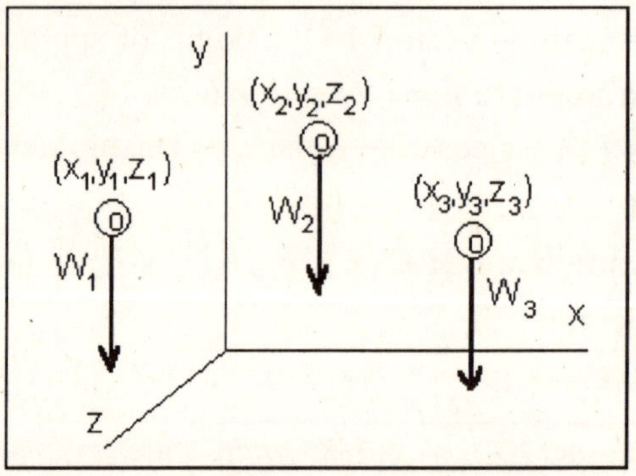

Solution: The single equivalent force must be equal to the sum of the forces: $\boxed{\vec{F} = (-W_1 - W_2 - W_3)\vec{j} = -\sum_{i=1}^{3} W_i}$

The moment about the x-axis is

$$\sum M_X = \vec{e}_X \cdot (\vec{r}_1 \times \vec{W}_1) + \vec{e}_X \cdot (\vec{r}_2 \times \vec{W}_2) + \vec{e}_X \cdot (\vec{r}_3 \times \vec{W}_3)$$

$$= \begin{vmatrix} 1 & 0 & 0 \\ x_1 & y_1 & z_1 \\ 0 & -W_1 & 0 \end{vmatrix} + \begin{vmatrix} 1 & 0 & 0 \\ x_2 & y_2 & z_2 \\ 0 & -W_2 & 0 \end{vmatrix} + \begin{vmatrix} 1 & 0 & 0 \\ x_3 & y_3 & z_3 \\ 0 & -W_3 & 0 \end{vmatrix} = \sum_{i=1}^{3} W_i z_i$$

. Similarly, the moment about the z-axis is $\sum M_Z = -\sum_{i=1}^{3} W_i x_i$. The moment arm is the ratio of the magnitude of the moment to the magnitude of the force, $\boxed{D_X = \dfrac{\sum_{i=1}^{3} W_i z_i}{\sum_{i=1}^{3} W_i}}$. $\boxed{D_Z = \dfrac{\sum_{i=1}^{3} W_i x_i}{\sum_{i=1}^{3} W_i}}$

Problem 4.140 Two forces act on the beam. If you represent them by a force $\vec{F}$ acting at C and a couple $\vec{M}$, what are $\vec{F}$ and $\vec{M}$?

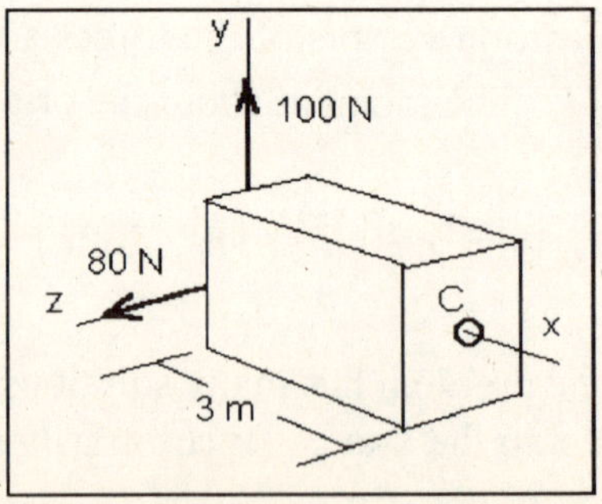

Solution: The equivalent force must equal the sum of forces: $\boxed{\vec{F} = 100\vec{j} + 80\vec{k}}$ The equivalent couple is equal to the moment about C:

$$\boxed{\sum \vec{M} = (3)(80)\vec{j} - (3)(100)\vec{k} = 240\vec{j} - 300\vec{k}}$$

Problem 4.141 An axial force of magnitude P acts on the beam. (a) It you represent it by a force $\vec{F}$ acting at the origin O and a couple $\vec{M}$, what are $\vec{F}$ and $\vec{M}$?

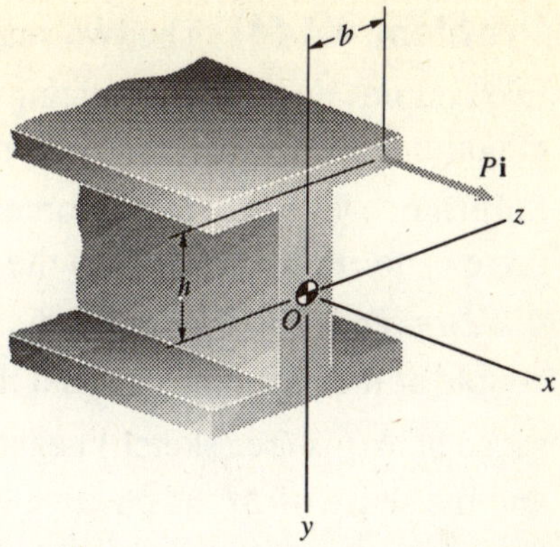

Solution: The equivalent force at the origin is equal to the applied force $\boxed{\vec{F} = P\,\vec{i}}$ The position vector of the applied force is $\vec{r} = -h\vec{j} + b\vec{k}$. The moment is

$$\vec{M} = (\vec{r} \times \vec{P}) = \begin{vmatrix} \vec{i} & \vec{j} & \vec{k} \\ 0 & -h & +b \\ P & 0 & 0 \end{vmatrix}$$

$\boxed{= bP\vec{j} + hP\vec{k}}$ This is the couple at the origin.

(Note that in the sketch the axis system has been rotated 180 about the x-axis; so that up is negative and right is positive for y and z.)

Problem 4.142 The brace is being used to remove a screw. (a) If you represent the forces acting one the brace by a force $\vec{F}$ acting at the origin O and a couple $\vec{M}$? (b) If you represent the forces acting on the brace by a force $\vec{F}'$ acting at a point P with coordinates (x_P, y_P, z_P) and a couple $\vec{M}'$, what are $\vec{F}'$ and $\vec{M}'$?

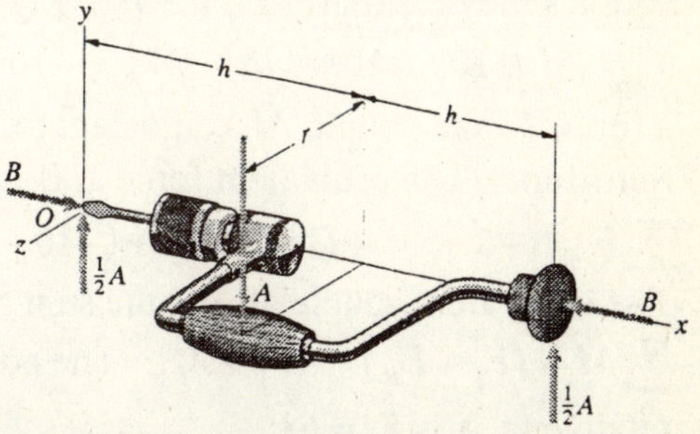

Solution: (a) Equivalent force at the origin O has the same value as the sum of forces, $\sum \vec{F}_X = (B - B)\vec{i} = 0$, $\sum \vec{F}_Y = (-A + \frac{1}{2}A + \frac{1}{2}A)\vec{j} = 0$, thus $\boxed{\vec{F} = 0}$. The equivalent couple moment has the same value as the moment exerted on the brace by the forces, $\sum \vec{M}_O = (rA)\vec{i}$. Thus the couple at O has the moment $\boxed{\vec{M} = rA\vec{i}}$. (b) The equivalent force at (x_P, y_P, z_P) has the same value as the sum of forces on the brace, and the equivalent couple at (x_P, y_P, z_P) has the same moment as the moment exerted on the brace by the forces: $\boxed{\vec{F} = 0}$, $\boxed{\vec{M} = rA\vec{i}}$.

Problem 4.143 Two forces and a couple act on the cube. If you represent them by a force $\vec{F}$ acting at point P and a couple $\vec{M}$, what are $\vec{F}$ and $\vec{M}$?

Solution: The equivalent force at P has the value of the sum of forces, $\sum \vec{F} = (2-1)\vec{i} + (1-1)\vec{j} + \vec{k}$, $\boxed{\vec{F}_P = \vec{i} + \vec{k}\ (kN)}$ The equivalent couple at P has the moment exerted by the forces and moment about P. The position vectors of the forces relative to P are : $\vec{r}_A = -\vec{i} - \vec{j} + \vec{k}$, and $\vec{r}_B = +\vec{k}$. The moment of the couple: $\sum \vec{M} = (\vec{r}_A \times \vec{F}_A) + (\vec{r}_B \times \vec{F}_B) + \vec{M}_C$

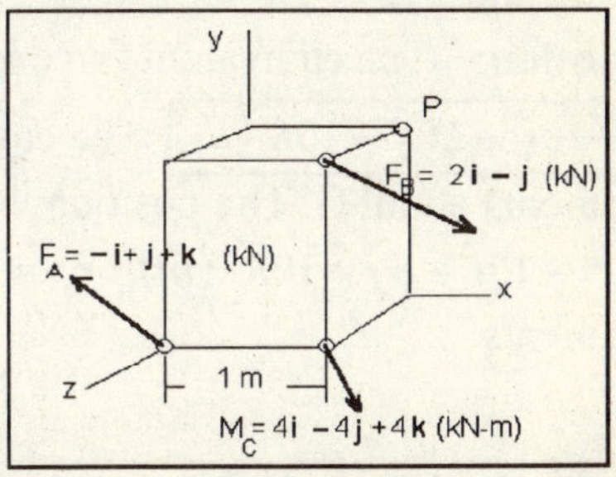

$$= \begin{vmatrix} \vec{i} & \vec{j} & \vec{k} \\ -1 & -1 & 1 \\ -1 & 1 & 1 \end{vmatrix} + \begin{vmatrix} \vec{i} & \vec{j} & \vec{k} \\ 0 & 0 & 1 \\ 2 & -1 & 0 \end{vmatrix} + \vec{M}_C = \boxed{3\vec{i} - 2\vec{j} + 2\vec{k}\ (kN\text{-}m)}$$

Problem 4.144 The two shafts are subjected to the torque (couples) shown. (a) If you represent the two couples by a force $\vec{F}$ acting at the origin O and a couple $\vec{M}$, what are $\vec{F}$ and $\vec{M}$? (b) What is the magnitude of the total moment exerted by the two couples?

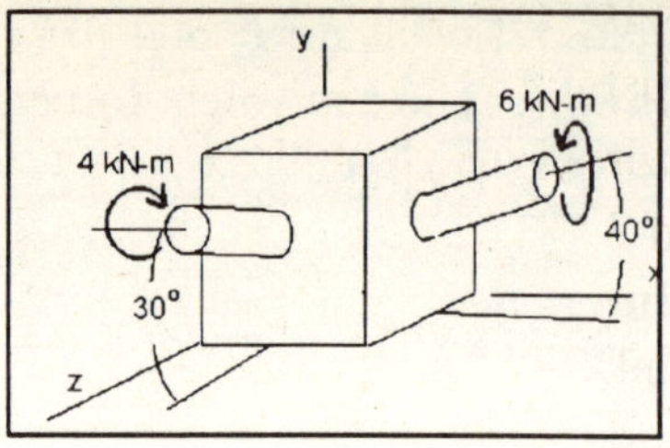

Solution: The equivalent force at the origin is zero, $\boxed{\vec{F}=0}$ since there is no resultant force on the system. Represent the couples of $4\ kN\text{-}m$ and $6\ kN\text{-}m$ magnitudes by the vectors $\vec{M}_1$ and $\vec{M}_2$. The couple at the origin must equal the sum : $\sum \vec{M} = \vec{M}_1 + \vec{M}_2$. The sense of $\vec{M}_1$ is (see sketch) negative with respect to both y and z, and the sense of $\vec{M}_2$ is positive with respect to both x and y.

$\vec{M}_1 = 4(-\vec{j}\sin 30^o - \vec{k}\cos 30^o) = -2\vec{j} - 3.464\vec{k}$, $\vec{M}_2 = 6(\vec{i}\cos 40^o + \vec{j}\sin 40^o) = 4.5963\vec{i} + 3.8567\vec{j}$.

Thus the couple at the origin is $\boxed{\vec{M}_o = 4.6\vec{i} + 1.86\vec{j} - 3.46\vec{k}\ \ (kN\text{-}m)}$ (b) The magnitude of the total moment exerted by the two couples is $|\vec{M}_o| = \sqrt{4.6^2 + 1.86^2 + 3.46^2}$ $\boxed{= 6.05\ \ (kN\text{-}m)}$

Problem 4.145 The persons A and B support a bar to which three dogs are tethered. The force and couples they exert are: $\vec{F}_A = -5\vec{i} + 15\vec{j} - 10\vec{k}\ \ (lb)$, $\vec{M}_A = 15\vec{j} + 10\vec{k}\ \ (ft\ lb)$, $\vec{F}_B = 5\vec{i} + 10\vec{j} - 10\vec{k}\ \ (lb)$, $\vec{M}_B = -10\vec{j} - 15\vec{k}\ \ (ft\ lb)$. If person B let go, person A would have to exert a force $\vec{F}$ and couple $\vec{M}$ equivalent to that both of them were exerting together. What are $\vec{F}$ and $\vec{M}$?

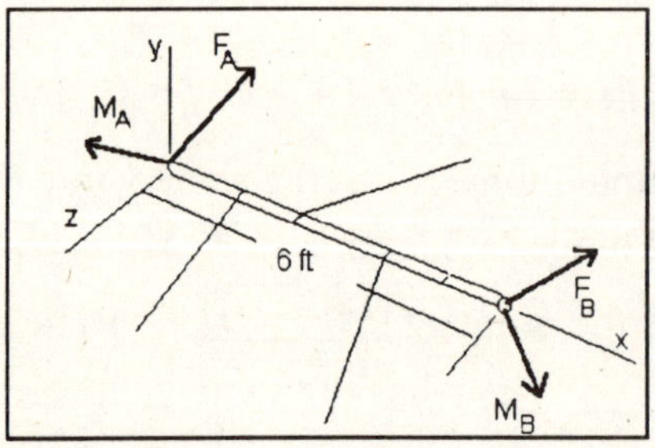

Solution: The equivalent force at B is the sum of the forces:

$\sum \vec{F} = (-5+5)\vec{i} + (15+10)\vec{j} + (-10-10)\vec{k}$ $\boxed{= 25\vec{j} - 20\vec{k}\ \ (lb)}$

The equivalent couple at A is the sum of the moments at A

$\sum \vec{M} = (\vec{r}_B \times \vec{F}_B) + \vec{M}_A + \vec{M}_B$ The position vector of B relative to A is $\vec{r}_B = 6\vec{i}$.

$$\text{Thus: } \sum \vec{M}_B = \begin{vmatrix} \vec{i} & \vec{j} & \vec{k} \\ 6 & 0 & 0 \\ 5 & 10 & -10 \end{vmatrix} + \vec{M}_A + \vec{M}_B$$

$$\sum \vec{M}_B = (60\vec{j} + 60\vec{k}) + (15\vec{j} + 10\vec{k}) + (-10\vec{j} - 15\vec{k}) \boxed{= 65\vec{j} + 55\vec{k}\ \ (ft\text{-}lb)}$$

Problem 4.146 Point G is at the center of the block. The forces are $\vec{F}_A = -20\vec{i} + 10\vec{j} + 20\vec{k}\ \ (lb)$, $\vec{F}_B = 10\vec{j} - 10\vec{k}\ \ (lb)$. If you represent the two forces by a force $\vec{F}$ acting at G and a couple $\vec{M}$, what are $\vec{F}$ and $\vec{M}$?

Solution: The equivalent force is the sum of the forces: $\sum \vec{F} = (-20)\vec{i} + (10+10)\vec{j} + (20-10)\vec{k}$ $\boxed{= -20\vec{i} + 20\vec{j} + 10\vec{k}\ (lb)}$. The equivalent couple is the sum of the moments about G. The position vectors are:

$\vec{r}_A = -15\vec{i} + 5\vec{j} + 10\vec{k}\ \ (in)$, $\vec{r}_B = 15\vec{i} + 5\vec{j} - 10\vec{k}$. The sum of the moments:

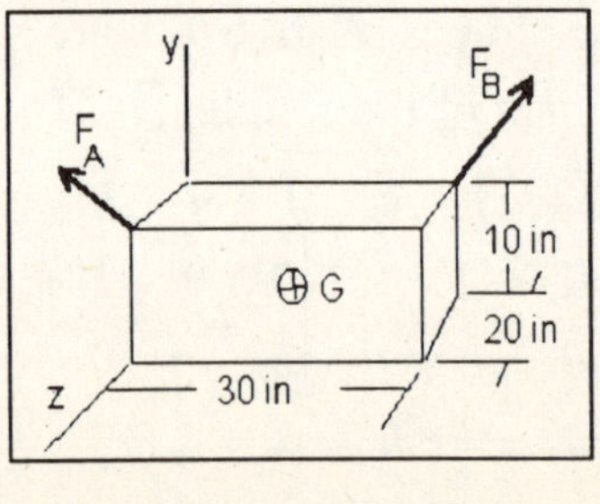

$$\sum \vec{M}_G = (\vec{r}_A \times \vec{F}_A) + (\vec{r}_B \times \vec{F}_B) = \begin{vmatrix} \vec{i} & \vec{j} & \vec{k} \\ -15 & 5 & 10 \\ -20 & 10 & 20 \end{vmatrix} + \begin{vmatrix} \vec{i} & \vec{j} & \vec{k} \\ 15 & 5 & -10 \\ 0 & 10 & -10 \end{vmatrix} =$$

$\boxed{50\vec{i} + 250\vec{j} + 100\vec{k}\ \ (in\ lb)}$

Problem 4.147 The engine above the airplane's fuselage exerts a thrust $T_O = 16\ kip$ and each engine below the fuselage exerts a thrust $T_U = 12\ kip$. The dimensions are $h = 8\ ft$, $c = 12\ ft$, and $b = 16\ ft$. If you represent the three thrust forces by a force $\vec{F}$ acting at the origin O and a couple $\vec{M}$, what are $\vec{F}$ and $\vec{M}$?

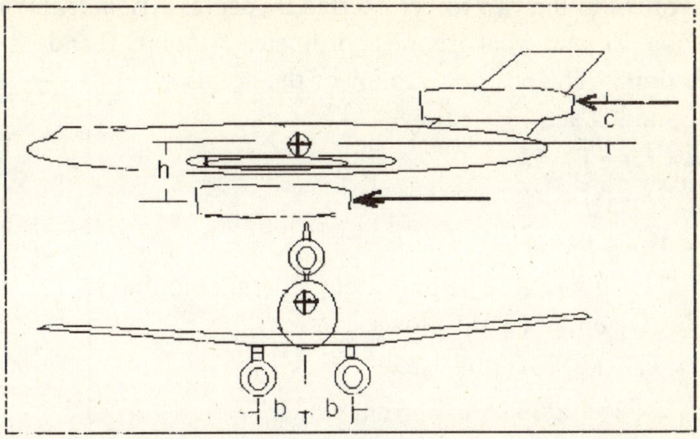

Solution: The equivalent thrust at the point G is equal to the sum of the thrusts: $\sum T = 16 + 12 + 12 \boxed{= 40\ kip}$

The sum of the moments about the point G is

$$\sum \vec{M} = (\vec{r}_{1U} \times \vec{T}_U) + (\vec{r}_{2U} \times \vec{T}_U) + (\vec{r}_O \times \vec{T}_O)$$

$$= (\vec{r}_{1U} + \vec{r}_{2U}) \times \vec{T}_U + (\vec{r}_O \times \vec{T}_O)$$

The position vectors are $\vec{r}_{1U} = +b\vec{i} - h\vec{j}$, $\vec{r}_{2U} = -b\vec{i} - h\vec{j}$, and $\vec{r}_O = +c\vec{j}$. For $h = 8\ ft$, $c = 12\ ft$, and $b = 16\ ft$, the sum of the moments is

$$\sum \vec{M} = \begin{vmatrix} \vec{i} & \vec{j} & \vec{k} \\ 0 & -16 & 0 \\ 0 & 0 & 12 \end{vmatrix} + \begin{vmatrix} \vec{i} & \vec{j} & \vec{k} \\ 0 & 12 & 0 \\ 0 & 0 & 16 \end{vmatrix} = (-192 + 192)\vec{i} = 0.$$

Thus the equivalent couple is $\boxed{\vec{M} = 0}$

Problem 4.148 Consider the airplane described in Problem 4.147, and suppose that the engine under the wing to the pilot's right loses thrust. (a) If you represent the two remaining thrust forces by a force $\vec{F}$ acting at the origin and a couple $\vec{M}$, what are $\vec{F}$ and $\vec{M}$? (c) If you represent the two remaining thrust forces by the force $\vec{F}$ alone, where does its line of action intersect the x-y plane?

Solution: The sum of the forces is now $\boxed{\sum \vec{F} = 12 + 16 = 28\vec{k}\ (kip)}$ The sum of the moments is now: $\sum \vec{M} = (\vec{r}_{2U} \times \vec{T}_U) + (\vec{r}_O \times \vec{T}_O)$. For $h = 8\ ft$, $c = 12\ ft$, and $b = 16\ ft$, using the position vectors for the engines given in Problem 4.147, the equivalent couple is

$$\vec{M} = \begin{vmatrix} \vec{i} & \vec{j} & \vec{k} \\ 16 & -8 & 0 \\ 0 & 0 & 12 \end{vmatrix} + \begin{vmatrix} \vec{i} & \vec{j} & \vec{k} \\ 0 & 12 & 0 \\ 0 & 0 & 16 \end{vmatrix} =$$

$\boxed{= 96\vec{i} - 192\vec{j}\ (ft\ kip)}$ (b) The moment of the single force is

$$\vec{M} = \begin{vmatrix} \vec{i} & \vec{j} & \vec{k} \\ x & y & z \\ 0 & 0 & 28 \end{vmatrix} = 28y\vec{i} - 28x\vec{j} = 96\vec{i} - 192\vec{j}$$

From which $\boxed{x = \frac{192}{28} = 6.86\ ft}$, and $\boxed{y = \frac{96}{28} = 3.43\ ft}$.

As to be expected, z can have any value, corresponding to any point on the line of action. Arbitrarily choose $z = 0$, so that the coordinates of the point of action are $(6.86, 3.43, 0)$.

Problem 4.149 The tension in cable AB is 100 *lb* and the tension in cable CD is 60 *lb*. Suppose that you want to replace these two cables by a single cable EF so that the force exerted on the wall at E is equivalent to the two forces exerted by cables AB and CD on the walls at A and C. What is the tension in cable EF and what are the coordinates of points E and F?

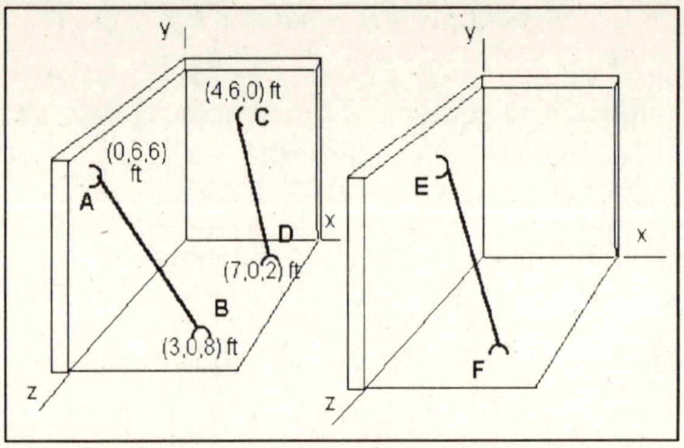

Solution: The position vectors of the points A, B, C, and D are

$\vec{r}_A = 0\vec{i} + 6\vec{j} + 6\vec{k}$

$\vec{r}_B = 3\vec{i} + 0\vec{j} + 8\vec{k}$

$\vec{r}_C = 4\vec{i} + 6\vec{j} + 0\vec{k}$

$\vec{r}_D = 7\vec{i} + 0\vec{j} + 2\vec{k}$ The unit vectors parallel to the cables are obtained as follows:

$\vec{r}_{AB} = \vec{r}_B - \vec{r}_A = 3\vec{i} - 6\vec{j} + 2\vec{k}$,

$|\vec{r}_{AB}| = \sqrt{3^2 + 6^2 + 2^2} = 7$, from which

$\vec{e}_{AB} = 0.4286\vec{i} - 0.8571\vec{j} + 0.2857\vec{k}$.

$\vec{r}_{CD} = \vec{r}_D - \vec{r}_C = 3\vec{i} - 6\vec{j} + 2\vec{k}$, $|\vec{r}_{CD}| = \sqrt{3^2 + 6^2 + 2^2} = 7$, from which

$\vec{e}_{CD} = 0.4286\vec{i} - 0.8571\vec{j} + 0.2857\vec{k}$. Since $\vec{e}_{AB} = \vec{e}_{CD}$, *the cables are parallel.* . To duplicate the force, the single cable EF must have the same unit vector.

The force on the wall at point A is $\vec{F}_A = 100\vec{e}_{AB} = 42.86\vec{i} - 85.71\vec{j} + 28.57\vec{k}$ (lb). The force on the wall at point C is $\vec{F}_C = 60\vec{e}_{CD} = 25.72\vec{i} - 51.43\vec{j} + 17.14\vec{k}$ (lb). The total force is

$\boxed{\vec{F}_{EF} = 68.58\vec{i} - 137.14\vec{j} + 45.71\vec{k}\ (lb),\ |\vec{F}_{EF}| = 160\ lb}$ For the systems to be equivalent, the moments about the origin must be the same. The moments about the origin are

$$\Sigma\vec{M}_O = (\vec{r}_A \times \vec{F}_A) + (\vec{r}_C \times \vec{F}_C) = \begin{vmatrix} \vec{i} & \vec{j} & \vec{k} \\ 0 & 6 & 6 \\ 42.86 & -85.71 & 28.57 \end{vmatrix} + \begin{vmatrix} \vec{i} & \vec{j} & \vec{k} \\ 4 & 6 & 0 \\ 25.72 & -51.43 & 17.14 \end{vmatrix}$$

$= 788.57\vec{i} + 188.57\vec{j} - 617.14\vec{k}$.

This result is used to establish the coordinates of the point E. For the one cable system, the end E is located at x=0. The moment is

$$\vec{M}_1 = \vec{r} \times \vec{F}_{EF} = \begin{vmatrix} \vec{i} & \vec{j} & \vec{k} \\ 0 & y & z \\ 68.58 & -137.14 & 45.71 \end{vmatrix} = (45.71y + 137.14z)\vec{i} + (68.58z)\vec{j} - (68.58y)\vec{k}$$

$= 788.57\vec{i} + 188.57\vec{j} - 617.14\vec{k}$, from above. From which $\boxed{y = \frac{617.14}{68.58} = 8.999... = 9\ ft}$

$\boxed{z = \frac{188.57}{68.58} = 2.75\ ft}$. Thus the coordinates of point E are $\boxed{E(0,9,2.75)\ ft}$ The coordinates of the point F are found as follows: Let L be the length of cable EF. Thus, from the definition of the unit vector,

$y_F - y_E = Le_y$ with the condition that $y_F = 0$, $L = \frac{9}{0.8571} = 10.5\ ft$. The other coordinates are

$x_F - x_E = Le_X$, from which $\boxed{x_F = 0 + 10.5(0.4286) = 4.5\ ft}$ $z_F - z_E = Le_Z$, from which

$z_F = 2.75 + 10.5(0.2857) = 5.75\ ft$ The coordinates of F are $\boxed{F(4.5,0,5.75)\ ft}$

==============================<>==============================

Problem 4.150 The distance $s = 4\ m$. If you represent the force and the 200 *N-m* couple by a force $\vec{F}$ acting at origin O and a couple $\vec{M}$, what are $\vec{F}$ and $\vec{M}$?

Solution: The equivalent force at the origin is

$\boxed{\vec{F} = 100\vec{i} + 20\vec{j} - 20\vec{k}}$.

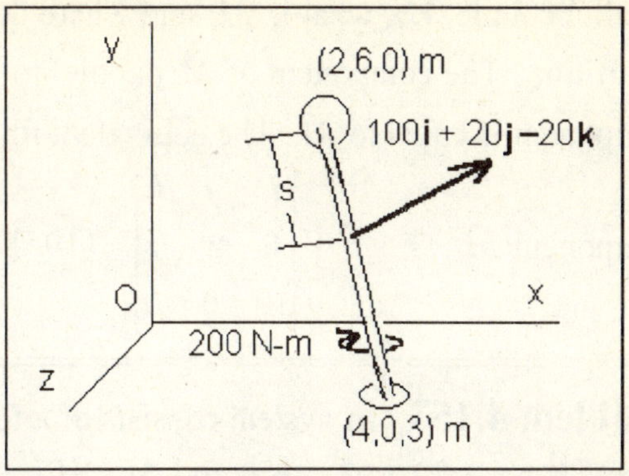

The strategy is to establish the position vector of the action point of the force relative to the origin O for the purpose of determining the moment exerted by the force about the origin. The position of the top of the bar is $\vec{r}_T = 2\vec{i} + 6\vec{j} + 0\vec{k}$. The vector parallel to the bar, pointing toward the base, is $\vec{r}_{TB} = 2\vec{i} - 6\vec{j} + 3\vec{k}$, with a magnitude of $|\vec{r}_{TB}| = 7$. The unit vector parallel to the bar is $\vec{e}_{TB} = 0.2857\vec{i} - 0.8571\vec{j} + 0.4286\vec{k}$. The vector from the top of the bar to the action point of the force is $\vec{r}_{TF} = s\vec{e}_{TB} = 4\vec{e}_{TB} = 1.1429\vec{i} - 3.4286\vec{j} + 1.7143\vec{k}$. The position vector of the action point from the origin is $\vec{r}_F = \vec{r}_T + \vec{r}_{TF} = 3.1429\vec{i} + 2.5714\vec{j} + 1.7143\vec{k}$. The moment of the force about the origin is

$$\vec{M}_F = \vec{r} \times \vec{F} = \begin{vmatrix} \vec{i} & \vec{j} & \vec{k} \\ 3.1429 & 2.5714 & 1.7143 \\ 100 & 20 & -20 \end{vmatrix} = -85.71\vec{i} + 234.20\vec{j} - 194.3\vec{k}$$

The couple is obtained from the unit vector and the magnitude. The sense of the moment is directed positively toward the top of the bar. $\vec{M}_C = -200\vec{e}_{TB} = -57.14\vec{i} + 171.42\vec{j} - 85.72\vec{k}$. The sum of the moments is $\boxed{\vec{M} = \vec{M}_F + \vec{M}_C = -142.86\vec{i} + 405.72\vec{j} - 280\vec{k}}$. This is the moment of the equivalent couple at the origin.

==============================<>==============================

Problem 4.151 The force $\vec{F}$ and couple $\vec{M}$ in System 1 are $\vec{F} = 12\vec{i} + 4\vec{j} - 3\vec{k}\ (lb)$, $\vec{M} = 4\vec{i} + 7\vec{j} + 4\vec{k}\ (ft)$. Suppose that you want to represent System 1 by a wrench (System 2). Determine the couple $\vec{M}_T$ and the coordinates x and z where the line of action of the force intersects the x-z plane.

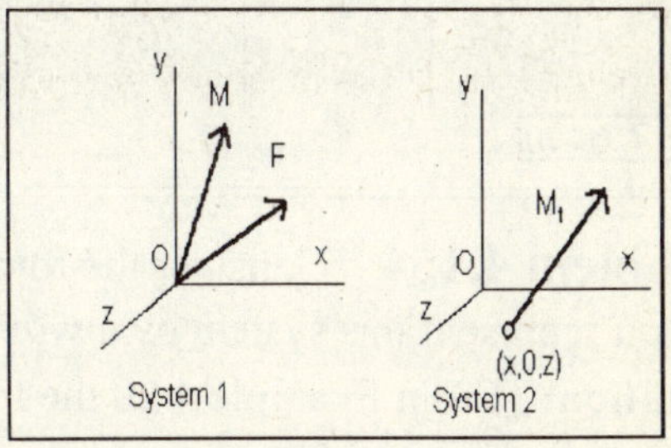

Solution: The component of $\vec{M}$ that is parallel to $\vec{F}$ is found as follows: The unit vector parallel to $\vec{F}$ is

$\vec{e}_F = \dfrac{\vec{F}}{|\vec{F}|} = 0.9231\vec{i} + 0.3077\vec{j} - 0.2308\vec{k}$. The component of $\vec{M}$ parallel to $\vec{F}$ is

$\boxed{\vec{M}_P = (\vec{e}_F \cdot \vec{M})\vec{e}_F = 4.5444\vec{i} + 1.5148\vec{j} - 1.1361\vec{k}\ (ft\text{-}lb)}$. The component of $\vec{M}$ normal to $\vec{F}$ is $\vec{M}_N = \vec{M} - \vec{M}_P = -0.5444\vec{i} + 5.4858\vec{j} + 5.1361\vec{k}\ (ft\text{-}lb)$. The moment of $\vec{F}$ must produce a moment equal to the normal component of $\vec{M}$. The moment is

$$\vec{M}_F = \vec{r} \times \vec{F} = \begin{vmatrix} \vec{i} & \vec{j} & \vec{k} \\ x & 0 & z \\ 12 & 4 & -3 \end{vmatrix} = -(4z)\vec{i} + (3x + 12z)\vec{j} + (4x)\vec{k}$$, from which $\boxed{z = \dfrac{-0.5444}{-4} = 0.1361\ ft}$

$\boxed{x = \dfrac{5.1362}{4} = 1.2840\ ft}$

==============================<>==============================

Problem 4.152 A system consists of a force $\vec{F}$ acting at the origin O and a couple moment $\vec{M}$, where $\vec{F} = 10\vec{i}$ (lb), $\vec{M} = 20\vec{j}$ $(ft\ lb)$. If you represent it by a wrench consisting of the force $\vec{F}$ and a parallel couple $\vec{M}_t$, what is $\vec{M}_t$ and where does the line of action intersect the y-z plane?

Solution: The component of $\vec{M}$ parallel to $\vec{F}$ is zero, since $\boxed{\vec{M}_P = (\vec{e}_F \cdot \vec{M})\vec{e}_F = 0}$. The normal component is equal to $\vec{M}$. The equivalent force must produce the same moment as the normal component $\vec{M} = \vec{r} \times \vec{F} = \begin{vmatrix} \vec{i} & \vec{j} & \vec{k} \\ 0 & y & z \\ 10 & 0 & 0 \end{vmatrix} = (10z)\vec{j} - (10y)\vec{k} = 20\vec{j}$, from which $\boxed{z = \frac{20}{10} = 2\ ft}$ and $\boxed{y = 0}$

Problem 4.153 A system consists of a force $\vec{F}$ acting at the origin O and a couple moment $\vec{M}$, where $\vec{F} = \vec{i} + 2\vec{j} + 5\vec{k}$ (N), and $\vec{M} = 10\vec{i} + 8\vec{j} - 4\vec{k}$ $(N\text{-}m)$. If you represent it by a wrench consisting of the force $\vec{F}$ and a parallel couple $\vec{M}_t$, (a) determine $\vec{M}_t$, and determine where the line of action of $\vec{F}$ intersects (b) the x-z plane, (c) the y-z plane.

Solution: The unit vector parallel to $\vec{F}$ is $\vec{e}_F = \frac{\vec{F}}{|\vec{F}|} = 0.1862\vec{i} + 0.3651\vec{j} + 0.9129\vec{k}$. (a) The parallel component of $\vec{M}_t$ is $\boxed{\vec{M}_P = (\vec{e}_F \cdot \vec{M})\vec{e}_F = 0.2\vec{i} + 0.4\vec{j} + 1.0\vec{k}\ (N\text{-}m)}$ The normal component is $\vec{M}_N = \vec{M} - \vec{M}_P = 9.8\vec{i} + 7.6\vec{j} - 5\vec{k}$. The moment of the force about the origin must be equal to the normal component of the moment. (b)The intersection with the x-z plane:

$$M_N = \vec{r} \times \vec{F} = \begin{vmatrix} \vec{i} & \vec{j} & \vec{k} \\ x & 0 & z \\ 1 & 2 & 5 \end{vmatrix} = -(2z)\vec{i} - (5x - z)\vec{j} + (2x)\vec{k} = 9.8\vec{i} + 7.6\vec{j} - 5\vec{k}\text{, from which}$$

$\boxed{z = \frac{9.8}{-2} = -4.9\ m}$, and $\boxed{x = \frac{5}{-2} = -2.5\ m}$ (c) The intersection with the y-z plane is

$$\vec{M}_N = \vec{r} \times \vec{F} = \begin{vmatrix} \vec{i} & \vec{j} & \vec{k} \\ 0 & y & z \\ 1 & 2 & 5 \end{vmatrix} = (5y - 2z)\vec{i} + (z)\vec{j} - (y)\vec{k} = 9.8\vec{i} + 7.6\vec{j} - 5\vec{k}\text{, from which } \boxed{y = 5\ m} \text{ and}$$

$\boxed{z = 7.6\ m}$

Problem 4.154 Consider the force $\vec{F}$ acting at origin O and the couple $\vec{M}$ given in Example 4.20. If you represent this system by a wrench, where does the line of action intersect the x-y plane ?

Solution: From Example 4.20 the force and moment are $\vec{F} = 3\vec{i} + 6\vec{j} + 2\vec{k}$ (N), and $\vec{M} = 12\vec{i} + 4\vec{j} + 6\vec{k}$ $(N\text{-}m)$. The normal component of the moment is $M_N = 7.592\vec{i} - 4.816\vec{j} + 3.061\vec{k}$ $(N\text{-}m)$. The moment produced by the force must equal the normal

$$\text{component: } \vec{M}_N = \vec{r} \times \vec{F} = \begin{vmatrix} \vec{i} & \vec{j} & \vec{k} \\ x & y & 0 \\ 3 & 6 & 2 \end{vmatrix} = (2y)\vec{i} - (2x)\vec{j} + (6x - 3y)\vec{k} = 7.592\vec{i} - 4.816\vec{j} + 3.061\vec{k}\ ,$$

from which $\boxed{x = \frac{4.816}{2} = 2.408\ m}$ and $\boxed{y = \frac{7.592}{2} = 3.796\ m}$

Problem 4.155 Consider the force $\vec{F}$ acting at the origin O and the couple $\vec{M}$ given in Example 4.20. If you represent this system by a wrench, where does the line of action intersect the plane $y = 3\ m$?

Solution: From Example 4.20 (see also Problem 4.154) the force is $\vec{F} = 3\vec{i} + 6\vec{j} + 2\vec{k}$, and the normal component of the moment is $\vec{M}_N = 7.592\vec{i} - 4.816\vec{j} + 3.061\vec{k}$. The moment produced by the force must be equal to the normal component: $\vec{M}_N = \vec{r} \times \vec{F} = \begin{vmatrix} \vec{i} & \vec{j} & \vec{k} \\ x & 3 & z \\ 3 & 6 & 2 \end{vmatrix} = (6-6z)\vec{i} - (2x-3z)\vec{j} + (6x-9)\vec{k}$

$= 7.592\vec{i} - 4.816\vec{j} + 3.061\vec{k}$, from which $\boxed{x = \dfrac{9+3.061}{6} = 2.01\ m}$ and $\boxed{z = \dfrac{6-7.592}{6} = -0.2653\ m}$

Problem 4.156 A wrench consists of a force of magnitude $100\ N$ acting at the origin O and a couple of magnitude $60\ N\,m$. The force and the couple point in the direction from O to the point $(1, 1, 2)\ m$. If you represent the wrench by a force $\vec{F}$ acting at point $(5, 3, 1)\ m$ and a couple $\vec{M}$, what are $\vec{F}$ and $\vec{M}$?

Solution: The vector parallel to the force is $\vec{r}_F = \vec{i} + \vec{j} + 2\vec{k}$, from which the unit vector parallel to the force is $\vec{e}_F = 0.4082\vec{i} + 0.4082\vec{j} + 0.8165\vec{k}$. The force and moment at the origin are $\boxed{\vec{F} = |\vec{F}|\vec{e}_{OF} = 40.82\vec{i} + 40.82\vec{j} + 81.65\vec{k}\ (N)}$, and $\vec{M} = 24.492\vec{i} + 24.492\vec{j} + 48.99\vec{k}\ (N\text{-}m)$. The force and moment are parallel. At the point $(5, 3, 1)\ m$ the equivalent force is equal to the force at the origin, given above. The moment of this force about the origin is

$\vec{M}_F = \vec{r} \times \vec{F} = \begin{vmatrix} \vec{i} & \vec{j} & \vec{k} \\ 5 & 3 & 1 \\ 40.82 & 40.82 & 81.65 \end{vmatrix} = 204.13\vec{i} - 367.43\vec{j} + 81.64\vec{k}$ For the moments to be equal in the two systems, the added equivalent couple must be

$\boxed{\vec{M}_C = \vec{M} - \vec{M}_F = -176.94\vec{i} + 391.92\vec{j} - 32.65\vec{k}\ (N\text{-}m)}$

Problem 4.157 System 1 consist of two forces and a couple. Suppose that you want to represent it by a wrench (System 2). Determine the force $\vec{F}$, the couple $\vec{M}_t$, and the coordinates x and z where the line of action of the wrench intersects the x-z plane.

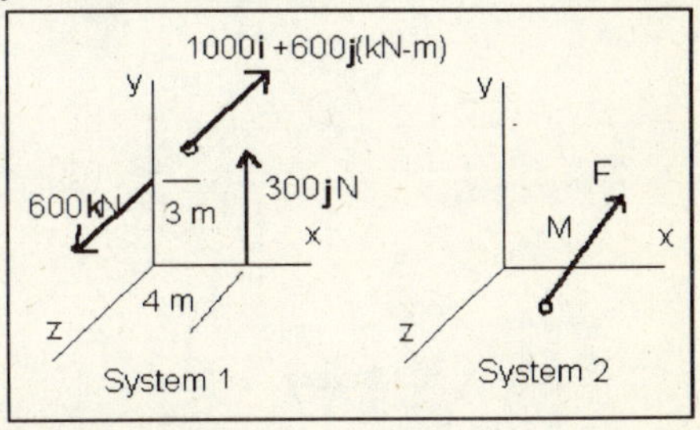

Solution: The sum of the forces in System 1 is $\boxed{\vec{F} = 300\vec{j} + 600\vec{k}\ (N)}$. The equivalent force in System 2 must have this value. The unit vector parallel to the force is $\vec{e}_F = 0.4472\vec{j} + 0.8944\vec{k}$ The sum of the moments in System 1 is $\vec{M} = 600(3)\vec{i} + 300(4)\vec{k} + 1000\vec{i} + 600\vec{j}$ $= 2800\vec{i} + 600\vec{j} + 1200\vec{k}\ (kN\,m)$. The component parallel to the force is $\vec{M}_P = 599.963\vec{j} + 1199.93\vec{k}\ (kN\text{-}m)$ $\boxed{= 600\vec{j} + 1200\vec{k}\ (kN\text{-}m)}$. The normal component is $\vec{M}_N = \vec{M} - \vec{M}_P = 2800\vec{i}$. The moment of the force

$\vec{M}_N = \begin{vmatrix} \vec{i} & \vec{j} & \vec{k} \\ x & 0 & z \\ 0 & 300 & 600 \end{vmatrix} = -300z\vec{i} - 600x\vec{j} + 300x\vec{k} = 2800\vec{i}$, from which $\boxed{x = 0}$, $\boxed{z = \dfrac{2800}{-300} = -9.333\ m}$

Problem 4.158 A plumber exerts the two forces shown to loosen a pipe. (a) What total moment does he exert about the axis of the pipe? (b) If you represent the two forces by a force $\vec{F}$ acting at O and a couple $\vec{M}$, what are $\vec{F}$ and $\vec{M}$? (c) If you represent the two forces by a wrench consisting of the force $\vec{F}$ and a parallel couple $\vec{M}$, what is $\vec{M}_t$ and where does the line of action of $\vec{F}$ intersect the x-y plane?

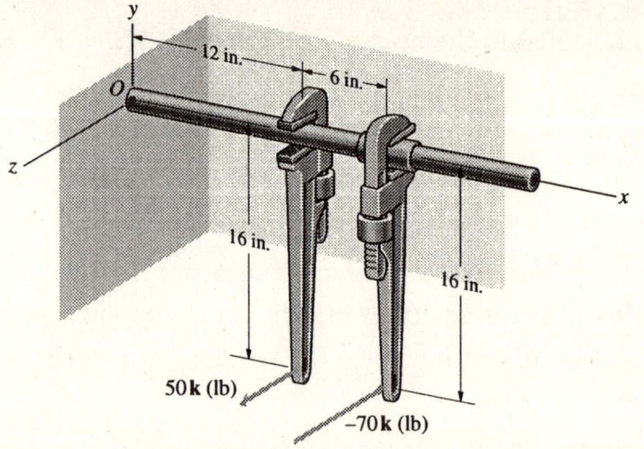

Solution: The sum of the forces is $\sum \vec{F} = 50\vec{k} - 70\vec{k} = -20\vec{k}$ (lb). (a) The total moment exerted on the pipe is $\boxed{\vec{M} = 16(20)\vec{i} = 320\vec{i}\ (ft\ lb)}$.

(b) The equivalent force at O is $\boxed{\vec{F} = -20\vec{k}}$. The sum of the moments about O is

$$\sum \vec{M}_O = (\vec{r}_1 \times \vec{F}_1) + (\vec{r}_2 \times \vec{F}_2) = \begin{vmatrix} \vec{i} & \vec{j} \\ 12 & -16 \\ 0 & 0 \end{vmatrix}$$

$\boxed{= 320\vec{i} + 660\vec{j}}$.

(c) The unit vector parallel to the force is $\vec{e}_F = \vec{k}$, hence the moment parallel to the force is $\boxed{\vec{M}_P = (\vec{e}_F \cdot \vec{M})\vec{e}_F = 0}$, and the moment normal to the force is $\vec{M}_N = \vec{M} - \vec{M}_P = 320\vec{i} + 660\vec{j}$. The force at the location of the wrench must produce this moment for the wrench to be equivalent. $\vec{M}_N = \begin{vmatrix} \vec{i} & \vec{j} & \vec{k} \\ x & y & 0 \\ 0 & 0 & -20 \end{vmatrix} = -20y\vec{i} + 20x\vec{j} = 320\vec{i} + 660\vec{j}$, from which

$\boxed{x = \frac{660}{20} = 33\ in}$, $\boxed{y = \frac{320}{-20} = -16\ in}$

Problem 4.159 Consider the system described in Problem 4.37. (a) Obtain a graph of the moment about A due to the force exerted by the spring on the circular bar at B for values of the angle α from zero to 90°. (b) Use the result of Part (a) to estimate the angle at which maximum moment occurs and the value of the moment.

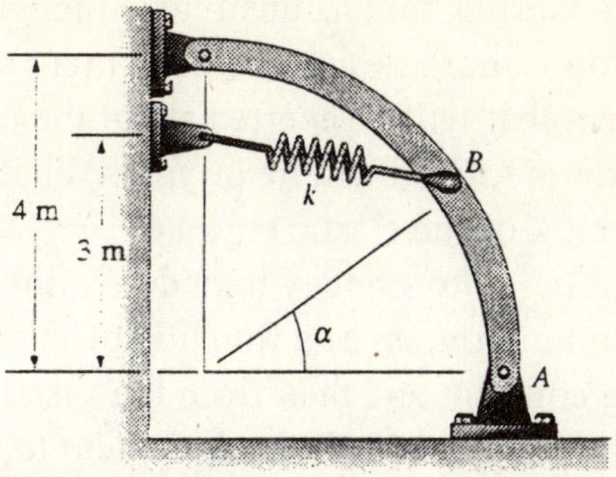

Solution: The unstretched length of spring is 1 m and the spring constant is $k = 20\ {}^{N}\!/_{m}$. Assume that the bar is a quarter circle, with a radius of 4 m. The stretched length of the spring is found from the Pythagorean Theorem: The height of the attachment point is $h = 4\sin\alpha\ m$, and the distance from the center is $4\cos\alpha$. The stretched length of the spring is $L = \sqrt{(3-h)^2 + (4\cos\alpha)^2}\ m$. The spring force is $F = (20)(L-1)\ N$. The angle that the spring makes with a vertical line parallel to A is $\beta = \tan^{-1}\left(\frac{3-h}{4\cos\alpha}\right)$. The horizontal component of the spring force is $F_X = F\cos\beta\ \ N$. The vertical component of the force is $F_Y = F\sin\beta\ \ N$. The displacement of the attachment point to the left of point A is $d = 4(1-\cos\alpha)\ m$, hence the action of the vertical component is negative, and the action of the horizontal component is positive. The moment about A is $\sum M_A = -dF_Y + hF_X$ Collecting terms and equations,

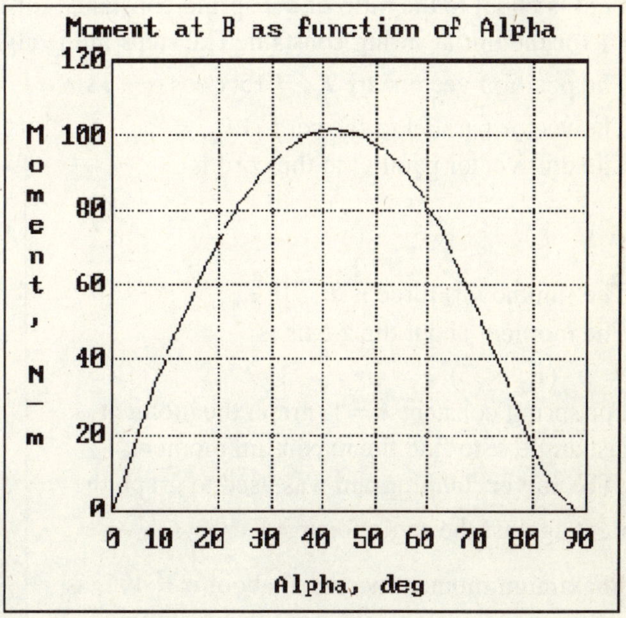

$h = 4\sin\alpha\ \ m,\ \ F_Y = F\sin\beta\ \ N,$

$F_X = F\cos\beta\ \ N,\ \ F = (20)(L-1)\ N,$

$L = \sqrt{(3-h)^2 + (4\cos\alpha)^2}\ \ m,\ \ \beta = \tan^{-1}\left(\frac{3-h}{4\cos\alpha}\right)$. A programmable calculator or a commercial package such as **TK Solver®** or **Mathcad®** is almost essential to the solution of this and the following problems. The commercial package **TK Solver 2.0** was used here to plot the graph of M against α. Using the graph as a guide, the following tabular values were taken about the maximum:

α, deg	Moment, N-m
41.5	101.463
42.0	101.483
42.5	101.472

The maximum value of the moment is estimated at $\boxed{\vec{M}_B = 101.49\ N\text{-}m}$, which occurs at approximately $\boxed{\alpha = 42.2^{o}}$

Problem 4.160 The exercise equipment shown is used by resting the elbow on the fixed pad and rotating the forearm to stretch the elastic cord AB. The cord behaves like a linear spring, and its unstretched length is 1 *ft*. Suppose that you want to design the equipment so that the maximum moment that will be exerted about the elbow joint E as the forearm is rotated will be 60 *ft lb*. What should the spring constant k of the elastic cord be?

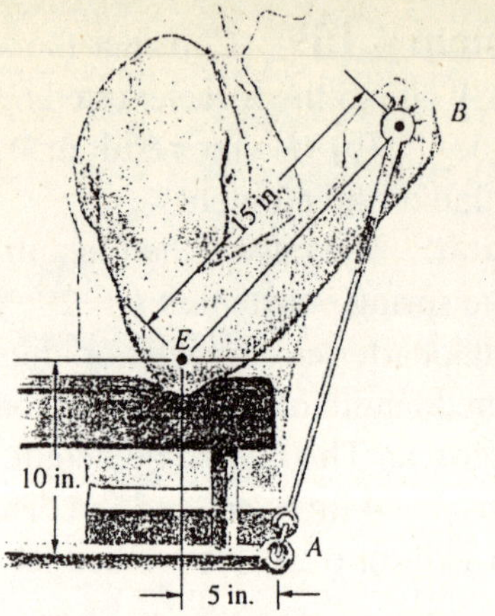

Solution: The strategy is to determine the position of maximum moment as a function of the angle α for a known spring constant and then from the system linearity use the fact that ratio of the desired moment to the actual maximum moment is equal to the ratio of the spring constants, for the same cord elongation. It is convenient to use $k_1 = 1$ for the initial spring constant. The steps in the algorithm are:

(1) The position vectors are $\vec{r}_{EB} = 15(\vec{i}\cos\alpha + \vec{j}\sin\alpha)$, and $\vec{r}_{EA} = 5\vec{i} - 10\vec{j}$.

(2) The vector parallel to the cord is $\vec{r}_{BA} = \vec{r}_{EA} - \vec{r}_{EB}$. ; its magnitude is the stretched length of the cord.

(3) The unit vector parallel to the cord is

$$\vec{e}_{BA} = \frac{\vec{r}_{BA}}{|\vec{r}_{BA}|}.$$

(4) The (unknown) force is $\vec{F} = |\vec{F}|\vec{e}_{BA}$.

(5) The moment about the z-axis is

$M_Z = \vec{k}\cdot(\vec{r}_{EB}\times\vec{F})$.

(6) For spring constant $k = 1$, graph the moment against angle α to find the maximum moment.

The **TK Solver 2.0** program was used to graph the moment against the angle for $k = 1\ {lb}/{ft}$.

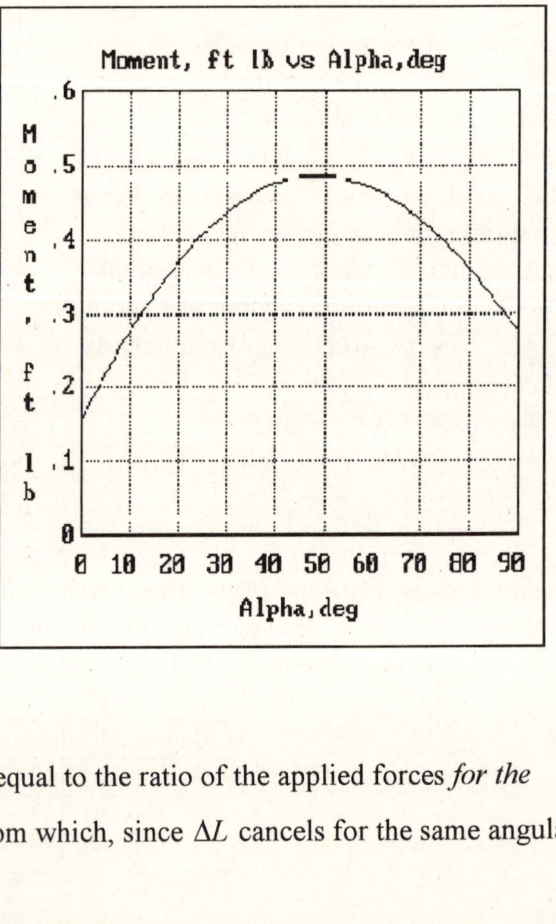

The maximum moment occurs at about $\alpha = 49^{\circ}$, as shown by the tabulated values at the maximum point, shown to four significant figures below.

α, deg	Moment, ft lb
48	0.4856
49	0.4856
50	0.4856

Since the system is linear, the ratio of the moments is equal to the ratio of the applied forces *for the same angular position*. Thus $\dfrac{60\ ft\ lb}{0.4856\ ft\ lb} = \dfrac{k_2\Delta L}{k_1\Delta L}$, from which, since ΔL cancels for the same angular position, and $k_1 = 1$, $\boxed{k_2 = \left(\dfrac{60}{0.4856}\right)k_1 = 123.54\ {lb}/{ft}}$.

Problem 4.161

Problem 4.161 The support cable extends from the top of the 3-m column A to a point B on the line L. The tension in the cable is 2 *kN*. The line L intersects the ground at the point (3, 0, 1) *m* and is parallel to the unit vector $\vec{e} = \frac{2}{7}\vec{i} + \frac{6}{7}\vec{j} - \frac{3}{7}\vec{k}$. The distance along L from the ground to point B is denoted s. What is the range of values of s for which the magnitude of the moment about O due to the force exerted by the cable at A exceeds 5.6 *kN-m*?

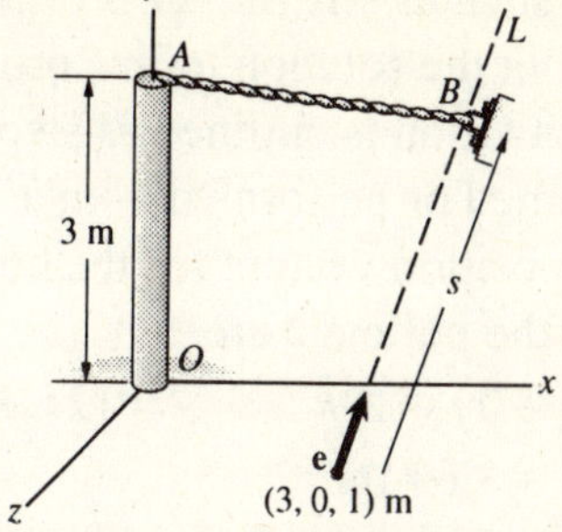

Solution: A programmable calculator or a commercial package such as **TK Solver®** or **Mathcad®** is almost essential in the solution to this problem. The **TK Solver 2.0** package was used here. The algorithm for computation is outlined as computational steps, which may be programmed or performed with a calculator:

(1) The position vectors are: $\vec{r}_{OG} = 3\vec{i} + 0\vec{j} + 1\vec{k}$, $\vec{r}_{OA} = 0\vec{i} + 3\vec{j} + 0\vec{k}$.

(2) The vector parallel to the line L is

$$\vec{r}_{GB} = s\vec{e}_{GB} = \left(\frac{2}{7}\vec{i} + \frac{6}{7}\vec{j} - \frac{3}{7}\vec{k}\right)s.$$

(3) The vector from O to B is the sum $\vec{r}_{OB} = \vec{r}_{OG} + \vec{r}_{GB}$.

(4) The vector parallel to AB is $\vec{r}_{AB} = \vec{r}_{OB} - \vec{r}_{OA}$.

(5) The unit vector parallel to AB is $\vec{e}_{AB} = \dfrac{\vec{r}_{AB}}{|\vec{r}_{AB}|}$.

(6) the tension at A is $\vec{T}_{AB} = 2\vec{e}_{AB}$.

(7) The moment about O due to the force applied at A is $\vec{M}_O = \vec{r}_{OA} \times \vec{T}_{AB}$. *Completes the computation for a given value of* s.

A plot of moment against s obtained using **TK Solver 2.0** is shown. The graph is relatively flat near the values of interest, which leads to inaccuracies in determining the limits on s. Using the graph as a guide, the following values were generated by **TK Solver.**

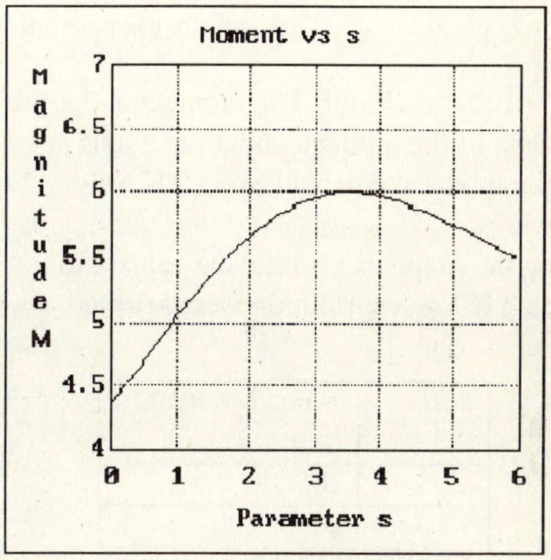

s, meters	Moment, kN m
1.89	5.596
1.90	5.601
5.57	5.602
5.58	5.5996

Thus the limits occur at about $\boxed{1.895 < s < 5.575\ m}$.

Problem 4.162 Consider Problem 4.94. Determine the distance d which causes the moment about the z axis due to the force exerted by the cable CD at point C to be a maximum. What is the maximum moment?

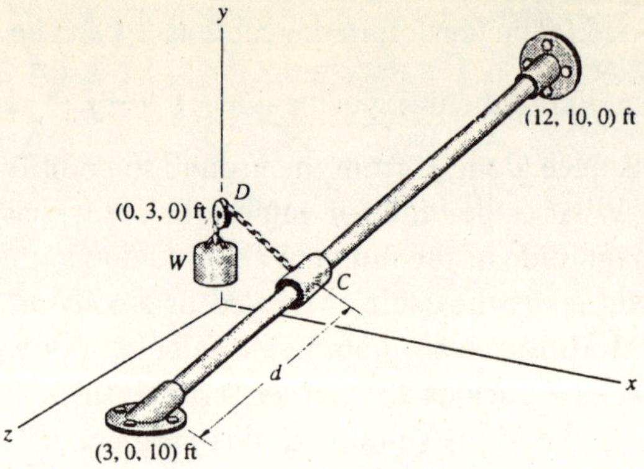

Solution: A programmable calculator or a commercial package such as **TK Solver®** or **Mathcad®** is almost essential in the solution to this problem. The algorithm for computation is outlined as steps that may be programmed or performed with a calculator:

(1) The position vectors for the bottom and top of the bar, and the pulley D are:

$\vec{r}_{OB} = 3\vec{i} + 0\vec{j} + 10\vec{k}$, $\vec{r}_{OT} = 12\vec{i} + 10\vec{j} + 0\vec{k}$

$\vec{r}_{OD} = 0\vec{i} + 3\vec{j} + 0\vec{k}$.

(2) The vector parallel to the bar, pointing to the top, is $\vec{r}_{BT} = \vec{r}_{OT} - \vec{r}_{OB}$. (3) The unit vector parallel to the bar is $\vec{e}_{BT} = \dfrac{\vec{r}_{BT}}{|\vec{r}_{BT}|}$. (4) The vector from the bottom of the bar to the point C is $\vec{r}_{BC} = d\vec{e}_{BT}$. (5) The vector position of the collar C is $\vec{r}_{OC} = \vec{r}_{OB} + \vec{r}_{BC}$. (6) The vector parallel to CD is $\vec{r}_{CD} = \vec{r}_{OD} - \vec{r}_{OC}$. (7) The unit vector parallel to CD is $\vec{e}_{CD} = \dfrac{\vec{r}_{CD}}{|\vec{r}_{CD}|}$. (8) The tension acting at the collar is $\vec{T}_{CD} = 100\vec{e}_{CD}\ lb$. (9) The moment about the z-axis is $\vec{M} = \vec{k} \cdot (\vec{r}_{OC} \times \vec{T}_{CD})$.

A graph of the moment about the z-axis against the parameter d obtained using **TK Solver 2.0** is shown.

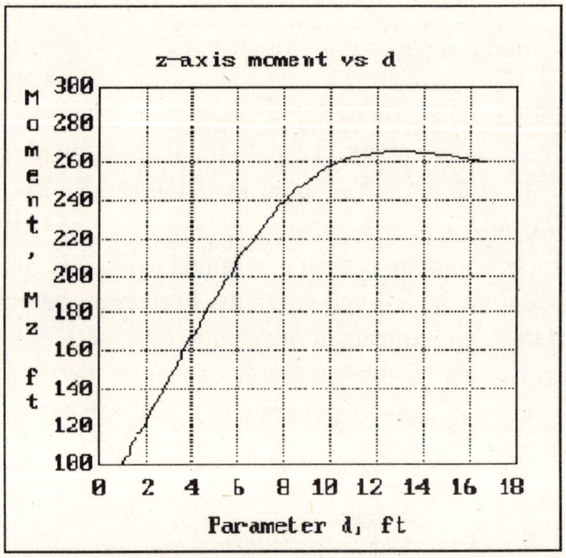

Using the graph as a guide, the values from a table in the **TK Solver** solution were selected:

d, ft	z-axis Moment, ft lb
12.8	265.38
13.0	265.41
13.2	265.38

The maximum appears to occur exactly at $\boxed{d = 13\ ft}$, with a maximum value of $\boxed{M_Z = 265.41\ ft\ lb}$

Problem 4.163 Consider Problem 4.95. The rod AB must exert a moment of magnitude 100 *ft lb* about the x-axis to support the hood of the car. Draw a graph of the magnitude of the force the rod must exert on the hood at B as a function of d for $1 \le d \le 4\ ft$. If you were designing the support AB, what value of d would you choose and what is the magnitude of the force it must exert on the hood?

Solution: The steps in an algorithm for computing the magnitude of the force as a function of the distance d are: (the detailed computations for one value of d are given in the solution to Problem 4.95.)

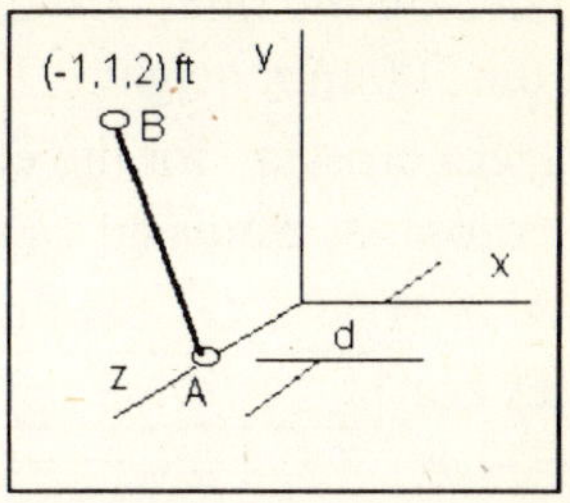

(1) The position coordinates of A and B are $\vec{r}_B = -1\vec{i} + 1\vec{j} + 2\vec{k}$, and $\vec{r}_A = 0\vec{i} + 0\vec{j} + d\vec{k}$.

(2) The vector parallel to AB is $\vec{r}_{AB} = \vec{r}_{OB} - \vec{r}_{OA}$.

(3) The unit vector parallel to AB is $\vec{e}_{AB} = \dfrac{\vec{r}_{AB}}{|\vec{r}_{AB}|}$. (4) The force exerted by the rod AB is $\vec{F} = |\vec{F}|\vec{e}_{AB}$. The moment about the x-axis is $M_X = \vec{i} \cdot (\vec{r}_{OA} \times \vec{F})$.

The commercial program **TK Solver 2.0** was used to graph the magnitude of the force as a function of the distance d. Using the graph as a guide, the following three values were taken from the table computed by **TK Solver.**

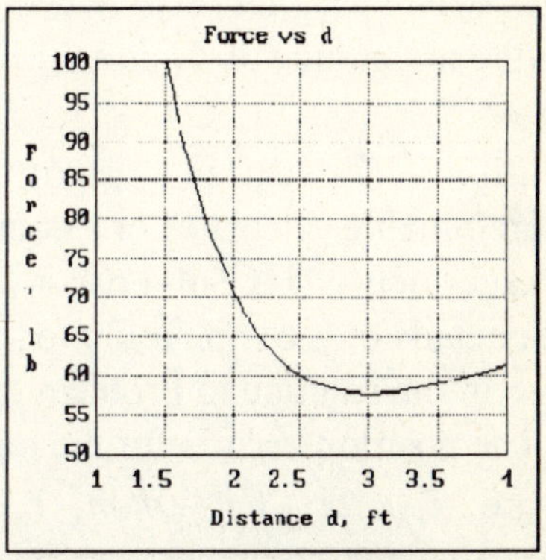

d, ft	Force, lb
2.8	58.02
3.0	57.74
3.2	57.96

The minimum force for the required moment occurs at $\boxed{d = 3\ ft}$, with a value of $\boxed{|\vec{F}| = 57.74\ lb}$

Problem 4.164 Consider the system shown in Problem 4.132. The forces exerted on the left post by cables AB and CD can be represented by a single force $\vec{F}$. Determine tensions in the cables so that $|\vec{F}| = 600\ N$ and the line of action intersects the y-axis at $y = 400\ mm$.

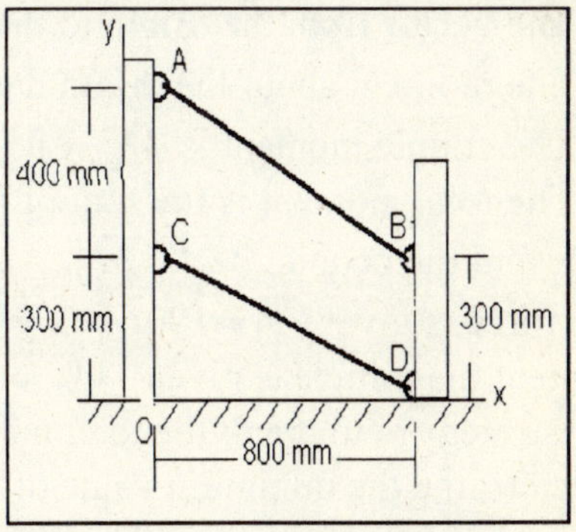

Solution: The strategy is to solve by iteration the simultaneous conditions on the forces and moments. Equating the horizontal and vertical components of the forces for the two systems:

$T_{AB} \cos\alpha + T_{CD} \cos\beta = 600\cos\theta$

$T_{AB} \sin\alpha + T_{CD} \sin\beta = 600\sin\theta$. where the angles are

$\alpha = -\tan^{-1}\left(\dfrac{400}{800}\right) = -26.6^{o}$ for cable AB, and

$\beta = -\tan^{-1}\left(\dfrac{300}{800}\right) = -20.6^{o}$ for cable CD., and the unknown angle θ applies to the single cable system. A guess at the unknown angle θ will yield a solution for the tensions which may not satisfy the equality of moments condition for the two systems. This difficulty is resolved as follows: denote the error in the equality of moments about the origin of the left post as $\varepsilon = 0.3(T_{AB}\cos\alpha) + 0.7(T_{CD}\cos\beta) - 0.4(600\cos\theta)$. Plot the error as a function of the angle θ over the allowed interval. For an angle θ at which $\varepsilon = 0$ the three conditions are satisfied.

Problem 4.164 continued on next page....

==================================<>==================================

continuation of solution to Problem 4.164...

The **TK Solver** software package was used to graph the error as a function of the angle. The range of allowed angles θ was determined by assuming that the single cable could be attached on the right post at any point between the ground and the 300 mm height.

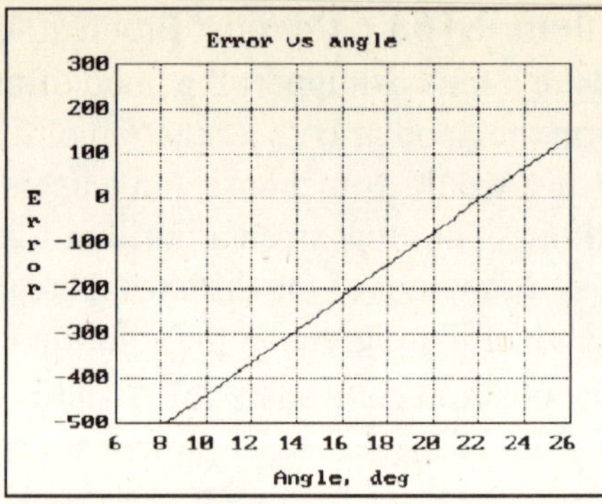

The zero crossing for the error occurs at about $\theta = 22^{o}$. A closer look at tabulated values near this value yields an angle of $\theta = 22.11^{o}$. At this angle the tensions in the two cables are $\boxed{T_{AB} = 155.4\ N}$, and $\boxed{T_{CD} = 445.26\ N}$

==================================<>==================================

Problem 4.165 Suppose that you want to represent the force and the 200 $N\,m$ couple in Problem 4.150 by a force $\vec{F}$ and a couple $\vec{M}$ and choose the distance s so that the magnitude of $\vec{M}$ is a minimum. Determine s, $\vec{F}$ and $\vec{M}$.

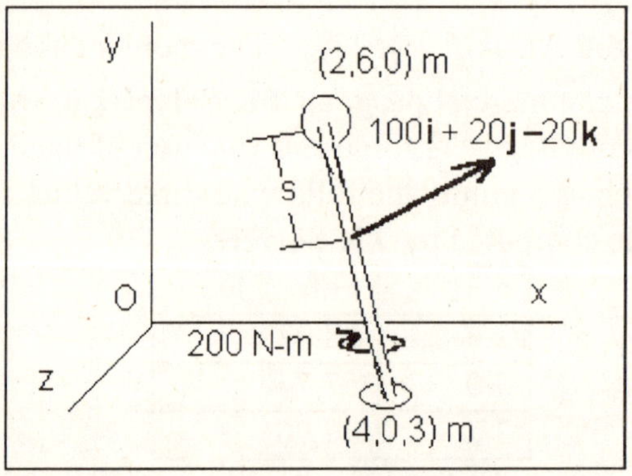

Solution: The steps in an algorithm for entering into a programmable calculator or a commercial software package such as **TK Solver®** or **Mathcad®** are given: (The details of a computation for one value of s are given in the solution to Problem 4.150)

(1) The position vectors for the top and bottom of the bar are : $\vec{r}_T = 2\vec{i} + 6\vec{j} + 0\vec{k}\ m,\ \vec{r}_B = 4\vec{i} + 0\vec{j} + 3\vec{k}\ \ m$.

(2) The vector parallel to the bar, pointing top to bottom, is $\vec{r}_{TB} = \vec{r}_B - \vec{r}_T = 2\vec{i} - 6\vec{j} + 3\vec{k}\ \ m$.

(3) The unit vector parallel to the bar, pointing top to bottom is $\vec{e}_{TB} = \dfrac{\vec{r}_{TB}}{|\vec{r}_{TB}|}$.

(4) The distance from the top of the bar to the point of application of the force is $\vec{r}_{TF} = s\vec{e}_{TB}\ m$. .

(5) The vector from the origin to the point of application of the force is $\vec{r}_{OF} = \vec{r}_T + \vec{r}_{TF}\ m$.

(6) The moment about the origin due to the action of the force is $\vec{M}_{OF} = \vec{r}_{OF} \times \vec{F}\ N\text{-}m$.

(7) The couple moment is $\vec{M}_C = M_C\vec{e}_{BT} = -M_C\vec{e}_{TB}\ \ N\text{-}m$.

(8) The total moment is the sum of the moments due to the force and the couple, $\vec{M}_T = \vec{M}_{OF} + \vec{M}_C$.

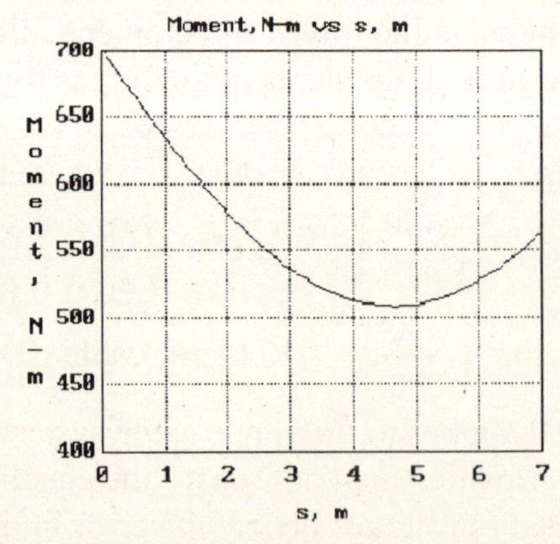

The force is given: $\vec{F} = 100\vec{i} + 20\vec{j} - 20\vec{k}\ \ N$. The couple moment magnitude is given: $M_C = 200\ N\text{-}m$.

The strategy is to graph the total moment as a function of s to determine the minimum value of the moment, and the value of s at the minimum. The **TK Solver 2.0** package was used to graph the total moment as a function of s. Using the graph as a guide, tabulated values near the minimum were examined. The minimum magnitude of the moment occurs at $\boxed{s = 4.66\ m}$. The equivalent force at the origin is $\boxed{\vec{F} = 100\vec{i} + 20\vec{j} - 20\vec{k}\ N}$. At s= 4.66 m the equivalent moment at the origin is $\boxed{\vec{M}_T = \vec{M}_{OF} + \vec{M}_C = -137.2\vec{i} + 437.77\vec{j} - 219.66\vec{k}\ \ N\text{-}m}$

==================================<>==================================

==============================<>==============================

Problem 4.166 Determine the moment of the 200 N force about A. (A) What is the two dimensional description of the moment? (b) Express the moment as a vector.

Solution: (a) The two dimensional description. By inspection, the perpendicular distance to the line of action from A is $d = (400-(-400))\ mm = 0.8\ m$. The moment is positive, since it is counterclockwise. Thus $\boxed{M_A = (0.8)(200) = 160\ N\text{-}m}$

(b) The vector description. The position vectors of A and the point of action are $\vec{r}_{OA} = 0.4\vec{i} - 0.2\vec{j}\ m$, and $\vec{r}_{OF} = -0.4\vec{i}\ m$. The distance from A to the point of action is $\vec{r}_{AF} = \vec{r}_{OF} - \vec{r}_{OA} = (-0.4-0.4)\vec{i} - (-0.2)\vec{j} = -0.8\vec{i} + 0.2\vec{j}\ m$.

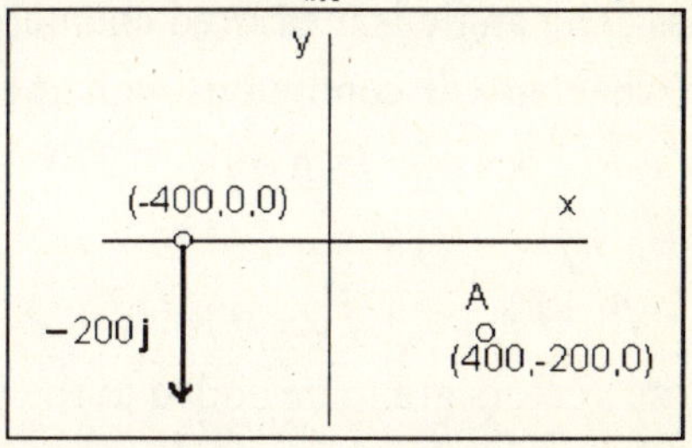

The moment is $\vec{M}_A = \vec{r}_{AF} \times \vec{F} = \begin{vmatrix} \vec{i} & \vec{j} & \vec{k} \\ -0.8 & 0.2 & 0 \\ 0 & -200 & 0 \end{vmatrix}$

$\boxed{= 160\vec{k}\ N\text{-}m}$

==============================<>==============================

Problem 4.167 The leaning tower of Pisa is approximately 180 ft tall and 23 ft in diameter. The horizontal displacement of the top of the tower from the vertical is approximately 15 ft. If you model the tower as a cylinder and assume that its weight is $7 \times 10^6\ lb$ and acts at the center, what is the magnitude of the moment exerted by the weight about the center of its base?

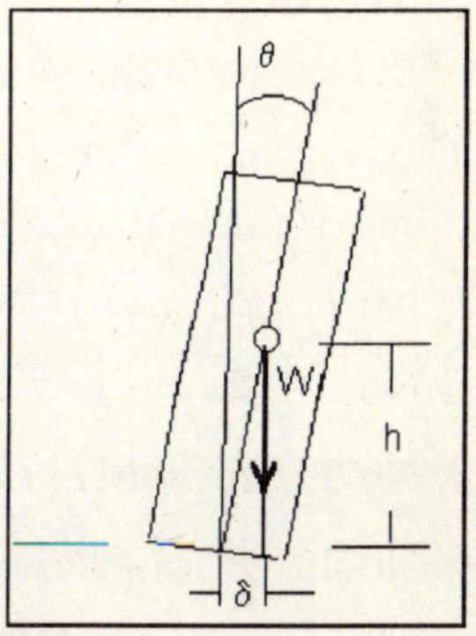

Solution: The angle of tilt is $\theta = \tan^{-1}\left(\frac{15}{180}\right) = 4.7636°$. The height of the center of gravity is $h = \frac{180}{2} = 90\ ft$. The horizontal distance at O from the line of action of the weight is $\delta = 90 \tan 4.7636° = 90\left(\frac{15}{180}\right) = 7.5\ ft$. (This result could have been got with greater simplicity using similar triangles.) The moment is negative (at least for this sketch representation). Thus $M_O = -7.5(W) = -52.5 \times 10^6$ $\boxed{= -5.25 \times 10^7\ ft\ lb}$

==============================<>==============================

Problem 4.168 The device shown has been suggested as a design for a perpetual motion machine. Determine the moment about the axis of rotation due to the four masses as a function of the angle as the device rotates 90° clockwise from the position shown and indicate whether gravity could cause rotation in that direction.

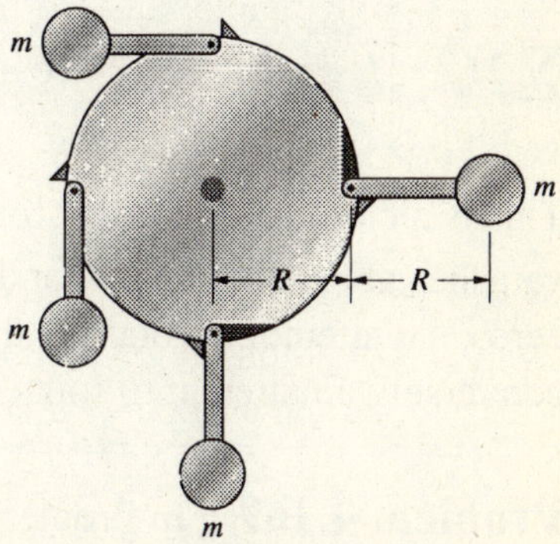

Solution: *Preliminary Remarks:* The problem is put in perspective by noting that the masses contribute zero moment in the configuration shown. The action line of the bottom weight passes through the wheel center, so that it contributes zero moment. The perpendicular distance from the wheel center to the line of action of the right hand weight is $2R$. Its moment about the wheel center due to the action of gravity is $\vec{M}_{RM} = -2Rmg\boldsymbol{j}$ The left hand and top weights are located at perpendicular distances R from the center. The sum of their moments is $\vec{M}_{LM} = 2Rmg\vec{j}$. The sum of these moments is zero: $\vec{M}_{RM} + \vec{M}_M = 0$. Thus *the wheel has no resulting moment to cause rotation* when in the configuration shown.

Problem 4.168 continued on next page....

=====================================<>=====================================

continuation of solution to Problem 4.168...

The problem, however, addresses another issue: what happens when the wheel is rotated from the current configuration through -90°? One can guess beforehand that the bottom weight will act like a pendulum weight, producing a restoring moment, that is, a moment opposed to the rotation. That this is indeed the case is shown in the following analysis.. *End of preliminary remarks.* The vector positions of the masses are the sum of two vectors: the position on the rim of the wheel, and the position of the mass. The angle is measured counterclockwise. The vector positions of the masses relative to the rim of the wheel are discontinuous with angle, with an abrupt discontinuity at $\alpha = 0^o$:

$$\vec{r}_m = R(\vec{i}\cos\alpha + \vec{j}\sin\alpha) \quad (270^o < \alpha < 360^o)$$

$$\vec{r}_m = -R\vec{j} \quad (180^o < \alpha < 270^o)$$

$$\vec{r}_m = R(-\vec{i}\sin\alpha + \vec{j}\cos\alpha) \quad (0^o < \alpha < 180^o)$$

These vectors are to be added to the wheel rim positions:

$$\vec{r}_W = R(\vec{i}\cos\alpha + \vec{j}\sin\alpha) \quad (0 < \alpha < 360^o).$$

Adding the vectors, the positions with respect to the center of the wheel for the three intervals are:

$$\vec{R}_1 = 2R(\vec{i}\cos\alpha + \vec{j}\sin\alpha) \quad (270^o < \alpha < 360^o)$$

$$\vec{R}_2 = R(\vec{i}\cos\alpha - \vec{j}(1-\sin\alpha)) \quad (180 < \alpha < 270^o)$$

$$\vec{R}_3 = R(\vec{i}(\cos\alpha - \sin\alpha) + \vec{j}(\cos\alpha + \sin\alpha)) \quad (0^o < \alpha < 180^o).$$

(a) The four masses are located 90° apart on the wheel. Start with α near but greater than zero. The positions of the masses are obtained by adding multiples of 90° to the angle.

Mass No 1: $\vec{R}_1 = 2R(\vec{i}\cos\alpha + \vec{j}\sin\alpha) \quad (\alpha \cong 0^o)$

Mass No 2: $\vec{R}_2 = R(\vec{i}\,sin\alpha - \vec{j}(1+cos\alpha)) \quad (\alpha \cong 0^o)$

Mass No 3: $\vec{R}_3 = R(\vec{i}(sin\,\alpha - cos\,\alpha) - \vec{j}(sin\,\alpha + cos\,\alpha)) \quad (\alpha \cong 0^o)$

Mass No 4: $\vec{R}_4 = -R(\vec{i}(sin\,\alpha + cos\,\alpha) + \vec{j}(sin\,\alpha - cos\,\alpha)) \quad (\alpha \cong 0^o)$

The weight vectors are $\vec{W} = -mg\vec{j}$. The sums of the moments:

$$\sum \vec{M}_O = (\vec{R}_1 \times \vec{W}) + (\vec{R}_2 \times \vec{W}) + (\vec{R}_3 \times \vec{W}) + (\vec{R}_4 \times \vec{W}) = (\vec{R}_1 + \vec{R}_2 + \vec{R}_3 + \vec{R}_4) \times \vec{W} = (\sum_1^4 \vec{R}_i) \times \vec{W}.$$

Collecting terms in the sum: $\sum_1^4 \vec{R}_i = +\vec{i}R\sin\alpha - \vec{j}R(1+\cos\alpha)$, from which

$$\sum \vec{M}_o = \begin{vmatrix} \vec{i} & \vec{j} & \vec{k} \\ R\sin\alpha & -R(1+\cos\alpha) & 0 \\ 0 & -mg & 0 \end{vmatrix} \boxed{= -\vec{k}mgR\sin\alpha}$$

This is a clockwise moment for $\alpha > 0^o$, and a counterclockwise moment for $\alpha < 0^o$. The moment vanishes at exactly $\alpha = 0^o$, and reverses sign on either side in a manner that tries to restore the angle to zero. An attempt to rotate in the negative direction will be opposed by a counterclockwise moment, and conversely an attempt to rotate in a clockwise manner will be opposed. This is analogous to a pendulum.

=====================================<>=====================================

Problem 4.169 In Problem 4.168, determine whether gravity could cause rotation in the counterclockwise direction.

Solution: The solution is contained in the discussion and solution to Problem 4.168. Gravity will not cause the system to rotate.

=====================================<>=====================================

Problem 4.170 Determine the moment of the 400 N force (a) about A, (b) about B.

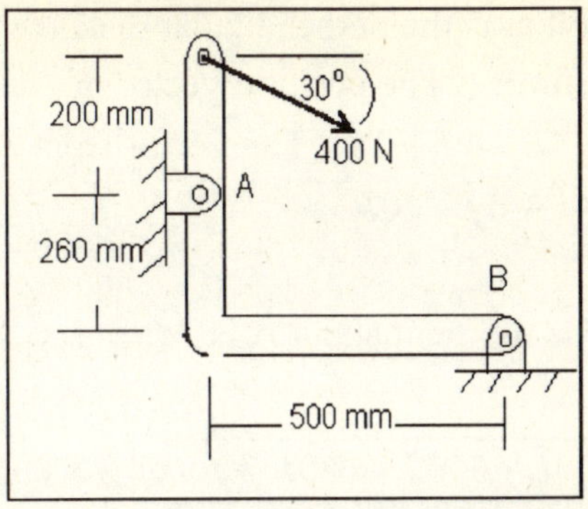

Solution: Use the two dimensional description of the moment. The vertical and horizontal components of the 200 N force are $F_Y = -400 \sin 30^o = -200\ N$, $F_X = +400\cos 30^o = 346.41\ N$.

(a) The moment arm from A to the line of action of the horizontal component is 0.22 m. The moment arm from A to the vertical component is zero. The moment about A is negative, $\boxed{M_A = -0.22(346.41) = -76.21\ N\text{-}m}$

(b) The perpendicular distances to the lines of action of the vertical and horizontal components of the force from B are $d_1 = 0.5\ m$, and $d_2 = 0.48\ m$. .The action of the vertical component is positive, and the action of the horizontal component is negative. The sum of the moments:

$\boxed{M_B = +0.5(200) - 0.48(346.41) = -66.28\ N\text{-}m}$

Problem 4.171 The vector sum of the forces acting on the beam is zero and the sum of the moments about A is zero. (a) What are the forces A_X, A_Y, and B? (a) What is the sum of the moments about B?

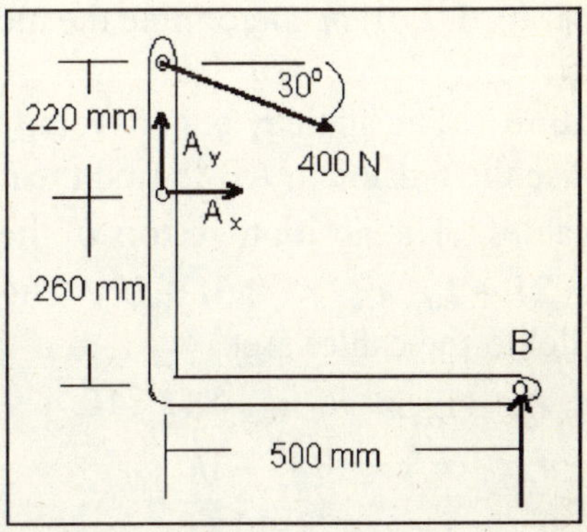

Solution: The vertical and horizontal components of the 400 N force are: $F_X = 400\cos 30^o = 346.41\ N$, $F_Y = 400 \sin 30^o = 200\ N$. The sum of the forces is

$\sum F_X = A_X + 346.41 = 0$, from which $\boxed{A_X = -346.41\ N}$

$\sum F_Y = A_Y + B - 200 = 0$. The sum of the moments about A is

$\sum M_A = 0.5B - 0.22(346.41) = 0$, from which $\boxed{B = 152.42\ N}$.

Substitute into the force equation to get

$\boxed{A_Y = 200 - B = 47.58\ N}$

(b) The moments about B are

$\boxed{M_B = -0.5A_Y - 0.48(346.41) - 0.26A_X + 0.5(200) = 0}$

Problem 4.172 To support the ladder, the force exerted at B by the hydraulic piston AB must exert a moment about C equal in magnitude to the moment about C due to the ladder's 450 lb weight. What is the magnitude of the force exerted at B?

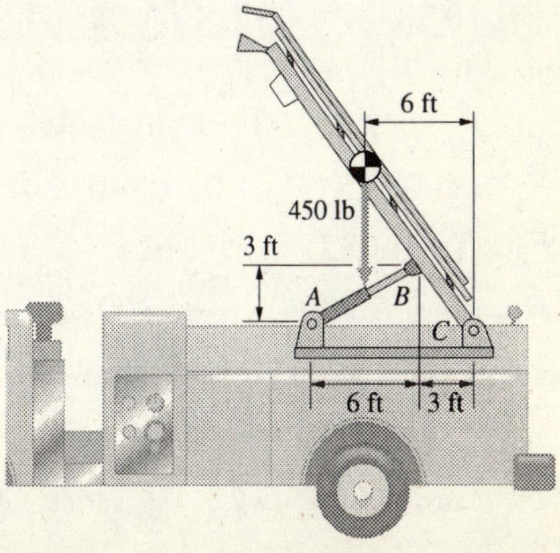

Solution: The moment about C exerted by the weight is $M_C = 450(6) = 2700\ ft\ lb$.

The ladder is at an elevation of 45° from the horizontal. The cylinder is at an angle $\theta = \tan^{-1}\left(\frac{3}{6}\right) = 26.56^o$. The vertical and horizontal components of the force at B due to the cylinder are

$F_X = F\cos 26.57^o = 0.8944F\ lb$

$F_Y = F\sin 26.57^o = 0.4472F\ lb$ The moment about C due to these forces is $M_C = -3(0.4472)F - 3(0.8944)F + 2700 = 0$.

Solving: $\boxed{F = \dfrac{2700}{4.0249} = 670.82\ lb}$

Problem 4.173 The force $\vec{F} = -60\vec{i} + 60\vec{j}\ (lb)$. (a) Determine the moment of $\vec{F}$ about point A. (b) What is the perpendicular distance from point A to the line of action of $\vec{F}$?

Solution: The position vector of A and the point of action are $\vec{r}_A = 8\vec{i} + 2\vec{j} + 12\vec{k}\ (ft)$, and $\vec{r}_F = 4\vec{i} - 4\vec{j} + 2\vec{k}$. The vector from A to F is $\vec{r}_{AF} = \vec{r}_F - \vec{r}_{OA} = (4-8)\vec{i} + (-4-2)\vec{j} + (2-12)\vec{k}$ $= -4\vec{i} - 6\vec{j} - 10\vec{k}$.

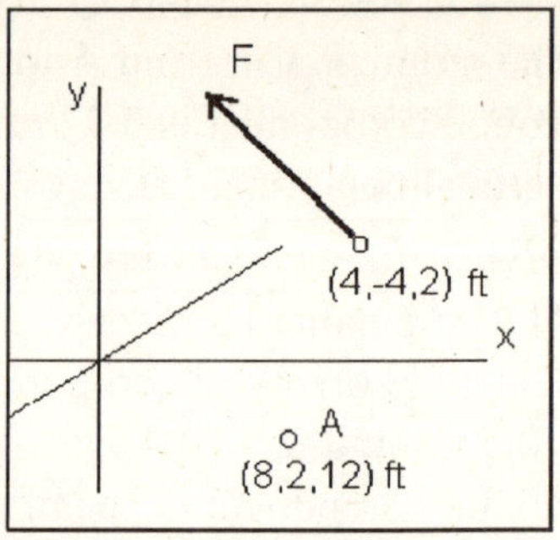

(a)The moment about A is $\vec{M}_A = \vec{r}_{AF} \times \vec{F} = \begin{vmatrix} \vec{i} & \vec{j} & \vec{k} \\ -4 & -6 & -10 \\ -60 & 60 & 0 \end{vmatrix}$

$\boxed{= 600\vec{i} - 600\vec{j} - 600\vec{k}\ (ft\ lb)}$ (b) The magnitude of the moment is $|\vec{M}_A| = \sqrt{600^2 + 600^2 + 600^2} = 1039.3\ ft\ lb$. The magnitude of the force is $|\vec{F}| = \sqrt{60^2 + 60^2} = 84.8528\ lb$. The perpendicular distance from A to the line of action is $\boxed{D = \frac{1039.3}{84.8528} = 12.25\ ft}$

Problem 4.174 The 20 kg mass is suspended by cables attached to three vertical 2 m posts. Point A is at (0, 1.2, 0) m. Determine the moment about the base E due to the force exerted on the post BE by cable AB.

Solution: The strategy is to develop the simultaneous equations in the unknown tensions in the cables, and use the tension in AB to find the moment about E. This strategy requires the unit vectors parallel to the cables. The position vectors of the points are: $\vec{r}_{OA} = 1.2\vec{j}$, $\vec{r}_{OB} = -0.3\vec{i} + 2\vec{j} + 1\vec{k}$, $\vec{r}_{OC} = 2\vec{j} - 1\vec{k}$, $\vec{r}_{OD} = 2\vec{i} + 2\vec{j}$, $\vec{r}_{OE} = -0.3\vec{i} + 1\vec{k}$. The vectors parallel to the cables are:

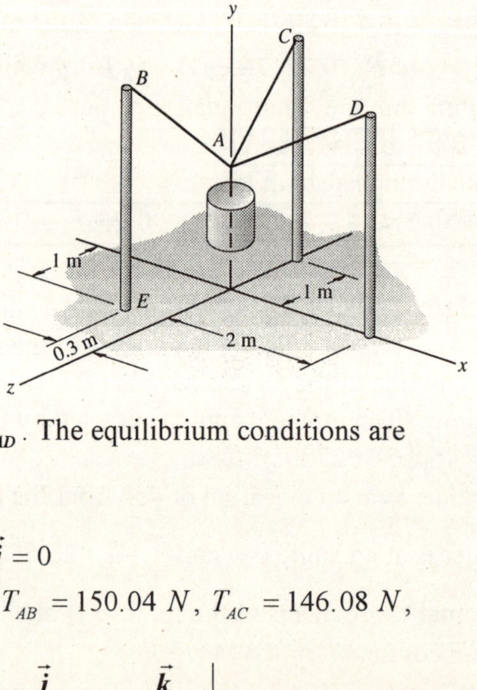

$\vec{r}_{AB} = \vec{r}_{OB} - \vec{r}_{OA} = -0.3\vec{i} + 0.8\vec{j} + 1\vec{k}$.,

$\vec{r}_{AC} = \vec{r}_{OC} - \vec{r}_{OA} = +0.8\vec{j} - 1\vec{k}$,

$\vec{r}_{AD} = \vec{r}_{OD} - \vec{r}_{OA} = +2\vec{i} + 0.8\vec{j}$.

The unit vectors parallel to the cables are:

$\vec{e}_{AB} = \frac{\vec{r}_{AB}}{|\vec{r}_{AB}|} = -0.2281\vec{i} + 0.6082\vec{j} + 0.7603\vec{k}$:

$\vec{e}_{AC} = 0\vec{i} + 0.6247\vec{j} - 0.7809\vec{k}$,

$\vec{e}_{AD} = +0.9848\vec{i} + 0.3714\vec{j} + 0\vec{k}$. The tensions in the cables are $\vec{T}_{AB} = T_{AB}\vec{e}_{AB}$, $\vec{T}_{AC} = T_{AC}\vec{e}_{AC}$, and $\vec{T}_{AD} = T_{AD}\vec{e}_{AD}$. The equilibrium conditions are $\vec{T}_{AB} + \vec{T}_{AC} + \vec{T}_{AD} = \vec{W}$. Collect like terms in $\vec{i}, \vec{j}, \vec{k}$:

$\Sigma\vec{F}_X = (-0.2281T_{AB} + 0T_{AC} + 0.9848T_{AD})\vec{i} = 0$

$\Sigma\vec{F}_Y = (+0.6082 \cdot T_{AB} + 0.6247 \cdot T_{AC} + 0.3714 \cdot T_{AD} - 196.2)\vec{j} = 0$

$\Sigma\vec{F}_Z = (+0.7603 \cdot T_{AB} - 0.7809 \cdot T_{AC} + 0 \cdot T_{AD})\vec{k} = 0$ Solve: $T_{AB} = 150.04\ N$, $T_{AC} = 146.08\ N$, $T_{AD} = 36.86\ N$. The moment about E is

$$\vec{M}_E = \vec{r}_{EB} \times (-T_{AB}\vec{e}_{AB}) = -T_{AB}(\vec{r}_{EB} \times \vec{e}_{AB}) = -150\begin{vmatrix} \vec{i} & \vec{j} & \vec{k} \\ 0 & 2 & 0 \\ -0.2281 & +0.6082 & +0.7603 \end{vmatrix} \boxed{= -228\vec{i} - 68.43\vec{k}\ (N\text{-}m)}$$

Problem 4.175

Problem 4.175 Three forces of equal magnitude are applied parallel to the sides of an equilateral triangle. (a) Show that the sum of the moments of the forces is the same about any point. (b) Determine the magnitude of the moment.

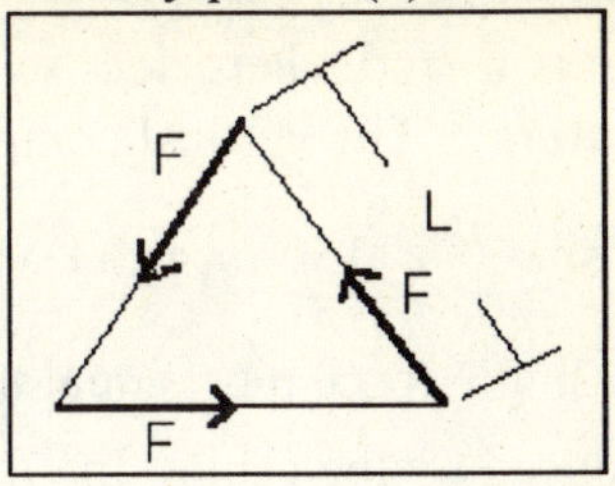

Solution: The interior angles of an equilateral triangle are 60°. Assume that the x-axis is coincident with the lower side, with the origin at the lower left corner. Denote the forces by labels 1,2, 3 counterclockwise beginning with the one coincident with the lower side, and label the corners 1,2,3 beginning with the lower left corner. The vectors to the lower corners are $\vec{r}_1 = 0$, and $\vec{r}_2 = L\vec{i}$. Let P(x,y) be any point in the space. The vector distances from P to the lower corners are $\vec{r}_{P1} = \vec{r}_1 - \vec{r}_P = -x\vec{i} - y\vec{j}$, $\vec{r}_{P2} = \vec{r}_2 - \vec{r}_p = (L-x)\vec{i} - y\vec{j}$. The unit vectors parallel to the forces are: $\vec{e}_1 = 1\vec{i}$, $\vec{e}_2 = -\frac{1}{2}\vec{i} + \frac{\sqrt{3}}{2}\vec{j}$, and $\vec{e}_3 = -\frac{1}{2}\vec{i} - \frac{\sqrt{3}}{2}\vec{j}$.. The sum of the moments about P is

$$\Sigma\vec{M} = (\vec{r}_{P1}\times\vec{F}_1) + (\vec{r}_{P1}\times\vec{F}_3) + (\vec{r}_{P2}\times\vec{F}_2) = F[\vec{r}_{P1}\times(\vec{e}_1+\vec{e}_3) + (\vec{r}_{P2}\times\vec{e}_2)]$$

$$\sum\vec{M}_P = F\begin{vmatrix}\vec{i} & \vec{j} & \vec{k}\\ -x & -y & 0\\ \frac{1}{2} & -\frac{\sqrt{3}}{2} & 0\end{vmatrix} + F\begin{vmatrix}\vec{i} & \vec{j} & \vec{k}\\ L-x & -y & 0\\ -\frac{1}{2} & \frac{\sqrt{3}}{2} & 0\end{vmatrix} = \vec{k}F\left(\frac{\sqrt{3}}{2}x + \frac{1}{2}y\right) + \vec{k}F\left(\frac{\sqrt{3}}{2}L - \frac{\sqrt{3}}{2}x - \frac{1}{2}y\right) = \boxed{\frac{\sqrt{3}}{2}FL\vec{k}}$$

Since the result is independent of the coordinates of P, the moment about any point is the same.

Problem 4.176

Problem 4.176 The bar AB supporting the lid of the grand piano exerts a force $\vec{F} = -6\vec{i} + 35\vec{j} - 12\vec{k}$ at B. The coordinates of B are (3, 4, 3) *ft*. What is the moment of the force about the hinge line of the lid (the x-axis)?

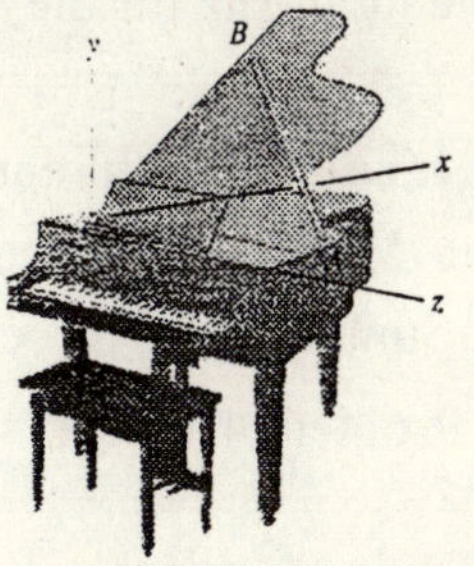

Solution: The position vector of point B is $\vec{r}_{OB} = 3\vec{i} + 4\vec{j} + 3\vec{k}$ The moment about the x-axis due to the force is $M_X = \vec{e}_X\cdot(\vec{r}_{OB}\times\vec{F}) = \vec{i}\cdot(\vec{r}_{OB}\times\vec{F})$

$$M_X = \begin{vmatrix}1 & 0 & 0\\ 3 & 4 & 3\\ -6 & 35 & -12\end{vmatrix} \boxed{= -153\ ft\ lb}$$

Problem 4.177

Problem 4.177 The system of cables and pulleys supports half of the 600 *lb* weight of the work platform. If you represent the upward force exerted at E by cable EF and the upward force exerted at G by cable GH by a single equivalent force $\vec{F}$, what is $\vec{F}$ and where does its line of action intersect the x-axis?

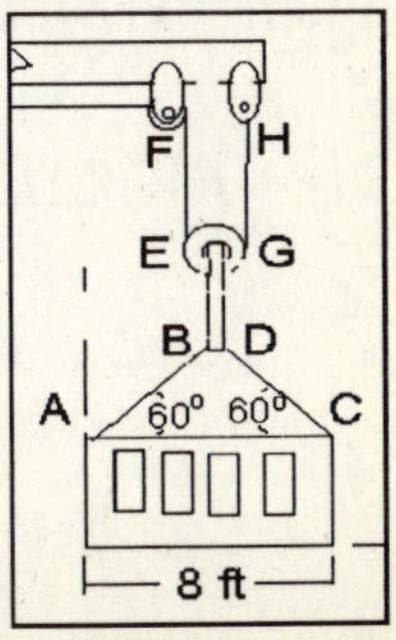

Solution: The cable-pulley combination does not produce a moment. Hence the equivalent force does not. The equivalent force is equal to the total supported weight, or $\boxed{\vec{F} = +\frac{600}{2}\vec{j} = 300\vec{j}\ (lb)}$ The force occurs at midpoint of the platform width, $\boxed{x = \frac{8}{2} = 4\ ft}$

=================================<>=================================

Problem 4.178 Consider the system in Problem 4.177. (a) What are the tensions in the cables AB and CD? (b) If you represent the forces exerted by the cables at A and C by a single equivalent force $\vec{F}$, what is $\vec{F}$ and where does its line intersect the x axis?

Solution: The vertical component of the tension is each cable must equal half the weight supported.

$T_{AB}\sin 60^o = 150\ lb$, from which $\boxed{T_{AB} = \dfrac{150}{\sin 60^o} = 173.2\ lb}$ By symmetry, the tension $\boxed{T_{CD} = 173.2\ lb}$.

The single force must equal the sum of the vertical components; since there is no resultant moment produced by the cables, the force is $\boxed{\vec{F} = 300\vec{j}\ lb}$ and it acts at the platform width midpoint $\boxed{x = 4\ ft}$.

=================================<>=================================

Problem 4.179 The systems are equivalent. Determine the forces A_X and A_y and the couple M_A.

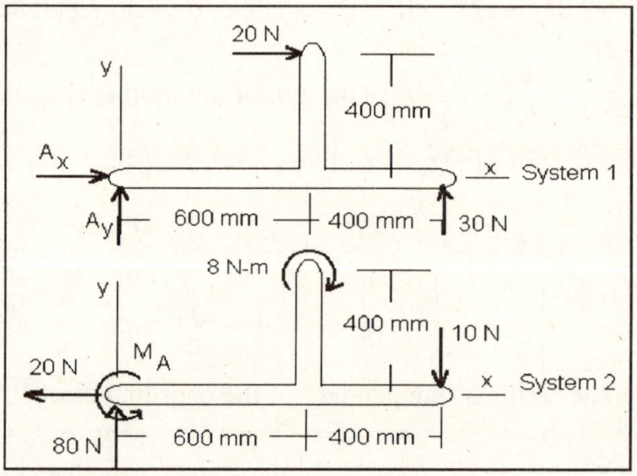

Solution: The sum of the forces for System 1 is $\sum \vec{F}_X = (A_X + 20)\vec{i}$,

$\sum \vec{F}_Y = (A_Y + 30)\vec{j}$ The sum of forces for System 2 is $\sum \vec{F}_X = (-20)\vec{i}$ and

$\sum \vec{F}_Y = (80 - 10)\vec{j}$ Equating the two systems:

$A_X + 20 = -20$ from which $\boxed{A_X = -40\ N}$

$A_Y + 30 = 80 - 10$ from which $\boxed{A_Y = 40\ N}$

The sum of the moments about the left end for System 1 is

$\sum M_1 = -(0.4)(20) + 30(1) = 22\ N\text{-}m$. The sum of moments about the left end for System 2 is $\sum M_2 = M_A - 10(1) - 8 = M_A - 18$.

Equating the moments for the two systems: $\boxed{M_A = 18 + 22 = 40\ N\text{-}m}$

=================================<>=================================

Problem 4.180 If you represent the two systems in Problem 4.179 by a force $\vec{F}$ acting at the origin and a couple $\vec{M}$, what are $\vec{F}$ and $\vec{M}$?

Solution: Summing the forces in System 1, $\vec{F} = (A_X + 20)\vec{i} + (A_Y + 30)\vec{j}$. Substituting from the solution in Problem 4.179, $\boxed{\vec{F} = -20\vec{i} + 70\vec{j}}$ The moment is $\boxed{\vec{M} = -20(0.4)\vec{k} + 30\vec{k} = 22\ \vec{k}\ (N\text{-}m)}$

=================================<>=================================

Problem 4.181 If you represent the equivalent systems in Problem 4.179 by a force $\vec{F}$ what is $\vec{F}$ and where does it cross the x-axis?

Solution: The force is $\boxed{\vec{F} = -20\vec{i} + 70\vec{j}}$ The moment to be represented is

$$\vec{M} = (\vec{r} \times \vec{F}) = 22\vec{k} = \begin{vmatrix} \vec{i} & \vec{j} & \vec{k} \\ x & 0 & 0 \\ -20 & 70 & 0 \end{vmatrix} = 70x\vec{k}, \text{ from which } \boxed{x = \frac{22}{70} = 0.3143\ m}$$

=================================<>=================================

Problem 4.182 The two systems are equivalent. If $\vec{F} = -100\vec{i} + 40\vec{j} + 30\vec{k}\ (lb)$, and $\vec{M}' = -80\vec{i} + 120\vec{j} + 40\vec{k}\ \ (in\ lb)$, determine $\vec{F}'$ and $\vec{M}$.

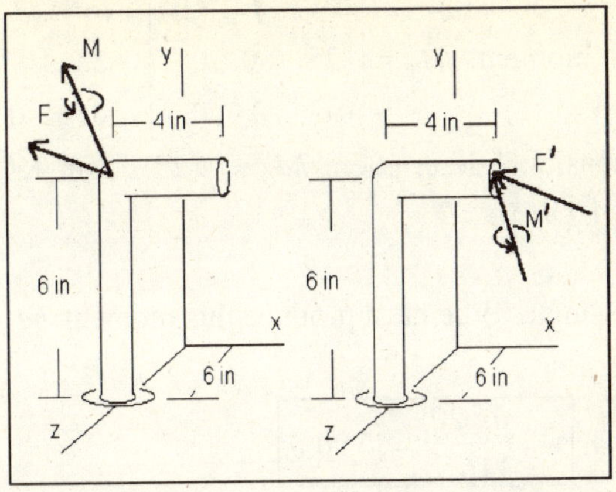

Solution: The sum of forces in the two systems must be equal, thus

$\boxed{\vec{F}' = \vec{F} = -100\vec{i} + 40\vec{j} + 30\vec{k}\ \ (lb)}$.

The moment for the unprimed system is $\vec{M}_T = \vec{r} \times \vec{F} + \vec{M}$. The moment for the primed system is $\vec{M}_T' = \vec{r}' \times \vec{F} + \vec{M}'$. The position vectors are $\vec{r} = 0\vec{i} + 6\vec{j} + 6\vec{k}$, and $\vec{r}' = 4\vec{i} + 6\vec{j} + 6\vec{k}$. Equating the moments and solving for the unknown moment $\vec{M} = \vec{M}' + (\vec{r}' - \vec{r}) \times \vec{F}$

$$= -80\vec{i} + 120\vec{j} + 40\vec{k} + \begin{vmatrix} \vec{i} & \vec{j} & \vec{k} \\ 4 & 0 & 0 \\ -100 & 40 & 30 \end{vmatrix} = -80\vec{i} + 120\vec{j} + 40\vec{k} - 120\vec{j} + 160\vec{k}\ \boxed{= -80\vec{i} + 200\vec{k}\ (in\text{-}lb)}$$

Problem 4.183 The tugboats A and B exert forces $|\vec{F}_A| = 1\ kN$ and $|\vec{F}_B| = 1.2\ \ kN$ on the ship. The angle $\theta = 30^o$. If you represent the two forces by a force $\vec{F}$ acting at the origin O and a couple $\vec{M}$, what are $\vec{F}$ and $\vec{M}$?

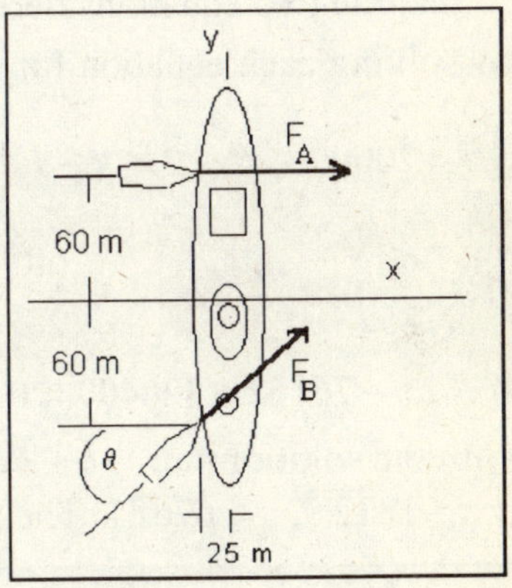

Solution: The sums of the forces are:

$\sum \vec{F}_X = (1 + 1.2\cos 30^o)\vec{i} = 2.0392\vec{i}\ \ (kN)$

$\sum \vec{F}_Y = (1.2\sin 30^o)\vec{j} = 0.6\vec{j}\ \ (kN)$. The equivalent force at the origin is

$\boxed{\vec{F}_{EQ} = 2.04\vec{i} + 0.6\vec{j}}$

The moment about O is $\vec{M}_O = \vec{r}_A \times \vec{F}_A + \vec{r}_B \times \vec{F}_B$. The vector positions are $\vec{r}_A = -25\vec{i} + 60\vec{j}\ \ (m)$, and $\vec{r}_B = -25\vec{i} - 60\vec{j}\ \ (m)$. The moment:

$$\vec{M}_O = \begin{vmatrix} \vec{i} & \vec{j} & \vec{k} \\ -25 & 60 & 0 \\ 1 & 0 & 0 \end{vmatrix} + \begin{vmatrix} \vec{i} & \vec{j} & \vec{k} \\ -25 & -60 & 0 \\ 1.0392 & 0.6 & 0 \end{vmatrix}\ \boxed{= -12.648\vec{k} = -12.6\vec{k}\ \ (kN\text{-}m)}$$

Check: Use a two dimensional description: The moment is

$M_O = -(25)F_B \sin 30^o + (60)(F_B \cos 30^o) - (60)(F_A) = 39.46F_B - 60F_A = -12.6\ kN\text{-}m$

Problem 4.184 The tugboats A and B in Problem 4.183 exert forces $|\vec{F}_A| = 600\ N$ and $|\vec{F}_B| = 800\ N$ on the ship. The angle $\theta = 45^o$. If you represent the two forces by a force $\vec{F}$, what is $\vec{F}$ and where does its line of action intersect the y-axis?

Solution: The equivalent force is $\vec{F} = (0.6 + 0.8\cos 45^o)\vec{i} + 0.8\sin 45^o\,\vec{j}$ $\boxed{= 1.1656\vec{i} + 0.5656\vec{j}\ \ (kN)}$

The moment produced by the two forces is $\vec{M}_O = \vec{r}_A \times \vec{F}_A + \vec{r}_B \times \vec{F}_B$. The vector positions are $\vec{r}_A = -25\vec{i} + 60\vec{j}\ \ (m)$, and $\vec{r}_B = -25\vec{i} - 60\vec{j}\ \ (m)$.

Problem 4.184 continued on next page..........

............*continuation of solution to Problem 4.184*.....

The moment: $\vec{M}_O = \begin{vmatrix} \vec{i} & \vec{j} & \vec{k} \\ -25 & 60 & 0 \\ 0.6 & 0 & 0 \end{vmatrix} + \begin{vmatrix} \vec{i} & \vec{j} & \vec{k} \\ -25 & -60 & 0 \\ 0.5656 & 0.5656 & 0 \end{vmatrix} = -16.20\vec{k}\ (kN\text{-}m)$ *Check:* Use a two dimensional description: $M_O = -(25)F_B \sin 45^o + (60)F_B \cos 45^o - 60F_A = 24.75F_B - 60F_A = -16.20\ kN\text{-}m$.

The single force must produce this moment. $\vec{M}_o = \begin{vmatrix} \vec{i} & \vec{j} & \vec{k} \\ 0 & y & 0 \\ 1.1656 & 0.5656 & 0 \end{vmatrix} = -1.1656y\vec{k} = -16.20\vec{k}$, from which $\boxed{y = \dfrac{16.20}{1.1656} = 13.90\ m}$

Problem 4.185 The tugboats A and B in Problem 4.183 want to exert two forces on the ship that are equivalent to a force $\vec{\boldsymbol{F}}$ acting at the origin of 2 *kN* magnitude. If $|\vec{\boldsymbol{F}}_A| = 800\ N$, determine the necessary values of $|\vec{\boldsymbol{F}}_B|$ and angle θ.

Solution: The equivalent force at the origin is $(F_A + F_B\cos\theta)^2 + (F_B \sin\theta)^2 = 2000^2$. The moment about the origin due to F_A and F_B must be zero: $\Sigma M_o = -60F_A + 60F_B\cos\theta - 25F_B \sin\theta = 0$. These are two equations in two unknowns $F_B \sin\theta$ and $F_B\cos\theta$. For brevity write $x = F_B\cos\theta$, $y = F_B\sin\theta$, so that the two equations become $x^2 + 2F_A x + F_A^2 + y^2 = 2000^2$ and $60x - 25y - 60F_A = 0$. Eliminate y by solving each equation for y^2 and equating the results:

$y^2 = 2000^2 - x^2 - 2F_A x - F_A^2 = \left(-\frac{60}{25}F_A + \frac{60}{25}x\right)^2$. Reduce to obtain the quadratic in x:

$\left[1+\left(\frac{60}{25}\right)^2\right]x^2 + 2F_A\left[1-\left(\frac{60}{25}\right)^2\right]x + \left[1+\left(\frac{60}{25}\right)^2\right]F_A^2 - 2000^2 = 0$. Substitute $F_A = 800\ N$ to obtain

$6.76x^2 - 7616x + 326400 = 0$. In canonical form: $x^2 + 2bx + c = 0$, where $b = -563.31$, and $c = 48284.0$, with the solutions $x = -b \pm \sqrt{b^2 - c} = 1082.0,\ = 44.62$. From the second equation, $y = -1812.9,\ = 676.81$. The force F_B has two solutions: Solve for F_B and θ : (1)

$\boxed{F_B = \sqrt{44.6^2 + 1812.9^2} = 1813.4\ N}$ at the angle $\boxed{\theta = \tan^{-1}\left(\frac{-1812.9}{44.6}\right) = -88.6^o}$, and (2)

$\boxed{F_B = \sqrt{676.8^2 + 1082.0^2} = 1276.2\ N}$, at the angle $\boxed{\theta = \tan^{-1}\left(\frac{676.8}{1082.0}\right) = 32.0^o}$

Problem 4.186 If you represent the forces exerted on the floor by the table legs by a force $\vec{F}$ acting at the origin and a couple $\vec{M}$, what are $\vec{F}$ and $\vec{M}$?

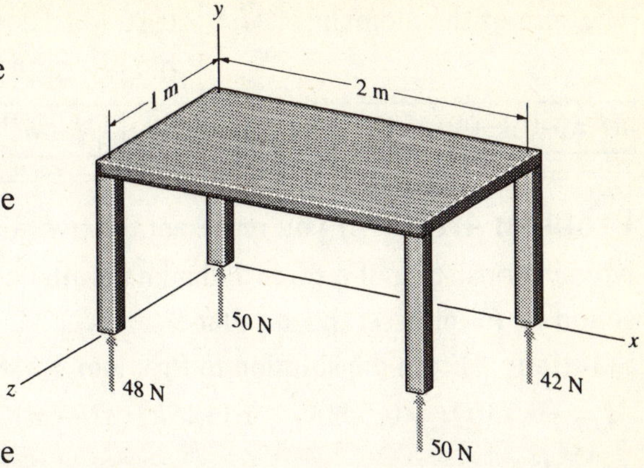

Solution: The sum of the forces is the equivalent force at the origin. $\boxed{\vec{F} = (50+48+50+42)\vec{j} = 190\vec{j}\ (N)}$ The position vectors of the legs are, numbering the legs counterclockwise from the lower left in the sketch: $\vec{r}_1 = +1\vec{k}$, $\vec{r}_2 = 2\vec{i} + 1\vec{k}$, $\vec{r}_3 = 2\vec{i}$,. $\vec{r}_4 = 0$. The sum of the moments about the origin is

$$\vec{M}_O = \begin{vmatrix} \vec{i} & \vec{j} & \vec{k} \\ 0 & 0 & 1 \\ 0 & 48 & 0 \end{vmatrix} + \begin{vmatrix} \vec{i} & \vec{j} & \vec{k} \\ 2 & 0 & 1 \\ 0 & 50 & 0 \end{vmatrix} + \begin{vmatrix} \vec{i} & \vec{j} & \vec{k} \\ 2 & 0 & 0 \\ 0 & 42 & 0 \end{vmatrix}$$

$\boxed{= -98\vec{i} + 184\vec{k}\ (N\text{-}m)}$ This is the couple that acts at the origin.

Problem 4.187 If you represent the forces exerted by the floor on the table legs in problem 4.186 by a force $\vec{F}$, what is $\vec{F}$ and where does its line of action intersect the x-z plane?

Solution: From the solution to Problem 4.186 the equivalent force is $\boxed{\vec{F} = 190\vec{j}}$. This force must produce the moment $\vec{M} = -98\vec{i} + 184\vec{k}$ obtained in Problem 4.186.

$$M = \begin{vmatrix} \vec{i} & \vec{j} & \vec{k} \\ x & 0 & z \\ 0 & 190 & 0 \end{vmatrix} = -190z\vec{i} + 190x\vec{k} = -98\vec{i} + 184\vec{k},$$ from which $\boxed{x = \frac{184}{190} = 0.9684\ m}$ and

$\boxed{z = \frac{98}{190} = 0.5158\ m}$

Problem 4.188 Two forces are exerted on the crankshaft by the connecting rods. The direction cosines of $\vec{F}_A$ are $\cos\theta_X = -0.182$, $\cos\theta_Y = 0.818$, and $\cos\theta_Z = 0.545$, and its magnitude is 4 *kN*. The direction cosines of $\vec{F}_B$ are $\cos\theta_X = 0.182$, $\cos\theta_Y = 0.818$, and $\cos\theta_Z = -0.545$, and its magnitude is 2 *kN*. If you represent the two forces by a force $\vec{F}$ acting at the origin and a couple $\vec{M}$, what are $\vec{F}$ and $\vec{M}$?

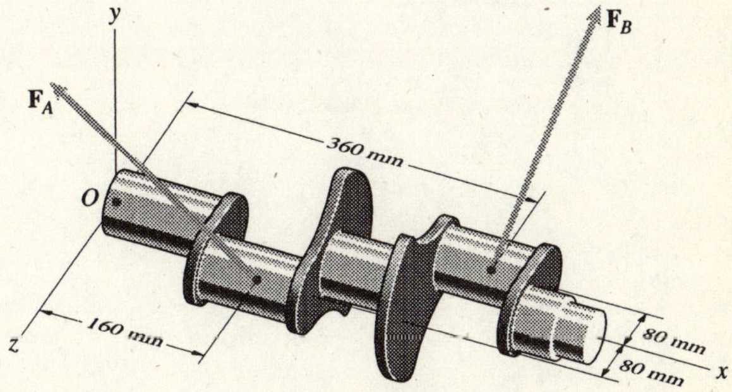

Solution: The equivalent force is the sum of the forces:

$\vec{F}_A = 4(-0.182\vec{i} + 0.818\vec{j} + 0.545\vec{k})$

$= -0.728\vec{i} + 3.272\vec{j} + 2.18\vec{k}\ (kN)$

$\vec{F}_B = 2(0.182\vec{i} + 0.818\vec{j} - 0.545\vec{k})$

$= 0.364\vec{i} + 1.636\vec{j} - 1.09\vec{k}\ (kN)$. The sum:

$\boxed{\vec{F}_A + \vec{F}_B = -0.364\vec{i} + 4.908\vec{j} + 1.09\vec{k}\ \ (kN)}$

The equivalent couple is the sum of the moments. $M = \vec{r}_A \times \vec{F}_A + \vec{r}_B \times \vec{F}_B$. The position vectors are: $\vec{r}_A = 0.16\vec{i} + 0.08\vec{k}$, $\vec{r}_B = 0.36\vec{i} - 0.08\vec{k}$.

Problem 4.188 continued on next page....

continuation of solution to Problem 4.188...

The sum of the moments: $\vec{M} = \begin{vmatrix} \vec{i} & \vec{j} & \vec{k} \\ 0.16 & 0 & 0.08 \\ -0.728 & 3.272 & 2.180 \end{vmatrix} + \begin{vmatrix} \vec{i} & \vec{j} & \vec{k} \\ 0.36 & 0 & -0.08 \\ 0.364 & 1.636 & -1.090 \end{vmatrix}$

$\boxed{\vec{M} = -0.1309\vec{i} - 0.0438\vec{j} + 1.1125\vec{k} \ (kN\text{-}m)}$

Problem 4.189 If you represent the two forces exerted on the crankshaft in Problem 4.188 by a wrench consisting of a force $\vec{F}$ and a parallel couple $\vec{M}_t$, what are $\vec{F}$ and $\vec{M}$ and where does the line of action of $\vec{F}$ intersect the x-z plane?

Solution: From the solution to Problem 4.188, $\vec{F} = -0.364\vec{i} + 4.908\vec{j} + 1.09\vec{k}\ (kN)$ and $\vec{M} = -0.1309\vec{i} - 0.0438\vec{j} + 1.1125\vec{k}\ (kN\text{-}m)$. The unit vector parallel to $\vec{F}$ is

$\vec{e}_F = \dfrac{\vec{F}}{|\vec{F}|} = -0.0722\vec{i} + 0.9737\vec{j} + 0.2162\vec{k}$. The moment parallel to the force is $\vec{M}_P = (\vec{e}_F \cdot \vec{M})\vec{e}_F$.

Carrying out the operations: $\boxed{\vec{M}_P = 0.2073\vec{e}_F = -0.01497\vec{i} + 0.2019\vec{j} + 0.0448\vec{k}\ (kN\text{-}m)}$. This is the equivalent couple parallel to $\vec{F}$.

The component of the moment perpendicular to $\vec{F}$ is $\vec{M}_N = \vec{M} - \vec{M}_P = -0.1159\vec{i} - 0.2457\vec{j} + 1.0688\vec{k}$. The force exerts this moment about the origin.

$$\vec{M}_N = \begin{vmatrix} \vec{i} & \vec{j} & \vec{k} \\ x & 0 & z \\ -0.364 & 4.908 & 1.09 \end{vmatrix} = (-4.908z)\vec{i} - (1.09x + 0.364z)\vec{j} + (4.908x)\vec{k}$$

$= -0.1159\vec{i} - 0.2457\vec{j} + 1.06884\vec{k}$. From which $\boxed{x = \dfrac{1.0688}{4.908} = 0.2178\ m}$, $\boxed{z = \dfrac{0.1159}{4.908} = 0.0236\ m}$

Problem 5.1 Draw free body diagrams of the objects a, b and c.

Solution:

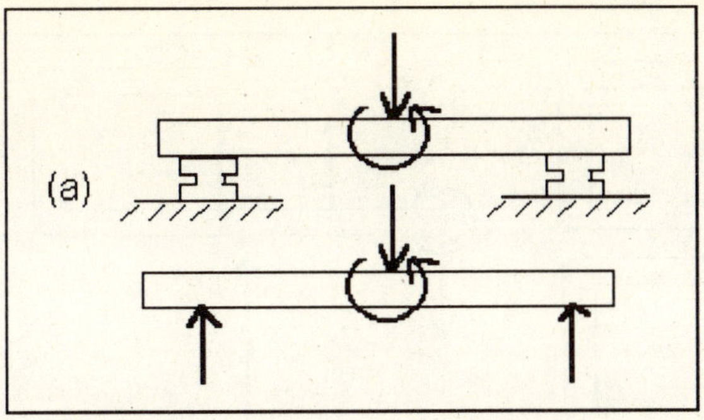

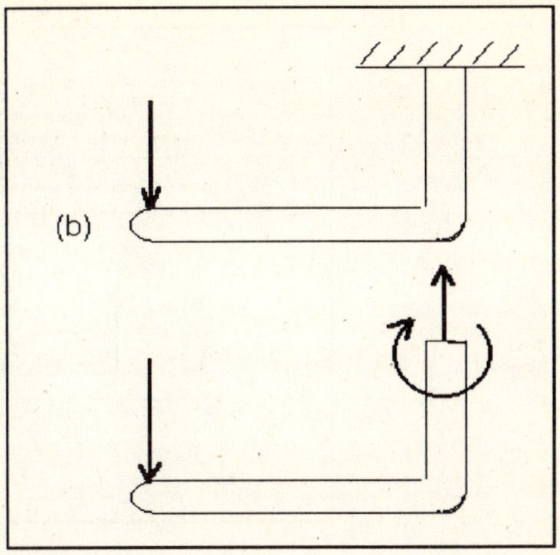

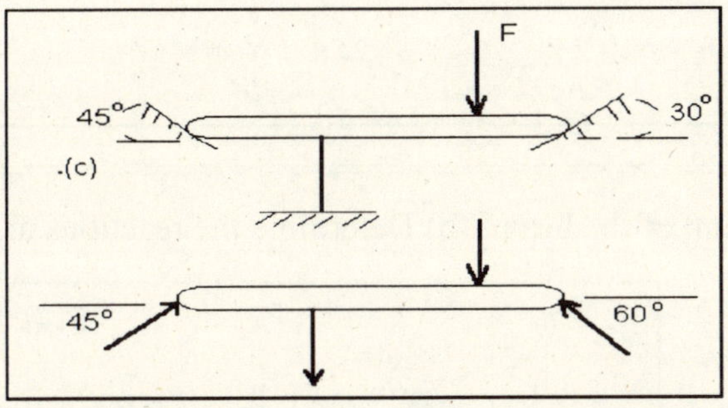

Problem 5.2 Draw free body diagrams of the objects a, b, and c.

Solution:

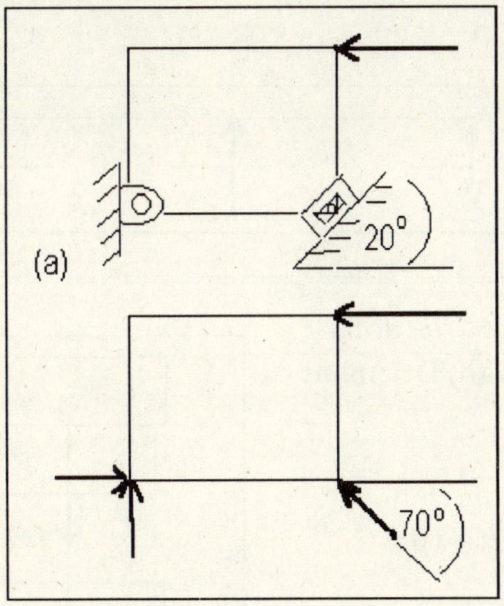

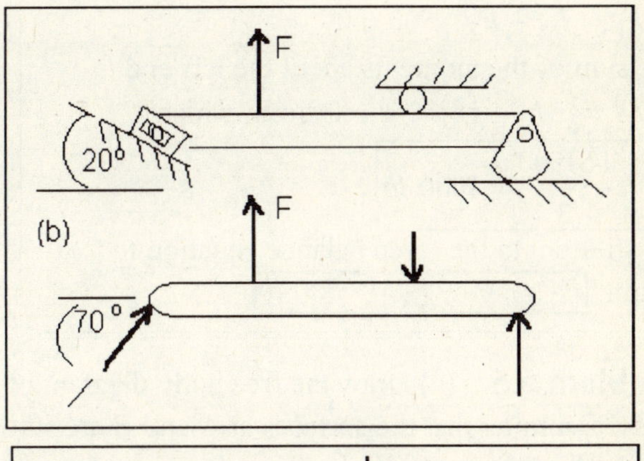

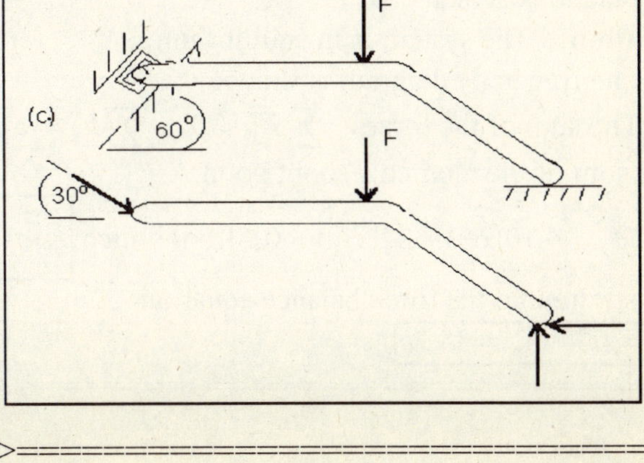

Problem 5.3 The tension in cable AB is T. Draw a free body diagram (a) of the bar and cable AB treating them as a single object, (b) of the bar not including cable AB, (c) of cable AB.

Solution:

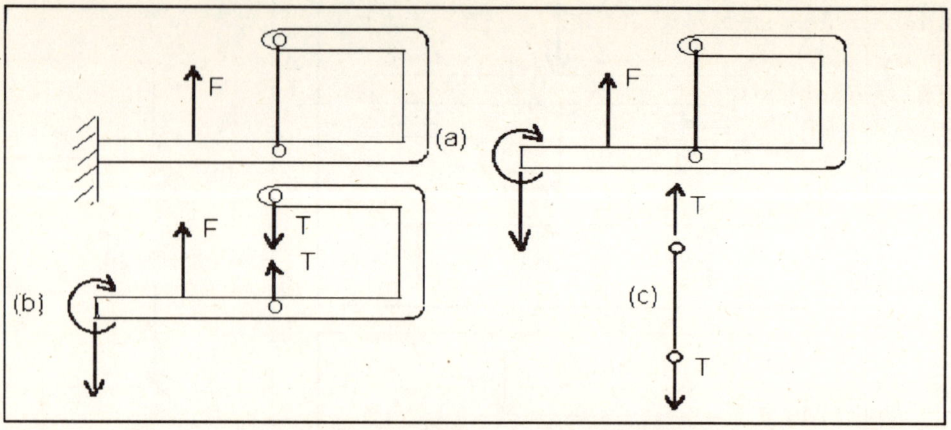

Problem 5.4 (a) Draw the free body diagram of the beam. (b) Determine the reactions at the supports.

Solution: The beam is in equilibrium, hence the sum of the forces and the sum of the moments both vanish.

The sums of the forces are

$\sum F_X = 0$

$\sum F_Y = F_A + F_B - 500 = 0$

The sum of the moments about the left end is

$\sum M = (3F_B - (6)500) = 0$ from which

$\boxed{F_B = \dfrac{(6)500}{3} = 1000\ lb}$

Substitute into the force balance equation to obtain: $\boxed{F_A = 500 - F_B = -500\ lb}$

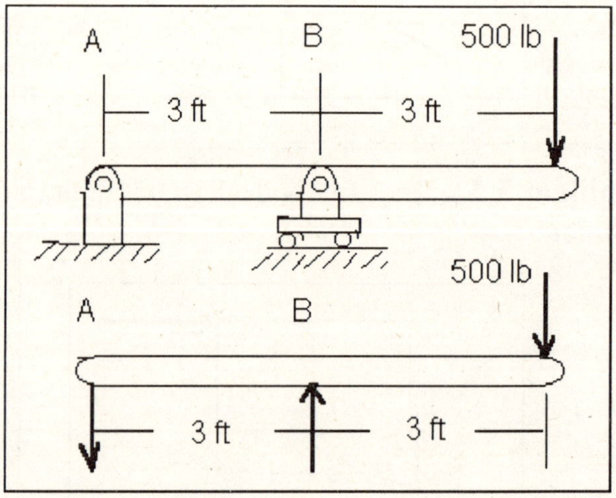

Problem 5.5 (a) Draw the free body diagram of the 60 lb drill press, assuming that the surfaces at A and B are smooth. (b) Determine the reactions at A and B.

Solution: The system is in equilibrium.

(a) The free body diagram is shown.

(b) The sum of the forces: $\sum F_X = 0$, $\sum F_Y = F_A + F_B - 60 = 0$

The sum of the moments about point A:

$\sum M_A = -10(60) + 24(F_B) = 0$, from which $\boxed{F_B = \dfrac{600}{24} = 25\ lb}$

Substitute into the force balance equation:

$\boxed{F_A = 60 - F_B = 35\ lb}$

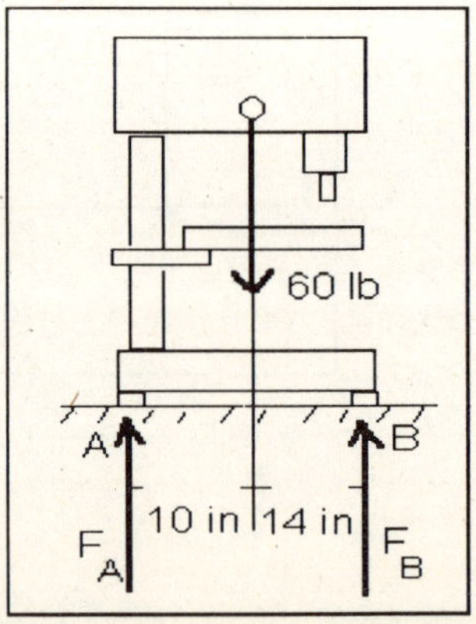

Problem 5.6 The mass of the person is 80 kg and the mass of the diving board is 45 kg. (a) Draw the free body diagram of the diving board. (b) Determine the reactions at the supports A and B,

Solution: The system is in equilibrium. The support at A is a pinned joint; the support at B is a roller support.

(a) The free body diagram is shown.

The weight of the person is $W_P = mg = 80(9.81) = 784.8\ N$

The weight of the diving board is

$W_D = mg = (45)(9.81) = 441.45\ N$

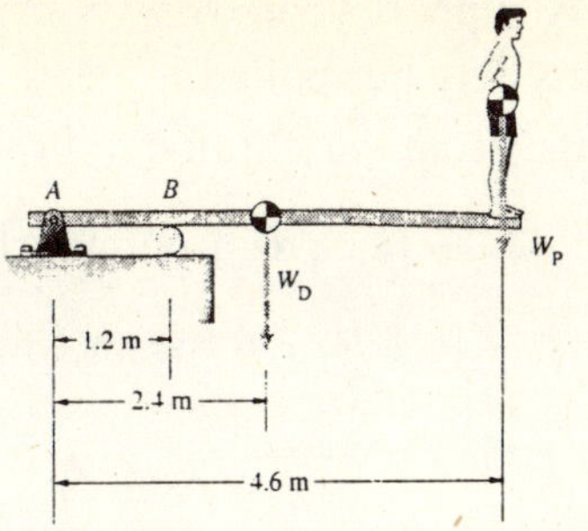

The sums of the forces are: $\boxed{\sum F_X = A_X = 0}$,

$\Sigma F_Y = -W_P - W_D + F_B + F_A = 0$

The sum of the moments about A is:

$\Sigma M_A = -4.6W_B - 2.4W_P + 1.2F_B = 0$ From which $\boxed{F_B = 3891.3\ N}$ Substituting into the force balance equation: $\boxed{F_A = -W_P - W_D + F_B = -2665\ N}$

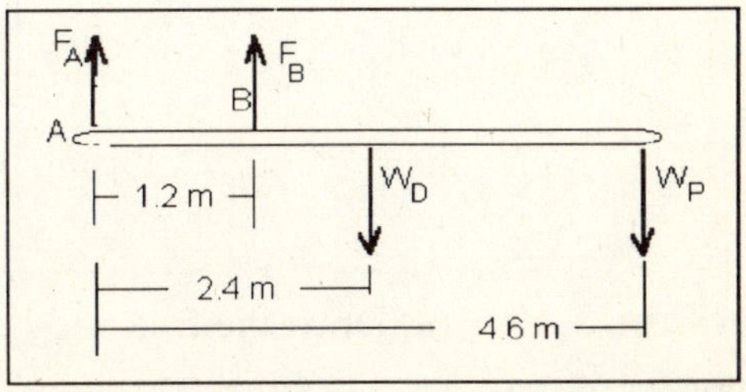

Problem 5.7 The ironing board has supports at A and B that can be modeled as roller supports. (a) Draw the free body diagram. (b) Determine the reactions at A and B.

Solution: The system is in equilibrium.

(a) The free-body diagram is shown.

(b) The sums of the forces are:

$\sum F_X = 0,\ \sum F_Y = F_A + F_B - 10 - 3 = 0$. The sum of the moments about A is

$\sum M_A = 12F_B - 22(10) - 42(3) = 0$, from which

$\boxed{F_B = \frac{346}{12} = 28.833\ lb}$. Substitute into the force balance equation: $\boxed{F_A = 13 - F_B = -15.833\ lb}$

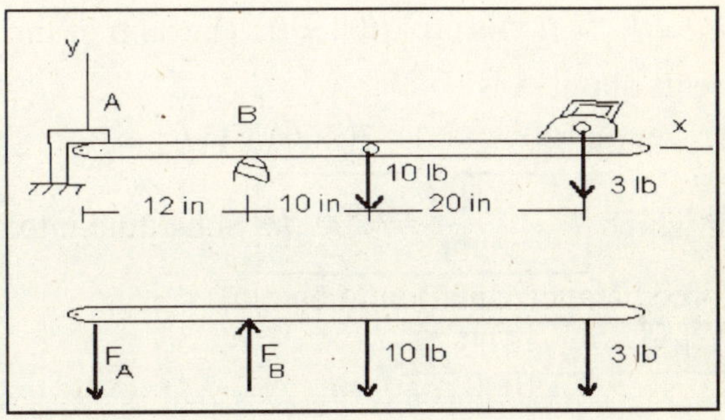

Problem 5.8 The distance $x = 2\ m$. (a) Draw the free body diagram of the beam. (b) Determine the reactions at the supports.

Solution: The system is in equilibrium. The point A is a pinned support. The point B is a roller support. (a) The free body diagram is shown. (b) The sums of the forces: $\boxed{\sum F_X = A_X = 0}$, $\sum F_Y = A_Y - 10 + F_B = 0$.

The sum of the moments about A is

$\sum M_A = -2(10) + 4F_B = 0$, from which

$\boxed{F_B = \frac{20}{4} = 5\ kN}$. Substitute into the force balance equation to obtain $\boxed{A_Y = 10 - F_B = 5\ kN}$

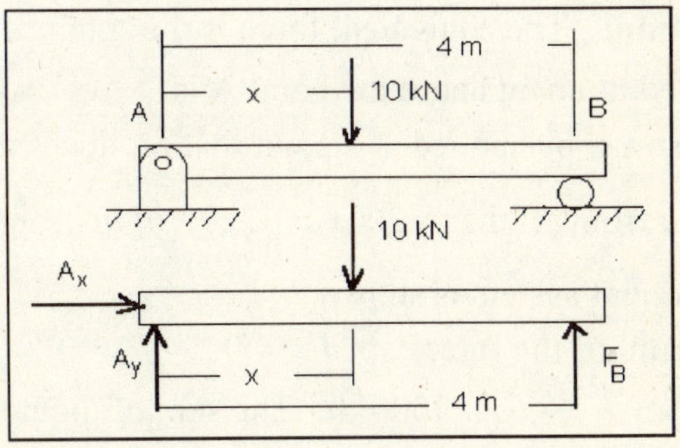

Problem 5.9 Consider the beam shown in Problem 5.8. An engineer determines that each support will safely support a force of 7.5 *kN*. What is the range of values of the distance x at which the 10 *kN* force can safely be applied?

Solution: From the solution to Problem 5.8 the equations for $\vec{F}_B$ and A_Y are:

$\sum M_A = 4F_B - 10x = 0$, from which $F_B = \frac{10x}{4} = 2.5x$.

$\sum F_Y = A_Y - 10 + F_B = 0$, from which $A_Y = 10 - F_B = 10 - 2.5x$ Solve for the value of *x*:

$x = \frac{F_B}{2.5}$, and $x = \frac{10 - A_Y}{2.5}$.

Let $F_B = 7.5\ kN$, then $A_Y = 2.5\ kN$ and $\boxed{x = \frac{7.5}{2.5} = 3\ m}$.

Let $A_Y = 7.5\ kN$, then $F_B = 2.5\ kN$, and $\boxed{x = \frac{10 - 7.5}{2.5} = 1\ m}$. Thus $\boxed{1 \le x \le 3\ m}$

Problem 5.10

(a) Draw the free body diagram of the beam. (b) Determine the reactions at the supports.

Solution: (a) Both supports are roller supports. The free body diagram is shown. (b) The sum of the forces: $\sum F_X = 0$, and

$\sum F_Y = F_A + F_B + 100 - 400 = 0$. The sum of the moments about A is

$\sum M_A = -3(100) + 900 - 7(400) + 11F_B = 0$

From which $\boxed{F_B = \frac{2200}{11} = 200\ lb}$ Substitute into the force balance equation to obtain

$\boxed{F_A = 300 - F_B = 100\ lb}$

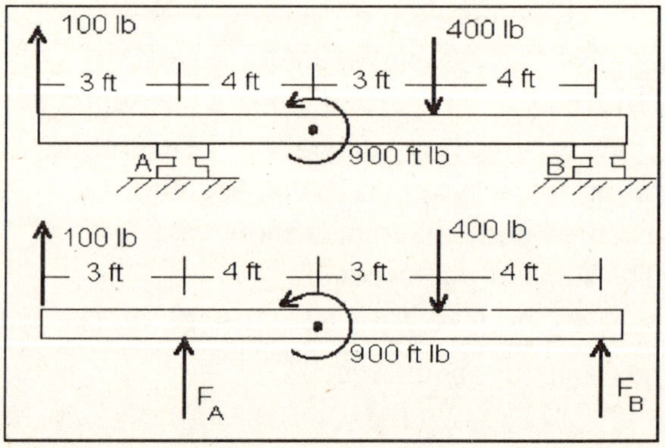

Problem 5.11

Consider the beam shown in Problem 5.10. First represent the loads (the 100 *lb* force, the 400 *lb* force and the 900 *ft lb* couple) by a single equivalent force, and then determine the reactions at the supports.

Solution: The equivalent force is the sum of the applied force: $\boxed{\sum F_Y = 100 - 400 = -300\ lb}$. The applied moment about the point A is $\sum M_A = -3(100) + 900 - 7(400) = -2200\ ft\ lb$. The equivalent force must be applied at a point that produces this moment about A. Let *x* be the distance to the line of action from A: then $-300x = -2200\ ft\ lb$ from which $x = \frac{2200}{300} = 7.33\ ft$ to the right of A. The equivalent system is shown.

The sum of the forces: $\sum F_X = 0$,

$\sum F_Y = F_A + F_B - 300 = 0$. The sum of moments about A is $\sum M_A = -(7.33)(300) + 11F_B = 0$, from which $\boxed{F_B = \frac{2200}{11} = 200\ lb}$. Substitute into the force balance equation to obtain: $\boxed{F_A = 300 - F_B = 100\ lb}$.

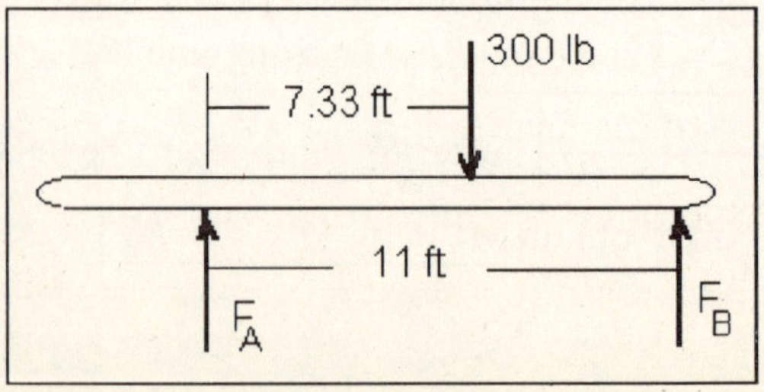

Problem 5.12 (a) Draw the free body diagram of the beam. (b) Determine the reactions at the support.

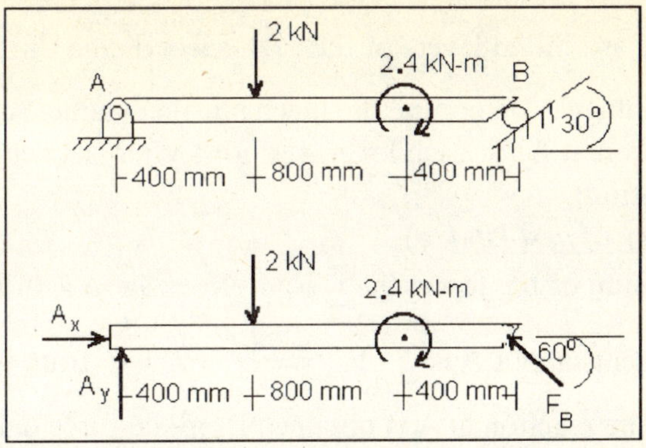

Solution: The support at A is a pinned support and the support at B is a roller support. (a) The free body diagram is shown. (b) The forces at point B are $F_{BX} = F_B \cos 120^o = -0.5F_B$ and $F_{BY} = F_B \sin 120 = 0.866F_B$. The sums of the forces are $\Sigma F_X = A_X - 0.5F_B = 0$, and $\Sigma F_Y = -2 + 0.866F_B + A_Y = 0$. The sum of the moments about A is $\Sigma M_A = -0.4(2) - 2.4 + 0.866F_B(1.6) = 0$. From which, $\boxed{F_B = \frac{3.2}{1.3856} = 2.3095\ kN}$ Substituting into the force balance equations: $\boxed{A_X = 0.5F_B = 1.1547\ kN}$ $\boxed{A_Y = 2 - 0.866F_B = 0\ \ kN}$

Problem 5.13 Consider the beam in Problem 5.12. First represent the loads (the 2 *kN* force and the 2.4 *kN* couple) by a single equivalent force, then determine the reactions at the supports.

Solution: The equivalent applied force is equal to the sum of the forces: $\boxed{F_{EQ} = 2\ kN}$. This force must be applied at a point that produces the applied moment about A: $M_{EQ} = -0.4(2) - 2.4 = -3.2\ kN\text{-}m$. Denote the distance from A to be x. Then the moment is $M_A = -2x = -3.2\ \ kN\text{-}m$, from which $x = \frac{-3.2}{-2} = 1.6\ m$. Thus the equivalent force is to be applied at point B. The equivalent system is shown. The sums of the forces are: $\Sigma F_X = A_X - 0.5F_B = 0$, and $\Sigma F_Y = -2 + 0.866F_B + A_Y = 0$. The sum of the moments: $\sum M_A = -1.6(2) + 1.6(0.866F_B) = 0$, from which $\boxed{F_B = 2.3095\ \ kN}$, $\boxed{A_X = 1.1547\ kN}$, and $\boxed{A_Y = 0}$

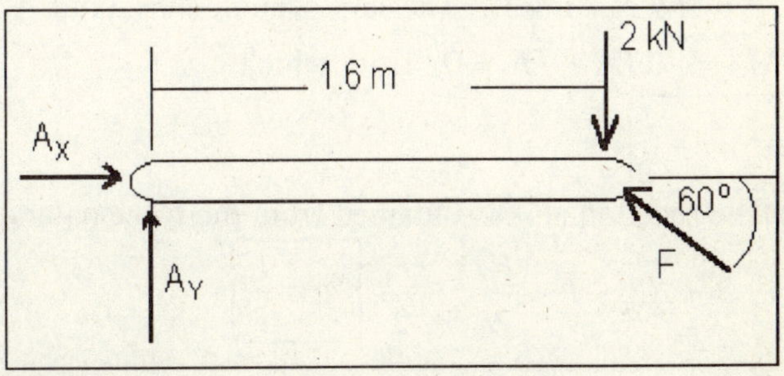

Problem 5.14 A man doing push-ups pauses in the position shown. His weight is $W = 175\ \ lb$. The distances $a = 10\ in,\ b = 29\ in$, and $c = 12\ in$. Determine the normal force exerted by the floor (a) on each hand, (b) on each foot.

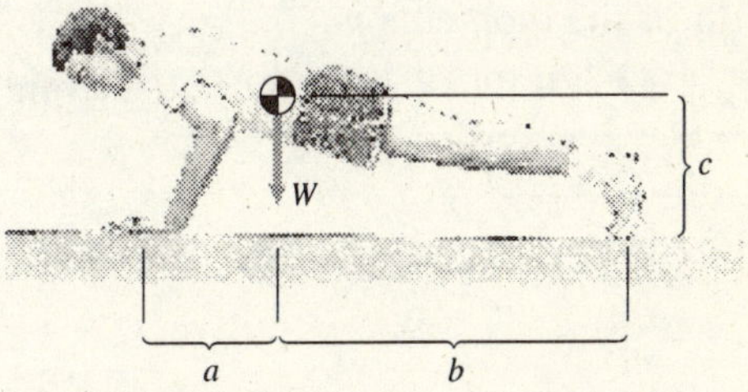

Solution: The sum of the forces is $\sum F_Y = -W + 2F_H + 2F_F = 0$. The sum of the moments about the hand position is $\sum W = -aW + (a+b)(2F_F) = 0$, from which $\boxed{F_F = \frac{aW}{2(a+b)} = \frac{10(175)}{2(39)} = 22.436\ lb}$ for each foot. Substitute into the force equation: $\boxed{F_H = \frac{W - 2F_F}{2} = 65.064\ \ lb}$ for each hand. *Check:* $2(65.064) + 2(22.436) = 175\ lb$, *Check.*

Problem 5.15 With each of the devices shown you can support a load R by applying a force F. They are called levers of the first, second, and third kinds. (a) The ratio $\frac{R}{F}$ is called the "mechanical advantage". Determine the mechanical advantage of each lever. (b) Determine the magnitude of the reaction at A for each lever. (Express your answer in terms of F.)

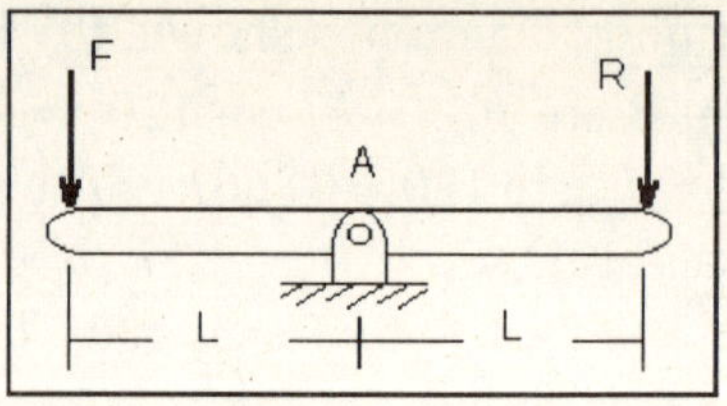

Solution:

Lever of first kind.(a)

The sum of the forces is $\sum F_Y = -F + A - R = 0$. The sum of the moments about A is $\sum M_A = FL - RL = 0$, from which $\boxed{\frac{R}{F} = \frac{L}{L} = 1}$

(b) The reaction at A is obtained from the force balance equation: $\boxed{A = R + F = 2F}$

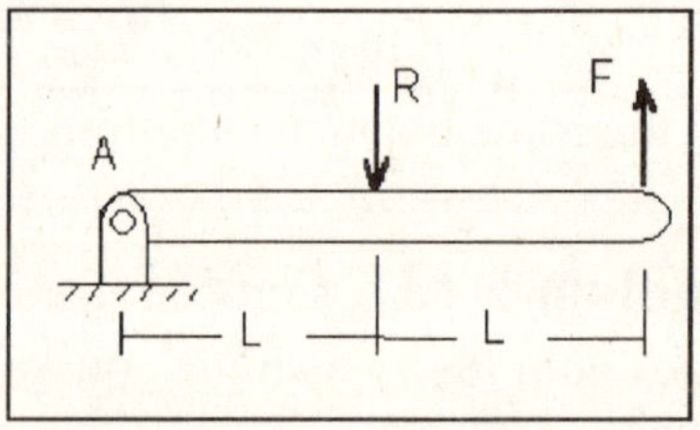

Lever of second kind. (a) The sum of forces is $\sum F_Y = A - R + F = 0$. The sum of the moments about A is $\sum M_A = -LR + 2LF = 0$, from which $\boxed{\frac{R}{F} = \frac{2L}{L} = 2}$

(b) The reaction at A is obtained from the force balance equation: $\boxed{A = -F + R = -F + 2F = F}$

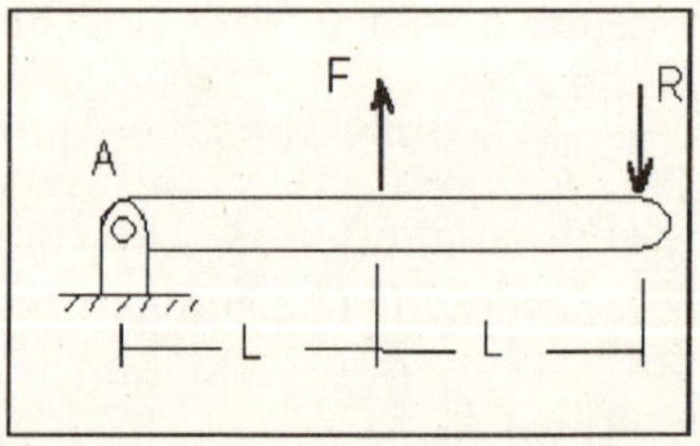

Lever of third kind. (a) The sum of forces is $\sum F_Y = A - R + F = 0$. The sum of moments about A is: $\sum M_A = -2LR + LF = 0$, from which: $\boxed{\frac{R}{F} = \frac{L}{2L} = \frac{1}{2}}$

(b) The reaction at A is obtained from the force balance equation

$A = -F + R = -F + \frac{F}{2} = -\frac{F}{2}$, $\boxed{|A| = \frac{F}{2}}$

Problem 5.16 (a) Draw the free body diagram of the beam. (b) Determine the reactions at the support.

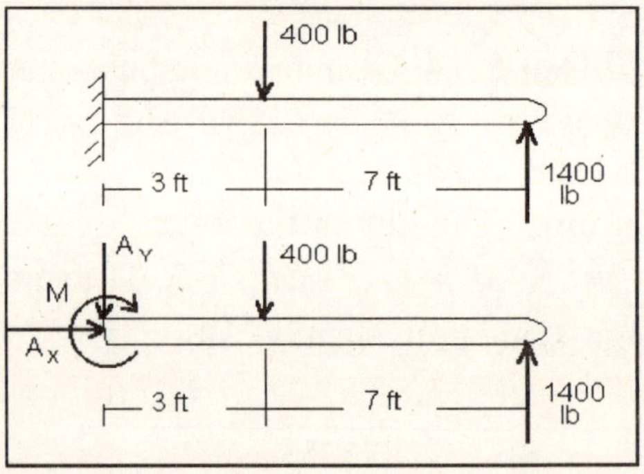

Solution: The sum of the forces is $\boxed{\sum F_X = A_X = 0}$

$\sum F_Y = A_Y - 400 + 1400 = 0$, from which $\boxed{A_Y = -1000 \ lb}$.

The sum of the moments is

$\sum M_A = M - 3(400) + 10(1400) = 0$, from which

$\boxed{M = -12800 \ ft \ lb}$

Problem 5.17 The force $F = 500\ lb$. Determine the reactions at the support.

Solution: The free body diagram is shown.

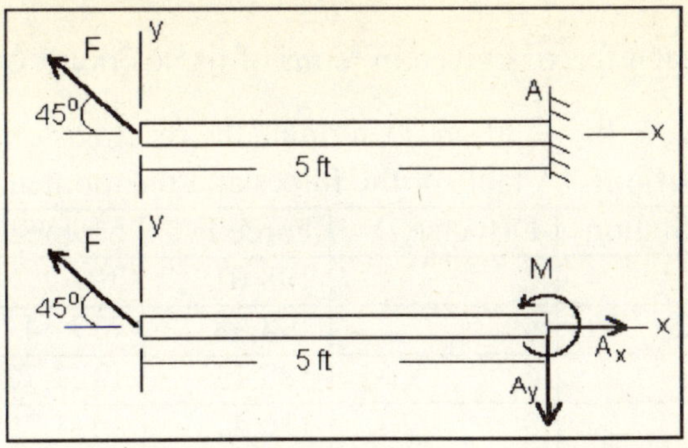

The sums of the forces:

$\sum F_X = -F\cos 45° + A_X = 0$, from which

$\boxed{A_X = F\cos 45° = 500(0.707) = 353.5\ lb}$

$\sum F_Y = F\cos 45° + A_Y = 0$, from which

$\boxed{A_Y = -F\cos 45° = -353.5\ lb}$

The sum of the moments about point A is

$\sum M_A = M - 5(F\cos 45°) = 0$, from which

$\boxed{M = 1767.5\ ft\ lb}$

Problem 5.18 Consider the beam in Problem 5.17. An engineer determines that the built-in support will fail if the magnitude of the total force exerted on the support exceeds 1000 *lb* or if the magnitude of the couple exerted on the support exceeds 3000 *ft lb.* Based on these criteria, what is the maximum force F that can be applied?

Solution: The strategy is to determine whether the load limitation is a force limitation or a moment limitation. The magnitude of the force at the built-in support is $F = 1000\ lb$. The moment limit is

$M_A = -F(5)(\cos 45°) = 3000\ ft\ lb$, from which $\boxed{F = \dfrac{3000}{5\cos 45°} = 848.66\ lb}$ which is the limit on F.

Problem 5.19 The mobile is in equilibrium. The object B weighs 27 *oz*. Determine the weights of the objects A, C, and D. (Neglect the weights of the crossbars.)

Solution: Denote the reactions at the supports by F_{AB}, F_{CD}, and F_{BCD} as shown. Start with the crossbar supporting the weights C and D. The sum of the forces is $\sum F_Y = -C - D + F_{CD} = 0$, from which $F_{CD} = C + D$.

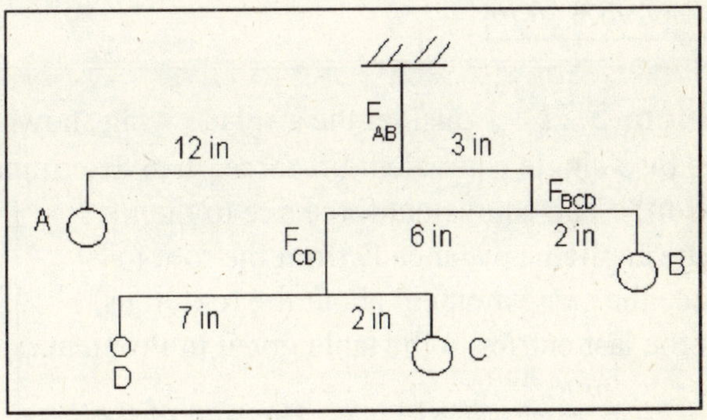

For the cross bar supporting the weight B, the sum of the forces is

$\sum F_Y = -B + F_{BCD} - F_{CD} = 0$, from which, substituting, $F_{BCD} = B + C + D$. For the crossbar supporting C and D, the sum of the moments about the support is $\Sigma M_{CD} = 7D - 2C = 0$, from which $D = \dfrac{2C}{7}$. For the crossbar supporting B, the sum of the moments is $\sum M_{BCD} = 6F_{CD} - 2B = 0$, from which, substituting from above

$F_{CD} = \dfrac{2B}{6} = C + D = C + \dfrac{2C}{7} = \dfrac{9C}{7}$, or $\boxed{C = \dfrac{7B}{27} = 7\ oz}$, and $\boxed{D = \dfrac{2C}{7} = 2\ oz}$ The sum of the moments about the crossbar supporting A is $\sum M_{AB} = 12A - 3F_{BCD} = 0$, from which, substituting from above,

$\boxed{A = \dfrac{3(B + C + D)}{12} = \dfrac{27 + 7 + 2}{4} = 9\ oz}$

Problem 5.20 The lift forces on an airplane's wing are represented by eight forces. The magnitude of each force is given in terms of its position x on the wing by $200[1-\left(\frac{x}{17}\right)^2]^{\frac{1}{2}}$ lb. The weight of the wing is $W = 400$ lb. Determine the reactions on the wing at the root R.

Solution: A table of the forces and the moments about the wing root is:

Station	Distance D	Force F	Moment
1	2	198.61	397.22
2	4	194.38	777.54
3	6	187.13	1122.78
4	8	176.47	1411.76
5	10	161.74	1617.38
6	12	141.67	1700.0
7	14	113.45	1588.37
8	16	67.58	1081.33
Sum		1241.03	9696.4

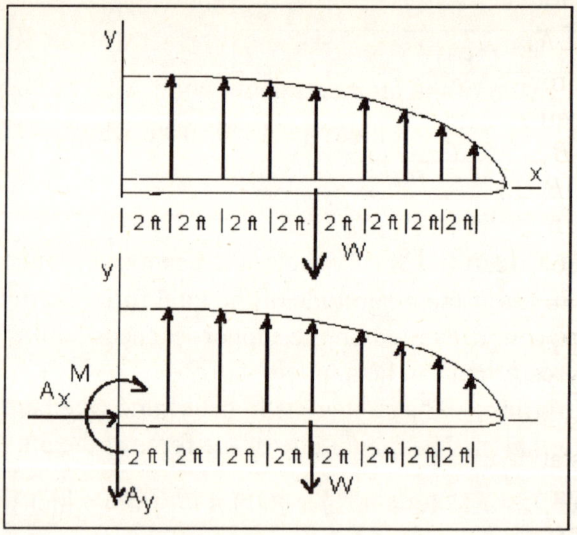

The sum of the lift forces is the last entry in the column of forces. The sum of the forces on the wing

$\boxed{\sum F_X = A_X = 0}$,

$\sum F_Y = A_Y + \sum F_L - W = 0$, from which

$\boxed{A_Y = -1241.03 + 400 = -841 \ lb}$

The sum of the moments about the root is $\sum M_R = M + \sum M_L - 8W = 0$, from which

$\boxed{M = -6496.3 \ ft\ lb}$

Problem 5.21 Consider the airplane wing shown in Problem 5.120. First represent the eight lift forces by a single equivalent lift force, then determine the reactions on the wing at the root R.

Solution: The equivalent force due to the lift must be applied a distance D from the root to produce the same moment about the root, thus, using the last entries in the table given in Problem 5.20, $D = \frac{9696.3}{1241} = 7.8133$ ft. The sum of the forces is unchanged from Problem 5.20,

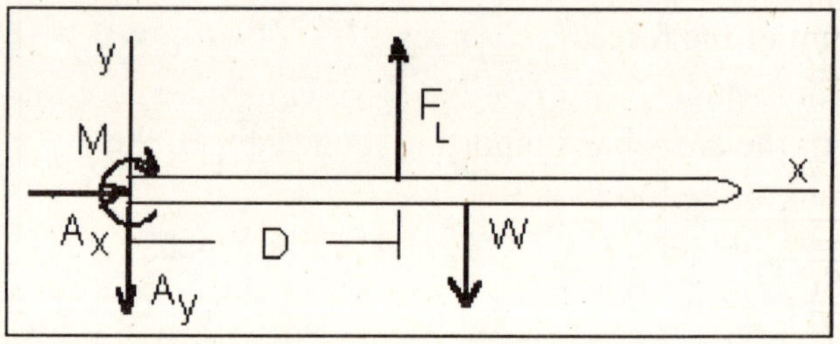

$\boxed{\sum F_X = A_X = 0}$,

$\sum F_Y = A_Y + \sum F_L - W = 0$ from which $\boxed{A_Y = -1241 + 400 = -841 \ lb}$.

The sum of the moments about the root is $\sum M_R = M + 7.8133(1241) - 8(400) = 0$, from which

$\boxed{M = -6496.3 \ ft\ lb}$

Problem 5.22 The 14.5 lb chain saw is subject to the loads at A by the log it cuts. Determine the reactions R, B_X, and B_Y that must be applied by the person using the saw to hold it in equilibrium.

Solution: The sum of the forces are

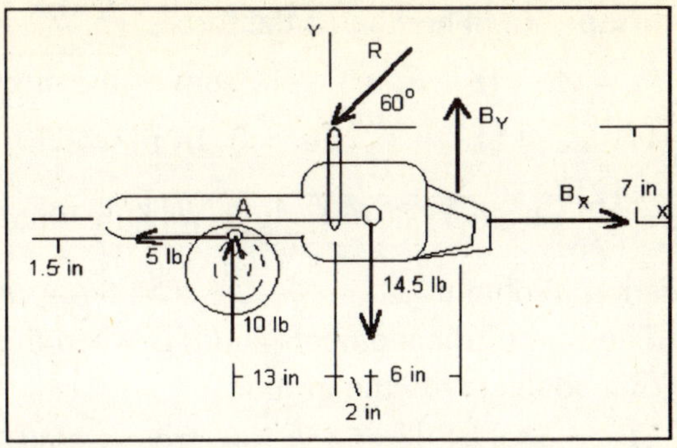

$\sum F_X = -5 + B_X - R\cos 60° = 0$.

$\Sigma F_Y = 10 - 14.5 + B_Y - R\sin 60° = 0$ The sum of the moments about the origin is

$\Sigma M_O = 7R\cos 60° + 8B_Y - 2(14.5) - 13(10) - 5(1.5) = 0$

From which $7R\cos 60° + 8B_Y - 166.5 = 0$. Collecting equations and reducing to 3 equations in 3 unknowns:

$B_X \ + 0B_Y \ - 0.5R = 5$

$0B_X \ + B_Y - 0.866R = 4.5$

$0B_X \ + 8B_Y + 3.5R = 166.5$.

Solving: $\boxed{B_X = 11.257\ lb}$, $\boxed{B_Y = 15.337\ lb}$, and $\boxed{R = 12.514\ lb}$

Problem 5.23 The mass of the trailer is 2.2 Mg (megagrams). The distances $a = 2.5\ m$ and $b = 5.5\ m$. The truck is stationary and the wheels of the trailer can turn freely, which means that the road exerts no horizontal force on them. (a) Draw the free body diagram of the trailer. (b) Determine the total normal force exerted on the rear tires at A and the reactions on the trailer at the pin support B,

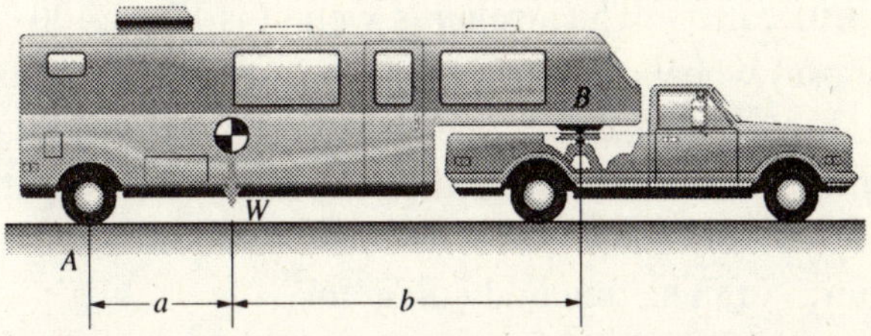

Solution: (a) The free body diagram is shown.

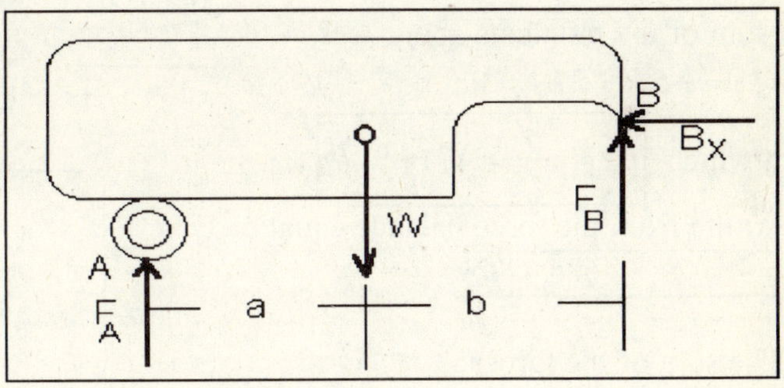

(b) The sum of forces: $\boxed{\sum F_X = B_X = 0}$.

$\sum F_Y = F_A - W + F_B = 0$. The sum of the moments about A:

$\sum M_A = -aW + (a+b)F_B = 0$, from which

$$F_B = \frac{aW}{a+b} = \frac{2.5(2.2 \times 10^3)(9.81)}{(2.5+5.5)}$$

$\boxed{= 6.744\ kN}$

Substitute into the force equation:

$\boxed{F_A = W - F_B = 14.838\ kN}$

Problem 5.24 The total weight of the wheelbarrow and its load is $W = 100\ lb$. (a) If $F = 0$, what are the vertical reactions at A and B? (b) What force is necessary to lift the support at A off the ground?

Solution: (a) The sum of the forces: $\boxed{\sum F_X = A_X = 0}$

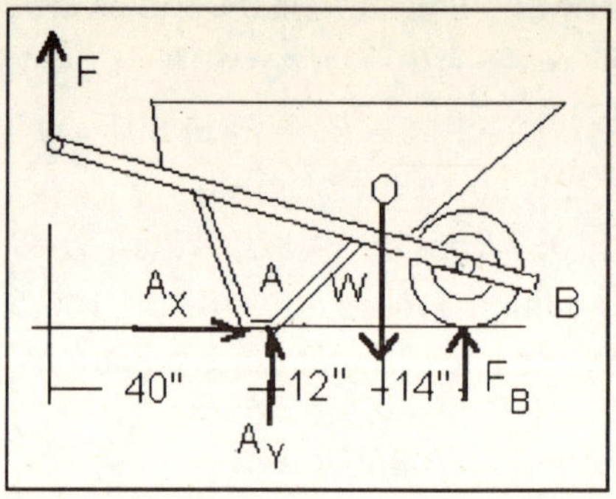

$\sum F_Y = A_Y - W + F_B = 0$. The sum of the moments about A is $\sum M_A = -W(12) + F_B(26) = 0$, from which

$F_B = \frac{12W}{26} = 46.1538\ lb$ $\boxed{= 46.2\ lb}$ Substitute into the force equation to obtain: $A_Y = W - F_B = 53.8462\ lb$ $\boxed{= 53.8\ lb}$

(b) The sum of the moments about B when the point A is not making contact with the ground:

$\sum M_B = (14)(100) - (66)F = 0$, from which

$F = \frac{(14)(100)}{66} = 21.2121$ $\boxed{= 21.2\ lb}$

Problem 5.25 The airplane's weight is $W = 2400\ lb$. Its brakes keep the rear wheels locked. The front (nose) wheel can turn freely, and so the ground exerts no horizontal force on it. The force T exerted by the airplane's propeller is horizontal. (a) Draw the free body diagram of the airplane. Determine the reaction exerted on the nose wheel and the total normal reaction on the rear wheels (b) when $T = 0$, and (c) when $T = 250\ lb$.

Solution: (a) The free body diagram is shown. (b) The sum of the forces:

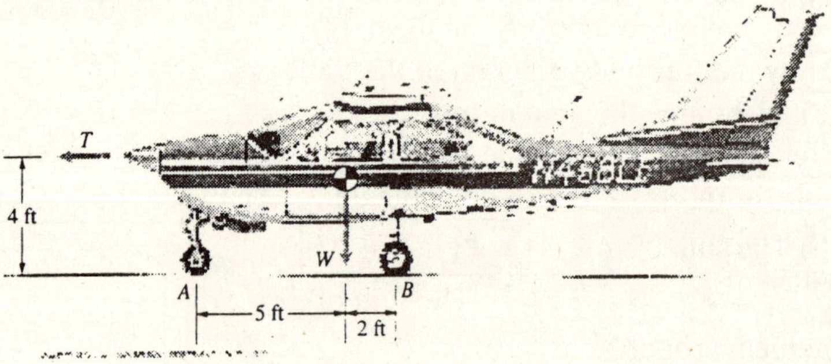

$\sum F_X = B_X = 0$

$\sum F_Y = A_Y - W + B_Y = 0$

The sum of the moments about A is

$\sum M_A = -5W + 7B_Y = 0$,

from which $\boxed{B_Y = \frac{5W}{7} = 1714.3\ lb}$

Substitute from the force balance equation:

$\boxed{A_Y = W - B_Y = 685.7\ lb}$

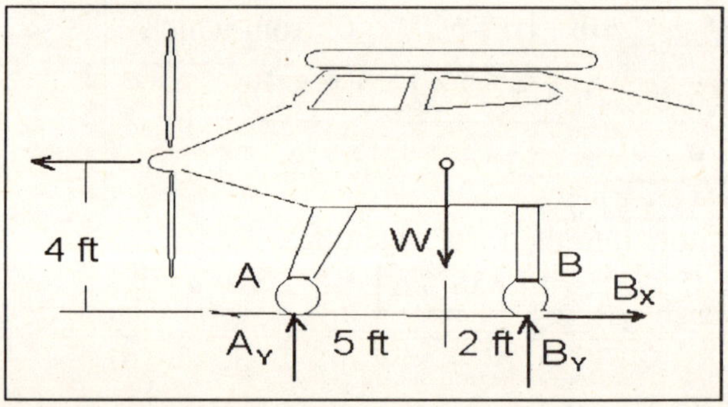

(c) The sum of the forces:

$\sum F_X = -250 + B_X = 0$, from which

$B_X = 250\ lb$

$\sum F_Y = A_Y - W + B_Y = 0$.

The sum of the moments about A:

$\sum M_A = (250)(4) - 5W + 7B_Y = 0$, from which $\boxed{B_Y = 1571.4\ lb}$ Substitute into the force balance equation to obtain: $\boxed{A_Y = 828.6\ lb}$

Problem 5.26 The fork lift truck is stationary. The front wheels are free to turn and the rear wheels are locked. The distances are $a = 1.25\ m$, $b = 0.50\ m$, and $c = 1.40\ m$. The weight of the load is $W_L = 2\ kN$ and the weight of the truck and operator is $W_F = 8\ kN$. What are the reactions at A and B?

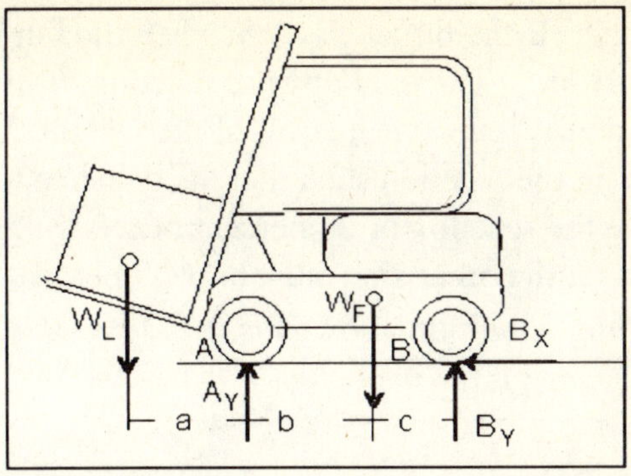

Solution: The sum of the forces: $\boxed{\sum F_X = B_X = 0}$ $\sum F_Y = A_Y - W_L - W_F + B_Y = 0$. The sum of the moments about A is $\sum M_A = +aW_L - bW_F + (b+c)B_Y = 0$, from which $\boxed{B_Y = \dfrac{bW_F - aW_L}{b+c} = 0.7895\ kN}$ Substitute into the force equation to obtain: $\boxed{A_Y = W_L + W_F - B_Y = 9.211\ kN}$

Problem 5.27 Consider the stationary fork-lift truck in Problem 5.26. The front wheels are free to turn and the rear wheels are locked. The distances $a = 45\ in$, $b = 20\ in$, and $c = 50\ in$. The weight of the truck and operator is $W_F = 3000\ lb$. For safety reasons a rule is established that the reaction at the rear wheels must be at least 400 lb. If the weight W_L of the load acts at the position shown, what is the maximum load?

Solution: From the solution to Problem 5.26, , $B_X = 0$, and the sum of the moments:

$\sum M_A = +aW_L - bW_F + (b+c)B_Y = 0$, from which $W_L = \dfrac{bW_F - (b+c)B_Y}{a} = 1333.33 - 1.5555B_Y$. Since B_Y is positive, the maximum load for $B_Y = 400\ lb$ is $\boxed{W_L = 711.1\ lb}$

Problem 5.28 The weight of the fan is $W = 20\ lb$. Its base has four equally spaced legs of length $b = 12\ in$ and $h = 36\ in$. What is the largest thrust T exerted by the fan's propeller for which the fan will remain in equilibrium?

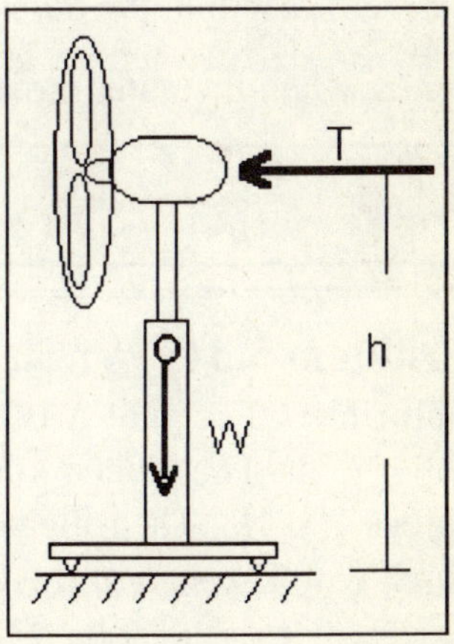

Solution: Each leg is assumed to be in contact with a rough surface, with (in two dimensions) two force components each.

The four equally spaced legs can be in two positions relative to the thrust line of action: In the first the distance to the center is b. In the second, the distance is $b \sin 45^o = 0.707b$. Tipping will occur when the leftmost (or rightmost) leg(s) has zero reaction on the floor.

For each position the sum of the moments about the center is:

$\sum M_T = -bW + Th = 0$, and $\sum M_T = -0.707bW + Th = 0$. From which the two tipping moment thrusts are:

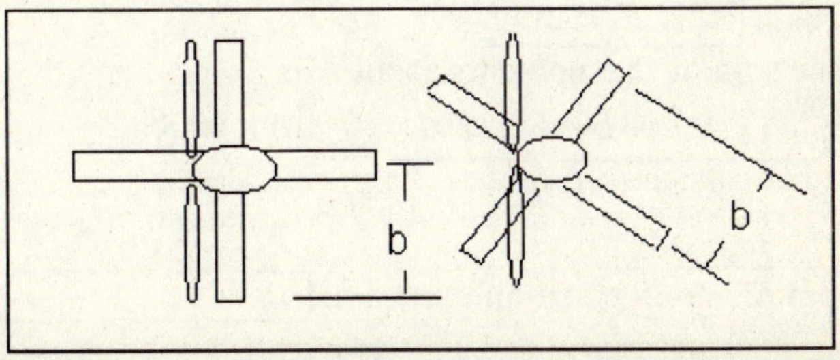

$T_1 = \dfrac{bW}{h} = 6.67\ lbs$,

$\boxed{T_2 = \dfrac{0.707bW}{h} = 4.71\ lb}$ which is the maximum thrust allowed.

==============================<>==============================

Problem 5.29 Consider the fan described in Problem 5.28. As a safety criterion, an engineer decides that the vertical reaction on any of the fans legs should not be less than 20 percent of the fan's weight. If the thrust T is 1 lb when the fan is on its highest speed, what is the maximum safe value of h?

Solution: The total upward reaction of the legs is equal to the weight of the fan, so that each leg normally bears one quarter of the weight. Under the condition of maximum tipping moment, with the legs in the position such that the distance to the center is $0.707b$, the legs in the outer position will each have the reaction of 20 percent of the weight, so that both will carry 40 percent of the weight. Thus the legs on the other side must bear 60 percent of the weight. The sum of the moments at the maximum tipping condition allowed is $\sum M_T = (0.707b)(0.4W) + Th - (0.707b)(0.6W) = 0$, from which:

$$\boxed{h = \frac{0.707b(0.2W)}{T} = 33.94\ in}$$

==============================<>==============================

Problem 5.30 To decrease costs, an engineer considers supporting a fan with three equally spaced legs instead of the four leg configuration shown in Problem 5.28. For the same values of b, h, and W, show that the largest thrust T for which the fan will remain in equilibrium with three legs is related to the value with four legs by $T_{three\ legs} = \frac{1}{\sqrt{2}} T_{four\ legs}$

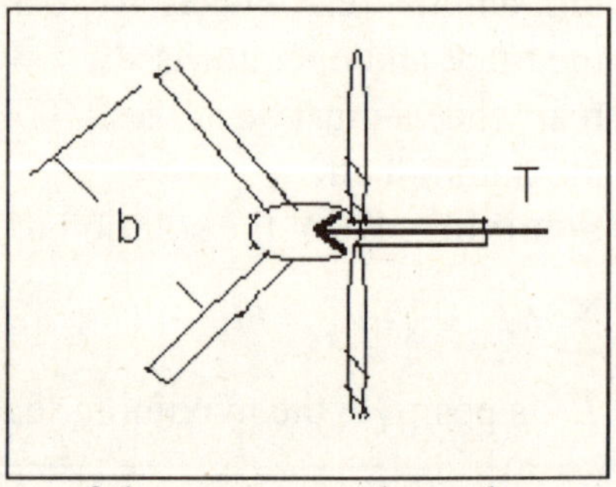

Solution: From the solution to Problem 5.28 the maximum thrust is $\boxed{T_{four\ legs} = \frac{bW\sin 45^o}{h} = \frac{bW}{\sqrt{2}h}}$. For three legs assume that the legs are in the position shown with respect to the line of action of the thrust, The distance to the center is $b\cos 60^o = \frac{b}{2}$. When the outermost leg has zero reaction, the other legs must bear the weight of the fan. The sum of the moments about the center when the outer most leg has zero reaction is $\sum M_T = -\frac{bW}{2} + Th = 0$, from which .

$$\boxed{T_{three\ legs} = \frac{bW}{2h} = \frac{1}{\sqrt{2}} T_{four\ legs.}}$$

==============================<>==============================

Problem 5.31 A force $F = 400\ N$ acts on the bracket. What are the reactions at A and B?

Solution: The joint A is a pinned joint; B is a roller joint. The pinned joint has two reaction forces A_X, A_Y. The roller joint has one reaction force B_X. The sum of the forces is $\sum F_X = A_X + B_X = 0$, $\Sigma F_Y = A_Y - F = 0$, from which $\boxed{A_Y = F = 400\ N}$

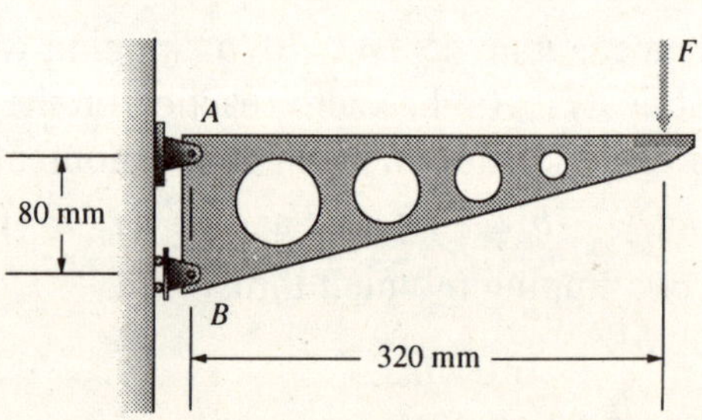

The sum of the moments about A is $\sum M_A = 0.08B_X - 0.320F = 0$, from which $\boxed{B_X = \frac{0.320(400)}{0.08} = 1600\ N}$ Substitute into the sum of forces equation to obtain: $\boxed{A_X = -B_X = -1600\ N}$

==============================<>==============================

Problem 5.32

The hanging sign exerts vertical 25 lb forces at A and B. Determine the tension in the cable and the reaction at the support C.

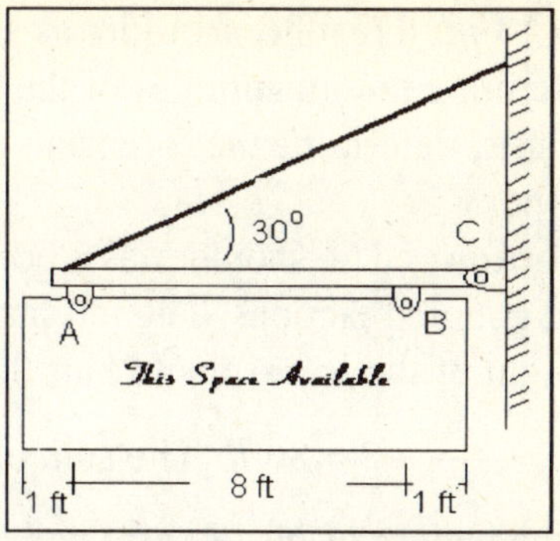

Solution: The joint C is pinned joint, with two forces, C_X, C_Y. The sum of the vertical forces is $\sum F_X = T_C \cos 30^o + C_X = 0$. $\sum F_Y = T_C \sin 30^o - A_Y - B_Y + C_Y = 0$. The sum of the moments about C is $\sum M_C = 9A_Y + 1B_Y - 10T_C \sin 30^o = 0$ from which $\boxed{T_C = \frac{9A_Y + B_Y}{10 \sin 30^o} = \frac{250}{5} = 50\ lb}$. Substitute into the vertical forces sum to obtain: $\boxed{C_Y = A_Y + B_Y - T_C \sin 30^o = 25\ lb}$ Substitute into the horizontal forces sum to obtain $\boxed{C_X = -T_C \cos 30^o = -43.3\ lb}$

Problem 5.33

This device, called a *swape* or *shadoof*, is used to help a person lift a heavy load. It was used in Egypt at least as early as 1550 B. C. and has continued to be used in various parts of the world to the present day. The distances $a = 12\ ft$ and $b = 4\ ft$. If the load being lifted weighs 100 lb and $W = 200\ lb$, determine the vertical force the person must exert to support the stationary load: (a) when the load is just above the ground (the position shown); (b) when the load is three feet above ground. (assume that the rope remains vertical.)

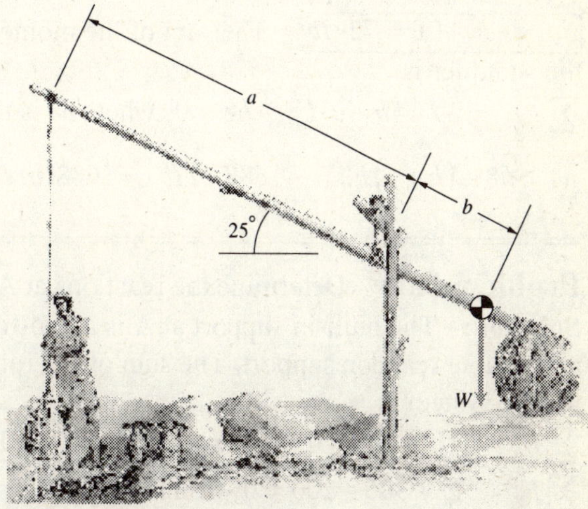

Solution: Denote the vertical force exerted by the laborer by F. (a) The sum of the moments about the fulcrum point is

$$\sum M_F = -aF \cos 25^o + aW_L \cos 25^o - bW \cos 25^o = 0$$

from which $F = \frac{(aW_L - bW)}{a}$, or

$\boxed{F = \frac{1200 - 800}{12} = 33.33\ lb}$. *the force exerted by the laborer is independent of the angle.*

(b) When the load is three feet above the ground, the swape is at a new angle. But since the force is independent of the angle, $\boxed{F = 33.33\ lb}$

Problem 5.34

This structure, called a *truss*, has a pin support at A a roller support at B, and is loaded by two forces. Determine the reactions at the supports.

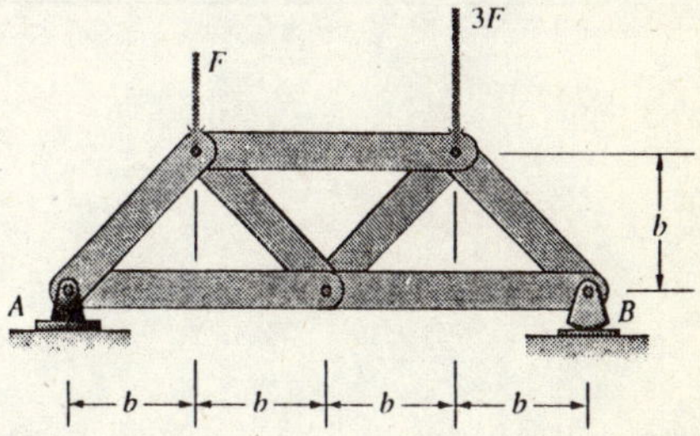

Solution: The truss can be treated as a single member. The pinned joint at A is a two force support; the roller support at B is a single force support. The sum of the forces: $\boxed{\sum F_X = A_X = 0}$. $\sum F_Y = A_Y + B_Y - F - 3F = 0$.. The sum of moments about A is

$\sum M_A = -bF - 3b(3F) + 4bB_Y = 0$, from which $\boxed{B_Y = \frac{10bF}{4} = 2.5F}$ From the vertical force sum, $\boxed{A_Y = 4F - 2.5F = 1.5F}$

Problem 5.35 An Olympic gymnast is stationary at the "iron cross" position. The weight of his left arm and the weight of his body *not including his arms* are shown. The distances $a = b = 9$ *in* and $c = 13$ *in*. Treat his shoulder as a built-in support and determine the magnitudes of the reactions at his shoulder. That is, determine the force and couple his shoulder must support.

Solution: The shoulder as a built-in joint has two-force and couple reactions. The left hand must support the weight of the left arm and half the weight of the body:

$F_H = \frac{144}{2} + 8 = 80\ lb$. The sum of the forces on the left arm is the weight of his left arm and the vertical reaction at the shoulder and hand:

$\sum F_X = S_X = 0$.

$\sum F_Y = F_H - S_Y - 8 = 0$, from which

$\boxed{S_Y = -F_H + 8 = 72\ lb}$. The sum of the moments about the shoulder is

$\sum M_S = M + (b+c)F_H - b8 = 0$, where M is the couple reaction at the shoulder. Thus

$$M = b8 - (b+c)F_H = -1688\ in\ lb = 1688(in\ lb)\left(\frac{1\ ft}{12\ in}\right) \boxed{= 140.67\ ft\ lb}$$

Problem 5.36 Determine the reactions at A.

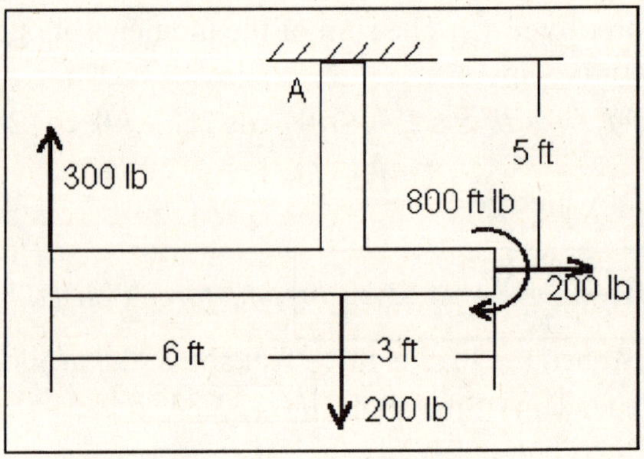

Solution: The built-in support at A is a two-force and couple reaction support. The sum of the forces for the system is

$\sum F_X = A_X + 200 = 0$, from which $\boxed{A_X = -200\ lb}$

$\sum F_Y = A_Y + 300 - 200 = 0$, from which

$\boxed{A_Y = -100\ lb}$

The sum of the moments about A:

$\sum M = -6(300) + 5(200) - 800 + M_A = 0$, from which $\boxed{M_A = 1600\ ft\ lb}$ which is the couple at A.

Problem 5.37 The car's brakes keep the rear wheels locked and the front wheels are free to turn. Determine the forces exerted on the front and rear wheels by the road when the car is parked (a) on an up-slope with $\alpha = 15^o$; b) on a down-slope with $\alpha = -15^o$.

Solution: The rear wheels are two force reaction support, and the front wheels are a one force reaction support. Denote the rear wheels by A and the front wheels by B, and define the reactions as being parallel to and normal to the road.. The sum of forces:

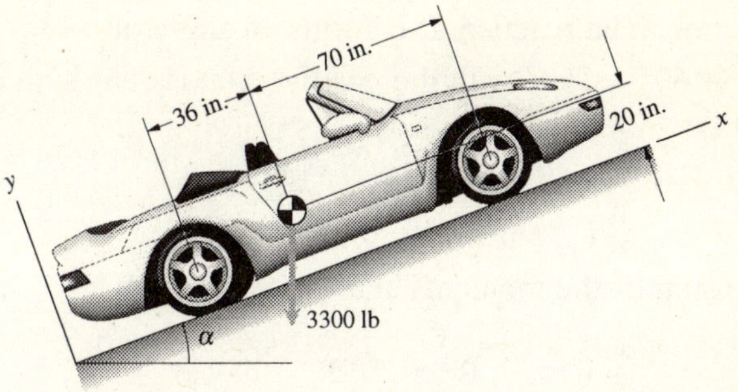

$\sum F_X = A_X - 3300 \sin 15^o = 0$, from which $\boxed{A_X = 854.1\ lb}$.

$\sum F_Y = A_Y - 3300 \cos 15^o + B_Y = 0$ Since the mass center of the vehicle is displaced above the point A, a component of the weight $(20W \sin\alpha)$ produces a positive moment about A, whereas the other component $(36W \cos\alpha)$ produces a negative moment about A. The sum of the moments about A: $\Sigma M_A = -36(3300 \cos 15^o) + 20(3300 \sin 15^o) + B_Y(106) = 0$, from which

$\boxed{B_Y = \frac{+97669}{106} = 921.4\ lb}$. Substitute into the sum of forces equation to obtain $\boxed{A_Y = 2266.1\ lb}$

(b) For the car parked down-slope the sum of the forces is $\sum F_X = A_X + 3300 \sin 15^o = 0$, from which $\boxed{A_X = -854\ lb}$ $\sum F_Y = A_Y - 3300 \cos 15^o + B_Y = 0$. The component $(20W \sin\alpha)$ now produces a negative moment about A. The sum of the moments about A is

$\sum M_A = -3300(36)\cos 15^o - 3300(20)\sin 15^o + 106B_Y = 0$, from which $\boxed{B_Y = \frac{131834}{106} = 1243.7\ lb}$

Substitute into the sum of forces equation to obtain $\boxed{A_Y = 1943.8\ lb}$

Problem 5.38 The weight W of the bar acts at its center. The surfaces are smooth. What is the tension in the horizontal string?

Solution: The surfaces are roller supports, with only one reaction force, which is normal to the contact surface. Denote the reaction at the top of the bar by B, the tension in the string by T, and the reaction at the base of the bar by A. The angle formed by the bar at the base is $\alpha = 45^o$, since the altitude and base of the triangle are equal. The reaction at the top of the bar forms the angle $(90 - \alpha)$ with the horizontal, and the reaction at the base is vertical. The sum of the forces is

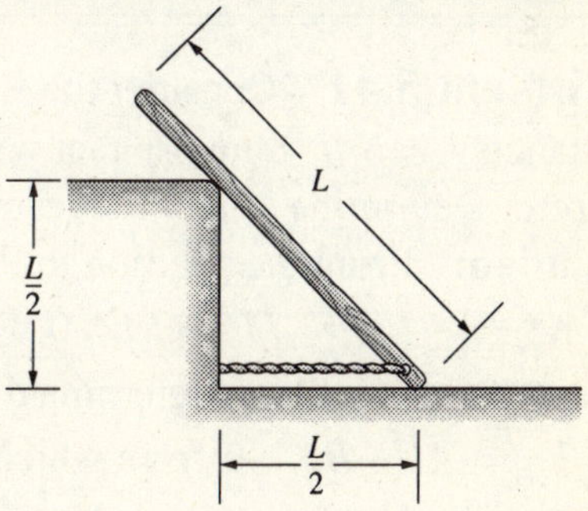

$\sum F_X = -T + B\cos(90 - \alpha) = -T + B \sin\alpha = 0$.

$\sum F_Y = -W + A_Y + B\cos\alpha = 0$. The sum of the moments about the lower end is

$\sum M_A = \left(\frac{WL}{2}\right)\cos\alpha - B\left(\frac{L}{\sqrt{2}}\right) = 0$, from which $B = \frac{W\cos\alpha}{\sqrt{2}}$. Substitute into the horizontal force equation to obtain the string tension $T = \frac{W}{\sqrt{2}} \sin\alpha \cos\alpha$ $\boxed{= \frac{W}{2\sqrt{2}} = 0.3536W}$

Problem 5.39 The plate is supported by a pin in a smooth slot at B. What are the reactions at the supports?

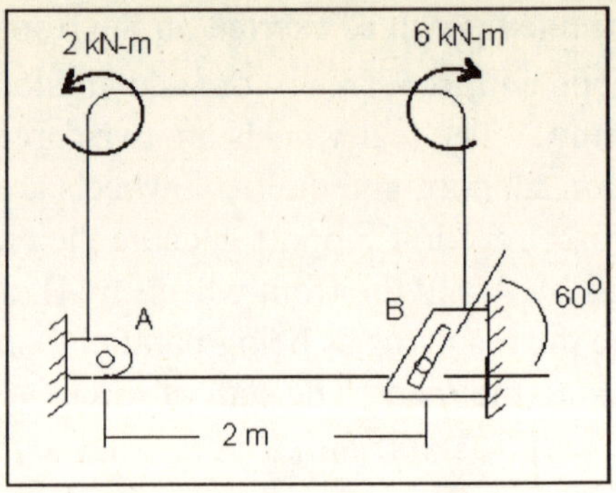

Solution: The pinned support is a two force reaction support. The smooth pin is a roller support, with a one force reaction. The reaction at B forms an angle of $90^o + 60^o = 150^o$ with the positive x-axis. The sum of the forces:

$\sum F_X = A_X + B\cos 150^o = 0$

$\sum F_Y = A_Y + B\sin 150^o = 0$

The sum of the moments about B is

$\Sigma M_B = -2A_Y + 2 - 6 = 0$, from which $\boxed{A_Y = -\frac{4}{2} = -2\ kN}$ Substitute into the force equations to obtain $B = \frac{A_Y}{\sin 150^o} = 4\ kN$, and $\boxed{A_X = -B\cos 150^o = 3.464\ kN}$ The horizontal and vertical reactions at B are $\boxed{B_X = 4\cos 150^o = -3.464\ kN}$, and $\boxed{B_Y = 4\sin 150^o = 2\ kN}$.

Problem 5.40 The force $F = 800\ N$ and the couple $M = 200\ N\text{-}m$. The distance $L = 2\ m$. What are the reactions at A and B?

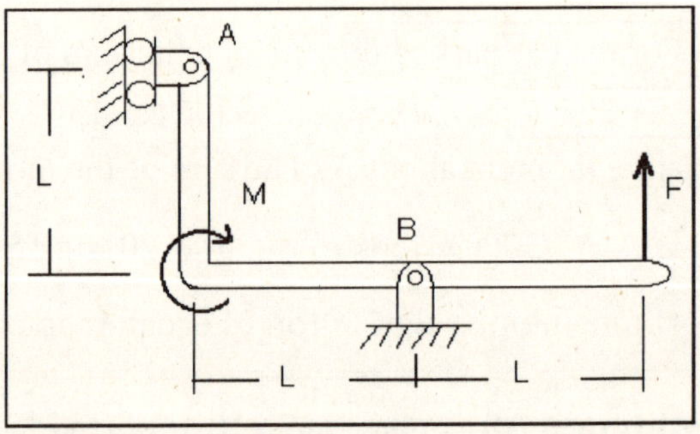

Solution: The sum of the forces:

$\sum F_X = A_X + B_X = 0$

$\sum F_Y = B_Y + F = 0$, from which $\boxed{B_Y = -F = -800\ N}$

The sum of the moments about B is

$\sum M_B = -LA_X + LF - M = 0$, from which

$\boxed{A_X = \frac{LF - M}{L} = 700\ N}$ Substitute into the force equations to obtain $\boxed{B_X = -700\ N}$.

Problem 5.41 Consider the L-shaped bar shown in Problem 5.40. The force $F = 300\ lb$ and the distance $L = 6\ ft$. Suppose that your design objectives require that you adjust M so that the reaction at A is zero. Determine M and the reactions at B.

Solution: From the solution to Problem 5.40 the sum of the moments about B are $\sum M_B = -LA_X + LF - M = 0$. For a zero reaction at A: $\boxed{M = LF = 6(300) = 1800\ ft\ lb}$.

The reactions at B are determined from the force equations:

$\sum F_X = A_X + B_X = 0$ from which $\boxed{B_X = 0}$.

$\sum F_Y = B_Y + F = 0$, from which $\boxed{B_Y = -F = -300\ lb}$

Problem 5.42 The weight $W = 1.2\ kN$. Determine the tension in the cable and the reactions at A.

Solution: The pinned support at A is a two force reaction support. Denote the tension in the cable by T. The sum of forces: $\sum F_X = A_X - T\cos 30^o = 0$

$\sum F_Y = A_Y + T\sin 30^o - W = 0$.

The sum of moments about the right end of the bar is $\sum M = 2W - 4A_Y = 0$, from which $\boxed{A_Y = \frac{W}{2} = 0.6\ kN}$. Substitute into the force equations to obtain $\boxed{T = \frac{W - A_Y}{\sin 30^o} = \frac{W}{2\sin 30^o} = W = 1.2\ kN}$, and

$\boxed{A_X = T\cos 30^o = 1.0392\ kN}$

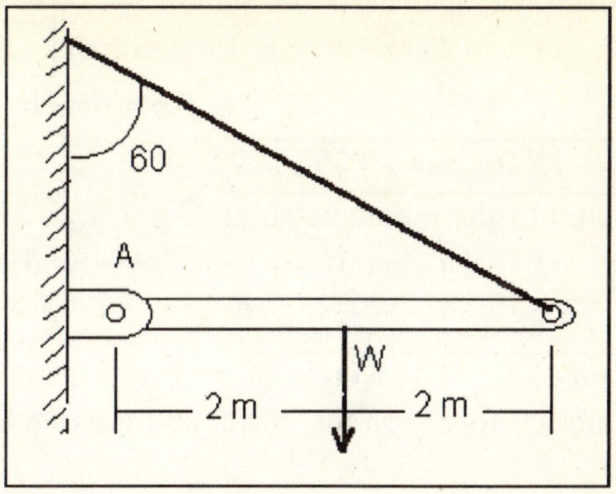

Problem 5.43 Consider the beam in Problem 5.42. An engineer determines that the cable will safely support a tension of 3 kN and that the support at A will safely support a force of magnitude 2.4 kN. Based on these criteria, what is the largest beam weight that can safely be supported?

Solution: The strategy is to determine the weight for the safe limit on cable tension, and then to test whether the reaction at A is within safe limits. From the solution to Problem 5.42 the sum of forces and the sum of moments leads to $A_X = T\cos 30^o$. For the safe limit of 3 kN on the tension, $A_X = 2.598\ kN$, which exceeds the safe limit of 2.4 kN for the reaction at A. *Thus the reaction at A limits the weight.* From the solution to Problem 5.42 the two equilibrium equations are:

$W = \quad T\sin 30^o + A_Y$

$0 \quad = -T\cos 30^o + A_X$

From Problem 5.42, $T = W$, from which $W(1 - \sin 30^o) = A_Y$, and $W\cos 30^o = A_X$. Square both equations, add, take the square root of the both sides, to obtain $W = \sqrt{A_Y^2 + A_X^2}$ .Substitute the safe limit for the magnitude of the reaction at A, from which $\boxed{W = 2.4\ kN}$.

Problem 5.44 The tension in cable BC is 100 lb. Determine the reactions at the built in support.

Solution: The cable does not exert an external force on the system, and can be ignored in determining reactions. The built-in support is a two-force and couple reaction support. The sum of forces:

$\boxed{\sum F_X = A_X = 0}$.

$\sum F_Y = A_Y - 200 = 0$, from which $\boxed{A_Y = 200\ lb}$.

The sum of the moments about A is

$\sum M = M_A - (3)(200) - 300 = 0$, from which

$\boxed{M_A = 900\ ft\ lb}$

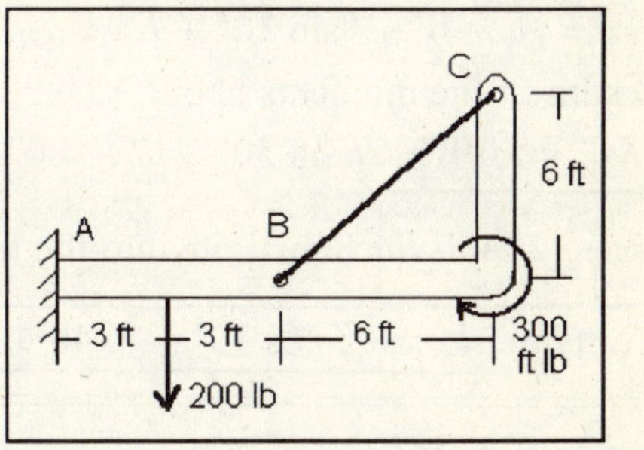

Problem 5.45 The tension in cable AB is 2 *kN*. What are the reactions at C in the two cases?

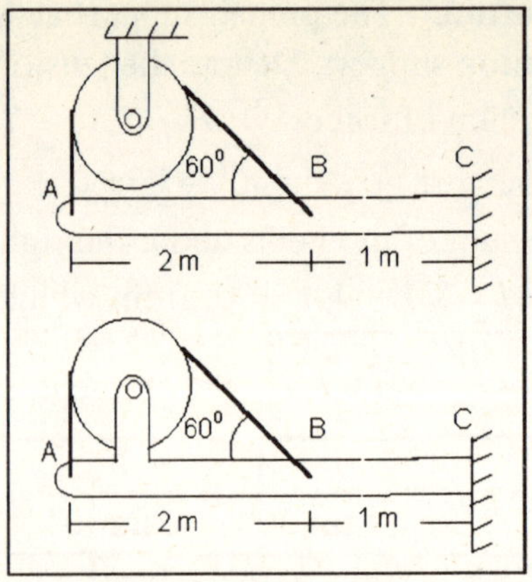

Solution: *First Case:* The sum of the forces:

$\sum F_X = C_X - T\cos 60^o = 0$, from which $\boxed{C_X = 2(0.5) = 1\ kN}$

$\sum F_Y = C_Y + T\sin 60^o + T = 0$, from which

$\boxed{C_Y = -1.866(2) = -3.732\ kN}$

The sum of the moments about C is

$\sum M = M_C - T\sin 60^o - 3T = 0$, from which

$\boxed{M_C = 3.866(2) = 7.732\ kN}$

Second Case: The weight of the beam is ignored, hence there are no external forces on the beam, and *the reactions at C are zero.*

Problem 5.46 Determine the reactions at the supports.

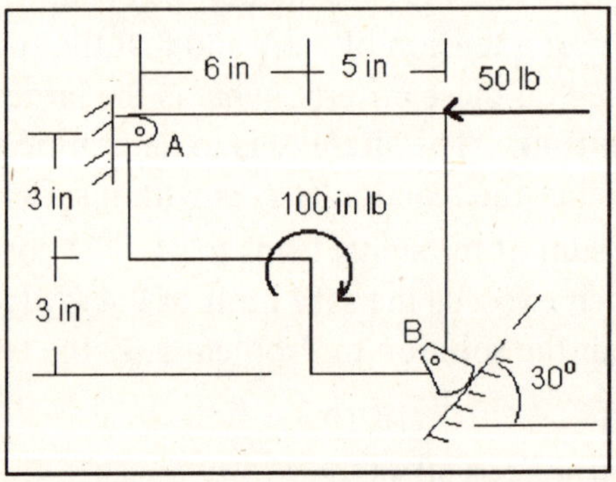

Solution: The reaction at A is a two-force reaction. The reaction at B is one-force, normal to the surface.

The sum of the forces:

$\sum F_X = A_X - B\cos 60^o - 50 = 0$. $\Sigma F_Y = A_Y + B\sin 60^o = 0$.

The sum of the moments about A is

$\sum M_A = -100 + 11B\sin 60^o - 6B\cos 60^o = 0$, from which

$\boxed{B = \dfrac{100}{(11\sin 60^o - 6\cos 60^o)} = 15.3\ lb}$. Substitute into the force equations to obtain $\boxed{A_Y = -B\sin 60^o = -13.3\ lb}$ and $\boxed{A_X = B\cos 60^o + 50 = 57.7\ lb}$

Problem 5.47 The weight $W = 1000\ lb$. Determine the tension in the cable and the reactions at A.

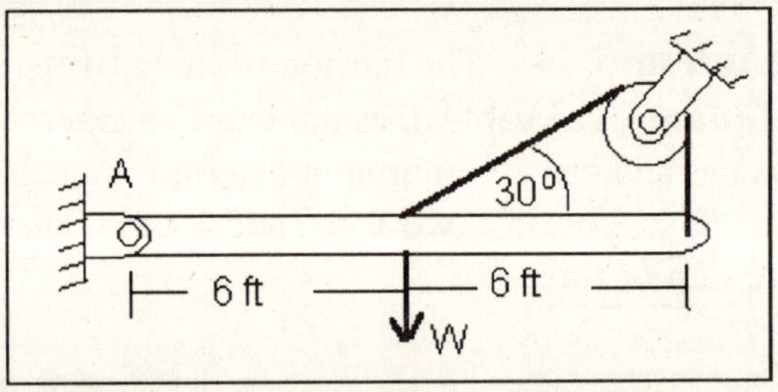

Solution: The sum of the forces:

$\sum F_X = A_X + T\cos 30^o = 0$

$\sum F_Y = A_Y - W + T\sin 30^o + T = 0$

The sum of the moments about A,

$\sum M_A = -6W + 6T\sin 30^o + 12T = 0$ from which

$\boxed{T = \dfrac{6W}{15} = 400\ lb}$. Substitute into the force equations top obtain $\boxed{A_X = -T\cos 30^o = -346.4\ lb}$, and $\boxed{A_Y = W - T\sin 30^o - T = 400\ lb}$

Problem 5.48 Consider the beam in Problem 5.47. An engineer determines that the cable will safely support a tension of 1000 *lb*. What is the largest beam weight *W* that can safely be supported?

Solution: From the sum of moments about A given in Problem 5.47,

$\sum M_A = -6W + 6T\sin 30^o + 12T = 0$, the weight is given by $\boxed{W = \dfrac{5T}{2} = 2500\ lb}$

==================================<>==================================

Problem 5.49 The distance $x = 3\ ft$. Determine the tension in the cable and the reactions at A and B.

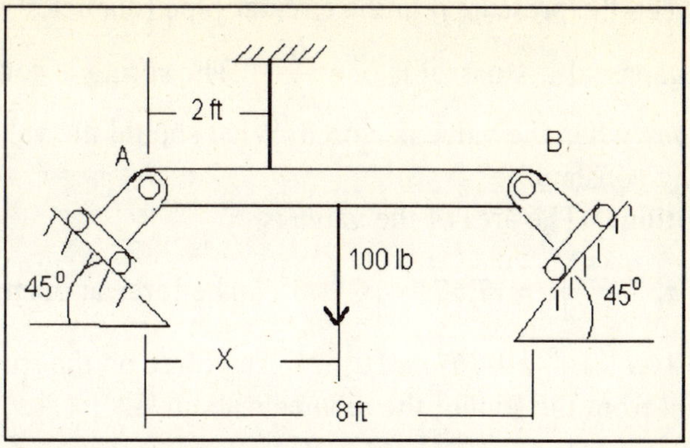

Solution: Denote the tension in the cable by T. The sum of the forces:

$\sum F_X = A\cos 45^o - B\cos 45^o = 0$, from which $A = B$.

$\sum F_Y = A\sin 45^o - W + T + B\sin 45^o = 0$

The sum of the moments about A is

$\sum M_A = 2T - xW + 8B\sin 45^o = 0$. Combine the equations to obtain the simultaneous equations:

$2T + 5.656B = 300$

$T + 1.414B = 100$

Solve: $\boxed{T = 50.0\ lb}$, and $B = 35.355\ lb$. From the force equations $\boxed{A = B = 35.4\ \ lb}$

==================================<>==================================

Problem 5.50 Consider the beam in Problem 5.49. Determine the range of values of the distance x for which the beam can be in equilibrium for the position shown.

Solution: From the sum of the moments about A in the solution to Problem 5.49,

$\sum M_A = 2T - xW + 8B\sin 45^o = 0$, the condition $A = B$, and the sum of the vertical forces:

$\sum F_Y = A\sin 45^o - W + T + B\sin 45^o = 0$, the two simultaneous equations in terms of x are obtained:

$2T + 4\sqrt{2}B = xW$

$T + \sqrt{2}B = W$

Eliminate the tension T by multiplying the second equation by 2 and subtracting from the first. The result: $2\sqrt{2}B = (x-2)W$. If $x \le 2$, then $B \le 0$, and the right hand reaction reverses sign, which makes no sense for a roller support. Similarly, if B is eliminated from the simultaneous equations by multiplying the second by 4 and subtracting, the result is $2T = (4-x)W$. If $x \ge 4$, then $T \le 0$, and the tension reverses sign: a zero or negative tension makes no sense for a cable support. Therefore the limits on x are $\boxed{2 \le x \le 4}$.

[*Check:* Take the sum of the moments about B, rather than about A as above:

$\sum M_B = (8-x)W - 6T - 8A\sin 45^o = 0$. Use the condition $A = B$, and the sum of the vertical forces:

$\sum F_Y = A\sin 45^o - W + T + B\sin 45^o = 0$. The two simultaneous equations in terms of x are obtained:

$6T + 4\sqrt{2}A = (8-x)W$

$T + \sqrt{2}A = W$

Eliminate T by multiplying the second equation by 6 and subtracting from the first. The result:

$2\sqrt{2}A = (x-2)W$. The reaction at A reverses sign for $x \le 2$. *Check.*

Eliminate A from the equations to obtain $2T = (4-x)W$, and the cable reaction changes sign at $x \ge 4$. *Check*].

==================================<>==================================

Problem 5.51 Suppose that you want to design the safety valve to open when the difference between the pressure p in the circular pipe (diameter =150 *mm*) and the atmospheric pressure is 10 *MPa* (Megapascals. A pascal is one $\frac{N}{m^2}$.) The spring is compressed 20 *mm* when the valve is closed. What should the value of the spring constant be?

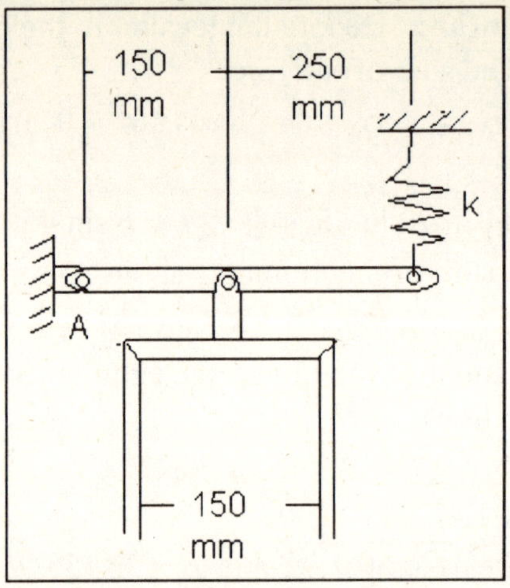

Solution: The area of the valve is

$a = \pi\left(\frac{0.15}{2}\right)^2 = 17.671 \times 10^{-3}\ m^2$. The force at opening is

$F = 10a \times 10^6 = 1.7671 \times 10^5\ N$. The force on the spring is found from the sum of the moments about A,

$\sum M_A = 0.15F - (0.4)k\Delta L = 0$. Solving,

$$k = \frac{0.15F}{(0.4)\Delta L} = \frac{0.15(1.7671 \times 10^5)}{(0.4)(0.02)} \boxed{= 3.313 \times 10^6\ \frac{N}{m}}$$

Problem 5.52 The rectangular plate is held in equilibrium by the horizontal force F. The weight W acts at the midpoint of the plate. Show that F is given by the equation: $F = \frac{(b\cos\alpha - h\sin\alpha)}{2(h\cos\alpha + b\sin\alpha)}W$

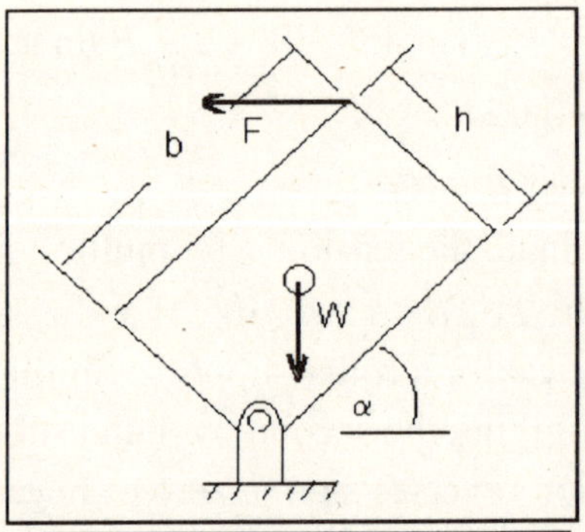

Solution: The moment about the pinned joint of the force is $M_F = HF$, where y is the vertical distance from the pinned joint to the action line of the force. This distance is the projection of the diagonal of the rectangle on the vertical axis, as shown in the second sketch: $H = \sqrt{b^2 + h^2}\sin(\alpha + \beta)$, where $\beta = \tan^{-1}\frac{h}{b}$. By a similar construction, the horizontal distance to the action line of the weight is $X = \left(\frac{1}{2}\right)\sqrt{b^2 + h^2}\cos(\alpha + \beta)$. The sum of the moments about the pinned joint is

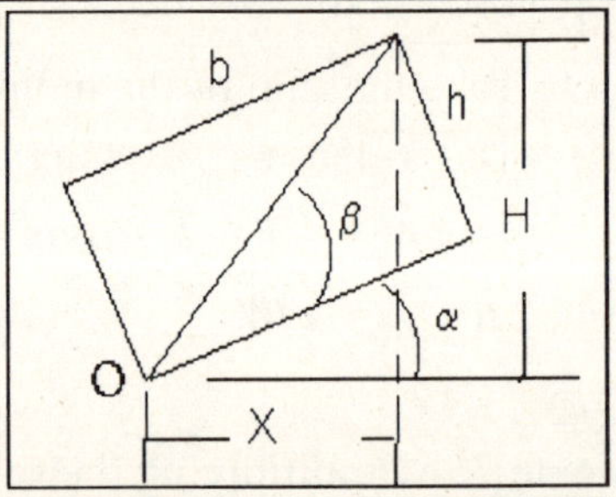

$$\sum M = F\left(\sqrt{b^2 + h^2}\right)\sin(\alpha + \beta) - \left(\frac{W}{2}\right)\left(\sqrt{b^2 + h^2}\right)\cos(\alpha + \beta) = 0.$$

Noting that $\cos\beta = \frac{b}{\sqrt{b^2 + h^2}}$, and $\sin\beta = \frac{h}{\sqrt{b^2 + h^2}}$, using the double angle formulas: $\cos(\alpha + \beta) = \cos\alpha\cos\beta - \sin\alpha\sin\beta$, and $\sin(\alpha + \beta) = \sin\alpha\cos\beta + \cos\alpha\sin\beta$, the moment equation reduces to

$$\sum M = F(b\sin\alpha + h\cos\alpha) - \left(\frac{W}{2}\right)(b\cos\alpha - h\sin\alpha) = 0,$$ from which $\boxed{F = \frac{(b\cos\alpha - h\sin\alpha)}{2(h\cos\alpha + b\sin\alpha)}W}$

Problem 5.53 The bar AB is of length L and weight W, and the weight acts at its midpoint. The angle $\alpha = 30^o$. What is the tension in the string?

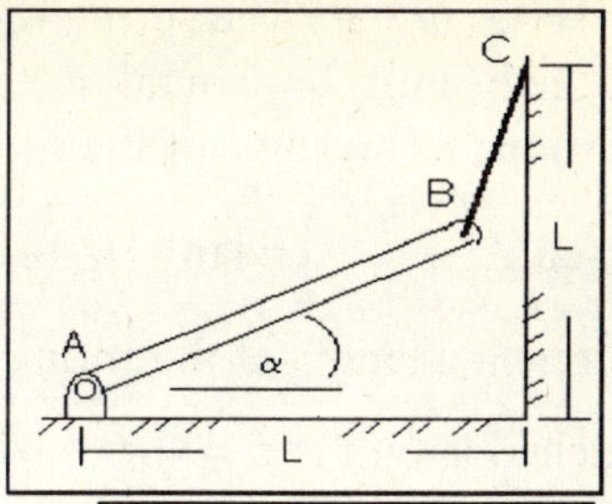

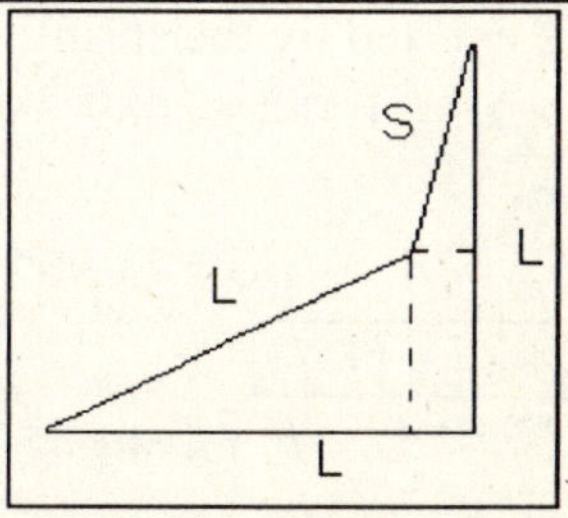

Solution: The strategy is to determine the angle formed by the string on the end of the rod (see sketch). The vertical distance from the pinned joint to the end of the rod is $D = L\sin\alpha$. The horizontal distance is $H = L\cos\alpha$. The sides of the small triangle formed by the string is $X = L - L\cos\alpha$ and $Y = L - L\sin\alpha$. The angle formed by the string with the horizontal is $\beta = \tan^{-1}\left(\dfrac{1-\sin\alpha}{1-\cos\alpha}\right) = 75^o$. The angle relative to the rod is $\theta = \beta - \alpha = 45^o$. The moment about the pinned joint is $\sum M = -\left(\dfrac{L}{2}\right)W\cos\alpha + TL\sin\theta = 0$, from which

$$T = \frac{W\cos\alpha}{2\sin\theta} = \frac{W(0.866)}{2(0.707)} = \frac{\sqrt{3}\,W}{2\sqrt{2}} \boxed{= 0.61237W}$$

Problem 5.54 The weight $W_1 = 1000\ lb$. Neglect the weight of the bar AB. The cable goes over a pulley at C. Determine the weight W_2 and the reactions at the pin support A.

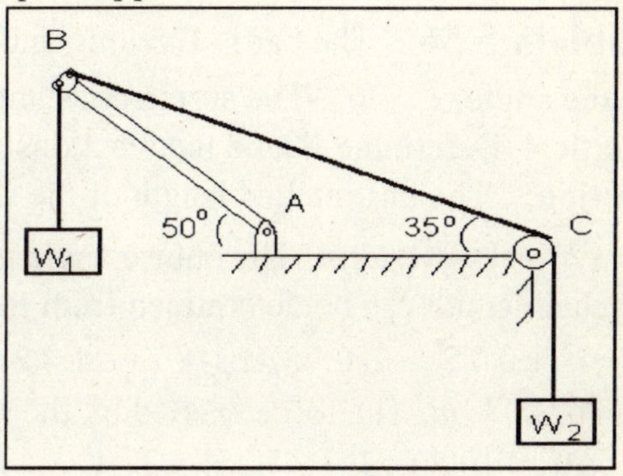

Solution: The strategy is to resolve the tensions at the end of bar AB into x- and y-components, and then set the moment about A to zero. The angle between the cable and the positive x axis is -35°. The tension vector in the cable is $\vec{T}_2 = W_2\left(\vec{i}\cos(-35^o) + \vec{j}\sin(-35^o)\right)$.

$= W_2\left(0.8192\vec{i} - 0.5736\vec{j}\right)\ (lb)$

Assume a unit length for the bar. The angle between the bar and the positive x-axis is $180^o - 50^o = 130^o$. The position vector of the tip of the bar relative to A is $\vec{r}_B = \vec{i}\cos(130^o) + \vec{j}\sin(130^o)$, $= -0.6428\vec{i} + 0.7660\vec{j}$. The tension exerted by W_1 is $\vec{T}_1 = -1000\vec{j}$. The sum of the moments about A is :

$$\sum \vec{M}_A = (\vec{r}_B \times \vec{T}_1) + (\vec{r}_B \times \vec{T}_2) = \vec{r}_B \times (\vec{T}_1 + \vec{T}_2) = L\begin{vmatrix} \vec{i} & \vec{j} & \vec{k} \\ -0.6428 & 0.7660 & 0 \\ 0.8191W_2 & -0.5736W_2 - 1000 & 0 \end{vmatrix} = 0\ .$$

$\sum \vec{M}_A = (-0.2587W_2 + 642.8)\vec{k} = 0$, from which $\boxed{W_2 = 2483.5\ lb}$

The sum of the forces:

$\sum \vec{F}_X = (A_X + W_2(0.8192))\vec{i} = 0$, from which $\boxed{A_X = -2034.4\ lb}$

$\sum \vec{F}_Y = (A_Y - W_2(0.5736) - 1000)\vec{j} = 0$, from which $\boxed{A_Y = 2424.5\ lb}$

==============================<>==============================

Problem 5.55 The distances $a = 6\ ft$ and $b = 3\ ft$. The couple $M = 200\ ft\ lb$. The spring constant $k = 1000\ lb/ft$. The spring would be unstretched if $h = 0$. The system is in equilibrium when $h = 4\ ft$ and the beam is horizontal. Determine the force F and the reactions at A.

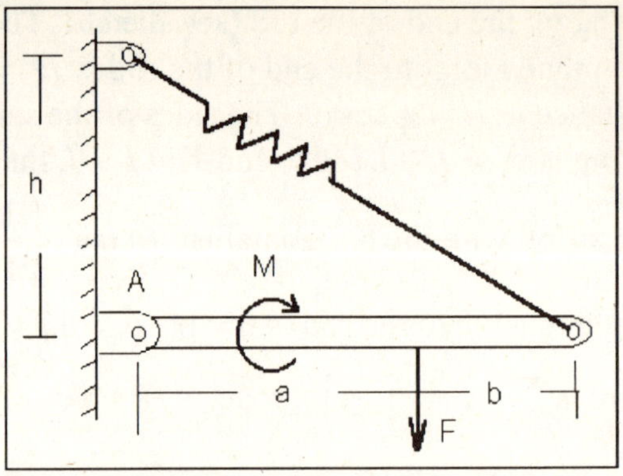

Solution: The angle of the spring relative to the beam is $\alpha = \tan^{-1}\left(\frac{h}{a+b}\right) = \tan^{-1}(0.4444) = 23.96°$. The unstretched length of the spring is $a + b = 9\ ft$. The stretched length is $L = \sqrt{h^2 + (a+b)^2} = 9.8489\ ft$. The force exerted by the spring is $T = k\Delta L = 1000(9.8489 - 9) = 848.86\ lb$. The sum of the forces:

$\sum F_X = A_X - T\cos 23.96° = 0$, from which $\boxed{A_X = 775.71\ lb}$.

$\sum F_Y = A_Y - F + T\sin 23.96° = 0$. The sum of the moments about A is

$\sum M_A = -M - aF + (a+b)T\sin 23.96° = 0$, from which $F = \frac{(a+b)T\sin 23.96° - M}{a} = \boxed{483.75\ lb}$.

Substitute into the force equation to obtain: $A_Y = F - T\sin 23.96° \boxed{= 139.0\ lb}$

==============================<>==============================

Problem 5.56 The bar is 1 m long and its weight W acts at its midpoint. The distance $b = 0.75\ m$ and the angle $\alpha = 30°$. The spring constant is $k = 100\ N/m$ and the spring is unstretched when the bar is vertical. Determine W and the reactions at A.

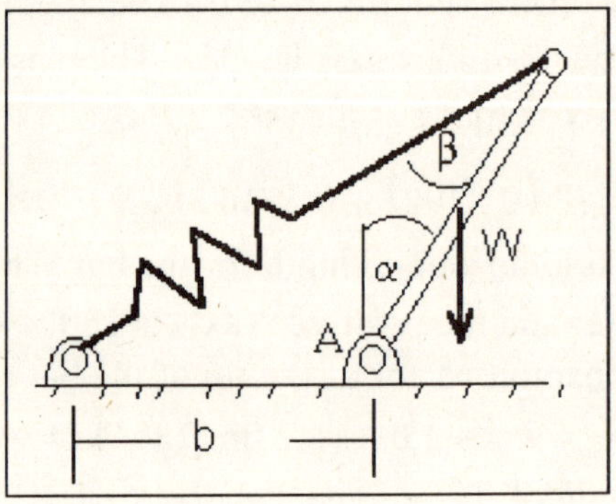

Solution: The unstretched length of the spring is $L = \sqrt{b^2 + 1^2} = 1.25\ m$. The obtuse angle is $90 + \alpha$, so the stretched length can be determined from the cosine law: $L_2^2 = 1^2 + 0.75^2 - 2(0.75)\cos(90 + \alpha) = 2.3125\ m^2$ from which $L_2 = 1.5207\ m$. The force exerted by the spring is $T = k\Delta L = 100(1.5207 - 1.25) = 27.1\ N$. The angle between the spring and the bar can be determined from the sine law:

$\frac{b}{\sin\beta} = \frac{1.5207}{\sin(90+\alpha)}$, from which $\sin\beta = 0.4271$, $\beta = 25.28°$.

The angle the spring makes with the horizontal is $180 - 25.28 - 90 - \alpha = 34.72°$. The sum of the forces:

$\sum F_X = A_X - T\cos 34.72° = 0$, from which $\boxed{A_X = 22.25\ N}$.

$\sum F_Y = A_Y - W - T\sin 34.72° = 0$. The sum of the moments about A is

$\sum M_A = T\sin 25.28° - \left(\frac{W}{2}\right)\sin\alpha = 0$, from which $\boxed{W = \frac{2T\sin 25.28°}{\sin\alpha} = 46.25\ N}$ Substitute into the force equation to obtain: $\boxed{A_Y = W + T\sin 34.72° = 61.66\ N}$

==============================<>==============================

==============================<>==============================

Problem 5.57 The *boom derrick* supports a suspended 15 *kip* load. The booms BC and DE are each 20 ft long. The distances $a = 15\,ft$ and $b = 2\,ft$ and the angle $\theta = 30^o$. Determine the tension in cable AB and reactions at the pin supports C and D.

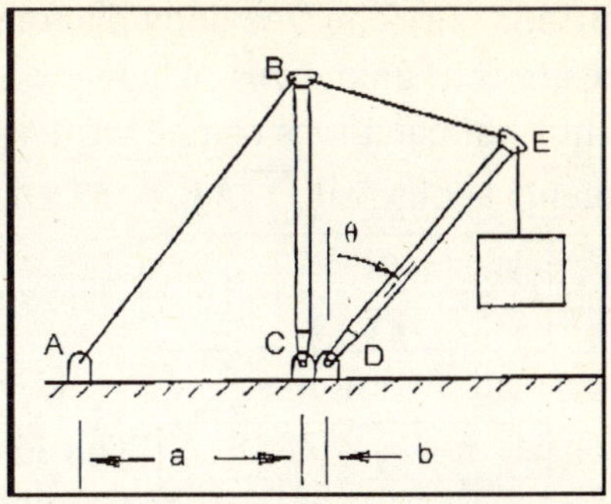

Solution Choose a coordinate system with origin at point C, with the y axis parallel to CB. The position vectors of the labeled points are: $\vec{r}_D = 2\vec{i}$

$\vec{r}_E = \vec{r}_D + 20(\vec{i}\sin 30^o + \vec{j}\cos 30^o) = 12\vec{i} + 17.3\vec{j}$,

$\vec{r}_B = 20\vec{j}$, $\vec{r}_A = -15\vec{i}$. The unit vectors:

$$\vec{e}_{DE} = \frac{\vec{r}_E - \vec{r}_D}{|\vec{r}_E - \vec{r}_D|} = 0.5\vec{i} + 0.866\vec{j},$$

$\vec{e}_{EB} = \dfrac{\vec{r}_B - \vec{r}_E}{|\vec{r}_B - \vec{r}_E|} = -0.976\vec{i} + 0.2179\vec{j}$. $\vec{e}_{CB} = \dfrac{\vec{r}_B - \vec{r}_C}{|\vec{r}_B - \vec{r}_C|} = 1\vec{j}$, $\vec{e}_{AB} = \dfrac{\vec{r}_A - \vec{r}_B}{|\vec{r}_A - \vec{r}_B|} = -0.6\vec{i} - 0.8\vec{j}$. *Isolate the juncture at E:* The equilibrium conditions are $\Sigma F_x = 0.5|\vec{D}| + 0.976|\vec{T}_{EB}| = 0$,

$\Sigma F_y = 0.866|\vec{D}| - 0.2179|\vec{T}_{EB}| - 15 = 0$, from which $\boxed{|\vec{D}| = 15.34\ kip}$ and $|\vec{T}_{EB}| = 7.86\ kip$. *Isolate the juncture at B:* The equilibrium conditions are: $\Sigma F_x = 0|\vec{C}| - 0.6|\vec{T}_{AB}| + 0.976|\vec{T}_{EB}|$, and

$\Sigma F_y = 1|\vec{C}| - 0.6|\vec{T}_{AB}| - 0.2179|\vec{T}_{EB}| = 0$, from which $\boxed{|\vec{T}_{AB}| = 12.79\ kip}$, and $|\vec{C}| = 11.94\ kip$. The components: $\boxed{D_x = 0.6|\vec{D}| = 7.67\ kip}$, $\boxed{D_y = 0.866|\vec{D}| = 13.287\ kip}$, and $\boxed{C_y = 1|\vec{C}| = 11.94\ kip}$

==============================<>==============================

Problem 5.58 The arrangement shown controls the elevators of an airplane. (The elevators are the horizontal control surfaces in the airplanes tail.) The elevators are attached to member EDG. Aerodynamic pressures on the elevators exert a clockwise couple of 120 *in lb*. Cable BG is slack and its tension can be neglected. Determine the force F and the reactions at pin support A.

Solution: Begin at the elevator. The moment arms at E and G are 6 *in*. The angle of the cable EC with the horizontal is $\alpha = \tan^{-1}\dfrac{12}{119.5}$ $= 5.734^o$. Denote the horizontal and vertical components of the force on point E by F_X and F_Y. The sum of the moments about the pinned support on the member EG is $\sum M_{EG} = 2.5F_Y + 6F_X - 120 = 0$. This is the tension in the cable EC. Noting that $F_X = T_{EC}\cos\alpha$, and $F_Y = T_{EC}\sin\alpha$,

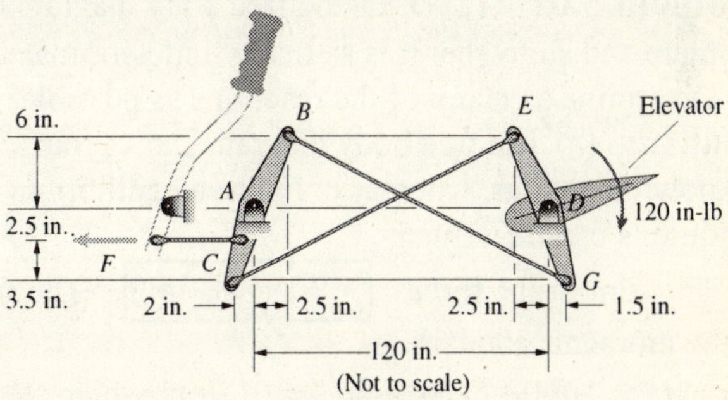

then $T_{EC} = \dfrac{120}{2.5\sin\alpha + 6\cos\alpha}$. The sum of the moments about the pinned support BC is

$\sum M_{BC} = -2T_{EC}\sin\alpha + 6T_{EC}\cos\alpha - 2.5F = 0$. Substituting: $F = \left(\dfrac{120}{2.5}\right)\left(\dfrac{6\cos\alpha - 2\sin\alpha}{6\cos\alpha + 2.5\sin\alpha}\right)$

$\boxed{= (48)(0.9277) = 44.53\ lb}$ The sum of the forces about the pinned joint A:

$\Sigma F_x = A_x - \boldsymbol{F} + T_{EC}\cos\alpha = 0$ from which $\boxed{A_x = 25.33\ lb}$, $\Sigma F_y = A_y + T_{EC}\sin\alpha = 0$ from which $\boxed{A_y = -1.93\ lb}$

==============================<>==============================

Problem 5.59 (a) Draw the free body diagram of the beam and show that it is statically indeterminate. (b) Determine as many of the reactions as possible.

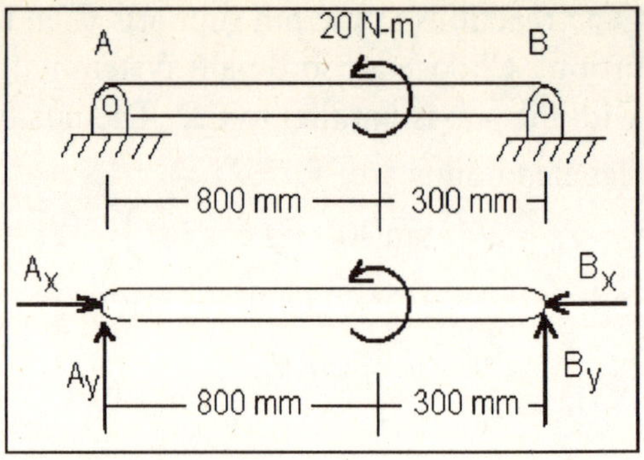

Solution: (a) The free body diagram shows that there are four unknowns, whereas only three equilibrium equations can be written. (b) The sum of moments about A is $\sum M_A = M + 1.1 B_Y = 0$, from which $\boxed{B_Y = -\frac{20}{1.1} = -18.18\ N}$. The sum of forces in the vertical direction is $\sum F_Y = A_Y + B_Y = 0$, from which $\boxed{A_Y = -B_Y = 18.18\ N}$. The sum of forces in the horizontal direction is $\sum F_X = A_X + B_X = 0$, from which the values of A_X and B_X are indeterminate.

Problem 5.60 Consider the beam in Problem 5.59. Choose supports at A and B such that it is not statically indeterminate. Determine the reactions at the supports.

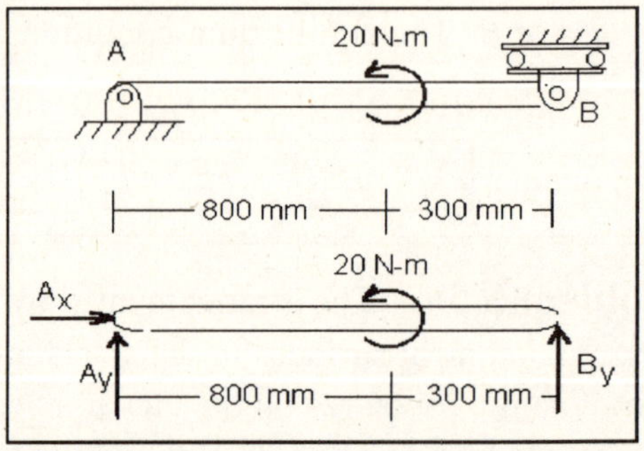

Solution: One possibility is shown: the pinned support at B is replaced by a roller support. The equilibrium conditions are:

$\boxed{\sum F_X = A_X = 0}$ The sum of moments about A is $\Sigma M_A = M + 1.1 B_Y = 0$, from which

$\boxed{B_Y = -\frac{20}{1.1} = -18.18\ N}$. The sum of forces in the vertical direction is $\sum F_Y = A_Y + B_Y = 0$, from which $\boxed{A_Y = -B_Y = 18.18\ N}$.

Problem 5.61 (a) Draw the free body diagram of the plate and show that it is statically indeterminate. (b) Determine as many of the reactions as possible.

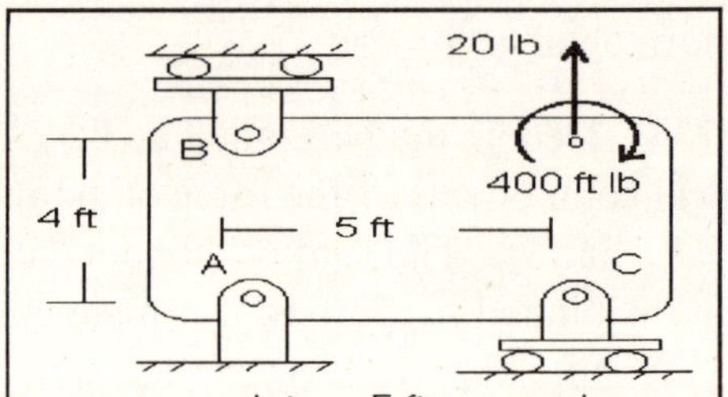

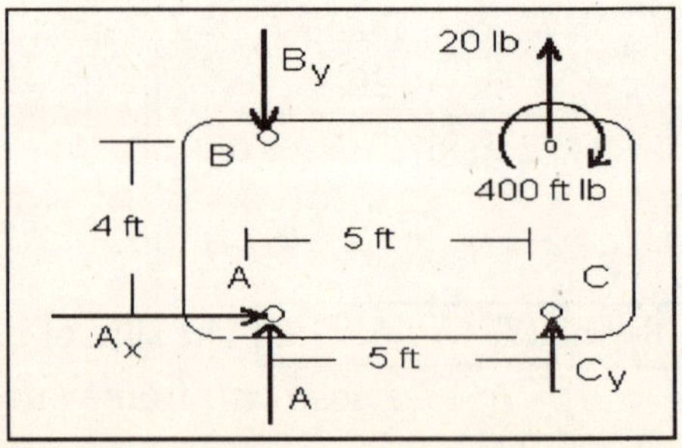

Solution: (a) The free body diagram shows four unknown reactions, whereas only three equilibrium conditions can be written

(b) The sum of the forces: $\boxed{\sum F_X = A_X = 0}$. . The sum of the moments about A is $\sum M_A = -400 + 5(20) + 5C_Y = 0$, from which $\boxed{C_Y = 60\ lb}$. The sum of the forces in the vertical direction is $\Sigma F_Y = -B_Y + A_Y + 20 = 0$, from which the forces A_Y and B_Y are indeterminate.

Problem 5.62 Consider the plate in Problem 5.61. Choose supports at A, B, and C so that is not statically indeterminate. Determine the reactions at the supports.

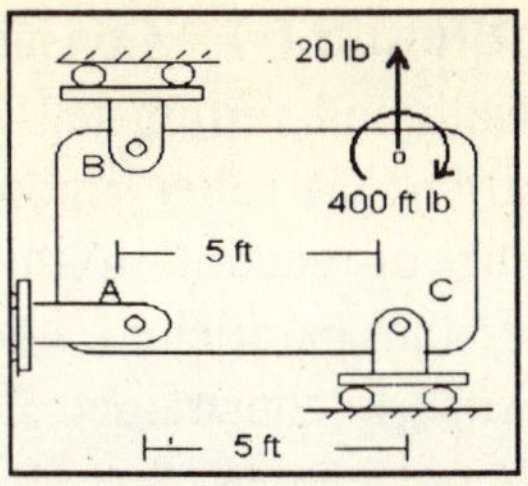

Solution: With three supports, each can have no more than one reaction. One solution is to replace the pin support at A with an x-direction roller support. The sum of forces: $\boxed{\sum F_X = A_X = 0}$

$\sum F_Y = C_Y - B_Y + 20 = 0$. The sum of the moments about A is

$\sum M_A = 5C_Y + 5F - M = 0$, from which $\boxed{C_Y = \frac{M}{5} - F = 60 \ lb}$, and

$\boxed{B_Y = 80 \ lb}$.

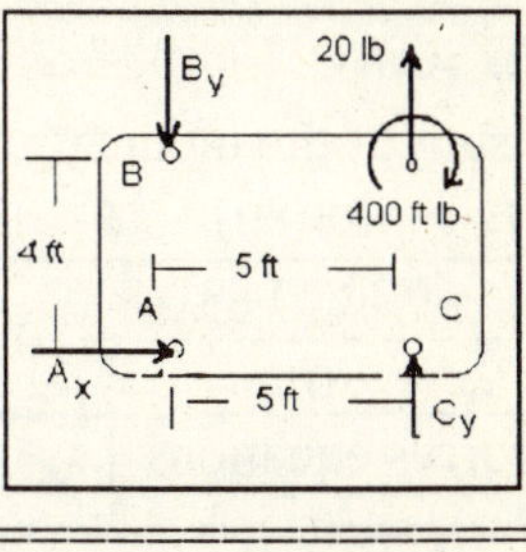

Problem 5.63 Draw the free body diagram of the L-shaped bar and show that it is statically indeterminate. Determine as many of the reactions as possible.

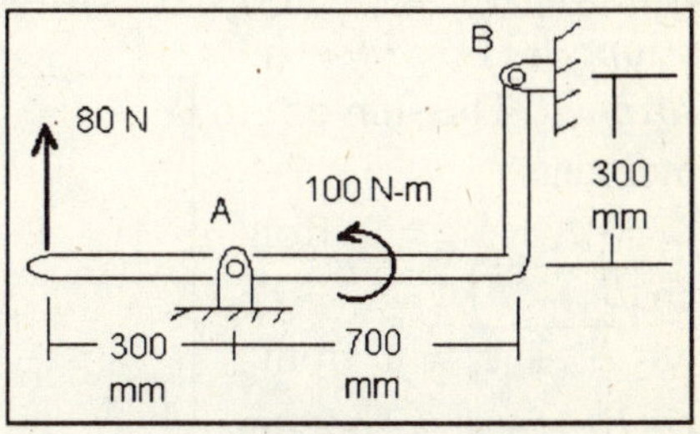

Solution: The free body diagram shows that there are four reactions, hence the system is statically indeterminate. The sum of the forces: $\sum F_X = (A_X + B_X) = 0$, and $\Sigma F_Y = A_Y + B_Y + F = 0$. A strategy for solving some statically indeterminate problems is to select a coordinate system such that the indeterminate reactions vanish from the sum of the moment equations. The choice here is to locate the x-axis on a line passing through both A and B, with the origin at A. Denote the reactions at A and B by A_N, A_P, B_N, and B_P, where the subscripts indicate the reactions are normal to and parallel to the new x-axis. Denote $F = 80 \ N$, $M = 100 \ N\text{-}m$

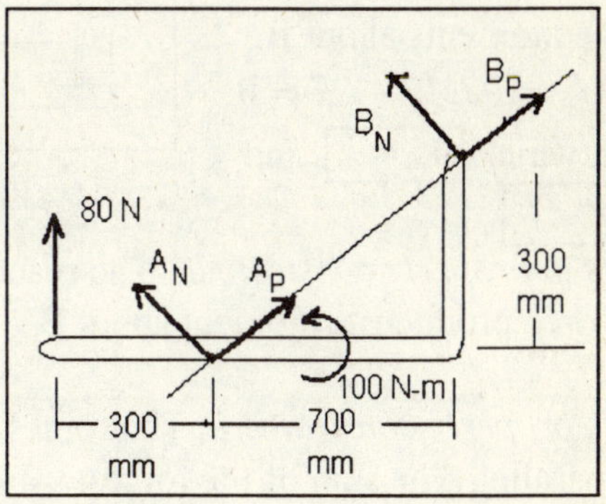

The length from A to B is $L = \sqrt{0.3^2 + 0.7^2} = 0.76157 \ m$.
The angle between the new axis and the horizontal is

$\theta = \tan^{-1}\left(\frac{0.3}{0.7}\right) = 23.2°$.

The moment about the point A is

$M_A = LB_N - 0.3F + M = 0$, from which

$\boxed{B_N = \frac{-M + 0.3F}{L} = \frac{-76}{0.76157} = -99.79 \ N}$, from which

The sum of the forces normal to the new axis is

$\sum F_N = A_N + B_N + F\cos\theta = 0$, from which $\boxed{A_N = -B_N - F\cos\theta = 26.26 \ lb}$

The reactions parallel to the new axis are i\r ̧determinate.

Problem 5.64

Consider the bar in Problem 5.63. Choose supports at A and B so that it is no statically indeterminate. Determine the reactions at the supports.

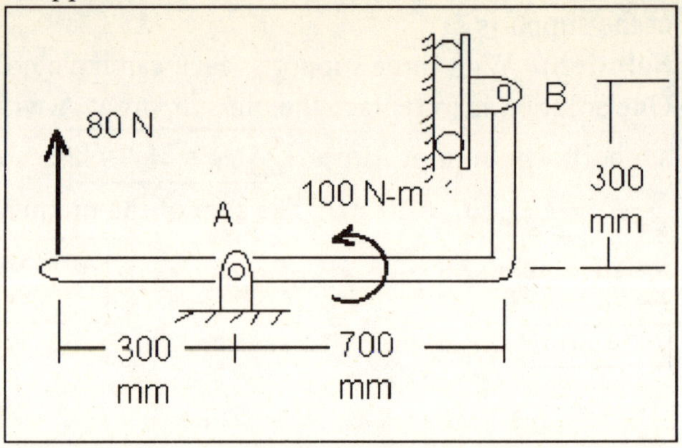

Solution: A roller support at B will reduce the number of reactions to three. The support reactions are neither parallel nor concurrent, so the bar will be properly constrained. The sum of the forces:

$\sum F_X = A_X + B_X = 0$.

$\sum F_Y = A_Y + F = 0$, from which $\boxed{A_Y = -80\ N}$

The sum of the moments about A:

$\Sigma M_A = -0.3(80) + M - 0.7B_X = 0$, from which

$\boxed{B_X = \dfrac{-0.3(80)+100}{0.7} = 108.57\ N}$. Substituting into the forces equations: $\boxed{A_X = 108.57\ N}$.

Problem 5.65

State whether each of the L-shaped bars shown is properly or improperly constrained. If a bar is properly constrained, determine the reactions at its supports.

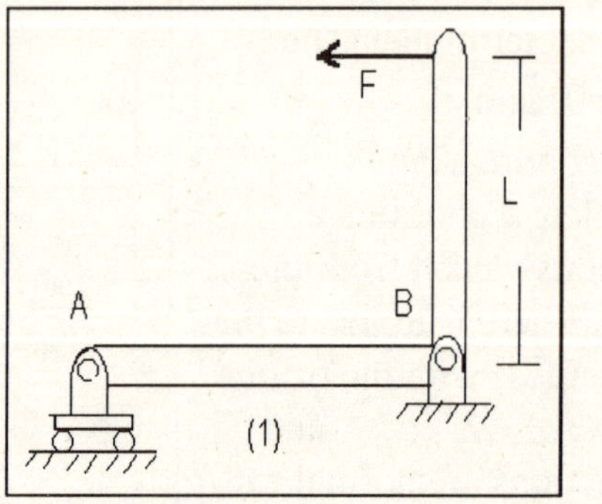

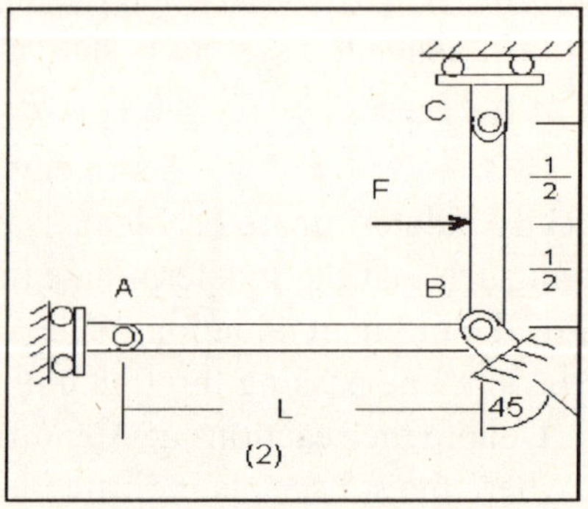

Solution: (1) *is properly constrained.* The sum of the moments:

$\Sigma F_X = -F + B_X = 0$, from which $\boxed{B_X = F}$.

$\Sigma F_Y = B_Y + A_y = 0$, from which $B_y = -A_y$. The sum of the moments about B:

$\sum M_B = -LA_Y + LF = 0$, from which $\boxed{A_Y = F}$, and $\boxed{B_{y\backslash} = -F}$

(2) *is improperly constrained.* The reactions intersect at B, while the force produces a moment about B.

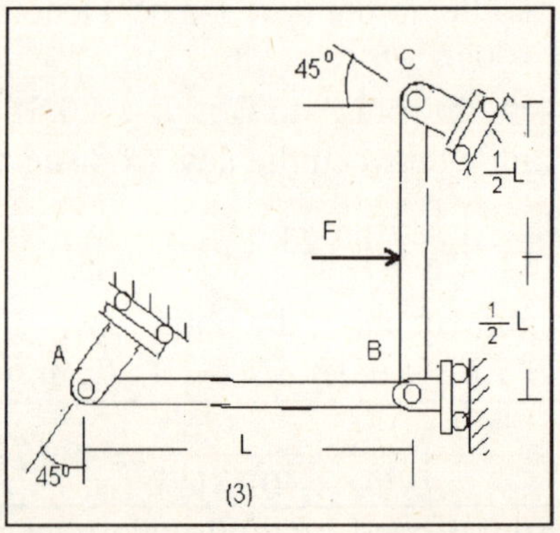

(3) *is properly constrained.* The forces are neither concurrent nor parallel. The sum of the forces:

$\Sigma F_X = -C\cos 45^o - B - A\cos 45^o + F = 0$.

$\sum F_Y = C\sin 45^o - A\sin 45^o = 0$ from which $A = C$

The sum of the moments about A:

$\sum M_A = -\dfrac{1}{2}LF + LC\cos 45^o + LC\sin 45^o = 0$, from which

$\boxed{C = \dfrac{F}{2\sqrt{2}}}$. Substituting and combining: $\boxed{A = \dfrac{F}{2\sqrt{2}}}$, $\boxed{B = \dfrac{F}{2}}$

Problem 5.66 State whether each of the L-shaped bars shown is properly or improperly constrained. If a bar is properly constrained, determine the reactions at its supports.

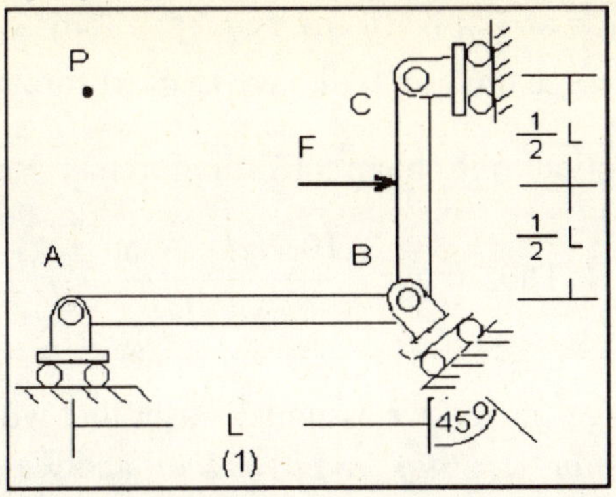

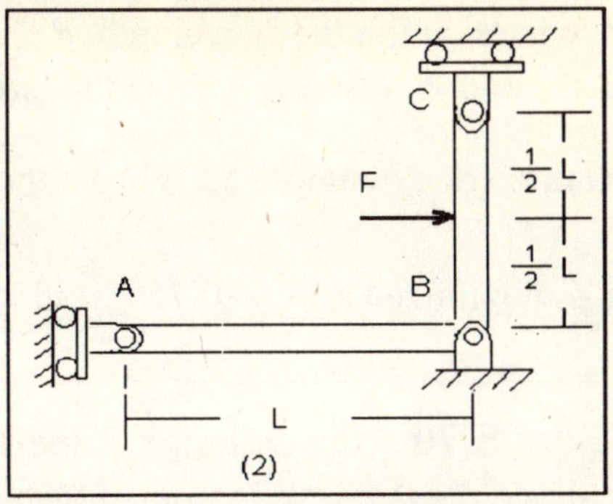

Solution: (1) *is improperly constrained.* The reactions intersect at a point P, and the force exerts a moment about that point.

(2) *is improperly constrained.* The reactions intersect at a point P and the force exerts a moment about that point.

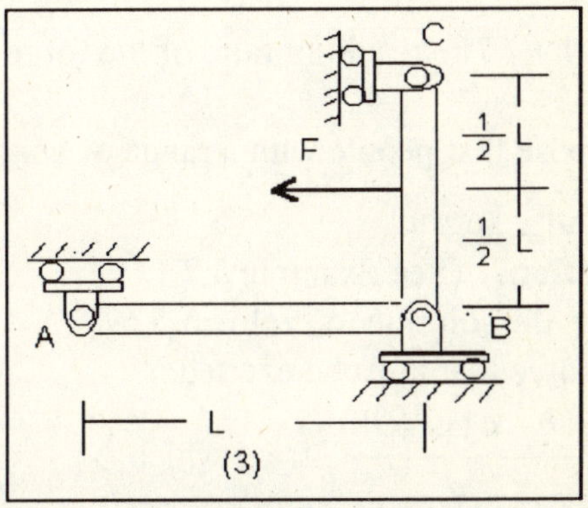

(3) *is properly constrained.* The sum of the forces:

$\sum F_X = C - F = 0$, from which $\boxed{C = F}$.

$\sum F_Y = -A + B = 0$, from which $A = B$. The sum of the moments about B: $LA + \frac{L}{2}F - LC = 0$, from which $\boxed{A = \frac{1}{2}F}$

$\boxed{B = \frac{1}{2}F}$

Problem 5.67 Suppose that $\alpha = 40^\circ$, $d = 1\ m$, $a = 200\ mm$, $b = 500\ mm$, $R = 75\ mm$, and the mass of the luggage is 40 *kg*. Determine *F* and *N*.

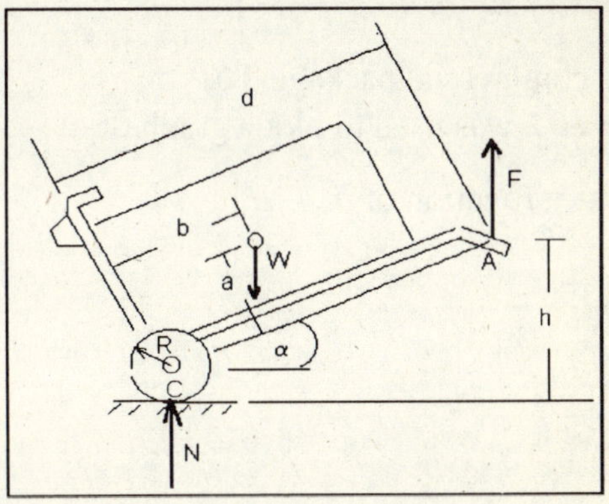

Solution: (See Example 5.7.)

The sum of the moments about the center of the wheel:

$\Sigma M_C = dF\cos\alpha + aW\sin\alpha - bW\cos\alpha = 0$, from which

$$F = \frac{(b - a\tan\alpha)W}{d} \boxed{= 130.35\ N}.$$

The sum of the forces:

$\sum F_Y = N - W + F = 0$, from which

$\boxed{N = 262.1\ N}$

Problem 5.68 Suppose that $\alpha = 35^\circ$, $d = 46\ in$, $a = 10\ in$, $b = 14\ in$, $R = 3\ in$, and you don't want the user to exert a force larger than 20 *lb*. What is the largest luggage weight that can be placed on the carrier?

Solution: (See Example 5.7.) From the solution to Problem 5.67, the force is $F = \frac{(b - a\tan\alpha)W}{d}$.

Solve for *W*: $W = \frac{Fd}{(b - a\tan\alpha)}$. For $F = 20\ lb$, $\boxed{W = 131.47 = 131.5\ lb}$

Problem 5.69 One of the difficulties in making design decisions is that you don't know how the user will place luggage on the carrier. Suppose that you assume that the point where the weight lies may be anywhere within the "envelope" $R \le a \le 0.75c$ and $0 \le b \le 0.75d$. If $\alpha = 30^\circ$, $c = 14\ in$, $d = 48\ in$, $R = 3\ in$, and $W = 80\ lb$, what is the largest force the user will have to exert for any luggage placement?

Solution: (See Example 5.7.) From the solution to Problem 5.67, the force is $F = \dfrac{(b - a\tan\alpha)W}{d}$. The force is maximized as $b \to 0.75d$, and $a \to R$. Thus $F_{MAX} = \dfrac{(0.75d - R\tan\alpha)}{d}$, $\boxed{F = 57.11\ lb}$

Problem 5.70 In our design of the luggage carrier, we assumed a user that will hold the handle at 36 *in* above the floor. We assumed that $R = 3\ in$, $a = 6\ in$, and $b = 12\ in$, and we chose the dimension $d = 4\ ft$. The resulting ratio of the force the user must exert to the weight of the luggage is $\dfrac{F}{W} = 0.132$.

Suppose that people with a range of heights use the carrier. Obtain a graph of $\dfrac{F}{W}$ as a function of *h* for $24 \le h \le 36\ in$.

Solution: (See Example 5.7.) From the solution to Problem 5.67, the force that must be exerted is $F = \dfrac{(b - a\tan\alpha)W}{d}$, from which $\dfrac{F}{W} = \dfrac{(b - a\tan\alpha)}{d}$. The angle α is given by $\alpha = \sin^{-1}\left(\dfrac{h-R}{d}\right)$.

The commercial package **TK Solver 2** was used to plot a graph of $\dfrac{F}{W}$ as a function of *h*.

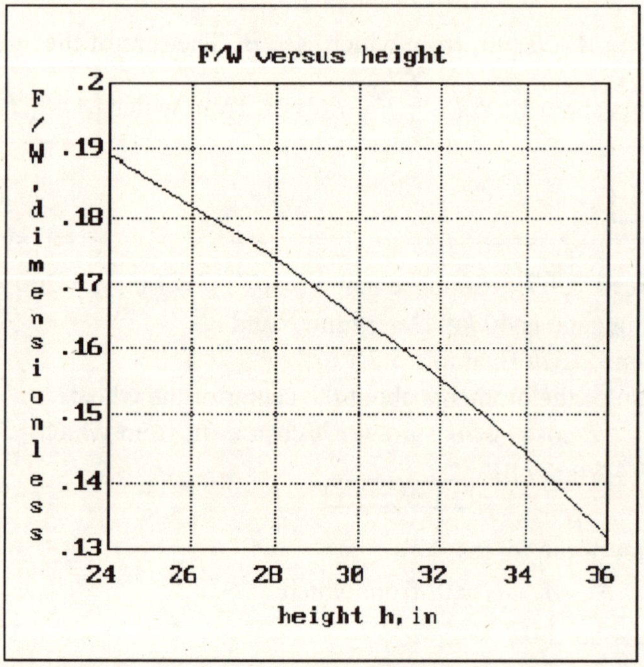

Problem 5.71 The bar ABC has a built-in support at A. It is loaded by forces $\vec{F}_B = 2\vec{i} + 6\vec{j} + 3\vec{k}\ (kN)$, $\vec{F}_C = \vec{i} - 2\vec{j} + 2\vec{k}\ (kN)$, and the couple: $\vec{M}_C = 2\vec{i} + \vec{j} - 2\vec{k}\ (kN\text{-}m)$.
(a) Draw the free body diagram of the bar. (b) Determine the reactions at A.

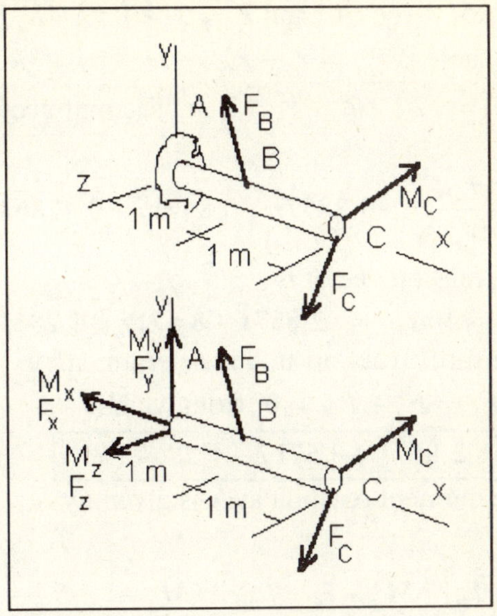

Solution: *Strategy*: Isolate the bar from the built-in support and show the loads and the reactions at the support. Then use Eqs (5.16)-(5.21) to determine the reactions. (Remember that you can sum the moments about any point.)

(a) The free body diagram shows the reactions $\vec{M}_X$, $\vec{M}_Y$, $\vec{M}_Z$, and $\vec{F}_X$, $\vec{F}_Y$, $\vec{F}_Z$ at the built in support.

(b) The sum of the forces:

$$\sum \vec{F}_X = (F_X + 2 + 1)\vec{i} = 0,$$

$$\sum \vec{F}_Y = (F_Y + 6 - 2)\vec{j} = 0$$

$\sum \vec{F}_Z = (F_Z + 3 + 2)\vec{k} = 0$, from which the force reaction at the built-in support is $\boxed{\vec{F}_A = F_X\vec{i} + F_Y\vec{j} + F_Z\vec{k} = -3\vec{i} - 4\vec{j} - 5\vec{k}\ (kN)}$.

The sum of the moments about A is

$$\sum \vec{M}_R = \vec{M}_A + \vec{M}_C + \vec{r}_B \times \vec{F}_B + \vec{r}_C \times \vec{F}_C = 0$$

$$\sum \vec{M}_R = \left(\vec{M}_A + \vec{M} + \begin{vmatrix} \vec{i} & \vec{j} & \vec{k} \\ 1 & 0 & 0 \\ 2 & 6 & 3 \end{vmatrix} + \begin{vmatrix} \vec{i} & \vec{j} & \vec{k} \\ 2 & 0 & 0 \\ 1 & -2 & 2 \end{vmatrix} \right) = 0.$$

$= (M_{AX} + 2)\vec{i} + (M_{AY} + 1 - 3 - 4)\vec{j} + (M_{AZ} - 2 + 6 - 4)\vec{k} = 0$, from which the moment reaction at the built-in support is $\boxed{\vec{M}_R = M_X^A\vec{i} + M_Y^A\vec{j} + M_Z^A\vec{k} = -2\vec{i} + 6\vec{j} + 0\vec{k}\ (kN\text{-}m)}$

Problem 5.72 The robotic manipulator is stationary. The weights of the arms AB and BC act at their midpoints. The direction cosines of the centerline of arm AB are $\cos\theta_X = 0.500$, $\cos\theta_Y = 0.866$, and $\cos\theta_Z = 0$. The direction cosines of arm BC are $\cos\theta_X = 0.707$, $\cos\theta_Y = 0.619$, $\cos\theta_Z = -0.342$. The support at A behaves like a built-in support. What are the reactions at A?

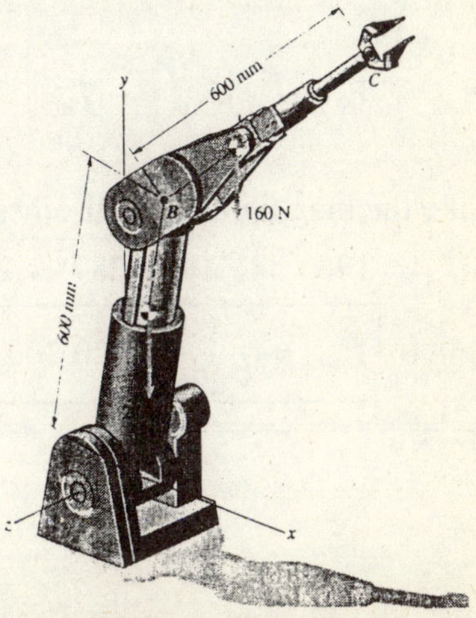

Solution: The position vectors of the arm centers are:

$$\vec{r}_{AB} = 0.3(0.5\vec{i} + 0.866\vec{j}),$$

$$\vec{r}_{AC} = 2\vec{r}_{AB} + 0.3(0.707\vec{i} + 0.619\vec{j} - 0.342\vec{k})$$

$= 0.5121\vec{i} + 0.7053\vec{j} - 0.1026\vec{k}$. The sum of the forces

$\sum \vec{F}_R = \vec{F}_A - 160\vec{j} - 200\vec{j} = 0$, from which the force reaction at A is $\boxed{\vec{F}_A = 0\vec{i} + 360\vec{j} + 0\vec{k}\ (N)}$. The sum of the moments about A

$$\Sigma \vec{M}_A = \vec{M}_A + \vec{r}_{AC} \times \vec{F}_{BC} + \vec{r}_{AB} \times \vec{F}_{AB} = 0$$

$$\Sigma \vec{M}_A = \vec{M}_A + \begin{vmatrix} \vec{i} & \vec{j} & \vec{k} \\ 0.15 & 0.2598 & 0 \\ 0 & -200 & 0 \end{vmatrix} + \begin{vmatrix} \vec{i} & \vec{j} & \vec{k} \\ 0.5121 & 0.7053 & -0.1026 \\ 0 & -160 & 0 \end{vmatrix} = 0$$

From which $\boxed{\vec{M}_A = +16.42\vec{i} + 0\vec{j} + 111.94\vec{k}\ (kN\text{-}m)}$

Problem 5.73 The bar AB has a built-in support at A. The tension in the cable is 10 *kN*. (a) Draw the free body diagram of the bar AB. (b) Determine the reactions at A.

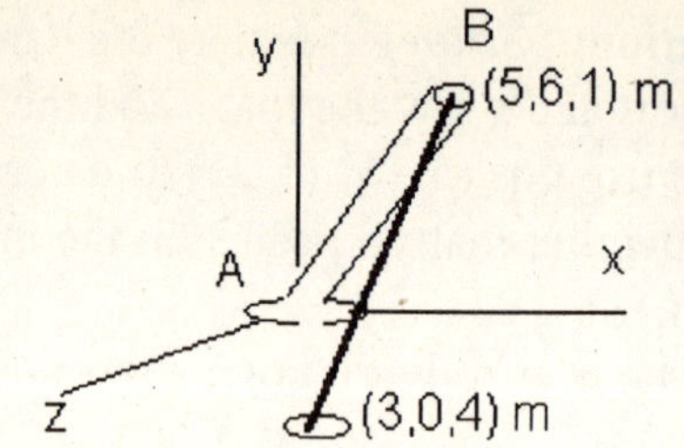

Solution: The position vectors of the ends of the cable are $\vec{r}_B = 5\vec{i} + 6\vec{j} + \vec{k}$, and $\vec{r}_C = 3\vec{i} + 0\vec{j} + 4\vec{k}$. The vector parallel to the cable is $\vec{r}_{BC} = \vec{r}_C - \vec{r}_B = -2\vec{i} - 6\vec{j} + 3\vec{k}$,

$|\vec{r}_{BC}| = \sqrt{2^2 + 6^2 + 3^2} = 7\ m$. The unit vector parallel to the cable is

$$\vec{e} = \frac{\vec{r}_{BC}}{|\vec{r}_{BC}|} = -0.2857\vec{i} - 0.8571\vec{j} + 0.4286\vec{k}.$$

The tension vector is

$\vec{T}_{BC} = 10\vec{e}_{BC} = -2.857\vec{i} - 8.571\vec{j} + 4.286\vec{k}$

The force reaction at A is determined by

$\sum \vec{F}_R = \vec{F}_A + \vec{T}_{BC} = 0$, from which

$$\boxed{\vec{F}_A = 2.857\vec{i} + 8.571\vec{j} - 4.286\vec{k}\ (kN)}$$

The moment reaction at A is given by

$$\sum \vec{M}_R = \vec{M}_A + \vec{r}_B \times \vec{T}_{BC} = \vec{M}_A + \begin{vmatrix} \vec{i} & \vec{j} & \vec{k} \\ 5 & 6 & 1 \\ -2.857 & -8.571 & 4.286 \end{vmatrix} = 0$$

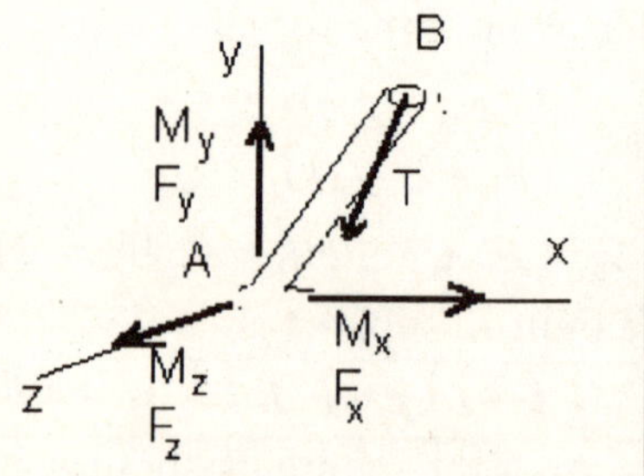

From which $\boxed{\vec{M}_A = -34.287\vec{i} + 24.287\vec{j} + 25.713\vec{k}\ (kN\text{-}m)}$

Problem 5.74 Consider the bar in Problem 5.73. The magnitude of the couple exerted on the bar by the built-in support is 100 *kN-m*. What is the tension in the cable BC?

Solution: From the solution to Problem 5.73, the moment reaction at A is the solution to the moment equilibrium equation: $\sum \vec{M}_R = \vec{M}_A + \vec{r}_B \times \vec{T}_{BC} = 0.$, from which $\vec{M}_A = -\vec{r}_B \times \vec{T}_{BC}$.

Noting that $\vec{T}_{BC} = |\vec{T}_{BC}|\vec{e}_{BC}$, then using the unit vector developed in the solution to Problem 5.73,

$\vec{e} = \frac{\vec{r}_{BC}}{|\vec{r}_{BC}|} = -0.2857\vec{i} - 0.8571\vec{j} + 0.4286\vec{k}$ and the position vector of the end of the bar:

$\vec{r}_B = 5\vec{i} + 6\vec{j} + \vec{k}$

$$\vec{M}_A = -\vec{r}_B \times |\vec{T}_{BC}|\vec{e}_{BC} = -|\vec{T}_{BC}| \begin{vmatrix} \vec{i} & \vec{j} & \vec{k} \\ 5 & 6 & 1 \\ -0.2857 & -0.8571 & 0.4286 \end{vmatrix} = -|\vec{T}_{BC}|(-3.4287\vec{i} + 2.4287\vec{j} + 2.5713\vec{k})$$

Take the magnitude of both sides

$$|\vec{M}_A| = 100 = |\vec{T}_{BC}|\sqrt{3.4287^2 + 2.4287^2 + 2.5713^2} = |\vec{T}_{BC}|(4.2961).$$

Solve: $\boxed{|\vec{T}_{BC}| = \frac{100}{4.2961} = 20.300\ kN}$

Problem 5.75 The boom AB is subjected to a force $\vec{F} = -1000\vec{j}\ (lb)$ at B and is supported by a ball and socket support at A and the cables CD and CE. (a) Draw the free body diagram of the boom. (b) Determine the tension in the cables and the reactions at A.

Solution: The ball and socket supports three force reactions. The strategy is to develop the unit vectors parallel to the cables so that the tension vectors can be used in the equilibrium conditions.

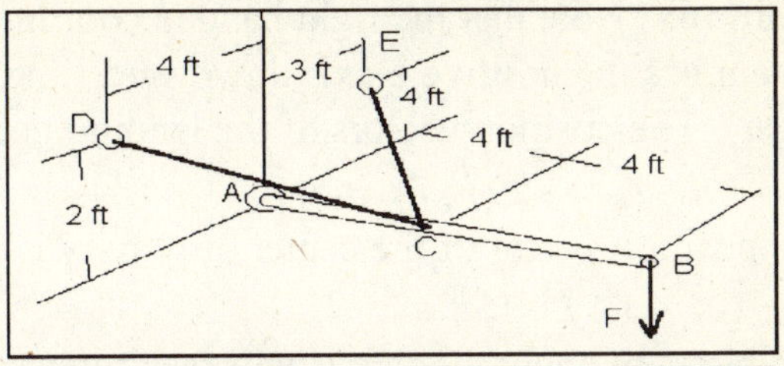

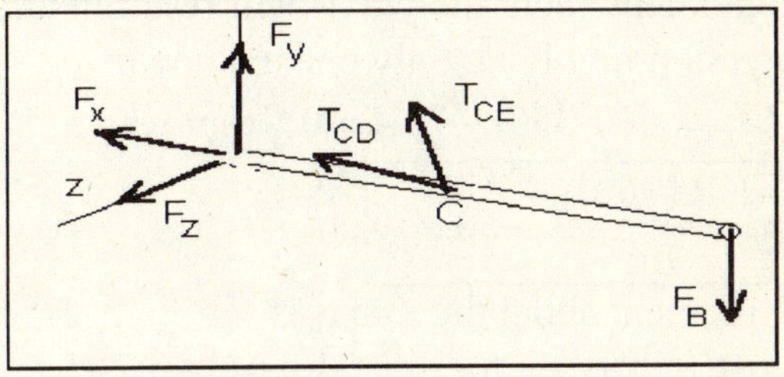

The position vectors at the ends of the cables are:
$\vec{r}_C = 4\vec{i}$, $\vec{r}_E = 0\vec{i} + 4\vec{j} - 3\vec{k}$, $\vec{r}_D = 0\vec{i} + 2\vec{j} + 4\vec{k}$. The vector parallel to cable CE is

$\vec{r}_{CE} = \vec{r}_E - \vec{r}_C = -4\vec{i} + 4\vec{j} - 3\vec{k}$,

$|\vec{r}_{CE}| = \sqrt{4^2 + 3^2 + 4^2} = 6.4031\ ft$

The vector parallel to the cable CD is

$\vec{r}_{CD} = \vec{r}_D - \vec{r}_C = -4\vec{i} + 2\vec{j} + 4\vec{k}$

$|\vec{r}_{CD}| = \sqrt{4^2 + 2^2 + 4^2} = 6\ ft$. The unit vectors are:

$\vec{e}_{CE} = -0.6247\vec{i} + 0.6247\vec{j} - 0.4685\vec{k}$

$\vec{e}_{CD} = -0.6667\vec{i} + 0.3333\vec{j} + 0.6667\vec{k}$

The tensions are $\vec{T}_{CE} = |\vec{T}_{CE}|\vec{e}_{CE}$ and $\vec{T}_{CD} = |\vec{T}_{CD}|\vec{e}_{CD}$.

The sum of the moments about A is

$$\sum \vec{M}_A = \vec{r}_C \times \vec{T}_{CE} + \vec{r}_C \times \vec{T}_{CD} + \vec{r}_B \times \vec{F}_B = 0$$

$$\sum \vec{M}_A = |\vec{T}_{CE}| \begin{vmatrix} \vec{i} & \vec{j} & \vec{k} \\ 4 & 0 & 0 \\ -0.6247 & 0.6247 & -0.4685 \end{vmatrix} + |\vec{T}_{CD}| \begin{vmatrix} \vec{i} & \vec{j} & \vec{k} \\ 4 & 0 & 0 \\ -0.6667 & 0.3333 & 0.6667 \end{vmatrix} + \begin{vmatrix} \vec{i} & \vec{j} & \vec{k} \\ 8 & 0 & 0 \\ 0 & -1000 & 0 \end{vmatrix} = 0$$

$\Sigma \vec{M}_A = |\vec{T}_{CE}|(0\vec{i} + 1.874\vec{j} + 2.499\vec{k}) + |\vec{T}_{CD}|(0\vec{i} - 2.667\vec{j} + 1.333\vec{k}) - 8000\vec{k} = 0$. Collecting terms, two simultaneous equations are obtained from the moment equilibrium condition:

$1.874|\vec{T}_{CE}| - 2.667|\vec{T}_{CD}| = 0$

$2.499|\vec{T}_{CE}| + 1.333|\vec{T}_{CD}| = 8000$. Solve: $\boxed{|\vec{T}_{CE}| = 2328.53\ lb}$ $\boxed{|\vec{T}_{CD}| = 1636.2\ lb}$ The sums of the forces:

$\sum \vec{F}_X = (A_X - 0.6247|\vec{T}_{CE}| - 0.6667|\vec{T}_{CD}|)\vec{i} = 0$

$\sum \vec{F}_Y = (A_Y + 0.6247|\vec{T}_{CE}| + 0.3333|\vec{T}_{CD}| - 1000)\vec{j} = 0$

$\sum \vec{F}_Z = (A_Z - 0.4685|\vec{T}_{CE}| + 0.6667|\vec{T}_{CD}|)\vec{k} = 0$. Substitute values for the tension, and solve:

$\boxed{A_X = 2545.5\ lb}$, $\boxed{A_Y = -1000\ lb}$, $\boxed{A_Z = 0}$

Problem 5.76 Consider the crane shown in Problem 5.75. The force $\vec{F}$ points downward (parallel to the y-axis). The cables CD and CE will each safely support a tension of 5000 *lb*. Based on this criterion, what is the largest safe magnitude of $\vec{F}$?

Solution: The problem is linear, thus the ratio of 5000 *lb* to the largest tension in the solution to Problem 5.75 is equal to the ratio of the forces: $\frac{|\vec{T}_{CD}|_{MAX}}{|\vec{T}_{CD}|} = \frac{5000}{2328.5} = \frac{|\vec{F}|_{MAX}}{|\vec{F}|} = \frac{|\vec{F}|_{MAX}}{1000}$ from which

$$\boxed{|\vec{F}|_{MAX} = \frac{5000}{2328.5}(1000) = 2147.3\ lb}$$

Problem 5.77 The airplane is at rest on the ground ($z = 0$ is at ground level). Its landing gear are at points A, B, and C. The coordinates of the point G where the 350 *kip* weight of the plane acts are $(10, 1.5, -15)$ *ft*.. What are the magnitudes of the normal reactions exerted by the ground on the plane's landing gear?

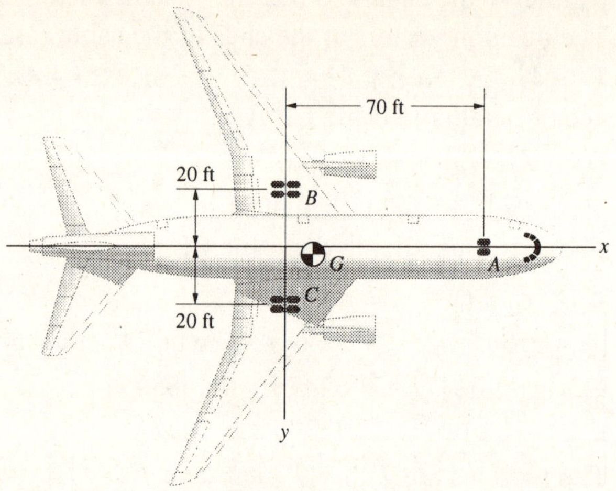

Solution: Note that the right handed coordinate system has the positive z-axis downward (into the paper). The position vectors of the landing gear are $\vec{r}_A = 70\vec{i}$, $\vec{r}_B = -20\vec{j}$, $\vec{r}_C = 20\vec{j}$.

The position vector of the center of gravity is $\vec{r}_G = 10\vec{i} + 1.5\vec{j} - 15\vec{k}$

The moment about the y-axis will result in an expression involving only one unknown:

$\sum M_Y = -10(350) - 70A = 0$, from which

$$\boxed{A = \frac{-10(350)}{70} = -50 \ kip}$$

The moment about the x-axis is

$\sum M_X = -20B + 20C + (350)(1.5) = 0$

The sum of the forces in the z-direction is

$\sum F_Z = B + C + 350 - 50 = 0$. The result is two simultaneous equations in the unknowns:

$-B + C = -26.25$

$B + C = -300$ These can be solved by (a) adding the two equations, (b) subtracting the two equations: $\boxed{C = \frac{-326.5}{2} = -163.1 \ kip}$ $\boxed{B = \frac{-273.75}{2} = -136.875 \ kip}$. (The negative sign indicates that the forces act in the negative z-direction, i.e., upward out of the paper.)

Problem 5.78 An engineer designs a system of pulleys to pull his model trains up out of the way when they aren't in use. What are the tensions in the three ropes when the system is in equilibrium?

Solution: Take the origin as the point C on the platform. The position vectors are $\vec{r}_A = -2\vec{i} + 8\vec{k}$, $\vec{r}_B = -4\vec{i}$, $\vec{r}_G = -1.6\vec{i} + 3.5\vec{k}$. The sum of the moments about C is

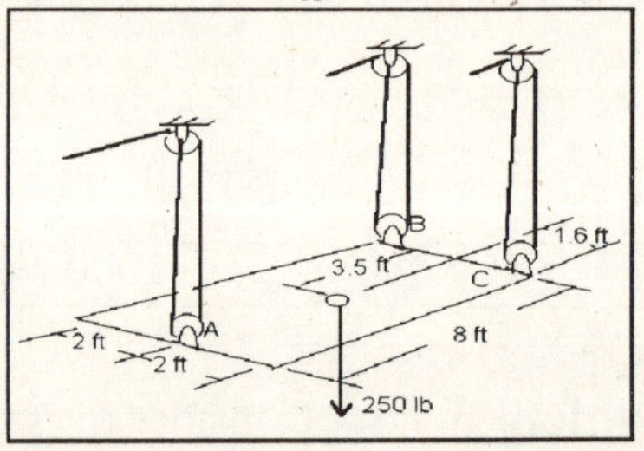

$$\sum \vec{M}_C = \begin{vmatrix} \vec{i} & \vec{j} & \vec{k} \\ -2 & 0 & 8 \\ 0 & A & 0 \end{vmatrix} + \begin{vmatrix} \vec{i} & \vec{j} & \vec{k} \\ -4 & 0 & 0 \\ 0 & B & 0 \end{vmatrix} + \begin{vmatrix} \vec{i} & \vec{j} & \vec{k} \\ -1.6 & 0 & 3.5 \\ 0 & -250 & 0 \end{vmatrix} = 0$$

$\Sigma \vec{M}_C = -(8A\vec{i} + 2A\vec{k}) - 4B\vec{k} + (875\vec{i} + 400\vec{k}) = 0$. Collecting like terms: $(-8A + 875)\vec{i} = 0$, $(-2A - 2B + 400)\vec{k} = 0$, from which $A = \frac{875}{8} = 109.375 \ lb$, $B = \frac{200 - A}{2} = 45.3125 \ lb$. The value of the reaction at C is found from the sum of the vertical forces: $\sum \vec{F}_Y = (-250 + A + B + C)\vec{j} = 0$, from which $C = 95.3125 \ lb$. Two pulley ropes support each point, so the tension in each rope is one-half the reaction at the point supported:

$$\boxed{T_A = \frac{A}{2} = 54.67 \ lb}, \boxed{T_B = \frac{B}{2} = 22.65 \ lb}, \boxed{T_C = \frac{C}{2} = 47.66 \ lb}$$

Problem 5.79 The L-shaped bar is supported by a bearing at A and rests on a smooth horizontal surface at B. The vertical force $F = 800\ lb$ and the distance $b = 1.5\ ft$. Determine reactions at A and B.

Solution: In general, a bearing can exert two couples and three force reactions. The smooth surface will exert only one force reaction. The position vectors of the force and the points A and B relative to the origin are $\vec{r}_A = 2\vec{i}$, $\vec{r}_F = b\vec{k}$, $\vec{r}_B = 3\vec{k}$. The vector distances from A are $\vec{r}_{AF} = \vec{r}_F - \vec{r}_A = -2\vec{i} + b\vec{k}$, and $\vec{r}_{AB} = \vec{r}_B - \vec{r}_A = -2\vec{i} + 3\vec{k}$. The sum of the moments about A is

$$\sum \vec{M}_A = \vec{M}_A + \vec{r}_{AB} \times \vec{B} + \vec{r}_{AF} \times \vec{F} = 0$$

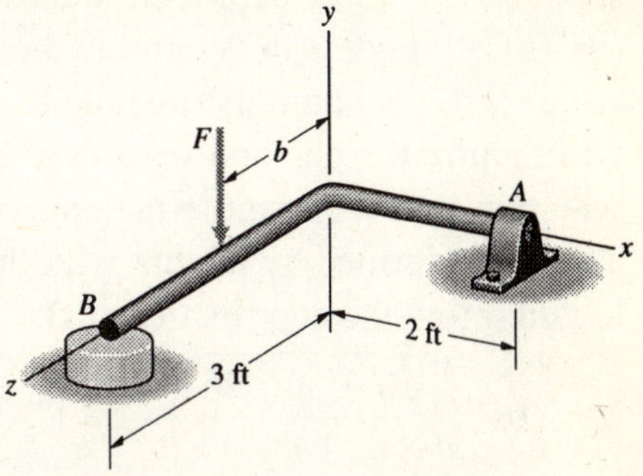

$$\sum \vec{M}_A = \vec{M}_A + \begin{vmatrix} \vec{i} & \vec{j} & \vec{k} \\ -2 & 0 & 3 \\ 0 & B & 0 \end{vmatrix} + \begin{vmatrix} \vec{i} & \vec{j} & \vec{k} \\ -2 & 0 & b \\ 0 & -F & 0 \end{vmatrix} = 0$$

$\sum \vec{M}_A = (M_X - 3B + Fb)\vec{i} + (M_Z - 2B + 2F)\vec{k} = 0$. From which $\boxed{M_Y^A = 0}$. The bearing at A cannot support a moment about the x-axis, hence $\boxed{M_X^A = 0}$. The result: $\boxed{B = \dfrac{bF}{3} = 400\ lb.}$, and $\boxed{M_Z^A = -1600 + 800 = -800\ ft\ lb}$. The sum of the forces: $\sum \vec{F}_X = A_X\vec{i} = 0$, from which $\boxed{A_X = 0}$. $\sum \vec{F}_Y = (B + A_Y - F)\vec{j} = 0$, from which $\boxed{A_Y = F - B = 400\ lb}$. Similarly $\sum \vec{F}_Z \boxed{= A_Z\vec{k} = 0}$.

Problem 5.80 Consider the system shown in Problem 5.79. The vertical force $F = 800\ lb$ and the distance $b = 1.5\ ft$. If you represent the reactions exerted at A and B by a single equivalent force, what is the force and where does its line of action intersect the centerline of the bar?

Solution: The equivalent force is equal to the applied force: $\vec{F}_{eq} = 800\vec{j}$. This force must produce the moment $\vec{M} = \vec{r}_A \times \vec{A} + \vec{r}_B \times \vec{B} + M_Z^A\vec{k}$. The sum of moment is

$$\sum \vec{M}_A = \begin{vmatrix} \vec{i} & \vec{j} & \vec{k} \\ x & 0 & z \\ 0 & 800 & 0 \end{vmatrix} = (-800z)\vec{i} + (800x)\vec{k} = \begin{vmatrix} \vec{i} & \vec{j} & \vec{k} \\ 2 & 0 & 0 \\ 0 & 400 & 0 \end{vmatrix} + \begin{vmatrix} \vec{i} & \vec{j} & \vec{k} \\ 0 & 0 & 3 \\ 0 & 400 & 0 \end{vmatrix} - 800\vec{k}.$$

$-(800z)\vec{i} + (800x)\vec{k} = 800\vec{k} - 1200\vec{i} - 800\vec{k}$, from which $x = 0$, and $\boxed{z = \dfrac{1200}{800} = 1.5\ ft}$

Problem 5.81 Consider the system in Problem 5.79. The vertical force $F = 800\ lb$. The bearing at A will safely support a force of 500 *lb* and a couple of 1000 *ft lb*. The surface at B will safely support a force of 500 *lb*. Based on these criteria, what is the allowable range of *b*?

Solution: From the solution to Problem 5.79 $\sum \vec{M}_A = (M_X - 3B + Fb)\vec{i} + (M_Z - 2B + 2F)\vec{k} = 0$, from which $b = \dfrac{3B}{F}$. Substitute the safe value for B : $b = \dfrac{1500}{800} = 1.875\ ft$. From the solution to Problem 5.79, $\sum \vec{F}_Y = (B + A_Y - F)\vec{j} = 0$, from which $A_Y = F - B = F - \dfrac{Fb}{3}$, or $b = \dfrac{3(F - A)}{F}$.

Substitute safe values for A: $b = 1.125\ ft$. Similarly, $M_z^A = 2(B - F) = 2(\dfrac{Fb}{3} - F)$, or

$b = \dfrac{3(M_Z^A + 2F)}{2F}$. Substitute the safe value of M_Z^A, $b = 4.875\ ft$. The limits are $\boxed{1.125 \le b \le 1.875\ ft}$

Problem 5.82 The 1.1 *m* bar is supported by a ball and socket support at A and the two smooth walls. The tension in the vertical cable CD is 1 *kN*. (a) Draw the free body diagram of the bar. (b) Determine the reactions at A and B.

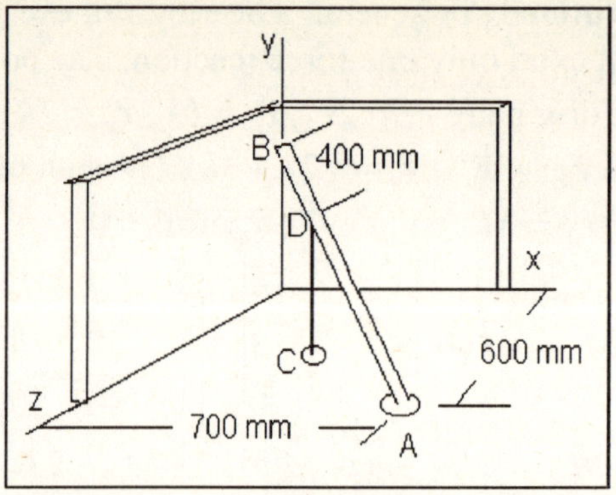

Solution: (a) The ball and socket cannot support a couple reaction, but can support a three force reaction. The smooth surface supports one-force normal to the surface. The cable supports one force parallel to the cable.

(b) The strategy is to determine the moments about A, which will contain only the unknown reaction at B. This will require the position vectors of B and D relative to A, which in turn will require the unit vector parallel to the rod. The angle formed by the bar with the horizontal is required to determine the coordinates of B:

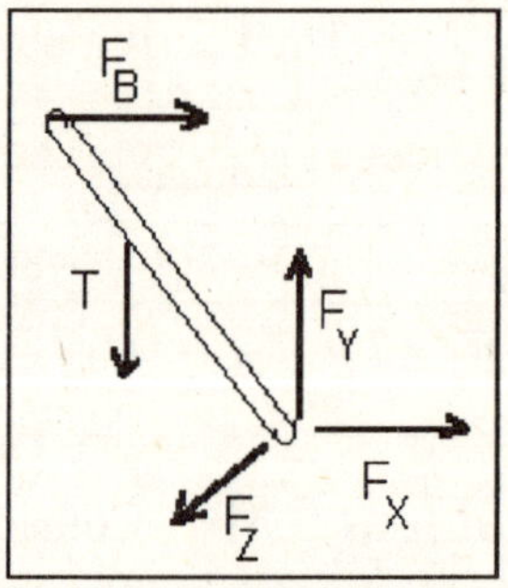

$\alpha = \cos^{-1}\left(\frac{\sqrt{0.6^2+0.7^2}}{1.1}\right) = 33.1°$. The coordinates of the points are:

$A(0.7, 0, 0.6)$, $B(0, 1.1(\sin 33.1°), 0) = (0, 0.6, 0)$, from which the vector parallel to the bar is $\vec{r}_{AB} = \vec{r}_B - \vec{r}_A = -0.7\vec{i} + 0.6\vec{j} - 0.6\vec{k}$ (m). The unit vector parallel to the bar is $\vec{e}_{AB} = \frac{\vec{r}_{AB}}{1.1} = -0.6364\vec{i} + 0.5455\vec{j} - 0.5455\vec{k}$.

The vector location of the point D relative to A is

$\vec{r}_{AD} = (1.1 - 0.4)\vec{e}_{AB} = 0.7\vec{e}_{AB} = -0.4455\vec{i} + 0.3819\vec{j} - 0.3819\vec{k}$.

The reaction at B is horizontal, with unknown x-component and z-components. The sum of the moments about A is

$$\sum \vec{M}_A = \vec{r}_{AB} \times \vec{B} + \vec{r}_{AD} \times \vec{D} = 0 = \begin{vmatrix} \vec{i} & \vec{j} & \vec{k} \\ -0.7 & 0.6 & -0.6 \\ B_X & 0 & B_Z \end{vmatrix} + \begin{vmatrix} \vec{i} & \vec{j} & \vec{k} \\ -0.4455 & 0.3819 & -0.3819 \\ 0 & -1 & 0 \end{vmatrix} = 0$$

Expand and collect like terms:

$$\sum \vec{M}_A = (0.6B_Z - 0.3819)\vec{i} - (0.6B_X - 0.7B_Z)\vec{j} + (-0.6B_X + 0.4455)\vec{k} = 0.$$

From which, $\boxed{B_Z = \frac{0.3819}{0.6} = 0.6365\ kN}$, $\boxed{B_X = \frac{0.4455}{0.6} = 0.7425\ kN}$.

The reactions at A are determined from the sums of the forces:

$\sum \vec{F}_X = (B_X + A_X)\vec{i} = 0$, from which $\boxed{A_X = -0.7425\ kN}$.

$\sum \vec{F}_Y = (A_Y - 1)\vec{j} = 0$, from which $\boxed{A_Y = 1\ kN}$.

$\sum \vec{F}_Z = (B_Z + A_Z)\vec{k} = 0$, from which $\boxed{A_Z = -0.6365\ kN}$

Problem 5.83 The 8 *ft* bar is supported by a ball and socket at A, the cable BD, and a roller support at C. Point B is the midpoint of the bar. The force $\vec{F} = -50\vec{k}\ (lb)$. Determine the tension in the cable BD and the reactions at A and C.

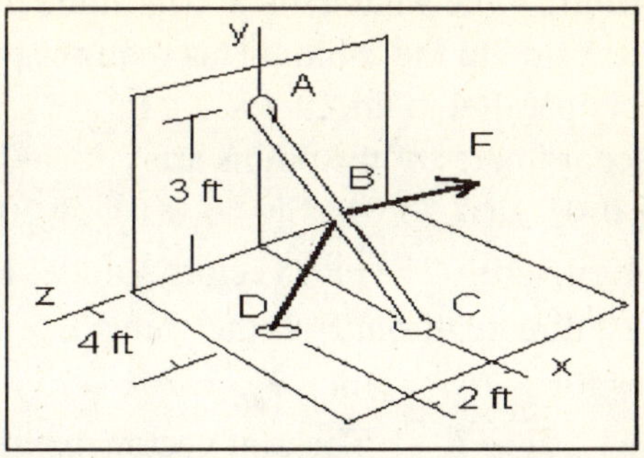

Solution: The strategy is to determine the sum of the moments about A, which will involve the unknown reactions at B and C. This will require the unit vectors parallel to the rod and parallel to the cable.

The angle formed by the rod is $\alpha = \sin^{-1}\left(\frac{3}{8}\right) = 22°$. The vector positions are: $\vec{r}_A = 3\vec{j}$, $\vec{r}_D = 4\vec{i} + 2\vec{k}$ and $\vec{r}_C = (8\cos 22°)\vec{i} = 7.4162\vec{i}$. The vector parallel to the rod is $\vec{r}_{AC} = \vec{r}_C - \vec{r}_A = 7.4162\vec{i} - 3\vec{j}$. The unit vector parallel to the rod is $\vec{e}_{AC} = 0.9270\vec{i} - 0.375\vec{j}$. The location of B is $\vec{r}_{AB} = 4\vec{e}_{AC} = 3.7081\vec{i} - 1.5\vec{j}$. The vector parallel to the cable is $\vec{r}_{BD} = \vec{r}_D - (\vec{r}_A + \vec{r}_{AB}) = 0.2919\vec{i} - 1.5\vec{j} + 2\vec{k}$. The unit vector parallel to the cable is $\vec{e}_{BD} = 0.1160\vec{i} - 0.5960\vec{j} + 0.7946\vec{k}$. The tension in the cable is $\vec{T} = |\vec{T}|\vec{e}_{BD}$. The reaction at the roller support C is normal to the x-z plane. The sum of the moments about A

$$\sum \vec{M}_A = \vec{r}_{AB} \times \vec{F} + \vec{r}_{AB} \times \vec{T} + \vec{r}_{AC} \times \vec{C} = 0$$

$$= \begin{vmatrix} \vec{i} & \vec{j} & \vec{k} \\ 3.7081 & -1.5 & 0 \\ 0 & 0 & -50 \end{vmatrix} + |\vec{T}| \begin{vmatrix} \vec{i} & \vec{j} & \vec{k} \\ 3.7081 & -1.5 & 0 \\ 0.1160 & -0.5960 & 0.7946 \end{vmatrix} + \begin{vmatrix} \vec{i} & \vec{j} & \vec{k} \\ 7.4162 & -3 & 0 \\ 0 & C_Y & 0 \end{vmatrix} = 0$$

$= 75\vec{i} + 185.4\vec{j} + |\vec{T}|(-1.192\vec{i} - 2.9466\vec{j} - 2.3840\vec{k}) + 7.4162C_Y\vec{k} = 0$, from which $\boxed{|\vec{T}| = \frac{75}{1.192} = 62.92\ lb}$

$\boxed{C_Y = \frac{2.3840|\vec{T}|}{7.4162} = 20.23\ lb}$ The reaction at A is determined from the sums of forces:

$\sum \vec{F}_X = (A_X + 0.1160|\vec{T}|)\vec{i} = 0$, from which $\boxed{A_X = -7.29\ lb}$, $\Sigma \vec{F}_Y = (A_Y - 0.5960|\vec{T}| + C_Y)\vec{j} = 0$, from which $\boxed{A_Y = +17.27\ lb}$ $\sum \vec{F}_Z = (A_Z + 0.7946|\vec{T}| - 50)\vec{k} = 0$, from which $\boxed{A_Z = 0\ lb}$

Problem 5.84 Consider the 8 *ft* bar in Problem 5.83. The force $\vec{F} = F_Y\vec{j} - 50\vec{k}\ (lb)$. What is the largest value of F_Y for which the roller support at C will remain on the floor?

Solution: From the solution to Problem 5.83, the sum of the moments about A is

$$\Sigma \vec{M}_A = \begin{vmatrix} \vec{i} & \vec{j} & \vec{k} \\ 3.7081 & -1.5 & 0 \\ 0 & F_Y & -50 \end{vmatrix} + |\vec{T}| \begin{vmatrix} \vec{i} & \vec{j} & \vec{k} \\ 3.7081 & -1.5 & 0 \\ 0.1160 & -0.5960 & 0.7946 \end{vmatrix} + \begin{vmatrix} \vec{i} & \vec{j} & \vec{k} \\ 7.4162 & -3 & 0 \\ 0 & C_Y & 0 \end{vmatrix} = 0$$

$= 75\vec{i} + 185.4\vec{j} + 3.7081F_Y\vec{k} + |\vec{T}|(-1.192\vec{i} - 2.9466\vec{j} - 2.384\vec{k}) + 7.4162C_Y\vec{k} = 0$, from which,

$|\vec{T}| = \frac{75}{1.192} = 62.92\ lb$. Collecting terms in $\vec{k}$, $3.7081F_Y + 2.384|\vec{T}| - 7.4162C_Y = 0$. For $C_Y = 0$,

$\boxed{F_Y = \frac{150.0}{3.708} = 40.453\ lb}$

Problem 5.85 The tower is 70 *m* tall. The tension in each cable is 2 *kN*. Treat the base of the tower as a built-in support. What are the reactions at A?

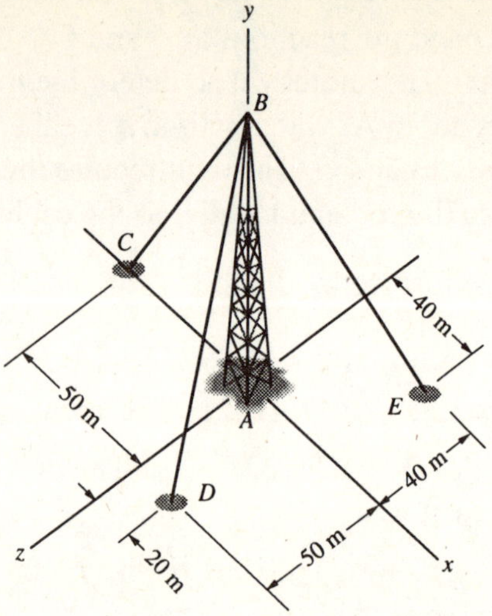

Solution: The strategy is to determine moments about A due to the cables. This requires the unit vectors parallel to the cables.
The coordinates of the points are:
$A(0,0,0)$, $B(0,70,0)$, $C(-50,0,0)$, $D(20,0,50)$ $E(40,0,-40)$. The unit vectors parallel to the cables, directed from B to the points E, D, and C $\vec{r}_{BE} = 40\vec{i} - 70\vec{j} - 40\vec{k}$, $\vec{r}_{BD} = 20\vec{i} - 70\vec{j} + 50\vec{k}$, $\vec{r}_{BC} = -50\vec{i} - 70\vec{j}$. The unit vectors parallel to the cables, pointing from B, are:
$\vec{e}_{BE} = 0.4444\vec{i} - 0.7778\vec{j} - 0.4444\vec{k}$,
$\vec{e}_{BD} = 0.2265\vec{i} - 0.7926\vec{j} + 0.5661\vec{k}$,
$\vec{e}_{BC} = -0.5812\vec{i} - 0.8137\vec{j} + 0\vec{k}$.
The tensions in the cables are:
$\vec{T}_{BD} = 2\vec{e}_{BD} = 0.4529\vec{i} - 1.5852\vec{j} + 1.1323\vec{k}\ (kN)$, $\vec{T}_{BE} = 2\vec{e}_{BE} = 0.8889\vec{i} - 1.5556\vec{j} - 0.8889\vec{k}\ (kN)$, $\vec{T}_{BC} = 2\vec{e}_{BC} = -1.1625\vec{i} - 1.6275\vec{j} - 0\vec{k}$. The sum of the moments about A is

$$\sum \vec{M}_A = \vec{M}^A + \vec{r}_{AB} \times \vec{T}_{BE} + \vec{r}_{AB} \times \vec{T}_{BD} + \vec{r}_{AB} \times \vec{T}_{BC} = 0 = \vec{M}^A + \vec{r}_{AB} \times (\vec{T}_{BE} + \vec{T}_{BC} + \vec{T}_{BD})$$

$$\sum \vec{M}_A = \vec{M}^A + \begin{vmatrix} \vec{i} & \vec{j} & \vec{k} \\ 0 & 70 & 0 \\ 0.1793 & -4.7682 & 0.2434 \end{vmatrix} = 0 = (M_X^A + 17.038)\vec{i} + (M_Y^A + 0)\vec{j} + (M_X^A - 12.551)\vec{k} = 0$$

from which $\boxed{M_X^A = -17.038\ kN\text{-}m}$, $\boxed{M_Y^A = 0}$, $\boxed{M_Z^A = 12.551\ kN\text{-}m}$ The force reactions at A are determined from the sums of forces. (Note that the sums of the cable forces have already been calculated and used above.) $\sum \vec{F}_X = (A_X + 0.17932)\vec{i} = 0$, from which $\boxed{A_X = -0.179\ kN}$, $\sum \vec{F}_Y = (A_Y - 4.7682)\vec{j} = 0$, from which $\boxed{A_Y = 4.768\ kN}$, $\sum \vec{F}_Z = (A_Z + 0.2434)\vec{k} = 0$, from which $\boxed{A_Z = -0.2434\ kN}$

Problem 5.86 Consider the tower in Problem 5.85. If the tension in cable BC is 2 *kN*, what do the tensions in cables BD and BC need to be if you want the couple exerted on the tower by the built in support at A to be zero? What are the resulting reactions at A?

Solution: From the solution to Problem 5.85, the sum of the moments about A is given by $\sum \vec{M}_A = \vec{M}^A + \vec{r}_{AB} \times (\vec{T}_{BE} + \vec{T}_{BC} + \vec{T}_{BD}) = 0$.. If the couple $\vec{M}^A = 0$, then the cross product is zero, which is possible only if the vector sum of the cable tensions is zero in the x and z directions.Thus, from Problem 5.85, $\vec{e}_x \cdot (\vec{T}_{BC} + |\vec{T}_{BE}|\vec{e}_{BE} + |\vec{T}_{BD}|\vec{e}_{BD}) = 0$, and $\vec{e}_z \cdot (\vec{T}_{BC} + |\vec{T}_{BE}|\vec{e}_{BE} + |\vec{T}_{BD}|\vec{e}_{BD}) = 0$. Two simultaneous equations in two unknowns result;
$0.4444|\vec{T}_{BE}| + 0.2265|\vec{T}_{BD}| = 1.1625$
$-0.4444|\vec{T}_{BE}| + 0.5661|\vec{T}_{BD}| = 0$. Solve: $\boxed{|\vec{T}_{BE}| = 1.868\ kN}$, $\boxed{|\vec{T}_{BD}| = 1.467\ kN}$. The reactions at A oppose the sum of the cable tensions in the x-, y-, and z- directions. $\boxed{A_X = 0}$, $\boxed{A_Y = 4.243\ kN}$, $\boxed{A_Z = 0}$. (These results are to be expected if there is no moment about A.)

Problem 5.87 The 40 *lb* door is supported by hinges at A and B. The y-axis is vertical. The hinges do not exert couples on the door and the hinge at B does not exert a force parallel to the hinge axis. The weight of the door acts at its midpoint. What are the reactions at A and B?

Solution: The position vector of the midpoint of the door:

$\vec{r}_{CM} = (2\cos 50^o)\vec{i} + 3.5\vec{j} + (2\cos 40^o)\vec{k} = 1.2856\vec{i} + 3.5\vec{j} + 1.532\vec{k}$.

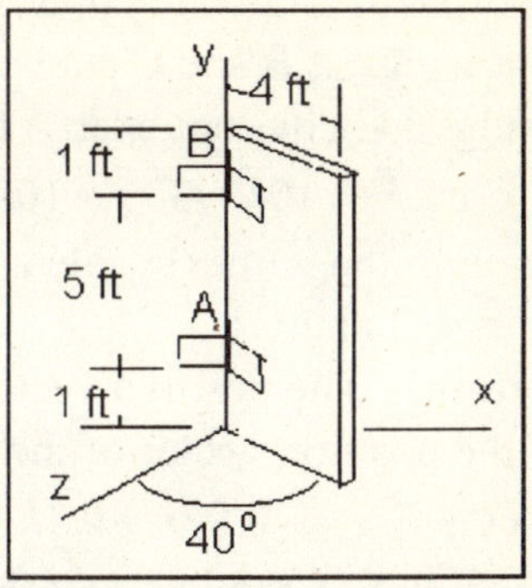

The position vectors of the hinges: $\vec{r}_A = \vec{j}$, $\vec{r}_B = 6\vec{j}$. The forces are: $\vec{W} = -40\vec{j}$, $\vec{A} = A_X\vec{i} + A_Y\vec{j} + A_Z\vec{k}$, $\vec{B} = B_X\vec{i} + B_Z\vec{k}$. The position vectors relative to A are $\vec{r}_{ACM} = \vec{r}_{CM} - \vec{r}_A = 1.2856\vec{i} + 2.5\vec{j} + 1.532\vec{k}$, $\vec{r}_{AB} = \vec{r}_B - \vec{r}_A = 5\vec{j}$. The sum of the moments about A

$$\sum \vec{M}_A = \vec{r}_{ACM} \times \vec{W} + \vec{r}_{AB} \times \vec{B} = \begin{vmatrix} \vec{i} & \vec{j} & \vec{k} \\ 1.2856 & 2.5 & 1.532 \\ 0 & -40 & 0 \end{vmatrix} + \begin{vmatrix} \vec{i} & \vec{j} & \vec{k} \\ 0 & 5 & 0 \\ B_X & 0 & B_Z \end{vmatrix} = 0$$

$\Sigma\vec{M}_A = (5B_Z + 40(1.532))\vec{i} + (-5B_X - 40(1.285))\vec{k} = 0$, from which $\boxed{B_Z = \frac{-40(1.532)}{5} = -12.256\ lb}$

$\boxed{B_X = \frac{-40(1.285)}{5} = -10.28\ lb}$. The reactions at A are determined from the sums of forces:

$\Sigma\vec{F}_X = (A_X + B_X)\vec{i} = 0$, from which $\boxed{A_X = 10.28\ lb}$,

$\Sigma\vec{F}_Y = (A_Y - 40)\vec{j} = 0$, from which $\boxed{A_Y = 40\ lb}$,

$\sum\vec{F}_Z = (A_Z + B_Z)\vec{k} = 0$, from which $\boxed{A_Z = 12.256\ lb}$

Problem 5.88 The vertical cable is attached at A. Determine the tension in the cable and the reactions at the bearing B due to the force $\vec{F} = 10\vec{i} - 30\vec{j} - 10\vec{k}\ (N)$.

Solution: The position vector of the point of application of the force is $\vec{r}_F = 0.2\vec{i} - 0.2\vec{k}$. The position vector of the bearing is $\vec{r}_B = 0.1\vec{i}$. The position vector of the cable attachment to the wheel is $\vec{r}_C = 0.1\vec{k}$. The position vectors relative to B are:

$\vec{r}_{BC} = \vec{r}_C - \vec{r}_B = -0.1\vec{i} + 0.1\vec{k}$,

$\vec{r}_{BF} = \vec{r}_F - \vec{r}_B = 0.1\vec{i} - 0.2\vec{k}$.

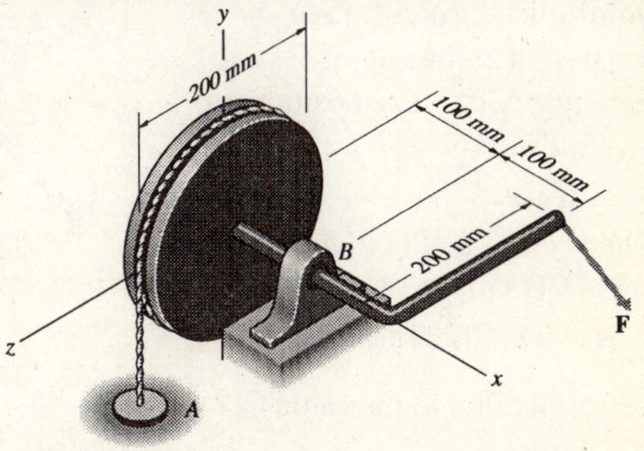

The sum of the moments about the bearing B is

$$\sum\vec{M}_B = \vec{M}^B + \vec{r}_{BF} \times \vec{F} + \vec{r}_{BC} \times \vec{C} = 0$$

$$\Sigma\vec{M}_B = \vec{M}^B + \begin{vmatrix} \vec{i} & \vec{j} & \vec{k} \\ 0.1 & 0 & -0.2 \\ 10 & -30 & -10 \end{vmatrix} + \begin{vmatrix} \vec{i} & \vec{j} & \vec{k} \\ -0.1 & 0 & 0.1 \\ 0 & -T & 0 \end{vmatrix}$$

$$= (-6 + 0.1T)\vec{i} + (M_Y^B - 1)\vec{j} + (M_Z^B - 3 + 0.1T)\vec{k} = 0,$$

from which $\boxed{T = \frac{6}{0.1} = 60\ N}$, $\boxed{M_Y^B = +1\ N\text{-}m}$, $\boxed{M_Z^B = -0.1T + 3 = -3\ N\text{-}m}$ The force reactions at the bearing are determined from the sums of forces: $\sum\vec{F}_X = (B_X + 10)\vec{i} = 0$, from which $\boxed{B_X = -10\ N}$.

$\sum\vec{F}_Y = (B_Y - 30 - 60)\vec{j} = 0$, from which $\boxed{B_Y = 90\ N}$.

$\sum\vec{F}_Z = (B_Z - 10)\vec{j} = 0$, from which $\boxed{B_Z = 10\ N}$

Problem 5.89 The device shown in Problem 5.88 is badly designed because of the couple that must be supported by the bearing at B, which would cause the bearing to "bind". (Imagine trying to open a door supported by only one hinge.) In this improved design, the bearings at B and C support no couples and the bearing at C does not exert a force in the x direction. If the force $\vec{F} = 10\vec{i} - 30\vec{j} - 10\vec{k}$ (N), what are the tension in the vertical cable and the reactions at the bearings B and C?

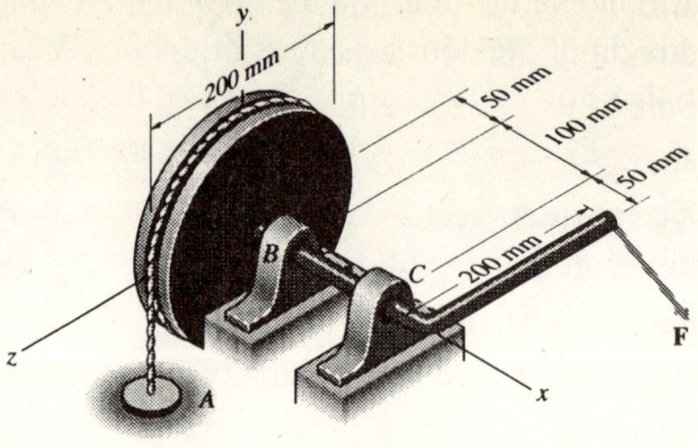

Solution: The position vectors relative to the bearing B are: the position vector of the cable attachment to the wheel is $\vec{r}_{BT} = -0.05\vec{i} + 0.1\vec{k}$. The position vector of the bearing C is: $\vec{r}_{BC} = 0.1\vec{i}$. The position vector of the point of application of the force is: $\vec{r}_{BF} = 0.15\vec{i} - 0.2\vec{k}$. The sum of the moments about B is

$$\sum \vec{M}_B = \vec{r}_{BT} \times \vec{T} + \vec{r}_{BC} \times \vec{C} + \vec{r}_{BF} \times \vec{F} = 0$$

$$\sum \vec{M}_B = \begin{vmatrix} \vec{i} & \vec{j} & \vec{k} \\ -0.05 & 0 & 0.1 \\ 0 & -T & 0 \end{vmatrix} + \begin{vmatrix} \vec{i} & \vec{j} & \vec{k} \\ 0.1 & 0 & 0 \\ 0 & C_Y & C_Z \end{vmatrix} + \begin{vmatrix} \vec{i} & \vec{j} & \vec{k} \\ 0.15 & 0 & -0.2 \\ 10 & -30 & -10 \end{vmatrix} = 0$$

$\sum \vec{M}_B = (0.1T - 6)\vec{i} + (-0.1C_Z + 1.5 - 2)\vec{j} + (0.05T + 0.1C_Y - 4.5)\vec{k} = 0$. From which: $\boxed{T = 60\ N}$, $\boxed{C_Z = \dfrac{-0.5}{0.1} = -5\ N}$, $\boxed{C_Y = \dfrac{4.5 - 0.05T}{0.1} = 15\ N}$ The reactions at B are found from the sums of forces: $\sum \vec{F}_X = (B_X + 10)\vec{i} = 0$, from which $\boxed{B_X = -10\ N}$. $\sum \vec{F}_Y = (B_Y + C_Y - T - 30)\vec{j} = 0$, from which $\boxed{B_Y = 75\ N}$. $\sum \vec{F}_Z = (B_Z + C_Z - 10)\vec{k} = 0$, from which $\boxed{B_Z = 15\ N}$

Problem 5.90 The crane's cable is attached to a stationary object at D. The crane is supported by the bearings E and F and the horizontal cable AB. The tension in cable AB is 8 *kN*. Determine the tension in the cable CD.

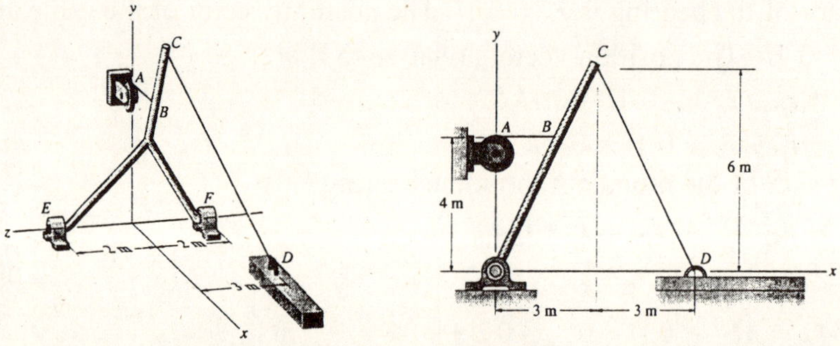

Solution: *Strategy*. (See discussion at the end of Example 5.10.) The position vectors are: $\vec{r}_B = 3\vec{i} + 4\vec{j}$, $\vec{r}_C = 3\vec{i} + 6\vec{j}$, $\vec{r}_D = 6\vec{i} - 3\vec{k}$. The vector parallel to the cable CD is $\vec{r}_{CD} = \vec{r}_D - \vec{r}_C$. $= 3\vec{i} - 6\vec{j} - 3\vec{k}$ The unit vector parallel to the cable CD is $\vec{e}_{CD} = \dfrac{\vec{r}_{CD}}{|\vec{r}_{CD}|} = 0.4082\vec{i} - 0.8165\vec{j} - 0.4082\vec{k}$. Noting $\vec{T}_{CD} = |\vec{T}_{CD}|\vec{e}_{CD}$

The sum of the moments about the z-axis is $\sum M_Z = \begin{vmatrix} 0 & 0 & 1 \\ 3 & 4 & 0 \\ -8 & 0 & 0 \end{vmatrix} + |\vec{T}_{CD}| \begin{vmatrix} 0 & 0 & 1 \\ 3 & 6 & 0 \\ 0.4082 & -0.8165 & -0.4082 \end{vmatrix}$

$= 32 + |\vec{T}_{CD}|(-3(0.8165) - 6(0.4082)) = 0$. From which $\boxed{|\vec{T}_{CD}| = \dfrac{32}{4.8987} = 6.5323\ kN}$

Problem 5.91 The crane in Problem 5.90 is supported by the horizontal cable AB and the bearings at E and F. The bearings do not exert couples and the bearing at F does not exert a force in the z-direction. The tension in cable AB is 8 *kN*. Determine the reactions at E and F.

Solution: The position vectors for the points B, C, E and F are: $\vec{r}_B = 3\vec{i} + 4\vec{j}$, $\vec{r}_C = 3\vec{i} + 6\vec{j}$, $\vec{r}_E = 2\vec{k}$, and $\vec{r}_F = -2\vec{k}$. The position vectors of the points B, C, and F relative the point E are: $\vec{r}_{EB} = \vec{r}_B - \vec{r}_E = 3\vec{i} + 4\vec{j} - 2\vec{k}$, $\vec{r}_{EC} = \vec{r}_C - \vec{r}_E = 3\vec{i} + 6\vec{j} - 2\vec{k}$, and $\vec{r}_{EF} = \vec{r}_F - \vec{r}_E = -4\vec{k}$. From the solution to Problem 5.90 the unit vector parallel to the cable CD is $\vec{e}_{CD} = 0.4082\vec{i} - 0.8165\vec{j} - 0.4082\vec{k}$. Noting that $\vec{T}_{CD} = |\vec{T}_{CD}|\vec{e}_{CD}$, the sum of the moments about E is $\sum \vec{M}_E = \vec{r}_{EB} \times \vec{B} + \vec{r}_{EC} \times \vec{T}_{CD} + \vec{r}_{EF} \times \vec{F} = 0$,

$$\sum \vec{M}_E = \begin{vmatrix} \vec{i} & \vec{j} & \vec{k} \\ 3 & 4 & -2 \\ -8 & 0 & 0 \end{vmatrix} + |\vec{T}_{CD}| \begin{vmatrix} \vec{i} & \vec{j} & \vec{k} \\ 3 & 6 & -2 \\ 0.4082 & -0.8165 & -0.4082 \end{vmatrix} + \begin{vmatrix} \vec{i} & \vec{j} & \vec{k} \\ 0 & 0 & -4 \\ F_X & F_Y & 0 \end{vmatrix} = 0$$

$\sum \vec{M}_E = \left(-4.0822|\vec{T}_{CD}| + 4F_Y\right)\vec{i} + \left(+0.4082|\vec{T}_{CD}| - 4F_X + 16\right)\vec{j} + \left(-4.8987|\vec{T}_{CD}| + 32\right)\vec{k} = 0$. From which:

$|\vec{T}_{CD}| = \dfrac{32}{4.8987} = 6.5323\ kN$, $\boxed{F_Y = \dfrac{4.0822|\vec{T}_{CD}|}{4} = 6.667\ kN}$, $\boxed{F_X = \dfrac{0.4082|\vec{T}_{CD}| + 16}{4} = 4.667\ kN}$

The reactions at E are determined from the sums of the forces:

$\sum \vec{F}_X = \left(E_X + F_X + |0.4082|\vec{T}_{CD}| - 8\right)\vec{i} = 0$, from which $\boxed{E_X = 0.667\ kN}$.

$\sum \vec{F}_Y = \left(E_Y + F_Y - 0.8165|\vec{T}_{CD}|\right)\vec{j} = 0$, from which $\boxed{E_Y = -1.333\ kN}$

$\sum \vec{F}_Z = \left(E_Z - 0.4082|\vec{T}_{CD}|\right)\vec{k} = 0$, from which $\boxed{E_Z = 2.667\ kN}$

Problem 5.92 The rocket launcher is supported by the hydraulic jack DE and the bearings A and B. The bearings lie on the x-axis and supports shafts parallel to the x-axis. The hydraulic cylinder DE exerts a force on the launcher that points along the line from D to E. The coordinates of D are (7,0,7) ft and the coordinates of E are (9,6,4) ft. The weight $W = 30\ kip$ acts at (4.5,5,2) ft. What is the magnitude of the reaction of the launcher at E?

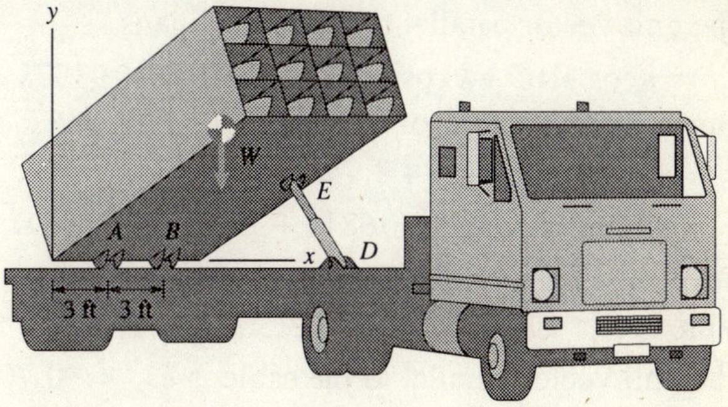

Solution: The position vectors of the points D, E and W are $\vec{r}_D = 7\vec{i} + 7\vec{k}$, $\vec{r}_E = 9\vec{i} + 6\vec{j} + 4\vec{k}\ (ft)$, $\vec{r}_W = 4.5\vec{i} + 5\vec{j} + 2\vec{k}\ (ft)$. The vector parallel to DE is $\vec{r}_{DE} = \vec{r}_E - \vec{r}_D = 2\vec{i} + 6\vec{j} - 3\vec{k}$. The unit vector parallel to DE is $\vec{e}_{DE} = 0.2857\vec{i} + 0.8571\vec{j} - 0.4286\vec{k}$

Since the bearings cannot exert a moment about the x-axis, the sum of the moments due to the weight and the jack force must be zero about the x-axis. The sum of the moments about the x-axis is:

$$\sum M_X = \begin{vmatrix} 1 & 0 & 0 \\ 4.5 & 5 & 2 \\ 0 & -30 & 0 \end{vmatrix} + |\vec{F}_{DE}| \begin{vmatrix} 1 & 0 & 0 \\ 9 & 6 & 4 \\ 0.2857 & 0.8571 & -0.4286 \end{vmatrix} = 0 = 60 - 6|\vec{F}_{DE}| = 0.$$ From which

$\boxed{|\vec{F}_{DE}| = \dfrac{60}{6} = 10\ kip}$

Problem 5.93 Consider the rocket launcher described in Problem 5.92. The bearings at A and B do not exert couples and the bearing B does not exert a force in the x-direction. Determine the reactions at A and B.

Solution: The position vectors to points A, B, E and W are: $\vec{r}_A = 3\vec{i}$, $\vec{r}_B = 6\vec{i}$, $\vec{r}_E = 9\vec{i} + 6\vec{j} + 4\vec{k}$, and $\vec{r}_W = 4.5\vec{i} + 5\vec{j} + 2\vec{k}$. The position vectors relative to point A are $\vec{r}_{AE} = \vec{r}_E - \vec{r}_A = 6\vec{i} + 6\vec{j} + 4\vec{k}$, $\vec{r}_{AW} = \vec{r}_W - \vec{r}_A = 1.5\vec{i} + 5\vec{j} + 2\vec{k}$, and $\vec{r}_{AB} = \vec{r}_B - \vec{r}_A = 3\vec{i}$. From the solution to Problem 5.92 the unit vector parallel to the jack DE is $\vec{e}_{DE} = 0.2857\vec{i} + 0.8571\vec{j} - 0.4286\vec{k}$. The sum of the moments about point A are: $\sum \vec{M}_A = \vec{r}_{AE} \times \vec{F}_E + \vec{r}_{AW} \times \vec{W} + r_{AB} \times \vec{B}$.

$$\sum \vec{M}_A = |\vec{F}_E| \begin{vmatrix} \vec{i} & \vec{j} & \vec{k} \\ 6 & 6 & 4 \\ 0.2857 & 0.8571 & -0.4286 \end{vmatrix} + \begin{vmatrix} \vec{i} & \vec{j} & \vec{k} \\ 1.5 & 5 & 2 \\ 0 & -30 & 0 \end{vmatrix} + \begin{vmatrix} \vec{i} & \vec{j} & \vec{k} \\ 3 & 0 & 0 \\ 0 & B_Y & B_Z \end{vmatrix} = 0$$

$= \left(-6|\vec{F}_E| - 60\right)\vec{i} + \left(3.7144|\vec{F}_E| - 3B_Z\right)\vec{j} + \left(3.4284|\vec{F}_E| - 45 + 3B_Y\right)\vec{k} = 0$. From which

$|\vec{F}_E| = \dfrac{60}{6} = 10\ kip$, $\boxed{B_Z = \dfrac{3.7144|\vec{F}_E|}{3} = 12.38\ kip}$, $\boxed{B_Y = \dfrac{-3.4284|\vec{F}_E| + 45}{3} = 3.572\ kip}$. The reaction at A is determined by the sums of forces:

$\sum \vec{F}_X = \left(A_X + 0.2857|\vec{F}_E|\right)\vec{i} = 0$, from which $\boxed{A_X = -2.857\ kip}$

$\sum \vec{F}_Y = \left(A_Y + B_Y + 0.8571|\vec{F}_E| - 30\right)\vec{j} = 0$, from which $\boxed{A_Y = 17.857\ kip}$

$\sum \vec{F}_Z = \left(A_Z + B_Z - 0.4286|\vec{F}_E|\right)\vec{k} = 0$, from which $\boxed{A_Z = -8.094\ kip}$

Problem 5.94 The plate is supported by hinges at A and B and the cable CE is loaded by the force at D. The hinges do not exert couples on the plate. What is the tension in the cable CE?

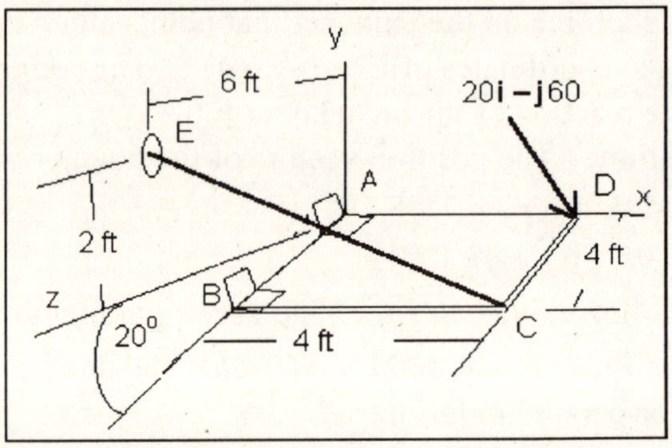

Solution: The hinges do not exert a moment about the hinge line, hence the sum of the moments about the hinge line is zero.

The unit vector parallel to the hinge line is $\vec{e}_H = \vec{j}\cos 110^o + \vec{k}\cos 20^o = -0.3420\vec{j} + 0.9397\vec{k}$.

The position vectors of the points B, C, D, and E are $\vec{r}_B = 4\vec{e}_H = -1.3681\vec{j} + 3.7588\vec{k}$, $\vec{r}_C = 4\vec{e}_H + 4\vec{i} = 4\vec{i} - 1.3681\vec{j} + 3.7588\vec{k}$, $\vec{r}_D = 4\vec{i}$, and $\vec{r}_E = 2\vec{j} + 6\vec{k}$. The position vector parallel to the cable is $\vec{r}_{CE} = \vec{r}_E - \vec{r}_C = -4\vec{i} + 3.3681\vec{j} + 2.2412\vec{k}$

The unit vector parallel to the cable is $\vec{e}_{CE} = -0.7031\vec{i} + .5920\vec{j} + 0.3939\vec{k}$. The sum of the moments about the hinge line is $\sum M_H = \vec{e}_H \cdot \left(\vec{r}_C \times \vec{T}\right) + \vec{e}_H \cdot \left(\vec{r}_F \times \vec{F}\right) = 0$

$$\sum M_H = |\vec{T}| \begin{vmatrix} 0 & -0.3420 & 0.9397 \\ 4 & -1.3681 & 3.7588 \\ -0.7031 & 0.5920 & 0.3939 \end{vmatrix} + \begin{vmatrix} 0 & -0.3420 & 0.9397 \\ 4 & 0 & 0 \\ 20 & -60 & 0 \end{vmatrix} = 0 = 2.7642|\vec{T}| - 225.53 = 0$$ From

which $\boxed{|\vec{T}| = \dfrac{225.53}{2.7642} = 81.59\ lb}$

Problem 5.95 Consider the plate in Problem 5.94. The hinges do not exert couples on the plate and the hinge at B does not exert a force parallel to the hinge axis. Determine the magnitudes of the forces at A and B.

Solution: From the solution to Problem 5.94, the unit vector parallel to the hinge line is $\vec{e}_H = \vec{j}\cos 110^o + \vec{k}\cos 20^o = -0.3420\vec{j} + 0.9397\vec{k}$. The position vectors of the points B, C, D, and E are $\vec{r}_B = 4\vec{e}_H = -1.3681\vec{j} + 3.7588\vec{k}$, $\vec{r}_C = 4\vec{e}_H + 4\vec{i} = 4\vec{i} - 1.3681\vec{j} + 3.7588\vec{k}$, $\vec{r}_D = 4\vec{i}$, and $\vec{r}_E = 2\vec{j} + 6\vec{k}$. The position vector parallel to the cable is $\vec{r}_{CE} = \vec{r}_E - \vec{r}_C = -4\vec{i} + 3.3681\vec{j} + 2.2412\vec{k}$ The unit vector parallel to the cable is $\vec{e}_{CE} = -0.7031\vec{i} + .5920\vec{j} + 0.3939\vec{k}$. The sum of the moments about the point A: $\sum \vec{M}_A = \vec{r}_C \times \vec{T} + \vec{r}_B \times \vec{B} + \vec{r}_F \times \vec{F} = 0$.

$$\Sigma \vec{M}_A = |\vec{T}|\begin{vmatrix} \vec{i} & \vec{j} & \vec{k} \\ 4 & -1.3681 & 3.7588 \\ -0.7031 & 0.5920 & 0.3939 \end{vmatrix} + \begin{vmatrix} \vec{i} & \vec{j} & \vec{k} \\ 0 & -1.3681 & 3.7588 \\ B_X & B_Y & B_Z \end{vmatrix} + \begin{vmatrix} \vec{i} & \vec{j} & \vec{k} \\ 4 & 0 & 0 \\ 20 & -60 & 0 \end{vmatrix} = 0$$

The three components lead to the equations:

$-2.7641|\vec{T}| - 3.759B_y - 1.368B_z = 0$, $-4.2184|\vec{T}| + 3.759B_x = 0$, and $1.4061|\vec{T}| + 1.3681B_x = 240$

Solve the last two equations: $|\vec{T}| = 81.596\ lb$, $B_x = 91.564\ lb$. *Check:* The tension value agrees with the solution to Problem 5.94, as it should. *check.* The first equation can be solved by defining a normal reaction at B: $B_n = B_y\cos 20^o$, and $B_n = B_z \sin 20^o$, from which $B_n = -56.4\ lb$.. The reactions at A are determined by the sums of the forces: $\Sigma\vec{F}_x = \left(A_x + B_x - 0.7031|\vec{T}| + 20\right)\vec{i} = 0$ from which $A_X = -54.2\ \ lb$. $\sum \vec{F}_Y = \left(A_Y + B_Y + 0.592|\vec{T}| - 60\right)\vec{j} = 0$, from which $A_Y = 64.7\ \ lb$.

$\Sigma\vec{F}_Z = \left(A_z + B_z + 0.3939|\vec{T}|\right)\vec{k} = 0$, from which $A_z = -12.9\ lb$. The magnitudes:

$|\vec{B}| = \sqrt{B_x^2 + B_n^2} = 107.5\ lb$, $|\vec{A}| = \sqrt{A_x^2 + A_y^2 + A_z^2} = 85.4\ \ lb$

Problem 5.96 The bar ABC is supported by ball and socket supports at A and C and the cable BD and loaded by the 200 *lb* suspended weight. That is the tension in the cable BD?

Solution: The strategy is to take the moments about A. Note that a ball and socket cannot support a couple reaction. The vector parallel to the cable is $\vec{r}_{BD} = -2i + 2\vec{j} - \vec{k}$. The unit vector parallel to the cable is $\vec{e}_{BD} = -0.6667\vec{i} + 0.6667\vec{j} - 0.3333\vec{k}$. The vector positions of the weight, point B and point C relative to point A are: $\vec{r}_{AW} = -4\vec{i}$, $\vec{r}_{AB} = -6\vec{i}$, and $\vec{r}_{AC} = -6\vec{i} + 4\vec{k}$. The sum of the moments about A is $\sum \vec{M}_A = \vec{r}_{AW} \times \vec{W} + \vec{r}_{AB} \times \vec{T} + \vec{r}_{AC} \times \vec{C} = 0$

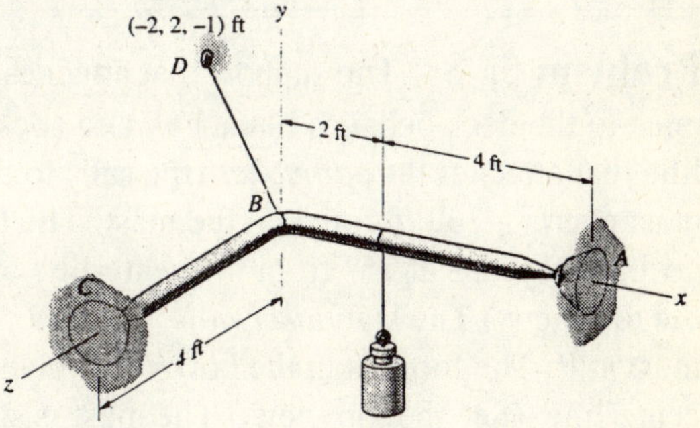

$$\sum \vec{M}_A = \begin{vmatrix} \vec{i} & \vec{j} & \vec{k} \\ -4 & 0 & 0 \\ 0 & -200 & 0 \end{vmatrix} + |\vec{T}|\begin{vmatrix} \vec{i} & \vec{j} & \vec{k} \\ -6 & 0 & 0 \\ -0.6667 & 0.6667 & -0.3333 \end{vmatrix} + \begin{vmatrix} \vec{i} & \vec{j} & \vec{k} \\ -6 & 0 & 4 \\ C_X & C_Y & C_Z \end{vmatrix} = 0$$

$= \left(-4C_Y\right)\vec{i} + \left(2|\vec{T}| + 6C_Z + 4C_X\right)\vec{j} + \left(-4|\vec{T}| - 6C_Y + 800\right)\vec{k} = 0$ From which: $C_Y = 0$, $|\vec{T}| = \dfrac{800}{4} = 200\ lb$

Problem 5.97 The bearings at A, B, and C do not exert couple on the bar and do not exert forces in the direction of the axis of the bar. Determine the reactions at the bearings due to the two forces on the bar.

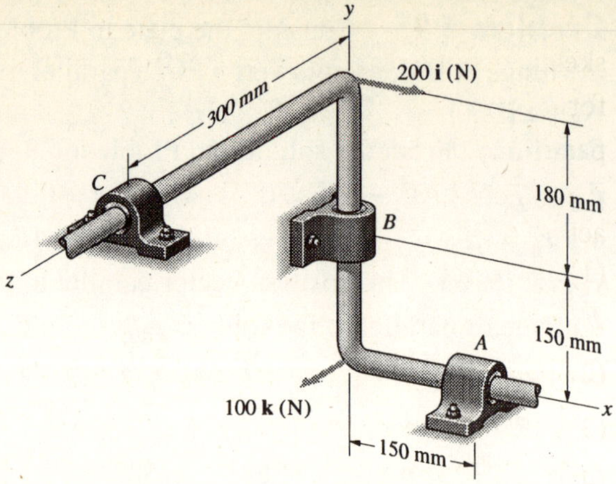

Solution: The strategy is to take the moments about A and solve the resulting simultaneous equations. The position vectors of the bearings relative to A are:

$\vec{r}_{AB} = -0.15\vec{i} + 0.15\vec{j}$, $\vec{r}_{AC} = -0.15\vec{i} + 0.23\vec{j} + 0.3\vec{k}$.

Denote the lower force by subscript 1, and the upper by subscript 2: $\vec{r}_{A1} = -0.15\vec{i}$, $\vec{r}_{A2} = -0.15\vec{i} + 0.23\vec{j}$.

The sum of the moments about A is:

$$\sum \vec{M}_A = \vec{r}_{A1} \times \vec{F}_1 + \vec{r}_{AB} \times \vec{B} + \vec{r}_{A2} \times \vec{F}_2 + \vec{r}_{AC} \times \vec{C} = 0$$

$$\sum \vec{M}_A = \begin{vmatrix} \vec{i} & \vec{j} & \vec{k} \\ -0.15 & 0 & 0 \\ 0 & 0 & 100 \end{vmatrix} + \begin{vmatrix} \vec{i} & \vec{j} & \vec{k} \\ -0.15 & 0.15 & 0 \\ B_X & 0 & B_Z \end{vmatrix} + \begin{vmatrix} \vec{i} & \vec{j} & \vec{k} \\ -0.15 & 0.33 & 0 \\ 200 & 0 & 0 \end{vmatrix} + \begin{vmatrix} \vec{i} & \vec{j} & \vec{k} \\ -0.15 & 0.33 & 0.3 \\ C_X & C_Y & 0 \end{vmatrix} = 0$$

$$\sum \vec{M}_A = (0.15B_Z - 0.3C_Y)\vec{i} + (15 + 0.15B_Z + 0.3C_X)\vec{j} + (-0.15B_X - 66 - 0.15C_Y - 0.33C_X)\vec{k} = 0$$

This results in three equations in four unknowns; an additional equation is provided by the sum of the forces in the x-direction (which cannot have a reaction term due to A)

$\sum \vec{F}_X = (B_X + C_X + 200)\vec{i} = 0$ The four equations in four unknowns:

$0B_X \ + 0.15B_Z \ + 0C_X \ - 0.3C_Z = 0$

$0B_X \ + 0.15B_Z \ + 0.3C_X \ + 0C_Y = -15$

$-0.15B_X \ + 0B_Z \ - 0.33C_X \ - 0.15C_Y = 66$

$B_X \ + 0B_Z \ + C_X \ + 0C_Z = -200$. (The HP-28S hand held calculator was used to solve these equations.) The solution: $\boxed{B_X = 750\ N}$, $\boxed{B_Z = 1800\ N}$, $\boxed{C_X = -950\ N}$, $\boxed{C_Y = 900\ N}$. The reactions at A are determined by the sums of forces:

$\sum \vec{F}_Y = (A_Y + C_Y)\vec{j} = 0$, from which $\boxed{A_Y = -C_Y = -900\ N}$

$\sum \vec{F}_Z = (A_Z + B_Z + 100)\vec{k} = 0$, from which $\boxed{A_Z = -1900\ N}$

Problem 5.98 The support that attaches the sailboat's mast to the deck behaves like a ball and socket support. The line that attaches the *spinnaker* (the sail) to the top of the mast exerts a 200 *lb* force on the mast. The force is in the horizontal plane at 15° from the centerline of the boat. (See the top view.) The *spinnaker pole* exerts a 50 *lb* force on the mast at P. The force is in the horizontal plane at 45° from the centerline. (See the top view.) The mast is supported by two cables, the *back stay* and the *port shroud* ACD. (The forestay AE and the starboard shroud AFG are slack and their tensions can be neglected.) Determine the tensions in the cables AB and CD and the reactions at the bottom of the mast.

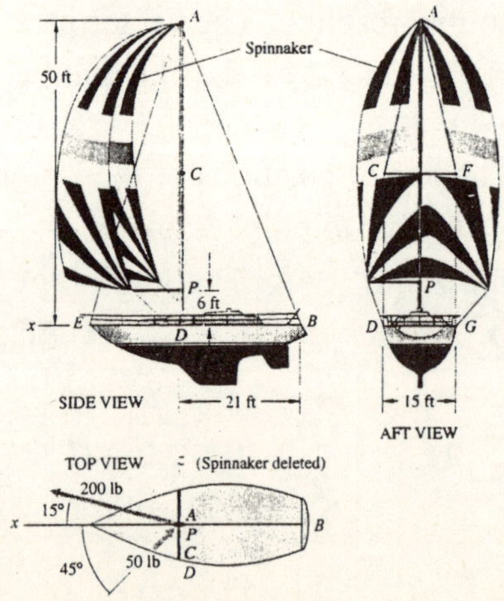

Solution: Although the dimensions are not given in the

Problem 5.98 continued on next page....

=================================<>=================================

continuation of solution to Problem 5.98...

sketch, assume that the point C is at the midpoint of the mast (25 ft above the deck).The position vectors for the points A, B, C, D, , and P are: $\vec{r}_A = 50\vec{j}$, $\vec{r}_B = -21\vec{i}$, $\vec{r}_P = 6\vec{j}$, $\vec{r}_C = 25\vec{j} - 7.5\vec{k}$. The vector parallel to the backstay AB is $\vec{r}_{AB} = \vec{r}_B - \vec{r}_A = -21\vec{i} - 50\vec{j}$. The unit vector parallel to backstay AB is $\vec{e}_{AB} = -0.3872\vec{i} - 0.9220\vec{j}$. The vector parallel to AC is $\vec{r}_{AC} = \vec{r}_C - \vec{r}_A = -25\vec{j} - 7.5\vec{k}$. The forces acting on the mast are:

(1) The force due to the spinnaker at the top of the mast:
$\vec{F}_A = 200\left(\vec{i}\cos 15^o + \vec{k}\cos 75^o\right) = 193.19\vec{i} + 51.76\vec{k}$.

(2) The reaction due to the backstay: $\vec{T}_{AB} = \left|\vec{T}_{AB}\right|\vec{e}_{AB}$

(3) The reaction due to the shroud: $\vec{T}_{AC} = \left|\vec{T}_{AC}\right|\vec{e}_{AC}$

(4) The force acting on the cross spar CE: $\vec{T}_{CE} = -\left(\vec{k}\cdot\vec{T}_{AC}\right)\vec{k} = 0.2873\left|\vec{T}_{AC}\right|\vec{k}$.

(5) The force due to the spinnaker pole: $\vec{F}_P = 50\left(-0.707\vec{i} + 0.707\vec{k}\right) - 35.35\vec{i} + 35.35\vec{k}$.

The sum of the moments about the base of the mast is

$\vec{M}_Q = \vec{r}_A \times \vec{F}_A + \vec{r}_A \times \vec{T}_{AB} + \vec{r}_A \times \vec{T}_{AC} + \vec{r}_C \times \vec{T}_{CE} + \vec{r}_P \times \vec{F}_P = 0$

$\sum\vec{M}_Q = \vec{r}_A \times \left(\vec{F}_A + \vec{T}_{AB} + \vec{T}_{AC}\right) + \vec{r}_C \times \vec{T}_{CE} + \vec{r}_P \times \vec{F}_P = 0$. From above,

$\vec{F}_A + \vec{T}_{AB} + \vec{T}_{AC} = F_{TX}\vec{i} + F_{TY}\vec{j} + F_{TZ}\vec{k} =$

$= \left(193.2 - (0.3872)\left|\vec{T}_{AB}\right|\right)\vec{i} + \left(-0.922\left|\vec{T}_{AB}\right| - 0.9578\left|\vec{T}_{AC}\right|\right)\vec{j} + \left(51.76 - 0.2873\left|\vec{T}_{AC}\right|\right)\vec{k}$

$$\sum\vec{M}_Q = \begin{vmatrix} \vec{i} & \vec{j} & \vec{k} \\ 0 & 50 & 0 \\ F_{TX} & F_{TY} & F_{TZ} \end{vmatrix} + \begin{vmatrix} \vec{i} & \vec{j} & \vec{k} \\ 0 & 25 & 0 \\ 0 & 0 & 0.2873\left|\vec{T}_{AC}\right| \end{vmatrix} + \begin{vmatrix} \vec{i} & \vec{j} & \vec{k} \\ 0 & 6 & 0 \\ -35.35 & 0 & 35.35 \end{vmatrix} = 0$$

$= \left(50F_{TZ} + (25)(0.2873)\left|\vec{T}_{AC}\right| + 212.1\right)\vec{i} + \left(-50F_{TX} + 212.1\right)\vec{k} = 0$. Substituting and collecting terms:

$(2800 - 7.1829\left|\vec{T}_{AC}\right|)\vec{i} + \left(-9447.9 + 19.36\left|\vec{T}_{AB}\right|\right)\vec{k} = 0$, from which $\left|\vec{T}_{AC}\right| = \dfrac{2800}{7.1829} = 389.81\ lb$,

$\boxed{\left|\vec{T}_{AB}\right| = 488.0\ lb}$ The tension in cable CD is the vertical component of the tension in AC,

$\boxed{\left|\vec{T}_{CD}\right| = \left|\vec{T}_{AC}\right|\left(\vec{j}\cdot\vec{e}_{AC}\right) = \left|\vec{T}_{AC}\right|(0.9578) = 373.37\ lb}$ The reaction at the base is found from the sums of the forces:

$\sum\vec{F}_X = \left(Q_X + 193.19 - 35.35 - \left|\vec{T}_{AB}\right|(0.3872)\right) = 0$, from which $\boxed{Q_X = 31.11\ lb}$

$\sum\vec{F}_Y = \left(Q_Y - 0.922\left|\vec{T}_{AB}\right| - 0.9578\left|\vec{T}_{AC}\right|\right)\vec{j} = 0$, from which $\boxed{Q_Y = 823.24\ lb}$

$\sum\vec{F}_Z = \left(Q_Z + 51.76 + \left(0.2873\left|\vec{T}_{AC}\right| - 0.2873\left|\vec{T}_{AC}\right| + 35.35\right)\right)\vec{k} = 0$, from which $\boxed{Q_Z = -87.11\ lb}$

Collecting the terms, the reaction is

$\boxed{\vec{Q} = 31.11\vec{i} + 823.24\vec{j} - 87.11\vec{k}\ (lb)}$

=================================<>=================================

Problem 5.99 The horizontal bar has a mass of 10 kg. Its weight acts at the midpoint of the bar and it is supported by a roller support at A and the cable BC. Use the fact that the bar is a three-force member to determine the angle α, the tension in the cable BC, and the magnitude of the reaction at A.

Solution: The roller support at A and the cable support at B are one-force supports. The reaction at A is $\vec{A} = A(\vec{i}\cos 60^o + \vec{j}\sin 60^o) = A(0.5\vec{i} + 0.855\vec{j})$. The reaction at B is $\vec{B} = B(\vec{i}\cos\alpha + \vec{j}\sin\alpha)$. The sum of the moments about B is $\sum M_B = +W(1) - A(0.866)(2) = 0$,

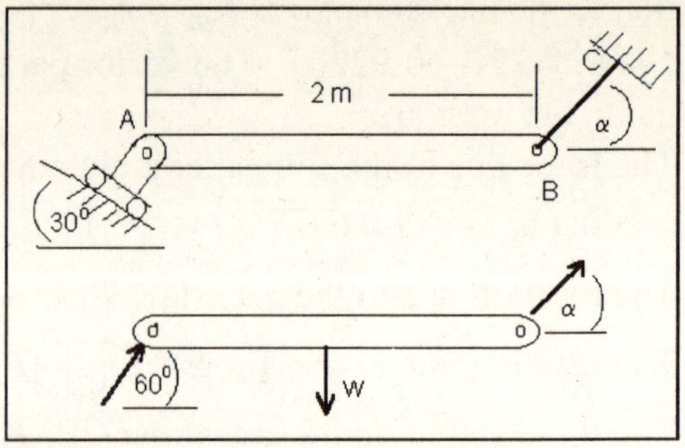

from which $\boxed{A = \frac{W}{(0.866)(2)} = \frac{(9.81)(10)}{1.732} = 56.64\ N}$

The sums of the forces:

$\sum F_X = A(0.5) + B(\cos\alpha) = 0$, from which $B\cos\alpha = -56.64(0.5) = -28.32$.

$\sum F_Y = A(0.866) - W + B\sin\alpha = 0$, from which $B\sin\alpha = 98.1 - 45.09 = 49.05$. Combining:

$\tan\alpha = \frac{49.05}{-28.32} = -1.732$, from which $\alpha = 120^o$ or $\alpha = 300^o$. Since $\cos\alpha \le 0$ and $\sin\alpha \ge 0$, the angle is in the second quadrant, hence $\boxed{\alpha = 120^o}$, and $\boxed{B = \frac{49.05}{\sin\alpha} = 56.64\ N}$

Problem 5.100 The horizontal bar is of negligible weight. Use the fact that the bar is a three-force member to determine the angle α necessary for equilibrium.

Solution: When the action lines of the reactions meet at a point, and the force does not produce a moment about that point, the system is in equilibrium. This situation occurs when all three action lines meet at the point P. Construct the two triangles shown. The hypotenuse of the left triangle is $h = \frac{4}{\cos 60^o} = 8$. The vertical distance to the point P is $D = \sqrt{8^2 - 4^2} = 6.9282$. The angle α is:

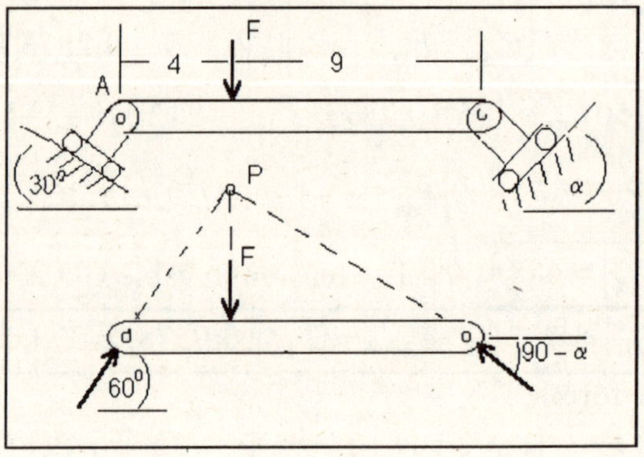

$90^o - \alpha = \tan^{-1}\left(\frac{6.9282}{9}\right)$, from which

$\boxed{\alpha = 90^o - 37.589^o = 52.4^o}$

Problem 5.101 The suspended load weighs 1000 lb. If you neglect the weight of the structure, it is a three force member. Use this fact to determine the magnitudes of the reactions at A and B.

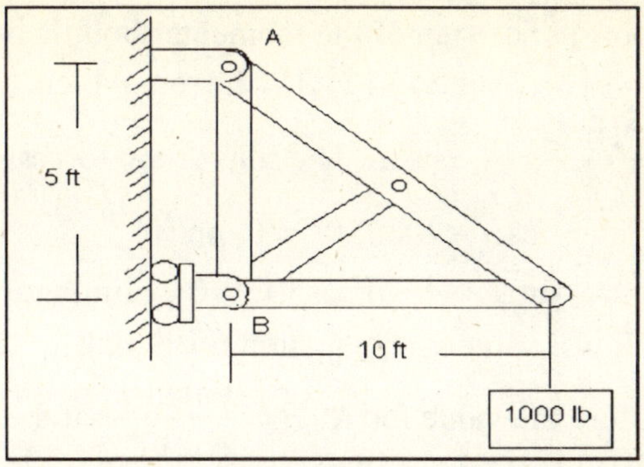

Solution: The pin support at A is a two-force reaction, and the roller support at B is a one force reaction. The moment about A is $M_A = 5B - 10(1000) = 0$, from which the magnitude at B is $\boxed{B = 2000\ lb}$. The sums of the forces: $\sum F_X = A_X + B = A_X + 2000 = 0$, from which $A_X = -2000\ lb$. $\sum F_Y = A_Y - 1000 = 0$, from which $A_Y = 1000\ lb$. The magnitude at A is

$\boxed{A = \sqrt{2000^2 + 1000^2} = 2236\ lb}$

Problem 5.102 The weight $W = 50\ lb$ of the circular disk acts at the center of the disk. Use the fact that the disk is a three force member to determine the tension in the cable and the magnitude of the reaction at the pin support.

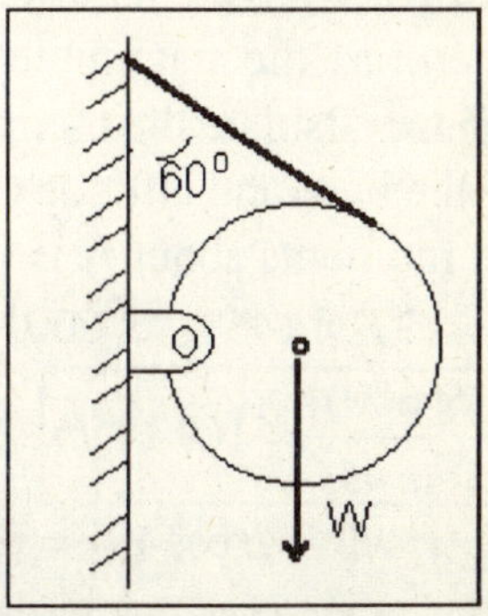

Solution: Denote the magnitude of the reaction at the pinned joint by B. t The sums of the forces are: $\sum F_X = B_X - T\sin 60° = 0$, and $\sum F_Y = B_Y + T\cos 60° - W = 0$. . The perpendicular distance to the action line of the tension from the center of the disk is the radius R. . The sum of the moments about the center of the disk is $M_C = -RB_Y + RT = 0$, from which $B_Y = T$. Substitute into the sum of the forces to obtain:

$T + T(0.5) - W = 0$, from which $\boxed{T = \frac{2}{3}W = 33.33\ lb}$ Substitute into the sum of forces to obtain $B_X = T\sin 60° = 28.86\ lb$. The magnitude of the reaction at the pinned joint is $\boxed{B = \sqrt{33.33^2 + 28.86^2} = 44.1\ lb}$

Problem 5.103 The weight $W = 40\ N$ of the circular disk acts at the center of the disk. What force is necessary to lift the disk off the floor?

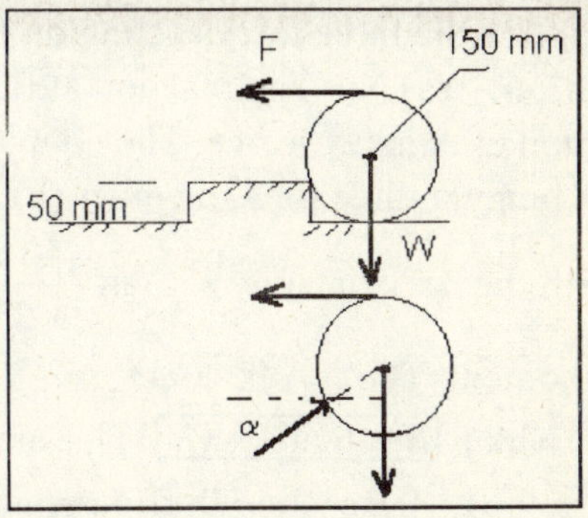

Solution: The reaction at the obstacle acts through the center of the disk (see sketch) Denote the contact point by B. When the moment is zero about the point B, the disk is at the verge of leaving the floor, hence the force at this condition is the force required to lift the disk. The perpendicular distance from B to the action line of the weight is $d = R\cos\alpha$, where α is given by (see sketch) $\alpha = \sin^{-1}\left(\frac{R-h}{R}\right) = \sin^{-1}\left(\frac{150-50}{150}\right) = 41.81°$. The perpendicular distance to the action line of the force is $D = 2R - h = 300 - 50 = 250\ mm$. The sum of the moments about the contact point is

$M_B = -(R\cos\alpha)W + (2R - h)F = 0$, from which $\boxed{F = \frac{(150\cos 41.81°)W}{250} = 0.4472W = 17.88\ N}$

Problem 5.104 Use the fact that the horizontal bar is a three force member to determine the angle α and the magnitudes of the reactions at A and B.

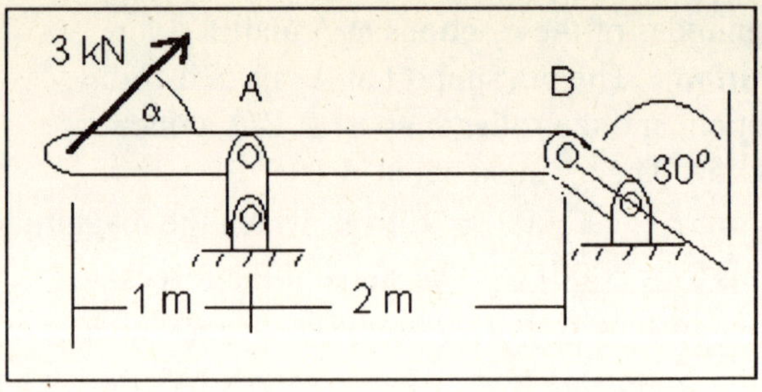

Solution: The sum of the moments about B is $M_B = -(3)(3\sin\alpha) + (2)A = 0$, from which $A = \dfrac{9\sin\alpha}{2} = 4.5\sin\alpha$. The sum of the forces: $\sum \vec{F}_X = 3\cos\alpha - B\sin 30^o = 0$, and $\sum F_Y = 3\sin\alpha - A + B\cos 30^o = 0$. Eliminate B as follows: solve each equation for B, and substitute the value for A: $B = \dfrac{1.5\sin\alpha}{\cos 30^o}$, and $B = \dfrac{3\cos\alpha}{\sin 30^o}$, from which $\dfrac{1.5\sin\alpha}{\cos 30^o} = \dfrac{3\cos\alpha}{\sin 30^o}$, or $\tan\alpha = \dfrac{3\cos 30^o}{1.5\sin 30^o} = 3.4641$, $\boxed{\alpha = 73.9^o}$. Substitute into the force equations to obtain $\boxed{B = 1.66\ kN}$, $\boxed{A = 4.32\ kN}$

Problem 5.105 The suspended load weighs 600 *lb.* Use the fact that ABC is a three-force member to determine the magnitudes of the reactions at A and B.

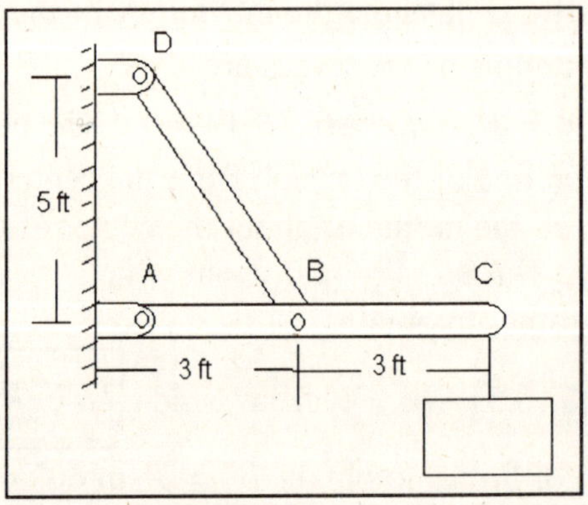

Solution: Isolate the member ABC. The angle ABD is 45° since the base and altitude of the triangle are equal·The sum of the moments about A is $M_B = +3B\sin 45^o - 6(600) = 0$ from which $\boxed{B = \dfrac{6(600)}{3\sin 45^o} = 1697.1\ lb}$. The sum of the forces $\sum F_X = A_X - B\cos 45^o = 0$, from which $A_X = 1199.8\ lb$. $\sum F_Y = A_Y + B\sin 45^o - 600 = 0$, from which $A_Y = 600 - 1199.82 = -599.82$. The magnitude at A is $\boxed{A = \sqrt{1199.8^2 + 599.8^2} = 1341.4\ lb}$

Problem 5.106 (a) Is the L-shaped bar a three-force member? (b) Determine the reactions at A and B. (c) Are the three forces acting on the L-shaped bar concurrent?

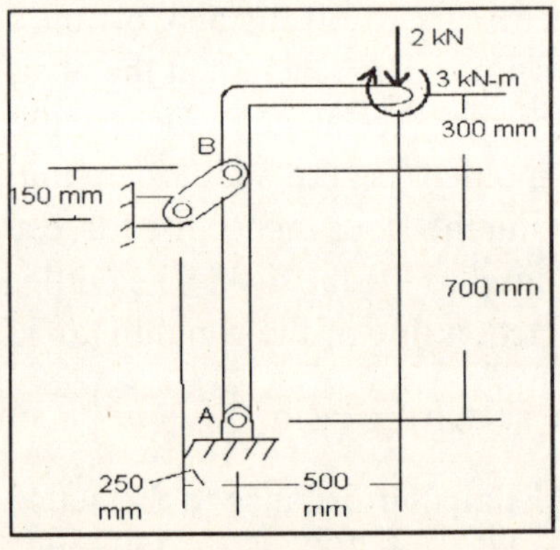

Solution: (a) No. The reaction at B is one-force, and the reaction at A is two-force. The couple keeps the L-shaped bar from being a three force member.(b) The angle of the member at B with the horizontal is $\alpha = \tan^{-1}\left(\dfrac{150}{250}\right) = 30.96^o$. The sum of the moments about A is $\sum M_B = -3 - 0.5(2) + 0.7B\cos\alpha = 0$, from which $\boxed{B = 6.6637\ kN}$. The sum of forces: $\sum F_X = A_X + B\cos\alpha = 0$, from which $A_X = -5.7143\ kN$. $\sum F_Y = A_Y - B\sin\alpha - 2 = 0$, from which $A_Y = 5.4281\ kN$. The magnitude at A: $\boxed{A = \sqrt{5.71^2 + 5.43^2} = 7.88\ kN}$ (c) No, by inspection.

Problem 5.107 The bucket of the excavator is supported by the two force member AB and the pin support at C. Its weight is $W = 1500\ lb$. What are the reactions at C?

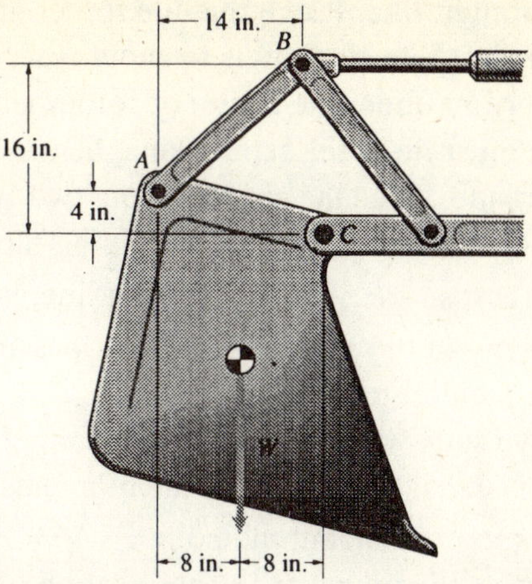

Solution: The angle of the member AB relative to the positive x axis is $\alpha = \tan^{-1}\left(\frac{12}{14}\right) = 40.6^o$. The moment about the point C is

$M_C = 4A\cos 40.6^o + 16A\sin 40.6^o + 8W = 0$, from which $A = -0.5948W = -892.23\ lb$. The sum of forces:

$\sum F_X = Cx - A\cos 40.6^o = 0$, from which:

$\boxed{C_X = -677.4\ lb}$

$\sum F_Y = C_Y - A\sin 40.6^o - W = 0$, from which

$\boxed{C_Y = 919.4\ lb}$

Problem 5.108 A rectangular plate is subjected to two forces A and B (Fig a). In Fig b, the two forces are resolved into components. By writing equilibrium equations in terms of the components A_X, A_Y, B_X, and B_Y, show that the two forces are equal in magnitude, opposite in direction, and directed along the line between their lines of action.

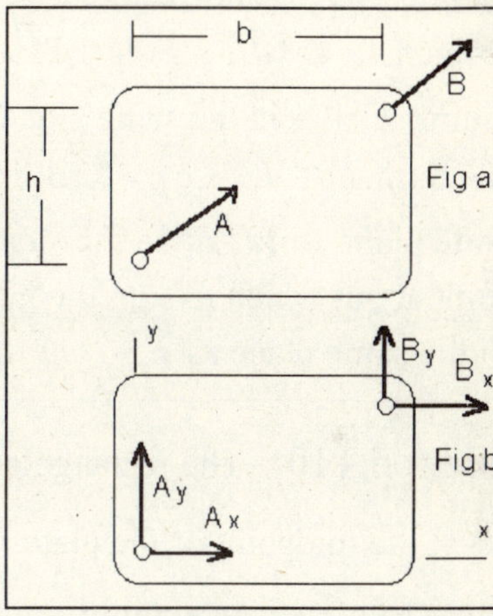

Solution: . The sum of forces:

$\sum F_X = A_X + B_X = 0$, from which $A_X = -B_X$

$\sum F_Y = A_Y + B_Y = 0$, from which $A_Y = -B_y$.

These last two equations show that A and B are equal and opposite in direction, since if the components are equal and opposite, the vectors are equal and opposite.

To show that the two vectors act along the line connecting the two points, determine the angle of the vectors relative to the positive x-axis. The sum of the moments about A is $M_A = bB_X - hB_y = 0$, from which the angle of direction of B is $\tan^{-1}\left(\frac{B_Y}{B_X}\right) = \tan^{-1}\left(\frac{h}{b}\right) = \alpha_B$. or $(180 + \alpha_B)$ Similarly, by substituting A: $\tan^{-1}\left(\frac{A_Y}{A_X}\right) = \tan^{-1}\left(\frac{h}{b}\right) = \alpha_A$, or $(180 + \alpha_A)$. But $\alpha = \tan^{-1}\left(\frac{h}{b}\right)$ describes direction of the line from A to B. The two vectors are opposite in direction, therefore the angles of direction of the vectors is one of two possibilities: B is directed along the line from A to B, or along the line B to A, and A is directed along the same line, oppositely to B.

==================================<>==================================

Problem 5.109 An object in equilibrium is subjected to three forces whose points of application lie on a straight line. Prove that the forces are coplanar.

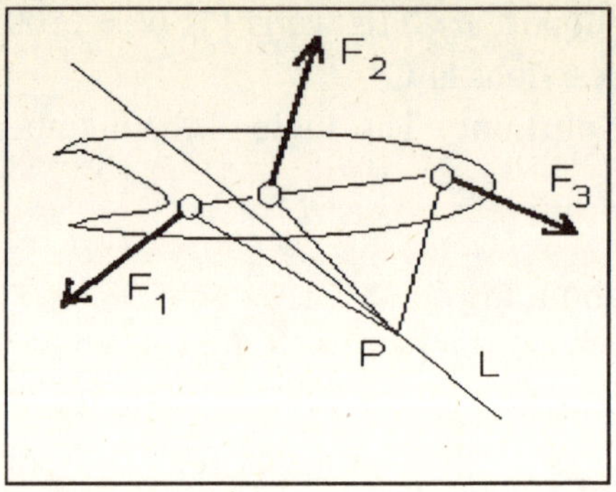

Solution: The strategy is to show that for a system in equilibrium under the action of forces alone, any two of the forces must lie in the same plane, hence all three must be in the same plane, since the choice of the two was arbitrary.
Let P be a point in a plane containing the straight line and one of the forces, say $\vec{F}_2$. Let L also be a line, not parallel to the straight line, lying in the same plane as $\vec{F}_2$, passing through P. Let $\vec{e}$ be a vector parallel to this line L..
First we show that the sum of the moments about any point in the plane is equal to the sum of the moments about one of the points of application of the forces. The sum of the moments about the point P: $\vec{M} = \vec{r}_1 \times \vec{F}_1 + \vec{r}_2 \times \vec{F}_2 + \vec{r}_3 \times \vec{F}_3 = 0$, where the vectors are the position vectors of the points of the application of the forces relative to the point P. (The position vectors lie in the plane.) Define $\vec{d}_{12} = \vec{r}_2 - \vec{r}_1$, and $\vec{d}_{13} = \vec{r}_3 - \vec{r}_1$. Then the sum of the moments can be rewritten, $\vec{M} = \vec{r}_1 \times (\vec{F}_1 + \vec{F}_2 + \vec{F}_3) + \vec{d}_{12} \times \vec{F}_2 + \vec{d}_{13} \times \vec{F}_3 = 0$. Since the system is in equilibrium, $\vec{F}_1 + \vec{F}_2 + \vec{F}_3 = 0$, and the sum of moments reduces to $\vec{M} = \vec{d}_{12} \times \vec{F}_2 + \vec{d}_{13} \times \vec{F}_3 = 0$, which is the moment about the point of application of $\vec{F}_1$. (The vectors $\vec{d}_{12}$, $\vec{d}_{13}$ are parallel to the line L.)The component of the moment parallel to the line L is $\vec{e} \cdot (\vec{d}_{12} \times \vec{F}_2)\vec{e} + \vec{e} \cdot (\vec{d}_{13} \times \vec{F}_3)\vec{e} = 0$, or $\vec{F}_2 \cdot (\vec{d}_{12} \times \vec{e})\vec{e} + \vec{F}_3 \cdot (\vec{d}_{13} \times \vec{e})\vec{e} = 0$. But by definition, $\vec{F}_2$ lies in the same plane as the line L, hence it is normal to the cross product $\vec{d}_{12} \times \vec{e} \neq 0$, and the term $\vec{F}_2 \cdot (\vec{d}_{12} \times \vec{e}) = 0$. But this means that $\vec{F}_3 \cdot (\vec{d}_{13} \times \vec{e})\vec{e} = 0$, which implies that $\vec{F}_3$ also lies in the same plane as $\vec{F}_2$, since $\vec{d}_{13} \times \vec{e} \neq 0$. Thus the two forces lie in the same plane. Since the choice of the point about which to sum the moments was arbitrary, this process can be repeated to show that $\vec{F}_1$ lies in the same plane as $\vec{F}_2$. *Thus all forces lie in the same plane.*

==================================<>==================================

Problem 5.110 The rectangular plate is held in equilibrium by the horizontal force $\vec{F}$. The weight W acts at the midpoint of the plate. The ratio $\frac{b}{h} = 4$. Determine the angle α at which the plate is in equilibrium for 5 values of the ratio $\frac{F}{W}$: 0, 0.5, 1.0, 1.5, and 2. (Assume that $0 \le \alpha \le 90^o$).

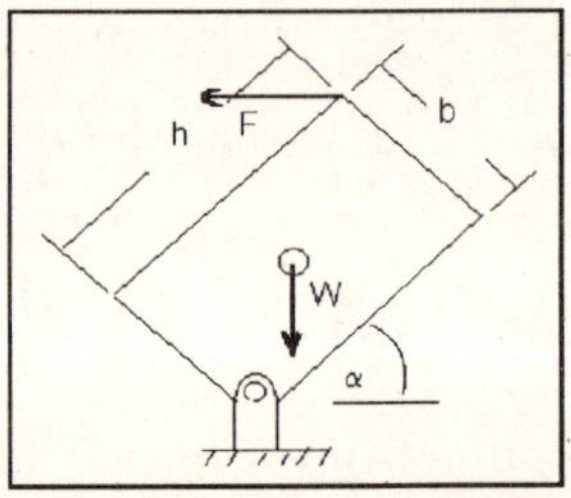

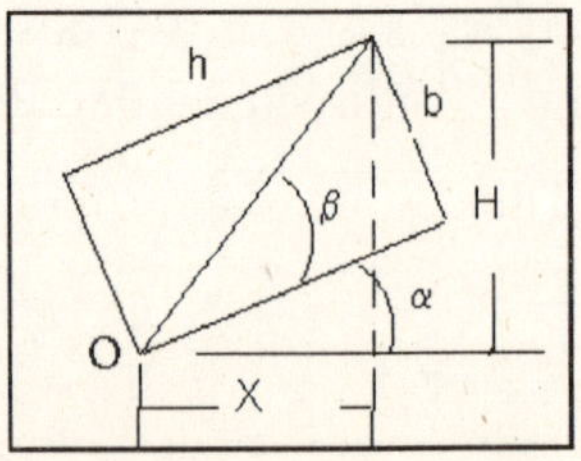

Solution: The moment about the pinned joint of the force is $M_F = HF$, where y is the vertical distance from the pinned joint to the action line of the force. This distance is the projection of the diagonal of the rectangle on the vertical axis, as shown in the second sketch: $H = \sqrt{b^2 + h^2} \sin(\alpha + \beta)$, where $\beta = \tan^{-1}\frac{h}{b}$. By a similar construction, the horizontal distance to the action line of the weight is $X = \left(\frac{1}{2}\right)\sqrt{b^2 + h^2} \cos(\alpha + \beta)$. The sum of the moments about the pinned joint is

Problem 5.110 continued on next page....

==================================<>==================================

continuation of solution to Problem 5.110...

$$\sum M = F\left(\sqrt{b^2+h^2}\right)\sin(\alpha+\beta) - \left(\frac{W}{2}\right)\left(\sqrt{b^2+h^2}\right)\cos(\alpha+\beta) = 0.$$

Noting that $\cos\beta = \dfrac{b}{\sqrt{b^2+h^2}}$, and $\sin\beta = \dfrac{h}{\sqrt{b^2+h^2}}$, using the double angle formulas:
$\cos(\alpha+\beta) = \cos\alpha\cos\beta - \sin\alpha\sin\beta$, and $\sin(\alpha+\beta) = \sin\alpha\cos\beta + \cos\alpha\sin\beta$, the moment equation reduces to $\sum M = F(b\sin\alpha + h\cos\alpha) - \left(\frac{W}{2}\right)(b\cos\alpha - h\sin\alpha) = 0$, from which

$$\frac{F}{W} = \frac{\left(\frac{b}{h}\cos\alpha - \sin\alpha\right)}{2\left(\cos\alpha + \frac{b}{h}\sin\alpha\right)} = \frac{\left(\frac{b}{h} - \tan\alpha\right)}{2\left(1 + \frac{b}{h}\tan\alpha\right)}.$$ Inverting: $\tan\alpha = \dfrac{\left(\frac{b}{h} - 2\frac{F}{W}\right)}{\left(2\frac{F}{W}\frac{b}{h} + 1\right)}$. Substituting $\frac{b}{h} = 4$, the results are for $\frac{F}{W} = 0, 0.5, 1.0, 1.5, 2.0,$ $\boxed{\alpha = 75.96^o, 30.96^o, 12.53^o, 4.40, 0}$

==================================<>==================================

Problem 5.111 The bar AB is of length L and weight W, and its weight acts at its midpoint. The tension in the string is $T = 1.2W$. What is the length of the string?

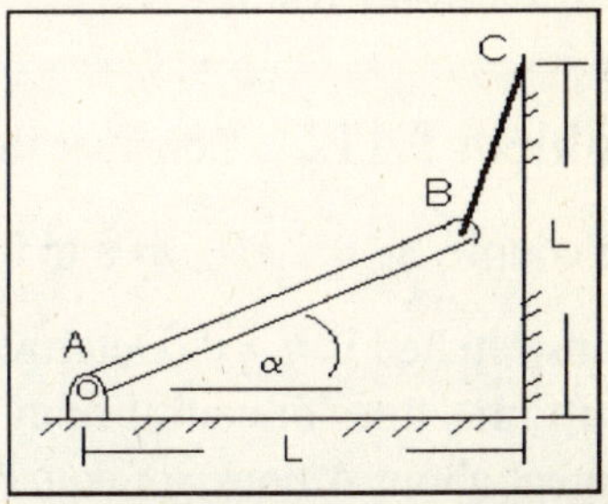

Solution: The strategy is to determine the angle formed by the string on the end of the rod (see sketch). The vertical distance from the pinned joint to the end of the rod is $D = L\sin\alpha$. The horizontal distance is $H = L\cos\alpha$. The sides of the small triangle formed by the string is $X = L - L\cos\alpha$ and $Y = L - L\sin\alpha$. The length of the string is $L_S = L\sqrt{(1-\cos\alpha)^2 + (1-\sin\alpha)^2}$ The angle formed by the string with the horizontal is $\beta = \tan^{-1}\left(\dfrac{1-\sin\alpha}{1-\cos\alpha}\right)$. The angle relative to the rod is $\theta = \beta - \alpha$. The moment about the pinned joint is

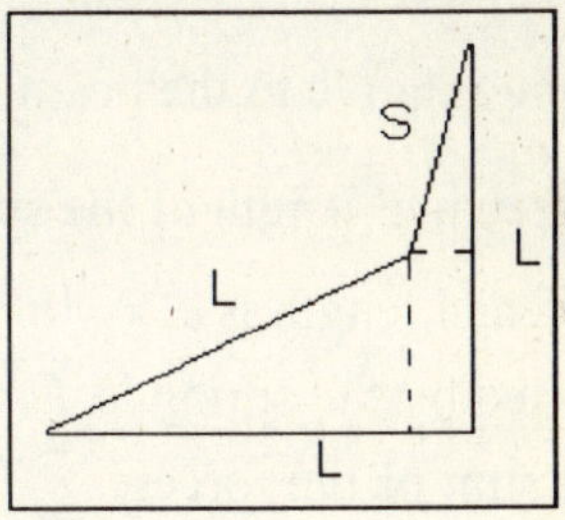

$\sum M = -\left(\frac{L}{2}\right)W\cos\alpha + LT\sin\theta = 0$, from which, substituting

$T = 1.2W$, $\cos\alpha = 2.4\sin\theta$

The angle α is a function of the length of the string: the length may be found from the zero crossing of the graph of the expression

$$f\left(\frac{Length}{L}\right) = 2.4\sin\theta - \sin\alpha$$

where $\theta = \beta - \alpha$, $\beta = \tan^{-1}\left(\dfrac{1-\sin\alpha}{1-\cos\alpha}\right)$, and the string length is $Length = L\sqrt{(1-\cos\alpha)^2 + (1-\sin\alpha)^2}$

The commercial package **TK Solver 2** was used to plot $f(Length/L)$ against $Length/L$. The procedure used was to graph the function on a coarse scale to find the approximate solution, and then to graph the function in the neighborhood of the solution.

Problem 5.111 continued on next page....

==================================<>==================================

continuation of solution to Problem 5.111...

From the graph, the length of the string is about $\boxed{Length = 0.4255L}$ at the zero crossing. At this value, the angle α is about

$\alpha = 40.3^o$

[*Explanatory Note:* The usual procedure would be to graph $f(\alpha)$ against α to determine the zero crossing value, and then to determine the length from this value of α. However, a feature of **TK Solver 2** is that it is unnecessary to express α explicitly in terms of the length, so that the length can be used directly as the independent variable. This feature offers advantages in the presentation of the solution, since the variable of interest is directly used. Since personal judgment is required in determining a zero crossing value, this elimination of an intermediate step also tends to improve precision in the result.]

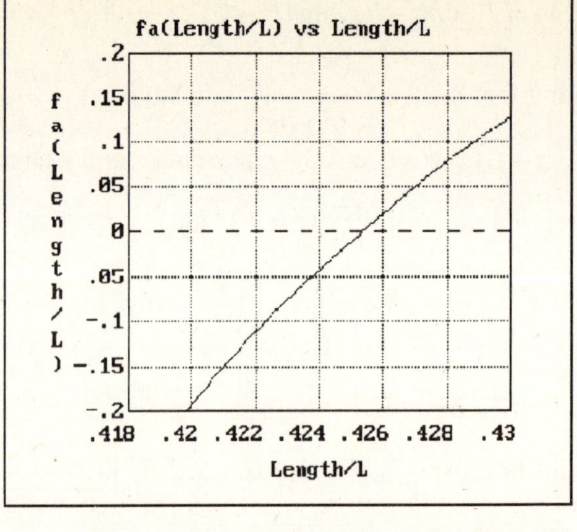

==================================<>==================================

Problem 5.112 Consider the system shown in Problem 5.55. The distances $a = 2\ m$ and $b = 1\ m$. The couple $M = 1\ kN\text{-}m$ and the force $F = 2\ kN$. The spring constant is $k = 3\ \dfrac{kN}{m}$. The spring would be unstretched if $h = 0$. Determine the value of h for equilibrium and the reactions at A.

Solution: Loss of equilibrium occurs when the moment about A does not equal zero. The angle of the spring relative to the beam is $\alpha = \tan^{-1}\left(\dfrac{h}{a+b}\right)$. The unstretched length of the spring is $a + b = 3\ m$. The stretched length is $L = \sqrt{h^2 + (a+b)^2}\ m$. The force exerted by the spring is $T = k\Delta L = 3(L - (a+b))\ kN$. The sum of the forces: $\sum F_X = A_X - T\cos\alpha = 0$, from which $A_X = T\cos\alpha$. $\sum F_Y = A_Y - F + T\sin\alpha = 0$, from which $A_Y = F - T\sin\alpha$. The sum of the moments about A is $\sum M_A = -M - aF + (a+b)T\sin\alpha = 0$, from which $F = \dfrac{(a+b)T\sin\alpha - M}{a}$. The threshold of equilibrium is determined by $f(h) = -M - aF + (a+b)T\sin\alpha \geq 0$. Graph $f(h)$ against h to determine the value of h at the zero crossing of $f(h)$, from which the other unknowns can be determined.

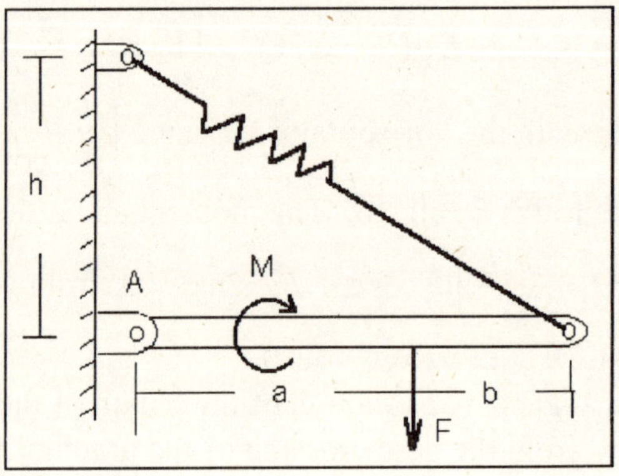

Problem 5.112 continued on next page....

continuation of solution to Problem 5.112...

The commercial package **TK Solver 2** was used to graph $f(h)$ against *h*. The equations are:

$f(h) = -M - aF + (a+b)T\sin\alpha \geq 0$,

$\alpha = \tan^{-1}\left(\dfrac{h}{a+b}\right)$, $L = \sqrt{h^2 + (a+b)^2}\ m$,

$T = k\Delta L = 3(L - (a+b))\ kN$, $A_X = T\cos\alpha$, and

$A_Y = F - T\sin\alpha$.

The value of *h* at the zero crossing is

$\boxed{h \geq 2.456\ m}$

The reactions are

$\boxed{A_X = 2.03\ kN}$

$\boxed{A_Y = 0.338\ kN}$

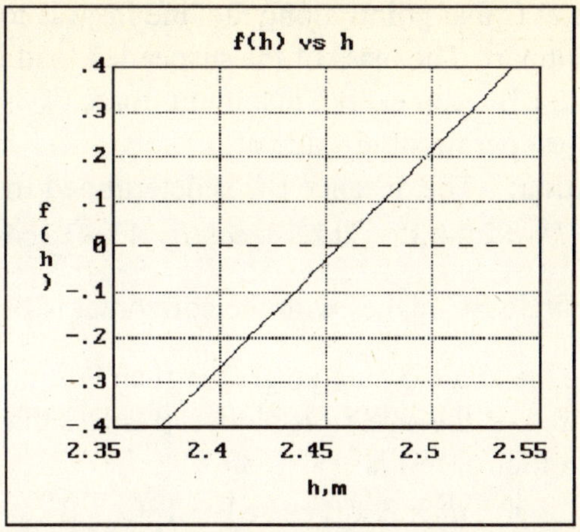

Problem 5.113 Consider the system shown in Problem 5.56. The bar is 1 *m* long and its weight $W = 35\ N$ acts at its midpoint. The distance $b = 0.75\ m$. The spring constant is $k = 100\ \dfrac{N}{m}$ and the spring is unstretched when the bar is vertical. Determine the angle α and the reactions at A.

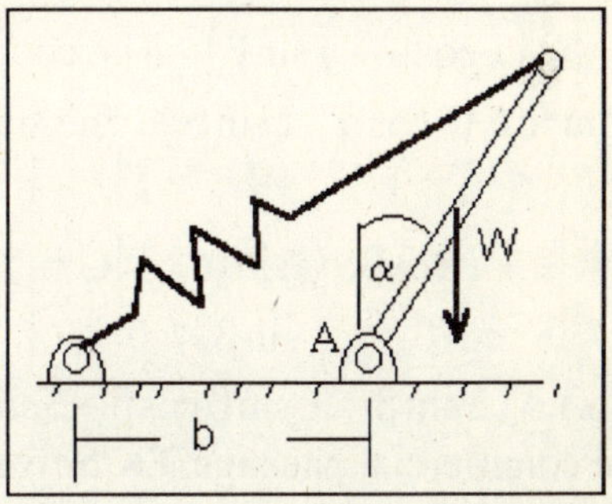

Solution: The unstretched length of the spring is $L = \sqrt{b^2 + 1^2} = 1.25\ m$. The obtuse angle is $90 + \alpha$, so the stretched length can be determined from the cosine law: $L_2^2 = 1^2 + 0.75^2 - 2(0.75)\cos(90+\alpha) = 1.5625 + 1.5\sin\alpha$ The force exerted by the spring is $T = k\Delta L = 100(L_2 - 1.25)\ N$. The angle between the spring and the bar can be determined from the sine law: $\dfrac{b}{\sin\beta} = \dfrac{L_2}{\sin(90+\alpha)}$, from which $\sin\beta = \dfrac{b\cos\alpha}{L_2}$.. The angle the spring makes with the horizontal is $\theta = 180 - \beta - 90 - \alpha = 90 - \beta - \alpha$. The sum of the forces: $\sum F_X = A_X - T\cos\theta = 0$, from which $A_X = T\cos\theta\ N$. $\sum F_Y = A_Y - W - T\sin\theta = 0$, from which $A_Y = W + T\sin\theta$. The sum of the moments about A is $\sum M_A = T\sin\beta - \left(\dfrac{W}{2}\right)\sin\alpha = 0$. The function

$f(\alpha) = T\sin\beta - \left(\dfrac{W}{2}\right)\sin\alpha$ is to be graphed against α to determine the value of α at the zero crossing. The commercial package **TK Solver 2** was used to graph the function. At the zero crossing $\boxed{\alpha = 44.1^o}$

$\boxed{A_X = 32.6\ N}$, and $\boxed{A_Y = 51.2\ N}$

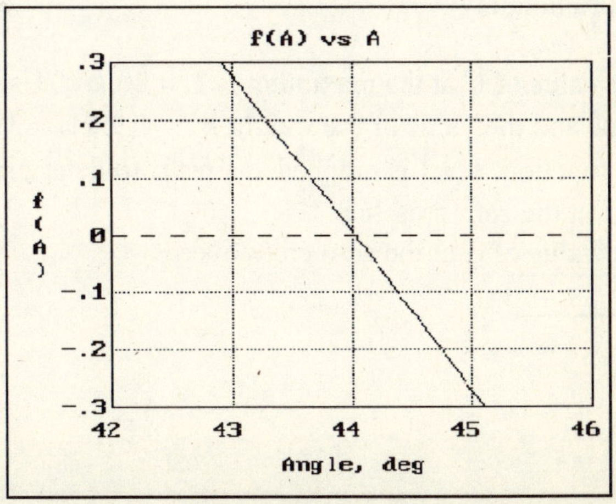

Problem 5.114 Consider the system shown in Problem 5.126. The hydraulic actuator BC exerts a force at C that points along the line from B to C. Treat A as a pin support. The mass of the suspended load is 4000 *kg*. If the actuator BC can exert a maximum force of 80 *kN* what is the smallest permissible value of α?

Solution: The distance BC is determined from the cosine law: $L^2 = 2^2 + 3^2 - 2(2)(3)\cos\alpha = 13 - 12\cos\alpha$.

The angle BC makes with the horizontal is $\beta = \sin^{-1}\left(\dfrac{3\sin\alpha}{L}\right)$.

Since the angle β may be greater than 90°, a quadrant check is required. If the angle γ is obtuse, β is less than or equal to 90°, thus a logical test is

if $\gamma \geq 90^o$ *then* $\beta = \beta$ *else* $\beta = 180^o - \beta$.

The angle γ is found as follows: the acute angle θ is found from the sine law: $\theta = \sin^{-1}\left(\dfrac{2\sin\alpha}{L}\right)$, and $\gamma = 180 - \alpha - \theta$.

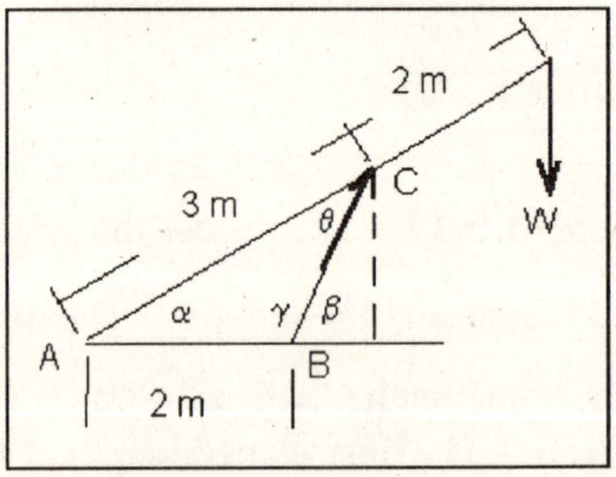

The coordinates of C with respect to an origin at A are $C(2 + L\cos\beta,\ 3\sin\alpha)$. The force exerted by the actuator: $\vec{C} = |\vec{C}|(\vec{i}\cos\beta + \vec{j}\sin\beta)$. The coordinates of the end of the boom are $(5\cos\alpha,\ 5\sin\alpha)$. The sum of the moments

$$\vec{M} = \begin{vmatrix} \vec{i} & \vec{j} & \vec{k} \\ 2 + L\cos\beta & 3\sin\alpha & 0 \\ \cos\beta & \sin\beta & 0 \end{vmatrix} C + \begin{vmatrix} \vec{i} & \vec{j} & \vec{k} \\ 5\cos\alpha & 5\sin\alpha & 0 \\ 0 & -W & 0 \end{vmatrix} = 0$$

$f(\alpha) = (2\sin\beta + L\sin\beta\cos\beta - 3\sin\alpha\cos\beta)C - 5W\cos\alpha = 0$

The commercial package **TK Solver 2** was used to graph the reaction at C and the angle β against α, as shown. The reaction decreases with an increase in α for two reasons: the perpendicular distance to the action line of the weight decreases, and the line BC tends toward a perpendicular to AC. Note that the angle β becomes greater than 90° for $\alpha > 48^o$. (approximately)..

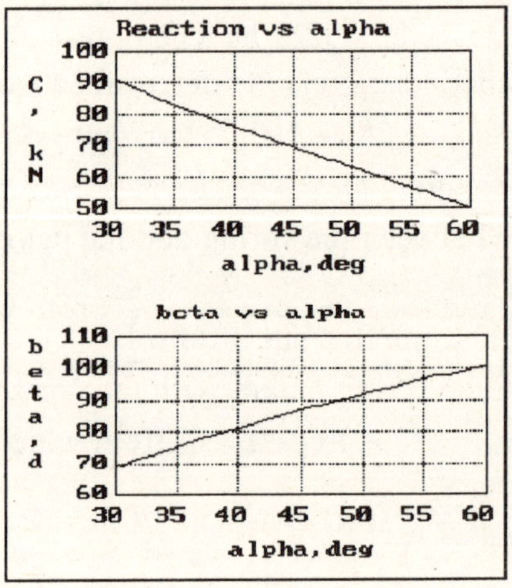

The value of C at the maximum is $C = 80\ kN$. Using this value and the value of the weight $W = 4(9.81) = 39.24\ kN$, the function $f(\alpha)$ is graphed against α to determine the value of α at the zero crossing.

The value of α at the zero cross over is

$\boxed{\alpha = 37.15^o}$

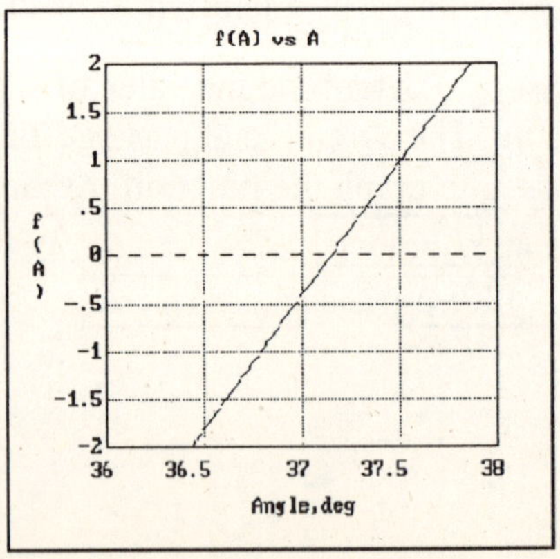

Problem 5.115 The beam is in equilibrium in the position shown. Each spring has an unstretched length of 1 m. Determine the distance b and the reactions at A.

Solution: The strategy is to determine the value of b that will cause the moment about A to equal zero. The angle of the top spring is $\alpha = \tan^{-1}\left(\frac{1}{3}\right) = 18.435^o$. The angle of the bottom spring is $\beta = \tan^{-1}\left(\frac{1}{b}\right)$. The stretched length of the top spring is $L_T = \sqrt{1^2 + 3^2} = 3.1623\ m$. The tension in the top spring is $T_T = (200)(3.1623 - 1) = 432.456\ N$. The tension in the bottom spring is

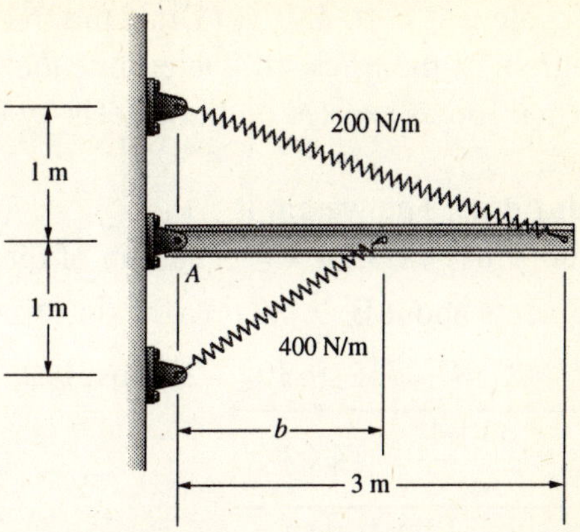

$T_B = 400\left(\sqrt{b^2 + 1} - 1\right) N$. The sum of the moments about the point A:

$M_A = -bT_B \sin\beta + 3(432.456)\sin 18.435^o = -bT_s$

. The sum of the forces:

$\sum F_X = A_X - T_B \cos\beta - 432.456 \cos 18.435 = 0$

, from which $A_X = T_B \cos\beta - 410.264\ N$.

$\sum F_Y = A_Y - T_B \sin\beta + 432.456 \sin 18.435^o = 0$,

from which $A_Y = T_B \sin\beta - 136.755\ N$.

The solution is obtained by graphing the sum of moments equation, $f(b) = -bT_B \sin\beta + 410.26$, against b to determine the value of b at the zero crossing.

The commercial package **TK Solver 2** was used to graph the function.

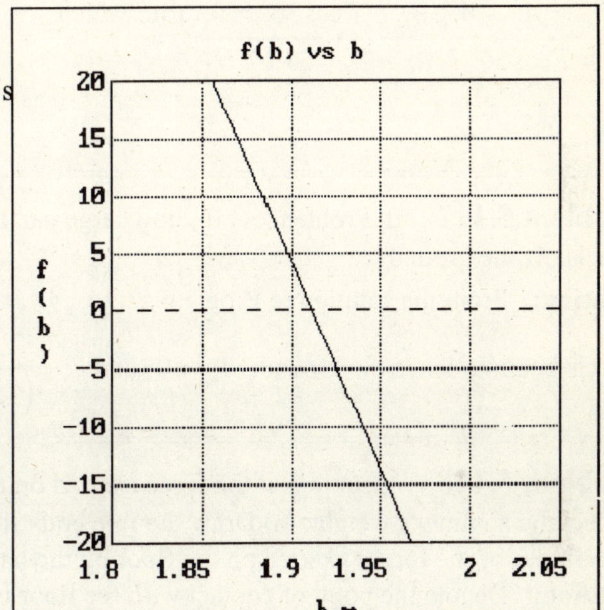

At the zero crossing the value of b is $\boxed{b = 1.91\ N}$ The reactions at A are: $\boxed{A_X = 820\ N}$, and $\boxed{A_Y = 77.8\ N}$

The values of A_X, A_Y are changing rapidly in the neighborhood of the zero crossing, so that the results are good only to three significant figures, at best.

Problem 5.116 The mass of the truck is 4 *Mg*. Its wheels are locked and the tension in its cable is $T = 10\ kN$. (a) Draw the freebody diagram of the truck. (b) Determine the normal forces exerted on the wheels by the road.

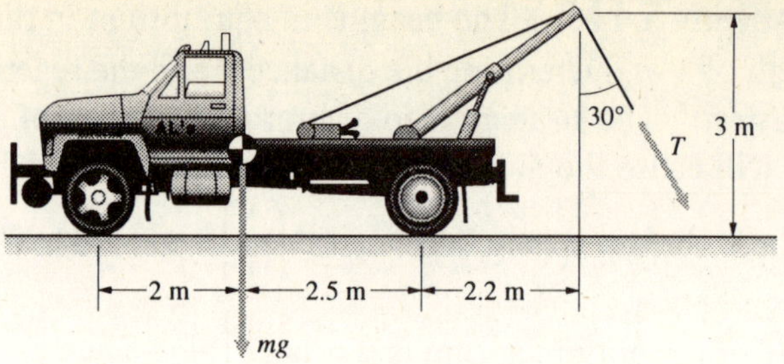

Solution: The weight is $4000(9.81) = 39.24\ kN$. The sum of the moments about B $\sum M_B = -3T\sin 30^o - 2.2T\cos 30^o + 2.5W - 4.5A_N = 0$ from which $A_N = 2.5W - T(3\sin 30^o + 2.2\cos 30^o)$,

$$\boxed{A_N = \frac{64.047}{4.5} = 14.23\ N}$$

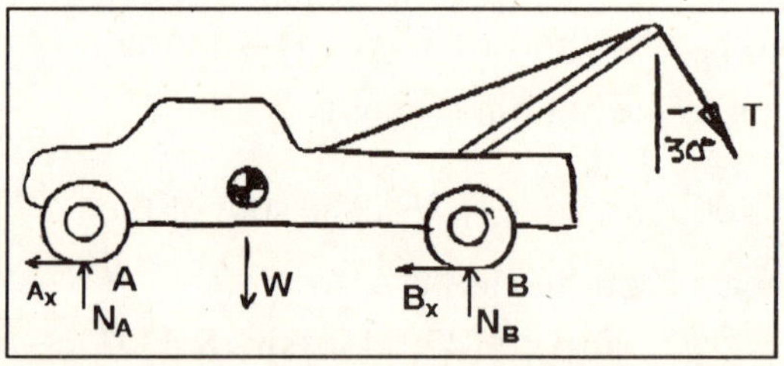

The sum of the forces:

$\sum F_Y = A_N - W + B_N - T\cos 30^o = 0$, from which

$$\boxed{B_N = T\cos 30^o - A_N + W = 33.67\ N}$$

Problem 5.117 In Problem 5.116, how large can the tension be without lifting the truck's front wheels off the ground?

Solution: From the solution to Problem 5.116, $A_N = 2.5W - T(3\sin 30^o + 2.2\cos 30^o)$. If the truck's wheels begin to leave the ground, $A_N = 0$, and $T = \dfrac{2.5W}{(3\sin 30^o + 2.2\cos 30^o)} = \dfrac{2.5(39.24)}{3.4053}\boxed{= 28.81\ kN}$

Problem 5.118 Assume that the force exerted on the head of the nail by the hammer is vertical and neglect the hammer's weight. (a) Draw the free body diagram of the hammer. (b) Determine the magnitudes of the forces exerted on the floor by the hammer.

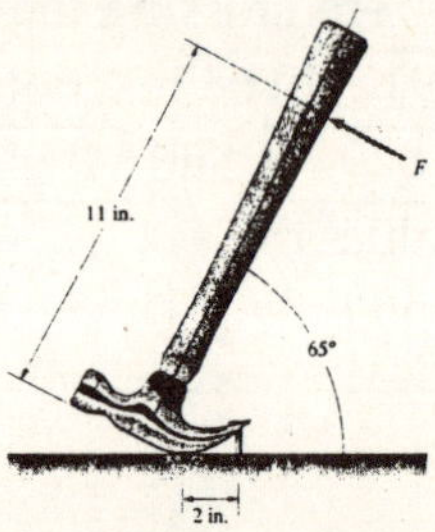

Solution: Denote the point of contact with the floor by B. The perpendicular distance from B to the line of action of the force is 11 in. The sum of the moments about B is $M_B = 11F - 2F_N = 0$, from which the force exerted by the nail head is $\boxed{F_N = \dfrac{11F}{2} = 5.5F}$. The sum of the forces: $\sum F_X = -F\cos 25 + H_x = 0$, from which the friction force exerted on the hammer is $\boxed{H_X = 0.9063F}$. $\sum F_Y = N_H - F_N + F\sin 25^o = 0$, from which the normal force exerted by the floor on the hammer is $\boxed{N_H = 5.077F}$

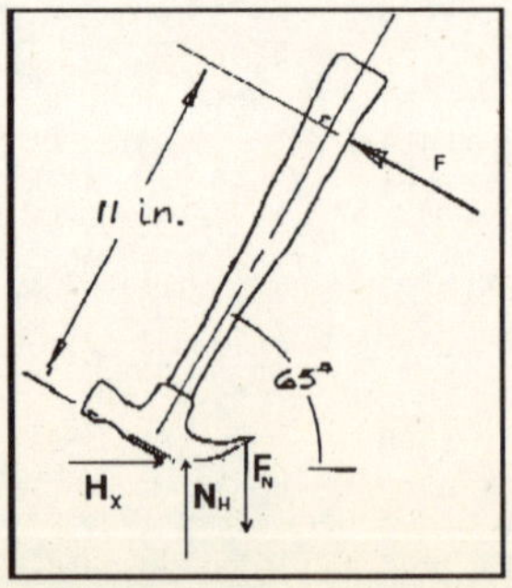

If the force on the handle is $F = 10\ lb$, then $\boxed{F_N = 55\ lb}$, $\boxed{H_X = 9.063\ lb}$, and $\boxed{N_H = 50.77\ lb}$

Problem 5.119 Scientists speculate that the Stegosaur could stand onit hind limbs for short periods to feed. Based on the free body diagram shown and assuming that $m = 2000\ kg$, determine the magnitudes of the forces B and C exerted by the ligament muscle brace and vertical column and determine the angle α.

Solution: Take the origin to be at the point of application of the force C. The position vectors of the points of application of the forces B and W are:

$\vec{r}_B = -415\vec{i} + 160\vec{j}\ (mm),\ \vec{r}_W = 790\vec{i} + 580\vec{j}\ (mm)$. The forces are

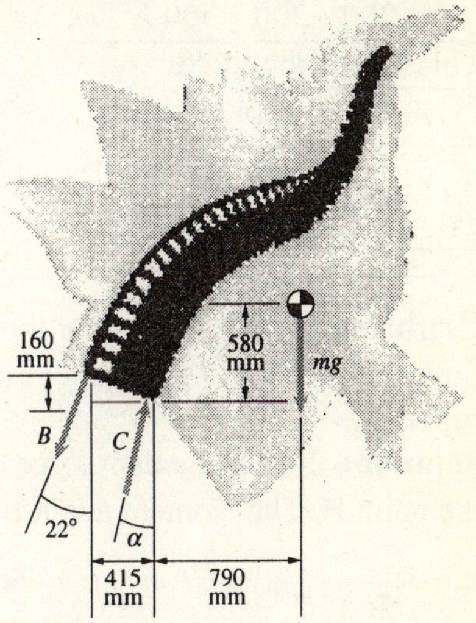

$\vec{C} = C(\vec{i}\cos(90^\circ - \alpha) + \vec{j}\sin(90^\circ - \alpha)) = C(\vec{i}\sin\alpha + \vec{j}\cos\alpha).$

$\vec{B} = B(\vec{i}\cos(270^\circ - 22^\circ) + \vec{j}\sin(270^\circ - 22^\circ))$

$= B(-0.3746\vec{i} - 0.9272\vec{j}).$

$\vec{W} = -2(9.81)\vec{j} = -19.62\vec{j}\ (kN)$. The moments about C,

$$\vec{M}_C = \begin{vmatrix} \vec{i} & \vec{j} & \vec{k} \\ -415 & 160 & 0 \\ -0.3746B & -0.9272B & 0 \end{vmatrix} + \begin{vmatrix} \vec{i} & \vec{j} & \vec{k} \\ 790 & 580 & 0 \\ 0 & -19.62 & 0 \end{vmatrix} = 0$$

$= 444.72B - 15499.8 = 0$, from which $\boxed{B = \frac{15499.8}{444.72} = 34.85\ kN}$

. The sums of the forces: $\sum \vec{F}_X = (C\sin\alpha - 0.3746B)\vec{i} = 0$, from which $C\sin\alpha = 13.06\ kN$.

$\sum \vec{F}_Y = (C\cos\alpha - 0.9272B - 19.62)\vec{j} = 0$, from which $C\cos\alpha = 51.93\ kN$.

The angle α is $\boxed{\alpha = \tan^{-1}\left(\frac{13.06}{51.93}\right) = 14.1^\circ}$. The magnitude of $\vec{C}$, $\boxed{C = \sqrt{13.06^2 + 51.93^2} = 53.55\ kN}$

Problem 5.120 (a) Draw the free body diagram of the 50 *lb* plate and explain why it is statically indetermate. (b) Determine as many of the reactions at A and B as possible.

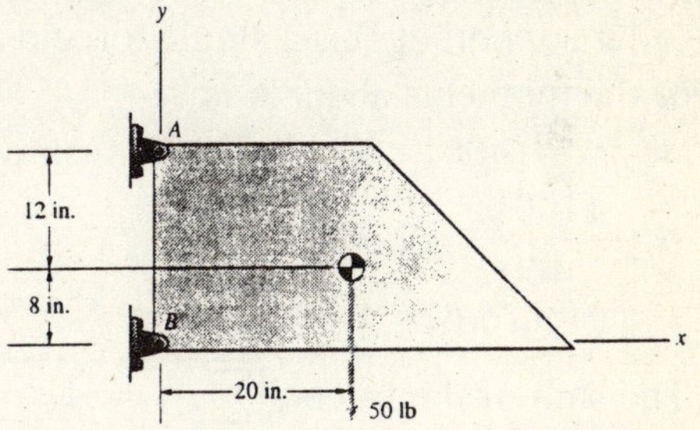

Solution: (a) The pin supports at A and B are two-force supports, thus there are four unknown reactions A_X, A_Y, B_X, and B_Y, but only three equilibrium equations can be written, two for the forces, and one for the moment. Thus there are four unknowns and only three equations, so the system is indeterminate.

(b) The sums of the forces:

$\sum F_X = A_X + B_X = 0$, from which $A_X = -B_X$

$\sum F_Y = A_Y + B_Y - 50 = 0$.

The sum of the moments about B

$M_B = 20A_X - 50(20) = 0$, from which $\boxed{A_X = 50\ lb}$, and from the sum of forces $\boxed{B_X = -50\ lb}$.

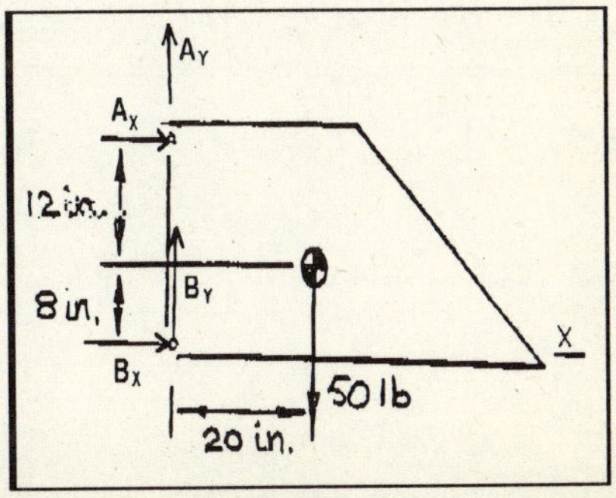

=================================<>=================================

Problem 5.121 (a) Draw the free body diagram of the beam. (b) Determine the reactions at the supports.

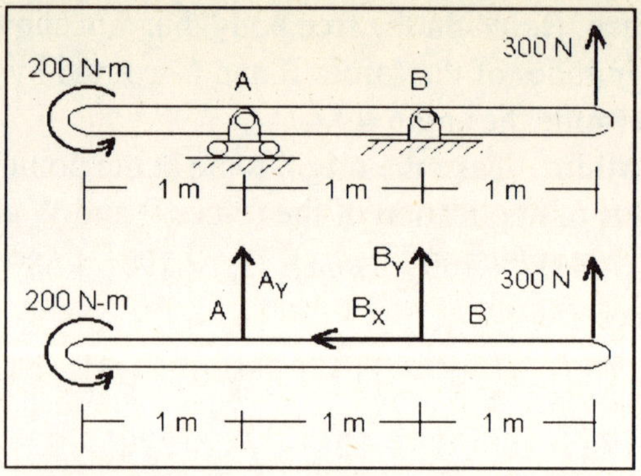

Solution: (a) The free body diagram is shown.

(b) The sum of the moments about B

$\sum M_B = +200-(1)A_Y+(1)300=0$, from which

$\boxed{A_Y = 200+300 = 500\ N\text{-}m}$

The sum of the forces:

$\boxed{\sum F_X = B_X = 0}$

$\sum F_Y = A_Y + B_Y + 300 = 0$, from which

$\boxed{B_Y = -A_Y - 300 = -800\ N}$

=================================<>=================================

Problem 5.122 Consider the beam shown in Problem 5.21. First represent the loads (the 300 *N* force and the 200 *N-m* couple) by a single equivalent force, and then determine the reactions at the supports.

Solution: The equivalent force is equal to the applied forces: $F = 300\ N$. Measure the distance x from the point B. The moment about B due to the loads: $M_B = 300x + 200 + 1(300) = 0$, from which $x = -\dfrac{500}{300} = -1.6667\ m$, or $1.6667\ m$ to the left of B. The reactions: The sum of the moments about B $M_B = 300(1.6667) - (1)A_Y = 0$ from which $\boxed{A_Y = 500\ N}$. The sum of the forces $\boxed{\sum F_X = B_X = 0}$, $\sum F_Y = A_Y + B_Y + 300 = 0$, from which $\boxed{B_Y = -A_Y - 300 = -800\ N}$

=================================<>=================================

Problem 5.123 The truss supports a 90-kg suspended object. What are the reactions at A and B?

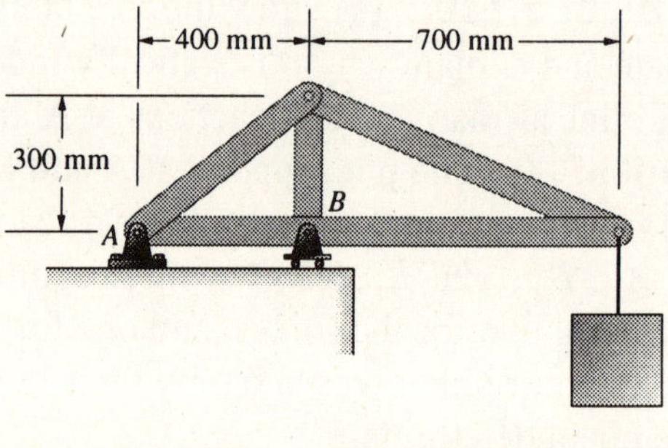

Solution: Treat the truss as a single element. The pin support at A is a two force reaction support; the roller support at B is a single force reaction. The sum of the moments about A is

$M_A = B(400) - W(1100) = 0$, from which

$B = \dfrac{1100W}{400} = 2.75W$

$B = 2.75(90)(9.81) = 2427.975\boxed{= 2.43\ kN}$

The sum of the forces: $\boxed{\sum F_X = A_X = 0}$

$\sum F_Y = A_Y + B - W = 0$, from which

$A_Y = W - B = 882.9 - 2427.975\ \boxed{= -1.545\ kN}$

=================================<>=================================

==================================<>==================================

Problem 5.124 The trailer is parked on a 15° slope. The hitch H behaves like a pin joint. Determine the reactions at A and H.

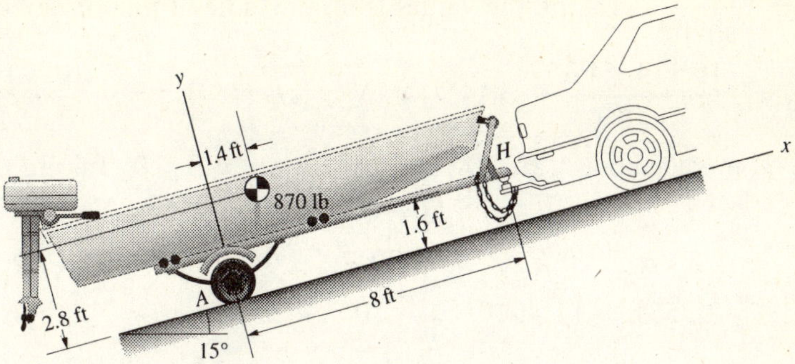

Solution: The coordinate system has the x-axis parallel to the road. The wheels are a one force reaction normal to the road, the pin H is a two force reaction. The position vectors of the points of the center of mass and H are:

$\vec{r}_W = 1.4\vec{i} + 2.8\vec{j}\ ft$

$\vec{r}_H = 8\vec{i} + 1.6\vec{j}$.

The angle of the weight vector realtive to the positive x-axis is $\alpha = 270^o - 15^o = 255^o$ The weight has the components $\vec{W} = W(\vec{i}\cos 255^o + \vec{j}\sin 255^o) = 870(-0.2588\vec{i} - 0.9659\vec{j}) = -225.173\vec{i} - 840.355\vec{j}\ (lb)$. The sum of the moments about H is $\vec{M}_H = (\vec{r}_W - \vec{r}_H) \times \vec{W} + (\vec{r}_A - \vec{r}_H) \times \vec{A}$,

$$\vec{M}_H = \begin{vmatrix} \vec{i} & \vec{j} & \vec{k} \\ -6.6 & 1.2 & 0 \\ -225.355 & -840.355 & 0 \end{vmatrix} + \begin{vmatrix} \vec{i} & \vec{j} & \vec{k} \\ -8 & -1.6 & 0 \\ 0 & A_Y & 0 \end{vmatrix} = 0 = 5816.55 - 8A_Y = 0, \text{ from which}$$

$\boxed{A_Y = \frac{5816.55}{8} = 727.1\ lb}$ The sum of the forces:

$\sum \vec{F}_X = (H_X - 225.173)\vec{i} = 0$, from which $\boxed{H_X = 225.2\ \ lb}$,

$\sum \vec{F}_Y = (A_Y + H_Y - 840.355)\vec{j} = 0$, from which $\boxed{H_Y = 113.3\ \ lb}$

==================================<>==================================

Problem 5.125 To determine the location of the point where the wqeight of a car acts (the *center of mass*), an engineer places the car on scales and measures the normal reactions at the wheels for two valuews of α, obtaining the following results:

α	A_Y, kN	B, kN
10°	10.134	4.357
20°	10.150	3.677

What are the distances *b* and *h* ?

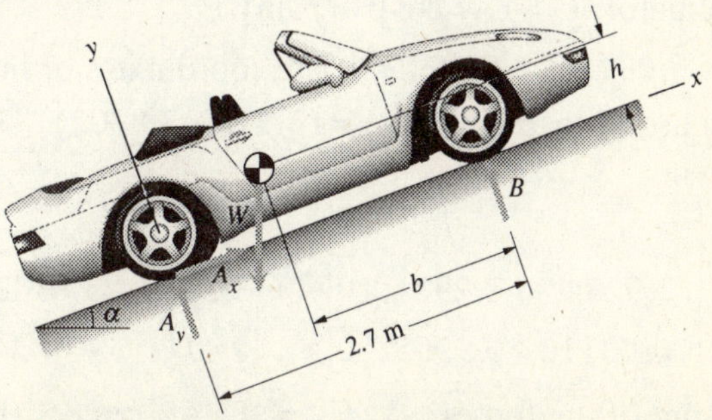

Solution: The position vectors of the cm and the point B are $\vec{r}_{CM} = (2.7 - b)\vec{i} + h\vec{j}$, $\vec{r}_B = 2.7\vec{i}$. . The angle between the weight and the positive x-axis is $\beta = 270 - \alpha$. The weight vector at each of the two angles is

$\vec{W}_{10} = W(\vec{i}\cos 260^o + \vec{j}\sin 260^o)$

$\vec{W}_{10} = W(-0.1736\vec{i} - 0.9848\vec{j})$

$\vec{W}_{20} = W(\vec{i}\cos 250^o + \vec{j}\sin 250^o)$

$\vec{W}_{20} = W(-0.3420\vec{i} - 0.9397\vec{j})$

Problem 5.125 continued on next page....

continuation of solution to Problem 5.125...

The weight W is found from the sum of forces: $\sum F_Y = A_Y + B_Y + W \sin\beta = 0$, from which $W_\beta = \dfrac{A_Y + B_Y}{\sin\beta}$. Taking the values from the table of measurements:

$$W_{10} = -\frac{10.134 + 4.357}{\sin 260^o} = 14.714\ kN\ ,\ [check:\ W_{20} = -\frac{10.150 + 3.677}{\sin 250^o} = 14.714\ kN\ \ check]$$

The moments about A are $\vec{M}_A = \vec{r}_{CM} \times \vec{W} + \vec{r}_B \times \vec{B} = 0$. Taking the values at the two angles:

$$\vec{M}_A^{10} = \begin{vmatrix} \vec{i} & \vec{j} & \vec{k} \\ 2.7-b & h & 0 \\ -2.5551 & -14.4910 & 0 \end{vmatrix} + \begin{vmatrix} \vec{i} & \vec{j} & \vec{k} \\ 2.7 & 0 & 0 \\ 0 & 4.357 & 0 \end{vmatrix} = 0 = 14.4903b + 2.5551h - 27.3618 = 0$$

$$\vec{M}_A^{20} = \begin{vmatrix} \vec{i} & \vec{j} & \vec{k} \\ 2.7-b & h & 0 \\ -5.0327 & -13.8272 & 0 \end{vmatrix} + \begin{vmatrix} \vec{i} & \vec{j} & \vec{k} \\ 2.7 & 0 & 0 \\ 0 & 3.677 & 0 \end{vmatrix} = 0\ \ 13.8272b + 5.0327h - 27.4054 = 0$$

These two simultaneous equations in two unknowns were solved using the HP-28S hand held calculator.
$\boxed{b = 1.80\ m}$, $\boxed{h = 0.50\ m}$

Problem 5.126

The hydraulic actuator BC exerts a force at C that points along the line from B to C. Treat point A as a pin support. The angle $\alpha = 50^o$. The mass of the suspended load is 4000 kg. Determine the force exerted at C and the reactions at A.

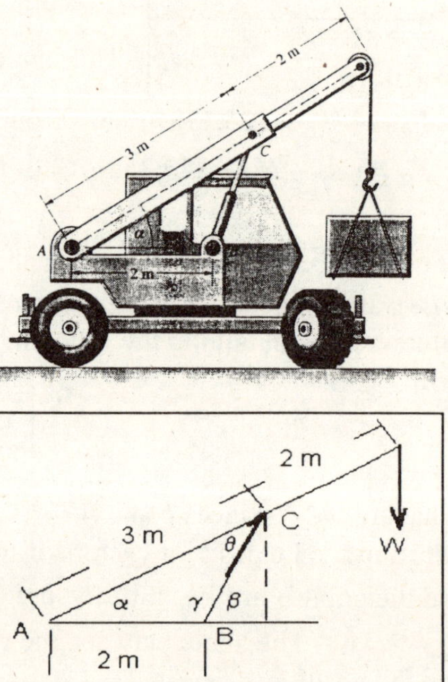

Solution: (See Problem 5.114 for discussion and graphs.)

The distance BC is determined from the cosine law: $L = \sqrt{2^2 + 3^2 - 2(2)(3)\cos\alpha} = \sqrt{5.2865} = 2.2992\ m$. The weight is $W = 4(9.81) = 39.24\ kN$. The angle BC makes with the horizontal is $\beta = \sin^{-1}\left(\dfrac{3\sin\alpha}{L}\right) = 88.215^o$ or $\beta = 91.785^o$.

If the angle γ is obtuse, β is less than or equal to 90°, thus a logical test is $if\ \gamma \geq 90^o\ then\ \beta = \beta\ \ else\ \beta = 180^o - \beta$.

where $\gamma = 180 - \alpha - \sin^{-1}\left(\dfrac{2\sin\alpha}{L}\right)$ from which $\beta = 91.785^o$

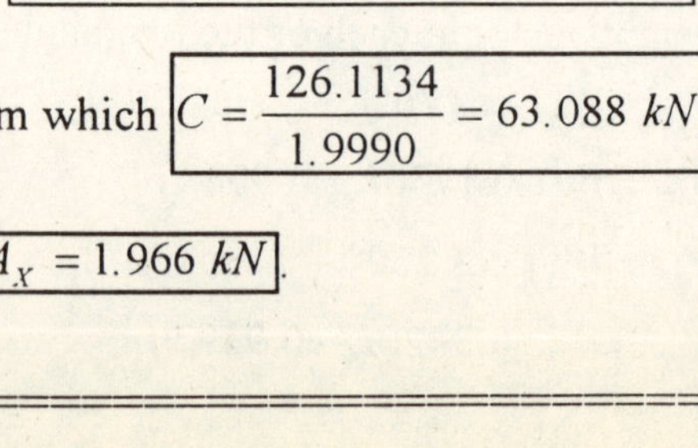

The coordinates of C with respect an origin at A are $(2 - L\cos\beta,\ 3\sin\alpha) = (1.9284, 2.2981)$. The force exerted by the actuator: $\vec{C} = C(\vec{i}\cos\beta + \vec{j}\sin\beta)$.
$= C(-0.03116\vec{i} + 0.9995\vec{j})$ The coordinates of the end of the boom are $(5\cos\alpha,\ 5\sin\alpha) = (3.2139, 3.8302)$. The sum of the moments

$$\vec{M} = \begin{vmatrix} \vec{i} & \vec{j} & \vec{k} \\ 1.9284 & 2.2981 & 0 \\ -0.03116 & 0.9995 & 0 \end{vmatrix} C + \begin{vmatrix} \vec{i} & \vec{j} & \vec{k} \\ 3.2139 & 3.8302 & 0 \\ 0 & -39.24 & 0 \end{vmatrix} = 0$$

, from which $\boxed{C = \dfrac{126.1134}{1.9990} = 63.088\ kN}$

The sum of the forces : $\sum F_X = A_X + C\cos\beta = 0$, from which $\boxed{A_X = 1.966\ kN}$.
$\sum F_Y = A_Y + C\sin\beta - W = 0$, from which $\boxed{A_Y = -23.817\ kN}$

================================<>================================

Problem 5.127 The horizontal rectangular plate weighs 800 N and is suspended by three vertical cables. The weight of the plate acts at its midpoint. What are the tensions in the cables?

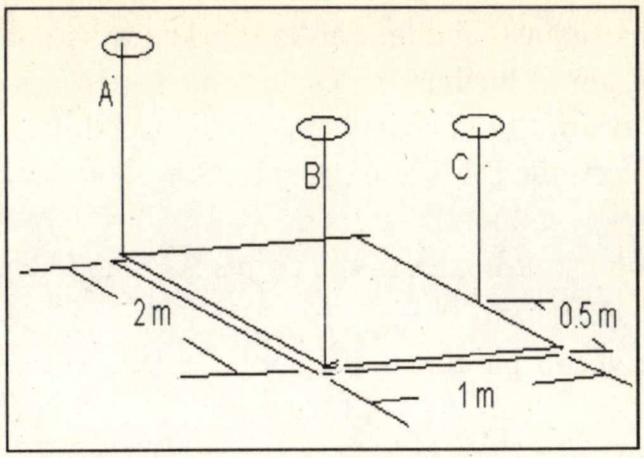

Solution: Choose an origin at A, with the x-axis coincident with the line AB, and the y-axis normal to the plate, positive upward. The position vectors of the points B, C, and the cm are

$\vec{r}_B = 2\vec{i}$, $\vec{r}_C = 1.5\vec{i} - 1\vec{k}$, $\vec{r}_{CM} = 1\vec{i} - 0.5\vec{k}$.

The sum of the moments about the point A is

$$\vec{M}_A = \vec{r}_B \times \vec{T}_B + \vec{r}_C \times \vec{T}_C + \vec{r}_{CM} \times \vec{W} = 0$$

$$\vec{M}_A = \begin{vmatrix} \vec{i} & \vec{j} & \vec{k} \\ 2 & 0 & 0 \\ 0 & T_B & 0 \end{vmatrix} + \begin{vmatrix} \vec{i} & \vec{j} & \vec{k} \\ 1.5 & 0 & -1 \\ 0 & T_C & 0 \end{vmatrix} + \begin{vmatrix} \vec{i} & \vec{j} & \vec{k} \\ 1 & 0 & -0.5 \\ 0 & -800 & 0 \end{vmatrix} = 0$$

$$\vec{M}_A = (T_C - 0.5(800))\vec{i} + (2T_B + 1.5T_C - 800)\vec{k} = 0$$

From which $\boxed{T_C = 400\ N}$, $\boxed{T_B = \frac{800 - 1.5T_C}{2} = 100\ N}$.

The reaction at A is found from the sum of forces:

$\sum F_Y = T_A + T_B + T_C - 800 = 0$, from which $\boxed{T_A = 800 - T_B - T_C = 400\ N}$

================================<>================================

Problem 5.128 Consider the suspended plate in Problem 5.127. The weight of the plate acts at its midpoint. If you represent the reactions exerted on the plate by the three cables by a single equivalent force, what is the force and where does its action line intersect the plate?

Solution: The equivalent force must equal the sum of the reactions: $F_{EQ} = T_A + T_B + T_C$ From the solution to Problem 5.127, $\boxed{F_{EQ} = 300 + 100 + 400 = 800\ N}$. The moment due to the action of the equaivalent force must equal the moment due to the reactions: The moment about A is

$$\vec{M}_A = \begin{vmatrix} \vec{i} & \vec{j} & \vec{k} \\ 2 & 0 & 0 \\ 0 & 100 & 0 \end{vmatrix} + \begin{vmatrix} \vec{i} & \vec{j} & \vec{k} \\ 1.5 & 0 & -1 \\ 0 & 400 & 0 \end{vmatrix} = \begin{vmatrix} \vec{i} & \vec{j} & \vec{k} \\ x & 0 & z \\ 0 & 800 & 0 \end{vmatrix}$$

$$\vec{M}_A = 400\vec{i} + 800\vec{k} = -(800z)\vec{i} + (800x)\vec{k},$$

from which $\boxed{z = -0.5\ m}$, and $\boxed{x = 1\ m}$., which corresponds to the midpoint of the plate. *Thus the equivalent force acts upward at the midpoint of the plate.*

================================<>================================

Problem 5.129 The 80 *lb* bar is supported by a ball and socket support at A, the smooth wall it leans against, and the cable AC. The weight of the bar acts at its midpoint. (a) Draw the free-body diagram of the bar. (b) Determine the tension in the cable and the reactions at A.

Solution: (a) The ball and socket is a three reaction force support; the cable and the smooth wall are each one force reaction supports.

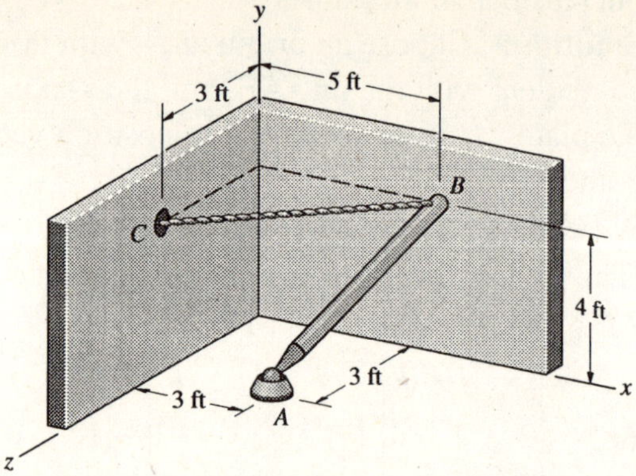

(b)The coordinates of the points A, B and C are $A(3,0,3)$, $B(5,4,0)$, and $C(0,4,3)$.

The vector parallel to the bar is

$\vec{r}_{AB} = \vec{r}_B - \vec{r}_A = 2\vec{i} + 4\vec{j} - 3\vec{k}$.

The length of the bar is $|\vec{r}_{AB}| = \sqrt{2^2 + 4^2 + 3^2} = 5.3852$.

The unit vector parallel to the bar is

$\vec{e}_{AB} = 0.3714\vec{i} + 0.7428\vec{j} - 0.5571\vec{k}$.

The vector parallel to the cable is

$\vec{r}_{BC} = \vec{r}_C - \vec{r}_B = -5\vec{i} + 3\vec{k}$.

The unit vector parallel to the cable is

$\vec{e}_{BC} = -0.8575\vec{i} + 0.5145\vec{k}$. The cable tension is $\vec{T} = |\vec{T}|\vec{e}_{BC}$.

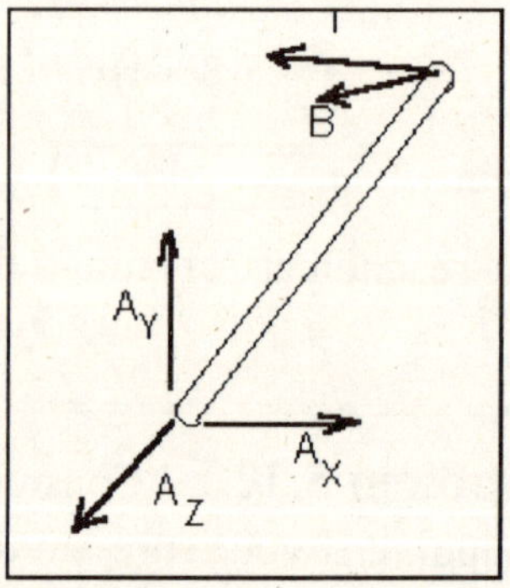

The point of application of the weight relative to A is $\vec{r}_{AW} = 2.6936\vec{e}_{AB}$

$\vec{r}_{AW} = 1.000\vec{i} + 2.000\vec{j} - 1.500\vec{k}$.

The reaction at B is $\vec{B} = |\vec{B}|\vec{k}$, since it is normal to a wall in the y-z plane. The sum of the moments about A is

$\vec{M}_A = \vec{r}_{AW} \times \vec{W} + \vec{r}_{AB} \times \vec{B} + \vec{r}_{AB} \times \vec{T} = 0$

$$\vec{M}_A = \begin{vmatrix} \vec{i} & \vec{j} & \vec{k} \\ 1 & 2 & -1.5 \\ 0 & -80 & 0 \end{vmatrix} + \begin{vmatrix} \vec{i} & \vec{j} & \vec{k} \\ 2 & 4 & -3 \\ 0 & 0 & |\vec{B}| \end{vmatrix} + \begin{vmatrix} \vec{i} & \vec{j} & \vec{k} \\ 2 & 4 & -3 \\ -0.8575 & 0 & 0.5145 \end{vmatrix} |\vec{T}| = 0$$

$\vec{M}_A = (-120 + 4|\vec{B}| + 2.058|\vec{T}|)\vec{i} - (2|\vec{B}| + 3.6015|\vec{T}|)\vec{j} + (3.43|\vec{T}| - 80)\vec{k} = 0$ Solve:

$\boxed{|\vec{T}| = \frac{80}{3.43} = 23.32\ lb}$ $\boxed{|\vec{B}| = \frac{120 - 2.058|\vec{T}|}{4} = 18.00\ lb}$ The reactions at A are found from the sums of forces:

$\sum F_X = A_X - |\vec{T}|(0.8575) = 0$ from which $\boxed{A_X = 20\ lb}$

$\sum F_Y = A_Y - 80 = 0$, from which $\boxed{A_Y = 80\ lb}$

$\sum F_Z = A_Z + |\vec{T}|(0.5145) + |\vec{B}| = 0$, from which $\boxed{A_Z = -30\ lb}$

Problem 5.130 The horizontal bar of weight W is supported by a roller support at A and the cable BC. Use the fact that the bar is a three-force member to determine the abgle α, the tension in the cable, and the magnitude of the reaction at A.

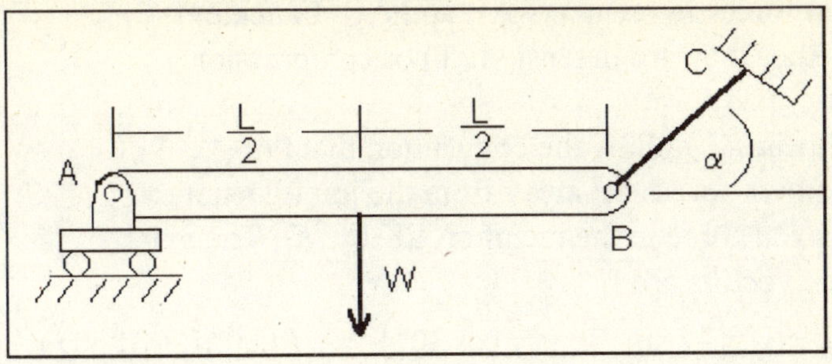

Solution: The sum of the moments about B is

$M_B = -LA_Y + \left(\frac{L}{2}\right)W = 0$, from which

$\boxed{A_Y = \frac{W}{2}}$. The sum of the forces:

$\sum F_X = T\cos\alpha = 0$, from which $\vec{T} = 0$ or $\cos\alpha = 0$. The choice is made from the sum of forces in the y-direction:

$\sum F_Y = A_Y - W + T\sin\alpha = 0$, from which $T\sin\alpha = W - A_Y = \frac{W}{2}$. This equation cannot be satisfied if $T = 0$, hence $\cos\alpha = 0$, or $\boxed{\alpha = 90^o}$, and $\boxed{T = \frac{W}{2}}$

Problem 5.131 The bicycle brake on the right is pinned to the bicycle's frame at A. Determine the force exerted by the brake pad on the wheel rim at B in terms of the cable tension T

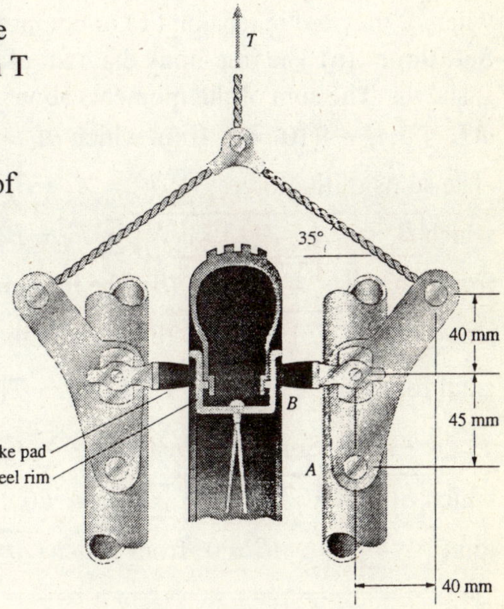

Solution: From the force balance equation for the cables: the force on the brake mechanism T_B in terms of the cable tension T is $T - 2T_B \sin 35^o = 0$, from which $T_B = \frac{T}{2\sin 35^o} = 0.8717T$

Take the origin of the system to be at A. The position vector of the point of attachment of B is $\vec{r}_B = 45\vec{j}\ (mm)$. The position vector of the point of attachement of the cable is $\vec{r}_C = 40\vec{i} + 85\vec{j}\ (mm)$. The force exerted by the brake pad is $\vec{B} = -B\vec{i}$. The force vector due the cable tension is $\vec{T}_B = T_B(\vec{i}\cos 145^o + \vec{j}\sin 145^o) = T_B(-0.8192\vec{i} + 0.5736\vec{j})$

The moment about A is

$\vec{M}_A = \vec{r}_B \times \vec{B} + \vec{r}_C \times \vec{T}_B = 0$

$$\vec{M}_A = \begin{vmatrix} \vec{i} & \vec{j} & \vec{k} \\ 0 & 45 & 0 \\ -B & 0 & 0 \end{vmatrix} + \begin{vmatrix} \vec{i} & \vec{j} & \vec{k} \\ 40 & 85 & 0 \\ -0.8192 & 0.5736 & 0 \end{vmatrix} T_B = 0$$

$\vec{M}_A = (45B + 92.576T_B)\vec{k} = 0$, from which $B = \frac{92.576T_B}{45} = 2.057T_B$. Substitute the expression for the cable tension: $\boxed{B = (2.057)(0.8717)T = 1.793T}$

Problem 6.1 The weight $W = 600\ lb$. Draw the free-body diagram of joint A. (b) Use the free-body diagram you obtained in Part (a) to determine the axial forces in members AB and AC. Indicate whether they are in tension (T) or compression (C).

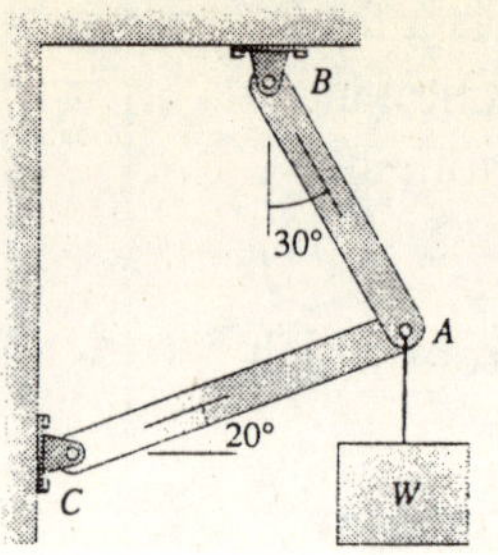

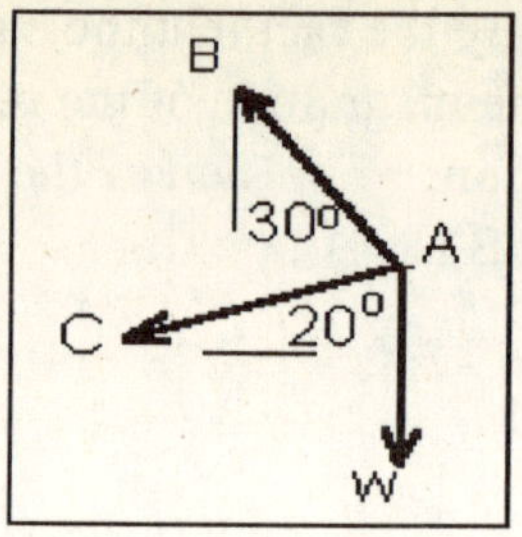

Solution: Follow the convention that the unknown forces act away from the joint. Denote the axial force in the member I,K by IK. The axial force vectors are

$$\vec{F}_{AB} = AB\left(-\vec{i}\sin 30^{\circ} + \vec{j}\cos 30^{\circ}\right) = AB(-0.5\vec{i} + 0.866\vec{j})$$

$$\vec{F}_{AC} = AC\left(-\vec{i}\cos 20^{\circ} - \vec{j}\sin 20^{\circ}\right) = AC\left(-0.9397\vec{i} - 0.3420\vec{j}\right)$$

$\vec{F}_{W} = -\vec{j}600\ (lb)$. The sum of forces: $\sum F_x = -0.5AB - 0.9397AC = 0$,

$\sum F_y = 0.866AB - 0.3420AC - 600 = 0$. Solve the two simultaneous equations: $\boxed{AB = 572.53\ lb\ (T)}$, $\boxed{AC = -304.64\ lb\ (C)}$

Problem 6.2 The truss supports a 10 kN load at C. (a) Draw the free body diagram of the entire truss and determine the reactions at its supports. (b) Determine the axial forces in its members. Indicate whether they are in tension (T) or compression (C).

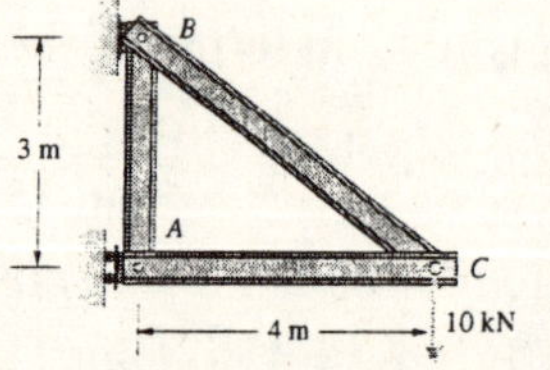

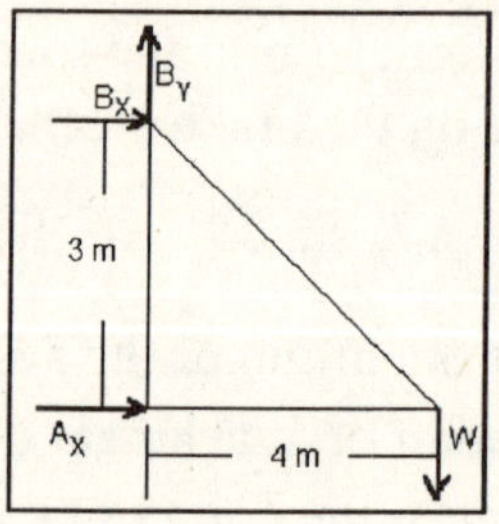

Solution: (a) The free-body diagram of the system is shown. The sum of the moments about B is: $M_B = 3A_x - 4(10) = 0$, from which $A_x = 13.33\ kN$. The sums of the forces: $\sum F_x = A_x + B_x = 0$, from which $\boxed{B_x = -A_x = -13.33\ kN}$. $\sum F_y = B_y - 10 = 0$, from which $\boxed{B_y = 10\ kN}$. (b) The interior angle ACB is $\alpha = \tan^{-1}(0.75) = 36.87^{\circ}$. (b) Assume that the unknown forces act away from the joint. Denote the axial force in the member I,K by IK. The axial forces are $\vec{F}_{CB} = BC\left(-\vec{i}\cos\alpha + \vec{j}\sin\alpha\right)$, and $\vec{F}_{CA} = -AC\vec{i}$. Summing the forces: $\sum F_y = BC\sin\alpha - 10 = 0$, from which $\boxed{BC = 16.67\ kN\ (T)}$. $\sum F_x = -BC\cos\alpha - AC = 0$, from which $\boxed{AC = -13.33\ kN\ (C)}$ For the joint A, $\sum F_y = AB = 0$, from which $\boxed{AB = 0}$

Problem 6.3 Consider the truss in Problem 6.2. An engineer decides that the members will safely support 30 kN axial forces in either tension or compression. What is the largest load the truss will safely support at C?

Solution: From the free body diagram for joint C (see the solution to Problem 6.2) the axial forces are $BC = 16.67\ kN\ (T)$ and $AC = -13.33\ kN\ (C)$. For a 30 kN axial force in BC, $\frac{30}{16.67} = \frac{W_{MAX\_BC}}{10}$., from which $W_{MAX\_BC} = 18\ kN$ For a 30 kN axial force in AC, $\frac{30}{13.33} = \frac{W_{MAX\_AC}}{10}$, from which $W_{MAX\_AC} = 22.5\ kN$. Thus the maximum load permitted is $\boxed{W_{MAX\_BC} = 18\ kN}$

Problem 6.4 text and solution:

Problem 6.4 The loads $F_1 = 60\ N$, and $F_2 = 40\ N$. (a) Draw the free body diagram of the entire truss and determine the reactions at its supports. (b) Determine the axial forces in its members. Indicate whether they are in compression (C) or tension (T).

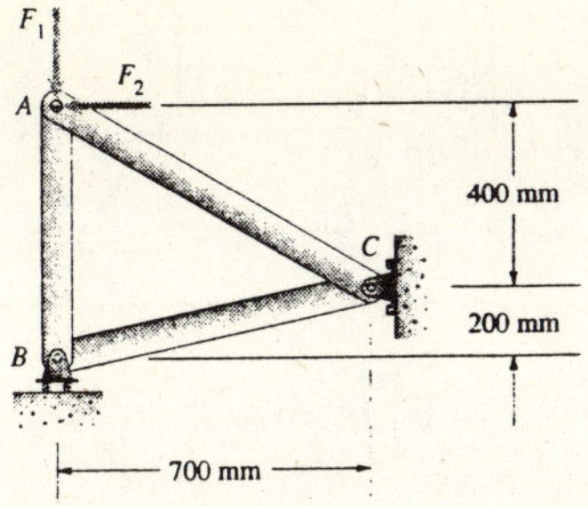

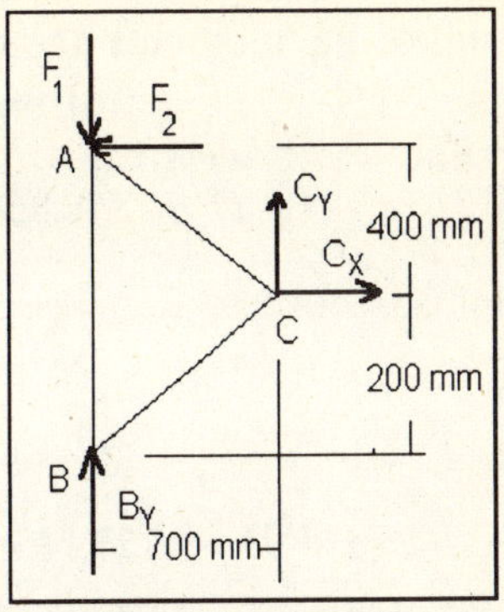

Solution: (a) The free-body diagram is shown. The sum of the moments about C is $M_C = -0.7B_y + 0.7F_1 + 0.4F_2 = 0$, from which

$$B_y = \frac{0.7(60) + 0.4(40)}{0.7}\ \boxed{B_y = 82.857\ N}.$$

The sums of forces:

$\sum F_x = C_x - F_2 = 0$, from which $\boxed{C_x = 40\ N}$.

$\sum F_y = B_y + C_y - F_1 = 0$, from which $\boxed{C_y = F_1 - B_y = -22.857 = -22.9\ N}$

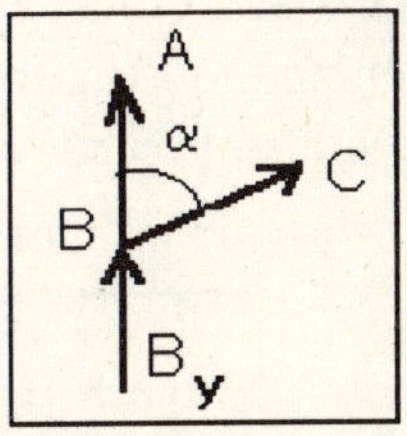

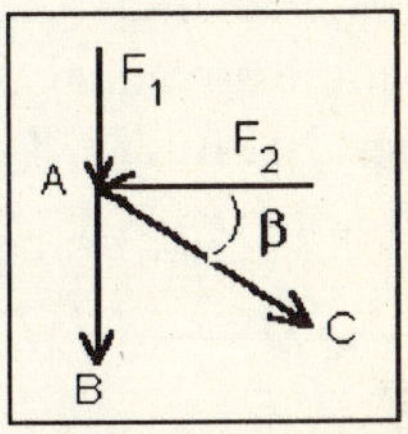

(b) The included angle CBA is $\alpha = 90^o - \tan^{-1}\left(\frac{200}{700}\right) = 74.05^o$. Denote the axial force in the member I,K by IK. The axial forces at joint B are $\vec{F}_{BC} = BC(\vec{i}\sin\alpha + \vec{j}\cos\alpha) = BC(0.9615\vec{i} + 0.2747\vec{j})$, $\vec{F}_{BA} = BA\vec{i}$, and the external force is $\vec{B} = B_y\vec{j}$.

The sums of the forces: $\sum F_x^B = 0.9615BC = 0$, from which $\boxed{BC = 0}$.

$\sum F_y^B = B_y + BA = 0$, from which $\boxed{BA = -B_y = -82.9\ N\,(C)}$ The magnitude of the angle formed by member AC with respect to the x-axis is $\beta = \tan^{-1}\left(\frac{0.4}{0.7}\right) = 29.74^o$. The sums of the axial forces:

$$\sum F_x^A = -F_2 + AC\cos\beta = 0, \text{ from which } \boxed{AC = \frac{F_2}{\cos\beta} = 46.1\ N\,(T)}$$

Problem 6.5 Consider the truss in Problem 6.4. The loads $F_1 = 440\ N$ and $F_2 = 160\ N$. Determine the axial forces in the members. Indicate whether they are in tension (T) or compression (C).

Solution: From the solution to Problem 6.4, the sum of the moments about C is $M_C = -0.7B_Y + 0.7F_1 + 0.4F_2 = 0$, from which $\boxed{B_y = \frac{0.7F_1 + 0.4F_2}{0.7} = 531.43\ N}$. Denote the axial force in the member I,K by IK. The axial loads at joint B are (see solution to Problem 6.4) $\boxed{AB = -B_y = -531.4\ N\,(C)}$. and $\boxed{BC = 0}$. Similarly, the sum of the forces at the joint A is

$$\sum F_x^A = -F_2 + AC\cos\beta = 0, \text{ from which } \boxed{AC = \frac{F_2}{\cos\beta} = 184.3\ N\,(T)}$$

Problem 6.6 The load $F = 600\ lb$. Determine the axial forces in the members.

Solution: The free body diagram of the system is shown. The sum of the moments about A is $M_A = -4F_l + 7D_Y = 0$, from which $D_y = \frac{4}{7}F_l$, $D_y = 342.86\ lb$

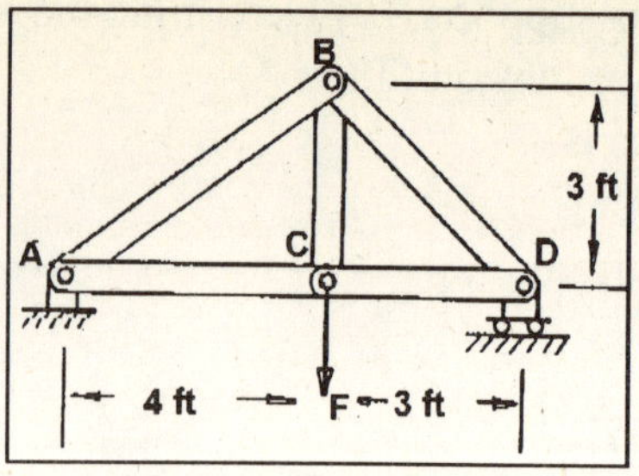

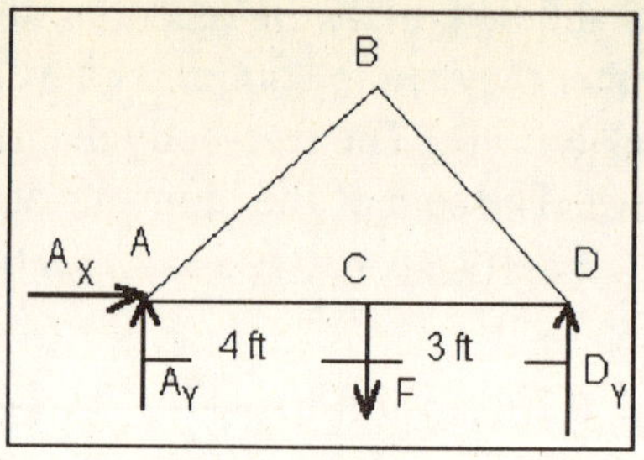

The sums of forces:

$\sum F_y = A_y = 0$, from which $A_y = 0$. $\sum F_y = A_y - F + D_y = 0$, from which $A_y = F - D_y = 257.14\ lb$.

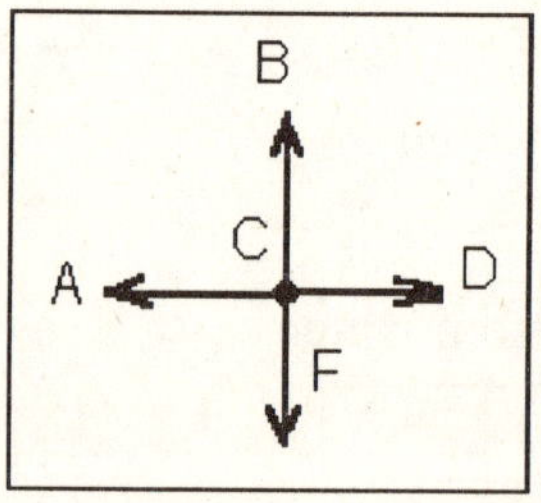

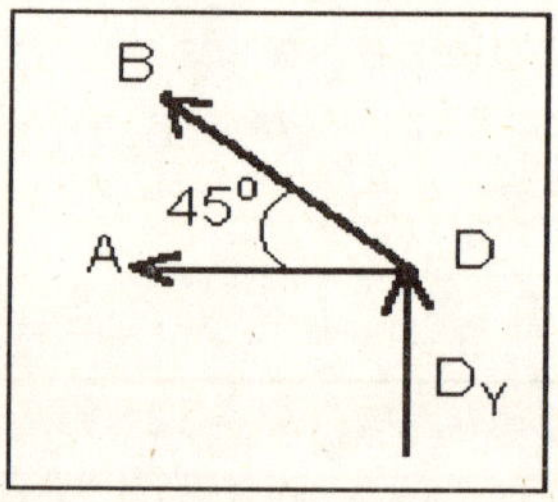

The included angle ADB is 45°, since the legs of the triangle are equal. Denote the axial force in the member I,K by IK. The sums of forces at D:

$\sum F_y^D = BD \sin 45^o + D_y = 0$, from which,

$BD = -\sqrt{2}D_y = -\sqrt{2}(342.86)$ $\boxed{BD = -484.9\ lb\ (C)}$.

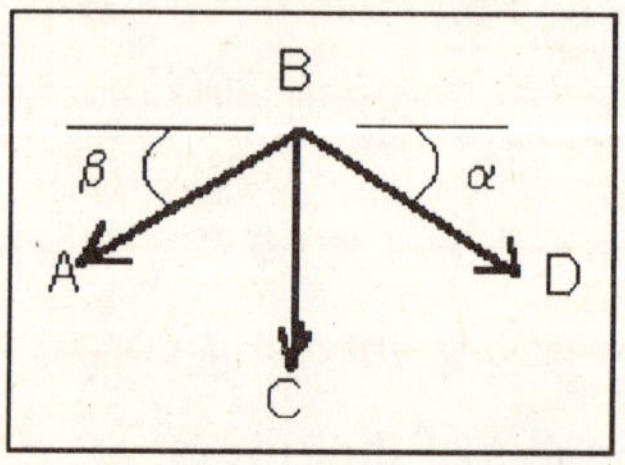

$\sum F_y^D = -CD - BD \cos 45^o = 0$, from which $CD = -BD(0.707)$, $\boxed{CD = 342.9\ lb\ (T)}$. At the joint C, the sums of forces: $\sum F_y^C = CD - AC = 0$, from which $\boxed{AC = CD = 342.9\ lb\ (T)}$ $\sum F_y^C = BC - F = 0$, from which $\boxed{BC = F = 600\ lb\ (T)}$. The angles at the joint B are $\alpha = 45^o$, and $\beta = \tan^{-1}\left(\frac{3}{4}\right) = 36.87^o$. The sums of forces: $\sum F_y^B = BD \cos\alpha - AB \cos\beta = 0$, from which $AB = \left(\frac{\cos\alpha}{\cos\beta}\right)BD = \boxed{-428.6\ lb\ (C)}$

Problem 6.7 Consider the truss shown in Problem 6.6. Each member will safely support a tensile force of 2000 lb and a compressive force of 1000 lb. Based on this criterion, determine the largest downward load F that can be safely applied at C.

Solution: From the solution to Problem 6.6, the reactions at D and A are: $D_y = \frac{4}{7}F$, and $A_y = \frac{3}{7}F$.

For compression members: $BD = \frac{4\sqrt{2}F}{7} > AB = \frac{8F}{7\cos\beta}$, from which the maximum force is limited by BD: $F_S = \frac{7(1000)}{4\sqrt{2}} = 1237\ lb$. For tension: $BC = F > AC = \frac{4}{7}F$, from which the maximum force is limited by BC: $F_S = 2000\ lb$. Thus the limit is $\boxed{F_S = 1237\ lb}$ [*Check*: The system is linear, so the ratio of applied forces is equal to the ratio of axial forces in each member. From Problem 6.6: for max compression $\frac{F_S}{600} = \frac{1000}{484.95}$, from which $F_S = 1237\ lb$. For the max tension $\frac{F_S}{600} = \frac{2000}{600}$, from which $F_S = 2000\ lb$. Thus the limit is in compression and $F_S = 1237\ lb$ *Check.*]

Problem 6.8 The load $F = 10\ kN$. Determine the axial forces in the members.

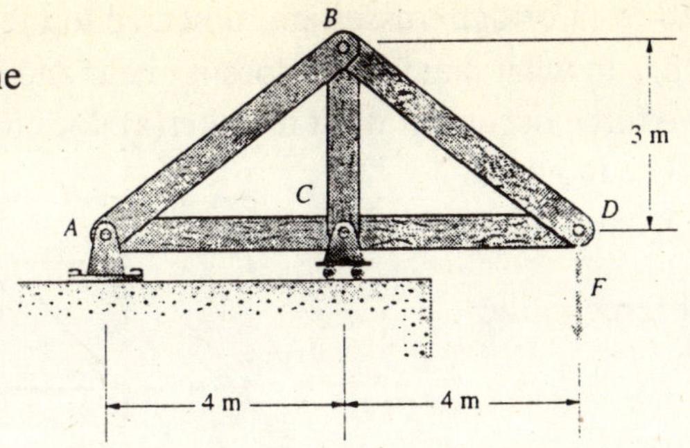

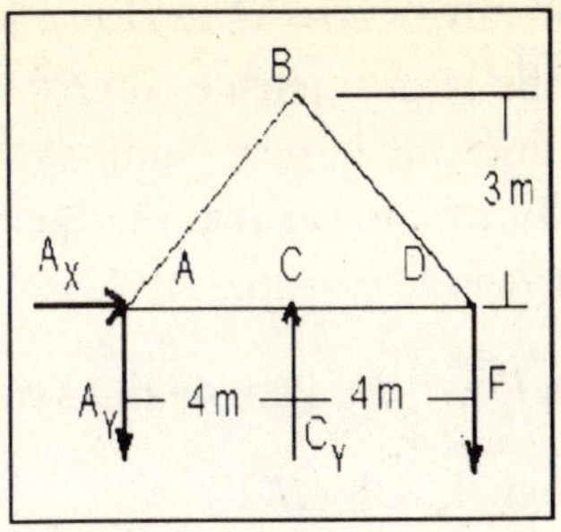

Solution: The sum of the moments about A is $M_A = 4C_y - 8F = 0$, from which $C_y = 2F = 20\ kN$. The included angle CDB is $\alpha = \tan^{-1}\left(\frac{3}{4}\right) = 36.87^o$. Denote the axial force in the member I, K by IK. The sum of forces in the joint D: $\sum F_y^D = BD\sin\alpha - F = 0$, from which

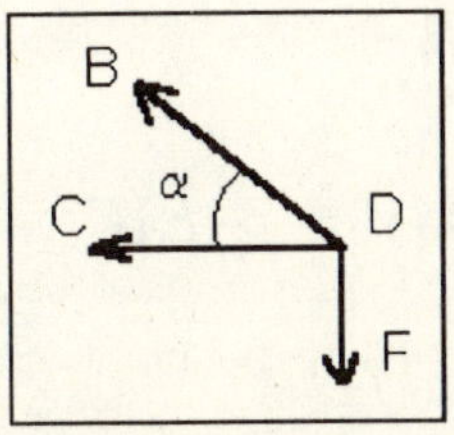

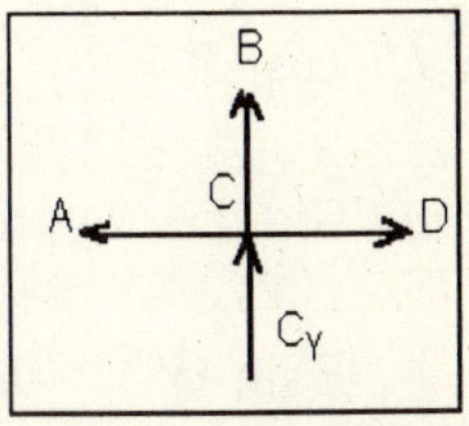

$$\boxed{BD = \frac{F}{\sin\alpha} = 16.67\ kN\,(T)}.$$

$\sum F_y^D = -BD\cos\alpha - CD = 0$, from which

$\boxed{CD = -BD\cos\alpha = -F\cot\alpha = -13.33\ kN\ (C)}$. For the joint C, the sum of forces: $\sum F_x^C = CD - AC = 0$, from which $\boxed{AC = CD = -F\cot\alpha = -13.33\ kN\ (C)}$. $\sum F_y^C = BC + C_y = 0$, from which $\boxed{BC = -C_y = -2F = -20\ kN\,(C)}$. For the angles at joint B, (see sketch), $\beta = \alpha$. The sum of forces: $\sum F_x^B = BD\cos\alpha - AB\cos\alpha = 0$, from

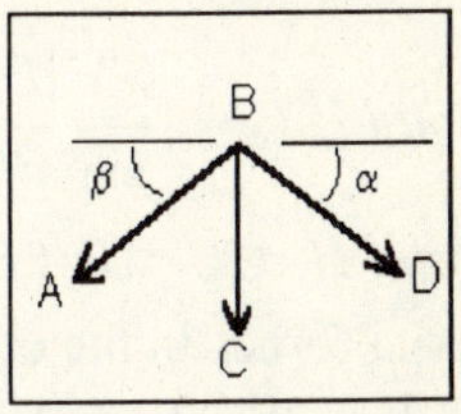

which $\boxed{AB = BD = \frac{F}{\sin\alpha} = 16.67\ kN\ (T)}$.

Problem 6.9 Consider the truss shown in Problem 6.8. Each member will safely support a tensile force of 150 *kN* and a compressive force of 30 *kN*. What is the largest downward load *F* that the truss will safely support?

Solution: From Problem 6.8 the magnitudes of the axial compression loads are : $BC = 2F > AC = F\cot\alpha = CD = F\cot\alpha$. Substitute the safe compressive load: $F = 0.5(30) = 15\ kN$.

The tension members are AB, BD. Take the magnitudes of the axial tension loads: $AB = BD = \frac{F}{\sin\alpha}$.

Substitute the safe tensile force: $F = 150(\sin\alpha) = 90\ kN$. Thus the limit is imposed by the compressive member BC and is $\boxed{F = 15\ kN}$. [*Check:* The member with the max compression load in the solution to Problem 6.8 is member BC. Since the system is linear, the ratio of applied forces is equal to the ratio of the axial forces: $\frac{F_S}{10} = \frac{30}{20}$, from which $F_S = 15\ kN$. *check*].

Problem 6.10 The Howe and Pratt bridge trusses are subjected to identical loads. (a) In which truss does the largest tensile force occur? In what member(s) does it occur and what is its value? (b) In which truss does the largest compressive force occur? In what member(s) does it occur and what is its value?

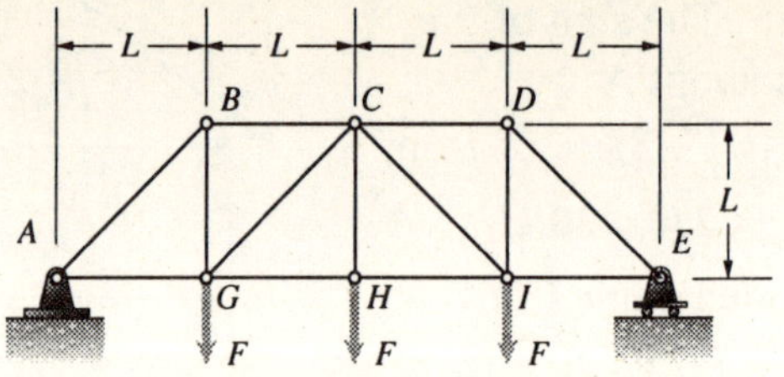

Howe

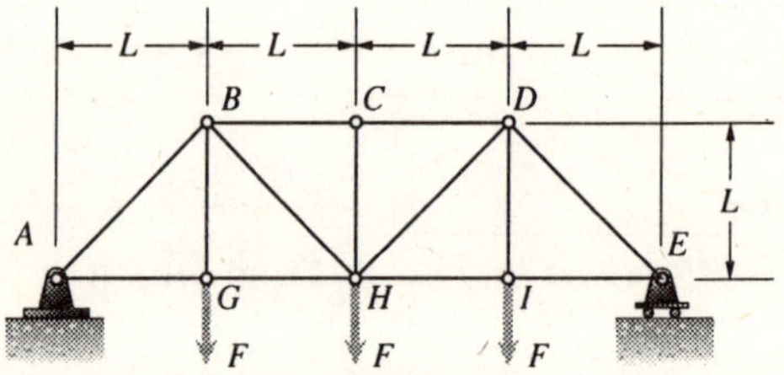

Pratt

Solution: (a) **Howe Bridge:** The moment about A is $M_A = -6F + 4E = 0$, from which $E = \frac{3}{2}F$. Denote the axial force in the member I, K by IK.

(1) *Joint E*: $DE = -\frac{E}{\sin 45^o} = -\frac{3\sqrt{2}}{2}F\ (C)$

$EI = DE\cos 45^o = \frac{3}{2}F\ (T)$

(2) *Joint D*: $CD = DE\cos 45^o = -\frac{3}{2}F\ (C)$,

$DI = -DE\sin 45^o = \frac{3}{2}F\ (T)$

(3) *Joint I*: $CI = \frac{F - |DI|}{\sin 45^o} = -\frac{\sqrt{2}}{2}F\ (C)$, $HI = EI - CI\sin 45^o = 2F\ (T)$

(4) *Joint H*: $CH = F\ (T)$, $GH = HI = 2F\ (T)$. By symmetry (the reaction at A has no x-component) the axial forces in the other members are $HG = HI$, $CG = CI$, $BG = DI$, $CD = BC$, $AG = EI$, and $AB = DE$. *In the Howe truss, the members HI and GH have the highest tensile force* $\boxed{GH = HI = 2F\ (T)}$ *and the members DE and AB have the highest compressive force*

$$\boxed{AB = DE = -\frac{3\sqrt{2}}{2}F\ (C)}$$

(a) **Pratt Bridge:** The moment about A is $M_A = -6F + 4E = 0$, from which $E = \frac{3}{2}F$.

(1) *Joint E:* $DE = -\frac{3\sqrt{2}}{2}F\ (C)$, $EI = \frac{-DE}{\sqrt{2}} = \frac{3}{2}F\ (T)$

(2) *Joint I*: $DI = F\ (T)$, $HI = EI = \frac{3}{2}F\ (T)$

(3) *Joint D*: $DH = -\sqrt{2}DI - DE = \frac{\sqrt{2}}{2}F\ (T)$, $DC = \frac{DE}{\sqrt{2}} - \frac{DH}{\sqrt{2}} = -2F\ (C)$

(4) *Joint C*: $BC = DC = -2F\ (C)$, $CH = 0$

(5) *Joint H*: $GH = HI = \frac{3}{2}F\ (T)$. The axial forces in the remaining members are determined from symmetry. *In the Pratt Bridge, the highest tensile force occurs in members* EI, HI, GH, and AG$\boxed{= \frac{3}{2}F\ (T)}$, *and the highest compressive force occurs in members* DE *and* AB$\boxed{= -\frac{3\sqrt{2}}{2}F\ (C)}$.
Thus (a) the Howe bridge has the highest tensile force in a member, and (b) the value of the compressive force is the same in members DE and AB $= \frac{3\sqrt{2}}{2}F\ (C)$ for both bridges.

Problem 6.11 The loads $F_1 = F_2 = 10\ kip$. Determine the axial forces in members BD, BE and BG.

Solution: *The complete structure as a free body:* The moments about A are $M_A = -12F_1 - 6F_2 + 16G = 0$, from which $G = \frac{180}{16} = 11.25\ kip$. Denote the axial force in a member joining two points I, K by IK.

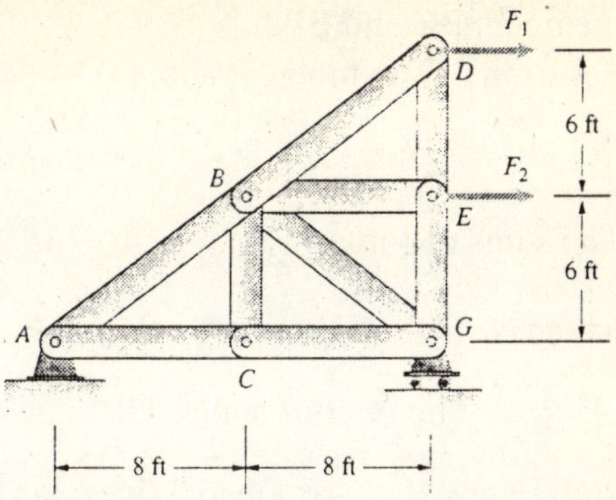

(1) Joint D: The interior angle DBE is $\alpha = \tan^{-1}\left(\frac{6}{8}\right) = 36.87^o$ $\sum F_y = -BD\cos\alpha + F_1 = 0$, from which $\boxed{BD = \frac{F_1}{\cos\alpha} = 12.5\ kip\ (T)}$. $\sum F_y = -BD\sin\alpha - DE = 0$, from which $DE = -BD\sin\alpha = -7.5\ kip(C)$

(2) Joint E: $\sum F_y = -BE + F_2 = 0$, from which $\boxed{BE = F_2 = 10\ kip\ (T)}$. $\sum F_y = DE - EG = 0$, from which $EG = DE = -7.5\ kip\ (C)$.

(3) Joint G: $\sum F_y = BG\sin\alpha + EG + G = 0$, from which $\boxed{BG = -\frac{EG + G}{\sin\alpha} = -6.25\ kip\ (C)}$

Problem 6.12 Consider the truss in Problem 6.11. The loads $F_1 = 60\ lb$ and $F_2 = 120\ lb$. Determine the axial forces in AB, BC, and BD.

Solution: The sum of the moments about A is $M_A = -12(60) - 6(120) + 16G = 0$, from which $G = 90\ lb$. The sum of forces: $\sum F_y = A_y + G = 0$, from which $A_y = -90\ lb$. Denote the axial force in a member joining two points I, K by IK

(1) *Joint D*: $\sum F_y = -BD\cos\alpha + F_! = 0$, from which $\boxed{BD = \frac{F_1}{\cos\alpha} = \frac{60}{0.8} = 75\ lb\ (T)}$

$\sum F_y = -BD\sin\alpha - DE = 0$, from which $DE = 45\ \ lb$

(2) *Joint E:* $\sum F_y = -BE + F_2 = 0$, from which $BE = 120\ \ lb$.

$\sum F_y = -DE + EG = 0$, from which $EG = DE = 45\ \ lb$

(3) *Joint G*: $\sum F_y = BG\sin\alpha + EG + G = 0$, from which $BG = -\frac{(EG + G)}{\sin\alpha} = -\frac{135}{0.6} = 225\ lb$.

$\sum F_y = -GC - BG\cos\alpha = 0$, from which $GC = -BG\cos\alpha = -225(0.8) = -180\ lb$

(4) *Joint C*: $\sum F_y = BC = 0$, from which $\boxed{BC = 0}$

$\sum F_x = AC - GC = 0$, from which $AC = 225\ lb$.

(5) *Joint A*: $\sum F_y = -A_y + AB\sin\alpha = 0$, from which $\boxed{AB = \frac{90}{0.6} = 150\ lb\ (T)}$

Problem 6.13 The truss supports loads at C and E. If $F = 3\ kN$, what are the axial forces in the members BC and BE?

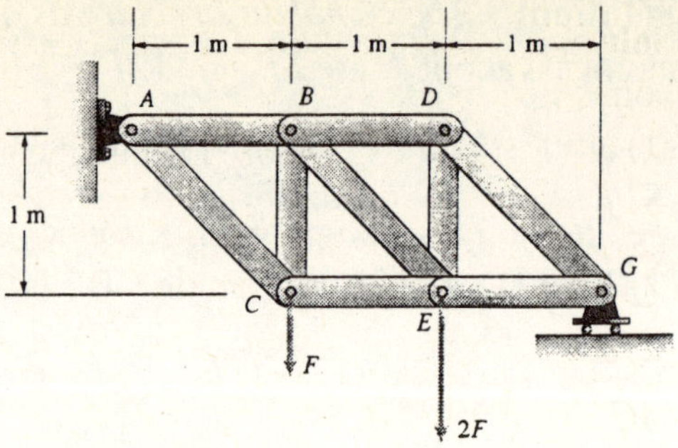

Solution: The moment about A is

$\sum M_A = -1F - 4F + 3G = 0$, from which $G = \frac{5}{3}F = 5\ kN$.

The sums of forces: $\sum F_Y = A_Y - 3F + G = 0$, from which $A_Y = -\frac{4}{3}F = -4\ kN$. $\sum F_X = A_X = 0$, from which $A_X = 0$. The interior angles GDE, EBC are 45°, from which $\sin\alpha = \cos\alpha = \frac{1}{\sqrt{2}}$. Denote the axial force in a member joining I, K by IK.

(1) *Joint G:* $\sum F_y = \frac{DG}{\sqrt{2}} + G = 0$, from which $DG = -\sqrt{2}G = -\frac{5\sqrt{2}}{3}F = -5\sqrt{2}\ kN\ (C)$.

$\sum F_x = -\frac{DG}{\sqrt{2}} - EG = 0$, from which $EG = -\frac{DG}{\sqrt{2}} = \frac{5}{3}F = 5\ kN\ (T)$.

(2) *Joint D:* $\sum F_y = -DE - \frac{DG}{\sqrt{2}} = 0$, from which $DE = \frac{5}{3}F = 5\ kN\ (T)$. $\sum F_x = -BD + \frac{DG}{\sqrt{2}} = 0$, from which $BD = -\frac{5}{3}F = -5\ kN\ (C)$.

(3) *Joint E:* $\sum F_y = \frac{BE}{\sqrt{2}} - 2F + DE = 0$, from which $\boxed{BE = 2\sqrt{2}F - \sqrt{2}DE = \frac{\sqrt{2}}{3}F = \sqrt{2}\ kN\ (T)}$.

$\sum F_x = -CE - \frac{BE}{\sqrt{2}} + EG = 0$, from which $CE = EG - \frac{BE}{\sqrt{2}} = \frac{4}{3}F = 4\ kN\ (T)$.

(4) *Joint A:* $\sum F_y = A_y - \frac{AC}{\sqrt{2}} = 0$, from which $AC = \frac{4\sqrt{2}}{3}F = 4\sqrt{2}\ kN\ (T)$. $\sum F_x = AB + \frac{AC}{\sqrt{2}} = 0$, from which $AB = -\frac{4}{3}F = -4\ kN\ (C)$.

(5) *Joint C:* $\sum F_y = BC + \frac{AC}{\sqrt{2}} - F = 0$, from which $\boxed{BC = F - \frac{AC}{\sqrt{2}} = -\frac{1}{3}F = -1\ kN\ (C)}$.

Problem 6.14 Consider the truss in Problem 6.13. Each member will safely support a tensile force of 28 *kN* and a compressive force of 12 *kN*. Based on this criterion, what is the ;largest safe (positive) value of F?

Solution: From the solution to Problem 6.14, the member with the largest tensile force is $EG = \frac{5}{3}F$, from which $F = \frac{3}{5}EG = 16.8\ kN$. The member with the largest compressive force is DG, $DG = \frac{-5\sqrt{2}}{3}F$, from which $\boxed{F = \frac{3}{5\sqrt{2}}DG = \frac{36}{5\sqrt{2}} = 5.09\ kN}$ is the largest safe value.

Problem 6.15

Problem 6.15 The truss attaches one end of a stretcher to a rescue helicopter. If the stretcher exerts 60 lb vertical loads at A and B, what are the axial forces in members CF, DF, and FG?

Solution: Denote the axial force in a member joining two points I, K by IK.

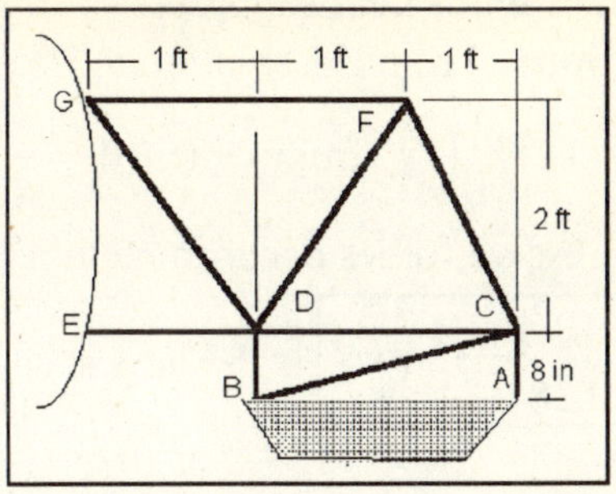

(1) *Joint B*: Denote the interior angle ABC as β.

$\sum F_x = BC\cos\beta = 0$, from which $BC = 0$.

$\sum F_y = BD + BC\sin\beta - F = 0$, from which $BD = F = 60\ lb$.

(2) *Joint A*: $\sum F_y = AC - F = 0$, from which $AC = F = 60\ lb\ (T)$

(3) *Joint C:* Denote the angle formed by the member CF with the vertical as $\alpha = \tan^{-1}(0.5) = 26.57^o$. $\sum F_y = AF\cos\alpha - AC = 0$, from which

$\boxed{CF = \dfrac{F}{\cos\alpha} = 1.118F = 67.1\ lb\ (T)}$ $\sum F_x = -CD - BC\cos\beta - CF\sin\alpha = 0$, from which $CD = -CF\sin\alpha = -0.45F = -27\ lb\ (C)$.

(4) *Joint F*: $\sum F_y = -CF\cos\alpha - DF\cos\alpha = 0$, from which $\boxed{DF = -CF = -1.118F = -67.1\ lb\ (C)}$

$\sum F_x = -FG + CF\sin\alpha - DF\sin\alpha = 0$, from which $\boxed{FG = F = 60\ lb\ (T)}$

Problem 6.16

Problem 6.16 Consider the truss in Problem 6.15. If the stretcher exerts 60 *lb* vertical loads at A and B, what are the axial forces in members AC, CD, and DG?

Solution: From the solution to Problem 6.15, $\boxed{AC = 60\ lb\ (T)}$, $\boxed{CD = -0.45F = -27\ lb\ (C)}$

Joint D: $\sum F_y = FG\cos\alpha + DG\cos\alpha - BD = 0$, from which

$\boxed{DG = \dfrac{BD}{\cos\alpha} - DF = 2.2362F = 134.2\ lb\ (T)}$

Problem 6.17

Problem 6.17 Determine the axial forces in the members in terms of the weight W.

Solution: Denote the axial force in a member joining two points I, K by IK. The angle between member DE and the positive *x* axis is $\alpha = \tan^{-1} 0.8 = 38.66^o$. The angle formed by member DB with the positive *x* axis is $90^o + \alpha$. The angle formed by member AB with the positive *x*-axis is α.

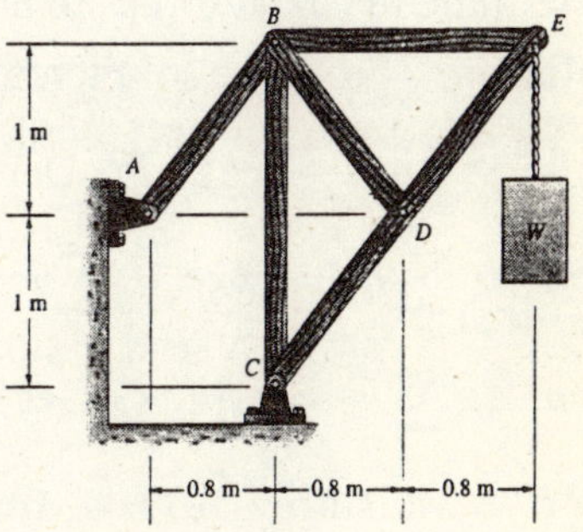

Joint E: $\sum F_y = -DE\cos\alpha - W = 0$, from which $\boxed{DE = -1.28W\ (C)}$. $\sum F_y = -BE - DE\sin\alpha = 0$, from which $\boxed{BE = 0.8W\ (T)}$

Joint D: $\sum F_x = DE\cos\alpha + BD\cos\alpha - CD\cos\alpha = 0$, from which $BD - CD = -DE$. $\sum F_y = -BD\sin\alpha + DE\sin\alpha - CD\sin\alpha = 0$, from which $BD + CD = DE$. Solving these two equations in two unknowns: $\boxed{CD = DE = -1.28W\ (C)}$, $\boxed{BD = 0}$

Joint B: $\sum F_x = BE - AB\sin\alpha - BD\sin\alpha = 0$, from which $\boxed{AB = \dfrac{BE}{\sin\alpha} = 1.28W\ (T)}$

$\sum F_y = -AB\cos\alpha - BC = 0$, from which $\boxed{BC = -AB\cos\alpha = -W\ (C)}$

Problem 6.18 Consider the truss shown in Problem 6.17. Each member will support a tensile force of 6 *kN* and a compressive force of 2 *kN*. Based on these criteria, determine the largest weight W that the truss will safely support.

Solution: From the solution to Problem 6.17, the largest tensile force is in member AB, $AB = 1.28W(T)$, from which $W = \frac{6}{1.28} = 4.69\ kN$ is the maximum safe load for tension. The largest compressive forces occur in members DE and CD, $DE = CD = 1.28W(C)$, from which $\boxed{W = \frac{2}{1.28} = 1.56\ kN}$ is the largest safe load for compression.

Problem 6.19 The loads $F_1 = 600\ lb$ and $F_2 = 300\ lb$. Determine the axial forces in members AE, BD, and CD.

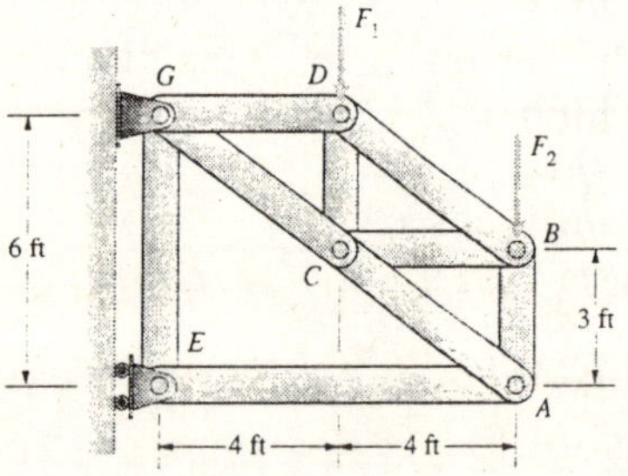

Solution: The reaction at E is determined by the sum of the moments about G: $M_G = +6E - 4F_1 - 8F_2 = 0$, from which $E = \frac{4F_1 + 8F_2}{6} = 800\ lb$. The interior angle EAG is $\alpha = \tan^{-1}\left(\frac{6}{8}\right) = 36.87^o$. From similar triangles this is also the value of the interior angles ACB, CBD, and CGD. *Method of joints:* Denote the axial force in a member joining two points I, K by IK. *Joint E*: $\sum F_y = E + AE = 0$, from which $\boxed{AE = -E = -800\ lb\ (C)}$. $\sum F_y = EG = 0$, from which $EG = 0$. *Joint A:* $\sum F_y = -AE - AC\cos\alpha = 0$, from which $AC = -\frac{AE}{0.8} = 1000\ lb\ (T)$. $\sum F_y = AC\sin\alpha + AB = 0$, from which $AB = -AC(0.6) = -600\ lb\ (C)$.

Joint B: $\sum F_y = BD\sin\alpha - AB - F_1 = 0$, from which $\boxed{BD = \frac{F_2 + AB}{0.6} = \frac{-300}{0.6} = -500\ lb\ (C)}$.

$\sum F_x = -BC - BD\cos\alpha = 0$, from which $BC = -BD(0.8) = 400\ lb\ (T)$.

Joint D: $\sum F_y = -BD\sin\alpha - CD - F_1 = 0$, from which $\boxed{CD = -F_1 - BD(0.6) = -300\ \ lb\ (C)}$

Problem 6.20 Consider the truss in Problem 6.19. The loads $F_1 = 450\ lb$ and $F_2 = 150\ lb$. Determine the axial forces in members AB, AC, and BC.

Solution: From the solution to Problem 6.19 the angle $\alpha = 36.87^o$ and the reaction at E is $E = \frac{4F_1 + 8F_2}{6} = 500\ lb$. Denote the axial force in a member joining two points I, K by IK.

Joint E: $\sum F_y = EG = 0$. $\sum F_x = AE + E = 0$, from which $AE = -E = -500\ lb\ (C)$.

Joint A: $\sum F_x = -AE - AC\cos\alpha = 0$, from which $\boxed{AC = -\frac{AE}{0.8} = 625\ lb\ (T)}$.

$\sum F_y = AC\sin\alpha + AB = 0$, from which $\boxed{AB = -AC(0.6) = -375\ lb\ (C)}$

Joint B: $\sum F_y = BD\sin\alpha - F_2 - AB = 0$, from which $BD = \frac{F_2 + AB}{0.6} = -375\ lb\ (C)$

$\sum F_x = -BC - BD\cos\alpha = 0$, from which $\boxed{BC = -BD(0.8) = 300\ lb\ (T)}$

Problem 6.21 Each member of the truss will safely support a tensile force of 4 kN and a compressive force of 1 kN. Determine the largest mass that can safely be suspended.

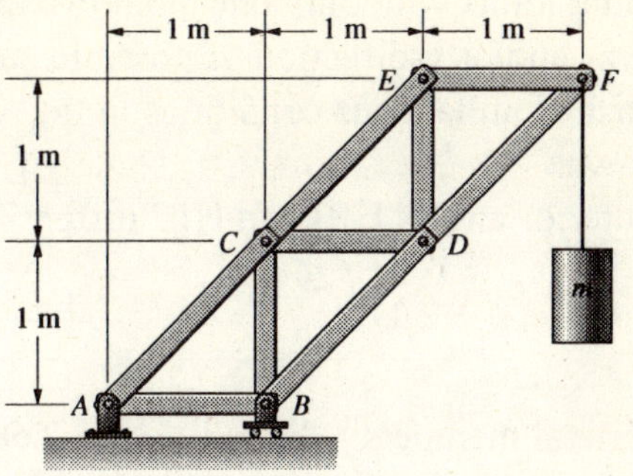

Solution: The common interior angle BAC=DCE=EFD=CDB is $\alpha = \tan^{-1}(1) = 45^o$.

Note $\cos\alpha = \sin\alpha = \frac{1}{\sqrt{2}}$ Denote the axial force in a member joining two points I, K by IK.

Joint F: $\sum F_y = -\frac{DF}{\sqrt{2}} - W = 0$, from which $DF = -\sqrt{2}\,W\,(C)$. $\sum F_x = -EF - \frac{DF}{\sqrt{2}} = 0$, from which $EF = W\,(T)$.

Joint E: $\sum F_x = -\frac{CE}{\sqrt{2}} + EF = 0$ from which $CE = \sqrt{2}\,W\,(T)$. $\sum F_y = -ED - \frac{CE}{\sqrt{2}} = 0$, from which $ED = -W\,(C)$.

Joint D: $\sum F_Y = ED + \frac{DF}{\sqrt{2}} - \frac{BD}{\sqrt{2}} = 0$, from which $BD = -2\sqrt{2}\,W\,(C)$.

$\sum F_X = \frac{DF}{\sqrt{2}} - \frac{BD}{\sqrt{2}} - CD = 0$, from which $CD = W\,(T)$

Joint C: $\sum F_x = -\frac{AC}{\sqrt{2}} + \frac{CE}{\sqrt{2}} + CD = 0$, from which $AC = 2\sqrt{2}\,W\,(T)$

$\sum F_y = -\frac{AC}{\sqrt{2}} + \frac{CE}{\sqrt{2}} - BC = 0$, from which $BC = -W\,(C)$

Joint B: $\sum F_x = -AB + \frac{BD}{\sqrt{2}} = 0$, from which $AB = -2W\,(C)$

This completes the determination of the axial forces in all nine members. The maximum tensile force occurs in member AC, $AC = 2\sqrt{2}\,W\,(T)$, from which the safe load is $W = \frac{4}{2\sqrt{2}} = \sqrt{2} = 1.414\ kN$. The maximum compression occurs in member BD, $BD = -2\sqrt{2}\,W\,(C)$, from which the maximum safe load is $W = \frac{1}{2\sqrt{2}} = 0.3536\ kN$. The largest mass m that can be safely supported is $\boxed{m = \frac{353.6}{9.81} = 36.0\ kg}$

Problem 6.22 The Howe truss helps support a roof. Model the supports at A and G as roller supports. Determine the axial forces in members AB, BC, and CD.

Solution: The strategy is to proceed from end A, choosing joints with only one unknown axial force in the x- and/or y-direction, if possible, and if not, establish simultaneous conditions in the unknowns.

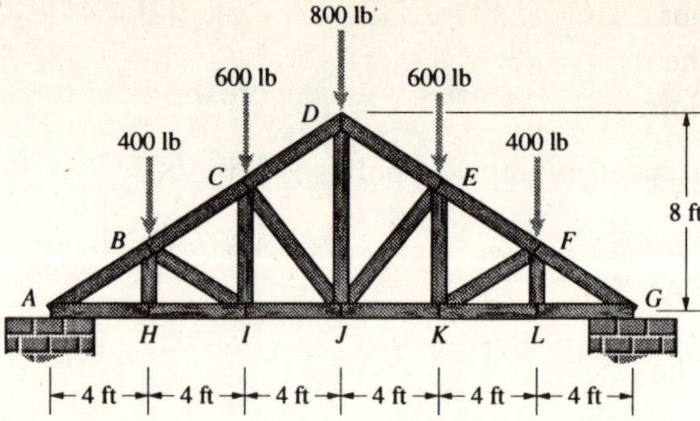

The interior angles HIB and HJC differ. The pitch angle is $\alpha_{Pitch} = \tan^{-1}\left(\frac{8}{12}\right) = 33.7^o$. The length of the vertical members: $\overline{BH} = 4\left(\frac{8}{12}\right) = 2.6667\ ft$, from which the angle $\alpha_{HIB} = \tan^{-1}\left(\frac{2.6667}{4}\right) = 33.7^o$.

$\overline{CI} = 8\frac{8}{12} = 5.3333\ ft$, from which the angle $\alpha_{IJC} = \tan^{-1}\left(\frac{5.333}{4}\right) = 53.1^o$.

The moment about G: $M_G = (4+20)(400) + (8+16)(600) + (12)(800) - 24A = 0$, from which $A = \frac{33600}{24} = 1400\ lb$. *Check:* The total load is 2800 *lb*. From left-right symmetry each support A, G supports half the total load. *check.*

The method of joints: Denote the axial force in a member joining two points I, K by IK.

Joint A: $\sum F_y = AB\sin\alpha_P + 1400 = 0$, from which $\boxed{AB = -\frac{1400}{\sin\alpha_P} = -2523.9\ lb\ (C)}$

$\sum F_x = AB\cos\alpha_{Pitch} + AH = 0$, from which $AH = (2523.9)(0.8321) = 2100\ lb\ (T)$

Joint H: $\sum F_y = BH = 0$, or, $BH = 0$. $\sum F_x = -AH + HI = 0$, from which $HI = 2100\ lb\ (T)$

Joint B: $\sum F_x = -AB\cos\alpha_{Pitch} + BC\cos\alpha_{Pitch} + BI\cos\alpha_{Pitch} = 0$, from which $BC + BI = AB$

$\sum F_y = -400 - AB\sin\alpha_{Pitch} + BC\sin\alpha_{Pitch} - BI\sin\alpha_{Pitch} = 0$, from which $BC - BI = AB + \frac{400}{\sin\alpha_{Pitch}}$.

Solve the two simultaneous equations in unknowns BC, BI: $BI = -\frac{400}{2\sin\alpha_{Pitch}} = -360.56\ lb\ (C)$, and

$\boxed{BC = AB - BI = -2163.3\ lb\ (C)}$

Joint I: $\sum F_x = -BI\cos\alpha_{Pitch} - HI + IJ = 0$, from which $IJ = 1800\ lb\ (T)$

$\sum F_y = +BI\sin\alpha_{Pitch} + CI = 0$, from which $CI = 200\ lb\ (T)$

Joint C: $\sum F_x = -BC\cos\alpha_{Pitch} + CD\cos\alpha_{Pitch} + CJ\cos\alpha_{IJC} = 0$, from which

$CD(0.8321) + CJ(0.6) = -1800$

$\sum F_y = -600 - CI - BC\sin\alpha_{Pitch} + CD\sin\alpha_{Pitch} - CJ\sin\alpha_{IJC} = 0$, from which

$CD(0.5547) - CJ(0.8) = -400$

Solve the two simultaneous equations to obtain $CJ = -666.67\ lb\ (C)$, and $\boxed{CD = -1682.57\ lb\ (C)}$

Problem 6.23 The plane truss forms part of the supports of a crane on an offshore oil platform. The crane exerts vertical 75 kN forces on the truss at B, C, and D. You can model the support at A as a pin support and model the support at E as a roller support which can exert a force normal to the dashed line but cannot exerts a force parallel to it. The angle $\alpha = 45^o$. Determine the axial forces in the members of the truss.

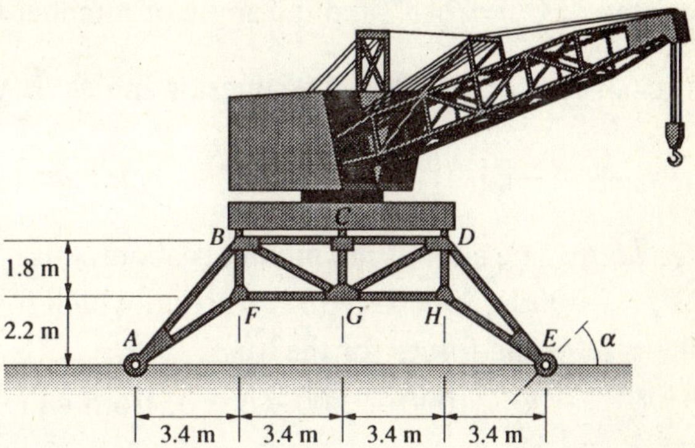

Solution: The included angles EAB and AED are each $\gamma = \tan^{-1}\left(\frac{4}{3.4}\right) = 49.64^o$. The included angles EAF and AEH are each $\beta = \tan^{-1}\left(\frac{2.2}{3.4}\right) = 32.91^o$. The included angles DBG and BDG are each $\theta = \tan^{-1}\left(\frac{1.8}{3.4}\right) = 27.9^o$. The angle between ED and the positive x- axis $\rho = 90^o + \alpha$.

The complete structure as a free body: The sum of the moments about A is $M_A = -(75)(3.4)(1+2+3) - (4)(3.4)E\sin\rho = 0$, from which $E = 159.1\ kN$. The projections of E on the x-and y-axis are: $E_x = -112.5\ kN$ and $E_y = 112.5\ kN$. The reactions at A are:

$\sum F_x^A = A_x + E_x = 0$, $BG = 80.1\ kN\,(T)$from which $A_X = -E_X = 112.5\ kN$.

$\sum F_y^A = A_y - 3(75) + E_y = 0$, from which $A_y = 112.5\ kN$. Thus the reactions at A and E are symmetrical about the truss center, which suggests that symmetrical truss members have equal axial forces.

The method of joints: Denote the axial force in a member joining two points I, K by IK.

Joint A: $\sum F_x = AB\cos\gamma + A_x + AF\cos\beta = 0$, $\sum F_y = AB\sin\beta + A_y + AF\sin\beta = 0$, from which two simultaneous equations are obtained. Solve: $\boxed{AF = -44.67\ kN\,(C)}$, and $\boxed{AB = -115.8\ kN\,(C)}$

Joint E: $\sum F_y = -DE\cos\gamma + E_x - EH\cos\beta = 0$. $\sum F_y = DE\sin\gamma + E_y + EH\sin\beta = 0$, from which two simultaneous equations are obtained. Solve: $\boxed{EH = -44.67\ kN\ (C)}$, and $\boxed{DE = -115.8\ kN\,(C)}$

Joint F: $\sum F_x = -AF\cos\beta + FG = 0$, from which $\boxed{FG = -37.5\ kN\ (C)}$

$\sum F_y = -AF\sin\beta + BF = 0$, from which $\boxed{BF = -24.26\ kN\,(C)}$

Joint H: $\sum F_x = EH\cos\beta - GH = 0$, from which $\boxed{GH = -37.5\ kN\ (C)}$

$\sum F_y = -EH\sin\beta + DH = 0$, from which $\boxed{DH = -24.26\ kN\ (C)}$

Joint B: $\sum F_y = -AB\sin\gamma - BF + BG\sin\theta - 75 = 0$, from which $\boxed{BG = 80.1\ kN\,(T)}$

$\sum F_x = -AB\cos\gamma + BC + BG\cos\theta = 0$, from which $\boxed{BC = -145.8\ kN\ (C)}$

Joint D: $\sum F_y = -DE\sin\gamma - DH - DG\sin\theta - 75 = 0$, from which $\boxed{DG = 80.1\ kN\ (T)}$

$\sum F_x = DE\cos\gamma - CD - DG\cos\theta = 0$, from which $\boxed{CD = -145.8\ kN\ (C)}$

Joint C: $\sum F_x = CD - BC = 0$, from which $\boxed{CD = BC}$ *Check.*

$\sum F_y = -CG - 75 = 0$, from which $\boxed{CG = -75\ kN\ (C)}$

Problem 6.24** (a) Design a truss attached to the supports A and B that supports the loads applied at points C and D. (b) Determine the axial forces in the members of the truss you designed in Part (a).

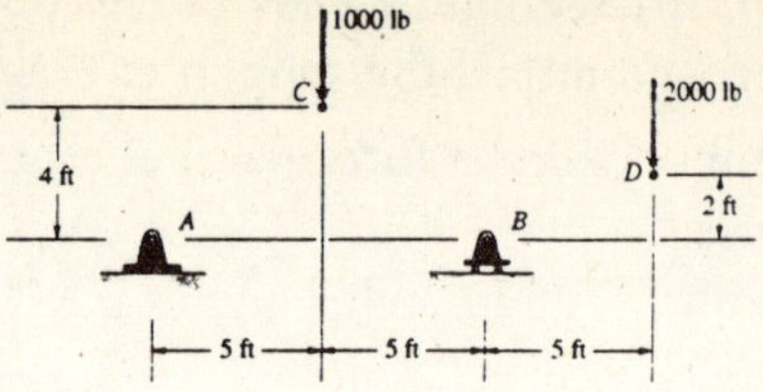

Solution: (a) An example truss is shown.
(b) Relative to the horizontal, the angle of member BD is $\alpha = \tan^{-1}\left(\frac{2}{5}\right) = 21.8^o$ and the angle of member CD is $\beta = \tan^{-1}\left(\frac{2}{10}\right) = 11.31^o$. The interior angles BAC and ABC are each $\gamma = \tan^{-1}\left(\frac{4}{5}\right) = 38.7^o$. *The complete structure as a free body:* The sum of the moments about A is $M_A = -5000 - 30000 + 10B = 0$, from which $B = 3500\ lb$. The sum of the forces for the truss

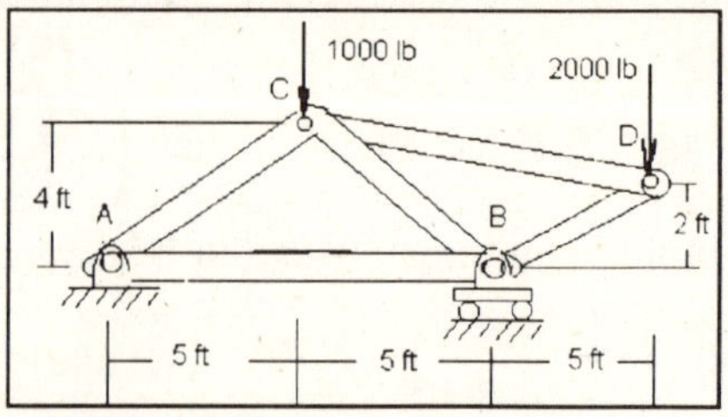

$\sum F_y^A = +A_y - 1000 - 3000 + B = 0$, from which $A_y = 500\ lb$. Similarly, $\sum F_x^A = A_x = 0$.
The method of joints: Denote the axial force in a member joining two points I, K by *IK*.
Joint D: $\sum F_x = -BD\cos\alpha - CD\cos\beta = 0$, $\sum F_y = -BD\sin\alpha + CD\sin\beta - 2000 = 0$, from which: $BD = -3590.1\ lb\ (C)$, and $\boxed{CD = 3399.3\ lb\ (T)}$. *Joint B:* $\sum F_y = BC\sin\gamma + BD\sin\alpha + B = 0$, from which $BC = -3468.4\ lb\ (C)$. $\sum F_x = -BC\cos\gamma + BD\cos\alpha - AB = 0$, from which $\boxed{AB = -625\ lb\ (C)}$.
Joint A: $\sum F_y = A_y + AC\sin\gamma = 0$, from which $\boxed{AC = -800\ lb\ (C)}$

Problem 6.25 (a) Design a truss attached to supports A and B that supports the loads applied at points C and D.

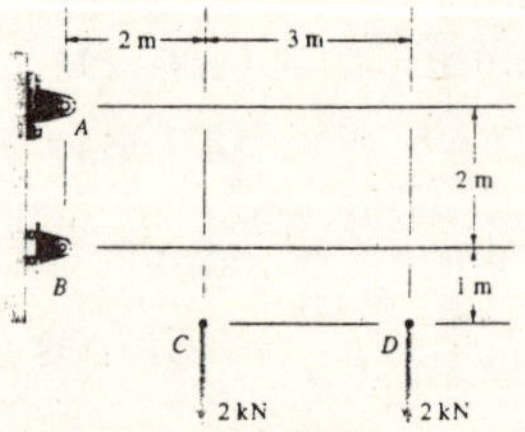

Solution: (a) An example design is shown.
(b) The angles are $\alpha = \tan^{-1}\left(\frac{3}{5}\right) = 30.96^o$, $\beta = \tan^{-1}\left(\frac{1}{2}\right) = 26.6^o$, $\gamma = \tan^{-1}\left(\frac{3}{2}\right) = 56.31^o$. *The complete structure as a free body:* The sum of the moments about A: $M_A = -2(2) - 5(2) + 2B = 0$, from which $B = 7\ kN$. $\sum F_y = A_y - 2 - 2 = 0$, from which $A_y = +4\ kN$. $\sum F_x = A_x + B = 0$, from which $A_x = -B = -7\ kN$.

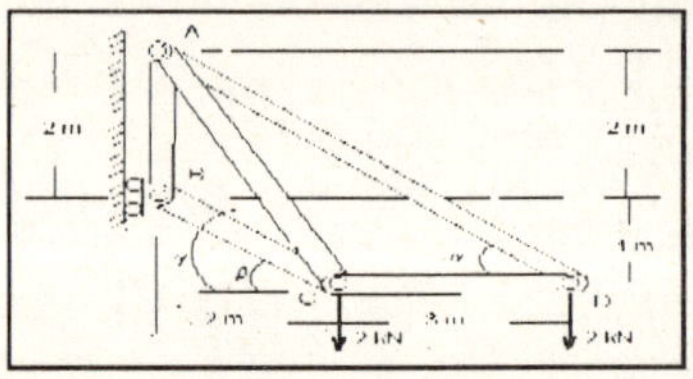

The method of joints: Denote the axial force in a member joining two points I, K by *IK*. *Joint D:* $\sum F_y = AD\sin\alpha - 2 = 0$, from which

$\boxed{AD = \frac{2}{\sin\alpha} = 3.89\ kN\ (T)}$. $\sum F_x = -AD\cos\alpha - CD = 0$, from which

$\boxed{CD = -AD\cos\alpha = -3.33\ kN\ (C)}$ *Joint A*: $\sum F_x = A_x + AD\cos\alpha + AC\cos\gamma = 0$, from which

$\boxed{AC = \frac{-AD\cos\alpha - A_x}{\cos\gamma} = 6.61\ kN\ (T)}$ $\sum F_y = A_y - AD\sin\alpha - AC\sin\gamma - AB = 0$, from which

$AB = AD\sin\alpha + AC\sin\gamma - A_Y = -3.5\ kN\ (C)$.

Joint B: $\sum F_y = AB - BC\sin\beta = 0$, from which $\boxed{BC = -\frac{AB}{\sin\beta} = -7.83\ kN\ (C)}$

Problem 6.26 Suppose that you want to design a truss supported at A and B (Fig. a) that provides support for a chinning bar at C that will subject the truss to a maximum downward load of 300 *lb*. If you use the simplest design (Fig. b), you obtain a 500 *lb* tensile force in member AC. Improve on this design by finding a truss for which the largest tensile force is less than 500 *lb*.

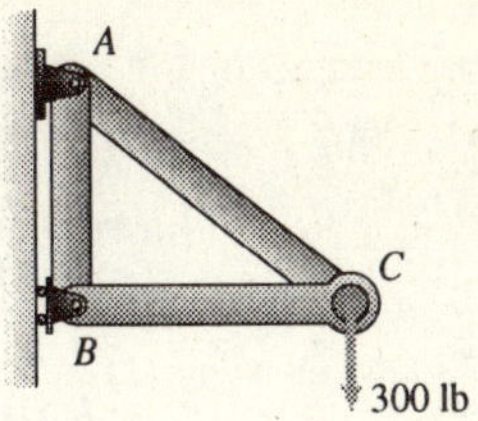

Solution: A simple truss that will reduce the tensile force is shown. The tensile force in member AC in the three member truss is replaced by a compressive force in member BD in the four member truss (which is not the best design idea, but given the constraints of the problem, it probably is the simplest way to reduce the maximum tensile stress).

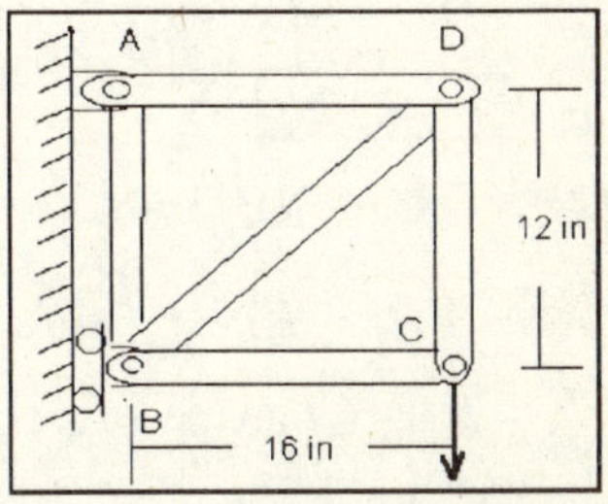

For the lower 4-member truss: The complete structure as a free body: The moment about A is $M_A = -300(16) + 12B = 0$, from which $B = 400\ lb$.

$\sum F_X^A = A_X + B = 0$, from which $A_X = -400\ lb$.

$\sum F_Y^A = A_Y - 300 = 0$, from which $A_Y = 300\ lb$.

The method of joints: Denote the axial force in a member joining two points I, K by IK.

Joint C: $\sum F_x = -CB = 0$, from which $CB = 0$ $\sum F_y = CD - 300 = 0$, from which $\boxed{CD = 300\ lb\ (T)}$

Joint A: $\sum F_x = A_x + AD = 0$, from which $AD = 400\ lb\ (T)$

$\sum F_y = A_y - AB = 0$, from which $\boxed{AB = 300\ lb\ (T)}$

Joint D: $\sum F_x = -AD - BD\cos\alpha = 0$, from which $\boxed{BD = \frac{-AD}{\cos\alpha} = -500\ lb\ (C)}$ Thus the maximum tensile force is 400 *lb* in member AD. The maximum compressive force is 500 *lb* in member BD.

Problem 6.27 The bars are of equal length. The spring is unstretched when $\alpha = 90^o$ and the horizontal surface is smooth. For what value of α in the range $0 < \alpha < 90^o$ will the system remain in equilibrium?

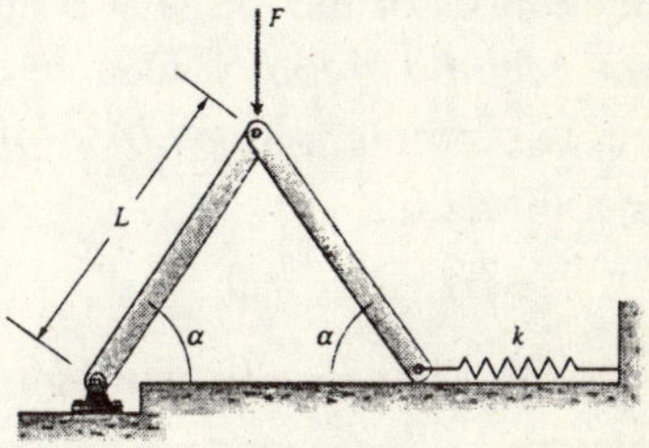

Solution: The stretched length of the spring is $\Delta L = 2L\cos\alpha$, and the force due to spring compression is $F_{spring} = -C_x = 2kL\cos\alpha$.

The complete structure as a free body: The sum of forces:

$\sum F_x = A_x - C_x = 0$, from which $A_x = -2kl\cos\alpha$. .Denote the axial force in a member joining two points I, K by IK.

Joint B: $\sum F_y = -AB\sin\alpha - BC\sin\alpha - F = 0$, $\sum F_y = -AB\cos\alpha + BC\cos\alpha = 0$, from which

$BC = \frac{-F}{2\sin\alpha}$, $AB = \frac{-F}{2\sin\alpha}$. *Joint C:* $\sum F_x - BC\cos\alpha - k\Delta L = 0$, from which $BC = -2kL$.

Substitute : $-2kL = -\frac{F}{2\sin\alpha}$, from which $\sin\alpha = \frac{F}{4kL}$, $\boxed{\alpha = \sin^{-1}\left(\frac{F}{4kL}\right)}$

Problem 6.28 Determine the axial forces in the Pratt truss in Figure 6.16 and confirm the values in Table 6.2

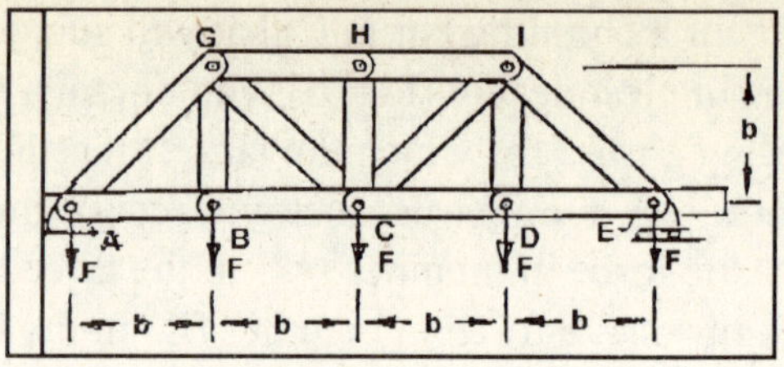

Solution: The common angle is $\alpha = \tan^{-1}\frac{b}{b} = 45^o$

The complete structure as a free body: The moment about A is $M_A = -bF(1+2+3+4) + 4bE = 0$, from which $E = 2.5F$. The reaction at A:

$\sum F_Y = A_Y - 4F + E = 0$, from which $A_Y = 2.5F$. $\sum F_X = A_X = 0$.

The method of joints: Joint E: $\sum F_y = E + EI\sin\alpha - F = 0$, from which $\boxed{EI = -2.12F}$.

$\sum F_x = -DE - EI\cos\alpha = 0$, from which $\boxed{DE = 1.5F}$. *Joint A:* $\Sigma F_y = AG\sin\alpha - F - A_y = 0$, from which $\boxed{AG = 2.12F}$. $\sum F_x = AG\sin\alpha + AB = 0$, from which $\boxed{AB = 1.5F}$.

Joint D: $\sum F_y = DI - F = 0$, from which $\boxed{DI = F}$. $\sum F_x = DE - CD = 0$, from which $\boxed{CD = 1.5F}$.

Joint I: $\sum F_y = -DI - CI\sin\alpha - EI\sin\alpha = 0$, from which $\boxed{CI = 0.707F}$.

$\sum F_x = -HI - CI\cos\alpha + EI\cos\alpha = 0$, from which $\boxed{HI = -2F}$.

Joint B: $\sum F_y = -BG - F = 0$, from which $\boxed{BG = F}$. $\sum F_x = -AB + BC = 0$, from which $\boxed{BC = 1.5F}$

Joint G: $\sum F_y = -AG\sin\alpha + CG\sin\alpha - BG = 0$, from which $\boxed{CG = 0.707F}$

$\sum F_x = -AG\cos\alpha + GH + CG\cos\alpha = 0$, from which $\boxed{GH = -2F}$

Joint H: $\boxed{\sum F_y = HC = 0}$. A term by term comparison confirms Table 6.2.

Problem 6.29 Determine the axial forces in the suspension bridge structure in Figure 6.18, including the reactions exerted on the towers, and confirm the values in Table 6.3

Solution: The roadway has pinned joints at H, I, and J, and at the towers. The strategy is to use the method of joints to show that the axial forces in members BH, CI and DF are each equal to F. An analysis of the joints at B and D yields the reaction at the towers. *Method of joints:* Denote the axial force in a member joining two points I, K, by IK. *Joints H, I, K.* The sum of forces in the x direction,

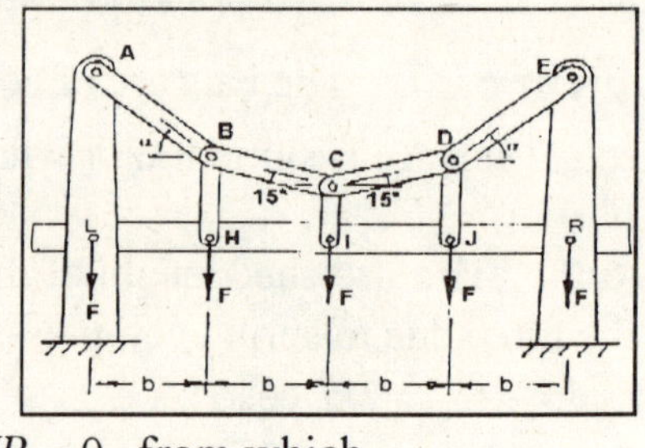

$\sum F_x^H = -HL + HI = 0$, $\sum F_x^I = -HI + IJ = 0$, $\sum F_x^J = -IJ + JR = 0$, from which $HL = HI = IJ = JL = 0$. The sum of the forces in the y-direction $\sum F_y^H = BH - F = 0$,

$\sum F_y^I = CI - F = 0$, $\sum F_y^J = DI - F = 0$, from which $\boxed{BH = F\ (T)}$, $\boxed{CI = F\ (T)}$, $\boxed{DJ = F\ (T)}$. *Joint C:* $\sum F_X = -BC\cos 15^o + CD\cos 15^o = 0$, from which $\boxed{CD = BC}$

$\sum F_y^B = CD\sin 15^o + BC\sin 15^o - F = 0$, from which $\boxed{CD = BC = 1.93F\ (T)}$ *Joint B:* The sum of forces: $\sum F_X = BC\cos 15^o - AB\cos a = 0$, and $\sum F_y = AB\sin a - BC\sin 15^o - F = 0$, from which .

$\boxed{\alpha = \tan^{-1}\left(\frac{BC\sin 15^o - F}{BC\cos 15^o}\right) = 38.79^o}$ and $\boxed{AB = BC\left(\frac{\cos 15^o}{\cos\alpha}\right) = 2.39F\ (T)}$ By symmetry, $\boxed{DE = AB}$

Problem 6.30 In Example 6.4, what are the axial forces in members CD and IJ?

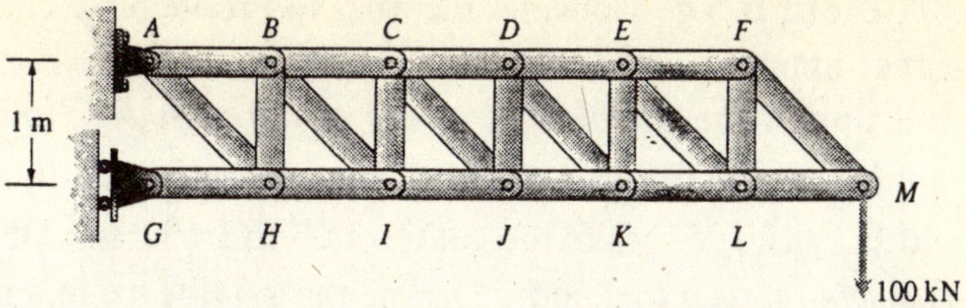

Solution: From Example 6.4, Figure 6.22, the horizontal and vertical members are each 1 m long. Make the cut as shown in Example 6.4 through members CD, CJ, and IJ. Consider the section to the right of the cut. Denote the axial force in a member joining I, K by *IK*.

The section as a free body: The sum of the moments about the joint J is $M_J = CD(1) - 100(3) = 0$, from which $\boxed{CD = 300\ kN(T)}$

The sum of the forces: $\sum F_y = CJ \sin 45^\circ - 100 = 0$, from which $CJ = 100\sqrt{2}\ kN\,(T)$

$\sum F_x = -IJ - 300 - CJ\cos 45^\circ = 0$, from which $\boxed{IJ = -300 - 100 = -400\ kN\,(C)}$

Problem 6.31 The truss supports loads at A and H. (a) Draw the free body diagram of the entire truss and determine the reactions at the supports B and F. (b) Obtain a section by cutting members CE, BE, and BD and draw its free body diagram. (c) Use the section you obtained in Part (b) to determine the axial forces in members CE,BE, and BD.

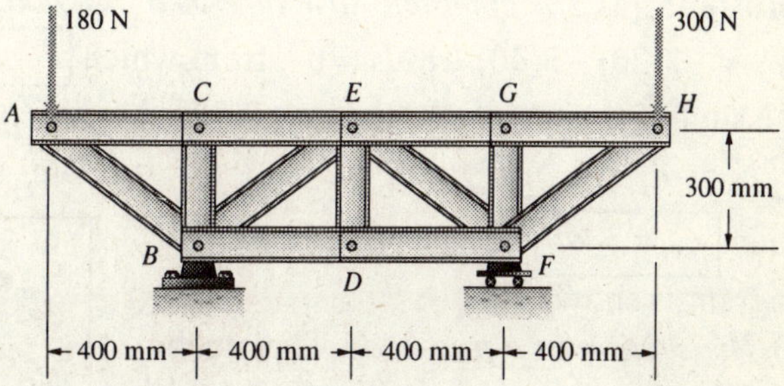

Solution: *The complete structure as a free body:* (a) The sum of the moments about B is $M_B = 180(0.4) - 300(1.2) + (0.8)F = 0$, from which $\boxed{F = \frac{-72+360}{0.8} = 360\ N}$

The reactions at B: $\boxed{\sum F_X = B_X = 0}$

$\sum F_Y = B_y - 180 - 300 + 360 = 0$, from which $\boxed{B_Y = 120\ N}$. (b) Make the cut as shown. The free-body diagram is shown.

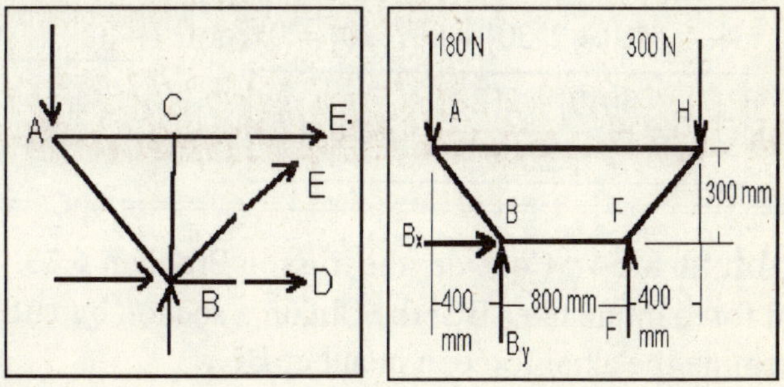

(c) *The section as a free body:* The interior angle DBE is $\alpha = \tan^{-1}\left(\frac{3}{4}\right) = 36.87^\circ$.

Denote the axial force in a member joining I, K by *IK*. The moments about B: $M_B = 180(0.4) - CE(0.3) = 0$, from which $\boxed{CE = \frac{72}{0.3} = 240\ N\,(T)}$. The sum of forces: $\sum F_y = -180 + B_y + BE\sin\alpha = 0$, from which

$\boxed{BE = \frac{180-120}{\sin\alpha} = 100\ N\,(T)}$. $\sum F_x = CE + BD + BE\cos\alpha = 0$, from which

$\boxed{BD = -240 - 100(0.8) = -320\ N\,(C)}$.

Problem 6.32 Consider the truss in Problem 6.31. Use the method of sections to determine the axial forces in members EG, EF, and DF.

Solution: *The complete structure as a free body:* From the dimensions given in Problem 6.31, the interior angle DFE is $\alpha = 36.87^\circ$, and from the solution to Problem 6.31, the reaction at B is $B_X = 0$ and $B_Y = 120\ N$. Make the cut through EG, EF, and DF. Consider the section to the left of the cut. *The section as a free body:* Denote the axial force in a member joining I, K by IK. The sum of the moments about E: $M_E = 180(0.8) + DF(0.3) - B_Y(0.4) = 0$, from which $\boxed{DF = -320\ N\ (C)}$. The sum of forces: $\sum F_y = -180 + B_y - EF \sin\alpha = 0$, from which $\boxed{EF = \dfrac{-60}{\sin\alpha} = -100\ N\ (C)}$.

$\sum F_x = EG + DF + EF\cos\alpha = 0$, from which $\boxed{EG = 320 + 80 = 400\ N\ (T)}$

Problem 6.33 The Howe bridge truss supports loads at F and G. (a) Draw the free body diagram of the entire truss and determine the reactions at the supports A and E. (b) Obtain a section by cutting members CD, CH, and GH and draw its free body diagram. (c) Use the section you obtained in Part (b) to determine the axial force in member GH.

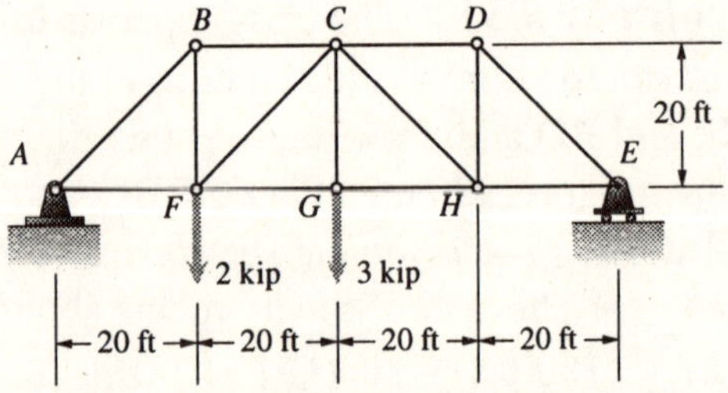

Solution: (a) *The complete structure as a free body:* $M_A = -2(20) - 3(40) + 80E = 0$, from which $\boxed{E = 2\ kip}$. The sum of the forces: $\sum F_x = A_x = 0$ $\sum F_y = A_y - 2 - 3 + E = 0$, from which $\boxed{A_y = 3\ kip}$. (b) The free-body diagram is shown.

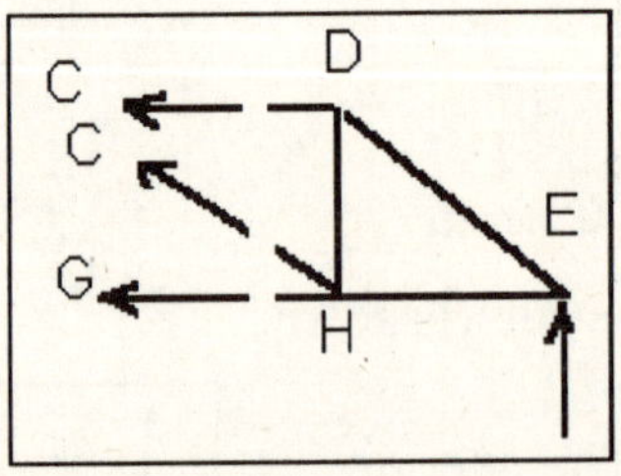

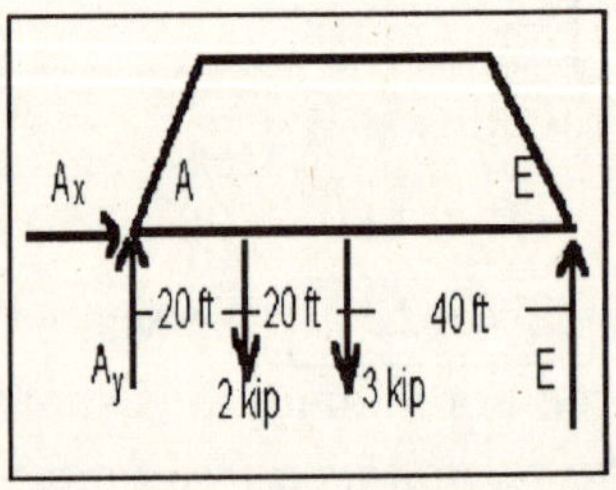

(c) *The section as a free body:* Denote the axial force in a member joining I, K by IK. The sum of the moments about C $M_C = -A_y(40) + 2(20) + GH(20) = 0$ from which $\boxed{GH = \dfrac{3(40) - 2(20)}{20} = 4\ kip\ (T)}$.

Problem 6.34 Consider the truss in Problem 6.33. (a) Use the method of joints to determine the axial force in member BC. (b) Obtain a section by cutting members BC, CF, and FG and use it to determine the axial force in member BC.

Solution: (a) *The complete structure as a free body:* Use the solution of Problem 6.33 to obtain $A_X = 0$ and $A_Y = 3\ kip$. Denote the axial force in a member joining I, K by IK. *The method of joints:* The interior angle FAB is $a = 45^o$. *Joint A:* $\sum F_Y = A_Y + AB\sin 45^o = 0$, from which $AB = -3\sqrt{2}\ kip\ (C)$. *Joint B:* $\sum F_X = -AB\cos 45^o + BC = 0$, from which $\boxed{BC = AB\cos 45^o = -3\ kip\ (C)}$. (b) The free-body diagram of the section is shown. (c) *The section as a free body:* The sum of the moments about F: $M_F = -A_Y(20) - BC(20) = 0$, from which $\boxed{BC = -A_Y = -3\ kip\ (C)}$.

Problem 6.35 The Pratt bridge truss supports five forces ($F = 75\ kip$). The dimension $L = 25\ ft$. Use the method of sections to determine the axial force in member JK.

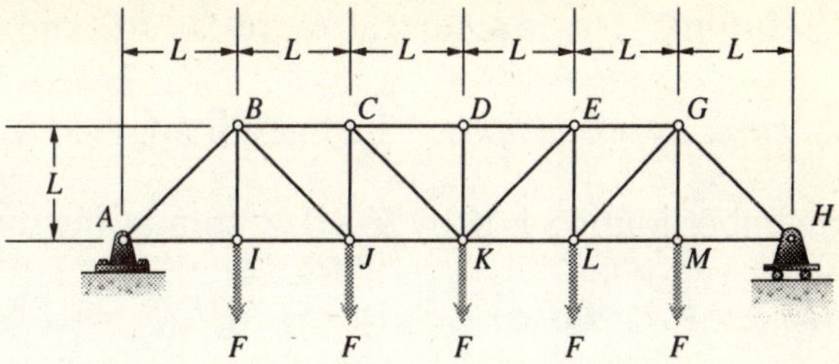

Solution: *The complete structure as a free body:* The sum of the moments about A $M_A = (L)F(1+2+3+4+5) - 6(L)H = 0$, from which $H = \frac{5}{2}F = 187.5\ kip$. The sum of the forces: $\sum F_x = A_x = 0$. $\sum F_y = A_y - 5F + H = 0$, from which $A_y = 187.5\ kip$. Make a cut through CD, CK, and JK. Denote the axial force in a member joining I, K by IK. *The section as a free body:* The sum of the moments about C is $M_C = -A_y(2L) + F(L) + JK(L) = 0$, from which $\boxed{JK = 2A_y - F = 300\ kip\ (T)}$.

Problem 6.36 Consider the bridge truss shown in Problem 6.35. The load $F = 40\ kip$ and the length $L = 30\ ft$. Use the method of sections to determine the axial force in member EK.

Solution: *The complete structure as a free body:* From the solution to Problem 6.35: $H = \frac{5}{2}F = 100\ kip$, and $A_Y = 5F - H = 100\ kip$. The interior angle DEK is 45°. Make a cut through DE, EK, and KL. Denote the axial force in a member joining I, K by IK.

The section as a free body: . The sum of the forces: $\sum F_y = A_y - 3F + EK\sin 45^o = 0$, from which

$\boxed{EK = \sqrt{2}(-A_y + 3F) = 28.28\ kip\ (T)}$.

Problem 6.37 The Warren truss supports a walkway between two buildings. The walkway exerts vertical 10 kip loads at B,D,F, and H. Model the supports at A and I as roller supports. (a) Use the method of joints to determine the axial force in member BC. (b) Use the method of sections to determine the axial force in member BC.

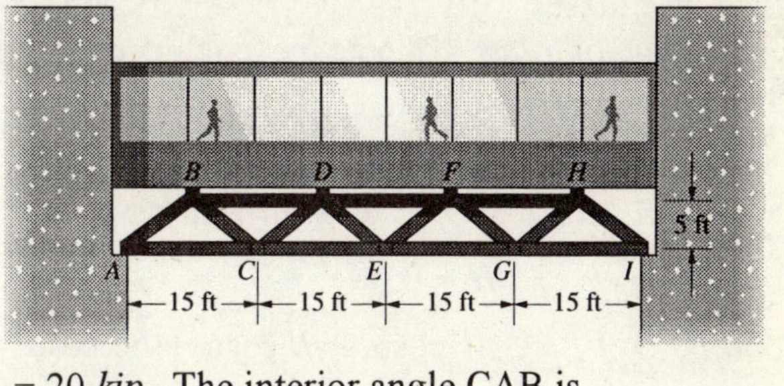

Solution: *The complete structure as a free body:* The moment about point A is $M_A = -10(7.5)(1+3+5+7) + 8(7.5)I_y = 0$, from which $I_y = \frac{160}{8} = 20\ kip$. The roller supports insure that $A_x = 0, I_x = 0$. The sum of the forces: $\sum F_y = A_y + I_y - 4F = 0$, from which $A_y = 4F - I_y = 20\ kip$. The interior angle CAB is $\alpha = \tan^{-1}\frac{5}{7.5} = 33.69^o$. *The method of joints:* Denote the axial force in a member joining I, K by IK.

Joint A: $\sum F_y = A_y + AB\sin\alpha = 0$, from which $AB = -\frac{A_y}{\sin\alpha} = -36.1\ kip\ (C)$.

Joint B: $\sum F_y = -AB\sin\alpha - BC\sin\alpha - F = 0$, from which $\boxed{BC = \frac{-AB\sin\alpha - F}{\sin\alpha} = 18\ kip\ (T)}$

.(b) Make a cut through BD, BC, and AC. *The section as a free body:* The sum of the forces:

$\sum F_y = A_y - F - BC\sin\alpha = 0$, from which $\boxed{BC = \frac{A_y - F}{\sin\alpha} = \frac{10}{0.5547} = 18\ kip\ (T)}$

Problem 6.38 Consider the truss described in Problem 6.37. Use the method of sections to determine the axial forces in members CE, DE, and DF.

Solution: The reactions at A are $A_x = 0$ and $A_y = 20\ kip$ Make a cut through CE, DE, and DF. *The section as a free body:* The interior angle FDE is $\alpha = \tan^{-1}\left(\frac{5}{7.5}\right) = 33.69^o$. Denote the axial force in a member joining I, K by IK. The sum of the moments about D is

$M_D = -A_y(3)(7.5) + F(2)(7.5) + CE(5) = 0$, from which $\boxed{CE = \frac{300}{5} = 60\ kip\ (T)}$. The sum of forces:

$\sum F_y = A_y - 2F - DE\sin\alpha = 0$, from which $\boxed{DE = \frac{A_y - 2F}{\sin\alpha} = 0}$.

$\sum F_x = DF + CE = 0$, from which $\boxed{DF = -CE = -60\ kip\ (C)}$.

Problem 6.39 The Howe truss helps support a roof. Model the supports A and G as roller supports. (a) Use the method of joints to determine the axial force in member BI. (b) Use the method of sections to determine the axial force in member BI.

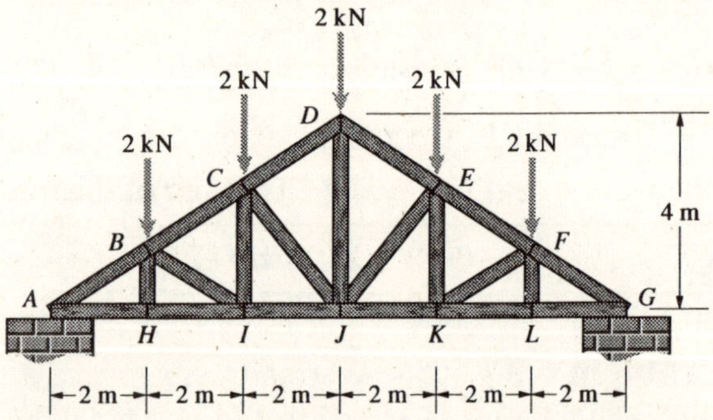

Solution: The pitch of the roof is $\alpha = \tan^{-1}\left(\frac{4}{6}\right) = 33.69^o$. This is also the value of interior angles HAB and HIB. *The complete structure as a free body:* The sum of the moments about A is $M_A = -2(2)(1+2+3+4+5) + 6(2)G = 0$, from which $G = \frac{30}{6} = 5\ kN$. The sum of the forces:

$\sum F_Y = A - 5(2) + G = 0$, from which $A = 10 - 5 = 5\ kN$.

The method of joints: Denote the axial force in a member joining I, K by IK.

(a) *Joint A:* $\sum F_y = A + AB\sin\alpha = 0$, from which $AB = \frac{-A}{\sin\alpha} = \frac{-5}{0.5547} = -9\ kN\ (C)$.

$\sum F_x = AB\cos\alpha + AH = 0$, from which $AH = -AB\cos\alpha = 7.5\ kN\ (T)$.

Joint H: $\sum F_y = BH = 0$.

Joint B: $\sum F_x = -AB\cos\alpha + BI\cos\alpha + BC\cos\alpha = 0$, $\sum F_y = -2 - AB\sin\alpha - BI\sin\alpha + BC\sin\alpha = 0$, from which the two simultaneous equations: $+BI + BC = -9$, and $-BI + BC = -5.41$. Solve:

$\boxed{BI = -1.803\ kN\ (C)}$, $BC = -7.195\ kN\ (C)$

(b) Make the cut through BC, BI and HI. *The section as a free body:* The sum of the moments about B:

$M_B = -A(2) + HI(2\tan\alpha) = 0$, from which $HI = \frac{3}{2}A = 7.5\ kN\ (T)$. The sum of the forces:

$\sum F_x = BC\cos\alpha + BI\cos\alpha + HI = 0$, $\sum F_y = A - F + BC\sin\alpha - BI\sin\alpha = 0$, from which the simultaneous equations: $BC + BI = -9$, $BC - BI = \frac{-A + F}{\sin\alpha} = -5.41$. Solve: $\boxed{BI = -1.803,kN(C)}$.

==================================<>==================================

Problem 6.40 Consider the truss described in Problem 6.39. Use the method of sections to determine the axial force in member EJ.

Solution: From the solution to Problem 6.39, the pitch angle is $\alpha = 36.69^o$, and the reaction $G = 5\ kN$. The length of member EK is $L_{EK} = 4\tan\alpha = \frac{16}{6} = 2.6667\ m$. The interior angle KJE is $\beta = \tan^{-1}\left(\frac{L_{EK}}{2}\right) = 53.13^o$. Make the cut through ED, EJ, and JK. Denote the axial force in a member joining I, K by IK. *The section as a free body:* The sum of the moments about E is $M_E = +4G - 2(F) - JK(2.6667) = 0$, from which $JK = \frac{20-4}{2.6667} = 6\ kN\ (T)$. The sum of the forces: $\sum F_x = -DE\cos\alpha - EJ\cos\beta - JK = 0$. $\sum F_y = DE\sin\alpha - EJ\sin\beta - 2F + G = 0$, from which the two simultaneous equations: $0.8321DE + 0.6EJ = -6$, $0.5547DE - 0.8EJ = -1$. Solve: $\boxed{EJ = -2.5\ kN\ (C)}$.

==================================<>==================================

Problem 6.41 Use the method of sections to determine the axial force in member EF.

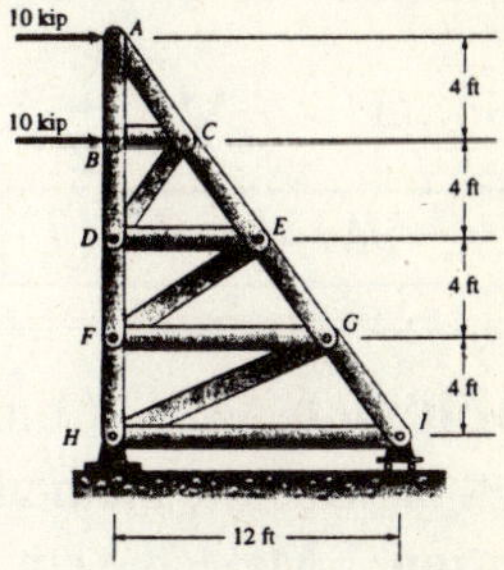

Solution: The included angle at the apex BAC is $\alpha = \tan^{-1}\left(\frac{12}{16}\right) = 36.87^o$. The interior angles BCA, DEC, FGE, HIG are $\gamma = 90^o - \alpha = 53.13^o$. The length of the member ED is $L_{ED} = 8\tan\alpha = 6\ ft$. The interior angle DEF is $\beta = \tan^{-1}\left(\frac{4}{L_{ED}}\right) = 33.69^o$. *The complete structure as a free body:* The moment about H is $M_H = -10(12) - 10(16) + I(12) = 0$, from which $I = \frac{280}{12} = 23.33\ kip$. The sum of forces: $\sum F_y = H_y + I = 0$, from which $H_y = -I = -23.33\ kip$. $\sum F_x = H_x + 20 = 0$, from which $H_x = -20\ kip$. Make the cut through EG, EF, and DE. Consider the upper section only. Denote the axial force in a member joining I, K by IK. *The section as a free body:* The sum of the moments about E is $M_E = -10(4) - 10(8) + DF(L_{ED}) = 0$, from which $DF = \frac{120}{6} = 20\ kip$. The sum of forces: $\sum F_y = -EF\sin\beta - EG\sin\gamma - DF = 0$, $\sum F_x = -EF\cos\beta + EG\cos\gamma + 20 = 0$, from which the two simultaneous equations: $0.5547EF + 0.8EG = -20$, and $0.8320EF - 0.6EG = 20$. Solve: $\boxed{EF = 4.0\ kip\ (T)}$

==================================<>==================================

Problem 6.42 Consider the truss shown in Problem 6.41. Use the method of sections to determine the axial force in member FG.

Solution: From the solution of Problem 6.41, the apex A included angle is $\alpha = 36.87^o$. The length of the member FG is $L_{FG} = 12\tan\alpha = 9\ ft$. Make the cut through EG, GF, and FH, and consider the upper section. Denote the axial force in a member joining I, K by IK. *The section as a free body:* The cut in EG is made very near the point G; the moment about this cut by $M_G = -(8+12)F + (L)FH = 0$ (where $F = 10\ kip$ from Problem 6.41), from which $FH = 22.22\ kip\ (T)$. The sum of the forces:, from which $EG = -27.78\ kip\ (C)$. $\sum F_x = -FG + 2F + EG\cos\beta = 0$, from which $\boxed{FG = -3.33\ kip\ (C)}$.

==================================<>==================================

Problem 6.43 The load $F = 20\ kN$ and the dimension $L = 2\ m$. Use the method of sections to determine the axial force in member HK.

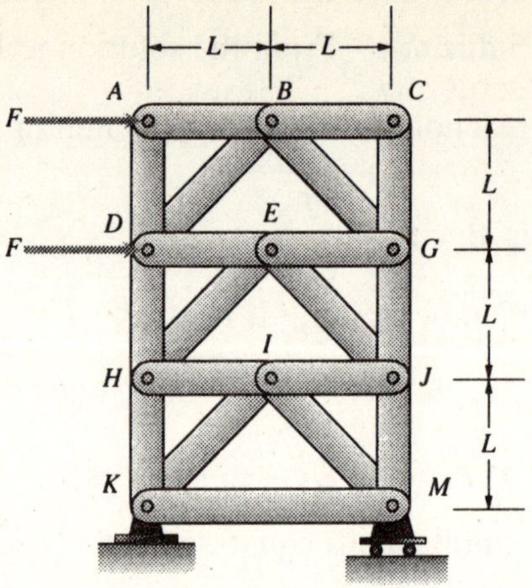

Solution: *Strategy*: Obtain a section by cutting members HK, HI, IJ, and JM. You can determine the axial forces in HK and JM even though the resulting free-body is statically indeterminate.

The complete structure as a free body: The sum of the moments about K is $M_K = -FL(2+3) + ML(2) = 0$, from which $M = \frac{5F}{2} = 50\ kN$. The sum of forces: $\sum F_Y = K_Y + M = 0$, from which $K_Y = -M = -50\ kN$. $\sum F_X = K_X + 2F = 0$, from which $K_X = -2F = -40\ kN$.

The section as a free body: Denote the axial force in a member joining I, K by IK. The sum of the forces: $\sum F_x = K_x - HI + IJ = 0$, from which $HI - IJ = K_x$. The sum of the moments about K is $M_K = M(L)(2) + JM(L)(2) - IJ(L) + HI(L) = 0$. Substitute $HI - IJ = K_x$, to obtain $JM = -M - \frac{K_x}{2} = -30\ kN\ (C)$. $\sum F_y = K_y + M + JM + HK = 0$, from which

$$\boxed{HK = -JM = 30\ kN\ (T)}$$

Problem 6.44 Consider the truss shown in Problem 6.43. The load $F = 50\ kN$ and the length $L = 2\ m$. Use the method of sections to determine the axial forces in members DH and GJ.

Solution: Make the cut through members DH, DE, EG, and GJ, and consider the upper section. Denote the axial force in a member joining I, K by IK

The section as a free body: The sum of the moments about D is $M_D = -F(L) - GJ(2L) = 0$, from which $\boxed{GJ = \frac{-F}{2} = -25\ kN\ (C)}$. The sum of the forces: $\sum F_y = -DH - GJ = 0$, from which

$$\boxed{DH = -GJ = 25\ kN\ (T)}$$

Problem 6.45 Consider the truss shown in Problem 6.21. The mass $m = 200\ kg$. Use the method of sections to determine the axial force in member CD.

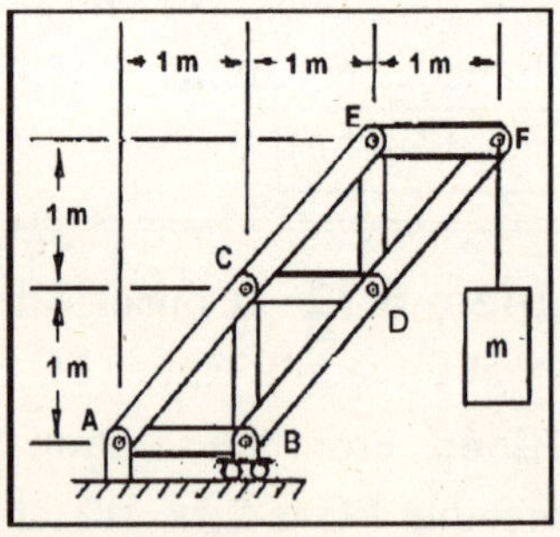

Solution: Make the cut through CE, CD, and BD, and consider the upper section. Denote the axial force in a member joining I, K by IK.

The section as a free body: The weight is $W = 9.81(200) = 1.962\ kN$.

The sum of the moments about D is $M_D = -W + \frac{EC}{\sqrt{2}} = 0$, from which

$EC = \sqrt{2}\,W = 2.775\ kN$. $\sum F_y = -\frac{EC}{\sqrt{2}} - \frac{BD}{\sqrt{2}} - W = 0$, from which

$BD = -2\sqrt{2}\,W = -2\sqrt{2}\,(1.962) = -5.55\ kN$. $\sum F_x = -\frac{EC}{\sqrt{2}} - \frac{BD}{\sqrt{2}} - CD = 0$, from which

$$\boxed{CD = -W + 2W = 1.96\ kN\ (T)}$$

Problem 6.46 The weight $W = 1000\ lb$. The cable passes over pulleys at A and D. (a) Determine the axial forces in member FG and HI. (b) By drawing free body diagrams of sections, explain why the axial forces in members FG and HI are equal.

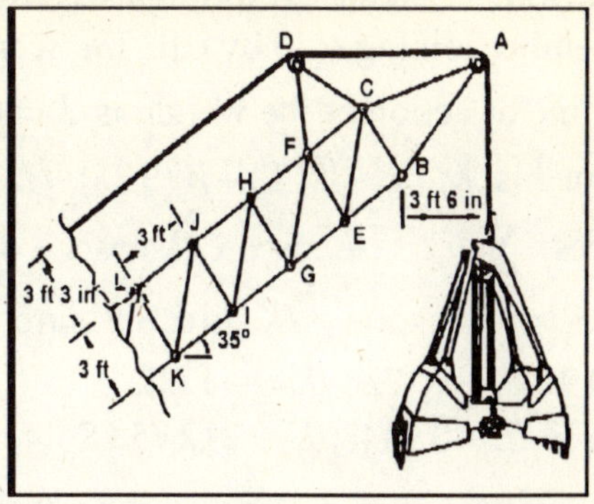

Solution: The truss is at angle $\alpha = 35^o$ relative to the horizontal. The angles of the members FG and HI relative to the horizontal are $\beta = 45^o + 35^o = 80^o$. (a) Make the cut through FH, FG, and EG, and consider the upper section. Denote the axial force in a member joining α, β by $\alpha\beta$.

The section as a free body: The perpendicular distance from point F is $L_{FW} = 3\sqrt{2}\sin\beta + 3.5 = 7.678\ ft$

The sum of the moments about F is

$M_F = -WL_{FW} + W(3.25) - |EG|(3) = 0$, from which

$EG = -1476.1\ lb\ (C)$. The sum of the forces:

$\sum F_Y = -FG\sin\beta - FH\sin\alpha - EG\sin\alpha - W\sin\alpha - W = 0$,

$\sum F_X = -FG\cos\beta - FH\cos\alpha - EG\cos\alpha - W\cos\alpha = 0$, from which the two simultaneous equations:

$-0.9848FG - 0.5736FH = 726.9$, and

$-0.1736FG - 0.8192FH = -389.97$. Solve: $\boxed{FG = -1158.5\ lb\ (C)}$,

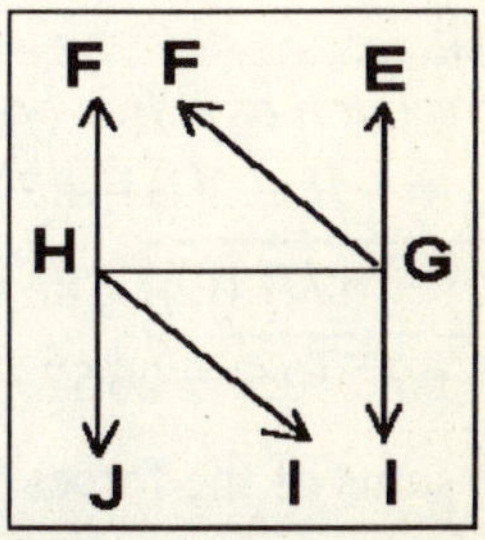

and $FH = 721.64\ lb\ (T)$. Make the cut through JH, HI, and GI, and consider the upper section. *The section as a free body:* The perpendicular distance from point H to the line of action of the weight is

$L_{HW} = 3\cos\alpha + 3\sqrt{2}\sin\beta + 3.5 = 10.135\ ft$. The sum of the moments about H is $M_H = -W(L) - |GI|(3) + W(3.25) = 0$, from which $|GI| = -2295\ lb\ (C)$.

$\sum F_Y = -HI\sin\beta - JH\sin\alpha - GI\sin\alpha - W\sin\alpha - W = 0$,

$\sum F_X = -HI\cos\beta - JH\cos\alpha - GI\cos\alpha - W\cos\alpha = 0$, from which the two simultaneous equations:

$-0.9848HI - 0.5736JH = 257.22$, and

$-0.1736HI - 0.8192JH = -1060.8$. Solve: $\boxed{HI = -1158.5\ lb\ (C)}$, and $\boxed{JH = 1540.6\ lb\ (T)}$.

(b) Choose a coordinate system with the y-axis parallel to JH. Isolate a section by making cuts through FH, FG,and EG, and through HJ, HI, and GI. The free section of the truss is shown. The sum of the forces in the x- and y- direction are each zero; since the only external x-components of axial force are those contributed by FG and HI, the two axial forces must be equal:

$\sum F_x = HI\cos 45^o - FG\cos 45^o = 0$, from which $HI = FG$

Problem 6.47 Consider the truss shown in Problem 6.46. The weight $W = 1000\ lb$. The cable passes over pulleys at A and D. Determine the axial forces in members IK and JL.

Solution: Make a cut through JL, JK, and IK, and consider the upper section. Denote the axial force in a member joining α, β by αβ. *The section as a free body:* The perpendicular distance from point J to the line of action of the weight is $L = 6\cos\alpha + 3\sqrt{2}\sin\beta + 3.5 = 12.593\ ft$. The sum of the moments about J is $M_J = -W(L) + W(3.25) - IK(3) = 0$, from which $\boxed{IK = -3114.4\ lb\ (C)}$. The sum of the forces: $\sum F_x = JL\cos\alpha - IK\cos\alpha - W\cos\alpha - JK\cos\beta = 0$, and

$\sum F_y = -JL\sin\alpha - IK\sin\alpha - W\sin\alpha - W - JK\sin\beta = 0$, from which two simultaneous equations:

$0.8192JL + 0.1736JK = -1732$

$0.5736JL + 0.9848JK = 212.75$. Solve: $\boxed{JL = 2360\ lb\ (T)}$, and $JK = -1158.5\ lb\ (C)$.

Problem 6.48 The truss supports loads at N, P and R. Determine the axial forces in members IL and KM.

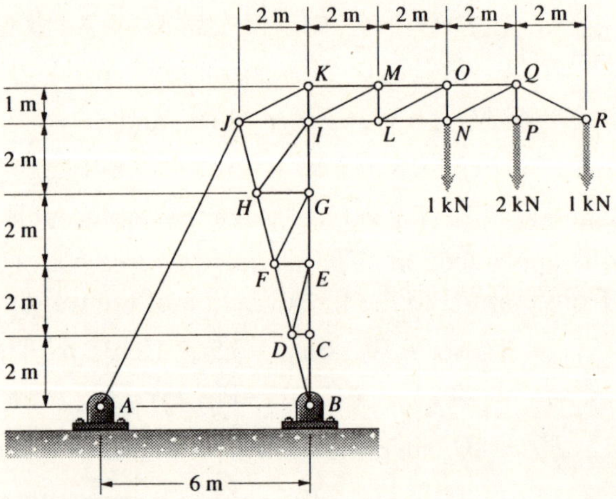

Solution: The strategy is to make a cut through KM, IM, and IL, and consider only the outer section. Denote the axial force in a member joining α, β by αβ.

The section as a free body: The moment about M is $M_M = -IL - 2(1) - 4(2) - 6(1) = 0$, from which $\boxed{IL = -16\ kN\ (C)}$. The angle of member IM is $\alpha = \tan^{-1}(0.5) = 26.57^o$

The sums of the forces: $\sum F_y = -IM\sin\alpha - 4 = 0$, from which

$IM = -\dfrac{4}{\sin\alpha} = -8.944\ kN\ (C)$.

$\sum F_x = -KM - IM\cos\alpha - IL = 0$, from which $\boxed{KM = 24\ kN\ (T)}$

Problem 6.49 Consider the truss shown in problem 6.48. Determine the axial forces in members HJ and GI.

Solution: The strategy is to make a cut through the four members AJ, HJ, HI, and GI, and consider the upper section. The axial force in AJ can be found by taking the moment of the structure about B.

The complete structure as a free body: The angle formed by AJ with the vertical is

$\alpha = \tan^{-1}\left(\dfrac{4}{8}\right) = 26.57^o$. The moment about B is $M_B = 6AJ\cos\alpha - 24 = 0$, from which

$AJ = 4.47\ kN\ (T)$.

The section as a free body: The angles of members HJ and HI relative to the vertical are

$\beta = \tan^{-1}\left(\dfrac{2}{8}\right) = 14.0^o$, and $\gamma = \tan^{-1}\left(\dfrac{1.5}{2}\right) = 28.2^o$, respectively. Make a cut through the four members AJ, HJ, HI, and GI, and consider the upper section. The moment about the point I is

$M_I = -24 + 2AJ\cos\alpha + 2HJ\cos\beta = 0$. from which $\boxed{HJ = 8.25\ kN\ (T)}$

Problem 6.49 continued on next page....

continuation of solution to Problem 6.49...

The sums of the forces: $\sum F_x = -AJ\sin\alpha + HJ\sin\beta - HI\sin\gamma = 0$, from which $HI = \dfrac{AJ\sin\alpha - HJ\sin\beta}{\sin\gamma} = \dfrac{2-2}{\sin\gamma} = 0$. $\sum F_Y = -AJ\cos\alpha - HJ\cos\beta - HI\cos\gamma - GI - 4 = 0$, from which $\boxed{GI = -16\ kN\ (C)}$

Problem 6.50 Consider the truss shown in Problem 6.48. By drawing free-body diagrams of sections, explain why the axial forces in members DE, FG, and HI are zero.

Solution: . Define α, β to be the interior angles BAJ and ABJ respectively. The sum of the forces in the x-direction at the base yields $A_X + B_X = 0$, from which $A_x = -B_x$. . Make a cut through AJ, BD and BC, from which the sum of forces in the x-direction , $A_x - BD\sin\beta = 0$. Since $A_x = AJ\sin\alpha$, then $AJ\sin\alpha - BD\sin\beta = 0$. A repeat of the solution to Problem 6.49 for each section shows that this result holds for each section, where BD is to be replaced by the member parallel to BD. For example: make a cut through AJ, FD, DE, and CE. Eliminate the axial force in member AJ as an unknown by taking the moment about A. Repeat the solution process in Problem 6.49, obtaining the result that $DE = \dfrac{AJ\sin\alpha - DF\sin\beta}{\cos\theta_{DE}} = 0$ where θ_{DE} is the angle of the member DE with the vertical. Similarly, a cut through AJ, FH, FG, and EG leads to $FG = \dfrac{AJ\sin\alpha - FH\sin\beta}{\cos\theta_{FG}} = 0$, and so on. Thus the explanation is that each member BD, DF, FH and HJ has equal tension, and that this tension balances the *x*-component in member AJ

Problem 6.51 The suspended weight $W = 600\ lb$. The bars AB and AC have ball and socket joints at each end. Determine the tension in the cable AD and the axial forces in members AB and AC.

Solution: The coordinates of the points are A(3,4,4), B(0,0,3), C(4,0,0), D(0,4,0). The position vectors parallel to each member are:

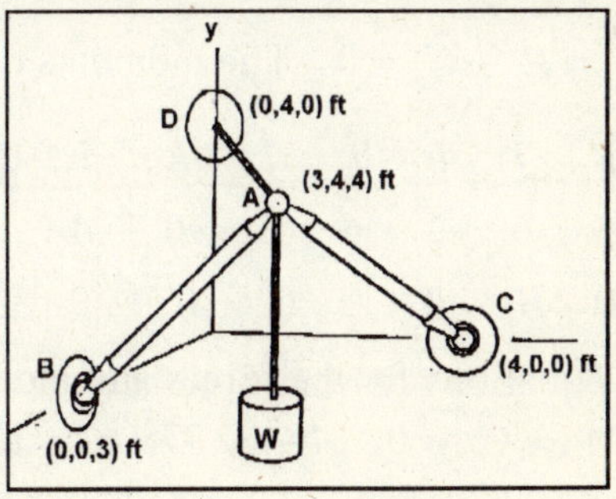

$\vec{r}_{AB} = -3\vec{i} - 4\vec{j} - \vec{k}$, $\vec{r}_{AC} = \vec{i} - 4\vec{j} - 4\vec{k}$,
$\vec{r}_{AD} = -3\vec{i} + 0\vec{j} - 4\vec{k}$. The unit vectors parallel to each member are $\vec{e}_{AB} = \dfrac{\vec{r}_{AB}}{|\vec{r}_{AB}|} = -0.5883\vec{i} - 0.7845\vec{j} - 0.1961\vec{k}$,
$\vec{e}_{AC} = 0.1741\vec{i} - 0.6963\vec{j} - 0.6963\vec{k}$, $\vec{e}_{AD} = -0.6\vec{i} - 0.8\vec{k}$. The unit vector parallel to the line of action of the weight is $\vec{e}_w = -\vec{j}$. The axial forces (tension in the cable) are given by $|\vec{T}_{AB}|\vec{e}_{AB}$, $|\vec{T}_{AC}|\vec{e}_{AC}$, and $|\vec{T}_{AD}|\vec{e}_{AD}$. The equilibrium conditions are

$$\sum\vec{F}_x = (-0.5883|\vec{T}_{AB}| + 0.1741|\vec{T}_{AC}| - 0.6|\vec{T}_{AD}|)\vec{i} = 0$$

$$\sum\vec{F}_y = (0.7845|\vec{T}_{AB}| - 0.6963|\vec{T}_{AC}| + 0|\vec{T}_{AD}| - 600)\vec{j} = 0$$

$$\sum\vec{F}_z = (-0.1961|\vec{T}_{AB}| - 0.6963|\vec{T}_{AC}| - 0.8|\vec{T}_{AD}|)\vec{k} = 0$$

Solve: $\boxed{|\vec{T}_{AB}| = -489.5\ lb\ (C)}$, $\boxed{|\vec{T}_{AC}| = -310.2\ lb\ (C)}$, $\boxed{|\vec{T}_{AD}| = 390\ lb\ (T)}$

Problem 6.52

Problem 6.52 The space truss supports a vertical 10 *kN* load at D. The reactions at the supports at points A, B, and C are shown. What are the axial forces in the members AD, BD, and CD?

Solution: Consider the joint D only. The position vectors parallel to the members from D are

$\vec{r}_{DA} = -4\vec{i} - 3\vec{j} - \vec{k}$,

$\vec{r}_{DB} = \vec{i} - 3\vec{j} + 2\vec{k}$, $\vec{r}_{DC} = 2\vec{i} - 3\vec{j} - \vec{k}$. The unit vectors parallel to the members from D are:

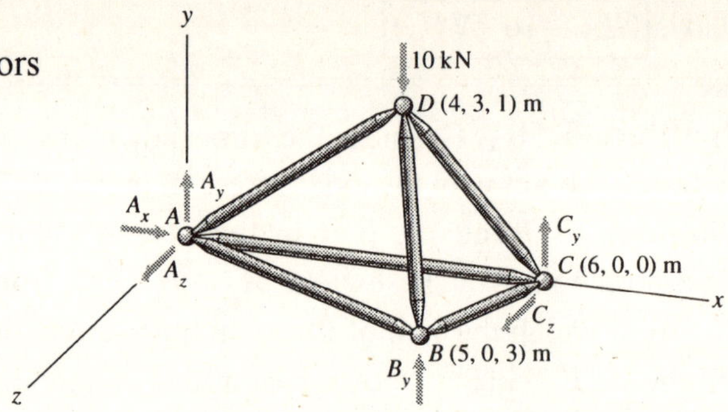

$$\vec{e}_{DA} = \frac{\vec{r}_{DA}}{|\vec{r}_{DA}|} = -0.7845\vec{i} - 0.5883\vec{j} - 0.1961\vec{k}$$

$$\vec{e}_{DB} = \frac{\vec{r}_{DB}}{|\vec{r}_{DB}|} = 0.2673\vec{i} - 0.8018\vec{j} + 0.5345\vec{k}$$

$$\vec{e}_{DC} = \frac{\vec{r}_{DC}}{|\vec{r}_{DC}|} = 0.5345\vec{i} - 0.8018\vec{j} - 0.2673\vec{k}$$

The equilibrium conditions for the joint D are $\sum \vec{F} = T_{DA}\vec{e}_{DA} + T_{DB}\vec{e}_{DB} + T_{DC}\vec{e}_{DC} - \vec{F}_D = 0$, from which

$\sum F_x = -0.7845T_{DA} + 0.2673T_{DB} + 0.5345T_{DC} = 0$

$\sum F_y = -0.5883T_{DA} - 0.8018T_{DB} - 0.8108T_{DC} - 10 = 0$

$\sum F_z = -0.1961T_{DA} + 0.5345T_{DB} - 0.2673T_{DC} = 0$. Solve: $\boxed{T_{DA} = -4.721\ kN\ (C)}$, $\boxed{T_{DB} = -4.157\ kN\ (C)}$

$\boxed{T_{DC} = -4.850\ kN\ (C)}$

Problem 6.53

Problem 6.53 Consider the space truss shown in Problem 6.52. The reactions at the supports at points A, B, and C are shown. What are the axial forces in members AB, AC, and AD?

Solution: The reactions at A are required for a determination of the equilibrium conditions at A.

The complete structure as a free body: The position vectors are $\vec{r}_{AB} = 5\vec{i} + 3\vec{k}$, $\vec{r}_{AC} = 6\vec{i}$, $\vec{r}_{AD} = 4\vec{i} + 3\vec{j} + \vec{k}$. The sum of the forces: $\sum F_x = A_x = 0$, $\sum F_y = A_y + C_c + B_y - 10 = 0$, and $\sum F_z = A_z + C_z = 0$. The moments due to the reactions: $\vec{M} = \vec{r}_{AB} \times \vec{F}_B + \vec{r}_{AC} \times \vec{F}_C + \vec{r}_{AD} \times \vec{F}_D = 0$

$$\vec{M} = \begin{vmatrix} \vec{i} & \vec{j} & \vec{k} \\ 5 & 0 & 3 \\ 0 & B_y & 0 \end{vmatrix} + \begin{vmatrix} \vec{i} & \vec{j} & \vec{k} \\ 6 & 0 & 0 \\ 0 & C_y & C_z \end{vmatrix} + \begin{vmatrix} \vec{i} & \vec{j} & \vec{k} \\ 4 & 3 & 1 \\ 0 & -10 & 0 \end{vmatrix} = 0 = (-3B_y + 10)\vec{i} - (6C_z)\vec{j} + (5B_y + 6C_y - 40)\vec{k} = 0.$$

These equations for the forces and moments are to be solved for the unknown reactions. The solution: $A_x = A_y = C_z = 0$, $A_y = 2.778\ kN$, $B_y = 3.333\ kN$, and $C_y = 3.889\ kN$

The method of joints: Joint A: The position vectors are given above. The unit vectors are: $\vec{e}_{AB} = 0.8575\vec{i} + 0.5145\vec{k}$, $\vec{e}_{AC} = \vec{i}$, $\vec{e}_{AD} = 0.7845\vec{i} + 0.5883\vec{j} + 0.1961\vec{k}$. The equilibrium conditions are:

$\sum \vec{F} = T_{AB}\vec{e}_{AB} + T_{AV}\vec{e}_{AC} + T_{AD}\vec{e}_{AD} + \vec{A} = 0$, from which

$\sum F_x = 0.8575T_{AB} + T_{AC} + 0.7845T_{AD} = 0$

$\sum F_y = 0T_{AB} + 0T_{AC} + 0.5883T_{AD} + 2.778 = 0$

$\sum F_z = 0.5145|T_{AB}| + 0|T_{AC}| + 0.1961|T_{AD}| = 0$ Solve: $\boxed{T_{AB} = 1.8\ kN\ (T)}$, $\boxed{T_{AC} = 2.16\ kN\ (T)}$

$\boxed{T_{AD} = -4.72\ kN\ (C)}$

Problem 6.54

Problem 6.54 Consider the space truss shown in Problem 6.52. The reactions at the supports at points A, B, and C are shown. What are the axial forces in members AB, BC, and BD?

Solution: From the solution of Problem 6.53, the reaction at B is $B_Y = 3.33\ kN$. The position vectors from point B to the other joints are: $\vec{r}_{BA} = -5\vec{i} - 3\vec{k}$, $\vec{r}_{BC} = \vec{i} - 3\vec{k}$, and $\vec{r}_{BD} = -\vec{i} + 3\vec{j} - 2\vec{k}$. The unit vectors are : $\vec{e}_{BA} = -0.8575\vec{i} - 0.5145\vec{k}$, $\vec{e}_{BC} = 0.3162\vec{i} - 0.9487\vec{k}$, $\vec{e}_{BD} = -0.2673\vec{i} + 0.8018\vec{j} - 0.5345\vec{k}$. The equilibrium conditions are

$\sum \vec{F} = T_{BA}\vec{e}_{BA} + T_{BC}\vec{e}_{BC} + T_{BD}\vec{e}_{BD} + \vec{B} = 0$, from which

$\sum F_x = -0.8575T_{AB} + 0T_{BC} - 0.5145T_{BD} = 0$

$\sum F_y = 0T_{BA} + 0T_{BC} + 0.8018T_{BD} + 3.333 = 0$

$\sum F_x = -0.5145T_{BA} - 0.9487T_{BC} - 0.5345T_{BD} = 0$. Solve: $\boxed{T_{BA} = 1.8\ kN\ (T)}$, $\boxed{T_{BC} = 1.366\ kN\ (T)}$, $\boxed{T_{BD} = -4.16\ kN\ (C)}$.

Problem 6.55

Problem 6.55 The space truss supports a vertical load F at A. Each member is of length L, and the truss has roller supports at B, C, and D. What are the axial forces in the members?

Solution: The strategy is to (a) develop the coordinates of the vertices, (b) solve for the reactions at each support, and (c) apply the equilibrium conditions at each joint.

The top view of the truss is shown. The distances in the sketch are $a = \frac{\sqrt{3}}{2}L$ and $b = \frac{L}{2}$. The bisectors of the angles in an equilateral triangle divide each other into parts with a 3-to-1 ratio, as demonstrated by the law of sines: $\frac{b}{\sin 60^o} = \frac{c}{\sin\ 30^o}$, from which $c = \frac{L}{2\sqrt{3}}$. The projected horizontal distances AB, AC, and AD are

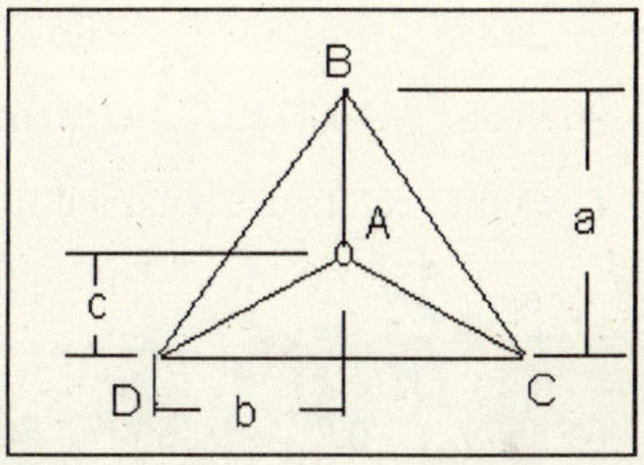

$d_{AB} = d_{AC} = d_{AD} = \frac{c}{\sin 30^o} = \frac{L}{\sqrt{3}}$. The height of the apex A above the base is $h_A = L\sqrt{1 - c^2} = \sqrt{\frac{2}{3}}\,L$.

Choose a coordinate system with the x-axis originating at B with positive direction parallel to the side BC, with the y-axis in the base plane. The coordinates in this system are $A\left(\frac{L}{2}, \frac{L}{2\sqrt{3}}, \sqrt{\frac{2}{3}}\,L\right)$, $B(0,0,0)$, $C(L, 0, 0)$, and $D\left(\frac{L}{2}, \frac{\sqrt{3}\,L}{2}, 0\right)$. The position vectors of the vertices A, B, C, D are

$\vec{r}_A = L(0.5\vec{i} + 0.2887\vec{j} + +0.8165\vec{k})$, $\vec{r}_B = L(0\vec{i} + 0\vec{j} + 0\vec{k})$, $\vec{r}_C = L(\vec{i} + 0\vec{j} + 0\vec{k})$, and $\vec{r}_D = L(0.5\vec{i} + 0.8660\vec{j} + 0\vec{k})$. *The complete structure as a free body:* Choose joint A. The position vectors of B ,C, and D relative to A are $\vec{r}_{AB} = \vec{r}_B - \vec{r}_A = L(-0.5\vec{i} - 0.2887\vec{j} - 0.8165\vec{k})$, $\vec{r}_{AC} = \vec{r}_C - \vec{r}_A = L(0.5\vec{i} - 0.2887\vec{j} - 0.8165\vec{k})$, and $\vec{r}_{AD} = \vec{r}_D - \vec{r}_A = L(0\vec{i} + 0.5774\vec{j} - 0.8165\vec{k})$.

The roller supports insure only vertical reactions at the points B, C, and D. The sum of forces $\sum F_Z = B + C + D - F = 0$ from which $\frac{B}{F} + \frac{C}{F} + \frac{D}{F} - 1 = 0$.

Problem 6.55 continued on next page....

================================<>================================

continuation of solution to Problem 6.55...

The sum of the moments about A:

$$\vec{M}_A = L\begin{vmatrix} \vec{i} & \vec{j} & \vec{k} \\ -0.5 & -0.2887 & -0.8165 \\ 0 & 0 & \frac{B}{F} \end{vmatrix} + L\begin{vmatrix} \vec{i} & \vec{j} & \vec{k} \\ 0.5 & -0.2887 & -0.8165 \\ 0 & 0 & \frac{C}{F} \end{vmatrix} + L\begin{vmatrix} \vec{i} & \vec{j} & \vec{k} \\ 0 & 0.5774 & -0.8165 \\ 0 & 0 & \frac{D}{F} \end{vmatrix} = 0$$

$\vec{M}_A = \left(-0.2887\frac{B}{F} - 0.2887\frac{C}{F} + 0.5774\frac{D}{F}\right)\vec{i} + \left(0.5\frac{B}{F} - 0.5\frac{C}{F}\right)\vec{j} + 0\vec{k} = 0$. The moment condition and the sum of vertical forces yields the simultaneous equations:

$-\frac{B}{F} - \frac{C}{F} + 2\frac{D}{F} = 0$, $\frac{B}{F} - \frac{C}{F} = 0$, $\frac{B}{F} + \frac{C}{F} + \frac{D}{F} = 1$. Solve: $\boxed{B = C = D = 0.3333F}$. *Check:* This answer is to be expected from symmetry, of course. *check.*

Joint A: Since L is the magnitude of the position vectors from A to the other vertices, the unit vectors are obtained directly from the position vectors: $\vec{e}_{AB} = -0.5\vec{i} - 0.2887\vec{j} - 0.8165\vec{k}$,
$\vec{e}_{AC} = 0.5\vec{i} - 0.2887\vec{j} - 0.8165\vec{k}$, and $\vec{e}_{AD} = 0\vec{i} + 0.5774\vec{j} - 0.8165\vec{k}$. The sum of the forces on Joint A are $\sum \vec{F} = T_{AB}\vec{e}_{AB} + T_{AC}\vec{e}_{AC} + T_{AD}\vec{e}_{AD} + \vec{F} = 0$, from which

$\sum F_x = -0.5T_{AB} + 0.5T_{AC} + 0T_{AD} = 0$

$\sum F_y = -0.2887T_{AB} - 0.2887T_{AC} + 0.5774T_{AD} = 0$

$\sum F_z = -0.8165T_{AB} - 0.8165T_{AC} + 0.8165T_{AD} - F = 0$. Solve: $\boxed{T_{AB} = T_{AC} = T_{AD} = -0.408F\ (C)}$

Joint B: The unit vectors are obtained from the position vectors: $\vec{e}_{BA} = 0.5\vec{i} + 0.2887\vec{j} + 0.8165\vec{k}$,
$\vec{e}_{BC} = 1\vec{i}$, $\vec{e}_{BD} = 0.5\vec{i} + 0.866\vec{j}$. The equilibrium conditions are

$\sum F_x = 0.5T_{BA} + T_{BC} + 0.5T_{BD} = 0$

$\sum F_y = 0.2887T_{BA} + 0T_{BC} + 0.8660T_{BD} = 0$

$\sum F_z = 0.8165T_{BA} + 0T_{BC} + 0T_{BD} = -0.3333F$. Solve: $T_{BA} = -0.408F\,(C)$, $\boxed{T_{BC} = T_{BD} = 0.136F\ (T)}$

Joint C: The dimensionless height of the apex $\frac{A}{L}$ is $h = \sqrt{\frac{2}{3}} = 0.8165$. The angle between member AC and the vertical is $\alpha = \cos^{-1}(0.8165) = 35.26^o$. The projection of the known axial force in AC on the horizontal plane is $F_H = T_{AC}\sin\alpha = 0.408F\sin\alpha$. The angle between members CB and CD and this projection of AC (an angle bisector) is $\pm 30^o$. The sum of the forces in the horizontal plane at joint C is $-T_{CB}\cos 30^o - T_{CD}\cos 30^o - T_{AC}\sin\alpha = 0$. But $T_{CB} = T_{BC}$ and $T_{CA} = T_{AC}$ are known, from which

$T_{CD} = -T_{AC}\left(\frac{\sin\alpha}{\cos 30^o}\right) - T_{BC} = 0.408F\left(\frac{0.5774}{0.8660}\right) - 0.136F$, from which $\boxed{T_{CD} = 0.136F\ (T)}$ This completes the determination of axial forces for the six members. *Discussion:* With assumptions about symmetry, the technique used for Joint C can be used entirely to solve this problem, by applying it to all joints. This provides a method for checking results. *End of Discussion.*

================================<>================================

Problem 6.56 The space truss has roller supports at B, C, and D and supports a vertical 800 *lb* load at A. What are the axial forces in members AB, AC, and AD?

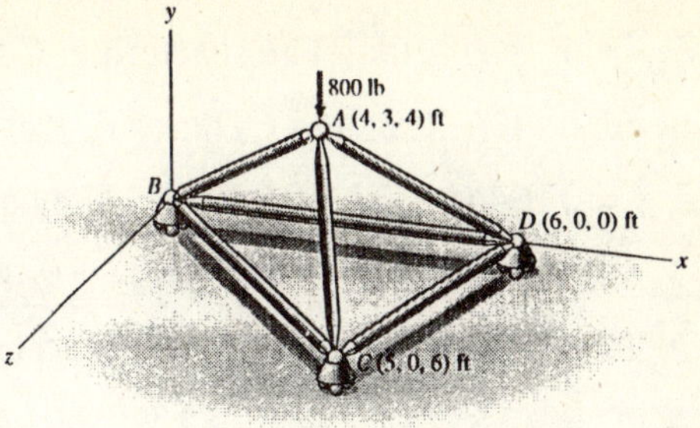

Solution: The position vectors of the points A, B, C, and D are $\vec{r}_A = 4\vec{i} + 3\vec{j} + 4\vec{k}$, $\vec{r}_C = 5\vec{i} + 6\vec{k}$, $\vec{r}_D = 6\vec{i}$. The position vectors from joint A to the vertices are:

$\vec{r}_{AB} = \vec{r}_B - \vec{r}_A = -4\vec{i} - 3\vec{j} - 4\vec{k}$, $\vec{r}_{AC} = \vec{r}_C - \vec{r}_A = 1\vec{i} - 3\vec{j} + 2\vec{k}$,

$\vec{r}_{AD} = \vec{r}_D - \vec{r}_A = 2\vec{i} - 3\vec{j} - 4\vec{k}$

Joint A: The unit vectors parallel to members AB, AC, and AD are $\vec{e}_{AB} = \dfrac{\vec{r}_{AB}}{|\vec{r}_{AB}|} = -0.6247\vec{i} - 0.4685\vec{j} - 0.6247\vec{k}$,

$\vec{e}_{AC} = \dfrac{\vec{r}_{AC}}{|\vec{r}_{AC}|} = 0.2673\vec{i} - 0.8018\vec{j} + 0.5345\vec{k}$, and $\vec{e}_{AD} = \dfrac{\vec{r}_{AD}}{|\vec{r}_{AD}|} = 0.3714\vec{i} - 0.5570\vec{j} - 0.7428\vec{k}$. The equilibrium conditions at point A: $\sum F_x = -0.6247T_{AB} + 0.2673T_{AC} + 0.3714T_{AD} = 0$

$\sum F_y = -0.4685T_{AB} - 0.8018T_{AB} - 0.5570T_{AD} - 800 = 0$

$\sum F_z = -0.6247T_{AB} + 0.5345T_{AC} - 0.7428T_{AD} = 0$. Solve: $\boxed{T_{AB} = -379.4\ lb\ (C)}$, $\boxed{T_{AC} = -665.2\ lb\ (C)}$, and $\boxed{T_{AD} = -159.6\ lb\ (C)}$

Problem 6.57 Consider the space truss described in Problem 6.56. What are the axial forces in members BC, BD, and CD?

Solution: . *The complete structure as a free body:* The position vectors relative to B are obtained directly from the coordinates, since B is at the origin: $\vec{r}_{BA} = 4\vec{i} + 3\vec{j} + 4\vec{k}$, $\vec{r}_{BC} = 5\vec{i} + 6\vec{k}$, $\vec{r}_{BD} = 6\vec{i}$. The reaction at B is required. The roller supports assure zero horizontal reactions. The sum of the forces in the vertical direction: $\sum F_y = B + D + C - 800 = 0$. The moments about point A:

$\vec{M}_B = (\vec{r}_{BA} \times \vec{F}) + (\vec{r}_{BC} \times \vec{C}) + (\vec{r}_{bD} \times \vec{D}) = 0$,

$$\vec{M}_B = \begin{vmatrix} \vec{i} & \vec{j} & \vec{k} \\ 4 & 3 & 4 \\ 0 & -800 & 0 \end{vmatrix} + \begin{vmatrix} \vec{i} & \vec{j} & \vec{k} \\ 5 & 0 & 6 \\ 0 & C & 0 \end{vmatrix} + \begin{vmatrix} \vec{i} & \vec{j} & \vec{k} \\ 6 & 0 & 0 \\ 0 & D & 0 \end{vmatrix} = 0 = (3200 - 6C)\vec{i} + 0\vec{j} + (-3200 + 5C + 6D)\vec{k} = 0.$$ The

solutions: $C = 533.3\ lb$, $D = 88.89\ lb$, and $B = 177.78\ lb$.

Joint B: The unit vectors are obtained from the position vectors relative to the origin:

$\vec{e}_{BA} = 0.6247\vec{i} + 0.4685\vec{j} + 0.6247\vec{k}$, $\vec{e}_{BC} = 0.6402\vec{i} + 0\vec{j} + 0.7682\vec{k}$, $\vec{e}_{BD} = 1\vec{i}$ The equilibrium conditions for joint B:

$\sum F_x = 0.6247T_{BA} + 0.6402T_{BC} + 1T_{BD} = 0$

$\sum F_y = 0.4865T_{BA} + 0T_{BC} + 0T_{BD} = -177.78$

$\sum F_z = 0.6247T_{BA} + 0.7682T_{BC} + 0T_{BD} = 0$. Solve: $T_{BA} = -379.4\ lb\ (C)$, $\boxed{T_{BC} = 308.56\ lb\ (T)}$,

$\boxed{T_{BD} = 39.51\ lb\ (T)}$

Problem 6.57 continued on next page....

continuation of solution to Problem 6.57...

Joint C: The position vectors relative to joint C are: $\vec{r}_{CA} = -1\vec{i} + 3\vec{j} - 2\vec{k}$, $\vec{r}_{CB} = -5\vec{i} + 0\vec{j} - 6\vec{k}$, and $\vec{r}_{CD} = 1\vec{i} + 0\vec{j} - 6\vec{k}$. The reaction at C (from above) is $C = 533.33\ lb$. The unit vectors parallel to the members CA, CB, and CD are: $\vec{e}_{CA} = -0.2673\vec{i} + 0.8018\vec{j} - 0.5345\vec{k}$, $\vec{e}_{CB} = -0.6402\vec{i} + 0\vec{j} - 0.7682\vec{k}$, $\vec{e}_{CD} = 0.1644\vec{i} + 0\vec{j} - 0.9864\vec{k}$. The equilibrium conditions for C:

$\sum F_x = -0.2673T_{AC} - 0.6402T_{BC} + 0.1644T_{CD} = 0$

$\sum F_y = 0.8018T_{AC} + 0T_{BC} + 0T_{CD} + 533.33 = 0$

$\sum F_z = -0.5345T_{AC} - 0.7682T_{BC} - 0.9864T_{CD} = 0$. Solve: $T_{AC} = -665.2\ lb\ (C)$, $T_{BC} = 308.6\ lb\ (T)$, and $\boxed{T_{CD} = 120.15\ lb\ (T)}$

Problem 6.58 The mirror housing of the telescope is supported by a 6-bar space truss. The mass of the housing is 3 Mg (megagrams) and its weight acts at G. The distance from the axis of the telescope to points A, B and C is 1 *m* and the distance from the axis to points D, E, and F is 2.5 *m*. If the telescope is vertical ($\alpha = 90^o$), what are the axial forces in the members of the truss?

Solution:

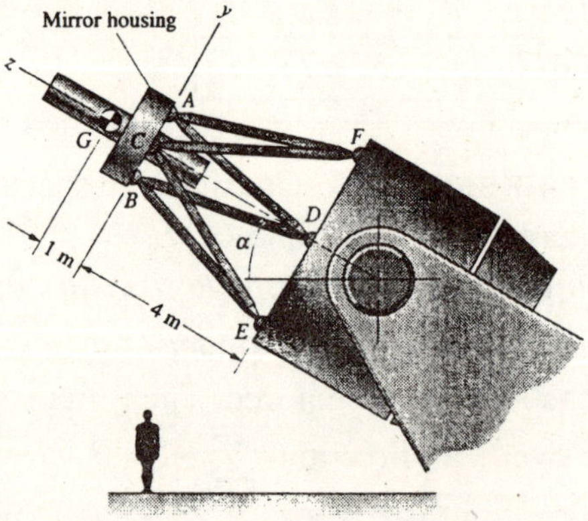

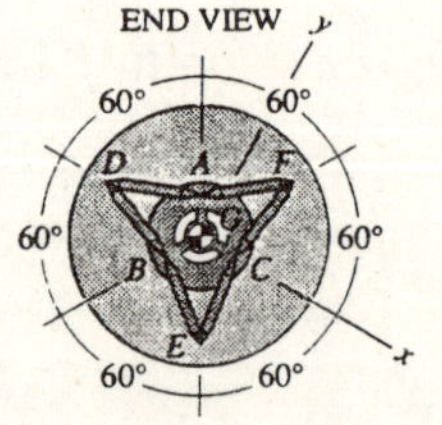

A cut through the 6-bar space truss leads to six equations in the unknowns (see Problem 6.59). However for this problem an alternate strategy based on reasonable assumptions about the equality of the tensions is used to get the reactions. Assume that each support carries one-third of the weight, which is equally divided between the two bars at the support. The coordinate system has its origin in the upper platform, with the x-axis passing though the point C. The coordinates of the points are:

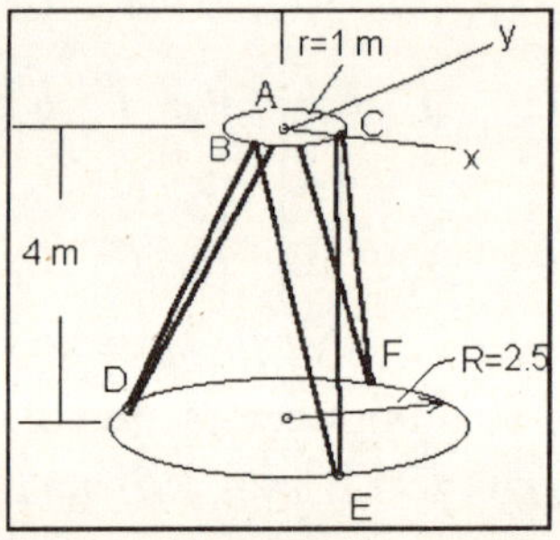

$A(-\cos 60^o,\ \sin 60^o, 0) = (-0.5, 0.866, 0)$,

$B(-\cos 60^o, -\sin 60^o, 0) = (-0.5, -0.866, 0)$, $C(1, 0, 0)$,

$D(-2.5, 0, -4)$,

$E(2.5\cos 60^o, -2.5\sin 60^o, -4) = (1.25, -2.165, -4)$,

$F(2.5\cos 60^o, 2.5\sin 60^o, -4) = (1.25, 2.165, -4)$.

Consider joint B in the upper housing. The position vectors of the points E and D relative to B are $\vec{r}_{BD} = -2\vec{i} + 0.866\vec{j} - 4\vec{k}$, $\vec{r}_{BE} = 1.75\vec{i} - 1.299\vec{j} - 4\vec{k}$. The unit vectors are $\vec{e}_{BD} = -0.4391\vec{i} + 0.1901\vec{j} - 0.8781\vec{k}$, and $\vec{e}_{BE} = 0.3842\vec{i} - 0.2852\vec{j} - 0.8781\vec{k}$. The weight is balanced by the *z* components: $\sum F_y = -\frac{W}{3} - (0.8781)T_{BD} - (0.8781)T_{BE} = 0$. Assume that the magnitude of the axial force is the same in both members BD and BE, $T_{BE} = T_{BD}$. The weight is $W = 3(9.81) = 29.43\ kN$. Thus the result: $\boxed{T_{BE} = T_{BD} = -5.5858\ kN\ (C)}$. From symmetry (and the assumptions made above) the axial force is the same in all members.

===================================<>===================================

Problem 6.59 Consider the telescope described in Problem 6.58. Determine the axial forces in the members of the truss if the angle α between the horizontal and the telescope axis is 20°.

Solution: .The coordinates of the points are,

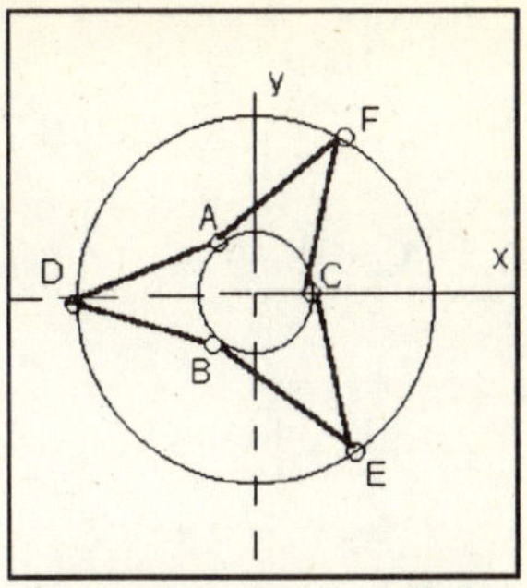

$A(-\cos 60^o,\ \sin 60^o, 0) = (-0.5, 0.866, 0)\ (m)$,

$B(-\cos 60^o, -\sin 60^o, 0) = (-0.5, -0.866, 0)\ (m)$, $C(1, 0, 0)\ (m)$,

$D(-2.5, 0, -4)\ (m)$,

$E(2.5\cos 60^o, -2.5\sin 60^o, -4) = (1.25, -2.165, -4)\ (m)$,

$F(2.5\cos 60^o, 2.5\sin 60^o, -4) = (1.25, 2.165, -4)\ (m)$. The coordinates of the center of gravity are $G(0, 0, 1)\ (m)$. Make a cut through the members just below the upper platform supports, such that the cut members have the same radial distance from the axis as the supports. Consider the upper section.

The section as a free body: The strategy is to sum the forces and moments to obtain six equations in the six unknown axial forces. The axial forces and moments are expressed in terms of unit vectors. The position vectors of the points E, D, and F relative to the points A ,B,and C are required to obtain the unit vectors parallel to the members. The unit vectors are obtained from these vectors. The vectors and their associated unit vectors are given in Table I. *Note:* While numerical values are shown below to four significant figures, the calculations were done with the full precision permitted (15 digits for **TK Solver 2.**)

:

Vector	x	y	z	Unit Vector	x	y	z
r_{AD}	-2	-0.866	-4	e_{AD}	-0.4391	-0.1901	-0.8781
r_{AF}	1.75	1.299	-4	e_{AF}	0.3842	0.2852	-0.8781
r_{BD}	-2	0.866	-4	e_{BD}	-0.4391	0.1901	-0.8781
r_{BE}	1.75	-1.299	-4	e_{BE}	0.3842	-0.2852	-0.8781
r_{CE}	0.25	-2.165	-4	e_{CE}	0.0549	-0.4753	-0.8781
r_{CF}	0.25	2.165	-4	e_{CF}	0.0549	-0.4753	-0.8781

Table I

The equilibrium condition for the forces is

$\left|\vec{T}_{AB}\right|\vec{e}_{AD} + \left|\vec{T}_{AF}\right|\vec{e}_{AF} + \left|\vec{T}_{BD}\right|\vec{e}_{BD} + \left|\vec{T}_{BE}\right|\vec{e}_{BE} + \left|\vec{T}_{CE}\right|\vec{e}_{CE} + \left|\vec{T}_{CF}\right|\vec{e}_{CF} + \vec{W} = 0$. This is three equations in six unknowns. The unit vectors are given in Table I. The weight vector is $\vec{W} = \left|\vec{W}\right|\left(-\vec{j}\cos\alpha - \vec{k}\sin\alpha\right)$, where α is the angle from the horizontal of the telescope housing. The remaining three equations in six unknowns are obtained from the moments:

$\vec{r}_A \times \left(\vec{T}_{AD} + \vec{T}_{AF}\right) + \vec{r}_B \times \left(\vec{T}_{BD} + \vec{T}_{BE}\right) + \vec{r}_C \times \left(\vec{T}_{CE} + \vec{T}_{CF}\right) + \vec{r}_G \times \vec{W} = 0$. Carry out the indicated operations on the moments to obtain the vectors defining the moments:

$$\vec{r}_A \times \vec{T}_{AD} = \left|\vec{T}_{AD}\right|\begin{vmatrix} \vec{i} & \vec{j} & \vec{k} \\ -0.5 & 0.866 & 0 \\ -0.4391 & -0.1901 & -0.8781 \end{vmatrix} = \left|\vec{T}_{AD}\right|\left(-0.7605\vec{i} - 0.4391\vec{j} + 0.4753\right)$$

$$= \left|\vec{T}_{AD}\right|\left(\vec{i}u_{ADx} + \vec{j}u_{ADy} + \vec{j}u_{ADz}\right)$$

Problem 6.59 continued on next page..........

==============================<>==============================

............continuation of solution to Problem 6.59.....

$$\vec{r}_A \times \vec{T}_{AF} = |\vec{T}_{AF}| \begin{vmatrix} \vec{i} & \vec{j} & \vec{k} \\ -0.5 & 0.866 & 0 \\ 0.3842 & 0.2852 & -0.8781 \end{vmatrix} = |\vec{T}_{AF}|(-0.7605\vec{i} - 0.4391\vec{j} - 0.4753\vec{k})$$

$$= |\vec{T}_{AF}|(\vec{i}u_{AFx} + \vec{j}u_{AFy} + \vec{k}u_{AFz})$$

$$\vec{r}_B \times \vec{T}_{BD} = |\vec{T}_{BD}| \begin{vmatrix} \vec{i} & \vec{j} & \vec{k} \\ -0.5 & -0.866 & 0 \\ -0.4391 & 0.1901 & -0.8781 \end{vmatrix} = |\vec{T}_{BD}|(0.7605\vec{i} - 0.4391\vec{j} - 0.4753\vec{k})$$

$$= |\vec{T}_{BD}|(\vec{i}u_{BDx} + \vec{j}u_{BDy} + \vec{k}u_{BDz})$$

$$\vec{r}_B \times \vec{T}_{BE} = |\vec{T}_{BE}| \begin{vmatrix} \vec{i} & \vec{j} & \vec{k} \\ -0.5 & -0.866 & 0 \\ 0.3842 & -0.2852 & -0.8781 \end{vmatrix} = |\vec{T}_{BE}|(0.7605\vec{i} - 0.4391\vec{j} - 0.4753\vec{k})$$

$$= |\vec{T}_{BE}|(\vec{i}u_{BEx} + \vec{j}u_{BEy} + \vec{k}u_{BEz})$$

$$\vec{r}_C \times \vec{T}_{BE} = |\vec{T}_{BE}| \begin{vmatrix} \vec{i} & \vec{j} & \vec{k} \\ 1 & 0 & 0 \\ 0.0549 & -0.4753 & -0.8781 \end{vmatrix} = |\vec{T}_{BE}|(0\vec{i} + 0.8781\vec{j} - 0.4753\vec{k})$$

$$= |\vec{T}_{CE}|(\vec{i}u_{CEx} + \vec{j}u_{CEy} + \vec{k}u_{CEz})$$

$$\vec{r}_C \times \vec{T}_{CF} = |\vec{T}_{CF}| \begin{vmatrix} \vec{i} & \vec{j} & \vec{k} \\ 1 & 0 & 0 \\ 0.0549 & 0.4753 & -0.8781 \end{vmatrix} = |\vec{T}_{CF}|(0\vec{i} + 0.8781\vec{j} + 0.4753\vec{k})$$

$$= |\vec{T}_{CF}|(\vec{i}u_{CFx} + \vec{j}u_{CFy} + \vec{k}u_{CFz})$$

$$\vec{r}_G \times \vec{W} = |\vec{W}| \begin{vmatrix} \vec{i} & \vec{j} & \vec{k} \\ 0 & 0 & 1 \\ -0 & -\cos\alpha & -\sin\alpha \end{vmatrix} = |\vec{W}|(\vec{i}\cos\alpha - \vec{j}0 + \vec{k}0) = (\vec{i}M_{Wx})$$

The six equations in six unknowns are:

$$|\vec{T}_{AD}|e_{ADx} + |\vec{T}_{AF}|e_{AFx} + |\vec{T}_{BD}|e_{BDx} + |\vec{T}_{BE}|e_{BEx} + |\vec{T}_{CE}|e_{CEx} + |\vec{T}_{CF}|e_{CFx} + W_x = 0$$

$$|\vec{T}_{AD}|e_{Ayx} + |\vec{T}_{AF}|e_{AFy} + |\vec{T}_{BD}|e_{BDy} + |\vec{T}_{BE}|e_{BEy} + |\vec{T}_{CE}|e_{CEy} + |\vec{T}_{CF}|e_{CFy} + W_y = 0$$

$$|\vec{T}_{AD}|e_{ADz} + |\vec{T}_{AF}|e_{AFz} + |\vec{T}_{BD}|e_{BDz} + |\vec{T}_{BE}|e_{BEz} + |\vec{T}_{CE}|e_{CEz} + |\vec{T}_{CF}|e_{CFz} + W_z = 0$$

$$|\vec{T}_{AD}|u_{ADx} + |\vec{T}_{AF}|u_{AFx} + |\vec{T}_{BD}|u_{BDx} + |\vec{T}_{BE}|u_{BEx} + |\vec{T}_{CE}|u_{CEx} + |\vec{T}_{CF}|u_{CFx} + M_{Wx} = 0$$

$$|\vec{T}_{AD}|u_{ADy} + |\vec{T}_{AF}|u_{Ayx} + |\vec{T}_{BD}|u_{BDy} + |\vec{T}_{BE}|u_{BEy} + |\vec{T}_{CE}|u_{CEy} + |\vec{T}_{CF}|u_{CFy} = 0,$$

$$|\vec{T}_{AD}|u_{ADz} + |\vec{T}_{AF}|u_{AFz} + |\vec{T}_{BD}|u_{BDz} + |\vec{T}_{BE}|u_{BEz} + |\vec{T}_{CE}|u_{CEz} + |\vec{T}_{CF}|u_{CFz} = 0$$

Problem 6.59 continued on next page..........

............continuation of solution to Problem 6.59.....

This set of equations was solved by iteration using **TK Solver 2** . For $\alpha = 20^o$ the results are: $\boxed{|\vec{T}_{AD}| = |\vec{T}_{BD}| = -1910.5\ N\ (C)}$, $\boxed{|\vec{T}_{AF}| = |\vec{T}_{CF}| = 16272.5\ N\ (T)}$, $\boxed{|\vec{T}_{BE}| = |\vec{T}_{CE}| = -19707\ N\ (C)}$. *Check:* For $\alpha = 90^o$, the solution is $|\vec{T}_{AD}| = |\vec{T}_{AF}| = |\vec{T}_{BD}| = |\vec{T}_{BE}| = |\vec{T}_{CE}| = |\vec{T}_{CF}| = -5585.8\ N$ (C), which agrees with the solution to Problem 6.58, obtained by another method. *check.*

Check: The solution of a six-by-six system by iteration has risks, since the matrix of coefficients may be ill-conditioned. As a reasonableness test for the solution process, **TK Solver 2** was used to graph the axial forces in the supporting bars over the range $-90^o < \alpha < 90^o$. The graph is shown. The negative values are compression, and the positive values are tension. When $\alpha = -90^o$, the telescope platform is pointing straight down, and the bars are in equal tension, as expected. When $\alpha = 90^o$ the telescope mount is upright and the supporting bars are in equal compression, as expected. The values of compression and tension at the two extremes are equal and opposite in value, and the values agree with those obtained by another method (see Problem 6.58), as expected. Since the axial forces go from tension to compression over this range of angles, all axial forces must pass through zero in the interval. *check.*

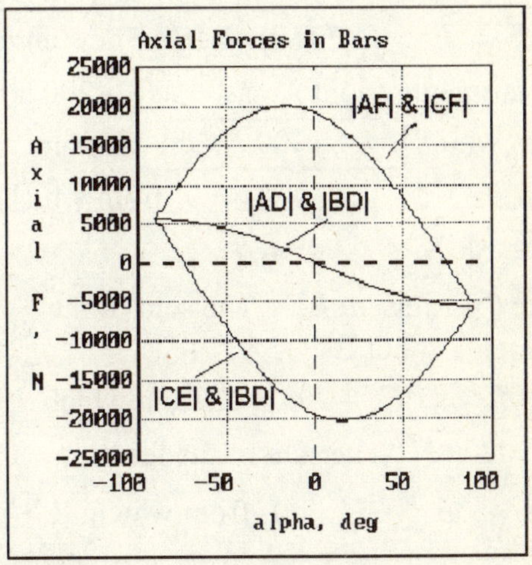

Problem 6.60 Draw the free-body diagram of (a) the entire frame, (b) the individual members AB and BC. (c) Determine the forces and couples on member AB

Solution: *Element BC:* The moment about B: the couple is positive, and the moment due to the 500 *N* force is negative.

$\sum M_B = C_y - 150 = 0$,

$C_y = 150\ N$

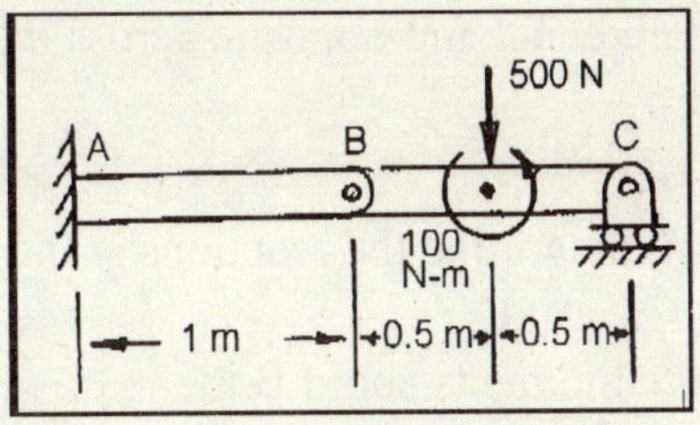

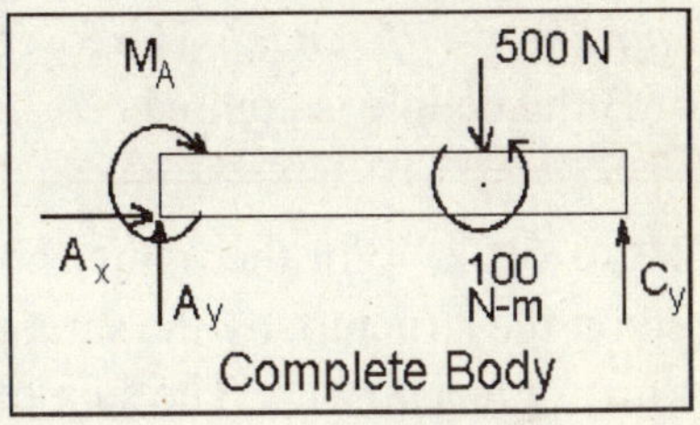

The sum of the forces:

$\sum F_y = B_y + C_y - 500 = 0$

from which $\boxed{B_y = 350\ N}$

$\sum F_x = B_x = 0$

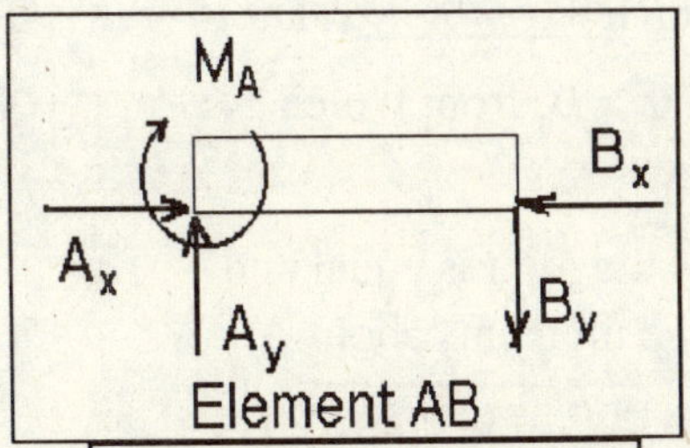

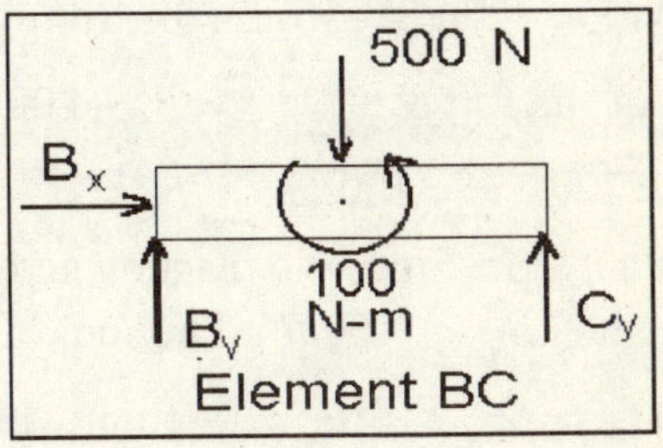

Element AB: The moment about A:

$\sum M_A = -M_A - (1)B_y = 0$. which $\boxed{M_A = -B_y = -350\ N\text{-}m}$,

The sums of forces $\sum F_y = A_y - B_y = 0$, from which $\boxed{A_y = B_y = 350\ N}$, $\boxed{\sum F_x = A_x = 0}$.

Problem 6.61 (a) Determine the forces and couples on member AB for Cases (a) and (b). (b) You know that the moment of a couple is the same about any point. Explain why the answers are not the same in Cases (a) and (b).

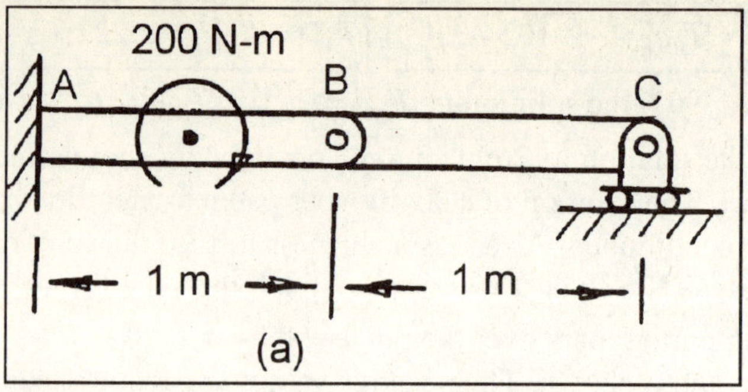

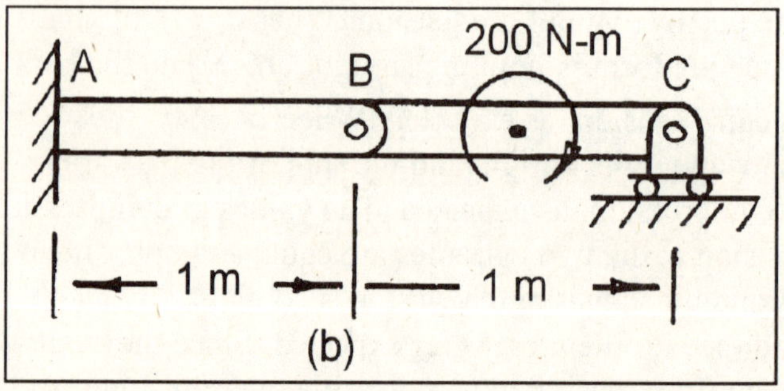

Solution: Case (a) *Element BC:* The moment about B is $M_B = (1)C_y = 0$, hence $\boxed{C_y = 0}$. The sum of the forces: $\sum F_y = B_y + C_y = 0$, from which $\boxed{B_y = 0}$. $\sum F_x = B_x = 0$. *Element AB:* The sum of the moments about A: $M = M_A - 200 = 0$, from which $\boxed{M_A = 200\ N\text{-}m}$. The sum of forces: $\sum F_y = B_y + A_y = 0$, from which $\boxed{A_y = 0}$, $\sum F_x = A_x = 0$

Case (b) *Element BC:* The sum of the moments about B:

$\sum M_B = (1)C_y - 200 = 0$, from which $C_y = 200\ N$. The sum of the forces:

$\sum F_y = B_y^{BC} + C_y = 0$, from which $B_y^{BC} = -200\ N$. $\sum F_x = B_x^{BC} = 0$. *Element AB:* The moments about A: $\sum M = M_A + (1)B_y^{AB} = 0$, from which, since the reactions across the joint are equal and opposite: $\boxed{B_y^{AB} = -B_y^{BC} = 200\ N}$, $\boxed{M_A = -200\ N\text{-}m}$. The sum of the forces: $\sum F_y = A_y + B_y^{AB} = 0$, from which $\boxed{A_y = -200\ N}$. $\sum F_x = A_x + B_x^{AB} = 0$, from which $\boxed{A_x = 0}$

Explanation of difference: The forces are equal and opposite across the joint B, so it matters on which side of B the couple is applied.

Problem 6.62 For the frame shown, determine the reactions at the built in support A and the force exerted on the structure by the smooth floor at C.

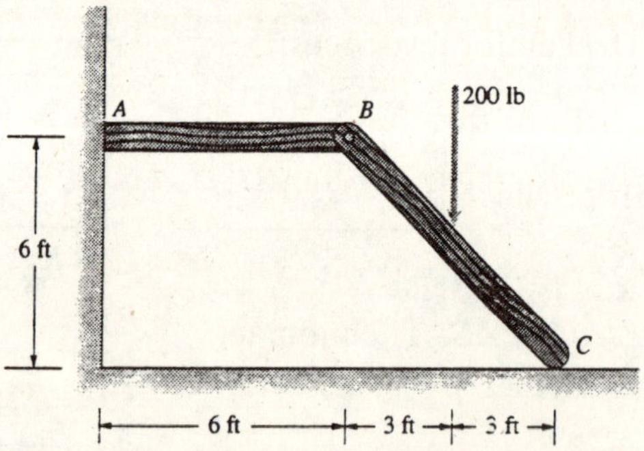

Solution: *Element BC:* The sum of the moments about B is $\sum M_B = -200(3) + (6)C = 0$, from which $\boxed{C = 100\ lb}$. The sum of the forces: $\sum F_y = B_y - 200 + C = 0$, from which $B_y = 200 - C = 100\ lb$. $\sum F_x = B_x = 0$

Element AB: Since the reaction across the joint is equal and opposite, $B_y = -100\ lb$. The sum of the moments about A:

$\sum M_R = M_A + (6)B_y = 0$, from which $\boxed{M_A = 600\ ft\ lb}$. The sum of the forces: $\sum F_y = A_y + B_y = 0$, from which $\boxed{A_y = 100\ lb}$. $\sum F_x = A_x + B_x = 0$, from which $\boxed{A_x = 0}$.

Problem 6.63 The force $F = 10\ kN$. Determine the forces on member ABC and present your answers as shown in Figure 6.35.

Solution: *The complete structure as a free body:*

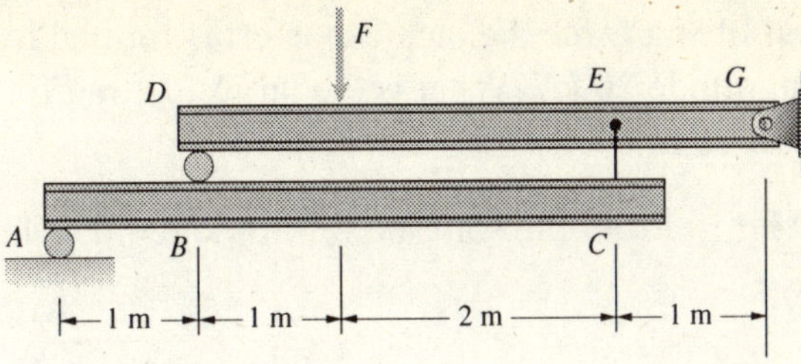

The sum of the moments about G:

$\sum M_G = +3F - 5A = 0$, from which

$A = \dfrac{3F}{5} = 6\ kN$ which is the reaction of the floor.

The sum of the forces: $\sum F_y = G_y - F + A = 0$,

from which $G_y = F - A = 10 - 6 = 4\ kN$.

$\sum F_x = G_x = 0$.

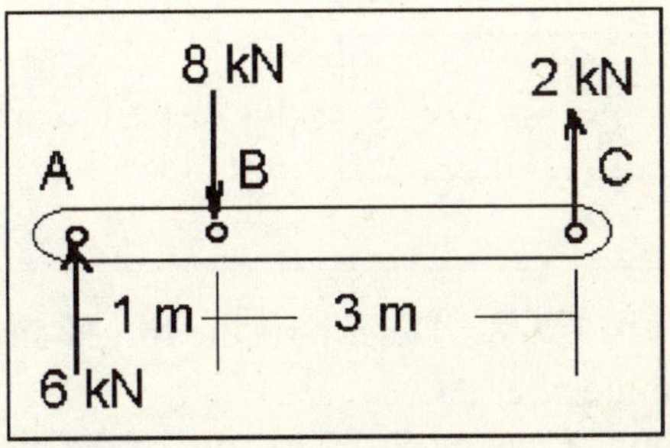

Element DEG: The sum of the moments about D

$\sum M = -F + 3E + 4G_y = 0$, from which

$E = \dfrac{F - 4G_y}{3} = \dfrac{10 - 16}{3} = -2\ kN$. The sum of the forces:

$\sum F_y = G_y - F + E + D = 0$, from which

$D = F - E - G_y = 10 + 2 - 4 = 8\ kN$.

Element ABC: Noting that the reactions are equal and opposite: $\boxed{B = -D = -8\ kN}$, and $\boxed{C = -E = 2\ kN}$. The sum of the forces: $\sum F_y = A + B + C = 0$, from which $\boxed{A = 8 - 2 = 6\ kN}$. *Check*

Problem 6.64 Consider the frame shown in Problem 6.63. The cable CE will safely support a tension of 10 *kN*. Based on this criterion, what is the largest downward force F that can be safely applied?

Solution: From the solution to Problem 6.63: $E = \dfrac{F - 4G_y}{3}$, $G_y = F - A$, and $A = \dfrac{3}{5}F$. Back substituting, $E = -\dfrac{F}{5}$ or $F = -5E$, from which, for $E = 10\ kN$, $\boxed{F = -50\ kN}$

Problem 6.65 The hydraulic actuator BD exerts a 6 *kN* force on member ABC. The force is parallel to BD and the actuator is in compression. Determine the forces on member ABC and present your answers as shown in Figure 6.35.

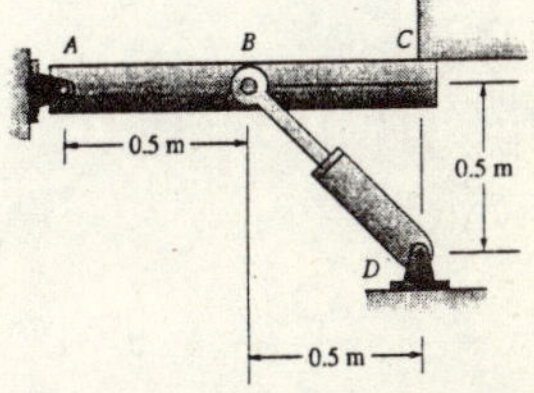

Solution: The surface at C is smooth.

Element ABC: The sum of the moments about A is

$\sum M = (0.5)B\sin 45^\circ + (1)C = 0$, from which

$\boxed{C = -3(0.707) = 2.121\ kN}$. The sum of the forces:

$\sum F_y = A_y + B\sin 45^\circ + C = 0$, from which

$\boxed{A_y = -B(0.707) - C = -2.121\ kN}$.

$\sum F_x = A_x - B\cos 45^\circ = 0$, from which $\boxed{A_x = 4.24\ kN}$

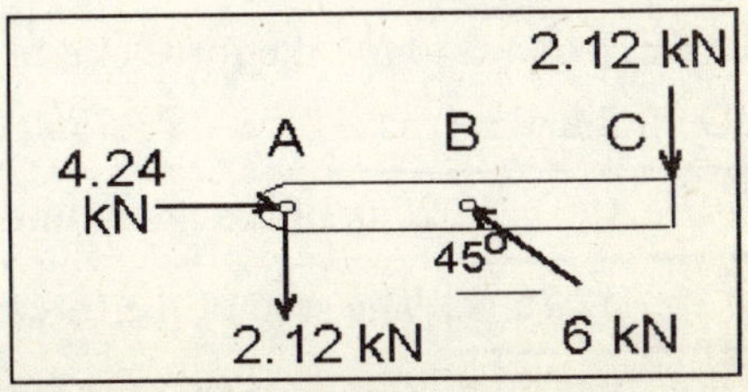

Problem 6.66 The simple hydraulic jack shown in Problem 6.65 is designed to exert a vertical force at point C. The hydraulic actuator BD exerts a force on the beam ABC that is parallel to BD. The largest lifting force the jack can exert is limited by the pin support A, which will safely support a force of magnitude 20 *kN*. What is the largest lifting force the jack can exert at C and what is the resulting axial force in the hydraulic actuator?

Solution: From the solution to Problem 6.65 $A_y = -\frac{B}{\sqrt{2}} - C$, $A_x = \frac{B}{\sqrt{2}}$, and $C = -\frac{B}{2\sqrt{2}}$.

Substituting, $A_y = -\frac{B}{2\sqrt{2}}$, and $|\vec{A}| = \sqrt{A_x^2 + A_y^2} = \frac{B}{\sqrt{2}}\sqrt{1^2 + \left(\frac{1}{2}\right)^2} = \frac{\sqrt{5}B}{2\sqrt{2}}$. For $|\vec{A}| = 20\ kN$,

$\boxed{B = \frac{2\sqrt{2}(20)}{\sqrt{5}} = 25.3\ kN}$ is the largest axial force, and $\boxed{C = -\frac{B}{2\sqrt{2}} = -8.944\ kN}$ is the largest lifting force.

Problem 6.67 Determine the forces on member BC and the axial force in member AC.

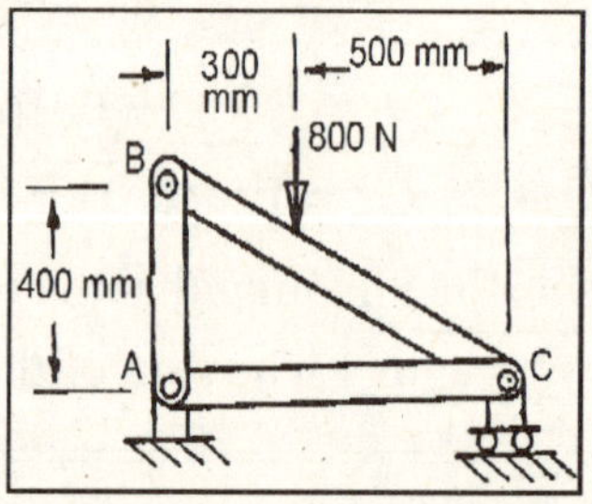

Solution: *Element BC:* The sum of the moments about B: $\sum M = -(0.3)800 + (0.8)C = 0$, from which $\boxed{C = 300\ N}$. The sum of the forces $\sum F_y = B - 800 + C = 0$, from which $\boxed{B = 500\ N}$. $\boxed{\sum F_x = C_x = 0}$ (The roller support prevents an x-direction reaction in C.) *Element AC:* The sum of the forces $\boxed{\sum F_x = A_x = 0}$

Problem 6.68 A football player works out with a "squat thrust" machine. To rotate the bar ABD, he must exert a vertical force at A such that the axial force in member BC is 250 *lb*. When bar ABD is on the verge of rotating, what force does he exert at A and what forces are exerted on member CDE at D?

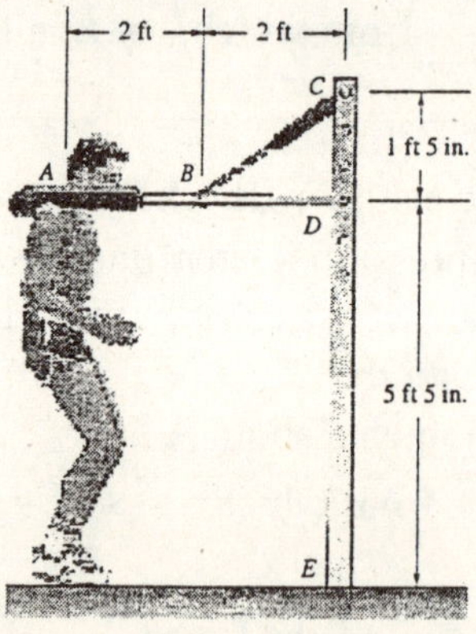

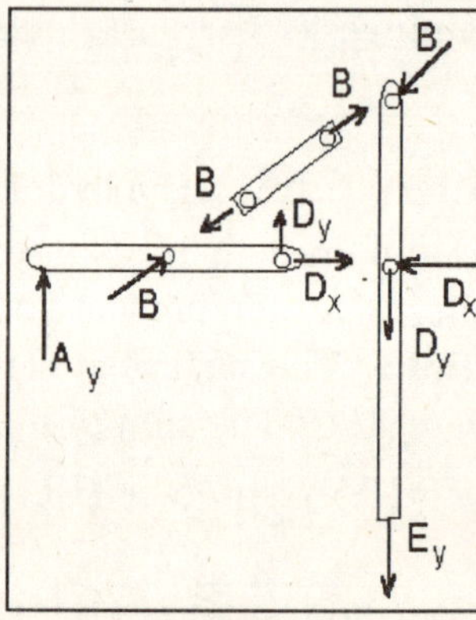

Solution: The angle member BC Makes with the vertical is

$$\alpha = \tan^{-1}\left(\frac{1.4167}{2}\right) = 35.31^o.$$

Element ABD: The sum of the moments about D is

$$\sum M_D = +(2)B\sin 35.31^o + (4)A_y^{AD} = 0,$$

from which

$$A_y^{BD} = -\frac{B\sin 35.31^o}{2} = -0.289B.$$ At an axial force $B = 250\ lb$, the force exerted by ABD on the player is $A_y^{BD} = -72.25\ lb$. The force exerted by the player is the reverse of this, $\boxed{A_y^P = -A_y^{BD} = 72.25\ lb}$ Take the moment about B $\sum M_B = -2A_y^{BD} + 2D_y^{BD} = 0$, from which $\boxed{D_y^{BD} = -72.25\ lb}$. The sum of the forces exerted by ABD, $\sum F_x = D_x^{BD} - B\cos\alpha = 0$, from which $\boxed{D_x^{BD} = 204\ lb}$ These are the forces exerted on CDE by element ABD at D.

Problem 6.69 The frame supports a 1 *kN* load. Determine the forces on member CDE.

Solution: The strategy is to demonstrate with free body diagrams of the structure, the element ABC, and the element BD that $D_x = 0$, $C_x = 0$, and $E_x = 0$, after which the other reactions can be determined from the free body diagram of CDE.

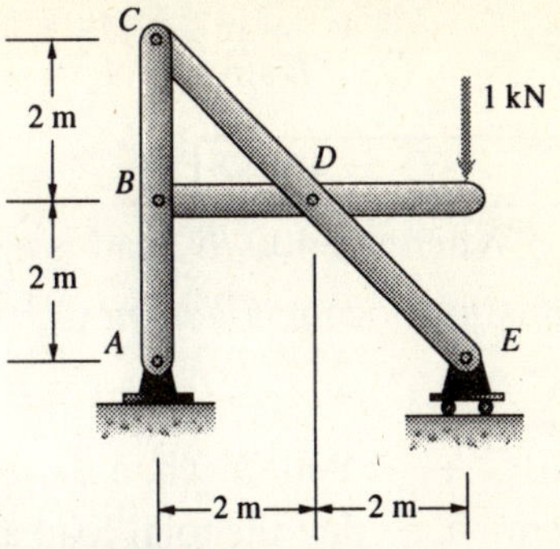

The complete structure as a free body: The sum of the moments about A: $\sum M_A = -4F + 4E_y = 0$, from which $\boxed{E_y = F = 1\ kN}$.,

$\sum F_y = A_y + E_y - F = 0$, from which $A_y = 0$

Element ABC: The sum of the moments about A:

$\sum M_A = 2B_x + 4C_x = 0$. The sum of the forces: $\sum F_x = B_x + C_x = 0$.

These have the solution $B_x = 0$, and $\boxed{C_x = 0}$.

Element BD: The sum of the forces $\sum F_x = B_x + D_x = 0$, from which $\boxed{D_x = -B_x = 0}$.

Element CDE: Since $D_x = 0$, and $E_x = 0$, they contribute nothing to the moment about C. The sum of the moments about C: $\sum M_C = 2D_y + 4E_y = 0$, from which $\boxed{D_y = -2E_y = -2\ kN}$ The sum of the forces $\sum F_y = C_y + D_y + E_y = 0$, from which $\boxed{C_y = -D_y - E_y = 1\ kN}$

Problem 6.70 Determine the forces on member ABC and present your answers as shown in Figure 6.35. Obtain the answer in two ways: (a) When you draw the free body diagrams of the individual members, place the 400 *lb* load on the free body diagram of member ABC. (b) When you draw the free body diagrams of the individual members, place the 400 *lb* load on the free body diagram of member CD.

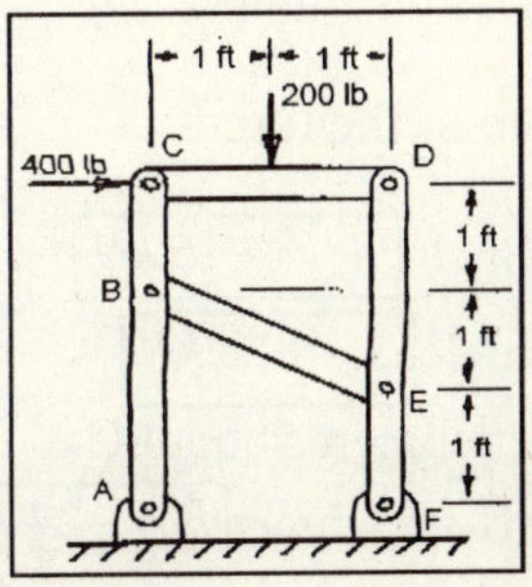

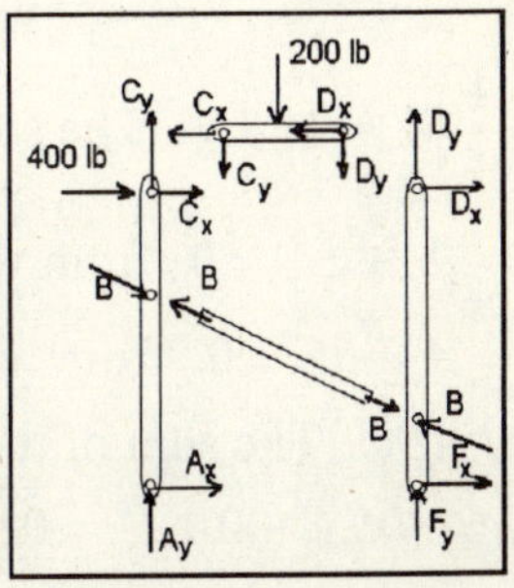

Solution: The angle of element BE relative to the horizontal is

$\alpha = \tan^{-1}\left(\frac{1}{2}\right) = 26.57^o$. *The complete structure as a free body:* The sum of the moments about A:

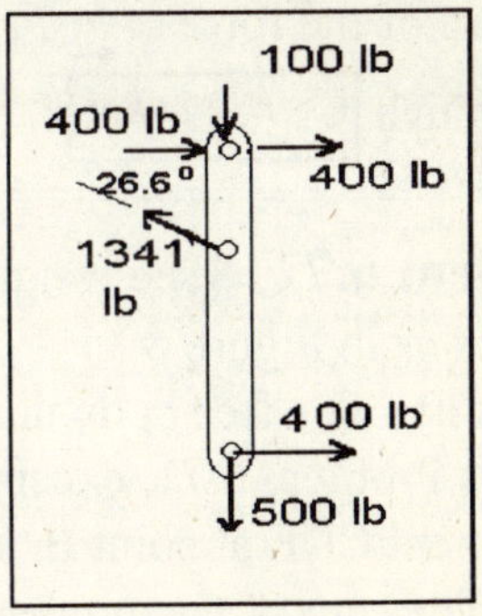

$\sum M_A = -3(400) - 1(200) + 2F_y = 0$

$\sum M_A = -3(400) - 1(200) + 2F_y = 0$, from which $F_y = 700\ lb$. The sum of forces: $\sum F_y = A_y + F_y - 200 = 0$, from which $\boxed{A_y = -500\ lb}$.

$\sum F_x = A_x + F_x + 400 = 0$.

(a) *Element CD:* The sum of the moments about D: $\sum M_D = 200 + 2C_y = 0$, from which $\boxed{C_y = -100\ lb}$. $\sum F_y = -D_y - C_y - 200 = 0$, from which $D_y = -100$. $\sum F_x = -C_x - D_x = 0$, from which $D_x = -C_x$.

Element DEF: The sum of the moments about F: $\sum M_F = -3D_x + B\cos\alpha = 0$, from which $D_x = -B\left(\frac{\cos\alpha}{3}\right)$. $\sum F_y = F_y + B\sin\alpha + D_y = 0$, from which $\boxed{B = \frac{-700 + 100}{\sin\alpha} = -1341.6\ lb}$, and $D_x = -400\ lb$

Problem 6.70 continued on next page....

continuation of solution to Problem 6.70...

Element ABC: $\sum M_A = -2B\cos\alpha - 3(400) - 3C_x = 0$. The sum of the forces $\sum F_y = C_y - B\sin\alpha + A_y = 0$, from which *Check:* $B = \dfrac{500+100}{\sin\alpha} = -1341.6\ lb$. *check.* From above: $\boxed{C_x = -D_x = 400\ lb}$. $\sum F_x = 400 + C_x + B\cos\alpha + A_x = 0$, from which $\boxed{A_x = 400\ lb}$.

(b) When the 400 lb load is applied to element CD instead, the following changes to the equilibrium equations occur: *Element CD:* $\sum F_x = -C_x - D_x + 400 = 0$, from which $C_x + D_x = 400$. *Element ABC:* $\sum F_x = C_x + A_x - B\cos\alpha = 0$. *Element DEF*: No changes. The changes in the solution for Element ABC $C_x = 800\ lb$ when the external load is removed, instead of $C_x = 400\ lb$ when the external load is applied, so that the total load applied to point C is the same in both cases.

Problem 6.71 The weight $W = 12\ kN$. Determine the forces on member ABC and present your answer as shown in Figure 6.35.

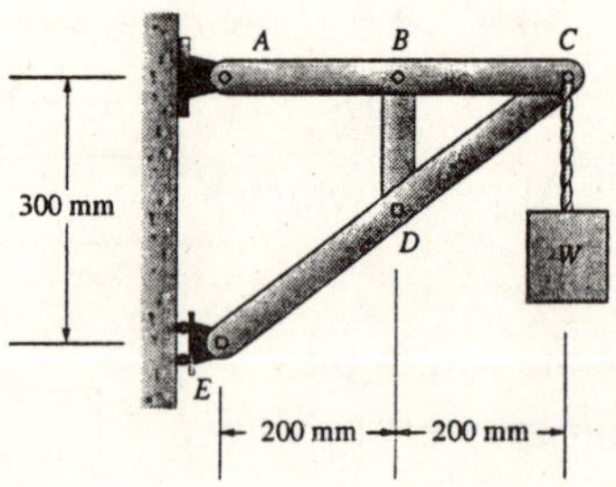

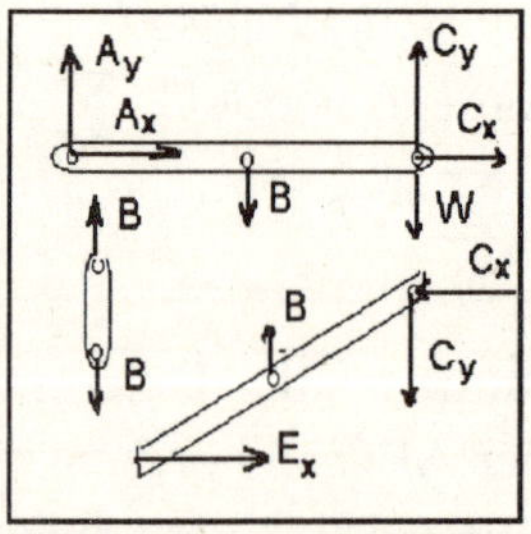

Solution: *The complete structure as a free body:* The sum of the moments about A: $\sum M_A = -0.4W + 0.3E_x = 0$, from which $E_x = \left(\dfrac{4}{3}\right)W = 16\ kN$. The sum of the forces:

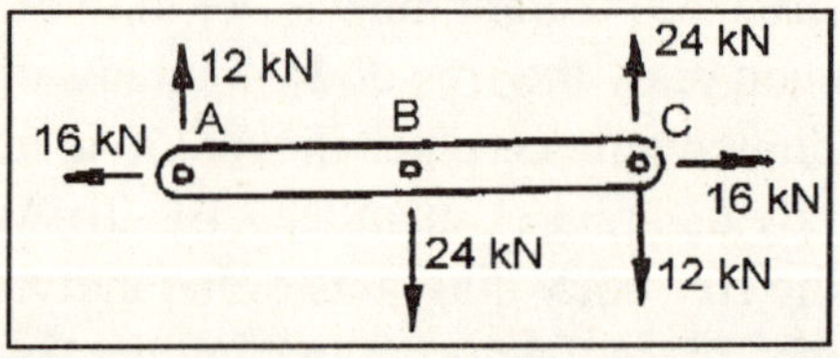

$\sum F_x = A_x + E_x = 0$, from which $\boxed{A_x = -16\ kN}$.
$\sum F_y = A_y - W = 0$, from which $\boxed{A_y = 12\ kN}$

Element ABC: The sum of the moments about C: $\sum M_C = -0.2B - 0.4A_y = 0$, from which $\boxed{B = -24\ kN}$.

The sum of the forces: $\sum F_x = A_x + C_x = 0$, from which $\boxed{C_x = 16\ kN}$. $\sum F_y = A_y + B + C_y - W = 0$, from which $\boxed{C_y = 24\ \ kN}$.

Problem 6.72 The weight $W = 12\ kN$. Determine the forces on member ABC and present your answers as in Figure 6.35.

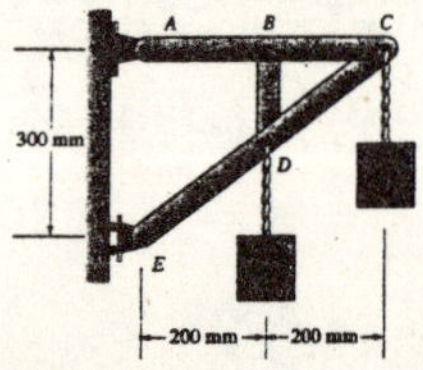

Solution: The free body diagrams of the elements are the same as Problem 6.71, except for the addition of a weight W on element CDE at point B. *The complete structure as a free body:* The sum of the moments about A:

$\sum M_A = -0.4W - 0.2W + 0.3E_x = 0$, from which $E_x = 2W = 24\ kN$. The sum of the forces:

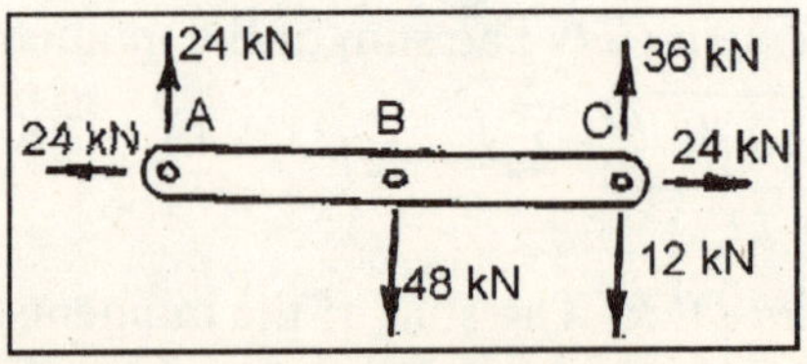

$\sum F_x = A_x + E_x = 0$, from which $\boxed{A_x = -24\ kN}$.
$\sum F_y = A_y - 2W = 0$, from which $\boxed{A_y = 24\ kN}$

Element ABC: The sum of the moments about C: $\sum M_C = -0.2B - 0.4A_y = 0$, from which $\boxed{B = -48\ kN}$. The sum of the forces: $\sum F_x = A_x + C_x = 0$, from which $\boxed{C_x = 24\ kN}$. $\sum F_y = A_y + B + C_y - 2W = 0$, from which $\boxed{C_y = 36\ kN}$.

Problem 6.73 The tension in cable BD is 500 lb. Determine the reactions at A for Cases (a) and (b).

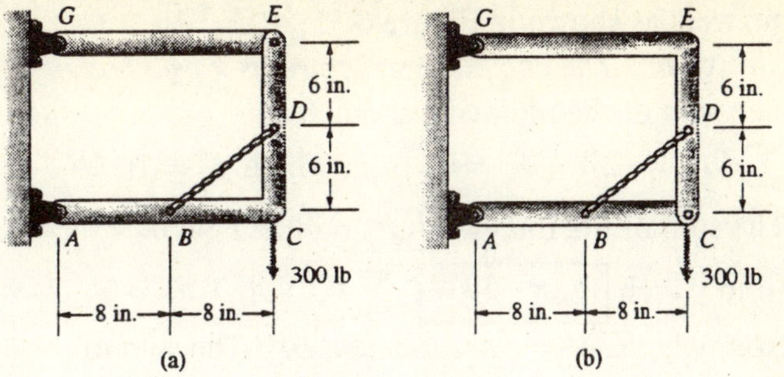

Solution: Case (a) *The complete structure as a free body:* The sum of the moments about G: $\sum M_G = -16(300) + 12A_x = 0$, from which $\boxed{A_x = 400\ lb}$. The sum of the forces: $\sum F_x = A_x + G_x = 0$, from which $G_x = -400\ lb$. $\sum F_y = A_y - 300 + G_y = 0$, from which $A_y = 300 - G_y$. *Element GE*: The sum of the moments about E: $\sum M_E = -16G_y = 0$, from which $G_y = 0$, and from above $\boxed{A_y = 300\ lb}$.

Case (b) .*The complete structure as a free body:* The sum of the moments about G: $\sum M_C = -16(300) + 12A_x = 0$, from which $\boxed{A_x = 400\ lb}$.

Element ABC.: The tension at the lower end of the cable is up and to the right, so that the moment exerted by the cable tension about point C is negative. The sum of the moments about C:

$\sum M_C = -8B\sin\alpha - 16A_y = 0$, noting that $B = 500\ lb$ and $\alpha = \tan^{-1}(\frac{6}{8}) = 36.87^o$, then $\boxed{A_y = -150\ lb}$.

Problem 6.74 Determine the forces on member ABCD and present your answers as shown in Figure 6.35.

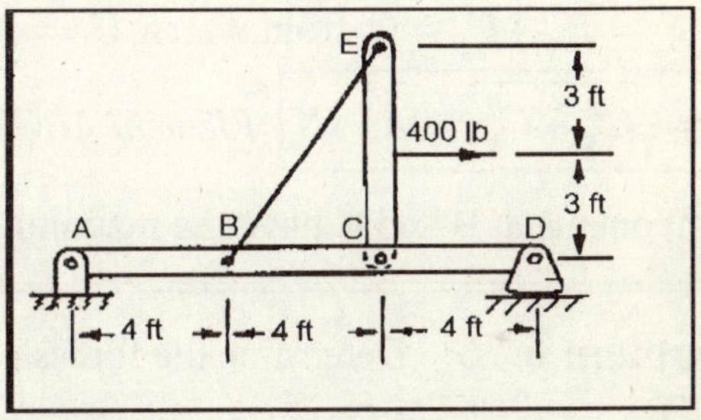

Solution: *The complete structure as a free body:* The sum of the moments about A: $\sum M_A = -400(3) + 12D_y = 0$, from which $\boxed{D_y = 100\ lb}$. The sum of the forces:

$\sum F_x = A_x + 400 = 0$, from which $\boxed{A_x = -400\ lb}$.

$\sum F_y = A_y + D_y = 0$, from which $\boxed{A_y = -100\ lb}$

Element EG: The sum of the moments about C:

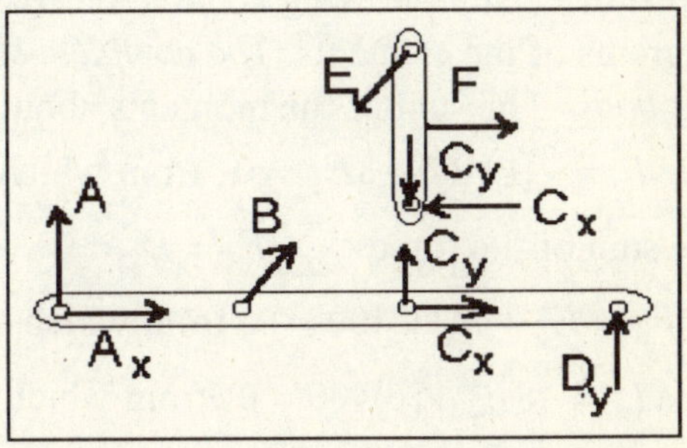

$\sum M_C = 6E\cos\alpha - 3(400) = 0$. The angle of cable element EB is $\alpha = \tan^{-1}\left(\frac{6}{4}\right) = 56.31^o$, from which the cable tension is $\boxed{E = 360.6\ lb}$. The sum of the forces:

$\sum F_x = C_x + 400 - E\cos\alpha = 0$, from which $C_x = -200\ lb$.

$\sum F_y = C_y - E\sin\alpha = 0$, from which $C_y = 300\ lb$.

Element ABCD: The tension in the cable acts on element ABCD with equal and opposite tension to the reaction on element EB, up and to the right at an angle of 56.31°, $\boxed{C_x = 200\ lb}$ to the right, and $\boxed{C_y = -300\ lb}$ downward.

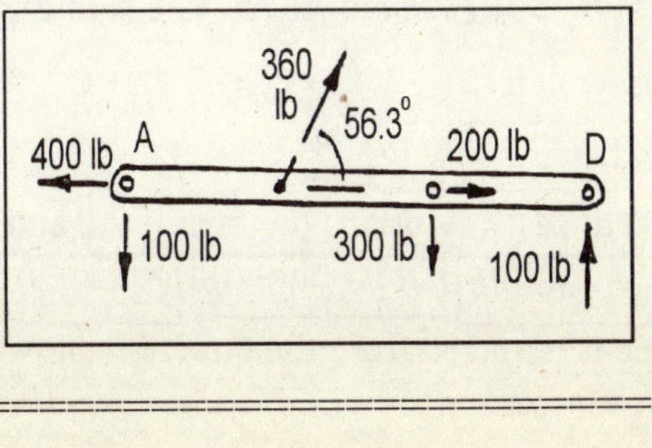

Problem 6.75

Problem 6.75 The weight $W = 5\ kN$. Determine the forces on member ABCD and present your answers as shown in Figure 6.35.

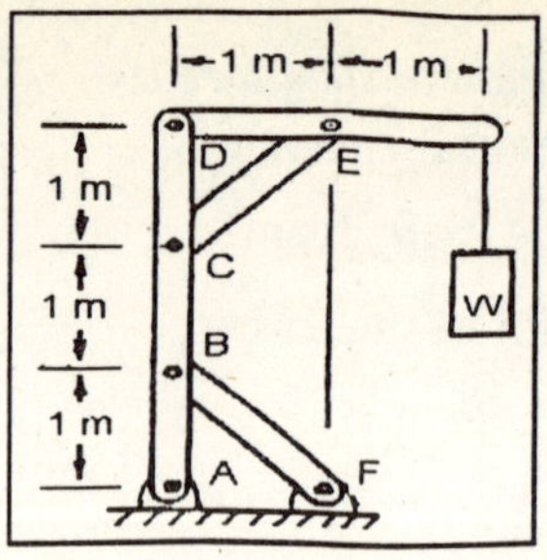

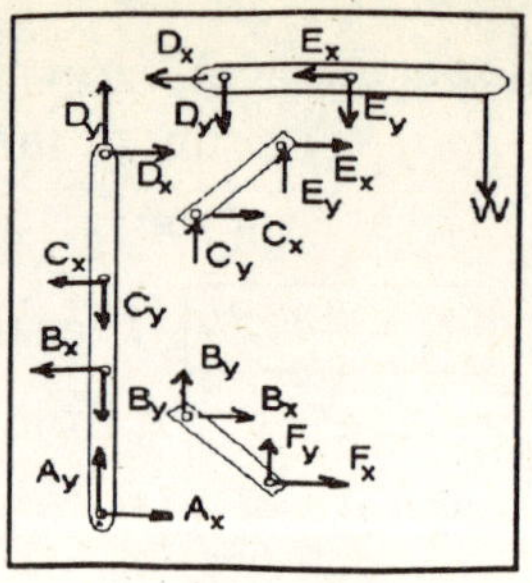

Solution: *The complete structure as a free body:* The sum of the moments about A: $\sum M_A = -2W + F_y = 0$, from which $F_y = 10\ kN$. The sum of the forces: $\sum F_y = A_y + F_y - W = 0$, from which $\boxed{A_y = -5\ kN}$, $\sum F_x = A_x + F_x = 0$, from which $A_x = -F_x$. *Element BF:* The sum of the moments about F: $\sum M_F = -B_y - B_y = 0$, from which $B_y = -B_x$. The sum of the forces: $\sum F_y = B_y + F_y = 0$, from which $B_y = -10\ kN$, and $B_x = 10\ kN$. $\sum F_x = B_x + F_x = 0$, from which $F_x = -10\ kN$, and from above, $\boxed{A_x = 10\ kN}$, from which $\boxed{A = \sqrt{A_x^2 + A_y^2} = 11.18\ kN}$. *Element DE:* The sum of the moments about D: $\sum M_D = -E_y - 2W = 0$, from which $E_y = -10\ kN$. The sum of the forces: $\sum F_y = -D_y - E_y - W = 0$, from which $\boxed{D_y = 5\ kN}$. $\sum F_x = -D_x - E_x = 0$, from which $D_x = -E_x$. *Element CE*: The sum of the moments about C: $\sum M_C = E_y - E_x = 0$, from which $E_x = -10\ kN$, and from above $\boxed{D_x = 10\ kN}$. $\sum F_y = E_y + C_y = 0$, from which $C_y = 10\ kN$. $\sum F_x = E_x + C_x = 0$, from which $C_x = 10\ kN$, and $\boxed{C = \sqrt{C_x^2 + C_y^2} = 14.1\ kN}$ *Element ABCD:* All reactions on ABCD have been determined above. The components at B and C have the magnitudes $\boxed{B = C = \sqrt{10^2 + 10^2} = 14.1\ kN}$, at .angles of 45°.

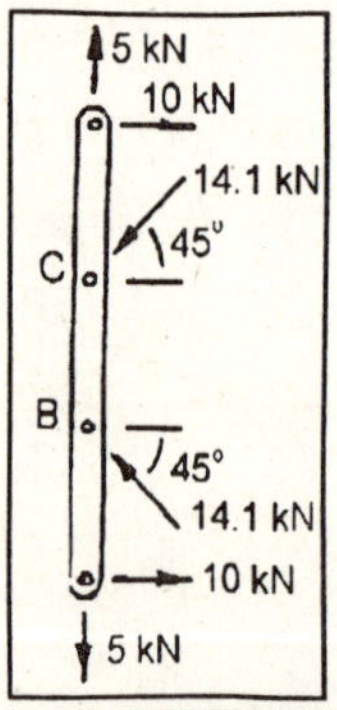

Problem 6.76

Problem 6.76 Determine the forces on member BCD.

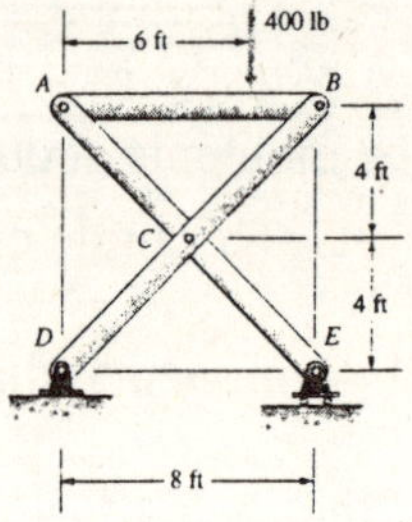

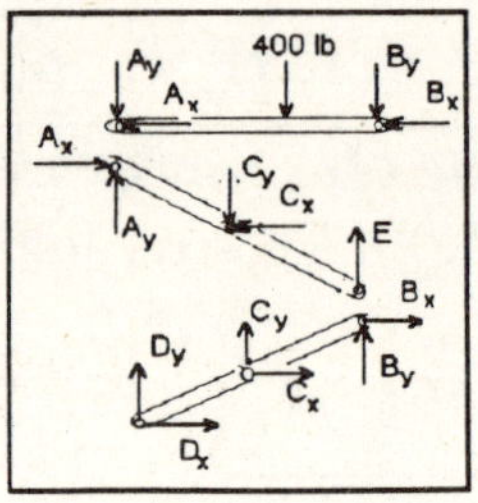

Solution: The following is based on free body diagrams of the elements: *The complete structure as a free body:* The sum of the moments about D: $\sum M_D = -(6)400 + 8E_y = 0$, from which $E_y = 300\ lb$. The sum of the forces: $\sum F_x = D_x = 0$. $\sum F_y = E_y + D_y - 400 = 0$, from which $D_y = 100\ lb$. *Element AB:* . The sum of the moments about A: $\sum M_A = -8B_y - (6)400 = 0$, from which $B_y = -300\ lb$. The sum of forces: $\sum F_y = -B_y - A_y - 400 = 0$, from which $A_y = -100\ lb$. $\sum F_x = -A_x - B_x = 0$, from which (1) $A_x + B_x = 0$ *Element ACE:* The sum of the moments about E: $\sum M_E = -8A_x + 4C_x - 8A_y + 4C_y = 0$, from which (2) $-2A_x + C_x - 2A_y + C_y = 0$. The sum of the forces: $\sum F_y = A_y + E_y - C_y = 0$, from which $\boxed{C_y = 200\ lb}$. $\sum F_x = A_x - C_x = 0$, from which (3) $A_x = C_x$. The three numbered equations are solved: $A_x = -400\ lb$, $\boxed{C_x = 400\ lb}$, and $\boxed{B_x = -400\ lb}$. *Element BCD:* The reactions are now known: $\boxed{B_y = -300\ lb}$, $\boxed{B_x = -400\ lb}$, $\boxed{C_y = 200\ lb}$, $\boxed{D_x = 0}$, $\boxed{D_y = 100\ lb}$, where negative sign means that the force is reversed from the direction shown on the free body diagram.

Problem 6.77 Determine the forces on member ABC.

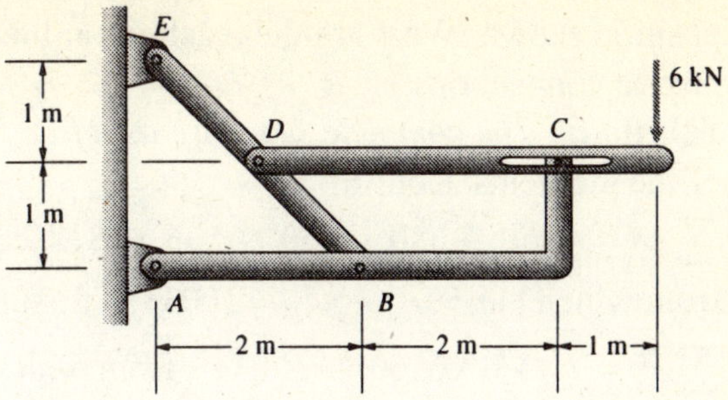

Solution: *The complete structure as a free body:* The sum of the moments about E: $\sum M_E = -5(6) + 2A_x = 0$, from which $\boxed{A_x = 15\ kN}$. The sum of the forces:

$\sum F_x = E_x + A_x = 0$, from which $E_x = -15\ kN$.

$\sum F_y = E_y + A_y - 6 = 0$, from which (1) $E_y + A_y = 6$

Element DC: The sum of the moments about D:

$\sum M_D = -3C_y - 4(6) = 0$, from which $\boxed{C_y = -8\ kN}$

The sum of the forces: $\sum F_y = -C_y - 6 - D_y = 0$, from which $\boxed{D_y = -2\ kN}$. $\boxed{\sum F_X = D_X = 0}$.

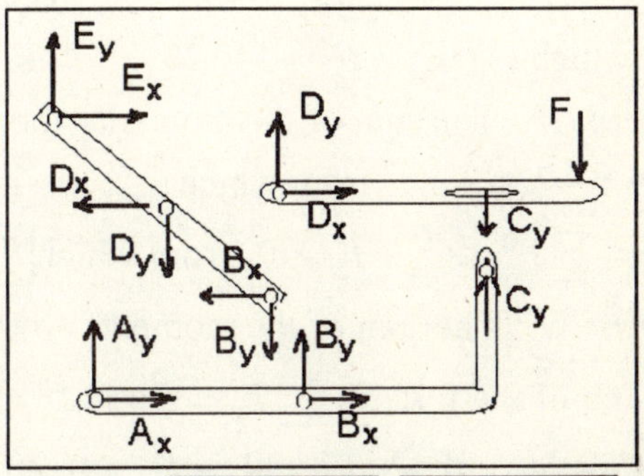

Element EDB: The sum of the moments about E:

$\sum M_E = -D_x - (1)D_y - 2B_y - 2B_x = 0$, from which, since $D_x = 0$, (2) $2B_y + 2B_x = -D_y$. The sum of forces:

$\sum F_y = -B_y - D_y + E_y = 0$, from which

(3) $B_y - E_y = D_y$: $\sum F_x = -B_x + E_x = 0$, from which

$\boxed{B_x = -15\ kN}$, and equations (1), (2) and (3) above are now solvable, $\boxed{B_y = 16\ kN}$, $\boxed{A_y = -8\ kN}$, and $\boxed{E_y = 14\ kN}$.

Element ABC: The reactions on Element ABC have been determined above:

$\boxed{A_x = 15\ kN,\ A_y = -8\ kN,\ B_x = -15\ kN,\ B_y = 16\ kN,\ C_y = -8\ kN}$

Problem 6.78 The weight $W = 80\ lb$. Determine the forces on member ABCD.

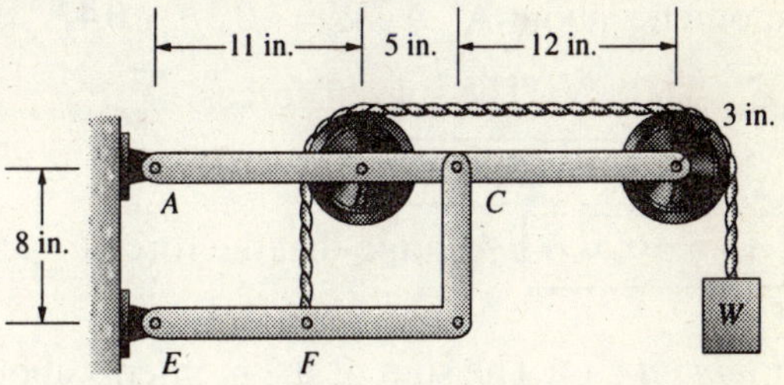

Solution: *The complete structure as a free body:* The sum of the moments about A: $\sum M_A = -31W + 8E_x = 0$, from which $E_x = 310\ lb$. The sum of the forces:

$\sum F_x = E_x + A_x = 0$, from which $\boxed{A_x = -310\ lb}$.

$\sum F_y = E_y + A_y - W = 0$, from which (1) $E_y + A_y = W$.

Element CFE: The sum of the forces parallel to x:

$\sum F_x = E_x - C_x = 0$, from which $\boxed{C_x = 310\ lb}$. The sum of the moments about E: $\sum M_E = 8F - 16C_y + 8C_x = 0$.

For frictionless pulleys, $F = W$, and thus $\boxed{C_y = 195\ lb}$.

The sum of forces parallel to y: $\sum F_y = E_y - C_y + F = 0$, from which $\boxed{E_y = 115\ lb}$. Equation (1) above is now solvable: $\boxed{A_y = -35\ lb}$.

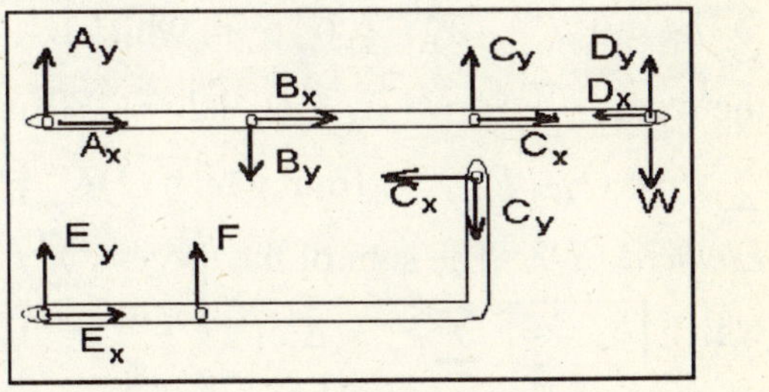

Element ABCD: The forces exerted by the pulleys on element ABCD are, by inspection:

$\boxed{B_x = W = 80\ lb}$, $\boxed{B_y = 80\ lb}$, $\boxed{D_x = 80\ lb}$, and $\boxed{D_y = -80\ lb}$, where the negative sign means that the force is reversed from the direction of the arrows shown on the free body diagram.

Problem 6.79 The man using the exercise machine is holding the 80-lb weight stationary in the position shown. What are the reactions at the built-in support E and the pin support F? (A and C are pinned connections.)

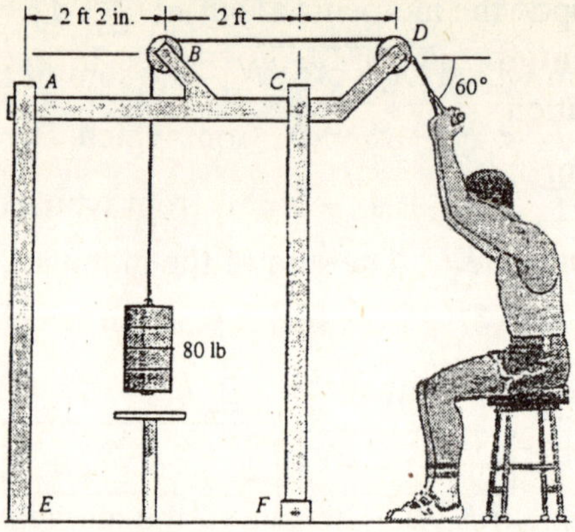

Solution: *The complete structure as a free body:* The sum of the moments about E:

$\sum M = -26W - 68W\sin 60^o + 50F_y - 81W\cos 60^o + M_E = 0$

from which (1) $50F_y + M_E = 10031$. The sum of the forces:

$\sum F_x = F_x + W\cos 60^o + E_x = 0$, from which

(2) $F_x + E_x = -40$. $\sum F_y = -W - W\sin 60^o + E_y + F_y = 0$,

from which (3) $E_y + F_y = 149.28$

Element CF: The sum of the moments about F:

$\sum M = -72C_x = 0$, from which $C_x = 0$. The sum of the forces: $\sum F_x = C_x + F_x = 0$, from which $\boxed{F_x = 0}$. From (2) above, $\boxed{E_x = -40\ lb}$

Element AE: The sum of the moments about E: $\sum M = M_E - 72A_x = 0$,. from which (4) $M_E = 72A_x$.

The sum of the forces : $\sum F_y = E_y + A_y = 0$, from which

(5) $E_y + A_y = 0$. $\sum F_x = A_x + E_x = 0$; from which $A_x = 40\ lb$, and from (4) $\boxed{M_E = 2880\ in\ lb = 240\ ft\ lb}$ From (1) $\boxed{F_y = 143.0\ lb}$, and from (2) $\boxed{E_y = 6.258\ lb}$. This completes the determination of the 5 reactions on E and F.

Problem 6.80 The force $F = 800\ N$. Determine the forces on member ABC.

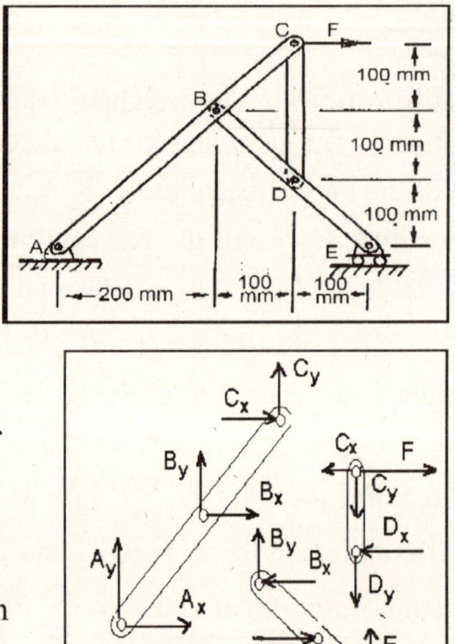

Solution: *The complete structure as a free body:* The moments about A: $\sum M = -0.3F + 0.4E_y = 0$, from which $E_y = 600\ N$. The sum of forces: $\sum F_X = A_X + F = 0$, from which $\boxed{A_x = -800\ N}$ $\sum F_y = A_y + E_y = 0$, from which $\boxed{A_y = -600\ N}$. Assume that the force F is applied to member CD.

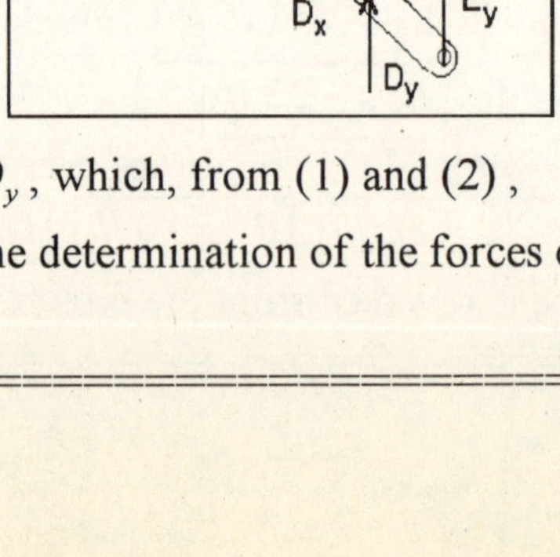

Element CD: The sum of the moments about D:

$\sum M = 0.2C_x - 0.2F = 0$, from which $\boxed{C_x = 800\ N}$. The sum of the forces: $\sum F_x = -C_x - D_x + F = 0$, from which $D_x = 0$.

$\sum F_y = C_y + D_y = 0$, from which (1) $C_y = -D_y$.

Element BDE: The sum of the forces: $\sum F_x = D_x - B_x = 0$, from which $\boxed{B_x = 0}$. $\sum F_y = E_y + D_y - B_y = 0$, from which (2) $B_y - D_y = E_y = 600$. The sum of the moments about E:

$\sum M_E = -0.1D_x - 0.1D_y + 0.2B_y + 0.2B_x = 0$, from which $B_y = 0.5D_y$, which, from (1) and (2), $D_y = -1200\ N$, $\boxed{B_y = -600\ N}$, and $\boxed{C_y = 1200\ N}$ This completes the determination of the forces on member ABC.

Problem 6.81 is marked by a separator line.

Problem 6.81 Consider the frame shown in Problem 6.80. Member CD will safely support a compressive load of 2 *kN*. Based on this criterion, what is the maximum force F that the frame will support?

Solution: From the solution of Problem 6.80, the equilibrium conditions on member CD leads to the relation: $C_y = -D_y$ where C_y is an axially compressive force on CD. From the equilibrium conditions on member BDE, $B_y - D_y = E_y$, and $B_y = 0.5D_y$ from which: $D_y = -2E_y$. From the equilibrium condition for the complete structure as a free body, $E_y = \frac{3}{4}F$. Substitute to obtain $C_y = \frac{3}{2}F$. For the maximum safe compressive force, $\boxed{F = \frac{2}{3}(2) = 1.33\ kN}$. *Check:* Since the system is linear, the ratio of the external forces to the axial loads on CD is a constant: $\frac{F}{2} = \frac{800}{1200}$, from which $F = 1.33\ kN$. *check.*

Problem 6.82 The unstretched length of the string is L_O. Show that when the system is in equilibrium, the angle α satisfies the relation $\sin\alpha = \frac{2\left(L_O - 2\frac{F}{k}\right)}{L}$

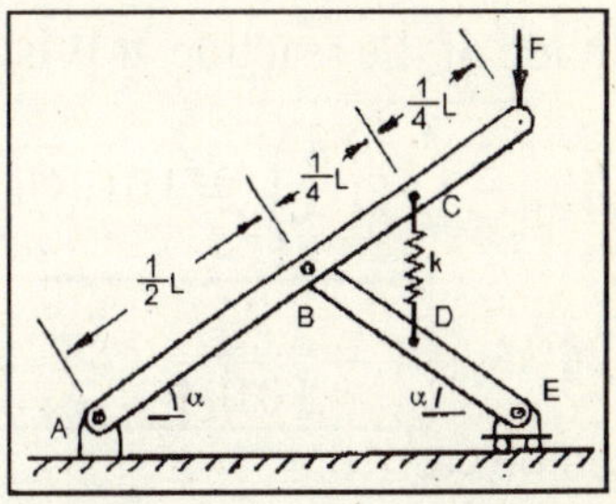

Solution: Since the action lines of the force F and the reaction E are co-parallel and coincident, the moment on the system is zero, and the system is always in equilibrium, for a non-zero force F. The object is to find an expression for the angle α for any non-zero force F.

The complete structure as a free body: The sum of the moments about A $\sum M_A = -FL\sin\alpha + EL\sin\alpha = 0$, from which $E = F$. The sum of forces: $\sum F_x = A_x = 0$, from which $A_x = 0$. $\sum F_y = A_y + E_y - F = 0$, from which $A_y = 0$, which completes a demonstration that F does not exert a moment on the system. *The spring C:* The elongation of the spring is $\Delta s = 2\frac{L}{4}\sin\alpha - L_O$, from which the force in the spring is $T = k\left(\frac{L}{2}\sin\alpha - L_O\right)$ *Element BE:* The strategy is to determine C_y, which is the spring force on BE. The moment about E is

$\sum M_E = -\frac{L}{4}C_y\cos\alpha - \frac{L}{2}B_y\cos\alpha - \frac{L}{2}B_x\cos\alpha = 0$, from which $\frac{C_y}{2} + B_y = -B_x$. The sum of forces: $\sum F_x = B_x = 0$, from which $B_x = 0$. $\sum F_y = C_y + B_y + E = 0$, from which $C_y + B_y = -E = -F$. The two simultaneous equations are solved: $C_y = -2F$, and $B_y = F$.

The solution for angle α: The spring force is $C_y = T = k\left(\frac{L}{2}\sin\alpha - L_O\right)$, from which

$k\left(\frac{L}{2}\sin\alpha - L_O\right) = -2F$. Solve: $\boxed{\sin\alpha = \frac{2\left(L_O - \frac{2F}{k}\right)}{L}}$

Problem 6.83 The pin support will safely support a force of 24 *kN* magnitude. Based on this criterion, what is the largest weight *W* that the frame will safely support?

Solution: *The complete structure as a free body:* Sum the forces in the x-direction: $\sum F_x = A_x = 0$, from which $A_x = 0$

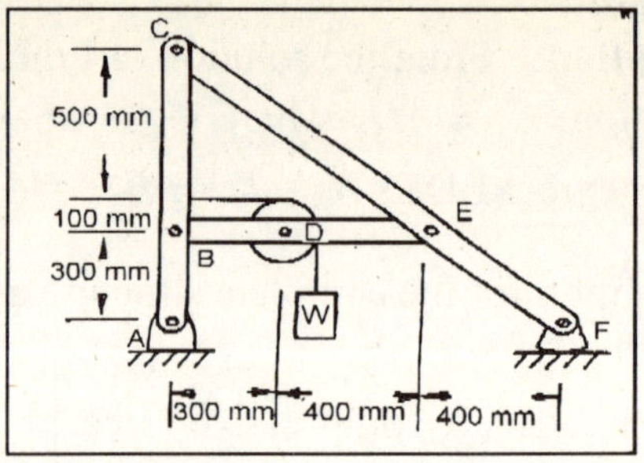

Element ABC: The sum of the moments about A:

$\sum M_A = +0.3B_x + 0.9C_x - 0.4W = 0$, from which (1) $0.3B_x + 0.9C_x = 0.4W$. The sum of the forces:

$\sum F_x = -B_x - C_x + W + A_x = 0$, from which (2) $B_x + C_x = W$.

Solve the simultaneous equations (1) and (2) to obtain

$$B_x = \frac{5}{6}W$$

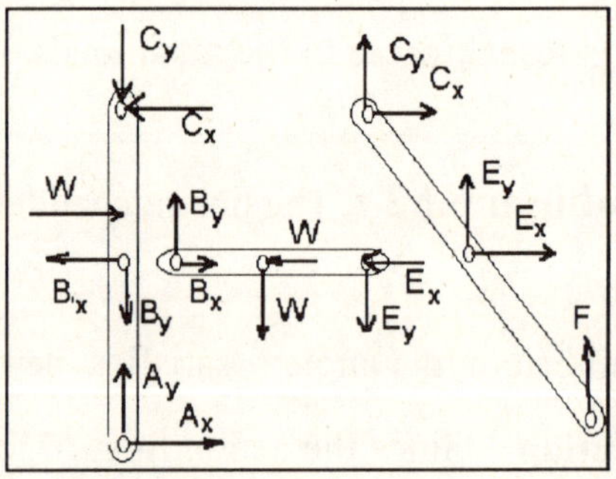

Element BE: The sum of the moments about E:

$\sum M_E = 0.4W - 0.7B_y = 0$, from which $B_y = \frac{4}{7}W$. The magnitude of the reaction at B is

$|\vec{\mathbf{B}}| = W\sqrt{\left(\frac{5}{6}\right)^2 + \left(\frac{4}{7}\right)^2} = 1.0104W$. For a safe value of

$|\vec{\mathbf{B}}| = 24\ kN$, $\boxed{W = \frac{24}{1.0104} = 23.752\ kN}$ is the maximum load that can be carried.

Problem 6.84 Determine the reactions at A and C.

Solution: *The complete structure as a free body:* The sum of the moments about A:

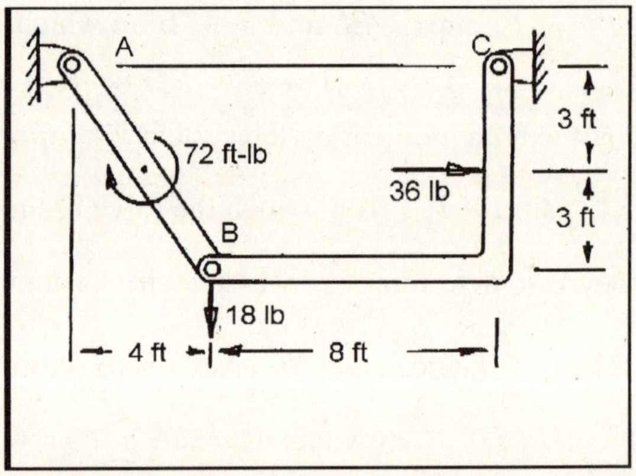

$\sum M_A = -4(18) + 3(36) + 12C_y - 72 = 0$, from which $\boxed{C_y = 3\ lb}$. The sum of the forces:

$\sum F_y = A_y + C_y - 18 = 0$, from which $\boxed{A_y = 15\ lb}$.

$\sum F_x = A_x + C_x + 36 = 0$, from which

(1) $C_x = -A_x - 36$

Element AB: The sum of the forces:

$\sum F_y = A_y - B_y - 18 = 0$, from which $B_y = -3\ lb$.

The sum of the moments: $\sum M_A = 6B_x - 4(18) - 4B_y - 72 = 0$, from which $B_x = 22\ lb$. The sum of the forces: $\sum F_x = A_x + B_x = 0$, from which $\boxed{A_x = -22\ lb}$ From equation (1) $\boxed{C_x = -14\ lb}$

Problem 6.85

Problem 6.85 Determine the forces on member AD.

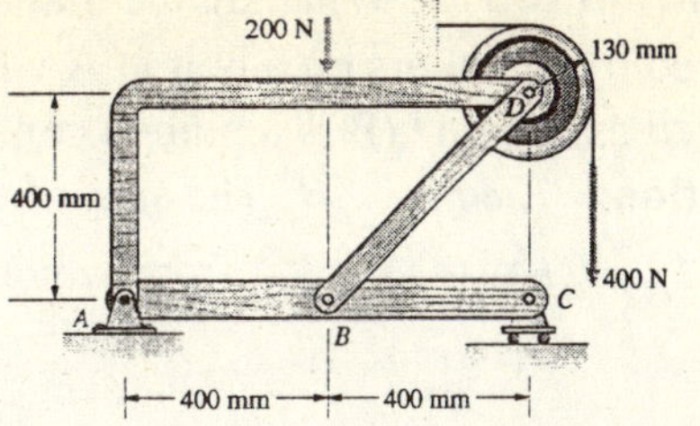

Solution: Denote the reactions of the support by R_x and R_y. *The complete structure as a free body:* $\sum F_x = R_x - 400 = 0$, from which $\boxed{R_x = 400\,N}$. The sum of moments: $\sum M_A = 800C - 400(800) - 400(400) - 400(200) = 0$, from which $C = 300\,N$. $\sum F_y = C + R_y - 400 - 200 = 0$, from which $R_y = 300\,N$. *Element ABC*: The sum of the moments: $\sum M_A = -4B_y + 8C = 0$, from which $B_y = 600\,N$. *Element BD*: The sum of the forces: $\sum F_y = B_y - D_y - 400 = 0$, from which $\boxed{D_y = 200\,N}$.

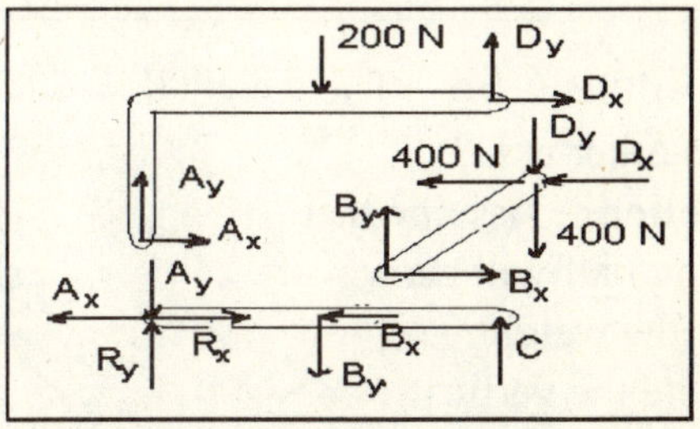

Element AD: The sum of the forces: $\sum F_y = A_y + D_y - 200 = 0$, from which $\boxed{A_y = 0}$: . *Element AD:* The sum of the forces: $\sum F_x = A_x + D_x = 0$ and $\sum M_A = -400(200) + 800D_y - 400D_x = 0$ $\boxed{A_x = -200\,N}$, and $\boxed{D_x = 200\,N}$. *Element BD:* The sum of forces: $\sum F_x = B_x - D_x - 400 = 0$ from which $B_x = 600\,N$. This completes the solution of the nine equations in nine unknowns, of which A_x, A_y, D_x, and D_y are the values required by the problem..

Problem 6.86

Problem 6.86 The frame shown is used to support high-tension wires. If $b = 3\,ft$, $\alpha = 30^o$, and $W = 200\,lb$, what is the axial force in member HJ?

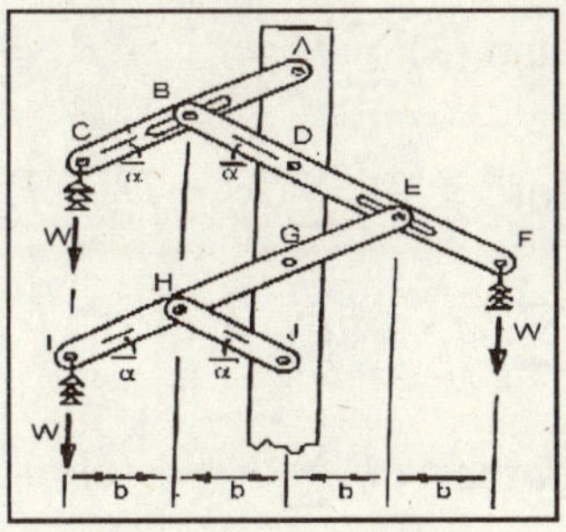

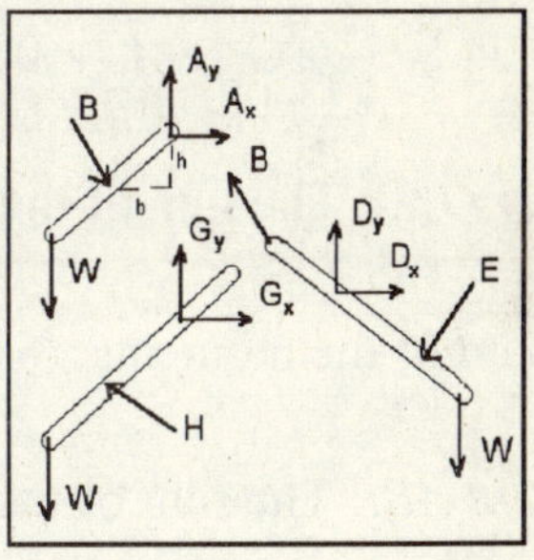

Solution: Joints B and E are sliding joints, so that the reactions are normal to AC and BF, respectively. Member HJ is supported by pins at each end, so that the reaction is an axial force. The distance $h = b\tan\alpha = 1.732\,ft$

Member ABC. The sum of the forces: $\sum F_x = A_x + B\sin\alpha = 0$, $\sum F_y = A_y - W - B\cos\alpha = 0$. The sum of the moments about B: $\sum M_B = bA_y - hA_x + bW = 0$. These three equations have the solution: $A_x = 173.21\,lb$, $A_y = -100\,lb$, and $B = -346.4\,lb$. *Member BDEF:* The sum of the forces: $\sum F_x = D_x - B\sin\alpha - E\sin\alpha = 0$, $\sum F_y = D_y - W + B\cos\alpha - E\cos\alpha = 0$. The sum of the moments about D: $\sum M_D = -2bW - bE\cos\alpha - hE\sin\alpha - bB\cos\alpha + hB\sin\alpha = 0$. These three equations have the solution: $D_x = -259.8\ lb$, $D_y = 350\ lb$, $E = -173.2\ lb$.

Member EGHI: The sum of the forces: $\sum F_x = G_x + E\sin\alpha - H\cos\alpha = 0$, $\sum F_y = G_y - W + E\cos\alpha + H\sin\alpha = 0$. The sum of the moments about H: $\sum M_H = bG_y - hG_x + bW + 2bE\cos\alpha - 2hE\sin\alpha = 0$. These three equations have the solution: $G_x = 346.4\,lb$, $G_y = 200\ lb$, and $\boxed{H = 300\ lb}$. This is the axial force in HJ.

Problem 6.87 What are the magnitudes of the forces exerted by the pliers on a bolt at A when 30 *lb* forces are applied as shown? (B is a pinned connection.)

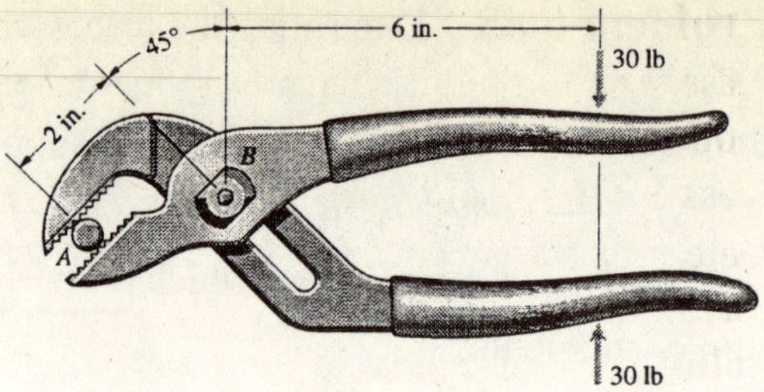

Solution: *Element AB:* The sum of the moments about B: $\sum M_B = 2F - (6)30 = 0$, from which $\boxed{F = 90\ lb}$.

Problem 6.88 The weight $W = 60\ kip$. What is the magnitude of the force the members exert on each other at D?

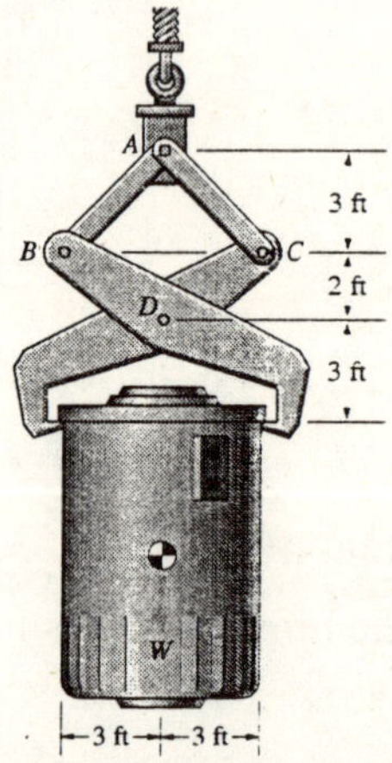

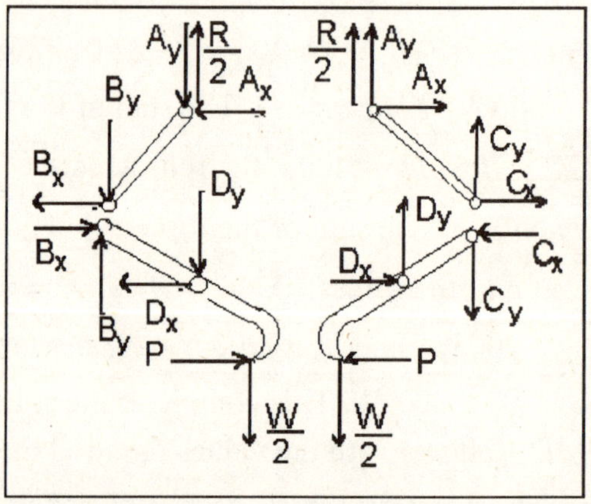

Solution: Assume that a tong half will carry half the weight, and denote the vertical reaction to the weight at A by R. *The complete structure as a free body:* The sum of the forces: $\sum F_y = R - W = 0$, from which $R = W$ *Tong-Half ACD: Element AC:* The sum of the moments about A: (1) $\sum M_A = 3C_y + 3C_x = 0$. The sum of the forces: (2) $\sum F_y = \frac{R}{2} + C_y + A_y = 0$, and (3) $\sum F_x = C_x + A_x = 0$.

Element CD: The sum of the forces: (4) $\sum F_x = D_x - P - C_x = 0$, and (5) $\sum F_y = D_y - C_y - \frac{W}{2} = 0$.

The sum of the moments: (6) $\sum M_D = 2C_x - 3C_y - 3P + \frac{3}{2}W = 0$

Element AB:. The sum of the forces: (7) $\sum F_y = -A_y + \frac{R}{2} - B_y = 0$, and (8) $\sum F_x = -A_x - B_x = 0$.

Element BD: The sum of the forces: (9) $\sum F_y = B_y - D_y - \frac{W}{2} = 0$, and (10) $\sum F_x = B_x - D_x + P = 0$.

These are ten equations in ten unknowns. These have the solution $R = 60\ kip$. *Check,* $A_x = -30\ kip$, $A_y = 0$, $B_x = 30\ kip$, $B_y = 30\ kip$, $C_x = 30\ kip$, $C_y = -30\ kip$, $D_x = 110\ kip$, $D_y = 0$, and $P = 80\ kip$. The magnitude of the force the members exert on each other at D is $\boxed{D = 110\ kip}$.

Problem 6.89 Figure (a) is a diagram of the bones and biceps muscle of a person's arm supporting a mass. Tension in the biceps muscle holds the forearm in the horizontal position. A simple mechanical model is shown in Figure (b). The weight of the forearm is 9 N and the mass $m = 2\ kg$.(a) Determine the tension in the biceps muscle AB. (b) Determine the magnitude of the force exerted on the upper arm by the forearm at the elbow joint C.

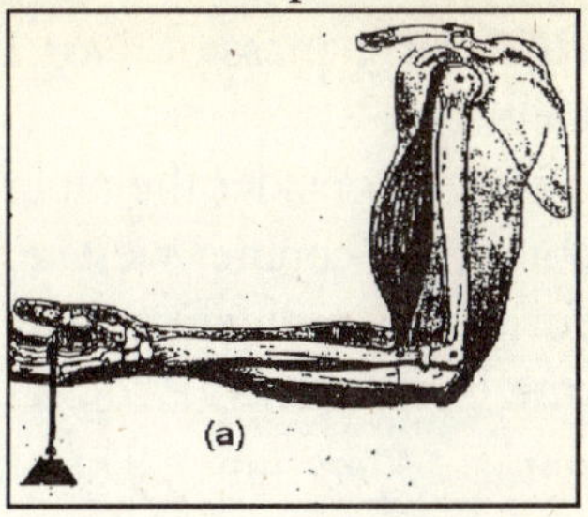

Solution: Make a cut through AB and BC just above the elbow joint C. The angle formed by the biceps muscle with respect to the forearm is $\alpha = \tan^{-1}\left(\frac{290}{50}\right) = 80.2^o$. The weight of the mass is $W = 2(9.81) = 19.62\ N$.

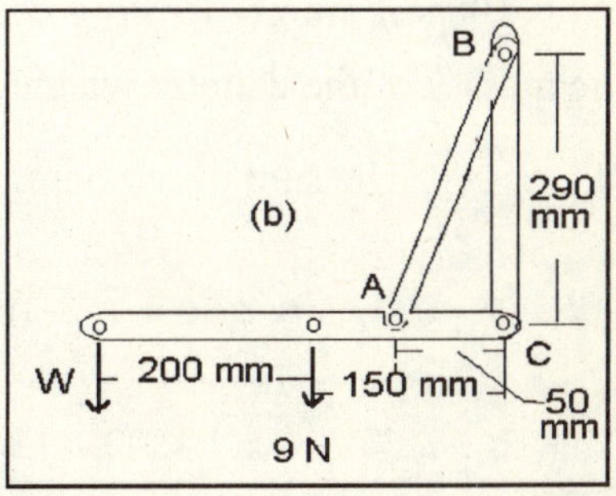

The section as a free body: The sum of the moments about C is $\sum M_C = 50T\cos\alpha - 150(9) - 350W = 0$, from which $\boxed{T = 166.76\ N}$ is the tension exerted by the biceps muscle AB.

The sum of the forces on the section is $\sum F_X = C_x + T\sin\alpha = 0$, from which $C_x = -28.33\ N$. $\sum F_Y = C_y + T\sin\alpha - 9 - W = 0$, from which $C_y = -135.72$. The magnitude of the force exerted by the forearm on the upper arm at joint C is $\boxed{F = \sqrt{C_x^2 + C_y^2} = 138.65\ N}$

Problem 6.90 The clamp presses two blocks of wood together. Determine the magnitude of the force the members exert on one another at C if the two blocks are pressed together with force of 200 N.

Solution: Consider the upper jaw only.

The section ABC as a free body:

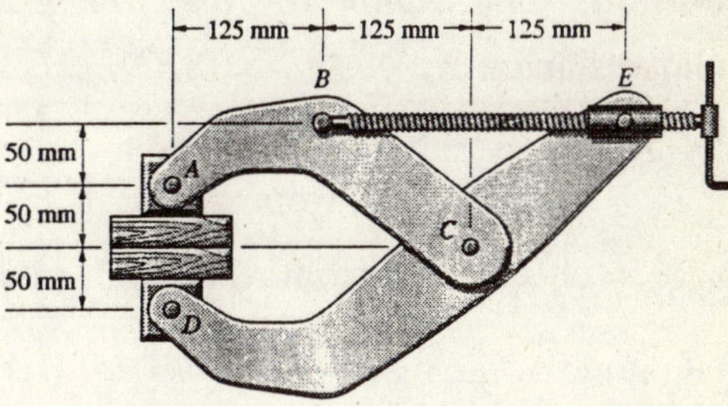

The sum of the moments about C is $\sum M_C = 100B - 250A = 0$, from which, for $A = 200\ N$, $B = 500\ N$. The sum of the forces: $\sum F_x = C_x - B = 0$, from which $C_x = 500\ N$, $\sum F_y = C_y + A = 0$, from which $C_y = -200\ N$. The magnitude of the reaction at C:

$\boxed{C = \sqrt{C_x^2 + C_y^2} = 538.52\ N}$

Problem 6.91

The mechanism is used to weigh mail. A package placed at A causes the weighted point to rotate through an angle α. Neglect the weight of the members except for the counter weight at B, which has a mass of 4 *kg*. If $\alpha = 20^o$, what is the mass of the package at A?

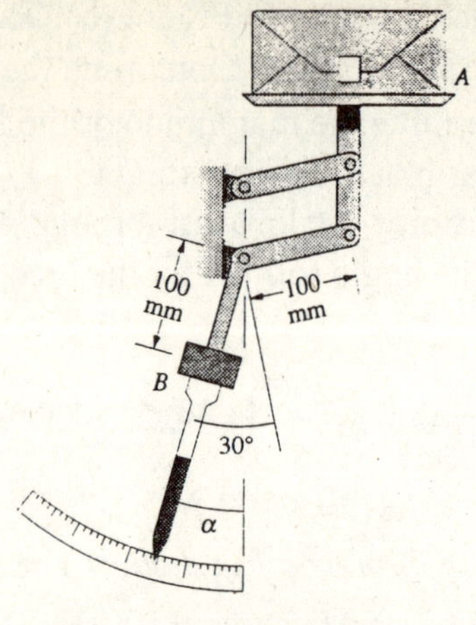

Solution: Consider the moment about the bearing connecting the motion of the counter weight to the motion of the weighing platform. The moment arm of the weighing platform about this bearing is $100\cos(30-\alpha)$. The restoring moment of the counter weight is $100mg\sin\alpha$. Thus the sum of the moments is $\sum M = 100m_B g\sin\alpha - 100m_A g\cos(30-\alpha) = 0$. Define the ratio of the masses of the counter weight to the mass of the package to be $R_M = \frac{m_B}{m_A}$. The sum of moments equation reduces to

$\sum M = R_M \sin\alpha - \cos(30-\alpha) = 0$, from which $R_M = \frac{\cos(30-\alpha)}{\sin\alpha} = 2.8794$, and the mass of the package is $\boxed{m_A = \frac{4}{R_M} = 1.3892 = 1.39\ kg}$

Problem 6.92

The scoop C of the front end loader is supported by two identical arms, one on each side of the loader. One of the two arms (ABC) is visible in the figure. It is supported by a pin support at A and the hydraulic actuator BD. The sum of the other loads on the arm, including its own weight, is $F = 1.6\ kN$. Determine the force BD and the reaction at A.

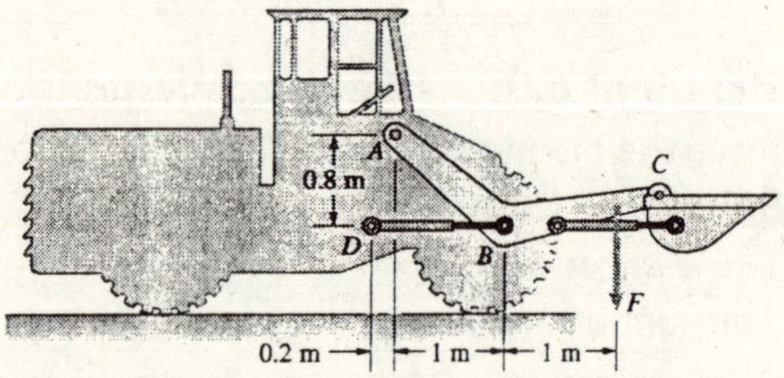

Solution: *The section ABC as a free body:* The sum of the moments about A: $\sum M_A = 0.8BD - 2F = 0$, from which $BD = 4\ kN$. The sum of the forces:

$\sum F_x = A_x + BD = 0$, from which $A_x = -4\ kN$.

$\sum F_y = A_y - F = 0$, from which $A_y = 1.6\ kN$. The magnitude of the reaction at A is $\boxed{A = \sqrt{A_x^2 + A_y^2} = 4.308\ kN}$

Problem 6.93

Determine the force exerted on the body by the bolt cutters and the magnitude of the force the members exert on each other at the pin connection A.

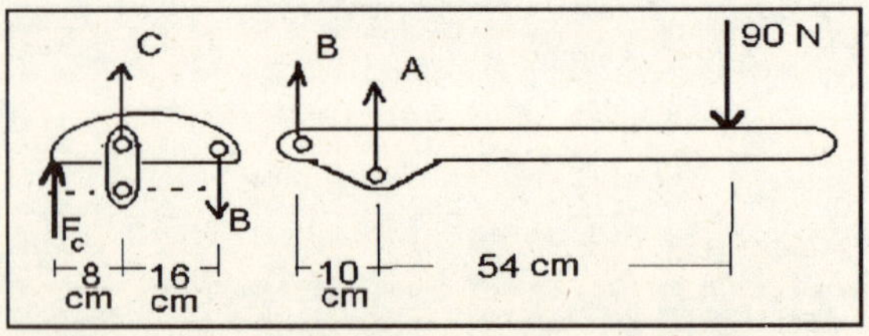

Solution: *Element AB:* The moment about A is $\sum M_A = -10B - 54F = 0$, where $F = 90\ N$. From which $B = -486\ N$. The sum of the forces: $\sum F_y = A + B - F = 0$, from which $\boxed{A = 576\ N}$

Element BC: The moment about C: $\sum M_C = -16B - 8F_C = 0$, from which the cutting force is $\boxed{F_C = 972\ N}$

Problem 6.94 The bones and tendons in a horse's leg are shown in Fig. (a). A mechanical model of the leg in shown in Fig. (b). If the horse is stationary and the normal force exerted by its leg by the ground is $N = 1200\ N$, determine the tensions in the superficial digit flexor BC and the patellar ligaments DF.

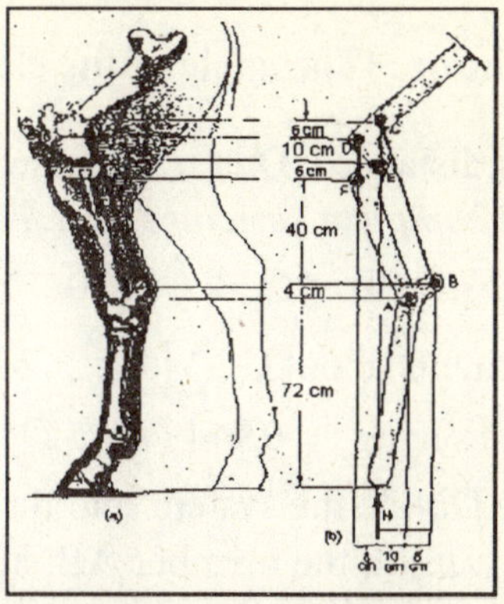

Solution: *The member AB as a free body:* To determine the tension in the flexor BC, the strategy is to take the moments about point A: The vertical distance from B to C is 62 *cm*, and the horizontal distance is 18 *cm*. The angle of the flexor BC with the positive *x* axis is $\alpha = 90^o + \tan^{-1}\left(\frac{18}{62}\right) = 106.2^o$. The unit vector parallel to the flexor BC is $\vec{e}_{BC} = \vec{i}\cos\alpha + \vec{j}\sin\alpha = -0.2788\vec{i} + 0.9603\vec{j} = e_X\vec{i} + e_Y\vec{j}$. The horizontal distance from A to B is 8 *cm*, and the vertical distance is 4 *cm*. The horizontal distance from A to the line of action of *N* is 10 *cm*. The moment about A is $\sum M_A = -4T_{BC}\cos\alpha + 8T_{BC}\sin\alpha - 1200(10) = 0$. Substitute and solve: $\boxed{T_{BC} = 1363.9\ N(T)}$. The reaction at A is needed for the patellar ligament tension solution: The sum of the forces: $\sum F_x = A_X + T_{BC}\cos\alpha = 0$, from which $A_x = 380.28\ N$. $\sum F_y = A_y + T_{BC}\sin a + 1200 = 0$, from which $A_y = -2509.86\ N$. *The member AEF as a free body:* The ligament FD is vertical, hence there is no horizontal component of tension. The horizontal component of distance from E to F is 8 *cm*. The horizontal component of distance from E to A is 10 *cm*. The vertical component of distance from E to A is 50 *cm*. Note that the reactions at A on AEF are equal and opposite to those at A on AB. The sum of the moments about E is $\sum M_E = -8T_{DF} - 50A_x - 10A_y = 0$, from which $\boxed{T_{DF} = 760.56\ N\ (T)}$

Problem 6.95 The device is designed to exert a large force on the horizontal bar at A for a stamping operation. If the hydraulic cylinder DE exerts an axial force of 800 *N* and $\alpha = 80^o$, what horizontal force is exerted on the horizontal bar at A?

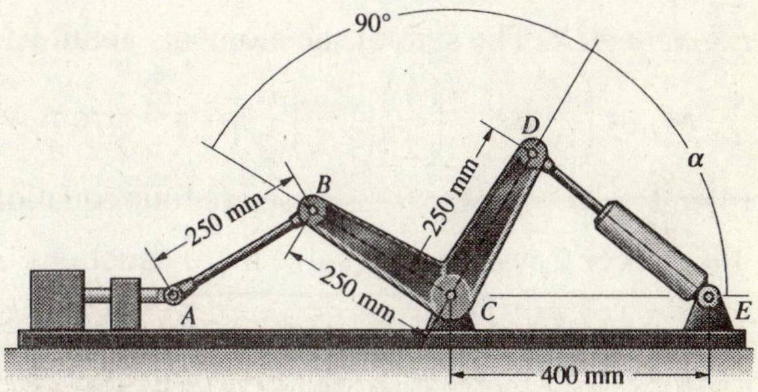

Solution: Define the x-y coordinate system with origin at C. The projection of the point D on the coordinate system is $R_y = 250\sin\alpha = 246.2\ mm$, and $R_x = 250\cos\alpha = 43.4\ mm$. The angle formed by member DE with the positive *x* axis is $\theta = 180 - \tan^{-1}\left(\frac{R_y}{400 - R_x}\right) = 145.38^o$. The components of the force produced by DE are $F_x = F\cos\theta = -658.3\ N$, and $F_y = F\sin\theta = 454.5\ N$. The angle of the element AB with the positive *x* axis is $\beta = 180 - 90 - \alpha = 10^o$, and the components of the force for this member are $P_x = P\cos\beta$ and $P_y = P\sin\beta$, where P is to be determined. The angle of the arm BC with the positive *x* axis is $\gamma = 90 + \alpha = 170^o$. The projection of point B is $L_x = 250\cos\gamma = -246.2\ mm$, and $L_y = 250\sin\gamma = 43.4\ mm$. The sum of the moments about C: $\sum M_C = R_xF_y - R_yF_x + L_xP_y - L_yP_x = 0$. Substitute and solve: $P = 2126.36\ N$, and $\boxed{P_x = P\cos\beta = 2094\ N}$ is the horizontal force exerted at A.

Problem 6.96 This device raises a load W by extending the hydraulic actuator DE. The bars AD and BC are 2 m long and the distances $b = 1.4\ m$ and $h = 0.8\ m$. If the load $W = 4\ kN$, what force must the actuator exert to hold the load in equilibrium?

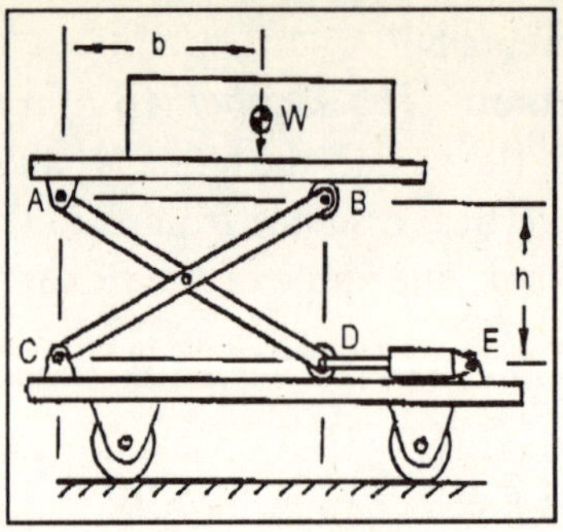

Solution: The angle of the element AD is $\alpha = \sin^{-1}\left(\frac{h}{2}\right) = 23.58^o$.

The distance CD is $d = 2\cos\alpha$.
The complete structure as a free body: The sum of the forces: $\sum F_y = -W + C_y + D_y = 0$. $\sum F_x = C_x + D_x = 0$. The sum of the moments about C: $\sum M_C = bW - dD_y = 0$. These have the solution: $C_y = 0.9449\ kN$, $D_y = 3.055\ kN$, and $C_x = -D_x$.
The Divide the system into three elements: the platform carrying the weight, the member AB, and the member BC.
The Platform: (See Free body diagram) The moments about the point A: $\sum M_A = -bW - dB = 0$. The sum of the forces: $\sum F_y = A + B + W = 0$. These have the solution: $B = -3.055\ kN$, and $A = -0.9449\ kN$.

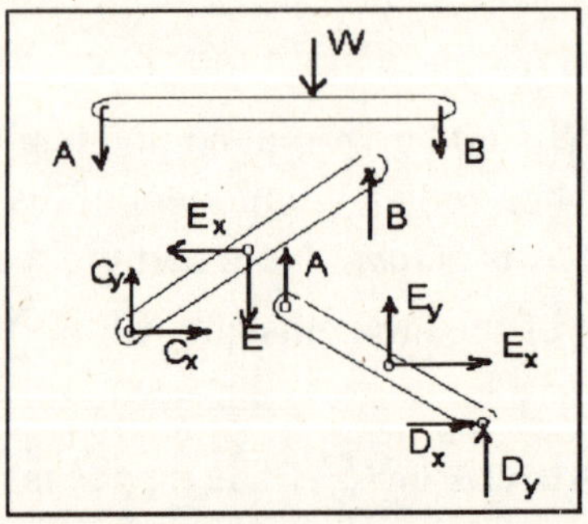

Element BC: The sum of the moments about E is

$$\sum M_C = -\left(\frac{h}{2}\right)C_y + \left(\frac{d}{2}\right)C_x + \left(\frac{d}{2}\right)B = 0\text{, from which}$$

(1) $dC_y - hC_x + dB = 0$ The sum of the forces:
$\sum F_x = -C_x + E_x = 0$, from which (2) $E_x - C_x = 0$,
$\sum F_y = C_y - E_y + B = 0$, from which (3) $C_y - E_y + B = 0$
Element AD: The sum of the moments about E:

$$\sum M_D = \left(\frac{d}{2}\right)D_y - \left(\frac{h}{2}\right)D_x - \left(\frac{d}{2}\right)A = 0\text{, from which}$$

(4) $dD_y - hD_x - dA = 0$.. These are four equations in the four unknowns: E_X, E_Y, D_x, C_X and D_X..
TK Solver 2 was used to solve these equations. The result: $E_x = 3.467\ kN$, $E_y = -2.110\ kN$, $C_x = 3.467\ kN$, $\boxed{D_X = 9.165 = 9.17\ kN}$ (*Note:* See Problem 11. 24.)

Problem 6.97 The linkage is in equilibrium under the action of the couples M_A and M_B. If $\alpha_A = 60^o$ and $\alpha_B = 70^o$, what is the ratio $\frac{M_A}{M_B}$?

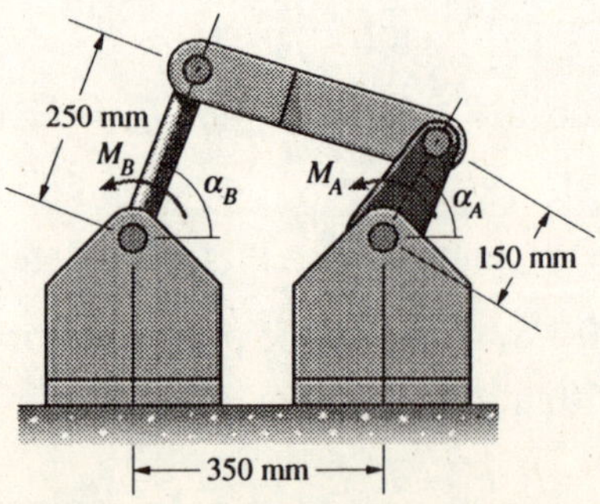

Solution: Make a cut through the linkage connecting the two cranks, and treat each system as a free body.. The equilibrium condition occurs when the reaction forces in the linkage are equal and opposite.
The position vector of the end of the system B crank is $\vec{r}_B = R_B(\vec{i}\cos\alpha_B + \vec{i}\sin\alpha_B) = 85.51\vec{i} + 234.92\vec{j}\ (mm)$.
The position vector at the end of the system A crank is $\vec{r}_A = R_A(\vec{i}\cos\alpha_A + \vec{j}\sin\alpha_A) = 75\vec{i} + 129.9\vec{j}\ (mm)$

Problem 6.97 continued on next page....

continuation of solution to Problem 6.97...

The angle of the linkage from the end of the system B crank with respect to the horizontal is

$$\beta = \tan^{-1}\left(\frac{y_A - y_B}{x_A - x_B + 350}\right) = -17.19^o.$$

The unit vector parallel to the linkage, originating at the B crank, is

$\vec{e}_{BA} = \vec{i}\cos\beta + \vec{j}\sin\beta = 0.9553\vec{i} - 0.2955\vec{j}$

The unit vector originating at A crank is $\vec{e}_{AB} = -\vec{e}_{BA}$. The components of the forces in the linkage are $|\vec{\mathbf{F}}|\mathbf{e}_{AB}$, and $|\vec{\mathbf{F}}|\mathbf{e}_{BA}$.

System B: When the system is in equilibrium, $M_B + |\vec{\mathbf{F}}|\begin{vmatrix} 0 & 0 & 1 \\ 85.5 & 234.9 & 0 \\ 0.9553 & -0.2955 & 0 \end{vmatrix} = 0$, from

which $M_B = 249.7|\vec{\mathbf{F}}|$.

System A: When the system is in equilibrium: $M_A + |\vec{\mathbf{F}}|\begin{vmatrix} 0 & 0 & 1 \\ 75 & 129.9 & 0 \\ -0.9553 & 0.2955 & 0 \end{vmatrix} = 0$, from

which $M_A = -146.27|\vec{\mathbf{F}}|$.

Complete system: Both systems are in equilibrium for the value $|\vec{\mathbf{F}}|$. Take the ratio of the two moments

to eliminate $|\vec{\mathbf{F}}|$. $\boxed{\frac{M_A}{M_B} = -\frac{146.27}{249.7} = -0.5858}$

Problem 6.98 Half of the load $W = 4\ kN$ of the front end loader is supported by the members ACG and the hydraulic actuator BC. Determine the reactions at A and the compressive axial force in the actuator BC.

Solution: Assume that half the load is 2 *kN* (that is, assume that W is the full load). Assume that the only external frame connection to the loader occurs at A and B.

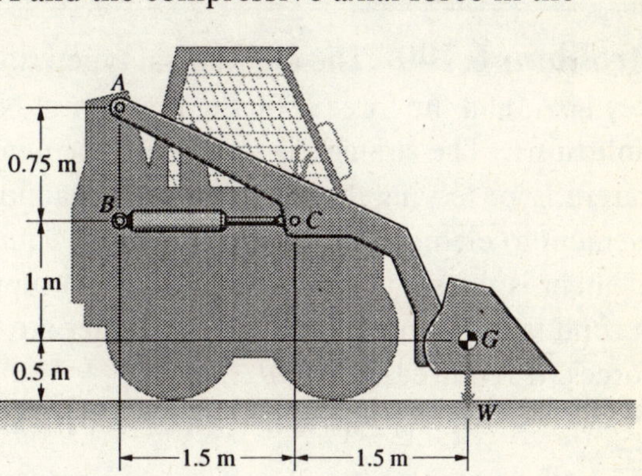

The sum of the moments about A is

$\sum M_A = 0.75BC - 3(2) = 0$, from which

$\boxed{BC = 8\ kN}$ is the axial force. The sum of the forces

$\sum F_X = A_X + BC = 0$, from which $\boxed{A_X = -8\ kN}$.

$\sum F_Y = A_Y - 2 = 0$, from which $\boxed{A_Y = 2\ kN}$

Problem 6.99 The weights $W_1 = 1500\ lb$ and $W_2 = 2000\ lb$. Determine the forces on member BCD at point B and D.

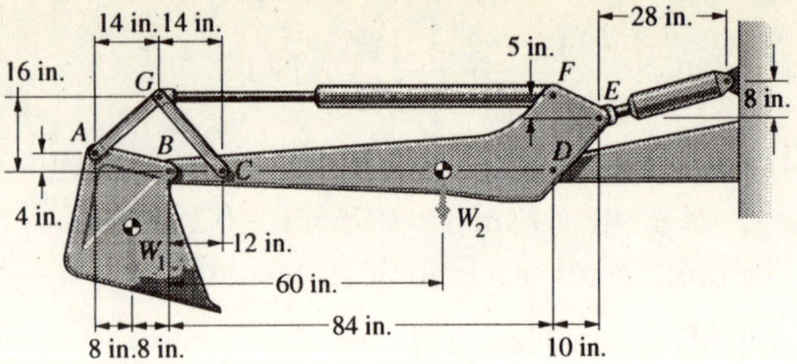

Solution: Assume that the only connection to an external frame is at D and E, where D is a pinned support, and E is an actuator oriented at an angle $\alpha = \tan^{-1}\left(\frac{8}{28}\right) = 15.95^o$ relative to the horizontal.

The complete structure as a free body: The horizontal distance between D and E is 10 *in.* The vertical distance between D and E is 11 *in.* The horizontal distances between D and W_1 and W_2 are 92 *in.* and 24 *in.* respectively. The sum of the moments about D is

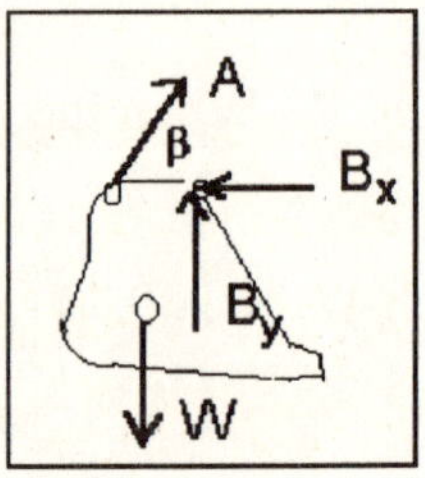

$\sum M_D = -(11)E\cos\alpha + (10)E\sin\alpha + (24)W_2 + (92)W_1 = 0$, from which $E = 23756\ lb$ is the force in the actuator element E. The sum of the forces:

$\sum F_x = D_x + E\cos\alpha = 0$, from which $\boxed{D_x = -22842\ lb}$.

$\sum F_y = D_y + E\sin\alpha - W_1 - W_2 = 0$, from which $\boxed{D_y = -3026.3\ lb}$

The bucket AB: The member AG is at an angle with the positive *x* axis $\beta = \tan^{-1}\left(\frac{12}{14}\right) = 40.6^o$. The moment about the point B is $\sum M_B = -(4)A\cos\beta - (16)A\sin\beta + 8W_1 = 0$, from which $A = 892.2\ lb$ is the reaction in member AG. The sum of the forces: $\sum F_x = A\cos\beta - B_x = 0$, , from which $B_x = 677.42\ lb$, and $\sum F_y = +A\sin\beta + B_y - W_1 = 0$, from which $B_y = 919.35\ lb$. The forces on member BCD are the reverse of this: $\boxed{B_y^{BCD} = -919.35\ lb}$ directed downward, and $\boxed{B_x^{BCD} = 677.42\ lb}$ directed to the right.

Problem 6.100 The mechanical typewriter is in equilibrium under the action of the 2 *N* force on the key at A, and the force *F* on the type bar at N. Determine the force F and the reactions at D, I, and M.

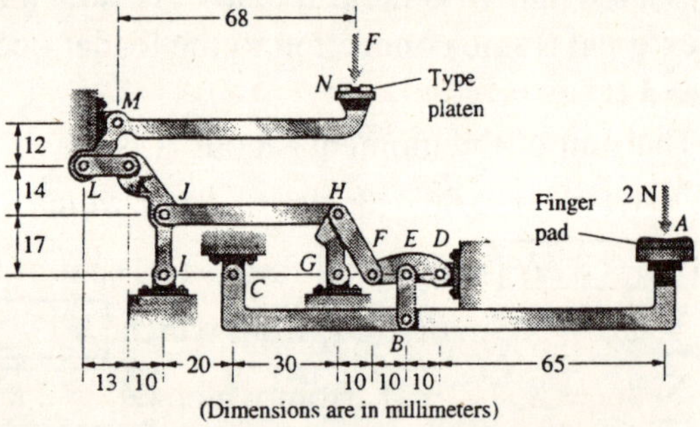

Solution: The strategy is to proceed element by element, carefully observing the reversal of the reactions from element to element. For brevity, the first equation for each element is the sum of moments about the support, the second is the solution, and the remainder are the sums of forces, if required. *Element ABC:*

$\sum M_C = 50B - (125)2 = 0, \therefore B = 5\ N$. *Element DEF:*

$\sum M_D = 10B - 20F\sin\alpha = 0$, $\alpha = \tan^{-1}(1.7) = 59.53^o$,

$\therefore E = 2.9\ N$. $\sum F_y = D_y - B + F\sin\alpha = 0$, $\boxed{D_y = 2.5\ N}$, $\sum F_x = D_x - F\cos\alpha = 0$, $\boxed{D_x = 1.471\ N}$.

Element GH: $\sum M_G = 17H - 17F\cos\alpha = 0$, $\therefore H = 1.47\ N$. *Element IJK:* $\sum M_I = -17H + 31K = 0$, $\therefore K = 0.8065\ N$. $\sum F_x = I_x - K + H = 0$, $\boxed{I_x = -0.6641\ N}$. $\sum F_y = I_y = 0$, $\boxed{I_y = 0}$. *Element LMN:*

$\sum M_M = -68F + 12K = 0$, $\boxed{F = 0.1423\ N}$. $\sum F_x = M_x + K = 0$, $\boxed{M_x = -0.8065\ N}$.

$\sum F_y = M_y - F = 0$, $\boxed{M_y = 0.1423\ N}$.

Problem 6.101 (a) For each member of the truss, obtain a graph of (axial force)/F as a function of x for $0 \le x \le 2\ m$. (b) If you were designing the truss, what value of x would you choose based on your results in Part (a)?

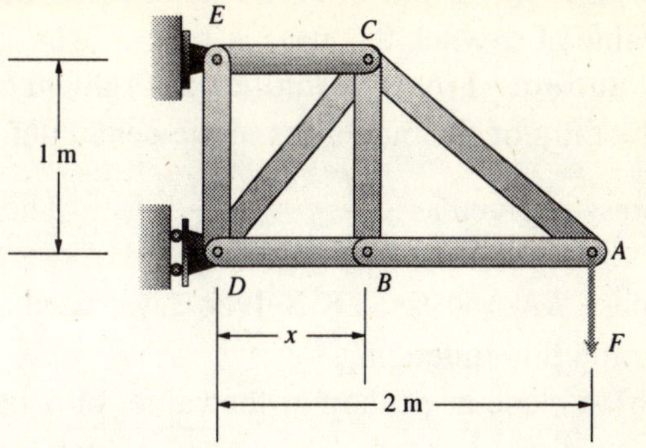

Solution: The angle of member CD with the positive x axis is $\beta = \tan^{-1}\left(\frac{1}{x}\right)$, which lies in the interval $90^o > \beta > 26.6^o$. The angle of member AC with the negative x axis is $\alpha = \tan^{-1}\left(\frac{1}{2-x}\right)$, which lies in the interval $26.57^o \le \alpha < 90^o$.

The complete structure as a free body: The sum of the moments about E is $\sum M_E = -2F + D = 0$, from which $D = 2F$. The sum of forces $\sum F_X = E_X + D = 0$, from which $E_X = -2F$. $\sum F_Y = E_Y - F = 0$, from which $E_Y = F$. These values are to be used as checks on the joint analysis.

The method of joints: Divide each axial force by the applied force F. (This is equivalent to adopting the value $F = 1$)

Joint A: The sums of forces: $\sum F_Y = AC \sin\alpha - 1 = 0$, and $\sum F_X = -AC\cos\alpha - AB = 0$

Joint B: The sums of forces: $\sum F_Y = BC = 0$, and $\sum F_X = -BD + AB = 0$

Joint C: The sums of forces: $\sum F_Y = -CD\sin\beta - AC\sin\alpha = 0$, and $\sum F_X = -CE + AC\cos\alpha - CD\cos\beta = 0$.

Joint E: The sums of forces:

$\sum F_X = E_X + CE = 0$ and

$\sum F_Y = E_Y - DE = 0$

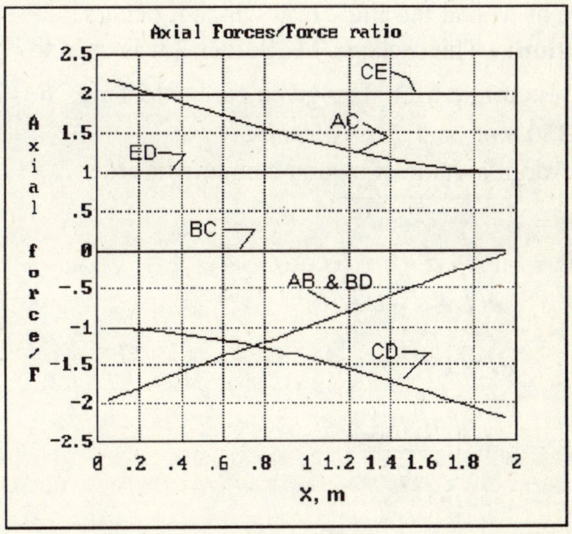

The **TK Solver 2** commercial package was used to plot the axial force ratios as a function of *x*.

(b) The minimum compressive axial forces occur at about $x = 0.8\ m$, so if compressive axial forces are a concern for safety reasons, this configuration offers advantages. However, note that as $\lim x \to 2$, the truss approaches a rectangle with CD as a cross brace. While the compressive force in CD is increased, the compressive force in BD is reduced (AB doesn't count, it no longer has any length) and the axial tension in AC is reduced. None of the axial tensions are increased. Thus this configuration offers the advantages of material saving, without safety penalties.

Problem 6.102 Consider the mechanism for weighing mail described in Problem 6.91.(a) Obtain a graph of the angle α as a function of the mass of the mail for values of the mass from 0 to 2 *kg*.
(b) Use the results of Part(a) to estimate the value of α when the mass is 1 *kg*.

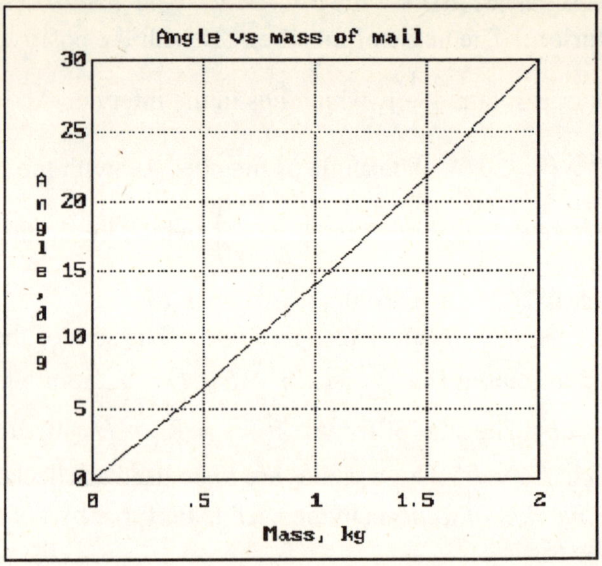

Solution: From the solution to Problem 6.91, the ratio of the mail mass to the pendulum bob mass is given as $\frac{M}{M_B} = \frac{\sin\alpha}{\cos(30^o - \alpha)}$. The bob mass $M_B = 4\ kg$. **TK Solver 2** was used to graph this function.

(b) A close inspection of the values of α near a mass of 1 yields an angle of $\boxed{\alpha = 13.9^o}$. *Check:* Substitute this value back into the expression above:

$$M = M_B\left(\frac{\sin 13.9^o}{\cos(30^o - 13.9^o)}\right) = 4(0.250035).$$

check.

Problem 6.103 Consider the system described in Example 6.11. (a) Obtain a graph of M as a function of α for values of α from 0° to 180°. (b) Use the results of Part (a) to estimate the maximum value of M and the angle α at which it occurs.

Solution: The package **TK Solver 2** was used to obtain the graph. The algorithm was taken directly from Example 6.11: The given parameters are: $R = 150\ mm$, $A = 0.02\ m^2$, $P_{atm} = 10^5\ Pa$, $b = 350\ mm$, and $L = 1050\ mm$

(1) Adopt a value of α, and compute in order:

(2) $q = \sqrt{b^2 + R^2\sin^2\alpha}$

(3) $V = A(L - d - q + R\cos\alpha)$

(4) $V_O = A(L - d - b - R)$

(5) $F = AP_{atm}(\frac{V_O}{V} - 1)$

(6) $\beta = \tan^{-1}\left(\frac{R\sin\alpha}{q}\right)$

(7) $Q = \frac{F}{\cos\beta}$

(8) $M = -QR\sin(\beta - \alpha)$

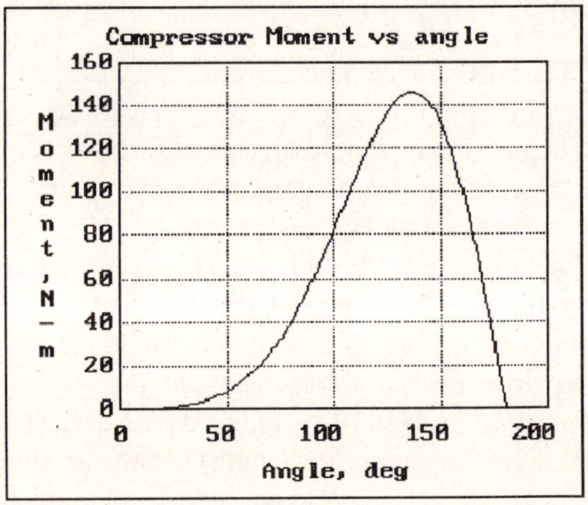

The graph is shown.

(b) A table of values of α vs M near maximum M shows that the maxima occurs between $136.70 \le \alpha \le 136.85$, at a value of $M = 145.75\ N\text{-}m$:

Problem 6.104 Consider the system described in Problem 6.94. The hydraulic cylinder DE exerts an axial force of 800 *N*. (a) Obtain a graph of the horizontal component of the force exerted on the horizontal bar at A by the rod AB for values of α from 45° to 85°. (b) Use the results of Part (a) to estimate the value of α for which the horizontal force is 2 *kN*.

Solution: The algorithm is taken directly from the solution to Problem 6.94 (which see). The parameters are: $F = 800\ N$, $b = 400\ mm$, $L_{CD} = 250\ mm$, $L_{BC} = 250\ mm$, and $L_{AB} = 250\ mm$.

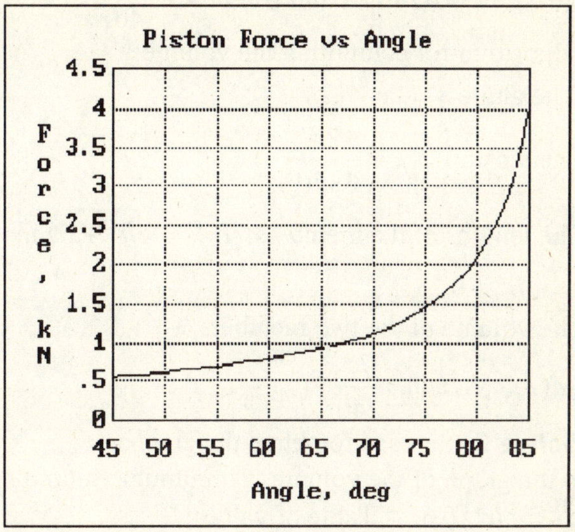

(1) Adopt a value of α in the interval, and compute in order:

(2) $R_y = L_{CD} \sin\alpha$, $R_x = L_{CD} \cos\alpha$

(3) $\theta = 180^o - \tan^{-1}\left(\dfrac{R_y}{b - R_x}\right)$

(4) $F_x = F\cos\theta$, $F_y = F\sin\theta$

(5) $\beta = 90^o - \alpha$

(6) $\gamma = 90^o + \alpha$

(7) $L_x = L\cos\gamma$, $L_y = L\sin\gamma$

(8) $P_x = \left(\dfrac{R_x F_y - R_y F_x}{L_y \cos\beta - L_x \sin\beta}\right)\cos\beta$

where P_x is the horizontal force.

(b) From an inspection of a Table of values, the force $P_x = 2\ kN$ occurs at an angle $\boxed{\alpha = 79.5^o}$

Problem 6.105 The suspended object weighs 10 *kN*. The two members have equal cross sectional areas A. Each member will safely support an axial force of $40A\ MN$, where *A* is in square meters. Determine the value of *h* that minimizes the total volume of material in the two members.

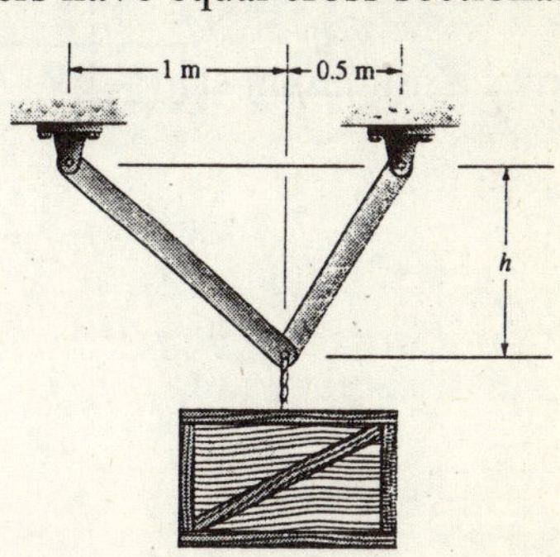

Solution: Assume that the maximum safe axial force is to be always imposed. The cross sectional area can always be modified to assure that this is true. However, one member will have a higher axial force than the other, and that member will determine the area for maximum safe axial force.

Make a cut near the pin supports. The sum of the moments about C is $\sum M_A = -W + 1.5B_y = 0$, from which $B_y = \left(\dfrac{2}{3}\right)W$. The sum of the forces $\sum F_y = C_y + B_y - W = 0$, from which $C_y = \left(\dfrac{1}{3}\right)W$. The vertical components of the forces: $C_y - T_{CD}\sin\alpha = 0$, from which $T_{CD}\sin\alpha = \left(\dfrac{1}{3}\right)W$. Similarly, $T_{BD}\sin\beta = \left(\dfrac{2}{3}\right)W$

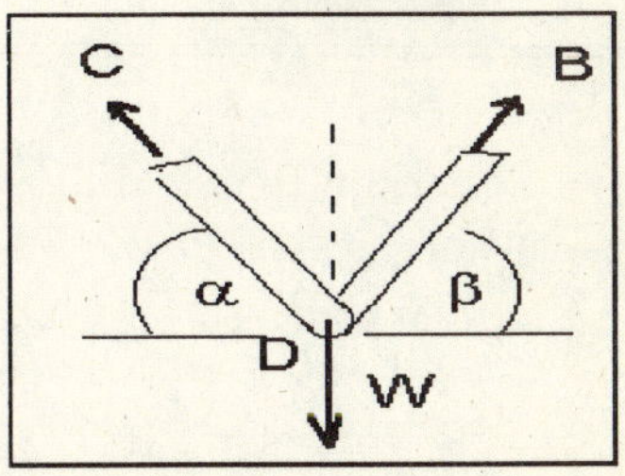

Problem 6.105 continued on next page....

continuation of solution to Problem 6.105...

The angle β is larger than the angle α, and both are in the first quadrant, hence $\sin\beta > \sin\alpha$, so it is not obvious from the expressions for the axial force which will be the larger. A graph of the tensions in the two members demonstrates that T_{BD} is the larger.

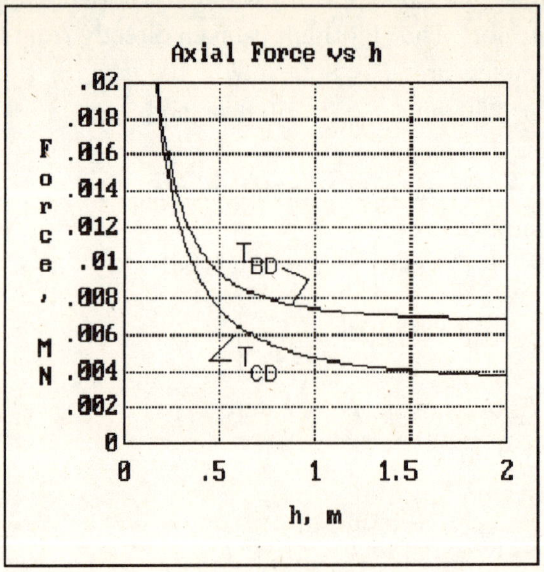

Thus, $T_{BD} = 40A\ MN$, from which $A = \frac{T_{BD}}{40}$.

The algorithm for computing the volume:

(1) The angle $\beta = \tan^{-1}(2h)$.

(2) The tension $T_{BD} = \frac{2W}{3\sin\beta}$

(3) The lengths of the members: $L_{CD} = \sqrt{h^2 + 1}$ and $L_{BD} = \sqrt{h^2 + 0.25}$

(4) The volume of the two members:

$$V = A(L_{CD} + L_{BD}) = \frac{T_{BD}}{40}(L_{CD} + L_{BD}).$$

TK Solver 2 was used to obtain the graphs.

From the graph of the volume, a minimum volume occurs for $h \cong 0.6$. A Table of values:

h,m	40V
0.59	0.16904
0.60	0.16898
0.61	0.16896
0.62	0.16898

shows a minimum at $\boxed{h = 0.61\ m}$

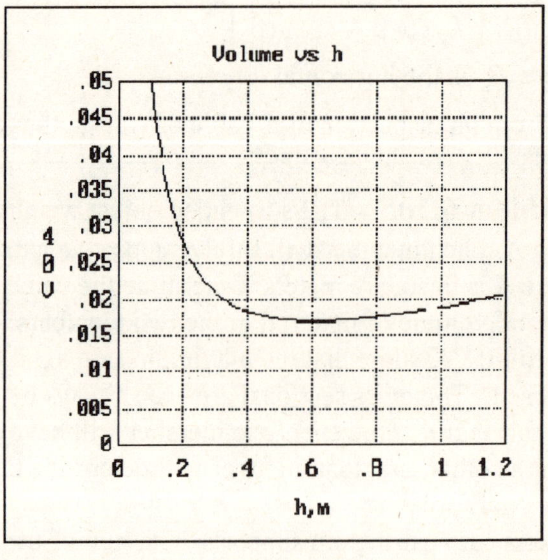

=================================<>=================================

Problem 6.106 Consider the device shown in Problem 6.96. The bars AD and BC are 2 *m* long, the distance $b = 1.4\ m$, and $W = 4\ kN$. If the largest force the hydraulic actuator DE can exert is 20 *kN*, what is the smallest height *h* at which the load can be supported?

Solution: The algorithm is taken directly from the solution to Problem 6.96. Divide the system into three elements: the platform carrying the weight, the member AB, and the member BC. (See free body diagrams in Problem 6.96) The algorithm for computing the values of D_x and *h* are taken directly from the solution to Problem 6.96.

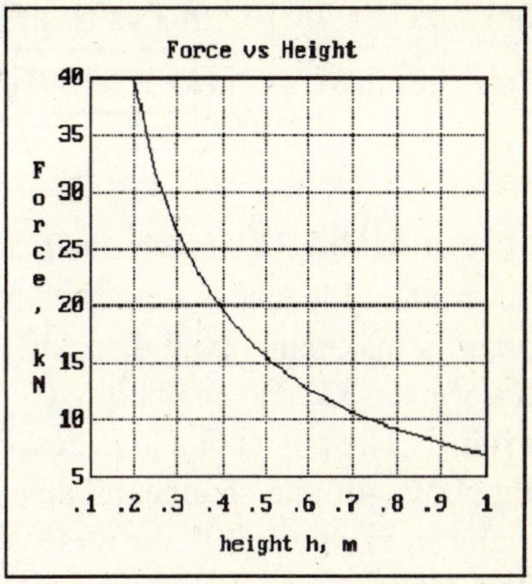

(1) Choose values of *h* in the interval $0 < h < 2\ m$

(2) Compute the angle: $\alpha = \sin^{-1}\left(\frac{h}{2}\right)$, and compute $d = 2\cos\alpha$

(3) Get the reactions of the complete structure by solving:

$-W + C_y + D_y = 0$,

$bW - dD_y = 0$

(3) Get the reactions for the platform by solving:

$-bW - dB = 0$,

$-W - A - B = 0$

(4) Solve the equilibrium conditions for *Element BEC*:

$E_x - C_x = 0$, $C_y - E_y + B = 0$, and

$\left(\frac{d}{2}\right)C_x - \left(\frac{h}{2}\right)C_y + \left(\frac{d}{2}\right)B = 0$.

(5) Solve the equilibrium condition for *Element AED:* $\left(\frac{d}{2}\right)D_y - \left(\frac{h}{2}\right)D_x - \left(\frac{d}{2}\right)A = 0$.

TK Solver 2 was used to graph the force D_x against the height *h*. From an inspection of the table from which the graph was done, at a force of 20 *kN*, $\boxed{h = 0.39\ m}$. *Check:* At $h = 0.8\ m$, the force is $D_X = 9.17\ kN$, as to be expected, since this was the solution to Problem 6.96. *check. Note:* The value of *h* can be obtained by a direct iteration of the equilibrium equations. The result checks the above.

=================================<>=================================

Problem 6.107 The linkage shown in Problem 6.97 is in equilibrium under the action of the couples M_A and M_B. When $\alpha_A = 60^o$, $\alpha_B = 70^o$. For the range of $0 \le \alpha_A \le 180^o$, estimate the maximum positive and negative values of $\frac{M_A}{M_B}$ and the values of α_A at which they occur.

Solution: The algorithm is taken from Problem 6.97, with the addition of the relationship between the angles. (See Figure with Problem 6.97 for dimensions.) . The algorithm is:

(a) For a value of α_A, $(0^o \le \alpha_A \le 180^o)$, solve for the angle α_B and the angle of the linkage connection β from the two simultaneous equations: (1) $R_A \cos\alpha_A - R_B \cos\alpha_B + 350 = L\cos\beta$, (2) $R_A \sin\alpha_A - R_B \sin\alpha_B = L\sin\beta$, where the length of the linkage connection L is

$$L = \sqrt{\left(R_A \sin(60^o) - R_B \sin(70^o)\right)^2 + \left(R_A \cos(60^o) - R_B \cos(70^o) + 350\right)^2} = 355.367\ mm$$

(b) Compute the positions of the crank ends: $x_A = R_A \cos\alpha_A$, $y_A = R_A \sin\alpha_A$, $x_B = R_B \cos\alpha_B$, $y_B = R_B \sin\alpha_B$.

(d) Get the components of unit vector parallel to the linkage: $e_{BAX} = \cos\beta$, $e_{BAY} = \sin\beta$.

Problem 6.107 continued on next page....

continuation of solution to Problem 6.107...

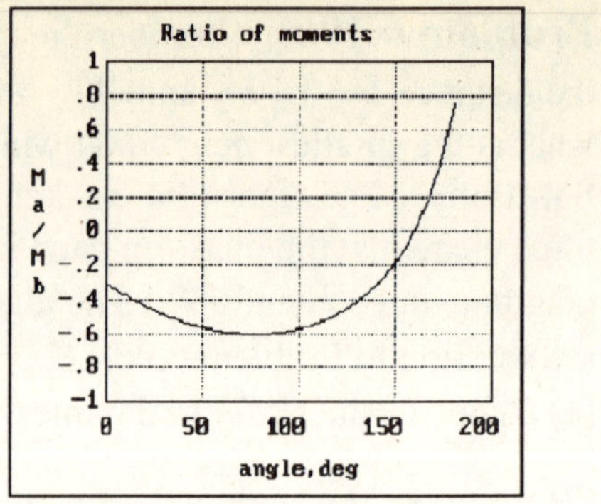

(e) Compute the moments: $M_B = -(x_B e_{ABY} - y_B e_{ABX})$, and $M_A = (x_A e_{ABY} - y_A e_{ABX})$

(f) Take the ratio: $R = \frac{M_A}{M_B}$. The graph is shown. The minimum occurs at $\boxed{\alpha_A = 78^o}$, with a value of $\boxed{R = -0.60289... = -0.603}$. The maximum occurs at $\boxed{\alpha_A = 180^o}$, with a value of $\boxed{R = 0.75}$

Problem 6.108 Consider the front end loader shown in Problem 6.98. Half of the load $W = 4\ kN$ of the front end loader is supported by the member ACG and the hydraulic actuator BC. If the actuator can exert a maximum axial force of 12 *kN*, what is the largest height above the ground at which the center of mass G can be supported?

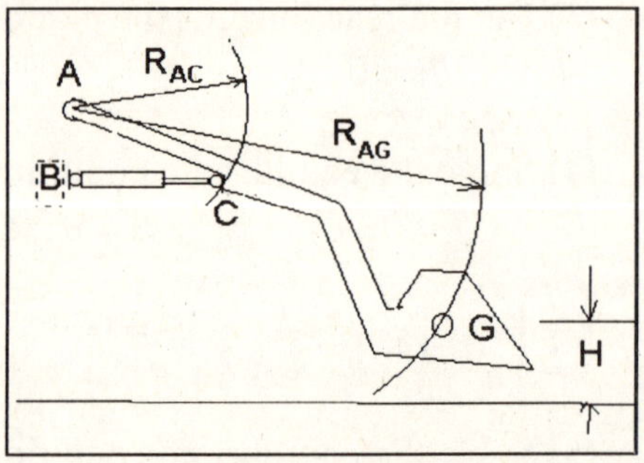

Solution: The points G and B describe arcs of circles as the height above ground changes. The radii of these circles (see sketch of loader in Problem 6.98 for dimensions) are

$R_{AG} = \sqrt{1.75^2 + 3^2} = 3.473\ m$, and

$R_{AC} = \sqrt{0.75^2 + 1.5^2} = 0.7649\ m$. The angles made by these radii from the horizontal line through point A when the point G is 0.5 *m* above the ground are $\gamma_{AC} = \tan^{-1}\left(\frac{0.75}{1.5}\right) = 26.57^o$, and

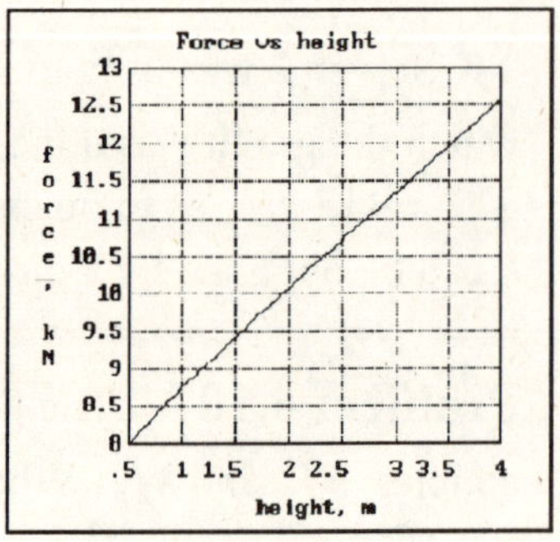

$\beta_{AG} = \tan^{-1}\left(\frac{1.75}{3}\right) = 30.25^o$. The algorithm for computing height as a function of the hydraulic actuator force is developed from the geometry and the equilibrium conditions: (1) The vertical distance from A to C as the height is increased: $h = -R_{AC}\sin(\gamma_{AC} - \alpha)$, where α is the angle of the element AG, ($0 < \alpha$), measured from the initial height of point G of 0.5 m, and a negative *h* means that h is below the point A. (2) The horizontal distance from A to C: $d = R_{AC}\cos(\gamma_{AC} - \alpha)$. (3) The height above the ground of point G: $H = 2.25 - R_{AG}\sin(\beta_{AG} - \alpha)$. (4) The angle of the hydraulic actuator BC with a horizontal line through B is $\theta = \tan^{-1}\left(\frac{0.75 + h}{d}\right)$. (5) The sum of the moments about point A is

$\sum M_A = hBC\cos\theta + dBC\sin\theta - \frac{W}{2} R_{AG}\cos(\beta_{AG} - \alpha) = 0$. *Check:* When $\alpha = 0$ these reduce to the algorithm used in the solution to Problem 6.98. *check.* The **TK Solver 2** commercial package was used to graph the force against height. The hydraulic actuator reaches the limit of $BC = 12\ kN$ when the height is $\boxed{H = 3.54\ m}$ *Note :* The height can be determined from a solution of the equilibrium equations; the graph is not necessary for a solution. The solution of the equilibrium equations for $BC = 12\ kN$, is (to three significant figures) $h = 0.724\ m$, $\theta = 44.3^o$, $\alpha = 52.1^o$, and $H = 3.54\ m$.

Problem 6.109 Consider the truss in Problem 6.23. The crane exerts vertical 75 *kN* forces on the truss at B, C, and D. You can model the support at A as a pin support and model the support at E as a roller support which can exert a force normal to the dashed line but cannot exert a force parallel to it. Determine value of the angle α for which the largest compressive force in any of the members is as small as possible. What are the resulting axial forces in the members?

Solution: The algorithm is taken directly from the solution to Problem 6.23, with the addition of a MAX function. The algorithm is as follows:

(1) Compute the angles of the members: AB and ED relative to the horizontal, $\gamma = \tan^{-1}\left(\frac{4}{3.4}\right) = 49.64^o$.

AF and EH relative to the horizontal: : $\beta = \tan^{-1}\left(\frac{2.2}{3.4}\right) = 32.91^o$. BG and DG relative to the horizontal:

$\theta = \tan^{-1}\left(\frac{1.8}{3.4}\right) = 27.9^o$. (See the sketch with Problem 6.23.)

(2) For a given value of α, get the unit vector components for the reaction angle at E:

$\vec{e}_E = \vec{i}\cos(90+\alpha) + \vec{j}\sin(90+\alpha) = -\vec{i}\sin\alpha + \vec{j}\cos\alpha$

(3) *The complete structure as a free body:* From the sum of moments: The x- and y- components:

$E_x = -E\sin\alpha$ *kN* and $E_y = 112.5$ *kN* (a constant). The reactions at A: :

$A_x = -E_x$. $A_y = 3(75) - E_y = 112.5$, from which $A_y = 112.5$ *kN*. (a constant)

Two equations are developed for each joint:

(d) *Joint A:* $AB\cos\gamma + A_x + AF\cos\beta = 0$, and $AB\sin\beta + A_y + AF\sin\beta = 0$.

(e) *Joint E:* $DE\cos\gamma - E_x + EH\cos\beta = 0$ and $DE\sin\gamma + E_y + EH\sin\beta = 0$

(f) *Joint F:* $-AF\cos\beta + FG = 0$, and $-AF\sin\beta + BF = 0$.

(g) *Joint H:* $EH\cos\beta - GH = 0$, and $-EH\sin\beta + DH = 0$

(h) *Joint B:* $-AB\sin\gamma - BF - BG\sin\theta - 75 = 0$, and $-AB\cos\gamma + BC + BG\cos\theta = 0$.

(i) *Joint D:* $-DE\sin\gamma - DH - DG\sin\theta - 75 = 0$, and $DE\cos\gamma - CD - DG\cos\theta = 0$

(j) *Joint C:* $CD = BC$ *Check.* $CG = -75$ *kN* (C)

(k) Use standard MAX function to get maximum compressive axial force:

$F_{MAX} = MAX(-AB, -AF, -FG, -BF, -BC, -BG, -CG)$,

where it is necessary to test only one of a matching value pair. The maximum axial compressive force for all members is graphed against the angle α. The min-max point occurs at about $\boxed{\alpha = 50.94...^o = 51^o}$, where

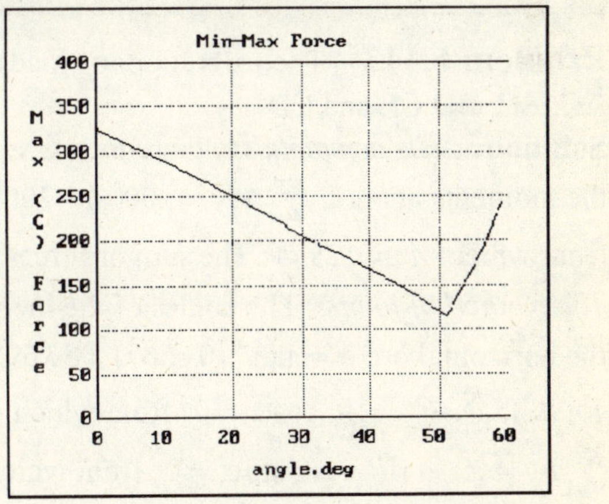

$AF = EH = 113.9$ *kN* (C),

$AB = DE = 66.5$ *kN* (C),

$BF = DH = 61.9$ *kN* (C),

$BG = DG = 80.2$ *kN* (T),

$BC = CD = 113.9$ *kN* (C)

$FG = GH = 95.6$ *kN* (C), and

$CG = 75$ *kN* (C). *Check.* The minimum maximum compressive stress at $\alpha = 51^o$ is $\boxed{R = 113.9 \text{ kN}}$ which occurs in elements AF, EH and BC, CD.

Problem 6.110 The Pratt bridge truss supports loads at F, G, and H. Determine the axial forces in members BC, BG, and FG.

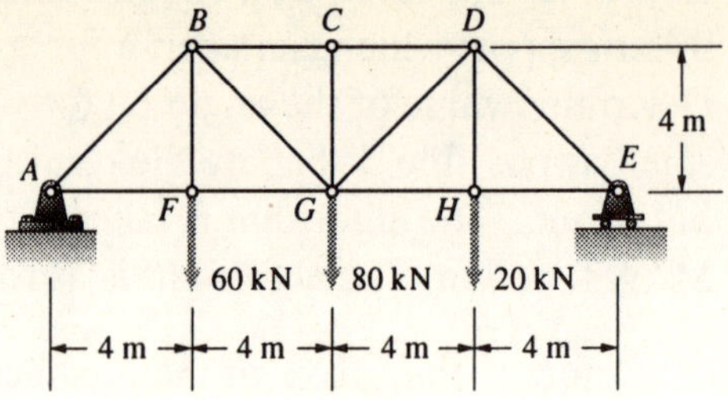

Solution: The angles of the cross-members are $\alpha = 45^o$.
The complete structure as a free body: The sum of the moments about A: $\sum M_A = -60(4) - 80(8) - 20(12) + 16E = 0$, from which $E = 70\ kN$. The sum of the forces: $\sum F_x = A_x = 0$.
$\sum F_y = A_y - 60 - 80 - 20 + E = 0$, from which $A_y = 90\ kN$
The method of joints:
Joint A: $\sum F_Y = A_y + AB\sin\alpha = 0$, from which $AB = -127.3\ kN\ (C)$, $\sum F_x = AB\cos\alpha + AF = 0$, from which $AF = 90\ kN\ (T)$. *Joint F:* $\sum F_x = -AF + FG = 0$, from which $\boxed{FG = 90\ kN\ (T)}$.
$\sum F_Y = BF - 60 = 0$, from which $BF = 60\ kN\ (C)$. *Joint B:* $\sum F_x = -AB\cos\alpha + BC + BG\cos\alpha = 0$, and $\sum F_y = -AB\sin\alpha - BF - BG\sin\alpha = 0$, from which: $-AB\sin\alpha - BF - BG\sin\alpha = 0$. Solve: $\boxed{BG = 42.43\ kN\ (T)}$, and $-AB\cos\alpha + BC + BG\cos\alpha = 0$, from which $\boxed{BC = -120\ kN\ (C)}$

Problem 6.111 Consider the truss in Problem 6.110. Determine the axial forces in members CD, GD, and GH.
Solution: Use the results of the solution of Problem 6.110: $BC = -120\ kN\ (C)$, $BG = 42.43\ kN\ (T)$, and $FG = 90\ kN\ (T)$. The angle of the cross-members with the horizontal is $\alpha = 45^o$. *Joint C:* $\sum F_x = -BC + CD = 0$, from which $\boxed{CD = -120\ kN\ (C)}$
$\sum F_Y = -CG = 0$, from which $CG = 0$. *Joint G:* $\sum F_y = BG\sin\alpha + GD\sin\alpha + CG - 80 = 0$, from which $\boxed{GD = 70.71\ kN\ (T)}$. $\sum F_y = -BG\cos\alpha + GD\cos\alpha - FG + GH = 0$, from which $\boxed{GH = 70\ kN\ (T)}$

Problem 6.112 The truss supports loads at F and H. Determine the axial forces in members AB, AC, BC, BD, CE and CD.

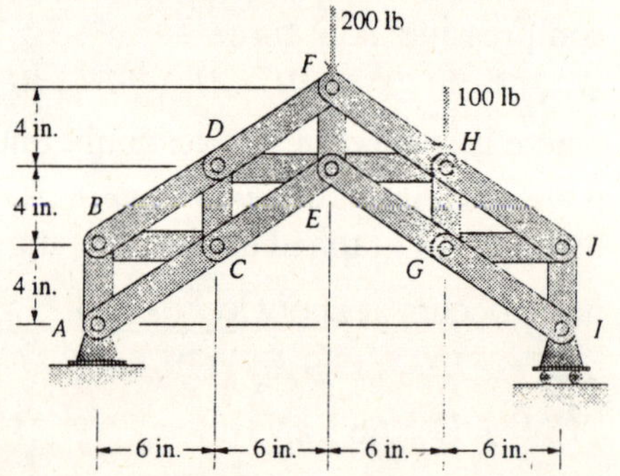

Solution: *The complete structure as a free body:* The sum of the moments about I: $\sum M_A = 100(6) + 200(12) - 24A_Y = 0$, from which $A_Y = 125\ lb$. The sum of forces: $\sum F_x = A_x = 0$.
The method of joints: The angles of the inclined members with the horizontal are $\alpha = \tan^{-1}(0.6667) = 33.69^o$
Joint A: $\sum F_x = AC\cos\alpha = 0$, from which $AC = 0$.
$\sum F_y = A_y + AB + AC\sin\alpha = 0$, from which $\boxed{AB = -125\ lb\ (C)}$
Joint B: $\sum F_{yt} = -AB + BD\sin\alpha = 0$, from which $\boxed{BD = -225.3\ lb\ (C)}$
$\sum F_x = BD\cos\alpha + BC = 0$, from which $\boxed{BC = 187.5\ lb\ (T)}$
Joint C: $\sum F_x = -BC - AC\cos\alpha + CE\cos\alpha = 0$, from which $\boxed{CE = 225.3\ lb\ (T)}$
$\sum F_y = -AC\sin\alpha + CD + CE\sin\alpha = 0$, from which $\boxed{CD = -125\ lb\ (C)}$

Problem 6.113

Consider the truss in Problem 6.112. Determine the axial forces in members EH and FH.

Solution: Use the results from the solution to Problem 6.112: $CE = 225.3\ lb\ (T)$, $CD = -125\ lb\ (C)$, $BD = -225.3\ lb\ (C)$.

The method of joints: The angle of inclined members with the horizontal is $\alpha = 33.69^o$.

Joint D: $\sum F_y = -BD\sin\alpha - CD + DF\sin\alpha = 0$, from which $DF = -450.7\ lb\ (C)$.

$\sum F_x = -DF\cos\alpha + DE - BD\cos\alpha = 0$, from which $DE = 187.5\ lb\ (T)$

Joint F: $\sum F_x = -DF\cos\alpha + FH\cos\alpha = 0$, from which $\boxed{FH = -450.7\ lb\ (C)}$

$\sum F_y = -200 - DF\sin\alpha - FH\sin\alpha - EF = 0$, from which $EF = 300\ lb\ (T)$

Joint E: $\sum F_y = -CE\sin\alpha + EF - EG\sin\alpha = 0$, from which $EG = 315\ lb\ (T)$

$\sum F_x = -DE + EH - CE\cos\alpha + EG\cos\alpha = 0$, from which $\boxed{EH = 112.5\ lb\ (T)}$

Problem 6.114

The truss supports a 400 N load at G. Determine the axial forces in members AC, CD, and CF

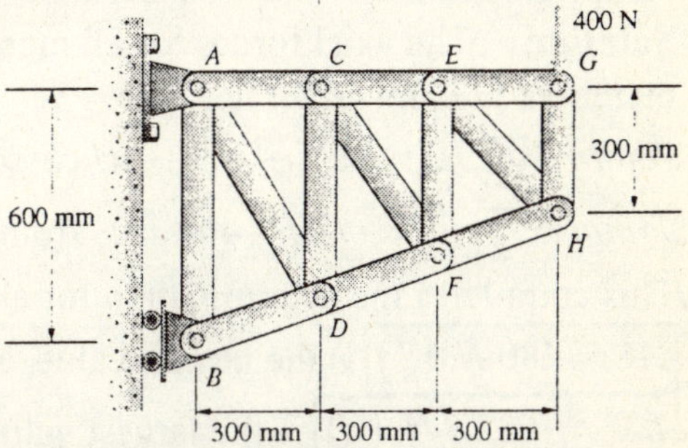

Solution: *The complete structure as a free body:* The sum of the moments about A: $\sum M_A = -900(400) + 600B = 0$, from which $B = 600\ N$. The sum of forces: $\sum F_x = A_x + B = 0$, from which $A_x = -600\ N$. $\sum F_y = A_y - 400 = 0$, from which $A_y = 400\ N$.

The method of joints: The angle from the horizontal of element BD is $\theta = \tan^{-1}\left(\frac{300}{900}\right) = 18.43^o$ The angle from the horizontal of element AD is

$$\alpha_{AD} = 90 - \tan^{-1}\left(\frac{300}{600 - 300\tan\theta}\right) = 59.04^o.$$

The angle from the horizontal of element CF is

$$\alpha_{CF} = 90 - \tan^{-1}\left(\frac{300}{600(1-\tan\theta)}\right) = 53.13^o.$$

Joint B: $\sum F_x = B + BD\cos\theta = 0$, from which $BD = -632.5\ N\ (C)$

$\sum F_y = AB + BD\sin\theta = 0$, from which $AB = 200\ N\ (T)$

Joint A: $\sum F_y = A_y - AD\sin\alpha_{AD} - AB = 0$, from which $AD = 233.2\ N\ (T)$

$\sum F_x = A_x + AC + AD\cos\alpha_{AD} = 0$, from which $\boxed{AC = 480\ N\ (T)}$

Joint D: $\sum F_x = -AD\cos\alpha_{AD} - BD\cos\theta + DF\cos\theta = 0$, from which $DF = -505.96\ N\ (C)$

$\sum F_y = AD\sin\alpha_{AD} + CD - BD\sin\theta + DF\sin\theta = 0$, from which $\boxed{CD = -240\ N\ (C)}$

Joint C: $\sum F_y = -CD - CF\sin\alpha_{CF} = 0$, from which $\boxed{CF = 300\ N\ (T)}$

==========================<>==========================

Problem 6.115 Consider the truss in Problem 6.114. Determine the axial forces in members CE, EF, and EH.

Solution: Use the results of the solution of Problem 6.114: $AC = 480\ N\ (T)$, $CF = 300\ N\ (T)$, $DF = -505.96\ N\ (C)$, $\theta = 18.4^{\circ}$, $\alpha_{CF} = 53.1^{\circ}$.

The method of joints: The angle from the horizontal of element EH is

$$\alpha_{EH} = 90 - \tan^{-1}\left(\frac{300}{600 - 900\tan\theta}\right) = 45^{o}$$

Joint C: $\sum F_x = -AC + CE + CF\cos\alpha_{CF} = 0$, from which $\boxed{CE = 300\ N\ (T)}$

Joint F: $\sum F_x = -CF\cos\alpha_{CF} - DF\cos\theta + FH\cos\theta = 0$, from which $FH = -316.2\ N\ (C)$

$\sum F_y = EF + CF\sin\alpha_{CF} - DF\sin\theta + FH\sin\theta = 0$, from which $\boxed{EF = -300\ N(C)}$

Joint E: $\sum F_y = -EH\sin\alpha_{EH} - EF = 0$, from which $\boxed{EH = 424.3\ N\ (T)}$

==========================<>==========================

Problem 6.116 Consider the truss in Problem 6.114. Which members have the largest tensile and compressive forces, and what are their values?

Solution: The axial forces for all members have been obtained in Problems 6.114 and 6.115 except for members EG and GH. These are:

Joint E: $\sum F_x = -CE + EG + EH\cos\alpha_{EH} = 0$, from which $EG = 0$

Joint G: $\sum F_y = -GH - 400 = 0$, from which $GH = -400\ N\ (C)$.

This completes the determination for all members. a comparison of tensile forces shows that $\boxed{AC = 480\ N\ (T)}$ is the largest value, and a comparison of compressive forces shows that $\boxed{BD = -632.5\ N\ (C)}$ is the largest value.

==========================<>==========================

Problem 6.117 A speaker system is suspended from the truss by cables attached at D and E. The mass of the speaker system is 130 *kg* and its weight acts at G. Determine the axial forces in members BC and CD.

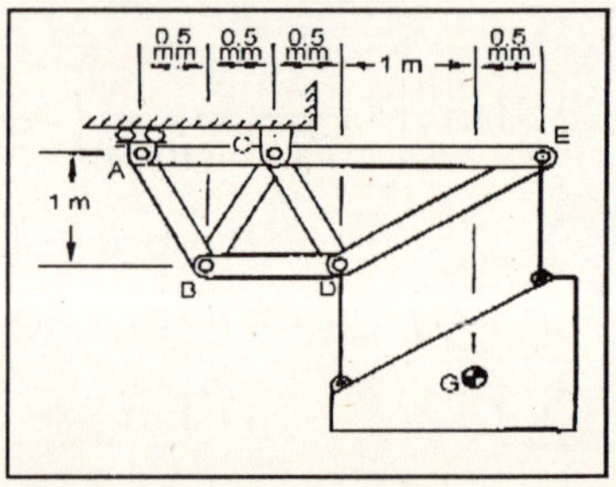

Solution: *The speaker as a free body:* The weight of the speaker is $W = 130(9.81) = 1275.3\ N$. Make a cut through the suspension cables D, E The sum of the moments about cable D is $\sum M_D = -(1)W + (1.5)E = 0$, from which $E = 850.2\ N$. The sum of the forces: $\sum F_y = D + E - W = 0$, from which $D = 425.1\ N$.

The structure as a free body: The sum of the moments about C is $\sum M_C = +(1)A - 0.5D - (2)E = 0$, from which $A = 1912.95\ N$. The sum of the forces: $\sum F_y = -A + C_y - W = 0$, from which $C_y = -637.65\ N$ .and $\sum F_x = C_x = 0$. *The method of joints:* The angle of member DE relative to the horizontal is $\alpha = \tan^{-1}\left(\frac{1}{1.5}\right) = 33.69^{o}$. The angles of members AB, BC, and CD are $\beta = 90 - \tan^{-1}(0.5) = 63.43^{o}$. *Joint E:* $\sum F_{yt} = -E - DE\sin\alpha = 0$, from which $DE = -1532.72\ N(C)$. $\sum F_x = -CE - DE\cos\alpha = 0$, from which $CE = 1275.3\ N\ (T)$

Problem 6.117 continued on next page....

continuation of solution to Problem 6.117....

Joint D: $\sum F_y = CD \sin\beta + DE \sin\alpha - D = 0$, from which $\boxed{CD = 1425.8\ N\,(T)}$

Joint C: $\sum F_y = -CD \sin\beta - BC \sin\beta + C_y = 0$, from which $\boxed{BC = 2138.7\ N\ (T)}$

Problem 6.118 Consider the system described in Problem 6.117. If each member of the truss will safely support a tensile force of 5 *kN* and a compressive force of 3 *kN*, what is the maximum safe value of the mass of the speaker system?

Solution: In the solution of Problem 1.117, the axial forces have been determined for four of the seven elements in the structure. The others are:

Joint D: $\sum F_x = -BD + DE \cos\alpha = 0$ from which $BD = -1275.7\ N\ (C)$

Joint C: $\sum F_x = -AC + CE - BC \cos\beta + CD \cos\beta = 0$, from which $AC = 956.5\ N\ (T)$

Joint B: $\sum F_y = BC \sin\beta + AB \sin\beta = 0$, from which $AB = -2138.7\ \ N(C)$

The values are tabulated:

Member	Axial F, kN	Type
DE	-1.533	(C)
CE	1.275	(T)
CD	1.425	(T)
BC	2.138	(T)
BD	-1.275	(C)
AC	0.956	(T)
AB	-2.138	(C)

The maximum tension occurs in element BC, and the maximum compression occurs in element AB. Since the system is linear, the ratio of the axial forces to two distinct mass loadings will be equal to the ratio of the masses: $\frac{T}{T_{SAFE}} = \frac{m}{m_T}$. Similarly, $\frac{C}{C_{SAFE}} = \frac{m}{m_C}$. The safe loading for max tensile forces $m_T = \left(\frac{T_{SAFE}}{T}\right)m = \left(\frac{5000}{BC}\right)130 = 303.9.. = 304\ kg$. Similarly for compressive forces,

$m_T = \left(\frac{C_{SAFE}}{C}\right)m = \left(\frac{-3000}{AB}\right)130 = 182.35... = 182.4\ kg$. The compressive force safety limit controls the loading, and the maximum safe mass is $\boxed{m = 182.4\ kg}$

Problem 6.119 Determine the forces on member ABC and present your answers as shown in Fig. 6.35.

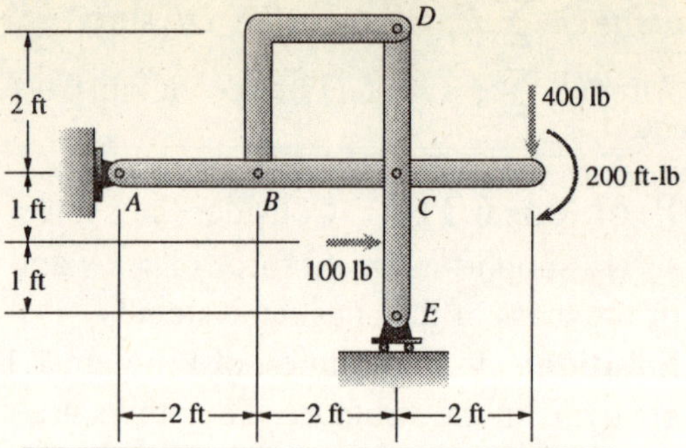

Solution: *The complete structure as a free body:* The sum of the moments: $\sum M_A = 100(1) - 400(6) - 200 + 4E = 0$, from which $E = 625\ lb$. The sum of the forces: $\sum F_y = A_y + E - 400 = 0$, from which $A_y = -225\ lb$. $\sum F_x = A_x + 100 = 0$, from which $A_x = -100\ lb$. These results are used as a check on the solution below.

Element ECD: (See the free body diagram.) The sum of the moments about E:

$\sum M_E = -4D_x - 2C_x - 100 = 0$, from which (1) $4D_x + 2C_x = -100$. The sum of the forces: $\sum F_x = D_x + C_x + 100 = 0$, from which (2) $D_x + C_x = -100$. $\sum F_y = E + C_y + D_y = 0$, thus (3) $D_y + C_y + E = 0$.

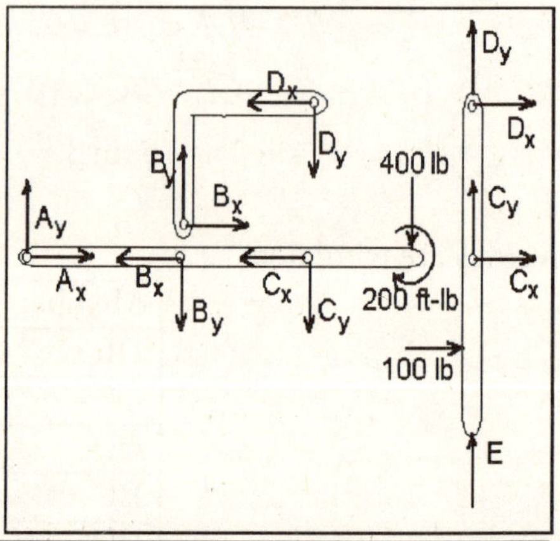

Element BD: The sum of the moments about B: $\sum M_B = 2D_x - 2D_y = 0$, from which (4) $D_x - D_y = 0$. The sum of the forces: $\sum F_x = B_x - D_x = 0$, from which (5) $B_x - D_x = 0$. $\sum F_y = B_y - D_y = 0$, from which (6) $B_y - D_y = 0$

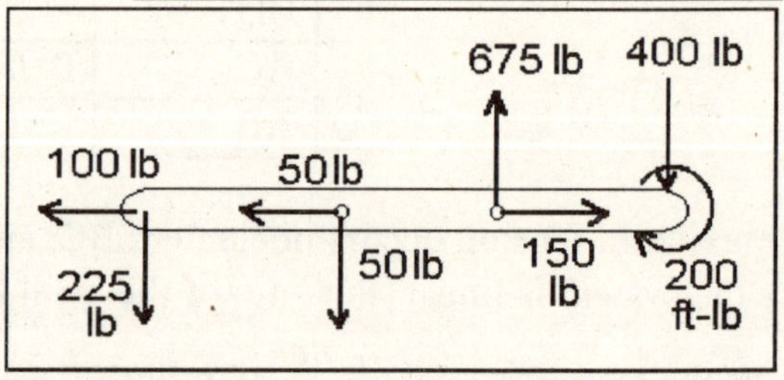

Element ABC: The sum of the moments about A: $\sum M_A = -2B_y - 4C_y - 200 - 6(400) = 0$, from which (7) $B_y + 2C_y = -1300$. The sum of the forces: $\sum F_x = A_x - B_x - C_x = 0$, from which (8) $A_x - B_x - C_x = 0$. $\sum F_y = A_y - B_y - C_y - 400 = 0$, from which (9) $A_y - B_y - C_y = 400$. These nine equations are solved for the nine reactions The reactions are tabulated:

Reaction	Value, lb
D_X	50
D_Y	50
C_X	-150
C_Y	-675
B_X	50
B_Y	50
A_X	-100 *Check.*
A_Y	-225 *Check.*
E	625 *Check.*

Observing the directions of the reactions on the free body diagram and the directions indicated by the tabulated values, the solution for the element ABC is shown as in Fig 6.35.

Problem 6.120 This structure supports a conveyor belt used in a lignite mining operation. The cables connected to the belt exert the force F at J. As a result of the counterweight $W = 8\ kip$, the reaction at E and the vertical reaction at D are equal. Determine F and the axial forces in members BG and EF.

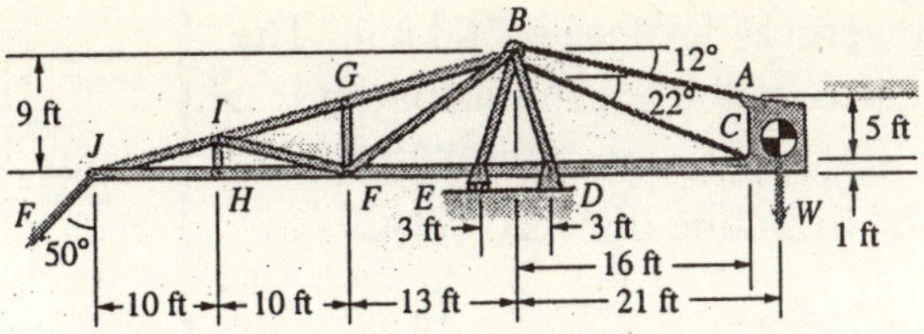

Solution: *The complete structure as a free body:* The moments about point J: using the fact that the vertical reactions at E and D are equal: $\sum M_J = -54W + 66D_y = 0$ from which $E = D_y = 6.545\ kip$. The sum of forces: $\sum F_y = -F\cos(50^o) + 2D_y - W = 0$, from which $\boxed{F = 7.920\ kip}$

The method of sections: Make a cut through BG, BF, and EF. Consider the section to the left of the cut.

The angle of member BG from the horizontal is $\alpha = \tan^{-1}\left(\frac{9}{33}\right) = 15.26^o$. The angle of member BF from the horizontal is $\beta = \tan^{-1}\left(\frac{9}{13}\right) = 34.7^o$. The length of member FG is $L_{FG} = 20\tan\alpha$.

The section as a free body: The sum of the moments about the point F:

$\sum M_F = 20F\cos(50^o) - L_{FG}BG\cos\alpha = 0$, from which $\boxed{BG = 19.35\ kip\ (T)}$

The sum of the forces: $\sum F_y = BF\sin\beta + BG\sin\alpha - F\cos(50^o) = 0$, from which $BF = 0$

$\sum F_y = EF + BF\cos\beta + BG\cos\alpha - F\sin(50^o) = 0$, from which $\boxed{EF = -12.6\ kip\ (C)}$

Problem 6.121 Consider the structure described in Problem 6.120. The counterweight $W = 8\ kip$ is pinned at D and is supported by cable ABC which passes over a pulley at A. What is the tension in the cable and what forces are exerted on the counterweight at D?

Solution: Make a cut through cable at AB, BC, and through CD near to the pinned joint at D. Consider the section to the right of the cut.

The section as a free body: The sum of the moments about the pinned joint at D is:

$\sum M_D = 13T\sin(22^o) + 1T\cos(22^o) + 6T\cos(12^o) + 13T\sin(12^o) - 18W = 0$, from which the cable tension is $\boxed{T = 10.022\ kip}$. The sum of forces:

$\sum F_y = T\sin(22^o) + T\sin(12^o) - W + D_y^{CD} = 0$, from which $\boxed{D_y^{CD} = 2.162\ kip}$

$\sum F_x = T\cos(22^o) + T\cos(12^o) + D_x^{CD} = 0$, from which $\boxed{D_x^{CD} = -19.095\ kip}$ This is the force exerted on the pinned joint at D by member CD.

Problem 6.122 The weights $W_1 = 4\ kN$ and $W_2 = 10\ kN$. Determine the forces on member ACDE at points A and E.

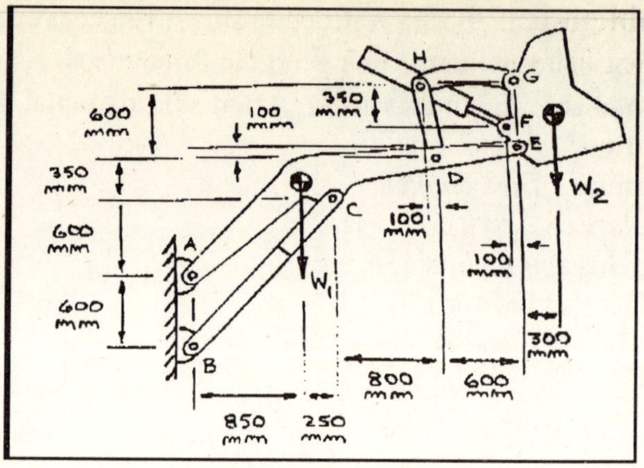

Solution: *The complete structure as a free body:* The horizontal distance from A to C is 1100 mm. The vertical distance is 600 mm. The horizontal distances to the action line of the weights W_1, W_2 are 850 mm and 2800 mm. The angle of the hydraulic lifter BC relative to the horizontal is $\alpha = \tan^{-1}\left(\frac{1200}{1100}\right) = 47.49^o$.

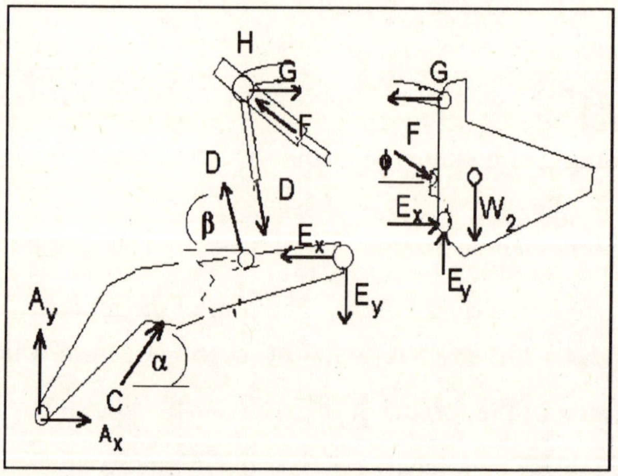

The sum of the moments about the point A is $\sum M_A = -850W_1 + 1100BC\sin\alpha - 2800W_2 = 0$ from which $BC = 77.45\ kN$. The sum of the forces: $\sum F_x = A_x + BC\cos\alpha = 0$, from which $A_x = -52.33\ kN$

$\sum F_y = A_y + BC\sin\alpha - W_1 - W_2 = 0$, from which $A_Y = -43.09\ kN$ These results are to be used as a check on the solution below.

The angle of member DH relative to the horizontal is $\beta = \tan^{-1}\left(\frac{600}{100}\right) = 80.54^o$. The angle of the force F relative to the horizontal is $\phi = \tan^{-1}\left(\frac{350}{600}\right) = 30.25^o$

The bucket as a free body: The sum of the moments about point E: $\sum M_E = 500G - 300W_2 - F(150\cos\phi - 100\sin\phi) = 0$. The sum of the forces on the bucket: $\sum F_x = -E_x + F\cos\phi - G = 0$, and $\sum F_y = -E_y - W_2 - F\sin\phi = 0$.

The linkage H as a free body: The sum of the forces: $\sum F_x = -F\cos\phi + G + D\cos\beta = 0$, and $\sum F_y = F\sin\phi - D\sin\beta = 0$.

The member ACDE as a free body: The sum of the moments about point A: $\sum M_A = -850W_1 - C(600\cos\alpha - 1100\sin\alpha) + D(1900\sin\beta + 950\cos\beta) + 2500E_y - 1050E_x = 0$. The sum of forces: $\sum F_x = A_x + C\cos\alpha - D\cos\beta + E_x = 0$, and $\sum F_y = A_y - W_1 + C\sin\alpha + D\sin\beta + E_y = 0$. These equilbrium conditions lead to 8 equations in 8 unknowns, which are solved by iteration using **TK Solver 2** The results: $\boxed{G = 7.53\ kN}$, $\boxed{D = 4.93\ kN}$, $\boxed{F = 9.65\ kN}$, $\boxed{E_x = 0.811\ kN}$, $\boxed{E_y = -14.9\ kN}$, $\boxed{A_x = -52.3\ kN}$, $\boxed{A_y = -43.1\ kN}$, and $\boxed{C = 77.4\ kN}$: The last three values check the results obtained from the solution for the complete structure as a free body.

Problem 7.1 Determine the x coordinate of the centroid of the area.

Strategy: The x-coordinate of the centroid is given by Eq. (7.6). You can use a differential element of area dA in the form of a vertical strip of width dx.

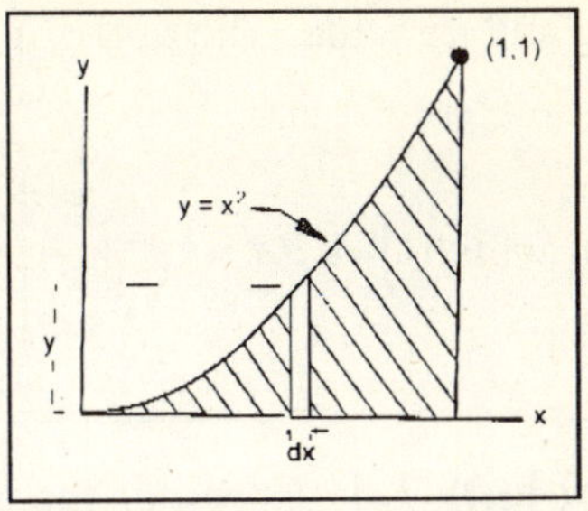

Solution: From Eq. (7.6) the x coordinate of the centroid is

$\bar{x} = \dfrac{\int_A x\,dA}{\int_A dA} = \dfrac{\int_A x\,dA}{A}$. The element of area is the vertical strip of height y and width dx. Thus $A = \int_0^1 y\,dx = \int_0^1 x^2 dx = \left[\dfrac{x^3}{3}\right]_0^1 = \dfrac{1}{3}$, and

$\bar{x}A = \int_A x\,dA = \int_0^1 x^3 dx = \left[\dfrac{x^4}{4}\right]_0^1 = \dfrac{1}{4}$, from which $\bar{x} = \dfrac{\int_A x\,dA}{A} = \boxed{\dfrac{3}{4}}$

Problem 7.2 Determine the y-coordinate of the centroid of the area in Problem 7.1.

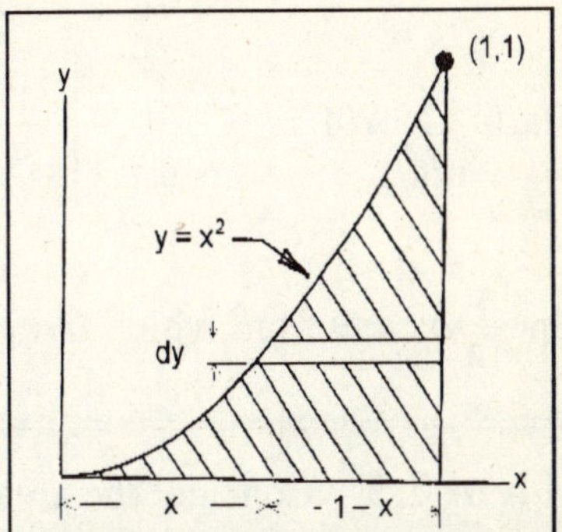

Solution: From Eq (7.7),

$\bar{y} = \dfrac{\int_A y\,dA}{\int_A dA} = \dfrac{\int_A y\,dA}{A}$. The element of area is the horizontal strip $(1-x)$ long and dy wide. Thus $A = \int_A dA = \int_0^1 (1-x)dy$

$= \int_0^1 (1 - y^{\frac{1}{2}})dx = \left[y - \dfrac{2y^{\frac{3}{2}}}{3}\right]_0^1 = \dfrac{1}{3}$, and

$\bar{y}A = \int_A y\,dA = \int_0^1 y\,(1-x)dy = \int_0^1 \left(y - y^{\frac{3}{2}}\right)dy = \left[\dfrac{y^2}{2} - \dfrac{2y^{\frac{5}{2}}}{5}\right]_0^1 = \dfrac{1}{10}$, from which $\boxed{\bar{y} = \dfrac{\int_A y\,dA}{A} = \dfrac{3}{10}}$

Problem 7.3 If $a = 2$, what are the coordinates of the centroid of the area?

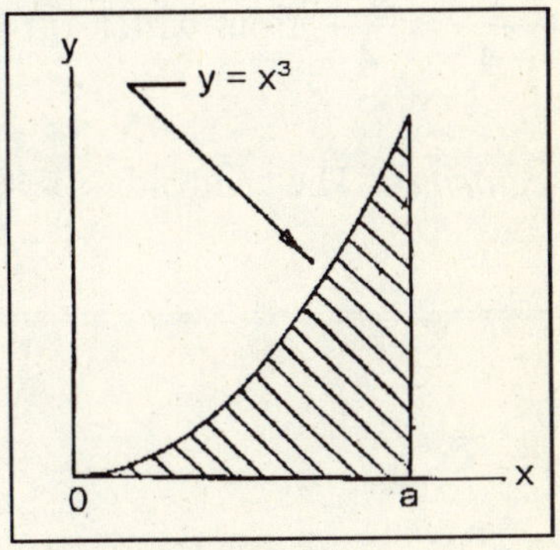

Solution: : *The area:* Take the element of area as the vertical strip dx wide. Then, $A = \int_0^2 y\,dx = \int_0^2 x^3 dx$, from which

$A = \left[\dfrac{x^4}{4}\right]_0^2 = \dfrac{2^4}{4} = 4$. *The x-coordinate:*

$\int_A x\,dA = \int_0^2 x\,y\,dx = \int_0^2 x^4 dx = \left[\dfrac{x^5}{5}\right]_0^2 = \dfrac{32}{5}$. Divide by the area:

$\boxed{\bar{x} = \dfrac{8}{5} = 1.6}$

Problem 7.3 continued on next page....

===================================<>===================================

continuation of solution to Problem 7.3...

The y-coordinate: . Take the element of area to be the product of the horizontal strip $(a-x)=(2-x)$ long and dy wide. The upper limit on the integration is $y=a^3=8$. $\bar{y}A=\int_A y\,dA=\int_0^8 y\left(2-y^{\frac{1}{3}}\right)dy$,

from which $\int_A y\,dA=\left[y^2\right]_0^8-\left[\frac{3y^{\frac{7}{3}}}{7}\right]_0^8=64-\frac{3(128)}{7}=9.1429$. $\boxed{\bar{y}=\frac{9.1429}{4}=2.2857}$

===================================<>===================================

Problem 7.4 Consider the area shown in Problem 7.3. If the y-coordinate of the centroid is $\bar{y}=5$, what is $\bar{x}$?

Solution: The strategy is to find the value of a from the value of the y-coordinate of the centroid, and then find the x-coordinate of the centroid. From the solution to Problem 7.3, $A=\int_0^a y\,dx=\left[\frac{x^4}{4}\right]_0^a=\frac{a^4}{4}$.

Similarly, $\int_A y\,dA=\left[\frac{ay^2}{2}\right]_0^{a^3}-\left[\frac{3y^{\frac{7}{3}}}{7}\right]_0^{a^3}=\frac{a^7}{2}-\frac{3a^7}{7}=\frac{1}{14}a^7$. Divide by the area and equate to $\bar{y}$ to obtain $\frac{4}{14}a^3=5$ from which $a=\left(\frac{70}{4}\right)^{\frac{1}{3}}=2.5962$. From the solution to Problem 7.3, $\boxed{\bar{x}=\frac{4}{5}a=2.0769}$

===================================<>===================================

Problem 7.5 Consider the area shown in Problem 7.3. The "center of the area" of an area is defined to be the point for which there is as much area to the right of the point as to the left of it, and as much area above the point as below it. If $a=4$, what are the x-coordinate of the center of area and the x coordinate of the centroid?

Solution: *Center of Area:* Let X be the coordinate of the center of the area. The area to the left of X is $A_L=\int_0^X x^3dx=\frac{X^4}{4}$. The area to the right is $A_R=\int_X^a x^3dx=\frac{a^4}{4}-\frac{X^4}{4}$. Equating the two areas, $\frac{a^4}{4}-\frac{X^4}{4}=\frac{X^4}{4}$, from which $a^4=2X^4$ or $X=a(2)^{-\frac{1}{4}}$. . For $a=4$, $\boxed{X=0.8408(4)=3.3636}$

The centroid: The centroid is $\bar{x}=\frac{\int_0^a x^4dx}{\int_0^a x^3dx}=\frac{4a}{5}$. For $a=4$, $\boxed{\bar{x}=\frac{16}{5}=3.200}$

===================================<>===================================

Problem 7.6

Problem 7.6 Suppose that an art student wants to paint a panel of wood as shown, with the horizontal and vertical lines passing through the centroid of the painted area, and asks you to determine the coordinates of the centroid. What are they?

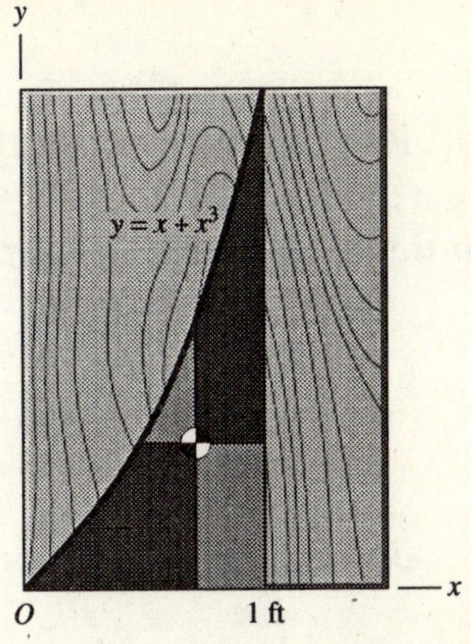

Solution: : *The area:* $A = \int_0^1 (x + x^3)dx = \left[\frac{x^2}{2} + \frac{x^4}{4}\right]_0^1 = \frac{3}{4}$.

The x-coordinate: $\int_0^1 x(x + x^3)dx = \left[\frac{x^3}{3} + \frac{x^5}{5}\right]_0^1 = \frac{8}{15}$. Divide

by the area: $\boxed{\bar{x} = \frac{32}{45} = 0.711}$

The y-coordinate: The element of area is $dA = (1 - x)dy$.

Note that $dy = (1 + 3x^2)dx$, hence $dA = (1 - x)(1 + 3x^2)dx$. Thus

$\bar{y}A = \int_A y\, dA = \int_0^4 (x + x^3)(1 - x)(1 + 3x^2)dx$, from which

$\int_0^1 (x - x^2 + 4x^3 - 4x^4 + 3x^5 - 3x^6)dx = \frac{1}{2} - \frac{1}{3} + \frac{4}{4} - \frac{4}{5} + \frac{3}{6} - \frac{3}{7} = 0.4381$ Divide by A $\boxed{\bar{y} = 0.5841}$

Problem 7.7

Problem 7.7 Determine the centroid of the area.

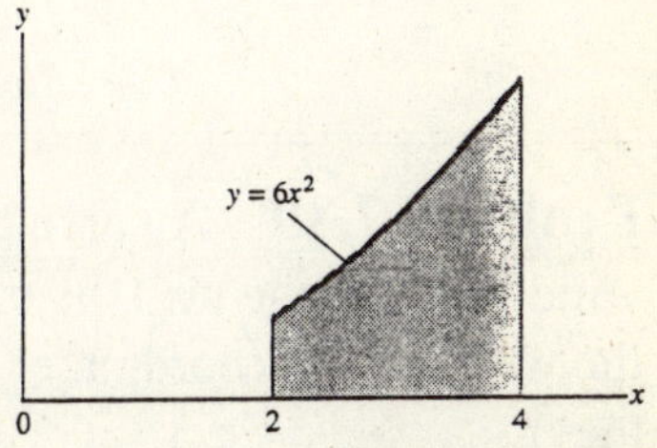

Solution: *The area:* $\int_2^4 6x^2 dx = [2x^3]_2^4 = 2(4^3 - 2^3) = 112$.

The x-coordinate: $\int_2^4 6x^3 dx = \left[\frac{3x^4}{2}\right]_2^4 = \frac{3}{2}(4^4 - 2^4) = 360$. Divide by the

area: $\boxed{\bar{x} = 3.2143}$ *The y-coordinate:* Divide the interval into two parts: $0 \le y \le 24$, and $24 \le y \le 96$. The elemental areas of the horizontal strips are $dA = 2dy$, and $dA = (4 - x)dy$ in the two intervals.

$\bar{y}A = \int_A y dA = \int_0^{24} 2y dy + \int_{24}^{96} y\left(4 - \sqrt{\frac{y}{6}}\right)dy = 24^2 + 2(96^2 - 24^2) - \frac{2}{5\sqrt{6}}\left(96^{\frac{5}{2}} - 24^{\frac{5}{2}}\right) = 3571.2$. Divide by

the area: $\boxed{\bar{y} = 31.8857}$

Problem 7.8

Problem 7.8 Determine the centroid of the area by letting *dA* be a vertical strip of width dx.

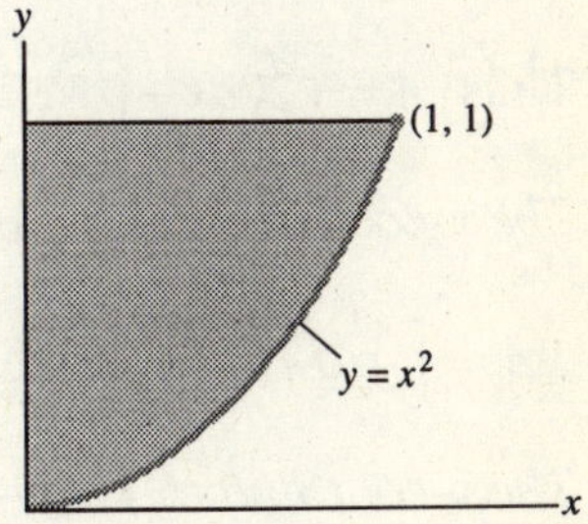

Solution: *The area:* The length of the vertical strip is $(1 - y)$, so that the elemental area is $dA = (1 - y)dx = (1 - x^2)dx$. The area:

$\int_A dA = \int_0^1 (1 - x^2)dx = \left[x - \frac{x^3}{3}\right]_0^1 = 1 - \frac{1}{3} = \frac{2}{3}$.

The x-coordinate: $\bar{x}A = \int_A x\, dA = \int_0^1 x(1 - x^2)dx = \left[\frac{x^2}{2} - \frac{x^4}{4}\right]_0^1 = \frac{1}{4}$: $\boxed{\bar{x} = \frac{3}{8}}$

The y-coordinate: The y-coodinate of the centroid of each element of area is located at the midpoint of the vertical dimension of the area element.

Problem 7.8 continued on next page....

continuation of solution to Problem 7.8...

$\bar{y} = y + \frac{1}{2}(1 - x^2)$. Thus $\int_A \bar{y} dA = \int_0 \left(x^2 + \left(\frac{1}{2}\right)(1 - x^2)(1 - x^2) \right) dx = \left(\frac{1}{2}\right)\left[x - \frac{x^5}{5} \right]_0^1 = \frac{2}{5}$. $\boxed{\bar{y} = \frac{3}{5}}$

Problem 7.9 Determine the centroid of the area in Problem 7.8 by letting dA be a horizontal strip of height dy.

Solution: *The area:* The length of the horizontal strip is x, hence the element of area is

$dA = x\,dy = y^{\frac{1}{2}}\,dy$. Thus $A = \int_0^1 y^{\frac{1}{2}}\,dy = \left[\frac{2y^{\frac{3}{2}}}{3} \right]_0^1 = \frac{2}{3}$ *Check:*

The x-coordinate: The x-coordinate of the centroid of each element of area is $\bar{x} = \frac{1}{2}x = \frac{1}{2}y^{\frac{1}{2}}$. Thus

$\int_A \left(\frac{1}{2}\right) y^{\frac{1}{2}}\,dA = \left(\frac{1}{2}\right)\int_0^1 y\,dy = \left(\frac{1}{2}\right)\left[\frac{y^2}{2}\right]_0^1 = \frac{1}{4}$. Divide by the area: $\boxed{\bar{x} = \frac{3}{8}}$

The y-coordinate: $\bar{y}A = \int_A y\,dA = \int_0^1 y\left(y^{\frac{1}{2}}dy\right) = \int_0^1 y^{\frac{3}{2}}\,dy = \left[\frac{2y^{\frac{5}{2}}}{5}\right]_0^1 = \frac{2}{5}$. Divide by the area: $\boxed{\bar{y} = \frac{3}{5}}$

Problem 7.10 An architect wants to build a wall with the profile shown. To estimate the effects of wind loads on the wall, he must determine the area of the wall and the coordinates of its centroid. What are they?

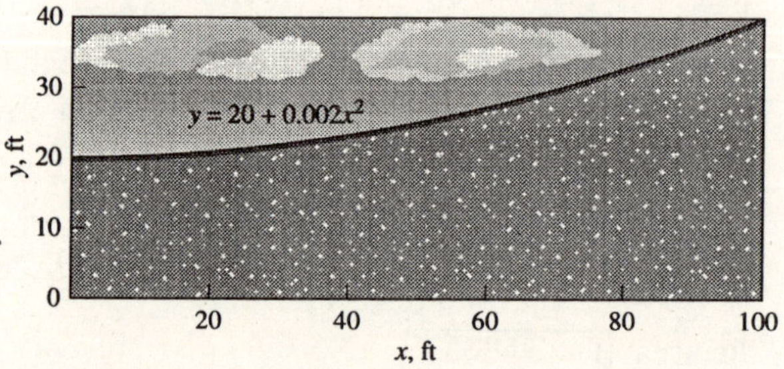

Solution: *The area:* For the element of area, take a vertical strip of width dx. Divide the integration interval into two parts: the first with a length of strip of 20 *ft*, and the second with a length of the strip $(y - 20)$. The element of area in the first part is $dA = 20dx$, with limits of integration of (0, 100) *ft*. The element of area in the second part is $dA = (y - 20)dx = (20 + 2\times10^{-3}x^2 - 20)dx = 2\times10^{-3}x^2 dx$. The limits of integration are (0, 100) *ft*.

Thus $A = \int_0^{100} 20dx + \int_0^{100}(2\times10^{-3}x^2)dx = \left[20x + 2\times10^{-3}\frac{x^3}{3}\right]_0^{100} = 2\frac{2}{3}\times10^3$ *sq ft* $\boxed{= 2666.7 \text{ sq ft}}$

The x-coordinate: Use the same elemental strips as above.

$\bar{x}A = \int_A x\,dA = \int_0^{100} 20xdx + \int_0^{100} 2\times10^{-3}x^3dx = \left[10x^2 + 1\times10^{-3}\frac{x^4}{2}\right]_o^{100} = 1.5\times10^5$ $\boxed{\bar{x} = 56.25 \text{ ft}}$

The y-coordinate: The y-coordinate of the centroid of each area element is $\bar{y} = 10$ for the first interval, and $y = 20 + \left(\frac{1}{2}\right)(y - 20) = 20 + \frac{y}{2} = 10 + 2\times10^{-3}\frac{x^2}{2}$ for the second interval.

$\int_0^{100} 200\,dx + (2\times10^{-3})\int_0^{100}(20x^2 + 2\times10^{-3}\frac{x^4}{2})dx = (2 + \frac{4}{3} + 0.4)\times10^4 = 37.3333\times10^3$. $\boxed{\bar{y} = 14 \text{ ft}}$

Problem 7.11 Determine the y-coordinate of the centroid of the area.

Solution: The x-intercepts are the roots of the quadratic equation $-x^2 + 12x - 27 = 0$, $x_{1,2} = 6 \pm \sqrt{36 - 27}$, from which $x_1 = 3$, and $x_2 = 9$. These roots establish the limits of integration.

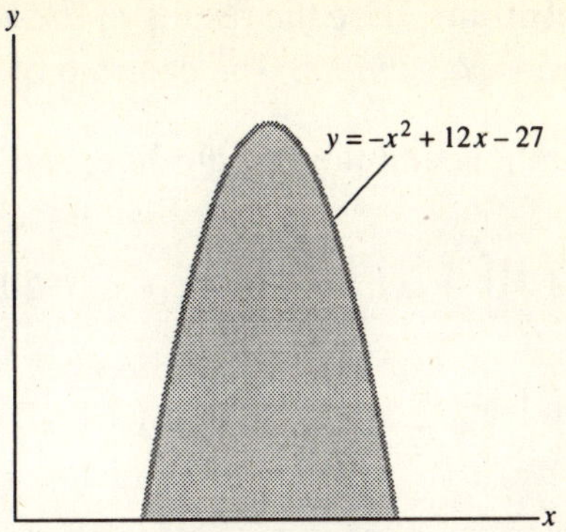

The area: Take an element of area as the vertical strip dx wide. The length of the strip is $y = -x^2 + 12x - 27$. Thus the element of area is $dA = (-x^2 + 12x - 27)dx$, and the area is

$$A = \int_3^9 \left(-x^2 + 12x - 27\right)dx = \left[-\frac{x^3}{3} + 12\frac{x^2}{2} - 27x\right]_3^9 = 36$$

The x-coordinate: $\bar{x}A = \int_A x\,dA = \int_3^9 (-x^3 + 12x^2 - 27x)dx = \left[-\frac{x^4}{4} + 4x^3 - \frac{27}{2}x^2\right]_3^9 = 216$.

$$\boxed{\bar{x} = \frac{216}{36} = 6}$$

Problem 7.12 Determine the y-coordinate of the centroid of the area in Problem 7.11.

Solution: Use the results of the solution to Problem 7.11 in the following.

The y-coordinate: The y-coordinate of the centroid of the elemental area is at the midpoint of the length, $\bar{y} = \left(\frac{1}{2}\right)\left(-x^2 + 12x - 27\right)$, from which $\int_A y\,dA = \int_3^9 \frac{1}{2}\left(-x^2 + 12x - 27\right)^2 dx$

$$= \int_3^9 \left(\frac{1}{2}\right)\left(x^4 - 24x^3 + 198x^2 - 648x + 729\right)dx = \left(\frac{1}{2}\right)\left[\frac{x^5}{5} - 6x^4 + \frac{198}{3}x^3 - 324x^2 + 729x\right]_3^9 = 129.6.$$

Divide by the area: $\boxed{\bar{y} = \frac{129.6}{36} = 3.6}$

Problem 7.13 Determine the x-coordinate of the centroid of the area.

Solution: The intercept of the straight line with the parabola occurs at the roots of the simultaneous equations: $y = x$, and $y = x^2 - 20$. This is equivalent to the solution of the quadratic $x^2 - x - 20 = 0$, $x_1 = -4$, and $x_2 = 5$. These establish the limits on the integration. *The area:* Choose a vertical strip dx wide. The length of the strip is $(x - x^2 + 20)$, which is the distance between the straight line $y = x$ and the parabola $y = x^2 - 20$. Thus the element of area is $dA = (x - x^2 + 20)dx$ and

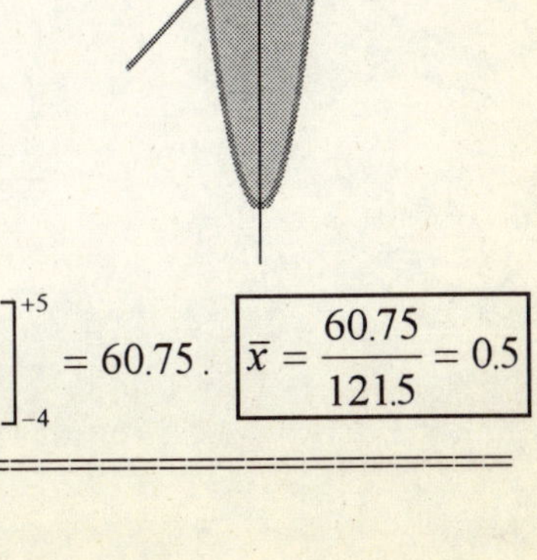

$$A = \int_{-4}^{+5} (x - x^2 + 20)dx = \left[\frac{x^2}{2} - \frac{x^3}{3} + 20x\right]_{-4}^{+5} = 121.5$$

The x-coordinate: $\bar{x}A = \int_A x\,dA = \int_{-4}^{+5} (x^2 - x^3 + 20x)dx = \left[\frac{x^3}{3} - \frac{x^4}{4} + 10x^2\right]_{-4}^{+5} = 60.75$. $\boxed{\bar{x} = \frac{60.75}{121.5} = 0.5}$

Problem 7.14 Determine the y-coordinate of the centroid of the area in Problem 7.13.

Solution: Use the results of the solution to Problem 7.13 in the following.

The y-coordinate: The centroid of the area element occurs at the midpoint of the strip enclosed by the parabola and the straight line, and the y-coordinate is: $\bar{y} = x - \left(\frac{1}{2}\right)(x - x^2 + 20) = \left(\frac{1}{2}\right)(x + x^2 - 20)$ ..

$$\bar{y}A = \int_A y\, dA = \left(\frac{1}{2}\right)\int_{-4}^{5} (x + x^2 - 20)\,(x - x^2 + 20)\, dx = \left(\frac{1}{2}\right)\int_{-4}^{+5} (-x^4 + 41x^2 - 400)dx$$

$$= \left(\frac{1}{2}\right)\left[-\frac{x^5}{5} + \frac{41x^3}{3} - 400x\right]_{-4}^{5} = -923.4\ .\ \boxed{\bar{y} = -\frac{923.4}{121.5} = -7.6}$$

Problem 7.15 Determine the y-coordinate of the centroid of the area.

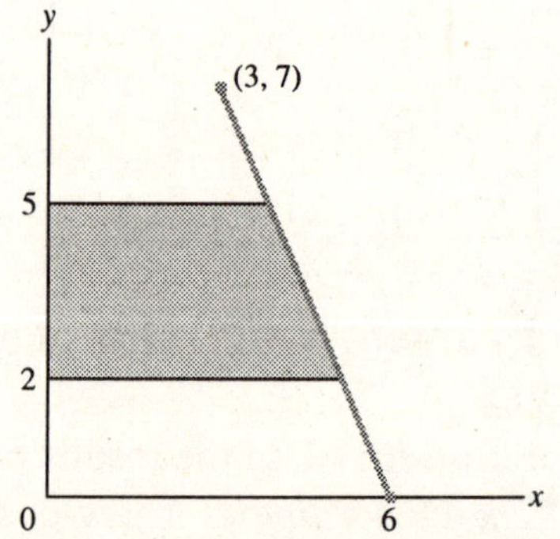

Solution: *The area:* The area element is the horizontal strip x-long and dy wide. The length is determined by the straight line, which has the equation $y = mx + b$, where $m = \frac{y_1 - y_2}{x_1 - x_2} = \frac{(7-0)}{(3-6)} = -2.3333$, and $b = y_1 - mx_1 = 7 - (-2.3333)3 = 14$. The length of the strip is $x = \left(\frac{1}{m}\right)(y - b)$. The element of area is $dA = \left(\frac{1}{m}\right)(y - b)dy$, from which

$$A = \left(\frac{1}{m}\right)\int_2^5 (y - b)dy = \left(\frac{1}{m}\right)\left[\frac{y^2}{2} - by\right]_2^5 = 13.5.$$

The y-coordinate: $\bar{y}A = \int_A y\, dA = \left(\frac{1}{m}\right)\int_2^5 y(y - b)dy = \left(\frac{1}{m}\right)\left[\frac{y^3}{3} - b\frac{y^2}{2}\right]_2^5 = 46.2857$. $\boxed{\bar{y} = 3.4286}$

Problem 7.16 Determine the x-coordinate of the centroid of the area in Problem 7.15.

Solution: From the solution to Problem 7.15, the area $A = 13.5$ is bounded by the straight line $y = mx + b$, where $m = -2.3333$, and $b = 14$. The horizontal strip is x-wide and dy high, where $x = \left(\frac{1}{m}\right)(y - b)$. The centroid of x is one half this strip, hence

$$\bar{x}A = \left(\frac{1}{2m^2}\right)\int (y - b)^2 dy = \left(\frac{1}{6m^2}\right)\left[(y - b)^3\right]_2^5 = 30.58, \text{ from which } \boxed{\bar{x} = \frac{30.58}{13.5} = 2.265}$$

Problem 7.17 An agronomist wants to measure the rainfall at the centroid of a plowed field between two roads. What are the coordinates of the point where the rain gauge should be placed?

Solution: *The area:* The element of area is the vertical strip $(y_t - y_b)$ long and dx wide, where $y_t = m_t x + b_t$ and $y_b = m_b x + b_b$ are the two straight lines bounding the area. , where

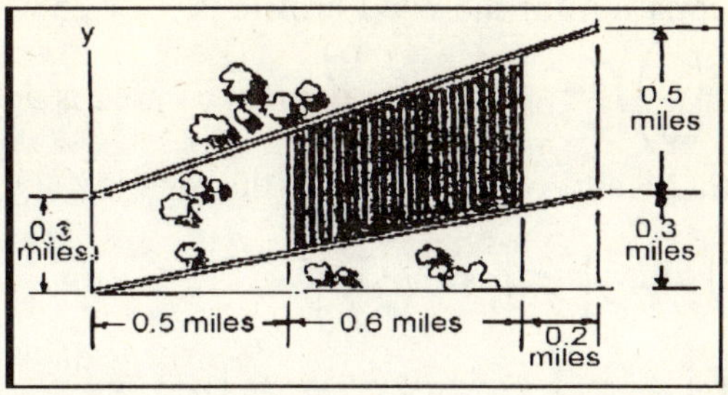

$$m_t = \frac{(0.8-0.3)}{(1.3-0)} = 0.3846, \text{ and } b_t = 0.8 - 1.3m_t = 0.3.$$

$$\text{Similarly}: m_b = \frac{(0.3-0)}{(1.3-0)} = 0.2308, \text{ and } b_b = 0$$

The element of area is

$dA = (y_t - y_b)\,dx = ((m_t - m_b)x + b_t - b_b)\,dx = (0.1538x + 0.3)dx$, from which

$$A = \int_{0.5}^{1.1} (0.1538x + 0.3)dx = \left[0.1538\frac{x^2}{2} + 0.3x\right]_{0.5}^{1.1} = 0.2538 \; sq\; mile$$

The x-coordinate: $\int_A x\,dA = \int_{0.5}^{1.1} (0.1538x + 0.3)x\,dx = \left[0.1538\frac{x^3}{3} + 0.3\frac{x^2}{2}\right]_{0.5}^{1.1} = 0.2058$. $\boxed{\bar{x} = 0.8109\; mi}$

The y-coordinate: The y-coordinate of the centroid of the elemental area is

$$\bar{y} = y_b + \left(\frac{1}{2}\right)(y_t - y_b) = \left(\frac{1}{2}\right)(y_t + y_b) = 0.3077x + 0.15. \text{ Thus,}$$

$$\bar{y}A = \int_A y\,dA = \int_{0.5}^{1.1} (0.3077x + 0.15)(0.1538x + 0.3)\,dx = \int_{0.5}^{1.1} (0.0473x^2 + 0.1154x + 0.045)dx$$

$$= \left[0.0471\frac{x^3}{3} + 0.1153\frac{x^2}{2} + 0.045x\right]_{0.5}^{1.1} = 0.1014. \text{ Divide by the area: } \boxed{\bar{y} = \frac{0.1014}{0.2538} = 0.3995\; mi}$$

Problem 7.18 The area shown is the cross section of an earth-fill dam. Determine the coefficients a and b so that the y-coordinate of the centroid of the cross section is 10 m.

Solution: *The area:* The elemental area is a vertical strip of length y and width dx, where $y = ax - bx^3$. Note that $y = 0$ at $x = 100$, thus $b = a \times 10^{-4}$.

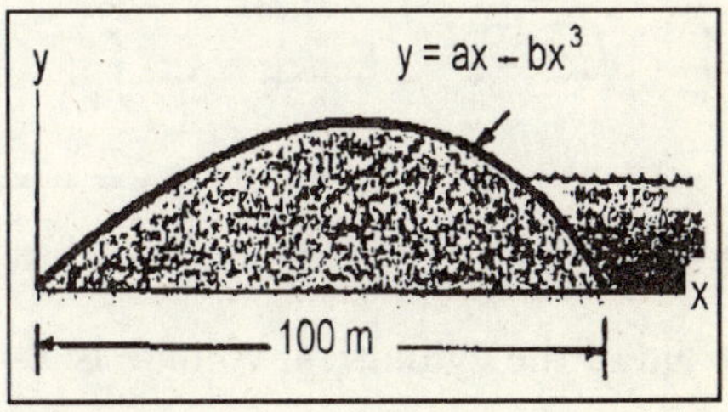

Thus $A = \int_A dA = a\int_0^{100} (x - (10^{-4})x^3)dx$

$= (0.5a)\left[x^2 - (0.5 \times 10^{-4})x^4\right]_0^{100} = 0.5a \times 10^4 - 0.25b \times 10^8$, and

the area is $A = 0.25a \times 10^4$.

The y-coordinate: The y-coordinate of the centroid of the elemental area is $\bar{y} = (0.5)(ax - bx^3) = (0.5a)(x - (10^{-4})x^3)$, from which

$$\bar{y}A = \int_A y\,dA = (0.5)a^2 \int_0^{100} (x - (10^{-4})x^3)^2 dx = (0.5)a^2 \int_0^{100} (x^2 - 2(10^{-4})x^4 + (10^{-8})x^6)\,dx$$

$$= (0.5a^2)\left[\frac{x^3}{3} - (10^{-4})\frac{2x^5}{5} + (10^{-8})\frac{x^7}{7}\right]_0^{100} = 3.81a^2 \times 10^4. \text{ Divide by the area:}$$

$$\bar{y} = \frac{3.810a^2 \times 10^4}{0.25a \times 10^4} = 15.2381a. \text{ For } \bar{y} = 10, \boxed{a = 0.6562}, \text{ and } \boxed{b = 6.562 \times 10^{-5}\; m^{-2}}$$

Problem 7.19 The airplane wing has an elliptical profile. Determine its centroid.

Solution: *The area:* When $x^2 = 0$, $y^2 = 1$, therefore $b^2 = 1$, $b = \pm 1$. When $x = 5$, $y = 0$, hence $a^2 = 25$, $a = \pm 5$.

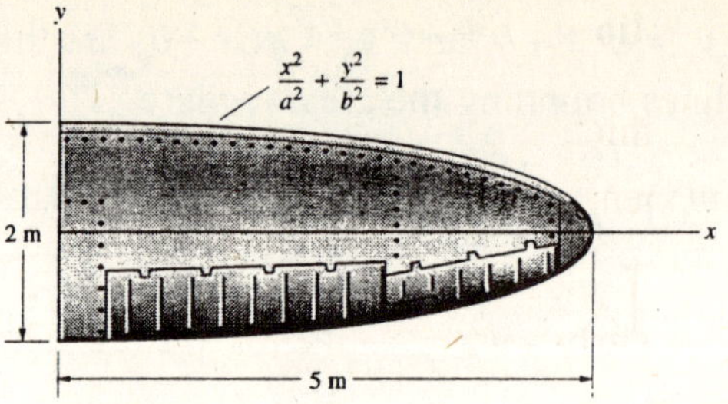

The equation for the wing profile is $x^2 + 25y^2 = 25$ or $y = \pm\left(\frac{1}{5}\right)\sqrt{25 - x^2}$. The element of area is a vertical strip of length $2y$ and width dx. Thus the area is

$$A = \int_A dA = \left(\frac{2}{5}\right)\int_0^5 \sqrt{5^2 - x^2}\,dx$$

$$= \left(\frac{2}{5}\right)\left[\frac{x\sqrt{5^2 - x^2}}{2} + \frac{5^2}{2}\sin^{-1}\left(\frac{x}{5}\right)\right]_0^5 = \frac{5}{2}\pi$$ where the principal value of the arcsine has been taken.

The x-coordinate: $\bar{x}A = \int_A x\,dA = \left(\frac{2}{5}\right)\int_0^5 x\sqrt{5^2 - x^2}\,dx = \left(\frac{2}{5}\right)\left[-\frac{(5^2 - x^2)^{\frac{3}{2}}}{3}\right]_0^5 = \frac{50}{3}$. $\boxed{\bar{x} = 2.122\ m}$ *The y-coordinate:* By symmetry, the y-coordinate of the centroid is $\boxed{\bar{y} = 0}$

Problem 7.20 Determine the centroid of the area.

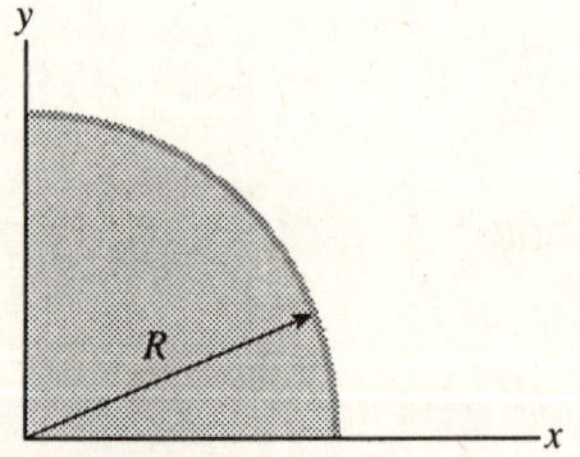

Solution: *The area:* The equation of the circle is $x^2 + y^2 = R^2$. Take the elemental area to be a vertical strip of height $y = \sqrt{R^2 - x^2}$ and width dx. , hence the element of area is $dA = \sqrt{R^2 - x^2}\,dx$. The area is $A = \frac{A_{circle}}{4} = \frac{\pi R^2}{4}$

The x-coordinate: $\bar{x}A = \int_A x\,dA = \int_0^R x\sqrt{R^2 - x^2}\,dx = \left[-\frac{(R^2 - x^2)^{\frac{3}{2}}}{3}\right]_0^R = \frac{R^3}{3}$: $\boxed{\bar{x} = \frac{4R}{3\pi}}$

The y-coordinate: The y-coordinate of the centroid of the element of area is at the midpoint:

$\bar{y} = \left(\frac{1}{2}\right)\sqrt{R^2 - x^2}$, hence $\bar{y}A = \int_A y\,dA = \left(\frac{1}{2}\right)\int_0^R (R^2 - x^2)\,dx = \left(\frac{1}{2}\right)\left[R^2 x - \frac{x^3}{3}\right]_0^R = \frac{R^3}{3}$ $\boxed{\bar{y} = \frac{4R}{3\pi}}$

Problem 7.21 Use the method described in Example 7.3 to confirm that the x-component of the centroid of the cylindrical volume is $\bar{x} = \frac{1}{2}L$

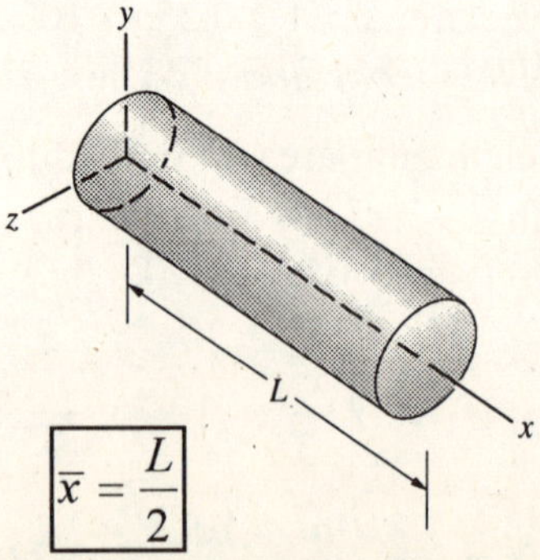

Solution: *The Volume:* T. Choose an element of volume to be the disk of radius R and thickness dx. The area of the disk is πR^2, and the element of volume is $dV = \pi R^2 dx$. Thus $V = \int_V dV = \int_0^L \pi R^2 dx = \pi R^2 L$.*The x-coordinate:* The centroid is defined by the equation $\bar{x}V = \int_V x\,dV = \pi R^2 \int_0^L x\,dx = \pi R^2\left[\frac{x^2}{2}\right]_0^L = \frac{\pi R^2 L^2}{2}$ $\boxed{\bar{x} = \frac{L}{2}}$

Problem 7.22 A volume of revolution is obtained by rotating the curve $y = x^{\frac{1}{2}}$ about the x axis. Determine its centroid.

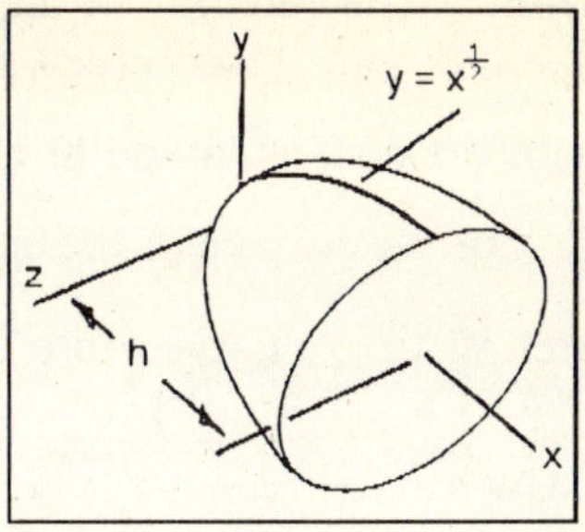

Solution: *The volume:* The element of volume is a disk of radius R and thickness dx. The radius is $R = x^{\frac{1}{2}}$, and the area of the disk element is $\pi R^2 = \pi x$. The volume element $dV = \pi x\, dx$, from which $V = \int_V dV = \pi \int_0^h x\, dx = \frac{\pi h^2}{2}$.

The x-coordinate: $\bar{x}V = \int_V x\, dV = \pi \int_0^h x^2 dx = \frac{\pi h^3}{3}$. $\boxed{\bar{x} = \frac{2}{3}h}$

The y-coordinate: By symmetry, the y-coordinate is zero. $\boxed{\bar{y} = 0}$

The z-coordinate: The argument applied to the y-coordinate also applies to the z-coordinate, since the figure is a volume of revolution about the axis normal to both z- and y- axes. $\boxed{\bar{z} = 0}$

Problem 7.23 A volume of revolution is obtained by revolving the curve $y = \frac{1}{4}x^2$ about the x-axis. Its volume is 10 m^3. Determine its centroid.

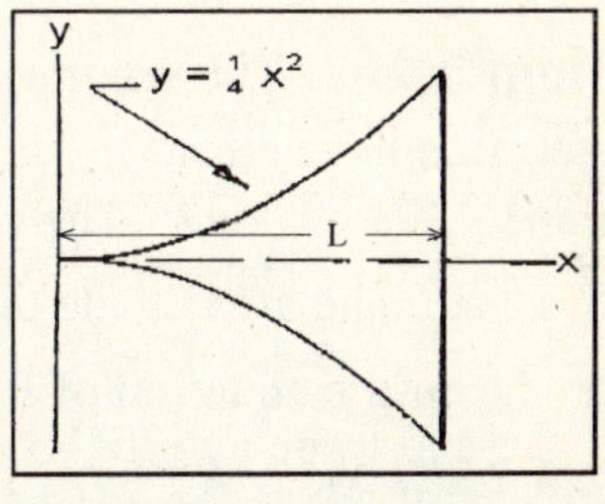

Solution: *The volume:* An element of volume is the disk of radius R and thickness dx. The radius is $R = \frac{1}{4}x^2$. The element of volume is $dV = \pi R^2 dx = \frac{\pi}{16}x^4 dx$, from which $V = \int_V dV = \frac{\pi}{16}\int_0^L x^4 dx = \frac{\pi}{80}\left[x^5\right]_0^L = \frac{\pi L^5}{80}$. The volume is given as 10 m^3, thus $L = \left(\frac{800}{\pi}\right)^{\frac{1}{5}} = 3.0282\ m$.

The x-coordinate: $\bar{x}V = \int_V x\, dV = \frac{\pi}{16}\int_0^L x^5 dx = \frac{\pi}{96}\left[x^6\right]_0^L = \frac{\pi L^6}{96}$. $\boxed{\bar{x} = \frac{5}{6}L = 2.5235\ m}$

The y- and z- coordinates: By symmetry, these coordinates are zero.

Problem 7.24 Determine the volume and centroid of the pyramid.

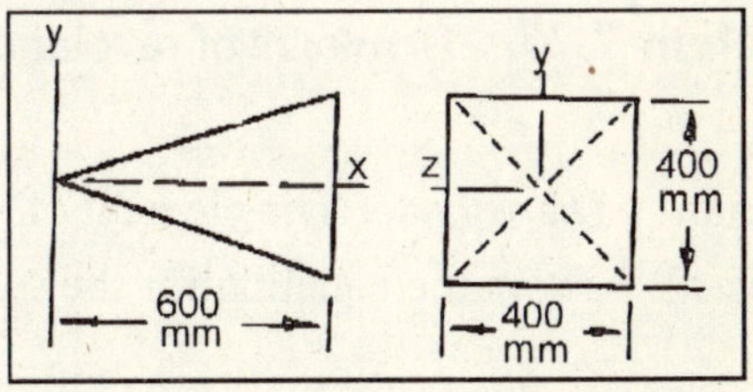

Solution: *The volume:* The element of volume is a square of thickness dx. The length of a side is a linear function of the height of the pyramid. Thus $L = Ax + B$. For $x = 0$, $L = 0$, and therefore $B = 0$. For $x = 600$, $L = 400$, therefore $A = \frac{2}{3}$. The area is $L^2 = \frac{4}{9}x^2$. The volume element is $dV = \frac{4}{9}x^2 dx$, from which $V = \int_V dV = \left(\frac{4}{9}\right)\int_0^{600} x^2 dx = \left(\frac{4}{27}\right)\left[x^3\right]_0^{600} = 3.2 \times 10^7\ mm^3$ $\boxed{= 0.032\ m^3}$

The x-coordinate: $\int_V x\, dV = \left(\frac{4}{9}\right)\int_0^{600} x^3 dx = \left(\frac{1}{9}\right)\left[x^4\right]_0^{600} = 1.44 \times 10^{10}$ .Divide by V: $\boxed{\bar{x} = 0.45\ m}$

By symmetry, the y- and z- coordinates of the centroid are zero.

Problem 7.25 Determine the centroid of the hemispherical volume.

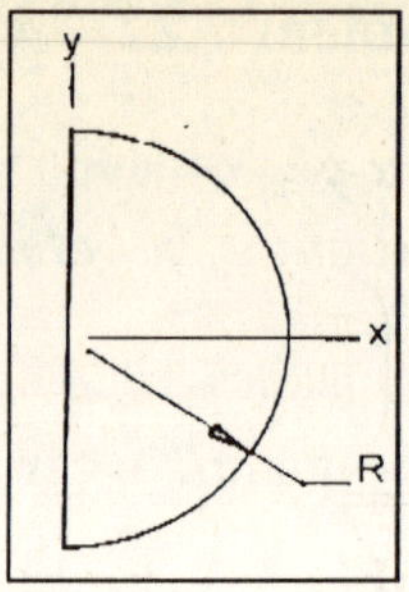

Solution: The equation of the surface of a sphere is $x^2 + y^2 + z^2 = R^2$.
The volume: The element of volume is a disk of radius ρ and thickness *dx*. The radius of the disk at any point within the hemisphere is $\rho^2 = y^2 + z^2$. From the equation of the surface of the sphere, $\rho^2 = (R^2 - x^2)$.

The area is $\pi\rho^2$, and therefore the element of volume is $dV = \pi(R^2 - x^2)dx$,

from which $V = \dfrac{V_{sphere}}{2} = \dfrac{2\pi}{3}R^3$. *The x-coordinate:*

$$\int_V x\,dV = \pi\int_0^R (R^2 - x^2)x\,dx = \pi\left[\frac{R^2x^2}{2} - \frac{x^4}{4}\right]_0^R = \frac{\pi}{4}R^4.$$ Divide by the volume:

$$\boxed{\bar{x} = \left(\frac{\pi R^4}{4}\right)\left(\frac{3}{2\pi R^3}\right) = \frac{3}{8}R}$$ By symmetry, the y- and z- coordinates of the centroid are zero.

Problem 7.26 The volume consists of a segment of a sphere of radius R. Determine its centroid.

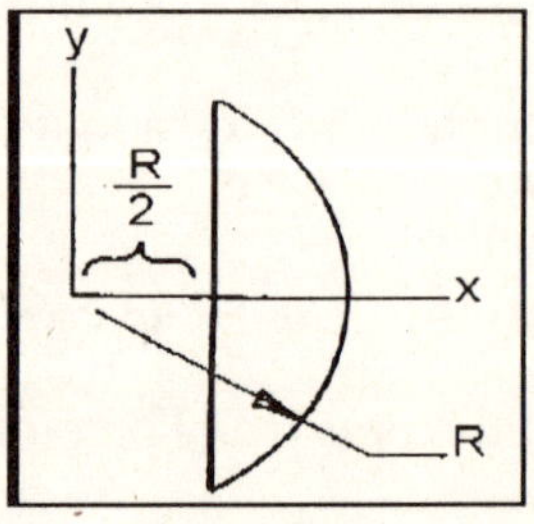

Solution: *The volume:* The element of volume is a disk of radius ρ and thickness *dx*. The area of the disk is $\pi\rho^2$, and the element of volume is $\pi\rho^2 dx$. From the equation of the surface of a sphere (see solution to Problem 7.25) $\rho^2 = R^2 - x^2$, from which the element of volume is

$dV = \pi(R^2 - x^2)dx$. Thus $V = \int_V dV = \pi\int_{\frac{R}{2}}^R (R^2 - x^2)dx = \pi\left[R^2x - \dfrac{x^3}{3}\right]_{\frac{R}{2}}^R$

$= \left(\dfrac{5\pi}{24}\right)R^3$. *The x-coordinate:* $\int_V x\,dV = \pi\int_{\frac{R}{2}}^R (R^2 - x^2)x\,dx = \pi\left[\dfrac{R^2x^2}{2} - \dfrac{x^4}{4}\right]_{\frac{R}{2}}^R = \dfrac{9\pi}{64}R^4$. Divide by

the volume: $\bar{x} = \left(\dfrac{9\pi R^4}{64}\right)\left(\dfrac{24}{5\pi R^3}\right) = \boxed{\dfrac{27}{40}R = 0.675R}$ By symmetry the y- and z-coordinates are zero.

Problem 7.27 A volume of revolution is obtained by revolving the curve $\dfrac{x^2}{a^2} + \dfrac{y^2}{b^2} = 1$ about the x-axis.

Solution: *The volume:* The element of volume is a disk of radius y and thickness *dx*. The area of the disk is πy^2. From the equation for the surface of the ellipse,

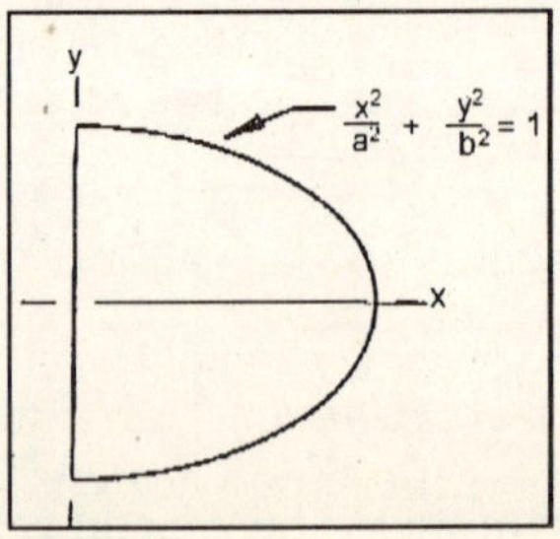

$\pi y^2 = \pi b^2(1 - \dfrac{x^2}{a^2})$ and $dV = \pi y^2 dx = \pi b^2\left(1 - \dfrac{x^2}{a^2}\right)dx$, from which

$$V = \int_V dV = \pi b^2\int_0^a \left(1 - \frac{x^2}{a^2}\right)dx = \pi b^2\left[x - \frac{x^3}{3a^2}\right]_0^a = \frac{2\pi b^2 a}{3}$$

Problem 7.27 continued on next page....

continuation of solution to Problem 7.27...

The x-coordinate: $\int_V x\,dV = \pi b^2 \int_0^a (1-\frac{x^2}{a^2})x\,dx = \pi b^2\left[\frac{x^2}{2}-\frac{x^4}{4a^2}\right]_0^a = \frac{\pi b^2 a^2}{4}$. Divide by volume:

$\bar{x} = \left(\frac{\pi b^2 a^2}{4}\right)\left(\frac{3}{2\pi b^2 a}\right) = \boxed{\left(\frac{3}{8}\right)a}$. By symmetry, the y- and z-coordinates of the centroid are zero.

Problem 7.28 The volume of revolution has a cylindrical hole of radius R. Determine its centroid.

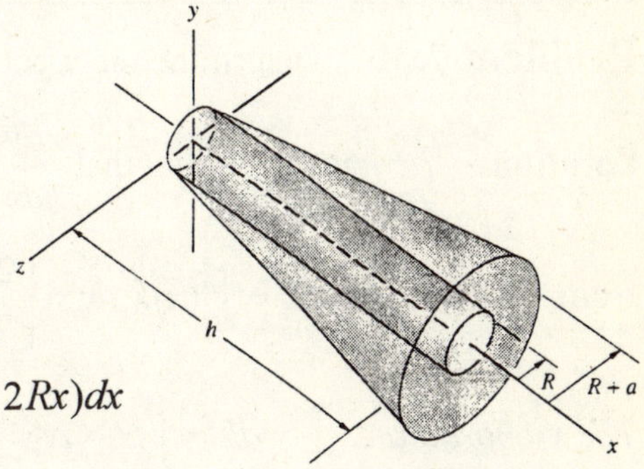

Solution: *The volume:* The element of volume is a disk of radius y and thickness dx The area of the disk is $\pi\,(y^2 - R^2)$

The radius is $y = \left(\frac{a}{h}\right)x + R$. The volume element is

$dV = \pi\left(\frac{a}{h}x + R\right)^2 dx - \pi R^2 dx$. Denote $m = \left(\frac{a}{h}\right)$,

$dV = \pi\left(m^2x^2 + 2mRx\right)dx$, from which $V = \int_V dV = \pi m\int_0^h (mx^2 + 2Rx)dx$

$= \pi m\left[m\frac{x^3}{3} + Rx^2\right]_0^h = \pi m h^2\left(\frac{mh}{3} + R\right)$

The x-coordinate: $\int_V x\,dV = \pi m\int_0^h \left(mx^3 + 2Rx^2\right)dx = \pi m\left[m\frac{x^4}{4} + \frac{2Rx^3}{3}\right]_0^h = \pi m h^3\left(m\frac{h}{4} + \frac{2R}{3}\right)$. Divide by

the volume: $\bar{x} = h\frac{\left(m\frac{h}{4} + \frac{2R}{3}\right)}{\left(m\frac{h}{3} + R\right)} = \boxed{h\frac{\left(\frac{a}{4} + \frac{2R}{3}\right)}{\left(\frac{a}{3} + R\right)}}$

By symmetry, the y-and z-coordinates of the centroid are zero.

Problem 7.29 Determine the y-coordinate of the centroid of the line in Example 7.4.

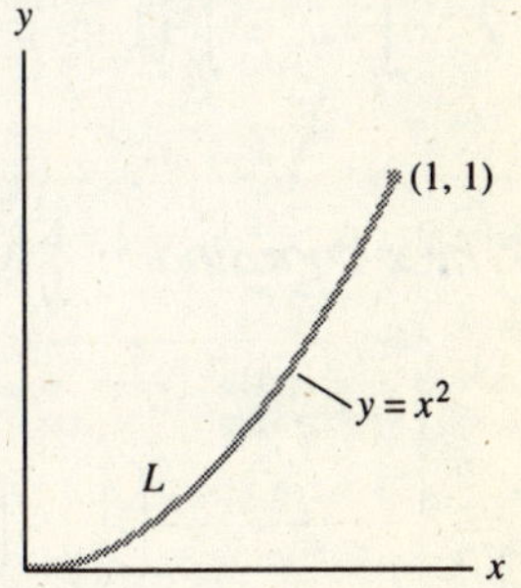

Solution: *The length of the line:* The elementary length of the line is $dL = \sqrt{1 + \left(\frac{dy}{dx}\right)^2}\,dx$. Noting that $\frac{dy}{dx} = 2x$, the element of length

is $dL = \left(1 + 4x^2\right)^{\frac{1}{2}} dx$. , from which

$$L = \int_L dL = \int_0^1 (1+4x^2)^{\frac{1}{2}} dx = \frac{1}{2}\left[\left(1+4x^2\right)^{\frac{1}{2}} x + \frac{1}{2}\log_e(2x + (1+4x^2)^{\frac{1}{2}})\right]_0^1 = \frac{\sqrt{5}}{2} + \frac{1}{4}\log_e(2+\sqrt{5}) = 1.4789$$

Problem 7.29 continued on next page....

continuation of solution to Problem 7.29...

The y-coordinate: The coordinate of the centroid of the length element is $\bar{y} = y = x^2$, from which

$$\bar{y} = \int_0^1 (1+4x^2)^{\frac{1}{2}} x^2 dx = \left[\frac{x(1+4x^2)^{\frac{3}{2}}}{16} - \frac{x(1+4x^2)^{\frac{1}{2}}}{32} - \frac{1}{64} log_e(2x + (1+4x^2)^{\frac{1}{2}}) \right]_0^1$$

$= \frac{1}{16}(5)^{\frac{3}{2}} - \frac{1}{32}(5)^{\frac{1}{2}} - \frac{1}{64}\log_e(2+\sqrt{5}) = 0.6063$ Divide by the length: $\boxed{\bar{y} = \frac{0.6063}{1.4789} = 0.410}$

Problem 7.30 Determine the x coordinate of the centroid of the line.

Solution: *The length:* Noting that $\frac{dy}{dx} = (x-1)^{\frac{1}{2}}$, the element of length is $dL = \sqrt{1 + \left(\frac{dy}{dx}\right)^2} dx = \sqrt{x}\, dx$

from which $L = \int_L dL = \int_1^5 (x)^{\frac{1}{2}} dx = \left[\frac{2}{3}(x)^{\frac{3}{2}}\right]_1^5 = 6.7869$

The x-coordinate: $\int_L x\, dL = \int_0^5 x^{\frac{3}{2}} dx = \left[\frac{2}{5}x^{\frac{5}{2}}\right]_1^5 = 21.961$. Divide by the

length: $\boxed{\bar{x} = \frac{21.961}{6.7869} = 3.2357}$

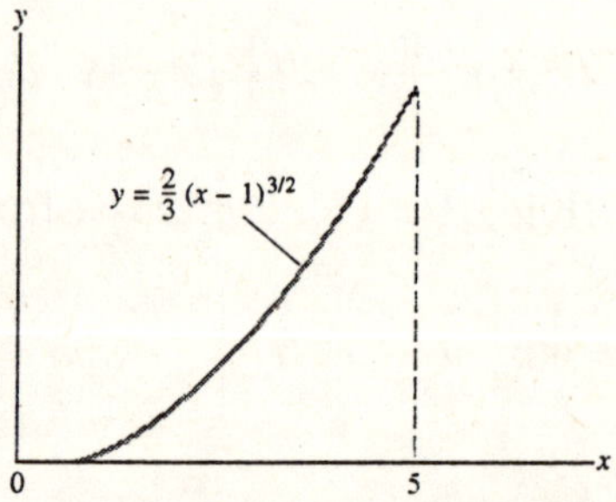

Problem 7.31 Determine the x-coordinate of the centroid of the line.

Solution: *The length:* Noting that $\frac{dy}{dx} = x^{\frac{1}{2}}$ the element of length is

$dL = \sqrt{1 + \left(\frac{dy}{dx}\right)^2} dx = \sqrt{1+x}\, dx$ from which

$$L = \int_L dL = \int_0^2 (1+x)^{\frac{1}{2}} dx = \left[\frac{2}{3}(1+x)^{\frac{3}{2}}\right]_0^2 = 2.7974$$

The x-coordinate: $\int_L x\, dL = \int_0^2 x(1+x)^{\frac{1}{2}} dx = 2\left[\frac{(1+x)^{\frac{5}{2}}}{5} - \frac{(1+x)^{\frac{3}{2}}}{3}\right]_0^2$

$= 2\left[\frac{3^{\frac{5}{2}}}{5} - \frac{3^{\frac{3}{2}}}{3} - \left(\frac{1}{5}\right) + \left(\frac{1}{3}\right)\right] = 3.0379$. Divide by the length: $\boxed{\bar{x} = 1.086}$

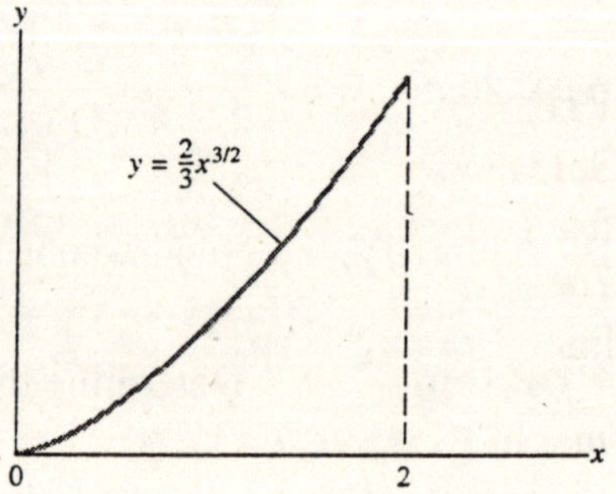

Problem 7.32 Determine the centroid of the circular arc.

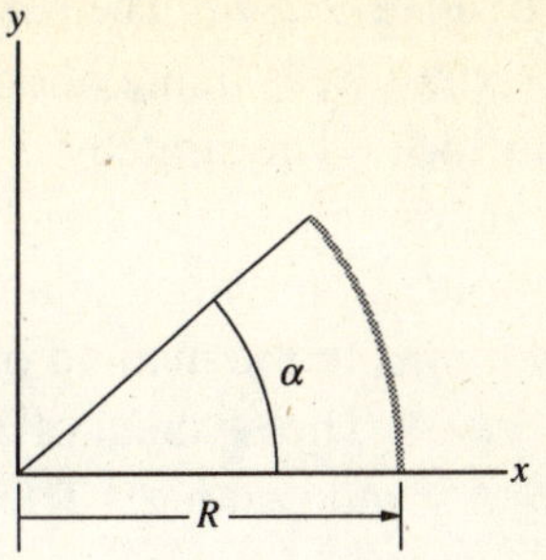

Solution: *The length:* From the equation for the circle, $y = \sqrt{R^2 - x^2}$ and $\frac{dy}{dx} = -(R^2 - x^2)^{-\frac{1}{2}} x$. The element of length

$$dL = \sqrt{1 + \left(\frac{dy}{dx}\right)^2}\, dx = R(R^2 - x^2)^{-\frac{1}{2}} dx\text{, from which}$$

$$L = \int_L dL = \int_{R\cos\alpha}^{R} R(R^2 - x^2)^{-\frac{1}{2}} dx$$

$= \left[R\sin^{-1}\left(\frac{x}{R}\right)\right]_{R\cos\alpha}^{R} = R\left(\frac{\pi}{2} - \sin^{-1}(\cos\alpha)\right) = R\left(\frac{\pi}{2} - \sin^{-1}\left(\sin\left(\frac{\pi}{2} - \alpha\right)\right)\right) = R\alpha$ *Check:* $L = R\alpha$ from the definition of α. *check.*

The x-coordinate: $\int_L x\, dL = R\int_{R\cos\alpha}^{R} x(R^2 - x^2)^{-\frac{1}{2}}\, dx = R\left[-(R^2 - x^2)^{\frac{1}{2}}\right]_{R\cos\alpha}^{R} = R^2 \sin\alpha$

Divide by the length: $\boxed{\bar{x} = \left(\frac{R}{\alpha}\right)\sin\alpha}$

The y-coordinate: The y-coordinate of the centroid of each element is $\bar{y} = y = \sqrt{R^2 - x^2}$. Hence $\int_L y\, dL = R\int_{R\cos\alpha}^{R} (R^2 - x^2)^{\frac{1}{2}} (R^2 - x^2)^{-\frac{1}{2}} dx = R\int_{R\cos\alpha}^{Rc} dx = R^2(1 - \cos\alpha)$. Divide by the length: $\boxed{\bar{y} = \left(\frac{R}{\alpha}\right)(1 - \cos\alpha)}$

Problem 7.33 The homogenous flat plate weighs 200 *lb*. What are the reactions at A and B?

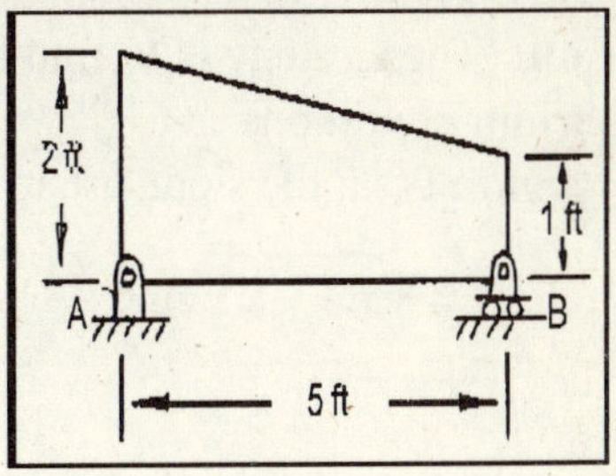

Solution: The strategy is to find the x-coordinate of the centroid from which to apply the equilibrium conditions.

The area: The plate is bounded on the upper side by the straight line, $y = mx + b$, where at $x = 0$, $y = b = 2$, and at $x = 5$, $y = 1$, $m = \frac{y - b}{x} = -0.2$. The element of area is a vertical strip of height y and width dx, $dA = (mx + b)dx$, from which

$$A = \int_A dL = \int_0^5 (mx + b)dx = \left[m\frac{x^2}{2} + bx\right]_0^5 = \left[\frac{25m}{2} + 5b\right] = 7.5 \text{ } sq\ ft$$

The x-coordinate: $\int_A x\, dA = \int_0^5 (mx + b)x\, dx = \left[\frac{mx^3}{3} + \frac{bx^2}{2}\right]_0^5 = 16.6667$. Divide by the area:

$\bar{x} = \frac{16.6667}{7.5} = 2.2222\ ft$. *The equilibrium conditions:* The sum of the moments about A:

$\sum M_A = -2.2222(200) + 5B = 0$, from which $\boxed{B = \frac{444.4}{5} = 88.89\ lb}$ The sum of the forces:

$\sum F_Y = A_Y + B - 200 = 0$, from which $\boxed{A_y = 111.1\ lb}$. $\boxed{\sum F_X = A_X = 0}$

=====================================<>=====================================

Problem 7.34 The suspended sign is a homogenous flat plate that has a mass of 130 *kg*. Determine the axial forces in members AD and CE.

Solution: The strategy is to determine the distance to the action line of the weight (x-coordinate of the centroid) from which to apply the equilibrium conditions to the method of sections.

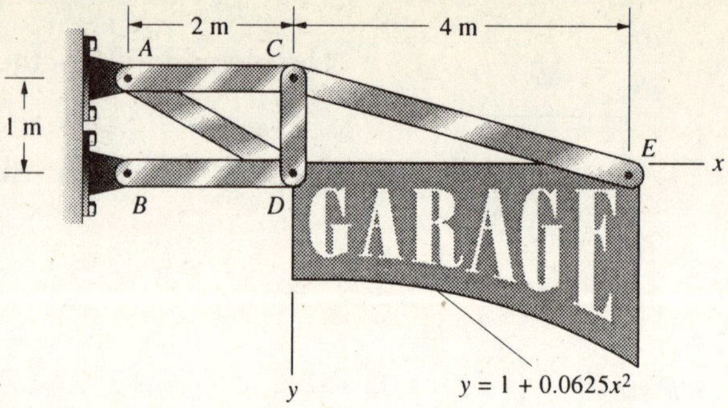

The area: The element of area is the vertical strip of length y and width *dx*. The element of area $dA = ydx = (1+ax^2)dx$, where $a = 0.0625$. Thus

$$A = \int_A dA = \int_0^4 (1+ax^2)dx = \left[x + \frac{ax^3}{3}\right]_0^4 = 5.3333\ sq\ ft .$$

The x-coordinate: $\int_A x\, dA = \int_0^4 x(1+ax^2)dx = \left[\frac{x^2}{2} + \frac{ax^4}{4}\right]_0^4 = 12$. Divide A: $\bar{x} = \frac{12}{5.3333} = 2.25\ ft$.

The equilibrium conditions: The angle of the member CE is $\alpha = \tan^{-1}\left(\frac{1}{4}\right) = 14.04^o$. The weight of the sign is $W = 130(9.81) = 1275.3\ N$. The sum of the moments about D is $\sum M_D = -2.2W + 4CE\sin\alpha = 0$, from which $\boxed{CE = 2957.7\ N(T)}$. *Method of sections:* Make a cut through members AC, AD and BD and consider the section to the right. The angle of member AD is $\beta = \tan^{-1}\left(\frac{1}{2}\right) = 26.57^o$. *The section as a free body:* The sum of the vertical forces: $\sum F_Y = AD\sin\beta - W = 0$ from which $\boxed{AD = 2851.7\ N(T)}$

=====================================<>=====================================

Problem 7.35 The circular slender bar has a mass of 20 *kg*. What are the reactions at A and B?

Solution: The strategy is to find the distance to the action line of the weight, from which to apply the equilibrium conditions.

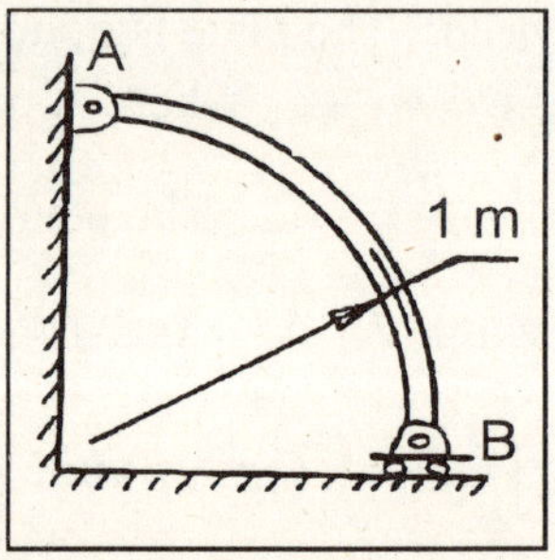

The length: Treat the slender bar as a mass line. from the equation for the circle $y = \sqrt{R^2 - x^2}$ and $\frac{dy}{dx} = (R^2 - x^2)^{-\frac{1}{2}}x$, and the element of line is $dL = \sqrt{1 + \left(\frac{dy}{dx}\right)^2} = R(R^2 - x^2)^{-\frac{1}{2}}dx$. The length is $L = \frac{2\pi R}{4} = \frac{\pi R}{2}$

The x- coordinate:

$$\int_L x\, dL = R\int_0^R x(R^2 - x^2)^{-\frac{1}{2}}dx = R\left[-(R^2 - x^2)^{\frac{1}{2}}\right]_0^R = R^2 .$$ Divide by length: $\bar{x} = \frac{2R}{\pi}$. *The equilibrium conditions:* The sum of the moments about A:

$\sum M_A = -\frac{2R}{\pi}W + RB = 0$, from which $B = \frac{2W}{\pi}$. The sum of the forces: $\sum F_y = A_y + B - W = 0$, from which $A_y = W\left(1 - \frac{2}{\pi}\right)$. Noting that $W = 20(9.81) = 196.2\ N$, then $\boxed{B = 124.9\ N}$, and $\boxed{A_y = 71.3\ N}$. The roller support guarantees that $\boxed{A_x = 0}$

=====================================<>=====================================

Problem 7.36 The $10\,ft$ horizontal cylinder with $1\,ft$ radius is supported at A and B. Its weight density is $\gamma = 100(1-0.002x^2)\ lb/ft^3$. What are the reactions at A and B?

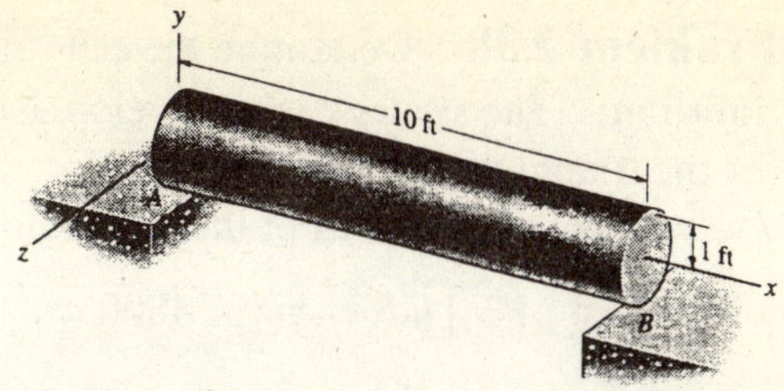

Solution: *The weight:* Denote $a = 0.002$. The element of volume is a disk of radius $R = 1\,ft$, thickness dx, and weight

$$dW = \gamma\pi R^2 dx, \text{ from which } W = \int dW = 100\pi R^2\int_0^L (1-ax^2)dx = 100\pi R^2\left[x - \frac{ax^3}{3}\right]_0^L = 2932.15\ lb$$

The x-coordinate of the mass center:

$$\int_W x\,dW = 100\pi R^2\int_0^L (1-ax^2)x dx = -\frac{25\pi R^2}{a}\left[(1-ax^2)^2\right]_0^L = 14137.17. \text{ Divide by W: } \bar{x} = 4.8214\,ft.$$

The equilibrium conditions: The sum of the moments about A is $\sum M_A = -W\bar{x} + 10B = 0$, from which $B = \frac{W\bar{x}}{L} = \frac{(2932.15)(4.8214)}{10}$ $\boxed{= 1413.7\ lb}$ The sum of the vertical forces: $\sum F_Y = A + B - W = 0$, from which $\boxed{A = 1518.4\ lb}$ The horizontal components of the reactions are zero: $\sum F_X = 0$

Problem 7.37 A horizontal cone with 800 mm length and 200 mm radius has a built-in support at A. Its mass density is $\rho = 6000(1+0.4x^2)\ kg/m^3$, where x is in meters. What are the reactions at A?

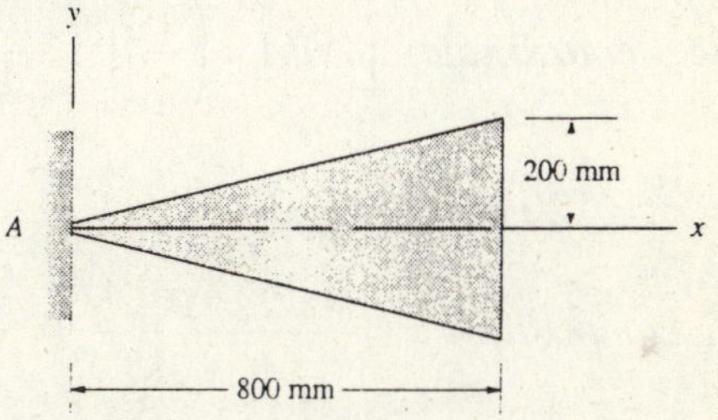

Solution: The strategy is to determine the distance to the line of action of the weight, from which to apply the equilibrium conditions.

The mass: The element of volume is a disk of radius y and thickness dx. y varies linearly with x: $y = 0.25x$. Denote $a = 0.4$. The mass of the disk is

$$dm = \rho\pi y^2 dx = 6000\pi(1+ax^2)(0.25x)^2 dx$$

$$= 375\pi(1+ax^2)x^2 dx, \text{ from which } m = 375\pi\int_0^{0.8}(1+ax^2)x^2 dx = 375\pi\left[\frac{x^3}{3} + a\frac{x^5}{5}\right]_0^{0.8} = 231.95\ kg$$

The x-coordinate of the mass center: $\int_m x\,dm = 375\pi\int_0^{0.8}(1+ax^2)x^3 dx$

$$= 375\pi\left[\frac{x^4}{4} + a\frac{x^6}{6}\right]_0^{0.8} = 141.23$$. Divide by the mass: $\bar{x} = 0.6089\ m$

The equilibrium conditions: The sum of the moments about A: $\sum M = M_A - mg\bar{x} = 0$, from which $M_A = mg\bar{x} = 231.94(9.81)(0.6089)$ $\boxed{= 1385.4\ N\text{-}m}$ The sum of the vertical forces: $\sum F_Y = A_Y - mg = 0$ from which $\boxed{A_Y = 2275.4\ N}$ The horizontal component of the reaction is zero, $\sum F_X = 0$.

====================================<>====================================

Problem 7.38 Determine the centroid.

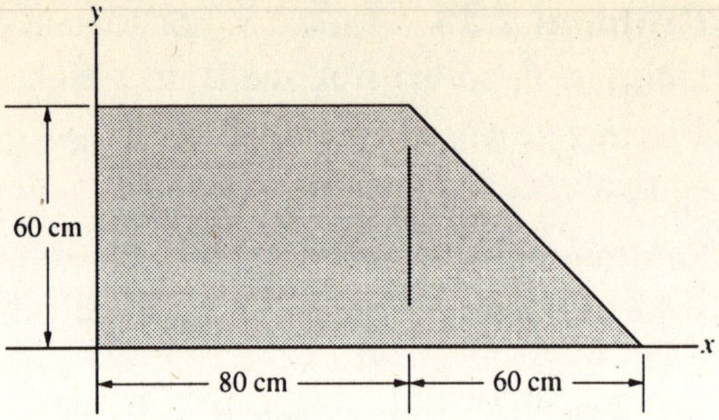

Solution: The strategy is to develop useful general results for the triangle and the rectangle.

The rectangle: The area of the rectangle of height h and width w is $A = \int_0^w h dx = hw = 4800\ cm^2$.

The x-coordinate: $\int_0^w hx dx = h\left[\frac{x^2}{2}\right]_0^w = \left(\frac{1}{2}\right)hw^2$. Divide by the area: $\bar{x} = \frac{w}{2}. = 40\ cm$

The y-coordinate: $\left(\frac{1}{2}\right)\int_0^w h^2 dx = \left(\frac{1}{2}\right)h^2 w$. Divide by the area: $\bar{y} = \left(\frac{1}{2}\right)h = 30\ cm$

The triangle: The area of the triangle of altitude a and base b is (assuming that the two sides a and b meet at the origin) $A = \int_o^b y(x)dx = \int_0^b \left(-\frac{a}{b}x + a\right)dx = \left[-\frac{ax^2}{2b} - ax\right]_0^b = \left[-\frac{ab}{2} + ab\right] = \frac{ab}{2}. = 1800\ cm^2$

Check: This is the familiar result. *check.*

The x-coordinate: $\int_0^b \left(-\frac{a}{b}x + a\right)x dx = \left[-\frac{ax^3}{3b} + \frac{ax^2}{2}\right]_0^b = \frac{ab^2}{6}$. Divide by the area: $\bar{x} = \frac{b}{3} = 20\ cm$

The y-coordinate: $\int_A y dA = \left(\frac{1}{2}\right)\int_0^b \left(-\frac{a}{b}x + a\right)^2 dx = -\frac{b}{6a}\left[\left(-\frac{a}{b}x + a\right)^3\right]_0^b = \frac{ba^2}{6}$. Divide by the area:

$\bar{y} = \frac{a}{3} = 20\ cm$

The composite: $\bar{x} = \frac{\bar{x}_R A_R + \bar{x}_T A_T}{A_R + A_T} = \frac{40(4800) + 100(1800)}{4800 + 1800}$ $\boxed{= 56.36\ cm}$

$\bar{y} = \frac{(30)(4800) + (20)(1800)}{4800 + 1800}$ $\boxed{= 27.27\ cm}$

====================================<>====================================

Problem 7.39 Determine the centroid.

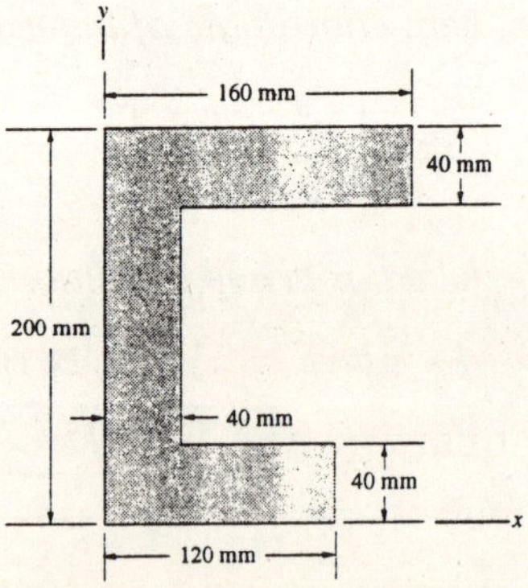

Solution: The strategy is to divide the figure into three rectangles, and to use the results of Problem 7.38 for the rectangles. The dimensions of the rectangles are:

(1) 40 mm height by 120 mm wide.

(2) 120 mm height by 40 mm wide.

(3) 40 mm high by 160 mm wide.

The areas: (1) 4800 sq mm, (2) 4800 sq mm. (3) 6400 sq mm.

The centroid locations: (1) $\bar{x}_1 = 60\ mm$, $\bar{y}_1 = 20\ mm$, (2) $\bar{x}_2 = 20\ mm$, $\bar{y}_2 = 100\ mm$, (3) $\bar{x}_3 = 80\ mm$, $\bar{y}_3 = 180\ mm$.

The composite: $\bar{x} = \frac{60(4800) + 20(4800) + 80(6400)}{4800 + 4800 + 6400}$ $\boxed{= 56\ mm}$

$\bar{y} = \frac{20(4800) + 100(4800) + 180(6400)}{4800 + 4800 + 6400}$ $\boxed{= 108\ mm}$

====================================<>====================================

Problem 7.40 Determine the centroid.

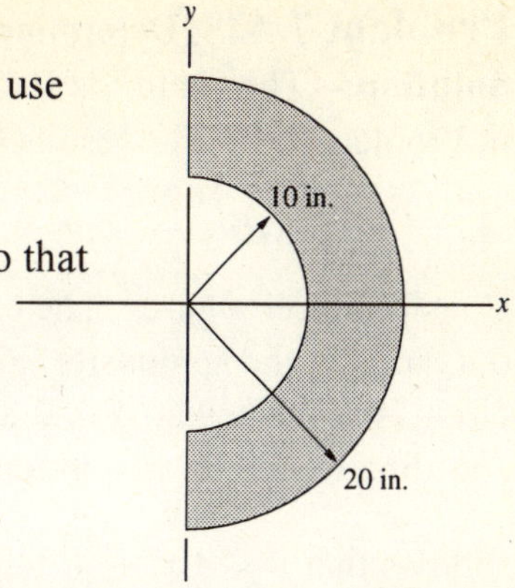

Solution: The strategy is to find the centroid for the half circle area, and use the result in the composite algorithm. *The area:* The element of area is a vertical strip y high and dx wide. From the equation of the circle, $y = \pm\sqrt{R^2 - x^2}$. The height of the strip will be twice the positive value, so that $dA = 2\sqrt{R^2 - x^2}\,dx$, from which

$$A = \int_A dA = 2\int_0^R (R^2 - x^2)^{\frac{1}{2}} dx = 2\left[\frac{x\sqrt{R^2 - x^2}}{2} + \frac{R^2}{2}\sin^{-1}\left(\frac{x}{R}\right)\right]_0^R = \frac{\pi R^2}{2}$$

The x-coordinate: $\int_A x\,dA = 2\int_0^R x\sqrt{R^2 - x^2}\,dx = 2\left[-\frac{(R^2 - x^2)^{\frac{3}{2}}}{3}\right]_0^R = \frac{2R^3}{3}$. Divide by A: $\bar{x} = \frac{4R}{3\pi}$

The y-coordinate: From symmetry, the y-coordinate is zero.

The composite: For a complete half circle $\bar{x}_1 = \frac{4(20)}{3\pi} = 8.488\ in$. For the inner half circle $\bar{x}_2 = 4.244\ in$. The areas are $A_1 = 628.32\ in^2$ and $A_2 = 157.08\ in.^2$. The area of the cut out half circle shown in the figure is $A_3 = 471.24\ in.^2$. The composite algorithm for the centroid of the cut-out segment is $\bar{x}_3 = \frac{x_1 A_1 - \bar{x}_2 A_2}{A_3} = \frac{(8.4883)(628.32) - (4.2441)(157.08)}{471.24}$ $\boxed{= 9.9\ in.}$

The y-coordinate of the centroid of the cut-out is zero.

Problem 7.41 Determine the centroid.

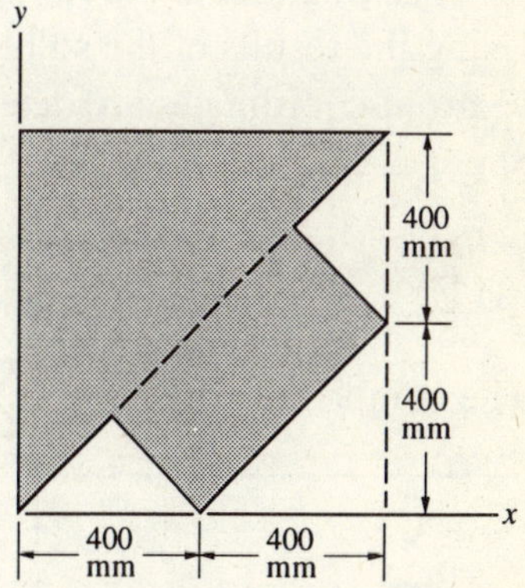

Solution: Divide the object into two areas:

(1) The triangle of altitude 800 mm and base 800 mm.

(2) the rectangle with sides $200\sqrt{2}$ mm and $400\sqrt{2}$ mm.

The areas and centroids of the areas are found by inspection.

(1) $A_1 = \left(\frac{1}{2}\right)(800)^2 = 3.2 \times 10^5\ mm^2$, $\bar{x}_1 = \frac{800}{3} = 266.67\ mm$,

$\bar{y}_1 = 800\left(\frac{2}{3}\right) = 533.33\ mm$.

(2) $A_2 = \left(200\sqrt{2}\right)\left(400\sqrt{2}\right) = 1.6 \times 10^5\ mm^2$, $\bar{y}_2 = \left(100\sqrt{2} + 200\sqrt{2}\right)\left(\frac{1}{\sqrt{2}}\right) = 300\ mm$.

$\bar{x}_2 = \bar{y}_2 + 200 = 500\ mm$

The composite algorithm:

$\boxed{\bar{x} = \frac{A_1\bar{x}_1 + A_2\bar{x}_2}{A_1 + A_2} = 344.4\ mm}$ $\boxed{\bar{y} = \frac{A_1\bar{y}_1 + A_2\bar{y}_2}{A_1 + A_2} = 455.6\ mm}$

Problem 7.42 Determine the centroid.

Solution: The centroids for the half-circle have been determined in Problem 7.40. The results for a half-circle of radius R:

$A_1 = \frac{\pi R^2}{2}$, $\bar{x}_1 = \frac{4R}{3\pi}$, $\bar{y}_1 = 0$.

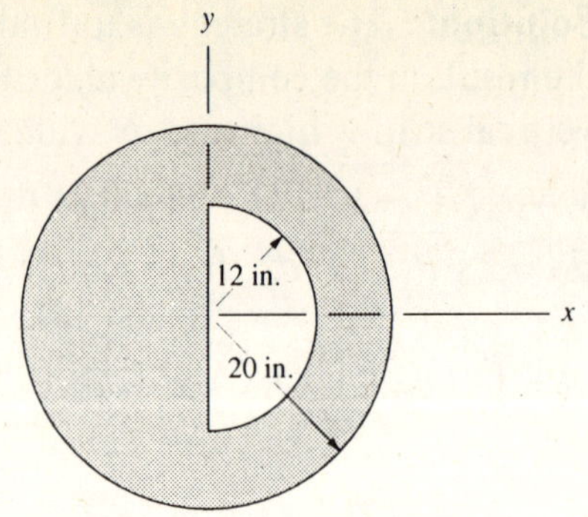

Consider three figures: The complete circle, (1) the half circle cut out, and (2) the composite figure. The centroid of the complete circle is at the origin; $\bar{x} = 0$ and $\bar{y} = 0$. The product of its centroid coordinates and its area is zero. From the composite algorithm, it follows that $0 = A_1\bar{x}_1 + A_2\bar{x}_2$, from which $\bar{x}_2 = -\left(\frac{A_1}{A_2}\right)\bar{x}_1$, and

$\bar{y}_2 = -\left(\frac{A_1}{A_2}\right)\bar{y}_1$. The areas: $A_1 = \frac{\pi R_1^2}{2}$, $A_2 = \pi R_2^2 - \frac{\pi R_1^2}{2}$. For $R_1 = 12$ *in.* and $R_2 = 20$ *in.*,

$A_1 = 226.2$ *in.*^{2}, and $A_2 = 1030.4$ *in.*^{2}, and $\bar{x}_1 = 5.093$ *in.*. Thus $\boxed{\bar{x}_2 = -\left(\frac{226.2}{1030.4}\right)(5.093) = -1.18 \text{ in.}}$,

and $\boxed{\bar{y}_2 = 0}$, since $\boxed{\bar{y}_1 = 0}$.

Problem 7.43 Determine the centroid.

Solution: Divide the object into three areas:

(1) A rectangle on the left, 100 mm by 60 mm.

(2) A rectangle at the lower right, 80 mm by 40 mm.

(3) A semi circle to the far lower right, radius 20 mm.

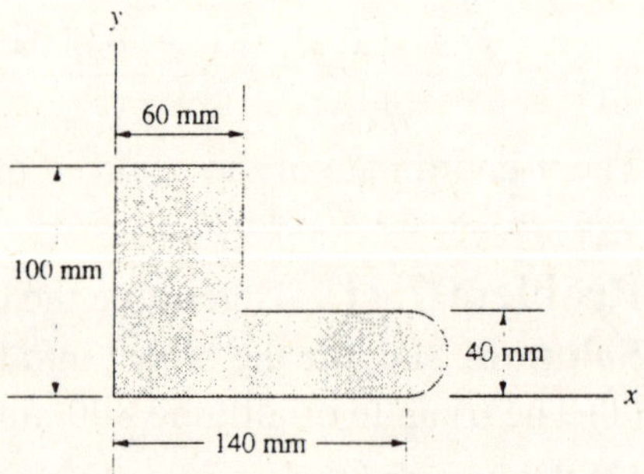

Using the results of the solutions previous problems, the areas and centroid coordinates are determined by inspection:

(1) $A_1 = 6 \times 10^3\ mm^2$, $\bar{x}_1 = 30\ mm$, $\bar{y}_1 = 50\ mm$

(2) $A_2 = 3.2 \times 10^3\ mm^2$, $\bar{x}_2 = 100\ mm$, $\bar{y}_2 = 20\ mm$

(3) $A_3 = 628.32\ mm^2$, $\bar{x}_3 = 148.49\ mm$, $\bar{y}_3 = 20\ mm$

The composite area is $A = \sum_1^3 A_i = 9.828 \times 10^3$ The centroid coordinates for the composite are

$\boxed{\bar{x} = \frac{\sum_1^3 A_i\bar{x}_i}{A} = 60.37\ mm}$, and $\boxed{\bar{y} = \frac{\sum_1^3 A_i\bar{y}_i}{A} = 38.31\ mm}$

=================================<>=================================

Problem 7.44 Determine the centroid.

Solution: Divide the object into four areas: (1) The rectangle 18 *in* by 18 *in*, (2) The triangle of altitude 18 *in* and base 6 *in*, and (3) the semi circle with radius 9 *in* and (4) The object itself.

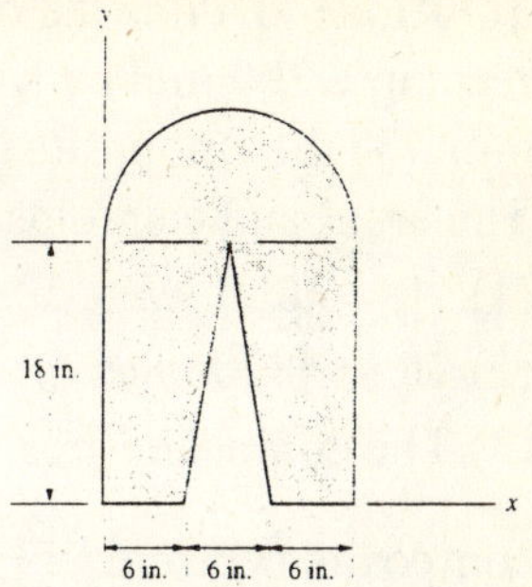

The areas and their centroids are determined by inspection:

(1) $A_1 = 18^2 = 324\ in.^2$, $\bar{x}_1 = 9\ in.$, $\bar{y}_1 = 9\ in.$

(2) $A_2 = \left(\frac{1}{2}\right)(18)(6) = 54\ in.^2$, $\bar{x}_2 = 9\ in.$, $\bar{y}_2 = 6\ in.$

(3) $A_3 = \frac{\pi 9^2}{2} = 127.2\ in.^2$, $\bar{x}_3 = 9\ in.$, $\bar{y}_3 = 18 + \frac{4(9)}{3\pi} = 21.8\ in.$

The composite area: $A = A_1 - A_2 + A_3 = 397.2\ in.^2$. The composite centroid:

$$\bar{x} = \frac{A_1\bar{x}_1 - A_2 x_2 + A_3 x_3}{A} = 9\ in.$$ *Check:* From symmetry, *check.* $$\bar{y} = \frac{A_1\bar{y}_1 - A_2\bar{y}_2 + A_3\bar{y}_3}{A} = 13.51\ in.$$

=================================<>=================================

Problem 7.45 Determine the centroid.

Solution: Divide the object into five areas (1) The triangle of altitude 12 in and base 8 in, (2) The rectangle 8 in by 12 in, (3) The semicircle of 6 in radius, (4) the circle of 6 in radius, and (5) the composite object itself. The areas and centroids are determined by inspection:

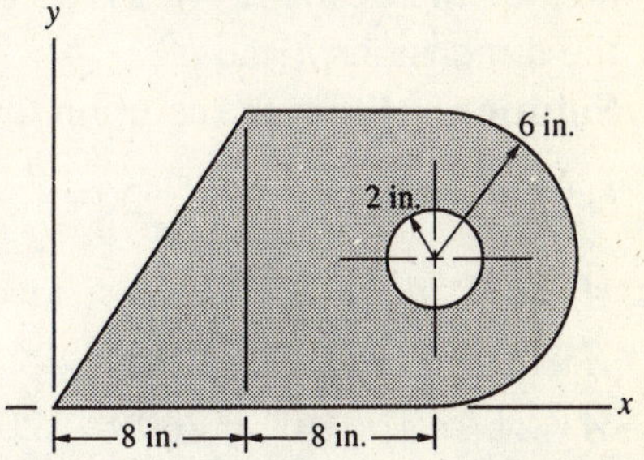

(1) $A_1 = \frac{(8)(12)}{2} = 48\ in.^2$, $\bar{x}_1 = \left(\frac{2}{3}\right)8 = 5.33\ in.$, $\bar{y}_1 = \left(\frac{1}{3}\right)12 = 4\ in.$,

(2) $A_2 = (8)(12) = 96\ in.$, $\bar{x}_2 = 12\ in.$, $\bar{y}_2 = 6\ in.$, (3) $A_3 = \frac{\pi 6^2}{2} = 56.55\ in.^2$, $\bar{x}_3 = 16 + \frac{4(6)}{3\pi} = 18.55\ in.$, $\bar{y}_3 = 6\ in.$, (4) $A_4 = \pi 2^2 = 12.57\ in.^2$, $\bar{x}_4 = 16\ in.$, $\bar{y}_4 = 6\ in.$. The composite area:

$A = A_1 + A_2 + A_3 - A_4 = 187.98\ in.^2$. The composite centroid: $$\bar{x} = \frac{A_1\bar{x}_1 + A_2\bar{x}_2 + A_3\bar{x}_3 - A_4\bar{x}_4}{A} = 12\ in.$$

$$\bar{y} = \frac{A_1\bar{y}_1 + A_2\bar{y}_2 + A_3\bar{y}_3 - A_4\bar{y}_4}{A} = 5.49\ in.$$

=================================<>=================================

Problem 7.46 Determine the centroid.

Solution: Divide the object into five areas: (1) The rectangle 80 mm by 80 mm, (2) The rectangle 120 mm by 80 mm, (3) the semicircle of radius 40 mm, (4) The circle of 20 mm radius, and (5) the composite object. The areas and centroids: (1) $A_1 = 6400\ mm^2$, $\bar{x}_1 = 40\ mm$, $\bar{y}_1 = 40\ mm$, (2) $A_2 = 9600\ mm^2$, $\bar{x}_2 = 120\ mm$, $\bar{y}_2 = 60\ mm$, (3) $A_3 = 2513.3\ mm^2$, $\bar{x}_3 = 120\ mm$, $\bar{y}_3 = 136.98\ mm$, (4) $A_4 = 1256.6\ mm^2$, $\bar{x}_4 = 120\ mm$, $\bar{y}_4 = 120\ mm$. (5) The composite area: $A = A_1 + A_2 + A_3 - A_4 = 17256.6\ mm^2$. The composite centroid:

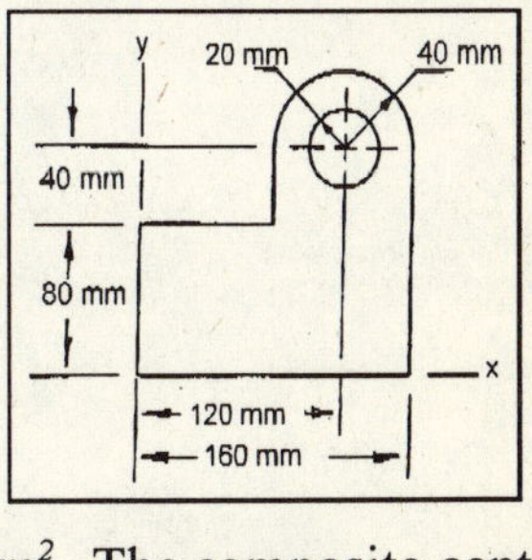

$$\bar{x} = \frac{A_1\bar{x}_1 + A_2\bar{x}_2 + A_3\bar{x}_3 - A_4\bar{x}_4}{A} = 90.3\ mm$$. $$\bar{y} = \frac{A_1\bar{y}_1 + A_2\bar{y}_2 + A_3\bar{y}_3 - A_4\bar{y}_4}{A} = 59.4\ mm$$

=================================<>=================================

Problem 7.47

The dimensions $b = 40\ mm$ and $h = 20\ mm$. Determine the y-coordinate of the centroid of the area.

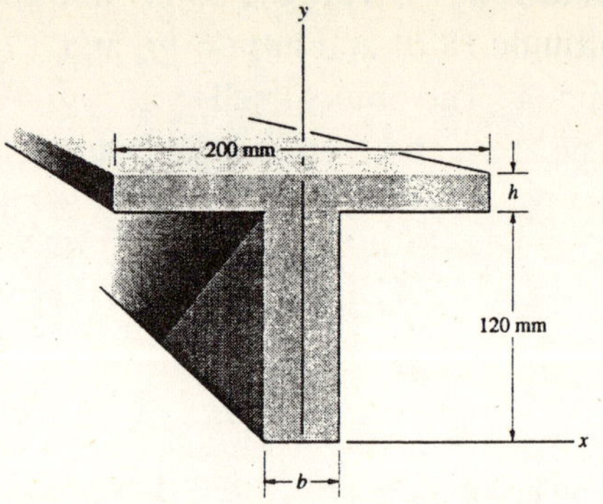

Solution: Divide the object into three areas (1) the upper rectangle 200 mm by h mm, (2) The bottom rectangle, 200 by b mm, (3) the composite object.

The areas and centroids are:

(1) $A_1 = 4000\ mm^2$, $\bar{x}_1 = 0$, $\bar{y}_1 = 130\ mm$

(2) $A_2 = 4800\ mm^2$, $\bar{x}_2 = 0$, $\bar{y}_2 = 60\ mm$

(3) The composite area: $A = A_1 + A_2 = 8800\ mm^2$. The composite centroid: $\bar{x} = 0$, $\boxed{\bar{y} = \frac{A_1\bar{y}_1 + A_2\bar{y}_2}{A} = 91.82\ mm}$

Problem 7.48

The area shown in Problem 7.47 is the cross section of a beam. If the cross sectional area of the beam $A = 0.01\ m^2$ and the y-coordinate of the centroid of the area is $\bar{y} = 80\ mm$, what are the dimensions b and h?

Solution: From the solution to Problem 7.47, the areas and centroids are:

(1) $A_1 = 200h$, $\bar{y}_1 = 120 + \frac{h}{2}$, (2) $A_2 = 120b$, $\bar{y}_2 = 60$, and the composite area is given: $A = 0.01\ m^2 = 10000\ mm^2$. Two simultaneous equations:

$A\bar{y} = A_1\bar{y}_1 + A_2\bar{y}_2 = 8 \times 10^5 = 2.4 \times 10^4 h + 10^2 h^2 + 7.2 \times 10^3 b$

$A = A_1 + A_2 = 10^4 = 200h + 120b$. Multiply the first by 120, and the second by 7.2×10^3, and subtract the second from the first. The result is a quadratic equation $h^2 + 120h - 2000 = 0$. This has the roots $h_{1,2} = (14.83, -134.8)$. Only the positive root makes sense, $\boxed{h = 14.83\ mm}$. Substitute this value into the equation for the area to obtain $\boxed{b = \frac{(10^4 - 200h)}{120} = 58.61\ mm}$

Problem 7.49-7.51

Three sails of a New York pilot schooner are shown The coordinates of the points are in feet.

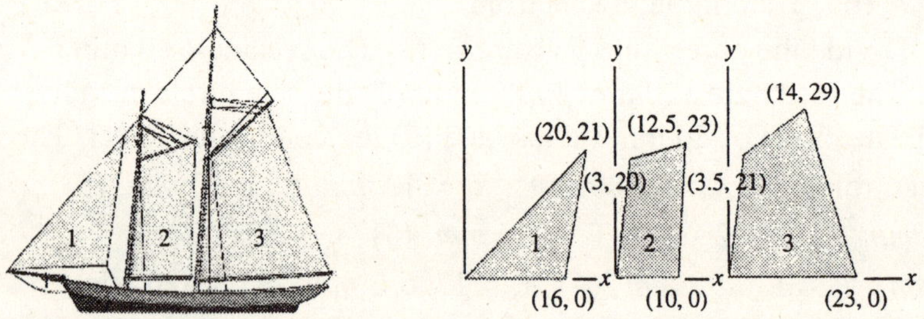

(a) (b)

Problems 7.49-7.51 continued on next page....

Problem 7.49 Determine the centroid of sail 1.

Solution: Divide the object into three areas: (1) The triangle with altitude 21 *ft* and base 20 *ft*. (2) The triangle with altitude 21 *ft* and base $(20-16)=4\ ft$, and (3) the composite sail. The areas and coordinates are: (1) $A_1 = 210\ ft^2$, $\bar{x}_1 = \left(\frac{2}{3}\right)20 = 13.33\ ft$, $\bar{y}_1 = \left(\frac{1}{3}\right)21 = 7\ ft$. (2) $A_2 = 42\ ft^2$, $\bar{x}_2 = 16 + \left(\frac{2}{3}\right)4 = 18.67\ ft$, $\bar{y}_2 = 7\ ft$. (3) The composite area: $A = A_1 - A_2 = 168\ ft^2$. The composite centroid: $\bar{x} = \dfrac{A_1\bar{x}_1 - A_2\bar{x}_2}{A} = 12\ ft$ $\bar{y} = \dfrac{A_1\bar{y}_1 - A_2\bar{y}_2}{A} = 7\ ft$

Problem 7.50 Determine the centroid of sail 2.

Solution: Divide the object into five areas: (1) a triangle on the left with altitude 20 *ft* and base 3 *ft*, (2) a rectangle in the middle 23 *ft* by 9.5 *ft*, (3) a triangle at the top with base of 9.5 *ft* and altitude of 3 *ft*. (4) a triangle on the right with altitude of 23 *ft* and base of 2.5 *ft*. (5) the composite sail. The areas and centroids are: (1) $A_1 = \frac{3(20)}{2} = 30\ ft^2$, $\bar{x}_1 = \left(\frac{2}{3}\right)3 = 2\ ft$, $\bar{y}_1 = \left(\frac{1}{3}\right)20 = 6.67\ ft$.

(2) $A_2 = (23)(9.5) = 218.5\ ft^2$, $\bar{x}_2 = 3 + \left(\frac{9.5}{2}\right) = 7.75\ ft$, $\bar{y}_2 = \frac{23}{2} = 11.5\ ft$

(3) $A_3 = \left(\frac{1}{2}\right)(3)(9.5) = 14.25\ ft^2$, $\bar{x}_3 = 3 + \left(\frac{1}{3}\right)9.5 = 6.167\ ft$, $\bar{y}_3 = 20 + \left(\frac{2}{3}\right)3 = 22\ ft$

(4) $A_4 = \left(\frac{1}{2}\right)(2.5)(23) = 28.75\ ft^2$, $\bar{x}_4 = 10 + \left(\frac{2}{3}\right)(2.5) = 11.67\ ft$, $\bar{y}_4 = \left(\frac{1}{3}\right)23 = 7.66\ ft$

(5) The composite area: $A = A_1 + A_2 - A_3 - A_4 = 205.5\ ft^2$. The composite centroid:

$\bar{x} = \dfrac{A_1\bar{x}_1 + A_2\bar{x}_2 - A_3\bar{x}_3 - A_4\bar{x}_4}{A} = 6.472\ ft$ $\bar{y} = \dfrac{A_1\bar{y}_1 + A_2\bar{y}_2 - A_3\bar{y}_3 - A_4\bar{y}_4}{A} = 10.603\ ft$

Problem 7.51 Determine the centroid of sail 3.

Solution: Divide the object into six areas: (1) The triangle Oef, with base 3.5 *ft* and altitude 21 *ft*. (2) The rectangle Oabc, 14 *ft* by 29 *ft*. (3) The triangle beg, with base 11.5 *ft* and altitude 8 *ft*. (4)The triangle bcd, with base 9 *ft* and altitude 29 *ft*. (5) The rectangle agef 3.5 *ft* by 8 *ft*. (6) The composite, Oebd. The areas and centroids are:

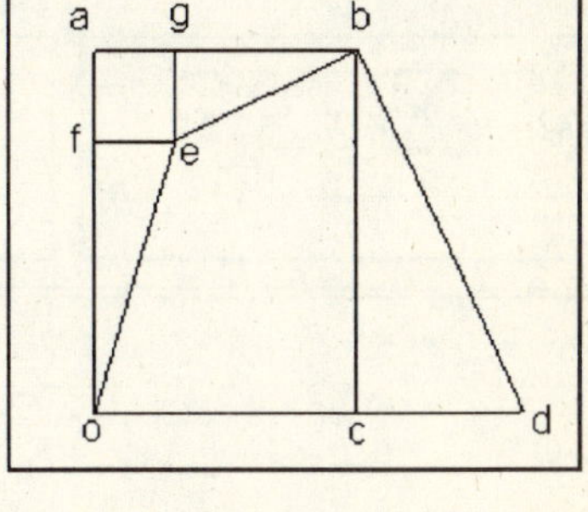

(1) $A_1 = 36.75\ ft^2$, $\bar{x}_1 = 1.167\ ft$, $\bar{y}_1 = 14\ ft$

(2) $A_2 = 406\ ft^2$, $\bar{x}_2 = 7\ ft$, $\bar{y}_2 = 14.5\ ft$

(3) $A_3 = 42\ ft^2$, $\bar{x}_3 = 7.\ ft$, $\bar{y}_3 = 26.33\ ft$

(4) $A_4 = 130.5\ ft^2$, $\bar{x}_4 = 17\ ft$, $\bar{y}_4 = 9.67\ ft$

(5) $A_5 = 28\ ft^2$, $\bar{x}_5 = 1.75\ ft$, $\bar{y}_5 = 25\ ft$

(6) The composite area: $A = -A_1 + A_2 - A_3 + A_4 - A_5 = 429.75\ ft^2$.

The composite centroid:

$$\bar{x} = \frac{-A_1\bar{x}_1 + A_2\bar{x}_2 - A_3\bar{x}_3 + A_4\bar{x}_4 - A_5\bar{x}_5}{A} = 10.877\ ft$$

$$\bar{y} = \frac{-A_1\bar{y}_1 + A_2\bar{y}_2 - A_3\bar{y}_3 + A_4\bar{y}_4 - A_5 y_5}{A} = 11.23\ ft$$

Problem 7.52 Determine the centroid of the volume.

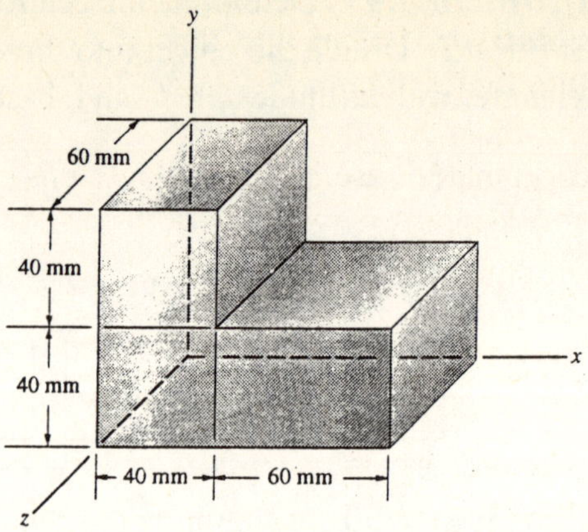

Solution: Divide the object into two volumes: (1) The left most volume with dimensions 40 by 80 by 60 *mm.* (2) The right most volume with dimensions 60 by 60 by 40 *mm.* The volumes and centroids are:

(1) $V_1 = 1.962 \times 10^5\ mm^3$, $\bar{x}_1 = 20\ mm$, $\bar{y}_1 = 40\ mm$, $\bar{z}_1 = 30\ mm$. (2) $V_2 = 1.44 \times 10^5\ mm^3$, $\bar{x}_2 = 70\ mm$, $\bar{y}_2 = 20\ mm$, $\bar{z}_2 = 30\ mm$. (3) The composite volume: $V = V_1 + V_2 = 3.36 \times 10^5\ mm^3$. The composite centroid:

$$\bar{x} = \frac{V_1\bar{x}_1 + V_2 x_2}{V} = 41.43\ mm$$

$$\bar{y} = \frac{V_1\bar{y}_1 + V_2\bar{y}_2}{V} = 31.43\ mm$$

$$\bar{z} = \frac{V_1\bar{z}_1 + V_2\bar{z}_2}{V} = 30\ mm$$

Problem 7.53 The holes are 3 *in* in diameter.

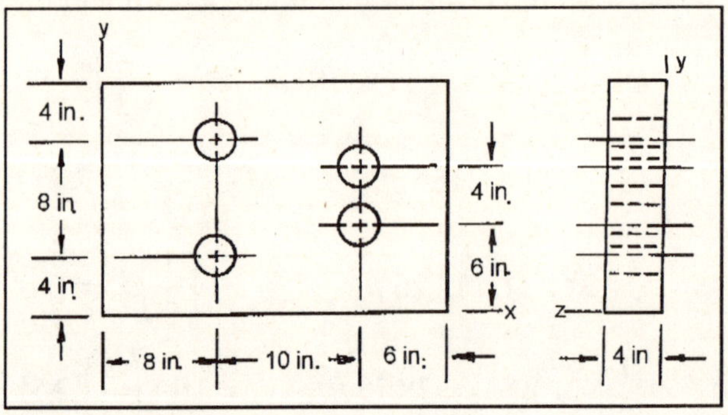

Solution: Divide the object into 6 volumes: (1) The block with dimensions 24 by 16 by 4 *in.* (2) The left top circular hole, (3) the left bottom hole,(4) the right top hole, (5) the right bottom hole, and (6) the composite volume.

The volumes and centroids are:

Volume	vol, cu in.	$\bar{x}$, in.	$\bar{y}$, in.	$\bar{z}$, in.
V1	1536	12	8	2
V2	28.27	8	12	2
V3	28.27	8	4	2
V4	28.27	18	10	2
V5	28.27	18	6	2

The composite volume is $V = V_1 - \sum_2^5 V_i = 1.4229 \times 10^3\ in^3$. The composite centroid:

$$\bar{x} = \frac{V_1\bar{x}_1 - \sum_2^5 V_i\bar{x}_i}{V} = 11.92\ in.$$

$$\bar{y} = \frac{V_1\bar{y}_1 - \sum_2^5 V_i\bar{y}_i}{V} = 8\ in.$$

$$\bar{z} = \frac{V_1\bar{z}_1 - \sum_2^5 V_i\bar{z}_i}{V} = 2\ in.$$

Problem 7.54 Find the centroid.

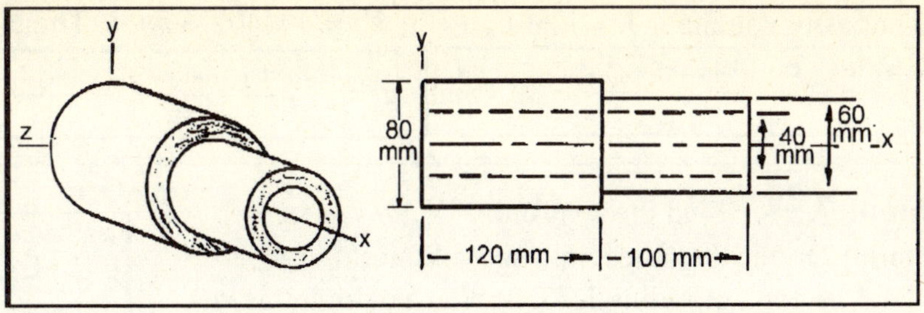

Solution: Divide the object into four volumes: (1) The left-most cylinder, with diameter 80 *mm* and length 120 *mm*, (2) the cylinder to the right, with diameter 60 *mm* and length 100 *mm*, (3) the cylinder bored through the center, with diameter 40 *mm* and length 220 *mm*, and (4) the composite cylinder. The volumes and centroids are: (1) $V_1 = 6.032 \times 10^5\ mm^3$, $\bar{x}_1 = 60\ mm$, $\bar{y}_1 = \bar{z}_1 = 0$, (2) $V_2 = 2.83 \times 10^5\ mm^3$, $\bar{x}_2 = 170\ mm$, $\bar{y}_2 = \bar{z}_2 = 0$, (3) $V_3 = 2.76 \times 10^5\ mm^3$, $\bar{x}_3 = 110\ mm$, $\bar{y}_3 = \bar{z}_3 = 0$. (4) The composite volume: $V = V_1 + V_2 - V_3 = 6.095 \times 10^5\ mm^3$. The composite centroid:

$$\boxed{\bar{x} = \frac{V_1\bar{x}_1 + V_2\bar{x}_2 - V_3\bar{x}_3}{V} = 88.4\ mm} \quad \boxed{\bar{y} = \bar{z} = 0}$$

Problem 7.55 Find the centroid.

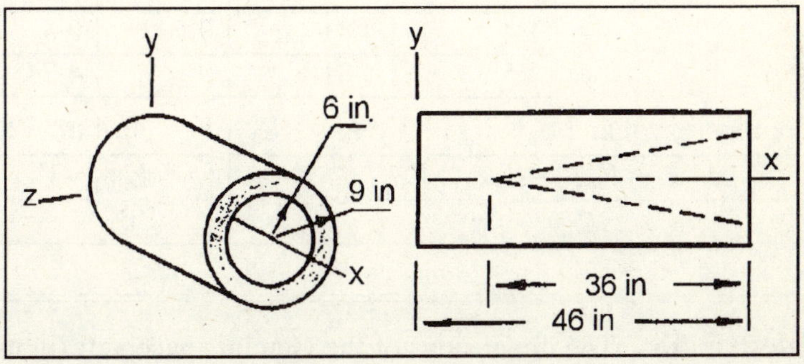

Solution: Divide the object into three volumes: (1) The cylinder with radius 9 *in.* and length 46 *in.*, (2) the cone of length 36 *in.* and base radius 6 *in.*, and (3) the composite volume. The volumes and centroids are:
(1) $V_1 = 11705.6\ in.^3$, $\bar{x}_1 = 23\ in.$, $\bar{y}_1 = \bar{z}_1 = 0$, (2) $V_2 = \left(\frac{1}{3}\right)\pi R^2 h = 1357.2\ in.^3$, $\bar{x}_2 = 10 + \left(\frac{3}{4}\right)h = 37\ in.$, $\bar{y}_2 = \bar{z}_2 = 0$. (3) The composite volume: $V = V_1 - V_2 = 10348.4\ in.^3$. The composite centroid: $\boxed{\bar{x} = \frac{V_1\bar{x}_1 - V_2\bar{x}_2}{V} = 21.16\ in.}$, $\boxed{\bar{y} = \bar{z} = 0}$

Problem 7.56 Find the centroid.

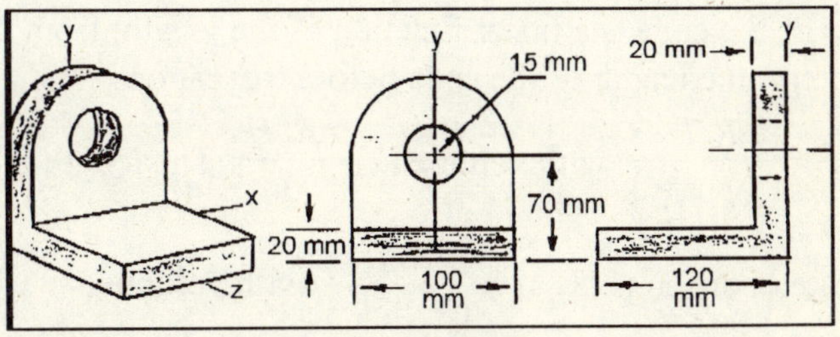

Solution: Divide the object into five volumes: (1) The base, 100 by 100 by 20 *mm*. (2) The vertical block, 100 by 50 by 20 *mm*. (3) The semicircular block, 20 *mm* thick with a 50 *mm* radius. (4) The circular cut out, 20 *mm* long with a radius of 15 *mm*. The volumes and centroids are:

Volume	Vol, cu mm	x	y	z
V1	240000	0	10	60
V2	100000	0	45	10
V3	78539.8	0	91.2	10
V4	14137.2	0	70	10

Problem 7.56 continued on next page....

continuation of solution to Problem 7.56...

The composite volume is $V = V_1 + V_2 + V_3 - V_4 = 404402.6\ mm^3$. The composite centroid: $\boxed{\bar{x} = 0}$

$$\boxed{\bar{y} = \frac{V_1\bar{y}_1 + V_2\bar{y}_2 + V_3\bar{y}_3 - V_4\bar{y}_4}{V} = 32.3\ mm} \quad \boxed{\bar{z} = \frac{V_1\bar{z}_1 + V_2\bar{z}_2 + V_3\bar{z}_3 - V_4\bar{z}_4}{V} = 39.7\ mm}$$

Problem 7.57 Find the centroid.

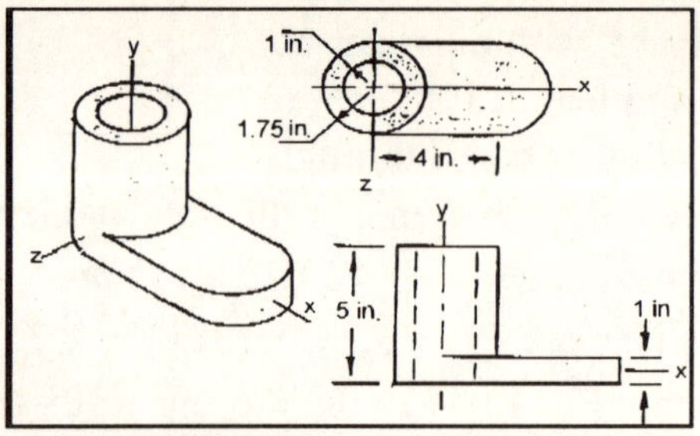

Solution: Divide the object into six volumes: (1) A cylinder 5 *in.* long of radius 1.75 *in.*, (2) a cylinder 5 *in.* long of radius 1 *in.*,(3) a block 4 *in.* long, 1 *in.* thick, and $2(1.75) = 3.5$ *in.* wide. (4) Semi-cylinder 1 *in* long with a radius of 1.75 *in.* , (5) a semi-cylinder 1 *in* long with a radius of 1.75 *in.* (6) The composite object. The volumes and centroids are:

Volume	Vol, cu in	$\bar{x}$, *in.*	$\bar{y}$, *in.*	$\bar{z}$, *in.*
V1	48.1	0	2.5	0
V2	15.7	0	2.5	0
V3	14	2	0.5	0
V4	4.81	0.743	0.5	0
V5	4.81	0	4.743	0

The composite volume is $V = V_1 - V_2 + V_3 - V_4 + V_5 = 46.4\ in^3$. The composite centroid:

$$\boxed{\bar{x} = \frac{V_1\bar{x}_1 - V_2\bar{x}_2 + V_3\bar{x}_3 - V_4\bar{x}_4 + V_5\bar{x}_5}{V} = 1.02\ in.} \quad \boxed{\bar{y} = \frac{V_1 y_1 - V_2 y_2 + V_3 y_3 - V_4\bar{y}_4 + V_5\bar{y}_5}{V} = 1.9\ in.},$$

$\boxed{\bar{z} = 0}$

Problem 7.58 The dimensions of the Gemini spacecraft (in meters) are $a = 0.70$, $b = 0.88$, $c = 0.74$, $d = 0.98$, $e = 1.82$, $f = 2.20$, $g = 2.24$, and $h = 2.98$. Determine the centroid of its volume.

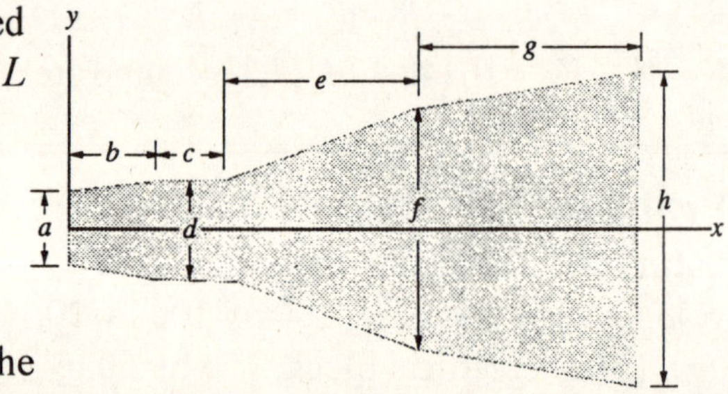

Solution: The spacecraft volume consists of three truncated cones and a cylinder. Consider the truncated cone of length L with radii at the ends R_1 and R_2, , where $R_2 > R_1$. Choose the origin of the x-y coordinate system at smaller end. The radius of the cone is a linear function of the length; from geometry, the length of the cone before truncations was (1) $H = \frac{R_2 L}{(R_2 - R_1)}$. with volume (2) $\frac{\pi R_2^2 H}{3}$. The length of the truncated portion is (3) $\eta = \frac{R_1 L}{(R_2 - R_1)}$.with volume (4) $\frac{\pi R_1^2 \eta}{3}$. The volume of the truncated cone is the difference of the two volumes, (5) $V = \frac{\pi L}{3}\left(\frac{R_2^3 - R_1^3}{R_2 - R_1}\right)$. The centroid of the removed part of the cone is (6) $\bar{x}_\eta = \left(\frac{3}{4}\right)\eta$, and the centroid of the complete cone is (7) $\bar{x}_h = \left(\frac{3}{4}\right)H$,.

Problem 7.58 continued on next page....

continuation of solution to Problem 7.58...

measured from the pointed end. From the composite theorem, the centroid of the truncated cone is

(8) $\bar{x} = \dfrac{V_h \bar{x}_h - V_\eta \bar{x}_\eta}{V} - \eta + x$., where x is the x-coordinate of the left hand edge of the truncated cone in the specific coordinate system. These eight equations are the algorithm for the determination of the volumes and centroids of the truncated cones forming the spacecraft.

Beginning from the left, the volumes are (1) a truncated cone, (2) a cylinder, (3) a truncated cone, and (4) a truncated cone. The algorithm and the data for these volumes were entered into **TK Solver 2** and the volumes and centroids determined. The volumes and x-coordinates of the centroids are:

Volume	Vol, cu m	$\bar{x}$, m
V1	0.4922	0.4884
V2	0.5582	1.25
V3	3.7910	2.752
V4	11.8907	4.8716
Composite	16.732	3.999

The last row is the composite volume and x-coordinate of the centroid of the composite volume. The total length of the spacecraft is 5.68 *m*, so the centroid of the volume lies at about 69% of the length as measured from the left end of the spacecraft. *Discussion:* The algorithm for determining the centroid of a system of truncated cones may be readily understood if it is implemented for a cone of known dimensions divided into sections, and the results compared with the known answer. Alternate algorithms (e.g. a Pappus-Guldinus algorithm) are useful for checking but arguably do not simplify the computations) *End discussion.*

Problem 7.59 Determine the centroid of the line.

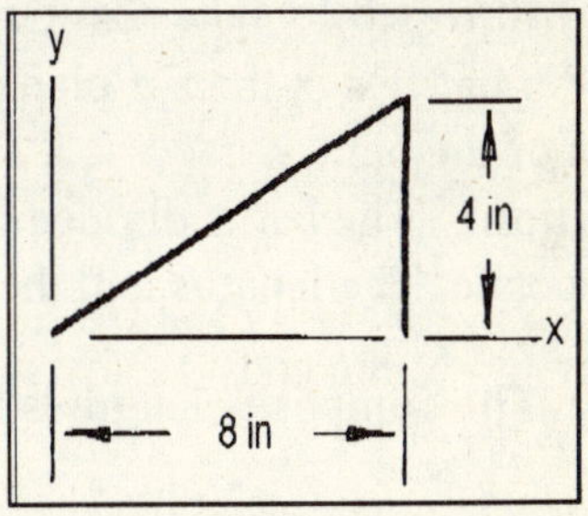

Solution: The object is divided into the two lines and the composite:(1) Length: $L_1 = \sqrt{8^2 + 4^2} = 8.9443\ in$, $\bar{x}_1 = 4\ in$, $\bar{y}_1 = 2\ in$. (2) Length: $L_2 = 4\ in$, $\bar{x}_2 = 8\ in$, $\bar{y}_2 = 2\ in$. The composite length: $L = 12.9443\ in$. The composite centroid:

$\bar{x} = \dfrac{L_1\bar{x}_1 + L_2\bar{x}_2}{L} = 5.236\ in$, $\bar{y} = \dfrac{L_1\bar{y}_1 + L_2\bar{y}_2}{L} = 2\ in$

Problem 7.60 Determine the centroid of the line.

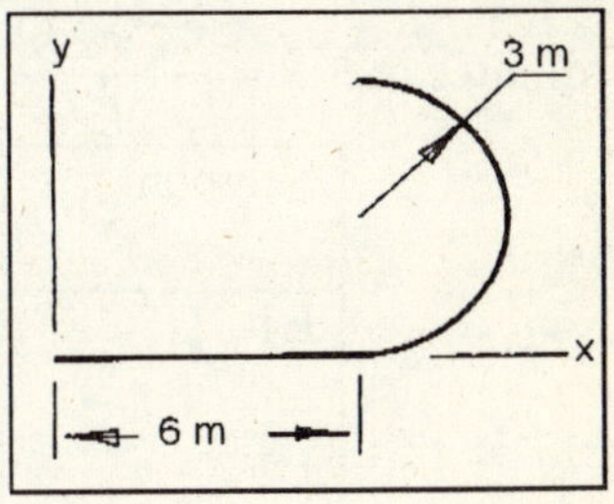

Solution: The object is divided into two lines and a composite.

(1) $L_1 = 6\ m$, $\bar{x}_1 = 3\ m$, $\bar{y}_1 = 0$. (2) $L_2 = 3\pi\ m$, $\bar{x}_2 = 6 + \dfrac{6}{\pi}\ m$, $\bar{y}_2 = 3$.

(3) The composite length: $L = 6 + 3\pi\ m$. The composite centroid:

$\bar{x} = \dfrac{L_1\bar{x}_1 + L_2\bar{x}_2}{L} = 6\ m$ $\bar{y} = \dfrac{3\pi}{2+\pi} = 1.83\ m$

Problem 7.61 Determine the centroid of the line.

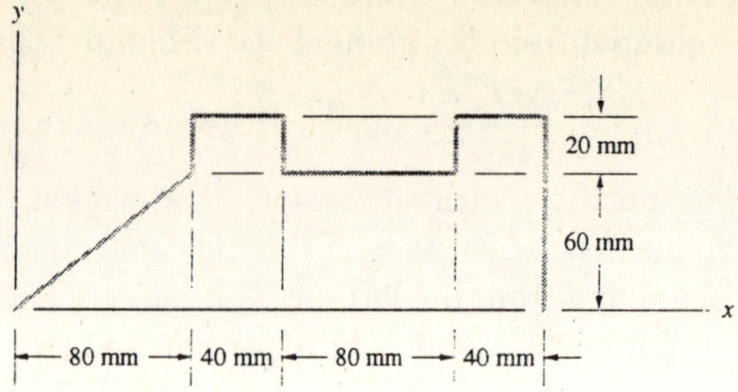

Solution: The line is divided into eight straight line segments. From right to left, the length and centroid of each segment are shown in the table. The last row is the composite length L and the centroid coordinates

The composite length is: $L = \sum_{i=1}^{8} L_i$. The composite coordinates are: $\bar{x} = \frac{\sum_{i=1}^{8} L_i \bar{x}_i}{L}$, and $\bar{y} = \frac{\sum_{i=1}^{8} L_i \bar{y}_i}{L}$.

Segment	Length, mm	x, mm	y, mm
L1	100	40	30
L2	20	80	70
L3	40	100	80
L4	20	120	70
L5	80	160	60
L6	20	200	70
L7	40	220	80
L8	80	240	40
Composite	400	142	54

Problem 7.62 The semicircular part of the homogenous, slender bar lies in the x-z plane. Determine the center of mass of the bar.

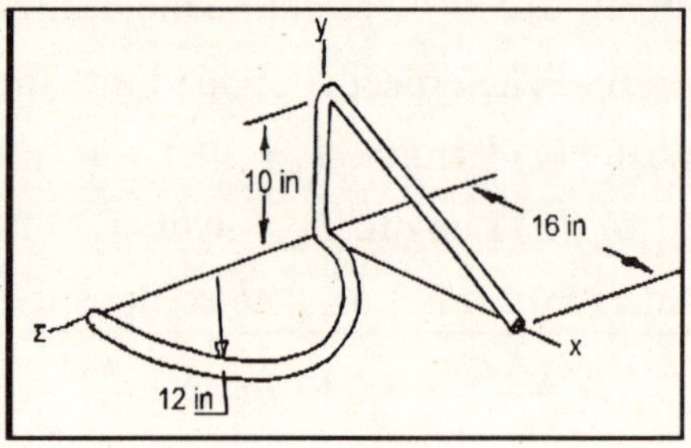

Solution: The bar is divided into three segments plus the composite. The lengths and the centroids are given in the table: The composite length is: $L = \sum_{i=1}^{3} L_i$. The composite coordinates are: $\bar{x} = \frac{\sum_{i=1}^{3} L_i \bar{x}_i}{L}$, and $\bar{y} = \frac{\sum_{i=1}^{3} L_i \bar{y}_i}{L}$

Segment	Length,in	$\bar{x}$, in.	$\bar{y}$, in.	$\bar{z}$, in.
L1	12π	$\frac{24}{\pi}$	0	12
L2	10	0	5	0
L3	18.868	8	5	0
Composite	66.567	6.594	2.168	6.796

Problem 7.63 The following theorem is *not* true. "The centroid of any area is coincident with the centroid of the line forming its boundary." Disprove it by finding a counterexample. That is, find an example for which is not true.

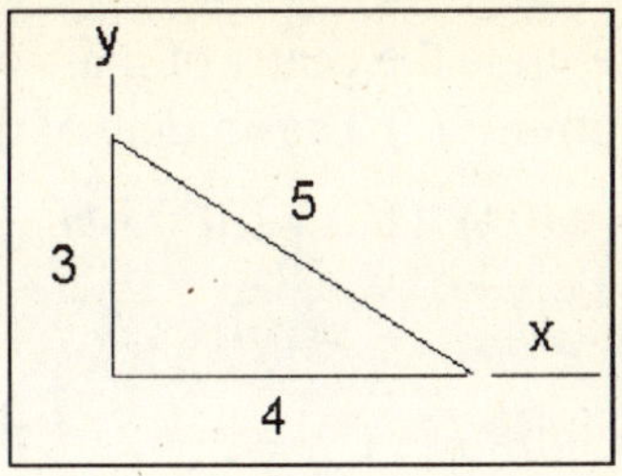

Solution: Consider the 3,4,5 triangle. The centroid of the area is $(x = \frac{4}{3}, y = 1)$ *Proof:* Consider only the x-coordinate. Choose a vertical strip of area ydx. The boundary is $y = -\frac{3}{4}x + 3$.

The area: $A = \int_0^4 (-\frac{3}{4}x + 3)dx = \left[-\frac{3}{8}x^2 + 3x\right]_0^4 = 6$. *Check* $A = \left(\frac{1}{2}\right)(3)(4) = 6$ *check.*

The x-coordinate: $\int_A x dA = \int_0^4 (-\frac{3}{4}x + 3)x\, dx = \left[-\frac{3}{12}x^3 + \frac{3}{2}x^2\right]_0^4 = 8$. Divide by the area: $\bar{x} = \frac{4}{3}$.

The Line: Divide the line into three segments, each corresponding to a side. (1) $L_1 = 3$, $\bar{x}_1 = 0$, $\bar{y}_1 = \frac{3}{2}$. (2) $L_2 = 4$, $\bar{x}_2 = 2$, $\bar{y}_2 = 0$. (3) $L_3 = 5$, $\bar{x}_3 = 2$, $\bar{y}_3 = \frac{3}{2}$. The composite length is $L = 3 + 4 + 5 = 12$. The centroid: $\bar{x} = \frac{(3)0 + (4)2 + (5)2}{12} = \frac{18}{12} = \frac{3}{2}$. *The theorem fails for the x-coordinate*. It is unnecessary to go further.

Problem 7.64 The positions of the centers of the three homogenous spheres are shown. The mass density of sphere 1 is ρ_0, the mass density of sphere 2 is $1.2\rho_0$, and the mass density of the sphere 3 is $1.4\rho_0$. (a) Determine the centroid of the volume of the three spheres. (b) Determine the center of mass of the three spheres.

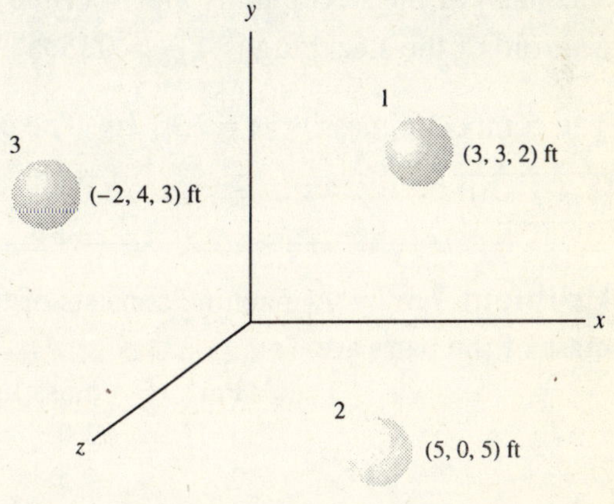

Solution: Assume that the spheres have equal volumes.

(a) *The volume centroid* is $\bar{x} = \frac{-2V + 3V + 5V}{3V} = 2\ ft$

$\bar{y} = \frac{4V + 3V + 0V}{3V} = 2.33\ ft$ $\bar{z} = \frac{3V + 2V + 5V}{3V} = 3.33\ ft$

(b) *The mass centroid:*

$$\bar{x}_m = \frac{-2(1.4\rho_0 V) + 3(\rho_0 V) + 5(1.2\rho_0 V)}{(1.4 + 1 + 1.2)\rho_0 V} = \frac{6.2}{3.6} = 1.722\ ft$$

$$\bar{y}_m = \frac{4(1.4\rho_0 V) + 3(\rho_0 V) + 0(1.2\rho_0 V)}{(1.4 + 1 + 1.2)\rho_0 V} = \frac{8.6}{3.6} = 2.39\ ft$$

$$\bar{z}_m = \frac{3(1.4\rho_0 V) + 2(\rho_0 V) + 5(1.2\rho_0 V)}{(1.4 + 1 + 1.2)\rho_0 V} = \frac{12.2}{3.6} = 3.39\ ft$$

Problem 7.65 The circular cylinder is made of aluminum (Al) with mass density $2700\ kg/m^3$ and iron (Fe) with mass density $7860\ kg/m^3$. (a) Determine the centroid of the volume of the cylinder. (b) Determine the center of mass of the cylinder.

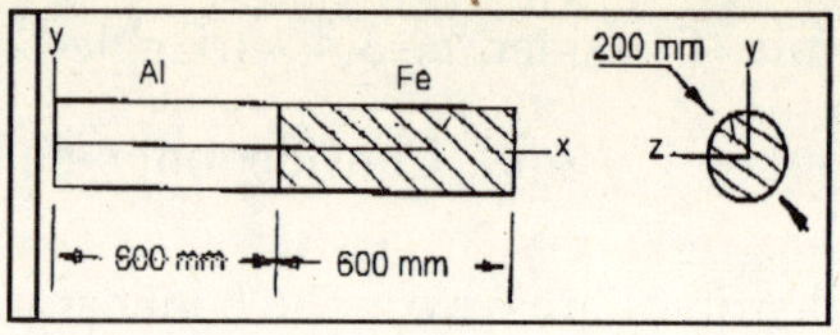

Solution: (a) The volume of the cylinder is $V = \pi(0.1)^2(1.2) = 0.0377\ m^3$. The volume of the parts: $V_{Al} = V_{Fe} = \dfrac{V}{2} = 0.0188\ m^3$. The centroid of the first part is $\bar{x}_{Al} = 0.3\ m$, $\bar{y}_{Al} = \bar{z}_{Al} = 0$. The centroid of the iron part is $\bar{x}_{Fe} = 0.6 + 0.3 = 0.9\ m$, $\bar{y}_{Fe} = \bar{z}_{Fe} = 0$.

The composite centroid $\boxed{\bar{x} = \dfrac{V_{Al}(0.3) + V_{Fe}(0.9)}{V} = \dfrac{1.2}{2} = 0.6\ m}$, $\boxed{\bar{y} = \bar{z} = 0}$.

(b) *The mass center:* The mass of the aluminum part is $m_{Al} = V_{Al}(2700) = 50.89\ kg$. The mass of the iron part is $m_{Fe} = V_{Fe}(7860) = 148.16\ kg$. The composite mass is $m = m_{Al} + m_{Fe} = 199.05\ kg$. The composite center of mass is $\boxed{\bar{x}_m = \dfrac{(50.89)(0.3) + (148.16)(0.9)}{199.05} = 0.7466\ m}$ $\boxed{\bar{y}_m = \bar{z}_m = 0}$

Problem 7.66 The cylindrical tube is made of aluminum with a mass density of $2700\ kg/m^3$. The cylindrical plug is made of steel with a mass density of $7800\ kg/m^3$. Determine the coordinates of the center of mass of the composite object.

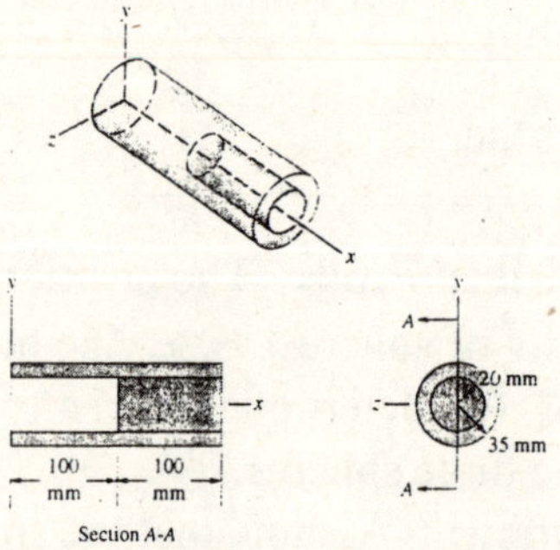

Solution: The volume of the aluminum tube is $V_{Al} = \pi(0.035^2 - 0.02^2)(0.2) = 5.18 \times 10^{-4}\ m^3$. The mass of the aluminum tube is $m_{Al} = (2700)V_{Al} = 1.4\ kg$ The centroid of the aluminum tube is $\bar{x}_{AL} = 0.1\ m$, $\bar{y}_{Al} = \bar{z}_{Al} = 0$.

The volume of the steel plug is $V_{Fe} = \pi(0.02)^2(0.1) = 1.26 \times 10^{-4}\ m^3$ The mass of the steel plug is $m_{Fe} = (7800)V_{Fe} = 0.9802\ kg$. The centroid of the steel plug is $\bar{x}_{Fe} = 0.15\ m, \bar{y}_{Fe} = \bar{z}_{Fe} = 0$.

The composite mass is $m = 2.38\ kg$. The composite centroid is $\boxed{\bar{x} = \dfrac{m_{Al}(0.1) + m_{Fe}(0.15)}{m} = 0.121\ m}$

$\boxed{\bar{y} = \bar{z} = 0}$

Problem 7.67 A machine consists of three parts. The masses and the locations of the centers of mass of the parts are:

Part	mass,kg	x, mm	y,mm	z,mm
1	2.0	100	50	-20
2	4.5	150	70	0
3	2.5	180	30	0

Determine the coordinates of the center of mass of the machine:

Solution: The composite mass is 9 kg. $\boxed{\bar{x} = \dfrac{2(100) + 4.5(150) + 2.5(180)}{9} = 147\ mm}$

$\boxed{\bar{y} = \dfrac{2(50) + 4.5(70) + 2.5(30)}{9} = 54.4\ mm}$ $\boxed{\bar{z} = \dfrac{2(-20)}{9} = -4.4\ mm}$

Problem 7.68 A machine consists of three parts. The masses and the locations of the centers of mass of two of the parts are:

Part	mass,kg	x,mm	y,mm	z,mm
1	2.0	100	50	-20
2	4.5	150	70	0

The mass of part 3 is 2.5 *kg*. The design engineer wants to position part 3 so that the center of mass of location of the machine is $\bar{x} = 120\ mm$, $\bar{y} = 80\ mm$, and $\bar{z} = 0$. Determine the necessary position of the center of mass of part 3.

Solution: The composite mass is $m = 2.0 + 4.5 + 2.5 = 9\ kg$. The location of the third part is

$$\boxed{\bar{x}_3 = \frac{120(9) - 2(100) - 4.5(150)}{2.5} = 82\ mm} \quad \boxed{\bar{y}_3 = \frac{80(9) - 2(50) - 4.5(70)}{2.5} = 122\ mm}$$

$$\boxed{\bar{z}_3 = \frac{2(20)}{2.5} = 16\ mm}$$

Problem 7.69 The homogenous plate weighs 500 N. Determine the reactions at the supports A and B.

Solution: Divide the object into three areas and the composite. Since the distance to the action line of the weight is the only item of importance, and since there is no horizontal component of the weight, it is unnecessary to determine any centroid coordinate other than the x-coordinate. The areas and the x-coordinate of the centroid are tabulated. The last row is the composite area and x-coordinate of the centroid.

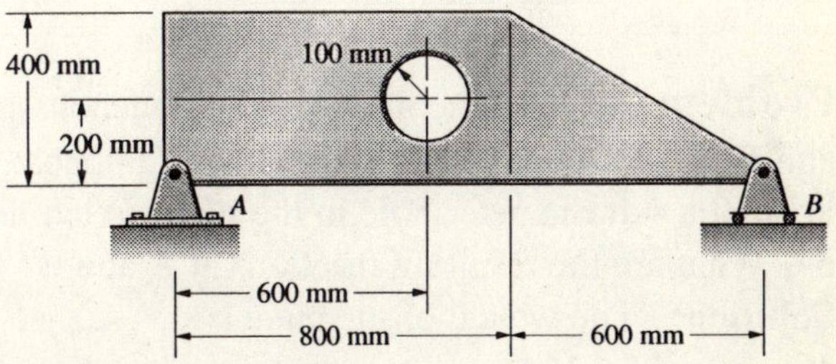

Area	A, sq mm	x
Rectangle	3.2×10^5	400
Circle	3.14×10^4	600
Triangle	1.2×10^5	1000
Composite	4.09×10^5	561

The composite area is $A = A_{rect} - A_{circ} + A_{triang}$. The composite x-coordinate of the centroid is

$$\bar{x} = \frac{A_{rect}\bar{x}_{rect} - A_{circ}\bar{x}_{circ} + A_{triang}\bar{x}_{triang}}{A}$$

. The sum of the moments about A:

$\sum M_A = -500(561) + 1400B = 0$, from which $\boxed{B = 200\ N}$ The sum of the forces:

$\sum F_y = A_y + B - 500 = 0$, from which $\boxed{A_y = 300\ N}$. $\boxed{\sum F_x = A_x = 0}$

Problem 7.70 The arrangement shown can be used to determine the position of the center of mass of a person. A horizontal board has a pin support at A and rests on a scale that measures weight at B. The distance from point A to point B is 90 *in*. When the person is not on the board, the scale weighs 20 *lb*. (a) When a 135 *lb* person is in position (1), the scale measures 98 *lb*. What is the x-coordinate of the center of mass? (b) When the person is in position (2) the scale measures 94 *lb*. What is the x-coordinate of the person's center of mass?

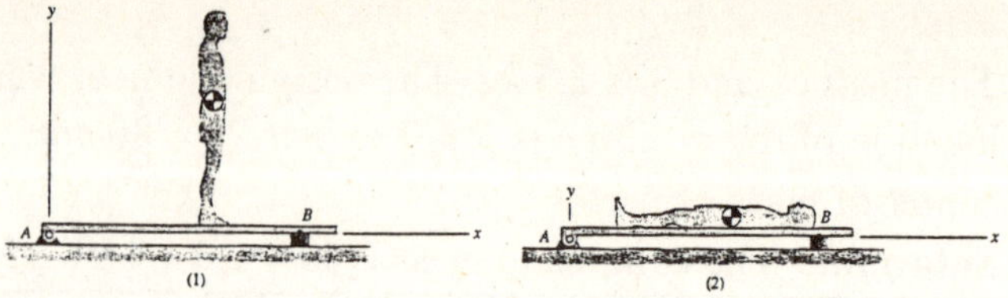

Solution: (a) Without the person the sum of the moments about A is $\sum M_A = -W_b\bar{x}_b + 90B = 0$, from which $W_b\bar{x}_b = 90B = 90(20) = 1800$ *in - lb*. With the person on the board, $\sum M_A = -(W_p + W_b)\bar{x} + 90B = 0$, from which $(W_p + W_b)\bar{x} = 90(98) = 8820$ *in - lb*. Noting that $(W_p + W_b)\bar{x} = W_p\bar{x}_p + W_b\bar{x}_b$, then

$W_p\bar{x}_p = 8820 - 1800 = 7020$ *in - lb*. For $W_p = 135$ *lb*, $\boxed{\bar{x}_p = \dfrac{7020}{135} = 52 \text{ in.}}$

(b) From above, $(W_p + W_b)\bar{x} = 90(94) = 8460$, from which $W_p\bar{x}_p = 8460 - 1800 = 6660$ *in - lb* For

$W_p = 135$ *lb* $\boxed{\bar{x}_p = \dfrac{6660}{135} = 49.3 \text{ in.}}$

Problem 7.71 The area of a homogenous plate is $10\,ft^2$. The vertical reactions on the plate at A and B are 80 *lb* and 84 *lb* respectively. Suppose that you want to equalize the reactions at A and B by drilling a 1-ft diameter hole in the plate. What horizontal distance from A should the center of the hole be? What are the resulting reactions at A and B?

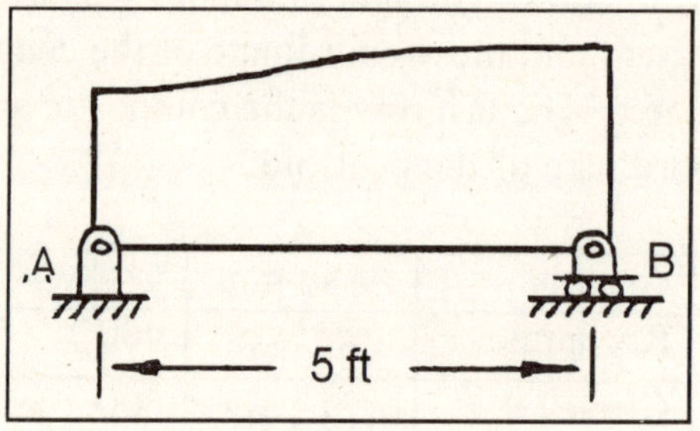

Solution: The weight of the plate is $W = 80 + 84 = 164$ *lb*. From the sum of moments about A, the centroid is $\bar{x} = \dfrac{84(5)}{W} = 2.56\,ft$. The weight density is $w = \dfrac{W}{10} = 16.4\ lb/ft^2$. The weight of the cutout is

$W_C = \pi(0.5^2)w = 12.88$ *lb*. The new weight of the plate is

$W_2 = W - W_C = 151.12$ *lb*. The new centroid must be at $\bar{x}_2 = \dfrac{5}{2} = 2.5\,ft$ for the reactions to be equal.

Therefore the x-coordinate of the center of the circle will be $\boxed{\bar{x}_C = \dfrac{W\bar{x} - W_2\bar{x}_2}{W_C} = 3.26\,ft}$ The reactions at

A and B will be $\boxed{A = B = \dfrac{W_2}{2} = 75.56\ lb}$

Problem 7.72 When the truck is unloaded, the total reactions at the front and rear wheels are $A = 12\ kip$ and $B = 8\ kip$. The load of gravel weighs $100\ lb/ft^3$. The dimension of the load in the z-direction is $10\ ft$ and its surface profile, given by the function shown, does not depend on z. What are the total reactions on the front and rear wheels?

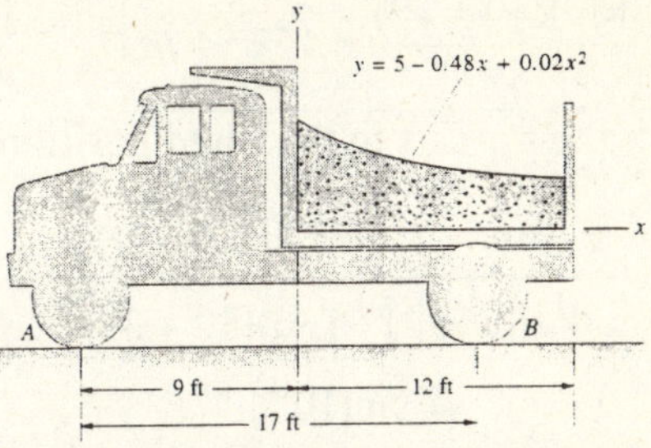

Solution: The weight of the empty truck is $W_t = (12+8)\times 10^3 = 20{,}000\ lb$. The center of mass of the empty truck is obtained from the sum of moments about A, $\sum M_A = -W_t\bar{x}_t + 17(8) = 0$, from which $\bar{x}_t = 6.8\ ft$.

The area of the gravel load is

$$A = \int_0^{12} dA = \int_0^{12} (5 - 0.48x + 0.02x^2)dx$$

$$= \left[5x - 0.24x^2 + \frac{0.02}{3}x^3\right]_0^{12} = 36.96\ sq\,ft$$. The weight of the gravel load is $W_g = 10A(100) = 36{,}960\ lb$. The center of mass of the load is

$\int_A x dA = \int_0^{12}(5 - 0.48x + 0.02x^2)xdx = = \left[2.5x^2 - 0.16x^3 + 0.005x^4\right]_0^{12} = 187.2$. Divide by the area to obtain $\bar{x} = 5.065\ ft$ from the front of the gravel load. From the front wheels, $\bar{x}_G = \bar{x}_g + 9 = 14.065\ ft$.

The sum of the moments about the front wheels: $\sum M_A = -W_t\bar{x}_t - W_g\bar{x}_G + 17B = 0$, from which $\boxed{B = 38{,}579\ lb}$ The sum of the forces: $\sum F_y = A - W_t - W_g + B = 0$, from which $\boxed{A = 18{,}381\ lb}$

Problem 7.73 A suitcase weighing 200 lb is put into the trunk of the car described in Example 7.14. The position of the center of mass of the suitcase is $\bar{x}_s = -21\ in.$, $\bar{y}_s = 30\ in.$, and $\bar{z}_s = -12\ in.$. If you consider the suitcase a part of the car, what is the position of the car's center of mass?

Solution: From Example 7.14, the center of mass of the car with empty trunk is $\bar{x}_c = 65\ in.$, $\bar{y}_c = 23\ in.$, $\bar{z}_C = 30.2\ in$. The weight of the car is given as $W_C = 3890\ lb$.

The composite weight $W = W_s + W_c = 3190\ lb$. The composite center of mass:

$\boxed{\bar{x} = \frac{W_c\bar{x}_c + W_s\bar{x}_s}{W} = 60.79\ in.}$ $\boxed{\bar{y} = \frac{W_c\bar{y}_c + W_s\bar{y}_s}{W} = 23.34\ in.}$ $\boxed{\bar{z} = \frac{W_c\bar{z}_c + W_s\bar{z}_s}{W} = 28.16\ in.}$

Problem 7.74 A group of engineering students construct a miniature device of the kind described in Example 7.14 and use it to determine the center of mass of a go-cart. The data they obtain is shown in the following table:

Wheelbase = 36 *in.*, Track = 30 *in.*	Measured Loads, lbs	
	α=0	α=10°
Left front Wheel, N_{LF}	35	32
Right front Wheel, N_{RF}	36	33
Left rear Wheel, N_{LR}	27	34
Right rear Wheel, N_{RR}	29	30

Determine the center of mass of the go-cart. Use the same coordinate system as in Example 7.14.

Problem 7.74 continued on next page....

continuation of solution to Problem 7.74...

The weight of the go-cart: $W = 35 + 36 + 27 + 29 = 127\ lb$. The sum of the moments about the z-axis

$\sum M_{z-axis} = (Wheelbase)(N_{LF} + N_{RF}) - \bar{x}W = 0$, from which $\boxed{\bar{x} = \dfrac{36(35+36)}{W} = 20.125\ in.}$

The sum of the moments about the x-axis: $\sum M_{x-axis} = \bar{z}W - (Track)(N_{RF} + N_{RR}) = 0$, from which

$$\boxed{\bar{z} = \frac{(30)(36+29)}{W} = 15.354\ in.}$$

With the go-cart in the tilted position, the sum of the moments about the z-axis

$\sum M_{z-axis} = (Wheelbase)(N_{LF} + N_{RF}) + \bar{y}W\sin(10^o) - \bar{x}W\cos\left(10^o\right) = 0$, from which

$$\boxed{\bar{y} = \frac{\bar{x}W\sin(10^o) - (36)(32+33)}{W\sin\left(10^o\right)} = 8.034\ in.}$$

Problem 7.75 With its engine removed, the mass of the car is 1100 *kg* and its center of mass is at C. The mass of the engine is 220 *kg*. (a) Suppose the you want to place the center of mass E of the engine so that the center of mass of the car is midway between the front wheels A and the rear wheels B. What is the distance b? (b) If the car is parked on a 15° slope facing up the slope, what total normal force is exerted by the road on the rear wheels B?

Solution: (a) The composite mass is $m = m_C + m_E = 1320\ kg$. The x-coordinate of the composite center of mass is given:

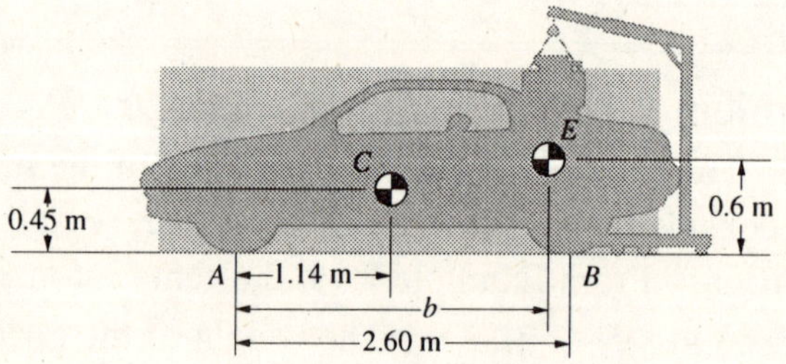

$\bar{x} = \dfrac{2.6}{2} = 1.3\ m$, from which the x-coordinate of the center of mass of the engine is

$\boxed{\bar{x}_E = b = \dfrac{(1.3m - 1.14m_C)}{m_E} = 2.1\ m}$ The y-coordinate of the composite center of mass is

$$\bar{y} = \frac{0.45m_C + 0.6m_E}{m} = 0.475\ m.$$

(b) Assume that the engine has been placed in the new position, as given in Part (a). The sum of the moments about B is $\sum M_A = 2.6A + \bar{y}mg\sin(15^o) - (2.6 - \bar{x})mg\cos(15^o) = 0$, from which $A = 5641.7\ N$. This is the normal force exerted by the road on A. The normal force exerted on B is obtained from; $\sum F_N = A - mg\cos(15^o) + B = 0$, from which $\boxed{B = 6866\ N}$

==================================<>==================================

Problem 7.76 The airplane is parked with its landing gear resting on scales. The weights measured at A, B, and C are 30 *kN*, 140 *kN*, and 146 *kN*, respectively. After a crate is loaded onto the plane, the weights measured at A, B, and C are 31 *kN*, 142 *kN*, and 147 *kN*, respectively. Determine the mass and the x and y coordinates of the center of mass of the crate.

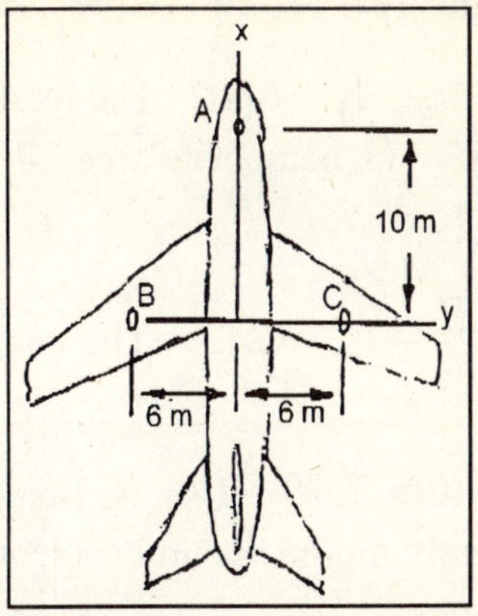

Solution: The weight of the airplane is $W_A = 30 + 140 + 146 = 316\ kN$. The center of mass of the airplane: $\sum M_{y-axis} = 30(10) - \bar{x}_A W_A = 0$, from which $\bar{x}_A = 0.949\ m$. $\sum M_{x-axis} = (140 - 146)(6) + \bar{y}_A W_A = 0$, from which $\bar{y}_A = 0.114\ m$. The weight of the loaded plane: $W = 31 + 142 + 147 = 320\ kN$. The center of mass of the loaded plane: $\sum M_{y-axis} = (31)10 - \bar{x}W = 0$, from which $\bar{x} = 0.969\ m$.

$\sum M_{x-axis} = (142 - 147)(6) + \bar{y}W = 0$, from which $\bar{y} = 0.0938\ m$.

The weight of the crate is $W_c = W - W_A = 4\ kN$. The center of mass of the crate:

$\boxed{\bar{x}_c = \frac{W\bar{x} - W_A\bar{x}_A}{W_c} = 2.5\ m}$, $\boxed{\bar{y}_c = \frac{W\bar{y} - W_A\bar{y}_A}{W_c} = -1.5\ m}$. The mass of the crate:

$\boxed{m_c = \frac{W_c \times 10^3}{9.81} = 407.75\ kg}$

==================================<>==================================

Problem 7.77 Use the Pappus-Guldinus theorems to determine the surface are A and volume V of a circular cylinder of length L and radius R.

Solution: For the area of the outside, choose the line to be the circle at the end of the cylinder. The centroid of the circle is (0,0), and the length of the line is $2\pi R$. The centroid moves a distance L, hence the surface area: $A = 2\pi RL$. For the area at the end of a cylinder, choose a radius line R. The centroid of the radius is $\left(\frac{R}{2}\right)$, and it moves a distance $d = 2\pi\left(\frac{R}{2}\right)$ and its length is R, thus the area of one end is $2\pi\left(\frac{R}{2}\right)R = \pi R^2$, from which the total area is $\boxed{A = 2\pi RL + 2\pi R^2}$. For the volume of the cylinder, choose an end with area πR^2. The centroid is (0,0), and it moves a distance L. Hence the volume of the cylinder is $\boxed{V = \pi R^2 L}$

==================================<>==================================

Problem 7.78 Use the second Pappus-Guldinus theorem to find the volume generated by revolving the line about the x-axis.

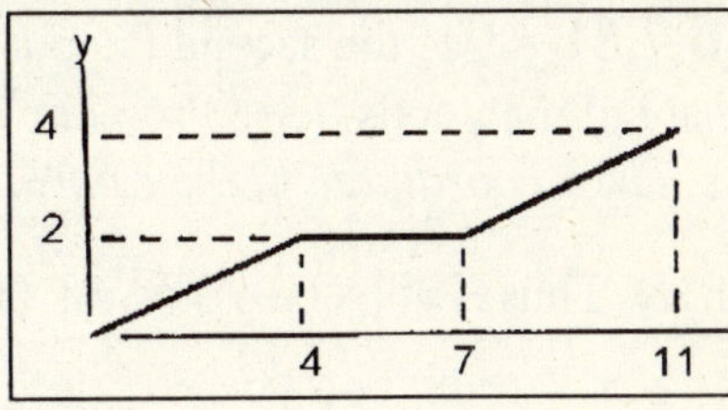

Solution: Divide the areas into four parts: two triangles and two rectangles.From left to right: (1) The area is $A_1 = \left(\frac{1}{2}\right)2(4) = 4$. The centroid y-coordinate:

$\bar{y}_1 = \left(\frac{1}{3}\right)2 = 0.6667$. The volume: $V_1 = 2\pi\bar{y}_1 A_1 = 16.755$. (2) The area is $A_2 = 2(3) = 6$. The centroid y-coordinate is $\bar{y}_2 = 1$, , the volume is $V_2 = 2\pi\bar{y}_2 A_2 = 37.7$.

Problem 7.78 continued on next page....

====================================<>====================================

continuation of solution to Problem 7.78...

(3) The area is $A_3 = 2(4) = 8$. The centroid y-coordinate : $\bar{y}_3 = 1$. The volume: $V_3 = 2\pi\bar{y}_3 A_3 = 50.27$.

(4) The area: $A_4 = \left(\frac{1}{2}\right)2(4) = 4$. The centroid y-coordinate is $\bar{y}_4 = 2 + \left(\frac{1}{3}\right)2 = 2.6667$. The volume is $V_4 = 2\pi\bar{y}_4 A_4 = 67.02$. The total volume: $V = V_1 + V_2 + V_3 + V_4 = 171.74$

Check: The composite area: $A = A_1 + A_2 + A_3 + A_4 = 22$. The composite centroid is

$$\bar{y} = \frac{\sum_1^4 A_i \bar{y}_i}{A} = 1.2424$$. The volume: $V = 2\pi\bar{y}A = 171.74$. *end check on arithmetic.*

====================================<>====================================

Problem 7.79 Use the first Pappus-Guldinus theorem to determine the area of the surface generated by revolving the line in Problem 7.78 about the x-axis.

Solution: Divide the line into three segments: (1) The length $L_1 = \sqrt{2^2 + 4^2} = 4.472$. The centroid y-coordinate is $\bar{y}_1 = 1$. The area is $A_1 = 2\pi\bar{y}_1 L_1 = 28.1$. (2) The length is $L_2 = 3$. The centroid y-coordinate is $\bar{y}_2 = 2$. The area $A_2 = 2\pi\bar{y}_2 L_2 = 37.7$. (3) The length: $L_3 = \sqrt{2^2 + 4^2} = 4.472$. The centroid y-coordinate is $\bar{y}_3 = 3$. The area: $A_3 = 2\pi\bar{y}_3 L_3 = 84.3$. The composite area:

$$\boxed{A = A_1 + A_2 + A_3 = 150.1}$$

====================================<>====================================

Problem 7.80 Use the second Pappus-Guldinus theorem to find the volume generated by revolving the curve about the x-axis.

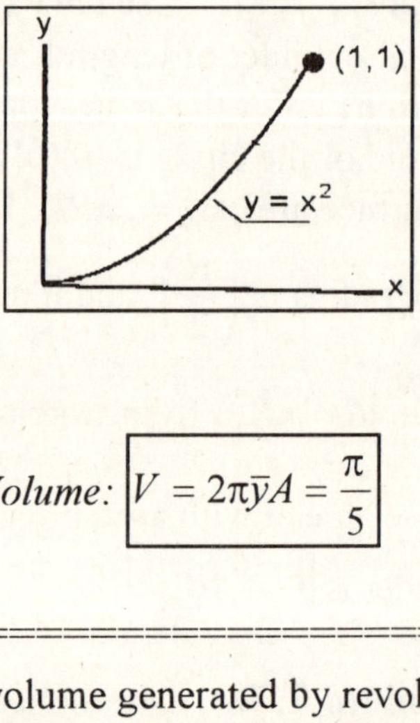

Solution: *The area:* The element of area is the vertical strip of height y and width dx. Thus $A = \int_0^1 y\,dx = \int_0^1 x^2 dx$. Integrating,

$A = \left[\frac{x^3}{3}\right]_0^1 = \frac{1}{3}$. *The y-coordinate:* $\int_A y\,dA = \int_0^1 y\,(1-x)dy =$

$\int_0^1 \left(y - y^{\frac{3}{2}}\right)dy = \left[\frac{y^2}{2} - \frac{2y^{\frac{5}{2}}}{5}\right]_0^1 = \frac{1}{10}$. Divide by the area: $\bar{y} = \frac{3}{10}$. *The Volume:* $\boxed{V = 2\pi\bar{y}A = \frac{\pi}{5}}$

====================================<>====================================

Problem 7.81 Use the second Pappus-Guldinus theorem to find the volume generated by revolving the curve about the y-axis. (See Problem 7.80.)

Solution: *The x coordinate of the centroid:* . The element of area is the vertical strip of height $(1-y)$ and width dx. Thus $A = \int_0^1 (1-y)dx = \int_0^1 (1-x^2)dx$. Integrating,

$A = \left[x - \frac{x^3}{3}\right]_0^1 = \frac{2}{3}$. $\int_A x\,dA = \int_0^1 (x - x^3)\,dx = \left[\frac{x^2}{2} - \frac{x^4}{4}\right]_o^1 = \frac{1}{4}$, divide by the area: $\bar{x} = \frac{3}{8}$. The volume is $\boxed{V = 2\pi\bar{x}A = \frac{\pi}{2}}$ (See solution to Problem 7.22.)

====================================<>====================================

Problem 7.82 The length of the curve is $L = 1.479$, and the area generated by rotating it about the x-axis is $A = 3.810$. Use the first Pappus-Guldinus therorm to determine the coordinate of the centroid of the curve. (See Problem 7.80)

Solution: The surface area is $A = 2\pi \bar{y} L$, from which $\boxed{\bar{y} = \frac{A}{2\pi L} = 0.41}$

Problem 7.83 Use the first Pappus-Guldinus theorem to determine the area of the surface generated by revolving the curve about the y-axis. (See Problem 7.80.)

Solution: The length of the line is given in Problem 7.82. $L = 1.479$. The elementary length of the curve is $dL = \sqrt{1 + \left(\frac{dy}{dx}\right)^2}\, dx$. Noting $\frac{dy}{dx} = 2x$, the element of line is $dL = \left(1 + 4x^2\right)^{\frac{1}{2}}$.

The x-coordinate: $\int_L x dL = \int_0^1 x\left(1 + 4x^2\right)^{\frac{1}{2}} dx = \frac{1}{12}\left[\left(1 + 4x^2\right)^{\frac{3}{2}}\right]_0^1 = \frac{5^{\frac{3}{2}} - 1}{12} = 0.8484$ Divide by the length to obtain $\bar{x} = 0.5736$ (See Example 7.4). The surface area is $\boxed{A = 2\pi \bar{x} L = 5.33}$

Problem 7.84 A nozzle for a large rocket engine is designed by revolving the function $y = \left(\frac{2}{3}\right)(x-1)^{\frac{3}{2}}$ about the y-axis. Use the first Pappus-Guldinus theorem to determine the surface area of the nozzle.

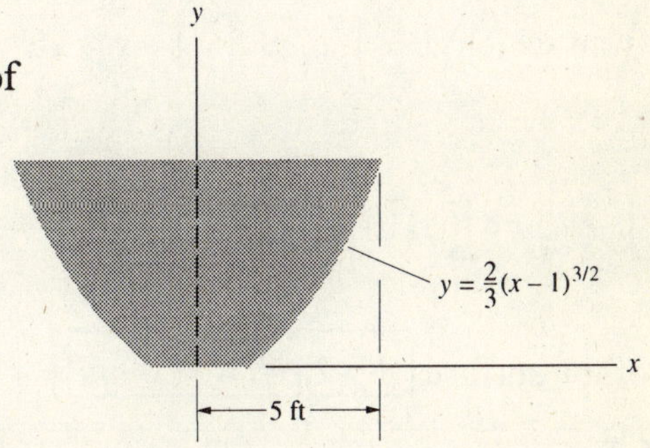

Solution: *The length:* Noting that $\frac{dy}{dx} = (x-1)^{\frac{1}{2}}$, the element of length is $dL = \sqrt{1 + \left(\frac{dy}{dx}\right)^2}\, dx = \sqrt{x}\, dx$ from which

$$L = \int_L dL = \int_1^5 (x)^{\frac{1}{2}}\, dx = \left[\frac{2}{3}(x)^{\frac{3}{2}}\right]_1^5 = 6.7869\ ft$$

The x-coordinate: $\int_L x\, dL = \int_1^5 x^{\frac{3}{2}}\, dx = \left[\frac{2}{5}x^{\frac{5}{2}}\right]_1^5 = 21.961$. Divide by the length: $\bar{x} = 3.2357$. The area

$\boxed{A = 2\pi \bar{x} L = 138\ ft^2}$

Problem 7.85 A volume of revolution is obtained by revolving the function $y=\left(\frac{2}{3}\right)x^{\frac{3}{2}}$ about the y-axis. Use the second Pappus-Guldinus theorem to determine its volume.

Solution: The element of area is the vertical strip between

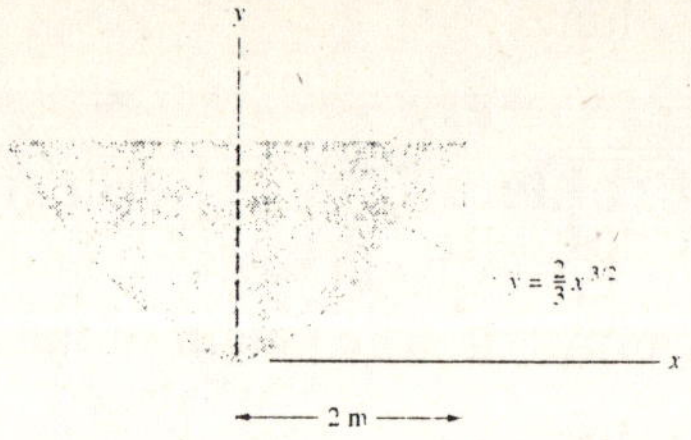

$y_1=\left(\frac{2}{3}\right)\left[x^{\frac{3}{2}}\right]_{x=2}=1.8856$ and y. $dA=(y_1-y)dx=(y_1-\left(\frac{2}{3}\right)x^{\frac{3}{2}})dx$.

The area is $\int_0^2(y_1-\left(\frac{2}{3}\right)x^{\frac{3}{2}})dx=\left[xy_1-\left(\frac{2}{3}\right)\left(\frac{2}{5}\right)x^{\frac{5}{2}}\right]_0^2=2.2627$. The x-coordinate of the centroid: $\int_0^2(y_1-\left(\frac{2}{3}\right)x^{\frac{3}{2}})xdx=\left[y_1\frac{x^2}{2}-\left(\frac{2}{3}\right)\left(\frac{2}{7}\right)x^{\frac{7}{2}}\right]_0^2=1.6162$. Divide by the area to obtain $\bar{x}=0.714\ m$. The volume is $\boxed{V=2\pi\bar{x}A=10.155\ m^3}$

Problem 7.86 Use the first Pappus-Guldinus theorem to determine the area of the curved surface of the volume of revolution in Problem 7.85.

Solution: *The length:* Noting that $\frac{dy}{dx}=x^{\frac{1}{2}}$ the element of length is $dL=\sqrt{1+\left(\frac{dy}{dx}\right)^2}\,dx=\sqrt{1+x}\,dx$ from which $L=\int_L dL=\int_0^2(1+x)^{\frac{1}{2}}\,dx=\left[\frac{2}{3}(1+x)^{\frac{3}{2}}\right]_0^2=2.7974\ m$. *The x-coordinate:*

$$\int_L x\,dL=\int_0^2 x(1+x)^{\frac{1}{2}}\,dx=2\left[\frac{(1+x)^{\frac{5}{2}}}{5}-\frac{(1+x)^{\frac{3}{2}}}{3}\right]_0^2=3.0379.$$

Divide by the length: $\bar{x}=1.086$ The area of the surface $\boxed{A=2\pi\bar{x}L=19.09\ m^2}$

Problem 7.87 The volume of revolution has a hole of radius R. (a) Use integration to determine its volume. (b) Use the second Pappus-Guldinus theorem to determine its volume.

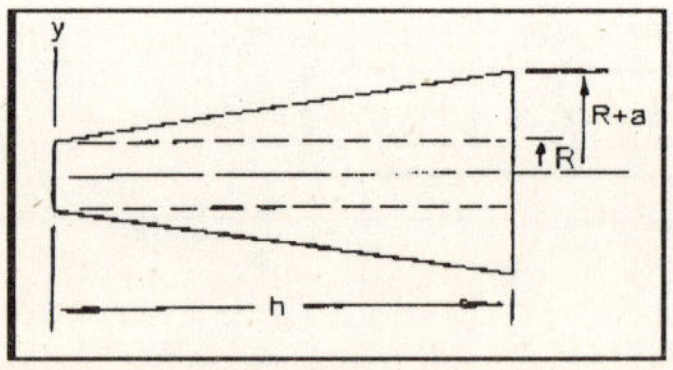

Solution: (a) The element of volume is a disk of radius y and thickness dx The area of the disk is $\pi\ (y^2-R^2)$ The radius is $y=\left(\frac{a}{h}\right)x+R$, from which $dV=\pi\left(\frac{a}{h}x+R\right)^2dx-\pi R^2dx$.

Denote $m=\left(\frac{a}{h}\right)$, $dV=\pi\left(m^2x^2+2mRx\right)dx$, from which $V=\int_V dV=\pi m\int_0^h(mx^2+2Rx)dx$

$$=\pi m\left[m\frac{x^3}{3}+Rx^2\right]_0^h=\pi mh^2\left(\frac{mh}{3}+R\right)=\pi ah\left(\frac{a}{3}+R\right).$$

Problem 7.87 continued on next page

continuation of Problem 7.87..........

continuation of Problem 7.87..........

(b) The area of the triangle is $A = \left(\frac{1}{2}\right)ah$. The y-coordinate of the centroid is $\bar{y} = R + \left(\frac{1}{3}\right)a$. The volume is $\boxed{V = 2\pi\bar{y}A = \pi ah\left(R + \left(\frac{1}{3}\right)a\right)}$

Problem 7.88 Determine the volume of the volume of revolution.

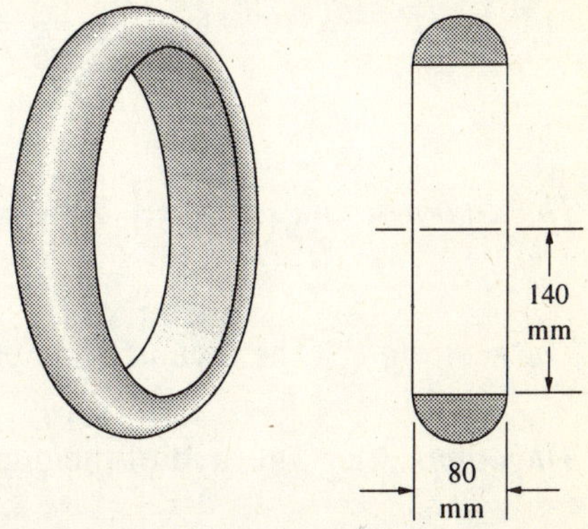

Solution: The area of the semicircle is $A = \frac{\pi r^2}{2}$. The centroid is $y = R + \frac{4r}{3\pi}$. The volume is

$V = 2\pi\left(\frac{\pi r^2}{2}\right)\left(R + \frac{4r}{3\pi}\right) = \pi^2 r^2 (R + \frac{4r}{3\pi})$. For $r = 40$ *mm* and $R = 140$ *mm*, $\boxed{V = 2.48 \times 10^{-3}\ m^3}$

Problem 7.89 Determine the surface area of the volume of revolution in Problem 7.88.

Solution: The length and centroid d of the semicircle is $L_o = \pi r$, $\bar{y} = R + \frac{2r}{\pi}$. The length and centroid of the inner line is $L_i = 2r$, and $\bar{y} = R$. $A = 2\pi(\pi r)\left(R + \frac{2r}{\pi}\right) + 2\pi(2r)(R) = 2\pi r(\pi R + 2r + 2R)$. For $r = 40$ *mm* and $R = 140$ *mm*, $\boxed{A = 0.201\ m^2}$

Problem 7.90 Determine the area of the surface of revolution obtained by revolving the line in Problem 7.61 about the x-axis.

Solution: From the solution to Problem 7.61, the composite length of the line is 400 mm, and the y-coordinate of the centroid is 54 mm. The area is $\boxed{A = 2\pi(400)(54) = 0.1357\ m^2}$

Problem 7.91 The mass of the Moon is 0.0123 times the mass of the Earth. If the Earth's center of mass is 238,000 miles from the center of mass of the moon, what is the distance of the center of mass of the Earth to the center of mass of the Earth-Moon system?

Solution: The composite mass is $M = M_e(1 + 0.0123)$. The centroid distance is

$$\boxed{\bar{x} = \frac{M_e 0 + 0.0123 M_e(238000)}{M_e(1 + 0.0123)} = 2891.8\ miles}$$

Problem 7.92

Problem 7.92 Determine the x-coordinate of the center of mass of the homogenous flat plate.

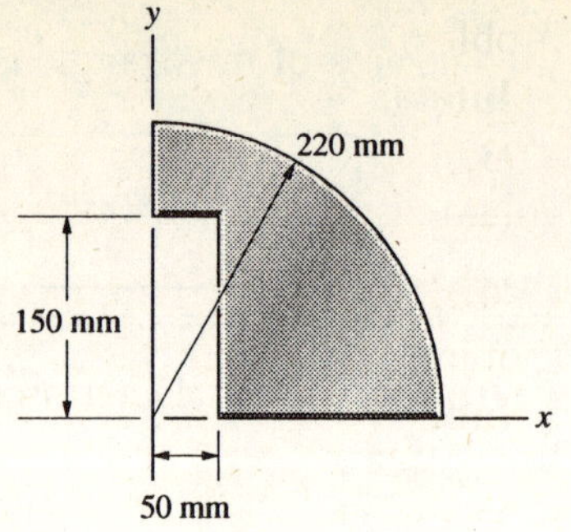

Solution: *The quarter circle:* The equation of the circle is $x^2 + y^2 = R^2$. Take the elemental area to be a vertical strip of height $y = \sqrt{R^2 - x^2}$ and width dx, hence the element of area is $dA = \sqrt{R^2 - x^2}dx$, and the area is

$$A = \int_0^R \sqrt{R^2 - x^2}dx = \left[\frac{x\sqrt{R^2 - x^2}}{2} + \frac{R^2}{2}\sin^{-1}\left(\frac{x}{R}\right)\right]_0^R = \frac{\pi R^2}{4}$$

The x-coordinate: $\bar{x}_C A = \int_A x\, dA = \int_0^R x\sqrt{R^2 - x^2}dx = \left[-\frac{(R^2 - x^2)^{\frac{3}{2}}}{3}\right]_0^R = \frac{R^3}{3}$ $\quad \bar{x}_C = \frac{4R}{3\pi}$

The rectangle: The area is $A = 50(150) = 7500\ mm^2$. The x-coordinate of the centroid is $\bar{x}_R = 25\ mm$.

The composite: The area of the quarter circle is $A_C = \frac{\pi(220)^2}{4} = 3.8013 \times 10^4\ mm$. The area of the rectangle is $A_R = 50(150) = 0.75 \times 10^4\ mm^2$. The composite area is $A = A_C - A_R = 3.0513 \times 10^4\ mm^2$.

The centroid: $\boxed{\bar{x} = \frac{A_C\bar{x}_C - A_R\bar{x}_R}{A} = 110\ mm}$

Problem 7.93

Problem 7.93 Determine the x-coordinate of the center of mass of the homogenous steel plate.

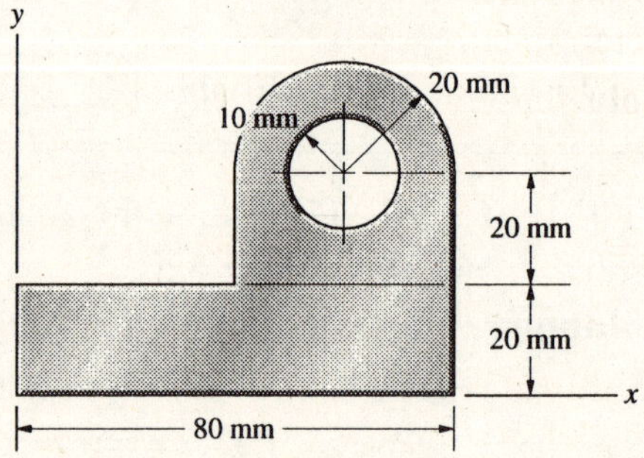

Solution: Divide the object into five areas: (1) The lower rectangle 20 by 80 *mm*, (2) an upper rectangle, 20 by 40 *mm*, (3) the semicircle of radius 20 *mm*, (4) the circle of radius 10 *mm*, and (5) the composite part. The areas and the centroids are tabulated. The last row is the composite and the centroid of the composite. The composite area is $A = \sum_1^3 A_i - A_4$. The centroid: $\bar{x} = \frac{\sum_1^3 A_i\bar{x}_i - A_4\bar{x}_4}{A}$, and $\bar{y} = \frac{\sum_1^3 A_i\bar{y}_i - A_4\bar{y}_4}{A}$. The following relationships were used for the centroids: For a rectangle: the centroid is at half the side and half the base. For a semicircle, the centroid is on the centerline and at $\frac{4R}{3\pi}$ from the base. For a circle, the centroid is at the center.

Area	A, sq mm	x, mm	y, mm
A1	1600	40	10
A2	800	60	30
A3	628.3	60	48.5
A4	314.2	60	40
Composite	2714	48.2	21.3

Problem 7.94 Determine the y-coordinate of the center of mass of the homogenous steel plate in Problem 7.93.

Solution: The solution is given in the last column and last row of the table in the solution to Problem 9.93.

Problem 7.95 Estimate the centroid of the of the volume of the Apollo Lunar return configuration (not including its rocket nozzle) by treating is as a cone and a cylinder.

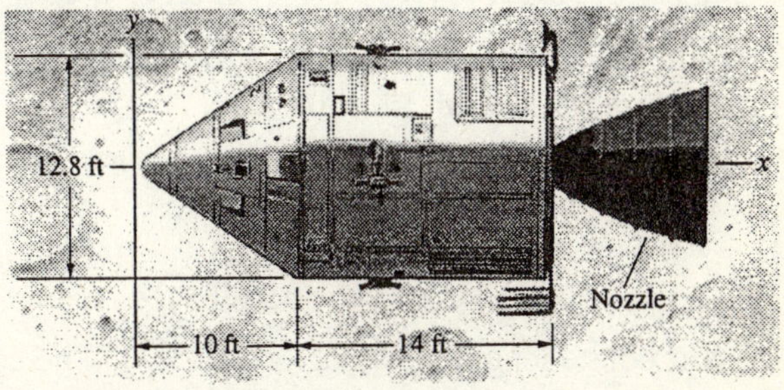

Solution: The volume of the cone is

$V_1 = \dfrac{\pi R^2 h}{3} = 428.93\ ft^3$. The x-coordinate of the centroid from the nose of the cone is

$\bar{x}_1 = \dfrac{3h}{4} = 7.5\ ft$. The volume of the cylinder is

$V_2 = \pi R^2 L = 1801.5\ ft^3$. The x-coordinate of the centroid from the nose of the cone is $\bar{x}_2 = h + \dfrac{L}{2} = 17\ ft$. The composite volume is

$V = V_1 + V_2 = 2230.4\ ft^3$. The x-coordinate of the composite centroid is $\boxed{\bar{x} = \dfrac{V_1 x_1 + V_2 x_2}{V} = 15.2\ ft}$. The y- and z-coordinates are zero, from symmetry.

Problem 7.96 The shape of the rocket nozzle of the Apollo Lunar return configuration is approximated by revolving the curve shown around the x-axis. In terms of the coordinate system shown, determine the centroid of the volume of the nozzle.

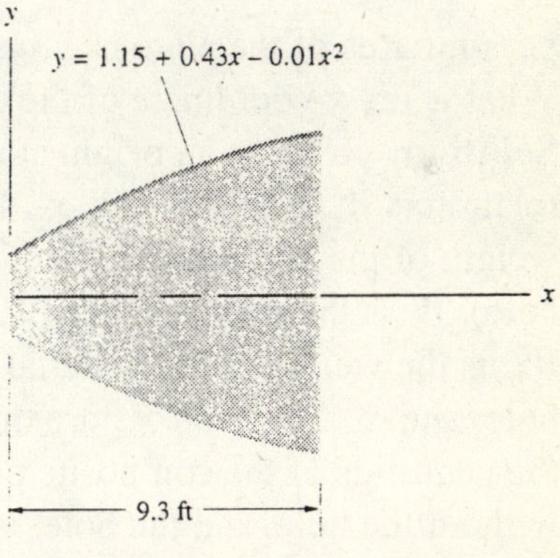

Solution: The element of volume is $dV = \pi y^2 dx$, from which

$V = \pi \int_0^{9.3} (1.15 + 0.43x - 0.01x^2)^2 dx$. Denote $a = 1.15$, $b = 0.43$, and $c = -0.01$. Expand the integrand :

$V = \pi \int_0^{9.3} (a^2 + 2abx + (2ac + b^2)x^2 + 2bcx^3 + c^2x^4) dx$.

$$V = \pi \left[a^2 x + abx^2 + \frac{(2ac + b^2)}{3} x^3 + \frac{bc}{2} x^4 + \frac{c^2}{5} x^5 \right]_0^{9.3}$$

$V = 263.2\ ft^3$.

$$\bar{x}V = \pi \int_0^{9.3} (a + bx + cx^2)^2\, x\, dx.$$

$$= \pi \int_0^{9.3} (a^2 x + 2abx^2 + (2abc + b^2)x^3 + 2bcx^4 + c^2 x^5) dx$$

$$= \pi \left[\frac{a^2x^2}{2} + \frac{2abx^3}{3} + \frac{(2ac + b^2)x^4}{4} + \frac{2bcx^5}{5} + \frac{c^2x^6}{6} \right]_0^{9.3} = 1621.9, \text{ from which } \boxed{\bar{x} = \frac{1621.9}{263.2} = 6.162\ ft}$$

$\bar{y} = \bar{z} = 0$ from symmetry.

Problem 7.97 If a string is tied to the slender bar at A and the bar is allowed to hang freely, what will be the angle between AB and the vertical?

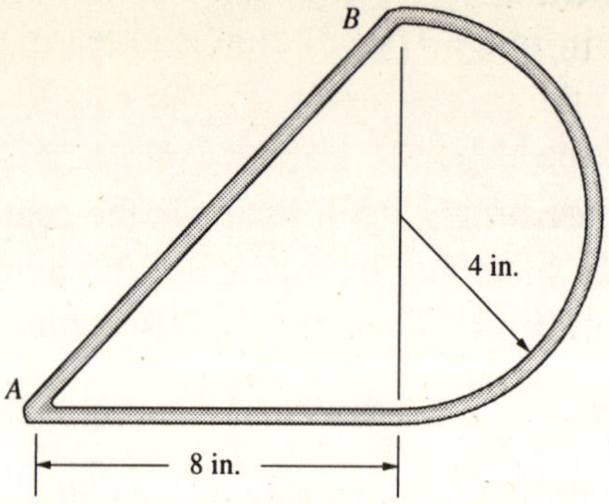

Solution: When the bar hangs freely, the action line of the weight will pass through the mass center. With a homogenous, slender bar, the mass center corresponds to the centroid of the lines making up the bar. Choose the origin at A, with the x-axis parallel to the lower bar. Divide the bar into three segments plus the composite: (1) The segment from A to the semi circle, (2) the segment AB, and (3) the semicircle.

(1) $L_1 = 8\ in,\ \bar{x}_1 = 4\ in,\ \bar{y}_1 = 0.$

(2) $L_2 = \sqrt{8^2 + 8^2} = 11.314\ in,\ \bar{x}_2 = 4,\ \bar{y}_2 = 4$

(3) $L_3 = 4\pi = 12.566\ in,\ \bar{x}_3 = 8 + \dfrac{2(4)}{\pi} = 10.546\ in,\ \bar{y}_3 = 4.$

The composite length $L = \sum_1^3 L_i = 31.88\ in$. The composite centroid: $\bar{x} = \dfrac{L_1\bar{x}_1 + L_2\bar{x}_2 + L_3\bar{x}_3}{L} = 6.58\ in$, $\bar{y} = \dfrac{L_1\bar{y}_1 + L_2\bar{y}_2 + L_3\bar{y}_3}{L} = 2.996\ in$. The angle from the point A to the centroid relative to the lower bar is $\alpha = \tan^{-1}\left(\dfrac{\bar{y}}{\bar{x}}\right) = 24.48^o$. The angle between AB and the lower bar is 45°, hence the angle between the line from A to the centroid and AB is $\boxed{\beta = 45 - \alpha = 20.52^o}$

Since the line from A to the centroid will be vertical, this is the angle between AB and the vertical.

Problem 7.98 The plate is of uniform thickness and is made of homogenous material whose mass per unit area of the plate is 2 kg/m^2. The vertical reactions at A and B are 6 N and 10 N respectively. What is the x-coordinate of the centroid of the hole?

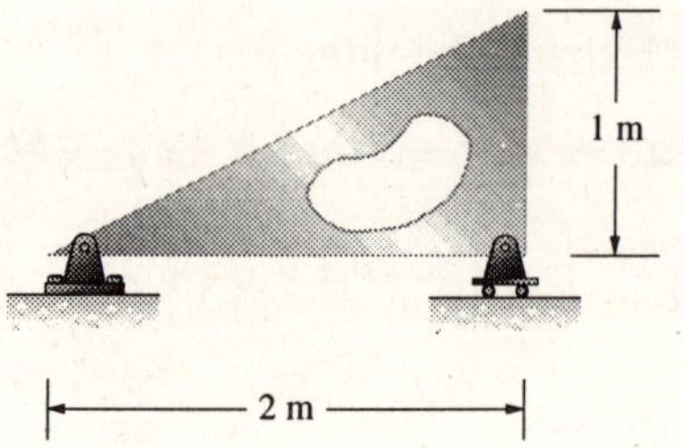

Solution: Choose an origin at A. The basic relation is $W_C\bar{x}_C = W\bar{x} - W_H\bar{x}_H$, where W_C is the weight of the composite plate (the one with the hole), W is the weight of the plate without the hole, W_H is the weight of the material removed from the hole, and $\bar{x}_C$, $\bar{x}$, and $\bar{x}_H$ are the x-coordinates of the centroids of the composite plate, the plate without the hole, and the hole, respectively.

The composite weight : $\sum F_Y = A + B - W_C = 0$, from which $W_C = 16\ N$. The x-coordinate of the centroid: $\sum M_A = -W_C\bar{x}_C + 2B = 0$, from which $\bar{x}_C = 1.25\ m$. The weight of the plate without the hole and the x-coordinate of the centroid: $W = \rho A g = \left(\dfrac{1}{2}\right)(2)(1)(2)(9.81) = 19.62\ N$, and

$\bar{x} = \left(\dfrac{2}{3}\right)2 = 1.3333\ m$. The weight of the material removed from the hole: $W_H = W - W_C = 3.62\ N$.

Solve: $\boxed{\bar{x}_H = \dfrac{W\bar{x} - W_C\bar{x}_C}{W_H} = 1.702\ m}$

Problem 7.99 Determine the center of mass of the homogenous sheet of metal.

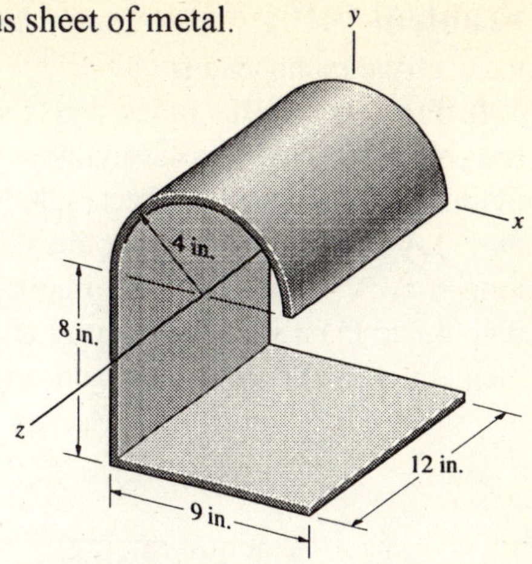

Solution: Divide the object into four parts: (1) The lower plate, (2) the left hand plate, (3) the semicircular plate, and (4) the composite plate. The areas and centroids are found by inspection:

(1) Area: $A_1 = 9(12) = 108\ in.^2$, $\bar{x}_1 = 0.5\ in.$, $\bar{y}_1 = -8\ in.$, $\bar{z}_1 = 6\ in.$

(2) $A_2 = 8(12) = 96\ in.^2$, $\bar{x}_2 = -4\ in.$, $\bar{y}_2 = -4\ in.$, $\bar{z}_2 = 6\ in.$

(3) $A_3 = \pi 4(12) = 150.8\ in.^2$, $\bar{x}_3 = 0, \bar{y} = \dfrac{2(4)}{\pi} = 2.546\ in.$, $z = 6\ in.$

The composite area is $A = \sum_1^3 A_i = 354.796\ in.^2$. The centroid for the composite: $\boxed{\bar{x} = \dfrac{\sum_1^3 A_i \bar{x}_i}{A} = -0.930\ in.}$ $\boxed{\bar{y} = \dfrac{\sum_1^3 A_i \bar{y}_i}{A} = -2.435\ in.}$ $\boxed{\bar{z} = \dfrac{\sum_1^3 A_i \bar{z}_i}{A} = 6\ in.}$

Problem 7.100 Determine the center of mass of the homogenous object.

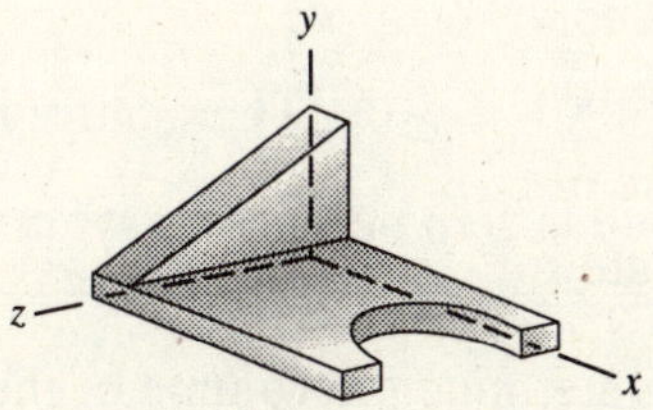

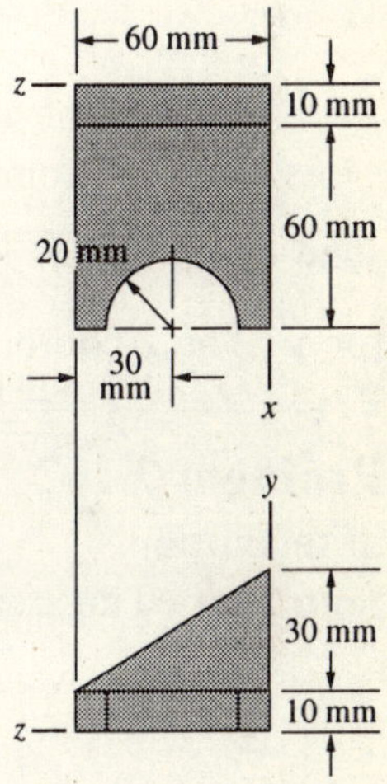

Solution: Divide the object into three parts and the composite: (1) A triangular solid 30 mm altitude, 60 mm base, and 10 mm thick. (2) A rectangle 60 by 70 mm by 10 mm. (3) A semicircle with radius 20 mm and 10 mm thick. The volumes and their centroids are determined by inspection:

(1) $V_1 = \left(\dfrac{1}{2}\right)(30)(60)(10) = 9000\ mm^3$, $\bar{x}_1 = 5\ mm$, $\bar{y}_1 = 10 + \dfrac{30}{3} = 20\ mm$, $\bar{z}_1 = \dfrac{60}{3} = 20\ mm$

(2) $V_2 = 60(70)(10) = 42000\ mm^3$, $\bar{x}_2 = 35\ mm$, $\bar{y}_2 = 5\ mm$, $\bar{z}_2 = 30\ mm$

(3) $V_3 = \dfrac{\pi 20^2}{2}(10) = 6283.2\ mm^3$, $\bar{x}_3 = 70 - \dfrac{4(20)}{3\pi} = 61.51\ mm$, $\bar{y}_3 = 5\ mm$, $\bar{z}_3 = 30\ mm$.

The composite volume is $V = V_1 + V_2 - V_3 = 44716.8\ mm^3$. The centroid is

$\boxed{\bar{x} = \dfrac{V_1\bar{x}_1 + V_2\bar{x}_2 - V_3\bar{x}_3}{V} = 25.237\ mm}$ $\boxed{\bar{y} = \dfrac{V_1\bar{y}_1 + V_2\bar{y}_2 - V_3\bar{y}_3}{V} = 8.019\ mm}$

$\boxed{\bar{z} = \dfrac{V_1\bar{z}_1 + V_2\bar{z}_2 - V_3\bar{z}_3}{V} = 27.99\ mm}$

Problem 7.101 Determine the center of mass of the homogenous object.

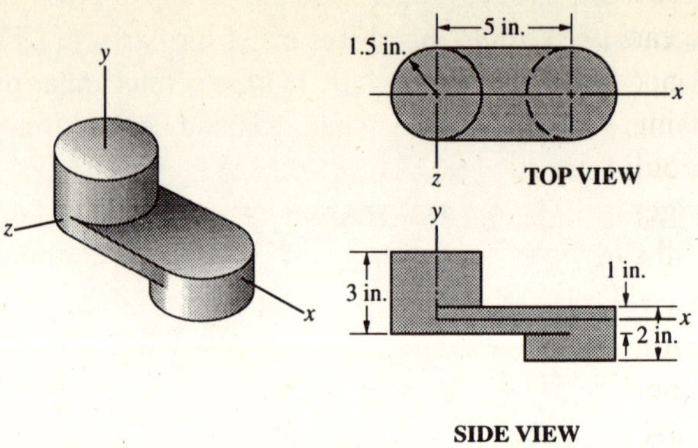

Solution: Divide the object into five parts plus the composite. (1) A solid cylinder with 1.5 *in.* radius, 3 *in.* long. (2) A rectangle 3 by 5 by 1 *in.* (3) A solid cylinder with radius 1.5 *in.*, 2 *in.* long. (4) A semicircle with radius 1.5 *in.* , 1 inch thick, (5) a semicircle with radius 1.5 *in.*, 1 inch thick. The volumes and centroids are determined by inspection. These are tabulated:

Part No	Vol,cu in.	$\bar{x}$, in.	$\bar{y}$, in.	$\bar{z}$, in.
V1	21.205	0	1	0
V2	15	2.5	0	0
V3	14.137	5	-0.5	0
V4	3.534	0.6366	0	0
V5	3.534	4.363	0	0
Composite	43.27	2.09	0.3267	0

The composite is $V = \sum_1^3 V_i - \sum_4^5 V_i$. The centroid: $\bar{x} = \dfrac{\sum_1^3 V_i\bar{x}_i - \sum_4^5 V_i\bar{x}_i}{V}$, with a corresponding expression for $\bar{y}$. The z-coordinate is zero because of symmetry.

Problem 7.102 Determine the volume of the volume of revolution.

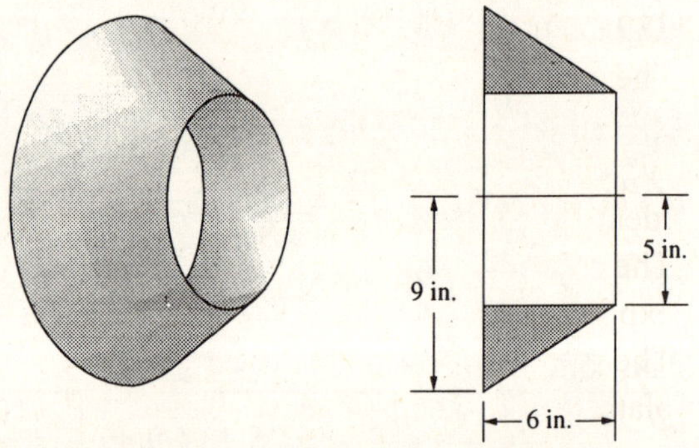

Solution: The area of the triangle is $A = \left(\frac{1}{2}\right)(4)(6) = 12\ in.^2$. The y-coordinate of the centroid is $\bar{y} = 5 + \left(\frac{1}{3}\right)4 = 6.3333\ in.$. The volume of revolution is $\boxed{V = 2\pi A\bar{y} = 477.52\ in.^3}$

Problem 7.103 Determine the surface area of the volume of revolution in Problem 7.102.

Solution: *The outer surface:* The length of the line is $L_1 = \sqrt{4^2 + 6^2} = 7.2111\ in.$. The y-coordinate of the centroid is $\bar{y} = 5 + 2 = 7\ in.$. The surface area is $A_1 = 2\pi L_1\bar{y}_1 = 317.16\ in.^2$.
The side surface: The length of the line is $L_2 = 4\ in$. The y-coordinate of the centroid is $\bar{y}_2 = 5 + 2 = 7\ in$. The surface area is $A_2 = 2\pi L_2\bar{y}_2 = 176.\ in^2$.
The inner surface: The length of the line is $L_3 = 6\ in$. The y-coordinate is $\bar{y}_3 = 5\ in$. The surface area is $A_3 = 2\pi L_3\bar{y}_3 = 188.5\ in.^2$. The total surface: $\boxed{A = A_1 + A_2 + A_3 = 681.6\ in.^2}$

==============================<>==============================

Problem 7.104 Engineering students participating in a charity benefit take part in an event in which a team of four students must stand on a circular platform with a ball and socket support and additional supports at the edges. Students A and B stand at the points shown. Knowing the coordinates of the points where A and B stand, the team must calculate the coordinates of points where students C and D should stand so that the center of mass of the team is at the center of the platform. The supports at the edges of the platform are then removed, and it the team has miscalculated, the platform will tilt and they will slide into a vat of mud. The students' weights are $W_A = 182\ lb$, $W_B = 122\ lb$, $W = 165\ lb$ and $W_D = 124\ lb$. Determine the coordinates of the points where C and D must stand.

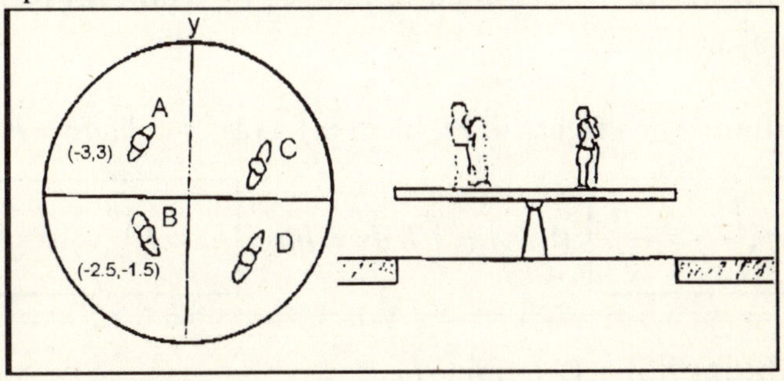

Solution: The strategy is to solve the problem in two stages: Assume that one person weighing the same as A and B is required to balance the platform. Solve for his position, and the determine the positions that C and D may take that are equivalent to his position and weight.

The sum of the moments about the y-axis: $\sum M_Y = W_A x_A + W_B x_B + Wx = 0$.

Substitute the weights and solve: $x = -\dfrac{(182(-3)+122(-2.5)}{289} = \dfrac{851}{289} = 2.94\ ft$. From the sum of moments about the x-axis, $y = -\dfrac{(182)(3)+(122)(-1.5)}{289} = -\dfrac{363}{289} = -1.256\ ft$. These are the coordinates for one (heavy) person to balance the platform.

The next step is to choose the coordinates the C and D must have to be equivalent to this one person. To be equivalent, the two must exert the same moment about the respective axes that the one person exerts. Thus, let (x_C, y_C) and (x_D, y_D) be the coordinates of C and D. The moments about the y- and x- axes:

$289(2.94) = 165x_C + 124x_D$,

$289(1.26) = 165y_C + 124y_D$

These may be written in the form of straight iine equations for convenience: rearranging,

$x_D = -1.33x_C + 6.85$,

$y_D = -1.33y_C - 2.294$.

Identical slopes means that the lines are parallel, although this has no apparent physical significance. The solution is given by these equations: given the coordinates (x_C, y_C) of C, the coordinates of D are explicitly specified. A solution exists only within the limits of the platform dimensions.

The proof that this is the solution follows from back substitution into the moment equations; the platform is indeed balanced for any allowed set of coordinates.

==============================<>==============================

Problem 8.1 Determine the I_y and k_y.

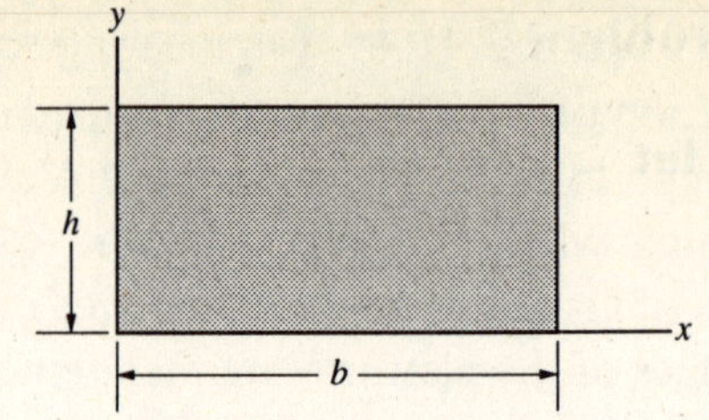

Solution: Let $dA = xdy = bdy$. $A = \int_0^h bdy = hb$.

$$\boxed{I_y = \int_A x^2 dA = h\int_0^b x^2 dx = h\left[\frac{x^3}{3}\right]_0^b = \frac{b^3 h}{3}}\quad \boxed{k_y = \sqrt{\frac{I_y}{A}} = \frac{b}{\sqrt{3}}}$$

Problem 8.2 Determine I_x and k_x by letting dA be (a) a horizontal strip of height dy; (b) a vertical strip of width dx.

Solution: (See figure in Problem 8.1.) (a) $A = \int_0^h bdy = hb$. $\boxed{I_x = \int_A y^2 dA = b\int_0^h y^2 dy = \frac{bh^3}{3}}$,

$\boxed{k_x = \sqrt{\frac{I_x}{A}} = \frac{h}{\sqrt{3}}}$ (b) $A = \int_0^b h\,dx = hb$ $\boxed{I_x = h\int_0^b y^2 dx = h\int_0^h y^2\left(\frac{b}{h}\right)dy = \frac{bh^3}{3}}$, $\boxed{k_x = \sqrt{\frac{I_x}{A}} = \frac{h}{\sqrt{3}}}$

Problem 8.3 Determine I_{xy}.

Solution: (See figure in Problem 8.1.) $A = \int_A dA = \int_0^b\int_0^h dx\,dy = hb$.

$$I_{xy} = \int_A xy\,dA = \int_0^b\int_0^h xy\,dx\,dy = \left[\frac{y^2}{2}\right]_0^h\left[\frac{x^2}{2}\right]_0^b \boxed{= \frac{h^2 b^2}{4}}$$

Problem 8.4 Determine I_x, k_x, I_y, and k_y.

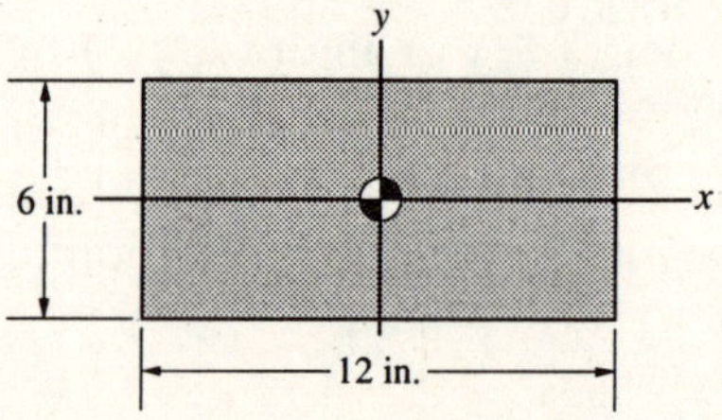

Solution:

$$A = \int_A dA = \int_{-3}^{+3}\int_{-6}^{+6} dy\,dx = [y]_{-3}^{+3}[x]_{-6}^{+6} = 6(12) = 72 \text{ in}^2$$

$$I_x = \int_A y^2 dA = \int_{-3}^{+3} y^2 dy\int_{-6}^{+6} dx = [x]_{-6}^{+6}\left[\frac{y^3}{3}\right]_{-3}^{+3}$$

$$\boxed{= \frac{24(27)}{3} = 216 \text{ in}^4}\quad \boxed{k_x = \sqrt{\frac{216}{72}} = \sqrt{3}\ in}$$

$$I_y = \int_A x^2 dA = \int_{-3}^{+3} dy\int_{-6}^{+6} x^2 dx = [y]_{-3}^{+3}\left[\frac{x^3}{3}\right]_{-6}^{+6}\quad \boxed{= \frac{12(216)}{3} = 864 \text{ in}^4}\quad \boxed{k_y = \sqrt{\frac{864}{72}} = 2\sqrt{3}\ in}$$

Problem 8.5 Determine I_{xy} and J_o for the area in Problem 8.4.

Solution: Use the solution of Problem 8.4. $I_{xy} = \int_A xydA = \int_{-3}^{+3} dy\int_{-6}^{+6} dy[xy] = \left[\frac{y^2}{2}\right]_{-3}^{+3}\left[\frac{x^2}{2}\right]_{-6}^{+6}$ $\boxed{= 0}$

$$\boxed{J_O = I_x + I_y = 216 + 864 = 1080\ in^4}$$

Problem 8.6 Determine I_y and k_y.

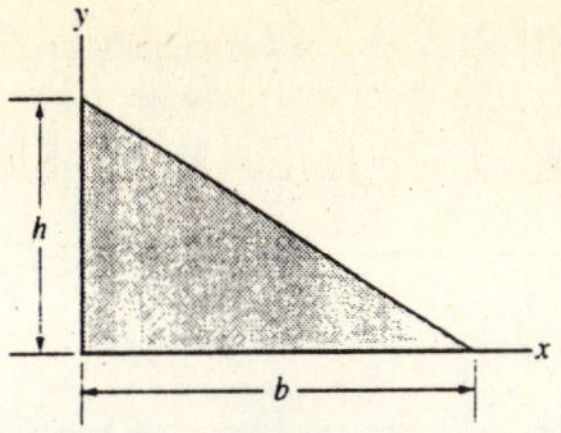

Solution: $y = -\left(\frac{h}{b}\right)x + h$, $dA = dydx$, therefore

$$A = \int_0^b dx \int_0^{-\frac{h}{b}x+h} dy = \int_0^b (-\frac{h}{b}x + h)dx = \left[-\frac{hx^2}{2b} + hx\right]_0^b = \frac{hb}{2}$$

$$I_y = \int_A x^2 dA = \int_0^b x^2 dx \int_0^{-\frac{h}{b}x+h} dy = \int_0^b \left(-\frac{h}{b}x^3 + hx^2\right)dx = \left[-\frac{hx^4}{4b} + \frac{hx^3}{3}\right]_0^b \boxed{= \frac{hb^3}{12}}$$

$$\boxed{k_y = \sqrt{\frac{I_y}{A}} = \frac{b}{\sqrt{6}}}$$

Problem 8.7 Determine J_o and k_o.

Solution: (See figure in Problem 8.6.) $y = -\left(\frac{h}{b}\right)x + h$, $dA = dxdy$

$$I_x = \int_A y^2 dA = \int_0^b dx \int^{-\frac{h}{b}x+h} y^2 dy = \frac{1}{3}\int_0^b \left(-\frac{h}{b}x + h\right)^3 dx = -\frac{b}{12h}\left[\left(-\frac{h}{b}x + h\right)^4\right]_0^b = \frac{bh^3}{12}$$

$$k_x = \sqrt{\frac{I_x}{A}} = \frac{h}{\sqrt{6}} \quad \boxed{J_o = I_x + I_y = \left(\frac{bh}{12}\right)(b^2 + h^2)} \quad \boxed{k_o = \sqrt{k_x^2 + k_y^2} = \sqrt{\frac{(b^2 + h^2)}{6}}}$$

Problem 8.8 Determine I_{xy}.

Solution: (See figure for Problem 8.6.) $I_{xy} = \int_A xy\, dA = \int_0^b x dx \int_0^{-\frac{h}{b}x+h} y dy = \frac{1}{2}\int_0^b x\left(-\frac{h}{b}x + h\right)^2 dx$

$$= \frac{1}{2}\left[\frac{h^2}{b^2}\frac{x^4}{4} - 2\frac{h^2}{b}\frac{x^3}{3} + h^2\frac{x^2}{2}\right]_0^b = \frac{h^2b^2}{2}\left[\frac{1}{4} - \frac{2}{3} + \frac{1}{2}\right] \boxed{= \frac{h^2b^2}{24}}$$

Problem 8.9 Determine I_Y and k_Y.

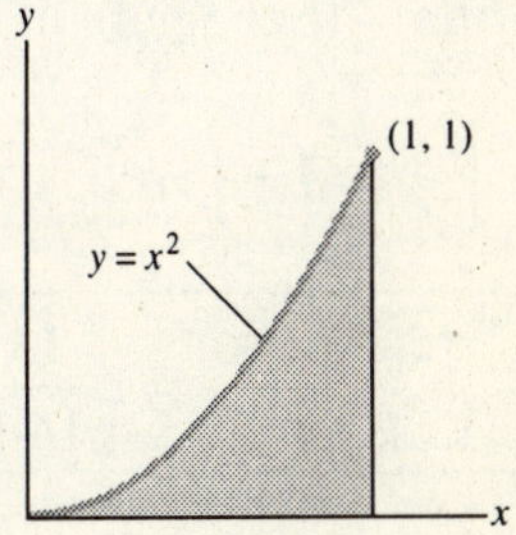

Solution: $dA = dxdy$ $A = \int_0^1 dx \int_0^{x^2} dy = \int_0^1 x^2 dx = \frac{1}{3}$.

$$I_y = \int_A x^2 dA = \int_0^1 x^2 dx \int_0^{x^2} dy = \int_0^1 x^4 dx \boxed{= \left[\frac{x^5}{5}\right]_0^1 = \frac{1}{5}}$$

$$\boxed{k_y = \sqrt{\frac{I_y}{A}} = \sqrt{\frac{3}{5}}}$$

Problem 8.10 Determine I_x and k_x.

Solution: (See figure in Problem 8.9.) $dA = dxdy$, $I_x = \int_A y^2 dA = \int_0^1 dx \int_0^{x^2} y^2 dy = \frac{1}{3}\int_0^1 x^6 dx$

$$= \left(\frac{1}{21}\right)\left[x^7\right]_0^1 = \frac{1}{21}$$ $$k_x = \sqrt{\frac{I_x}{A}} = \frac{1}{\sqrt{7}}$$

Problem 8.11 Determine J_o and k_o.

Solution: (See figure in Problem 8.9.) $J_o = I_x + I_y = \frac{1}{5} + \frac{1}{21} = \frac{26}{105}$,

$$k_o = \sqrt{k_x^2 + k_y^2} = \sqrt{\frac{3}{5} + \frac{1}{7}} = \sqrt{\frac{26}{35}}$$

Problem 8.12 Determine I_{xy}.

Solution: (See figure in Problem 8.9.) $dA = dxdy$

$$I_{xy} = \int_A xy\, dA = \int_0^1 xdx \int_0^{x^2} ydy = \frac{1}{2}\int_0^1 x^5 dx = \frac{1}{12}\left[x^6\right]_0^1 = \frac{1}{12}$$

Problem 8.13 Determine I_y and k_y.

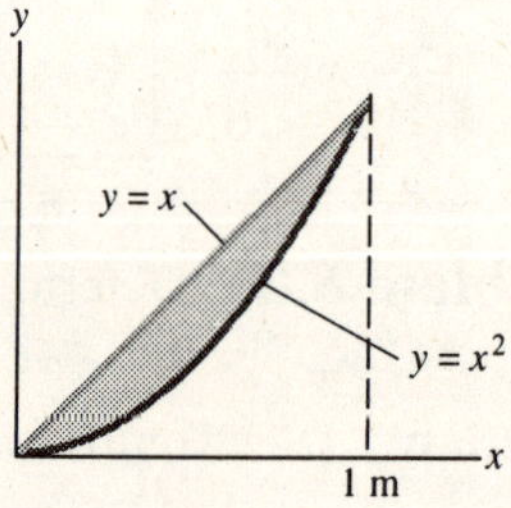

Solution: $dA = dxdy$ $A = \int_0^1 dx \int_{x^2}^{x} dy = \int_0^1 (x - x^2) dx = \left[\frac{x^2}{2} - \frac{x^3}{3}\right]_0^1 = \frac{1}{6}$

$$I_y = \int_A x^2 dy = \int_0^1 x^2 dx \int_{x^2}^{x} dy = \int_0^1 x^2 (x - x^2) dx = \left[\frac{x^4}{4} - \frac{x^5}{5}\right]_0^1 = \frac{1}{20} \text{ m}^4$$

$$k_y = \sqrt{\frac{I_y}{A}} = \sqrt{\frac{6}{20}} = \sqrt{\frac{3}{10}}\ m$$

Problem 8.14 Determine I_x and k_x.

Solution: (See figure in Problem 8.13.) $dA = dxdy$.

$$I_x = \int_A y^2 dA = \int_0^1 dx \int_{x^2}^{x} y^2 dx = \int_0^1 dx \left[\frac{y^3}{3}\right]_{x^2}^{x} = \frac{1}{3}\int_0^1 (x^3 - x^6) dx = \frac{1}{3}\left[\frac{x^4}{4} - \frac{x^7}{7}\right]_0^1 = \frac{1}{28} \text{ m}^4$$

$$k_x = \sqrt{\frac{I_x}{A}} = \sqrt{\frac{6}{28}} = \sqrt{\frac{3}{14}}\ m$$

Problem 8.15 Determine J_o and k_o.

Solution: (See figure in Problem 8.13.) $J_o = I_x + I_y = \frac{1}{28} + \frac{1}{20} = \frac{3}{35}\ m^4$

$$k_o = \sqrt{k_x^2 + k_y^2} = \sqrt{\frac{3}{10} + \frac{3}{14}} = \sqrt{\frac{18}{35}}\ m$$

Problem 8.16 Determine I_{xy}.

Solution: (See figure in Problem 8.13.)

$$I_{xy} = \int_A xy\,dA = \int_0^1 x\,dx\int_{x^2}^{x} y\,dy = \frac{1}{2}\int_0^1 x\left(x^2 - x^4\right)dx = \boxed{\frac{1}{2}\left[\frac{x^4}{4} - \frac{x^6}{6}\right]_0^1 = \frac{1}{24}\ \text{m}^4}$$

Problem 8.17 Determine k_y.

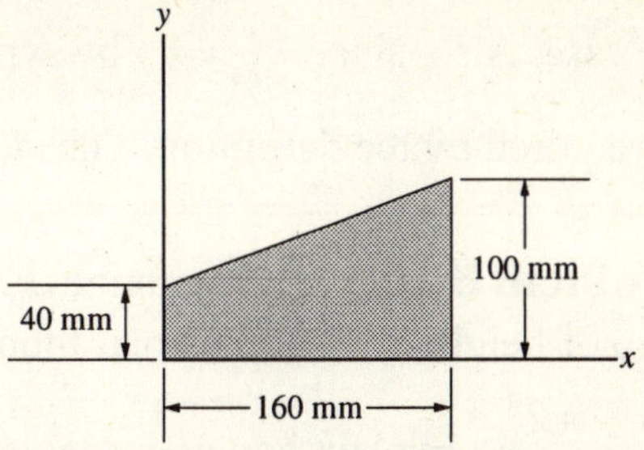

Solution: The upper boundary is a straight line,

$$y = \left(\frac{60}{160}\right)x + 40 = \frac{3}{8}x + 40. \quad dA = dx\,dy$$

$$A = \int_A dA = \int_0^{0.160} dx\int_0^{\frac{3}{8}x+0.040} dy = \int_0^{0.160}\left(\frac{3}{8}x + 0.040\right)dx$$

$$= \left[\frac{3}{16}x^2 + 0.040x\right]_0^{0.160} = 0.0112\ \text{m}^2 \qquad I_y = \int_A x^2 dA = \int_0^{0.160} x^2 dx\int_0^{\frac{3}{8}x+0.040} dy = \int_0^{0.160} x^2\left(\frac{3}{8}x + 0.040\right)dx$$

$$= \left[\frac{3}{8}\frac{x^4}{4} + 0.040\frac{x^3}{3}\right]_0^{0.160} = 1.16\times10^{-4}\ \text{m}^4 \quad \boxed{k_y = \sqrt{\frac{I_y}{A}} = \sqrt{\frac{1.16\times10^{-4}}{1.12\times10^{-2}}} = 0.1018\ m}$$

Problem 8.18 Determine k_o for the area in Problem 8.17.

Solution: From the solution to problem 8.17, $y = \frac{3}{8}x + 0.04$ m, and $A = 1.12\times10^{-2}\ \text{m}^2$.

$$I_x = \int_A y^2 dA = \int_0^{0.16} dx\int_0^{\frac{3}{8}x+0.04} y^2 dy = \frac{1}{3}\int_0^{0.16}\left(\frac{3}{8}x + 0.04\right)^3 dx = \frac{2}{9}\left[\left(\frac{3}{8}x + 0.04\right)^4\right]_0^{0.16} = 2.222\times10^{-5}\ \text{m}^4$$

$$k_x = \sqrt{\frac{I_x}{A}} = 4.454\times10^{-2}\ m. \quad \boxed{k_o = \sqrt{k_x^2 + k_y^2} = 0.1109\ m}$$

Problem 8.19 (a) Determine I_y and k_y by letting dA be a vertical strip of width dx. (b) The polar moment of inertia of a circular area with its center at the origin is $J_o = \frac{\pi R^4}{2}$. Explain how you can use this information to confirm your answer in Part (a).

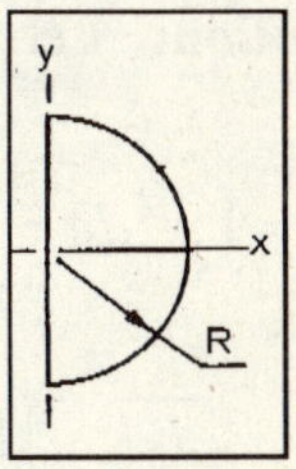

Solution: The equation of the circle is $x^2 + y^2 = R^2$, from which $y = \pm\sqrt{R^2 - x^2}$. The strip dx wide and y long has the elemental area $dA = 2\sqrt{R^2 - x^2}\,dx$. The area of the semicircle is $A = \frac{\pi R^2}{2}$ $I_y = \int_A x^2 dA = 2\int_0^R x^2\sqrt{R^2 - x^2}\,dx$

$$= 2\left[-\frac{x(R^2 - x^2)^{\frac{3}{2}}}{4} + \frac{R^2 x\left(R^2 - x^2\right)^{\frac{1}{2}}}{8} + \frac{R^4}{8}\sin^{-1}\left(\frac{x}{R}\right)\right]_0^R = \boxed{\frac{\pi R^4}{8}} \quad \boxed{k_y = \sqrt{\frac{I_y}{A}} = \frac{R}{2}}$$

Problem 8.19 continued on next page

continuation of Problem 8.19..........

(b) If the integration were done for a circular area with the center at the origin, the limits of integration for the variable x would be from $-R$ to R, doubling the result. Hence, doubling the answer above, $I_y = \frac{\pi R^4}{4}$. By symmetry, $I_x = I_y$, and the polar moment would be $J_o = 2I_y = \frac{\pi R^4}{2}$, which is indeed the case. Also, since $k_x = k_y$ by symmetry for the full circular area, $k_o = \sqrt{\frac{I_x}{A} + \frac{I_y}{A}} = \sqrt{2\frac{I_y}{A}} = \sqrt{\frac{J_o}{A}}$ as required by the definition. Thus the result checks.

Problem 8.20 (a) Determine I_x and k_x for the area in Problem 8.19 by letting dA be a horizontal strip of height dy. (b) The polar moment of inertia of a circular area with its center at the origin is $J_o = \frac{\pi R^4}{2}$. Explain how you can use this information to confirm your answer to Part (a).

Solution: Use the results of the solution to Problem 8.19, $A = \frac{\pi R^2}{2}$. The equation for the circle is $x^2 + y^2 = R^2$, from which $x = \pm\sqrt{R^2 - y^2}$. The horizontal strip is from 0 to R, hence the element of area is $dA = \sqrt{R^2 - y^2}\,dy$. $I_x = \int_A y^2 dA = \int_{-R}^{+R} y^2 \sqrt{R^2 - y^2}\,dy$

$$= \left[-\frac{y(R^2 - y^2)^{\frac{3}{2}}}{4} + \frac{R^2 y(R^2 - y^2)^{\frac{1}{2}}}{8} + \frac{R^4}{8}\sin^{-1}\left(\frac{y}{R}\right)\right]_{-R}^{R} \boxed{= \left[\frac{R^4}{8}\frac{\pi}{2} + \frac{R^4}{8}\frac{\pi}{2}\right] = \frac{\pi R^4}{8}} \quad \boxed{k_x = \sqrt{\frac{I_x}{A}} = \frac{R}{2}} \text{ (b)}$$

If the area were circular, the strip would be twice as long, and the moment of inertia would be doubled: $I_x = \frac{\pi R^4}{4}$. By symmetry $I_y = I_x$, and $J_o = 2I_x = \frac{\pi R^4}{2}$, which is indeed the result. Since $k_x = k_y$ by symmetry for the full circular area, the $k_o = \sqrt{\frac{I_x}{A} + \frac{I_y}{A}} = \sqrt{2\frac{I_x}{A}} = \sqrt{\frac{J_o}{A}}$ as required by the definition. This checks the answer.

Problem 8.21 Determine J_o and k_o.

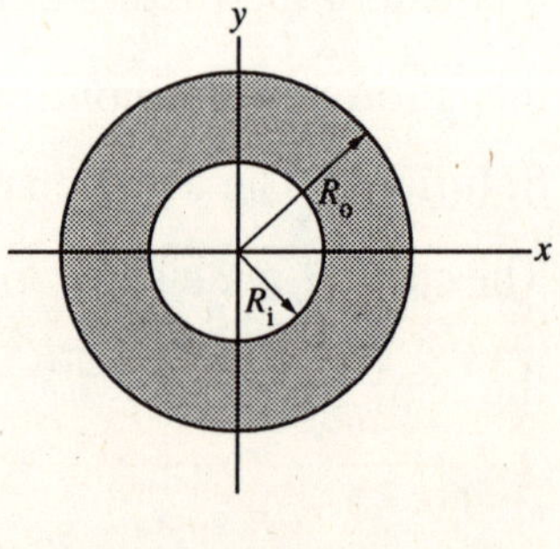

Solution: Let $dA = (r d\theta)dr = r dr d\theta$. $A = \int_0^{2\pi} d\theta \int_{R_i}^{R_o} r dr = \pi(R_o^2 - R_i^2)$

$$J_o = \int_A r^2 dA = \int_0^{2\pi} d\theta \int_{R_i}^{R_o} r^3 dr = 2\pi\left[\frac{r^4}{4}\right]_{R_i}^{R_o} \boxed{= \frac{\pi}{2}(R_o^4 - R_i^4)}$$

$$k_o = \sqrt{\frac{J_o}{A}} = \sqrt{\frac{1}{2}\frac{(R_o^4 - R_i^4)}{(R_o^2 - R_i^2)}} = \sqrt{\frac{1}{2}\frac{(R_o^2 + R_i^2)(R_o^2 - R_i^2)}{(R_o^2 - R_i^2)}} \boxed{= \sqrt{\frac{1}{2}(R_o^2 + R_i^2)}}$$

Problem 8.22 A vertical plate of area A is beneath the surface of a stationary body of water. The pressure of the water subjects each element dA of the surface of the plate to a force $(p_o + \gamma y)dA$, where p_o is the pressure at the surface of the water and γ is the weight density of the water. Show that the magnitude of the moment about the x-axis due to the pressure on the front face of the plate is $M_{(x\text{-axis})} = p_o \bar{y} A + \gamma I_X$

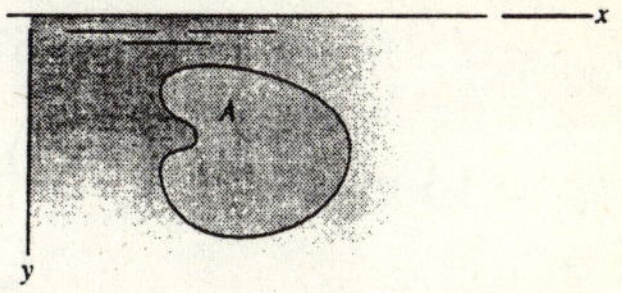

Solution: The moment about the x-axis is $dM = y(p_o + \gamma y)dA$. integrating over the surface of the plate: $M = \int_A (p_o + \gamma y) y\, dA$. Noting that p_o and γ are constants over the area, $M = p_o \int_A y dA + \gamma \int y^2 dA$. By definition,

$\bar{y} = \frac{\int_A y dA}{A}$ and $I_x = \int_A y^2 dA$, then $\boxed{M = p_o \bar{y} A + \gamma I_X}$, which demonstrates the result.

Problem 8.23 Determine I_y and k_y for the composite area in Example 8.4.

Solution:

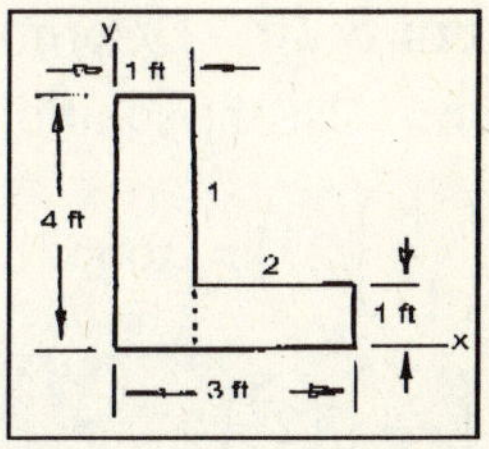

	dx, ft	A, sq ft	$I_{y'}$	$I_{y'} + d_x^2 A$
Part 1	0.5	4	0.333	1.333
Part 2	2	2	0.6667	8.6667
Iy		6		$10\ (ft^4)$
ky				$1.29\ (ft)$

The I_x and k_x are shown in the last column of the last two rows.

Problem 8.24 Determine I_x, k_x and I_{xy} for the composite area by dividing it into rectangles 1 and 2 as shown and compare your results to those of Example 8.4.

Solution:

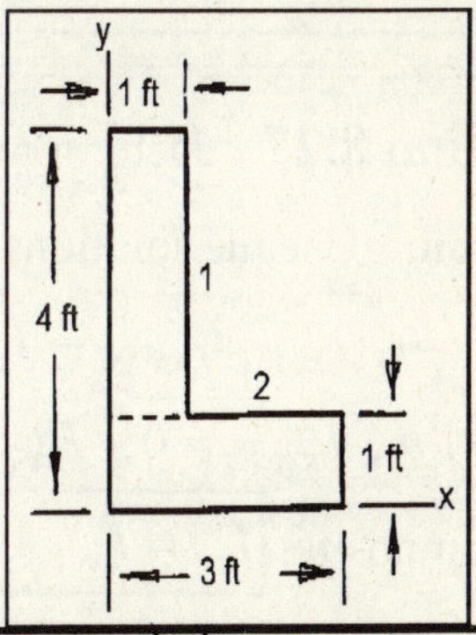

	$d_x\ (ft)$	$d_y\ (ft)$	$A\ (ft^2)$	$I_x\ (ft^4)$	$I_{xy}\ (ft^4)$	$I_{xy} + d_x d_y A$ (ft^4)	$I_x + d_y^2 A$ (ft^4)
Part 1	0.5	2.5	3	2.25	0	3.75	21
Part 2	1.5	0.5	3	0.25	0	2.25	1
						$I_{xy} = 6\ (ft^4)$	$I_x = 22\ (ft^4)$
			6				$k_x = 1.915\ (ft)$

Problem 8.25 Determine I_y and k_y.

Solution: Divide the area into two parts: (1) the part to the left of the y-axis, and (2) the part to the right of the y-axis.

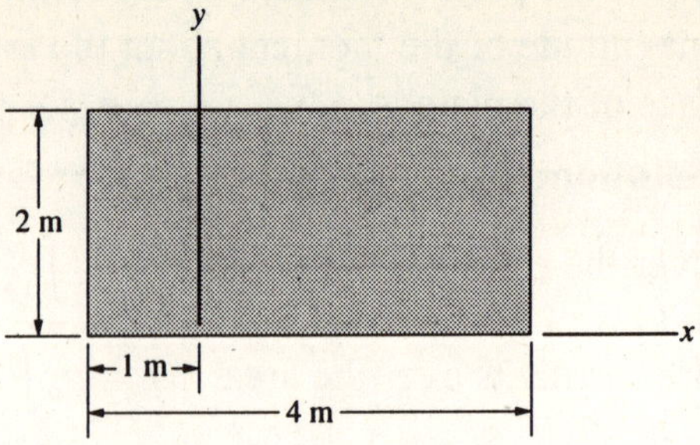

Part(1): The area

$A_1 = 2 \, . \, d_X = -0.5 \, m \, . \, I_{yy1} = \left(\frac{1}{12}\right) 2(1^3) = 0.1667 \, m^4.$

$I_{y1} = d_x^2 A_1 + I_{yy1} = 0.6667$. Part(2): The area: $A_2 = 6 \, m^2$,

$d_{x2} = 1.5 \, m$. $I_{yy2} = \left(\frac{1}{12}\right) 2(3^3) = 4.5 \, m^4$.

$I_{y2} = d_{x2}^2 A_2 + I_{yy2} = 18 \, m^4$.

The composite: $\boxed{I_y = I_{y1} + I_{y2} = 18.67 \, m^4}$ $\boxed{k_y = \sqrt{\frac{I_y}{(A_1 + A_2)}} = \sqrt{\frac{18.6667}{8}} = 1.528 \, m}$

Problem 8.26 Determine J_o and k_o.

Solution: Use the solution to Problem 8.25. Part (1): $A_1 = 2 \, m^2$ $d_{y1} = 1 \, m$,

$I_{xx1} = \left(\frac{1}{12}\right)(2^3) = 0.6667$, $I_{x1} = d_{y1}^2 A_1 + I_{xx1} = 2.6667 \, m^4$. Part (2): $A_2 = 6 \, m^2$,. $d_{y2} = 1 \, m$,

$I_{xx2} = \left(\frac{1}{12}\right) 3(2^3) = 2 \, m^4$, $I_{x2} = d_{y2}^2 A_2 + I_{xx2} = 8 \, m^4$.

The composite: $I_x = I_{x1} + I_{x2} = 10.6667 \, m^4$. $k_x = \sqrt{\frac{10.667}{(A_1 + A_2)}} = 1.155 \, m$.

$\boxed{J_o = I_y + I_x = 18.6667 + 10.6667 = 29.33 \, m^4}$ $\boxed{k_o = \sqrt{k_y^2 + k_x^2} = 1.915 \, m}$

Problem 8.27 Determine I_{xy}.

Solution: Use the solutions to Problems 8.25 and 8.26. Part (1): $A_1 = 2 \, m^2$, $d_{x1} = -0.5$, $d_{y1} = 1 \, m$, $I_{xxyy1} = 0$, $I_{xy1} = I_{xxyy1} + d_{x1} d_{y1} A_1 = 0 + (-0.5)(1)2 = -1 \, m^4$. Part (2): $A_2 = 6 \, m^2$, $d_{x2} = 1.5 \, m$, $d_{y2} = 1 \, m$, $I_{xxyy2} = 0$, $I_{xy2} = I_{xxyy2} + d_{x2} d_{y2} A_2 = 0 + 1.5(1)(6) = 9 \, m^4$

The composite: $\boxed{I_{xy} = I_{xy1} + I_{xy2} = -1 + 9 = 8 \, m^4}$

Problem 8.28 Determine I_y, k_y, J_o, and k_o.

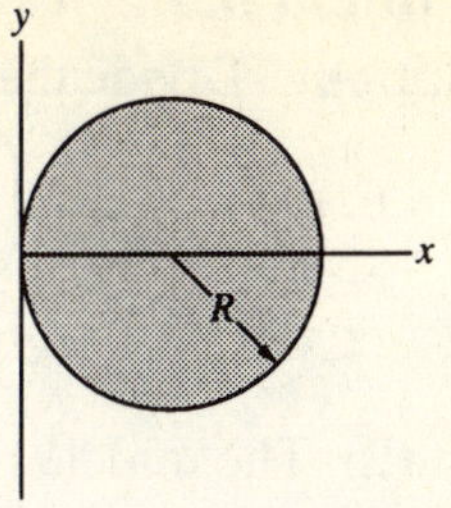

Solution: For a circle about the center: $I_{yy} = I_{xx} = \left(\frac{\pi}{4}\right)R^4$.

$$I_y = R^2(\pi R^2) + \left(\frac{\pi}{4}\right)R^4 = \left(\frac{5}{4}\right)\pi R^4$$ $$I_x = \frac{\pi R^4}{4}.$$

$$k_y = \sqrt{\frac{I_y}{\pi R^2}} = \frac{\sqrt{5}}{2}R.$$ $$J_o = I_y + I_x = \frac{5}{4}\pi R^4 + \frac{1}{4}\pi R^4 = \frac{3}{2}\pi R^4$$

$$k_o = \sqrt{\frac{J_o}{\pi R^2}} = \sqrt{\frac{3}{2}}R$$

Problem 8.29 Determine I_y and k_y.

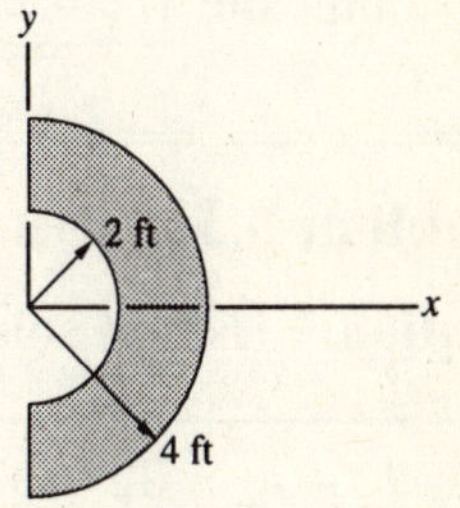

Solution: For a semicircle about a diameter: $I_{yy} = I_{xx} = \left(\frac{1}{8}\right)\pi R^4$.,

$$I_y = \left(\frac{1}{8}\right)\pi(4)^4 - \left(\frac{1}{8}\right)\pi(2)^4 = \frac{\pi}{8}(4^4 - 2^4) = 94.25\ ft^4$$

$$k_y = \sqrt{\frac{2I_y}{\pi(4^2 - 2^2)}} = 2.236\ ft$$

Problem 8.30 Determine I_y and k_y.

Solution: For a semicircle: $I_{yy} = I_{xx} = \left(\frac{1}{8}\right)\pi R^4$. $I_x = \frac{\pi}{8}(4^4 - 2^4) = 94.248\ ft^4$

$$k_x = \sqrt{\frac{2I_x}{\pi(4^2 - 2^2)}} = 2.236\ f\ t$$

Problem 8.31 Determine J_o and k_o.

Solution: Use the solutions to Problems 8.29 and 8.30. $J_o = I_x + I_y = 2(94.248) = 188.5\ ft^4$

$$k_o = \sqrt{\frac{2J_o}{\pi(4^2 - 2^2)}} = 3.16\ ft$$

Problem 8.32 Determine I_x and k_x .

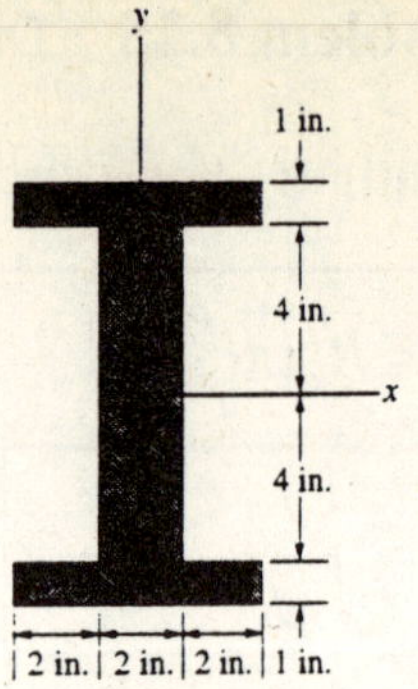

Solution: Divide the beam into three parts: Part (1) the top rectangle:

$$A_1 = 6 \ in^2 \ , \ I_{xx1} = \left(\frac{1}{12}\right)(1^3)(6) = 0.5 \ in^4 \ , \ I_{yy1} = \left(\frac{1}{12}\right)(1)(6^3) = 18 \ in^4 \ ,$$

$$d_{y1} = 4.5 \ in \ . I_{x1} = d_{y1}^2 A_1 + I_{xcx1} = 122 \ in^4 \ , I_{y1} = I_{yy1} = 18 \ in^4 \ .$$

Part (2): The middle rectangle: $A_2 = 16 \ in^2$,

$$I_{xx2} = \left(\frac{1}{12}\right)(8^3)(2) = 85.3333 \ in^4 \ , \ I_{yy2} = \left(\frac{1}{12}\right)(8)(2^3) = 5.3333 \ in^4 \ .,$$

$I_{x2} = I_{xx2} = 85.3333 \ in^4$, $I_{y2} = I_{yy2} = 5.3333 \ in^4$. Part (3) The lower rectangle: by symmetry, the area and moments of inertia are identical to Part (1).

The composite: $\boxed{I_x = 2I_{x1} + I_{x2} = 244 + 85.3333 = 329.33 \ in^4}$ $\boxed{k_x = \sqrt{\frac{I_x}{(2A_1 + A_2)}} = 3.43 \ in}$

Problem 8.33 Determine J_o and k_o for the area in Problem 8.32.

Solution: Use the solution in Problem 8.32. $I_y = 2I_{y1} + I_{y2} = 36 + 5.3333 = 41.3333 \ in^4$.

$\boxed{J_o = I_x + I_y = 329.3333 + 41.3333 = 370.67 \ in^4}$ $\boxed{k_o = \sqrt{\frac{J_o}{(2A_1 + A_2)}} = \sqrt{\frac{370.6667}{28}} = 3.64 \ in}$

Problem 8.34 If you design the beam cross section so that $I_x = 6.4 \times 10^5 \ mm^4$, what are the resulting values of I_y and J_o ?

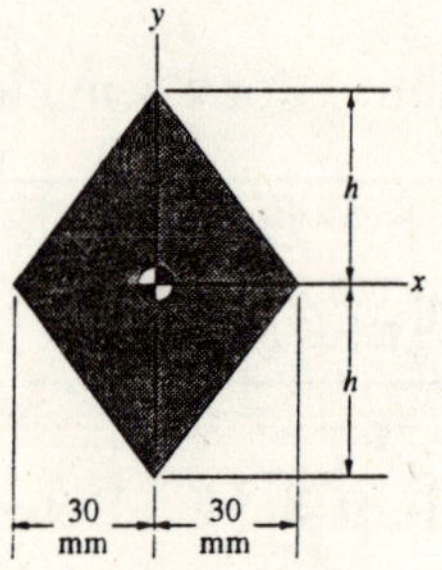

Solution: The area moment of inertia for a triangle about the base is

$$I_{xx} = \left(\frac{1}{12}\right)bh^3 \ , \text{from which } I_x = 2\left(\frac{1}{12}\right)(60)h^3 = 10h^3 \ mm^4 \ ,$$

$$I_x = 10h^3 = 6.4 \times 10^5 \ mm^4 \ , \text{from which } h = 40 \ mm .$$

$$I_{yt} = 2\left(\frac{1}{12}\right)(2h)(30^3) = \left(\frac{1}{3}\right)h(30^3) \ \text{ from which}$$

$\boxed{I_y = \left(\frac{1}{3}\right)(40)(30^3) = 3.6 \times 10^5 \ mm^4}$ and $\boxed{J_o = I_x + I_y = 3.6 \times 10^5 + 6.4 \times 10^5 = 1 \times 10^6 \ mm^4}$

Problem 8.35 Determine I_y and k_y.

Solution: Divide the area into three parts: Part (1): The top rectangle. $A_1 = 160(40) = 6.4 \times 10^3 \; mm^2$.,

$d_{x1} = \frac{160}{2} = 80 \; mm$.. $I_{yy1} = \left(\frac{1}{12}\right)(40)(160^3) = 1.3653 \times 10^7 \; mm^4$. From which

$I_{y1} = d_{x1}^2 A_1 + I_{yy1} = 5.4613 \times 10^7 \; mm^4$. Part (2): The middle rectangle:

$A_2 = (200 - 80)(40) = 4.8 \times 10^3 \; mm^2$, $d_{x2} = 20 \; mm$, $I_{yy2} = \left(\frac{1}{12}\right)(120)(40^3) = 6.4 \times 10^5 \; mm^4$. From

which, $I_{y2} = d_{x2}^2 A_2 + I_{yy2} = 2.56 \times 10^6 \; mm^4$. Part (3) The bottom rectangle:

$A_3 = 120(40) = 4.8 \times 10^3 \; mm^2$, $d_{x3} = \frac{120}{2} = 60 \; mm$, $I_{yy3} = \left(\frac{1}{12}\right)40(120^3) = 5.76 \times 10^6 \; mm^4$ From

which $I_{y3} = d_{X3}^2 A_3 + I_{yy3} = 2.304 \times 10^7 \; mm^4$. The composite:

$I_y = I_{y1} + I_{y2} + I_{y3} = 8.0213 \times 10^7 \; mm^4$ $\quad k_y = \sqrt{\frac{I_y}{(A_1 + A_2 + A_3)}} = 70.8 \; mm$

Problem 8.36 Determine I_x and k_x.

Solution: Use the solution to Problem 8.35. Divide the area into three parts: Part (1): The top rectangle. $A_1 = 6.4 \times 10^3 \; mm^2$. $d_{y1} = 200 - 20 = 180 \; mm$, $I_{xx1} = \left(\frac{1}{12}\right)(160)(40^3) = 8.533 \times 10^5 \; mm^4$.

From which $I_{x1} = d_{y1}^2 A_1 + I_{xx1} = 2.082 \times 10^8 \; mm^4$ Part (2) : The middle rectangle:

$A_2 = 4.8 \times 10^3 \; mm^2$, $d_{y2} = \frac{120}{2} + 40 = 100 \; mm$, $I_{xx2} = \left(\frac{1}{12}\right)(40)(120^3) = 5.76 \times 10^6 \; mm^4$ from which

$I_{x2} = d_{y2}^2 A_2 + I_{xx2} = 5.376 \times 10^7 \; mm^4$

Part (3) The bottom rectangle: $A_3 = 4.8 \times 10^3 \; mm^2$, $d_{y3} = 20 \; mm$,

$I_{xx3} = \left(\frac{1}{12}\right)120(40^3) = 6.4 \times 10^5 \; mm^4$ and $I_{x3} = d_{y3}^2 A_3 + I_{xx3} = 2.56 \times 10^6 \; mm^4$. The composite:

$I_x = I_{x1} + I_{x2} + I_{x3} = 2.645 \times 10^8 \; mm^4$ $\quad k_x = \sqrt{\frac{I_x}{(A_1 + A_2 + A_3)}} = 128.6 \; mm$

Problem 8.37 Determine I_{xy}.

Solution: (See figure in Problem 8.35). Use the solutions in Problems 8.35 and 8.36. Divide the area in three parts: Part (1): $A_1 = 160(40) = 6.4 \times 10^3 \; mm^2$., $d_{x1} = \frac{160}{2} = 80 \; mm$.. $d_{y1} = 200 - 20 = 180 \; mm$.

$I_{xxyy1} = 0$, from which $I_{xy1} = d_{x1} d_{y1} A_1 + I_{xxyy1} = 9.216 \times 10^7 \; mm^4$.

Part (2) $A_2 = (200 - 80)(40) = 4.8 \times 10^3 \; mm^2$,, $d_{x2} = 20 \; mm$, $d_{y2} = \frac{120}{2} + 40 = 100 \; mm$, from which

$I_{xy2} = d_{x2} d_{y2} A_2 = 9.6 \times 10^6 \; mm^4$. Part (3): $A_3 = 120(40) = 4.8 \times 10^3 \; mm^2$, $d_{x3} = \frac{120}{2} = 60 \; mm$,

Problem 8.37 continued on next page

continuation of Problem 8.37..........

$d_{y3} = 20\ mm$, from which $I_{xy3} = d_{x3}d_{y3}A_3 = 5.76 \times 10^6$. The composite:

$$\boxed{I_{xy} = I_{xy1} + I_{xy2} + I_{xy3} = 1.0752 \times 10^8\ mm^4}$$

Problem 8.38 The origin of the coordinate system is at the centroid of the composite area. Determine the I_x and k_x.

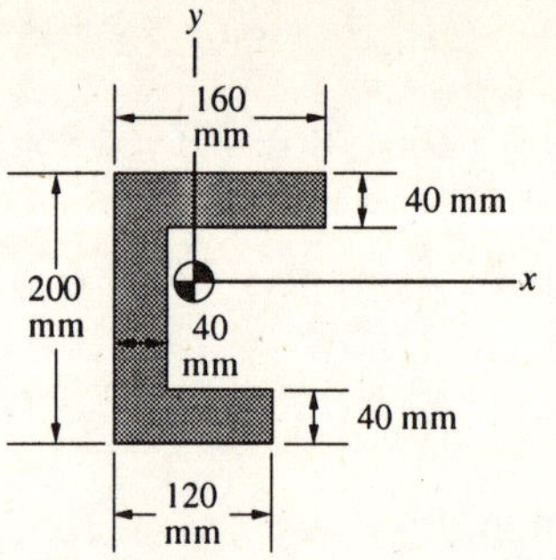

Solution: The strategy is to use the relationship $I_x = d^2 A + I_{xc}$, where I_{xc} is the area moment of inertia about the centroid. From this $I_{xc} = -d^2 A + I_x$. Use the solutions to Problems 8.35, 8.36, and 8.37. Divide the area into three parts and locate the centroid relative to the coordinate system in the Problems 8.35, 8.36, and 8.37.

Part (1) $A_1 = 6.4 \times 10^3\ mm^2, .. d_{y1} = 200 - 20 = 180\ mm$.

Part (2) $A_2 = (200 - 80)(40) = 4.8 \times 10^3\ mm^2$, $d_{x1} = \frac{160}{2} = 80\ mm$, $d_{x2} = 20\ mm$,

$d_{y2} = \frac{120}{2} + 40 = 100\ mm$, Part (3) $A_3 = 120(40) = 4.8 \times 10^3\ mm^2$, $d_{x2} = \frac{120}{2} = 60\ mm$,

$d_{y3} = 20\ mm$. The total area is $A = A_1 + A_2 + A_3 = 1.6 \times 10^4\ mm^2$. The centroid coordinates are

$\bar{x} = \frac{A_1 d_{x1} + A_2 d_{x2} + A_3 d_{x3}}{A} = 56\ mm$, $\bar{y} = \frac{A_1 d_{y1} + A_2 d_{y2} + A_3 d_{y3}}{A} = 108\ mm$. from which

$$\boxed{I_{xc} = -\bar{y}^2 A + I_x = -1.866 \times 10^8 + 2.645 \times 10^8 = 7.788 \times 10^7\ mm^4} \quad \boxed{k_{xc} = \sqrt{\frac{I_{xc}}{A}} = 69.77\ mm}$$

Problem 8.39 Determine I_y and k_y.

Solution: The strategy is to use the relationship $I_y = d^2 A + I_{yc}$, where I_{yc} is the area moment of inertia about the centroid. From this $I_{yc} = -d^2 A + I_y$. Use the solution to Problem 8.38. The centroid coordinates are $\bar{x} = 56\ mm$, $\bar{y} = 108\ mm$, from which

$$\boxed{I_{yc} = -\bar{x}^2 A + I_y = -5.0176 \times 10^7 + 8.0213 \times 10^7 = 3.0 \times 10^7\ mm^4}, \quad \boxed{k_{yc} = \sqrt{\frac{I_{yc}}{A}} = 43.33\ mm}$$

Problem 8.40 Determine I_{xy}.

Solution: Use the solution to Problem 8.38. . The centroid coordinates are $\bar{x} = 56\ mm$, $\bar{y} = 108\ mm$, from which $\boxed{I_{xyc} = -\bar{x}\bar{y}A + I_{xy} = -9.6768 \times 10^7 + 1.0752 \times 10^8 = 1.0752 \times 10^7\ mm^4}$

Problem 8.41 Determine I_x and k_x.

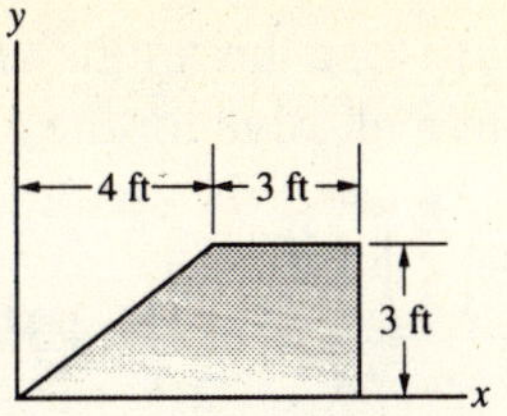

Solution: Divide the area into two parts: Part (1): a triangle and Part (2): a rectangle. The area moment of inertia for a triangle about the base is $I_x = \left(\frac{1}{12}\right)bh^3$. The area moment of inertia about the base for a rectangle is $I_x = \left(\frac{1}{3}\right)bh^3$.

Part (1) $I_{x1} = \left(\frac{1}{12}\right)4(3^3) = 9\ ft^2$. Part (2) $I_{x2} = \left(\frac{1}{3}\right)3(3^3) = 27$. The composite:

$\boxed{I_x = I_{x1} + I_{x2} = 36\ ft^4}$. The area: $A = \left(\frac{1}{2}\right)4(3) + 3(3) = 15\ ft^2$. $\boxed{k_x = \sqrt{\frac{I_x}{A}} = 1.549\ ft}$.

Problem 8.42 Determine J_o and k_o.

Solution: (See Figure in Problem 8.41.) Use the solution to Problem 8.41. Part (1): The area moment of inertia about the centroidal axis parallel to the base for a triangle is

$I_{yc} = \left(\frac{1}{36}\right)bh^3 = \left(\frac{1}{36}\right)3(4^3) = 5.3333\ ft^4$, from which $I_{y1} = \left(\frac{8}{3}\right)^2 A_1 + I_{yc} = 48\ ft^4$. where $A_1 = 6\ ft^2$.

Part (2): The area moment of inertia about a centroid parallel to the base for a rectangle is

$I_{yc} = \left(\frac{1}{12}\right)bh^3 = \left(\frac{1}{12}\right)3(3^3) = 6.75\ ft^4$, from which $I_{y2} = (5.5)^2 A_2 + I_{yc} = 279\ ft^4$, where $A_2 = 9\ ft^2$..

The composite: $I_y = I_{y1} + I_{y2} = 327\ ft^4$, from which, using a result from Problem 8.41,

$\boxed{J_o = I_x + I_y = 327 + 36 = 363\ ft^4}$ and $\boxed{k_o = \sqrt{\frac{J_o}{A}} = 4.92\ ft}$

Problem 8.43 Determine I_{xy}.

Solution: (See Figure in Problem 8.41.) Use the results of the solutions to Problems 8.41 and 8.42. The area cross product of the moment of inertia about centroidal axes parallel to the bases for a triangle is $I_{x'y'} = \frac{1}{72}b^2h^2$, and for a rectangle it is zero. Therefore: $I_{xy1} = \left(\frac{1}{72}\right)(4^2)(3^2) + \left(\frac{8}{3}\right)\left(\frac{3}{3}\right)A_1 = 18\ ft^4$

and $I_{xy2} = (1.5)(5.5)A_2 = 74.25\ ft^4$, $\boxed{I_{xy} = I_{x'y'1} + I_{xy2} = 92.25\ ft^4}$

Problem 8.44 The origin of the coordinate system is at the centroid of the composite area. Determine I_x and k_x.

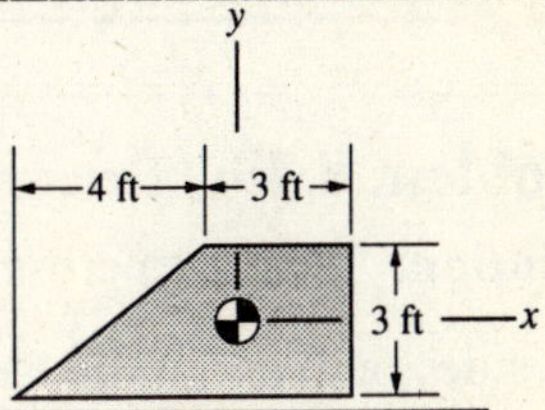

Solution: Use the results of Problems 8.41, 8.42, and 8.43. The strategy is to use the parallel axis theorem and solve for the area moment of inertia about the centroidal axis. The centroidal coordinate

$\bar{y} = \frac{A_1(1) + A_2(1.5)}{A} = 1.3\ ft$. From which $\boxed{I_{xc} = -\bar{y}^2 A + I_x = 10.65\ ft^4}$ and $\boxed{k_{xc} = \sqrt{\frac{I_{xc}}{A}} = 0.843\ ft}$

Problem 8.45 Determine J_O and k_O

Solution: Use the results of Problems 8.41,8.42, and 8.43. The strategy is to use the parallel axis theorem and solve for the area moment of inertia about the centroidal axis. The centroidal coordinate :

$$\bar{x} = \frac{A_1\left(\frac{8}{3}\right) + A_2(5.5)}{A} = 4.3667\ ft$$, from which $I_{YC} = -\bar{x}^2 A + I_Y = 40.98\ ft^4$. Using a result from

Problem 8.44, $\boxed{J_O = I_{XC} + I_{YC} = 10.65 + 40.98 = 51.63\ ft^4}$ and $\boxed{k_O = \sqrt{\frac{J_O}{A}} = 1.855\ ft}$

Problem 8.46 Determine I_{XY}.

Solution: Use the results of Problems 8.41-8.45 The strategy is to use the parallel axis theorem and solve for the area moment of inertia about the centroidal axis. Using the centroidal coordinates determined in Problems 8.44 and 8.45, $\boxed{I_{xyc} = -\bar{x}\bar{y}A + I_{xy} = -85.15 + 92.25 = 7.1\ ft^4}$

Problem 8.47 Determine I_x and k_x .

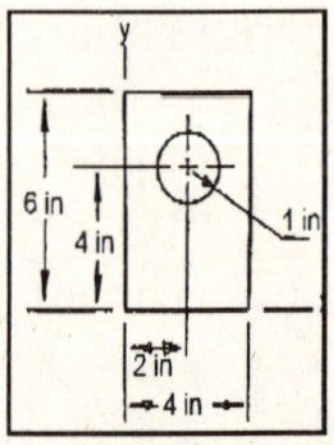

Solution: Divide the area into two parts: Part (1) the 4- by 6 in rectangle, and Part (2), the circle of 1 in. radius. Part (1): The area moment of inertia of a rectangle about a base is $I_{x1} = \left(\frac{1}{3}\right)bh^3 = \left(\frac{1}{3}\right)4\left(6^3\right) = 288\ in^4$. Part (2) The area moment of inertia of a circle about a centroidal axis is $I_{x2c} = \left(\frac{1}{4}\right)\pi R^4 = 0.7854\ in^4$, from which $I_{x2} = 4^2\pi R^2 + 0.7854 = 51.05\ ft^4$. The composite: $\boxed{I_x = I_{x1} - I_{x2} = 236.95\ in^4}$ The composite area: $A = A_1 - A_2 = 24 - \pi R^2 = 20.86\ in^2$, from which $\boxed{k_x = \sqrt{\frac{I_x}{A}} = 3.37\ in}$

Problem 8.48 Determine J_o and k_o .

Solution: Use the solution to Problem 8.47. Part (1) $I_{y1} = \left(\frac{1}{3}\right)6\left(4^3\right) = 128\ in^4$. Part (2)

$I_{y2} = 2^2\pi R^2 + 0.7854 = 13.35\ in^4$. The composite: $I_y = I_{y1} + I_{y2} = 128 - 13.35 = 114.65\ in^4$, from which $\boxed{J_o = I_x + I_y = 236.95 + 114.65 = 351.6\ in^4}$ $\boxed{k_o = \sqrt{\frac{J_o}{A}} = 4.11\ in}$

Problem 8.49 Determine I_{xy}.

Solution: The area cross product of inertia is zero about the centroidal axes of a rectangle parallel to the sides, and about the centroidal axes of a circle. Therefore $\boxed{I_{xy} = 3(2)A_1 - 4(2)\pi R^2 = 118.867\ in^4}$

Problem 8.50 The origin of the coordinate system is at the centroid of the composite area. Determine I_x and k_x.

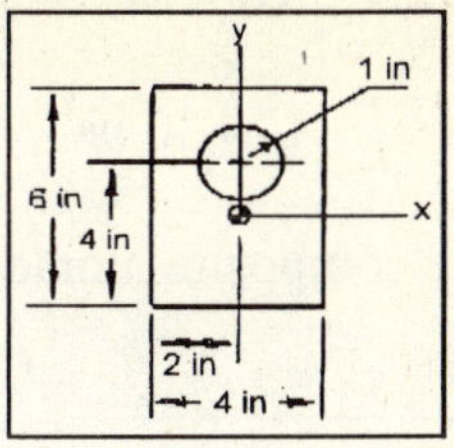

Solution: Use the results in the solutions to Problems 8.47-8.49.. The strategy is to use the parallel axis theorem, and solve for the area moment of inertia about the centroid. The centroid coordinate:

$\bar{y} = \dfrac{A_1(3) - A_2(4)}{A} = \dfrac{24(3) - \pi(4)}{20.858} = 2.85\ in$, from which

$\boxed{I_{xc} = -\bar{y}^2 A + I_x = -169.34 + 236.95 = 67.6\ in^4}$ and $\boxed{k_{xc} = \sqrt{\dfrac{I_{xc}}{A}} = 1.80\ in}$

Problem 8.51 Determine J_o and k_o.

Solution: Use the results in the solutions to Problems 8.47-8.50.. The strategy is to use the parallel axis theorem, and solve for the area moment of inertia about the centroid. The centroid coordinate:

$\bar{x} = \dfrac{A_1(2) - A_2(2)}{A} = 2\ in$, from which $I_{yc} = -\bar{x}^2 A + I_y = -83.43 + 114.65 = 31.21\ in^4$. Using a result from Problem 8.50, $\boxed{J_{oc} = I_{xc} + I_{yc} = 67.6 + 31.2 = 98.8\ in^4}$ and $\boxed{k_{oc} = \sqrt{\dfrac{Joc}{A}} = 2.176\ in}$

Problem 8.52 Determine I_{xy}.

Solution: Use the results in the solutions to Problems 8.47-8.51. The strategy is to use the parallel axis theorem, and solve for the area moment of inertia about the centroid. Use the results from Problems 8.49, 8.50 and 8.51: $\boxed{I_{xyc} = -\bar{x}\bar{y}A + I_{xy} = -118.87 + 118.87 = 0}$

Problem 8.53 The origin of the coordinate system is at the centroid of the composite area. Determine I_x and k_x.

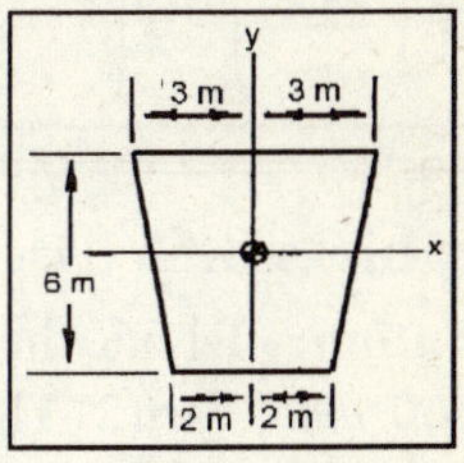

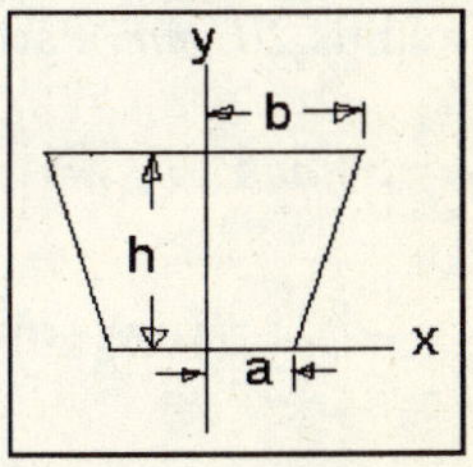

Solution: The strategy is to determine the location of the centroid relative to the base of the figure, from which the area moment of inertia can be determined. Break the area into three parts: Part (1) The rectangle with base $2a$ and altitude h; Part (2) The triangle on the right with base $(b-a)$ and altitude h, and Part (3) The triangle on the left with base $(b-a)$ and altitude h. Part (1) The area is $A_1 = 2ah = 24\ m^2$. The centroid is $\bar{x}_1 = 0$ and $\bar{y}_1 = \dfrac{h}{2} = 3\ m$. The area moment of inertia about the centroid is

$I_{xx1} = \left(\dfrac{1}{12}\right)(2a)h^3 = \left(\dfrac{1}{6}\right)ah^3 = 72\ m^4$, and

$I_{yy1} = \left(\dfrac{1}{12}\right)h(2a)^3 = \left(\dfrac{2}{3}\right)ha^3 = 32\ m^4$. Part (2): $A_2 = \left(\dfrac{1}{2}\right)h(b-a) = 3\ m^2$, $\bar{x}_2 = a + \dfrac{b-a}{3} = 2.3333\ m$,

$\bar{y}_2 = \left(\dfrac{2}{3}\right)h = 4\ m$, $I_{yy2} = \left(\dfrac{1}{36}\right)h(b-a)^3 = 0.1667\ m^4$, $I_{xx2} = \left(\dfrac{1}{36}\right)(b-a)h^3 = 6\ m^4$.

Problem 8.53 continued on next page

continuation of Problem 8.53..........

Part (3): $A_3 = A_2$, $\bar{x}_3 = -\bar{x}_2$, $\bar{y}_3 = \bar{y}_2$, $I_{xx3} = I_{xx2}$, $I_{yy3} = I_{yy2}$. The composite area is

$A = \sum_1^3 A_i = 30\ m^2$. The centroid: $\bar{x} = \dfrac{A_1\bar{x}_1 + A_2\bar{x}_2 + A_3\bar{x}_3}{A} = 0$, $\bar{y} = \dfrac{A_1\bar{y}_1 + A_2\bar{y}_2 + A_2\bar{y}_2}{A} = 3.2\ m$.

The composite moment of inertia about the centroid is

$$I_x = (\bar{y} - \bar{y}_1)^2 A_1 + I_{xx1} + (\bar{y} - \bar{y}_2)^2 A_2 + I_{xx2} + (\bar{y} - \bar{y}_3)^2 A_3 + I_{xx3},\quad \boxed{I_x = 88.8\ m^4}$$

$$\boxed{k_x = \sqrt{\frac{I_x}{A}} = \sqrt{\frac{88.8}{30}} = 1.720\ m}$$

Problem 8.54 Determine I_y and k_y.

Solution: Use the results of the solution to Problem 8.53.

$$I_y = (\bar{x} - \bar{x}_1)^2 A_1 + I_{yy1} + (\bar{x} - \bar{x}_2)^2 A_2 + I_{yy2} + (\bar{x} - \bar{x}_3)^2 A_3 + I_{yy3},\quad \boxed{I_y = 65\ m^4}$$

$$\boxed{k_y = \sqrt{\frac{I_y}{A}} = \sqrt{\frac{65}{30}} = 1.472\ m}$$

Problem 8.55 Suppose that you want to decrease I_x by twenty percent by placing the a circular cutout with its center at the centroid of the area. What is the radius of the cutout?

Solution: Use the results of the solution to Problem 8.54. The decrease is $I_{co} = 0.2I_x = 17.76\ m^4$. The area moment of inertia of a circle about a diameter is $I_{xxc} = \left(\frac{1}{4}\right)\pi R^4$, from which

$$\boxed{R = \left(\frac{4I_{co}}{\pi}\right)^{\frac{1}{4}} = \left(\frac{4(17.76)}{\pi}\right)^{\frac{1}{4}} = 2.181\ m}$$

Problem 8.56 Determine I_y and k_y.

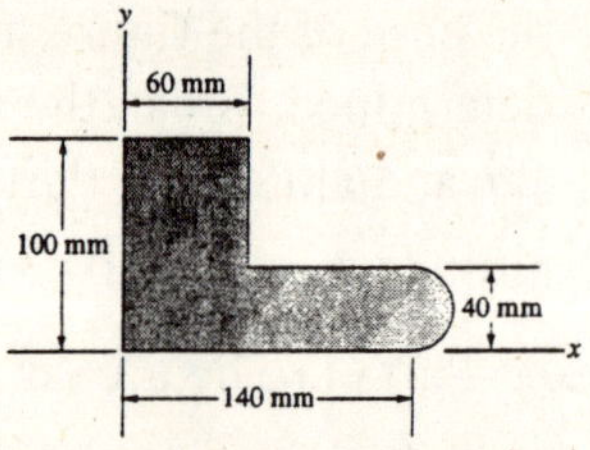

Solution: Divide the object into three parts: Part (1): The rectangle 100 mm by 60 mm, Part(2) The rectangle 80 mm by 40 mm. Part (3): The semicircle of radius 20 mm. Part (1). $A_1 = 100(60) = 6000\ m^2$, $\bar{x}_1 = 30\ mm$,

$\bar{y}_1 = 50\ mm$, $I_{xx1} = \left(\frac{1}{12}\right)(60)(100)^3 = 5 \times 10^6\ mm^4$,

$I_{yy1} = \left(\frac{1}{12}\right)(100)(60)^3 = 1.8 \times 10^6\ mm^4$. Part (2) $A_2 = 80(40) = 3200\ mm^2$, $\bar{x}_2 = 100\ mm$,

$\bar{y}_2 = 20\ mm$. $I_{xx2} = \left(\frac{1}{12}\right)80(40)^3 = 4.2667 \times 10^5\ mm^4$, $I_{yy2} = \left(\frac{1}{12}\right)40(80)^3 = 1.7067 \times 10^6\ mm^4$. Part

(3): $A_3 = \dfrac{\pi(20)^2}{2} = 628.32\ mm^2$, $\bar{x}_3 = 140 + \dfrac{4(20)}{3\pi} = 148.49\ mm$, $\bar{y}_3 = 20\ mm$.

Problem 8.56 continued on next page

continuation of Problem 8.56..........

$I_{xx3} = \left(\frac{1}{8}\right)\pi R^4 = 62831.8\ mm^4$, $I_{yy3} = \left(\frac{1}{8}\right)\pi R^4 = 62831.8\ mm^4$. The area: $A = \sum_1^3 A_i = 9828.32\ mm^2$.

$I_y = \bar{x}_1 A_1 + I_{yy1} + \bar{x}_2 A_2 + I_{yy2} + \bar{x}_3 A_3 + I_{yy3}$. $\boxed{I_y = 5.482 \times 10^7\ mm^4}$ $\boxed{k_y = \sqrt{\frac{I_y}{A}} = 74.69\ mm}$

Problem 8.57 Determine I_x and k_x.

Solution: Use the results of the solution to Problem 8.56.

$I_x = \bar{y}_1^2 A_1 + I_{xx1} + \bar{y}_2^2 A_2 + I_{xx2} + \bar{y}_3^2 A_3 + I_{xx3}$ $\boxed{I_x = 2.202 \times 10^7\ mm^4}$. $\boxed{k_x = \sqrt{\frac{I_x}{A}} = 47.33\ mm}$

Problem 8.58 Determine I_{xy}.

Solution: Use the solutions to Problems 8.56 and 8.57. The area cross product of inertia for each of the areas about their centroids is zero. $\boxed{I_{xy} = \bar{x}_1\bar{y}_1 A_1 + \bar{x}_2\bar{y}_2 A_2 + \bar{x}_3\bar{y}_3 A_3 = 1.727 \times 10^7\ mm^4}$

Problem 8.59 The origin of the coordinate system is at the centroid of the composite area. Determine I_y and k_y.

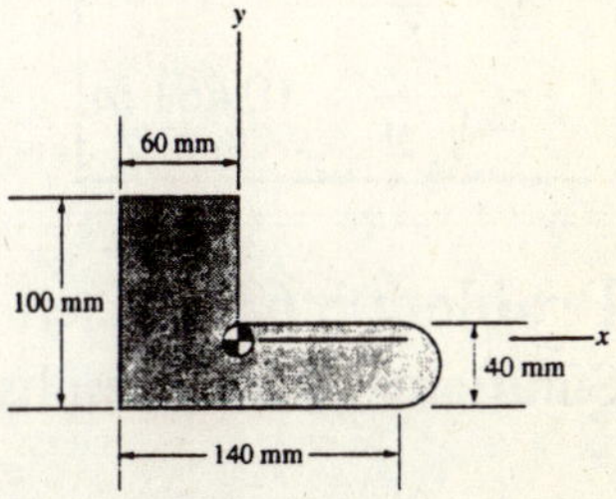

Solution: Use the solutions to Problems 8.56-8.58. The location of the centroid is

$$\bar{x} = \frac{A_1\bar{x}_1 + A_2\bar{x}_2 + A_3 x_3}{A} = \frac{1.8(10^5) + 3.2(10^5) + 9.33(10^4)}{9.828(10^3)} = 60.37\ mm$$

The area moment of inertia about the composite centroid:

$\boxed{I_{yc} = -\bar{x}^2 A + I_y = -3.58 \times 10^7 + 5.48 \times 10^7 = 1.9 \times 10^7\ mm^4}$ $\boxed{k_{yc} = \sqrt{\frac{I_{yc}}{A}} = 43.98\ mm}$

Problem 8.60 Determine I_x and k_x.

Solution: Use the solutions to Problems 8.56-8.58. The location of the centroid is

$\bar{y} = \frac{A_1\bar{y}_1 + A_2\bar{y}_2 + A_3\bar{y}_3}{A} = \frac{3(10^5) + 6.4(10^4) + 1.26(10^4)}{9.828(10^3)} = 38.3\ mm$. The area moment of inertia about the centroid is $\boxed{I_{xc} = -\bar{y}^2 A + I_x = -1.443 \times 10^7 + 2.202 \times 10^7 = 7.593 \times 10^6\ mm^4}$

$\boxed{k_{xc} = \sqrt{\frac{I_{xc}}{A}} = 27.8\ mm}$

Problem 8.61 Determine I_{xy}.

Solution: Use the results of the solutions to Problems 8.56-8.60.

$\boxed{I_{xyc} = -\bar{x}\bar{y}A + I_{xy} = -2.27 \times 10^7 + 1.726 \times 10^7 = -5.47 \times 10^6\ mm^4}$

Problem 8.62 Determine I_y and k_y.

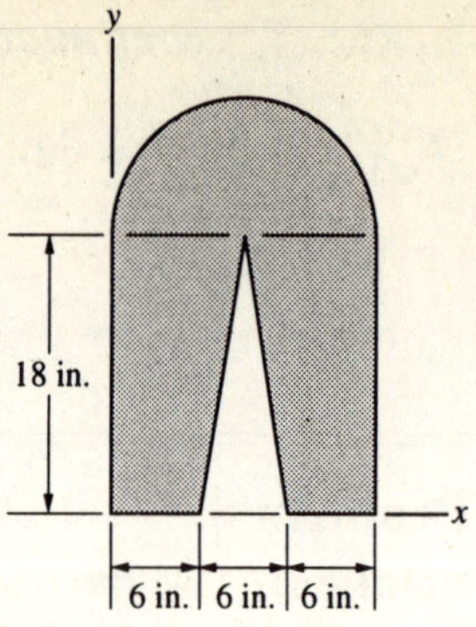

Solution: Divide the area into three parts: Part (1) The rectangle 18 by 18 inches; Part (2) The triangle with base 6 in and altitude 18 in; Part (3) The semicircle of 9 in radius. Part (1): $A_1 = 18(18) = 324\ in^2$, $\bar{x}_1 = 9\ in$,

$\bar{y}_1 = 9\ in$, $I_{xx1} = \left(\frac{1}{12}\right)18\left(18^3\right) = 8748\ in^4$, $I_{yy1} = \left(\frac{1}{12}\right)18\left(18^3\right) = 8748\ in^4$.

Part (2): $A_2 = \left(\frac{1}{2}\right)18(6) = 54\ in^2$, $\bar{x}_2 = 9\ in$, $\bar{y}_2 = \left(\frac{1}{3}\right)18 = 6\ in$,

$I_{xx2} = \left(\frac{1}{36}\right)6\left(18^3\right) = 972\ in^4$, $I_{yy2} = (1/18)18\left(3^3\right) = 27\ in^4$. Part (3) $A_3 = \frac{\pi\left(9^2\right)}{2} = 127.23\ in^2$,

$\bar{x}_3 = 9\ in$, $\bar{y}_3 = 18 + \left(\frac{4(9)}{3\pi}\right) = 21.82\ in$, $I_{xx3} = \left(\frac{1}{8}\right)\pi\left(9^4\right) - \left(\frac{4(9)}{3\pi}\right)^2 A_3 = 720.1\ in^4$,

$I_{yy3} = \left(\frac{1}{8}\right)\pi\left(9^4\right) = 2576.5\ in^4$. The composite area: $A = \sum_1^3 A_i = 397.23\ in^2$. The area moment of inertia: $I_y = \bar{x}_1^2 A_1 + I_{yy1} - \bar{x}_2^2 A_2 - I_{yy2} + \bar{x}_3^2 A_3 + I_{yy3}$, $\boxed{I_y = 4.347 \times 10^4\ in^4}$,

$\boxed{k_y = \sqrt{\frac{I_y}{A}} = 10.461\ in}$

Problem 8.63 Determine I_x and k_x.

Solution: Use the results of the solution to Problem 8.62.

$I_x = \bar{y}_1^2 A_1 + I_{XX1} - \bar{y}_2^2 A_2 - I_{XX2} + \bar{y}_3^2 A_3 + I_{XX3}$, $\boxed{I_x = 9.338 \times 10^4\ in^4}$, $\boxed{k_x = \sqrt{\frac{I_x}{A}} = 15.33\ in}$

Problem 8.64 Determine I_{xy}.

Solution: Use the results of the solutions to Problems 8.62 and 8.63. $I_{xy} = \bar{x}_1\bar{y}_1 A_1 - \bar{x}_2\bar{y}_2 A_2 + \bar{x}_3\bar{y}_3 A_3$

$\boxed{I_{xy} = 4.8313 \times 10^4\ in^4}$

Problem 8.65 Determine I_y and k_y.

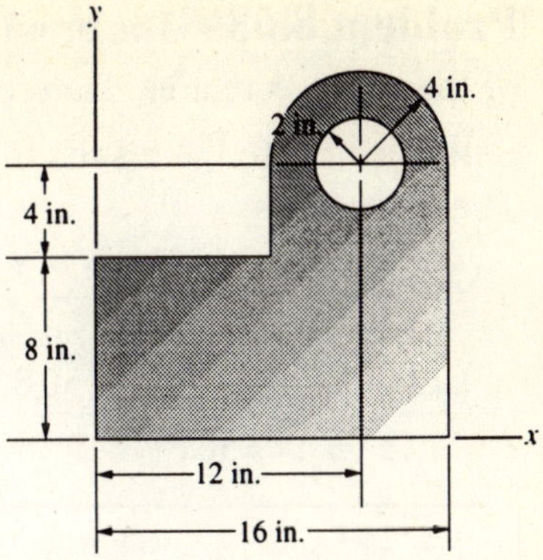

Solution: Divide the area into four parts: Part (1) The rectangle 8 in by 16 in. Part (2): The rectangle 4 in by 8 in. Part (3) The semicircle of radius 4 in, and Part (4) The circle of radius 2 in.

Part (1): $A_1 = 16(8) = 128\ in^2$, $\bar{x}_1 = 8\ in$, $\bar{y}_1 = 4\ in$,

$$I_{xx1} = \left(\frac{1}{12}\right)16(8^3) = 682.67\ in^4,\ I_{yy1} = \left(\frac{1}{12}\right)8(16^3) = 2730.7\ in^4.$$

Part (2): $A_2 = 4(8) = 32\ in^2$, $\bar{x}_2 = 12\ in$, $\bar{y}_2 = 10\ in$,

$$I_{xx2} = \left(\frac{1}{12}\right)8(4^3) = 42.667\ in^4,\ I_{yy2} = \left(\frac{1}{12}\right)4(8^3) = 170.667\ in^4.$$

Part (3): $A_3 = \dfrac{\pi(4^2)}{2} = 25.133\ in^2$, $\bar{x}_3 = 12\ in$. $\bar{y}_3 = 12 + \left(\dfrac{4(4)}{3\pi}\right) = 13.698\ in$. The area moments of inertia about the centroid of the semicircle are $I_{yy3} = \left(\dfrac{1}{8}\right)\pi(4^4) = 100.53\ in^4$,

$$I_{xx3} = \left(\frac{1}{8}\right)\pi(4^4) - \left(\frac{4(4)}{3\pi}\right)^2 A_3 = 28.1\ in^4.\ \textit{Check: } I_{xx3} = 0.1098(R^4) = 28.1\ in^4.\ \textit{check.}$$

Part (4): $A_4 = \pi(2^2) = 12.566\ in^2$, $\bar{x}_4 = 12\ in$, $\bar{y}_4 = 12\ in$, $I_{xx4} = \left(\dfrac{1}{4}\right)\pi(2^4) = 12.566\ in^4$,

$I_{yy4} = I_{xx4} = 12.566\ in^4$. The composite area: $A = \sum_1^3 A_i - A_4 = 172.566\ in^2$. The area moment of inertia: $I_y = \bar{x}_1^2 A_1 + I_{yy1} + \bar{x}_2^2 A_2 + I_{yy2} + \bar{x}_3^2 A_3 + I_{yy3} - \bar{x}_4^2 A_4 - I_{yy4}$ $\boxed{I_y = 1.76 \times 10^4\ in^4}$,

$$\boxed{k_y = \sqrt{\frac{I_y}{A}} = 10.1\ in}$$

Problem 8.66 Determine I_x and k_x.

Solution: Use the results in the solution to Problem 8.65.

$I_x = \bar{y}_1^2 A_1 + I_{xx1} + \bar{y}_2^2 A_2 + I_{xx2} + \bar{y}_3^2 A_3 + I_{xx3} - \bar{y}_4^2 A_4 - I_{xx4}$ $\boxed{I_x = 8.89 \times 10^3\ in^4}$

$$\boxed{k_x = \sqrt{\frac{I_x}{A}} = 7.18\ in}$$

Problem 8.67 Determine I_{xy}.

Solution: Use the results in the solution to Problem 8.65. $I_{xy} = \bar{x}_1\bar{y}_1 A_1 + \bar{x}_2\bar{y}_2 A_2 + \bar{x}_3\bar{y}_3 A_3 - \bar{x}_4\bar{y}_4 A_4$,

$$\boxed{I_{xy} = 1.0257 \times 10^4\ in^4}$$

Problem 8.68 The origin of the coordinate system is at the centroid of the composite area. Determine I_y and k_y.

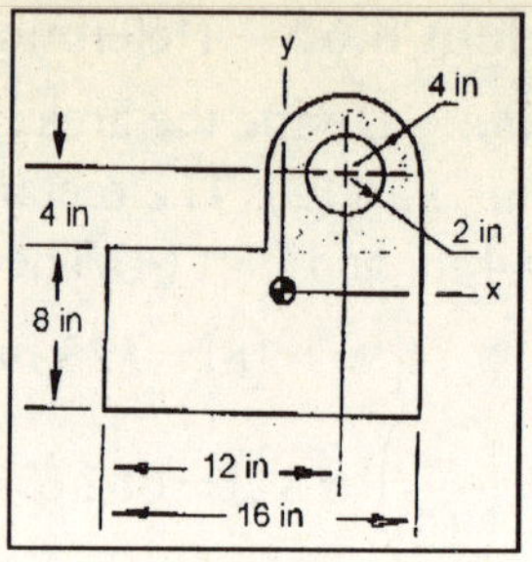

Solution: : Use the results in the solutions to Problems 8.65 to 8.67. The centroid is

$$\bar{x} = \frac{\bar{x}_1 A_1 + \bar{x}_2 A_2 + \bar{x}_3 A_3 - \bar{x}_4 A_4}{A}$$

$$= \frac{1024 + 384 + 301.6 - 150.8}{172.567} = 9.033 \; in., \text{ from which}$$

$$\boxed{I_{yc} = -\bar{x}^2 A + I_y = -1.408 \times 10^4 + 1.7598 \times 10^4 = 3518.2 \; in^4} \quad \boxed{k_{yc} = \sqrt{\frac{I_{yc}}{A}} = 4.52 \; in}$$

Problem 8.69 Determine I_x and k_x.

Solution: Use the results in the solutions to Problems 8.65 to 8.67. The centroid is

$$\bar{y} = \frac{\bar{y}_1 A_1 + \bar{y}_2 A_2 + \bar{y}_3 A_3 - \bar{y}_4 A_4}{A} = 5.942 \; in, \text{ from which}$$

$$\boxed{I_{xc} = -\bar{y}^2 A + I_x = -6092.9 + 8894 = 2801 \; in^4} \quad \boxed{k_{xc} = \sqrt{\frac{I_{xc}}{A}} = 4.03 \; in}$$

Problem 8.70 Determine I_{xy}.

Solution: Use the results in the solutions to Problems 8.65 to 8.69.

$$\boxed{I_{xyc} = -\bar{x}\bar{y}A + I_{xy} = -9.263 \times 10^3 + 1.0257 \times 10^4 = 994.5 \; in^4}$$

Problem 8.71 The dimensions of the concrete masonry unit are in inches. What is I_y?

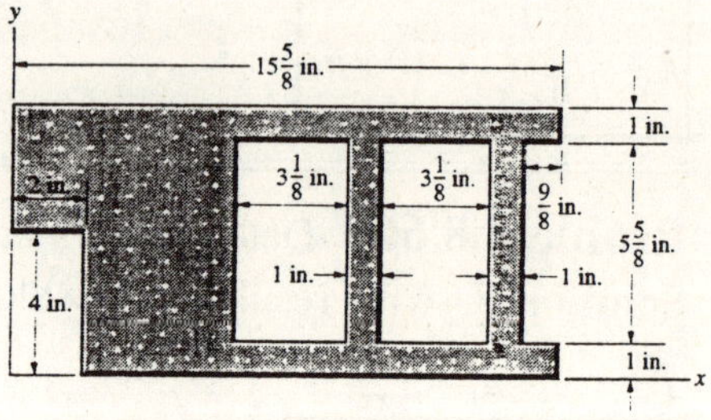

Solution: Divide the unit into 5 parts: Part (1): The rectangle $15\frac{5}{8}$ *in* by $7\frac{5}{8}$ in, Part (2): The cut-out to the far right, $15\frac{5}{8}$ *in* by $\frac{9}{8}$ *in*, Part (3) The rectangular cut out $5\frac{5}{8}$ *in* by $3\frac{1}{8}$ *in* to the far right, Part (4) The rectangular cut out $5\frac{5}{8}$ *in* by $3\frac{1}{8}$ *in* in the center, and Part (5) The cut out 4 *in* by 2 *in* on the lower left.

Part (1) $A_1 = (15.625)(7.625) = 119.14 \; in^2$,

$$\bar{x}_1 = \frac{15.625}{2} = 7.8125 \; in, \; \bar{y}_1 = \frac{7.625}{2} = 3.8125 \; in, \; I_{xx1} = \left(\frac{1}{12}\right)(15.625)(7.625)^3 = 577.24 \; in^4,$$

$$I_{yy1} = \left(\frac{1}{12}\right)(7.625)\left(15.625^3\right) = 2423.9 \; in^4. \text{ Part (2): } A_2 = (1.125)(5.625) = 6.328 \; in^2.$$

$$\bar{x}_2 = 15.625 - \frac{1.125}{2} = 15.0625 \; in, \; \bar{y}_2 = 1 + \frac{5.625}{2} = 3.8125 \; in, \; I_{xx2} = \left(\frac{1}{12}\right)(1.125)\left(5.625^3\right) = 16.685 \; in^4,$$

$$I_{yy2} = \left(\frac{1}{12}\right)(5.625)\left(1.125^3\right) = 0.667 \; in^4. \text{ Part (3): } A_3 = (3.125)(5.625) = 17.578 \; in^2,$$

Problem 8.71 continued on next page

continuation of Problem 8.71..........

$\bar{x}_3 = 15.625 - 1.125 - 1 - \frac{3.125}{2} = 11.9375\ in$, $\bar{y}_3 = 1 + \frac{5.625}{2} = 3.8125\ in$,

$I_{xx3} = \left(\frac{1}{12}\right)(3.125)(5.625^3) = 46.3486\ in^4$, $I_{yy3} = \left(\frac{1}{12}\right)(5.625)(3.125^3) = 14.305\ in^4$. Part (4):

$A_4 = A_3 = 17.578\ in^2$, $\bar{x}_4 = 15.625 - 1.125 - 1 - 3.125 - 1 - \frac{3.125}{2} = 7.8125\ in$, $\bar{y}_4 = \bar{y}_3 = 3.8125\ in$,

$I_{xx4} = I_{xx3} = 46.3486\ in^4$, $I_{yy4} = I_{yy3} = 14.305\ in^4$. Part (5): $A_5 = 2(4) = 8\ in^2$, $\bar{x}_5 = 1\ in$, $\bar{y}_5 = 2\ in$

$I_{xx5} = \left(\frac{1}{12}\right)2(4^3) = 10.667\ in^4$, $I_{yy5} = \left(\frac{1}{12}\right)4(2^3) = 2.667\ in^4$.

The results are tabulated:

Area	base, in	height, in	Area, in^2	$\bar{x}$, in	$\bar{y}$, in	I_{XX}, in^4	I_{YY} in^4
1	15.625	7.625	119.14	7.8125	3.8125	577.24	2423.9
2	1.125	5.625	6.3281	15.063	3.8125	16.685	0.6667
3	3.125	5.625	17.578	11.938	3.8125	46.349	14.305
4	3.125	5.625	17.578	7.8125	3.8125	46.349	14.305
5	2	4	8	1	2	10.667	2.6667

The area moment of inertia about the y-axis:

$I_y = \bar{x}_1^2 A_1 + I_{yy1} - \bar{x}_2^2 A_2 - I_{yy2} - \bar{x}_3^2 A_3 - I_{yy3} - \bar{x}_4^2 A_4 - I_{yy4} - \bar{x}_5^2 A_5 - I_{yy5}$ $\boxed{I_y = 4641.9\ in^4}$

Problem 8.72 Determine I_X for the concrete masonry unit in Problem 8.71.

Solution: Use the results of the solution to Problem 8.71.

$I_x = \bar{y}_1^2 A_1 + I_{xx1} - \bar{y}_2^2 A_2 - I_{xx2} - \bar{y}_3^2 A_3 - I_{xx2} - \bar{y}_4^2 A_4 - I_{xx4} - \bar{y}_5^2 A_5 - I_{xx5}$, $\boxed{I_x = 1553.94\ in^4}$

Problem 8.73 The area $A = 2 \times 10^4\ mm^2$. Its moment of inertia about the y-axis is $I_y = 3.2 \times 10^8\ mm^4$. Determine its moment of inertia about the $\hat{y}$ axis.

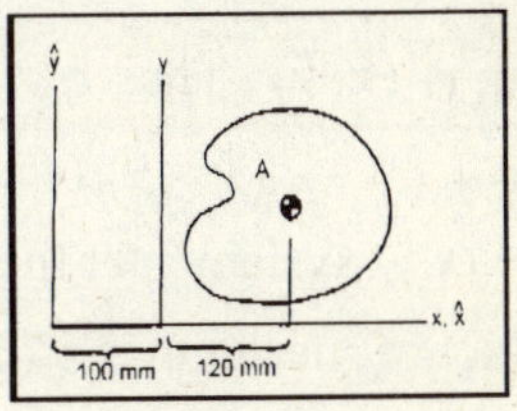

Solution: Use the parallel axis theorem. The moment of inertia about the centroid of the figure is

$I_{yc} = -x^2 A + I_y = -(120^2)(2 \times 10^4) + 3.2 \times 10^8 = 3.20 \times 10^7\ mm^4$. The moment of inertia about the $\hat{y}$ axis is $I_{\hat{y}} = \bar{x}^2 A + I_{yc}$ $\boxed{I_{\hat{y}} = (220^2)(2 \times 10^4) + 3.2 \times 10^7 = 1 \times 10^9\ mm^4}$

Problem 8.74 The area $A = 100\ in^2$ and is symmetric about the x' axis. The moments of inertia $I_{x'} = 420\ in^4$, $I_{y'} = 580\ in^4$, $J_o = 11000\ in^4$, and $I_{xy} = 4800\ in^4$. What are I_x and I_y?

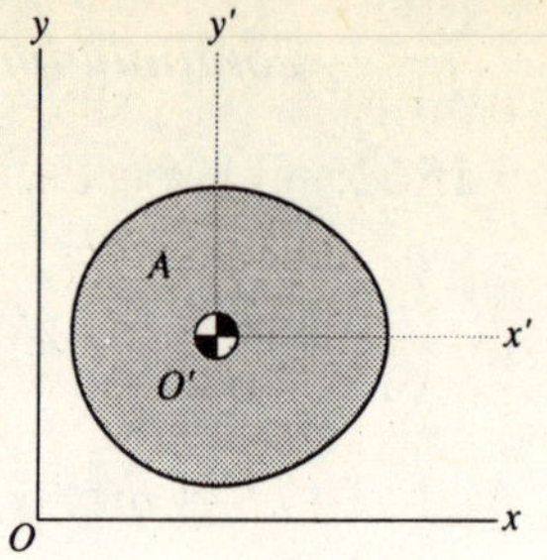

Solution: The basic relationships: (1) $I_x = y^2 A + I_{xc}$, (2) $I_y = x^2 A + I_{yc}$, (3) $J_o = Ar^2 + J_c$, (4) $J_o = I_x + I_y$, (5) $J_c = I_{xc} + I_{yc}$, and (6) $I_{xy} = Axy + I_{xyc}$, where the subscript c applies to the primed axes, and the others to the unprimed axes. The x,y values are the displacement of the primed axes from the unprimed axes. The steps in the demonstration are:

(i) From symmetry about the x_c axis, the product of inertia $I_{xyc} = 0$.

(ii) From (3): $r^2 = \dfrac{J_o - J_c}{A} = 100\ in^2$, from which $r^2 = x^2 + y^2 = 100\ in^2$

(iii) From (6) and $I_{xyc} = 0$, $y = \dfrac{I_{xy}}{Ax}$, from which $x^2 r^2 = x^4 + \left(\dfrac{I_{xy}}{A}\right)^2$. From which: $x^4 - 100x^2 + 2304 = 0$. (iv) The roots: $x_1^2 = 64$, and $x_2^2 = 36$. The corresponding values of y are found from $y = \sqrt{r^2 - x^2}$ from which $(x_1, y_1) = (8,6)$, and $(x_2, y_2) = (6,8)$. (v) Substitute these pairs to obtain the possible values of the area moments of inertia: $\boxed{I_{x1} = Ay_1^2 + I_{xc} = 4020\ in^4}$, $\boxed{I_{y1} = Ax_1^2 + I_{yc} = 6980\ in^4}$ $\boxed{I_{x2} = Ay_2^2 + I_{xc} = 6820\ in^4}$, $\boxed{I_{y2} = Ax_2^2 + I_{yc} = 4180\ in^4}$

Problem 8.75 Derive the parallel axis theorem for the product of inertia, Eq (8.12) by using the same procedures we used to derive Eqs (8.10) and (8.11).

Solution: By definition, $I_{xy} = \int_A xy\, dA$. Choose a coordinate system originating at the centroid, so that $\int_A x dA = \int_A y dA = 0$. Choose a new system (ξ, η) such that the distance to the centroid from the origin is $(\hat{x}, \hat{y})$. Thus $\xi = \hat{x} - x$, and $\eta = \hat{y} - y$. The area product of inertia in the new system is $I_{\xi\eta} = \int_A (\hat{x} - x)(\hat{y} - y) dA = \int_A \hat{x}\hat{y} dA - \int_A \hat{y} x dA - \int_A \hat{x} y dA + \int_A xy dA$. Since the $(\hat{x}, \hat{y})$ are constant numbers, if the *(x,y)* system is at the centroid of the area, then the second and third integrals vanish, so that the product of inertia in the (ξ, η) coordinate system is $I_{\xi\eta} = \hat{x}\hat{y}\int_A dA + I_{XY} = \hat{x}\hat{y}A + I_{XY}$, which is the parallel axis theorem for the area product of inertia.

Problem 8.76 Derive the parallel axis theorem for the polar moment of inertia, Eq (8.13), (a) by using the same procedures we used to derive Eqs (8.10) and (8.11); (b) by using Eqs (8.10) and (8.11).

Solution: (a) By definition $J_o = \int_A r^2 dA$. Define coordinates (ξ, η) such that $r^2 = \xi^2 + \eta^2$ Suppose that the polar moment of inertia is defined about the centroid of the area, such that $\int_A \xi \, dA = \int_A \eta \, dA = 0$. Choose a new system such that the origins are separated by a distance $(\hat{x}, \hat{y})$. In the new system: $\hat{J} = \int_A \rho^2 \, dA$. Note that $\rho^2 = (\hat{x} + \xi)^2 + (\hat{y} + \eta)^2$,

$\hat{J} = \int_A \rho^2 \, dA = \int_A (\hat{x} + \xi)^2 dA + \int_A (\hat{y} + \eta)^2 dA =$

$\hat{J} = \int_A \hat{x}^2 dA + 2\int_A \hat{x}\xi \, dA + \int_A \xi^2 dA + \int_A \hat{y}^2 dA + 2\int \hat{y}\eta \, dA + \int_A \eta^2 \, dA$ The integrals of first powers of ξ and η vanish, from which $\hat{J} = \int_A \left(\hat{x}^2 + \hat{y}^2\right) dA + \int_A \left(\xi^2 + \eta^2\right) dA$. Since $\hat{x}$ and $\hat{y}$ are constant numbers, $\hat{J} = d^2 A + \int r^2 dA = d^2 A + J_o$, where d is the distance from the origin of the new system to the centroid of the area. This is the parallel axis theorem for the area polar moment of inertia.

(b) Eqs (8.10) and (8.11) are $I_x = I_{xc} + d_y^2 A$ and $I_y = I_{yc} + d_x^2 A$, where $I_{xc} = \int_A x^2 \, dA$ and $I_{yc} = \int_A y^2 dA$ are area moments of inertia about axes passing through the area centroid, and parallel to the new axes located at a distance (d_x, d_y) from the centroid. Define $\rho^2 = d_X^2 + d_Y^2$ and $r^2 = x^2 + y^2$. Add Eqs (8.10) and (8.11) and combine terms. The result, $\boxed{J = \rho^2 A + \int_A r^2 \, dA}$, which is the parallel axis theorem for the polar moment of inertia.

Problems 8.77-8.80 are related to Example 8.6.

Problem 8.77 Determine the moment of inertia of the beam cross section about the x-axis. Compare your result with the moment of inertia of a solid cross section of equal area and confirm the ratio shown in Figure 8.16.

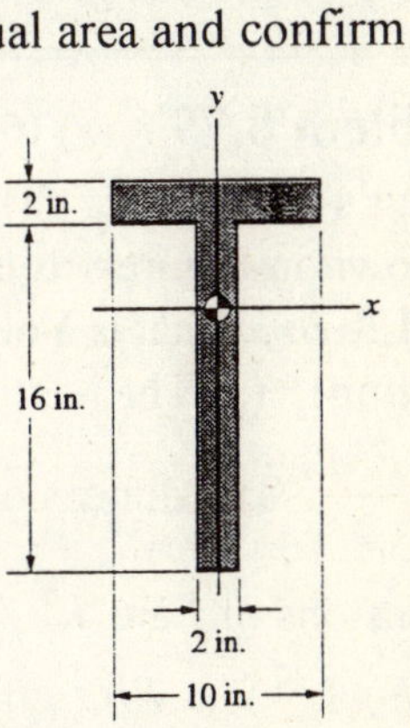

Solution: Divide the beam into two parts: Part (1) The upper rectangle 2 *in* by 10 *in*. Part (2): The lower rectangle 2 *in* by 16 *in*. Compute the centroidal distance from the base of the lower rectangle, and use the parallel axis theorem to transfer to the composite centroid.

Part (1) $A_1 = 2(10) = 20 \; in^2$, $\bar{x}_1 = 0$, $\bar{y}_1 = 16 + 1 = 17 \; in$,

$I_{xx1} = \left(\frac{1}{12}\right) 10\left(2^3\right) = 6.6667 \; in^4$, $I_{yy1} = \left(\frac{1}{12}\right) 2\left(10^3\right) = 166.6667 \; in^4$.

Part (2) $A_2 = 2(16) = 32 \; in^2$, $\bar{x}_2 = 0$, $\bar{y}_2 = 8 \; in$,

$I_{xx2} = \left(\frac{1}{12}\right) 2\left(16^3\right) = 682.67 \; in^4$, $I_{yy2} = \left(\frac{1}{12}\right) 16\left(2^3\right) = 10.667 \; in^4$ The area moments of inertia about the base: $I_x = \bar{y}_1^2 A_1 + I_{xx1} + \bar{y}_2^2 A_2 + I_{xx2} = 8517.3 \; in^4$.

Problem 8.77 continued on next page....

=============================<>=============================

continuation of solution to Problem 8.77...

$I_y = \bar{x}_1^2 A_1 + I_{yy1} + \bar{x}_2^2 A_2 + I_{yy2} = 177.33\ in^4$. The composite centroidal coordinates:

The composite area is $A = A_1 + A_2 = 52\ in^2$. The centroid coordinates: $\bar{x} = \frac{A_1\bar{x}_1 + A_2\bar{x}_2}{A} = 0$,

$\bar{y} = \frac{A_1\bar{y}_1 + A_2\bar{y}_2}{A} = 11.462\ in$. The moment of inertia about the composite centroid is

$I_{xc} = -\bar{y}^2 A + I_X = 1686.26\ in^4$

For a square cross section of equal area, each side is $b = \sqrt{A} = \sqrt{52} = 7.2111\ in$. The area moment of inertia about the centroid is $I_x = \left(\frac{1}{12}\right)(7.211)(7.211^3) = 225.33\ in^{4}$. The ratio: $\boxed{R = \frac{1686.26}{225.33} = 7.4833}$ which confirms the ratio given in Figure 8.16.

=============================<>=============================

Problem 8.78 The area of the beam cross section is $5200\ mm^2$. Determine the moment of inertia of the beam cross section about the x-axis. Compare your result with the moment of inertia of a solid section of equal area and confirm the ratio shown in Figure 8.16.

Solution: Let the outside dimension be *b mm*, then the inside dimension is $b - 40\ mm$. The cross section is $A = b^2 - (b-40)^2 = 5200\ mm^2$. Solve: $b = 85\ mm$. Divide the beam cross section into two parts: the inner and outer squares. Part (1)

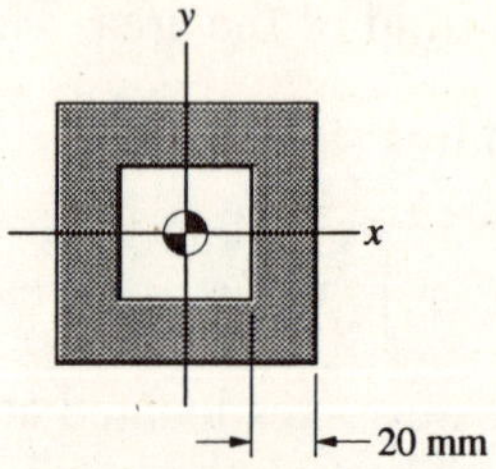

$A_1 = 85^2 = 7225\ mm^2$, $I_{xx1} = \left(\frac{1}{12}\right)85(85^3) = 4.35 \times 10^6$. Part (2)

$A_2 = 45^2 = 2025\ mm^2$. $I_{xx2} = \left(\frac{1}{12}\right)45(45^3) = 3.417 \times 10^5$. The composite moment of inertia about the centroid is $I_x = I_{xx1} - I_{xx2} = 4.008 \times 10^6\ mm^4$ For a square cross section of the same area, $h = \sqrt{5200} = 72.111\ mm$. The area moment of inertia is

$I_{xb} = \left(\frac{1}{12}\right)72.111(72.111^3) = 2.253 \times 10^6\ in^4$. The ratio: $\boxed{R = \frac{4.008 \times 10^6}{2.253 \times 10^6} = 1.7788 = 1.78}$ which confirms the value given in Figure 8.16.

=============================<>=============================

Problem 8.79 (a) If I_x is expressed in in^4, R is in *in.*, and M is in *in-lb*, what are the dimensions c Young's modulus E? (b) A beam with the cross section shown is subjected to couples $M = 1500\ in\ lb$ as shown in Figure 8.15b and it bends into a circular arc with radius $R = 120\ in$. What is Young's modulus for the material?

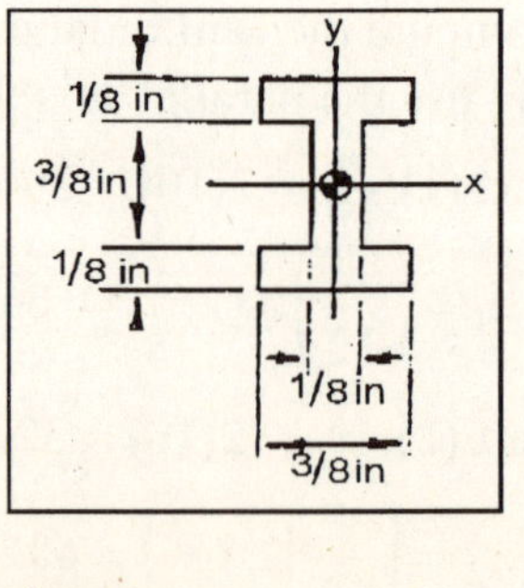

Solution: (a) The relationship between E and the curvature of a beam is $E = \frac{RM}{I}$. The dimensions of R are L^1, of M are L^1F^1, and the dimensions of I are L^4. Thus E has the dimensions of $L(LF)(L^{-4}) \rightarrow FL^{-2}$. If F is in *lb* and L is in *in.,* then E has the dimensions of lb / in^2

Problem 8.79 continued on next page....

============================<>============================

continuation of solution to Problem 8.79...

(b) Divide the beam cross section into three parts: Part (1) is the upper rectangle, Part (2) is the middle rectangle, and Part (3) is the lower rectangle. Part (1): $A_1 = \left(\frac{1}{8}\right)\left(\frac{3}{8}\right) = \frac{3}{64}\ in^2$, $\bar{y}_1 = \frac{1}{4}\ in$,

$I_{xx1} = \left(\frac{1}{12}\right)\left(\frac{3}{8}\right)\left(\frac{1}{8}\right)^3 = 6.1 \times 10^{-5}\ in^4$. Part (2): $A_2 = A_1$, $\bar{y}_2 = 0$,

$I_{xx2} = \left(\frac{1}{12}\right)\left(\frac{1}{8}\right)\left(\frac{3}{8}\right)^3 = 5.49 \times 10^{-4}\ in^4$. Part (3) $A_3 = A_1$, $\bar{y}_3 = -\bar{y}_1$, $I_{xx3} = I_{xx1}$. The composite:

$I_x = \bar{y}_1^2 A_1 + I_{xx1} + \bar{y}_2^2 A_2 + I_{xx2} + \bar{y}_3^2 A_3 + I_{xx3}$, from which $I_x = 6.5 \times 10^{-3}\ in^4$. Young's modulus is $\boxed{E = \frac{RM}{I_x} = \frac{(120)(1500)}{6.5 \times 10^{-3}} = 2.76 \times 10^7\ lb/in^2}$

============================<>============================

Problem 8.80 Suppose that you want to design a beam made of material whose density is $8000\ kg/m^3$. The beam is to be $4\ m$ in length and have a mass of $320\ kg$. Design a cross section for the beam such that $I_x = 3 \times 10^{-5}\ m^4$

Solution: The strategy is to determine the cross sectional area, and then use the ratios given in Figure 8.16 to design a beam. The volume of the beam is $V = AL = 4A\ m^3$. The mass of the beam is $m = V(8000) = 32000A = 320\ kg$, from which $A = 0.01\ m^2$. The moment of inertia for a beam of square cross section with this area is $I_{xxb} = \left(\frac{1}{12}\right)(0.1)(0.1^3) = 8.333 \times 10^{-6}\ m^4$. The ratio is

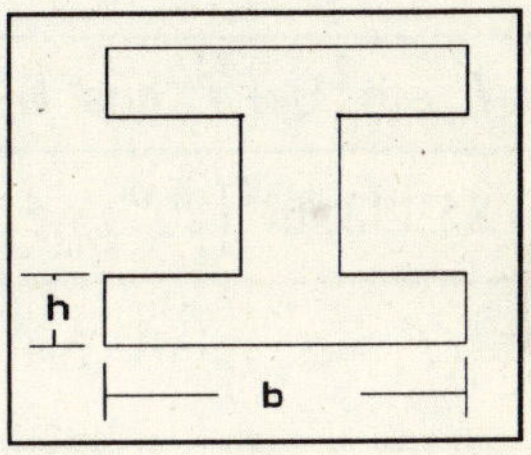

$R = \frac{3 \times 10^{-5}}{8.333 \times 10^{-6}} = 3.6$. From Figure 8.16, this ratio suggests an I-beam of the form shown in the sketch. Choose an I-beam made up of three equal area rectangles, of dimensions b by $h\ m$ in section. The moment of inertia about the centroid is $I_x = \bar{y}_1^2 A_1 + I_{xx1} + \bar{y}_2^2 A_2 + I_{xx2} + \bar{y}_3^2 A_3 + I_{xx3}$.

Since all areas are equal, $A_1 = A_2 = A_3 = bh$, and $\bar{y}_1 = \frac{b+h}{2}$, $\bar{y}_2 = 0$, and $\bar{y}_3 = -\bar{y}_1$, this reduces to

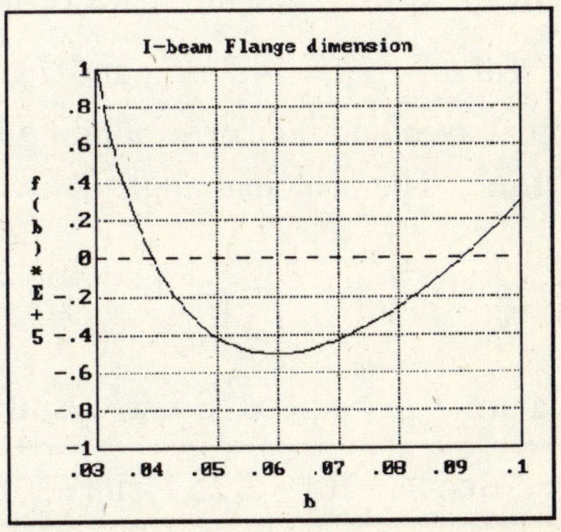

$I_x = \left(\frac{1}{6}\right)bh^3 + 2\left(\frac{b+h}{2}\right)^2 hb + \left(\frac{1}{12}\right)hb^3$. Note that $bh = \frac{A}{3}$, where A is the known total cross section area. These are two equations in two unknowns. Plot the function

$f(b) = \left(\frac{1}{6}\right)bh^3 + 2\left(\frac{b+h}{2}\right)^2 bh + \left(\frac{1}{12}\right)hb^3 - I_x$ subject to the condition that $hb = \frac{A}{3}$. The function was graphed using **TK Solver 2.** The graph crosses the zero axis at approximately $b = 0.0395\ m$. and $b = 0.09\ m$. The lower value is an allowable value for h and the greater value corresponds to an allowable value of b. Thus the I beam design has the flange dimensions, $\boxed{b = 90\ mm}$ and $\boxed{h = 39.5\ mm}$.

============================<>============================

Problem 8.81 Determine $I_{x'}$, $I_{y'}$, and $I_{x'y'}$.

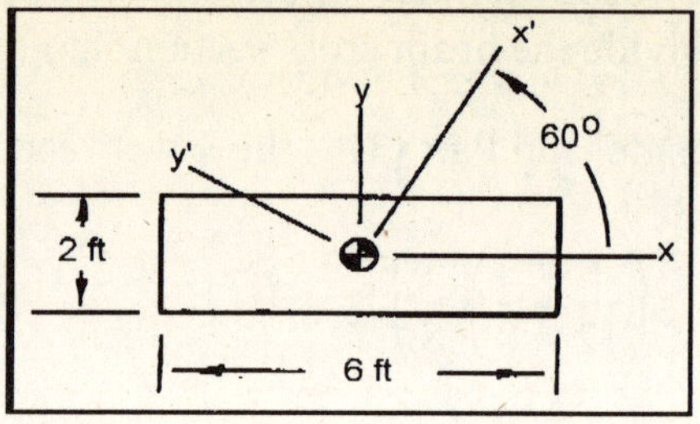

Solution: The area moments of inertia for the unrotated coordinates are $I_x = \left(\frac{1}{12}\right)6\left(2^3\right) = 4\ ft^4$, $I_y = \left(\frac{1}{12}\right)2\left(6^3\right) = 36\,ft^4$. The rotation angle is 60°. The area moments in the rotated system are

$I_{x'} = I_x \cos^2\theta - 2I_{xy}\sin\theta\cos\theta + I_y \sin^2\theta$ $\;I_{y'} = I_x \sin^2\theta + 2I_{xy}\sin\theta\cos\theta + I_y\cos^2\theta$, and

$I_{x'y'} = \left(I_x - I_y\right)\sin\theta\cos\theta + \left(\cos^2\theta - \sin^2\theta\right)I_{xy}$. By symmetry, $I_{xy} = 0$, so that

$\boxed{I_{x'} = 4\cos^2 60 + 36\sin^2 60 = 28\,ft^4}$, $\boxed{I_{y'} = 4\sin^2 60 + 36\cos^2 60 = 12\,ft^4}$, and

$\boxed{I_{x'y'} = (4-36)\sin 60\cos 60 = -13.856\,ft^4}$

Problem 8.82 Determine $I_{x'}$, $I_{y'}$, and $I_{x'y'}$. (Do not use Mohr's circle.)

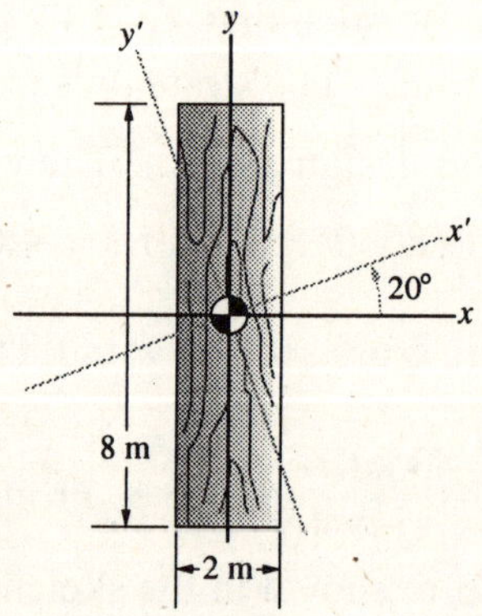

Solution: The area moments of inertia in the unrotated system are $I_x = \left(\frac{1}{12}\right)2\left(8^3\right) = 85.33\,m^4$, $I_y = \left(\frac{1}{12}\right)\left(2^3\right)(8) = 5.333\,m^4$. The rotation angle is 20°. By symmetry, $I_{xy} = 0$. The area moment of inertia for the rotated system is $I_{x'} = I_x\cos^2\theta + I_y\sin^2\theta$.

$\boxed{I_{x'} = 85.333\cos^2\left(20^o\right) + 5.333\sin^2\left(20^o\right) = 75.975\,ft^4}$

$I_{y'} = I_x\sin^2\theta + I_y\cos^2\theta$.

$\boxed{I_{y'} = 85.333\sin^2\left(20^o\right) + 5.333\cos^2\left(20^o\right) = 14.692\,ft^4}$.

$I_{x'y'} = (I_x - I_y)\sin\theta\cos\theta$. $\boxed{I_{x'y'} = (85.333 - 5.333)\sin\left(20^o\right)\cos\left(20^o\right) = 25.712\,ft^4}$

Problem 8.83 The moments of inertia of the rectangular area are $I_x = 76.0\,m^4$, $I_y = 14.7\,m^4$, and $I_{xy} = 25.7\,m^4$. Determine a set of principal axes and the corresponding principal moments of inertia.

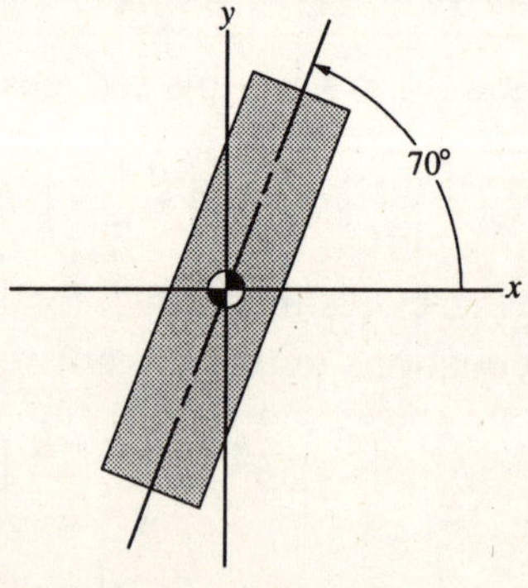

Solution: The principal angle is given by

$\theta = \left(\frac{1}{2}\right)\tan^{-1}\left(\frac{2I_{xy}}{I_y - I_x}\right) = --19.99 = -20^o$. The principal moments of inertia are $I_{xP} = I_x\cos^2\theta - 2I_{xy}\sin\theta\cos\theta + I_y\sin^2\theta$,

$\boxed{I_{xP} = 76\cos^2(-20) - 2(25.7)\sin(-20)\cos(-20) + 14.7\sin^2(-20) = 85.35\,m^4}$,

$I_{yP} = I_x\sin^2\theta + 2I_{xy}\sin\theta\cos\theta + I_y\cos^2\theta$.

$\boxed{I_{yP} = (76)\sin^2(-20) + 2(25.7)\sin(-20)\cos(-20) + 14.7\cos^2(-20) = 5.35\,m^4}$

Problem 8.84 Determine the moments of inertia $I_{x'}$, $I_{y'}$, and $I_{x'y'}$ if $\theta = 15^o$. (Do not use Mohr's circle.)

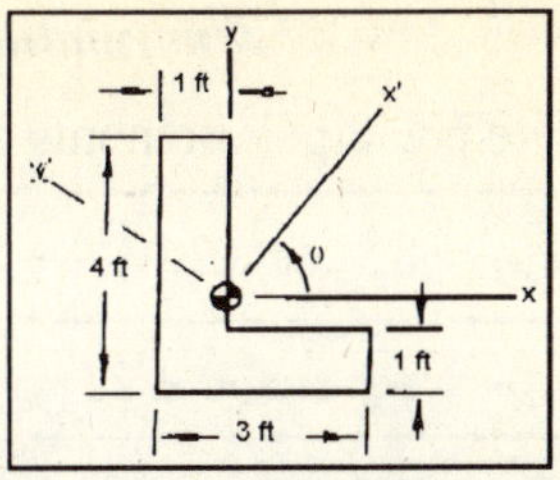

Solution: The strategy is to determine the moments of inertia about the centroid for the unrotated system, from which to determine the moments of inertia for the rotated system. The determination of the moments of inertia about the centroid requires the determination of the location of the centroid of the composite. Divide the area into two parts: Part (1) is the vertical rectangle 1 *ft* by 4 *ft;* Part (2) is the horizontal rectangle 2 *ft* by 1 *ft*. The centroid is computed from the lower left corner of Part (1).

Part(1): $A_1 = 4\,ft^2$, $\bar{x}_1 = 0.5\,ft$, $\bar{y}_1 = 2\,ft$, $I_{xx1} = \left(\frac{1}{12}\right)1(4^3) = 5.333\,ft^4$, $I_{yy1} = \left(\frac{1}{12}\right)4(1^3) = 0.3333\,ft^4$.

Part (2): $A_2 = 2\,ft^2$, $\bar{x}_2 = 1 + 1 = 2\,ft$, $\bar{y}_2 = 0.5\,ft$ $I_{xx1} = \left(\frac{1}{12}\right)2(1^3) = 0.1667\,ft^4$,

$I_{yy2} = \left(\frac{1}{12}\right)1(2^3) = 0.6667\,ft^4$. The composite area is $A = A_1 + A_2 = 6\,ft^2$. The moment of inertia about the lower left corner is $I_x = \bar{y}_1^2 A_1 + I_{xx1} + \bar{y}_2^2 A_2 + I_{xx2} = 22\,ft^4$,

$I_y = \bar{x}_1^2 A_1 + I_{yy1} + \bar{x}_2^2 A_2 + I_{yy2} = 10\,ft^4$, $I_{xy} = \bar{x}_1\bar{y}_1 A_1 + \bar{x}_2\bar{y}_2 A_2 = 6\,ft^4$. The composite centroid relative to the lower left corner is $\bar{x} = \frac{\bar{x}_1 A_1 + \bar{x}_2 A_2}{A} = 1\,ft$, $\bar{y} = \frac{\bar{y}_1 A_1 + \bar{y}_2 A_2}{A} = 1.5\,ft$. The moments of inertia about the centroid: $I_{xc} = -\bar{y}^2 A + I_x = 8.5\,ft^4$, $I_{yc} = -\bar{x}_2^2 A + I_y = 4\,ft^4$,

$I_{xyc} = -\bar{x}\bar{y}A + I_{xy} = -3\,ft^4$. The moments of inertia for the rotated system are

$I_{x'} = I_{xc}\cos^2(15) - 2I_{xyc}\sin(15)\cos(15) + I_{yc}\sin^2(15)$.

$$I_{x'} = 8.5\cos^2(15^o) - 2(-3)\sin(15^o)\cos(15^o) + 4\sin^2(15^o) = 9.7\,ft^4$$

$I_{y'} = I_{xc}\sin^2(15) + 2I_{xyc}\sin(15)\cos(15) + I_{yc}\cos^2(15)$

$$I_{y'} = 8.5\sin^2(15^o) + 2(-3)\sin(15^o)\cos(15^o) + 4\cos(15^o) = 2.8\,ft^4$$

$I_{x'y'} = (I_{xc} - I_{yc})\sin(15)\cos(15) + (\cos^2(15) - \sin^2(15))I_{xyc}$

$$I_{x'y'} = (8.5 - 4)\sin(15^o)\cos(15^o) + \left(\cos^2(15^o) - \sin^2(15^o)\right)(-3) = -1.47\,ft^4$$

Problem 8.85 For the area and coordinate systems in Problem 8.84, determine a set of principal axes and the corresponding principal moments of inertia. (Do not use Mohr's circle.)

Solution: Use the results of the solution to Problem 8.84: $I_{xc} = 8.5\,ft^4$, $I_{yc} = 4\,ft^4$, and $I_{xyc} = -3\,ft^4$.

The principal angle is $\theta = \left(\frac{1}{2}\right)\tan^{-1}\left(\frac{2I_{xyc}}{I_{yc} - I_{xc}}\right) = 26.57^o$

Problem 8.85 continued on next page

---

continuation of Problem 8.85..........

The principal moments of inertia are $I_{xP} = I_{xc}\cos^2\theta - 2I_{xyc}\sin\theta\cos\theta + I_{yc}\sin^2\theta$

$$\boxed{I_{xP} = 8.5\cos^2(26.6^o) - 2(-3)\sin(26.6^o)\cos(26.6^o) + 4\sin^2(26.6^o) = 10\ ft^4}$$

$$I_{yP} = I_{xc}\sin^2\theta + 2I_{xyc}\sin\theta\cos\theta + I_{yc}\cos^2\theta$$

$$\boxed{I_{yP} = 8.5\sin^2(26.6^o) + 2(-3)\sin(26.6^o)\cos(26.6^o) + 4\cos^2(26.6^o) = 2.5\ ft^4}$$

Problem 8.86 Solve the Problem 8.81 by using Mohr's circle.

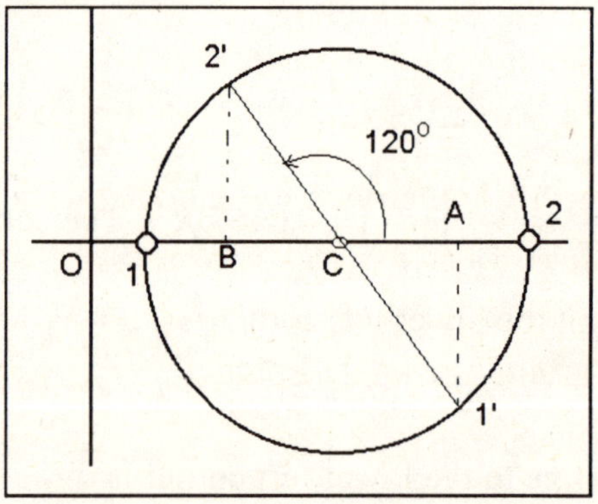

Solution: Use the results of the solution in Problem 8.81: the moments of inertia of the unrotated system are $I_x = 4\ ft^4$, $I_y = 36\ ft^4$, and $I_{xy} = 0$. The circle is constructed as follows: The point 1 is located at the coordinates $(I_x, I_{xy}) = (4,0)$, and labeled. The point 2 is located at $(I_y, -I_{xy}) = (36,0)$ and labeled. The center of the circle is located at $OC = \frac{I_x + I_y}{2} = 20$. The circle is drawn about the center such that it passes through points 1 and 2. A straight line drawn through the center and the points 1 and 2 coincides with the axis, as shown. The point $1'$ is located on the circle at an angle of 120° counterclockwise from point 1 and labeled. A straight line drawn though point $1'$ and the center of the circle defines the point $2'$. The new moments of inertia are taken from the graph as follows: $\boxed{I_{x'} = \overline{OA} = 28\ ft^4}$. $\boxed{I_{y'} = \overline{OB} = 12\ ft^4}$, and the product of inertia is $\boxed{I_{x'y'} = \overline{A1'} = -13.8\ ft^4}$ It is negative because the point $1'$ has the coordinates $(I_{x'}, I_{x'y'}) = (28, -13.8)$.

Problem 8.87 Solve Problem 8.82 by using Mohr's circle.

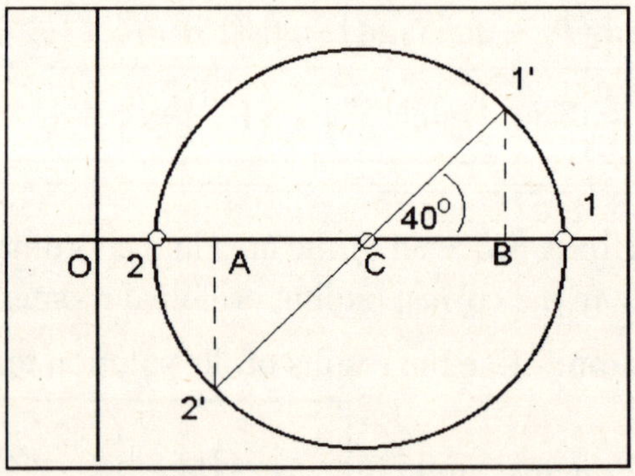

Solution: From the solution to Problem 8.82, the area moments of inertia for the unrotated system are $I_x = 85.33\ m^4$, $I_y = 5.333\ m^4$, and $I_{xy} = 0$. The center of the circle is located on the x-axis at $\overline{OC} = \frac{I_x + I_y}{2} = 45.3\ m^4$. The points 1 and 2 have the coordinates $(I_x, I_{xy}) = (85.3,0)$ and $(I_y, -I_{xy}) = (5.3,0)$ respectively. The circle is drawn through these two points about the center. The point $1'$ is located on the perimeter of the circle at an angle of 40° counterclockwise from point 1. Point $2'$ is located on the perimeter by drawing a straight line though $1'$ and the center. The new moments of inertia are taken from the graph as follows:

$\boxed{I_{x'} = \overline{OB} = 76\ m^4}$, $\boxed{I_{y'} = \overline{OA} = 14.7\ ft^4}$, and $\boxed{I_{x'y'} = \overline{B1'} = 25.7\ ft^4}$

Problem 8.88 Solve Problem 8.83 by using Mohr's circle.

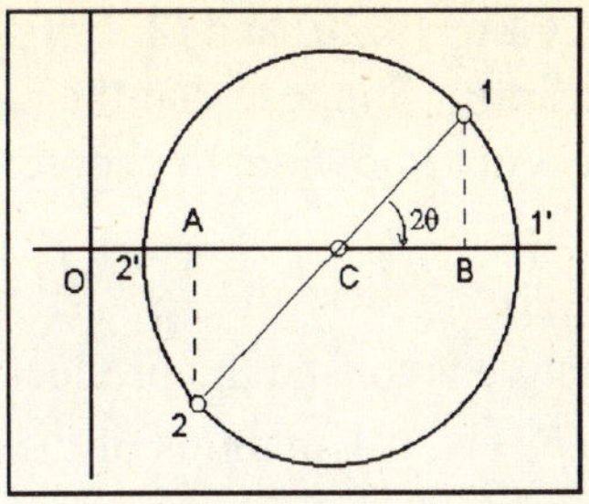

Solution: The area moments of inertia are given: $I_x = 76\ m^4$, $I_y = 14.7\ m^4$, and $I_{xy} = 25.7\ m^4$. The objective is to determine the principal axes and the principal moments of inertia. The center of the circle lies on the x-axis at $\overline{OC} = \dfrac{I_x + I_y}{2} = 45.4\ m^4$. The points 1 and 2 are located at the coordinates $(I_x, I_{xy}) = (76, 25.7)$, and $(I_y, -I_{xy}) = (14.7, -25.7)$, respectively. The circle is drawn through these two points about the center. The point $1'$ lies on the periphery of the circle at the intersection with the x-axis. The point $2'$ is on the perimeter of the circle at the intersection of a straight line drawn through point $1'$ and the center of the circle. The angle of rotation is negative, and is half the measured angle on the circle, $\boxed{\theta = -20^o}$. The moments of inertia are taken from the graph: $\boxed{I_x = \overline{OB} = 85.4\ m^4}$, $\boxed{I_y = \overline{OA} = 5.4\ m^4}$, and $\boxed{I_{xy} = 0}$.

Problem 8.89 Solve Problem 8.84 using Mohr's circle.

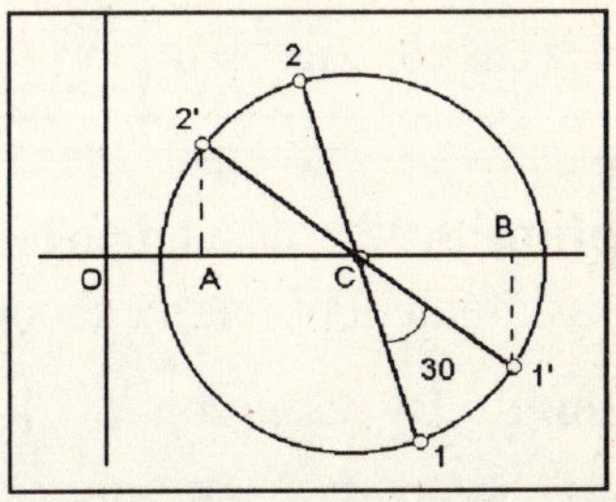

Solution: From the solution to Problem 8.84 the area moments of inertia about the centroid are $I_{xc} = 8.5\,ft^4$, $I_{yc} = 4\,ft^4$, and $I_{xyc} = -3\,ft^4$. The center of the circle lies on the x-axis at $\overline{OC} = \dfrac{I_x + I_y}{2} = 6.25\,ft^4$. The points 1 and 2 have the coordinates $(I_{xc}, I_{xyc}) = (8.5, -3)$, and $(I_{yc}, -I_{xyc}) = (4,3)$, respectively. The circle is drawn through these points about the center. The point $1'$ lies on the perimeter of the circle at an angle of 30° counterclockwise from point 1. The point $2'$ is defined by the intersection of the perimeter and a straight line drawn through $1'$ and the center of the circle. The new moments of inertia are taken from the graph: $\boxed{I_{x'} = \overline{OB} = 9.7\,ft^4}$, $\boxed{I_{y'} = \overline{OA} = 2.8\,ft^4}$, and $\boxed{I_{x'y'} = \overline{B1'} = -1.5\,ft^4}$

Problem 8.90 Solve Problem 8.85 using Mohr's circle.

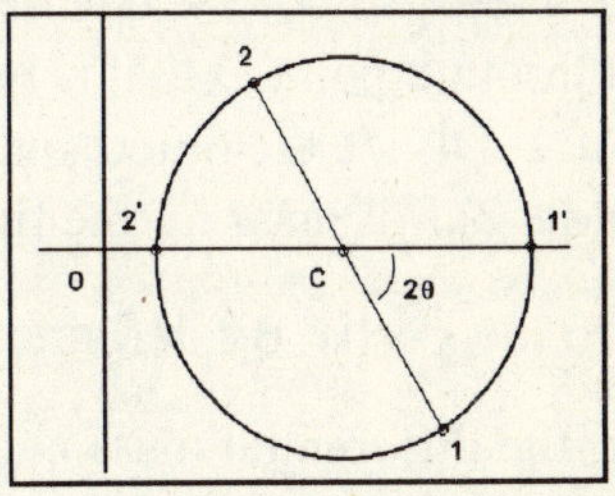

Solution: From the solution to Problems 8.85 and 8.89, $I_x = 8.5\,ft^4$, $I_y = 4\,ft^4$, and $I_{xy} = -3\,ft^4$. The objective is to locate the principal axes and to determine the principal moments of inertia. The center of the circle is at $\overline{OC} = \dfrac{8.5 + 4}{2} = 6.25\,ft^4$. The points 1 and 2 are located at $(I_x, I_{xy}) = (8.5, -3)$, $(I_y, -I_{xy}) = (4, -3)$, respectively. The principal axes coincide with the x-axis: point $1'$ lies on the perimeter at an angle $2\theta = 2(26.6^o)$ counterclockwise from point 1. Thus the angle to the principal axes is $\boxed{\theta = 26.6^o}$ and the principal moments of inertia are $\boxed{I_{xP} = \overline{O1'} = 10\,ft^4}$ and $\boxed{I_{yP} = \overline{O2'} = 2.5\,ft^4}$

Problem 8.91 Derive Eq (8.22) for the product of inertia by using the same procedure we used to derive Eqs (8.20) and (8.21).

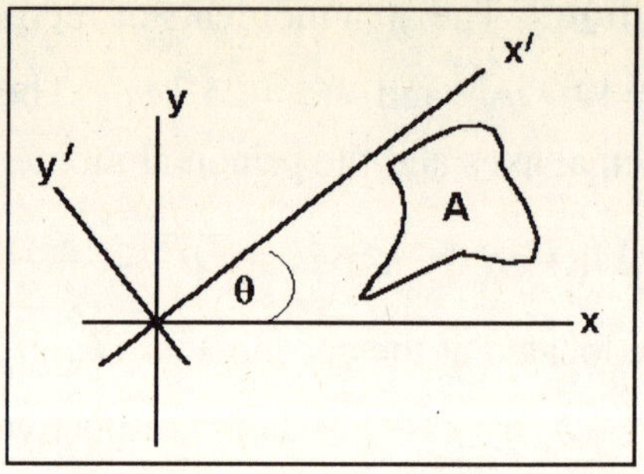

Solution: Suppose that the area moments of inertia of the area A are known in the coordinate system (x, y), $I_x = \int_A y^2 dA$, $I_y = \int_A x^2 dA$, and $I_{xy} = \int_A xyA$. The objective is to find the product of inertia in the new coordinate system (x', y') in terms of the known moments of inertia. The new (x', y') system is formed from the old (x, y) system by rotation about the origin through a counterclockwise angle θ.

By definition, $I_{x'y'} = \int_A x'\, y'\, dA$. From geometry, $x' = x\cos\theta + y\sin\theta$, and $y' = -x\sin\theta + y\cos\theta$. The product is $x'\, y' = xy\cos^2\theta - xy\sin^2\theta + y^2\cos\theta\,\sin\theta - x^2\cos\theta\,\sin\theta$. Substitute into the definition:

$$I_{x'y'} = \left(\cos^2\theta - \sin^2\theta\right)\int_A xy\, dA + \left(\cos\theta\,\sin\theta\right)\left(\int_A y^2 dA - \int_A x^2\, dA\right),$$ from which

$$\boxed{I_{x'y'} = \left(\cos^2\theta - \sin^2\theta\right)I_{xy} + \left(I_x - I_y\right)\sin\theta\,\cos\theta},$$ which is the expression required.

Problem 8.92 The homogenous, slender bar has mass m and length l. Use integration to determine its mass moment of inertia about the perpendicular axis L_o.

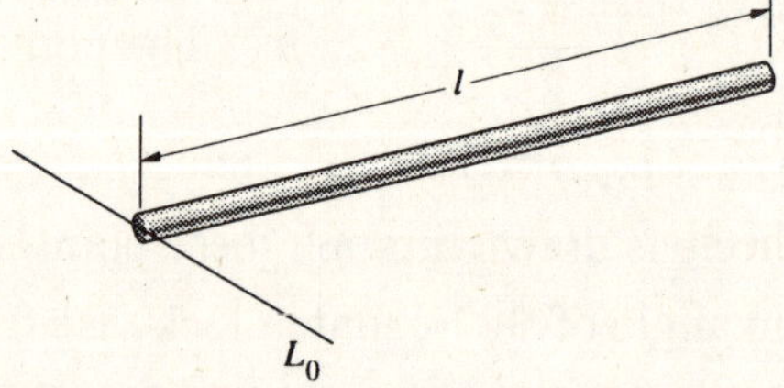

Solution: By definition $I = \int_0^l r^2 dm$. The differential mass for a slender homogenous bar is $dm = \rho A dr$, where ρ in the mass density per unit volume and A is the cross section area. Thus

$$\boxed{I = \rho A\int_0^l r^2 dr = \rho A\left[\frac{r^3}{3}\right]_0^l = \frac{\rho A l^3}{3} = \frac{ml^2}{3}}$$

Problem 8.93 Two homogenous, slender bars, each of mass m and length l, are welded together to form the T-shaped object. Use integration to determine the mass moment of inertia of the object about the axis through point O that is perpendicular to the bars.

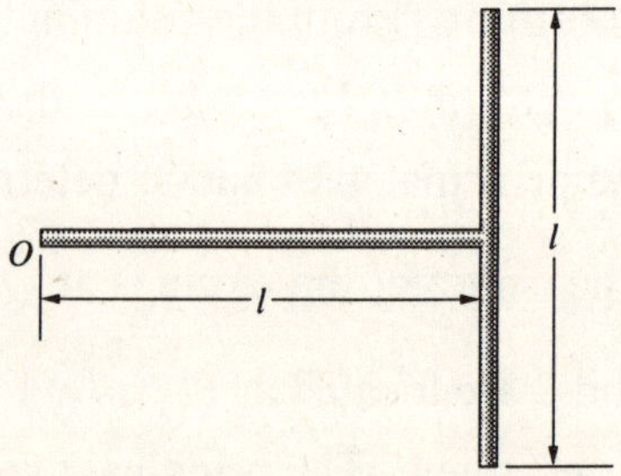

Solution: Divide the object into two pieces, each corresponding to a slender bar of mass m; the first parallel to the y-axis, the second to the x-axis. By definition $I = \int_0^l r^2 dm + \int_m r^2 dm$. For the first bar, the differential mass is $dm = \rho A dr$. Assume that the second bar is very slender, so that the mass is concentrated at a distance l from O. Thus $dm = \rho A dx$, where x lies between the limits $-\frac{l}{2} \le x \le \frac{l}{2}$. The distance to a differential dx is $r = \sqrt{l^2 + x^2}$. Thus the definition becomes

$$I = \rho A\int_0^l r^2 dr + \rho A\int_{-\frac{l}{2}}^{\frac{l}{2}} (l^2 + x^2)dx \quad I = \rho A\left[\frac{r^3}{3}\right]_0^l + \rho A\left[l^2 x + \frac{x^3}{3}\right]_{-\frac{l}{2}}^{\frac{l}{2}} \boxed{= ml^2\left(\frac{1}{3} + 1 + \frac{1}{12}\right) = \frac{17}{12}ml^2}$$

Problem 8.94 The homogenous, slender bar has mass m and length l. Use integration to determine the mass moment of inertia about the axis L.

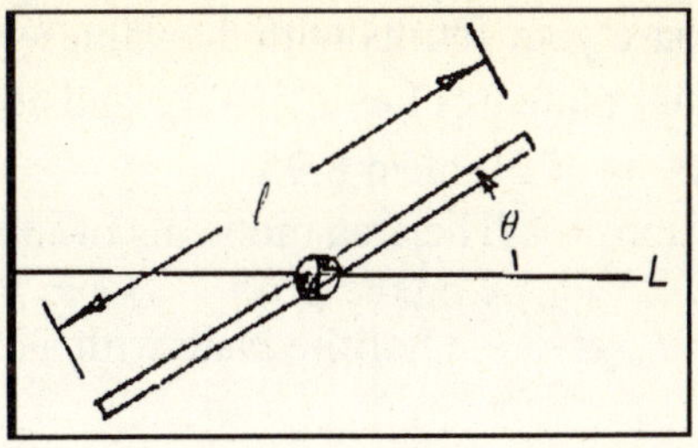

Solution: By definition, $I = \int_m r^2 dm$. Let the differential mass be $dm = \rho A dx$, where x is measured along the bar from $x = -\frac{l}{2}$ to $x = \frac{l}{2}$. The distance from the line L to the bar is r, where $r = x\sin\theta$. Substitute to obtain

$$I = \rho A \sin^2\theta \int_{-\frac{l}{2}}^{\frac{l}{2}} x^2 dx = \rho A \sin^2\theta \left[\frac{x^3}{3}\right]_{-\frac{l}{2}}^{\frac{l}{2}} \boxed{= \frac{ml^2}{12}\sin^2\theta}$$

Problem 8.95 A homogenous, slender bar is bent into a circular ring of mass m and radius R. Determine mass moment of inertia of the ring (a) about the axis through its center of mass that is perpendicular to the ring; (b) about the axis L.

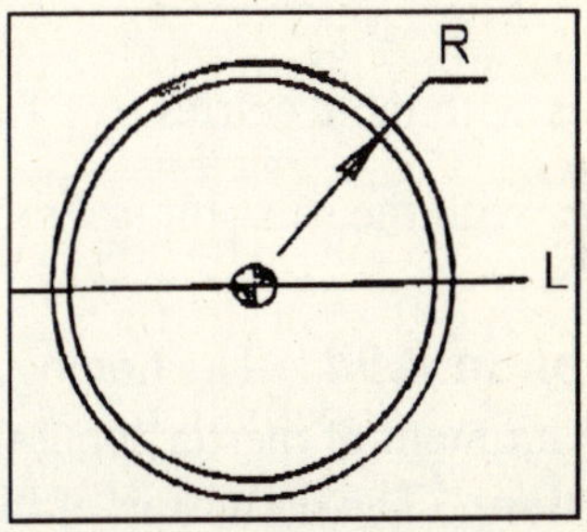

Solution: The definition is $I = \int_m r^2 dm$.

(a) The differential mass of the ring is $dm = \rho A R d\theta$, where A is the cross sectional area of the bar. The distance to the ring from the center of mass is a constant, thus $I = \rho A R^3 \int_0^{2\pi} d\theta = 2\pi\rho A R^3 = mR^2$.

(b) The equation of a slender ring is $x^2 + y^2 = R^2$. The distance to the ring from the line L is y, thus the definition becomes $I = 2\int_m y^2 dm$, where each half of the ring contributes. The differential element of mass is $dm = \rho A R d\theta$. Substitute into the definition $I = 2\rho A R \int_0^{\pi} y^2 d\theta$. Noting that $y^2 = R^2 - x^2$ and $x = R\sin\theta$, $I = 2\rho A R^3 \int_0^{\pi} \sin^2\theta d\theta$

$$\boxed{= 2\rho A R^3 \left[\frac{\theta}{2} - \frac{\sin 2\theta}{4}\right]_0^{\pi} = \frac{mR^2}{2}}$$

Problem 8.96 The homogenous thin plate is of uniform thickness and mass m. Determine its mass moments of inertia about the x, y, and z axes.

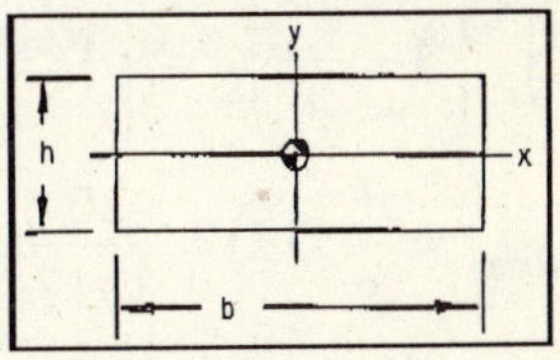

Solution: *Strategy:* The mass moments of inertia of a thin plate of arbitrary shape are given by Eqs (8.30)-(8.32) in terms of the area moments of inertia of the cross sectional area of the plate. The area moments of inertia can be obtained from Appendix B.

The area moments of inertia are $I_x = \left(\frac{1}{12}\right)bh^3$, $I_y = \left(\frac{1}{12}\right)hb^3$,

$I_z = I_x + I_y$. The area mass density of the plate is $\frac{m}{A} = \frac{m}{bh}$, hence the ,mass moments of inertia are

$$I_{(x-axis)} = \frac{m}{bh} I_x = \frac{m}{12}h^2, \boxed{I_{(y-axis)} = \frac{m}{bh} I_y = \frac{m}{12}b^2} \; I_{(z-axis)} = \frac{m}{12}\left(h^2 + b^2\right)$$

Problem 8.97 The homogenous thin plate is of uniform thickness and mass *m*. (a) Determine its mass moments of inertia about the x and z axes. (b) Let $R_i = 0$ and compare your results with the values given in Appendix C for a thin circular plate. (c) Let $R_i \rightarrow R_o$ and compare your results with the solutions of Problem 8.95.

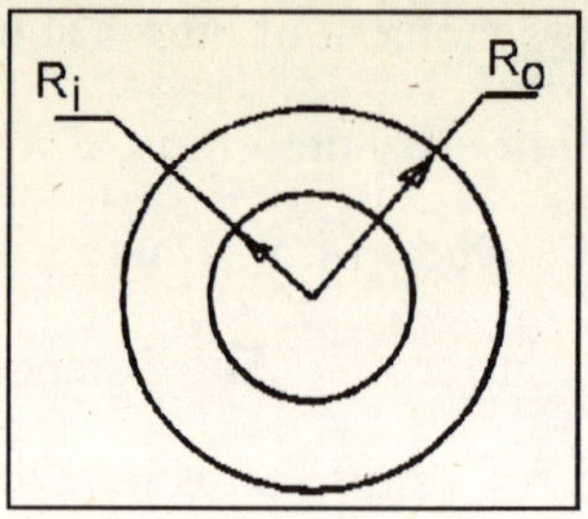

Solution: (a)The area moments of inertia for a circular area are $I_x = I_y = \frac{\pi R^4}{4}$. For the plate with a circular cutout, $I_x = \frac{\pi}{4}\left(R_o^4 - R_i^4\right)$ The area mass density is $\frac{m}{A}$, thus for the plate with a circular cut, $\frac{m}{A} = \frac{m}{\pi\left(R_o^2 - R_i^2\right)}$, from which the mass moments of inertia $I_{(x-axis)} = \frac{m\left(R_o^4 - R_i^4\right)}{4\left(R_o^2 - R_i^2\right)} = \frac{m}{4}\left(R_o^2 + R_i^2\right)$ $I_{(z-axis)} = 2I_{(x-axis)} = \frac{m}{2}\left(R_o^2 + R_i^2\right)$. (b) Let $R_i = 0$, to obtain $I_{x-axis} = \frac{m}{4}R_o^2$, $I_{(z-axis)} = \frac{m}{2}R_o^2$, which agrees with table entries. (c) Take the limit as $R_i \rightarrow R_o$, $I_{(x-axis)} = \frac{m}{2}R_o^2$, $I_{(z-axis)} = mR_o^2$, which agrees with the solution to Problem 8.95.

Problem 8.98 The homogenous thin plate is of uniform thickness and weighs 20 *lb*. Determine its mass moment of inertia about the *y* axis.

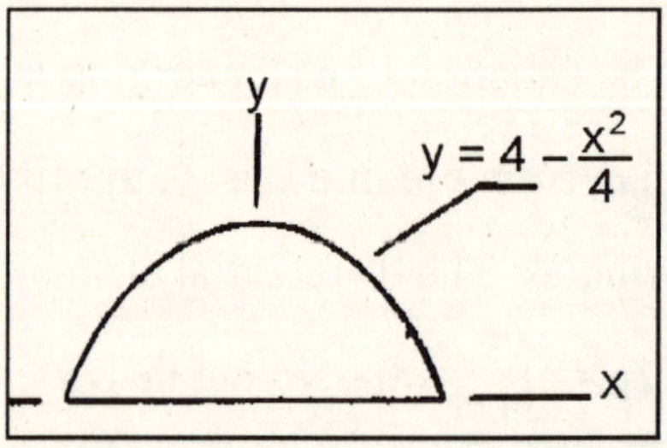

Solution: The definition of the mass moment of inertia is $I = \int_m r^2 dm$. The distance from the y-axis is *x*, where *x* varies over the range $-4 \le x \le 4$. Let $\tau = \frac{m}{A} = \frac{W}{gA}$ be the area mass density. The mass of an element *ydx* is $dm = \frac{W}{gA} ydx$. Substitute into the definition: $I_{y-axis} = \frac{W}{gA}\int_{-4}^{4} x^2\left(4 - \frac{x^2}{4}\right)dx = \frac{W}{gA}\left[\frac{4x^3}{3} - \frac{x^5}{20}\right]_{-4}^{+4} = \frac{W}{gA}[68.2667]$. The area is

$A = \int_{-4}^{4}\left(4 - \frac{x^2}{4}\right)dx = \left[4x - \frac{x^3}{12}\right]_{-4}^{4} = 21.333\ ft^2$ The mass moment of inertia about the y-axis is

$I_{(y-axis)} = \frac{W}{g}(3.2) = \frac{20}{32.17}(3.2) = 1.989\ slug\text{-}ft^2$

Problem 8.99 Determine the mass moment of inertia of the plate in Problem 8.98 about the x-axis.

Solution: The differential mass is $dm = \frac{W}{gA} dydx$. The distance of a mass element from the x-axis is y, thus $I = \frac{W}{gA}\int_{-4}^{+4} dx \int_{0}^{4-\frac{x^2}{4}} y^2 dy = \frac{W}{3gA}\int_{-4}^{+4}\left(4 - \frac{x^2}{4}\right)^3 dx$

$$= \frac{W}{3gA}\left[64x - 4x^3 + \frac{3}{20}x^5 - \frac{x^7}{448}\right]_{-4}^{4} = \frac{W}{3gA}[234.057]$$. From the solution to Problem 8.98,

$A = 21.333\ ft^2$. Thus the mass moment of inertia about the x-axis is

$$\boxed{I_{x-axis} = \frac{W}{3g}\frac{(234.057)}{(21.333)} = \frac{W}{g}(3.657) = 2.27\ slug\text{-}ft^2}.$$

Problem 8.100 The mass of the object is 10 *kg*. Its mass moment of inertia about L_1 is $10\ kg\text{-}m^2$ What is its mass moment of inertia about L_2? The three axes are in the same plane.)

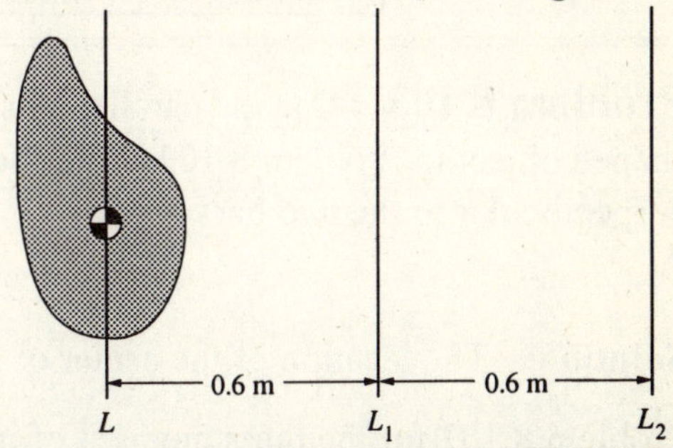

Solution: The strategy is to use the data to find the mass moment of inertia about L, from which the mass moment of inertia about L_2 can be determined.

$I_L = -(0.6)^2(10) + 10 = 6.4\ m^2$, from which

$$\boxed{I_{L2} = (1.2)^2(10) + 6.4 = 20.8\ m^2}$$

Problem 8.101 An engineer gathering data for the design of a maneuvering unit determines that an astronaut's center of mass is at $x = 1.01\ m$, $y = 0.16\ m$ and that his mass moment of inertia about the z axis is $105.6\ kg\text{-}m^2$. His mass is 81.6 *kg*. What is his mass moment of inertia about the z' axis through his center of mass?

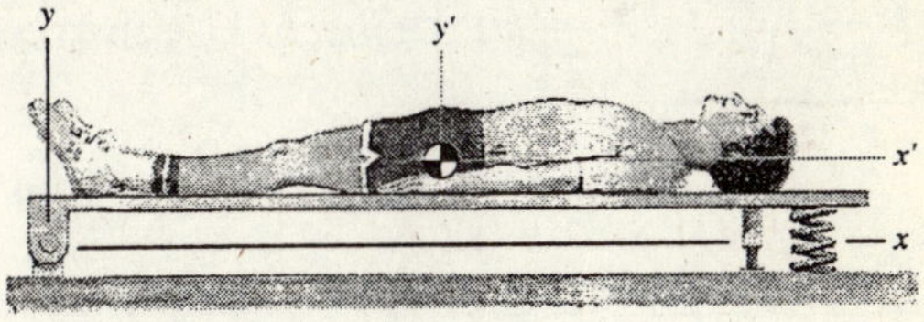

Solution: The distance from the z' axis to the z axis is $d = \sqrt{x^2 + y^2} = 1.02257\ m$. The mass moment of inertia about the z' axis is $\boxed{I_{z'-axis} = -d^2 m + I_{z-axis} = -(1.0457)(81.6) + 105.6 = 20.27\ kg\text{-}m^2}$

Problem 8.102 Two homogenous, slender bars, each of mass m and length l, are welded together to form the T-shaped object. Use the parallel axis theorem to determine the mass moment of inertia of the object about the axis through point O that is perpendicular to the bars.

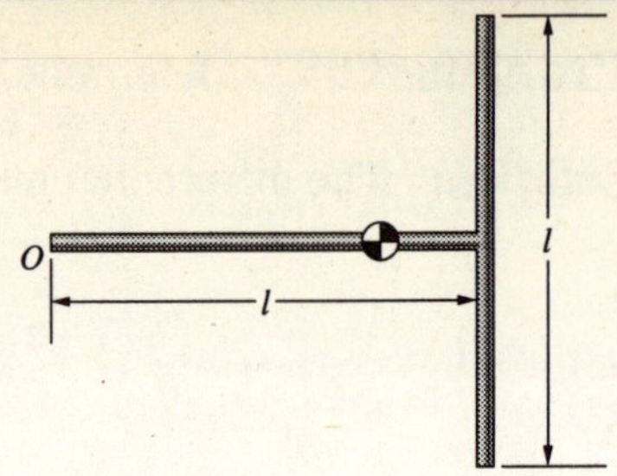

Solution: Divide the object into two pieces, each corresponding to a bar of mass m. By definition $I = \int_0^l r^2 dm$. For the first bar, the differential mass is $dm = \rho A dr$, from which the mass moment of inertia about one end is

$$I_1 = \rho A \int_0^l r^2 dr = \rho A \left[\frac{r^3}{3}\right]_0^l = \frac{ml^2}{3}.$$ For the second bar $$I_2 = \rho A \int_{-\frac{l}{2}}^{\frac{l}{2}} r^2 dr = \rho A \left[\frac{r^3}{3}\right]_{-\frac{l}{2}}^{\frac{l}{2}} = \frac{ml^2}{12}$$ is the mass moment of inertia about the center of the bar. From the parallel axis theorem, the mass moment of inertia about O is $\boxed{I_o = \frac{ml^2}{3} + l^2 m + \frac{ml^2}{12} = \frac{17}{12} ml^2}$

Problem 8.103 Use the parallel axis theorem to determine the mass moment of inertia of the T-shaped object in Problem 8.102 about the axis through the center of mass of the object that is perpendicular to the two bars.

Solution: The location of the center of mass of the object is $\bar{x} = \dfrac{m\left(\frac{l}{2}\right) + lm}{2m} = \frac{3}{4}l$. Use the results of Problem 8.102 for the mass moment of inertia of a bar about its center. For the first bar,

$I_1 = \left(\frac{l}{4}\right)^2 m + \frac{ml^2}{12} = \frac{7}{48}ml^2$. For the second bar, $I_2 = \left(\frac{l}{4}\right)^2 m + \frac{ml^2}{12} = \frac{7}{48}ml^2$. The composite:

$\boxed{I_c = I_1 + I_2 = \frac{7}{24}ml^2}$ *Check:* Use the results of Problem 8.102:

$$I_c = -\left(\frac{3l}{4}\right)^2 (2m) + \frac{17}{12}ml^2 = \left(\frac{-9}{8} + \frac{17}{12}\right)ml^2 = \frac{7}{24}ml^2. \textit{ check.}$$

Problem 8.104 The mass of the homogenous, slender bar is 20 *kg*. Determine the mass moment of inertia about the z axis.

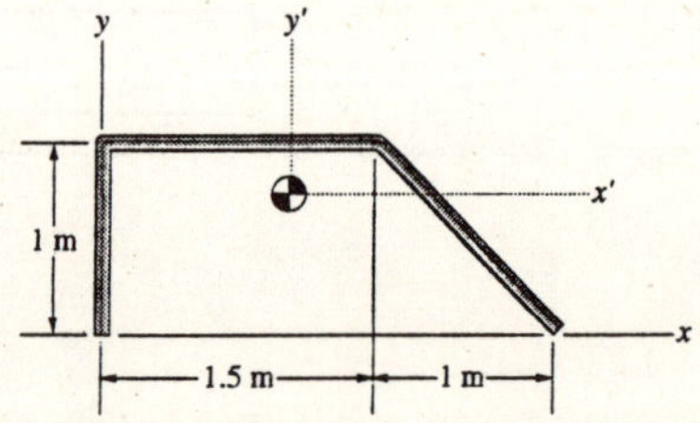

Solution: Divide the object into three segments. Part (1) is the 1 m bar on the left, Part (2) is the 1.5 m horizontal segment, and Part (3) is the segment on the far right. The mass density is

$\rho = \frac{m}{L} = \frac{20}{(1 + 1.5 + \sqrt{2})} = 5.1\ kg/m$ The mass moments of inertia about the centers of mass and the distances to the centers of mass from the z-axis are:

Part (1) $I_1 = \rho\left(\frac{l_1^3}{12}\right) = m_1\frac{l_1^2}{12} = 0.425\ kg\text{-}m^2$, $m_1 = \rho 1 = 5.1\ kg$, $d_1 = 0.5\ m$,

Part (2), $I_2 = \rho\frac{l_2^3}{12} = m_2\frac{l_2^2}{12} = 1.437\ kg\text{-}m^2$, $m_2 = 7.665\ kg$, $d_2 = \sqrt{0.75^2 + 1^2} = 1.25\ m$

Problem 8.104 continued on next page....

continuation of solution to Problem 8.104...

Part (3) $I_3 = \rho \frac{l_3^3}{12} = m_3 \frac{(\sqrt{2})^2}{12} = 1.204\ kg\text{-}m^2$, $m_3 = 7.225\ kg$, $d_3 = \sqrt{2^2 + 0.5^2} = 2.062\ m$. The composite: $\boxed{I = d_1^2 m_1 + I_1 + d_2^2 m_2 + I_2 + d_3^2 m_3 + I_3 = 47.02\ kg\text{-}m^2}$

Problem 8.105 Determine the mass moment of inertia of the bar in Problem 8.104 about the z' through its center of mass.

Solution: The center of mass: $\bar{x} = \frac{x_1 m_1 + x_2 m_2 + x_3 m_3}{20} = \frac{0 + 0.75(7.665) + 2(7.225)}{20} = 1.01\ m$.

$\bar{y} = \frac{0.5 m_1 + 1 m_2 + 0.5 m_3}{20} = \frac{0.5(5.1) + 1(7.665) + 0.5(7.225)}{20} = 0.6914\ m$. The distance from the z-axis to the center of mass is $d = \sqrt{\bar{x}^2 + \bar{y}^2} = 1.221\ m$. The mass moment of inertia about the center o mass:

$\boxed{I_c = -d^2(20) + I_o = 17.19\ kg\text{-}m^2}$

Problem 8.106 The rocket is used for atmospheric research. Its weight and its mass moment of inertia about the z-axis through its center of mass (including its fuel) are $10\ kip$ and $10200\ slug\text{-}ft^2$ respectively. The rocket's fuel weighs 6000 *lb*, its center of mass is located at $x = -3\ ft$, $y = 0$, and $z = 0$, and the mass moment of inertia of the fuel about the axis through the fuel's center of mass parallel to z is 2200 *slug-ft²*. When the fuel is exhausted, what is the rocket's mass moment of inertia about the axis through its new center of mass parallel to z?

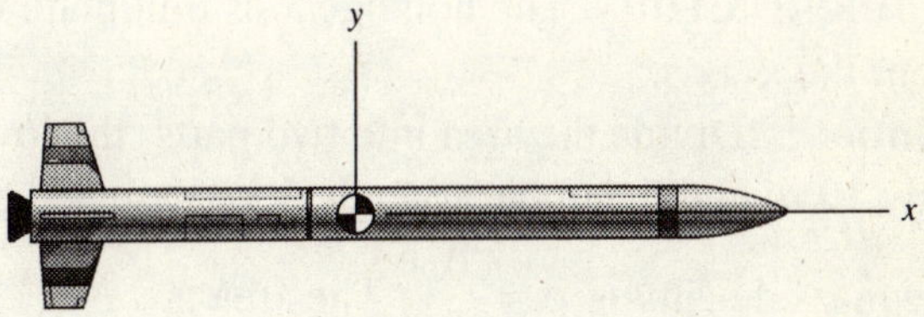

Solution: Denote the mass moment of inertia of the empty rocket as I_E about a center of mass x_E, and the mass moment of inertia of the fuel as I_F about a mass center x_F. Using the parallel axis theorem, the mass moment of inertia of the filled rocket is $I_R = I_E + x_E^2 m_E + I_F + x_F^2 m_F$, about a mass center at the origin ($x_R = 0$). Solve: $I_E = I_R - x_E^2 m_E - I_F - x_F^2 m_F$. The objective is to determine values for the terms on the right from the data given. Since the filled rocket has a mass cente at the origin, the mass center of the empty rocket is found from $0 = m_E x_E + m_F x_F$, from which $x_E = -\left(\frac{m_F}{m_E}\right) x_F$. Using a value of $g = 32\ ft/s^2$, $m_F = \frac{W_F}{g} = \frac{6000}{32} = 186.34\ slug$,

$m_E = \frac{(W_R - W_F)}{g} = \frac{10000 - 6000}{32} = 124.23\ slug$. From which $x_E = -\left(\frac{186.335}{124.224}\right)(-3) = 4.5\ ft$ is the new location of the center of mass. Substitute:

$\boxed{I_E = I_R - x_E^2 m_E - I_F - x_F^2 m_F = 10200 - 2515.5 - 2200 - 1677.01 = 3807.5\ slug\text{-}ft^2}$

Problem 8.107

Problem 8.107 The mass of the homogenous thin plate is 36 kg. Determine its mass moment of inertia about the x- axis.

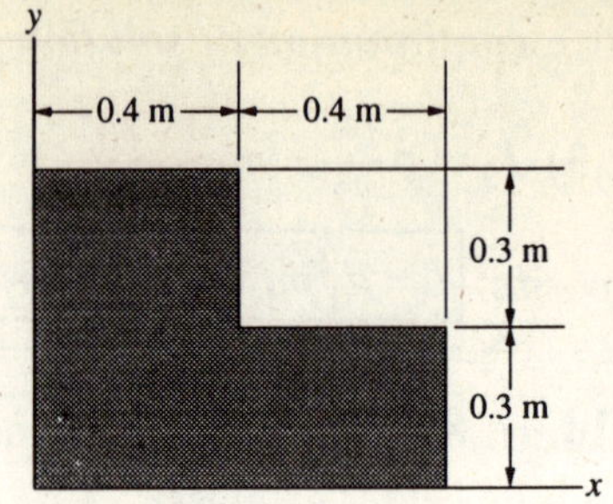

Solution: Divide the plate into two areas: the rectangle 0.4 m by 0.6 m on the left, and the rectangle 0.4 m by 0.3 m on the right. The mass density is $\rho = \frac{m}{A}$. The area is

$A = (0.4)(0.6) + (0.4)(0.3) = 0.36\ m^2$, from which

$\rho = \frac{36}{0.36} = 100\ kg/m^2$. The mass moment of inertia about the x-axis is

$$\boxed{I_{x-axis} = \rho\left(\frac{1}{3}\right)(0.4)(0.6^3) + \rho\left(\frac{1}{3}\right)(0.4)(0.3)^3 = 3.24\ kg\text{-}m^2}$$

Problem 8.108

Problem 8.108 Determine the mass moment of inertia of the plate in Problem 8.107 about the z-axis.

Solution: The basic relation to use is $I_{z-axis} = I_{x-axis} + I_{y-axis}$. The value of I_{x-axis} is given in the solution of Problem 8.107. The mass moment of inertia about the y-axis using the same divisions as in Problem 8.107 and the parallel axis theorem is

$I_{y-axis} = \rho\left(\frac{1}{3}\right)(0.6)(0.4)^3 + \rho\left(\frac{1}{12}\right)(0.3)(0.4)^3 + (0.6)^2\rho(0.3)(0.4) = 5.76\ kg\text{-}m^2$, from which

$$\boxed{I_{z-axis} = I_{x-axis} + I_{y-axis} = 3.24 + 5.76 = 9\ kg\text{-}m^2.}$$

Problem 8.109

Problem 8.109 The homogenous thin plate weighs 10 lb. Determine its mass moment of inertia about the x-axis.

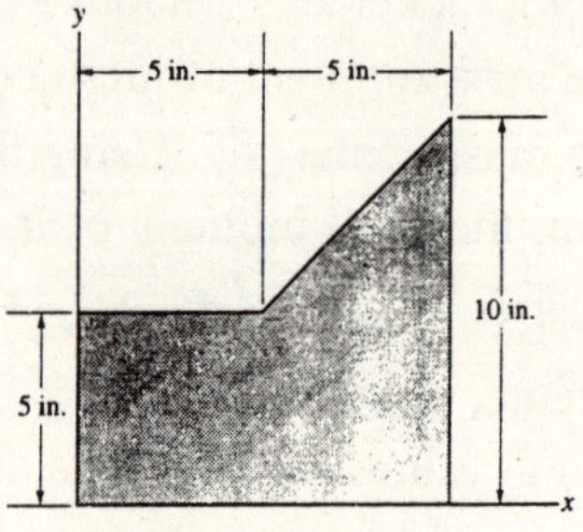

Solution: Divide the area into two parts: the lower rectangle 5 in by 10 in and the upper triangle 5 in base and 5 in altitude. The mass density is $\rho = \frac{W}{gA}$. The area is

$A = 5(10) + \left(\frac{1}{2}\right)5(5) = 62.5\ in^2$. Using $g = 32\ ft/s^2$, the mass density is $\rho = \frac{W}{gA} = 0.005\ slug/in^2$. Using the parallel axis theorem, the mass moment of inertia about the x-axis is $I_{x-axis} = \rho\left(\frac{1}{3}\right)(10)(5)^3 + \rho\left(\frac{1}{36}\right)(5)(5^3) + \rho(5 + 5/3)^2\left(\frac{1}{2}\right)(5)(5) = 4.948\ slug\text{-}in^2$

$$\boxed{I_{x-axis} = 0.03436\ slug\text{-}ft^2}$$

Problem 8.110

Problem 8.110 Determine the mass moment of inertia of the plate in Problem 8.109 about the y-axis.

Solution: Use the results of the solution in Problem 8.109 for the area and the mass density.

$$I_{y-axis} = \rho\left(\frac{1}{3}\right)5(10^3) + \rho\left(\frac{1}{36}\right)5(5^3) + \rho(5 + 10/3)^2\left(\frac{1}{2}\right)5(5) = 12.76\ slug\text{-}in^2 \boxed{= 0.0886\ slug\text{-}ft^2}$$

Problem 8.111 The thermal radiator (used to eliminate excess heat from a satellite) can modeled as a homogenous, thin, rectangular plate. Its mass is 5 slugs. Determine its mass moment of inertia about the x- y- and z- axes.

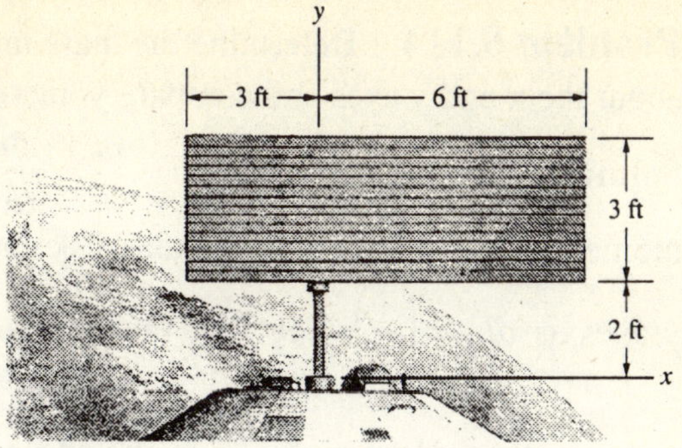

Solution: The area is $A = 9(3) = 27\ ft^2$. The mass density is $\rho = \frac{m}{A} = \frac{5}{27} = 0.1852\ slugs/ft^2$. The mass moment of inertia about the centroid of the rectangle is

$I_{xc} = \rho\left(\frac{1}{12}\right)9(3^3) = 3.75\ slug\text{-}ft^2$, $I_{yc} = \rho\left(\frac{1}{12}\right)3(9^3) = 33.75\ slug\text{-}ft^2$. Use the parallel axis theorem: $\boxed{I_{x-axis} = \rho A(2+1.5)^2 + I_{xc} = 65\ slug\text{-}ft^2}$, $\boxed{I_{y-axis} = \rho A(4.5-3)^2 + I_{yc} = 45\ slug\text{-}ft^2}$.

$\boxed{I_{z-axis} = I_{x-axis} + I_{y-axis} = 110\ slug\text{-}ft^2}$

Problem 8.112 The mass of the homogenous, thin plate is 2 *kg*. Determine its mass moment of inertia about the x-axis L_o through point O that is perpendicular to the plate.

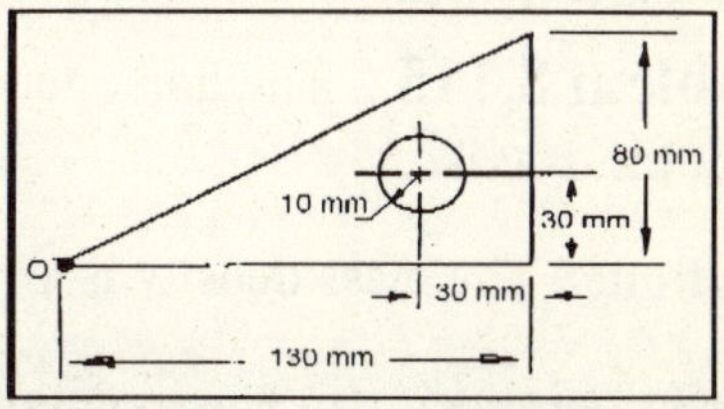

Solution: The strategy is to determine I_x and I_y about point O, from which to determine I_z. The area is

$A_1 = \left(\frac{1}{2}\right)80(130) = 5200\ mm^2$, $A_2 = \pi(10^2) = 314.16\ mm^2$,

$A = A_1 - A_2 = 4885.84\ mm^2$. The mass density is $\rho = \frac{m}{A} = \frac{2}{A} = 0.0004093\ kg/mm^2$. The mass moments of inertia about the centroid of the circle are $I_{xc} = I_{yc} = \rho\left(\frac{1}{4}\right)\pi(10^4) = 2.849\ \ kg\text{-}mm^2$. Use the parallel axis theorem to determine the mass moments of inertia about the x- and y- axes.

$$I_{y-axis} = \rho\left(\frac{1}{36}\right)80(130^3) + \rho A_1\left(\frac{2}{3}130\right)^2 - \rho A_2(100^2) - I_{yc} = 16697.5\ kg\text{-}mm^2$$

$$I_{x-axis} = \rho\left(\frac{1}{12}\right)130(80^3) - \rho A_2(30^2) - I_{xc} = 2151.5\ kg\text{-}mm^2.$$

$I_{z-axis} = I_{x-axis} + I_{y-axis} = 18849\ kg\text{-}mm^2$ $\boxed{I_{z-axis} = 0.018849\ kg\text{-}m^2}$

Problem 8.113 The homogenous cone is of mass *m*. Determine its mass moment of inertia about the z-axis and compare your result with the value in Appendix C.

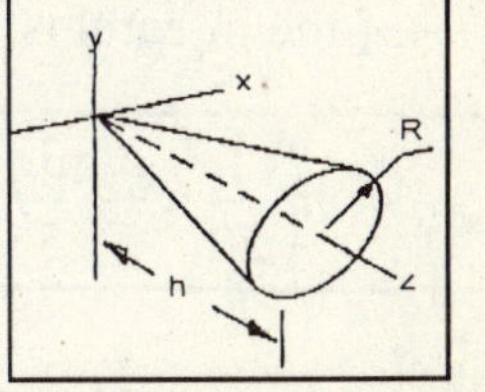

Solution: The differential mass $dm = \left(\frac{m}{V}\right)\pi r^2 dz = \frac{3m}{R^2 h} r^2\ dz$. The moment of inertia of this disk about the z-axis is $\frac{1}{2}mr^2$. The radius varies with z, $r = \left(\frac{R}{h}\right)z$, from which $\boxed{I_{z-axis} = \frac{3mR^2}{2h^5}\int_0^h z^4 dz = \frac{3mR^2}{2h^5}\left[\frac{z^5}{5}\right]_0^h = \frac{3mR^2}{10}}$

Problem 8.114 Determine the mass moments of inertia of the homogenous cone in Problem 8.113 about the x and y axes and compare your results with the values in Appendix C.

Solution: The mass density is $\rho = \frac{m}{V} = \frac{3m}{\pi R^2 h}$ The differential element of mass is $dm = \rho\pi r^2 dz$.. The moment of inertia of this elemental disk about an axis through its center of mass, parallel to the x- and y-axes, is $dI_x = \left(\frac{1}{4}\right) r^2 dm$. Use the parallel axis theorem, $I_x = \int_m \left(\frac{1}{4}\right) r^2 dm + \int_m z^2 dm$. Noting that $r = \frac{R}{h} z$, then $r^2 dm = \rho\left(\frac{\pi R^4}{h^4}\right) z^4 dz$, and $z^2 dm = \rho\left(\frac{\pi R^2}{h^2}\right) z^4 dz$. Substitute:

$I_x = \rho\left(\frac{\pi R^4}{4h^4}\right)\int_0^h z^4 dz + \rho\left(\frac{\pi R^2}{h^2}\right)\int_0^h z^4 dz$. Integrating and collecting terms:

$$I_x = \left(\frac{3mR^2}{4h^5} + \frac{3m}{h^3}\right)\left[\frac{z^5}{5}\right]_0^h = m\left(\frac{3}{20}R^2 + \frac{3}{5}h^2\right)$$

By symmetry, $I_y = I_x$

Problem 8.115 The homogenous pyramid is of mass m. Determine its mass moment of inertia about the z-axis.

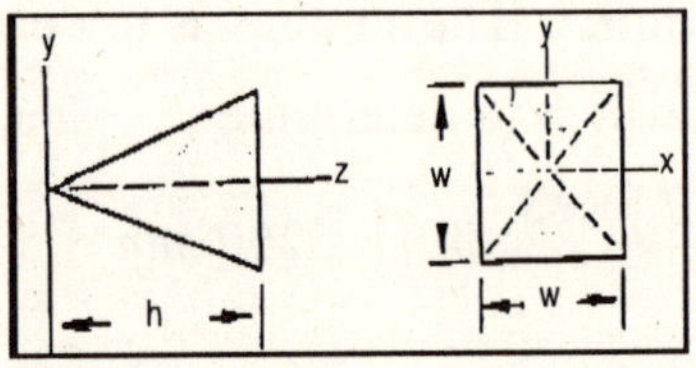

Solution: The mass density is $\rho = \frac{m}{V} = \frac{3m}{w^2 h}$. The differential mass is $dm = \rho\omega^2 dz$. The moment of inertia of this element about the z-axis is $dI_z = \left(\frac{1}{6}\right)\omega^2 dm$. Noting that $\omega = \frac{wz}{h}$, then

$dI_z = \rho\left(\frac{w^4}{6h^4}\right) z^4 dz = \frac{mw^2}{2h^5} z^4 dz$. Integrating: $I_{z-axis} = \left(\frac{mw^2}{2h^5}\right)\int_0^h z^4 dz = \frac{1}{10} mw^2$

Problem 8.116 Determine the mass moment of inertia of the homogenous pyramid in Problem 8.115 about the x- and y-axes.

Solution: Use the results of the solution of Problem 8.115 for the mass density. The elemental disk is $dm = \rho\omega^2 dz$. The moment of inertia about an axis through its center of mass parallel to the x- axis is $dI_x = \left(\frac{1}{12}\right)\omega^2 dm$. Use the parallel axis theorem: $I_{x-axis} = \left(\frac{1}{12}\right)\int_m \omega^2 dm + \int_m z^2 dm$. Noting that $\omega = \frac{w}{h} z$, the integral is $I_{x-axis} = \frac{\rho w^4}{12h^4}\int_0^h z^4 dz + \frac{\rho w^2}{h^2}\int_0^h z^4 dz$. Integrating and collecting terms

$$I_{x-axis} = m\left(\frac{1}{20}w^2 + \frac{3}{5}h^2\right)$$

By symmetry, $I_{y-axis} = I_{x-axis}$

Problem 8.117 The homogenous parallelepiped is of mass m. Determine its mass moment of inertia about the x, y and z axes and compare your results with the values in Appendix C.

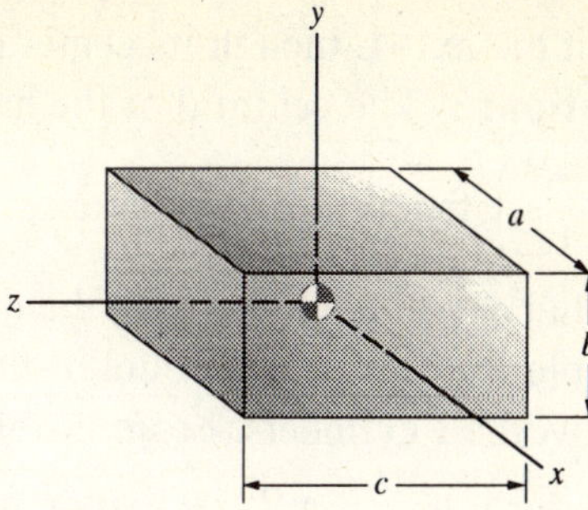

Solution: Consider a rectangular slice normal to the x- axis of dimensions b by c and mass dm. The area density of this slice is $\rho = \frac{dm}{bc}$. From Eqs (8.30) and (8.31), the mass moment of inertia about the y axis of the centroid of a thin plate is the product of the area density and the area moment of inertia of the plate: $dI_y = \rho\left(\frac{1}{12}\right)bc^3$., from which $dI_y = \left(\frac{1}{12}\right)c^2 dm$ By symmetry, the mass moment of inertia about the z axis is $dI_z = \left(\frac{1}{12}\right)b^2 dm$. Since the labeling of the x- y- and z- axes is arbitrary, $dI_x = dI_z + dI_y$, where the x-axis is normal to the area of the plate. Thus $dI_x = \left(\frac{1}{12}\right)(b^2 + c^2)dm$, from which $\boxed{I_x = \left(\frac{1}{12}\right)(b^2 + c^2)\int_m dm = \frac{m}{12}(b^2 + c^2)}$. By symmetry, the argument can be repeated for each coordinate, to obtain $\boxed{I_y = \frac{m}{12}(a^2 + c^2)}$ $\boxed{I_z = \frac{m}{12}(b^2 + a^2)}$ *Discussion:* The parallel axis theorem does not apply above because the mass moment of inertia about the y and z axes are about *an origin at the centroid of the thin slice* and not at the origin of the coordinate system at the centroid of the block. Thus the mass moment of inertia about the x-axis of the slice $bc\,dx$ is independent of the location on the x-axis. *End of discussion.*

Problem 8.118 The homogenous ring consists of steel of density $\rho = 15\ slug/ft^3$. Determine the its mass moment of inertia about the axis L through its center of mass.

Solution: Consider a solid cylinder of radius R and height h.. Its mass density is $\rho = \frac{m}{V} = \frac{m}{\pi R^2 h}$. Consider a thin slice normal to the axis L. The mass moment of inertia of this slice about the axis L is the product of the area density and the area moment of inertia:

$dI_L = (\rho\, h)r^2 dA = (\rho\, h)(2\pi)r^3 dr$, from which

$I_L = (2\rho\, h\pi)\int_0^R r^3 dr = \left(\frac{1}{2}\right)mR^2$. From which, for the outer ring $I_{Lo} = \left(\frac{1}{2}\right)m_o R_o^{\ 2}$. For the inner ring $I_{Li} = \left(\frac{1}{2}\right)m_i R_i^2$, from which

$I_L = I_{Lo} - I_{Li} = \left(\frac{1}{2}\right)(m_o R_o^2 - m_i R_i^2)$. The mass of the outer cylinder is

$m_o = \rho\pi R_o^2 h = 15\pi(0.5^2)(1/3) = 3.927\ slugs$. The mass of the inner cylinder is

$m_i = \rho\pi R_i^2 h = 15\pi(0.25^2)(1/3) = 0.982\ slugs$, from which $\boxed{I_L = 0.46\ slug\text{-}ft^2}$

Problem 8.119 The homogenous half cylinder is of mass *m*. Determine its mass moment of inertia about the axis L though its center of mass.

Solution: : The centroid of the half cylinder is located a distance of $\left(\frac{4R}{3\pi}\right)$ from the edge diameter. The strategy is to use the parallel axis theorem to treat the mass moment of inertia of a complete cylinder as the sum of the mass moments of inertia for the two half cylinders.

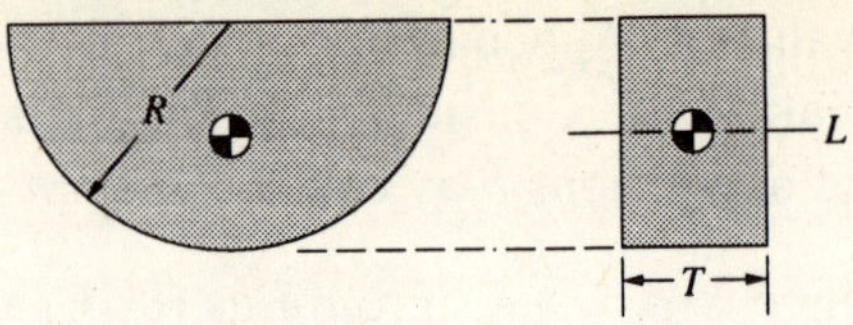

From Problem 8.118, the mass moment of inertia about the geometric axis for a cylinder is $I_{cL} = mR^2$, where *m* is one half the mass of the cylinder. By the parallel axis theorem,

$$I_{cL} = 2\left(\left(\frac{4R}{3\pi}\right)^2 m + I_{hL}\right). \text{ Solve } I_{hL} = \left(\frac{I_{cL}}{2} - \left(\frac{4R}{3\pi}\right)^2 m\right) = \left(\frac{mR^2}{2} - \left(\frac{16}{9\pi^2}\right)mR^2\right) = mR^2\left(\frac{1}{2} - \frac{16}{9\pi^2}\right) =$$

$$\boxed{= mR^2\left(\frac{1}{2} - \frac{16}{9\pi^2}\right) = 0.31987mR^2 = 0.32mR^2}$$

Problem 8.120 The object shown consists of steel of density $\rho = 7800\ kg/m^3$ of width $w = 40\ mm$. Determine the mass moment of inertia about the axis L_O.

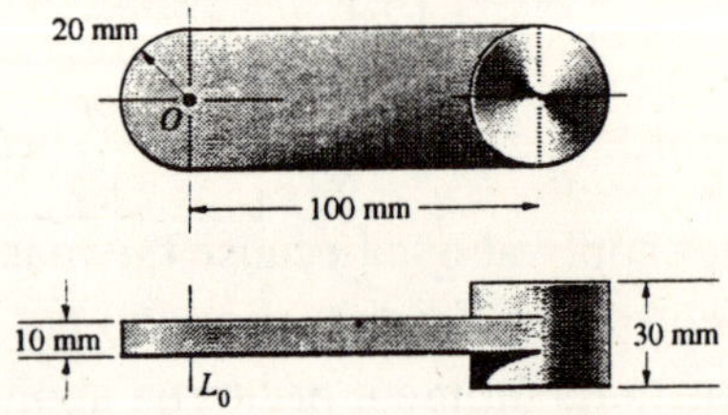

Solution: Divide the object into four parts: Part (1) The semi-cylinder of radius $R = 0.02\ m$, height $h_1 = 0.01\ m$. Part (2): The rectangular solid $L = 0.1\ m$ by $h_2 = 0.01\ m$ by $w = 0.04\ m$. Part (3): The semi-cylinder of radius $R = 0.02\ m$, $h_1 = 0.01\ m$ Part (4) The cylinder of radius $R = 0.02\ m$, height $h = 0.03\ m$.

Part (1) $m_1 = \frac{\rho\pi R^2 h_1}{2} = 0.049\ kg$, $I_1 = \frac{m_1 R^2}{4} = 4.9 \times 10^{-6}\ kg\text{-}m^2$, Part (2): $m_2 = \rho w L h_2 = 0.312\ kg$,

$$I_2 = \left(\frac{1}{12}\right)m_2(L^2 + w^2) + m_2\left(\frac{L}{2}\right)^2 = 0.00108\ kg\text{-}m^2.$$

Part (3) $m_3 = m_1 = 0.049\ kg$, $I_3 = -\left(\frac{4R}{3\pi}\right)^2 m_2 + I_1 + m_3\left(L - \frac{4R}{3\pi}\right)^2 = 0.00041179\ kg\ m^2$. Part (4)

$m_4 = \rho\pi R^2 h = 0.294\ kg$, $I_4 = \left(\frac{1}{2}\right)m_4\left(R^2\right) + m_4 L^2 = 0.003\ kg\ m^2$. The composite:

$$\boxed{I_{Lo} = I_1 + I_2 - I_3 + I_4 = 0.003674\ kg\ m^2}$$

Problem 8.121 Determine the mass moment of inertia of the object in Problem 8.120 about the axis through the center of mass of the object parallel to L_O.

Solution: The center of mass is located relative to L_O

$$\bar{x} = \frac{m_1\left(-\frac{4R}{3\pi}\right) + m_2(0.05) - m_3\left(0.1 - \frac{4R}{3\pi}\right) + m_4(0.1)}{m_1 + m_2 - m_3 + m_4} = 0.066\ m,$$

$$\boxed{I_c = -\bar{x}^2 m + I_{Lo} = -0.00265 + 0.00367 = 0.00102\ kg\ m^2}$$

Problem 8.122
The thick plate consists of steel with a density of $\rho = 15\ slug/ft^2$. Determine its mass moment of inertia about the z- axis .

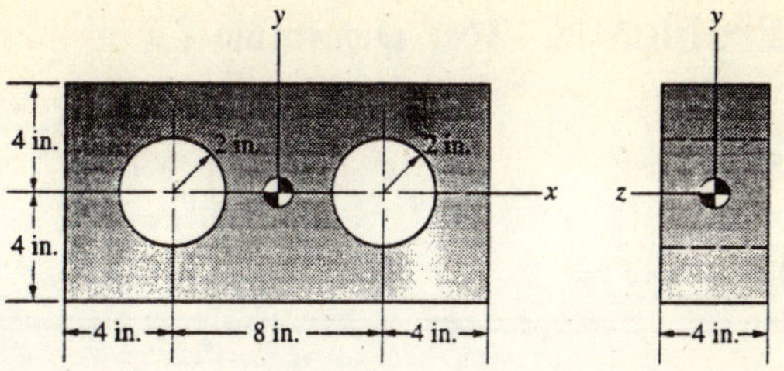

Solution: Divide the object into three parts: Part (1) the rectangle 8 *in* by 16 *in* , Parts (2) & (3) the cylindrical cut outs. Part (1): $m_1 = \rho 8(16)(4) = 4.444\ slugs$.

$I_1 = (\frac{1}{12})m_1(16^2 + 8^2) = 118.52\ slug\ in^2$. Part (2): $m_2 = \rho\pi(2^2)(4) = 0.4363\ slug$,

$I_2 = \frac{m_2(2^2)}{2} + m_2(4^2) = 7.854\ slug\ in^2$. Part (3): $m_3 = m_2 = 0.4363\ slugs$, $I_3 = I_2 = 7.854\ slug\text{-}in^2$.

The composite: $I_{z-axis} = I_1 - 2I_2 = 102.81\ slug\text{-}in^2$ $\boxed{I_{z-axis} = 0.714\ slug\text{-}ft^2}$

Problem 8.123
. Determine the mass moment of inertia of the object in Problem 8.122 about the x-axis.

Solution: Use the same divisions of the object as in Problem 8.122. Part (1):

$$I_{1x-axis} = \left(\frac{1}{12}\right)m_1(8^2 + 4^2) = 28.63\ slug\text{-}in^2,$$

Part (2): $I_{2x-axis} = \left(\frac{1}{12}\right)m_2(3(2^2) + 4^2) = 1.018\ slug\text{-}in^2$. The composite:

$I_{x-axis} = I_{1x-axis} - 2I_{2x-axis} = 27.59\ slug\ in^2$ $\boxed{= 0.1916\ slug\ ft^2}$

Problem 8.124
Determine I_y and k_y.

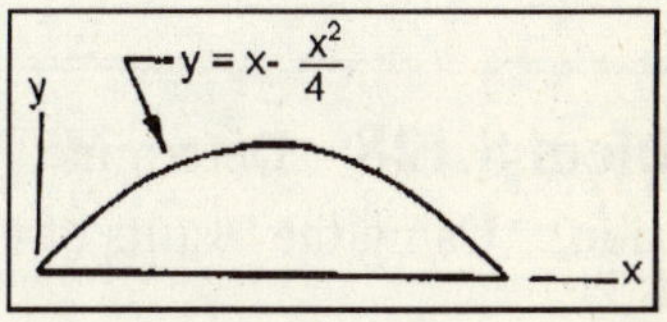

Solution: By definition, $I_y = \int_A x^2 dA$. The element of area is $dA = dx\,dy$. The limits on the variable x are $0 \le x \le 4$. The area is

$$A = \int_0^4 dx \int_0^{X-\frac{X^2}{4}} dy = \left[\frac{x^2}{2} - \frac{x^3}{12}\right]_0^4 = 2.6667\ \ I_y = \int_0^4 x^2 dx \int_0^{x-\frac{x^2}{4}} dy = \int_0^4 (x - \frac{x^2}{4})x^2 dx.$$

$$\boxed{= \left[\frac{x^4}{4} - \frac{x^5}{20}\right]_0^4 = 12.8}\text{ from which } \boxed{k_y = \sqrt{\frac{I_y}{A}} = 2.19}$$

Problem 8.125
Determine I_x and k_x.

Solution: By definition, $I_x = \int_A y^2 dA$, from which $I_x = \int_0^4 dx \int_0^{x-\frac{x^2}{4}} y^2 dy = \left(\frac{1}{3}\right)\int_0^4 \left(x - \frac{x^2}{4}\right)^3 dx$

$$\boxed{I_x = \left(\frac{1}{3}\right)\left[\frac{x^4}{4} - \frac{3}{20}x^5 + \frac{3}{96}x^6 - \frac{x^7}{448}\right]_0^4 = 0.6095}. \text{ From Problem 8.124, } A = 2.667, \boxed{k_x = \sqrt{\frac{I_x}{A}} = 0.4781}$$

Problem 8.126 Determine I_{xy} .

Solution: $I_{xy} = \int_A xy\, dA, = \int_0^4 x\, dx \int_0^{x-\frac{x^2}{4}} y dy = \left(\frac{1}{2}\right) \int_0^4 (x - \frac{x^2}{4})^2 x dx = \left(\frac{1}{2}\right)\left[\frac{x^4}{4} - \frac{x^5}{10} + \frac{x^6}{96}\right]_0^4$

$\boxed{= 2.1333}$

Problem 8.127 The origin of the x', y' coordinate system is at the centroid of the area. Determine $I_{y'}$ and $k_{y'}$.

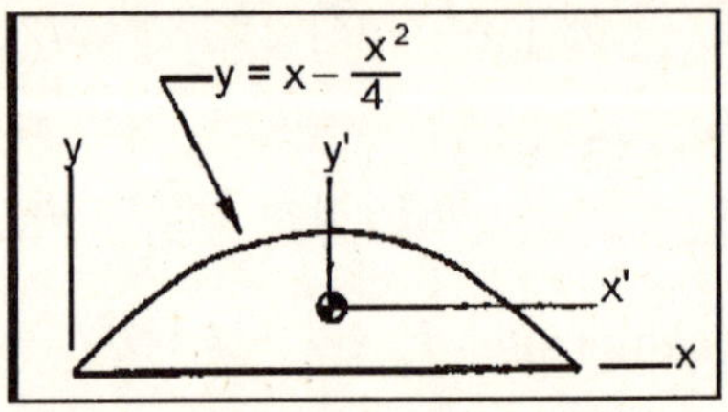

Solution: The limits on the variable x are $0 \le x \le 4$. By definition,

$$A\bar{y} = \int_A y dA = \int_0^4 dx \int_0^{X-\frac{X^2}{4}} y dy = \left(\frac{1}{2}\right) \int_0^4 (x - \frac{x^2}{4})^2 dx$$

$= \left(\frac{1}{2}\right)\left[\frac{x^3}{3} - \frac{x^4}{8} + \frac{x^5}{80}\right]_0^4 = 1.06667$. From Problem 8.124 the area is $A = 2.667$, from which

$\bar{y} = 0.3999 = 0.4$. Similarly, $A\bar{x} = \int_0^4 x dx \int_0^{X-\frac{X^2}{4}} dy = \int_0^4 x\left(x - \frac{x^2}{4}\right) dx = \left[\frac{x^3}{3} - \frac{x^4}{16}\right]_0^4 = 5.3333$, from

which $\bar{x} = 1.9999 = 2$. The area moment of inertia is $I_{yy} = -\bar{x}^2 A + I_y$. Using the result of Problem 8.24, $I_y = 12.8$, from which the area mass moment of inertia about the centroid is

$\boxed{I_{y'} = -10.6666 + 12.8 = 2.133}$ and $\boxed{k_{y'} = \sqrt{\frac{I_{y'}}{A}} = 0.8944}$

Problem 8.128 Determine $I_{x'}$ and $k_{x'}$.

Solution: Using the results of Problems 8.125 and 8.127, $I_x = 0.6095$ and $\bar{y} = 0.4$. The area mass moment of inertia about the centroid is $\boxed{I_{x'} = -\bar{y}^2 A + I_x = 0.1828}$ and $\boxed{k_{x'} = \sqrt{\frac{I_{x'}}{A}} = 0.2618}$

Problem 8.129 Determine $I_{x'y'}$.

Solution: From Problems 8.126 and 8.127, $I_{xy} = 2.133$ and $\bar{x} = 2$, $\bar{y} = 0.4$. The product of the moment of inertia about the centroid is $\boxed{I_{x'y'} = -\bar{x}\bar{y}A + I_{xy} = -2.133 + 2.133 = 0}$

Problem 8.130 Determine I_y and k_y.

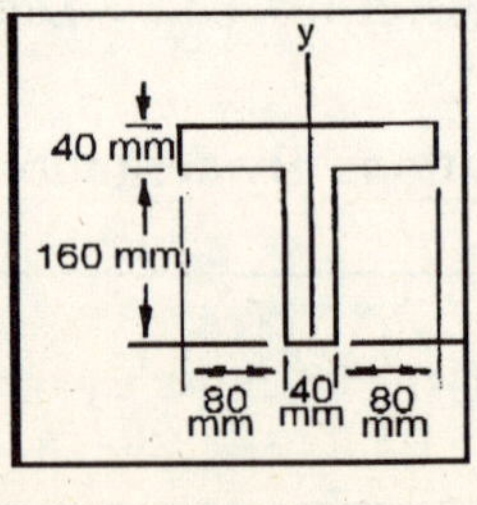

Solution: Divide the section into two parts: Part (1) is the upper rectangle 40 *mm* by 200 *mm*, Part (2) is the lower rectangle, 160 *mm* by 40 *mm*.

Part (1) $A_1 = 0.040(0.200) = 0.008\ m^2$, $\bar{y}_1 = 0.180\ m \backslash$, $\bar{x}_1 = 0$,

$I_{y1} = \left(\frac{1}{12}\right) 0.04(0.2)^3 = 2.6667 \times 10^{-5} m^4$.

Problem 8.130 continued on next page

continuation of Problem 8.130..........

Part(2): $A_2 = (0.04)(0.16) = 0.0064\ m^2$, $\bar{y}_2 = 0.08\ m$, $\bar{x}_2 = 0$, $I_{y2} = \left(\frac{1}{12}\right)(0.16)(0.04)^3 = 8.5 \times 10^{-7}\ m^4$.

The composite: $A = A_1 + A_2 = 0.014\ m^2$, $I_y = I_{y1} + I_{y2}$, $\boxed{I_y = 2.752 \times 10^{-5}\ m^4 = 2.752 \times 10^7\ mm^4}$,

and $\boxed{k_y = \sqrt{\frac{I_y}{A}} = 0.0437\ m = 43.7\ mm}$

Problem 8.131 Determine I_x and k_x for the area in Problem 8.130.

Solution: Use the results in the solution to Problem 8.130. Part (1) $A_1 = 0.040(0.200) = 0.008\ m^2$,

$\bar{y}_1 = 0.180\ m$, $I_{x1} = \left(\frac{1}{12}\right)0.2\left(0.04^3\right) + (0.18)^2 A_1 = 2.603 \times 10^{-4}\ m^4$.

Part(2): $A_2 = (0.04)(0.16) = 0.0064\ m^2$, $\bar{y}_2 = 0.08\ m$,

$I_{x2} = \left(\frac{1}{12}\right)(0.04)(0.16)^3 + (0.08)^2 A_2 = 5.461 \times 10^{-5}\ m^4$,. The composite: $A = A_1 + A_2 = 0.0144\ m^2$,

The area moment of inertia about the x axis is $\boxed{I_x = I_{x1} + I_{x2} = 3.15 \times 10^{-4}\ m^4 = 3.15 \times 10^8\ mm^4}$, and

$\boxed{k_x = \sqrt{\frac{I_x}{A}} = 0.1479\ m = 147.9\ mm}$

Problem 8.132 The origin of the coordinate system is at the centroid of the composite area. Determine I_x and k_x.

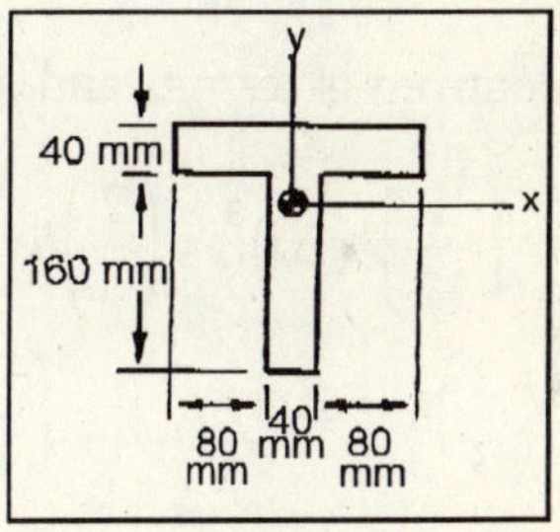

Solution: Use the results of the solutions to Problems 8.130-8.131. The centroid is located relative to the base at $\bar{x}_c = \frac{\bar{x}_1 A_1 + \bar{x}_2 A_2}{A} = 0$,

$\bar{y}_c = \frac{\bar{y}_1 A_1 + \bar{y}_2 A_2}{A} = 0.1356\ m$. The moment of inertia about the x-axis is

$\boxed{I_{xc} = -\bar{y}_C^2 A + I_X = 5.0396 \times 10^{-5}\ m^4 = 5.04 \times 10^7\ mm^4}$

and $\boxed{k_{xc} = \sqrt{\frac{I_{xc}}{A}} = 0.05909\ m = 59.1\ mm}$

Problem 8.133 Determine J_o and k_o for the area in Problem 8.132.

Solution: Use the results of the solutions to Problems 8.130-8.131. The area moments of inertia about the centroid are $I_{xc} = 5.028 \times 10^{-5}\ m^4$ and $I_{yc} = I_y = 2.752 \times 10^{-5}\ m^4$, from which

$\boxed{J_o = I_{xc} + I_{yc} = 7.78 \times 10^{-5}\ m^4 = 7.78 \times 10^7\ mm^4}$ and $\boxed{k_o = \sqrt{\frac{J_o}{A}} = 0.0735\ m = 73.5\ mm}$

Problem 8.134 Determine I_x and k_x.

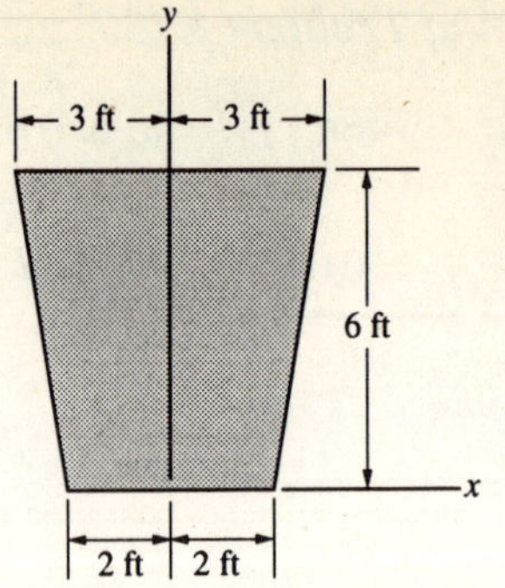

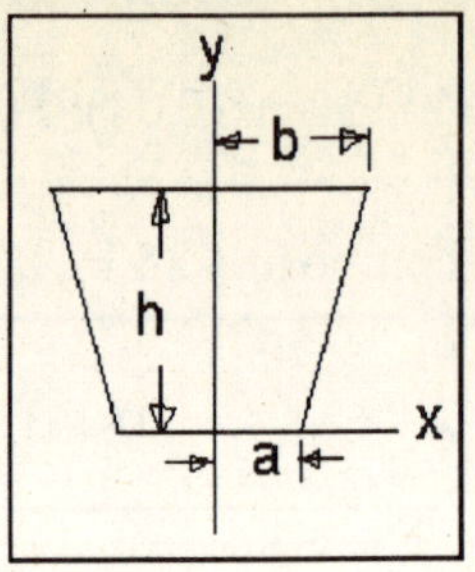

Solution: Break the area into three parts: Part (1) The rectangle with base $2a$ and altitude h; Part (2) The triangle on the right with base $(b-a)$ and altitude h, and Part (3) The triangle on the left with base $(b-a)$ and altitude h. Part (1) The area is $A_1 = 2ah = 24\ ft^2$. The centroid is $\bar{x}_1 = 0$ and $\bar{y}_1 = \dfrac{h}{2} = 3\ ft$. The area moment of inertia about the centroid is $I_{xc1} = \left(\dfrac{1}{12}\right)(2a)h^3 = \left(\dfrac{1}{6}\right)ah^3 = 72\ ft^4$. Part (2): $A_2 = \left(\dfrac{1}{2}\right)h(b-a) = 3\ ft^2$, $\bar{x}_2 = a + \dfrac{b-a}{3} = 2.3333\ ft$, $\bar{y}_2 = \left(\dfrac{2}{3}\right)h = 4\ ft$, $I_{xc2} = \left(\dfrac{1}{36}\right)(b-a)h^3 = 6\ ft^4$. Part (3): $A_3 = A_2$, $\bar{x}_3 = -\bar{x}_2$, $\bar{y}_3 = \bar{y}_2$, $I_{xc3} = I_{xc2}$,. The composite area is $A = A_1 + A_2 + A_2 = 30\ ft^2$ The composite moment of inertia $I_x = (\bar{y}_1)^2 A_1 + I_{xc1} + (\bar{y}_2)^2 A_2 + I_{xc2} + (\bar{y}_3)^2 A_3 + I_{xc3}$, $\boxed{I_x = 396\ ft^4}$

$$\boxed{k_x = \sqrt{\frac{I_x}{A}} = \sqrt{\frac{396}{30}} = 3.633\ ft}$$

Problem 8.135 Determine I_y and k_y for the area in Problem 8.134.

Solution: Divide the area as in the solution to Problem 8.134. Part (1) The area is $A_1 = 2ah = 24\ ft^2$. The centroid is $\bar{x}_1 = 0$ and $\bar{y}_1 = \dfrac{h}{2} = 3\ ft$. The area moment of inertia about the centroid is $I_{yc1} = \left(\dfrac{1}{12}\right)h(2a)^3 = \left(\dfrac{2}{3}\right)ha^3 = 32\ ft^4$ Part (2): $A_2 = \left(\dfrac{1}{2}\right)h(b-a) = 3\ ft^2$, $\bar{x}_2 = a + \dfrac{b-a}{3} = 2.3333\ ft$, $\bar{y}_2 = \left(\dfrac{2}{3}\right)h = 4\ ft$, $I_{yc2} = \left(\dfrac{1}{36}\right)h(b-a)^3 = 0.1667\ ft^4$. Part (3): $A_3 = A_2$, $\bar{x}_3 = -\bar{x}_2$, $\bar{y}_3 = \bar{y}_2$, , $I_{yc3} = I_{yc2}$. The composite area is $A = A_1 + A_2 + A_2 = 30\ ft^2$ The composite moment of inertia, $I_y = \bar{x}_1^2 A_1 + I_{yc1} + \bar{x}_2^2 A_2 + I_{yc2} + \bar{x}_3^2 A_3 + I_{yc3}$, $\boxed{I_y = 65\ ft^4}$ and $\boxed{k_y = \sqrt{\dfrac{I_y}{A}} = 1.472\ ft}$

Problem 8.136 The moments of inertia of the area are $I_x = 36\ m^4$, $I_y = 145\ m^4$, and $I_{xy} = 44.25\ m^4$. Determine a set of principal axes and the principal moment of inertia.

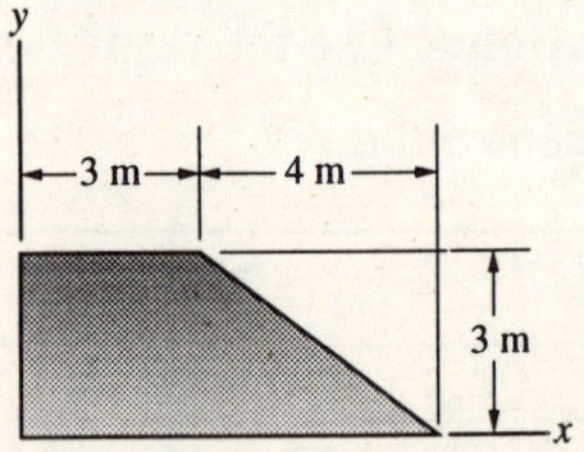

Solution: The principal angle is $\boxed{\theta = \left(\dfrac{1}{2}\right)\tan^{-1}\left(\dfrac{2I_{xy}}{I_y - I_x}\right) = 19.54^o}$

The principal moments of inertia are $\boxed{I_{xP} = I_x \cos^2\theta - 2I_{xy}\sin\theta\cos\theta + I_y \sin^2\theta = 20.298 = 20.3\ m^4}$

$$\boxed{I_{yP} = I_x \sin^2\theta + 2I_{xy}\sin\theta\cos\theta + I_y \cos^2\theta = 160.70\ m^4}$$

Problem 8.137 The mass moment of the 31-oz bat about a perpendicular axis through point B is 0.093 *slug-ft*^{2}. What is the bat's mass moment of inertia about a perpendicular axis through point A? (Point A is the bat's "instantaneous center" or center of rotation, at the instant shown.)

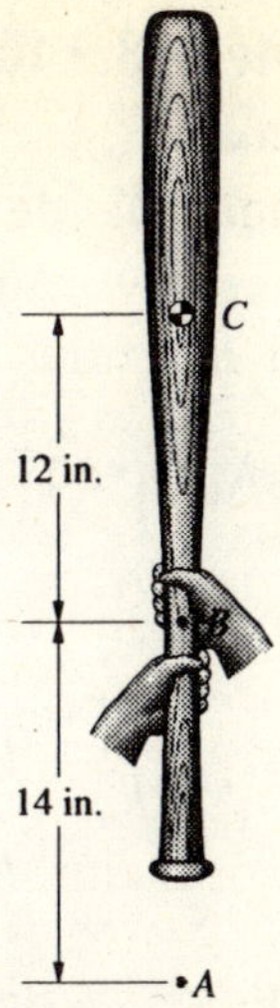

Solution: The mass of the bat is $m = \dfrac{31}{16(32.17)} = 0.06023\ slugs$. Use the parallel axis theorem to obtain the mass moment of inertia about the center of mass C, and then use the parallel axis theorem to translate to the point A.:

$$\boxed{I_A = \left(\frac{12+14}{12}\right)^2 m + 0.093 = 0.3155\ slug\text{-}ft^2}$$

Problem 8.138 The mass of the thin, homogenous plate is 4 *kg*. Determine its mass moment of inertia about the y axis.

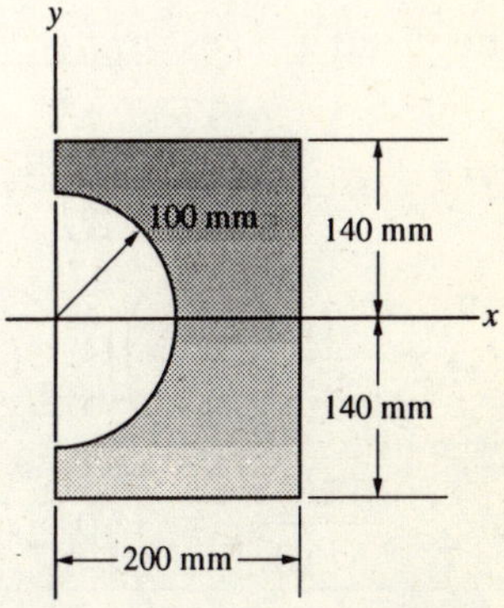

Solution: Divide the object into two parts: Part (1) is the semicircle of radius $100\,mm$, and Part (2) is the rectangle 200 *mm* by 280 *mm*. The area of Part (1) $A_1 = \dfrac{\pi R^2}{2} = 15708.\ mm^2$. The area of Part (2) is $A_2 = 280(200) = 56000\ mm^2$. The composite area is $A = A_2 - A_1 = 40292\ mm^2$. The area mass density is $\rho = \dfrac{4}{A} = 9.9275\times 10^{-5}\ kg/mm^2$ For Part (1) $\bar{x}_1 = \bar{y}_1 = 0$, $I_{y1} = \rho\left(\dfrac{1}{8}\right)\pi R^4 = 3898.5\ kg\text{-}mm^2$. For Part (2) $\bar{x}_2 = 100\ mm$.

$I_{y2} = \bar{x}_2^2 \rho A_2 + \rho\left(\dfrac{1}{12}\right)(280)(200^3) = 74125.5\ kg\text{-}mm^2$. The composite:

$$\boxed{I_y = I_{y2} - I_{y1} = 70226\ kg\text{-}mm^2 = 0.070226\ kg\text{-}m^2}$$

Problem 8.139 Determine the mass moment of inertia of the plate in Problem 8.138 about the z-axis.

Solution: Use the same division of the parts and the results of the solution to Problem 8.138. For Part (1), $I_{x1} = \rho\left(\dfrac{1}{8}\right)\pi R^4 = 3898.5\ kg\text{-}mm^2$. For Part (2) $I_{x2} = \rho\left(\dfrac{1}{12}\right)(200)(280^3) = 36321.5\ kg\text{-}mm^2$. The composite: $I_x = I_{x2} - I_{x1} = 32423\ kg\text{-}mm^2$, from which, using the result of the solution to Problem 8.138, $\boxed{I_z = I_x + I_y = 32422 + 70226 = 102649\ kg\text{-}mm^2 = 0.10265\ kg\text{-}m^2}$

Problem 8.140 The homogenous object weighs 400 *lb*. Determine its mass moment of inertia about the x-axis.

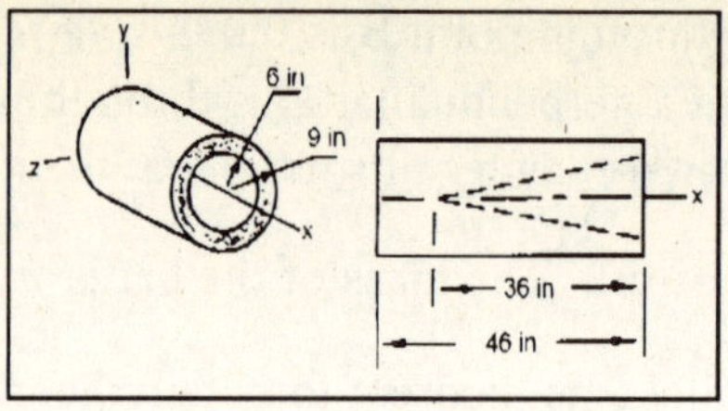

Solution: Divide the object into two volumes: Part (1): The cylinder 9 *in.* radius and height 46 *in.* Part (2) The cone with base 6 *in.* radius and 36 *in.* height. The volumes are:

$V_1 = \pi R_1^2 h_1 = 11705.6\ in^3$, and $V_2 = \left(\frac{1}{3}\right)\pi R_2^2 h_2 = 1357.2\ in^3$.

The composite volume: $V = V_1 - V_2 = 10348.4\ in^3$. The volume density is

$\rho = \frac{m}{V} = \frac{400}{gV} = 0.0012\ slugs/in^3$, using $g = 32.17\ ft/s^2$. Part (1): Consider a thin slice normal to the x-axis of radius R_1 and thickness dx. The moments of inertia about y'- and z'- axes <u>*originating at the centroid of the thin slice*</u> are $dI_y = \left(\frac{1}{4}\right)R^2 dm$ and $dI_z = \left(\frac{1}{4}\right)R^2 dm$, from which

$dI_x = dI_y + dI_z = \left(\frac{1}{2}\right)R^2 dm$. Integrating: $I_x = \left(\frac{1}{2}\right)R^2 \int_m dm = \frac{\rho}{2}\pi h_1 R_1^4 = \frac{\rho V_1}{2}R^2 = 569.6\ slug\text{-}in^2$.

Part (2): A thin slice normal to the x-axis of radius r and thickness dx has the moments of inertia about the y and z- axes <u>*originating at the centroid of the thin slice*</u> of $dI_y = dI_z = \left(\frac{1}{4}\right)r^2 dm$, from which

$dI_x = dI_y + dI_z = \left(\frac{1}{2}\right)r^2 dm = \left(\frac{\rho\pi}{2}\right)r^4 dx$. Noting that $r = \frac{R_2}{h_2}(x + h_2 - h_1)$, the integral becomes:

$$I_{x2} = \left(\frac{R_2}{h_2}\right)^4\left(\frac{\rho\pi}{2}\right)\int_{h_1-h_2}^{h_1}(x + h_2 - h_1)^4 dx = \left(\frac{R_2}{h_2}\right)^4\left(\frac{\rho\pi}{10}\right)\left[(x + h_2 - h_1)^5\right]_{h_1-h_2}^{h_1}$$

$I_{x2} = \frac{\rho\pi R_2^4 h_2}{10} = \frac{3}{10}\rho V_2 R_2^2 = 17.61\ slug\text{-}in^2$ The composite:

$$\boxed{I_x = I_{x1} - I_{x2} = 552.0\ slug\ in^2 = 3.83\ slug\ ft^2}$$

Problem 8.141 Determine the mass moments of inertia of the object in Problem 8.140 about the y and z axes.

Solution: Use the same division into parts that was used in Problem 8.140. Part (1) The mass moment of inertia of a cylinder about its centroid is $I_{yc1} = \left(\frac{1}{12}\right)\rho V_1(3R_1^2 + h_1^2)$. Use the principal axis theorem:

$I_{y1} = \left(\frac{1}{12}\right)\rho V_1\left(3R_1^2 + h_1^2\right) + \left(\frac{h_1}{2}\right)^2 \rho V_1 = \rho V_1\left(\frac{R_1^2}{4} + \frac{h_1^2}{3}\right) = 10205\ slug\ in^2$. Part (2): The moment of inertia about the centroid of a cone is $I_{yc2} = \left(\frac{3}{80}\right)\rho V_2\left(4R_2^2 + h_2^2\right) = 88.06\ slug\ in^2$. Use the principal axis theorem, and note that the centroid of the cone is located $\bar{x} = 10 + (3/4)(36) = 37\ in.$ from the origin. Thus $I_{y2} = I_{yc2} + (37)^2 \rho V_2 = 2320.5\ slug\ in^2$. For the composite:

$\boxed{I_y = I_{y1} - I_{y2} = 7884.6\ slug\ in^2 = 54.75\ slug\ ft^2}$ By symmetry, $\boxed{I_z = I_y}$

Problem 8.142 Determine the mass moment of inertia of the 14 *kg* flywheel about the axis L.

Solution: Divide the flywheel into three hollow cylinders. Part (1): The outer cylinder. Part (2): The middle cylinder Part (3): The inner cylinder. Part(1) The volume:

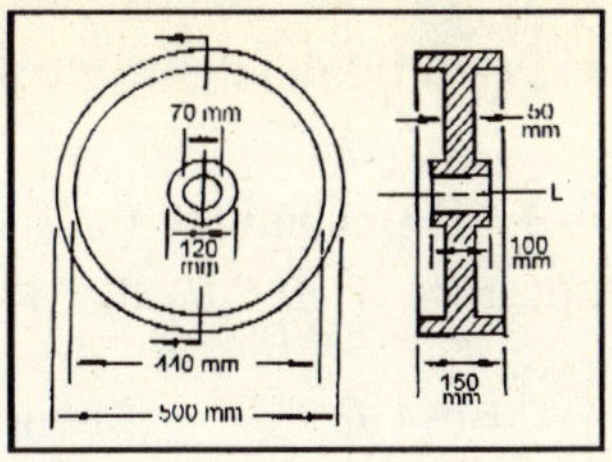

$V_1 = \pi\left(R_{1o}^2 - R_{1i}^2\right)h_1 = \pi\left(0.250^2 - 0.220^2\right)(0.150) = 0.006644\ m^3$,

Part (2):

$V_2 = \pi\left(R_{2o}^2 - R_{2i}^2\right)h_2 = \pi\left(0.220^2 - 0.060^2\right)(0.05) = 0.00704\ m^3$.

Part (3) $V_3 = \pi\left(R_{3o}^2 - R_{3i}^2\right)h_3 = \pi\left(0.06^2 - 0.035^2\right)0.1 = 0.000746\ m^3$ The composite volume is

$V = V_1 + V_2 + V_3 = 0.014428\ m^3$. The volume mass density is $\rho = \frac{m}{V} = \frac{14}{V} = 970.35\ kg/m^3$.

The mass moment of inertia of a hollow cylinder about the geometric axis is $I = \frac{m}{2}(R_o^2 + R_i^2)$. The composite is $I_1 = \frac{\rho V_1}{2}\left(0.25^2 + 0.22^2\right) + \frac{\rho V_2}{2}(0.22^2 + 0.06^2) + \frac{\rho V_3}{2}(0.06^2 + 0.035^2)$ $\boxed{= 0.5368\ kg\ m^2}$

==================================<>==================================

Problem 9.1 The value of the distributed load w at $x = 0$ is 36 *lb/ft*. (a) Determine the function w. (b) Use Eq (9.1) to determine the total force exerted by the distributed load. (c) Determine the total force exerted by the distributed load by calculating the "area " between the loading curve and the x-axis.

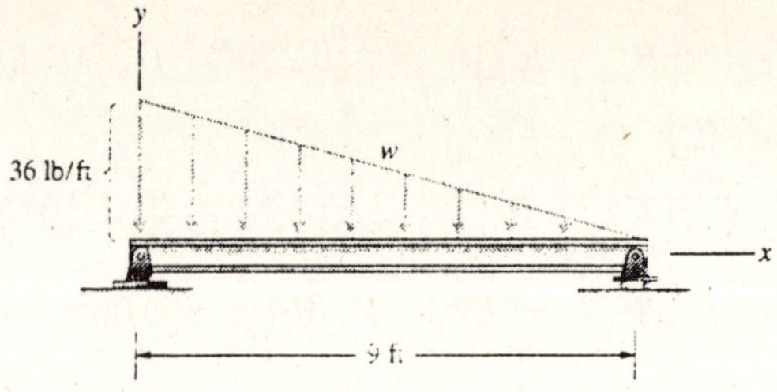

Solution (a) The load distribution is a straight line, with intercept $w = 36\ lb/ft$ at $x = 0$, and a slope of $-\left(\frac{36}{9}\right) = -4\ lb/ft^2$. . , hence $\boxed{w = -4x + 36\ lb/ft}$. (b)

From Eq (9.1) the force is $F = \int_L w\,dx$, from which the force is $\boxed{F = \int_0^9 (-4x + 36)\,dx = \left[-2x^2 + 36x\right]_0^9 = 162\ lb}$ (c) The area of a triangle is $\boxed{F = \left(\frac{1}{2}\right)ba = \left(\frac{1}{2}\right)(9)(36) = 162\ lb}$

==================================<>==================================

Problem 9.2 In preliminary design studies for a pedestrian bridge, an engineer models the weight of the bridge and its maximum expected load by the distributed load shown. What reactions are exerted on the supports of the bridge at A and B?

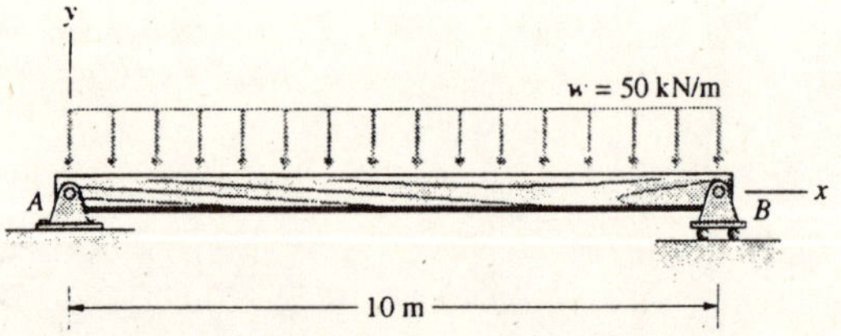

Solution From Eq (9.2) the moment due to a distributed load is $M = \int_L w\,x\,dx$. The sum of the moments about A is $\sum M = -\int_0^{10} (50)x dx + 10B_y = 0$, from which $\boxed{B_y = \frac{50}{10}\left[\frac{x^2}{2}\right]_0^{10} = 250\ kN}$. The sum of the forces is

$\sum F_y = A_y - \int_0^{10} 50\,dx + B_y = 0$, from which $\boxed{A_y = 500 - 250 = 250\ kN}$. and $\boxed{\sum F_x = A_x = 0}$

==================================<>==================================

Problem 9.3 Determine the reactions at A and B.

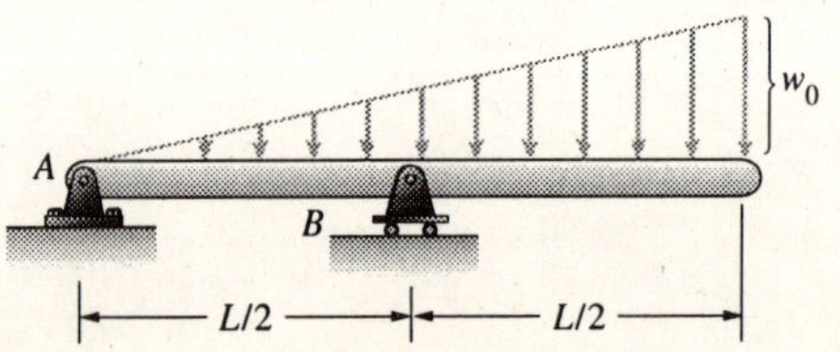

Solution The load distribution is a straight line with intercept $w = 0$ at $x = 0$, and slope $\left(\frac{w_o}{L}\right)$. The sum of the moments about A is

$\sum M_A = -\int_0^L \left(\frac{w_o}{L}\right)x^2 dx + \left(\frac{L}{2}\right)B_y = 0$, from which

$\boxed{B_y = \left(\frac{2}{L}\right)\left(\frac{w_o}{L}\right)\left[\frac{x^3}{3}\right]_0^L = \left(\frac{2}{3}\right)w_o L}$. The sum of the forces is $\sum F_y = A_y - \int_0^L \left(\frac{w_o}{L}\right)x\,dx + B_y = 0$,

from which $\boxed{A_y = \left(\frac{w_o}{L}\right)\left[\frac{x^2}{2}\right]_0^L - \left(\frac{2}{3}\right)w_o L = -\left(\frac{1}{6}\right)w_o L}$ and $\boxed{\sum F_x = A_x = 0}$

==================================<>==================================

Problem 9.4 The cantilever beam is subjected to a triangular distributed load. What are the reactions at A?

Solution The load distribution is a straight line with intercept $w = 200\ N/m$ at $x = 0$, and slope $-\left(\frac{200}{10}\right) = -20\ N/m^2$. The sum of the moments is

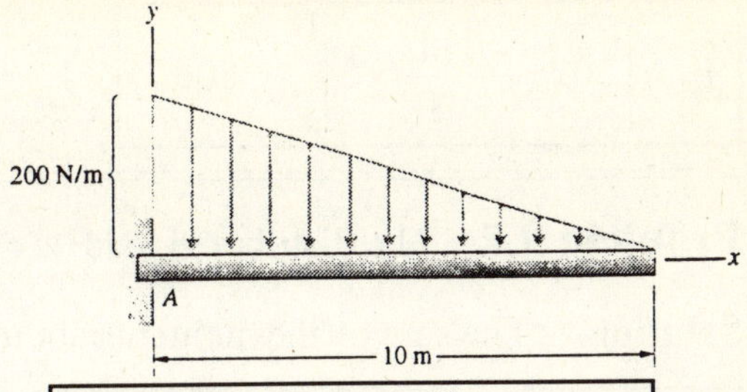

$\sum M = M_A - \int_0^{10} (-20x + 200)x\, dx = 0$, from which

$\boxed{M_A = \left[-\frac{20}{3}x^3 + 100x^2\right]_0^{10} = 3333.3\ N\,m}$. The sum of the forces: $\sum F_y = A_y - \int_0^{10} (-20x + 200)\, dx = 0$, from which $\boxed{A_y = \left[-10x^2 + 200x\right]_0^{10} = 1000\ N}$, and $\boxed{\sum F_x = A_x = 0}$

Problem 9.5 An engineer measures the forces exerted by the soil on a 10 m section of a building foundation and finds that they are described by the distributed load $w = -10x - x^2 + 0.2x^3\ kN/m$.

(a) Determine the magnitude of the total force exerted on the foundation by the distributed load.

(b) Determine the magnitude of the moment about A due to the distributed load.

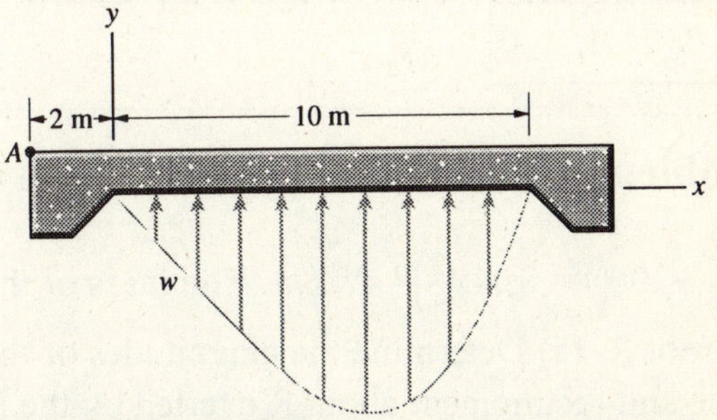

Solution (a) The total force is

$$F = -\int_0^{12} \left(10x + x^2 - 0.2x^3\right) dx$$

$= \left[-5x^2 - \frac{x^3}{3} + \frac{0.2}{4}x^4\right]_0^{10}$ $\boxed{|F| = 333.3\ kN}$ (b) The moment about the origin is

$M = -\int_0^{10} \left(10x + x^2 - 0.2x^3\right)x\, dx = \left[-\frac{10}{3}x^3 - \frac{1}{4}x^4 + \frac{0.2}{5}x^5\right]_0^{10}$, $|M| = 1833.33\ kN$. The distance from the origin to the equivalent force is $d = \frac{|M|}{F} = 5.5\ m$, from which $\boxed{|M_A| = (d + 2)F = 2500\ kN\,m}$.

Problem 9.6 The beam is subjected to a distributed load $w = 100x - cx^2\ lb/ft$, where c is a constant. The moment about point A due to the distributed load is zero. What are the reactions at A?

Solution The moment about A is

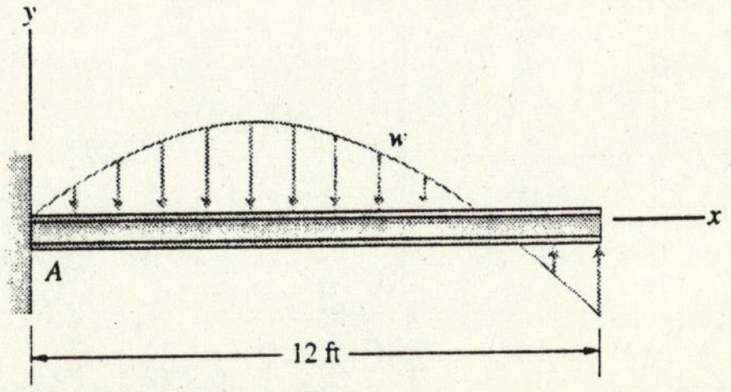

$\sum M = M_A - \int_0^{12} \left(100x - cx^2\right)x\, dx = 0$. Noting that $M_A = 0$, then $\left[\frac{100}{3}x^3 - \frac{c}{4}x^4\right]_0^{12} = 0$,

$= 57600 - 5184c = 0$ from which $c = 11.11$.

Problem 9.6 continued on next page

continuation of Problem 9.6..........

The sum of the forces: $\sum F_y = A_y - \int_0^{12}(100x - 11.11x^2)dx = 0$, from which

$$\boxed{A_y = \left[50x^2 - \frac{11.11}{3}x^3\right]_0^{12} = 800\ lb} \quad \boxed{\sum F_x = A_x = 0}$$

Problem 9.7 The distributed load $w = 10x^2\ lb/ft$. Determine the reactions at A and B.

Solution The sum of the moments about A is $\sum M_A = -\int_0^{12}(10x^2)x\,dx + 8B_y = 0$, from which

$$\boxed{B_y = \frac{10}{8}\int_0^{12} x^3 dx = \frac{10}{32}\left[x^4\right]_0^{12} = 6480\ lb}$$

The sum of the forces: $\sum F_y = A_y - \int_0^{12} 10x^2 dx + B_y = 0$,

from which $\boxed{A_y = \left[\frac{10}{3}x^3\right]_0^{12} - B_y = -720\ lb}$.

$$\boxed{\sum F_x = A_x = 0}$$

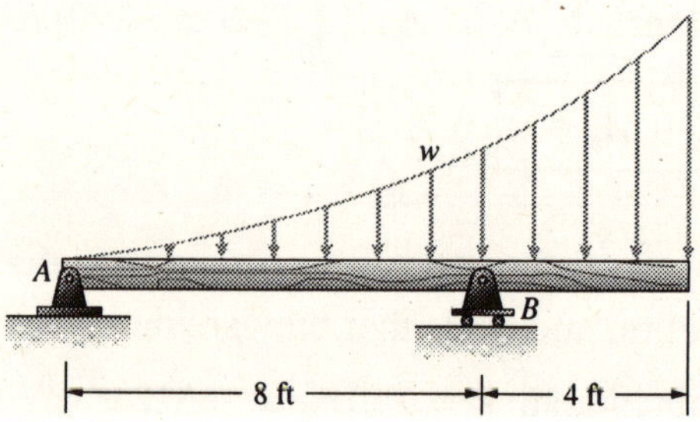

Problem 9.8 The aerodynamic lift of the wing is described by the distributed load $w = -300(1 - 0.04x^2)^{\frac{1}{2}}\ N/m$. The mass of the wing is 27 kg and its center of mass is located 2 m from the root R. (a) Determine the magnitudes of the force and the moment about R exerted by the lift of the wing. (b) Determine the reactions on the wing at R.

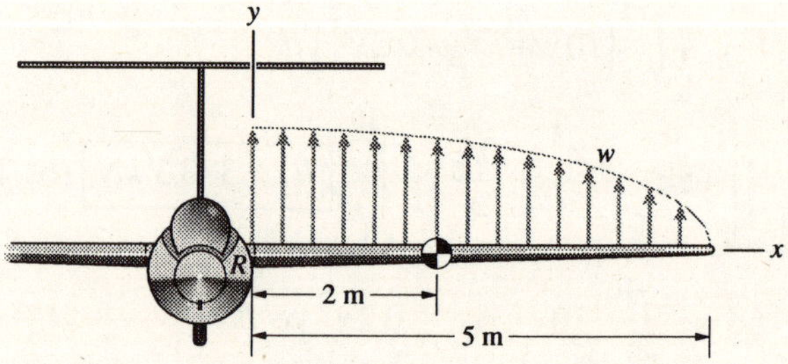

Solution (a) The force due to the lift is

$$F = -w = \int_0^5 300(1 - 0.04x^2)^{\frac{1}{2}} dx,$$

$$F = \frac{300}{5}\int_0^5 (25 - x^2)^{\frac{1}{2}} dx \quad F = 60\left[\frac{x\sqrt{25 - x^2}}{2} + \frac{25}{2}\sin^{-1}\left(\frac{x}{5}\right)\right]_0^5 = 375\pi\ N, \ \boxed{|F| = 1178.1\ N}.$$

The moment about the root due to the lift is $M = 300\int_0^5 (1 - 0.04x^2)^{\frac{1}{2}} x\,dx$,

$$M = -60\left[\frac{(25 - x^2)^{\frac{3}{2}}}{3}\right]_0^5 = \frac{60(25)^{\frac{3}{2}}}{3} = 2500 \quad \boxed{|M| = 2500\ N\,m}.$$

(b) The sum of the moments about the root: $\sum M = M^R + 2500 - 27g(2) = 0$, from which $\boxed{M^R = -1970\ N\text{-}m}$ The sum of forces $\sum F_y = F_R + 1178.1 - 27g = 0$, from which $\boxed{F_R = -1178.1 + 27g = -913.2\ N}$

Problem 9.9 If $w_B = 100\ lb/ft$, what are the reactions at A and B?

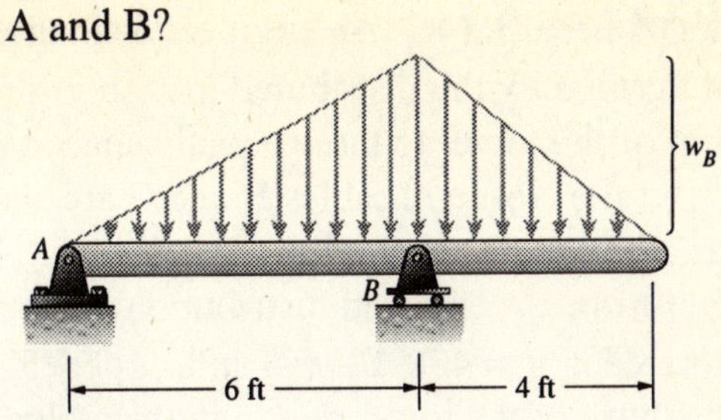

Solution The load distribution is two straight lines, the first over the interval $0 \le x \le 6$ has the intercept $w = 0$ at $x = 0$, and a slope of $\left(\frac{100}{6}\right) = 16.67\ ft/lb^2$. The second over the interval $6 \le x \le 10$ has the intercept $w = 250$ at $x = 0$ and a slope of -25. The sum of the moments about A is $\sum M = -\int_0^6 16.67x^2 dx - \int_6^{10}(-25x^2 + 250x)dx + 6B_y = 0$, from which

$B_y = \left(\frac{1}{6}\right)\left(\frac{16.67}{3}\right)\left[x^3\right]_0^6 + \left(\frac{1}{6}\right)\left[-\frac{25}{3}x^3 + \frac{250}{2}x^2\right]_6^{10}$ $\boxed{= 200 + 244.4 = 444.4\ lb}$. The sum of the forces

$\sum F_y = A_y - \int_0^6 16.67x\,dx - \int_6^{10}(-25x + 250)dx + B_y = 0$, from which

$A_y = \left(\frac{16.67}{2}\right)\left[x^2\right]_0^6 + \left[\frac{-25}{2}x^2 + 250x\right]_6^{10}$ $\boxed{A_y = 300.06 + 200 - 444.4 = 55.57\ lb}$ *Check:* From the area of the triangles: $F = F_1 + F_2 = \left(\frac{1}{2}\right)(6)(100) + \left(\frac{1}{2}\right)(4)(100) = 300 + 200 = 500\ lb$. The centroids of the triangles are $d_1 = \left(\frac{2}{3}\right)6 = 4\ ft$, $d_2 = 6 + \left(\frac{1}{3}\right)4 = 7.333\ ft$. The sum of the moments about A:

$\sum M = -300d_1 - 200d_2 + 6B_y = 0$, from which $B_y = \left(\frac{1}{6}\right)(300)(4) + \left(\frac{1}{6}\right)(200)(7.33) = 444.4\ lb$, *check.*

The sum of the forces: $A_y = 500 - 444.4 = 55.6\ lb$ *check.* $\boxed{\sum F_x = A_x = 0}$

Problem 9.10 If the supports of the beam in Problem 9.9 will each safely support a vertical load of 1000 lb, what is the largest safe value of w_B?

Solution Use the areas under the triangles from Problem 9.9: The centroidal distance of the first triangle is 4 *ft,* of the second: 7.333 *ft*. The sum of the moments:

$\sum M = -\left(\frac{1}{2}\right)w_B(6)(4) - \left(\frac{1}{2}\right)w_B(4)(7.33) + 6B_y = 0$, from which $B_y = 4.44w_B$. The sum of the forces: $\sum F_y = A_y - \left(\frac{1}{2}\right)6w_B - \left(\frac{1}{2}\right)4w_B + B_y = 0$, from which $A_y = 5w_B - 4.44w_B = 0.5567w_B$. The limiting load is that on B, $\boxed{w_B = \frac{1000}{4.44} = 225\ lb/ft}$

Problem 9.11.. An engineer estimates that the maximum wind load on the 40 *m* tower in Fig (a) is described by the distributed load in Fig (b). The tower is supported by three cables A, B, and C from the top of the tower to three equally spaced points 15 *m* from the bottom of the tower. If the wind blows from the west and cables B and C are slack, what is the tension in cable A? (Model the base of the tower as a ball and socket joint.)

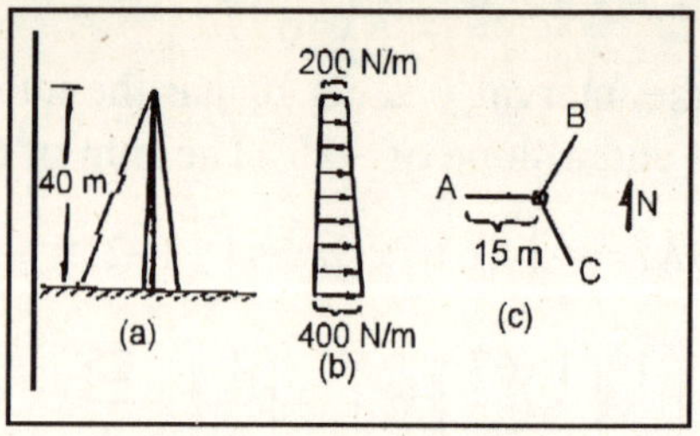

Solution The load distribution is a straight line with the intercept $w = 400\ N/m$, and slope -5. The moment about the base of the tower due to the wind load is

$$M_W = \int_0^{40} (-5y + 400)y\, dy,$$

$$M_W = \left[-\frac{5}{3}y^3 + 200y^2\right]_0^{40} = 213.33\ kN\text{-}m$$, clockwise about the base, looking North. The angle formed by the cable with the horizontal at the top of the tower is $\theta = 90^o - \tan^{-1}\left(\frac{15}{40}\right) = 69.44^o$. The sum of the moments about the base of the tower is

$$\sum M = -M_W + 40T_A \cos\theta = 0,$$ from which $\boxed{T_A = \left(\frac{1}{40\cos\theta}\right)M_W = 15.19\ kN}$

Problem 9.12 If the wind in Problem 9.11 blows from the east and cable A is slack, what are the tensions in cables B and C?

Solution From Problem 9.11, the moment about the base of the tower is $M_W = 213.33\ kN\text{-}m$, counterclockwise if the wind is from the east and the observer is looking North. The angle in the horizontal plane between the cables and the east is 60°. The sum of the moments about the base is $\sum M = M_W - 40T_B \cos\theta \cos 60^o - 40T_C \cos\theta \cos 60^o = 0$. From symmetry, the tensions in the two cables are equal, from which $\boxed{T_B = T_C = \left(\frac{1}{80\cos\theta \cos 60^o}\right)M_W = 15.19\ kN}$

Problem 9.13 What is the axial load in member BD of the frame?

Solution The distributed load is two straight lines: Over the interval $0 \le y \le 5$ the intercept is $w = 0$ at $y = 0$ and the slope is $+\frac{100}{5} = 20$. Over the interval $5 \le y \le 10$, the load is a constant $w = 100\ N/m$. The moment about the origin E due to the load is $M_E = \int_0^5 (20y)y\, dy + \int_5^{10} 100y\, dy$, from which

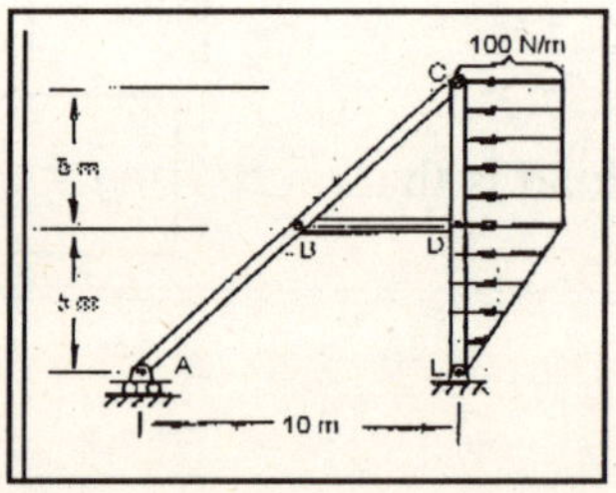

$$M_E = \left[\frac{20}{3}x^3\right]_0^5 + \left[\frac{100}{2}x^2\right]_5^{10} = 4583.33\ N\text{-}m.$$ *Check:* The area of the triangle is $F_1 = \left(\frac{1}{2}\right)(5)(100) = 250\ N$. The area of the rectangle: $F_2 = 500\ N$. The centroid distance for the triangle is $d_1 = \left(\frac{2}{3}\right)5 = 3.333\ m$.

Problem 9.13 continued on next page..........

............continuation of solution to Problem 9.13.....

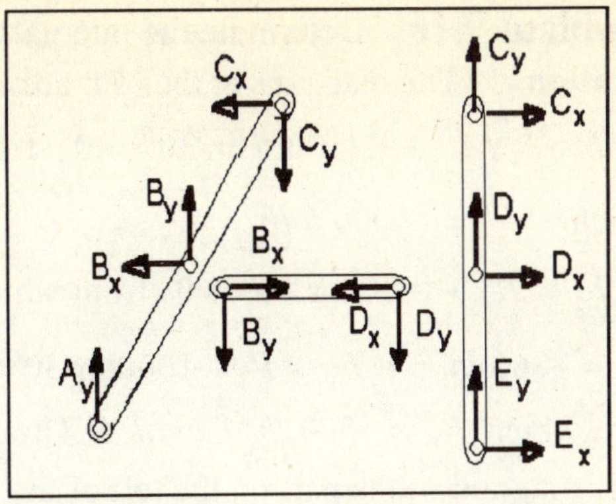

The centroid distance of the rectangle is $d_2 = 7.5\ m$. The moment about E is $M_E = d_1 F_1 + d_2 F_2 = 4583.33\ N\,m$. *check.*

The Complete Structure: The sum of the moments about E is $\sum M = -10A_R + M_E = 0$, where A_R is the reaction at A, from which $A_R = 458.33\ N$.

The element ABC: . Element BD is a two force member, hence $B_y = 0$. The sum of the moments about C: $\sum M_C = -5B_x - 10A_y = 0$, where A_Y is equal and opposite to the reaction of the support, from which $B_x = -2A_y = 2A_R = 916.67\ N$. Since the reaction in element BD is equal and opposite, $\boxed{B_x = -916.67\ N}$, *which is a tension in BD.*

Problem 9.14 Determine the internal forces and moment at A, B and C.

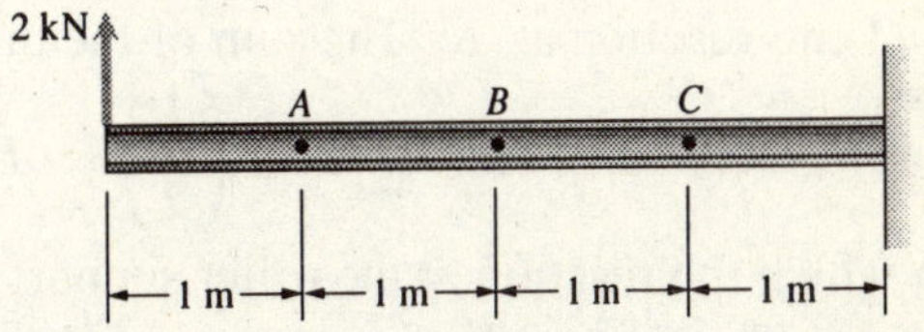

Solution *Strategy:* In this case you don't need to determine the reactions at the built-in support. Cut the beam at the point where you want to determine the internal forces and moment and draw the free body diagram of the part to the left of your cut. Remember the P, V, and M must be in their defined positive directions in your free body diagrams.

A typical cut (for this problem) is shown.

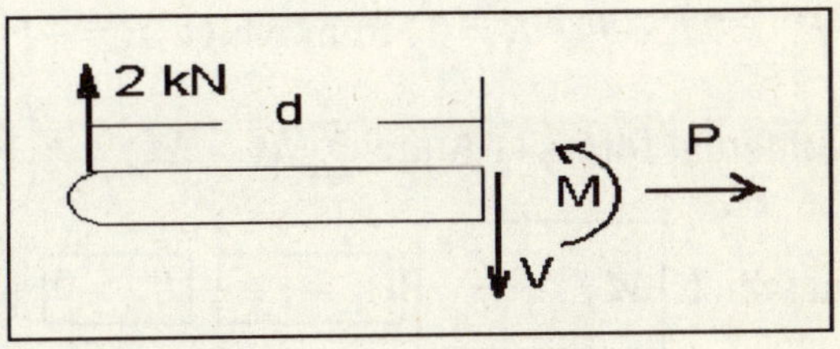

Cut A: $d = 1\ m$. From the sum of forces: $\boxed{V_A = 2\ kN}$ $\boxed{P_A = 0}$. From the sum of moments $\boxed{M_A = 2\ kN\ m}$

Cut B: $d = 2\ m$. From the sum of forces: $\boxed{V_B = 2\ kN}$ $\boxed{P_B = 0}$. From the sum of moments: $\boxed{M_B = 4\ kN\ m}$.

Cut C: $d = 3\ m$. From the sum of forces, $\boxed{V_C = 2\ kN}$, $\boxed{P_C = 0}$. From the sum of moments: $\boxed{M_C = 6\ kN\ m}$

Problem 9.15 Determine the internal forces and moment at A, B, and C.

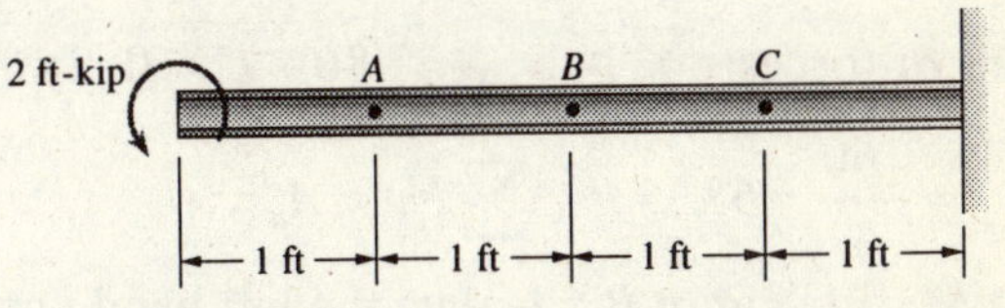

Solution A cut at A is shown. Since no forces act on the beam to the left, the shear and axial forces at each cut is zero, $\boxed{P_A = P_B = P_C = 0}$, and $\boxed{V_A = V_B = V_C = 0}$. From the sum of the moments

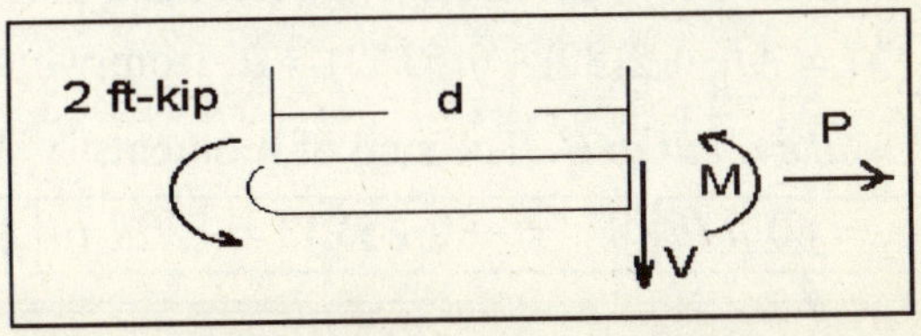

$M_A + 2 = 0$, from which $\boxed{M_A = -2\ ft\ kip}$

Since the effect of a couple is independent of the distance, the moment is the same for each cut:

$\boxed{M_A = M_B = M_C = -2\ ft\text{-}kip}$

Problem 9.16 Determine the internal forces and moments at A.

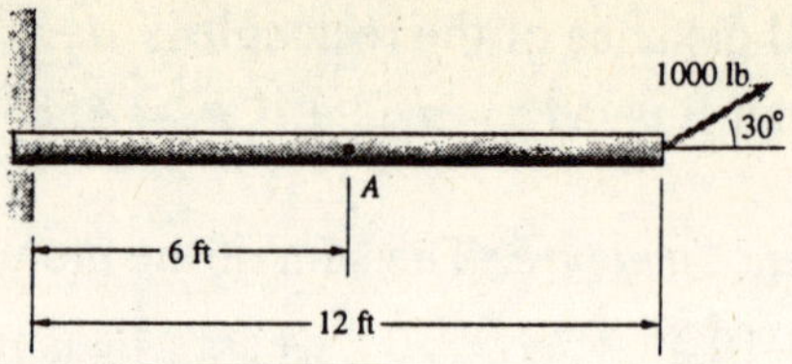

Solution The reactions at the root are:

$\sum M_R = M^R + 12(1000)\sin 30^o = 0$, from which $M^R = -6000\,ft\,lb$.

$\sum F_y = R_y + 1000\cos 30^o = 0$, from which $R_x = -866\ lb$. $\sum F_y = R_y + 1000\sin 30^o = 0$, from which $R_y = -500\ lb$. *Cut at A:* The free body diagram of the part on the left of the cut is shown. The sum of the moments about A:

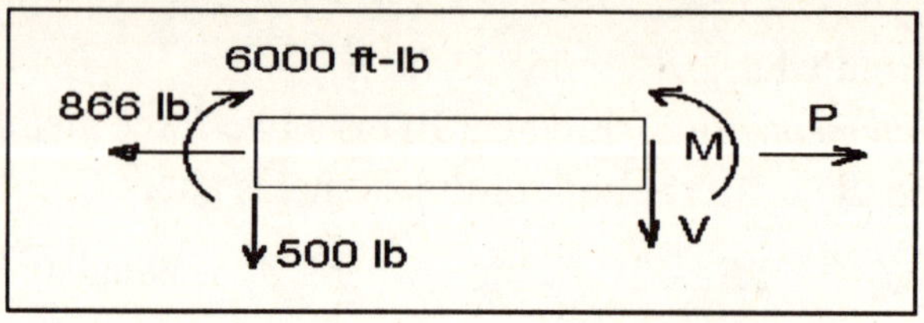

$\sum M = M^A + 6(500) - 6000 = 0$, from which

$\boxed{M_A = 3000\,ft\,lb}$, $\boxed{V_A = -500\ lb}$, The sum of the forces: , $\sum F_x = P_A - 866 = 0$, $\boxed{P_A = 866\ lb}$

Problem 9.17 Determine the internal forces and moments at A.

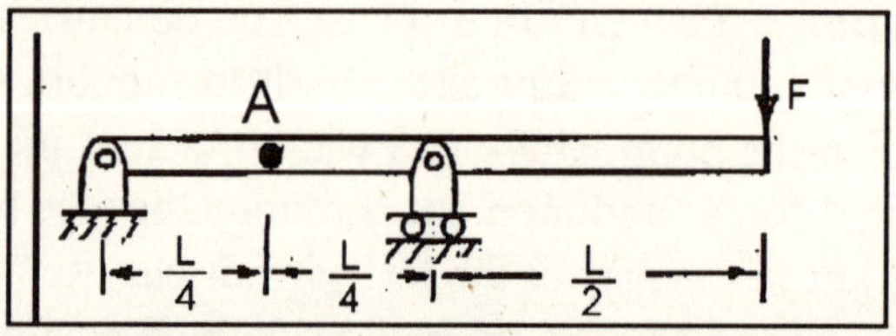

Solution Denote the roller reaction as B and the pinned end reaction as R. The sum of the moments about the left pinned end: $\sum M = \left(\frac{L}{2}\right)B - LF = 0$, from which the reaction at the roller support is $B = 2F$. The sum of the forces:

$\sum F_y = 2F - F + R = 0$, from which $R = -F$.

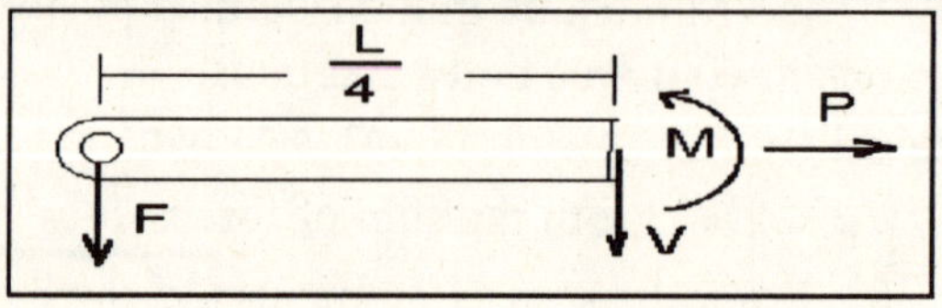

The internal forces at A are $\sum M = M_A + F\left(\frac{L}{4}\right) = 0$,

from which $\boxed{M_A = -\frac{FL}{4}}$ $\boxed{V_A = -F}$, $\boxed{P_A = 0}$

Problem 9.18 Determine the internal forces and moment (a) at B; (b) at C.

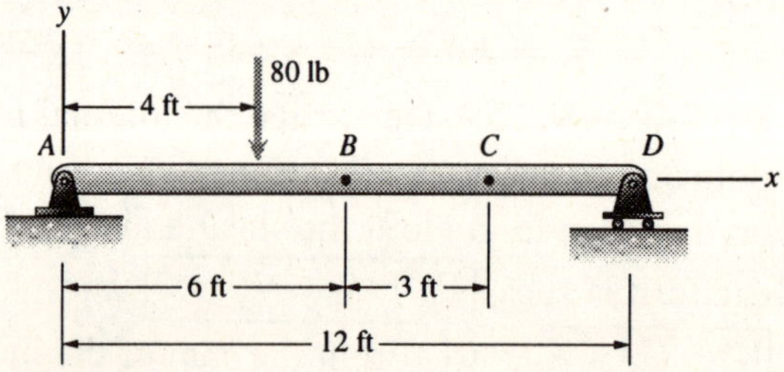

Solution The sum of the moments about A is

$\sum M_A = -4(80) + 12D = 0$, from which $D = \frac{80}{3}\ lb$. The sum of forces: $\sum F_y = A_y - 80 + D = 0$, from which

$A_y = \frac{160}{3} = 53.33\ lb$. $\sum F_x = A_x = 0$

(a) *Make a cut at B.* Isolate the left hand part. The sum of moments

$\sum M = M_B + 2(80) - 6(53.33) = 0$, from which $\boxed{M_B = 160\,ft\,lb}$. $\boxed{V_B = 53.33 - 80 = -26.7\ lb}$. $\boxed{P_B = 0}$

(b) *Make a cut at C:* The sum of moments $\sum M = M_C + 80(5) - 9(53.33) = 0$, from which

$\boxed{M_C = 80\,ft\,lb}$. $\boxed{V_C = -80 + 53.33 = -26.7\ lb}$. $\boxed{P_C = 0}$

Problem 9.19 Determine the internal forces and moment at B (a) if $x = 250\ mm$; (b) if $x = 750\ mm$.

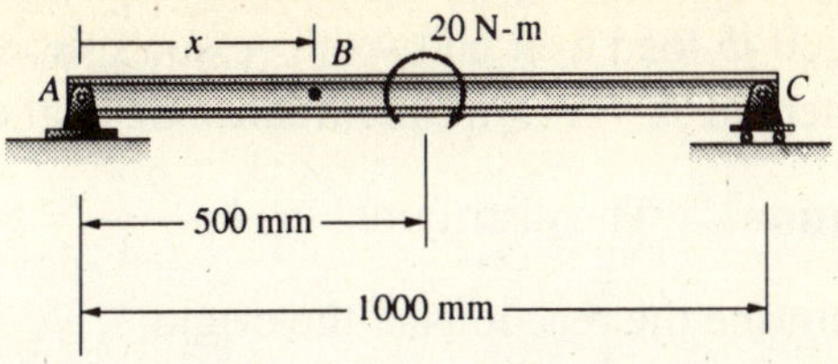

Solution (a) The sum of the moments about A $\sum M_A = -20 + 1C = 0$, from which $C = 20\ N$. The sum of forces $\sum F_y = A_y + C = 0$, from which $A_y = -C = -20\ N$. $\sum F_x = A_x = 0$

(a) Make a cut at B. Isolate the left hand part. The sum of moments $\sum M = M_B + 0.25(20) = 0$, from which $\boxed{M_B = -5\ N m}$. $\boxed{V_B = -20\ N}$, $\boxed{P_B = 0}$.

(b) *Make a cut at B*: Isolate the left hand part. The sum of moments: $\sum M = M_B - 20 + 0.75(20) = 0$, from which $\boxed{M_B = 5\ N\text{-}m}$. $\boxed{V_B = -20\ N}$, $\boxed{P_B = 0}$.

Problem 9.20 Determine the internal forces and moment at A for each loading.

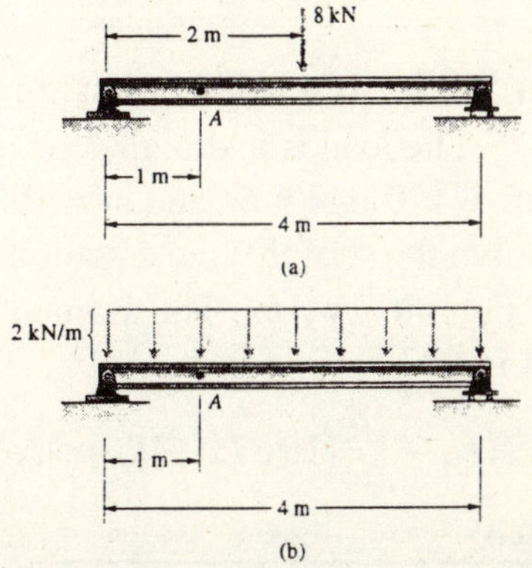

Solution (a) Denote the reaction at the pinned left end by R, and the reaction at the roller support by B. The reaction at B: $\sum M = -2(8) + B(4) = 0$, from which $B = 4\ kN$. The reaction at R: $\sum F_y = R_y - 8 + B = 0$, from which $R_y = 4\ kN$. $\sum F_x = R_x = 0$. *Make a cut at A*: Isolate the left hand part. The sum of moments: $\sum M = M_A - 4(1) = 0$, from which $\boxed{M_A = 4\ kN\ m}$ $\boxed{V_A = 4\ kN}$ $\boxed{P_A = 0}$

(b) Determine the reaction at B: The sum of the moments about R: $\sum M_R = -\int_0^4 2x\,dx + 4B = 0$, from which $B = \left(\frac{1}{4}\right)\left[2\frac{x^2}{2}\right]_0^4 = \frac{16}{4} = 4\ kN$. The reaction at R: $\sum F_y = R_y - \int_0^4 2\,dx + B = 0$, from which $R_y = 8 - 4 = 4\ kN$, $\sum F_x = R_x = 0$. *Make a cut at A:* Isolate the left hand part. The sum of moments:

$\sum M = M_A - (1)R_y + \int_0^1 2x\,dx = 0$, from which $\boxed{M_A = R_y - 1 = 3\ kN\ m}$.

$\boxed{V_A = R_y - \int_0^1 2\,dx = 4 - 2 = 2\ kN}$ $\boxed{P_A = 0}$.

Problem 9.21 Model the ladder rung as a simply supported (pin supported) beam and assume that the 200 *lb* load exerted by the person's shoes is uniformly distributed. Determine the internal forces and moment at A. (A is in the middle side of the shoe, at $x = 10$ *in*.)

Solution The distributed load is $\frac{200}{4} = 50\ lb/in$.

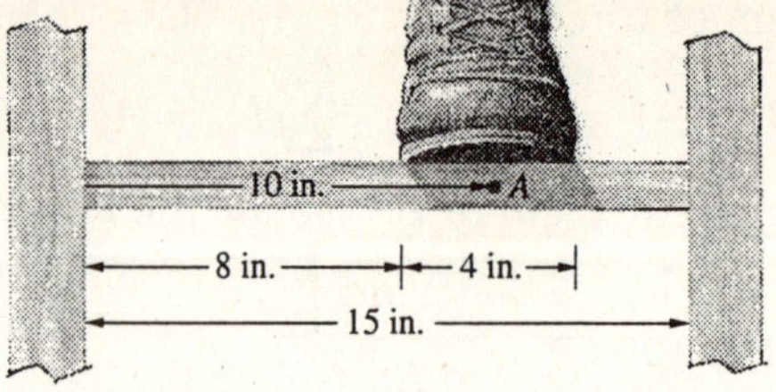

Determine the reactions at the origin:

$\sum M = -\int_8^{12} 50x\,dx + 15B_y = 0$, from which

$B_y = \left(\frac{1}{15}\right)\left[50\frac{x^2}{2}\right]_8^{12} = 133.33\ lb$. $\sum F_y = R_y - 100 + B_y = 0$,

from which $R_y = 100 - 133.33 = -33.33\ lb$.

Make a cut at A: Isolate the left hand part: The sum of moments $\sum M = M_A - 10(33.33) + 100(9) = 0$, from which $\boxed{M_A = 566.67\ in\text{-}lb}$ $\boxed{V_A = -33.33\ lb}$, $\boxed{P_A = 0}$

Problem 9.22 If $x = 3\ m$, what are the internal forces and moment at A?

Solution The load is a straight line defined at two points: $w = 0$ at $x = 6$, and $w = 100$ at $x = 12$. Solve the straight line equation to obtain $w = 16.67x - 100\ kN/m$. The moment reaction at the root is

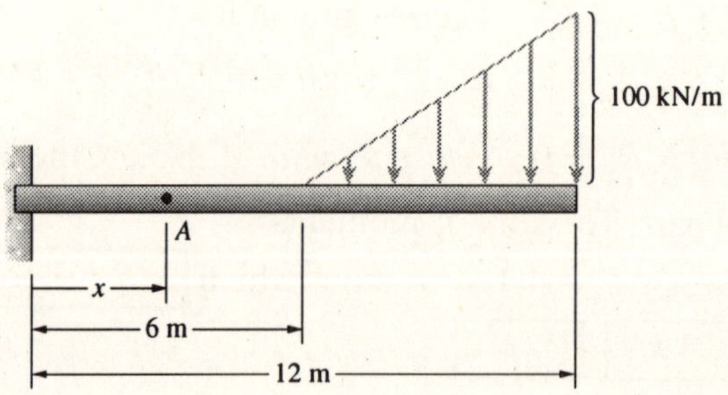

$$\sum M_R = M_A - \int_6^{12} (16.67x - 100)x\,dx = 0$$

From which

$$\boxed{M_A = 100\left[\frac{x^3}{18} - \frac{x^2}{2}\right]_6^{12} = 3000\ kN\,m}.$$

Check: The area under the triangle is $F = \left(\frac{1}{2}\right)(100)(6) = 300\ kN$. The centroid of the triangle is at $\bar{x} = 6 + \left(\frac{2}{3}\right)6 = 10\ m$. The moment about the root is $M_R = F\bar{x} = 300(10) = 3000\ kN\,m$. *check.* The sum of forces is $\sum F_y = R_y - 300 = 0$, from which $R_y = 300\ kN$. $\sum F_x = R_x = 0$.

Make a cut at A: Isolate the left hand part. The moment at A: $\sum M_A = M_A + M_R - 3R_y = 0$, from which $\boxed{M_A = -3000 + 900 = -2100\ kN\,m}$. $\boxed{V_A = 300\ kN}$, $\boxed{P_A = 0}$

Problem 9.23 If $x = 9\ m$ in Problem 9.22, what are the internal forces and moments at A?

Solution Use the results of the solution to Problem 9. 22: The loading is $w = 16.67x - 100\ kN/m$ over the interval $6 \le x \le 12$. The reaction at the root is $M^R = 3000\ kN\ m$ and $R_y = 300\ kN$, $R_x = 0$. *Make a cut at A:* Isolate the left hand part. The area under the triangle from 6 m to 9 m is

$F = \left(\frac{1}{2}\right)50(3) = 75\ kN$. The centroid distance from A is $\bar{x} = \left(\frac{1}{3}\right)3 = 1\ m$. From the sum of moments:

$\boxed{M^A = -3000 + 9(300) - 1(75) = -375\ kN\ m}$.. $\boxed{V_A = 300 - 75 = 225\ kN}$, $\boxed{P_A = 0}$

Problem 9.24 Determine the internal forces and moments at A.

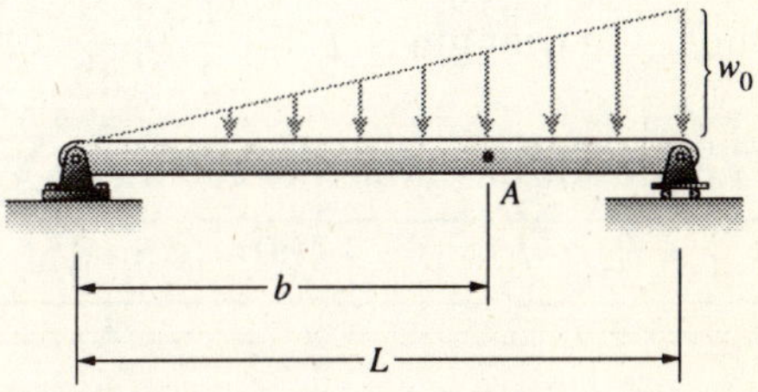

Solution The loading is a straight line, with slope $\frac{w_o}{L}$ and intercept $w = 0$. The area under the triangle is $F = \left(\frac{1}{2}\right)w_o L$. The centroidal distance from the left end is $\bar{x} = \left(\frac{2}{3}\right)L$. The sum of the moments about the left end of the beam is $\sum M = -\bar{x}F + BL = 0$, from which $B = \frac{w_o L}{3}$. The sum of the forces:

$\sum F_y = R_y - F + B = 0$, from which $R_y = F - B = \left(\frac{1}{2}\right)w_o L - \left(\frac{1}{3}\right)w_o L = \left(\frac{1}{6}\right)w_o L$. *Make a cut at A:*

Isolate the left end of the beam. The area under the triangle from 0 to b is $F_1 = \left(\frac{1}{2}\right)b\left(\frac{w_o b}{L}\right) = \left(\frac{b^2 w_o}{2L}\right)$.

The centroidal distance from A is $\bar{x}_1 = \left(\frac{b}{3}\right)$. The sum of the moments about A is

$\sum M = M^A - bR_y + \bar{x}_1 F_1 = 0$, from which $\boxed{M^A = b\left(\frac{1}{6}\right)w_o L - \left(\frac{1}{6}\right)\frac{b^3 w_o}{L} = \frac{b w_o}{6}\left(L - \frac{b^2}{L}\right)}$,

$\boxed{V_A = R_y - F_1 = \left(\frac{1}{6}\right)w_o L - \left(\frac{1}{2}\right)\frac{w_o b^2}{L} = \frac{w_o}{2}\left(\frac{L}{3} - \frac{b^2}{L}\right)}$, $\boxed{P_A = 0}$

Problem 9.25 If $x = 3\ ft$, what are the internal forces and moment at A?

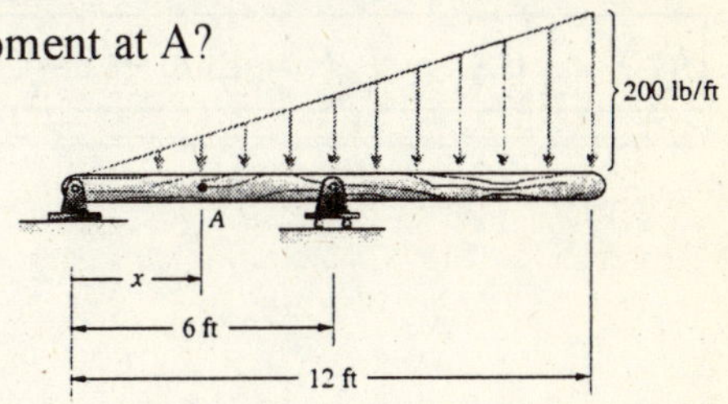

Solution The area under the triangle is

$F = \left(\frac{1}{2}\right)(200)(12) = 1200\ lb$. The centroidal distance:

$\bar{x} = \left(\frac{2}{3}\right)12 = 8\ ft$. The reaction at the roller:

$\sum M = -\bar{x}F + 6B = 0$, from which $B = 1600\ lb$. The reaction at the origin: $\sum F_y = R_y - F + B = 0$, from which $R_y = 1200 - 1600 = -400\ lb$.

Problem 9.25 continued on next page

continuation of Problem 9.25...........

Make a cut at A: Isolate the left hand part. The area under the triangle $F_1 = \left(\frac{1}{2}\right)3\left(\frac{3}{12}200\right) = 75\ lb$. The centroid distance from A: $\bar{x}_1 = 1$. The moment: $\sum M = M^A - 3R_y + 75 = 0$, from which

$\boxed{M^A = -1200 - 75 = -1275\ ft\ lb}$ $\boxed{V_A = R_y - 75 = -475\ lb}$ $\boxed{P_A = 0}$

Problem 9.26 If $x = 9\ ft$ in Problem 9.25, what are the internal forces and moment at A?

Solution Use the solution to Problem 9.25: The reaction at the roller support $B = 1600\ lb$. The reaction at the left end $R_y = -400\ lb$, and $R_x = 0$. *Make a cut at A:* Isolate the left hand part. The area under the triangle is $F = \left(\frac{1}{2}\right)9\left(\frac{9}{12}200\right) = 675\ lb$. The centroidal distance from A is $\bar{x} = \left(\frac{1}{3}\right)9 = 3\ ft$.

The sum of the moments about A: $\sum M = M^A - 9R_y + \bar{x}F - 3B = 0$, from which $\boxed{M^A = -825\ ft\ lb}$

$\boxed{V_A = R_y - F + B = -400 - 675 + 1600 = 525\ lb}$, $\boxed{P_A = 0}$

Problem 9.27 Determine the internal forces and moment at the midpoint of the beam in Problem 9.6.

Solution From the solution to Problem 9.6, the distributed load is $w = 100x - cx^2\ lb/ft$, where c is a constant, determined from the condition that the moment about A is zero, $c = 11.11$. The reaction at A is $A_y = 800\ lb$. *Make a cut at the midpoint of the beam.* At the midpoint of the beam the area under the curve is

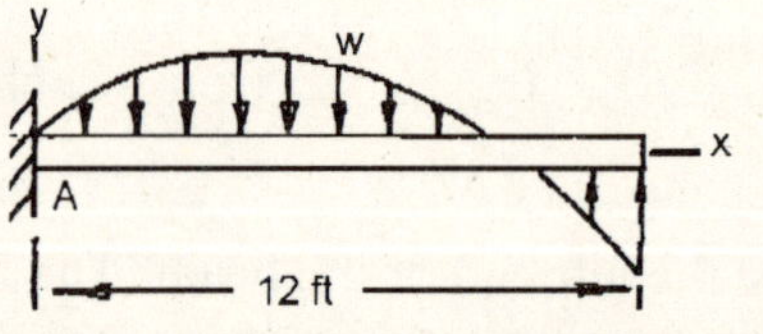

$F = \int_0^6 (100x - cx^2)dx = \left[50x^2 - \left(\frac{c}{3}\right)x^3\right]_0^6 = 1000\ lb$, acting downward.. The moment about the origin is $M = \int_0^6 (100x - cx^2)x dx = \left[\frac{100}{3}x^3 - \frac{c}{4}x^4\right]_0^6 = 3600\ ft\ lb$, acting clockwise. The centroidal distance from point A is $\bar{x} = \frac{M}{F} = 3.6\ ft$. The distance of the centroid from the midpoint is $\bar{x}_A = 6 - \bar{x} = 2.4\ ft$.

The moment about the midpoint is $\sum M_{MP} = M^{MP} - 6A_y + \bar{x}_A F = 0$, from which

$\boxed{M^{MP} = 6A_y - \bar{x}_A F = 4800 - (2.4)(1000) = 2400\ ft\ lb}$. $\boxed{V_{MP} = A_y - F = -200\ lb}$ $\boxed{P_{MP} = 0}$

Problem 9.28 Determine the internal forces and moment at the point midway between points A and B of the beam in Problem 9.7.

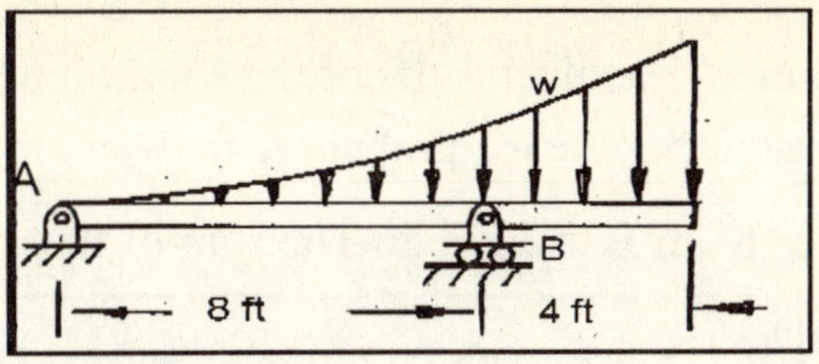

Solution From the solution to Problem 9.7, the distributed load is $w = 10x^2\ lb/ft$; the reaction at A is $A_y = -720\ lb$. *Make a cut midway between A and B:* Isolate the left hand part. The area under the curve between A and the cut is

$F = \int_0^4 10x^2 dx = \left[\frac{10}{3}x^3\right]_0^4 = 213.33\ lb$. The moment relative to A is

$M = \int_0^4 10x^3 dx = \left[\frac{10}{4}x^4\right]_0^4 = 640\ ft\ lb$. The centroidal distance from A is $\bar{x} = \frac{M}{F} = 3\ ft$. The distance of this centroid from the midpoint is $\bar{x}_{MP} = 4 - 3 = 1\ ft$. The sum of the moments about the midpoint: $\sum M = M^{MP} - 4A_y + \bar{x}_A F = 0$, from which

$\boxed{M^{MP} = 4A_y - \bar{x}_{MP}F = -(4)(720) - 1(213.33) = -3093.33\ ft\ lb}$

$\boxed{V_{MP} = A_y - F = -720 - 2133.33 = -913.33\ lb}$, $\boxed{P_{MP} = 0}$

Problem 9.29 Determine the internal forces and moment at A.

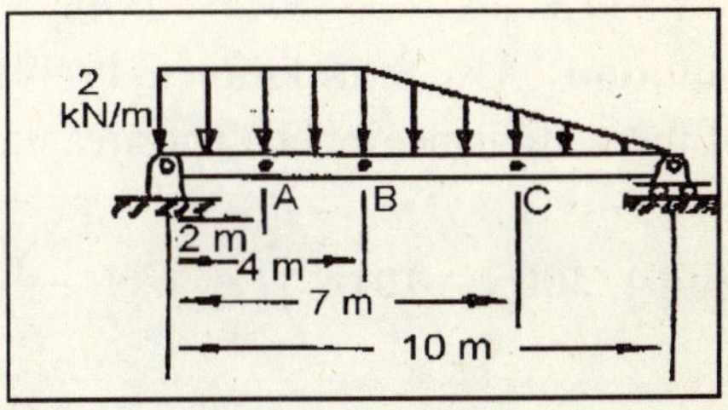

Solution The area under the load curve is

$F = 4(2) + \left(\frac{1}{2}\right)(2)6 = 14\ kN$. The centroids relative to the left end are $\bar{x}_1 = 2\ m$ and $\bar{x}_2 = 4 + \left(\frac{1}{3}\right)6 = 6\ m$. The moments about the left end due to the loading are

$M_L = (4)(2)(2) + (6)\left(\frac{1}{2}\right)(2)(6) = 52\ kN\ m$. Denote the reaction at the right end by D. The sum of the moments about the left end $\sum M = -M_L + 10D = 0$, from which $D = 5.2\ kN$. The sum of the forces: $\sum F_y = R_y - F + D = 0$, from which $R_y = F - D = 8.8\ kN$

Make a cut at A: Isolate the left hand part. The area under the loading curve between the left end and A is $F_1 = (2)(2) = 4\ kN$. The centroidal distance from A is $\bar{x}_A = 2 - 1 = 1\ m$. The sum of the moments about A is $\sum M = M^A - 2A_y + (1)F_1 = 0$, from which $\boxed{M^A = 2R_y - F_1 = 13.6\ kN\ m}$

$\boxed{V_A = R_y - F_1 = 4.8\ kN}$, $\boxed{P_A = 0}$

==================================<>==================================

Problem 9.30 Determine the internal forces and moment about point B of the beam in Problem 9.29.

Solution Use the results of the solution to Problem 9.29. *Make a cut at B:* Isolate the left hand part. The area under the loading curve between the left end and B is $F = (2)(4) = 8\ kN$. The centroid distance from B is $x_B = 2$. The sum of the moments about B: $\sum M = M^B - 4R_y + \bar{x}_B F = 0$, from which $\boxed{M^B = 4R_y - \bar{x}_B F = 4(8.8) - 2(8) = 19.2\ kN\,m}$ $\boxed{V_B = R_y - F = 0.8\ kN}$ $\boxed{P_B = 0}$

==================================<>==================================

Problem 9.31 Determine the internal forces and moment at point C of the beam in Problem 9.29.

Solution Use the solution to Problem 9.29. The area under the loading curve between C and the left end is made up of two rectangles and a triangle: $F = 2(4) + \left(\frac{1}{2}\right)(3)(2-1) + (3)(2-1) = 12.5\ kN$. The centroid distances from C are $\bar{x}_1 = 3 + 2 = 5$, $\bar{x}_2 = (2/3)(3) = 2$, $\bar{x}_3 = 1.5\,ft$. The sum of the moments about C: $\sum M = M^C - 7R_y + 8(5) + 2\left(\frac{1}{2}\right)(3) + 1.5(3) = 0$, from which $\boxed{M^C = 7R_y - 47.5 = 14.1\ kN\,m}$

$\boxed{V_C = R_y - F = -3.7\ kN}$ $\boxed{P_C = 0}$

==================================<>==================================

Problem 9.32 Neglecting the weight of the wing in Problem 9.8, determine the internal forces and moment at the wing's center of mass.

Solution Use the solution to Problem 9.8: $M^R = -2500\ N\,m$ and $F_R = -1178.1\ N$, when the weight of the wing is neglected. The area under the lift distribution between the root and the center of mass is

$$F = \int_0^2 300\left(1 - 0.04x^2\right)^{\frac{1}{2}} dx = 60\left[\frac{x\sqrt{25 - x^2}}{2} + \frac{25}{2}\sin^{-1}\left(\frac{x}{5}\right)\right]_0^2 = 583.59\ N,$$ acting upward. The moment

about the root is $$M = 300\int_0^2 (1 - 0.04x^2)^{\frac{1}{2}} x\, dx = -60\left[\frac{\left(25 - x^2\right)^{\frac{3}{2}}}{3}\right]_0^2 = +575.32\ N\,m,$$ acting

counterclockwise. The centroid distance from the root: $\bar{x} = \frac{M}{F} = 0.9858\ m$. The distance of the centroid from the center of mass is $\bar{x}_{CM} = 2 - 0.9858 = 1.014\ m$. Make a cut at the *CM* and consider the left hand part. The sum of the moments about the center of mass:

$\sum M_{CM} = M^{CM} + 2|F_R| - 583.6(1.014) - |M_R| = 0$, from which

$\boxed{M^{CM} = -(2)(1178.1) + 583.6(1.014) + 2500 = 735.6\ N\,m}$,

$\boxed{V_{CM} = -F_R + F = -1178.1 + 583.6 = -594.5\ N}$ $\boxed{P_{CM} = 0}$

==================================<>==================================

Problem 9.33 Determine the internal forces an moment at A.

Solution: *The complete structure as a free body:* .

The sum of the moments about the right end:

$\sum M = 3(10) - 5R = 0$, from which

$R = \frac{30}{5} = 6\ kN$. The sum of forces about the right

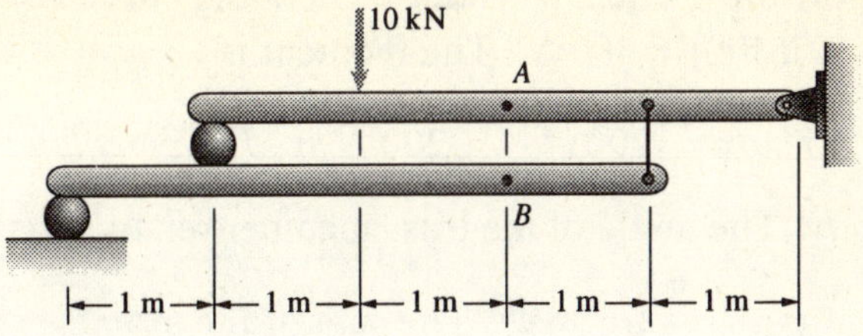

end: $\sum F_y = R_y + C_y - 10 = 0$, from which

$C_y = 4\ kN$ *The element CA as a free body:* The sum of the moments about C:

$\sum M_C = -4F_1 + 10(3) - F_2 = 0$. The sum of the forces: $\sum F_y = C_y + F_2 - 10 + F_1 = 0$. Solve the simultaneous equations: $F_1 = 8\ kN$, $F_2 = -2\ kN$. *Make a cut at A:* Isolate the left end of CA. The sum of the moments about A: $\sum M = M_A - 2F_1 + 10 = 0$, from which $\boxed{M_A = -10 + 16 = 6\ kN\,m}$

$\boxed{V_A = 8 - 10 = -2\ kN}$, $\boxed{P_A = 0}$

Problem 9.34 Determine the internal forces and moments about point B in Problem 9.33.

Solution: Use the solutions to Problem 9.33. *Make a cut at point B:* Isolate the left part. The sum of the moments about B: $\sum M_B = M_B + 2F_1 - 3R_y = 0$, from which $\boxed{M_B = -16 + 18 = 2\ kN\,m}$

$\boxed{V_B = R_y - F_1 = 6 - 8 = -2\ kN}$ $\boxed{P_B = 0}$

Problem 9.35 Determine the internal forces and moment at A.

Solution: *The complete structure as a free body:* The sum of the moments about the left end:

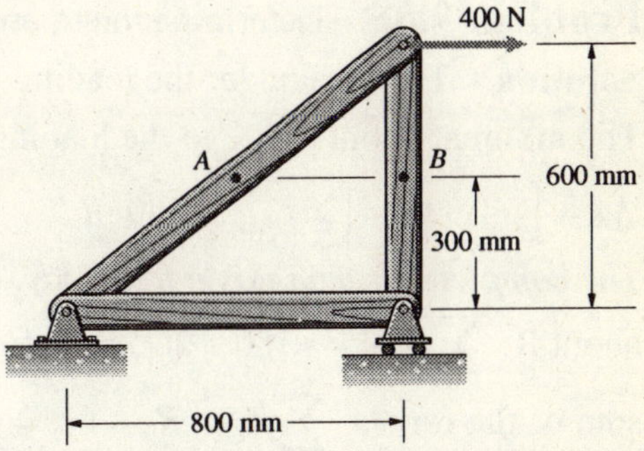

$\sum M_R = -600(400) + 800C_y = 0$, from which $C_y = 300\ N$.

The sum of the forces: $\sum F_y = R_y + 300 = 0$, from which

$R_y = -300\ N$. $\sum F_x = R_x + 400 = 0$, from which

$R_x = -400\ N$. *Make a cut at A:* Isolate the part to the left of A: The resultant force reaction at R is

$R = \sqrt{400^2 + 300^2} = 500\ N$ at an angle of $\theta = 180 + \tan^{-1}\frac{3}{4} = 216.9^\circ$. The angle element RA relative to the horizontal is $\alpha = \tan^{-1}\left(\frac{3}{4}\right) = 36.9^\circ$. Thus the angle of RA is equal and 180 deg opposite to the reaction at R. Since the roller support prevents the horizontal member of the structure from exerting an x- component of load, the member RA bears the full reaction as an axial load. Thus $\boxed{M^A = 0}$ $\boxed{V_A = 0}$ and $\boxed{P_A = 500\ N}$.

Problem 9.36 Determine the internal forces and moment at point B of the truss in Problem 9.35.

Solution: Use the solution to Problem 9.35. *Make a cut at B:* Isolate the lower portion. The reaction at the roller support is vertical, hence the reaction in the vertical member is an axial load, hence $\boxed{M_B = 0}$,

$\boxed{V_B = 0}$, and $\boxed{P_B = -300\ N}$.

Problem 9.37 Determine the internal forces and moment at A.

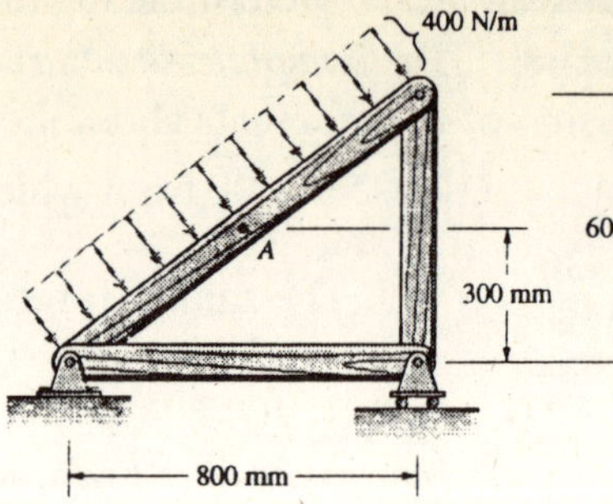

Solution: The area under the loading curve is $F = 1(400) = 400\ N$. The moment is $\int_0^1 400x dx = \left[200x^2\right]_0^1 = 200\ N\,m$. *The complete structure as a free body:* The angle of the truss member relative to the horizontal is $\alpha = \tan^{-1}\left(\frac{3}{4}\right) = 36.9^o$. The sum of the moments about the left end is $\sum M_R = -200 + 0.8B = 0$, from which $B = 250\ N$. The sum of the forces: $\sum F_y = R_y - F\cos\alpha + B = 0$, from which $R_y = F\cos\alpha - B = 70\ N$, $\sum F_x = R_x + F\sin\alpha = 0$, from which $R_x = -F\sin\alpha = -240\ N$. The angle of the reaction relative to the positive x-axis: $\theta = 180^o + \tan^{-1}\left(\frac{70}{240}\right) = 163.74^o$. *Make a cut at A:* Isolate the section RA. The area under the curve from R to A: $F_{RA} = 400(0.5) = 200\ N$. The moment due to the loading to the left of A: $M_{RA} = \int_0^{0.5} 400x\,dx = \left[200x^2\right]_0^{0.5} = 50\ N\,m$. The sum of the moments to the left of A: $\sum M_{RA} = 0.5R\sin(\alpha - \theta) + 50 + M_A = 0$, from which $\boxed{M_A = 50\ N\,m}$. $\boxed{V_A = R\sin(\alpha - \theta) + F_{RA} = 0}$. $\sum F_P = P_A + R\cos(\alpha - \theta) = 0$, from which $\boxed{P_A = 150\ N}$

Problem 9.38 Determine the internal forces and moment at A.

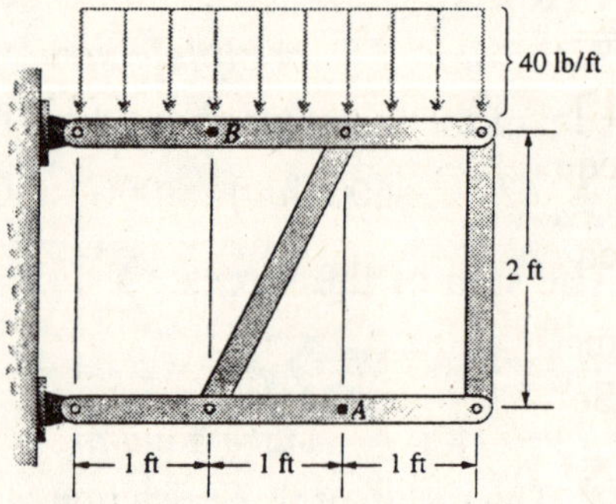

Solution The area under the loading curve: $F = 40(3) = 120\ lb$. The moment about R due to the loading is $M = \int_0^3 40x\,dx = \left[20x^2\right]_0^3 = 180\ ft\,lb$.

The complete structure as a free body: The sum of the moments about R: $\sum M_R = -180 + 2C_x = 0$, from which $C_x = 90\ lb$.. The sum of the forces: $\sum F_x = R_x + C_x = 0$, from which $R_x = -90\ lb$.

The element CA as a free body: The angle of the cross brace relative to the positive x- axis is $\alpha = \tan^{-1}(2) = 63.4^o$. The pinned joint at the right end cannot exert a force along the x- axis, so that the sum of the forces: $\sum F_x = C_x - F_1\cos\alpha = 0$, from which $F_1 = 201.2\ lb$. The sum of the moments about the right end: $\sum M = -3C_y + 2F_1\sin\alpha = 0$, from which $C_y = \left(\frac{2}{3}\right)F_1\sin\alpha = 120\ lb$. *Make a cut at A:* Isolate the section CA. The sum of the moments about A $\sum M_{CA} = M_A - 2C_y + F_1\sin\alpha = 0$, from which $\boxed{M_A = 60\ ft\,lb}$ $\boxed{V_A = C_y - F_1\sin\alpha = -60\ lb}$, $\boxed{P_A = F_1\cos\alpha - C_x = 0}$

Problem 9.39 Determine the internal forces and moment at point B of the frame is Problem 9.38.
Solution Use the solution to Problem 9.38. *The element RB as a free body.* The sum of forces $\sum F_y = R_y - 120 + C_y = 0$, from which $R_y = 0$. The area under the curve between B and R is $F_{RB} = 1(40) = 40\ lb$. The moment is $M_{RB} = 20(1)^2 = 20\ ft\ lb$. *Make a cut at B*: The sum of the moments about B is $\sum M_{RB} = M_B - 20 = 0$, from which $\boxed{M_B = 20\ ft\ lb}$. $\boxed{V_B = R_y - 40 = -40\ lb}$, $\boxed{P_A = -R_x = 90\ lb}$ (tension).

Problem 9.40 (a) Determine the internal forces and moment as functions of x. (b) Draw the shear force and bending moment diagrams. *Strategy:* You don't have to determine the reactions at the built in support. Cut the beam at an arbitrary position x and draw the free body diagrams.

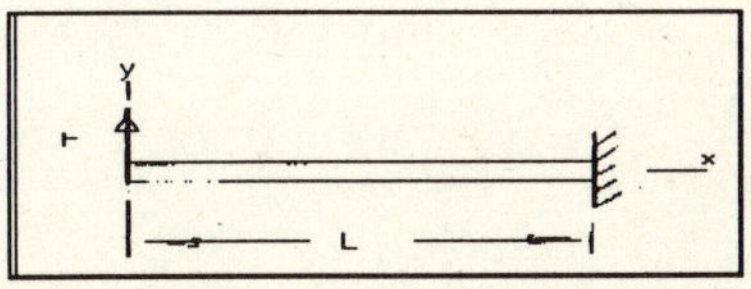

Solution *Make a cut a point x units from the left end.* By definition, the shear is the sum of the forces to the left: $\boxed{V_x = F}$.

The free body diagram of the of the left part: $M(x) - Fx = 0$, from which $M(x) = Fx$ which is a straight line with slope $F\ ft\ lb / ft$. The axial forces are zero $\boxed{P(x) = 0}$

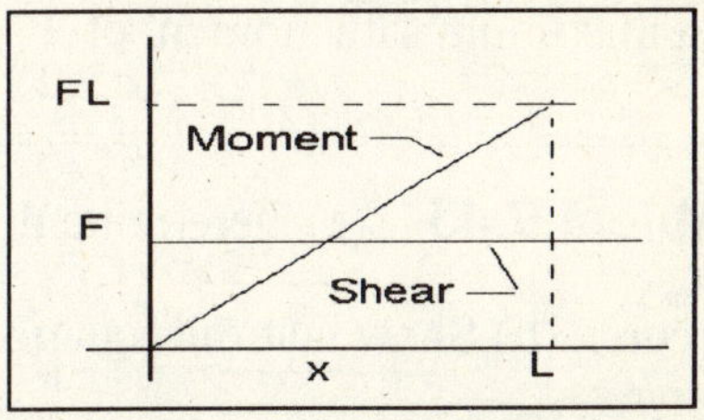

Problem 9.41 (a) Determine the internal forces and moment as functions of x. (b) Show that the equations for V and M as functions of x satisfy the equation $V = \frac{dM}{dx}$. (c) Draw the shear force and bending moment diagrams.

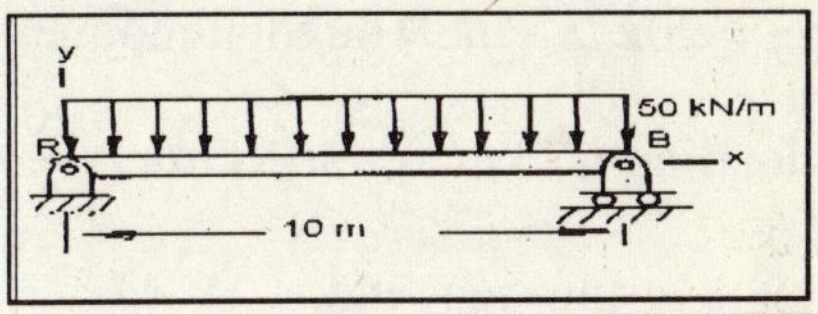

Solution The sum of the moments about the left end is $\sum M = -\int_0^{10} 50x\,dx + 10B = 0$, from which $B = 250\ kN$. The sum of the forces: $\sum F_y = R_y - \int_0^{10} 50dx + B = 0$, from which $R_Y = 500 - 250 = 250\ kN$

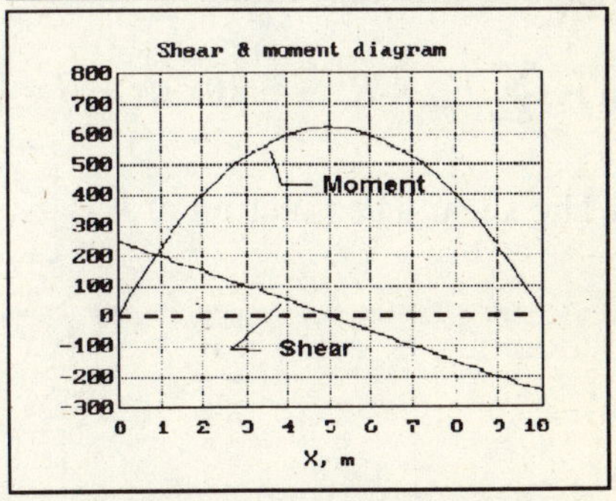

(a) Make a cut at x. By definition, for a simple beam the shear is the sum of the forces to the left of x.

$\boxed{V(x) = \sum F = +R_y - \int_0^x 50dx = 250 - 50x\ (kN)}$, which is straight line with intercept $250\ kN$ and slope $-50\ kN/m$. The free body diagram for the moment $\sum M^x = M(x) - R_y x + \int_0^x 50x dx = 0$, from which $\boxed{M(x) = 250x - 25x^2\ \ kN\ m}$. The axial force is zero, $\boxed{P(x) = 0}$. Note that $\frac{dM}{dx} = \frac{d(250x - 25x^2)}{dx} = 250 - 50x = V(x)$. (c) The graphs are shown.

Problem 9.42 The beam in Problem 9.41 will safely support a bending moment of 1 MN m (1 meganewton-meter) at any cross section. Based on this criteria, what is the maximum safe value of the distributed load?

Solution The sum of the moments about the left end is $\sum M = -\int_0^{10} wx\,dx + 10B = 0$, from which the reaction at the right end is $B = \left(\frac{1}{10}\right)\left[w\frac{x^2}{2}\right]_0^{10} = 5w\ kN$. The sum of the forces:

$\sum F_y = R_y - \int_0^{10} w\,dx + B = 0$, from which the reaction at the left end is $R = 10w - 5w = 5w\ kN$. The moment as a function of x is $M(x) = \int_0^x 5w dx - \int_0^x wx\,dx = 5wx - \frac{w}{2}x^2$. The maximum moment occurs at $\frac{dM}{dx} = 0 = 5w - wx = 0$, or $x = 5\ m$. Thus the maximum moment is $M_{MAX} = 25w - 25\left(\frac{w}{2}\right) = 12.5w$.

For a maximum safe moment of $1 \times 10^3\ kN\text{-}m$, $\boxed{w = \frac{10^3}{12.5} = 0.08(10^3) = 80\ kN/m}$

Problem 9.43 (a) Determine the maximum bending moment in the beam and the value of x where it occurs. (b) Show that the equations for V and M satisfy the equation $V = \frac{dM}{dx}$.

Solution The distributed load is a straight line function of x, with zero intercept and slope $\frac{360}{18} = 20\ lb/ft^2$, from which $w(x) = 20x\ lb/ft$. The total moment about the left end due to the loading is $M = \int_0^{18} w(x)xdx = \frac{20}{3}\left[x^3\right]_0^{18} = 38880\ ft\ lb$. The

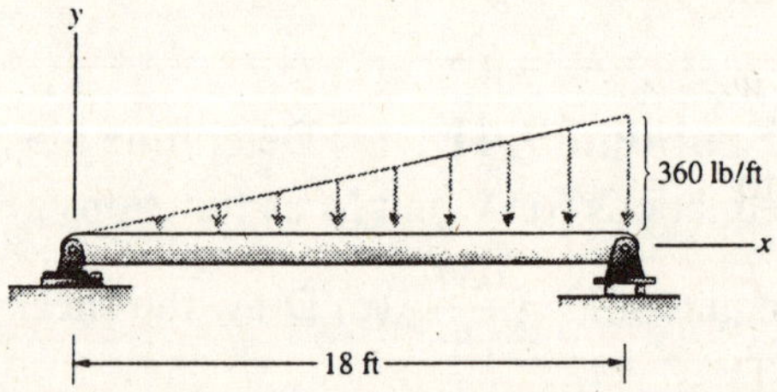

reaction at the right end is $\sum M = -38880 + 18B = 0$, from which $B = 2160\ lb$. The reaction at the left end is $\sum F_y = R_y - \int_0^{18} 20x\,dx + B = 0$, from which $R_y = \left[10x^2\right]_0^{18} - 2160 = 1080\ lb$.

(a) The shear as a function of x is $\boxed{V(x) = \sum F = R_y - \int_0^x 20x\,dx = 1080 - 10x^2\ lb}$. The moment as a function of x is $M(x) = R_y x - \int_0^x 10x^2 dx = 1080x - \left(\frac{10}{3}\right)x^3\ ft\ lb$. The maximum moment occurs where $\frac{dM}{dx} = 0 = 1080 - 10x^2 = 0$, or at $\boxed{x = \sqrt{\frac{1080}{10}} = 10.39\ ft}$. The maximum moment is

$$\boxed{M_{MAX} = 1080(10.39) - \frac{10}{3}(10.39)^3 = 7482.5\ ft\ lb}$$

(b) The shear is $\frac{dM}{dx} = \frac{d(1080x - \left(\frac{10}{3}\right)x^3)}{dx} = 1080 - 10x^2\ lb$.

Problem 9.44 Draw the shear and bending moment diagrams for the beam in Problem 9.43.

Solution The shear and moment diagrams were graphed using **TK Solver 2.** The graph is shown.

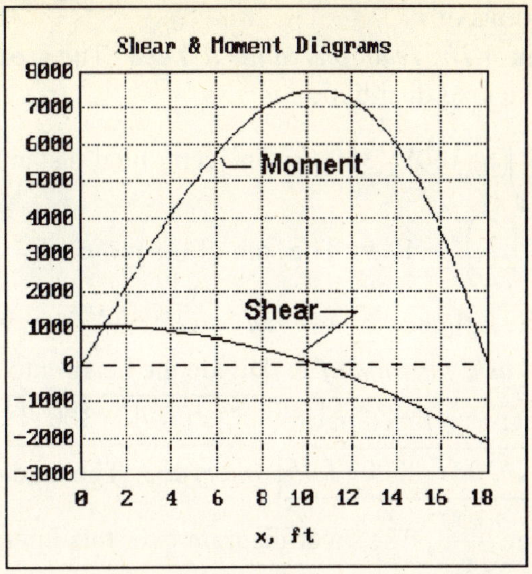

Problem 9.45 Determine the internal forces and moment as functions of x for the beam in Problem 9.4.

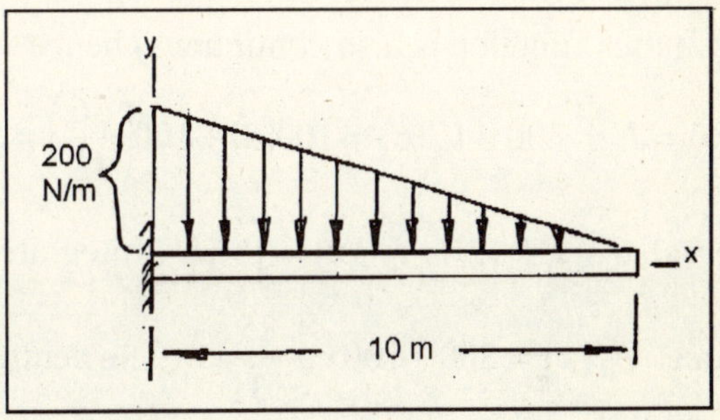

Solution *The reactions at the left end:* The distributed load is a straight line with intercept $w = 200\ N/m$ and slope $-\frac{200}{10} = -20\ N/m^2$. The total force exerted by the distributed load is

$$F = \int_0^{10}(200 - 20x)dx = \left[200x - 10x^2\right]_0^{10} = 1000\ N$$

The sum of the moments about the left end:

$$\sum M = M^A - \int_0^{10}(200 - 20x)x\,dx = 0, \text{ from which}$$

$$M^A = \int_0^{10}\left(200x - 20x^2\right)dx = \left[100x^2 - \frac{20}{3}x^3\right]_0^{10} = \frac{10^4}{3}\ N\,m.$$ *The shear and moment as functions of x:*

The shear is $V(x) = \sum F = F - \int_0^x (200 - 20x)dx = 1000 - 200x + 10x^2 \boxed{= 10(x-10)^2\ N}$. The moment as a function of x.

$$M(x) = Fx + \int_0^x (200 - 20x)xdx = 1000x + \int_0^x (200 - 20x)xdx = 1000x - 100x^2 - \left(\frac{20}{3}\right)x^3.$$ Reducing:

$\boxed{M(x) = \left(\frac{10}{3}\right)(x-10)^3}$ *Check:* The internal moment at $x = 0$, $\sum M = M(0) + M^A = 0$, from which $M(0) = -M^A$. *check.* The axial forces are zero $\boxed{P(x) = 0}$

Problem 9.46 Determine the shear and moment as functions of x:

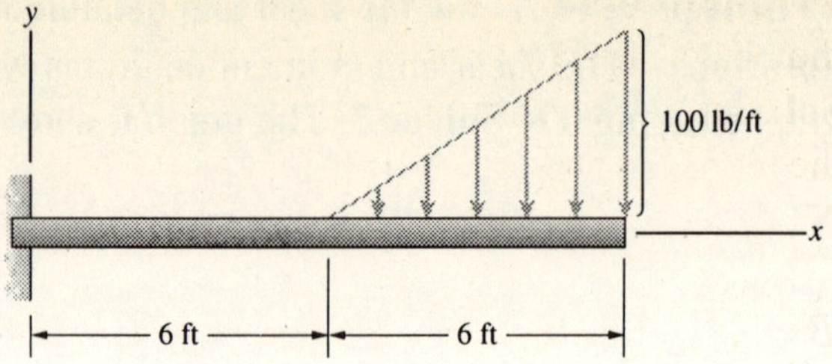

Solution *The reactions at the left end:* The area under the load distribution is $F = \left(\frac{1}{2}\right)(6)(100) = 300\ lb$. The centroidal distance is $\bar{d} = 6 + \left(\frac{2}{3}\right)6 = 10\ ft$. The sum of the moments about the left end: $\sum M = M^A - \bar{d}F = 0$, from which $M^A = 10(300) = 3000\ ft\ lb$. *The shear and moment as a function of x:* Divide the beam into two intervals: $0 \le x \le 6$, and $6 < x \le 12$. *Interval 1:* The shear as a function of x: $\boxed{V_1(x) = F = 300\ lb}$. The moment: $M_1(x) = Fx - M^A$, from which $\boxed{M_1(x) = 300x - 3000\ ft\ lb}$. *Interval 2:* The load curve is a straight line, with intercept $-100\ lb/ft$, and slope $\frac{100}{6}$. The shear diagram over this interval is

$$\boxed{V_2(x) = F - \int_6^x (-100 + \frac{100}{6}x)dx = F - 100\left[-x + \frac{1}{12}x^2\right]_6^x = 100\left(x - \frac{x^2}{12}\right) lb}$$

Check: The load curve is continuous at $x = 6$, hence the shear diagram must be continuous at $x = 6$, (since the integral of a continuous function is also continuous) hence $V_1(6) = V_2(6)$

$V_1(6) = F = 300 = V_2(6) = 100(6) - 100\left(\frac{36}{12}\right) = 300$ *check.* The force due to the distributed load in the interval $6 \le x \le 12$ is $F_2(x) = \int_6^x w dx$. Integrate and reduce: $F_2(x) = 300 - 100x + \frac{25}{3}x^2$. The centroid distance from x is $\bar{d}_2 = \frac{(x-6)}{3}$. The moment about x is $M_2(x) = Fx - M^A - \bar{d}_2 F_2$. Substitute and reduce: $\boxed{M_2(x) = -2400 + 50x^2 - \left(\frac{25}{9}\right)x^3\ ft\ lb}$. *Check:* The moment must be zero at $x = 12$. *check. Check:* The moment must be continuous at at $x = 6$, $M_1(6) = M_2(6)$, from which $M_1(6) = -1200$, and $M_2(6) = -1200$ *check.* The axial forces are zero, $P(x) = 0$

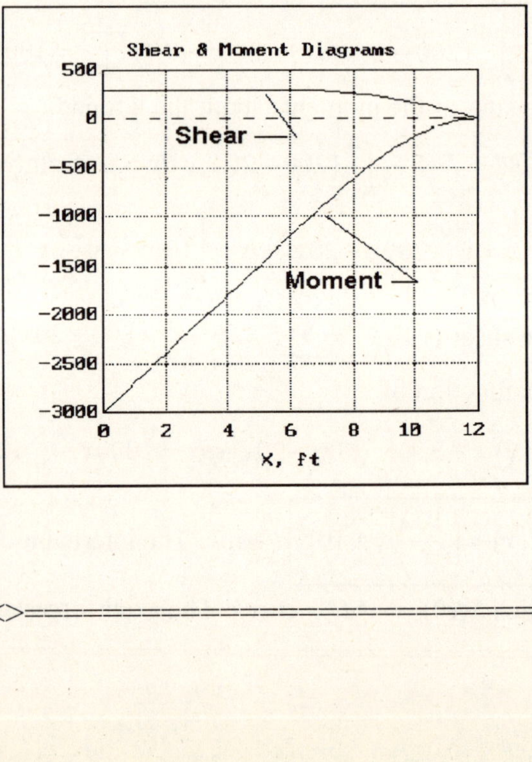

(b) The graph was drawn with **TK Solver 2.**

Problem 9.47 The loads $F = 200\ N$ and $C = 800\ N\ m$. (a) Determine the internal forces and moment as functions of x. (b) Draw the shear and moment diagrams.

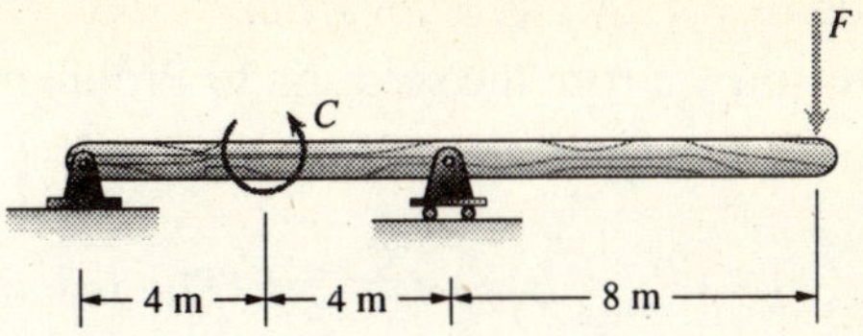

Solution *The reactions at the supports:* The sum of the moments about the left end:

$\sum M = C + 8B - 16F = 0$, from which

$$B = \left(\frac{1}{8}\right)(-C + 16F) = \left(\frac{1}{8}\right)(-800 + 16(200)) = 300\ N.$$

The sum of the forces: $\sum F_y = R_y + B - F = 0$, from which $R_y = -100\ N$. *The intervals as free bodies:* Divide the interval into three parts: $0 \le x \le 4$, $4 < x \le 8$, and $8 < x \le 16$. *Interval I.* The shear is $\boxed{V_1(x) = R_y = -100\ N}$. The moment is $\boxed{M_1(x) = +R_y x = -100x\ N\ m}$. *Interval 2.* The shear is $\boxed{V_2(x) = R_y = -100\ N}$. The sum of the moments is $\sum M^x = M_2(x) + C - R_y x = 0$, from which, $\boxed{M_2(x) = -100x - 800\ N\ m}$. *Interval 3:* The shear is $\boxed{V_3(x) = R_Y + B = -100 + 300 = 200\ N}$. The sum of the moments is $\sum M^x = M_3(x) + C - R_y x - B(x - 8) = 0$, from which $\boxed{M_3(x) = -800 - 100x + 300x - 2400 = 200x - 3200\ N\ m}$ The axial forces are zero, $\boxed{P(x) = 0}$ in all intervals.

(b) The diagrams are shown.

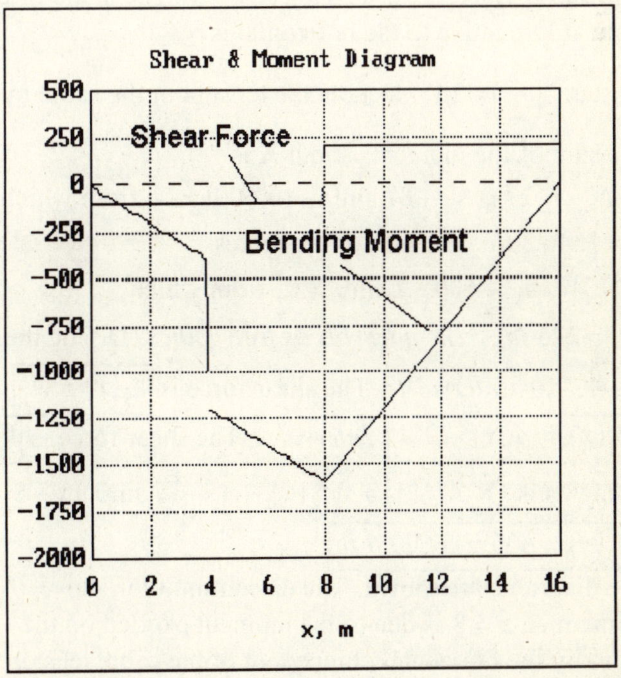

Problem 9.48

Problem 9.48 The beam in Problem 9.47 will safely support shear forces and bending moments of magnitudes 2 *kN* and 6.5 *kN m* respectively. Based on this criterion, can it safely be subjected to the loads $F = 1\ kN$, $C = 1.6\ kN\,m$?

Solution From the solution to Problem 9.47, the shear and the moments in the intervals are *Interval 1:* $V_1 = R_y$, $M_1(x) = R_y x$, *Interval 2:* $V_2(x) = R_y$, $M_2(x) = R_y x + C$, *Interval 3:* $V_3(x) = R_y + B$, $M_3(x) = (R_y + B)x + C - 8B$. The reactions are $B = \left(\frac{1}{8}\right)(16F - C)$, and $R_y = F - B$. The maximum shears in each interval have the magnitude rank: $|V_1(x)| = |V_2(x)| \le |V_3(x)|$, so that the largest shear for a force $F = 1\ kN$ is $V_3(x) = R_y - B = F - B + B = F = 1\ kN$, which can be safely supported. The maximum moment magnitudes in each interval have the rank: $|M_1(x)| \le |M_2(x)| \le |M_3(x)|$. The maximum moment magnitude occurs in the third interval:

$M_3(x) = (R_y + B)x + C - 8B = Fx + C - (16F - C) = Fx - 16F$. The maximum magnitude occurs at $x = 8$, $\boxed{|M_3(8)| = 8F = 8\ kN\,m}$ and it exceeds the safe limit by 2.5 $kN\,m$.. $\boxed{NO}$

Problem 9.49

Problem 9.49 Draw the shear and bending moment diagrams for beam ABC.

Solution *The structure as a free body:* The angle of the cable at D relative to the horizontal is $\theta = \tan^{-1}\left(\frac{1}{4}\right) = 14^o$. Denote the tension in the cable by T.

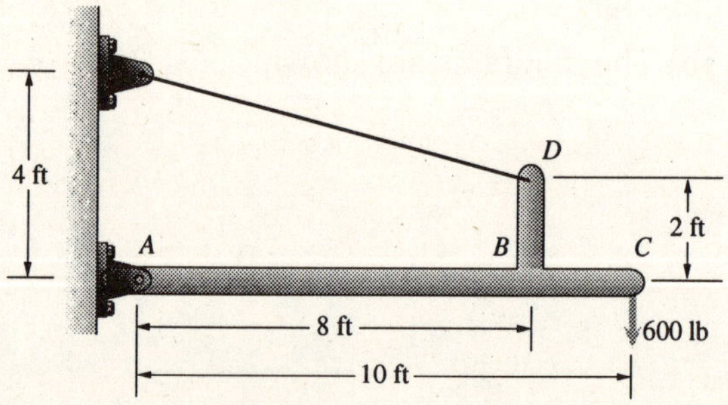

The sum of the moments about A is $\sum M_A = 2T\cos\theta + 8T\sin\theta - 10(600) = 0$, from which $T = 1546.2\ lb$. The sum of the forces: $\sum F_y = A_y - 600 + T\sin\theta = 0$, from which $A_y = 225\ lb$. *The intervals as free bodies:* Divide the beam into two intervals: $0 \le x < 8$, and $8 \le x \le 10$. *Interval 1:* The shear force is $\boxed{V_1(x) = A_y}$. The internal bending moment is $\boxed{M_1(x) = A_y x = 225x}$. *Interval 2*: The shear force is $\boxed{V_2(x) = A_y + T\sin\theta = 600\ lb}$. The sum of the moments is $\sum M^x(x) = M_2(x) - A_y - (T\sin\theta)(x-8) + 2T\cos\theta = 0$, from which $\boxed{M_2(x) = 600x - 6000\ ft\ lb}$.

The diagrams are shown. The discontinuity in the moment at $x = 8$ is due to the moment exerted on the beam by the horizontal component of the cable tension. at D. This horizontal component of tension exerts a moment that is independent of the distance x along the beam ABC. In this sense, it behaves like a couple.

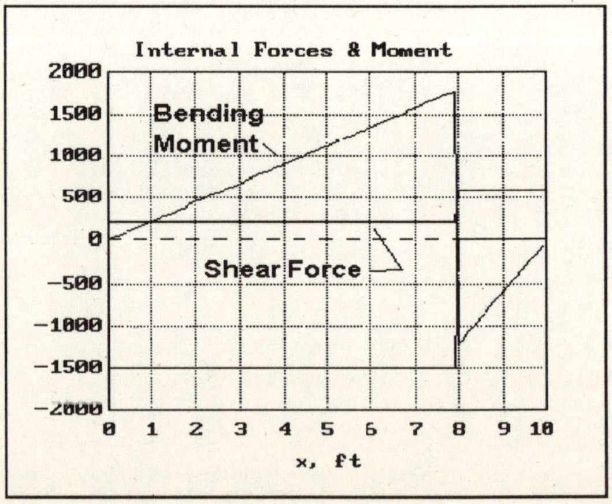

Problem 9.50 Draw the shear and moment diagrams for beam ABC.

Solution *The structure as a free body.* The sum of the moments about A: $\sum M_A = -4(4) + 6C = 0$, from which $C = \frac{16}{6} = \frac{8}{3}$ kN. The sum of the forces

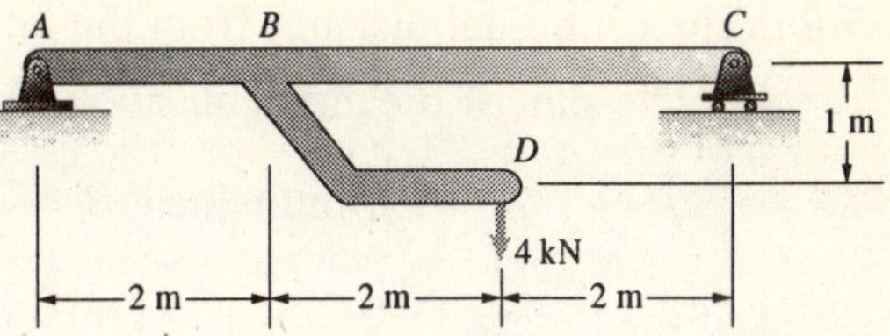

$\sum F_y = A_y - 4 + C = 0$, from which $A_y = 4 - \frac{8}{3} = \frac{4}{3}$ kN. *The intervals as free bodies::* Divide the beam into two intervals, $0 \le x < 2$, and $2 \le x < 6$. *Interval 1:* The shear is

$$\boxed{V_1(x) = A_y = \frac{4}{3}\ kN}$$

The moment is

$$\boxed{M_1(x) = A_y x = \frac{4}{3} x\ kN\ m}$$

Interval 2: The shear is

$$\boxed{V_2(x) = A_y - 4 = \frac{4}{3} - 4 = -\frac{8}{3}\ kN}$$

The moment at B is $B = -2(4) = -8\ kN\ m$. The internal moment is

$M_2(x) = A_y x + B - 4(x - 2)\ kN\ m$,

$$\boxed{M_2(x) = -\frac{8}{3}x + 16\ kN\ m}$$

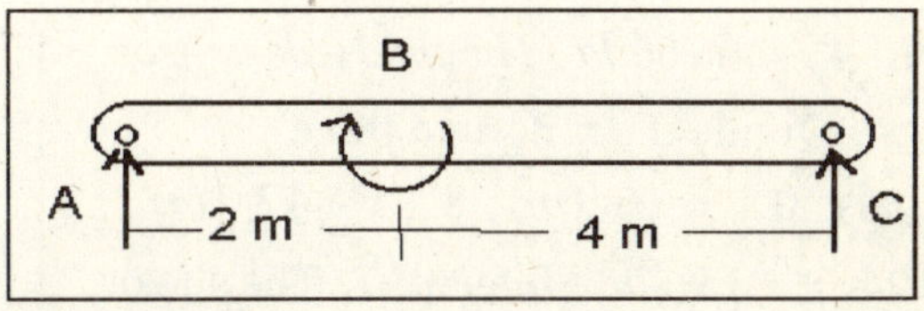

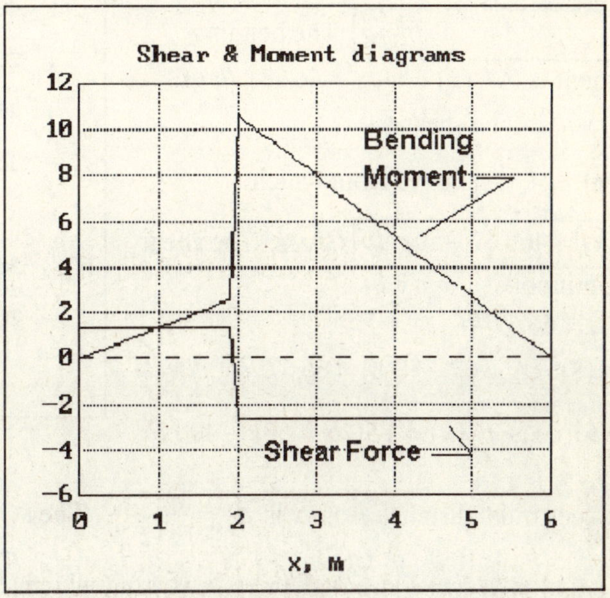

The diagrams are shown. The discontinuity in the moment at $x = 2$ is due to the moment exerted at B by the force at D. This moment acts like a couple, in that its magnitude is independent of the distance x along the beam ABC.

Problem 9.51 Model the ladder rung as a simply supported (pin supported) beam and assume that the 200 lb load exerted by the person's shoe is uniformly distributed. Draw the shear force and bending moment diagrams.

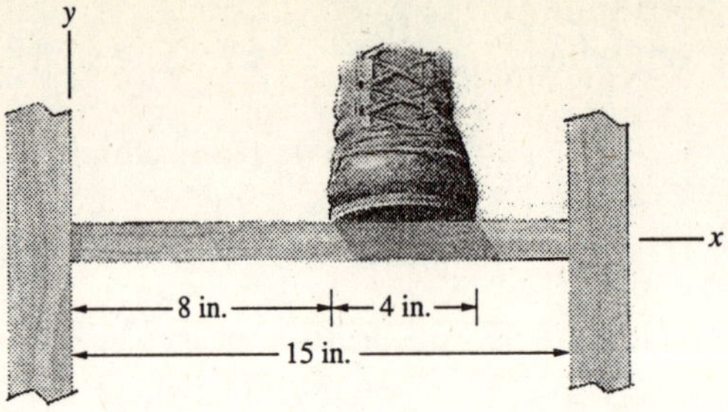

Solution *The structure as a free body.* The load density is 50 *lb/in.* The centroidal distance from the left end is $\bar{d} = 10\ in$. The sum of the moments about the left end is $\sum M_A = -\bar{d}F + 15B = 0$, from which $B = \frac{2000}{15} = 133.33\ lb$. The sum of the forces: $\sum F_y = A_y - 200 + B = 0$, from which $A_y = 66.66\ lb$. *The intervals as free bodies:*. Divide the rung into three intervals: $0 \le x < 8\ (in)$, $8 \le x < 12\ (in)$, and $12 \le x < 15\ (in)$. *Interval 1:* The shear is $\boxed{V_1(x) = A_y = 66.67\ lb}$. The bending moment is $M_1(x) = A_y x = 66.67x\ lb\ in.$. *Interval 2:* The shear is $V_2(x) = A_y - \int_8^x 50dx$, from which $\boxed{V_2(x) = 466.67 - 50x\ lb}$ *Check:* The shear is continuous at $x = 8\ in.$. $V_1(8) = 66.67\ lb$ $= V_2(8) = 466.67 - 50(8) = 66.67\ lb$, *check.* The force is $F_2(x) = \int_8^x 50dx = 50x - 400\ lb$.

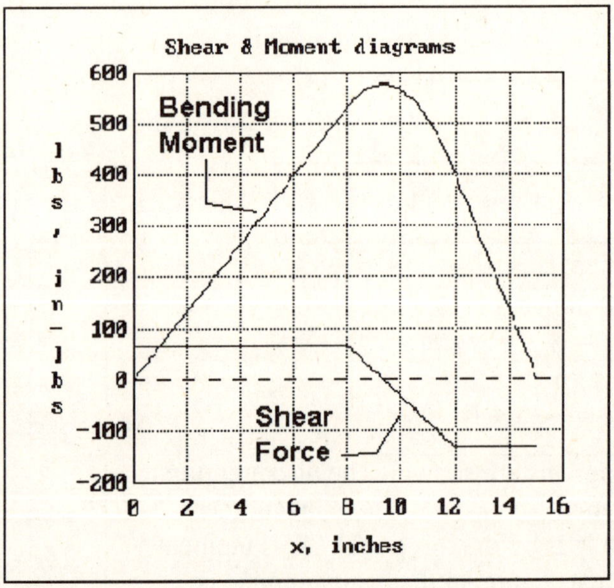

The centroidal distance from x is $\bar{d}_2 = \frac{(x-8)}{2}$. The sum of the moments is $\sum M(x) = M_2(x) - A_y x + F_2(x)\bar{d}_2 = 0$ from which $\boxed{M_2(x) = 466.67x - 25x^2 - 1600\ in\ lb}$. *Check:* The moment is continuous at $x = 8$. $M_1(8) = 66.67(8) = 533.33\ lb\ in.$. $M_2(8) = 466.67(8) - 25(8^2) - 1600 = 533.33\ in\ lb$. *check. Interval 3*: The shear is $\boxed{V_3(x) = A_y - \int_8^{12} 50dx = -133.33\ lb}$. The total load is 200 lb. The centroid distance from x is $\bar{d} = (x-10)$. The moment is $M_3(x) = A_y x - 200(x-10)$, from which $\boxed{M_3(x) = -133.33x + 2000\ in\ lb}$.

Check: The moment at $x = 12$ is continuous. $M_2(12) = 466.67(12) - 25(12^2) - 1600 = 400\ in.\ lb$. $M_3(12) = -133.33(12) + 2000 = 400\ in.\ lb$. *check. Check:* The moment is zero at $x = 15\ in.$. $M_3(15) = -133.33(15) + 2000 = 0$, *check.* The diagrams are shown.

===<>===

Problem 9.52 What is the maximum bending moment in the ladder rung in Problem 9.51 and where does it occur?

Solution From the solution to Problem 9.51, the maximum bending moment occurs in the second interval (under the shoe). The moment is given by $M_2(x) = 466.67x - 25x^2 - 1600$ *in lb*. The maximum occurs at $\frac{dM_2(x)}{dx} = 0 = 466.67 - 50x = 0$, or $\boxed{x = \frac{466.67}{50} = 9.33 \text{ in}}$. The value of the moment: $\boxed{M_{MAX} = 466.7(9.33) - 25(9.33)^2 - 1600 = 577.8 \text{ in lb}}$

===<>===

Problem 9.53 Assume that the surface that the beam rests on exerts a uniformly distributed load on it. Draw the shear force and bending moment diagrams.

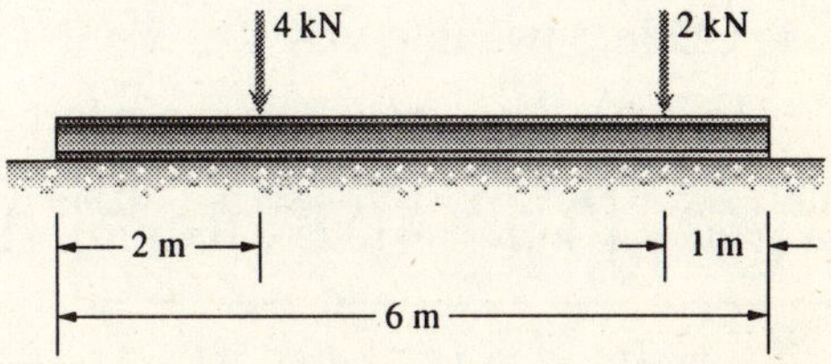

Solution The load density is $w = \frac{6}{6} = 1$ kN/m. *The intervals as free bodies:* Divide the beam into three intervals: $0 \le x < 2$ (m), $2 \le x < 5$ (m), and $5 \le x \le 6$ (m). *Interval 1*: The shear force is $V_1(x) = \int_0^x w dx = x$ kN. The force to the left is $F_1(x) = \int_0^x w dx = x$ kN. The centroid distance from x is $\bar{d}_1 = \frac{x}{2}$. The moment is $\boxed{M_1(x) = F_1(x)\bar{d}_1 = \frac{x^2}{2} \text{ kN m}}$.

Interval 2: The shear force is $\boxed{V_2(x) = V_1(x) - 4 = x - 4 \text{ kN}}$. The moment is $\boxed{M_2(x) = M_1(x) - 4(x-2) = \frac{x^2}{2} - 4x + 8 \text{ kN m}}$.

Interval 3: The shear force is $\boxed{V_3(x) = x - 4 - 2 = x - 6 \text{ kN}}$.

The bending moment is $\boxed{M_3(x) = M_2(x) - 2(x-5) = \frac{x^2}{2} - 6x + 18 \text{ kN}}$.

The shear and moment diagrams are shown.

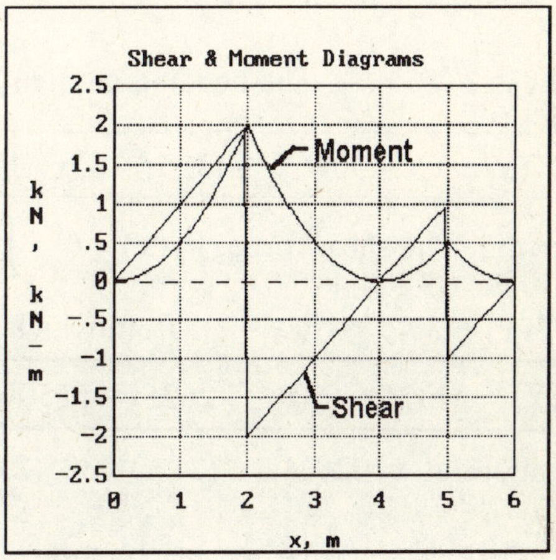

===<>===

==================================<>==================================

Problem 9.54 The homogenous beams AB and CD weight 600 *lb* and 500 *lb*, respectively. Draw the shear force and moment diagrams for beam CD.

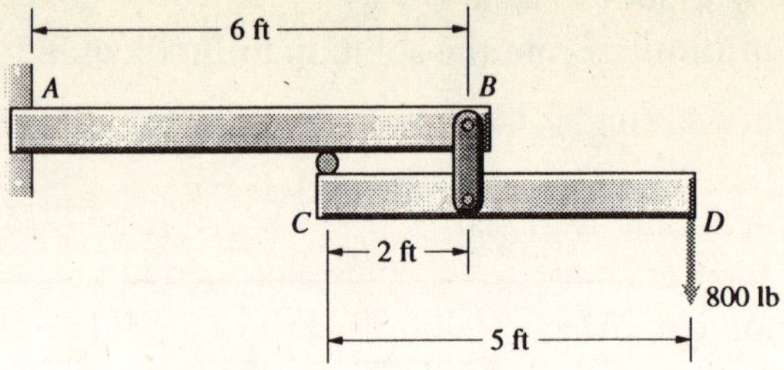

Solution *The element CD as a free body.* The sum of the moments about point C:

$\sum M_C = -500(2.5) + 2B - 5(800) = 0$, from which $B = 2625\ lb$. The sum of the forces:

$\sum F_y = C + B - 800 - 500 = 0$, from which $C = 800 + 500 - 2625 = -1325\ lb$. The weight density $w = \dfrac{500}{5} = 100\ lb/ft$. *The intervals as free bodies:* Divide CD into two intervals: $0 \le x < 2\ (ft)$, and $2 \le x < 5\ (ft)$. *Interval 1*: The shear force

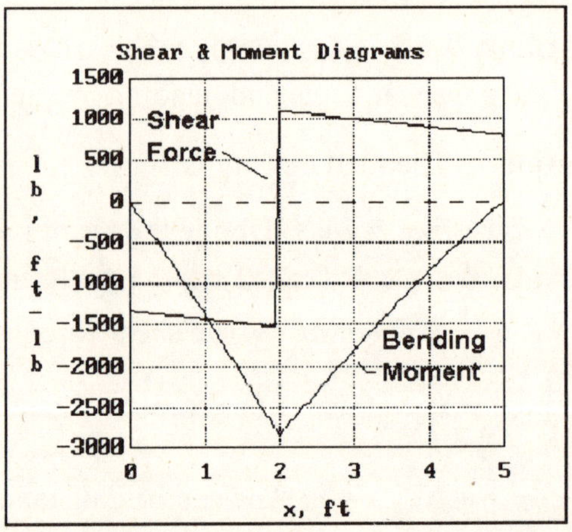

$\boxed{V_1(x) = C - \int_0^x w\,dx = -1325 - 100x\ lb}$. The distributed force is $F_1^{CD} = \int_0^x w dx = 100x$. The centroidal distance from x is $\bar{d}_1^{CD} = \dfrac{x}{2}$. The bending moment:

$\boxed{M_1^{CD}(x) = Cx - 50x^2 = -1325x - 50x^2\ lb\,ft}$.

Interval 2: The shear force: $V_2(x) = C - \int_0^X \rho dx + B = 1300 - 100x\ lb$. The moment

$M_2^{CD}(x) = M_1^{CD} + B(x-2)$, from which

$\boxed{M_2(x) = -1325x - 50x^2 + 2625x - 5250 = 1300x - 50x^2 - 5250\ lb\,ft}$ *Check:* The moment is continuous at $x = 2$. $M_1^{CD}(2) = -1325(2) - 50(2^2) = -2850\ lb\,ft$.

$M_2^{CD} = 1300(2) - 50(2^2) - 5250 = -2850\ lb\,ft$. *check. Check:* The moment is zero at $x = 5$.

$M_2^{CD}(5) = 1300(5) - 50(5^2) - 5250 = 0$. *check.*

==================================<>==================================

Problem 9.55

Problem 9.55 Draw the shear force and bending moment diagrams for beam AB in Problem 9.54.

Solution Use the solution to Problem 9.54. The reactions at C and B are $C = 1325\ lb$ and $B = -2625\ lb$. *The member AB as a free body*: The sum of the moments about A is $\sum M_A = M_A + 4C + 6B - 600(3) = 0$, from which $M_A = 12250\ ft\ lb$. $\sum F_Y = A_y + C + B - 600 = 0$, from which $A_y = 1900\ lb$ The weight density is $w = \frac{600}{6} = 100\ ft/lb$. *The intervals as free bodies:* Divide the beam into the intervals: $0 \le x < 4\ (ft)$ and $4 \le x < 6\ (ft)$. *Interval 1:* The shear force: $V_1^{AB}(x) = A_y - \int_0^x w\,dx = 1900 - 100x\ lb$. The distributed force is $F_1^{AB} = \int_0^x w dx = 100x$. The centroid distance from x is $\bar{d}_1^{AB} = \frac{x}{2}$. The moment is

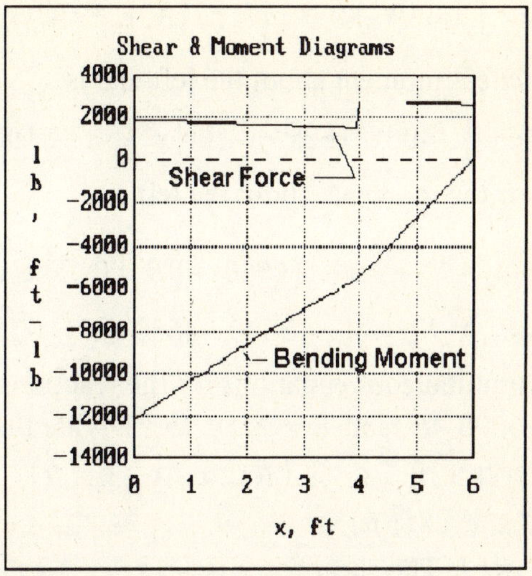

$M_1^{AB}(x) = -M_A + A_y x - F_1^{AB}(x)\bar{d}_1^{AB}$, from which $\boxed{M_1^{AB}(x) = -12250 + 1900x - 50x^2\ ft\ lb}$. *Interval 2:* The shear force is $\boxed{V_2^{AB}(x) = V_1^{AB}(x) + C = 3225 - 100x\ lb}$. The moment is $M_2^{AB}(x) = M_1^{AB}(x) + C(x-4)$, $\boxed{M_2^{AB}(x) = -17550 + 3225x - 50x^2\ ft\ lb}$. *Check:* The moment is continuous at $x = 4$. $M_1^{AB}(4) = -12250 + 1900(4) - 50(4^2) = -5450\ lb\ ft$.

$M_2^{AB}(4) = -17550 + 3225(4) - 50(4^2) = -5450\ lb\ ft$. *check. Check:* The moment is zero at $x = 6$.

$M_2^{AB}(6) = -17550 + 3225(6) - 50(6^2) = 0$. *check.*

Problem 9.56

Problem 9.56 The load $F = 4650\ lb$. Draw the shear force and moment diagrams.

Solution *The structure as a free body:* Denote the left support reaction by A and the right support reaction by B. The equivalent force exerted by the load is $F_L = -\int_0^3 (400x^2)dx = -\left(\frac{400}{3}\right)\left[x^3\right]_0^3 = -3600\ lb$. The equivalent moment about the left end is $M_L = -\int_0^3 (400x^2)x\,dx = -100\left[x^4\right]_0^3 = -8100\,ft\ lb$.. The sum of the moments about the left end:

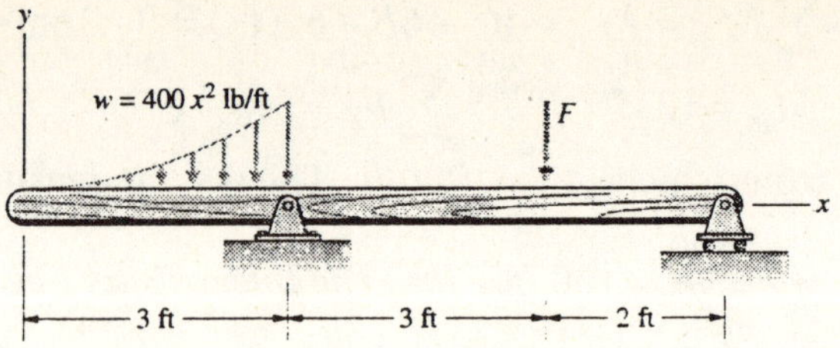

$\sum M = -M + 3A - 6(4650) + 8B = 0$. The sum of the forces: $\sum F_y = A_y - 3600 - 4650 + B = 0$. Solve these two simultaneous equations for the reactions: $A = 6000\ \ lb$, and $B = 2250\ \ lb$. Divide the beam into 3 intervals: $0 \le x < 3\ \ (ft)$, $3 \le x < 6\ \ (ft)$, and $6 \le x < 8\ \ (ft)$.

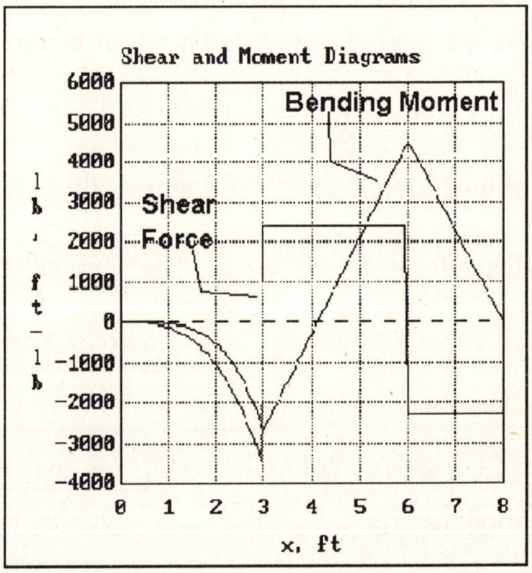

Interval 1: The shear force is $\boxed{V_1(x) = -\int_0^x 400x^2 dx = -\left(\frac{400}{3}\right)x^3\ \ lb}$. The distributed force is $F_1(x) = \int_0^x w dx = \left(\frac{400}{3}\right)x^3\ lb$. The centroid distance from x is $\bar{d}_1 = x - \frac{\int_0^x wxdx}{\int_0^x wdx} = x - 0.75x = 0.25x$

The moment is $\boxed{M_1(x) = -F_1\bar{d}_1 = -\frac{100}{3}x^4\ lb\ ft}$.

Interval 2: The shear force is $\boxed{V_2(x) = -\left(\frac{400}{3}\right)(3^3) + A = 2400\ lb}$. The total distributed force is $F_1(3) = -\frac{400}{3}(3^3) = -3600\ lb$. The centroid distance from x is $\bar{d}_2 = x - (0.75)(3) = x - 2.25$. The bending moment is $\boxed{M_2(x) = F_1(3)\bar{d}_2 + A(x-3) = 2400x - 9900\ lb\ ft}$.

Interval 3: The shear force is $\boxed{V_3(x) = V_2(x) - F = 2400 - 4650 = -2250\ \ lb}$. The bending moment is $\boxed{M_3(x) = M_2(x) - F(x-6) = -2250x + 18000\,ft\ lb}$. *Check:* The moments are continuous at $x = 3$ and $x = 6$. $M_1(3) = -\frac{100}{3}(3^4) = -2700\ lb\,ft$. $M_2(3) = 2400(3) - 9900 = -2700\ lb\,ft$ *.check.* $M_2(6) = 2400(6) - 9900 = 4500\ lb\,ft$. $M_3(6) = -2250(6) - 18000 = 4500\ lb\,ft$. *check. Check:* The moment is zero at $x = 8$. $M_3(8) = -2250(8) + 18000 = 0$ *check*

==================================<>==================================

Problem 9.57 If the load $F = 2150\ lb$ in Problem 9.56, what are the maximum and minimum shear forces and bending moments and at what values of x do they appear?

Solution Use the solutions and the solution procedures used in Problem 9.56. *The structure as a free body:*. From the solution to Problem 9.56, the equivalent force due to the distributed load is $F_L = -3600\ lb$. The moment about the left end due to the distributed load is $M = -8100\ ft\ lb$. Denote the left support reaction by A, and the right support reaction by B. The sum of the moments about the left end is $\sum M = -8100 + 3A - 6(2150) + 8B = 0$. The sum of the forces $\sum F_Y = -3600 + A - 2150 + B = 0$. Solve these simultaneous equations to obtain $A = 5000\ lb$, and $B = 750\ lb$. Divide the beam into 3 intervals: $0 \le x < 3\ (ft)$, $3 \le x < 6\ (ft)$, and $6 \le x < 8\ (ft)$.

Interval 1: The shear force is $\boxed{V_1(x) = -\left(\frac{400}{3}\right)x^3\ lb}$. The moment is $\boxed{M_1(x) = -\left(\frac{100}{3}\right)x^4}$. *Interval 2:* The shear force is $\boxed{V_2(x) = -3600 + A = 1400\ lb}$ The moment is $M_2(x) = F_1(3)\bar{d}_2 + A(x-3)$, where from Problem 9.56 $F_1(3) = 3600\ lb$ and $\bar{d}_2 = x - 2.25$. Substitute and reduce: $\boxed{M_2(x) = 1400x - 6900\ ft\ lb}$.

Interval 3: The shear force is $\boxed{V_3(x) = V_2(x) - F = -750\ lb}$. The moment is $\boxed{M_3(x) = M_2(x) - F(x-6) = -750x + 6000\ lb\ ft}$.

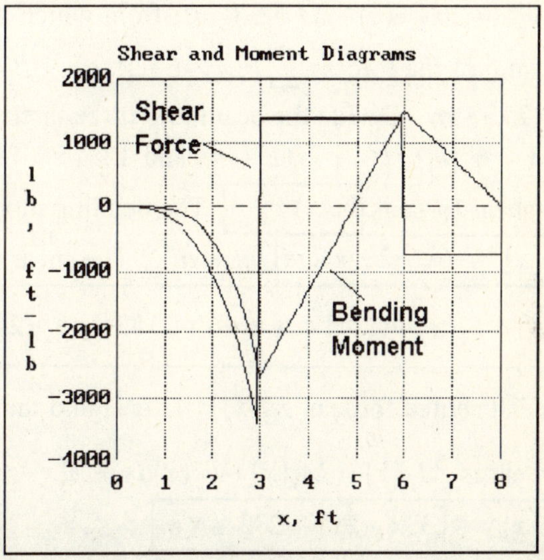

From inspection of the shear force and moment diagrams, the minimum value of the shear force occurs at $x = 3\ ft$, $\boxed{V_{Min} = -3600\ lb}$, and the maximum occurs in the second interval, $3 \le x < 6\ (ft)$, $\boxed{V_{MAX} = V_2(x) = 1400\ lb}$. The minimum value of the bending moment occurs at $x = 3\ ft$, $\boxed{M_{Min} = M_2(3) = 1400(3) - 6900 = -2700\ ft\ lb}$. The maximum value of the moment occur at $x = 6\ ft$, $\boxed{M_{Max} = 1400(6) - 6900 = 1500\ ft\ lb}$

==================================<>==================================

Problem 9.58 Draw the shear force and bending moment diagrams.

Solution *The structure as a free body:* Denote the reaction at the left support by A and the reaction at the right support by B. The equivalent force due to the load distribution is the area: $F = 4(6) + \left(\frac{1}{2}\right)4(6) = 36\ kN$. The moment about the left end of the distribution is the sum of products of the force and the centroidal distances

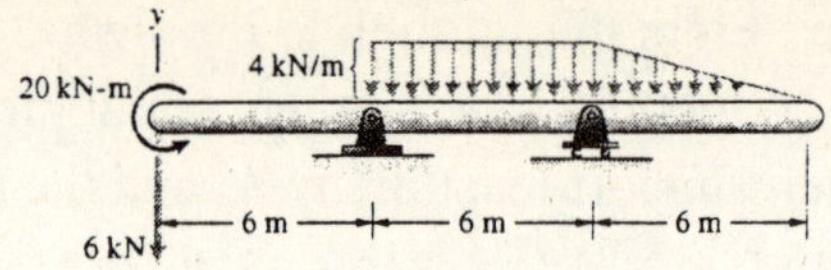

$M = 4(6)(3) + \left(\frac{1}{2}\right)4(6)(6 + \frac{6}{3}) = 168\ kN\,m$ The sum of the moments about the left support is

$\sum M_A = 20 + 6(6) - M + 6B = 0$, from which $B = 18.67\ kN$.

The sum of the forces: $\sum F_Y = -6 + A - F + B = 0$, from which $A = 23.33\ kN$. Divide the beam into three intervals: $0 \le x < 6\ (m)$, $6 \le x < 12\ (m)$, and $12 \le x < 18\ (m)$. *Interval 1:* The shear force is $\boxed{V_1(x) = -6}$ The bending moment is $\boxed{M_1(x) = -6x - 20\ kN\,m}$. *Interval 2:* The shear force is

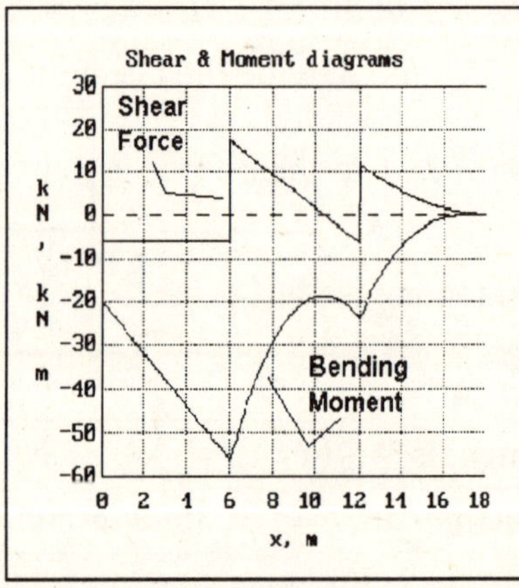

$$\boxed{V_2(x) = V_1(x) + A - \int_6^x 4dx = -6 + 23.33 - 4x + 24 = 41.33 - 4x\ kN}$$

The distributed force is $F_2(x) = 4x\ kN$, and the centroid distance from x is $\bar{d}_2 = \frac{x-6}{2}$. The bending moment is $M_2(x) = M_1(x) - F_2(x)\bar{d}_2 + A(x-6)$. Substitute and reduce:

$$\boxed{M_2(x) = 41.33x - 2x^2 - 232\ kN\,m}$$

Interval 3: .The shear force is $V_3(x) = V_2(12) + B - \int_{12}^{x}(-\frac{2}{3}x + 12)\,dx$. $\boxed{V_3(x) = \left(\frac{1}{3}\right)x^2 - 12x + 108\ kN}$.

The distributed force in the 3rd interval is $F_3(x) = \int_{12}^{x} w_3 dx$. The centroid distance from x is

$\bar{d}_3 = x - \frac{\int_{12}^{x} w_3 x dx}{\int_{12}^{x} w_3 dx}$, from which $F_3(x)\bar{d}_3 = xF_3(x) - \int_{12}^{x} w_3 x dx$ is the moment. The total moment is

$M_3(x) = V_2(12)(x-12) + F_3(x)\bar{d}_3 + B(x-12)$. The load density: $w_3(x) = -\frac{2}{3}x + 12\ kN$. Substitute and reduce: $\boxed{M_3(x) = \left(\frac{1}{9}\right)x^3 - 6x^2 + 108x - 648\ kN\,m}$. *Check:* The moments are continuous at $x = 6$ and $x = 12$. $M_1(6) = -6(6) - 20 = -56\ kN\,m$. $M_2(6) = 41.33(6) - 2(6^2) - 232 = -56\ kN\,m$ *check.*

$M_2(12) = 41.33(12) - 2(12^2) - 232 = -24\ kN\,m$, $M_3(12) = \left(\frac{1}{9}\right)(12^3) - 6(12^2) + 108(12) - 648 = -24$.

check. Check: The moment vanishes at $x = 18$. $M_3(18) = \left(\frac{1}{9}\right)(18^3) - 6(18^2) + 108(18) - 648 = 0$. *check.*

The shear force and moment diagram are shown for the complete beam.

Problem 9.59 In Problem 9.58, what are the maximum and minimum shear forces and bending moments and where do they occur?

Solution Use the solutions to Problem 9.58. From an inspection of the graph in Problem 9.58, the minimum shear force occurs at $x = 12\ m$, $\boxed{V_{MIN} = V_2(12) = 41.33 - 4(12) = -6.67\ kN}$. The maximum shear force occurs at $x = 6\ m$, $\boxed{V_{MAx} = V_2(6) = 41.33 - 4(6) = 17.33\ kN}$. The minimum bending moment occurs at $x = 6\ m$, $\boxed{M_{MIN} = M_1(6) = M_2(6) = -6(6) - 20 = -56\ kN\, m}$. The maximum moment is zero at the right end of the beam, $\boxed{x = 18\ m}$.

Problem 9.60 Draw the shear force diagram for the airplane wing in Problem 9.8.(Neglect the weight of the wing.)

Solution Use the solutions of Problem 9.8. The reaction at the root is $F_R = -1178.1\ N$, when the weight of the wing is neglected. From Problem 9.8, the load distribution is

$w = -300(1 - 0.04x^2)^{\frac{1}{2}} = 60\left(25 - x^2\right)^{\frac{1}{2}}$. The shear force is

$$V(x) = F_R - \int_0^x w dx = F_R + 60\int_0^x \left(25 - x^2\right)^{\frac{1}{2}} dx.$$

$$V(x) = F_R + 60\left[\frac{x\sqrt{25 - x^2}}{2} + \frac{25}{2}\sin^{-1}\left(\frac{x}{5}\right)\right]_0^x,$$

$$V(x) = -1178.1 + 30\left(x\sqrt{25 - x^2} + 25\sin^{-1}\left(\frac{x}{5}\right)\right)$$

The shear distribution as a function of the distance from the root is shown.

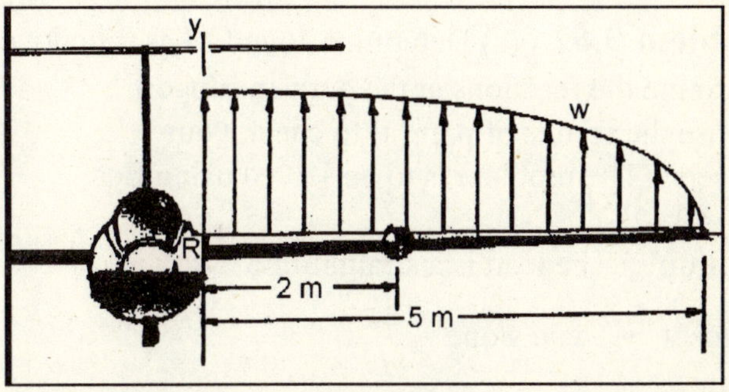

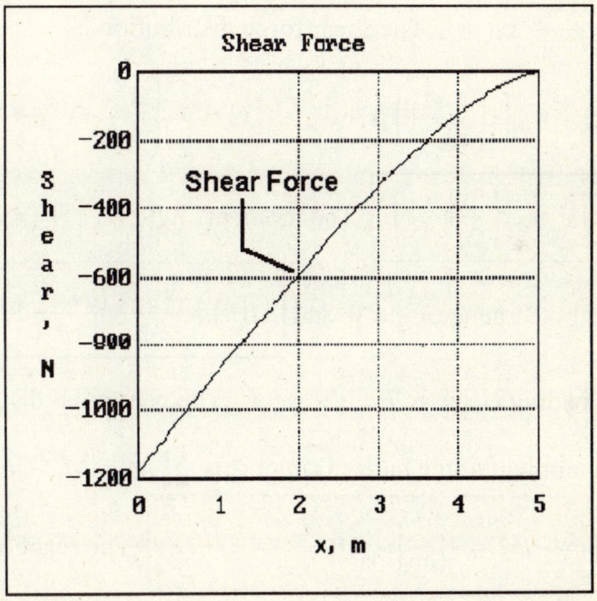

Problem 9.61 Determine V and M as functions of x. (The problem is to be solved using Eqs (9.7)-(9.12).)

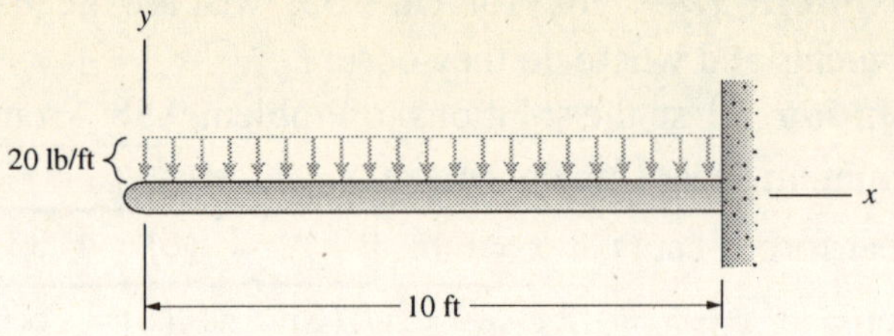

Solution The shear distribution is

$V(x) = -\int 20dx + C_1 = -20x + C_1$ The shear is zero at $x = 0$, hence $C_1 = 0$, and

$\boxed{V(x) = -20x\ lb}$. The moment is $M(x) = \int V(x)dx + C_2 = -10x^2 + C_2$. The moment is zero at $x = 0$, hence $C_2 = 0$, and $\boxed{M(x) = -10x^2\ ft\ lb}$

Problem 9.62 (a) Determine V and M as functions of x. (b) Use the equilibrium conditions to determine the reactions at the built-in support and use the results of part (a) to check your answers. (The problem is to be solved using Eqs (9.7)-(9.12).)

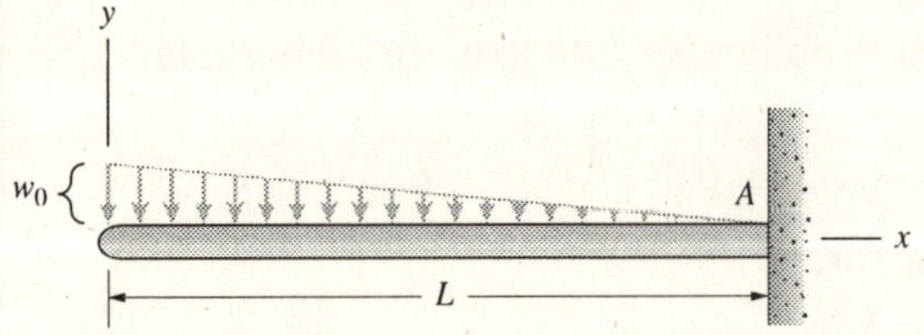

Solution The load is a straight line with intercept w_o and slope $-\frac{w_o}{L}$,

$w = -\frac{w_o}{L}x + w_o$. The shear force distribution is

$V(x) = w_o \int \left(\frac{x}{L} - 1\right)dx + C_1$ $V(x) = w_o\left(\frac{x^2}{2L} - x\right) + C_1$. At $x = 0$, $V(0) = 0$, from which $C_1 = 0$, and

$\boxed{V(x) = w_o x\left(\frac{x}{2L} - 1\right)}$ The moment: $M(x) = \int V(x)dx + C_2 = \frac{w_o x^2}{2}\left(\frac{x}{3L} - 1\right) + C_2$. At $x = 0$, $M(0) = 0$, hence $C_2 = 0$ and $\boxed{M(x) = \frac{w_o x^2}{2}\left(\frac{x}{3L} - 1\right)}$ (b) The equilibrium conditions: The area under the loading curve is $F = \left(\frac{1}{2}\right)w_o L$. The centroidal distance is $\bar{d} = L - \left(\frac{L}{3}\right) = \frac{2L}{3}$ from the right end. The moment at the built in support is $\sum M = M_A + F\bar{d} = 0$, from which

$\boxed{M_A = -F\bar{d} = -\left(\frac{1}{2}\right)w_o L\left(\frac{2L}{3}\right) = -\frac{w_o L^2}{3}}$. The reaction at the right end is $\sum F_y = A_y - F = 0$, from which $\boxed{A_y = F = \left(\frac{1}{2}\right)w_o L}$. *Check:* Use the shear force to determine the reaction at the right end:

$V(x) \xrightarrow[X=L]{} w_o L\left(\frac{L}{2L} - 1\right) = -\frac{w_o L}{2}$. The reaction is equal and opposite to the shear, hence

$A = \left(\frac{1}{2}\right)w_o L$. *check.* The bending moment distribution is $M(x) \xrightarrow[X=L]{} \frac{w_o L^2}{2}\left(\frac{L}{3L} - 1\right) = -\frac{w_o L^2}{3}$.

check. $\boxed{\sum F_x = A_x = 0}$

Problem 9.63 Determine V and M as functions of x. (The problem is to be solved using Eqs (9.7)-(9.12).)

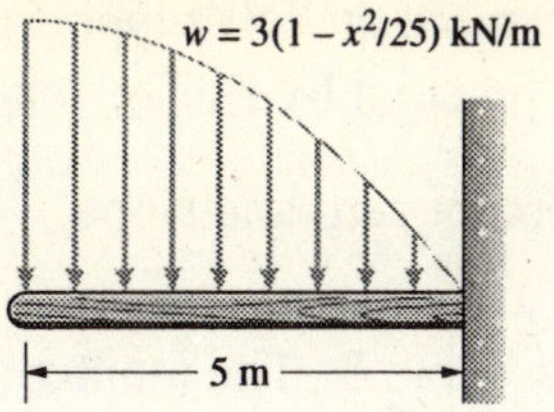

Solution The shear force is

$$V(x) = -\int 3\left(1 - \frac{x^2}{25}\right)dx + C_1 = -3\left(x - \frac{x^3}{75}\right) + C_1.$$

At $x = 0$, $V(0) = 0$, hence $C_1 = 0$, and

$$\boxed{V(x) = -3\left(x - \frac{x^3}{75}\right)\ kN}$$

The moment is

$M(x) = \int V(x)dx + C_2 = 3\left(\frac{x^2}{2} - \frac{x^4}{300}\right) + C_2$. At $x = 0$, $M(0) = 0$, hence $C_2 = 0$, and

$$\boxed{M(x) = -3\left(\frac{x^2}{2} - \frac{x^4}{300}\right)\ kN\ m}$$

Problem 9.64 Determine V and M as functions of x for the beam in Problem 9.41. (The problem is to be solved using Eqs (9.7)-(9.12).)

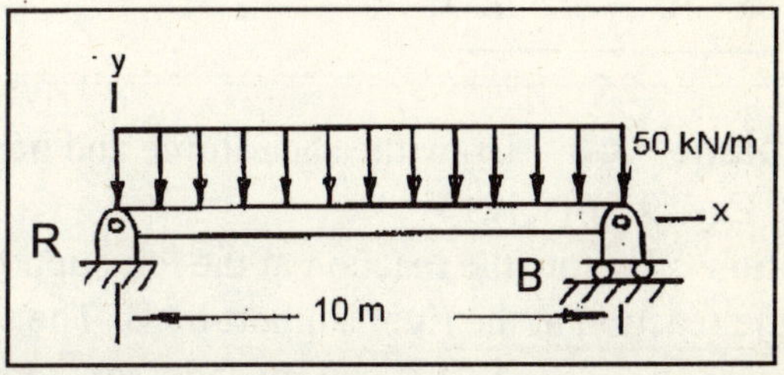

Solution This is a repeat of Part (a) of Problem 9.41. The sum of the moments about the left end is

$\sum M = -\int_0^{10} 50x\,dx + 10B = 0$, from which

$B = 250\ kN$. The sum of the forces:

$\sum F_y = R_y - \int_0^{10} 50dx + B = 0$, from which

$R_y = 500 - 250 = 250\ kN$

(a) By definition, for a simple beam the shear is the integral of the force density to the left:

$V(x) = -\int 50\,dx + C = -50x + C$. At $x = 0$, $V(0) = 250\ kN$, hence the constant of integration $C = 250\ kN$, and

$\boxed{V(x) = 250 - 50x\ kN}$, which is straight line with intercept $250\ kN$ and slope $-50\ kN/m$.

By definition, for a simple beam the moment is the integral of the shear to the left,

$M(x) = \int V(x)dx + C = 250x - 25x^2 + C$. The moment is zero at $x = 0$, hence the constant of integration $C = 0$. Thus

$\boxed{M(x) = 250x - 25x^2\ kN\ m}$.

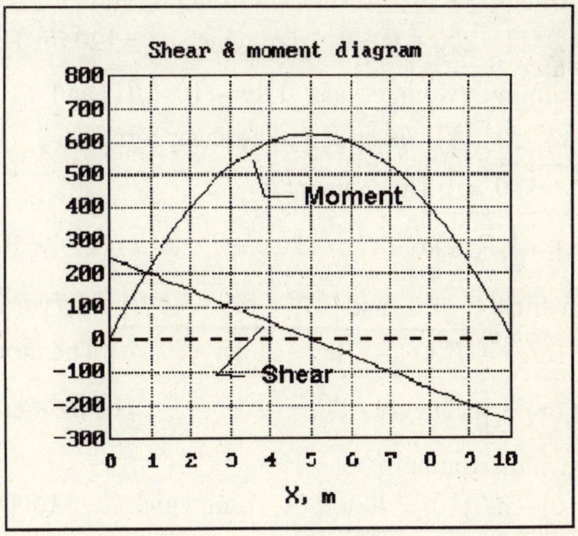

Problem 9.65 Determine V and M as functions of x. (The problem is to be solved using Eqs (9.7)-(9.12).)

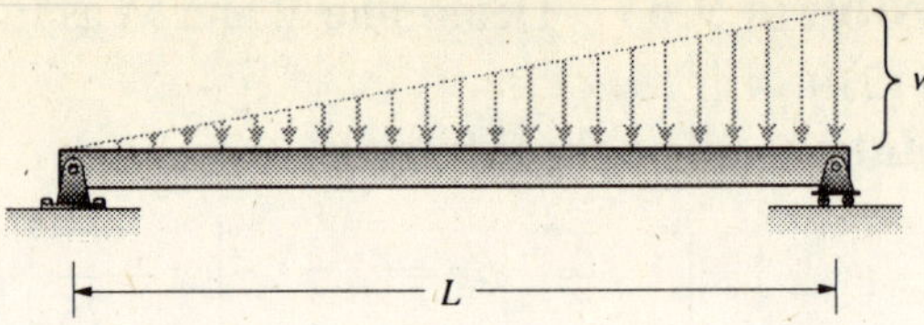

Solution Denote the left support reaction by A and the right support reaction by B. The load curve is a straight line with intercept zero and slope $\frac{w_o}{L}$. The area under the curve is $F = \left(\frac{1}{2}\right)w_o L$. The moment about the left support is $\sum M_A = -\int_0^L \left(\frac{w_o}{L}\right)x^2 dx + BL = 0$, from which $B = \left(\frac{1}{L}\right)\left(\frac{w_o L^2}{3}\right) = \frac{w_o L}{3}$. The sum of the forces $\sum F_y = A - \left(\frac{1}{2}\right)w_o L + \frac{w_o L}{3} = 0$, from which $A = \frac{w_o L}{6}$. The shear force is $V(x) = \frac{w_o L}{6} - \int \left(\frac{w_o}{L}\right)x\,dx = \frac{w_o L}{6} - \frac{w_o x^2}{2L}$. $\boxed{= \frac{w_o}{2}\left(\frac{L}{3} - \frac{x^2}{L}\right)}$ The moment is $M(x) = \int V(x)\,dx + C_1 = \frac{w_o x}{6}\left(L - \frac{x^2}{L}\right) + C_1$. At $x = 0$, $M(0) = 0$, hence $C_1 = 0$, and

$$\boxed{M(x) = \frac{w_o x}{6}\left(L - \frac{x^2}{L}\right)}$$

Problem 9.66 Draw the shear force and bending moment diagrams. (The problem is to be solved using Eqs (9.7)-(9.12).)

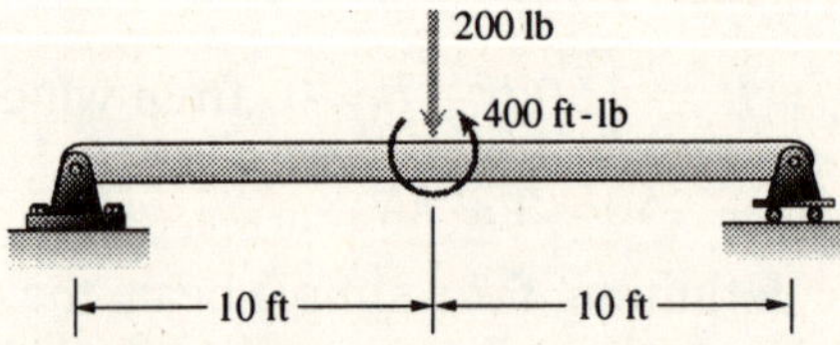

Solution Denote the reaction at the left support by A, and the reaction at the right support by B. The sum of the moments about A: $\sum M_A = 400 - 200(10) + 20B = 0$, from which $B = 80\ lb$. The sum of the forces: $\sum F_y = A - 200 + B = 0$, from which $A = 120\ lb$. Divide the beam into two intervals: $0 \le x < 10\ (ft)$, and $10 \le x < 20\ (ft)$. *Interval 1:* The shear force is $\boxed{V_1(x) = 120\ lb}$. The moment is $M_1(x) = \int V_1(x)dx + C_1 = 120x + C_1$. At $x = 0$, $M_1(0) = 0$, from which $C_1 = 0$, and $\boxed{M_1 = 120x\ \ lb\ ft}$. *Interval 2:* The shear force is $\boxed{V_2(x) = V_1(x) - 200 = -80\ \ lb}$. The moment is $M_1(x) = \int V_2(x)dx + C_2 = -80x + C_2$. The moment has a known discontinuity, $x = 10$, $M_1(10) - M_2(10) = 400\ ft\ lb$, from which $C_2 = 1600$, and $\boxed{M_2(x) = -80x + 1600\ ft\ lb.}$ *Check:* The moment is zero at $x = 20\ ft$. $M_2(20) = -18(20) + 1600 = 0$. *check.*

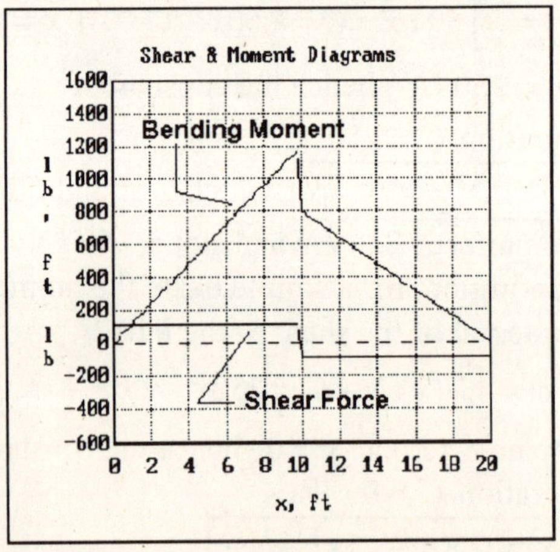

Problem 9.67 Draw the shear force and moment diagrams for the beam in Problem 9.47. (The problem is to be solved using Eqs (9.7)-(9.12).)

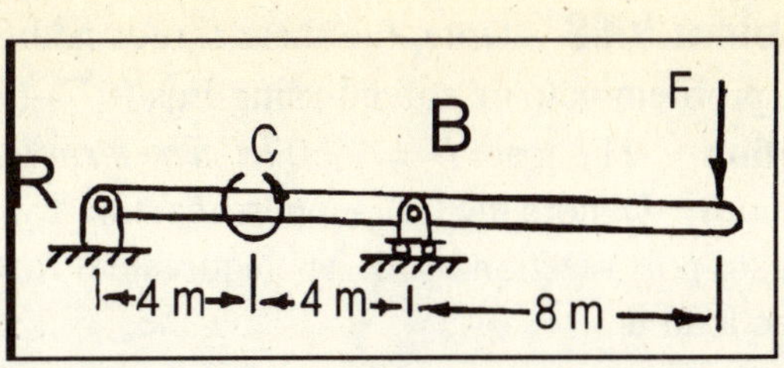

Solution The loads $F = 200\ N$ and $C = 800\ N m$. *The reactions at the supports:* Denote the reaction at the left support by R and the reaction at the right support by B. The sum of the moments about the left end: $\sum M = C + 8B - 16F = 0$, from which

$B = \left(\frac{1}{8}\right)(-C + 16F) = \left(\frac{1}{8}\right)(-800 + 16(200)) = 300\ N$. The sum of the forces: $\sum F_y = R_y + B - F = 0$,

from which $R_y = -100\ N$. Divide the interval into three parts: $0 \le x \le 4$, $4 < x \le 8$, and $8 < x \le 16$.

Interval I. The shear is $\boxed{V_1(x) = R_y = -100\ N}$.

The moment is

$M_1(x) = \int V_1(x)dx + C_1 = R_y x + C_1$.

At $x = 0$, $M_1(0) = 0$, from which $C_1 = 0$, and $\boxed{M_1(x) = -100x}$. *Interval 2.* The shear is $\boxed{V_2(x) = V_1(x) = R_y = -100\ N}$. The moment is $M_2(x) = \int V_2(x)dx + C_2$. There is a known discontinuity in the moment at $x = 4\ m$, $M_1(4) - M_2(4) = 800\ N\ m$, from which $C_2 = -800$, and $\boxed{M_2(x) = -100x - 800\ N\ m}$. *Interval 3:* The shear is $\boxed{V_3(x) = V_2(x) + B = -100 + 300 = 200\ N}$. The moment is

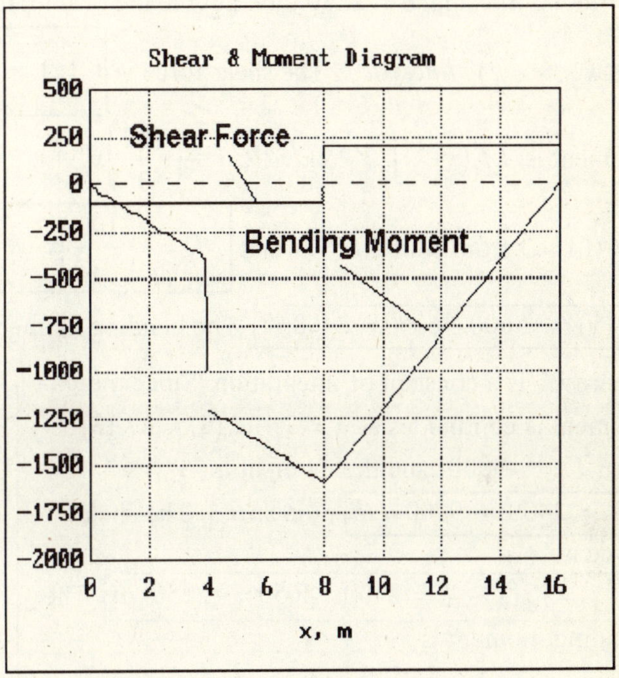

$M_3(x) = \int_0^x V_3(x)dx + C_3 = 200x + C_3$. The moment is continuous at $x = 8$, $M_2(8) = M_3(8)$ from which $C_3 = -3200$, and

$\boxed{M_3(x) = 200x - 3200\ N m}$

Problem 9.68 Draw the shear force and bending moment diagrams for the beam in Problem 9.56. (The problem is to be solved using Eqs (9.7)-(9.12).)

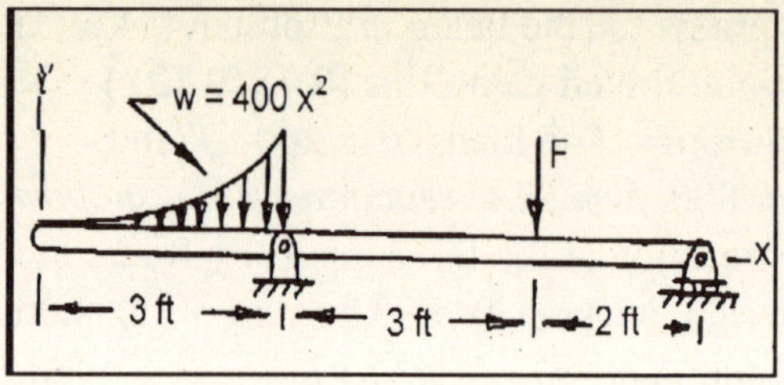

Solution The load $F = 4650\ lb$. *The structure as a free body:* Denote the left support reaction by A and the right support reaction by B. The equivalent force exerted by the load is

$F_L = -\int_0^3 (400x^2)dx = -\left(\frac{400}{3}\right)\left[x^3\right]_0^3 = -3600\ lb$. The equivalent moment about the left end is $M_L = -\int_0^3 (400x^2)x\,dx = -100\left[x^4\right]_0^3 = -8100\ lb\,ft$.. The sum of the moments about the left end: $\sum M = -M + 3A - 6(4650) + 8B = 0$. The sum of the forces: $\sum F_y = A_y - 3600 - 4650 + B = 0$. Solve these two simultaneous equations for the reactions: $A = 6000\ lb$, and $B = 2250\ lb$. Divide the beam into 3 intervals: $0 \le x < 3\ (ft)$, $3 \le x < 6\ (ft)$, and $6 \le x < 8\ (ft)$. *Interval 1*: The shear force is $\boxed{V_1(x) = -\int_0^X 400x^2 dx = -\left(\frac{400}{3}\right)x^3\ lb}$. The bending moment is $M_1(x) = \int_0^x V_1(x)dx + C = -\left(\frac{100}{3}\right)x^4 + C$, where C is a constant of integration. Note $M_1(0) = 0$ from which $C = 0$ and $\boxed{M_1 = -\left(\frac{100}{3}\right)x^4\ ft\ lb}$. *Interval 2:* The shear force is $\boxed{V_2(x) = -3600 + 6000 = 2400\ lb}$. The bending moment is $M_2(x) = \int V_2(x)dx + C_1 = 2400x + C_1$, where C_1 is a constant of integration. Since the moment is continuous at $x = 3$, $M_1(3) = M_2(3)$, then $C_1 = -9900$, and the moment is $\boxed{M_2 = 2400x - 9900\ ft\ lb}$. *Interval 3*: The shear force is $\boxed{V_3(x) = V_2(x) - F = 2400 - 4650 = -2250\ lb}$. The bending moment is

$M_3(x) = \int V_3(x)dx + C_2 = -2250x + C_2$ .Since the moment is continuous at $x = 6$, $M_2(6) = M_3(6)$, $C_2 = 18000$, and the moment is $\boxed{M_3(x) = -2250x + 18000\ lb\,ft}$

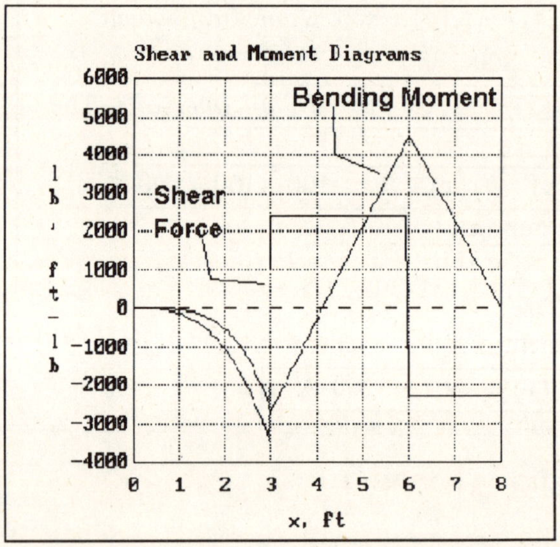

Problem 9.69 Draw the shear force and moment diagrams for the beam in Problem 9.58.

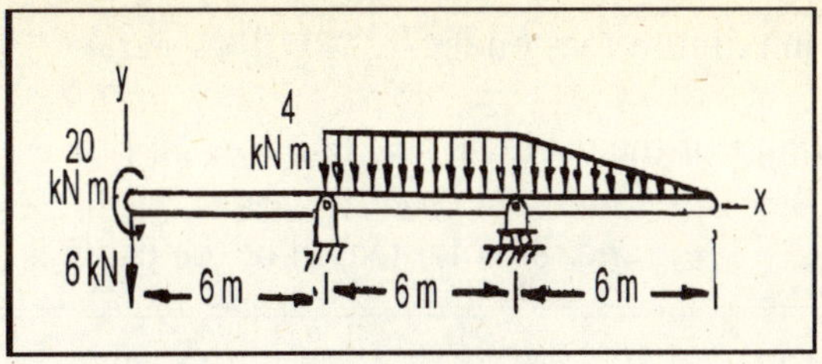

Solution Denote the reaction at the left support by A and the reaction at the right support by B. Use the results of the solution to Problem 9.58. From these results: the equivalent force due to the load distribution is $F = 36\ kN$ The moment about the left end of the distribution is $M = 168\ ft\ lb$ The reaction at the right support is $B = 18.67\ \ kN$. The reaction at the left support is $A = 23.33\ kN$. Divide the beam into three intervals: $0 \le x < 6\ (ft)$, $6 \le x < 12\ (ft)$, and $12 \le x < 18\ (ft)$. *Interval 1:* The shear force is $\boxed{V_1(x) = -6}$ The bending moment is $M_1(x) = \int V_1(x)dx + C$. At $x = 0$, $M_1(0) = -20\ kN\ m$, hence $\boxed{M_1(x) = -6x - 20\ \ kN\ m}$. *Interval 2:* The shear force is $\boxed{V_2(x) = V_1(x) + A - \int_6^x w dx = 41.33 - 4x\ kN}$

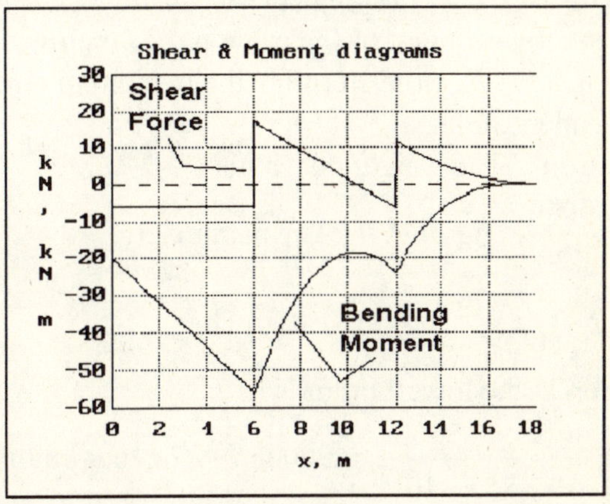

The bending moment is $M_2(x) = \int V_2(x)dx + C$. The moment is continuous at $x = 6$, hence $C = -232$, and $\boxed{M_2(x) = 41.33x - 2x^2 - 232\ \ kN\ m}$ *Interval 3:* The shear force is $V_3(x) = V_2(x) - \int_{12}^{x}\left(-\frac{2}{3}x + 12\right) + B$, from which $\boxed{V_3(x) = \left(\frac{1}{3}\right)x^2 - 12x + 108\ kN}$. The moment is $M_3 = \int V_3(x)dx + C$. The moment is continuous at $x = 12$, hence $C = -648$, and

$$\boxed{M_3(x) = \left(\frac{1}{9}\right)x^3 - 6x^2 + 108x - 648\ kN\ m}$$

The shear force and moment diagram are shown for the complete beam.

Problem 9.70 The cable supports a uniformly distributed load $w = 1\ \ kN/m$. (a) What is the maximum tension? (b) What is the length of the cable? *Strategy:* You know the coordinates of the attachment points of the cable relative to its lowest point, so you can use Eq. (9.15) to determine the coefficient a, then use $a = \frac{w}{T_o}$ to determine the tension at the lowest point.

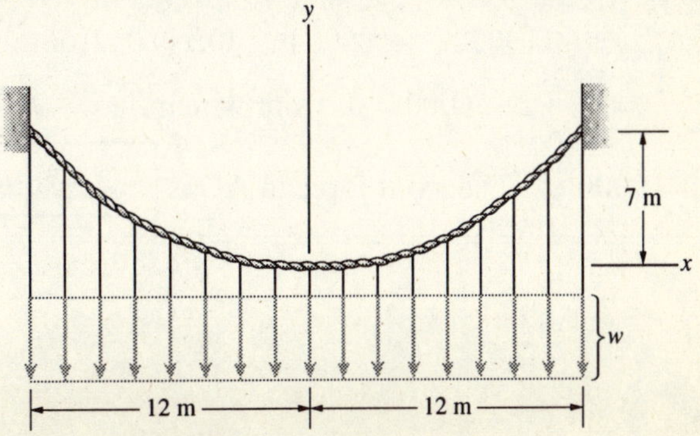

Solution (a) The equation of the parabola is $y = \frac{1}{2}ax^2$, from which $a = 2\left(\frac{7}{12^2}\right) = 0.09722$.

The tension at the lowest point is $T_o = \frac{w}{a} = \frac{1}{0.09722} = 10.286\ \ kN$.

Problem 9.70 continued on next page

continuation of Problem 9.70...........

The maximum tension is $T_{MAX} = T_o\sqrt{1+a^2x^2}$, which for $x = 12$, $\boxed{T_{MAX} = 15.8\ kN}$ The length of the right half of the cable is $s = (1/2)\left(x\sqrt{1+a^2x^2} + \left(\frac{1}{a}\right)\ln\left(ax + \sqrt{1+a^2x^2}\right)\right)$, from which, for $x = 12$, $s = 14.33\ m$, and the total length of the cable is $\boxed{L = 2s = 28.66\ m}$

Problem 9.71 Engineers use a cable to suspend a pipeline above a river. The towers are 120 *ft* apart and the lowest point of the cable is 10 *ft* below the tops of the towers. The total weight of the suspended pipe is 6000 *lb*. (a) Determine the maximum tension in the cable. (b) Determine the length of the suspended cable.

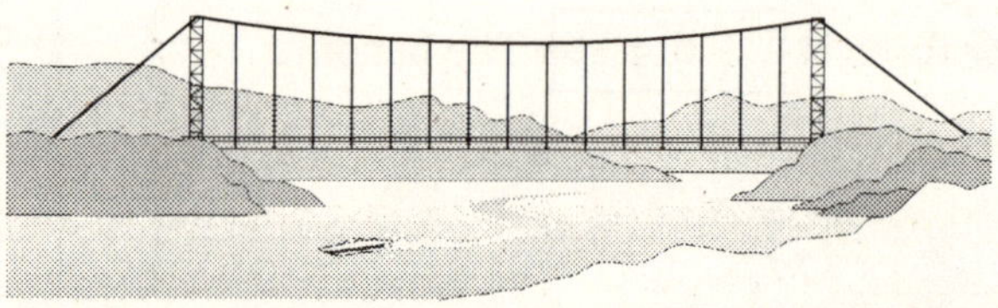

Solution The weight density is $w = \frac{6000}{120} = 50\ lb/ft$. The parameter $a = 2\frac{y}{x^2} = \frac{20}{60^2} = 5.55 \times 10^{-3}$. The tension at the lowest point is $T_o = \frac{w}{a} = \frac{50}{5.55 \times 10^{-3}} = 9000\ lb$. The maximum tension is $T_{MAX} = T_o\sqrt{1+a^2x^2}$, which for $x = 60$ $\boxed{T_{MAX} = 9486.8\ lb}$ The half-length of the cable is $s = (1/2)\left(x\sqrt{1+a^2x^2} + \left(\frac{1}{a}\right)\ln\left(ax + \sqrt{1+a^2x^2}\right)\right)$, which, for $x = 60$, $s = 61\ ft$, and the total length of the cable is $\boxed{L = 2s = 122.2\ ft}$

Problem 9.72 The cable supports a uniformly distributed load $w = 100\ lb/ft$. The lowest point of the cable is 20 *ft* below the attachment points C and D. Determine the axial loads in the truss members AC and BC.

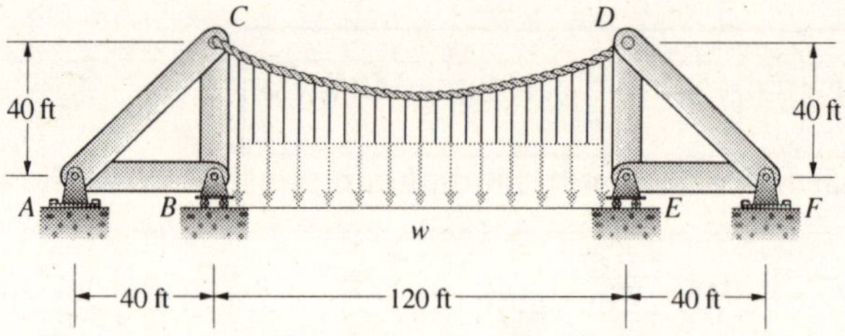

Solution The parameter $a = 2\left(\frac{y}{x^2}\right) = 2\frac{20}{60^2} = 0.0111$. The tension at the lowest point is $T_o = \frac{w}{a} = \frac{100}{0.01111} = 9000\ lb$. The components of the cable tension at point C are $T\cos\theta = T_o$, and $T\sin\theta = wx = 6000\ lb$. The sum of the moments about point A is $\sum M_A = -40(9000) - 40(6000) + 40B = 0$, from which $\boxed{B = 15000\ lb}$. The sum of forces $\sum F_y = A_y + B - 6000 = 0$, from which $A_y = -9000\ lb$. $\sum F_x = A_x + 9000 = 0$, from which $A_x = 9000\ lb$. The axial force in AC is $\boxed{F_{AC} = \sqrt{9000^2 + 9000^2} = 12728\ lb\ (T)}$

Problem 9.73 The cable supports a railway bridge between two tunnels. The distributed load is $w = 1\ MN/m$ and $h = 40\ m$. (a) What is the maximum tension in the cable? (b) What is the length of the cable?

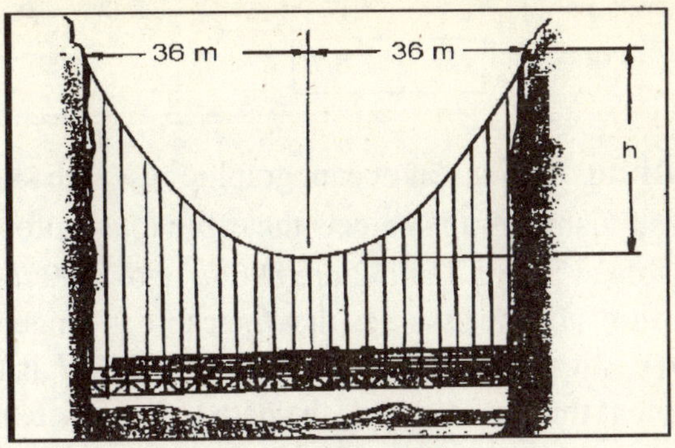

Solution The parameter $a = 2\frac{y}{x^2} = 2\frac{40}{36^2} = 0.06173$. The tension at the lowest point: $T_o = \frac{w}{a} = \frac{1000}{a} = 16200\ kN$. The maximum tension: $T_{MAX} = T_o\sqrt{1+a^2x^2}$, which, for $x = 36\ m$, $\boxed{T_{MAX} = 39477\ kN}$. The cable length is

$$s(x) = \left(x\sqrt{1+a^2x^2} + \frac{1}{a}\ln\left(ax + \sqrt{1+a^2x^2}\right)\right),$$ which, for $x = 36\ m$, $\boxed{L = 112.66\ m}$.

Problem 9.74 The cable in Problem 9.73 will safely support a tension of 40 MN. What is the shortest cable that can be used, and what is the corresponding value of h?

Solution The tension at the lowest point is $T_o = \frac{w}{a}$. The maximum tension is $T_{MAX} = T_o\sqrt{1+a^2x^2}$.

Square both sides, substitute and reduce algebraically: $T_o^2 = T_{MAX}^2 - w^2x^2$. The terms on the right are known: $T_{MAX}^2 = 40^2(10^6)$, and $w^2x^2 = 36^2(10^6)$. Solve for the parameter a,

$$a^2 = \frac{10^6}{(40^2 - 36^2)(10^6)} = 3.29(10^3),$$ from which $a = 0.0574$. The height is

$$\boxed{h = \left(\frac{1}{2}\right)ax^2 = \left(\frac{1}{2}\right)(0.0574)(36^2) = 37.165\ m}$$. The length is

$$s(x) = \left(x\sqrt{1+a^2x^2} + \frac{1}{a}\ln\left(ax + \sqrt{1+a^2x^2}\right)\right),$$ which, for $x = 36\ m$, $\boxed{L = 108.26\ m}$

Problem 9.75 At A the main cable of the suspension bridge is horizontal and its tension is $1 \times 10^8\ lb$. (a) Determine the distributed load acting on the cable. (b) What is the tension at B?

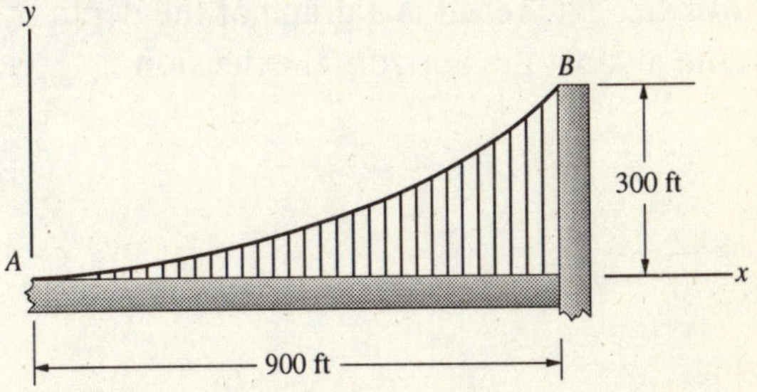

Solution (a) The parameter $a = 2\frac{y}{x^2} = 2\frac{300}{900} = 7.4074 \times 10^{-4}$. The distributed load is

$$\boxed{w = T_o a = (1 \times 10^8)(7.4 \times 10^{-4}) = 7.4074 \times 10^4\ lb/ft}$$

(b) The tension at B is

$$\boxed{T_B = T_o\sqrt{1 + a^2 900^2} = 1.2 \times 10^8\ lb}$$

Problem 9.76 Determine the length of the main cable in Problem 9.75.

Solution The cable length is $s(x) = \left(\frac{1}{2}\right)\left(x\sqrt{1+a^2x^2} + \frac{1}{a}\ln\left(ax + \sqrt{1+a^2x^2}\right)\right)$. For $a = 7.4074 \times 10^{-4}$ and $x = 900\ ft$, $\boxed{L = 962.8\ ft}$

Problem 9.77 An oceanographic research ship tows an instrument package from a cable. Hydrodynamic drag subjects the cable to a uniformly distributed force $w = 2\ lb/ft$. The tensions in the cable at points 1 and 2 are 800 *lb* and 1300 *lb*, respectively. Determine the distance h.

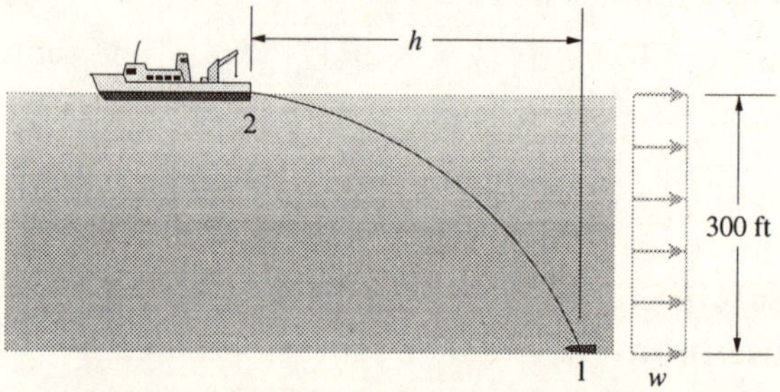

Solution If one assumes that the cable is tangent to the vertical at the point 1, so that the 800 *lb* is the tension at the lowest point, the data is inconsistent; therefore the point 1 must be at a distance x_1 from the lowest point. There are three unknowns in the problem: the distance x_1, the tension at the lowest point T_o, and the parameter *a*. The three equations that define these unknowns are:

(1) $800 = T_o\sqrt{1+a^2x_1^2}$, (2) $T_o = \frac{w}{a} = \frac{2}{a}$, (3) $1300 = T_o\sqrt{1+a^2(x_1+300)^2}$. These are reduced to two equations in two unknowns: (1) $800 = \left(\frac{2}{a}\right)\sqrt{1+a^2x_1^2}$, (2) $1300 = \left(\frac{2}{a}\right)\sqrt{1+a^2(x_1+300)^2}$ and solved by iteration using **TK Solver 2**. The result: $a = 3.596 \times 10^{-3}$, $x_1 = 287.5\ ft$. Using these values, the distance is $\boxed{h = \left(\frac{1}{2}\right)a(x_1+300)^2 - \left(\frac{1}{2}\right)ax_1^2 = 471.94\ ft}$

Problem 9.78 Draw a graph of the shape of the cable in Problem 9.77.

Solution The following equations are graphed:

(1) $IF\ d > x_1\ THEN\ w = h - \left(\frac{1}{2}\right)ad^2 + \left(\frac{1}{2}\right)ax_1^2$,

(2) $z = 300 + x_1 - d$, where $h = 471.9\ ft$, $a = 3.59573 \times 10^{-3}$, $x_1 = 287.5\ ft$. The value *w* is plotted on the abscissa, and *z* is plotted on the ordinate. The result is a graph of the depth of the cable against the horizontal extension.

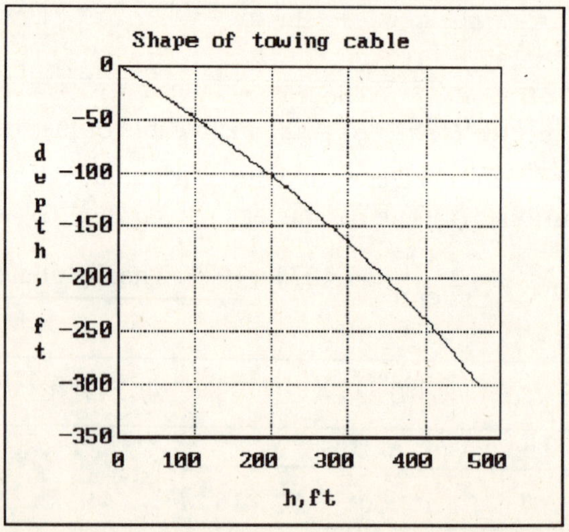

Problem 9.79 The rope weighs 1 *N/m*. The tension at its lowest point is 5 *N*. (a) What is the maximum tension? (b) What is the length of the rope?

Solution The rope satisfies the condition for a catenary. The tension at the lowest point is $T_o = \frac{w}{a}$, from which the parameter $a = 0.2$. (a) The maximum tension is

$$\boxed{T = T_o \cosh(ax) = 5\cosh(6/5) = 9.05\ N}$$

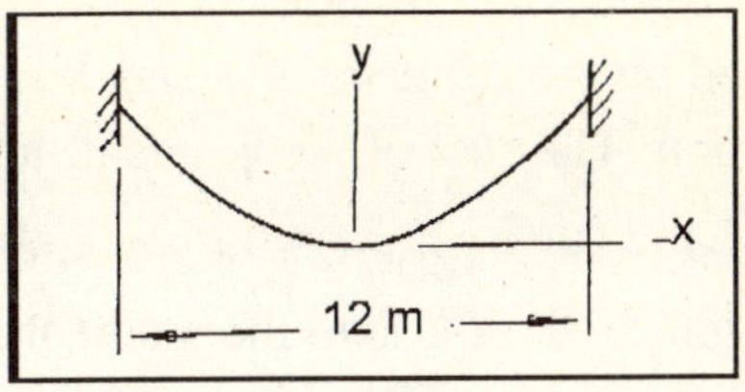

(b) The length is $\boxed{L = 2\left(\frac{\sinh(ax)}{a}\right) = 15.09\ m}$

Problem 9.80 The mass per unit length of lines AB and BC is 2 *kg/m*. The tension at the lowest point of cable AB is 1.8 *kN*. The two lines exert equal horizontal forces at B. (a) Determine the sags h_1 and h_2. (b) Determine the maximum tensions in the two lines.

Solution The lines meet the condition for a catenary. (a) *The line AB*. The weight density is $w = 2(9.81) = 19.62\ N/m$. The parameter

$$a_1 = \frac{w}{T_{AB}} = \frac{19.62}{1800} = 0.0109.$$ The sag is

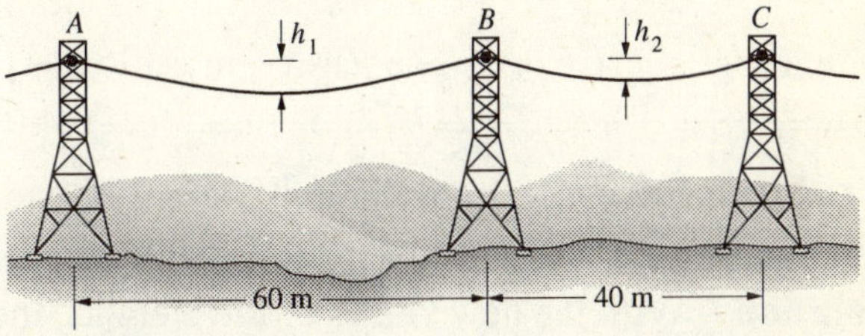

$$\boxed{h_1 = \left(\frac{1}{a_1}\right)\left(\cosh(30a_1) - 1\right) = 4.949\ m}$$. *The line BC*. The horizontal component of the tension at B is $T_{AB} = 1.8\ kN$. Thus the tension at the lowest point in BC is 1.8 *kN*., and the parameter *a* for line BC is equal to a_1. The sag is $\boxed{h_2 = \left(\frac{1}{a_1}\right)\left(\cosh(20a_1) - 1\right) = 2.189\ m}$. (b) *The line AB*. The maximum tension is

$$\boxed{T_{MAX}^{AB} = T_{AB}\cosh(30a_1) = 1897.1\ N}$$ *The line BC*. The maximum tension is

$$\boxed{T_{MAX}^{BC} = T_{AB}\cosh(20a_1) = 1842.9\ N}$$.

Problem 9.81 The rope is loaded by 10 *lb* weights suspended at 1 *ft* intervals along its length. Neglect the weight of the rope. If the tension at the lowest point of the rope is 45 *lb*, determine *h*, the maximum tension in the rope, and the length of the rope. *Strategy:* Obtain an approximate answer by modeling the discrete loads exerted on the rope as a uniformly distributed load.

Solution The rope satisfies the conditions for a catenary. The load distribution is $w = 10\ lb/ft$. The parameter $a = \frac{10}{45} = 0.2222$. The distance *h* is

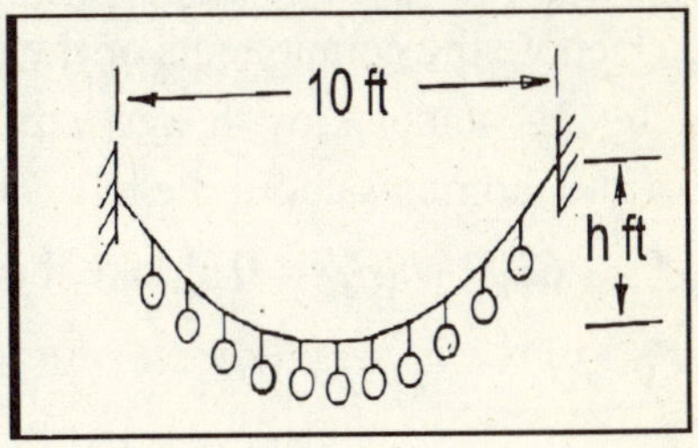

$$\boxed{h = \left(\frac{1}{a}\right)\left(\cosh(5a) - 1\right) = 3.076\ ft}$$. The maximum tension is

$\boxed{T_{MAX} = 45\cosh(5a) = 75.76\ lb}$ The length is $\boxed{L = 2\left(\frac{\sinh(5a)}{a}\right) = 12.188\ ft}$

Problem 9.82 In Example 9.12, what are the tensions in cable segments 1 and 3? ($W_1 = 100\ N$, $W_2 = 200\ N$, and $h_1 = 1\ m$.)

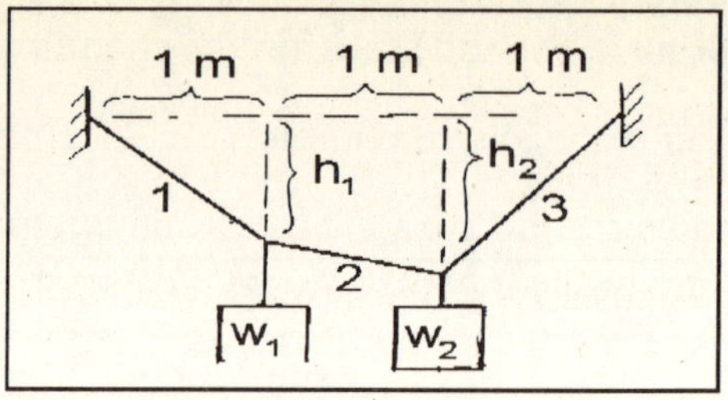

Solution Make a cut to right of the left attachment, and to the right of W_1. Denote the components of the force exerted by the string by F_H and F_V, with sign indicating direction. The sum of the moments about the left of the right hand cut is $\sum M = +F_H - F_V = 0$. Make the cut to the right of W_2 and sum the moments about the part to the left of that point: $\sum M = h_2 F_H - 2F_V + W_1 = 0$. Make a cut at the right attachment and sum the moments about the left part: $\sum M = -3F_V + 2W_1 + W_2 = 0$. Solve these simultaneous equations for the unknowns to obtain: $F_V = F_H = 133.33\ N$ and $h_2 = 1.25\ m$. The tension in cable 1 is $\boxed{T_1 = \sqrt{F_V^2 + F_H^2} = 188.56\ N}$. Take the sum of the forces for the complete system: $\sum F_x = -F_H + F_{3H} = 0$, from which $F_{3H} = F_H = 133.33\ N$. $\sum F_y = F_V - W_1 - W_2 + F_{3V} = 0$, from which $F_{3V} = 166.67\ N$. The tension in cable 3 is $\boxed{T_3 = \sqrt{F_{3H}^2 + F_{3v}^2} = 213.44\ N}$

Problem 9.83 If the weights $W_1 = 240\ N$ and $W_2 = 400\ N$ in Example 9.12, what is the vertical distance h_2 and what is the tension in cable segment 3?

Solution With the new values of the weights, the three simultaneous equations in Problem 9.82 have the solutions: $F_H = F_V = 293.33\ N$, and $\boxed{h_2 = 1.182\ m}$. The sums of the forces for the complete system are $\sum F_x = -F_H + F_{3H} = 0$ and $\sum F_y = F_V - W_1 - W_2 + F_{3V} = 0$, from which $F_{3H} = F_V = 293.33\ N$, $F_{3V} = 346.67\ N$, and the tension in cable 3 is $\boxed{T_3 = \sqrt{F_{3H}^2 + F_{3V}^2} = 454.12\ N}$

Problem 9.84 Two weights are suspended from a string. The weights $W_1 = W_2 = 50\ lb$. The vertical distance $h_1 = 4\ ft$. (a) Determine the vertical distance h_2. (b) What is the maximum tension in the cable?

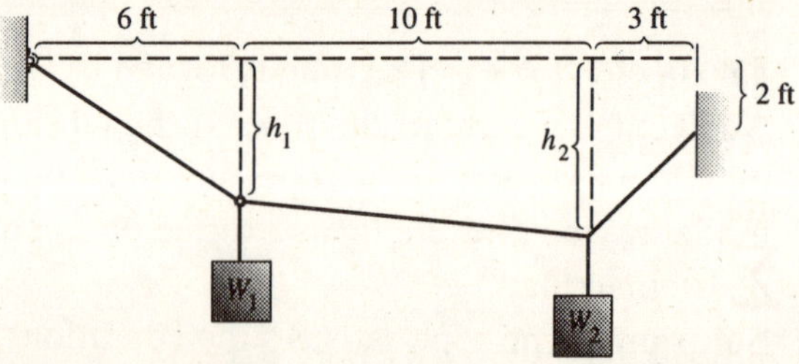

Solution The strategy is to make cuts along the string and sum the moments to the left of each cut. The three simultaneous equations are then solved for the unknowns. Make a cut at left attachment and to the right of W_1. Denote the components of the force exerted by the string by F_H and F_V, with sign indicating direction. The sum of the moments about the left part: $\sum M = -6F_V - h_1 F_H = 0$. Make the cut to the right of W_2. The sum of the moments about the left part: $\sum M = -h_2 F_H - 16F_V + 10W_1 = 0$. Make the cut at the right attachment point and sum the moments to the left: $\sum M = -19F_V - h_2 F_H + 13W_1 + 3W_2 = 0$. Solve: $\boxed{h_2 = 4\ ft}$, $F_V = 50\ lb$, and $F_H = -75\ lb$. From the sum of the forces for the complete string the tension in the right support string is equal to the tension in the string. Thus the maximum tension is $\boxed{T = \sqrt{F_V^2 + F_H^2} = 90.14\ lb}$.

Problem 9.85 Three weights are suspended from a string. The weights are $W = 10\ N$. (a) Determine the vertical distances h_1 and h_3 and sketch the configuration of the string. (b) What are the tensions in cable segments 1 and 2?

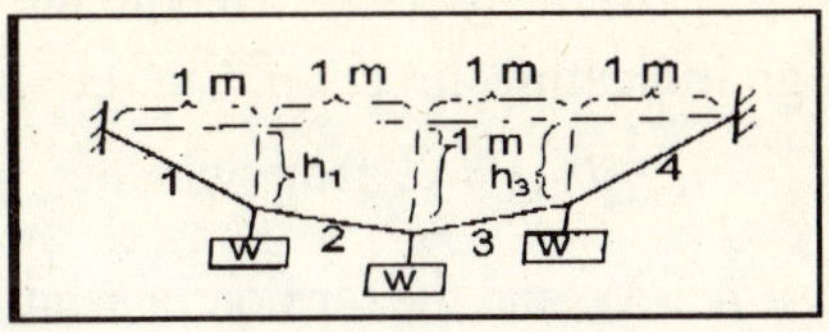

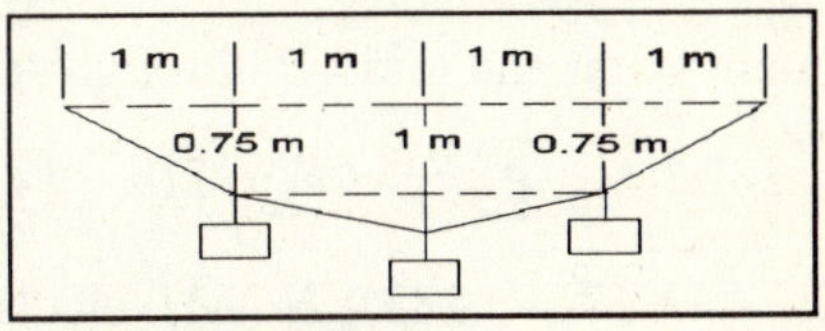

Solution The strategy is to make cuts in the string and sum the moments to the left of the cut. The resulting simultaneous equations are solved for the unknowns. (a) Make a cut at the left attachment point and to the right of the leftmost weight. Denote the components of the force exerted by the string by F_H and F_V, with sign indicating direction. The sum of the moments: $\sum M = -h_1 F_H - 1F_V = 0$. Make a cut at the left attachment and to the right of the second weight. The sum of the moments: $\sum M = -1F_H - 2F_V + 1W = 0$. Make a cut at the left attachment point and to the right of the third weight. The sum of the moments: $\sum M = -h_3 F_H - 3F_V + 1W + 2W = 0$. Make a cut at the right and left attachment points. The sum of the moments: $\sum M = -4F_V + 3W + 2W + 1W = 0$. Solve: $\boxed{h_1 = 0.75\ m}$, $\boxed{h_3 = 0.75\ m}$, $F_H = -20\ N$, and $F_V = 15\ N$. (b) The tension in segment 1 is $\boxed{T_1 = \sqrt{F_H^2 + F_V^2} = 25\ N}$. Make a cut at the left attachment point and at the midpoint of segment 2. Sum the forces to the left: $\sum F_x = F_H + F_{2H} = -20 + F_{2H} = 0$, from which $F_{2H} = 20\ N$. $\sum F_y = F_V + F_{2V} - W = 0$, from which $F_{2V} = 10 - 15 = -5\ N$. The tension in segment 2: $\boxed{T_2 = \sqrt{5^2 + 20^2} = 20.62\ N}$

Problem 9.86 Three weights are suspended from a string. The weight $W = 100\ N$. The vertical distance $h_1 = 400\ mm$. Determine the vertical distances h_2 and h_3.

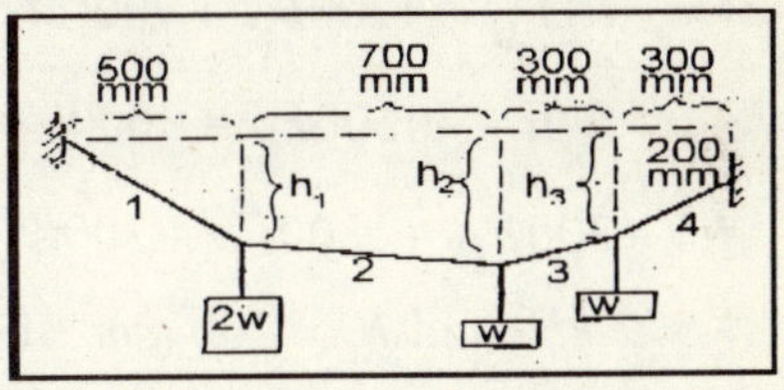

Solution Make a cut at the left attachment point and to the right of the leftmost weight and sum moments to the left: Denote the components of the force exerted by the string by F_H and F_V, with sign indicating direction. $\sum M = -h_1 F_H - 500F_V = 0$. Make a cut at the left attachment point, to the right of the middle weight and sum moments to the left: $\sum M = -h_2 F_H - 1200F_V + 700(2W) = 0$. Make a cut at the left attachment point, to the right of the rightmost weight, and sum the moments to the left: $\sum M = -h_3 F_H - 1500F_V + 300W + 1000(2W) = 0$. Make a cut at both left and right attachment points and sum the moments to the left: $\sum M = -200F_H - 1800F_V + 300W + 600W + 1300(2W) = 0$. Solve these simultaneous equations for the unknowns: $\boxed{h_2 = 464\ mm}$, $\boxed{h_3 = 385.1\ mm}$.

Problem 9.87 In Problem 9.86, what is the maximum tension in the cable and in what segment does it occur?

Solution The tension in segment 1 is known from the solution to Problem 9.86: $T = \sqrt{F_V^2 + F_H^2} = 361.47\ N$. Denote the components of the force exerted by the string by F_H and F_V, with sign indicating direction. Cut the string at both attachment points and sum forces: $\sum F_x = F_H + F_{4H} = 0$, from which $F_{4H} = 282.3\ N$. $\sum F_y = F_V - 4W + F_{4V} = 0$, from which $F_{4V} = 174.2\ N$, and the tension in segment 4 is $T_4 = \sqrt{F_{4H}^2 + F_{4V}^2} = 331.7\ N$. Cut the string at the left attachment point and midway through segment 2 and sum forces: $\sum F_x = F_H + F_{2H} = 0$, from which $F_{2H} = 282.3\ N$. $\sum F_y = F_V - 2W + F_{2V} = 0$, from which $F_{2V} = -25.8\ N$, and the tension is $T_2 = \sqrt{F_{2V}^2 + F_{2H}^2} = 283.4\ N$. Make cuts at the midpoint of segments 2 and 3 and sum forces: $\sum F_x = F_{2H} + F_{3H} = 0$, from which $F_{3H} = -282.3\ N$. $\sum F_y = F_{2V} - W + F_{3V} = 0$, from which $F_{3V} = 74.2\ N$, and $T_3 = \sqrt{F_H^2 + F_V^2} = 292\ N$. Thus the maximum tension occurs in segment 1, $\boxed{T_{MAX} = 361.5\ N}$.

Problem 9.88 The cable in Problem 9.86 will safely support a tension of 15 kN. Is $W = 2\ kN$ a safe load if $h_1 = 200\ mm$?

Solution Use the procedure used in the solution to Problem 9.86 to write the following simultaneous equations in the unknowns: Denote the components of the force exerted by the string by F_H and F_V, with sign indicating direction.

$\sum M = -h_1 F_H - 500 F_V = 0$ (Sum of moments, first cut.)

$\sum M = -h_2 F_H - 1200 F_V + 700(2W) = 0$ (Sum of moments, second cut.)

$\sum M = -h_3 F_H - 1500 F_V + 300W + 1000(2W) = 0$ (Sum of moments, third cut.)

$\sum M = -200 F_H - 1800 F_V + 300W + 600W + 1300(2W) = 0$ (Sum of moments, fourth cut.)

For $W = 2\ kN$ and $h_1 = 200\ mm$, the solution is $h_2 = 272\ mm$, $h_3 = 258.3\ mm$, $F_H = -13.46\ kN$, and $F_V = 5.38\ kN$. To find the maximum tension: In segment 1: $T = \sqrt{F_V^2 + F_H^2} = 14.5\ kN$. Cut the string at both attachment points and sum forces: $\sum F_x = F_H + F_{4H} = 0$, from which $F_{4H} = 13.46\ kN$. $\sum F_y = F_V - 4W + F_{4V} = 0$, from which $F_{4V} = 2.62\ kN$, and the tension in segment 4 is $T_4 = \sqrt{F_{4H}^2 + F_{4V}^2} = 13.71\ kN$. Cut the string at the left attachment point and midway through segment 2 and sum forces: $\sum F_x = F_H + F_{2H} = 0$, from which $T_{2H} = 13.46\ kN$. $\sum F_y = F_V - W + F_{2V} = 0$, from which $F_{2V} = -1.38\ kN$, and the tension is $T_2 = \sqrt{F_{2V}^2 + F_{2H}^2} = 13.53\ kN$. Make cuts at the midpoint of segments 2 and 3 and sum forces: $\sum F_x = F_{2H} + F_{3H} = 0$, from which $F_{3H} = -13.46\ kN$. $\sum F_Y = T_{2V} - 2W + T_{3V} = 0$, from which $T_{3V} = 5.38\ kN$, and $T_3 = 14.498\ kN$. Thus the maximum tension occurs in segment 1, $\boxed{T_{MAX} = 14.5\ kN}$, and the load can be safely carried.

Problem 9.89 The cable is 10 m long and subjected to a distributed load $w = 200\ N/m$. What is the maximum tension? *Strategy:* By setting $x = 4\ m$ in Eq (9.17) and obtaining a graph of s as a function of the parameter *a*, you can determine the value of a corresponding to $s = 5\ m$ and so determine $T_O = \frac{w}{a}$. Then you can determine the maximum tension from Eq (9.16).

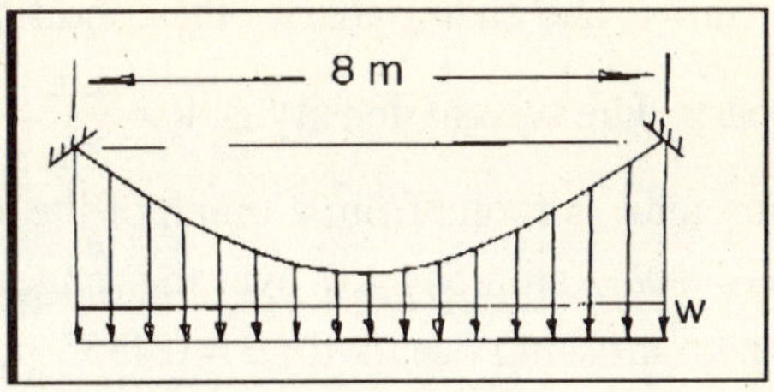

Solution Graph the equation

$f(a) = 4\sqrt{1+16a^2} + \left(\frac{1}{a}\right)\ln\left(4a + \sqrt{1+16a^2}\right) - 10$ over the range $0.3 < a \le 0.4$ and determine the value of a at the zero crossing of the function. The graph crosses zero at a value of $a = 0.337$, approximately. The tension at the lowest point: $T_O = \frac{w}{a} = 593.4\ N$. The maximum tension is

$$T_{MAX} = T_O\sqrt{1+16a^2} = 996.1\ N$$

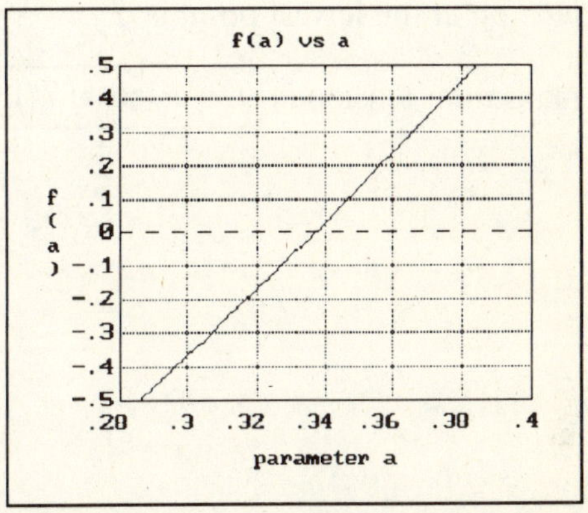

Problem 9.90 The rope weighs 1 *N/m* and is 16 *m* in length. (a) What is the maximum tension? (b) What is the vertical distance from the attachment points to the lowest point of the rope?

Solution The rope satisfies the condition for a catenary.

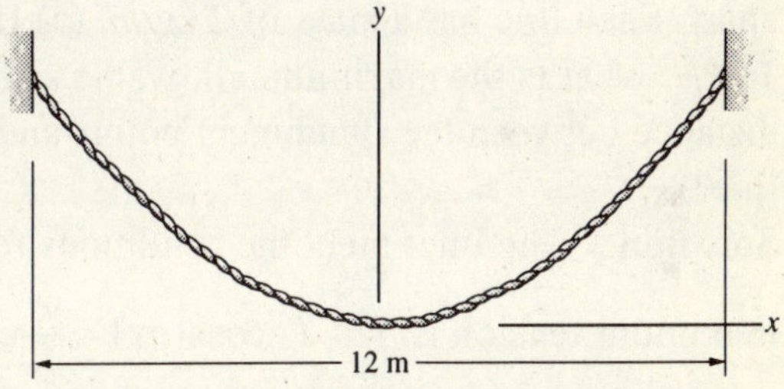

The parameter a is found from $s = \left(\frac{\sinh(ax)}{a}\right)$, from which $f(a) = 8a - \sinh(6a) = 0$. A graph of the function *f(a)* over the range $0.1 < a \le 0.3$ shows a zero crossing for a value of $a = 0.2252$ approximately. The tension at the lowest point is $T_o = \frac{w}{a} = 4.44\ N$.

(a) The maximum tension is

$$T_{MAX} = T_O\cosh(6a) = 9.15\ N$$

(b) The sag is $h = \left(\frac{1}{a}\right)(\cosh(6a) - 1) = 4.709\ m$

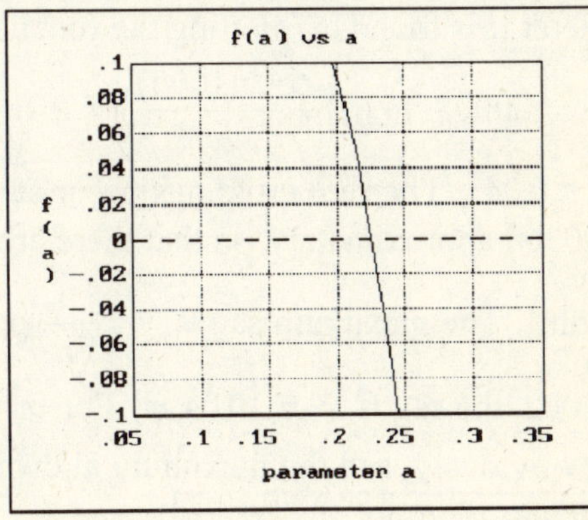

Problem 9.91 A chain weighs 20 *lb* and is 20 *ft* long. It is suspended from two points that are at the same height and 10 *ft* apart. (a) Determine the maximum tension in the chain. (b) Draw a sketch of the shape of the chain.

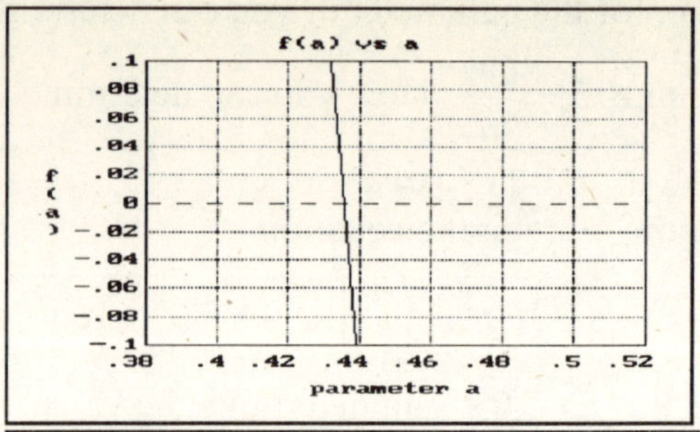

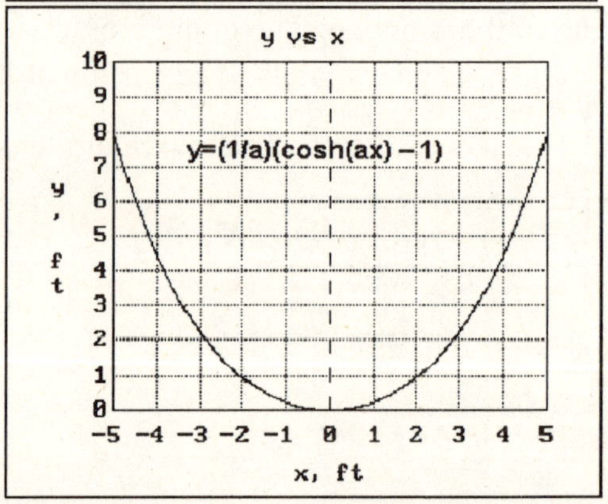

Solution The chain satisfies the condition for a catenary. The weight density is $w = \frac{20}{20} = 1\ lb/ft$. The parameter a is found from a graph of the function $f(a) = 10a - \sinh(5a) = 0$ over the range $0.4 < a \le 0.5$. The zero crossing occurs for $a = 0.4355$, approximately.

The tension at the lowest point is $T_o = \frac{w}{a} = 2.97\ lb$.

The maximum tension is $\boxed{T = T_o \cosh(5a) = 10.26\ lb}$

Problem 9.92 An engineer wants to suspend high-voltage power lines between poles 200 meters apart. Each line has a mass of 2 *kg/m*. (a) If he doesn't want to subject the lines to a tension greater than 10 *kN*, what is the maximum allowable sag between poles? That is, what is the largest allowable vertical distance between the attachment points and the lowest point of the line? (b) What is the length of each line?

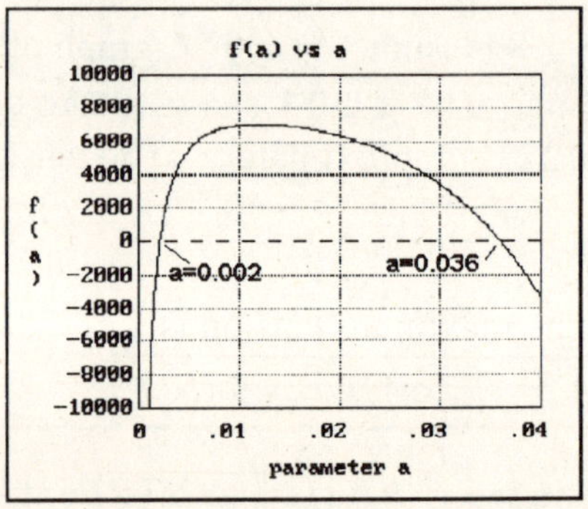

Solution The lines meet the conditions for a catenary. The maximum tension is $T = T_o \cosh(ax) = \frac{w}{a}\cosh(100a)$. The parameter a is found by plotting the function $f(a) = 10000 - 19.62\left(\frac{\cosh(100a)}{a}\right) = 0$ over the range $0 < a \le 0.04$. The zero crossings occur at values $a = 0.002$ and $a = 0.036$, approximately, so that there are two possible solutions. The maximum sag is $y = \left(\frac{1}{a}\right)(\cosh(100a) - 1)$. For $a = 0.002$ the sag is $y_1 = 10.04\ m$ For $a = 0.036$, $y_2 = 481.93\ m$.(a) The second value is not a practical value, obviously, and the maximum allowable sag is $\boxed{h = 10.04\ m}$. (b) The length of the line is

$$\boxed{L = 2\left(\frac{\sinh(100a)}{a}\right) = 201.34\ m}$$

Problem 9.93 The mass per unit length of lines AB and BC is 2 *kg/m*. The length of line AB is 62 *m*. The two lines have equal horizontal forces at B. (a) Determine the sags h_1 and h_2. (b) Determine the maximum tensions in the two lines.

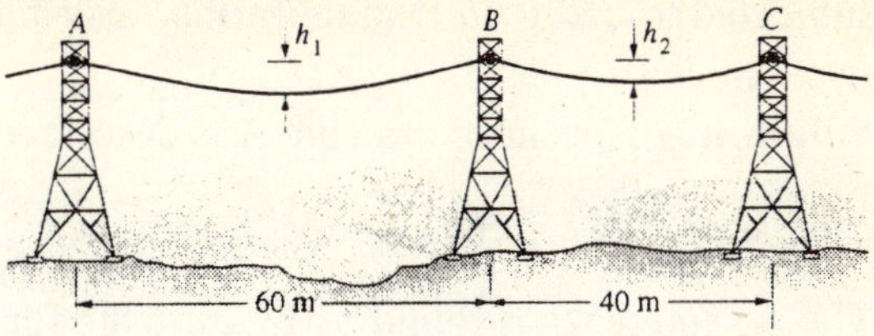

Solution The lines meet the conditions for a catenary. The weight density is $w = 2(9.81) = 19.62\ N/m$

Line AB: The length of line AB is $s = \dfrac{\sinh(30a)}{a}$. The parameter a is found by graphing the function $f(a) = 31 - \dfrac{\sinh(30a)}{a} = 0$. The zero crossing occurs at a value $a = 0.0148$, approximately. The tension at the lowest point is $T_o = \dfrac{w}{a} = \dfrac{19.62}{0.0148} = 1322.68$ The sag is $\boxed{h_1 = \left(\dfrac{1}{a}\right)(\cosh(30a) - 1) = 6.786\ m}$ The maximum tension is $\boxed{T_{MAX}^{AB} = T_o \cosh(30a) = 1455.8\ N}$

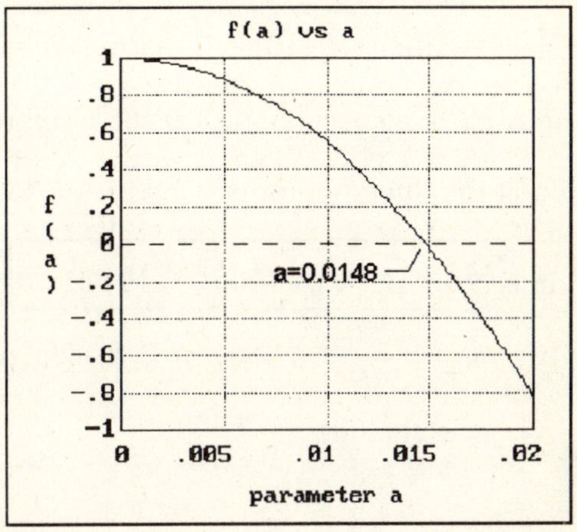

Line BC: The horizontal component of the force is the same for both lines, T_o. The weight density is the same for both lines, hence the parameter a is the same for lines. The sag is $\boxed{h_2 = (1/a)(\cosh(20a) - 1) = 2.99\ m}$. The max tension is $\boxed{T_{MAX}^{BC} = T_o \cosh(20a) = 1381.3\ N}$

Problem 9.94 The mass per unit length of the lines AB and BC in Problem 9.93 is 2 *kg/m*. The sag $h_1 = 4.5\ m$. The two lines exert equal horizontal forces at B. (a) Determine the sag h_2. (b) Determine the maximum tensions in the two lines.

Solution The weight density is $w = 2g = 19.62\ N/m$. The sag $h_1 = \left(\dfrac{1}{a}\right)(\cosh(ax) - 1)$. Solve

$f(a) = 4.5 - \dfrac{\cosh(30a) - 1}{a} = 0$ by iteration, to obtain $a = 0.009926$. The tension at the lowest point is

$T_o = \dfrac{w}{a} = 1976.5\ N$. This is also the tension at the lowest point in BC, and

(a) $\boxed{h_2 = \dfrac{\cosh(20a) - 1}{a} = 1.992\ m}$. (b) The maximum tensions are $\boxed{T_{AB} = T_o \cosh(30a) = 2064.8\ N}$, $\boxed{T_{BC} = T_o \cosh(20a) = 2015.6\ N}$

Problem 9.95 Two 30 ft cables A and B are each suspended from points that are at the same height and 20 ft apart. Cable A is subjected to a 200 lb load uniformly distributed horizontally. Cable B is subjected to a 200 lb load uniformly distributed along its length. What are the maximum tensions in the two cables?

Solution The shape of cable A is described by a parabola; the shape of cable B is described by a catenary. *Cable A:* The weight density is $w_A = \frac{200}{20} = 10 \ lb/ft$.

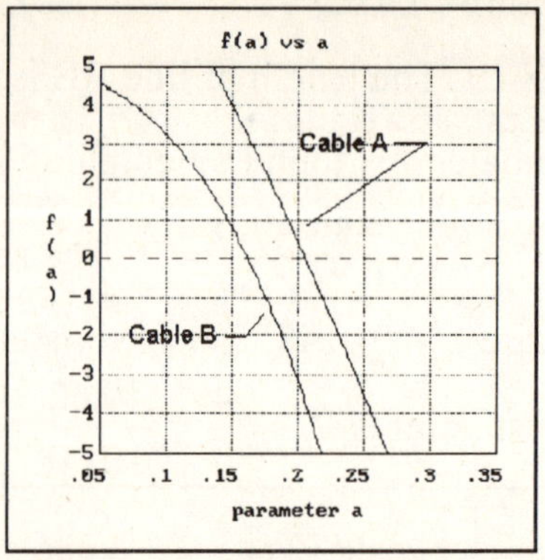

The parameter a is found by plotting the function

$$f(a) = 30 - 10\sqrt{1+a^2(10)^2} - \left(\frac{1}{a}\right)\ln\left(10a + \sqrt{1+a^2(10)^2}\right) = 0.$$

The zero crossing occurs at $a = 0.2055$, approximately. The tension at the lowest point is $T_{oA} = \frac{w_A}{a} = 48.65 \ lb$. The maximum tension is $\boxed{T_A = T_{oA}\sqrt{1+100a^2} = 111.21 \ lb}$ *Cable B*: The weight density is $w_B = \frac{200}{30} = 6.6667 \ lb/ft$. The parameter a is found by plotting the function

$f(a) = 15 - \frac{\sinh(10a)}{a} = 0$. The zero crossing occurs at $a = 0.1622$. The tension at the lowest point is

$T_{oB} = \frac{w_B}{a} = 41.098 \ lb$. The maximum is $\boxed{T_B = T_{oB}\cosh(10a) = 108.1 \ lb}$

Problem 9.96 Draw a graph of the two cables in Problem 9.95 comparing their shapes.

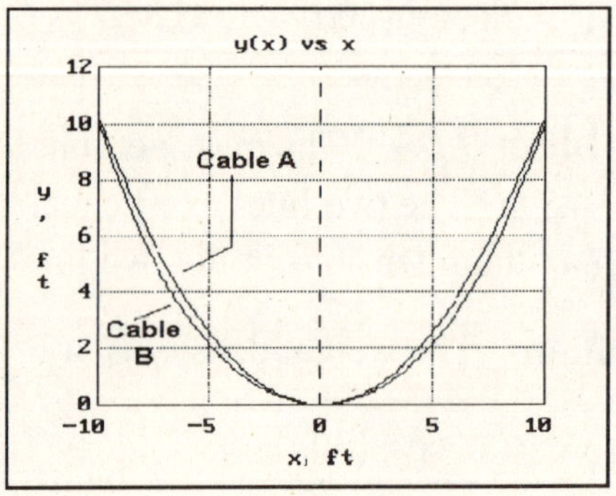

Solution The functions $y_A = \left(\frac{1}{2}\right)ax^2$ and

$y_B = \left(\frac{1}{a}\right)(\cosh(ax) - 1)$ are graphed over the range $-10 \le x \le +10$. The lower graph is the shape of Cable B.

Problem 9.97 The weights $W_1 = 100\ N$ and $W_2 = 200\ N$ and the total length L of the three segments of the rope is 5 *m*. (a) What are h_1 and h_2? (b) What is the maximum tension in the rope? *Strategy:* If you choose a value of h_1, you can determine h_2 and then L. By obtaining a graph of L as a function of h_1, you can determine the value of h_1 that corresponds to $L = 5\ m$.

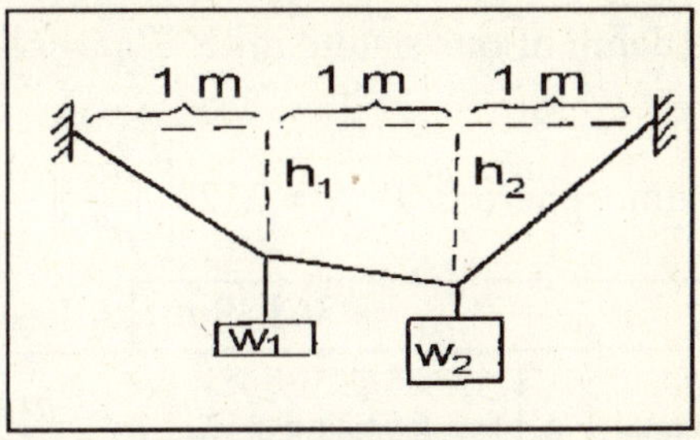

Solution Make a cut at the left support and to the right of W_1. Denote the components of the forces associated with the tensions by F_H and F_V, with the sign indicating direction. The sum of the moments to the left: $\sum M = -1F_V + h_1 F_H = 0$, from which $h_1 = \dfrac{F_V}{F_H}$. Make a cut at the left support and to the right of W_2. The sum of the moments to the left: $\sum M = -2F_V + h_2 F_H + 1W_1 = 0$, from which $h_2 = \dfrac{2F_V - W_1}{F_H}$ The ratio of the two values: $\left(\dfrac{h_2}{h_1}\right) = \left(\dfrac{2F_V - W_1}{F_V}\right)$. Make cuts at the left and right supports. The sum of the moments to the left: $\sum M = -3F_V + 1W_2 + 2W_1 = 0$. Solve the last equation: $F_V = \left(\dfrac{1}{3}\right)(2W_1 + W_2) = 133.33\ N$, from which $\left(\dfrac{h_2}{h_1}\right) = \left(\dfrac{2F_V - W_1}{F_V}\right) = 1.25$. The length of the rope in terms

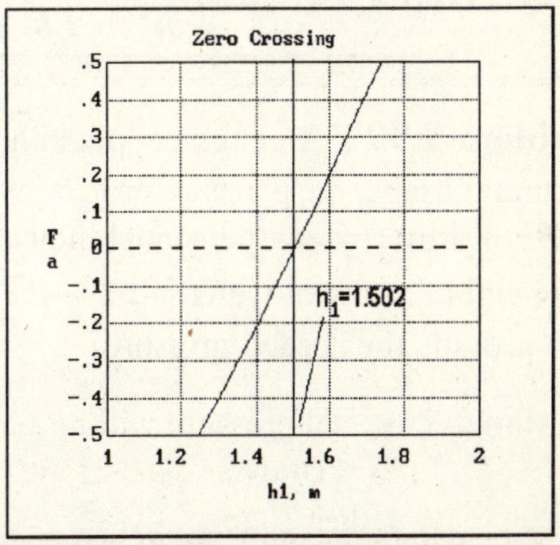

of h_1 and h_2 is $L = \sqrt{1+h_1^2} + \sqrt{1+(h_2-h_1)^2} + \sqrt{1+h_2^2}$. Substitute $h_2 = 1.25h_1$, $L = \sqrt{1+h_1^2} + \sqrt{1+(0.25h_1)^2} + \sqrt{1+(1.25h_1)^2}$. This can be solved by iteration for a value of h_1 that yields $L = 5\ m$, or by graphing L against h_1. A graph of $(L-5)$ against h_1 has a zero crossing at $\boxed{h_1 = 1.5\ m}$, approximately. From this $\boxed{h_2 = 1.25h_1 = 1.88\ m}$. *The tension in the rope*: From the sum of the forces for the system, the horizontal tension component in the rope has the same magnitude everywhere: $|F_H| = 88.77\ N$. The tension in segment 1 is $\boxed{T_1 = \sqrt{F_V^2 + F_H^2} = 160.18\ N}$. In the 3rd segment $\sum F_y = F_V - W_1 - W_2 + F_{3V} = 0$, from which $F_{3V} = 166.67\ N$, and the tension in segment 3 is $\boxed{T_3 = \sqrt{F_{3H}^2 + F_{3v}^2} = 188.8\ N}$ Make a cut at the midpoint of segment 2 and at the left attachment point and sum the vertical forces: $\sum F_y = F_V - W_1 + F_{2V} = 0$, from which $F_{2V} = -33.33\ N$, and the tension in segment 2 is $\boxed{T_2 = \sqrt{F_{2H}^2 + F_{2V}^2} = 94.8\ N}$. Thus the maximum tension occurs in segment 3, $\boxed{T_{MAX} = 188.8\ N}$. *Check:* The angle in segment 1 is $\alpha_1 = \tan^{-1}(h_1) = 56.34^o$, and the tension is $T_1 = \dfrac{F_H}{\cos\alpha_1} = 160.18\ N$. The angle in segment 2 is $\alpha_2 = \tan^{-1}(h_2 - h_3) = 20.6^o$, and the tension is $T_2 = \dfrac{F_H}{\cos\alpha_2} = 94.8\ N$. For segment 3, $\alpha_3 = \tan^{-1}(h_2) = 62^o$, and $T_3 = \dfrac{F_H}{\cos\alpha_3} = 188.8\ N$ *.check.*

Problem 9.98 A deep submersible research vehicle operates near the bottom of the Indian Ocean at a depth of one mile. The weight density of the water is approximately $\gamma = 64\ lb/ft^3$ and the atmospheric pressure is $p_{atm} = 2120\ lb/ft^2$. Determine the pressure on the vehicle's surface a) in psi; (b) in Pa.

Solution The pressure due to the sea water at a depth x is $P_{SW} = \gamma x$, where the units are consistent. At a depth of one statute mile: $P_{SW} = 64\left(\frac{lb}{ft^3}\right)\left(\frac{5280\ ft}{1\ mile}\right)\left(\frac{1\ ft}{12\ in}\right)^2 = 2346.67\ \left(\frac{lb}{in^2}\right)$. The pressure due to the atmosphere is $P_{ATM} = 2120\left(\frac{lb}{ft^2}\right)\left(\frac{1\ ft}{12\ in}\right)^2 = 14.722\ \left(\frac{lb}{in^2}\right)$. The sum of the pressures: $\boxed{P_{PSI} = P_{SW} + P_{ATM} = 2361.39\ psi}$. A Pascal is defined to be one Newton per square meter. The pressure in Pascals is $\boxed{Pa = 2361.39\left(\frac{lb}{in^2}\right)\left(\frac{12\ in}{1\ ft}\right)^2\left(\frac{1\ ft}{0.3048\ m}\right)^2\left(\frac{4.448\ N}{1\ lb}\right) = 16.28\ MPa}$

Problem 9.99 An engineer planning a water system for a new community estimates that at expected maximum usage, the pressure drop between the central system and the farthest planned fire hydrant will be 25 *psi*. Fire fighting personnel indicate that a gage pressure of 40 *psi* at the fire hydrant is required. The weight density of water is $62.4\ lb/ft^3$. How tall would a water tower at the central system have to be to provide the needed pressure?

Solution The total pressure will be $65\ \left(\frac{lb}{in^2}\right)$. A tower of height h will produce a pressure of $P = \gamma\ h$, where the units are to be consistent. The tower height is $h = \left(\frac{P}{\gamma}\right)$. Let the tower height be in feet: $\boxed{h\ ft = \left(\frac{1}{62.4}\right)\left(\frac{ft^3}{lb}\right)(65)\left(\frac{lb}{in^2}\right)\left(\frac{12\ in}{1\ ft}\right)^2 = 150\ ft}$

Problem 9.100 A cube of material is suspended below the surface of a liquid of weight density γ. By calculating the forces exerted on the faces of the cube by pressure, show that their sum is an upward force of magnitude $\gamma\ b^3$.

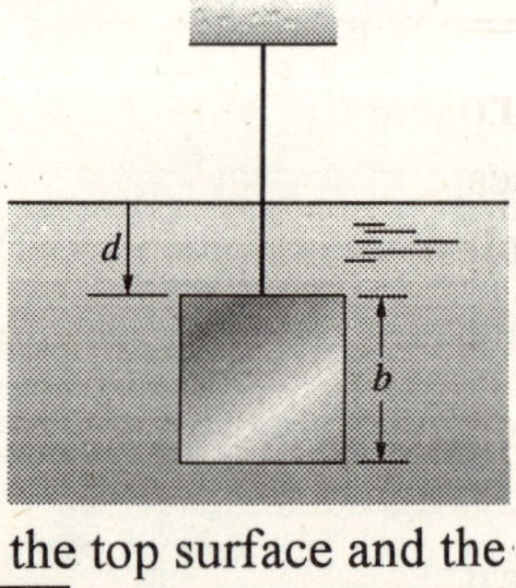

Solution Neglect the pressures on the supporting wire. The force on the top surface is $F_{TOP} = \gamma\ db^2$. The force on the sides is $F_S = \gamma\,(d + \frac{b}{2})b^2$, where $(d + \frac{b}{2})$ is the distance of the centroid of the area from the surface. The force on the bottom surface is $F_{BOTTOM} = \gamma\,(d + b)b^2$. The forces on the sides cancel by symmetry. The difference between the downward force on the top surface and the upward force on the bottom surface is the resultant: $\boxed{R = \gamma\,(d + b)b^2 - \gamma db^2 = \gamma b^3}$.

Problem 9.101 The area is subjected to a uniform pressure $p_{ATM} = 1 \times 10^5 \, Pa$. (a) What is the total force exerted on the area by the pressure? (b) What is the moment about the y-axis due to the pressure?

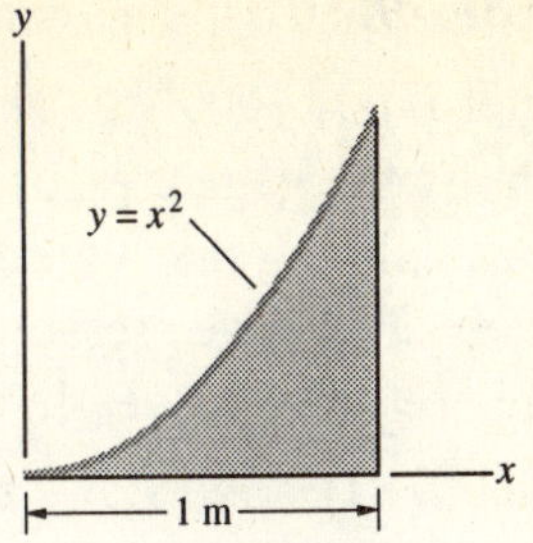

Solution The element of area is $dA = dxdy$. (a) The element of force is $dF = pdA$. Integrating:

$$F = \int_A pdA = (10^5)\int_0^1 dx \int_0^{x^2} dy = (10^5)\int_0^1 x^2 dx$$

$$\boxed{F = \left(\frac{10^5}{3}\right) N = 3.33 \times 10^4 \; N}$$

(b) The element of moment about the y-axis is $dM = pxdA$. Integrating:

$$M_{y-axis} = \int_A pxdA = p\int_0^1 xdx \int_0^{x^2} dy = p\int_0^1 x^3 dx, \quad \boxed{M_{y-axis} = \left(\frac{10^5}{4}\right) = 2.5 \times 10^4 \; Nm}$$

Problem 9.102 Determine the coordinates of the center of pressure in Problem 9.101.

Solution The center of pressure

$$\boxed{\bar{x} = \frac{\int_A pxdA}{p\int_A dA} = \frac{2.5 \times 10^4}{3.33 \times 10^4} = 0.75 \; m}. \quad \bar{y} = \frac{\int_A pydA}{\int_A pdA} = \frac{p\int dx \int_0^{x^2} ydy}{3.33 \times 10^4} = \frac{\frac{p}{2}\int_0^1 x^4 dx}{3.33 \times 10^4}, \quad \boxed{\bar{y} = \frac{\frac{p}{10}}{3.333 \times 10^4} = 0.3 \; m}$$

Problem 9.103 The top of the rectangular plate is $2 \, ft$ below the surface of a lake. Atmospheric pressure $p_{atm} = 2120 \; lb/ft^2$ and the weight density of water is $\gamma = 62.4 \; lb/ft^2$. (a) What is the maximum pressure exerted on the plate by the water? (b) Determine the force exerted on a face of the plate by the pressure of the water.

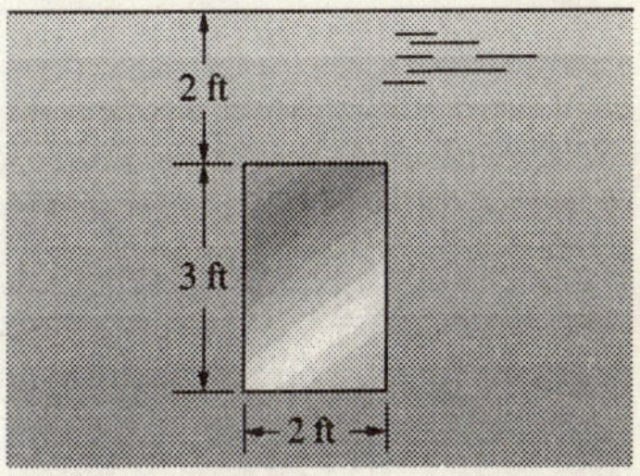

Solution (a) The maximum pressure occurs at the maximum depth: $\boxed{p = p_{atm} + \gamma 5 = 2120 + 312 = 2432 \; lb/ft^2}$

(b) The force is $F = \int_A pdA = \int_2^5 (p_{atm} + \gamma x)2dx = 6p_{atm} + \gamma\left[x^2\right]_2^5$ $\boxed{F = 12720 + 62.4(21) = 14030.4 \; lb}$

Problem 9.104 How far from the top of the plate in Problem 9.103 is the center of pressure located?

Solution The center of pressure from the surface:

$$\bar{x} = \frac{\int_A pxdA}{\int_A pdA} = \frac{\int_2^5 (p_{atm} + \gamma x)2xdx}{\int_A pdA} = \frac{2120(21) + 62.4(\frac{2}{3})(125 - 8)}{14030.4} = 3.52 \; ft \text{ from the surface.. From the top}$$

of the plate: $\boxed{d = \bar{x} - 2 = 1.52 \; ft}$

Problem 9.105 The width of the dam (the dimension into the page) is 100 *m*. The mass density of the water is $\rho = 1000 \ kg/m^3$. Determine the force exerted on the dam by the gage pressure of the water (a) by integration, (b) by calculating the "volume" of the pressure distribution.

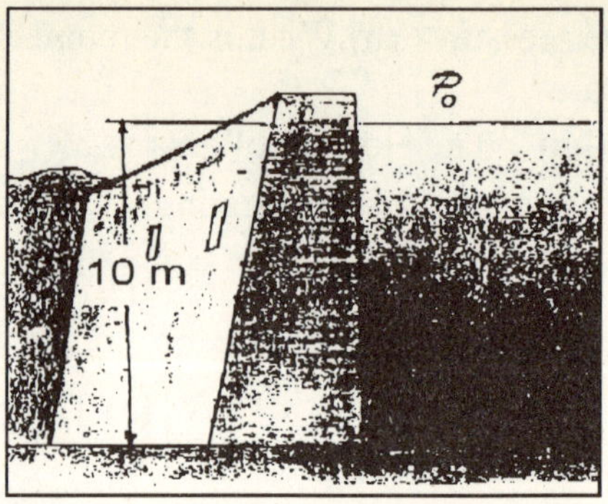

Solution For gage pressure, we neglect the pressure of the atmosphere. (a) The weight density of the water is $\gamma = 1000(9.81) = 9810 \ N/m^3$. The force on the dam is $F = \int_A pdA = 100\gamma \int_0^{10} xdx = 5000\gamma \ \boxed{= 49.05 \times 10^6 \ N}$ (b) The pressure distribution is a triangle, with base 10γ and altitude $10 \ m$. The length dimension of this "solid" is 100 *m*. The volume of this triangular "solid" is

$$\boxed{V = \left(\frac{1}{2}\right)100(10)(9810) = 49.05 \times 10^6 \ N}$$

Problem 9.106 In Problem 9.106, how far down from the surface of the water is the center of pressure due to the gage pressure of the water on the dam?

Solution The center of pressure is the centroid of the pressure distribution, which is a triangle of altitude 10 *m* and base 10γ. The centroid of a triangle is $\left(\frac{1}{3}\right)$ of the altitude, or, from the surface,

$$\boxed{\bar{d} = \left(\frac{2}{3}\right)10 = 6.67 \ m}$$

Problem 9.107 The width of the gate (the dimension into the page) is 4 *ft*, and it weighs 200 *lb*. The weight density of the water is $\gamma = 62.4 \ lb/ft^3$ and the atmospheric pressure $p_{atm} = 2120 \ lb/ft^2$. Determine the reactions at A.

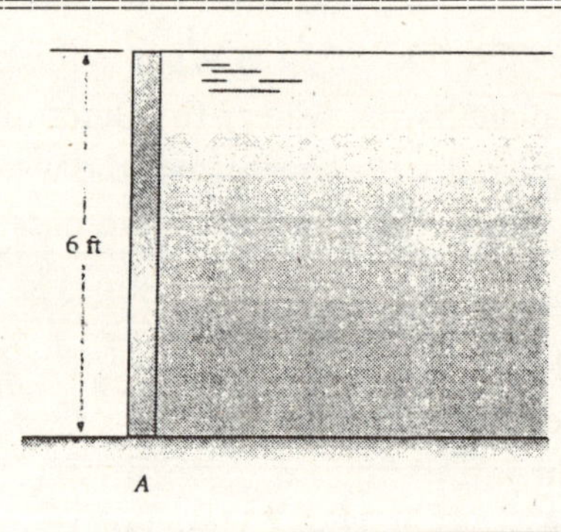

Solution Since the atmospheric pressure applies to both sides of the gate, it is ignored. The strategy is to use the "volume" of the pressure to calculate the force and moment. The pressure distribution is a triangle with base 6γ and altitude 6 *ft*. The "volume" is $F = \left(\frac{1}{2}\right)62.4(6)(6)(4) = 4492.8 \ lb$. The reaction is equal and opposite: $\boxed{A_x = 4492.8 \ lb}$ to right. The centroid of the pressure "volume" is $\bar{d} = \left(\frac{1}{3}\right)6 = 2 \ ft$ from the bottom, hence the moment acting on A is $\boxed{M = 2F = 8985.6 \ lb\text{-}ft}$ clockwise. The full weight of the gate rests on A, so that the vertical component of the reaction is $\boxed{A = 200 \ lb}$ upward.

==================================<>==================================

Problem 9.108 The homogenous gate weighs 100 *lb* and its width (the dimension into the page) is 3 *ft*. The weight density of the water is $\gamma = 62.4\ lb/ft^3$, and the atmospheric pressure is $2120\ lb/ft^2$. Determine the reactions at A and B.

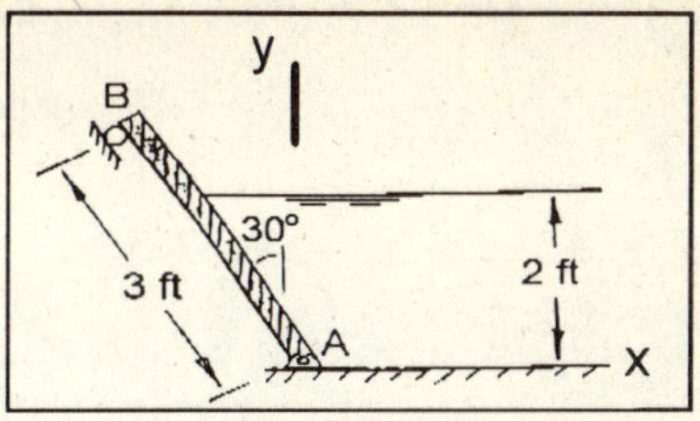

Solution The atmospheric pressure acts on both sides of the gate, so it is ignored. The strategy is use the "volume" of the pressure distribution to compute the force acting on the face of the gate. The pressure distribution is a triangle with base 2γ. The pressure acts over an area $\left(\dfrac{6}{\cos(30)}\right) = 6.92\ ft^2$

Thus the "volume" is $F = \left(\dfrac{1}{2}\right)(62.4)(2)(6.92) = 432.32\ lb$. This force acts normally to the surface of the gate, or at an angle of $\theta = 210^o$ relative to the positive x-axis. The centroid of the pressure is $\bar{d} = \left(\dfrac{2}{3}\right)\left(\dfrac{1}{\cos 30^o}\right) = 0.7698\ ft$ along the inner face of the gate from A. The sum of the moments about A is $\sum M_A = \bar{d}F + W(1.5)\sin 30^o - 3B = 0$, from which $\boxed{B = 135.933\ lb}$, normal to the inner face of the gate. The sum of the forces normal to the gate surface is $\sum F_N = +A_N - F + B - W\sin 30^o = 0$,

from which $A_N = 346.3\ lb$ at an angle $\alpha = 30^{o^o}$ relative to the positive x axis. The sum of the forces acting parallel to the gate surface is $\sum F_P = A_P - W\cos(30^o) = 0$, from which $A_P = 86.6\ lb$ at an angle of 120°. Thus the components of the reaction at A are

$\boxed{A_x = -346.3\cos(210^o) + 86.6\cos(120^o) = 256.7\ lb}$, to the right, and

$\boxed{A_y = -346.3\sin(210^o) + 86.6\sin(120^o) = 248.2\ lb}$ upward.

==================================<>==================================

Problem 9.109 The gate has water of depth $d = 1\ m$ on one side. The width of the gate (the dimension into the page) is 2 *m*, and the gate weighs 800 *N*. The mass density of the water is $\rho = 1000\ kg/m^3$ and the atmospheric pressure $p_{atm} = 10^5\ Pa$. Determine the reactions on the gate at A and B. (The support at B exerts only a horizontal reaction on the gate.)

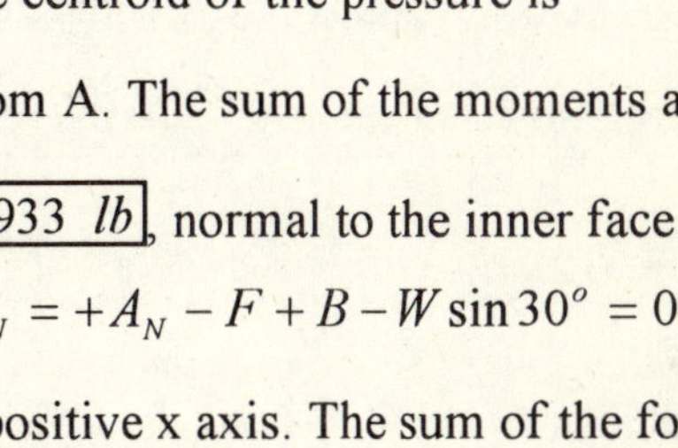

Solution Since atmospheric pressure acts on both faces of the gate, it is ignored. The weight density of the water is $\gamma = \rho g = 9810\ N/m^3$. The strategy is to use the "volume" of the pressure distribution to determine the reactions. The force on the gate surface above A is $F_1 = (d - 0.5)^2\gamma = 2452.5$. The moment about A due to the force above point A $M_1 = \left(\dfrac{1}{3}\right)(d-0.5)^3\gamma = 408.75\ Nm$. The force on the gate surface below A is $F = (d-0.5)\gamma + (0.5^2)\gamma = 7357.5\ N$. *Check:* The sum of the forces is $F_1 + F_2 = d^2\gamma = 9810\ N$. *check.*

Problem 9.109 continued on next page

continuation of Problem 9.109..........

The moment about A due to the force below A is due to a rectangle and a triangle,

$M_2 = 0.25(d-0.5)(\gamma) + (0.5^2)\left(\frac{\gamma}{3}\right) = 2043.75\ N\,m$. The sum of the moments about A:

$\sum M_A = +M_1 - M_2 + 0.5B = 0$, from which $\boxed{B = 3270\ N}$. The sum of the forces:

$\sum F_x = A_x - F_1 - F_2 + B = 0$, from which $\boxed{A_x = 6540\ N}$ to the right. The vertical reaction at A is the weight of the gate: $\boxed{A_y = 800\ N}$ upward.

Problem 9.110 When the water depth d in Problem 9.109 exceeds a certain value, the gate will open releasing the water. What is this value?

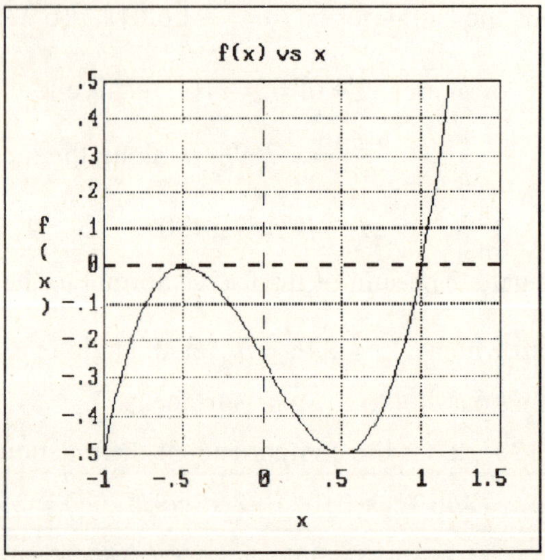

Solution The force closing the gate will be zero when the moments about A equal zero. From the solution to Problem 109, $M_1 = \left(\frac{1}{3}\right)(d-0.5)^3\gamma$, and

$M_2 = 0.25(d-0.5)(\gamma) + (0.5^2)\left(\frac{\gamma}{3}\right)$. Equating and reducing: $(d-0.5)^3 - \left(\frac{3}{4}\right)(d-0.5) - \left(\frac{1}{4}\right) = 0$. Let $x = d - 0.5$, from which $x^3 - \left(\frac{3}{4}\right)x - \left(\frac{1}{4}\right) = 0$.. This equation can be solved by iteration or by graphing the function $f(x) = x^3 - \left(\frac{3}{4}\right)x - \left(\frac{1}{2}\right)$ to find the zero crossings. Since there is more than one real root, the safest procedure is to graph the function. The graph is shown. The osculating contact at $x = -0.5$ corresponds to $d = 0$. This is a valid but trivial solution: it corresponds to zero water behind the gate. The root at $x = 1$ corresponds to a depth of $\boxed{d = 1.5\ m}$, at which point the closing forces are neutral.

Problem 9.111 The dam has water of depth 4 ft on one side. The width of the dam (the dimension into the page) is 8 ft. The weight density of the water is $\gamma = 62.4\ lb/ft^3$, and the atmospheric pressure $p_{atm} = 2120\ lb/ft^2$. If you neglect the weight of the dam, what are the reactions at A and B?

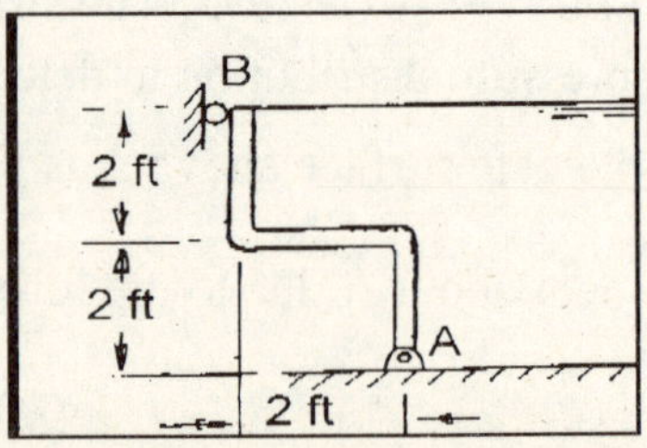

Solution The atmospheric pressure acts on both faces of the dam, so it is ignored. The strategy is to use the "volume" of the pressure distribution to the determine the reactions. The pressure distribution is a triangle of base 4γ and altitude $4\ ft$. The force on the vertical faces of the dam is $F_1 = \left(\frac{1}{2}\right)(4)(4)(\gamma)8 = 3993.6\ lb$. The moment about A due to the force on the vertical faces is $M_1 = \left(\frac{4}{3}\right)F_1 = 5324.8\ ft\,lb$. The force on the horizontal face of the dam is $F_2 = (2)(\gamma)(2)(8) = 1996.8\ lb$.

Problem 9.111 continued on next page

continuation of Problem 9.111..........

The moment about A due to the force on the horizontal face is $M_2 = 1F_2 = 1996.8\ ft\ lb$. The sum of the moments about A: $\sum M_A = M_1 + M_2 - 4B = 0$, from which $\boxed{B = 1830.4\ lb}$. The sum of the forces: $\sum F_x = A_x - F_1 + B = 0$, from which $\boxed{A_x = 2163.3\ lb}$ to the right. $\sum F_y = A_y - F_2 = 0$, from which $\boxed{A_y = 1996.8\ lb}$ upward.

Problem 9.112 The spherical tank of 400 *mm* inner radius is full of water ($\rho = 1000\ kg/m^3$). The pressure of the water at the top of the tank is $4 \times 10^5\ Pa$. (a) What is the pressure of the water of the water at the bottom of the tank? (b) What is the total force exerted on the inner surface of the tank by the pressure of the water?

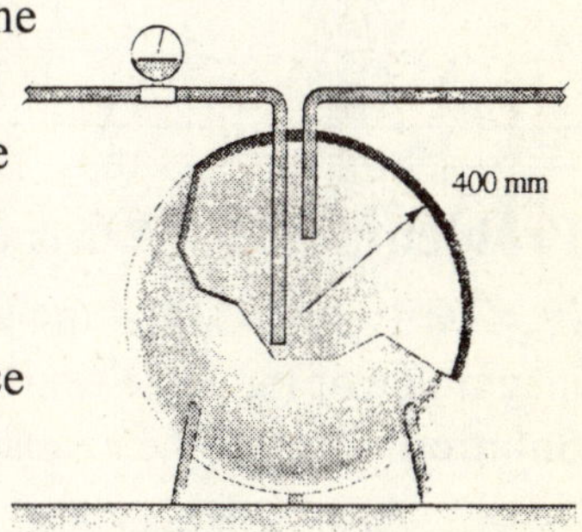

Solution The weight density is $\gamma = \rho g = 9810\ N/m^3$. ((a) The pressure distribution is $P(x) = p_O + \gamma x$.) The pressure at the bottom of the tank is $\boxed{P(0.8) = p_o + \gamma(0.5) = 4 \times 10^5 + (9810)(0.8) = 4.0785 \times 10^5\ Pa}$.

(b) From the free body diagram of the sphere of water, the unbalanced force is the weight of water, acting downward: $\boxed{W = \left(\frac{4}{3}\right)\pi R^3 \gamma = 2629.9\ N}$

Problem 9.113 Consider a plane, vertical area *A* below the surface of a liquid. Let p_O be the pressure at the surface. (a) Show that the force exerted on the area is $F = \bar{p}A$, where $\bar{p} = p_O + \gamma\bar{x}$ is the pressure of the liquid at the centroid of the area. (b) Show that the x-coordinate of the center of pressure is $x_P = \bar{x} + \frac{\gamma I_{y'}}{\bar{p}A}$, where $I_{y'}$ is the moment of interia about the y' axis through its centroid.

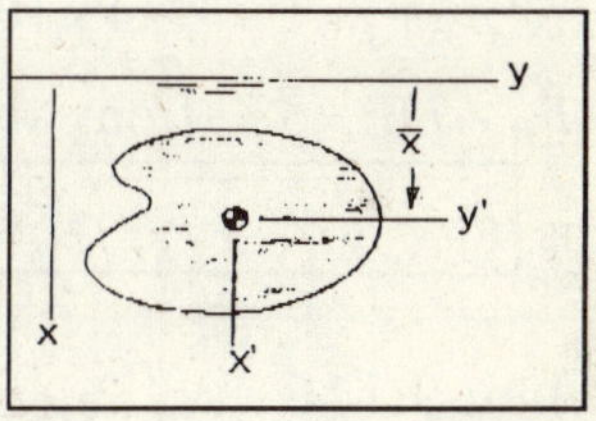

Solution (a) The definition of the centroid of the area is $\bar{x}A = \int_A x\,dA$. . The force on the plate is

$F = \int_A p\,dA = p_O \int_A dA + \gamma \int_A x\,dA = p_O A + \gamma\bar{x}A$. But by definition, the pressure at the centroid of the area is $\bar{p} = p_O + \gamma\bar{x}$, hence $\boxed{F = \bar{p}A}$. (b) The moment about the y axis is $M_{y\text{-}axis} = \int_A px\,dA = p_O \int_A x\,dA + \gamma \int_A x^2\,dA$. The moment of inertia is defined: $I_y A = \int_A x^2\,dA$, from which the moment is $M_{y\text{-}axis} = p_O\bar{x}A + \gamma I_y$. But the moment is also given by $M_{y\text{-}axis} = Fx_P$, where F is the force on the plate and x_P is the distance to the center of pressure. Thus $\bar{p}Ax_P = p_O\bar{x}A + \gamma I_y$,

from which $x_P = \left(\frac{p_O}{\bar{p}}\right)\bar{x} + \frac{\gamma I_y}{\bar{p}A}$. From the parallel axis theorem, $I_y = \bar{x}^2 A + I_{y'}$. Substitute and reduce:

$\boxed{x_P = \frac{(p_O + \gamma\bar{x})\bar{x}}{\bar{p}} + \frac{\gamma I_{y'}}{\bar{p}A} = \bar{x} + \frac{\gamma I_{y'}}{\bar{p}A}}$, which demonstrates the required result.

==================================<>==================================

Problem 9.114 A circular plate of 1 m radius is below the surface of a stationary pool of water. Atmospheric pressure $p_{atm} = 100\ kPa$ and the mass density of the water is $\rho = 1000\ kg/m^3$. Determine (a) the force exerted on the face of the plate by the pressure of the water, and (b) the x coordinate of the center of pressure. (See Problem 9.113)

Solution (a) From Problem 9.113, the pressure on the face of the plate is $F = \bar{p}A$, where $\bar{p}$ is the pressure at the centroid of the area.

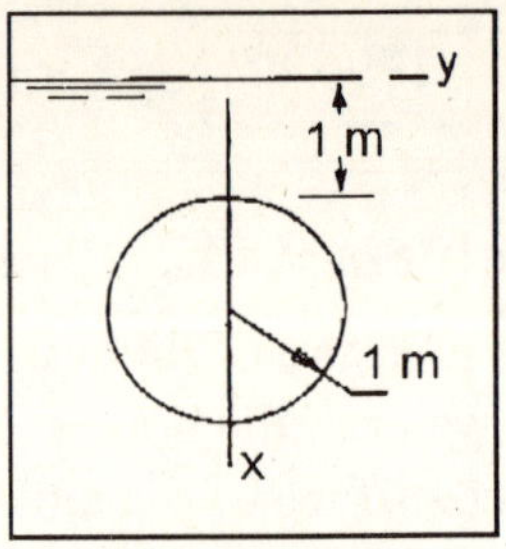

$\boxed{F = (p_O + \gamma 2)(\pi 1^2) = 375.8\ kN}$. (b) From Problem 9.113,

$$\boxed{x_P = \bar{x} + \frac{\gamma I_{y'}}{\bar{p}A} = 2 + \frac{9810\left(\dfrac{\pi 1^4}{4}\right)}{F} = 2.0205\ m}$$

==================================<>==================================

Problem 9.115 A tank consists of a cylinder with hemispherical ends. It is filled with water ($\gamma = 62.4\ lb/ft^3$) and the pressure of the water at the top of the tank is $3000\ lb/ft^2$. Determine the magnitude of the force exerted by the pressure of the water on one of the hemispherical ends.

Solution Isolate the hemisphere as a free body. . The weight of the water is $W = \left(\frac{2}{3}\right)\pi R^3\gamma = 1.045 \times 10^6\ lb$, acting downward. The force due to the pressure is obtained by integrating over the circular face on the left of the hemisphere:

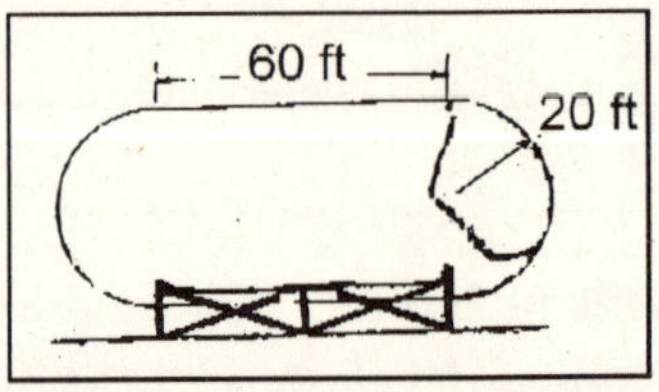

$F = \int_A p dA = p_O \int_A dA + \gamma \int_A x dA = p_O A + \gamma \bar{x} A$. acting to the right. Substituting,

$F = \pi R^2 p_o + \gamma \pi R^3 = 5.34 \times 10^6\ lb$ The sum of the forces acting on the hemisphere:

$0 = -\vec{F}_P + \vec{i}W + \vec{j}F$, from which $\vec{F}_P = \vec{i}W + \vec{j}F = 1.04 \times 10^6 \vec{i} + 5.34 \times 10^6 \vec{j}\ lb$. The magnitude of the force is $\boxed{F_P = \sqrt{W^2 + F^2} = 5.4396 \times 10^6\ lb}$

==================================<>==================================

Problem 9.116 An object of volume V and weight W is suspended below the surface of the stationary liquid of weight density γ. Show that the tension in the cord is $W - V\gamma$. The result is due to Archimedes (287-212 B.C.)

Solution The result follows from the free body diagram of the space occupied by the object and the development of the force exerted by the pressure in terms of the volume of the object. Let A be the area bounding the volume V. The force exerted on a surface A is

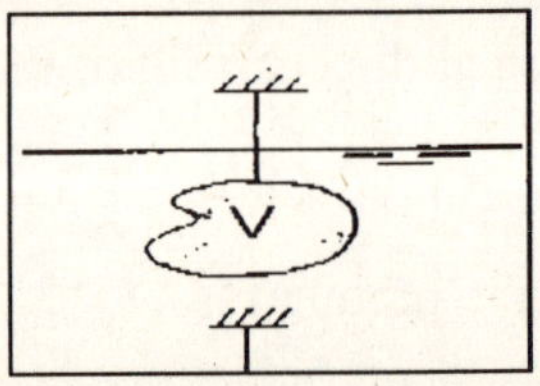

$\vec{F} = -\int_A \vec{n} p dA$, where $\vec{n}$ is a unit vector normal to the elemental surface dA, positive outward from the surface, and the negative sign comes from the equilibrium condition that the reaction force acts oppositely to the unit vector. Choose an x, y, z coordinate system such that the elemental forces are $\vec{n} p dA = \vec{i} p dydz + \vec{j} p dxdz + \vec{k} p dxdy$. The force becomes

$\vec{F} = -\int_A \vec{i} p dydz - \int_A \vec{j} p dxdz - \int_A \vec{k} p dxdy$. But these integrals can also be written as

$\vec{F} = -\vec{i}\int_V \frac{\partial p}{\partial x} dxdydz - \vec{j}\int_V \frac{\partial p}{\partial y} dxdydz - \vec{k}\int_V \frac{\partial p}{\partial z} dxdydz$.

Problem 9.116 continued on next page

---

continuation of Problem 9.116...........

This set of integrals can be collapsed into $\vec{F} = -\int_A \vec{n} p dA = -\int_V \nabla p \, dV$,
where the volume V is bounded by the surface A,, ∇p is a shorthand notation for
$\nabla p = \vec{i}\frac{\partial p}{\partial x} + \vec{j}\frac{\partial p}{\partial y} + \vec{k}\frac{\partial p}{\partial z}$, and $dV = dxdydz$. The pressure $p = p_o + \gamma x$, from which $\nabla p = \vec{i}\gamma$, and the integral becomes $\vec{F} = -\int_A \vec{n} p dA = -\vec{i}\gamma \int_V dV = -\vec{i}\gamma V$. The weight of the object acts in the positive x direction, so the resultant force is $\boxed{\vec{F}_R = \vec{i}(W - \gamma V)}$

Problem 9.117 Determine the internal forces and moments at x (a) if $x = 250\ mm$; (b) if $x = 750\ mm$.

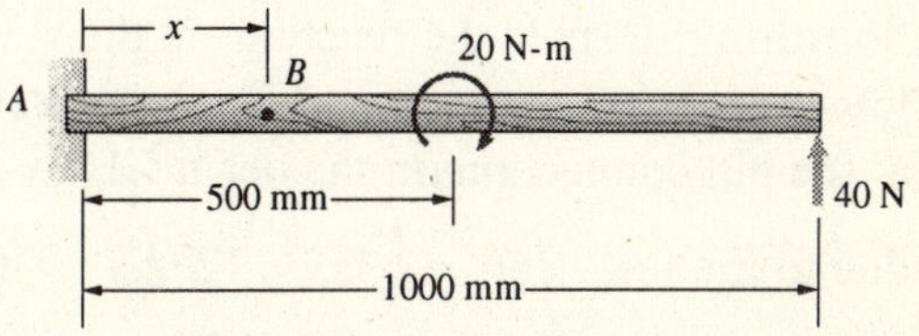

Solution *The complete beam:* The sum of the moments about A is $\sum M_A = M_A - 20 + 1(40) = 0$, from which $M_A = -20\ N m$. The sum of the forces:
$\sum F_y = A_y + 40 = 0$, from which $A_y = -40\ N$.
$\sum F_x = A_x = 0$ *The internal forces at* $x = 250\ mm$.The shear is $\boxed{V_1(x) = -40\ N}$. The moment is $M_1(x) = \int V_1(x)dx + C_1 = -40x + C_1$ At $x = 0$, $M_1(0) = -M_A = 20\ N m$, from which $C_1 = 20\ kN$ .Thus the moment is $\boxed{M(0.25) = -40(0.25) + 20 = 10\ Nm}$. *The internal forces at* $x = 750\ mm$. Divide the beam into two segments: $0 \le x < 0.5\ (m)$, and $0.5 \le x < 0.75\ (m)$. The shear in the second segment is $\boxed{V_2(x) = -40\ N}$ The moment is $M_2(x) = \int V_2(x)dx + C_2 = -40x + C_2$. A known discontinuity exists in the moment at $x = 0.5\ m$, $M_1(0.5) - M_2(0.5) = -20\ N m$, from which $C_2 = 40$ and the moment is $M_2(x) = -40x + 40\ Nm$. At $x = 0.75$, $\boxed{M_2(0.75) = -40(0.75) + 40 = 10\ N m}$. The axial forces $\boxed{P(x) = 0}$ everywhere.

Problem 9.118 The homogenous beam weighs 1000 *lb*. What are the internal forces and bending moments at its midpoint?

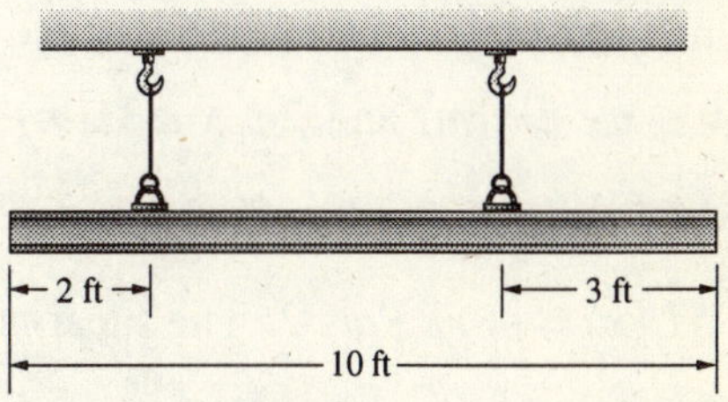

Solution Denote the support reactions by A and B. The load distribution is $w = 1000/10 = 100\ lb/ft$. The moment about the left end due to the load of the beam is

$M_L = \int_0^{10} wxdx = 100\left[\frac{x^2}{2}\right]_0^{10} = 5000\ ft\ lb$. The sum of the moments about the left end is $\sum M = -M_L + 2A + 7B = 0$ The sum of the forces:
$\sum F_y = A + B - 1000 = 0$. Solve the two simultaneous equations to obtain: $A = 400\ lb$, $B = 600\ lb$. $\sum F_x = 0$. Divide the left half of the beam into two parts: $0 \le x < 2\ (ft)$ and $2 \le x < 5\ (ft)$.

Problem 9.118 continued on next page

continuation of Problem 9.118..........

Part 1: The shear is $V_1(x) = -100x$. The moment is $M_1(x) = \int V_1(x)dx + C_1 = -50x^2 + C_1$. At $x = 0$, $M_1(0) = 0$, thus $C_1 = 0$, and the moment is $M_1(x) = -50x^2$. *Part 2:* The shear is $V_2(x) = V_1(x) + 400\ lb$. The moment is $M_2(x) = \int V_2(x)dx + C_2 = -50x^2 + 400x + C_2$. The moment is continuous at $x = 2$, $M_1(2) = M_2(2)$, from which $C_2 = -800$, and the moment is $M_2(x) = -50x^2 + 400x - 800$. At the midpoint, $x = 5\ ft$, the shear is $\boxed{V_2(5) = -100(5) + 400 = -100\ lb}$ and the moment is $\boxed{M_2(5) = -50(25) + 400(5) - 800 = -50\ ft\ lb}$. The axial force is zero, $\boxed{P(5) = 0}$

Problem 9.119 Determine the internal forces and moments at A.

Solution Denote the reactions at the slide support of BC by D, and the reactions at the upper and lower pin supports by E and G respectively, and the reaction at the pin connection on the beam GB by H. .The total force due to the load is $F = \int_0^4 3dx = 12\ kN$. The moment about the left end of BC is $M = \int_0^4 3xdx = 24\ kN\,m$. The sum of the moments about the point B is $\sum M_B = -M + 3D = 0$, from which $D = 8\ kN$. The sum of the forces: $\sum F_y = B - F + D = 0$, from which $B = 4\ kN$ is the vertical reaction at B. The centroid distance of the load from the left end of BC is $d = \frac{M}{F} = \frac{24}{12} = 2\ m$. The sum of the moments about the upper pin support is $\sum M = -(1+2)F + 2G_x = 0$, from which $G_x = 18\ kN$. With these known quantities, the following equations can be written: (See figure to right for assumed positive directions.)

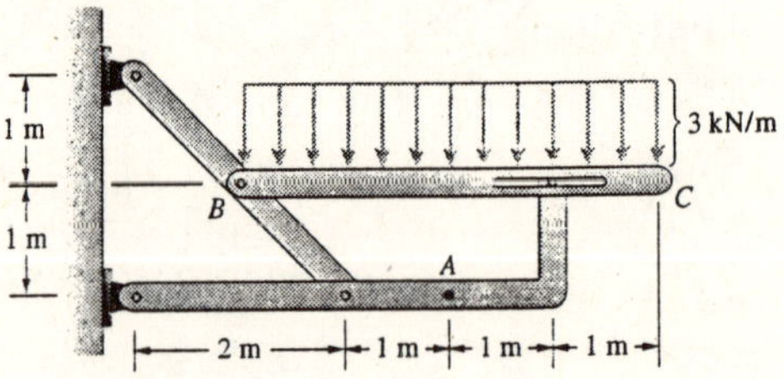

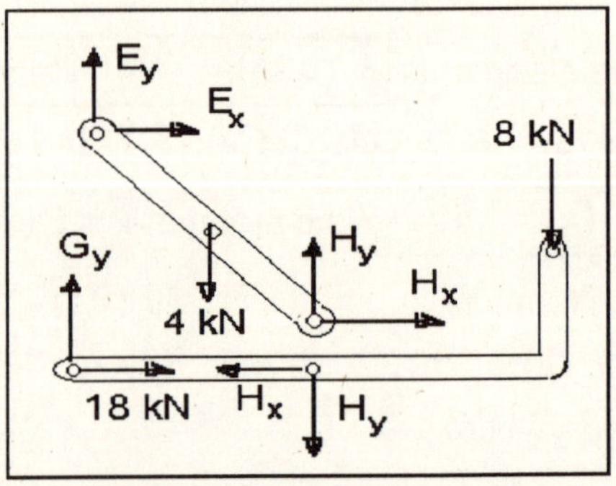

Sum of moments about H for lower member yields $G_y = -8\ kN$. The sum of the forces for the lower member yields: $H_y = -16\ kN$ and $H_x = 18\ kN$. Divide the lower member into two parts $0 \le x < 2\ (m)$ and $2 \le x < 3\ (m)$. *Part 1:* The shear is $V_1(x) = -8\ kN$. The moment is $M_1(x) = \int V_1(x)dx + C_1 = -8x + C_1$. The moment at $x = 0$, $M_1(0) = 0$, from which $C_1 = 0$, and $M_1(x) = -8x$. *Part 2:* The shear is $V_2(x) = V_1(x) + 16 = 8\ kN$. The moment is $M_2(x) = \int V_2(x)dx + C_2 = 8x + C_2$. The moment is continuous at $x = 2$, $M_1(2) = M_2(2)$, from which $C_2 = -32$, and the moment is $M_2(x) = 8x - 32\ kN\,m$. Thus at point A, $x = 3$, the internal forces and bending moment are $\boxed{V_2(3) = 8\ kN}$, $\boxed{P(3) = 0}$, $\boxed{M_2(3) = 24 - 32 = -8\ kN\,m}$

Problem 9.120 Draw the shear force and bending moment diagrams for the beam in Problem 9.117.

Solution From the solution to Problem 9.117 , the shear and bending moment are $V_1(x) = -40\ N$, $M_1(x) = -40x + 20\ N\,m\ (0 \le x < 0.5\ m)$, and $V_2(x) = -40\ N$, $M_2(x) = -40x + 40\ N\,m$, $(0.5 \le x < 1\ m)$

The shear force and bending moment diagrams are shown.

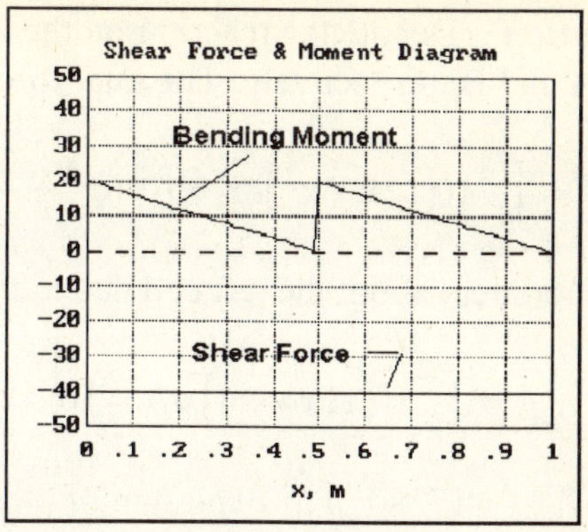

Problem 9.121 Draw the shear force and bending moment diagrams for the beam in Problem 9.118.

Solution From the solution for Problem 9.118, the shear force and bending moment are: (1) $V_1(x) = -100x\ lb$, $M_1(x) = -50x^2\ ft\,lb, (0 \le x < 2\ ft)$. (2) $V_2(x) = -100x + 400\ lb$, $M_2(x) = -50x^2 + 400x - 800\ ft\,lb, (2 \le x < 7\ ft)$. (3) $V_3(x) - 100x + 1000\ lb$, $M_3(x) = -50x^2 + 1000x - 5000\ ft\,lb$, $(7 \le x < 10\ ft)$

The shear force and bending moment diagrams are shown.

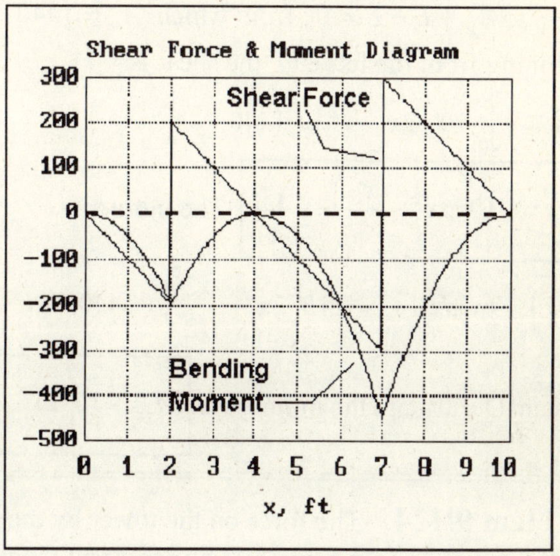

Problem 9.122 Draw the shear force and bending moment diagrams for the beam in Problem 9.119.

Solution From the solution to Problem 9.119, the shear force and bending moment are: (1) $V_1(x) = -8\ kN$, $M_1(x) = -8x\ kN\,m$, $(0 \le x < 2\ m)$, (2) $V_2(x) = 8\ kN$, $M_2(x) = 8x - 32\ kN\,m$, $(2 \le x < 4\ m)$

The shear force and bending moment diagrams are shown.

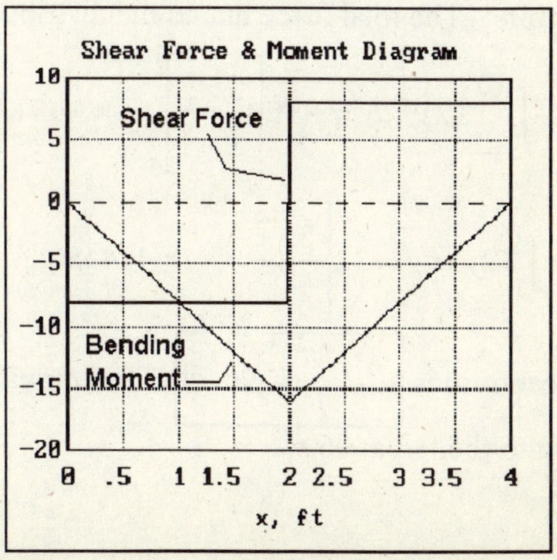

Problem 9.123 Use Eqs (9.7)-(9.12) to determine the shear force and bending moment diagrams for the beam.

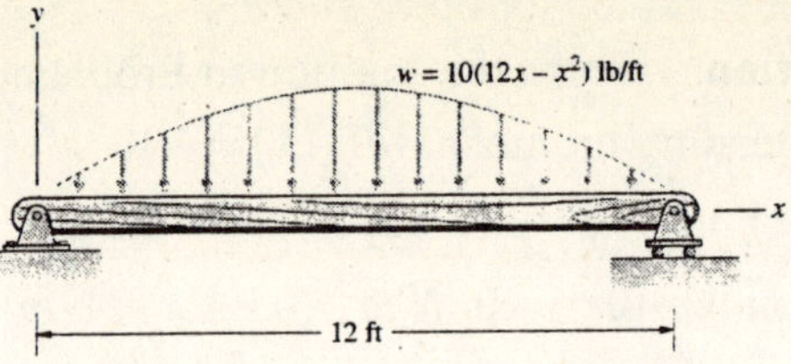

Solution Denote the reactions at the left and right ends by A and B, respectively. The total force due to the load is $F = \int_0^{12} 10(12x - x^2)dx = 10\left[6x^2 - \frac{x^3}{3}\right]_0^{12} = 2880\ lb$.

The moment about the left end due to the load is

$$M_{LOAD} = \int_0^{12} 10(12x - x^2)xdx = 10\left[4x^3 - \frac{x^4}{4}\right]_0^{12}$$

$M_{LOAD} = 17280\ ft\ lb$. The sum of the moments about the left end: $\sum M_A = -M + 12B = 0$, from which $B = 1440\ lb$. The sum of the forces:

$\sum F_y = A_y + B - F = 0$, from which $A_y = 1440\ lb$.

Beginning from the left end, the shear is

$$V(x) = -\int_0^x 10(12x - x^2) + 1440$$

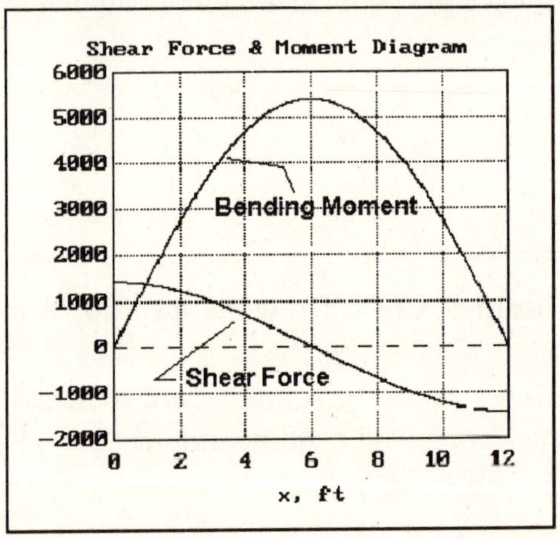

$\boxed{V(x) = -10\left(6x^2 - \frac{x^3}{3}\right) + 1440}$ The moment is

$M = \int V(x)dx + C = -10\left(2x^3 - \frac{x^4}{12}\right) + 1440x + C$. The moment is zero at $x = 0$, hence the constant $C = 0$, and the moment is $\boxed{M = -10\left(2x^3 - \frac{x^4}{12}\right) + 1440x}$

Problem 9.124 The force on the tower by the wind is described by the distributed force shown. The wind is parallel to the x axis. If you represent the wind force by a single equivalent force ***F***, what is ***F*** and where does its line of action intersect the y-axis?

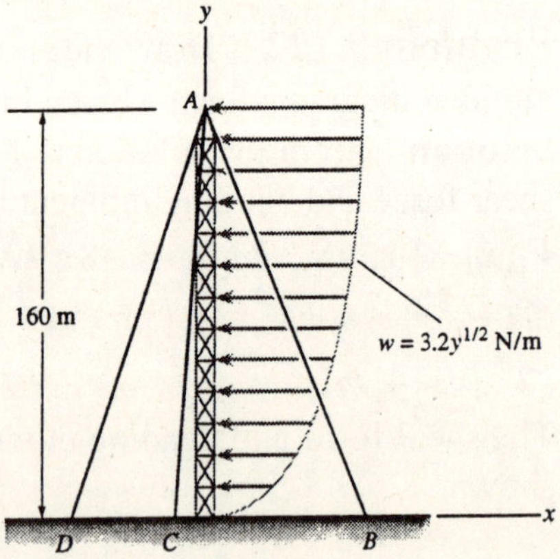

Solution The total force due to the distributed load is

$$F = -\int_0^{160} 3.2y^{\frac{1}{2}}dy = -3.2\left[\frac{2}{3}y^{\frac{3}{2}}\right]_0^{160} = -4317.56\ N.$$ The moment is

$$M = \int_0^{160} 3.2y^{\frac{3}{2}}dy = 3.2\left[\frac{2}{5}y^{\frac{5}{2}}\right]_0^{160} = 414486\ N\,m,$$ counterclockwise.

The centroid is $\boxed{\bar{d} = \frac{M}{F} = 96\ m}$. The equivalent force is $\boxed{4.318\ kN}$ acting to the left at 96 m.

Problem 9.125 If the base of the tower in Problem 9.124 can be modeled as a ball and socket support and the tension in cable AD is 1 *kN*, what are the tensions in cable AB and AC.

Solution The reaction at the base is found from the sum of the moments about A: $\sum M_A = -(160-96)F + 160E = 0$, from which $\vec{E} = +1727\vec{i}\ N$. Assume that all cables are in tension. Then the horizontal components of the tension are

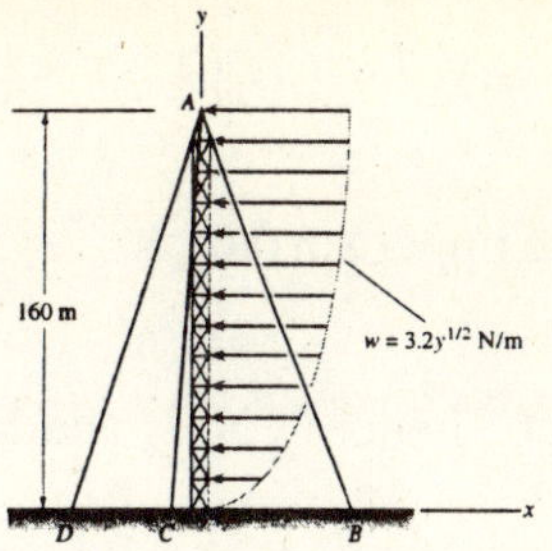

$$\vec{B}_H = |\vec{B}_H|(\vec{i}\cos 20^o - \vec{k}\sin 20^o) = |\vec{B}_H|(0.9397\vec{i} - 0.3420\vec{k})$$

$$\vec{C}_H = |\vec{C}_H|(\vec{i}\cos 260^o - \vec{k}\sin 260^o) = |\vec{C}_H|(-0.1736\vec{i} + 0.9848\vec{k})$$

$$\vec{D}_H = |\vec{D}_H|(\vec{i}\cos 140^o - \vec{k}\sin 140^o) = |\vec{D}_H|(-0.7660\vec{i} - 0.6428\vec{k})$$

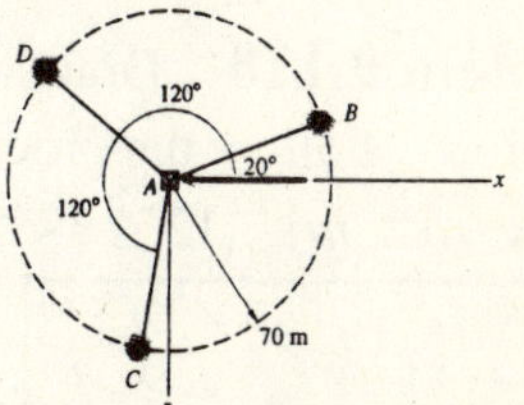

The sum of the forces: $\sum \vec{F} = \vec{B}_H + \vec{C}_H + \vec{D}_H + \vec{F} + \vec{E} = 0$

$$\sum F_x = -4317.6 - 0.1736|\vec{C}_H| + 0.9397|\vec{B}_H| - 0.7660|\vec{D}_H| + 1727 = 0$$

$$\sum F_z = -0.9848|\vec{C}_H| + 0.3420|\vec{B}_H| + 0.6428|\vec{D}_H| = 0$$

The angle from the horizontal is $\theta = \tan^{-1}\left(\frac{160}{70}\right) = 66.37^o$. $|\vec{D}_H| = 1000(\cos 66.37^o) = 400.8\ N$. Solve:

$|\vec{C}_H| = 1423..9\ N$, $|\vec{B}_H| = 3346.7\ N$, from which $\boxed{|\vec{C}| = \frac{|\vec{C}_H|}{\cos 66.37^o} = 3552.5\ N}$,

$$\boxed{|\vec{B}| = \frac{|\vec{B}_H|}{\cos 66.37^o} = 8349.6\ N}$$

Problem 9.126 Determine the maximum bending moment in the tower in Problem 9.24 and the position on the y-axis where is occurs.

Solution From the solution to Problem 9.125, the reaction at the base of the tower is 1727 *N*. The shear force is $V(y) = -\int_0^Y 3.2y^{\frac{1}{2}}dy + 1727 = -2.133y^{\frac{3}{2}} + 1727$. The bending moment is

$M = \int V(y)dy + C = -0.8533y^{\frac{5}{2}} + 1727y + C$. The bending moment is zero at $y = 0$, hence $C = 0$, and

the bending moment is $M = -0.8533y^{\frac{5}{2}} + 1727y\ Nm$. The maximum bending moment occurs at

$V(y) = 0$, from which $-2.133y^{\frac{3}{2}} + 1727 = 0$. Solve: $\boxed{y = 86.86\ m}$, and $\boxed{M_{MAX} = 90004.9\ Nm}$

Problem 9.127 The steel plate weighs 234 *lb*. Obtain an equation for the shear force V as a function of x from $x = 0$ to $x = 12$.

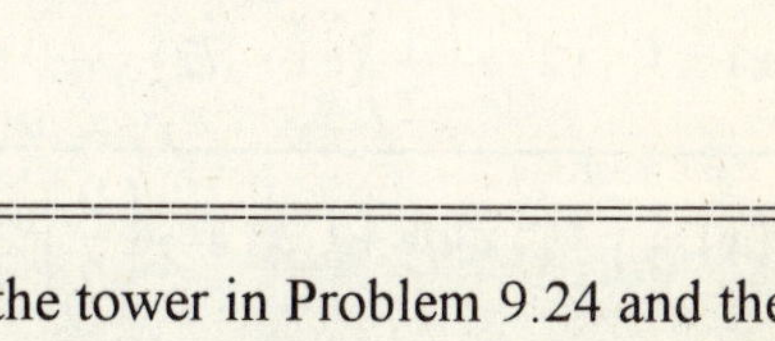

Solution The strategy is to obtain an equation for the area of the plate as a function of *x*. The area of the plate is $A = 2(34) + 4(10) + 4(12) = 156\ in^2$. The centroid of the plate relative to the left end is $\bar{x} = 17\ in$, from symmetry. The sum of the moments about the left end is $\sum M = -234(17) + 34B = 0$.

Problem 9.127 continued on next page..........

.............*continuation of solution to Problem 9.127....*

From which $B = 117\ lb$. From the sum of the forces the reaction at $A = 117\ lb$, as expected from symmetry. *Part 1*: $(0 \le x < 12\ in)$. The density per unit area is $\rho = \frac{234}{156} = \left(\frac{3}{2}\right)\ lb/in^2$. The equation for the upper surface is $y = \frac{x}{3} + 2$. The area is $A_1(x) = \int y dx + C_1 = \frac{x^2}{6} + 2x + C_1$. The area is zero at $x = 0$, hence $C_1 = 0$, and the shear is $\boxed{V_1(x) = V_1(0) - \left(\frac{3}{2}\right)\left(\frac{x^2}{6} + 2x\right) = 117 - \frac{x^2}{4} - 3x\ lb}$.

Problem 9.128 Draw the shear force and bending moment diagrams for the plate in Problem 9.127.

Solution Follow the procedure begun in the solution to Problem 9.127. Divide the plate into intervals: $(0 \le x < 12\ in)$, $(12 \le x < 22\ in)$, and $(22 \le x < 34\ in)$. *Part 1:* The shear is given in Problem 9.127: $\boxed{V_1(x) = 117 - \frac{x^2}{4} - 3x\ \ lb}$. The moment is $M_1(x) = \int V_1(x)dx + C_1 = 117x - \frac{x^3}{12} - \frac{3x^2}{2} + C_1$, The moment is zero at $x = 0$, hence $C_1 = 0$, and $\boxed{M_1(x) = 117x - \frac{x^3}{12} - \frac{3x^2}{2}\ \ in\ lb}$ *Part 2:* The area is $A_2(x) = \int 6dx + C_2 = 6x + C_2$. The area is zero at $x = 12$, hence $C_2 = -72$ The shear is

$\boxed{V_2(x) = V_1(12) - \left(\frac{3}{2}\right)(6x - 72) = -9x + 153\ \ lb}$. The moment is

$M_2(x) = \int V_2(x)dx + C_{2M} = -\left(\frac{9}{2}\right)x^2 + 153x + C_{2M}$. The moment is continuous at $x = 12$, $M_1(12) = M_2(12) = 1044\ ft\ lb$, hence $C_{2M} = -144$ and the moment is $\boxed{M_2(x) = -\frac{9x^2}{2} + 153x - 144\ \ in\ lb}$

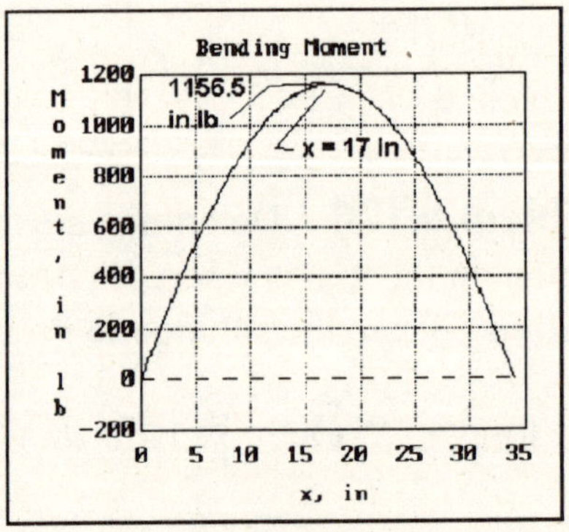

Part 3: The equation for the upper surface is $y = -\frac{x}{3} + \frac{40}{3}$. The area is $A_3(x) = \int y dx + C_3 = -\frac{x^2}{6} + \frac{40x}{3} + C_3$. The area is zero at $x = 22$, hence $C_3 = -212.67$. The shear is $V_3(x) = V_2(22) - \left(\frac{3}{2}\right)A_3(x)$, $\boxed{V_3(x) = \frac{x^2}{4} - 20x + 274\ lb}$. The moment is $M_3(x) = \int V_3(x)dx + C_3 = \frac{x^3}{12} - 10x^2 + 274x + C_3$.

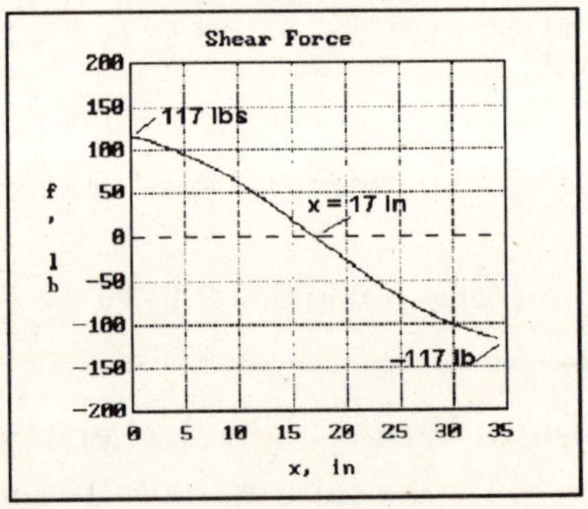

The moment is continuous at $x = 22$, $M_2(22) = M_3(22) = 1044$, hence $C_3 = -1031.33$, and

$\boxed{M_3(x) = \frac{x^3}{12} - 10x^2 + 274x - 1031.33\ \ in\ lb}$

The shear force and bending moment diagrams are shown.

=================================<>=================================

Problem 9.129 The power line has a mass of 1.4 *kg/m*. If the line will safely support a tension of 5 *kN,* determine whether it will safely support an ice accumulation of 4 *kg/m*.

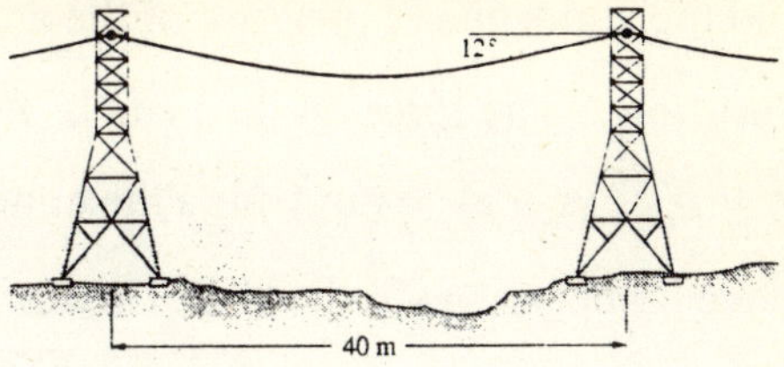

Solution The power line meets the conditions for a catenary. The weight density with an ice load is $\gamma = (1.4 + 4)(9.81) = 52.974 \ N/m$. The angle at the attachment point is related to the length and the parameter *a* by (use Eq (9.21)) $sa = \tan\theta$. But $sa = \sinh(ax)$, and x is known. Thus the parameter can be found from

$a = \dfrac{\sinh^{-1}(\tan\theta)}{x} = \dfrac{\sinh^{-1}(0.2126)}{20} = 0.01055$. The tension at the lowest point is

$T_o = \dfrac{w}{a} = 502153 = 5.02 \ kN$. The maximum tension is $\boxed{T = T_o \cosh(ax) = 5133.7 = 5.133 \ kN}$. Thus the line will not sustain the load.

=================================<>=================================

Problem 9.130 The gate has water of 2 *m* depth on one side. The width of the gate (the dimension into the paper) is 4 *m*, and the mass of the gate is 160 *kg.* The mass density of the water is $\rho = 1000 \ kg/m^3$ and the atmospheric pressure is $p_{atmos} = 10^5 \ Pa$. Determine the reactions on the gate at A and B. (The support B exerts only a horizontal reaction on the gate.)

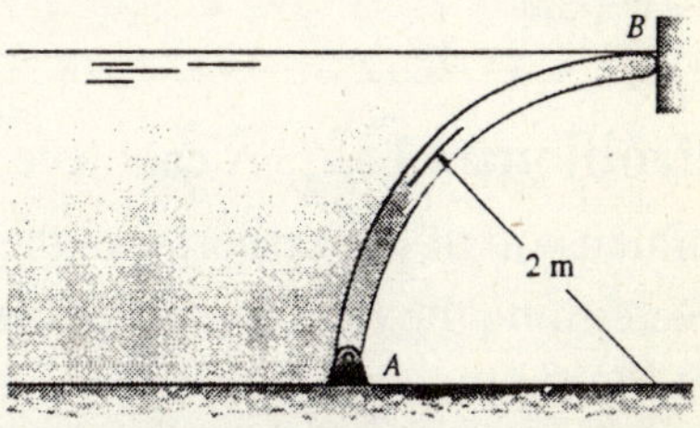

Solution The weight density of the water is $\gamma = 9810 \ N/m^3$. Since atmospheric pressure acts on both sides of the gate, it is ignored. Choose a reference at point B, with *y* positive downward and *x* positive to the right. The strategy is to determine the equivalent forces and moments exerted by the water. Divide the cross section areas (parallel to the paper) into the (1) the area below the gate (2) The area above the gate and (3) the area (into the paper) of a vertical plane passing through A. The area below the gate is

$A_1 = \dfrac{\pi R^2}{4} = 1.571 \ m^2$. The area above the gate is $A_2 = R^2 - A_1 = \dfrac{3\pi R^2}{4} \ 0.8584 \ m^2$. The weight of the water above the gate is $Q = \gamma A_2 4 = 3\gamma\pi R^2 = 33683.9 \ N$. The composite centroid of A_1 and A_2 is $\dfrac{R}{2}$,

from which $-\dfrac{R}{2} = \dfrac{-\left(\dfrac{4\pi R}{3}\right)A_1 + \bar{x}_1 A_2}{A_1 + A_2}$, and $\bar{x}_1 = -R\left(\dfrac{2}{3(4-\pi)}\right) = -1.553 \ m$. The force exerted by the water on the vertical plane at A is $F = \dfrac{\gamma 4R}{2} = 78480 \ N$, acting at $\bar{y}_1 = \dfrac{2R}{3} = 1.333 \ m$. The weight of the gate is $W = 160(9.81) = 1569.6 \ N$, acting at $\bar{x}_2 = \dfrac{2R}{\pi} = -1.273 \ m$. The sum of the forces:

$\sum F_x = A_x + F - B = 0$, $\sum F_y = -A_y + W + Q = 0$. The sum of moments about B:

$\sum M_B = |\bar{x}_1|Q + |\bar{x}_2|W + |\bar{y}_1|F + RA_x - RA_y = 0$. Solve: $\boxed{A_x = -44225.7 \ N}$ (toward left), $\boxed{A_y = 35253.5 \ N}$ (upward) $\boxed{B = 34254.3 \ N}$ (toward left).

=================================<>=================================

Problem 9.131 The water depth at the center of the elliptical aquarium window is 20 *ft*. Determine the magnitude of the net force exerted on the window by the pressure of the sea water ($\gamma = 64 \ lb/ft^3$) and the atmospheric pressure of the air on the opposite side. (See Problem 9.113.)

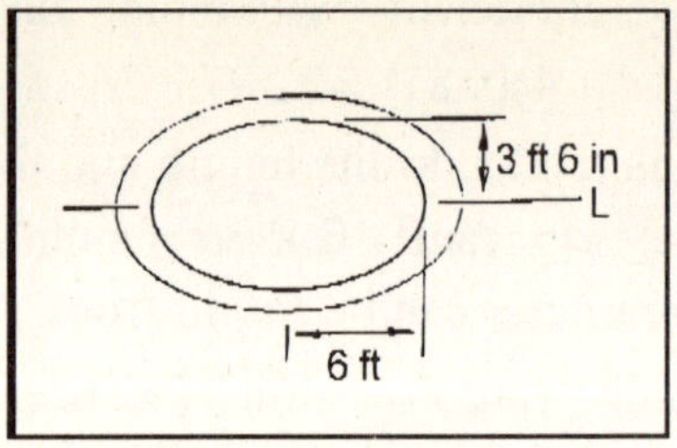

Solution The force on the plate is $F = \int_A p dA$. The pressure is $p = p_o + \gamma x - p_o = \gamma x$, where the atmospheric pressure cancels, since it appears on both sides. The force: $F = \gamma \int_A x dA = \gamma \bar{x} A$, hence the force on the window is $F = 1280 A \ lb$. The area of the ellipse is $A = \pi ab = 21\pi$, and the force is $\boxed{F = 84445 \ lb}$

Problem 9.132 In Problem 9.131, determine the magnitude of the net moment exerted on the window about the horizontal axis L by the pressure of the sea water ($\gamma = 64 \ lb/ft^3$) and the atmospheric pressure of the air on the other side.

Solution The moment is $M = \int pxdA = \gamma \int x^2 dA = \gamma I_y$. The moment of inertia is

$I_y = \dfrac{\pi ab^3}{4} = 202 \ ft^4$, and the moment is $\boxed{M = \gamma I_y = 12930.8 \ ft \ lb}$

Problem 9.133 A cantilevered carport roof has a wedge shaped accumulation of snow. The dimension of the carport into the page is 20 ft. The pressure due to the snow at A is $80 \ lb/ft^2$. Determine the total force and the moment about A due to the snow.

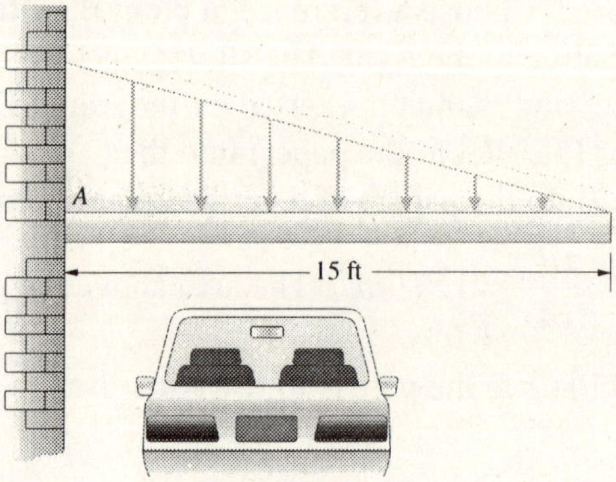

Solution The "volume" of pressure is

$\boxed{F = \left(\frac{1}{2}\right)(80)(15)(20) = 12000 \ lb}$ The centroid of the volume is $d = \left(\frac{1}{3}\right)15 = 5 \ ft$ from the left end. the moment is $\boxed{M = -5(12) = -60000 \ ft \ lb}$

Problem 10.1 The weight of the box is $W = 30\ lb$, $\alpha = 20^o$, and the force F is perpendicular to the inclined surface. The coefficient of friction between the box and the inclined surface is $\mu_S = 0.2$. (a) If $F = 30\ lb$, what is the magnitude of the friction force exerted on the box? (b) If $F = 10\ lb$, show that the box cannot remain at rest on the inclined surface.

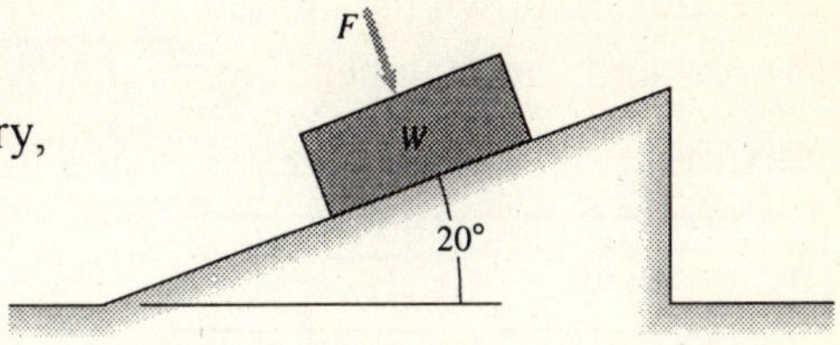

Solution The maximum friction force is defined to be $f = \mu_S N$, where N is the normal force. (a) The box is stationary, hence the friction force is equal to the force acting to move the box down the plane: $\sum F_P = f - W_P = 0$, from which $\boxed{f = W_P = W\sin\alpha = 10.26\ lb}$

(b) The component of force parallel to the surface is $W_P = W\sin\alpha = 10.26\ lb$ acting to move the box down the plane. The friction force is $f = \mu_s(10 + 30\cos\alpha) = 7.638\ lb$, acting to hold the box in place. Since $W_P > f$, the box will move.

Problem 10.2 In Problem 10.1, what is the smallest force necessary to hold the box stationary on the inclined surface?

Solution At impending slip, the sum of the forces parallel to the surface is $\sum F_P = f - W_P = 0$, from which $f = W_P$. The friction force is $f = \mu_S(F + W\cos\alpha)$, and $W_P = W\sin\alpha$. Equate and solve:

$$\boxed{F = W\left(\frac{\sin\alpha}{\mu_S} - \cos\alpha\right) = 30\left(\frac{\sin 20^o}{0.2} - \cos 20^o\right) = 23.1\ lb}$$

Problem 10.3 The truck weighs 18000 lb. (a) If the truck is stationary on the incline, and $\alpha = 10^o$, what is the magnitude of the friction force exerted on it by the road? (b) If the coefficient of friction between the truck's tires and the road is $\mu_s = 0.6$, what is the largest value of α for which the truck can remain stationary? (c) If the incline is icy and $\mu_s = 0.1$, what is the largest value of α for which the truck can remain stationary?

Solution (a) Since the truck is stationary, the friction force balances the force acting to move the truck down the incline: $\boxed{f = 18000\sin 10^o = 3125.67\ lb}$ (b) At impending slip, $f = \mu_s(W\cos\alpha) = W\sin\alpha$, from which $\tan a = \mu_s$, and $\boxed{\alpha = \tan^{-1}(0.6) = 30.96^o}$. (c) The angle is $\boxed{\alpha = \tan^{-1}(0.1) = 5.71^o}$

Problem 10.4 The 50 N box is stationary on the inclined surface. (a) What is the magnitude of the friction force on the box? (b) What is the smallest value of the coefficient of static friction that will permit the box to remain stationary?

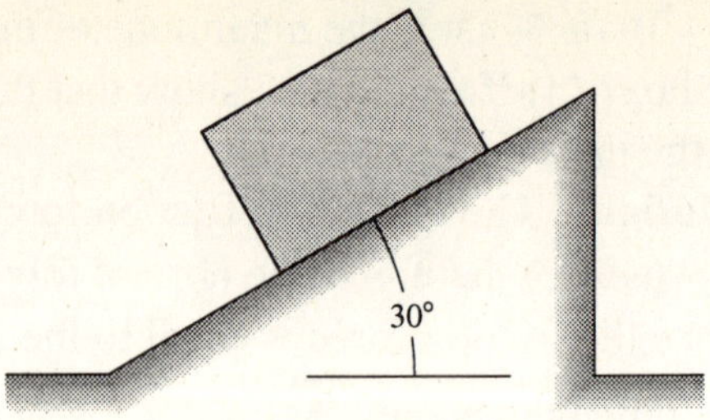

Solution (a) The friction force balances the force acting to slide the box down the incline. $\boxed{f = 50\sin 30^o = 25\ N}$

(b) Assume slip is impending, so that the friction force is maximum: $f = \mu_s(50\cos 30^o) = 50\sin 30^o$, from which

$\boxed{\mu_s = \tan(30^o) \doteq 0.577... = 0.58}$

Problem 10.5 You can measure the coefficient of static friction between two materials as follows. Place a sheet of one of the materials on a flat board. Bond a sheet of the other material to a block A. Place the block on the board and slowly increase the tilt α of the board until the block slips. Show that the coefficient of static friction between the two materials is related to the angle α by $\mu_s = \tan\alpha$.

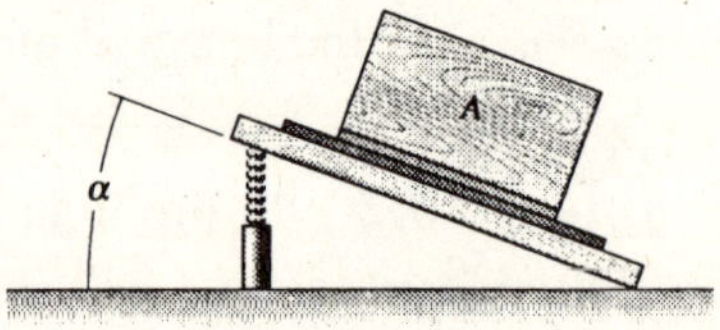

Solution At impending slip, the friction force balances the force acting to slide the block down the incline. Let W be the weight of the block. Then $\sum F = \mu_S(W\cos\alpha) - W(\sin\alpha) = 0$, from which $\boxed{\mu_s = \tan\alpha}$.

Problem 10.6 The mechanism shown is designed to position pieces of luggage on a ramp. The suitcase S weights 20 lb. The mechanism exerts a force on the suitcase in the direction parallel to the ramp. The coefficients of friction between the suitcase and the ramp are $\mu_s = 0.20$ and $\mu_k = 0.18$. (a) Will the suitcase remain stationary on the ramp when the mechanism exerts no force on it? (b) Determine the force the mechanism must exert on the suitcase to start it moving up the ramp. (c) Determine the force that the mechanism must exert on the suitcase to keep it moving up the ramp once it has started moving.

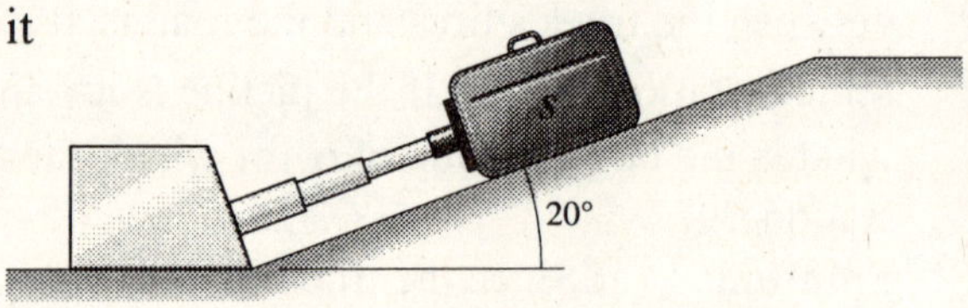

Solution (a) The force acting to move the suitcase down the ramp is $F_P = W\sin\alpha = 20\sin 20^o = 6.84\ lb$. The maximum static friction force acting to keep the suitcase stationary is $f = \mu_s W\cos\alpha = 0.2(20)\cos 20^o = 3.76\ lb$. Since $F_P > f$, the suitcase will slide down the ramp. (b) The mechanism must exert a force to overcome static friction and the action of gravity. Thus $\boxed{F_M = \mu_s W\cos\alpha + W\sin\alpha = 20(0.2\cos 20^o + \sin 20^o) = 10.6\ lb}$. (c) The force required to keep it moving is $\boxed{F_M = \mu_k W\cos\alpha + W\sin\alpha = 20(0.18\cos 20^o + \sin 20^o) = 10.22\ lb}$

Problem 10.7 The weight of the box is 30 *N*. The force *F* is horizontal, and the angle $\alpha = 20^o$. The coefficient of static friction between the box and the inclined surface is $\mu_s = 0.2$. For what range of values of *F* will the box remain stationary?

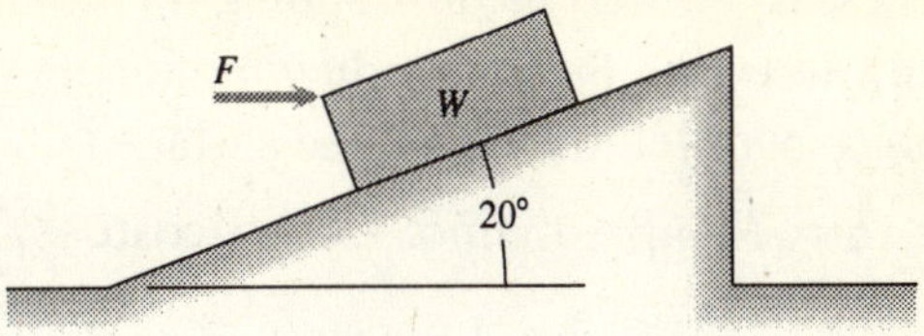

Solution The range of *F* is determined by the value at which the box begins to slide down the incline, and the value at which the box begins to move up the incline. The component of *F* that acts parallel to the plane is $F\cos\alpha$, and the component that acts normally to the plane is $F\sin\alpha$. Gravity acts to slide the box downward. At impending motion up the incline:

$F_{MAX}\cos\alpha - W\sin\alpha - \mu_s(W\cos\alpha + F_{MAX}\sin\alpha) = 0$. Solve: $\boxed{F_{MAX} = \dfrac{W(\mu_s\cos\alpha + \sin\alpha)}{\cos\alpha - \mu_s\sin\alpha} = 18.25\ N}$.

. At impending motion down the incline $F\cos\alpha - W\sin\alpha + \mu_s(W\cos\alpha + F\sin\alpha) = 0$. Solve:

$$\boxed{F_{MIN} = W\left(\frac{\sin\alpha - \mu_s\cos\alpha}{\cos\alpha + \mu_s\sin\alpha}\right) = 4.585\ N}$$

Problem 10.8 Box A weighs 100 *lb* and box B weighs 30 *lb*. The coefficients of friction between Box A and the ramp are $\mu_s = 0.30$ and $\mu_k = 0.28$. What is the magnitude of the friction force exerted on box A by the ramp?

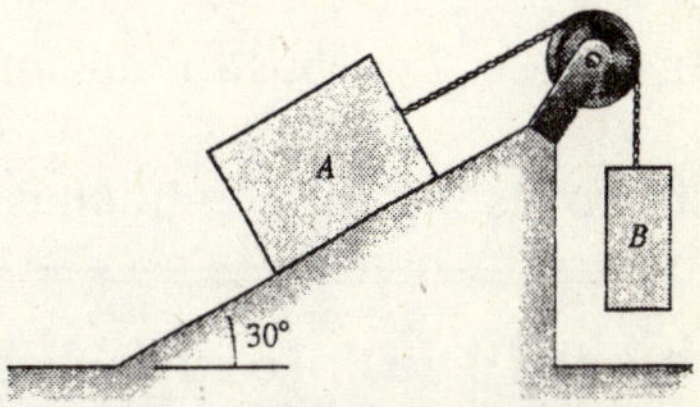

Solution The sum of the forces parallel to the inclined surface is $\sum F = -A\sin\alpha + B + f = 0$, from which

$$\boxed{f = A\sin\alpha - B = 100\cos 30^o - 30 = 20\ lb}$$

Problem 10.9 In Problem 10.8, box A weighs 100 *lb* and the coefficients of friction between box A and the ramp are $\mu_s = 0.30$ and $\mu_k = 0.28$. For what value of the weights of the box B will the system be stationary?

Solution The upper and lower limits on the range are determined by the weight required to move the box up the ramp, and the weight that will allow the box to slip down the ramp. Assume impending slip. The friction force opposes the impending motion. For impending motion up the ramp the sum of forces parallel to the ramp are $\sum F = A\sin\alpha - B_{MAX} + \mu_S A\cos\alpha = 0$, from which

$\boxed{B_{MAX} = A(\sin\alpha + \mu_s\cos\alpha) = 100(\sin 30^o + 0.3\cos 30^o) = 75.98\ lb}$ For impending motion down the ramp: $\sum F = A\sin\alpha - B_{MIN} - \mu_s A\cos\alpha = 0$, from which

$$\boxed{B = A(\sin\alpha - \mu_s\cos\alpha) = 100(\sin 30^o - 0.3\cos 30^o) = 24.02\ lb}$$

Problem 10.10 The box is stationary on the inclined surface. The coefficient of static friction is μ_s. (a) Determine the force T necessary to start the box sliding up the surface. (b) Show that the force necessary to start the box sliding up the surface is a minimum when $\tan\beta = \mu_s$.

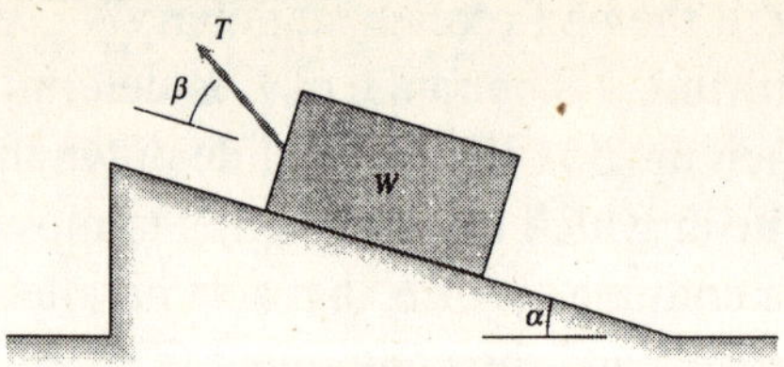

Solution (a) For impending motion up the ramp, the sum of forces parallel to the inclined surface is

$\sum F = T\cos\beta - W\sin\alpha - \mu_s(W\cos\alpha - T\sin\beta) = 0$, from which $T = W\left(\dfrac{\sin\alpha + \mu_s\cos\alpha}{\cos\beta + \mu_s\sin\beta}\right)$. (b) The minimum value of $T(\beta)$ occurs when $\dfrac{dT}{d\beta} = 0 = -W(\sin\alpha + \mu_s\cos\alpha)\left(\dfrac{-\sin\beta + \mu_s\cos\beta}{(\cos\beta + \mu_s\sin\beta)^2}\right) = 0$. This condition is satisfied when $-\sin\beta + \mu_s\cos\beta = 0$, from which $\boxed{\mu_s = \dfrac{\sin\beta}{\cos\beta} = \tan\beta}$. That this is indeed a minimum (and not a maximum) can be found from the substitution into the second derivative:

$\dfrac{d^2T}{d\beta^2} = -W(\sin\alpha + \mu_s\cos\alpha)\left\{\dfrac{-(\cos\beta + \mu_s\sin\beta)}{(\cos\beta + \mu_s\sin\beta)^2} - 2\dfrac{(-\sin\beta + \mu_s\cos\beta)^2}{(\cos\beta + \mu_s\sin\beta)^3}\right\}$. Substitute $\mu_s = \tan\beta$ to obtain $\dfrac{d^2T}{d\beta^2} = T$. Thus for all non zero values of the magnitude T the second derivative is positive, which proves that the condition is a minimum for non zero values of T.

Problem 10.11 To explain observations of ship launchings at the port of Rochefort in 1779, Coulomb analyzed the system in Problem 10.10 to determine the minimum force T necessary to hold the box stationary on the inclined surface. Show that the result is $T = W\dfrac{(\sin\alpha - \mu_s\cos\alpha)}{(\cos\beta - \mu_s\sin\beta)}$.

Solution The force necessary to hold the box stationary occurs when the impending motion is down the incline. The friction force is in opposition to the direction of impending motion. The sum of forces parallel to the inclined surface is $\sum F = T\cos\beta - W\sin\alpha + \mu_s(W\cos\alpha - T\sin\beta) = 0$, from which

$$\boxed{T = W\left(\dfrac{\sin\alpha - \mu_s\cos\alpha}{\cos\beta - \mu_s\sin\beta}\right)}$$

Problem 10.12 Two sheets of plywood A and B lie on the bed of a truck. They have the same weight W, and the coefficient of static friction between the two sheets of wood and between the wood and the truck bed is μ_s. (a) If you apply a horizontal force to sheet A and apply no force to sheet B, can you slide sheet A off the truck without causing sheet B to move? What force is necessary to cause sheet A to start moving? (b) If you prevent sheet A from moving by applying a horizontal force on it, what horizontal force on sheet B is necessary to start it moving?

Solution (a) The friction force exerted by sheet A on B at impending motion is $f_{AB} = \mu_s W$. The friction force exerted by sheet B on the bed of the truck is $f_{BT} = \mu_s(2W)$, since the normal force is due to the weight of both sheets. Since $f_{BT} > f_{AB}$, the top sheet will begin moving before the bottom sheet. Yes The force required to start sheet A to move is $\boxed{F = f_{AB} = \mu_s W}$ (b) The force on B is the friction between A and B and the friction between B and the truck bed. Thus the force required to start B in motion is $\boxed{F_B = f_{AB} + f_{BT} = 3\mu_s W}$.

Problem 10.13 Suppose that the truck in Problem 10.12 is loaded with N sheets of plywood, each of the same weight W, labeled (from the top) sheets 1, 2, ... N. The coefficient of static friction between the sheets of wood and between the bottom sheet and the truck bed is μ_s. If you apply a horizontal force to the sheets above it to prevent them from moving, can you pull out the i *th* sheet, $1<i<N$, without causing any of the sheets below it to move? What force must you apply to the *i th* sheet to cause it to start moving?

Solution . The force holding the sheets below the *i th* sheet from moving is the friction force between the bed of the truck and the bottom sheet. The weight is NW, hence the force opposing motion of the sheets below the *i th* sheet is $F_N = \mu_s NW$. The force causing motion to start is the friction between the *i th* sheet and those below it, which is $F_B = \mu_s(iW)$. The resultant force is $F_R = \mu_s W(N-i)$. The bottom sheets will begin to move when the resultant is zero, which can only occur for the last sheet, $i = N$. Thus the *i th* sheet can be extracted without the sheets below it moving. Yes The force required to move the *i th* sheet is the force required to overcome the friction force due to sheet above it and the sheet below it. The weight of the sheets above is $(i-1)\,W$ and the weight on the sheet below it is iW. The maximum total friction force opposing motion is

$$\boxed{F_i = \mu_s(iW + (i-1)W) = \mu_s W(2i-1)}$$

Problem 10.14 The coefficient of static friction between the two boxes and between the lower box and the inclined surface is μ_s. What is the largest angle α for which the lower box will not slip?

Solution At impending slip the force opposing motion of the lower box is the friction force between the two boxes and the friction force between the lower box and the inclined surface. The force inducing the lower box to move is the component of weight parallel to the inclined surface. The sum of the forces parallel to the inclined surface is

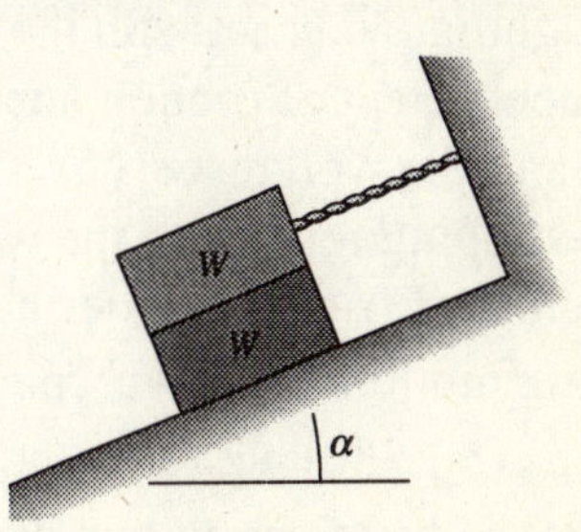

$\sum F_P = W\sin\alpha - \mu_s(2W\cos\alpha) - \mu_s(W\cos\alpha) = 0$, from which

$\mu_s = \left(\frac{1}{3}\right)\tan\alpha$, or $\boxed{\alpha = \tan^{-1}(3\mu_s)}$

Problem 10.15 The coefficient of static friction between the two boxes and between the lower box and the inclined surface is μ_s. What is the largest force F that will not cause the boxes to slip?

Solution At impending motion, the sum of the forces parallel to the inclined surface for the upper box is

$\sum F_P = F - \mu_s W \cos\alpha + W \sin\alpha - T = 0$, where T is the tension in the string, and the friction force opposes impending motion in the direction of F. For the lower box,

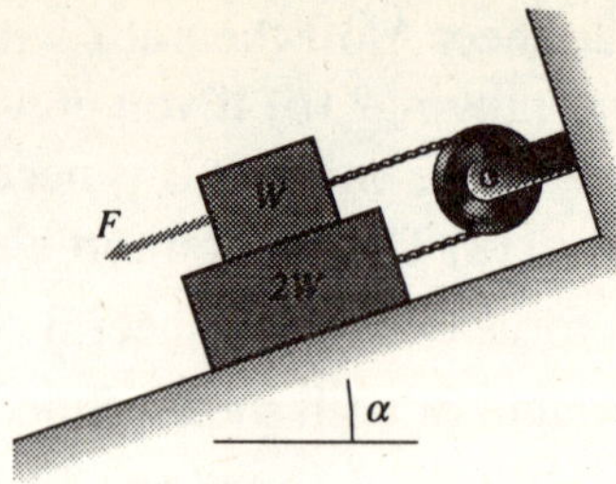

$\sum F = 2W \sin\alpha + \mu_s(3W \cos\alpha) + \mu_s W \cos\alpha - T = 0$, where the friction force opposes impending motion in the direction of T. Combining:

$$\boxed{F = \mu_s(5W \cos\alpha) + W \sin\alpha = W(5\mu_s \cos\alpha + \sin\alpha)}$$

Problem 10.16 Consider the system shown in Problem 10.15. The coefficient of static friction between the two boxes and between the lower box and the inclined surface is μ_s. If $F = 0$, the lower box will slip down the inclined the inclined plane. What is the smallest force F for which the boxes will not slip?

Solution The solution is obtained by the same procedure as in Problem 10.15, with the exception that the friction forces now oppose impending motion in the direction of T for the upper box, and impending motion down the surface, for the lower box. The sums of forces parallel to the inclined surface for the two boxes are: $\sum F_P = F + \mu_s W \cos\alpha + W \sin\alpha - T = 0$ and

$\sum F = 2W \sin\alpha - \mu_s(3W \cos\alpha) - \mu_s W \cos\alpha - T = 0$. Combining:

$$\boxed{F = -\mu_s W \cos\alpha + W \sin\alpha - 4\mu_s W \cos\alpha = W(\sin\alpha - 5\mu_s \cos\alpha)}$$

Problem 10.17 A sander consists of a rotating disk with sandpaper bonded to the edge surface. The normal force exerted on the workpiece A by the sander is 30 *lb*. The workpiece A weighs 50 *lb*. The coefficients of friction between the sander and the workpiece A are $\mu_s = 0.65$ and $\mu_k = 0.60$. The coefficients of friction between the workpiece A and the table are $\mu_s = 0.35$ and $\mu_k = 0.30$. Will the workpiece remain stationary while it is being sanded?

Solution Two possible situations can cause impending motion: (1) before the contact surface has begun slipping with respect to the workpiece (that is, as the sanding wheel is brought into contact with the work piece), so that the static coefficient of friction applies between the sanding wheel and the workpiece. (2) After the sanding wheel has begun slipping relative to the workpiece, when the kinetic coefficient of friction applies between the sander and the workpiece. In the first situation, the force inducing motion of the workpiece is $F = \mu_s 30 = 19.5$ *lb*. The force resisting motion is $F_A = \mu_s(30 + 50) = 28$ *lb*. Thus the workpiece will not slip. The second situation is less severe, since the force inducing motion is $F = 30\mu_k = 18$ *lb*, and the force opposing motion is the same. The workpiece will remain stationary.

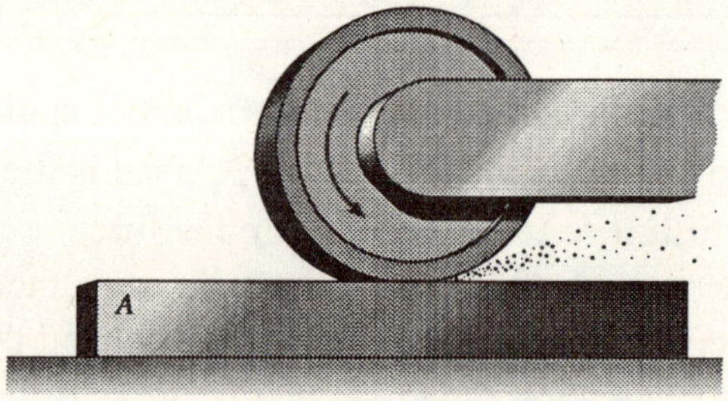

Problem 10.18 Blocks A and B are connected by a horizontal bar. The coefficient of static friction between the inclined surface and the 400 *lb* block A is 0.3. The coefficient of static friction between the surface and the 300 *lb* block B is 0.5. What is the smallest force that will prevent the blocks from slipping down the surface?

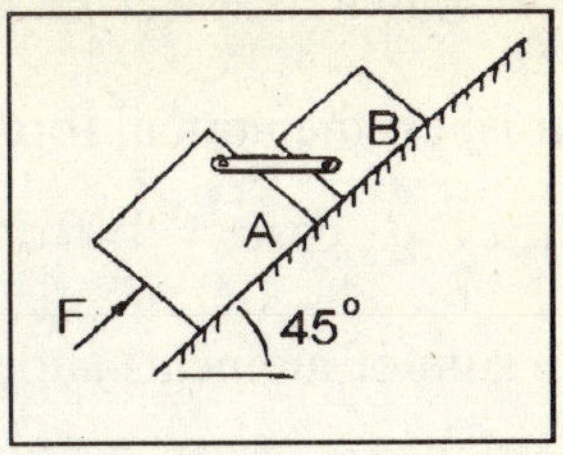

Solution The (horizontal) connecting bar exerts a component of force normal to the inclined surface. This force increases the normal force exerted on B by the inclined plane, and reduces the normal force exerted on A by the inclined plane. Isolate B. Denote the component of the linkage force parallel to the surface by f_{link}. The equilibrium conditions on B are

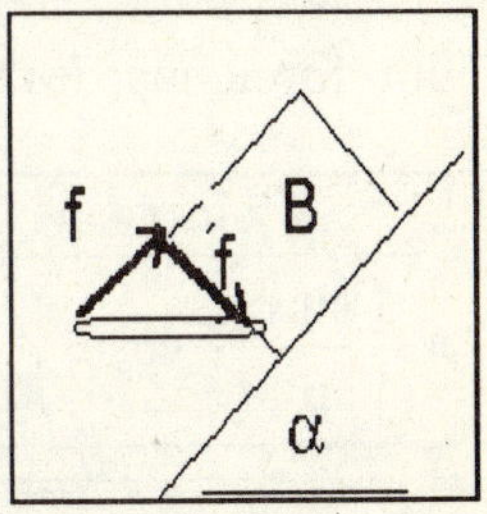

$\sum F = -f_{link} + B\sin\alpha - \mu_{sB} N_B = 0$, where $N_B = B\cos\alpha + f_{link}$, (where f_{link} forms the sides of a right triangle), from which

$$f_{link} = \frac{B}{(1+\mu_{sB})}(\sin\alpha - \mu_{sB}\cos\alpha) = 70.71\ lb.$$

Isolate A. The equilibrium conditions are $\sum F = f_{link} - F + A\sin\alpha - \mu_{sA} N_A = 0$, where $N_A = A\cos\alpha - f_{link}$, from which $F - f_{link} = A(\sin\alpha - \mu_{sA}\cos\alpha) - \mu_{sA} f_{link} = 219.2\ lb$. The total force required to keep the blocks from slipping is $\boxed{F = 70.71 + 219.2 = 289.91\ lb}$

Problem 10.19 What force F is necessary to cause the blocks in Problem 9.18 to move up the plane?

Solution The friction forces oppose impending motion up the plane. Use the results of the solution to Problem 10.18 with the friction forces reversed. *Isolate B*. The linkage force is

$$f_{link} = \frac{B}{(1-\mu_{sB})}(\sin\alpha + \mu_{sB}\cos\alpha) = 636.4\ lb.$$

Isolate A. The resultant force on A is

$F - f_{ink} = A(\sin\alpha + \mu_{sA}\cos\alpha) - \mu_{sA} f_{link} = 176.78\ lb$. The resultant force required to cause the blocks to start to move up the plane is $\boxed{F = 636.4 + 176.8 = 813.2\ lb}$

Problem 10.20 Suppose that you want the bar of length L to act as a simple brake that will allow the workpiece A to slide to the left but will not allow it to slide to the right no matter how large a horizontal force is applied to it. The weight of the bar is W and the coefficient of static friction between it and the workpiece A is μ_s. You can neglect friction between the workpiece and the surface it rests on. (a) What is the largest angle α for which the bar will prevent the workpiece from moving to the right? (b) If α has the value determined in Part (a), what horizontal force is required to start the workpiece A in motion to the left?

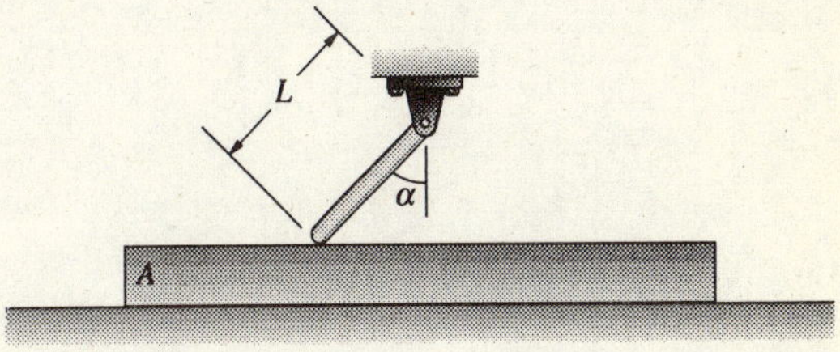

Solution (a) To resist motion to the right no matter how large the horizontal force requires a very large friction force. For impending motion to the right, the sum of the moments about the bar hinge is

$$\sum M = +\frac{W}{2}L\sin\alpha - F_N L\sin\alpha + fL\cos\alpha = 0,$$ where

F_N is the normal force and f is the friction force resisting the impending motion.

Problem 10.20 continued on next page

continuation of Problem 10.20..........

Noting that $f = \mu_s F_N$, the sum of moments yields $F_N = \dfrac{W \sin a}{2(\sin\alpha - \mu_s \cos\alpha)}$. If the denominator vanishes, the normal force and hence the friction force become as large as required. Thus $\sin\alpha - \mu_s \cos\alpha = 0$, from which $\mu_s = \tan\alpha$, or $\boxed{\alpha = \tan^{-1}(\mu_s)}$ (b) For impending motion to the left, the sum of moments about the bar hinge is $\sum M = +W\left(\dfrac{L}{2}\right)\sin\alpha - F_N L \sin\alpha - fL\cos\alpha = 0$, where the friction force opposes the impending motion. Noting $F_N = \dfrac{f}{\mu_s}$, then

$$\boxed{f = \frac{W \sin\alpha}{2\left(\dfrac{\sin a}{\mu_s} + \cos\alpha\right)} = \frac{W \sin\alpha}{4\cos\alpha} = \frac{\mu_s W}{4}}$$ is the force required to start motion to the left.

Problem 10.21 The coefficient of friction between the 20 *lb* bar and the floor is $\mu_s = 0.3$. Neglect friction between the bar and the wall. (a) If $\alpha = 20^o$, what is the magnitude of the friction force exerted on the bar by the floor? (a) What is the maximum value of α for which the bar will not slip?

Solution The sum of the moments about the upper end of the bar is

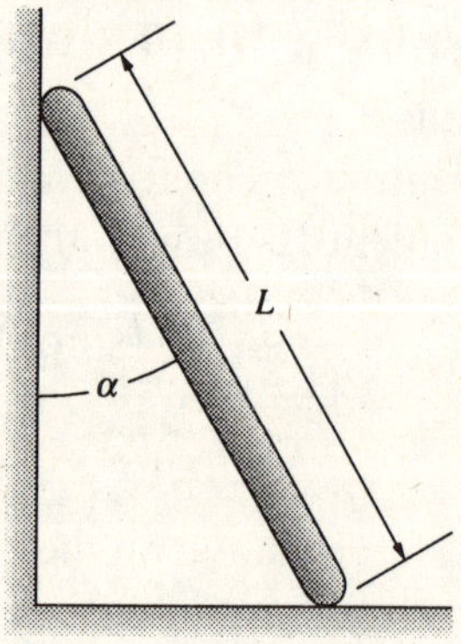

$\sum M = -\dfrac{WL}{2}\sin\alpha + F_N L \sin\alpha - fL\cos\alpha = 0$, from which

$f = \dfrac{\left(-\dfrac{W}{2} + F_N\right)\sin\alpha}{\cos\alpha}$. The sum of the forces in the vertical direction

$\sum F_Y = -W + F_N = 0$, from which $F_N = W$. Substitute:

$\boxed{f = \dfrac{W \sin\alpha}{2\cos\alpha} = 10\tan 20^o = 3.64 \ lb}$ (b) At impending slip,

$f = \mu_s F_N = \mu_s W$, from which, substituting above, $\mu_s = \left(\dfrac{1}{2}\right)\tan\alpha$, or

$\alpha = \tan^{-1}(2\mu_s) = \tan^{-1}(0.6) = 31^o$

Problem 10.22 The masses of the ladder and person are 18 *kg* and 90 *kg* respectively. The center of mass of the 4 *m* ladder is at its midpoint. If $\alpha = 30^o$, what is the minimum coefficient of static friction between the ladder and the floor necessary for the person to climb to the top of the ladder? Neglect friction between the ladder and the wall.

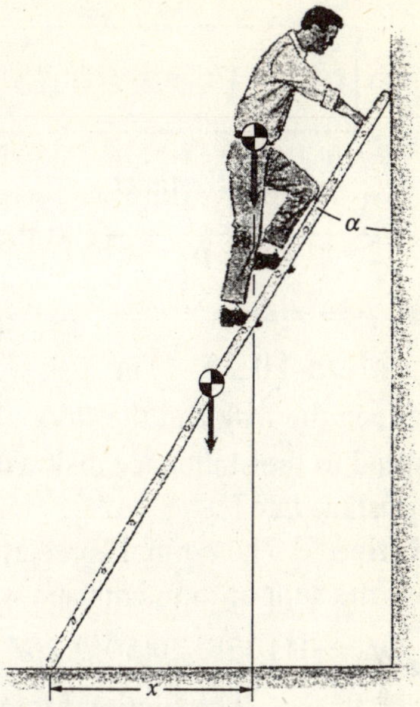

Solution The weight of the ladder is $W = 18g = 176.58\ N$. The weight of the person is $P = 90g = 882.9\ N$. Let h be the distance along the ladder of the person's center of mass, and L be the length of the ladder. The horizontal distance is. The sum of the moments about the top of the ladder:

$$\sum M = P(L\sin\alpha - x) - F_N L\sin\alpha + W\frac{L}{2}\sin\alpha + fL\cos\alpha = 0.$$

From the sum of the forces, $\sum F_Y = F_N - W - P = 0$, from which the normal force at the foot of the ladder is $F_N = W + P$. Substitute, solve for the friction force, and reduce algebraically:

$f = \left(\frac{h}{L}P + \frac{1}{2}W\right)\tan\alpha$. At the top of the ladder, $\frac{h}{L} = 1$, hence

$f = \left(P + \frac{W}{2}\right)\tan\alpha = (883 + 88.3)(0.5774) = 560.72\ N$ At

impending slip, $f = \mu_s F_N = \mu_s(P + W)$, from which $\boxed{\mu_s = \frac{f}{P+W} = \frac{560.72}{1059.48} = 0.5292}$

Problem 10.23 The coefficient of static friction between the ladder in Problem 10.22 and the floor is $\mu_s = 0.6$. The masses of the ladder and the person are 18 *kg* and 100 *kg* respectively. The center of mass of the 4 *m* ladder is at its midpoint. What is the maximum value of α for which the person can climb to the top of the ladder? Neglect friction between the wall and the ladder.

Solution The solution in Problem 10.22 for the friction force is $f = \left(\frac{h}{L}P + \frac{1}{2}W\right)\tan\alpha$

and $f = \mu_s(W + P)$. At impending slip $\tan\alpha = \frac{\mu_s(W+P)}{\left(\frac{h}{L}P + \frac{1}{2}W\right)}$. At the top of the ladder $\frac{h}{L} = 1$, and

$\tan\alpha = 0.6495$, or $\boxed{\alpha = \tan^{-1}(0.6495) = 33^o}$

Problem 10.24 The coefficient of static friction between the ladder in Problem 10.22 and the floor is $\mu_s = 0.6$ and $\alpha = 35^o$. The center of mass of the 4 *m* ladder is at the midpoint, and its mass is 18 *kg*. (a) If a football player weighing 140 *kg* attempts to climb the ladder, what maximum value of x will he reach? (b) What minimum friction coefficient would be required for him to reach the top?

Solution The weight of the football player is $P = 140g = 1373.4\ N$, and the weight of the ladder is $W = 176.58\ N$ From the solution to Problem 10.22, $f = \left(\frac{h}{L}P + \frac{1}{2}W\right)\tan\alpha$ and $f = \mu_s(W + P)$. .

Problem 10.24 continued on next page..........

==<>==

............continuation of solution to Problem 10.24.....

Substitute and solve for $\frac{h}{L} = \frac{\mu_s(W+P) - \frac{W}{2}\tan\alpha}{P\tan\alpha} = 0.90277$. From Problem 9.22, $x = h\sin\alpha$, from which $\boxed{x = \left(\frac{h}{L}\right)L\sin\alpha = 2.07\ m}$ (b) Solve the above for the static friction coefficient:

$\mu_s = \frac{\left(\frac{h}{L}P + \frac{1}{2}W\right)\tan\alpha}{(W+P)}$. At the top of the ladder, $\frac{h}{L} = 1$, and $\mu_s = \frac{\left(P + \frac{1}{2}W\right)\tan\alpha}{(P+W)}$, $\boxed{\mu_s = 0.66}$

==<>==

Problem 10.25 The disk weighs 50 *lb*. Neglect the weight of the bar. The coefficients of friction between the disk and the floor are $\mu_s = 0.6$ and $\mu_k = 0.4$. What is the largest moment that can be applied to the stationary disk without causing it to rotate? What moment is necessary to rotate the disk at a constant rate?

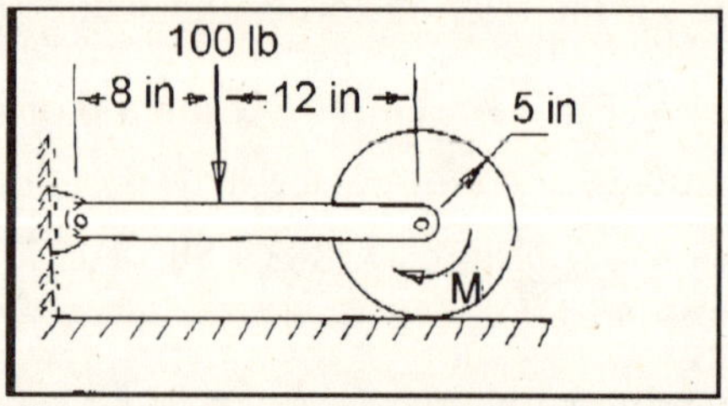

Solution The normal force at the point of contact is found from the sum of moments about the pin support.

$\sum M = -8(100) - 20(50) + 20F_N = 0$, from which $F_N = 90\ lb$. .The friction force is $f = \mu_s F_N$ The moment exerted by the friction force is

$\boxed{M_F = \mu_s R F_N = 0.6(5)(90) = 270\ in\ lb}$ This is the moment to be overcome at impending slip.

(b) The moment required to rotate the disk at a constant rate is $\boxed{M_K = \mu_k R F_N = 0.4(5)(90) = 180\ in\ lb}$

==<>==

Problem 10.26 The cylinder has weight *W*. The coefficient of static friction between the cylinder and the floor and the cylinder and the wall is μ_s. What is the largest moment that can be applied to the stationary cylinder without causing it to rotate?

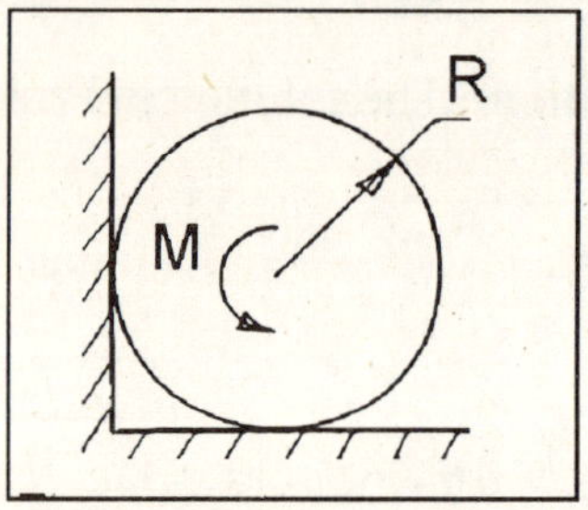

Solution Assume impending slip. The force opposing rotation is the sum of the friction force at the wall and at the floor. Denote the normal force at the wall by F_{NW} and the normal force on the floor by F_{NF}. From the sum of forces: $\sum F_y = \mu_s F_{NW} + F_{NF} - W = 0$, and $\sum F_x = F_{NW} - \mu_s F_{NF} = 0$. Solve these two simultaneous equations to obtain: $F_{NF} = \frac{W}{1+\mu_s^2}$, and $F_{NW} = \frac{\mu_s W}{1+\mu_s^2}$. The sum of moments about the center of the cylinder is $\sum M_C = M_{app} - \mu_s R F_{NW} - \mu_s R F_{NF} = 0$. Substitute and solve:

$\boxed{M_{app} = \mu_s R W\left(\frac{1+\mu_s}{1+\mu_s^2}\right)}$. At impending slip, this is the maximum moment that can be applied to the cylinder.

==<>==

Problem 10.27 The cylinder has weight W. The coefficient of static friction between the cylinder and the floor and between the cylinder and the wall is μ_s. What is the largest moment M that can be applied to the stationary cylinder without causing it to rotate?

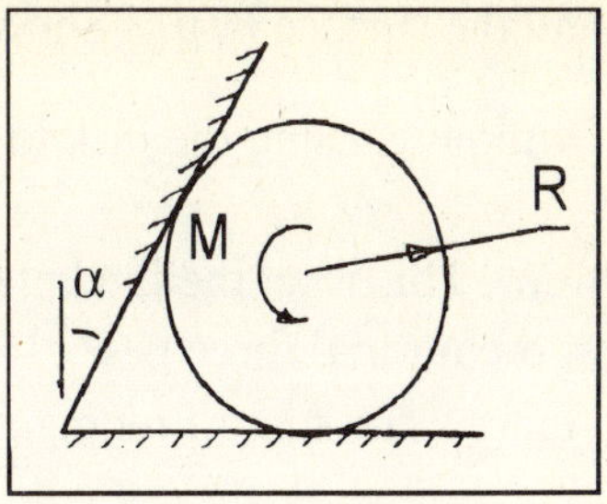

Solution Assume impending slip. Denote the normal force at the wall by F_{NW} and the normal force at the floor by F_{NF}. The projection of the friction force at the wall on an x-y coordinate system is

$\vec{f}_W = \mu_s |\vec{F}_{NW}|(\vec{i}\cos(90-\alpha)+\vec{j}\sin(90-\alpha))$

$= \mu_s |\vec{F}_{NW}|(\vec{i}\sin\alpha + \vec{j}\cos\alpha)$. The projection of the normal force at the wall on an x-y coordinate system is $\vec{F}_{NW} = |\vec{F}_{NW}|(\vec{i}\cos\alpha - \vec{j}\sin\alpha)$. The sum of the forces:

$\sum F_y = |\vec{F}_{NW}|(\mu_s\cos\alpha - \sin\alpha) + |\vec{F}_{NF}| - |\vec{W}| = 0$, and $\sum F_x = |\vec{F}_{NW}|(\cos\alpha + \mu_s\sin\alpha) - \mu_s|\vec{F}_{NF}| = 0$.

Solve these simultaneous equations to obtain: $|\vec{F}_{NF}| = \dfrac{|\vec{W}|(\cos\alpha + \mu_s\sin\alpha)}{(1+\mu_s^2)\cos\alpha}$, and $|\vec{F}_{NW}| = \dfrac{\mu_s|\vec{W}|}{(1+\mu_s^2)\cos\alpha}$.

The sum of the moments about the center of the cylinder is $\sum M_C = M_{app} - \mu_s R|\vec{F}_{NF}| - \mu_s R|\vec{F}_{NW}| = 0$.

Substitute and reduce algebraically: $\boxed{M_{app} = \dfrac{\mu_s RW(\cos\alpha + \mu_s\sin\alpha + \mu_s)}{(1+\mu_s^2)\cos\alpha}}$. This is the maximum applied moment at impending slip. *Check:* This reduces to the solution of Problem 10.26 when $\alpha = 0$, as it should. *check.*

Problem 10.28 Suppose that $\alpha = 30^o$ in Problem 10.27 and a moment $M = 0.5RW$ is required to turn the cylinder at a constant rate. What is the coefficient of kinetic friction?

Solution Substitute the angle and the moment into the solution of Problem 10.27 to obtain:

$$0.5 = \frac{\mu_k\left(\frac{\sqrt{3}}{2}+\frac{\mu_k}{2}+\mu_k\right)}{(1+\mu_k^2)\frac{\sqrt{3}}{2}} = \frac{\mu_k(1+\sqrt{3}\mu_k)}{(1+\mu_k^2)}$$

This reduces to the quadratic equation $\mu_k^2 + 2b\mu_k - c = 0$, where $b = c = \dfrac{1}{(2\sqrt{3}-1)}$. The solution is $\mu_k = -b \pm \sqrt{b^2+c}$. Substitute numerical values:

$\boxed{\mu_k = 0.3495}$, or $\mu_k = -1.1612$. The negative value has no meaning here.

Problem 10.29 The disk of weight W and radius R is held in equilibrium on the circular surface by a moment M. The coefficient of static friction between the disk and the surface is μ_s. Show that the largest value M can have without causing the disk to slip is $M = \frac{\mu_s RW}{\sqrt{1+\mu_s^2}}$.

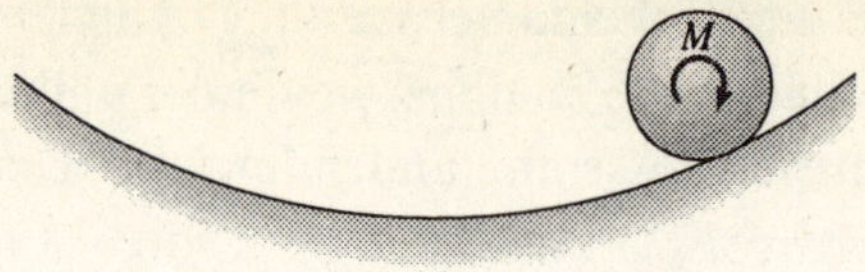

Solution This is an inclined plane problem. Let α be the angle at the point of contact. From the sum of forces: the normal force is $F_N = W\cos\alpha$, and the friction force is $f = \mu_s F_N = W\sin\alpha$, from which $\mu_s = \tan\alpha$. The sum of moments about the center of the disk yields $M = fR = \mu_s RW\cos\alpha$. Noting that $\cos\alpha = \frac{1}{\sqrt{1+\tan^2\alpha}}$, then $\boxed{M = \frac{\mu_s RW}{\sqrt{1+\mu_s^2}}}$, which is the moment at impending slip.

Problem 10.30 The coefficient of static friction between the jaws of the pliers and the gripped object is μ_s. What is the largest value of the angle α for which the gripped object will not slip? (Neglect the object's weight.)

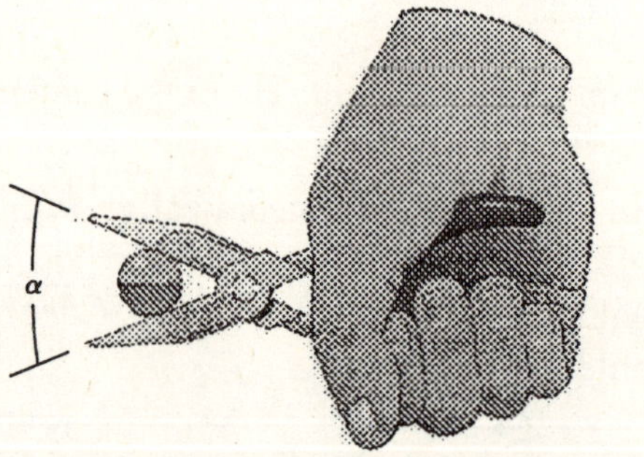

Solution Choose an x-y coordinate system such that the x- axis bisects the angle α. Define $\beta = \frac{\alpha}{2}$ The projection of the normal forces on the x-y system is, for the top: $\vec{F}_{NT} = |\vec{F}_{NT}|(-\vec{i}\sin\beta - \vec{j}\cos\beta)$. For the bottom: $\vec{F}_{NB} = |\vec{F}_{NB}|(-\vec{i}\sin\beta + \vec{j}\cos\beta)$. The projection of the friction forces on the x-y system is, for the top: $\vec{f}_T = \mu_s|\vec{F}_{NT}|(\vec{i}\cos\beta - \vec{j}\sin\beta)$. For the bottom: $\vec{f}_B = \mu_s|\vec{F}_{NB}|(\vec{i}\cos\beta + \vec{j}\sin\beta)$. The forces tending to expel the gripped object are the components of the normal forces in the negative x direction, and the components tending to retain the gripped object are the friction forces in the positive x-direction. These must balance: $\sum F_x = -|\vec{F}_{NB}|\sin\beta - |\vec{F}_{NT}|\sin\beta + \mu_s|\vec{F}_{NT}|\cos\beta + \mu_s|\vec{F}_{NB}|\cos\beta = 0$, from which $|\vec{F}_{NB}|(-\sin\beta + \mu_s\cos\beta) + |\vec{F}_{NT}|(-\sin\beta + \mu_s\cos\beta) = 0$. From symmetry, $|\vec{F}_{NB}| = |\vec{F}_{Nt}|$, since the weight of the object is neglected. For non-trivial values of the normal forces, $-\sin\beta + \mu_s\cos\beta = 0$, from which $\mu_s = \tan\beta$, or $\beta = \tan^{-1}(\mu_s)$. Noting $\beta = \frac{\alpha}{2}$, $\boxed{\alpha = 2\tan^{-1}(\mu_s)}$

==================================<>==================================

Problem 10.31 The stationary disk has a radius of 300 *mm* and is attached to a pin support at D. The disk is held in place by the brake ABC in contact with the disk at C. The hydraulic actuator BE exerts a horizontal 400 N force on the brake at B. The coefficients of friction between the disk and the brake are $\mu_s = 0.6$ and $\mu_k = 0.5$. What moment must be applied to the stationary disk to cause it to slip in the counterclockwise direction?

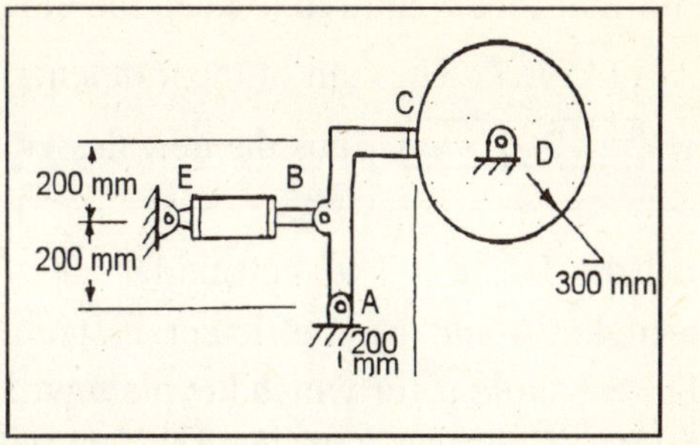

Solution Assume impending slip. For counterclockwise motion the friction force $f = \mu_s F_N$ opposes the impending slip, so that it acts on the brake in a downward direction, producing a negative moment (clockwise) about A.. The sum of the moments about A:

$\sum M_A = -0.2(400) + (0.4 - 0.2\mu_s)F_N$, from which $F_N = 285.7\ N$. The sum of the moments about the center of the disk: $\sum M_D = M - 0.3(\mu_s)F_N = 0$, from which $\boxed{M = 51.43\ Nm}$.

==================================<>==================================

Problem 10.32 What moment must be applied to the stationary disk in Problem 10.31 to cause if to rotate in a clockwise direction?

Solution Assume impending slip. For clockwise motion the friction force $f = \mu_S F_N$ opposes the impending slip, so that it acts on the brake in an upward direction, producing a positive moment (counterclockwise) about A. The sum of the moments about A:

$\sum M_A = -0.2(400) + (0.4 + 0.2\mu_s)F_N = 0$, from which $F_N = 153.85\ N$. The sum of the moments about the center of the disk: $\sum M_D = M - 0.3\mu_s F_N = 0$, from which $\boxed{M = 27.69\ Nm}$

==================================<>==================================

Problem 10.33 Block B weighs 80 N. The coefficient of static friction between the surfaces of the clamp and the block is $\mu_s = 0.2$. When the clamp is aligned as shown, what minimum force must the spring exert to prevent the block from slipping out?

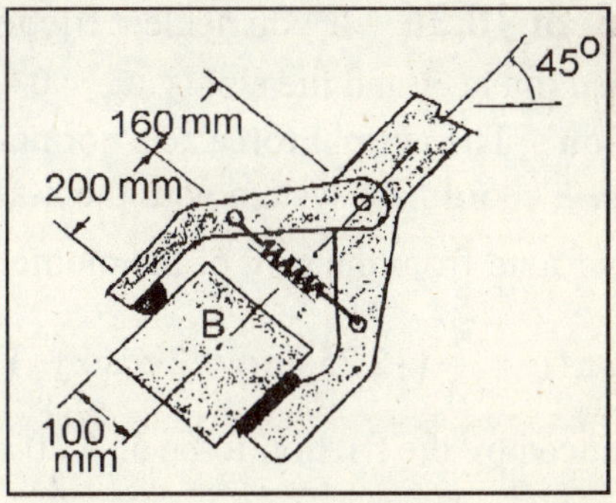

Solution The force tending to make the block slip is the component of the lift parallel to the jaws of the clamp: $F = W\sin\alpha$. The two contacts each exert a friction force on the block that must balance this force: For the top contact the normal force is $F_N = F_T$, and for the bottom contact the normal force is $F_N = F_T + W\cos\alpha$. The sum of the friction forces is $\sum f = \mu_s F_T + \mu_s(F_T + W\cos\alpha)$. The friction forces exactly balances the gravity force at impending slip $2\mu_s F_T + \mu_s W\cos\alpha - W\sin\alpha = 0$, from which $F_T = 113.14\ N$. The friction force opposes impending slip, so that it exerts a positive moment on the clamp jaw pin support. The sum of the moments about the clamp jaw pin support is $\sum M_H = 0.16F_S - 0.36F_T + 0.1\mu_s F_T = 0$, from which $F_S = 240.42\ N$ is the force exerted by the spring on the jaw clamp at impending slip.

==================================<>==================================

Problem 10.34 By altering its dimensions, redesign the clamp in Problem 10.33 so that the minimum force the spring must exert to prevent the block from slipping is 180 N. Draw a sketch of your new design.

Solution One alternative is to increase the 160 mm dimension between the pin support and the spring. The normal force required to keep the block from slipping (from the solution of Problem 10.33) is $F_N = 113.14\ N$. The sum of the moments $\sum M = 180x - (x + 0.2)113.14 + 0.1\mu_S(113.14) = 0$, from which $\boxed{x = 305\ mm}$. Thus the new design has the 160 mm dimension increased to 305 mm.

Problem 10.35 The rectangular 100 lb plate is supported by the pins A and B. If friction can be neglected at A and the coefficient of static friction between the pin at B and the slot is $\mu_s = 0.4$, what is the largest angle α for which the plate will not slip?

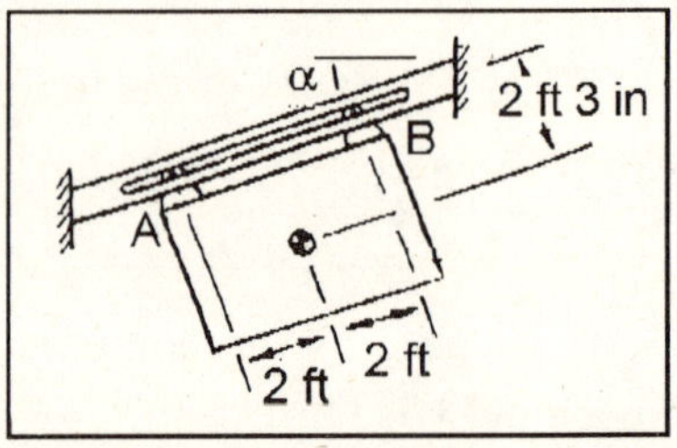

Solution Choose a coordinate system with the x- axis parallel to the rail. The sum of the moments about A is $\sum M_A = -2W\cos\alpha - 2.25W\sin\alpha + 4B = 0$, from which $B = \frac{W}{4}(2\cos\alpha + 2.25\sin\alpha)$. The component of weight causing the plate to slide is $F = W\sin\alpha$. This must be balanced by the friction force: $0 = -W\sin\alpha + \mu_s B$, from which $\frac{W\sin\alpha}{\mu_s} = \frac{W}{4}(2\cos\alpha + 2.25\sin\alpha)$. Reduce algebraically to obtain

$$\boxed{\alpha = \tan^{-1}\left(\frac{1}{\frac{2}{\mu_s} - 1.125}\right) = 14.47^o}$$

Problem 10.36 If you neglect friction at B in Problem 10.35 and the coefficient of static friction between pin at A and the slot is $\mu_s = 0.4$, what is the largest angle α for which the plate will not slide?

Solution The normal force acts normally to the slots, and the friction force acts parallel to the slot. Choose a coordinate system with the x-axis parallel to the slots. The normal component of the reaction at A is found from the sum of the moments about B: $\sum M_B = -2.25W\sin\alpha + 2W\cos\alpha - 4A = 0$, from which $A_N = \frac{W}{4}(-2.25\sin\alpha + 2\cos\alpha)$. The force tending to make the plate slide is $F = -W\sin\alpha$. This is balanced by the friction force at A, $0 = -W\sin\alpha + \mu_s A_N$, from which $\frac{W\sin\alpha}{\mu_s} = \frac{W}{4}(-2.25\sin\alpha + 2\cos\alpha)$. Reduce algebraically to obtain $\boxed{\alpha = \tan^{-1}\left(\frac{1}{\frac{2}{\mu_s} + 1.125}\right) = 9.27^o}$

Check: The normal reactions at A and B are unequal: as the slots are inclined from the horizontal, the parallel component of the gravity force reduces the normal force at A, and increases the normal force at B. *check*. . *Check:* The sum of the reactions at A and B are $A_N + B_N = W\cos\alpha$. *check.* The magnitude $\sqrt{(A_N + B_N)^2 + (\mu_s A_N)^2} = W$, hence the system is in equilibrium at impending slip. *check.*

Problem 10.37 The shelf is designed so that it can be placed at any height on the vertical beam. It is supported by friction between the two horizontal cylinders and the vertical beam. The combined weight of the shelf and camera is W. If the coefficient of static friction between the vertical beam and the horizontal cylinders is μ_s, what is the minimum distance b necessary for the shelf to stay in place?

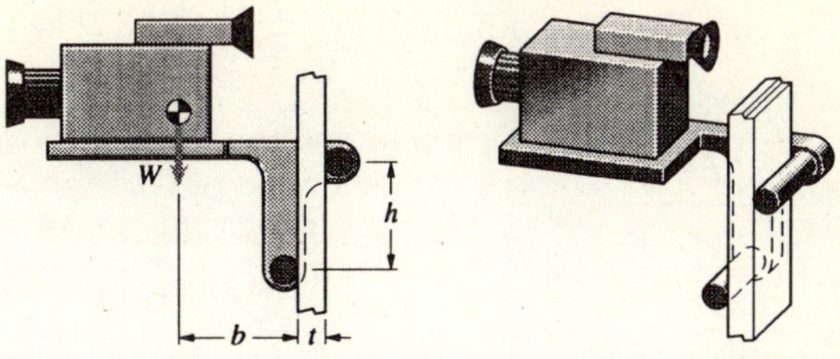

Solution Take the sum of the moments about the lower cylinder

$\sum M_{LC} = +bW - hF_{NU} + t\mu_s F_{NU} = 0$. The sum of the forces $\sum F_y = F_{NU} - F_{NL} = 0$,

from which the normal forces at the two cylinders are equal, and $F_N = \dfrac{bW}{(h - \mu_s t)}$. The force causing slippage is the weight, which is balanced by the friction force: $0 = -W + 2\mu_s F_N = -W + \dfrac{2\mu_s bW}{(h - t\mu_s)} = 0$,

from which $\boxed{b = \dfrac{h - t\mu_s}{2\mu_s} = \left(\dfrac{1}{2}\right)\left(\dfrac{h}{\mu_s} - t\right)}$

Problem 10.38 The refrigerator weighs 350 lb. The distances $h = 60\ in$ and $b = 14\ in$. The coefficient of static friction at A and B is $\mu_s = 0.24$. (a) What force is necessary for impending slip? (b) Will the refrigerator tip over before it slips?

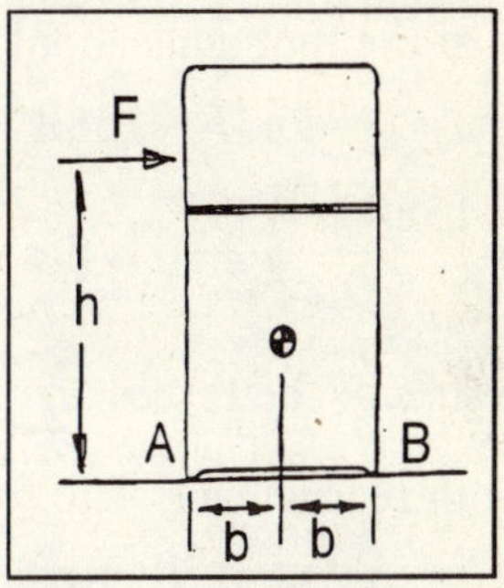

Solution The normal forces on the right and left supports are found: The sum of moments about support A: $\sum M_A = -bW - hF + 2bB = 0$, from which $B = \dfrac{bW + hF}{2b}$. The sum of forces: $\sum F_Y = A + B - W = 0$, from which $A = W - B = \dfrac{bW - hF}{2b}$. (a) The friction forces must balance the applied force at impending slip:

$0 = F - \mu_s A - \mu_s B = F - \mu_s\left(\dfrac{bW - hF}{2b}\right) - \mu_s\left(\dfrac{bW + hF}{2b}\right)$, from which $\boxed{F = \mu_s W = 0.24(350) = 84\ lb}$.

(b) If the normal force at A approaches zero before motion occurs, the refrigerator will start to tip:

$F_{tip} = \left(\dfrac{b}{h}\right)W = \left(\dfrac{14}{60}\right)350 = 81.67\ lb$. Since $F_{tip} < F_{move}$, the refrigerator will tip. Yes.

Problem 10.39 If you want the refrigerator in Problem 10.38 to slip before it tips, what is the minimum height h at which you can push it?

Solution From the solution to Problem 10.38, the tipping force must be equal to or greater than the moving force, $F_{mpve} = 84\ lb$. Thus when the normal force at A approaches zero, the tipping force must equal or exceed 84 lb, from which $h \le b\left(\dfrac{W}{F_{move}}\right) = 14\left(\dfrac{350}{84}\right) \le 58.33\ in$

==================================<>==================================

Problem 10.40 The refrigerator shown in Problem 10.38 weighs 350 *lb*. The distances $h = 45$ *in* and $b = 15$ *in*. The coefficients of static friction at A and B are 0.24 and 0.30 respectively. (a) What force F is necessary for impending slip? (b) Will the refrigerator tip over before it slips?

Solution Use the solution to Problem 10.38. The reactions at A and B are $A = \dfrac{bW - hF}{2b}$ and $B = \dfrac{bW + hF}{2b}$. The friction forces and the moving force are in balance at impending slip:

$$0 = F - \mu_{sA}A - \mu_{sB}B = F - \mu_{sA}\left(\frac{bW - hF}{2b}\right) - \mu_{sB}\left(\frac{bW + hF}{2b}\right)$$ from which

$$F = \frac{bW(\mu_{sA} + \mu_{sB})}{2b + h(\mu_{sA} - \mu_{sB})}$$.(*Check:* See results of Problem 10.44.) $\boxed{F = \dfrac{15(350)(0.24 + 0.3)}{2(15) + 45(0.24 - 0.3)} = 103.8\ lb}$.

Check: If $\mu_{sA} = \mu_{sB} = 0.24$, $h = 60$, $b = 14$, then $F = 84\ lb$, which is the solution to Problem 10.38, as it should be. *check.* (b) The tipping force is determined by $A = 0$, or

$F_{tip} = \left(\dfrac{b}{h}\right)W = \left(\dfrac{15}{45}\right)W = 116.7\ \ lb$. The tipping force is greater than the moving force, $116.7 > 84\ lb$ so the refrigerator will not tip.

==================================<>==================================

Problem 10.41 The mass of the Mercedes-Benz 600SL is 2020 *kg* and all four wheels are locked. The coefficient of static friction between the tires and the road is $\mu_s = 0.8$. Determine the force F that must be exerted on the rear bumper to cause the car to slip.

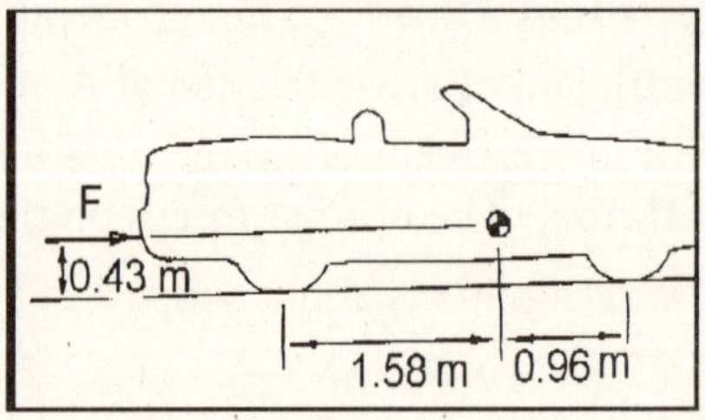

Solution The weight of the car is $W = 2020g = 19816\ N$. The sum of the moments about the rear wheels is

$\sum M_R = -0.43F - 1.58W + 2.54B$, from which

$B = \dfrac{1.58W + 0.43F}{2.54}$ is the vertical reaction of both front wheels.

The sum of the forces: $\sum F_Y = A + B - W = 0$, from which $A = W - B = \dfrac{0.96W - 0.43F}{2.54}$ is the reaction of both rear wheels. The force must be balanced by the friction forces at impending motion: $0 = F - \mu_s A - \mu_s B$. Reduce and solve: $\boxed{F = \mu_s W = 0.8(19816) = 15853\ N}$

==================================<>==================================

Problem 10.42 Determine the force F that must be exerted on the car in Problem 10.41 to cause it to slip if (a) The rear wheels are locked but the front wheels are free to roll. (b) The front wheels are locked but the rear wheels are free to roll. (The wheels that are free to roll exert no friction force.)

Solution (a) Use the solution to Problem 10.41.The reaction of both rear wheels is $A = \dfrac{0.96W - 0.43F}{2.54}$. The sum of forces: $0 = F - \mu_s A = F - \dfrac{0.96\mu_s W}{2.54} + \dfrac{0.43\mu_s F}{2.54}$. Reduce and solve:

$\boxed{F = \dfrac{0.96\mu_s W}{(2.54 + 0.43\mu_s)} = \dfrac{0.96(0.8)(19816)}{(2.54 + 0.8(0.43))} = 5277\ N}$ (b) The reaction of both front wheels is

$B = \dfrac{1.58W + 0.43F}{2.54}$. The sum of forces:, $0 = F - \mu_s B$. Solve: $\boxed{F = \dfrac{1.58\mu_s W}{(2.54 - 0.43(0.8))} = 11406\ N}$

==================================<>==================================

Problem 10.43 The coefficient of static friction between the 3000 *lb* car's tires and the road is $\mu_s = 0.5$. Determine the steepest grade (the largest angle α) the car can drive up at constant speed if the car has (a) rear wheel drive; (b) front wheel drive; (c) four wheel drive.

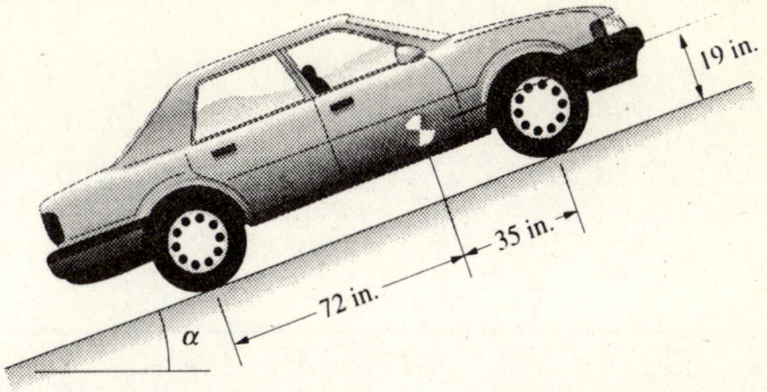

Solution The friction force acts parallel to the incline, and the normal force is normal to the incline. Choose a coordinate system with the x-axis parallel to the incline. The component of the weight that acts parallel to the incline is $W\sin\alpha$, and the component acting normally to the incline is $W\cos\alpha$. (a) *For rear wheel drive:* The moment about the point of contact of the front wheels: $\sum M_{FW} = 35W\cos\alpha + 19W\sin\alpha - 107R = 0$, from which the normal reaction of the two rear wheels is $R = \frac{W}{107}(35\cos\alpha + 19\sin\alpha)$. The force causing impending slip is $W\sin\alpha$, which is balanced by the friction force: $0 = W\sin\alpha - \mu_s R$, from which $\frac{W\sin\alpha}{\mu_s} = \frac{W}{107}(35\cos\alpha + 19\sin\alpha)$. Reduce and solve: $\boxed{\alpha = \tan^{-1}\left(\frac{35}{\frac{107}{\mu_s} - 19}\right) = 10.18^o}$ is the maximum angle at impending slip. (b) *For front wheel drive*: The moments about the point of contact of the rear wheels is $\sum M_{RW} = -72W\cos\alpha + 19W\sin\alpha + 107F = 0$, from which the normal reaction of the two front wheels is $F = \frac{W}{107}(72\cos\alpha - 19\sin\alpha)$. The friction force balances the component of gravity parallel to the incline: $0 = -W\sin\alpha + \mu_s F$, from which $\frac{W\sin\alpha}{\mu_s} = \frac{W}{107}(72\cos\alpha - 19\sin\alpha)$. Reduce and solve: $\alpha = \tan^{-1}\left(\frac{72}{\frac{107}{\mu_s} + 19}\right) = 17.17^o$ (c) *For four wheel drive:* Use the reactions of the front and rear wheels obtained in Parts (a) and (b). The sum of the forces parallel to the incline is $\sum F_X = -W\sin\alpha + \mu_s R + \mu_s F = 0$, from which $\frac{W\sin\alpha}{\mu_s} = \frac{W}{107}(35\cos\alpha + 19\sin\alpha + 72\cos\alpha - 19\sin\alpha)$. Reduce and solve: $\boxed{\alpha = \tan^{-1}(\mu_s) = 26.57^o}$

Check: This result is the same as if the Mercedes with four wheel drive were a box on an incline, as it should be. See Problems 10.4 & 10.5.

Problem 10.44 The stationary cabinet has weight W. Determine the force F that must be exerted to move it if (a) the coefficient of static friction at A and B is μ_s, (b) if the coefficient of static friction at A is μ_{sA} and the coefficient of static friction at B is μ_{sB}.

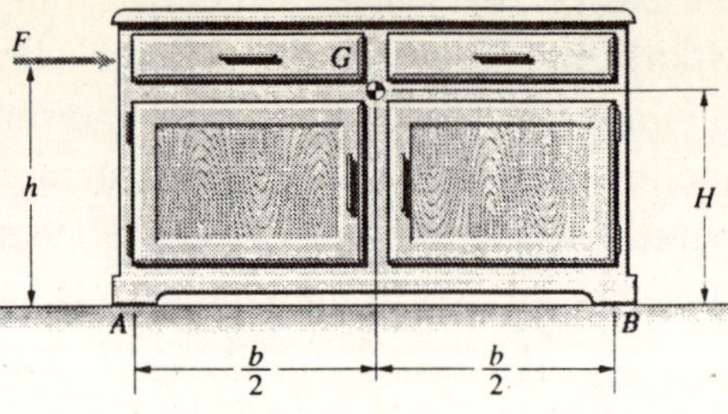

Solution (a) The sum of the moments about B is

$\sum M_B = -hF + \left(\frac{b}{2}\right)W - bA = 0$, from which $A = \frac{W}{2} - \left(\frac{h}{b}\right)F$.

The sum of forces: $\sum F_y = -W + A + B = 0$, from which

$B = W - A = \frac{W}{2} + \left(\frac{h}{b}\right)F$. $\sum F_x = F - \mu_s A - \mu_s B = 0$, from

which $\boxed{F = \mu_s(\frac{W}{2} + \left(\frac{h}{b}\right)F + \frac{W}{2} - \left(\frac{h}{b}\right)F) = \mu_s W}$ (b) Use the normal reactions found in Part (a). From

the sum of forces parallel to the floor, $F = \mu_{sA}A + \mu_{sB}B = \mu_{sA}\left(\frac{W}{2} - \left(\frac{h}{b}\right)F\right) + \mu_{sB}\left(\frac{W}{2} + \left(\frac{h}{b}\right)\right)$. Reduce

and solve: $\boxed{F = \dfrac{\frac{W}{2}(\mu_{sA} + \mu_{sB})}{\left(1 + \frac{h}{b}(\mu_{sA} - \mu_{sB})\right)}}$

Problem 10.45 The 3 m bar weighs 200 N. It will slip if the angle α is larger than 15°. What is the coefficient of static friction between the ends of the bar and the circular surface? *Strategy:* Assume that slip impends at both ends of the bar.

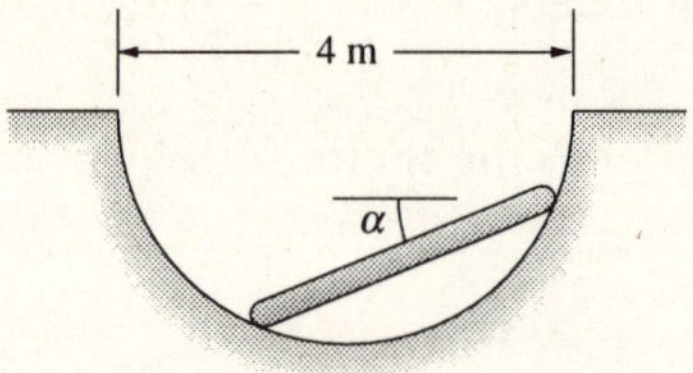

Solution Assume that the circular surface is a circle of radius $R = 2\ m$. Choose a coordinate system in which the x-axis is parallel to the bar. (See Figure.) The normal force is normal to the curved surface, and the friction forces are tangent to the surface. The strategy is to project these forces onto the coordinate system. This requires that the values of the angles be determined. Begin with the center of the circle O, which is also the origin of the coordinate system. Construct a bisector O-C' through the midpoint of the bar M. The line M-M' is parallel to x-x', and from geometry the angle formed by O-O' and O-C is $\alpha = 15°$. The angle formed by O-C and O-B is $\beta = \sin^{-1}\left(\frac{L}{2R}\right) = 48.6^o$, where $L = 3\ m$ is the length of the bar. The angle formed by O-C and O-A is also β, by symmetry. The angle $\theta = 90 - (\beta - \alpha) = 56.4^o$. The angle $\gamma = 90 - (\beta + \alpha) = 26.41^o$. The normal force at A is directed A-O, and is $\vec{A} = |\vec{A}|(\vec{i}\cos\theta + \vec{j}\sin\theta)$. The friction force at A opposes slip, and is normal to O-A: $\vec{f}_A = \mu_s|\vec{A}|(\vec{i}\sin\theta - \vec{j}\cos\theta)$.

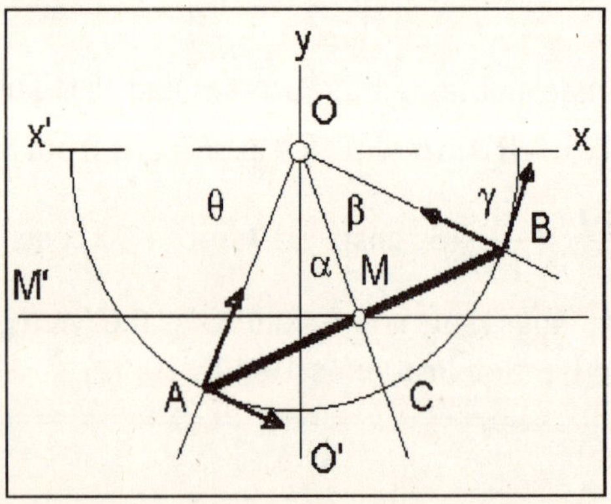

Problem 10.45 continued on next page..........

............continuation of solution to Problem 10.45.....

The normal force at B is directed B-O, and is $\vec{B} = |\vec{B}|(-\vec{i}\cos\gamma + \vec{j}\sin\gamma)$. The friction force at B opposes the impending slip, and is normal to O-B, $\vec{f}_B = \mu_s|\vec{B}|(\vec{i}\sin\gamma + \vec{j}\cos\gamma)$. The weight of the bar is directed downward, $\vec{W} = -\vec{j}|\vec{W}|$. The sum of the forces is

$$\sum F_x = |\vec{A}|(\cos\theta + \mu_s\sin\theta) + |\vec{B}|(-\cos\gamma + \mu_s\sin\gamma) = 0$$

$$\sum F_y = |\vec{A}|(\sin\theta - \mu_s\cos\theta) + |\vec{B}|(\sin\gamma + \mu_s\cos\gamma) - |\vec{W}| = 0$$

. The sum of the moments about the center of the circle is $\sum M_O = +R\mu_s|\vec{A}| + R\mu_s|\vec{B}| - R|\vec{W}|\cos\beta\sin\alpha = 0$, where the moment about the origin due to the weight is $-R\cos\beta(|\vec{W}|\sin\alpha)$. These are three equations in three unknowns. Solve by iteration (using **TK Solver 2**) to obtain $|\vec{A}| = 168\ N$, $|\vec{B}| = 129.2\ N$, and $\mu_S = 0.1152$. *Check:* The three simultaneous equations have a one degree of non linearity in μ_s so that there is a risk that another solution exists that may have been missed by the simple brute force iteration. To avoid the risk, solve the three equations analytically for μ_s as follows: Write the first two equations for the force equilibrium conditions as $a_{11}|\vec{A}| + a_{12}|\vec{B}| = 0$, $a_{21}|\vec{A}| + a_{22}|\vec{B}| = |\vec{W}|$. Multiply the first by a_{21} and the second by a_{11} and subtract to eliminate $|\vec{A}|$. The result: $|\vec{B}| = \dfrac{a_{11}|\vec{W}|}{\det}$, where $\det = a_{11}a_{22} - a_{12}a_{21}$ is the determinant of the two simultaneous equations. Similarly, $|\vec{B}|$ can be eliminated to obtain

$|\vec{A}| = -\dfrac{a_{12}|\vec{W}|}{\det}$. Substitute these into the third equation for the moment equilibrium condition and reduce algebraically: The result: $\mu_S(a_{11} - a_{12}) - \det(\cos\beta\sin\alpha) = 0$. This expression can be reduced algebraically: reducing term by term: $\det = (1+\mu_s^2)\sin(\theta+\gamma) = (1+\mu_s^2)(0.9922)$, and

$\mu_s(a_{11} - a_{12}) = \mu_s(\cos\theta + \cos\gamma) + \mu_s^2(\sin\theta - \sin\gamma) = \mu_s(1.4489) + \mu_s^2(0.3882)$, so that the expression for μ_s becomes, with substitution of numerical values, $\mu_s^2 + 2(3.317)\mu_s - 0.7778 = 0$. This quadratic expression has two solutions: $\mu_s = -3.317 \pm \sqrt{3.317^2 + 0.7778} = -6.75, = 0.1152$. The negative solution has no meaning here, so $\boxed{\mu_S = 0.1152}$, which agrees with the iterative solution. (See Problem 10.128 for an alternative formulation of the solution to this problem.)

Problem 10.46 The coefficient of static friction between the right bar and the surface at A is $\mu_s = 0.6$. Neglect the weights of the bars. If $\alpha = 20°$, what is the magnitude of the friction force exerted at A?

Solution Note that the condition of impending slip does not necessarily apply. The moments about the left pin support:

$\sum M = -F\cos\alpha + 2A\sin\alpha = 0$, from which $A = \dfrac{F}{2\tan\alpha}$.

Problem 10.46 continued on next page

continuation of Problem 10.46..........

Isolate the right bar and take moments about the upper pin joint: $\sum M = AL\sin\alpha - fL\cos\alpha = 0$, from which $f = A\tan\alpha$. Substitute for A: $\boxed{f = A\tan\alpha = \frac{F\tan\alpha}{2\tan\alpha} = \frac{F}{2}}$

Problem 10.47 Consider the system shown in Problem 10.46. The coefficient of static friction between the right bar and the surface at A is $\mu_s = 0.6$. Neglect the weight of the bars. What is the largest angle α at which the truss will remain stationary without slipping?

Solution From the solution to Problem 10.46, $f = A\tan\alpha$. Since $f = \mu_S A$ at impending slip, $\mu_s = \tan\alpha$, from which $\boxed{\alpha = \tan^{-1}(0.6) = 30.96^o}$

Problem 10.48 The coefficient of static friction between the box and the floor is μ_S. Neglect the weights of the bars. What force F will cause the box to slip?

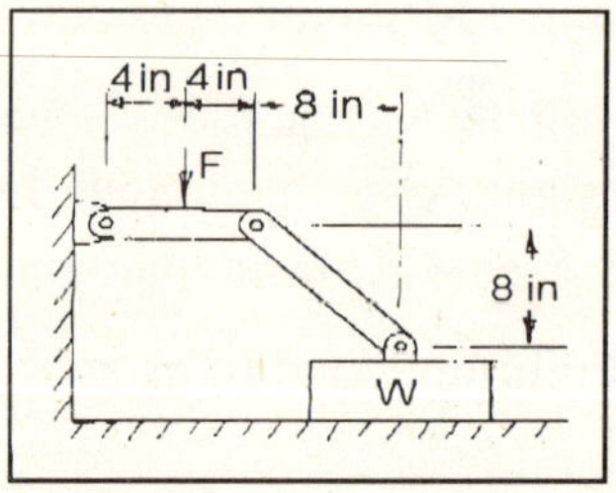

Solution Denote the normal force (acting upward) exerted by the surface on the box by A, and the friction force (acting to the left) exerted by the surface on the box by f. The sum of the moments about the leftmost pin support is $\sum M = -4F - 16W + 16A - 8f = 0$ *Isolate the right frame element*. The sum of the moments about the upper pin support is $\sum M = -8W + 8A - 8f = 0$, from which (1) $f = A - W$. Substitute into the first moment equation to obtain (2) $A = W + \frac{F}{2}$. Since by definition $f = \mu_S A$, then from (1) $A = (1 - \mu_S)W$. Substitute this into (2) to obtain $\boxed{F = \frac{2W\mu_S}{(1-\mu_S)}}$

Problem 10.49 The structure is supported at B by the normal and friction forces exerted on the plate by the wall. Neglect the weights of the bars. (a) What is the magnitude of the friction force? (b) What is the minimum coefficient of static friction necessary for the structure to remain in place?

Solution Assume that each bar is of the same length L. (a) Note that the condition for impending slip does not necessarily apply for Part (a). The sum of the moments about the upper pin support is

$$\sum M = -WL\sin 30^o + BL(\cos 30^o + \sin 8^o) - fL(\cos 8^o - \sin 30^o) = 0$$

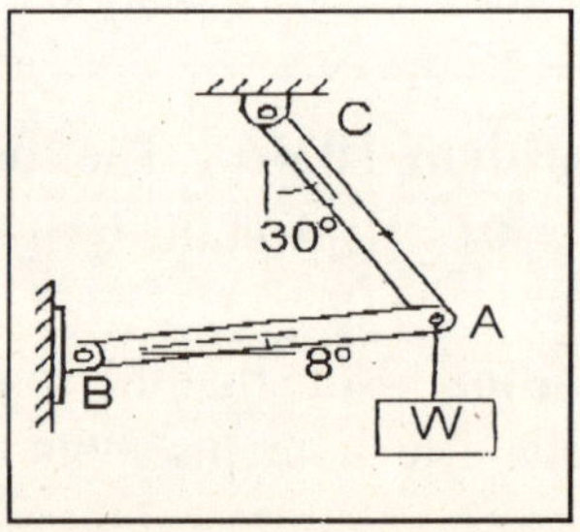

Isolate the lower frame member. The sum of the moments about the right pin support is $\sum M = -fL\cos 8^o + BL\sin 8^o = 0$, from which (1) $f = B\tan 8^o$. Substitute this into the first moment equation to obtain

(2) $B = \frac{W\sin 30^o}{(\cos 30^o + \sin 30^o \tan 8^o)}$. Substitute (2) into (1) to obtain

$\boxed{f = 0.0751W}$ (b) At impending slip $f = \mu_S B$. From (1) above, $\boxed{\mu_S = \tan 8^o = 0.1405}$

Problem 10.50 Suppose that the bars AB and AC in Problem 10.49 are of equal length L and each has a weight of $\frac{W}{2}$. (a) What is the magnitude of the friction force at B? (b) What is the minimum coefficient of static friction necessary for the structure to remain in place?

Solution The sum of the moments about C is

$\sum M = -WL(\frac{7}{4})\sin 30^o + BL(\cos 30^o + \sin 8^o) - fL(\cos 8^o - \sin 30^o) + \frac{W}{4}\cos 8^o = 0$. Isolate the lower frame element. The sum of the moments about the right pin support is

$\sum M = -(f - \frac{W}{4})L\cos 8^o + BL\sin 8^o = 0$, from which (1) $f = B\tan 8^o + \frac{W}{4}$. Substitute into the first moment equation to obtain (2) $B = \frac{1.5W\sin 30^o}{(\cos 30^o + \sin 30^o \tan 8^o)} = 0.801W$. (a) From (1)

$\boxed{f = B\tan 8^o + \frac{W}{4} = 0.3626W}$. (b) At impending slip $f = \mu_s B$. From (1) $\boxed{\mu_s = \tan 8^o + \frac{W}{4B} = 0.4526}$

Problem 10.51 The frame is supported by the normal and friction forces exerted on the plates at A and G by the fixed surfaces. The coefficient of static friction at A is $\mu_s = 0.6$. Will the frame slip when it is subjected to the forces shown?

Solution The strategy is to write the equilibrium equations and solve for the unknown. *The complete structure as a free body:* Denote the normal forces as A and G, and the friction forces as f_A and f_G. The sum of forces: (1)

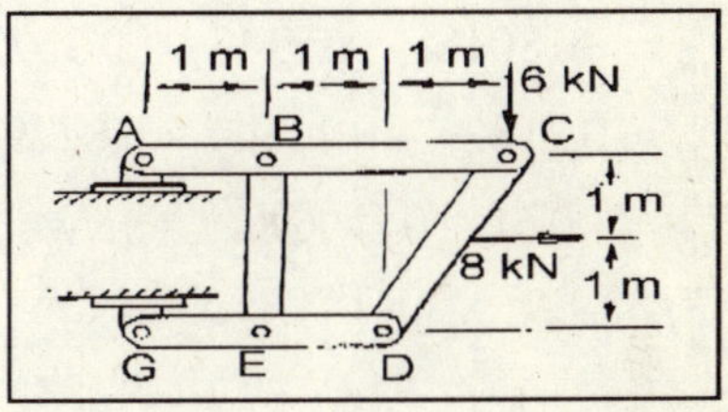

$\sum F_x = -f_A + f_G + 8 = 0$, (2) $\sum F_y = -A + G - 6 = 0$. *The elements as free bodies: Element ABC:* (See Figure)

(3) $\sum M_B = +1A + 2C_y - 12 = 0$. (4) $\sum F_x = -f_A + C_x = 0$,

(5) $\sum F_y = -A - B + C_y - 6 = 0$. *Element BE:*

(6) $\sum F_y = B - E = 0$. *Element CD:*(7)

$\sum M_C = +2D_x - D_y + 8 = 0$, (8) $\sum F_x = -C_x + D_x + 8 = 0$

(9) $\sum F_y = D_y - C_y = 0$. *Element DEG:*

(10) $\sum M_D = -2G - E = 0$

(11) $\sum F_x = f_G - D_x = 0$

(12) $\sum F_y = G - D_y + E = 0$ These twelve equations in ten unlknowns can be solved by iteration or by back substitution. The results in detail: $A = -24$ kN, $f_A = 13$ kN, $C_y = 18$ kN, $C_x = 13$ kN, $B = 36$ kN, $D_x = 5$ kN, $D_y = 18$ kN, $E = 36$ kN, $G = -18$ kN, $f_G = 5$ kN. The assumed directions are shown in the Figure; a negative sign means that the result is opposite to the assumed direction. The magnitude of the coefficient of static friction for the reaction at A required to hold the frame in equilibrium is

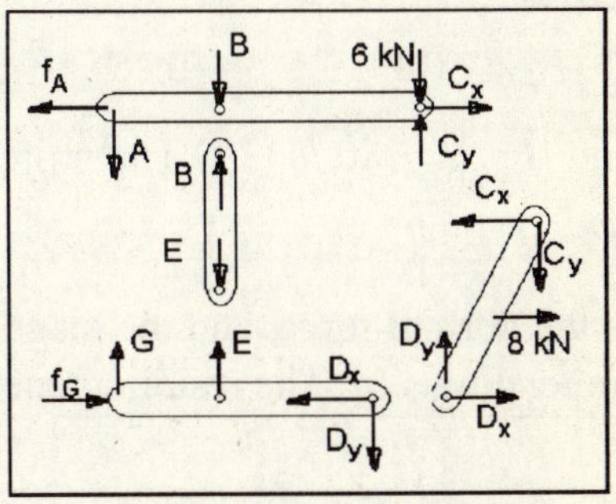

$\boxed{\mu_{sA} = \frac{f_A}{A} = 0.5417}$. Since this is less than the known value, $\mu_s = 0.6$, the frame will not slip at A.

=============================<>=============================

Problem 10.52 The frame in Problem 10.51 is supported by the normal and friction forces exerted on the plates at A and G by the fixed surfaces. The coefficient of static friction at G is $\mu_s = 0.25$. Will the frame slip at G when it is subjected to the loads shown?

Solution From the solution to Problem 10.51 the normal and friction force reactions at G required to hold the frame in equilibrium are $G = 18\ kN$, and $f_G = 5\ kN$. The coefficient of static friction is $\boxed{\mu_{sB} = \frac{5}{18} = 0.2778}$. Since this is greater than the known value $\mu_s = 0.25$, the frame will slip at G.

=============================<>=============================

Problem 10.53 The suspended weight $W = 600\ lb$. The bars AB and AC have ball and socket supports at each end. Suppose that you want the ball and socket at B to be held in place by the normal and friction forces between the support and the wall. What minimum coefficient of friction is required?

Solution The strategy is to develop the unit vectors parallel to the bars and the rope, project the forces onto these unit vectors, and solve the associated equilibrium equations. The vector positions are $\vec{r}_A = 3\vec{i} + 2\vec{j} + 4\vec{k}$, $\vec{r}_B = 0\vec{i} + 0\vec{j} + 3\vec{k}$, $\vec{r}_C = 4\vec{i} + 0\vec{j} + 0\vec{k}$, and $\vec{r}_O = 0\vec{i} + 2\vec{j} + 0\vec{k}$. The unit vectors are defined: $\vec{e}_{AB} = \frac{\vec{r}_B - \vec{r}_A}{|\vec{r}_B - \vec{r}_A|}$. Apply this definition to obtain:

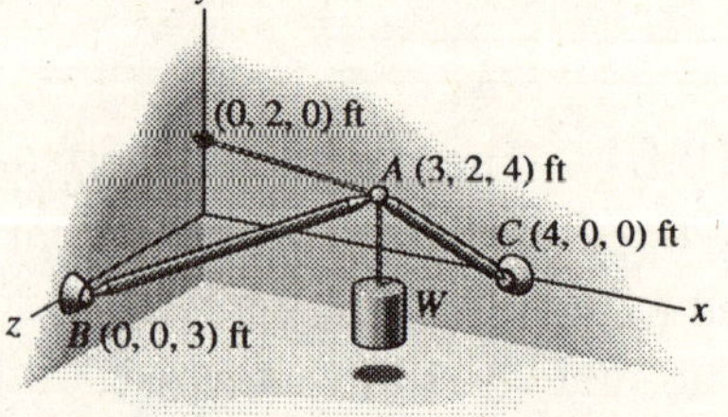

$\vec{e}_{AB} = -0.8018\vec{i} - 0.5345\vec{j} - 0.2673\vec{k}$,

$\vec{e}_{AC} = 0.2182\vec{i} - 0.4364\vec{j} - 0.8729\vec{k}$,

$\vec{e}_{AO} = -0.6\vec{i} + 0\vec{j} - 0.8\vec{k}$. The weight vector is $\vec{W} = -600\vec{j}$. The equilibrium conditions:

$\sum F_x = -0.2182|\vec{F}_{AC}| - 0.8018|\vec{F}_{AB}| - 0.6|\vec{F}_{AO}| = 0$,

$\sum F_y = -0.4364|\vec{F}_{AC}| - 0.5345|\vec{F}_{AB}| + 0|\vec{F}_{AO}| - |\vec{W}| = 0$

$\sum F_z = -0.8729|\vec{F}_{AC}| - 0.2673|\vec{F}_{AB}| - 0.8|\vec{F}_{AO}| = 0$ Solve: $|\vec{F}_{AC}| = -494.92\ lb\ (C)$, $|\vec{F}_{AB}| = -718.4\ lb\ (C)$, and $|\vec{F}_{AO}| = 780\ lb\ (T)$. The projection of $|\vec{F}_{AB}|$ onto the unit vector is the force exerted on the wall at B: $\vec{F}_{AB} = |\vec{F}_{AB}|(-0.8018\vec{i} - 0.5345\vec{j} - 0.2673\vec{k}) = +576\vec{i} + 384\vec{j} + 192\vec{k}$. The magnitude of the *x*-component is the normal force, and the magnitude of the resultant of the y- and z- component is the friction force at impending slip. The resulting coefficient of static friction at impending slip is $\boxed{\mu_s = \frac{\sqrt{384^2 + 192^2}}{576} = 0.7454}$. This is the minimum coefficient of static friction required to prevent slip.

=============================<>=============================

Problem 10.54 In Problem 10.53 what friction force is exerted on the support at B by the wall?

Solution From the solution to Problem 10.53, the magnitude of the friction force is

$\boxed{f_B = \sqrt{384^2 + 192^2} = 429.3\ lb}$

=============================<>=============================

Problem 10.55 A force $F = 200\ N$ is necessary to raise A at a constant rate. Between all contacting surfaces, $\mu_s = 0.25$ and $\mu_k = 0.20$. The weight of the wedge is negligible. What is the weight of A?

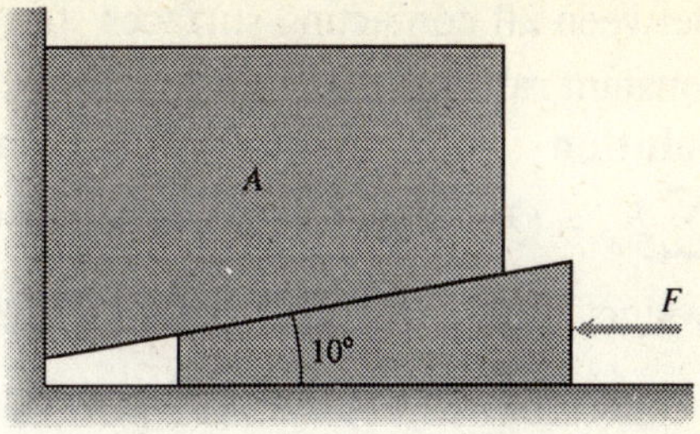

Solution The contacting surfaces are on the wedge and between the load A and the wall. Denote the normal force on the upper surface of the wedge by *N*, and the normal force on the lower surface by *P*. Denote the normal force between the wall and A by *Q*. The equilibrium conditions for the load A moving at a constant rate are:

$\sum F_x = Q - N\sin\alpha - \mu_k N\cos\alpha = 0$. $\sum F_y = N\cos\alpha - \mu_k N\sin\alpha - \mu_k Q - W_A = 0$. The equilibrium conditions for the wedge are $\sum F_x = N\sin\alpha + \mu_k N\cos\alpha + \mu_k P - F = 0$.

$\sum F_y = P - N\cos\alpha + \mu_k N\sin\alpha = 0$. These four simultaneous equations in four unknowns may be solved by Gaussian multiplication and subtraction to eliminate unknowns, or by iteration. The solution: $Q = 132.2\ N$, $N = 356.7\ N$, $P = 338.94\ N$ and $\boxed{W_A = 312.49\ N}$ *Check:* Use Eq (10.7) to obtain:

$$W_A = \frac{(1-\mu_k^2) - 2\mu_k\tan\alpha}{(1-\mu_k^2)\tan\alpha + 2\mu_k}F = \left(\frac{0.8895}{0.5693}\right)(200) = 312.49\ N. \quad \textit{check.}$$

Problem 10.56 Suppose that A in Problem 10.55 weighs 1000 *lb*. Between all contacting surfaces $\mu_s = 0.25$ and $\mu_k = 0.20$. Neglect the weight of the wedge. (a) What force is necessary to start the wedge moving to the left? (b) Will the wedge remain in place if the force is removed?

Solution (a) From the solution to Problem 10.55, the equilibrium conditions at impending slip are: for the load A: $\sum F_x = Q - N\sin\alpha - \mu_s N\cos\alpha = 0$, $\sum F_y = N\cos\alpha - \mu_s N\sin\alpha - \mu_s Q - W_A = 0$. For the wedge: $\sum F_x = N\sin\alpha + \mu_s N\cos\alpha + \mu_s P - F = 0$, $\sum F_y = P - N\cos\alpha + \mu_s N\sin\alpha = 0$, where the coefficient of static friction is used. These four simultaneous equations in four unknowns are solved to obtain: $Q = 501.95\ lb$, $N = 1195.6\ lb$, $P = 1125.4\ lb$ and $\boxed{F = 783.3\ lb}$. (b) The strategy is to solve the equilibrium equations for μ_s when the force is zero. The equilibrium conditions for the wedge when the force is zero and slip is impending are $\sum F_x = N\sin\alpha - \mu_s N\cos\alpha - \mu_s P = 0$,

$\sum F_y = P - N\cos\alpha - \mu_s N\sin\alpha = 0$. Eliminate *P* to obtain: $(1-\mu_s^2)\sin\alpha - 2\mu_s\cos\alpha = 0$. The result is a quadratic: $\mu_s^2 + 2\mu_s\cot\alpha - 1 = 0$, with the solution $\mu_s = 0.087,\ = -11.43$. The negative value has no meaning here, so the coefficient of static friction required to maintain the wedge in place is $\boxed{\mu_s = 0.087}$, so that the wedge will not slip out.

Problem 10.57 Suppose that A in Problem 10.55 weighs 1000 *lb* and the wedge weighs 200 *lb*. Between all contacting surfaces, $\mu_s = 0.25$ and $\mu_k = 0.20$. What force is necessary to raise the box at a constant rate?

Solution The equilibrium conditions for the wedge in steady motion are, for the box: $\sum F_x = Q - N\sin\alpha - \mu_k N\cos\alpha = 0$, $\sum F_y = N\cos\alpha - \mu_k N\sin\alpha - \mu_k Q - W_A = 0$, and for the wedge: $\sum F_x = N\sin\alpha + \mu_k N\cos\alpha + \mu_k P - F = 0$, $\sum F_y = P - N\cos\alpha + \mu_k N\sin\alpha - W_W = 0$.

Solve: $Q = 423.1\ lb$, $N = 1141.6\ lb$, $P = 1284.6\ lb$, and $\boxed{F = 680\ lb}$. *Check*: Use Eq (10.7)

$$F = \mu_k W_W + \frac{(1-\mu_k^2)\tan\alpha + 2\mu_k}{(1-\mu_k^2) - 2\mu_k\tan\alpha} W = 40 + \frac{0.5693}{0.8895}(1000) = 680\ lb\ \ check.$$

Problem 10.58 The stationary blocks A, B, and C each weigh 2 *kN*. Between all contacting surfaces, $\mu_s = 0.6$. What force is necessary to start B moving downward?

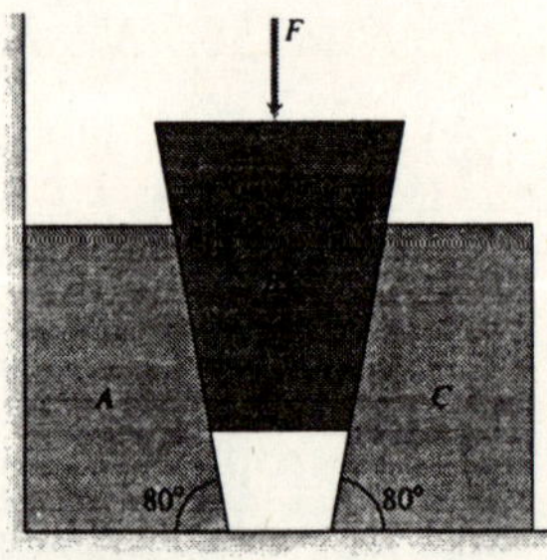

Solution The wedge angle is 10° for each side. The block A cannot move, hence the friction contact surfaces are the wedge surfaces plus the bottom surface of block C. The equilibrium conditions are (See Free body diagrams) *for the wedge*

$$\sum F_y = -F - W + 2\mu_s N\cos\alpha + 2N\sin\alpha = 0$$

For the block:

$$\sum F_y = -W - N\sin\alpha - \mu_s N\cos\alpha + P = 0$$

$$\sum F_x = N\cos\alpha - \mu_s N\sin\alpha - \mu_s P = 0$$

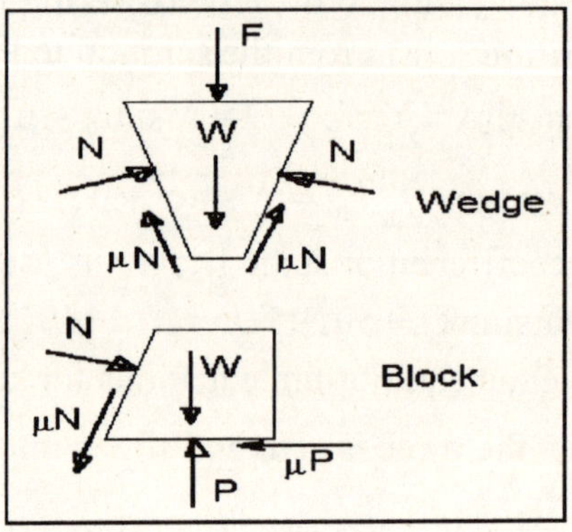

These three equations in three unknowns are solved by iteration: $N = 2.84\ kN$, $P = 4.17\ kN$, and $\boxed{F = 2.349\ kN}$ *Check:* The three equations can be solved by back substitution: define $a_1 = \sin\alpha + \mu_s\cos\alpha = 0.7645$, and $a_2 = \cos\alpha - \mu_s\sin\alpha = 0.8806$. From the last equation:

$P = N\left(\frac{a_2}{\mu_s}\right)$. Substitute into the second equation to obtain

$N = \frac{\mu_s W}{a_2 - \mu_s a_1} = 2.84\ kN$. Substitute into the first equation to obtain

$$F = W\left(\frac{3\mu_s a_1 - a_2}{a_2 - \mu_s a_1}\right) = 2\left(\frac{0.4955}{0.4219}\right) = 2.349\ kN, \text{ and } P = 4.17\ kN.\ check.$$

Problem 10.59 Suppose that the force $F = 1\ kN$ in Problem 10.58 and A, B, and C each weigh 2 kN. The coefficient of static friction between all surfaces is μ_s. What is the smallest value of μ_s for which all blocks will remain stationary?

Solution From the solution to Problem 10.58, the equilibrium conditions are: *For the wedge* $\sum F_y = -F - W + 2\mu_s N\cos\alpha + 2N\sin\alpha = 0$. *For the block:*

$\sum F_y = -W - N\sin\alpha - \mu N\cos\alpha + P = 0$

$\sum F_x = N\cos\alpha - \mu_s N\sin\alpha - \mu_s P = 0$. Substitute $\alpha = 10^o$, $F = 1\ kN$, and $W = 2\ kN$ and solve the three simultaneous equations in the three unknowns by iteration. The result: $N = 2.124\ kN$, $P = 3.5\ kN$, and $\boxed{\mu_s = 0.5407}$ *Check:* From the solution to Problem 10.58 for the force: $F = W\left(\dfrac{3\mu_s a_1 - a_2}{a_2 - \mu_s a_1}\right)$. Substitute and reduce algebraically to obtain the quadratic:

$\mu_s^2 + 2\left(\dfrac{F+2W}{F+3W}\right)\mu_s \tan\alpha - \left(\dfrac{F+W}{F+3W}\right) = 0$, from which $\mu_s^2 + 2(0.1259)\mu_s - 0.4286 = 0$. This has the solution $\mu_s = -0.1259 \pm \sqrt{0.1259^2 + 0.4286} = -0.792,\ \ 0.5407$. The negative value has no meaning here, so $\mu_s = 0.5407$ which agrees with the iterative solution. *check.*

Problem 10.60 The object A is held in place by two wedges. The coefficient of static friction between all surfaces is μ_s. Neglect the weights of the wedges. What is the largest angle α for which all wedges will remain in place?

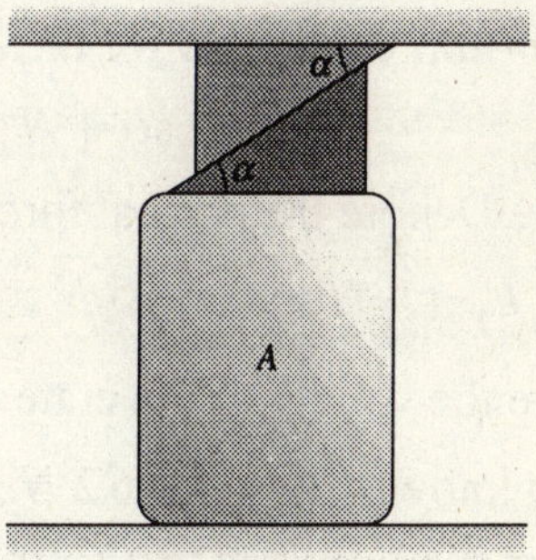

Solution The equilibrium equations for the two wedges are: *For the top wedge:* Denote the normal force acting downward on the top wedge by *P*, and the normal force acting on the lower surface by *N*. The friction force acts to oppose slip.

$\sum F_{y1} = -P + N\cos\alpha + \mu_s N\sin\alpha = 0$,

$\sum F_{x1} = \mu_s P + \mu_s N\cos\alpha - N\sin\alpha = 0$ *For the bottom wedge:* Denote the normal force acting upward on the lower surface by Q.

$\sum F_{y2} = -N\cos\alpha - \mu_s N\sin\alpha + Q = 0$

$\sum F_{x2} = N\sin\alpha - \mu_s N\sin\alpha - \mu_s Q = 0$. Define: $a_1 = \sin\alpha - \mu_s\cos\alpha$ and $a_2 = \cos\alpha + \mu_s\sin\alpha$.

From the last equation: $Q = \dfrac{Na_1}{\mu_s}$. Substitute into the third equation to obtain $a_2 - \dfrac{a_1}{\mu_s} = 0$, from which

$\mu_s^2 + 2\mu_s\cot\alpha - 1 = 0$. Solve: $\tan\alpha = \dfrac{2\mu_s}{(1-\mu_s^2)}$, or $\boxed{\alpha = \tan^{-1}\left(\dfrac{2\mu_s}{1-\mu_s^2}\right)}$., if the equilibrium condition of no slip is to be satisfied for the lower wedge. *For the upper wedge:* From the last equation $P = \dfrac{Na_1}{\mu_s}$.

Substitute into the first equation to obtain $a_2 - \dfrac{a_1}{\mu_s} = 0$, which leads to the result obtained for the lower wedge for the angle α.

Problem 10.61 The crate A weighs 600 *lb*. Between all contacting surfaces, $\mu_s = 0.32$ and $\mu_k = 0.30$. Neglect the weight of the wedges. What force F is required to move A to the right at a constant rate?

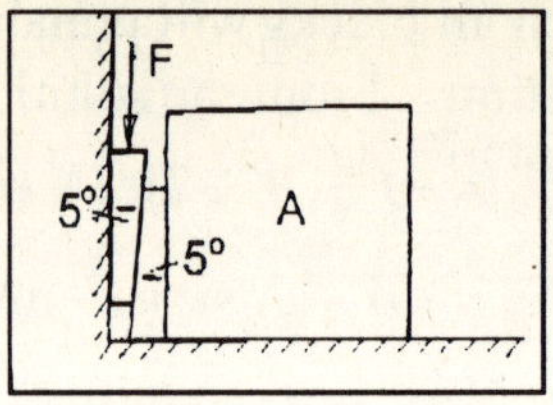

Solution The active sliding contact surfaces are between the wall and the left wedge, between the wedges, between the floor and the bottom of the right wedge, and between the crate and the floor. *Leftmost wedge:* Denote the normal force exerted by the wall by Q, and the normal force between the wedges by N. The equilibrium conditions for the left wedge moving at a constant rate are: $\sum F_y = -F + \mu_k N\cos\alpha + N\sin\alpha + \mu_k Q = 0$.

$\sum F_x = Q - N\cos\alpha + \mu_k N\sin\alpha = 0$. *For the right wedge:* Denote the normal force exerted by the crate by A, and the normal force exerted by the floor by P. $\sum F_y = -N\sin\alpha - \mu_k N\cos\alpha + P = 0$.

$\sum F_x = N\cos\alpha - \mu_k N\sin\alpha - \mu_k P - A = 0$. *For the crate:* Denote the weight of the crate by W.

$\sum F_x = A - \mu_k W = 0$. These five equations are solved for the five unknowns by iteration:

$Q = 204.4$ *lb*, $N = 210.7$ *lb* $P = 81.34$ *lb*, $A = 180$ *lb*, and $\boxed{F = 142.66 \ lb}$

Problem 10.62 Suppose that between all contacting surfaces in Problem 10.61, $\mu_s = 0.32$ and $\mu_k = 0.30$. Neglect the weight of the 5° wedges. If a force $F = 800$ N is required to move A to the right at a constant rate, what is the weight of A?

Solution From the solution to Problem 10.61, the equilibrium equations are: *Leftmost wedge:* Denote the normal force exerted by the wall by Q, and the normal force between the wedges by N. The equilibrium conditions for the left wedge moving at a constant rate are:

$\sum F_y = -F + \mu_k N\cos\alpha + N\sin\alpha + \mu_k Q = 0$. $\sum F_x = Q - N\cos\alpha + \mu_k N\sin\alpha = 0$. *For the right wedge:* Denote the normal force exerted by the crate by A, and the normal force exerted by the floor by P. $\sum F_y = -N\sin\alpha - \mu_k N\cos\alpha + P = 0$. $\sum F_x = N\cos\alpha - \mu_k N\sin\alpha - \mu_k P - A = 0$. *For the crate:* Denote the weight of the crate by W. $\sum F_x = A - \mu_k W = 0$. These five equations are solved for the five unknowns: $Q = 1146.2$ N, $N = 1181.6$ N, $P = 456.1$ N, $\boxed{W = 3364.7 \ N}$, and $\boxed{A = 1009 \ N}$.

Problem 10.63 Box A weighs 800 N and the wedge B weighs 400 N. Between all contacting surfaces, $\mu_s = 0.15$ and $\mu_k = 0.12$. What force is required to raise A at a constant rate?

Solution The active surfaces are between the wall and the box, and between the box and the upper surface of the wedge, and between the lower surface of the wedge and the base. *For the box A*: Denote the normal force exerted by the wall by Q, and the normal force exerted by the wedge by N. The friction forces oppose motion.

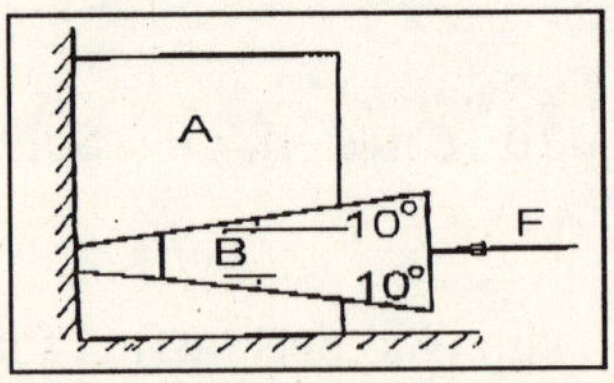

$\sum F_y = -W + N\cos\alpha - \mu_k N\sin\alpha - \mu_k Q = 0$,

$\sum F_x = -\mu_k N\cos\alpha - N\sin\alpha + Q = 0$. *For the wedge B.* Denote the normal force on the lower surface by P.

Problem 10.63 continued on next page

continuation of Problem 10.63..........

$\sum F_x = -F + \mu_k N\cos\alpha + \mu_k P\cos\alpha + P\sin\alpha + N\sin\alpha = 0$.

$\sum F_y = -N\cos\alpha + P\cos\alpha + \mu_k N\sin\alpha - \mu_k P\sin\alpha - W_w = 0$. These four equations in four unknowns are solved: $Q = 251.3\ N$, $N = 861.2\ N$, $P = 1276.1\ N$ and $\boxed{F = 623.7\ N}$

Problem 10.64 Suppose that in Problem 10.63, A weighs 800 *lb* and B weighs 400 *lb*. The coefficients of friction between all contacting surfaces are $\mu_s = 0.15$ and $\mu_k = 0.12$. Will B remain in place if the force F is removed?

Solution The equilibrium conditions are: *For the box A*: Denote the normal force exerted by the wall by Q, and the normal force exerted by the wedge by N. The friction forces oppose motion.

$\sum F_y = -W + N\cos\alpha + \mu_s N\sin\alpha + \mu_s Q = 0$, $\sum F_x = +\mu_s N\cos\alpha - N\sin\alpha + Q = 0$.

For the wedge B. Denote the normal force on the lower surface by P.

$\sum F_x = -\mu_s N\cos\alpha - \mu_s P\cos\alpha + P\sin\alpha + N\sin\alpha = 0$.

$\sum F_y = -N\cos\alpha + P\cos\alpha - \mu_s N\sin\alpha + \mu_k P\sin\alpha - W_w = 0$. (A comparison with the equilibrium conditions for Problem 10.63 will show that the friction forces are reversed, since for slippage the box A will move downward, and the wedge B to the right.) The strategy is to solve these equations for the required μ_s to keep the wedge B in place when $F = 0$. The solution $Q = 0$, $N = 787.8\ lb$, $P = 1181.8\ lb$ and $\boxed{\mu_s = 0.1763}$. Since the value of μ_s required to hold the wedge in place is greater than the value given, the wedge will slip out. *Check:* The value of μ_s can be obtained directly from one of the equilibrium equations for the wedge, namely,

$\sum F_x = -\mu_s N\cos\alpha - \mu_s P\cos\alpha + P\sin\alpha + N\sin\alpha = 0$, from which $(\sin\alpha - \mu_s\cos\alpha)(P + N) = 0$. If $P > 0$ and $N > 0$, then $\mu_s = \tan\alpha = 0.1763$, which is the result for a box on an incline. See Problems 10.4 & 10.5. *check.*

Problem 10.65 Between A and B, $\mu_s = 0.20$, and between B and C, $\mu_s = 0.18$. Between C and the wall, $\mu_s = 0.30$. The weights $W_B = 20\ lb$ and $W_C = 80\ lb$. What force F is required to start C moving upward?

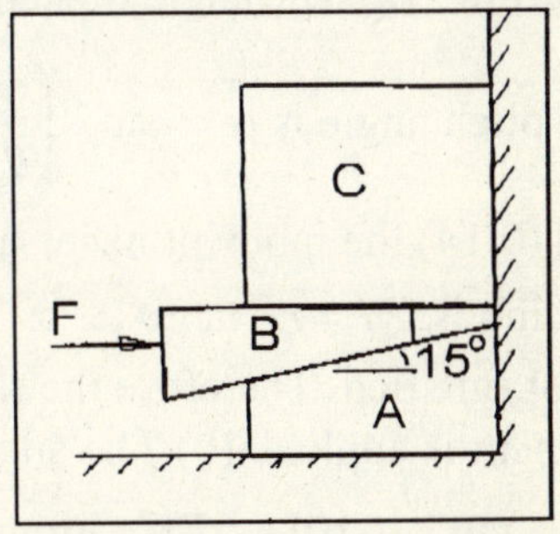

Solution The active contact surfaces are between the wall and C, between the wedge B and C, and between the wedge B and A. *For the weight C*: Denote the normal force exerted by the wall by Q, and the normal force between B and C by N. Denote the several coefficients of static friction by subscripts. The equilibrium conditions are:

$\sum F_y = -W_C + N - \mu_{CW} Q = 0$, $\sum F_x = -Q + \mu_{BC} N = 0$. *For the wedge B*: Denote the normal force between A and B by P.

$\sum F_y = -N + P\cos\alpha - \mu_{AB} P\sin\alpha - W_B = 0$. $\sum F_x = F - \mu_{BC} N - \mu_{AB} P\cos\alpha - P\sin\alpha = 0$. These four equations in four unknowns are solved: $Q = 15.2\ lb$, $N = 84.6\ lb$, $P = 114.4\ lb$, and $\boxed{F = 66.9\ lb}$

Problem 10.66 The masses of A, B, and C are 8 *kg*, 12 *kg*, and 80 *kg*, respectively. Between all contacting surfaces, $\mu_s = 0.4$. What force F is required to start C moving upward?

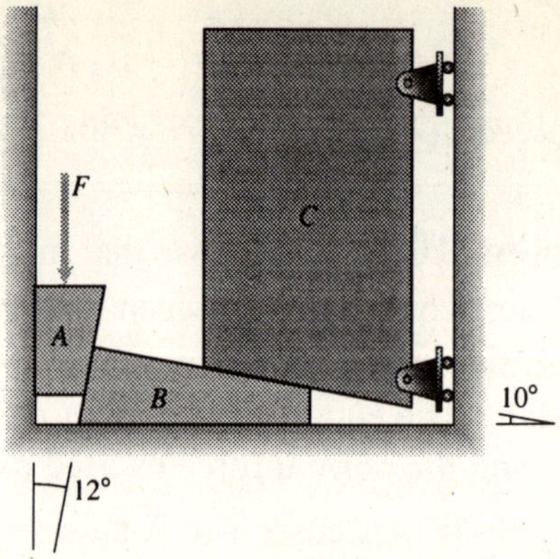

Solution The active contact surfaces are between A and B, between A and the wall, between B and the floor, and between B and C. Assume that the roller supports between C and the wall exert no friction forces. *For the wedge A:* Denote the normal force exerted by the wall as Q and the normal force between A and B as N. The weight is $W_A = 8g = 78.48\ N$. The equilibrium conditions:

$\sum F_y = -F + \mu_s Q + \mu_s N\cos\alpha + N\sin\alpha - W_A = 0$

$\sum F_y = Q - N\cos\alpha + \mu_s N\sin\alpha = 0$. *For wedge B*: Denote the normal force exerted on B by the floor by P, and the normal exerted by the weight C as S. The weight of B is $W_B = 12g = 117.72\ N$. The equilibrium conditions:

$\sum F_y = -N\sin\alpha - S\cos\beta + P - \mu_s N\cos\alpha + \mu_s S\sin\beta - W_B = 0$.

$\sum F_x = N\cos\alpha - \mu_s N\sin\alpha - \mu_s P - \mu_s S\cos\beta - S\sin\beta = 0$. *For the weight C:* The weight is $W_C = 80g = 784.8\ N$. The equilibrium conditions: $\sum F_y = -W_C + S\cos\beta - \mu_s S\sin\beta = 0$. These five equations in five unknowns are solved: $Q = 1157.6\ N$, $N = 1293.5\ N$, $S = 857.4\ N$, $P = 1677.5$, and $\boxed{F = 1160\ N}$

Problem 10.67 The vertical threaded shaft fits into a mating groove in the tube C. The pitch of the threaded shaft is $p = 0.1\ in$. and the mean radius of the thread is $r = 0.5\ in$. The coefficients of friction between the thread and the mating groove are $\mu_s = 0.15$ and $\mu_k = 0.10$. The weight $W = 200\ lb$. Neglect the weight of the threaded shaft. (a) Will the stationary threaded shaft support the weight if no moment is applied to the shaft? (b) What moment must be applied to the threaded shaft to raise the weight at a constant rate?

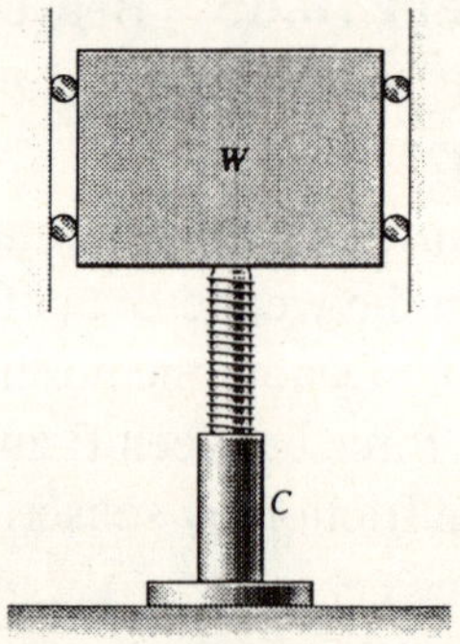

Solution (a) The angle of static friction is $\theta_s = \tan^{-1}(0.15) = 8.53^o$.

The pitch angle is $\alpha = \tan^{-1}\left(\dfrac{p}{2\pi r}\right) = \tan^{-1}\left(\dfrac{0.1}{2\pi(0.5)}\right) = 1.82^o$. From Eq (10.14) the moment necessary for the shaft to be on the verge of rotating is $M = rF\tan(\theta_s - \alpha)$.. For a zero moment, $\theta_s = \alpha$, which is not satisfied. Therefore the shaft will support the weight when no moment is applied. (b) The angle of kinetic friction is $\theta_k = \tan^{-1}(0.10) = 5.71^o$. From Eq (10.11) the moment required to raise the weight at a constant rate is $\boxed{M = rW\tan(\theta_k + \alpha) = 0.5(200)\tan(7.533) = 13.2\ in\ lb}$

Problem 10.68 Suppose that in Problem 10.67 the pitch of the threaded shaft is $p = 2\ mm$ and the mean radius of the thread is $r = 20\ mm$. The coefficients of friction between the thread and the mating groove are $\mu_s = 0.22$, and $\mu_k = 0.20$ The weight $W = 500\ N$. Neglect the weight of the threaded shaft. What moment must be applied to the threaded shaft to lower the weight at a constant rate?

Solution The angle of kinetic friction is $\theta_k = \tan^{-1}(0.2) = 11.31^o$. The angle of pitch is $\alpha = \tan^{-1}\left(\frac{p}{2\pi r}\right) = \tan^{-1}\left(\frac{2}{2\pi(20)}\right) = 0.9118^o$. The moment required to lower the weight at a constant rate is $\boxed{M = 0.02(500)\tan(11.31 - 0.9118) = 1.835\ N\,m}$

Problem 10.69 The position of the horizontal beam can be adjusted by turning the machine screw A. Neglect the weight of the beam. The pitch of the screw is $p = 1\ mm$ and the mean radius of the thread is $r = 4\ mm$. The coefficients of friction between the thread and the mating groove are $\mu_s = 0.20$ and $\mu_k = 0.18$. If the system is initially stationary, determine the moment that must be applied to the screw to start moving (a) upward; (b) downward.

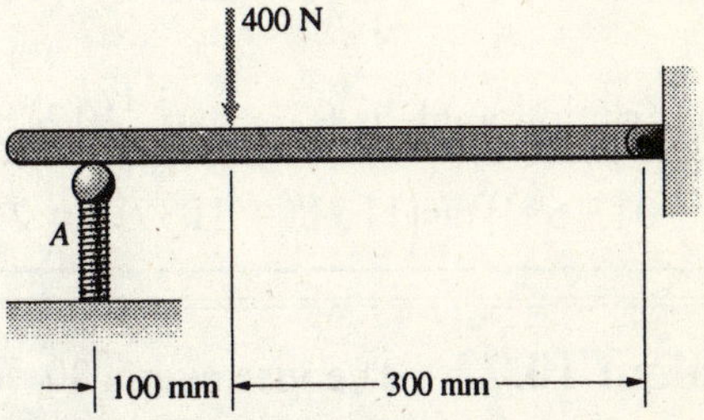

Solution The sum of the moments about the pin support is $\sum M = -0.4F + (0.3)400 = 0$, from which the force exerted by the screw is $F = 300\ N$. The pitch angle is $\alpha = \tan^{-1}\left(\frac{1}{2\pi(4)}\right) = 2.28^o$. The static friction angle is $\theta_s = \tan^{-1}(0.2) = 11.31^o$. (a) The moment required to start motion upward is $\boxed{M = 0.004(300)\tan(11.31^o + 2.28^o) = 0.29\ N\,m}$ (b) The moment required to start motion downward is $\boxed{M = 0.004(300)\tan(11.31^o - 2.28^o) = 0.19\ N\,m}$

Problem 10.70 Suppose that in Problem 10.69, the pitch of the machine screw is $p = 1\ mm$ and the mean radius of the thread is $r = 4\ mm$. What minimum value of the coefficient of static friction between the thread and the mating groove is necessary for the beam to remain in the position shown with no moment applied to the screw?

Solution From the solution to Problem 10.69 the force applied to the screw is $F = 300\ N$. The pitch angle is $\alpha = \tan^{-1}\left(\frac{1}{2\pi(4)}\right) = 2.28^o$. The moment required to start motion downward is

$M = 0.004(300)\tan(\theta_s - \alpha)$. For $M = 0$, $\tan(\theta_s - \alpha) = 0$, from which $\theta_s = \alpha = 2.279^o$, and

$\boxed{\mu_s = \tan(2.279^o) = 0.0398}$

Problem 10.71

The block A weighs 600 N. Neglect the weight of the 5° wedge. The coefficient of kinetic friction between the contacting surfaces of the block A, the wedge, the table and the wall is $\mu_k = 0.4$. The pitch of the threaded shaft is 5 mm, the mean radius of the threads is 15 mm, and the coefficient of kinetic friction between the threads and the mating groove is 0.2. What moment must be exerted on the threaded shaft to raise the block at a constant rate?

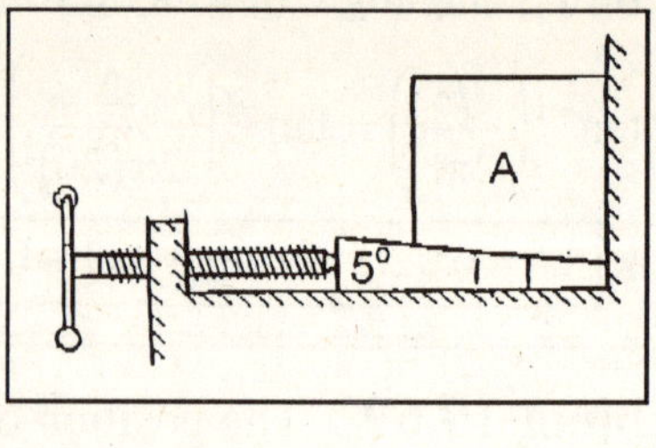

Solution Denote the wedge angle by $\beta = 5^o$ and the normal force on the top by N and on the lower surface by P. *For the wedge*:

$\sum F_y = -N\cos\beta + P + \mu_k N\sin\beta = 0$

$\sum F_x = F - \mu_k N\cos\beta - \mu_k P - N\sin\beta = 0$. *For the block A*: the normal force at the wall is Q.: $\sum F_y = -W + N\cos\beta - \mu_k N\sin\beta - \mu_k Q = 0$,

$\sum F_x = \mu_k N\cos\beta - Q + N\sin\beta = 0$. These four equations are solved to obtain: $Q = 379.9\ N$,

$N = 782.2\ N$, $P = 751.9\ N$, and $F = 680.6\ N$. The pitch angle is $\alpha = \tan^{-1}\left(\frac{5}{2\pi(15)}\right) = 3.037^o$, and the kinetic friction angle is $\theta_k = \tan^{-1}(0.2) = 11.31^o$. The required moment is

$M = 0.015(681)\tan(11.31^o + 3.037^o) = 2.61\ N\ m$

Problem 10.72

The vise exerts 80 lb forces on block A. The threaded shafts are subjected only to axial loads by the jaws of the vise. The pitch of their threads is $p = \frac{1}{8}\ in$, the mean radius of the threads is $r = 1\ in$, and the coefficient of static friction between the threads and the mating grooves is 0.2. Suppose that you want to loosen the vise by turning one of the shafts. Determine the moment you must apply (a) to shaft B; (b) to shaft A.

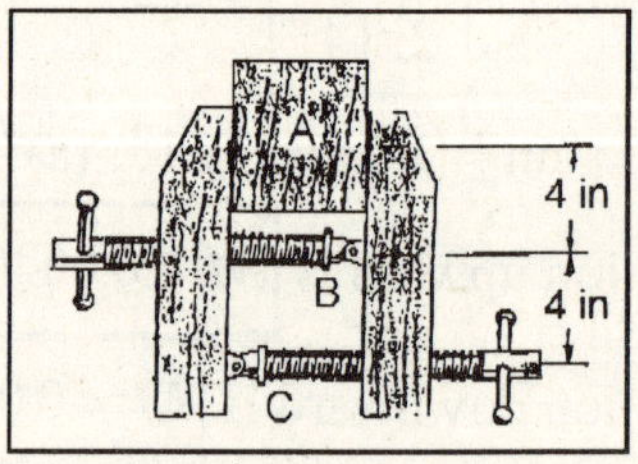

Solution Isolate the left jaw. The sum of the moments about C: $\sum M_C = -4B + 8(80) = 0$, from which $B = 160\ lb\ (T)$. The sum of the forces: $\sum F_x = -80 + B - C = 0$, from which $C = 80\ lb\ (C)$.

The pitch angle is $\alpha = \tan^{-1}\left(\frac{1}{16\pi}\right) = 1.14^o$. The static friction angle is $\theta_s = \tan^{-1}(0.2) = 11.31^o$. The moments required to loosen the vise are $M_B = \left(\frac{1}{12}\right)(160)\tan(11.31^o - 1.14^o) = 2.39\ ft\ lb$, and

$M_C = rC\tan(\theta_s - \alpha) = 1.2\ ft\ lb$.

Problem 10.73 Suppose that you want to tighten the vise in Problem 10.72 by turning one of the shafts. Determine the moment you must apply (a) to shaft B; (c) to shaft C.

Solution Use the solution to Problem 10.72. (a) The moment on shaft B required to tighten the vise is $M_B = rB\tan(\theta_s + \alpha)$. Noting that $r = \frac{1}{12}$, $B = 160\ lb$, $\alpha = \tan^{-1}\left(\frac{1}{16\pi}\right) = 1.14^o$ and $\theta_s = \tan^{-1}(0.2) = 11.31^o$, then $\boxed{M_B = 2.94\ ft\ lb}$ (b) For shaft C, $M_C = rC\tan(\theta_s + \alpha)$, where $C = 80\ lb$, $\boxed{M_C = 1.74\ ft\ lb}$

Problem 10.74 The clamp presses two pieces of wood together. The pitch of the threads is $p = 2\ mm$ and the mean radius of the threads is $r = 8\ mm$, and the coefficient of kinetic friction between the threads and the mating groove is 0.24. What moment must be exerted on the threaded shaft to press the pieces of wood together with a force of 200 *N*?

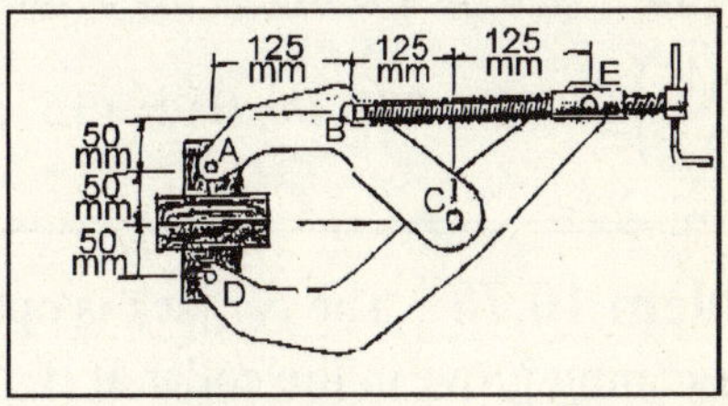

Solution *Isolate the upper jaw.* The sum of the moments about the pin support is: $\sum M = 0.1F - (0.25)(200) = 0$, from which $F = 500\ N$. The pitch angle is $\alpha = \tan^{-1}\left(\frac{2}{2\pi(8)}\right) = 2.28^o$. The kinetic friction angle is $\theta_k = \tan^{-1}(0.24) = 13.5^o$. The moment required to exert the 200 *N* on the block is

$$\boxed{M = (0.008)(500)\tan\left(13.5^o + 2.28^o\right) = 1.13\ N\ m}$$

Problem 10.75 Suppose that in Problem 10.74, the coefficient of static friction between the threaded shaft and the mating grooves is 0.24. If the clamp presses the pieces of wood together with a force of 200 *N*, what moment must be exerted on the threaded shaft to begin loosening the clamp?

Solution Use the solution to Problem 10.74. The angle of the coefficient of static friction is $\theta_s = \tan^{-1}(0.3) = 16.7^o$. The moment required is $\boxed{M = (0.008)(500)\tan(16.7 - 2.28) = 1.03\ N\,m}$

Problem 10.76 The threaded shaft has a ball and socket support at B. The 400 *lb* load A can be raised or lowered by rotating the threaded shaft, causing the threaded collar at C to move relative to the shaft. Neglect the weights of the members. The pitch of the shaft is $p = \frac{1}{4}\ in$, the mean radius of the thread is $r = 1\ in$, and the coefficient of static friction between the thread and the mating groove is 0.24. If the system is stationary in the position shown, what moment is necessary to start the shaft rotating to raise the load?

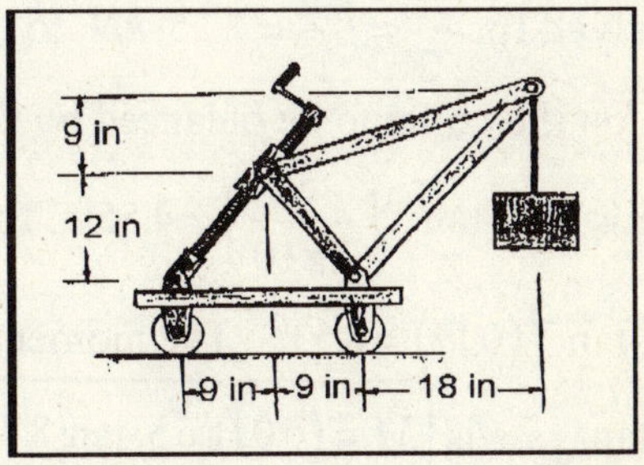

Solution Denote the lower right pin support by D. The length of the connecting member CD is $L_{CD} = \sqrt{9^2 + 12^2} = 15\ in$. The angle between the threaded shaft and member CD is $\beta = 2\tan^{-1}\left(\frac{9}{12}\right) = 73.74^o$.

Problem 10.76 continued on next page

............continuation of solution to Problem 10.76.....

The sum of the moments about D is $\sum M_D = L_{CD} F \cos(90-\beta) - 18W = 0$, from which $F = 500\ lb$.

The pitch angle is $\alpha = \tan^{-1}\left(\frac{p}{2\pi r}\right) = 2.28^o$. The angle of static friction is $\theta_s = \tan^{-1}(0.24) = 13.5^o$. The moment required to start the threaded collar in motion is

$$\boxed{M = rF\tan(\theta_s + \alpha) = \left(\frac{1}{12}\right)(500)\tan\left(13.5^o + 2.28^o\right) = 11.77\ ft\ lb}$$

Problem 10.77 In Problem 10.76, if the system is stationary in the position shown, what moment is necessary to start the shaft rotating to lower the load?

Solution Use the results of the solution to Problem 10.76. The moment is $M = rF\tan(\theta_s - a)$, where $r = \left(\frac{1}{12}\right)\ ft$, $F = 500\ lb$, $\theta_s = 13.5^o$, and $\alpha = 2.28^o$, from which $\boxed{M = 8.26\ ft\ lb}$

Problem 10.78 The car jack is operated by turning the threaded shaft at A. The threaded shaft fits into a mating grove in the collar at B, causing the collar to move relative to the shaft as the shaft turns. As a result, points B and D move closer together or farther apart, causing point C (where the jack is in contact with the car) to move up or down. The pitch of the threaded shaft is $p = 5\ mm$, the mean radius of the thread is $r = 10\ mm$, and the coefficient of kinetic friction between the thread and the mating groove is 0.15. What moment is necessary to turn the shaft at a constant rate and raise the jack if it is in the position shown and $F = 6.5\ kN$?

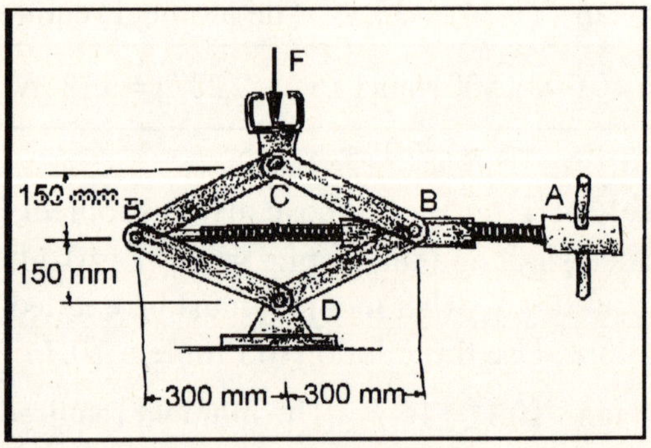

Solution Isolate members BC and BD. Assume that half the car load is carried by these members. The equilibrium conditions for member BC are:

$\sum F_x = C_x - B_x = 0$, $\sum M_B = 0.3\left(\frac{F}{2}\right) - 0.15 C_x = 0$ These equations are solved for $\frac{F}{2} = \frac{6.5}{2} = 3.25\ kN$ to obtain $B_x = 6.5\ kN$, which is the force on the collar to be balanced by the rotating threaded shaft. The pitch angle is $\alpha = \tan^{-1}\left(\frac{5}{2\pi 10}\right) = 4.55^o$. The angle of kinetic friction is

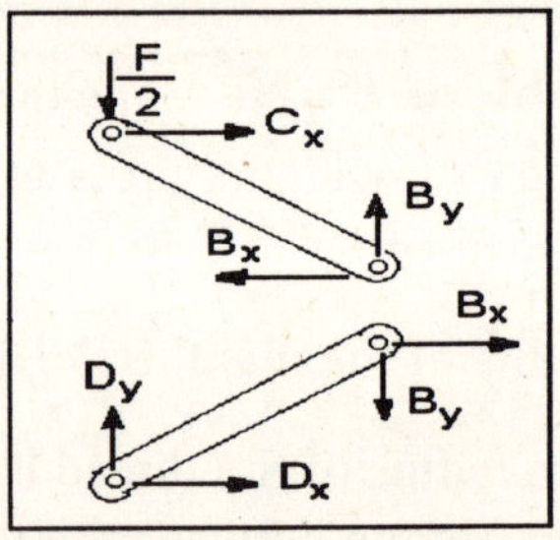

$\theta_s = \tan^{-1}(0.15) = 8.53^o$. The moment required to rotate the shaft at a constant rate is $\boxed{M = (0.01)(6.5)\tan\left(8.53^o + 4.55^o\right) = 0.0151\ kN\ m = 15.1\ N\ m}$

Problem 10.79 In Problem 10.78, what moment is necessary to turn the threaded shaft at a constant rate and lower the jack when it is in the position shown if the force $F = 6500\ N$?

Solution Use the results of the solution of Problem 10.78. The moment required to lower the jack at a constant rate is $M = rB\tan(\theta_k - \alpha)$, where $r = 0.01\ m$, $B = 6500\ N$, $\theta_k = 8.53^o$, $\alpha = 4.55^o$, from which $\boxed{M = 4.52\ N\text{-}m}$

Problem 10.80 A turnbuckle is used to adjust the tension in a bar or cable. The central link is threaded at both ends. Rotating it draws the threaded ends of the bar together or moves them apart. Suppose that the pitch of the threads is $p = \frac{1}{8}\ in$, their mean radius is $r = 1\ in$, and the coefficient of static friction between the threads and the mating grooves is 0.24. If $T = 200\ lb$, what moment must be exerted on the turnbuckle to start tightening it?

Solution The pitch angle is

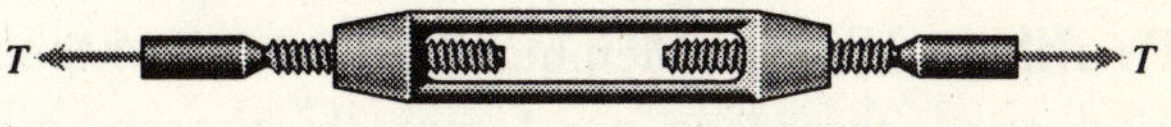

$\alpha = \tan^{-1}\left(\frac{1}{8(2\pi)(1)}\right) = 1.14^o$. The static friction angle is $\theta_s \tan^{-1}(0.24) = 13.5^o$. The moment required for one threaded shaft is $M = rT\tan(\theta_s + \alpha) = \left(\frac{1}{12}\right)(200)\tan\left(13.5^o + 1.14^o\right) = 4.35\ ft\ lb$. The moment for both threaded shafts is $\boxed{M = 8.70\ ft\ lb}$

Problem 10.81 Member BE of the frame has a turnbuckle. (See Problem 10.80.) The threads have pitch $p = \frac{1}{32}\ in$, the mean radius of the threads is $r = \frac{1}{4}\ in$, and the coefficient of static friction between the threads and the mating grooves is 0.2. What moment must be exerted on the turnbuckle to start loosening it?

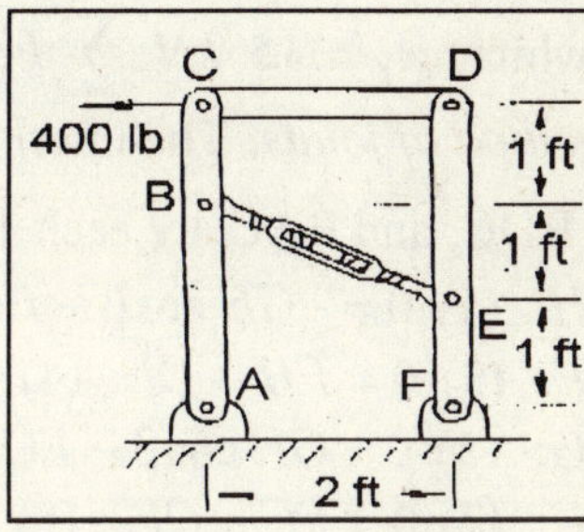

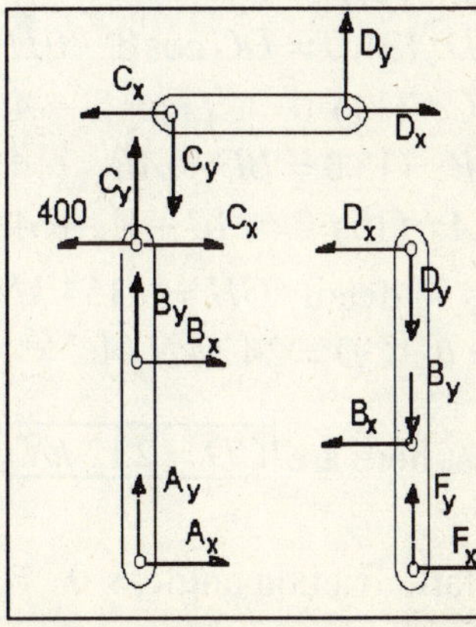

Solution The strategy is to determine the tension in the turnbuckle from the frame member equilibrium conditions. *The complete structure as a free body:* The sum of moments about A: $\sum M_A = 3(400) + 2F_y = 0$, from which $F_y = -600\ lb$. The sums of the forces: $\sum F_y = A_y + F_y = 0$, from which $A_y = 600\ lb$

(1) $\sum F_x = 400 + A_x + F_x = 0$.

Make a cut through the turnbuckle cable BE.

The member AC as a free body:

(2) $\sum F_x = A_x + B_x - 400 + C_x = 0$

(3) $\sum F_y = A_y + B_y + C_y = 0$,

(4) $\sum M_A = -2B_x + 3(400) - 3C_x = 0$.

Problem 10.81 continued on next page

continuation of Problem 10.81..........

The member CD as a free body: (5) $\sum F_x = -C_x + D_x = 0$, (6) $\sum F_y = -C_y + D_y = 0$

The member DF as a free body: (7) $\sum F_x = -D_x - B_x + F_x = 0$, (8) $\sum F_y = -D_y - B_y + F_y = 0$, (9) $\sum M_F = +1B_x + 3D_y = 0$. These nine equations in six unknowns are solved to obtain $B_x = 1200\ lb$ and $B_y = -600\ lb$. The tension on the turnbuckle is

$T = \sqrt{B_y^2 + B_x^2} = \sqrt{600^2 + 1200^2} = 1341.6\ lb$. The pitch angle is $\alpha = \tan^{-1}\left(\frac{4}{32(2\pi)(1)}\right) = 1.14^o$. The angle of static friction is $\theta_s = \tan^{-1}(0.2) = 11.31^o$. The moment required to loosen each threaded shaft is

$M = \left(\frac{1}{48}\right)(1341.6)\tan\left(11.31^o - 1.14^o\right) = 5.014\ ft\ lb$. The total moment is $\boxed{M_t = 2M = 10.028\ ft\ lb}$

Problem 10.82 Members CD and DG of the truss have turnbuckles. (See Problem 10.80.) The pitch of the threads is $p = 4\ mm$, their mean radius is $r = 10\ mm$, and the coefficient of static friction between the threads and the mating grooves is 0.18. What moment must be exerted on the turnbuckle to start loosening it?

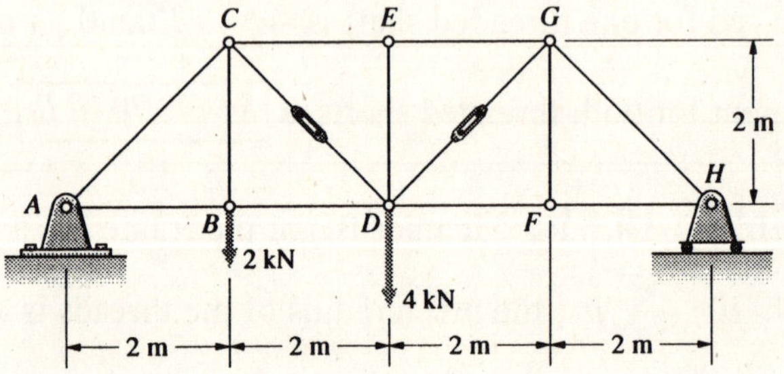

Solution *The complete structure as a free body:* The equilibrium conditions:

$\sum M_A = -2(2) - 4(4) + 8H = 0$, from which

$H = \frac{20}{8} = 2.5\ kN$. $\sum F_y = A_y + H - 2 - 4 = 0$, from which $A_Y = 3.5\ kN$. $\sum F_x = A_x = 0$.

The method of joints: The interior angles GHF, DFG, BDC, and BAC are each $\beta = 45^o$.

Joint H: (1) $0 = -GH\cos\beta - FH$, (2) $0 = GH\sin\beta + H$, .

Joint F: (3) $0 = FH - DF$, (4) $GF = 0$,

Joint G: (5) $0 = GH\cos\beta - EG - DG\cos\beta$, (6) $0 = -GH\sin\beta - DG\sin\beta$.

Joint E: (7) $0 = EG - CE$, (8) $0 = DE$,

Joint D: (9) $0 = DG\cos\beta - CD\cos\beta + DF - DB$,(10) $0 = DG\sin\beta + CD\sin\beta + DE - 4$.

Joint C: (11) $0 = CD\cos\beta - AC\cos\beta + CE$, (12) $0 = -CD\sin\beta - AC\sin\beta - BC$.

Joint B: (130 $= BD - AB$, (14) $0 = BC - 2$.

Joint A: (15) $0 = AB + A_x + AC\cos\beta$, (16) $0 = A_y + AC\sin\beta$. These equations are solved: The results in detail: $GH = -3.54\ kN$, $FH = 2.5\ kN$, $DF = 2.5\ kN$, $GF = 0$, $EG = -5\ kN$, $DG = 3.54\ kN$, $DE = 0$, $CD = 2.12\ kN$ $AC = -4.95\ kN$, $BC = 2\ kN$, $BD = 3.5\ kN$, $AB = 3.5\ kN$. The members of interest here are $\boxed{CD = 2.12\ kN}$ and $\boxed{DG = 3.54\ kN}$. The pitch angle is $\alpha = \tan^{-1}\left(\frac{4}{(2\pi)(10)}\right) = 3.64^o$.

The static friction angle is $\theta_s = \tan^{-1}(0.18) = 10.20^o$. The moment required to loosen the turnbuckle is

$$\boxed{M = 2(0.01)(2.12)\tan\left(10.2^o - 3.64^o\right) = 0.00488\ kN\,m = 4.88\ N\,m}$$

==================================<>==================================

Problem 10.83 In Problem 10.82, what moment must be exerted on the turnbuckle on member DG to start loosening it?

Solution Use the results of the solution of Problem 10.82. The moment required to loosen the turnbuckle is $M = 2rT\tan(\theta_s - \alpha)$, where $r = 0.01\ m$, $T = DG = 3.54\ kN$, $\theta_s = 10.2^o$, and $\alpha = 3.64^o$

$$\boxed{M = 2(0.01)(3.54)\tan\left(10.2^o - 3.64^o\right) = 0.00813\ kN\,m = 8.13\ N\,m}$$

==================================<>==================================

Problem 10.84 The load $W = 800\ N$ can be raised or lowered by rotating the threaded shaft. The distances $h = 200\ mm$, and $b = 75\ mm$. The pinned bars are each 300 mm in length. The pitch of the threaded shaft is $p = 5\ mm$, the mean radius of the thread is $r = 15\ mm$, and the coefficient of kinetic friction between the thread and the mating groove is 0.2. When the system is in the position shown, what moment must be exerted on the threaded shaft to raise the load at a constant rate?

Solution The vertical distances HE, BE, AD, DG are 100 *mm*.

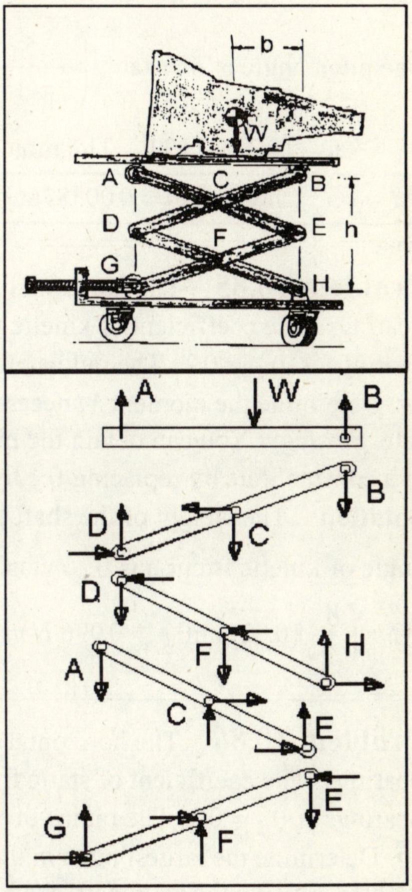

The included angle ABC is $\beta = \sin^{-1}\left(\dfrac{50\ mm}{150\ mm}\right) = 19.47^o$. The distance $L = 300\cos\beta\ mm$. Isolate the members and write the equilibrium equations, beginning at the top: *Isolate the frame AB which supports the load W*. The equilibrium conditions are (1) $\sum M_B = -AL + Wb = 0$,

(2) $\sum F_y = A - W + B_y = 0$, $\sum F_x = B_x = 0$

Isolate BD: (3) $\sum F_y = -B_y - C_y + D_y = 0$,

(4) $\sum F_x = -C_x + D_x = 0$,

(5) $\sum M_D = \left(\dfrac{h}{4}\right)C_x - \left(\dfrac{L}{2}\right)C_y - LB = 0$

Isolate DH: (6) $\sum F_y = -D_y - F_y + H_y = 0$,

(7) $\sum F_x = -D_x - F_x + H_x = 0$,

(8) $\sum M_H = \left(\dfrac{h}{4}\right)F_x + \left(\dfrac{L}{2}\right)F_y + LD_y + \left(\dfrac{h}{2}\right)D_x = 0$

Isolate AE (9) $\sum F_y = -A + C_y + E_y = 0$,

(10) $\sum F_x = C_x + E_x = 0$,

(11) $\sum M_E = -\left(\dfrac{h}{4}\right)(C_x) - \left(\dfrac{L}{2}\right)C_y + LA = 0$.

Isolate EG (12) $\sum F_y = -E_y + F_y + G_y = 0$,

(13) $\sum F_x = -E_x + F_x + G_x = 0$,

(14) $\sum M_G = -\dfrac{h}{4}F_x + \left(\dfrac{L}{2}\right)F_y - L\,E_y + \left(\dfrac{h}{2}\right)E_x = 0$. These 14 equations in 14 unknowns are to be solved to determine the reaction G_x, which is the force that the threaded shaft must overcome to raise

Problem 10.84 continued on next page

continuation of Problem 10.84..........

the load at a constant rate.. An analytic solution is obtained as follows:

From (1) $A = \frac{Wb}{L}$. From (2) $B = W\left(1 - \frac{b}{L}\right)$. From (8) and (11) $C_y = W\left(\frac{2b}{L} - 1\right)$, $C_x = \frac{2WL}{h}$.

From (9) $E_y = W(1 - \frac{b}{L})$. From (10) $E_x = -\frac{2}{h}WL$. From (4) $D_x = \frac{2}{h}WL$

From (3) $D_y == \frac{Wb}{L}$. From (8) and (14) $F_y = W\left(1 - \frac{2b}{L}\right)$, $F_x = \frac{6WL}{h}$. From (12) $G_y = \frac{Wb}{L}$ *Check:*

This value of G_y is expected from the overall equilibrium conditions. *check.* From (13) $\boxed{G_x = \frac{4WL}{h}}$

Substitute: $G_x = \frac{1200W}{h}\cos\beta\ N$. Note that $\cos\beta = \sqrt{1 - \left(\frac{h}{600}\right)^2}$, from which $\boxed{G_x = 2W\frac{\sqrt{600^2 - h^2}}{h}}$

The pitch angle is $\alpha = \tan^{-1}\left(\frac{5}{2\pi(15)}\right) = 3.037^o$. The angle of kinetic friction is

$\theta_k = \tan^{-1}(0.2) = 11.31^o$. The moment required to raise the load at a constant rate:

$\boxed{M = rG_x \tan(\theta_k + \alpha) = 0.003836G_x = 17.36\ N\,m}$

Problem 10.85 The horizontal shaft is supported by two journal bearings. The coefficient of kinetic friction between the shaft and the bearings is $\mu_k = 0.2$. The radius of the shaft is 20 *mm* and its mass is 5 *kg*. Determine the moment M necessary to rotate the shaft at a constant rate. *Strategy:* You can obtain the moment necessary to rotate the shaft at a constant rate by replacing θ_s by θ_k in Eq. (10.15).

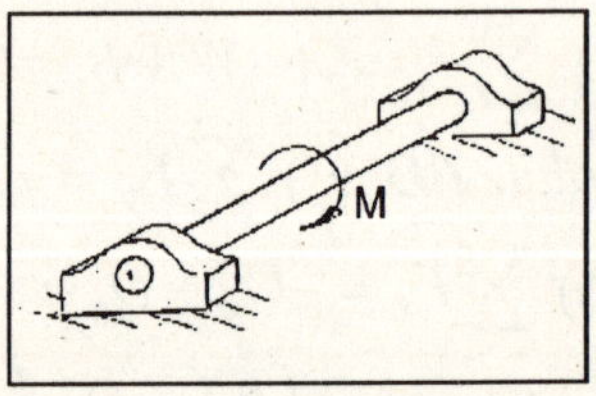

Solution The weight of the shaft is $W = mg = 5(9.81) = 49\ N$, divided between two bearings. The angle of kinetic friction is $\theta_k = \tan^{-1}(0.2) = 11.31^o$. The moment per bearing is

$M = \left(\frac{W}{2}\right)(0.02)\sin\theta_k = 0.096\ N\,m$. The total moment is $\boxed{M_t = 0.192\ N\,m}$

Problem 10.86 The horizontal shaft is supported by two journal bearings. The coefficient of static friction between the shaft and the bearings is $\theta_s = 0.3$. The radius of the shaft is 20 *mm* and its mass is 5 *kg*. Determine the largest mass *m* that can be suspended as shown without causing the stationary shaft to slip in the bearings.

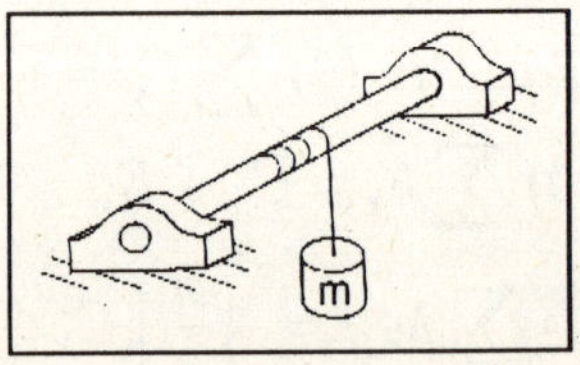

Solution The weight of the shaft is $W = mg = 5(9.81) = 49\ N$. This weight is divided between two bearings. The angle of static friction is $\theta_s = \tan^{-1}(0.3) = 16.7^o$. The load per bearing is $F = \frac{W + W_m}{2}$,

and the moment required to start rotation is $M_m = (W + W_m)r\sin\theta_s$ where W_m is the suspended weight,

$W_m = \frac{M_m}{r}$.. From which $W_m r = (W + W_m)r\sin\theta_k$, from which $\boxed{m = \frac{W_m}{g}\left(\frac{\sin\theta_s}{1 - \sin\theta_s}\right) = 2.02\ \ kg}$.

Problem 10.87 Suppose that in Problem 10.86 the mass $m = 8\ kg$ and the coefficient of kinetic friction between the shaft and bearing is $\mu_k = 0.26$. What moment must be applied to the shaft to raise the mass at a constant rate?

Solution From the solution to Problem 10.86 the moment required is $M_{applied} = (W + W_m) r \sin\theta_k$, where $W_m = mg$ is the weight of the suspended mass. The moment required to raise the suspended mass is $M_m = W_m r$. The total moment is the sum of the moment required to turn the shaft and the moment required to raise the mass:

$$M_{total} = (W + mg) r \sin\theta_k + W_m r = (49 + 78.5)(0.02)\sin(14.6^o) + 1.57 = 2.21\ N\,m$$

Problem 10.88 The pulley is mounted on a horizontal shaft supported by journal bearings. The coefficient of kinetic friction between the shaft and the bearings is $\mu_k = 0.3$. The radius of the shaft is 20 *mm* and the radius of the pulley is 150 *mm.* The mass $m = 10\ kg$. Neglect the masses of the pulley and shaft. What force T must be applied to the cable to move the mass upward at a constant rate?

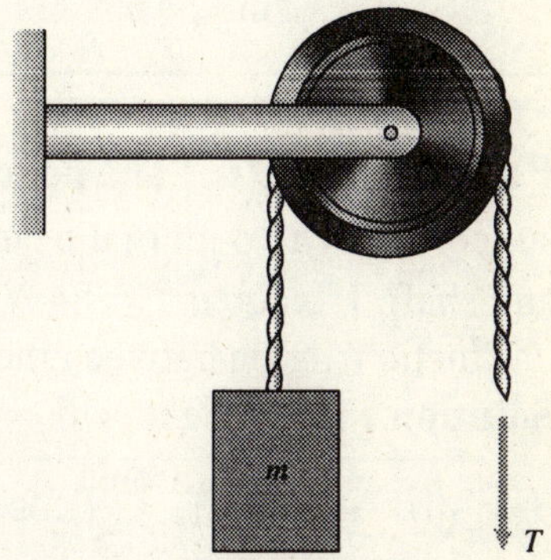

Solution The angle of kinetic friction is $\theta_k = \tan^{-1}(\mu_k) = 16.7^o$. The moment required to turn the shaft is $M = (mg + T) r \sin\theta_k$. The applied moment is $M = (T - mg)R$ where R is the radius of the pulley. Equating and reducing:

$$T = mg\left(\frac{1 + \frac{r}{R}\sin\theta_k}{1 - \frac{r}{R}\sin\theta_k}\right) = (98.1)\left(\frac{1.0383}{0.9617}\right) = 105.92\ N$$

Problem 10.89 In Problem 10.88, what force must be applied to the cable to lower the mass at a constant rate?

Solution Form the solution to Problem 10.88, $\theta_k = \tan^{-1}(\mu_k) = 16.7^o$, and $M = (mg + T) r \sin\theta_k$. The applied moment is $M = (mg - T)R$. Substitute and reduce:

$$T = mg\left(\frac{1 - \frac{r}{R}\sin\theta_k}{1 + \frac{r}{R}\sin\theta_k}\right) = (98.1)\left(\frac{0.9617}{1.0383}\right) = 90.86\ N$$

Problem 10.90 The pulley of 8 *in* radius is mounted on a shaft of 1 *in* radius. The shaft is supported by two journal bearings. The coefficient of static friction between the bearings and the shaft is $\mu_s = 0.15$. Neglect the weights of the pulley and shaft. The 50 *lb* block A rests on the floor. If sand is slowly poured into the bucket B, what do the bucket and sand weigh when the shaft slips in the bearings?

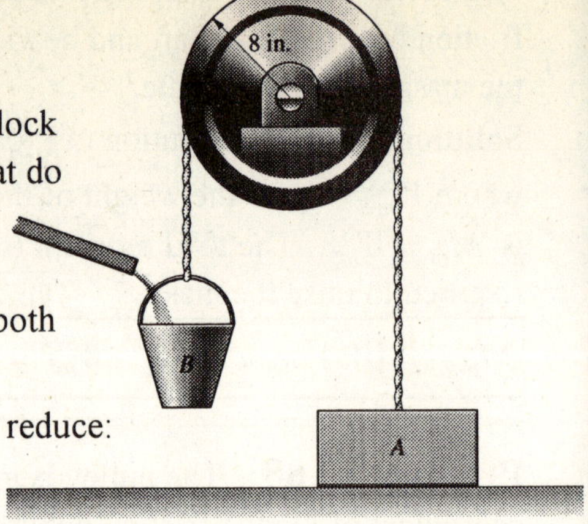

Solution (See Problem 10.88.) The angle of static friction is $\theta_s = \tan^{-1}(\mu_s) = 8.53^o$. The moment required to start rotation for both bearings is $M = r(B+W)\sin\theta_s$. The applied moment is $M = (B-W)R$, where R is the radius of the pulley. Substitute and reduce:

$$\boxed{B = W\left(\frac{1+\frac{r}{R}\sin\theta_s}{1-\frac{r}{R}\sin\theta_s}\right) = (50)\left(\frac{1.0185}{0.9815}\right) = 51.9\ lb}$$

Problem 10.91 The pulley of 50 *mm* radius is mounted on a shaft of 10 *mm* radius. The shaft is supported by two journal bearings. The mass of the block A is 8 *kg*. Neglect the weights of the pulley and shaft. If a force $T = 84\ N$ is necessary to raise the block A at a constant rate, what is the coefficient of kinetic friction between the shaft and the bearings?

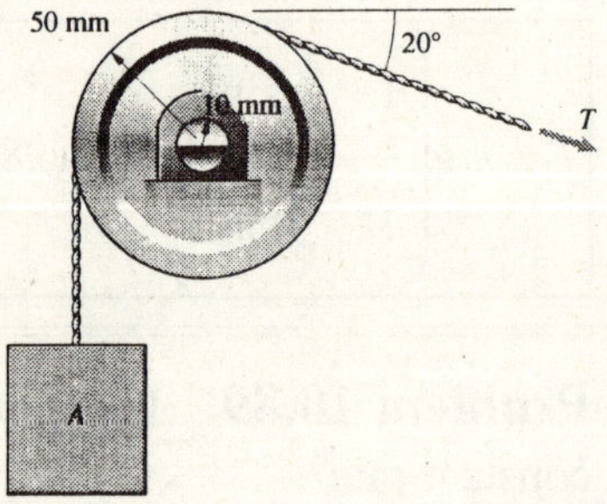

Solution The weight is $W = mg = 78.5\ N$. The force on the pulley is $F = \sqrt{(W+T\sin\alpha)^2 + (T\cos\alpha)^2}$, where $\alpha = 20^o$.

$F = \sqrt{107.2^2 + 78.9^2} = 133.13\ N$. The moment required to raise the mass at constant rate for both bearings is $M = rF\sin\theta_k = 1.33\sin\theta_k$. The applied moment is $M = (T-W)R = 0.276\ N\,m$. Substitute and reduce:

$\sin\theta_k = \frac{(T-W)R}{rF} = \frac{0.276}{1.33} = 0.2073$, from which $\theta_k = 11.96^o$ and

$$\boxed{\mu_k = \tan(11.96^o) = 0.2119}$$

Problem 10.92 Suppose that in Problem 10.91, the coefficient of kinetic friction between the shaft and the bearings is $\mu_k = 0.18$. What force *T* is required to raise the block T at a constant rate?

Solution The angle of kinetic friction is $\theta_k = \tan^{-1}(0.18) = 10.2^o$. From the solution to Problem 10.91 the force on the bearings is $F = \sqrt{(W+T\sin\alpha)^2 + (T\cos\alpha)^2}$. The moment required to raise the mass (for both bearings) is $M = rF\sin\theta_k$. The applied moment is $M = (T-W)R$. Equate the moments: $(T-W)R = r\sqrt{(W+T\sin\alpha)^2 + (T\cos\alpha)^2}\sin\theta_k$. Square both sides and reduce to obtain the quadratic

equation: $T^2 - 2WT\left(\frac{1+\left(\frac{r}{R}\right)^2\sin^2\theta_k\sin\alpha}{1-\left(\frac{r}{R}\right)^2\sin^2\theta_k}\right) + W^2 = 0$ or $T^2 - 2(78.6)T + 6159.1 = 0$.

Problem 10.92 continued on next page

continuation of Problem 10.92..........

This has the solutions $T_{1,2} = 78.6 \pm \sqrt{78.6^2 - 6159.1} = 83.2\ N,\ \ 74.1\ N$. A heuristic argument for choosing the greater root for the tension when the block is being raised is based on a "loss" of tension in the pulley because of the moment required to turn the pulley on its shaft. A more rigorous argument is: The applied moment required to raise the block A is $M_{applied}^{raise} = +(T-W)R$, and the applied moment required to lower the block is $M_{applied}^{raise} = -(T-W)R$, however, when the quantity is squared, the distinction is lost, since $(M_{applied}^{raise})^2 = (M_{applied}^{lower})^2$. Thus the solution contains both results. But note from the condition $M_{applied} = r\sqrt{(W + T\sin\alpha)^2 + (T\cos\alpha)^2}\ \sin\theta_k$ that $M_{applied} \ge 0$ always since no term on the right can be negative. The tension required to raise or lower the block is $T_{raise} = \dfrac{M_{applied}^{raise}}{R} + W$ and $T_{lower} = -\dfrac{M_{applied}^{lower}}{R} + W$. Since as shown above $M_{applied}^{raise} \ge 0$, $M_{applied}^{lower} \ge 0$, and also since $W \ge 0$, then always $T_{raise} \ge T_{lower}$. It follows that the greatest root of the quadratic is the tension required to raise the block, and the lesser root is the tension required to lower the block, $\boxed{T_{raise} = 83.2\ N}$

Problem 10.93 Suppose that in Problem 10.91, the coefficient of kinetic friction between the shaft and the bearings is $\mu_k = 0.24$. What force is necessary to lower the block at a constant rate?

Solution The angle of kinetic friction is $\theta_k = \tan^{-1}(0.24) = 13.5^o$. From the solution to Problem 10.92, $T^2 - 2WT\left(\dfrac{1+\left(\dfrac{r}{R}\right)^2 \sin^2\theta_k \sin\alpha}{1-\left(\dfrac{r}{R}\right)^2 \sin^2\theta_k}\right) + W^2 = 0$, $T^2 - 2(78.7)T + 6159 = 0$, from which

$T_{1,2} = 84.72\ N,\ \ 72.7\ N$ From the argument given in the solution to Problem 10.92, the tension required to lower the block is the lesser root, $\boxed{T_{lower} = 72.7\ N}$

Problem 10.94 The two pulleys have a radius of 4 *in* and are mounted on shafts of 1 *in* radius supported by journal bearings. Neglect the weight of the pulleys and shafts. The tension in the spring is 40 *lb*. The coefficient of kinetic friction between the shaft and bearings is $\mu_k = 0.3$. What moment is required to turn the left pulley at constant rate?

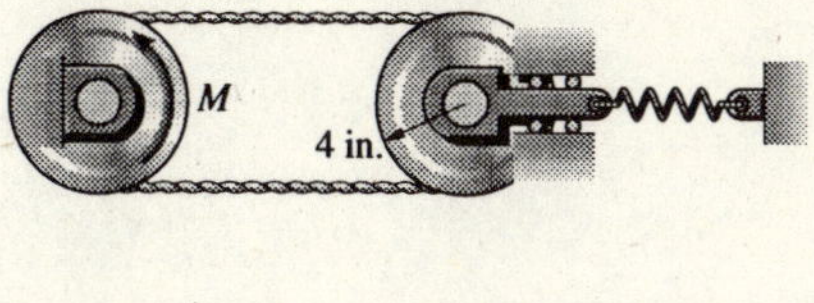

Solution The angle of kinetic friction is $\theta_k = \tan^{-1}(0.3) = 16.7^o$. The load on the bearings is $F = 40\ lb$. The moment required to turn both pulleys at constant rate is $M = 2rF\sin\theta_k$. This is equal to the applied moment,

$$\boxed{M_{applied} = 2rF\sin\theta_k = (2)\left(\frac{1}{12}\right)40\sin\left(16.7^o\right) = 1.92\ ft\ lb}$$

Problem 10.95 The two pulleys have a radius of 6 *in* and are mounted on shafts of 1 *in.* radius supported by journal bearings. Neglect the weight of the pulleys and shafts. The coefficient of kinetic friction between the shafts and the bearings is $\mu_k = 0.2$. If a force $T = 200\ lb$ is required to raise the man at a constant rate, what is his weight?

Solution Denote the tension in the horizontal portion of the cable by H. The angle of kinetic friction is $\theta_k = \tan^{-1}(0.2) = 11.31^o$

Consider the right Pulley: The force on the right pulley is $F = \sqrt{T^2 + H^2}$ The magnitude of the moment required to turn the shaft in the bearing is $M_{right} = r\sqrt{T^2 + H^2}\sin\theta_k$. The applied moment is $M_{applied} = (T - H)R$, from which $(T - H)R = r\sqrt{T^2 + H^2}\sin\theta_k$. Square both sides and reduce to obtain the quadratic: $H^2 - 2TH\left(\dfrac{1}{1 - \left(\dfrac{r}{R}\right)^2 \sin^2\theta_k}\right) + T^2 = 0$, or

$H^2 - 2(200.124)H + 40000 = 0$. This has the solutions:

$H = 200.124 \pm \sqrt{200.124^2 - 40000} = 209.38,\quad 190.87$. By an argument exactly analogous to that in the solution to Problem 10.92, the lesser root corresponds to the horizontal tension, $H = 190.97.. = 191\ lb$. (The argument may also be used that this corresponds to the "loss" in tension, when the man is being raised, because of the moment required to turn the pulley.)

Consider the left Pulley: The force on the pulley is $F_{left} = \sqrt{W^2 + H^2}$. The applied moment is $M_{applied} = -(H - W)R$, from which $(H - W)R = r\sqrt{W^2 + H^2}\sin\theta_k$. Square both sides and reduce to the quadratic: $W^2 - 2\left(\dfrac{H}{1 - \left(\dfrac{r}{R}\right)^2 \sin^2\theta_k}\right)W + H^2 = 0$, or $W^2 - 2(191.1)W + 36431.9 = 0$. This has the solutions: $W_{1,2} = 191.166 \pm \sqrt{191.166^2 - 36431.9} = 199.9\ lb,\quad 182.25\ lb$. By an analogous argument to that used in Problem 10.92, the lesser root corresponds to the weight of the man, $\boxed{W_{raised} = 182.3\ lb}$

Problem 10.96 If the man in Problem 10.95 weighs 160 *lb*, what force *T* is necessary to lower him at a constant rate?

Solution Use the solution to Problem 10.96, with $W = 160\ lb$. *Begin with the left pulley*: the quadratic relation between the weight and the horizontal tension is $H^2 - 2\left(\dfrac{W}{1-\left(\dfrac{r}{R}\right)^2 \sin^2\theta_k}\right)H + W^2 = 0.$, or

$H^2 - 2(160.171)H + 25600 = 0$. This has the solutions:

$H_{1,2} = 160.171 \pm \sqrt{160.171^2 - 25600} = 167.57\ lb,\quad 152.77\ lb$. Since the man is being lowered, his weight "absorbs" the loss due to the moment required to turn the pulley (see also the argument in the solution to Problem 10.92), and the lesser root corresponds to the horizontal tension: $H = 152.77\ lb$

Consider the right pulley: The quadratic relation between the tension and the horizontal tension is

$T^2 - 2\left(\dfrac{H}{1-\left(\dfrac{r}{R}\right)^2 \sin^2\theta_k}\right)T + H^2 = 0$, or $T^2 - 2(152.93)T + 23338.4 = 0$. This has the solutions:

$T_{1,2} = 152.93 \pm \sqrt{153.93^2 - 23338.4} = 160\ lb,\quad 145.87\ lb$. By previous arguments, the lesser root corresponds to the tension when the man is being lowered at a constant rate, $\boxed{T_{lower} = 145.9\ lb}$.

Problem 10.97 Each pulley has a radius of 100 *mm* and a mass of 2 *kg*. They are mounted on shafts of 5 *mm* radius supported by journal bearings. The coefficient of kinetic friction between the shafts and the bearings is $\mu_k = 0.18$. The mass of A is 14 *kg*. What force *T* is required to raise A at a constant rate?

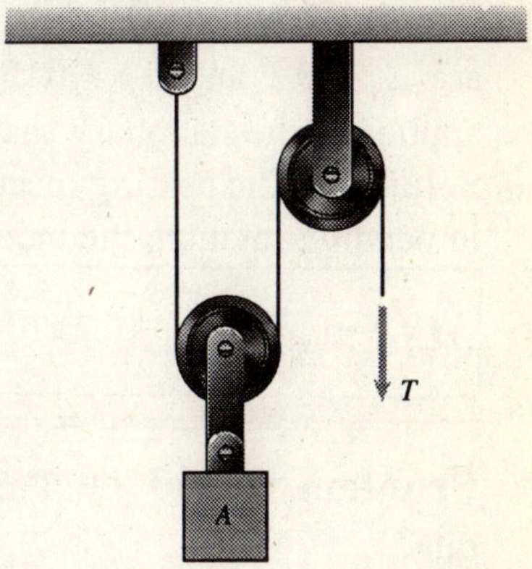

Solution The angle of kinetic friction is $\theta_k = \tan^{-1}(0.18) = 10.2^o$. The weight of A is $W_A = 14g = 137.34\ N$. The weight of each pulley is $W_P = 2g = 19.62\ N$. Denote the tension in the left cable by *P* and the tension in the middle cable by *Q*.

Consider the left Pulley: The force on the pulley is $W = W_A + W_P = 156.96\ N$. The moment required to turn the pulley on its bearing is $M = rW\sin\theta_k$. The applied moment is $M_{applied} = (Q-P)R$, from which $(Q-P)R = rW\sin\theta_k$. The sum of the tensions in the cables: $P + Q = W$, from which

$Q = \dfrac{W}{2}\left(1 + \left(\dfrac{r}{R}\right)\sin\theta_k\right) = 79.18\ N$.

Consider the right Pulley: The force on the pulley is $F = Q + T + W_P$. The moment required to turn the pulley on its bearing is $M = r(Q + T + W_P)\sin\theta_k$.

Problem 10.97 continued on next page

continuation of Problem 10.97...........

The applied moment is $M_{applied} = (T-Q)R$, from which $(T-Q)R = r(T+Q+W_P)\sin\theta_k$. Reduce algebraically:

$$T = \left(\frac{Q(1+\left(\frac{r}{R}\right)\sin\theta_k) + W_P\left(\frac{r}{R}\right)\sin\theta_k}{1-\left(\frac{r}{R}\right)\sin\theta_k} \right) = 80.77\ N$$

Problem 10.98 In Problem 10.97, what force T is necessary to lower A at a constant rate?

Solution Use the results of the solution to Problem 10.97. The angle of friction is $\theta_k = 10.2^o$. *Left Pulley:* The moment required to turn the bearing is $M = rW\sin\theta_k$. The applied moment is $M_{applied} = -(Q-P)R$, from which $-(Q-P)R = rW\sin\theta_k$. The sum of the tensions in the cables: $P+Q=W$, from which, $Q = \frac{W}{2}\left(1-\left(\frac{r}{R}\right)\sin\theta_k\right) = 77.785\ N$. *Right Pulley:* The force on the pulley is $F = Q+T+W_P$. The moment required to turn the pulley on its bearing is $M = r(T+Q+W_p)\sin\theta_k$. The applied moment is $M_{applied} = -(T-Q)R$, from which $-(T-Q)R = r(T+Q+W_p)\sin\theta_k$. Reduce algebraically:

$$T = \left(\frac{Q(1-\left(\frac{r}{R}\right)\sin\theta_k) - W_P\left(\frac{r}{R}\right)\sin\theta_k}{1+\left(\frac{r}{R}\right)\sin\theta_k} \right) = 76.25\ N$$

Problem 10.99 The circular flat ended shaft is pressed into the thrust bearing by an axial load of 100 N. Neglect the weight of the shaft. The coefficients of friction between the end of the shaft and the bearing are $\mu_s = 0.2$ and $\mu_k = 0.15$. What is the largest moment M that can be applied to the stationary shaft without causing it to rotate in the bearing?

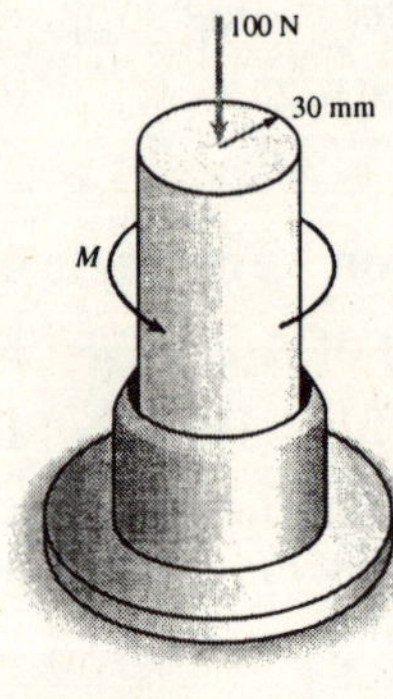

Solution The bearing meets the conditions for Eq. (10.17). For impending rotation, the moment is

$$M = \frac{2}{3}\mu_s Fr = \left(\frac{2}{3}\right)(0.2)(100)(0.03) = 0.4\ N\,m$$

Problem 10.100 In Problem 10.99, what moment M is required to rotate the shaft at a constant rate?

Solution The bearing meets the conditions for Eq. (10.17). The moment required to sustain a constant rate of rotation is $M = \left(\frac{2}{3}\right)\mu_k Fr = \left(\frac{2}{3}\right)(0.15)(100)(0.03) = 0.3\ N\,m$

Problem 10.101

Suppose that the end of the shaft in Problem 10.99 is supported by a thrust bearing of the type shown in Fig 10.102, where $r_o = 30\ mm$, $r_1 = 10\ mm$, $\alpha = 30^o$, and $\mu_k = 0.15$. What moment M is required to rotate the shaft at a constant rate?

Solution The bearing meets the conditions for Eq (10.16). The moment required to sustain a constant rate of rotation is $\boxed{M = \left(\frac{2\mu_k F}{3\cos\alpha}\right)\left(\frac{r_o^3 - r_1^3}{r_o^2 - r_1^2}\right) = 0.3753\ N\,m}$

Problem 10.102

The disk D is rigidly attached to the vertical shaft. The shaft has flat ends supported by thrust bearings. The disk and the shaft together weigh 500 *lb* and the diameter of the shaft is 2 *in*. The vertical force exerted on the end of the shaft by the upper bearing is 100 *lb*. The coefficient of kinetic friction is $\mu_k = 0.25$. What moment is required to rotate the shaft at a constant rate?

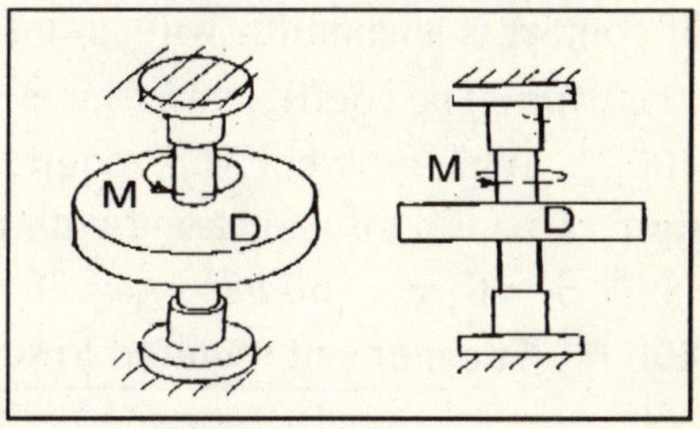

Solution The moment required to rotate the upper bearing is

$$M_{upper} = \left(\frac{2}{3}\right)\mu_k Fr = \left(\frac{2}{3}\right)(0.25)\left(\frac{100}{12}\right) = 1.389\ ft\ lb.$$

The moment required to rotate the lower bearing is

$$M_{lower} = \left(\frac{2}{3}\right)\mu_k Fr = \left(\frac{2}{3}\right)(0.25)\left(\frac{600}{12}\right) = 8.333\ ft\ lb$$

The total moment is $\boxed{M = M_{uppwr} + M_{lower} = 9.72\ ft\ lb}$

Problem 10.103

Suppose that the ends of the shaft in Problem 10.102 are supported by thrust bearings of the type shown in Figure 10.102, where $r_o = 1\ in$, $r_1 = \frac{1}{4}\ in$, $\alpha = 45^o$, and $\mu_k = 0.25$. What moment is required to rotate the shaft at a constant rate?

Solution For the combined bearings: $\boxed{M = \left(\frac{2\mu_k}{3\cos\alpha}\right)\left(\frac{r_o^3 - r_1^3}{r_o^2 - r_1^2}\right)(100 + 600) = 14.44\ ft\ lb}$

Problem 10.104

The shaft is supported by thrust bearings that subject it to an axial load of 800 *N*. The coefficient of kinetic friction between the shaft and the left and right bearings are $\mu_k = 0.20$ and $\mu_k = 0.26$, respectively. What moment is required to rotate the shaft at a constant rate?

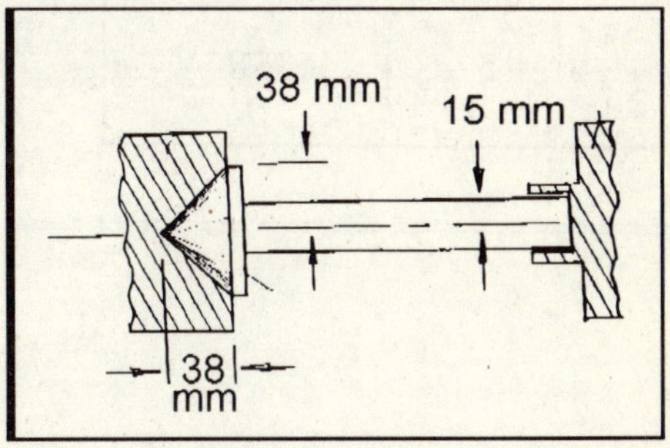

Solution *The left bearing:* The parameters are $r_o = 38\ mm$, $r_i = 0$, $\alpha = 45^o$, $\mu_k = 0.2$, and $F = 800\ N$. The moment required to sustain a constant rate of rotation is $M_{left} = \frac{2\mu_k F}{3\cos\alpha}\left(\frac{r_o^3 - r_i^3}{r_o^2 - r_i^2}\right) = 5.73\ N\,m$

The right bearing: This is a flat end bearing. The parameters are $\mu_k = 0.26$, $r = 15\ mm$, and $F = 800\ N$.

The moment required to sustain a constant rate of rotation is $M_{right} = \frac{2\mu_k Fr}{3} = 2.08\ N\,m$. The sum of the moments: $\boxed{M = 5.73 + 2.08 = 7.81\ N\,m}$

Problem 10.105 A motor is used to rotate a paddle for mixing chemicals. The shaft of the motor is coupled to the paddle wheel using a friction clutch of the type shown in Figure 10.24a. The radius of the clutch is 120 *mm* and the coefficient of static friction between the disks is 0.6. If the motor transmits a maximum moment of 15 *N m* to the paddle, what is the minimum normal force between the plates of the clutch necessary to prevent slipping?

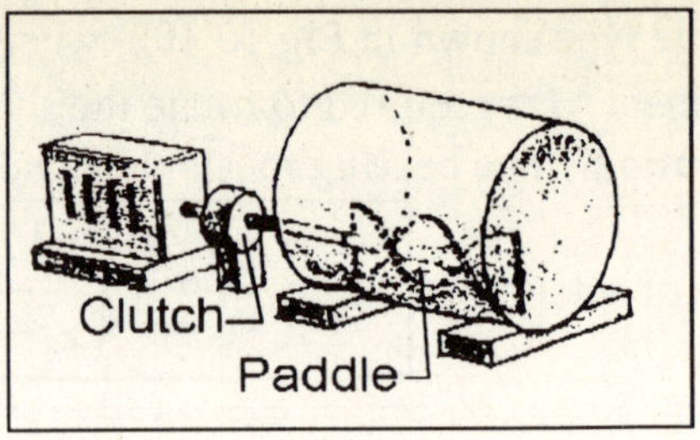

Solution The moment necessary to prevent slipping is

$$M = \frac{2\mu_s Fr}{3} = \frac{2(0.6)(0.12)F}{3} = 15\ N\,m. \text{ Solve: } \boxed{F = 312.5\ N}$$

Problem 10.106 The thrust bearing is supported by contact of the collar C with a fixed plate. The area of contact is an annulus with an inside diameter $D_1 = 40\ mm$ and an outside diameter $D_2 = 120\ mm$. The coefficient of kinetic friction between the collar and the plate is $\mu_k = 0.3$. The force $F = 400\ N$. What moment M is required to rotate the shaft at a constant rate?

Solution This is a thrust bearing with parameters $\mu_k = 0.3$, $\alpha = 0$, $r_o = 60\ mm$, $r_i = 20\ mm$, and $F = 400\ N$. The moment required to sustain rotation at a constant rate is $\boxed{M = \frac{2\mu_k F}{3}\left(\frac{r_o^3 - r_i^3}{r_o^2 - r_i^2}\right) = 5.2\ N\,m}$

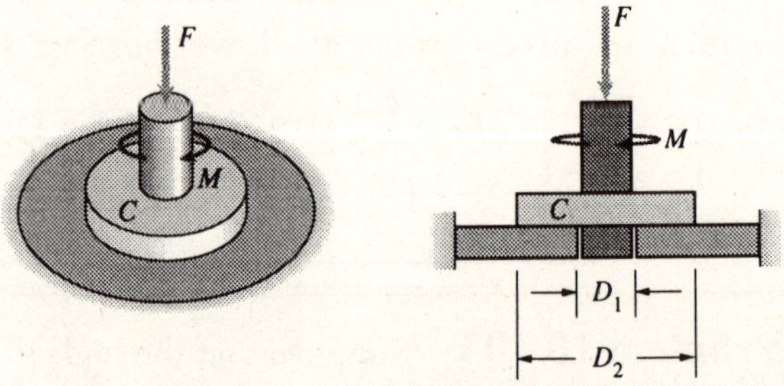

Problem 10.107 An experimental automobile brake design works by pressing an annular plate against the rotating wheel. If $\mu_k = 0.6$, what force F pressing the plate against the wheel is necessary to exert a moment of 200 *N m* on the wheel?

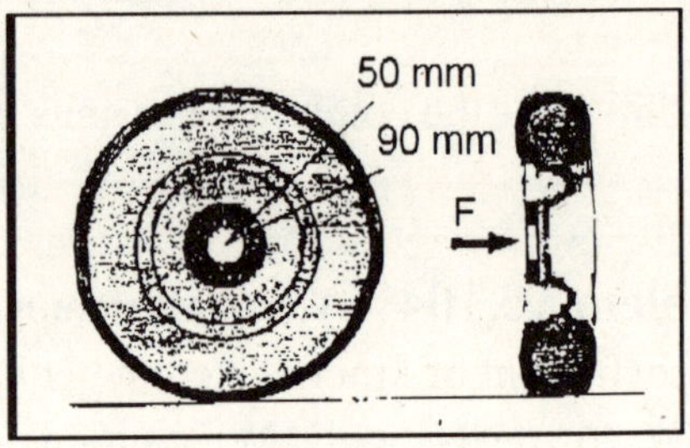

Solution This is a thrust bearing with parameters $\mu_k = 0.6$, $\alpha = 0$, $r_o = 90\ mm$, $r_i = 50\ mm$, and $M = 200\ N\,m$. The moment is $M = \frac{2\mu_k F}{3}\left(\frac{r_o^3 - r_i^3}{r_o^2 - r_i^2}\right)$. Solve: .

$$\boxed{F = \frac{3M}{2\mu_k}\left(\frac{r_o^2 - r_i^2}{r_o^3 - r_i^3}\right) = 4635.8\ N\text{-}m}$$

Problem 10.108 In Problem 10.107, suppose that $\mu_k = 0.65$ and the force pressing against the wheel is $F = 2\ kN$. (a) What moment is exerted on the wheel? (b) What percentage increase in the moment exerted on the wheel is obtained if the outer radius of the brake is increased from 90 *mm* to 100 *mm*.?

Solution Use the results of the solution to Problem 10.107, with parameters $\mu_k = 0.65$, $F = 2\ kN$.

(a) The moment is $M = \frac{2\mu_k F}{3}\left(\frac{r_o^3 - r_i^3}{r_o^2 - r_i^2}\right) = 0.0935\ kN\text{-}m = 93.5\ N\text{-}m$ (b) The new moment is

$M = \frac{2\mu_k F}{3}\left(\frac{r_o^3 - r_i^3}{r_o^2 - r_i^2}\right) = 0.1011\ kN\text{-}m = 101.1\ N\text{-}m$ The percentage increase is

$$\Delta M\% = \left(\frac{101.1 - 93.5}{93.5}\right)100 = 8.17\%$$

Problem 10.109 The coefficient of static friction between the plates of the car's clutch is $\mu_s = 0.8$. If the plates are pressed together with a force $F = 600\ lb$, what is the maximum torque the clutch will support without slipping?

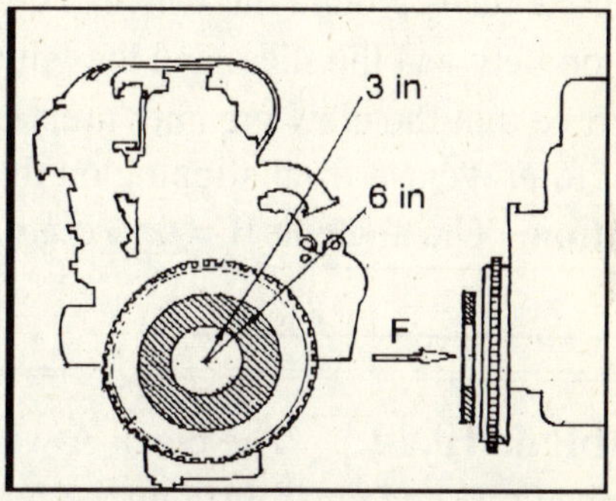

Solution This is the equivalent of a thrust bearing with parameters $\mu_s = 0.8$, $\alpha = 0$, $r_o = 6\ in$, $r_i = 3\ in$, and $F = 600\ lb$. The moment is

$$M = \frac{2\mu_s F}{3}\left(\frac{r_o^3 - r_i^3}{r_o^2 - r_i^2}\right) = 186.67\ ft\ lb$$

Problem 10.110 The "Morse" taper is used to support the workpiece on a machinist's lathe. The taper is driven into the spindle and is held in place by friction. If the spindle exerts a uniform pressure $p = 15\ psi$ on the taper and $\mu_s = 0.2$, what moment must be exerted about the axis of the taper to loosen it?

Solution The outer radius of the taper is $r_o = 1\ in.$, and the inner radius is $r_i = 0.625\ in$. The angle of the taper is $\alpha = 90 - \tan^{-1}\left(\frac{r_o - r_i}{L}\right) = 90 - \tan^{-1}\left(\frac{1 - 0.625}{9}\right) = 87.6^o$.

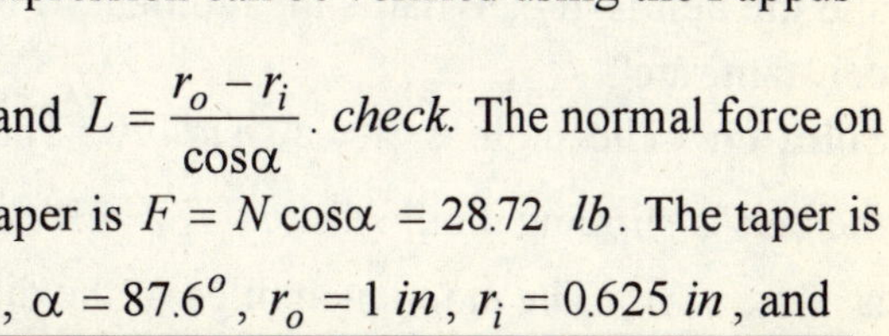

The active area of contact of the taper is the area of a truncated cone: $A = \frac{\pi(r_o^2 - r_i^2)}{\cos\alpha} = 45.99\ in^2$. *Check:* This expression can be verified using the Pappus-Guldinus Theorem (see Example 7.15) where $\bar{y} = \frac{r_o + r_i}{2}$, and $L = \frac{r_o - r_i}{\cos\alpha}$. *check.* The normal force on the taper is $N = pA = 689.8\ lb$, and the axial force on the taper is $F = N\cos\alpha = 28.72\ lb$. The taper is equivalent to a thrust bearing with the parameters $\mu_s = 0.2$, $\alpha = 87.6^o$, $r_o = 1\ in$, $r_i = 0.625\ in$, and $F = 28.72\ lb$. The moment required to initiate slip is $M = \frac{2\mu_s F}{3\cos\alpha}\left(\frac{r_o^3 - r_i^3}{r_o^2 - r_i^2}\right) = 114.1\ in\ lb = 9.51\ ft\ lb$

Problem 10.111 Suppose that you want to lift a 50 *lb* crate off the ground by using a rope looped over a tree limb as shown The coefficient of static friction between the rope and the limb is 0.4 and the rope is wound 120° around the limb. What force must you exert to lift the crate?

Solution This meets the conditions for Eq (10.22). The angle in radians is $\beta = (120^o)\left(\frac{\pi}{180}\right) = 2.094 \ radians$. The force required is

$$\boxed{T = 50e^{\mu_s \beta} = 115.6 \ lb}$$

Problem 10.112 "Winches" are used on sailboats to help support the ropes ("sheets") that position the sails. The simple winch shown is a post that will rotate in the clockwise direction (as seen from above) but not in the counter clockwise direction. The sail exerts an 800 *N* force on the sheet. The sheet is wrapped three complete turns around the winch. The coefficient of static friction between the sheet and the winch is $\mu_s = 0.2$. Determine the force that the crew member must exert on the end of the sheet to prevent it from slipping on the winch.

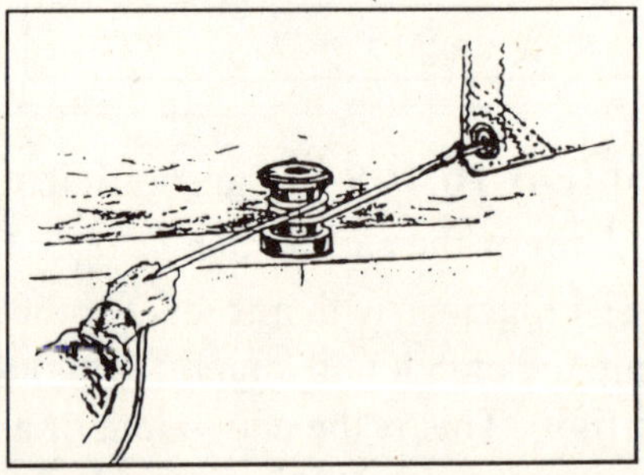

Solution The angle is $\beta = 6\pi \ radians$. The tension required is $\boxed{T = 800e^{-\mu_s \beta} = 18.44 \ N}$

Problem 10.113 The block A weighs 30 *lb*. The rope is wrapped one and one fourth turns around the fixed post. The coefficient of friction between the fixed post and the rope are $\mu_s = 0.2$ and $\mu_k = 0.15$. (a) What is the smallest force the person must exert to support the stationary block? (b) What force must the person exert to raise the block at a constant rate?

Solution The angle is $\frac{5\pi}{2} \ radians$. (a) The force required to hold the block stationary is $\boxed{F = 30e^{-\mu_s \beta} = 6.236 \ lb}$ (b) The force required to raise the block at a constant rate is

$$\boxed{F = 30e^{\mu_k \beta} = 97.45 \ lb}$$

Problem 10.114 The weight of the block A is *W*. The disk is supported by a smooth bearing. The coefficient of kinetic friction between the disk and the belt is μ_k. What is the moment M necessary to turn the disk at a constant rate?

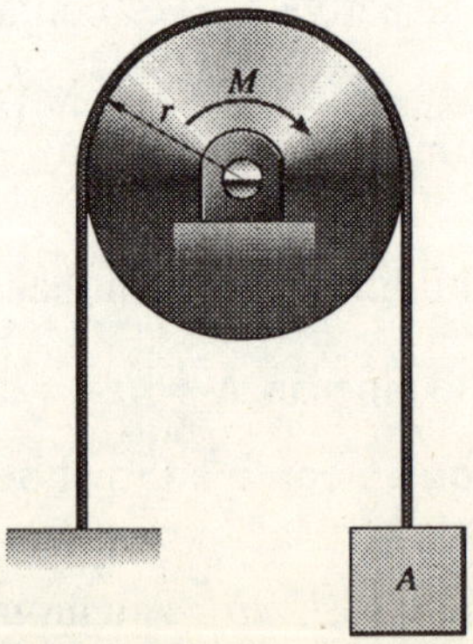

Solution The angle is $\beta = \pi \ radians$. The tension in the left belt when the belt is slipping on the disk is $T_{left} = We^{\mu_k \beta}$. The tension in the right belt is $T_{right} = W$. The moment applied to the disk is

$\boxed{M = R(T_{left} - T_{right}) = R(We^{\mu_k \beta} - W) = RW(e^{\mu_k \pi} - 1)}$ This is the moment that is required to rotate the disk at a constant rate.

Problem 10.115 The moment required to turn the wheel of the exercise bicycle is adjusted by changing the weight W. The coefficient of kinetic friction between the wheel and the belt is μ_k. Assume that the wheel turns clockwise. (a) Show that the moment M required to turn the wheel is $M = WR\left(1 - e^{-3.4\mu_k}\right)$. (b) If $W = 40$ *lb* and $\mu_k = 0.2$, what force will the scale S indicate when the bicycle is in use?

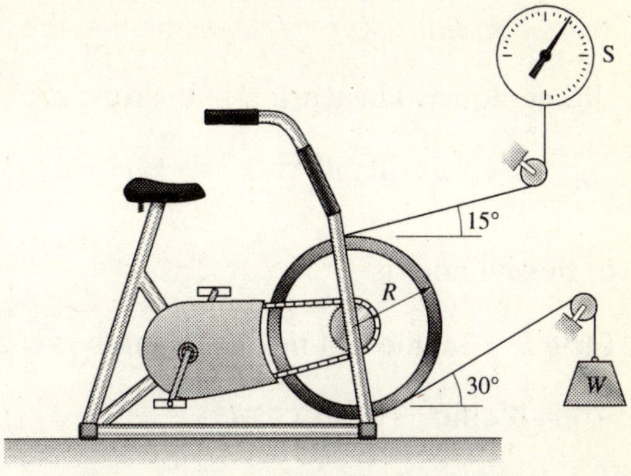

Solution Let β be the angle in radians of the belt contact with wheel. . The tension in the top belt when the belt slips is $T_{upper} = We^{-\mu_k \beta}$. The tension in the lower belt is $T_{lower} = W$. The moment applied to the wheel is $M = R\left(T_{lower} - T_{upper}\right) = RW(1 - e^{-\mu_k \beta})$. This is the moment required to turn the wheel at a constant rate. The angle β in radians is $\beta = \pi + (30 - 15)\left(\frac{\pi}{180}\right) = 3.40$ *radians*, from which $\boxed{M = RW(1 - e^{-3.4\mu_k})}$. (b) The upper belt tension is $\boxed{T_{upper} = 40e^{-3.4(0.2)} = 20.26\ lb}$. This is also the reading of the scale S.

Problem 10.116 The box B weighs 50 *lb*. The coefficient of friction between the cable and the fixed round supports are $\mu_s = 0.4$ and $\mu_k = 0.3$. (a) What is the minimum force F required to support the box? (b) What force F is required to move the box upward at a constant rate?

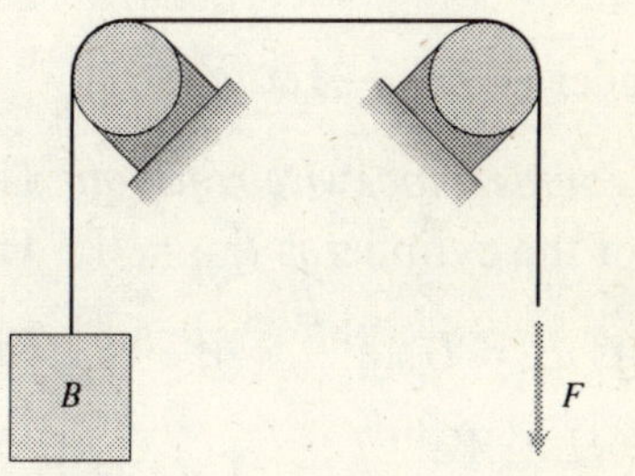

Solution The angle of contact between the cable and each round support is $\beta = \frac{\pi}{4}$ *radians*.(a) Denote the tension in the horizontal part of the cable by H. The tension in H is $H = We^{-\mu_s \beta}$. The force F is $F = He^{-\mu_s \beta} = We^{-2\mu_s \beta}$, from which $\boxed{F = 14.23\ lb}$ is the force necessary to hold the box stationary. (b) As the box is being raised, $H = We^{\mu_k \beta}$, and $F = He^{\mu_k \beta} = We^{2\mu_k \beta}$, from which $\boxed{F = 128.32\ lb}$

Problem 10.117 In Example 10.8, the coefficient of kinetic friction is 0.16 between the rope and the left cylinder and 0.34 between the rope and the right cylinder. The mass of the crate is 50 *kg*. What force would the woman have exert to raise the 50 *kg* crate at a constant rate? Would she need help?

Solution The weight is $W = mg = 490.5\ N$. The angle of contact of the rope and each cylinder is $\beta = \frac{\pi}{2}$ *radians*. Denote the tension in the horizontal part of the rope by H. As the crate is raised, the horizontal tension is $H = We^{0.16\beta}$. The force exerted is $F = He^{0.34\beta} = We^{(0.16+0.34)\beta} = We^{0.5\beta}$, from which $\boxed{F = 1075.8\ N}$. This force corresponds roughly to lifting 240 *lb*. Most folks would welcome help.

Problem 10.118 The block A weighs 200 N. The coefficient of static friction between the block and the inclined surface is 0.3 and the coefficient of static friction between the cable and the fixed cylinder is 0.2. For what range of the tension T will the block remain stationary?

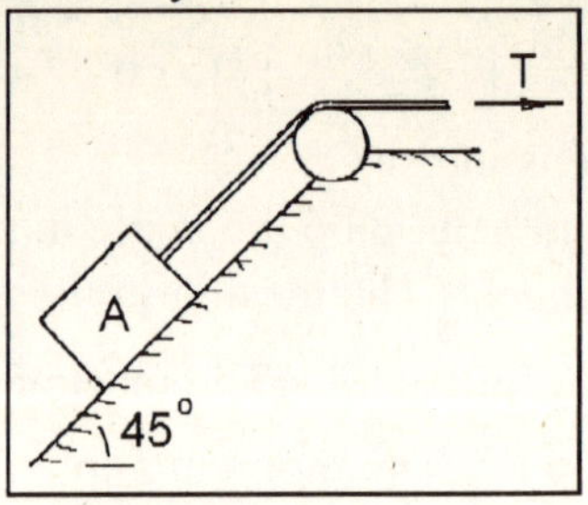

Solution The angle of contact between the rope and the cylinder is $\beta = \frac{\pi}{4}$ *radians*. *Case 1:* Assume that the box is on the verge of sliding down. The tension in the rope above the box is $T_A = W\sin\alpha - \mu_{sA} W\cos\alpha = \frac{W}{\sqrt{2}}(1-\mu_{sA})$. The tension on the right of the cylinder is $T_C = T_A e^{-\mu_{sc}\beta} = \frac{W}{\sqrt{2}}(1-\mu_{sA})e^{-\mu_{sc}\beta} = 84.6\ N$.

Case 2: Assume that the box is on the verge of sliding upward. The tension in the rope above the box is $T_A = W\sin\alpha + \mu_{sA} W\cos\alpha = \frac{W}{\sqrt{2}}(1+\mu_{sA})$. The tension in the rope to the right of the cylinder is $T_C = T_A e^{\mu_{sc}\beta} = \frac{W}{\sqrt{2}}(1+\mu_{sA})e^{\mu_{sc}\beta} = 215.1\ lb$. Thus the range of tensions is $\boxed{84.6 \le T_C \le 215.1\ (lb)}$

Problem 10.119 The block A weighs 30 lb. The coefficient of kinetic friction between the rope and the cylinder is 0.2. If the cylinder is rotated at a constant rate, first in the counterclockwise direction and then in the clockwise direction, the difference in the height of the block A is one foot. What is the spring constant k?

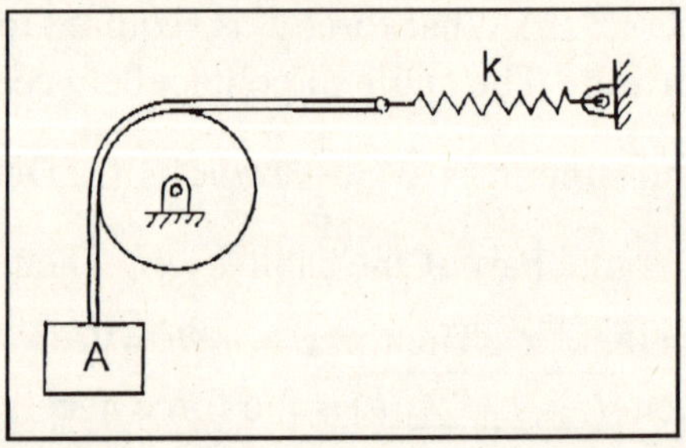

Solution The angle of contact between the rope and the cylinder is $\beta = \frac{\pi}{2}$ *radians*.

For counterclockwise rotation: The tension in the belt to the left of the cylinder is $T_{left} = W$. The tension in the horizontal part is $H = T_{left}e^{\mu_k\beta} = We^{\mu_k\beta}$. The elongation of the spring is $s_1 = \frac{H}{k} = \frac{We^{\mu_k\beta}}{k}$. *For clockwise rotation:* The tension in the belt to the left is $T_{left} = W$. The tension in the horizontal part is $H = T_{left}e^{-\mu_k\beta} = We^{-\mu_k\beta}$. The elongation of the spring is $s_2 = \frac{H}{k} = \frac{We^{-\mu_k\beta}}{k}$. The difference in elongations is $s_1 - s_2 = \frac{W}{k}\left(e^{\mu_k\beta} - e^{-\mu_k\beta}\right) = \frac{2W}{k}\sinh(\mu_k\beta) = 1$, from which

$$\boxed{k = 2W\sinh(\mu_k\beta) = 60\sinh(0.2\frac{\pi}{2}) = 19.16\ lb/ft}$$

Problem 10.120 If the force F in Example 10.9 is increased to 400 N, what are the largest values of the couples M_A and M_B for which the belt will not slip?

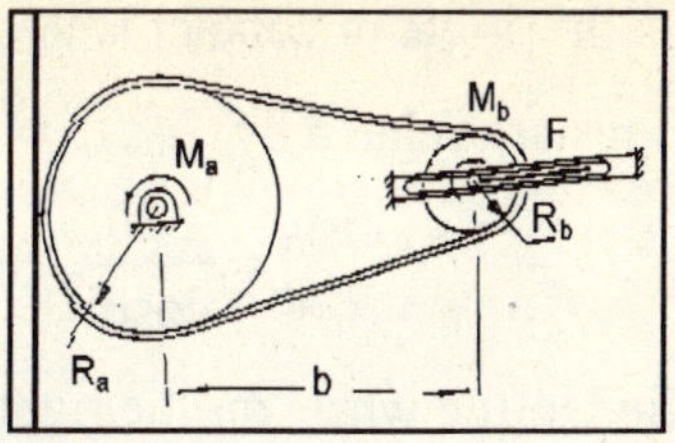

Solution From Example 10.9, $b = 500\ mm$, $\mu_s = 0.8$, $R_a = 200\ mm$, $R_b = 100\ mm$. The angle of contact for pulley A is $\beta_a = \pi + 2\alpha$. The angle of contact for pulley B is $\beta_b = \pi - 2\alpha$, where $\alpha = \sin^{-1}\left(\frac{R_a - R_b}{b}\right) = \sin^{-1}\left(\frac{0.1}{0.5}\right) = 0.2014\ radians$. The belt contact is less for pulley B, so it is most likely to slip first. The couples are in opposition so that the tension in the upper belt is greater than the tension in the lower belt: For belt B:

$T_{upper} = T_{lower}e^{\mu_s\beta_b} = 8.945T_{lower}$. The force is $F = \left(T_{upper} + T_{lower}\right)\cos\alpha$, from which

$T_{lower} = \frac{F}{\left(1 + e^{\mu_s\beta_b}\right)\cos\alpha} = 41.05\ N$, and $T_{upper} = 8.945T_{lower} = 367.19\ N$. The couples are

$\boxed{M_b = R_b\left(T_{upper} - T_{lower}\right) = 32.61\ N\text{-}m}$. $\boxed{M_a = R_a\left(T_{upper} - T_{lower}\right) = 65.23\ N\text{-}m}$

Problem 10.121 If the belt in Example 10.9 is a V-belt with angle $\gamma = 45^o$, what are the largest values of the couples M_A and M_B for which the belt will not slip?

Solution From Example 10.9, $b = 500\ mm$, $\mu_s = 0.8$, $R_a = 200\ mm$, $R_b = 100\ mm$, and $F = 200\ N$. Use the results of Problem 10.120: The contact angle for belt B is $\beta_b = \pi - 2\alpha = 2.7468\ radians$, where $\alpha = 0.1974\ radians$. For brevity, denote $\Psi = \frac{\mu_s\beta_b}{\sin\frac{\gamma}{2}} = 5.7422$ The tensions are related by

$T_{upper} = T_{lower}e^{\Psi} = 311.74T_{lower}$. The force $F = 200 = \left(T_{upper} + T_{lower}\right)\cos\alpha$, from which

$T_{lower} = \frac{F}{(1 + 311.74)\cos\alpha} = 0.652\ N$, and $T_{upper} = 203.3\ N$. The couples are

$\boxed{M_a = R_a\left(T_{upper} - T_{lower}\right) = 40.53\ N\,m}$, $\boxed{M_b = R_b\left(T_{upper} - T_{lower}\right) = 20.27\ N\,m}$

Problem 10.122 The spring exerts a 320 N force on the left pulley. The coefficient of static friction between the flat belt and the pulleys is $\mu_s = 0.5$. The right pulley cannot rotate. What is the largest moment M that can be exerted on the left pulley without causing the belt to slip?

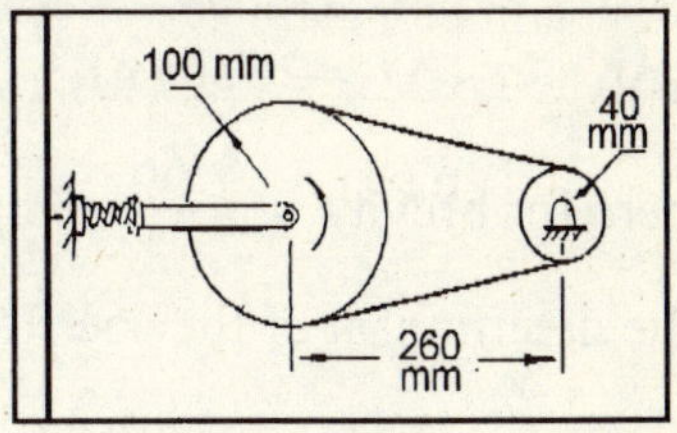

Solution The angle of of the belt relative to the horizontal is is $\alpha = \sin^{-1}\left(\frac{100 - 40}{260}\right) = 0.2329\ radians$. .For the right pulley the angle of contact is $\beta_{right} = \pi - 2\alpha = 2.676\ radians$. The sum of the horizontal components of the tensions equals the force exerted by the spring: $F = (T_{upper} + T_{lower})\cos\alpha = 320\ N$.

Problem 10.122 continued on next page

continuation of *Problem 10.122..........*

Since the angle of contact is less on the right pulley, it should slip there first. At impending slip, the tensions are related by $T_{upper} = T_{lower}e^{\mu_s\beta_{right}} = 3.811T_{lower}$. Substitute and solve:

$T_{lower}\left(1+e^{0.5(2.676)}\right) = \frac{320}{\cos\alpha}$, from which $T_{lower} = 68.34\ N$, and $T_{upper} = 260.48\ N$. The moment applied to the wheel on the right is $\boxed{M_{applied} = R\left(T_{upper} - T_{lower}\right) = 0.1(192.16) = 19.22\ N\text{-}m}$

Problem 10.123 Suppose that the belt in Problem 10.122 is a V-belt with angle $\gamma = 30^o$. What is the largest moment *M* that can be exerted on the left pulley without causing the belt to slip?

Solution Use the results of the solution to Problem 10.122, as applicable: The angle of contact with the right pulley is $\beta_{right} = 2.676\ radians$. The sum of the horizontal components of the tensions equals the force exerted by the spring: $F = (T_{upper} + T_{lower})\cos\alpha = 320\ N$. Since the angle of contact is less on the right pulley, it will slip there. At impending slip, the tensions are related by $T_{upper} = T_{lower}e^{\Psi}$, where for brevity $\Psi = \frac{\mu_s\beta_{right}}{\sin\left(\frac{\gamma}{2}\right)} = 5.169$ has been substituted. Substitute and solve: $T_{lower}\left(1+e^{\Psi}\right) = \frac{320}{\cos\alpha}$,

from which $T_{lower} = 1.86\ N$, and $T_{upper} = 327.01\ N$. The moment applied to the wheel on the right is

$\boxed{M_{applied} = R\left(T_{upper} - T_{lower}\right) = 0.1(325.16) = 32.52\ N\text{-}m}$

Problem 10.124 Beginning with Eqs. (10.26), derive Eq. (10.27):

$T_2 = T_1 e^{\frac{\mu_s\beta}{\sin\left(\frac{\gamma}{2}\right)}}$.

Solution The Eqs. (10.26) are:

$\sum F_{tngential} = 2\mu_s\Delta N + T\cos\left(\frac{\Delta\alpha}{2}\right) - (T+\Delta T)\cos\left(\frac{\Delta\alpha}{2}\right) = 0$

$\sum F_{normal} = 2\Delta N\sin\left(\frac{\gamma}{2}\right) - (T+\Delta T)\sin\left(\frac{\Delta\alpha}{2}\right) - T\sin\left(\frac{\Delta\alpha}{2}\right) = 0$. These can be written as two simultaneous equations in two unknowns:

$a_{11}\Delta N + a_{12}\Delta T = 0$,

$a_{21}\Delta N + a_{22}\Delta T = 2T\sin(\eta)$,

where for brevity $\eta = \frac{\Delta\alpha}{2}$. These have the solution $\Delta T = \frac{2T\sin(\eta)a_{11}}{\det}$, where $\det = a_{11}a_{22} - a_{12}a_{21}$ is the determinant of the coefficients. Reducing term by term: $a_{11} = 2\mu_s$, and

$\det = -2\left(\mu_s\sin(\eta) - \sin\left(\frac{\gamma}{2}\right)\cos(\eta)\right)$.

Problem 10.124 continued on next page

=================================<>=================================

continuation of Problem 10.124...........

The angle $\Delta\alpha$ is small, so $\eta << 1$, and from an expansion in a Taylor's series about $\eta = 0$, $\sin(\eta) = \eta - \frac{\eta^3}{3!} + ...$, $\cos(\eta) = 1 - \frac{\eta^2}{2} + ...$ For small η,. $\eta << 1$, only the first terms in the expansion apply, so that $a_{11}\sin\eta = 2\mu_s\eta$, and $\det = -2\left(\mu_s\eta - \sin\left(\frac{\gamma}{2}\right)\right)$. The solution: $\Delta T = \dfrac{-\mu_s T(\Delta\alpha)}{\left(\mu_s\left(\frac{\Delta\alpha}{2}\right) - \sin\frac{\gamma}{2}\right)}$.

Divide both sides by $\Delta\alpha$ and take the limit:

$$\lim\left(\frac{\Delta T}{\Delta\alpha}\right)_{\Delta\alpha\to 0} = \frac{dT}{d\alpha} = \lim\left(\frac{\mu_s T}{\left(\mu_s\Delta\alpha - \sin\frac{\gamma}{2}\right)}\right)_{\Delta\alpha\to 0} = \frac{\mu T}{\sin\frac{\gamma}{2}},$$

from which $\dfrac{dT}{d\alpha} = \dfrac{\mu_s T}{\sin\frac{\gamma}{2}}$. This first order differential equation can be solved by the standard separation of variables procedure:

$\int_{T_1}^{T_2}\frac{dT}{T} = \int_{\alpha_1}^{\alpha_2}\frac{\mu_s}{\sin\frac{\gamma}{2}}d\alpha$, from which $\log_e(T_2) - \log_e(T_1) = \frac{\mu_s}{\sin\frac{\gamma}{2}}(\alpha_2 - \alpha_1)$. Define $\beta = \alpha_2 - \alpha_1$, and

simplify: $\log_e\left(\frac{T_2}{T_1}\right) = \left(\frac{\mu_s\beta}{\sin\frac{\gamma}{2}}\right)$, or alternatively $\boxed{T_2 = T_1 e^{\frac{\mu_s\beta}{\sin\frac{\gamma}{2}}}}$

=================================<>=================================

Problem 10.125 The mass of the block is 20 *kg* and the coefficient of static friction between the block and the floor is $\mu_s = 0.3$. The spring constant $k = 1\ kN/m$ and the spring is unstretched when it is vertical. How far can the slider B be moved to the right without causing the block to slip?

Solution Let A_o be the original position with unstretched spring, and let B be the new position. The distance to the right is *d*. The angle $\alpha = 10^o$. The new length of the spring is $L = \sqrt{(1 + d\tan\alpha)^2 + d^2}$. The spring force is $kL - 1$. The angle θ is given by $\theta = \sin^{-1}\left(\frac{d}{L}\right)$,

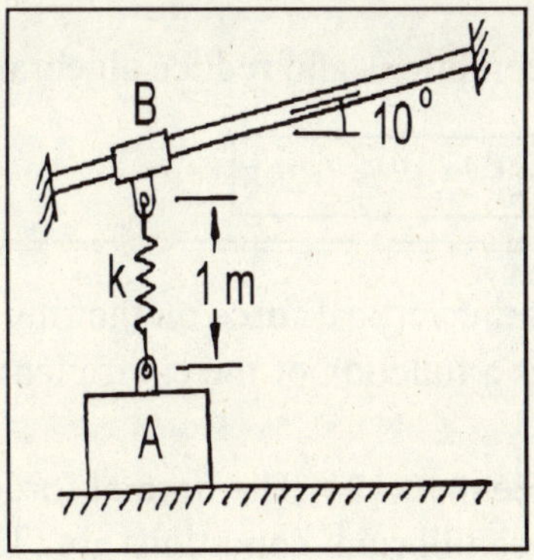

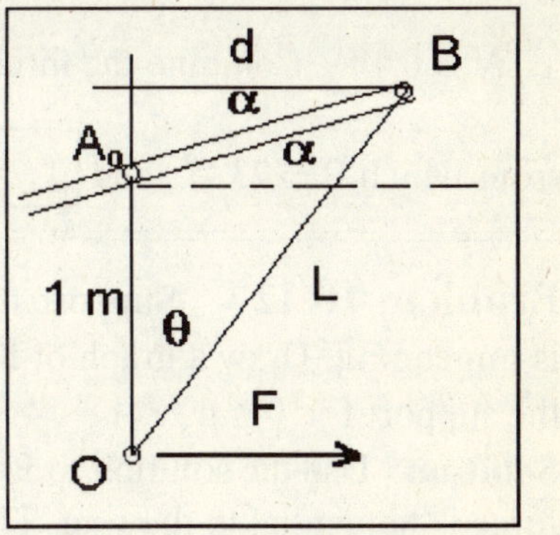

and the horizontal component of the force in the block is $F = (kL - 1)\sin\theta = k(L-1)\left(\frac{d}{L}\right) = k\frac{d(L-1)}{L}$.

Problem 10.125 continued on next page

continuation of Problem 10.125..........

The normal force exerted by the floor on the block is $N = W - k(L-1)\cos\theta$. The equilibrium condition at impending slip is $F - \mu_s N = 0$, from which

$$k\frac{d(L-1)}{L} - \mu_s(W - k(L-1)\cos\theta) = 0.$$ Graph the equation

$$f(d) = k\frac{d(L-1)}{L} - \mu_s(W - k(L-1)\cos\theta),$$ where

$L = \sqrt{(1 + d\tan\alpha)^2 + d^2}$, and $\theta = \sin^{-1}\left(\frac{d}{L}\right)$, and determine the zero crossing point. The zero crossing point corresponds to $d = 0.317\ m$, which is the horizontal distance to the right. The distance along the sliding bar is $\boxed{D = \frac{d}{\cos 10^o} = 0.322\ m}$

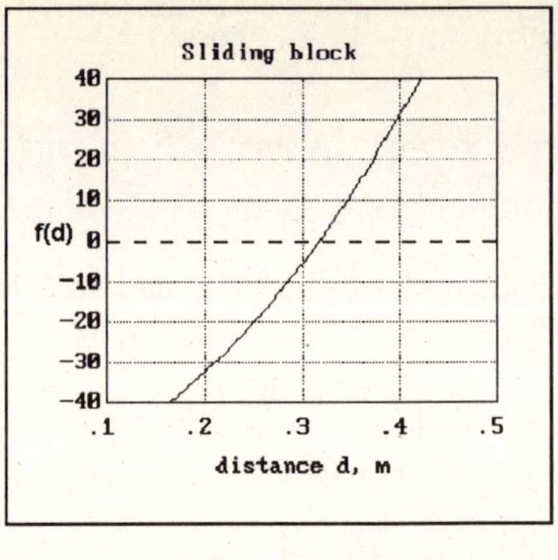

Problem 10.126 The slender circular ring of weight W is supported by normal and friction forces at A. If slip is impending when the vertical force $F = 0.4W$, what is the coefficient of static friction between the ring and the support?

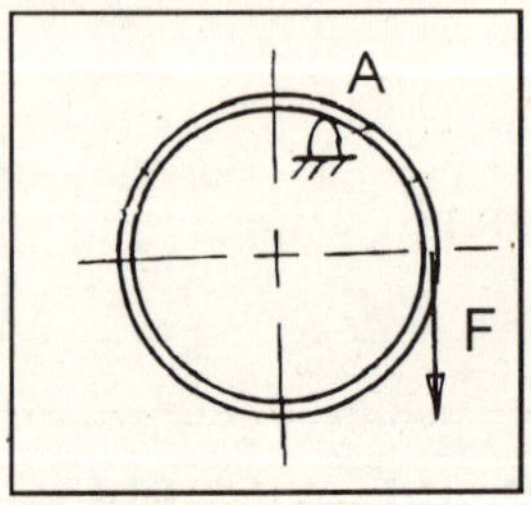

Solution Let θ be the angle between the x-axis and point A. The normal force is normal to the ring, and the friction force is tangential to the ring. The sum of forces: $\sum F_x = N\cos\theta - \mu_s N\sin\theta = 0$,

$\sum F_y = N\sin\theta + \mu_s N\cos\theta - W - F = 0$. The sum of the moments about the center of the ring: $\sum M_c = +R\mu_s N - RF = 0$. Substitute $F = 0.4W$ and reduce each equation to obtain the three equations: $\mu_s = \cot\theta$, $N = \frac{1.4W}{\sin\theta + \mu_s\cos\theta}$, and

$\mu_s N = 0.4W$. Combine the three conditions and reduce algebraically to obt ain $\cos\theta = \left(\frac{0.4}{1.4}\right) = 0.2857$,

from which $\theta = 73.4^o$, and $\boxed{\mu_s = \cot 73.4^o = 0.298}$

Problem 10.127 Suppose that the vertical force on the ring in Problem 10.126 is $F = KW$ and slip is impending. Draw a graph of K as a function of the coefficient of static friction between the ring and the support for $0 \le \mu_s \le 1$.

Solution Use the solution to Problem 10.126. The normal force is normal to the ring, and the friction force is tangential to the ring. The equilibrium conditions are: The sum of forces:

$\sum F_x = N\cos\theta - \mu_s N\sin\theta = 0$, $\sum F_y = N\sin\theta + \mu_s N\cos\theta - W - F = 0$. The sum of the moments about the center of the ring: $\sum M_c = +R\mu_s N - RF = 0$, where θ is the angle between the y-axis and point A. Substitute the condition $F = KW$ and reduce to obtain the three intermediate equations:

$\mu_s = \cot\theta$, $N = \frac{(1+K)W}{\sin\theta + \mu_s\cos\theta}$, and $\frac{\mu_s(1+K)}{\sin\theta + \mu_s\cos\theta} = K$.

Problem 10.127 continued on next page

=====================================<>=====================================

continuation of Problem 10.127..........

Combine and reduce to obtain: $\cos\theta = \dfrac{K}{1+K}$., or $K = \dfrac{\cos\theta}{1-\cos\theta}$.

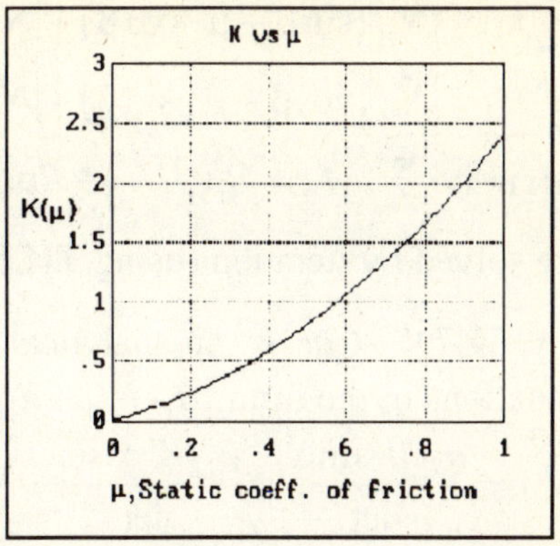

Since $0 \le \mu_s \le 1$, then from $\mu_s = \cot\theta$, and $45^o \le \theta \le 90^o$ it is permissible to use the relation $\sin\theta = \sqrt{1-\cos^2\theta}$ with a positive sign. Thus $\mu_s = \cot\theta = \dfrac{\cos\theta}{\sqrt{1-\cos^2\theta}}$, $\cos\theta = \dfrac{\mu_s}{\sqrt{1+\mu_s^2}}$,

and $\boxed{K = \dfrac{\mu_s}{\sqrt{1+\mu_s^2}-\mu_s}}$. This is graphed over the range $0 \le \mu_s \le 1$. The limits on K are $0 \le K \le \dfrac{1}{\sqrt{2}-1} \cong 2.414$. *Check.*

$$\lim_{\mu_s \to 0} K = \left[\frac{\mu_s}{\sqrt{1+\mu_s^2}-\mu_s}\right]_{\mu_s \to 0} \to 0.. \quad \lim_{\mu_s \to 1} K = \left[\frac{\mu_s}{\sqrt{1+\mu_s^2}-\mu_s}\right]_{\mu_s \to 1} \to \frac{1}{\sqrt{2}-1} = 2.4142. \ check.$$

=====================================<>=====================================

Problem 10.128 The 3 m bar weighs 200 N and the coefficient of static friction between the ends of the bar and the circular surface is $\mu_s = 0.3$. What is the largest value of the angle for which the bar will not slip? *Strategy:* Assume that slip impends at the both ends of the bar.

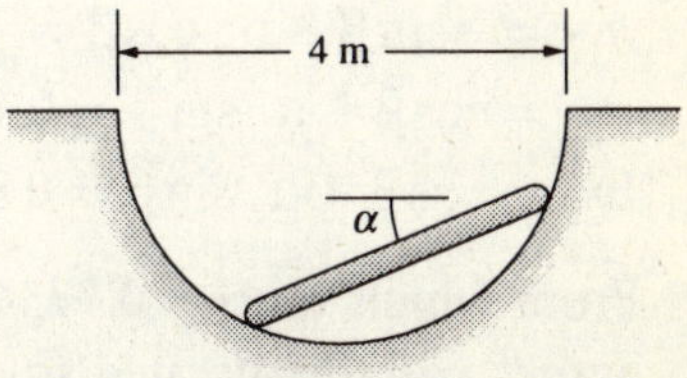

Solution The strategy is to write the equilibrium equations in a form in which the angle α is isolated. This can be done by choosing a coordinate system fixed to the bar, in which α appears only in the direction of the gravity vector. Choose a coordinate system with the origin at the center of the circle forming the circular bowl, with the y-axis perpendicular to the bar. Construct the radii O-A and O-B through the ends of the bar. The circular bowl has a radius $R = 2\ m$. The angle $\beta = \sin^{-1}\left(\dfrac{L}{2R}\right) = 48.6^o$.

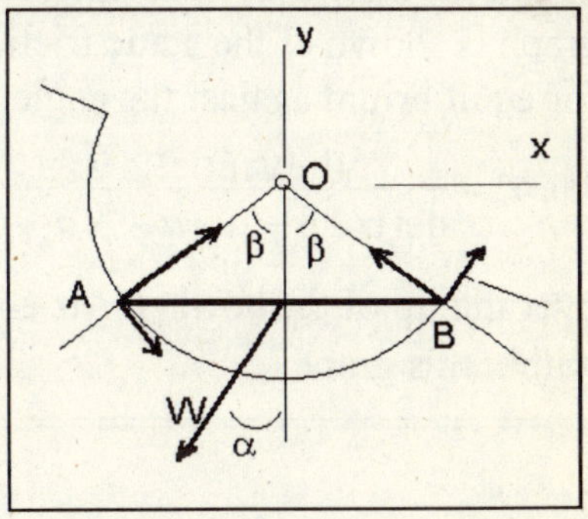

The normal force at A is directed along the radii A-O, with the angle relative to the positive x-axis of $90^o - \beta$, and the projection $\vec{N}_A = |\vec{N}_A|(\vec{i}\sin\beta + \vec{j}\cos\beta)$. The friction force at A opposes slip, and is tangent to the radii O-A ,with the components $\vec{f}_A = \mu_s|\vec{N}_A|(\vec{i}\cos\beta - \vec{j}\sin\beta)$

The normal force at B is directed along the radii B-O, with the angle relative to the positive x-axis of $90^o + \beta$, and the projection $\vec{N}_B = |\vec{N}_B|(-\vec{i}\sin\beta + \vec{j}\cos\beta)$ The friction force at B opposes slip, and is perpendicular to the radii O-B, with the components $\vec{f}_B = \mu_s|\vec{N}_B|(\vec{i}\cos\beta + \vec{j}\sin\beta)$. The weight vector acts through the midpoint of the bar, and has the components $\vec{W} = |\vec{W}|(-\vec{i}\sin\alpha - \vec{j}\cos\alpha)$.

Problem 10.128 continued on next page

continuation of Problem 10.128..........

The equilibrium conditions are:

$$\sum F_x = |\vec{N}_A|(\sin\beta + \mu_s\cos\beta) + |\vec{N}_B|(-\sin\beta + \mu_s\cos\beta) - |\vec{W}|\sin\alpha = 0$$

$$\sum F_y = |\vec{N}_A|(\cos\beta - \mu_s\sin\beta) + |\vec{N}_B|(\cos\beta + \mu_s\sin\beta) - |\vec{W}|\cos\alpha = 0$$

The sum of the moments about the origin is: $\sum M_O = R\mu_s|\vec{N}_A| + R\mu_s|\vec{N}_B| - R|\vec{W}|\cos\beta\sin\alpha = 0$. These three equations in three unknowns are solved by iteration (using **TK Solver 2).** The results: $|\vec{N}_A| = 181.1\ N$, $|\vec{N}_B| = 89.2\ N$, and $\alpha = 37.79^o$. *Check:* An analytical solution for α can be obtained as follows: write the first two equations in the form: $a_{11}N_A + a_{12}N_B = W\sin\alpha$, $a_{21}N_A + a_{22}N_B = W\cos\alpha$. These have the solution $N_A = \frac{a_{22}W\sin\alpha}{\det} - \frac{a_{12}W\cos\alpha}{\det}$, $N_B = \frac{-a_{21}W\sin\alpha}{\det} + \frac{a_{11}W\cos\alpha}{\det}$., where $\det = a_{11}a_{22} - a_{21}a_{12}$ is the determinant of the coefficients. Substitute these into the third equation and reduce:

$\mu_s(a_{22} - a_{21})\sin\alpha + \mu_s(a_{11} - a_{12})\cos\alpha = (\det)\cos\beta\sin\alpha$, from which

$$\tan\alpha = \frac{\mu_s(a_{11} - a_{12})}{\det\cos\beta - \mu_s(a_{22} - a_{21})}.$$

The terms on the right hand side are known quantities involving only β and μ_s:

$a_{11} = \sin\beta + \mu_s\cos\beta = 0.9484$.

$a_{12} = -\sin\beta + \mu_s\cos\beta = -0.5516$,

$a_{21} = \cos\beta - \mu_s\sin\beta = 0.4364$, and

$a_{22} = \cos\beta + \mu_s\sin\beta = 0.8864$, $(\det)\cos\beta = 0.7153$.

From which $\tan\alpha = 0.7754$, and $\boxed{\alpha = 37.79^o}$, which agrees with the iterated solution. *#2 Check:* For $\alpha = 15^o$ the expression can be solved by iteration to obtain the answer to Problem 10.45, as it should. A graph is shown of the static friction coefficient required for equilibrium against the angle.

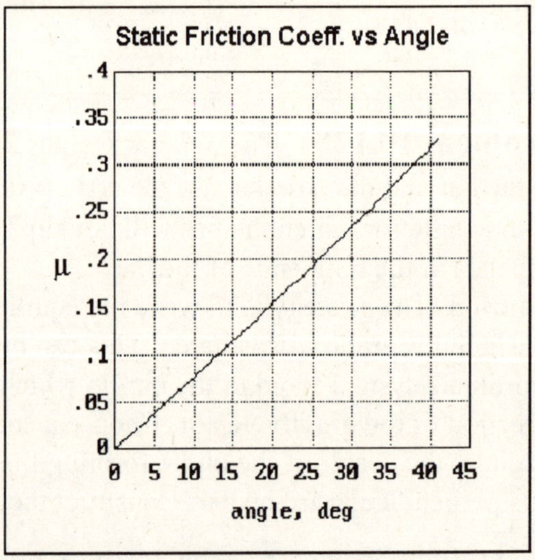

The graph was computed from the expression

$$\tan\alpha = \frac{\mu_s(a_{11} - a_{12})}{\det\cos\beta - \mu_s(a_{22} - a_{21})}.$$

over the allowed range of angles such that the bar does not extend over the lip of the bowl, $0 \le \alpha \le 90^o - \beta = 41.4^o$. Note that the solution to Problem 10.45 is contained within this graph.

Problem 10.129 In Problem 10.84, the load $W = 800\ N$ can be raised or lowered by rotating the threaded shaft. The distances $b = 75\ mm$ and $h = 200\ mm$. The pinned bars are each 300 *mm* in length. The pitch of the threaded shaft is $p = 5\ mm$, the mean radius of the thread is $r = 15\ mm$, and the coefficient of kinetic friction between the thread and the mating groove is 0.2. Draw a graph of the moment that must be exerted on the threaded shaft to raise the load at a constant rate as a function of the height *h* from $h = 100\ mm$ to $h = 400\ mm$.

Solution Assume that the threaded shaft is to be turned at a constant rate. Use the solution to Problem 10.84, as applicable. The vertical distances HE, BE, AD, DG are $y = \frac{h}{2}\ mm$.. The included angle BAC is $\beta = \sin^{-1}\left(\frac{h}{600}\right)$

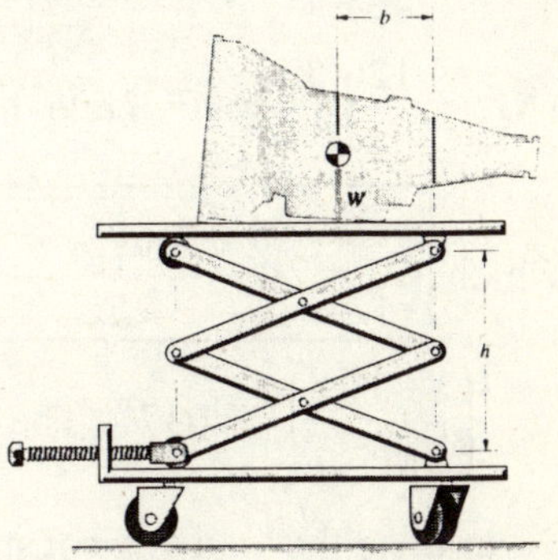

The distance AB is $L = 300\cos\beta\ \ mm$. Isolate the members and write the equilibrium equations, beginning at the top: *Isolate the frame AB which supports the load W.* The equilibrium conditions are (1) $\sum M_B = -AL + Wb = 0$,

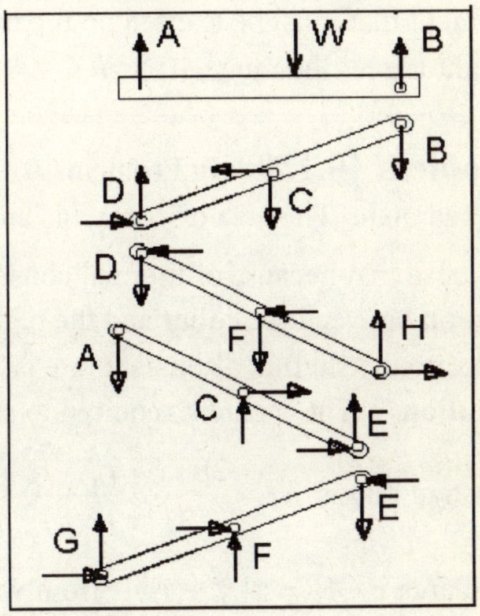

(2) $\sum F_y = A - W + B_y = 0$, $\sum F_x = B_x = 0$

Isolate BD: (3) $\sum F_y = -B_y - C_y + D_y = 0$,

(4) $\sum F_x = -C_x + D_x = 0$,

(5) $\sum M_D = \left(\frac{h}{4}\right)C_x - \left(\frac{L}{2}\right)C_y - LB = 0$

Isolate DH:(6) $\sum F_y = -D_y - F_y + H_y = 0$,

(7) $\sum F_x = -D_x - F_x + H_x = 0$,

(8) $\sum M_H = \left(\frac{h}{4}\right)F_x + \left(\frac{L}{2}\right)F_y + LD_y + \left(\frac{h}{2}\right)D_x = 0$

Isolate AE (9) $\sum F_y = -A + C_y + E_y = 0$,

(10) $\sum F_x = C_x + E_x = 0$,

(11) $\sum M_E = -\left(\frac{h}{4}\right)(C_x) - \left(\frac{L}{2}\right)C_y + LA = 0$.

Isolate EG (12) $\sum F_y = -E_y + F_y + G_y = 0$,

(13) $\sum F_x = -E_x + F_x + G_x = 0$,

(14) $\sum M_G = -\frac{h}{4}F_x + \left(\frac{L}{2}\right)F_y - L\,E_y + \left(\frac{h}{2}\right)E_x = 0$. These 14 equations in 14 unknowns are to be solved to determine the reaction G_x, which is the force that the threaded shaft must overcome to raise the load at a constant rate.. An analytic solution is obtained as follows:

From (1) $\boxed{A = \frac{Wb}{L}}$. From (2) $\boxed{B = W\left(1 - \frac{b}{L}\right)}$. From (8) and (11) $\boxed{C_y = W\left(\frac{2b}{L} - 1\right)}$, $\boxed{C_x = \frac{2WL}{h}}$.

From (9) $\boxed{E_y = W\left(1 - \frac{b}{L}\right)}$. From (10) $\boxed{E_x = -\frac{2}{h}WL}$. From (4) $\boxed{D_x = \frac{2}{h}WL}$

Problem 10.129 continued on next page..........

............*continuation of solution to Problem 10.129.....*

From (3) $\boxed{D_y = \frac{Wb}{L}}$. From (8) and (14) $F_y = W\left(1 - \frac{2b}{L}\right)$, $\boxed{F_x = \frac{6WL}{h}}$. From (13) $\boxed{G_x = \frac{4WL}{h}}$. From (12) $\boxed{G_y = \frac{Wb}{L}}$

Check: This value of G_y is expected from the equilibrium conditions for the complete structure. *check.* Substitute to obtain $G_x = \frac{1200W}{h}\cos\beta$. Note that $\cos\beta = \sqrt{1 - \left(\frac{h}{600}\right)^2}$, from which $\boxed{G_x = 2W\frac{\sqrt{600^2 - h^2}}{h}}$ The pitch angle is

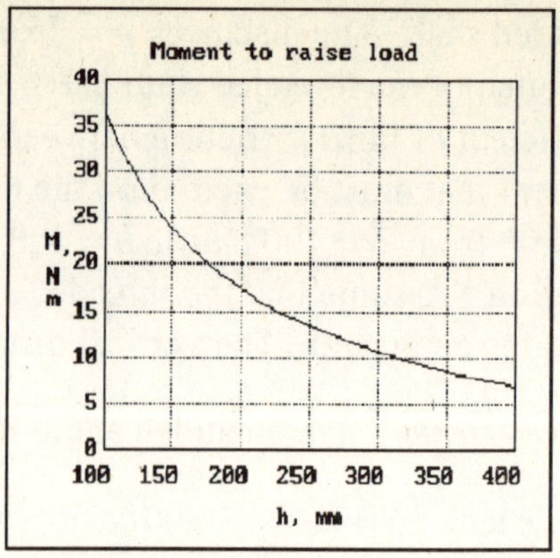

$\alpha = \tan^{-1}\left(\frac{5}{2\pi(15)}\right) = 3.037^o$. The angle of kinetic friction is $\theta_k = \tan^{-1}(0.2) = 11.31^o$. The moment required to raise the load at a constant rate: $M = rG_x \tan(\theta_k + \alpha) = 0.003836G_x$. The graph of the moment that must be exerted on the threaded shaft to raise the 800 N load is shown as a function of the height h over the range $100 \le h \le 400$ (mm).

Problem 10.130 In Problem 10.106, The thrust bearing is supported by contact of the collar C with a fixed plate. The area of contact is an annulus with an inside diameter D_1 and an outside diameter D_2. Suppose that because of thermal constraints, you want the area to be $0.02\ m^2$. The coefficient of kinetic friction between the collar and the plate is $\mu_k = 0.3$. The force $F = 600\ N$ and the moment M required to rotate the shaft at a constant rate is 10 $N\,m$. What are the diameters D_1 and D_2?

Solution . The moment required to sustain rotation at a constant rate is $M = \frac{2\mu_k F}{3}\left(\frac{r_o^3 - r_i^3}{r_o^2 - r_i^2}\right)$ Note that the total area is $A_T = \pi\left(r_o^2 - r_i^2\right)$, from which

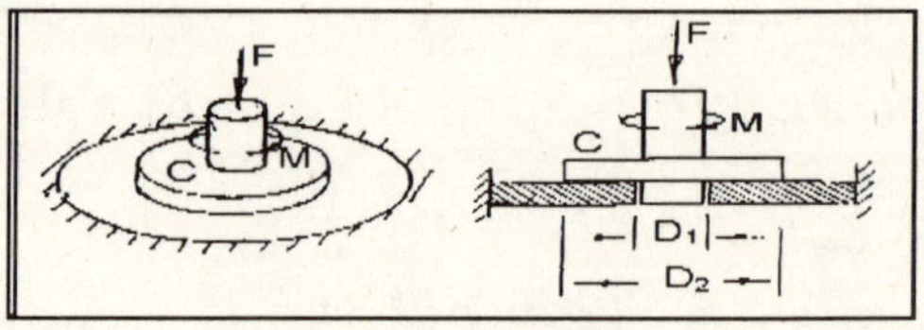

$M = \frac{\pi\mu_k F}{12A_T}\left(D_2^3 - D_1^3\right)$. Solve, and note that $D_2^2 = \frac{4A_T}{\pi} + D_1^2$, from which $\left(\frac{4A_T}{\pi} + D_1^2\right)^{\frac{3}{2}} - D_1^3 = \frac{12A_T M}{\pi\mu_k F}$. From which $\left(D_1^2 + 0.02546\right)^{\frac{3}{2}} - D_1^3 - 0.00424 = 0$. The value of D_1 is determined by graphing the function

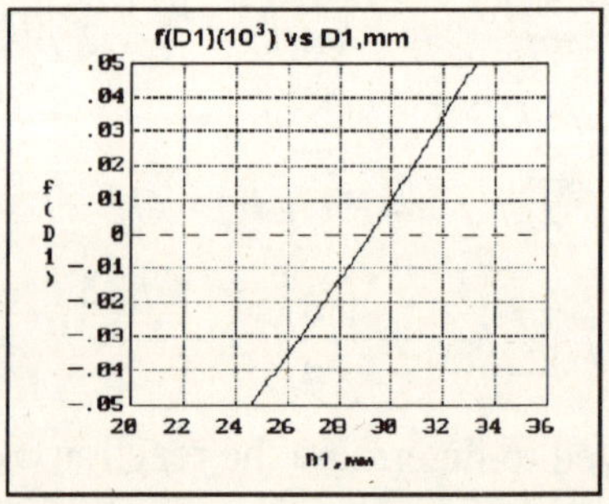

$f(D_1) = \left(D_1^2 + 0.02546\right)^{\frac{3}{2}} - D_1^3 - 0.00424$ to determine the zero crossings. The zero crossing occurs at $\boxed{D_1 = 29.169 = 29.2\ mm}$, from which the outside diameter is $\boxed{D_2 = \sqrt{\frac{4A_T}{\pi} + D_1^2} = 162.2\ mm}$

Problem 10.131 In Problem 10.119, suppose that the block A weighs 30 *lb* and the spring constant $k = 30\ lb/ft$. If the cylinder is rotated at a constant rate, first in the counterclockwise direction and then in the clockwise direction, the difference in the height of the block is two feet. What is the coefficient of friction between the rope and the cylinder?

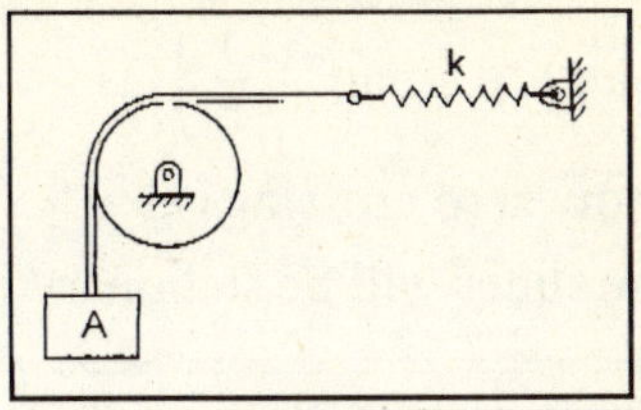

Solution The angle of contact between the rope and the cylinder is $\beta = \frac{\pi}{2}$ *radians*. *For counterclockwise rotation:* The tension in the belt to the left of the cylinder is $T_{left} = W$. The tension in the horizontal part is $H = T_{left}e^{\mu_k\beta} = We^{\mu_k\beta}$. The elongation of the spring is $s_1 = \frac{H}{k} = \frac{We^{\mu_k\beta}}{k}$. *For clockwise rotation:* The tension in the belt to the left is $T_{left} = W$. The tension in the horizontal part is $H = T_{left}e^{-\mu_k\beta} = We^{-\mu_k\beta}$. The elongation of the spring is $s_2 = \frac{H}{k} = \frac{We^{-\mu_k\beta}}{k}$. The difference in elongations is

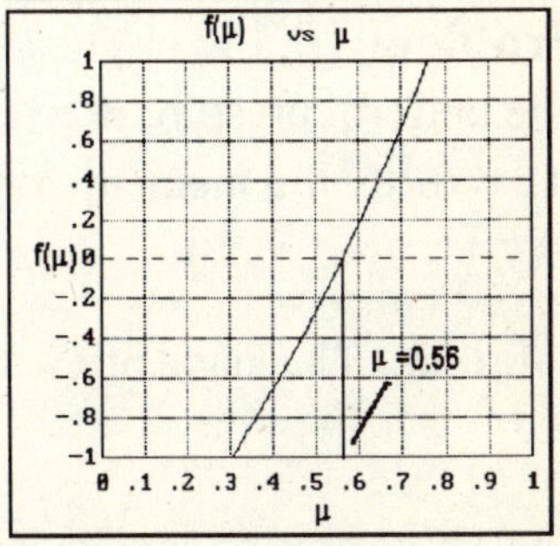

$s_1 - s_2 = \frac{W}{k}\left(e^{\mu_k\beta} - e^{-\mu_k\beta}\right) = 2\ ft$. Graph the function $f(\mu_k) = \frac{W}{k}(e^{\mu_k\beta} - e^{-\mu_k\beta}) - 2$ against μ_k to find the zero crossings. The crossing occurs at $\mu_k = 0.56$, approximately. *Check:* An analytic solution is obtained using the well known relation: $\frac{W}{k}(e^{\mu_k\beta} - e^{-\mu_k\beta}) = \frac{2W}{k}\sinh(\mu_k\beta) = 2\ ft$, from which

$$\mu_k = \frac{\sinh^{-1}\left(\frac{W}{k}\right)}{\beta} = \frac{2\sinh^{-1}(1)}{\pi} = 0.561$$

Problem 10.132 The coefficient of static friction between the 1 *kg* slider and the vertical bar is $\mu_s = 0.6$. The constant of the spring is $k = 20\ N/m$ and its unstretched length is 1 *m*. Determine the range of values of *y* at which the slider will remain stationary on the bar.

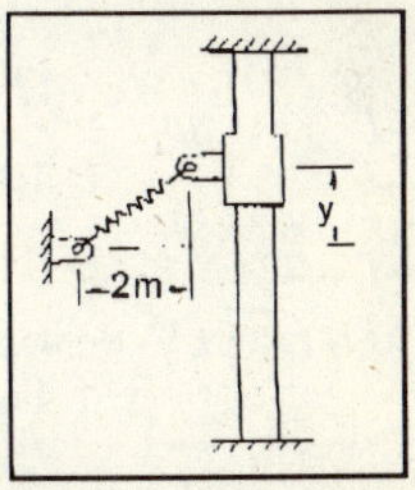

Solution The weight of the slider is $W = 9.81\ N$. The spring length is $L = \sqrt{y^2 + 4}$. The elongation of the spring is $\Delta L = \sqrt{y^2 + 4} - 1$. The force exerted by the spring is $F = k\Delta L$. The angle formed by the spring is $\beta = \tan^{-1}\left(\frac{y}{2}\right)$. The normal force on the slider is $N = F\cos\beta$. The friction force parallel to the bar is $f = \mu_s N$. The friction force opposes slip, so that it is up when β is positive, and down when β is negative. Thus the sum of the forces parallel to the bar is $\sum F_y = -W - F\sin\beta + \mu_s N \operatorname{sgn}(\beta) = 0$, where $\operatorname{sgn}(\beta)$ is the standard signature function (called signum) with unit value and the sign of β. .

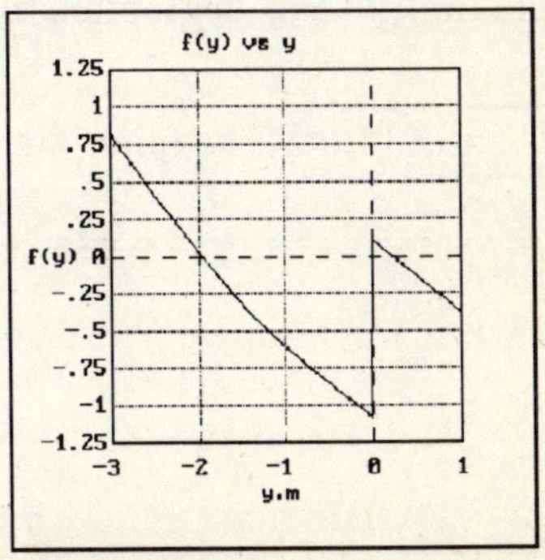

Problem 10.132 continued on next page

continuation of Problem 10.132..........

Substitute and reduce: $f(y) = -\frac{W}{k} - (\sqrt{y^2+4} - 1)(\sin\beta - \mu_s \operatorname{sgn}(\beta)\cos\beta) = 0$. The function $f(y)$, where $\beta = \tan^{-1}\left(\frac{y}{2}\right)$, is graphed as a function of y to determine the zero crossings. Tabulated values show zero crossings at $y = -1.9625 = -1.963\ m$ and $y = 0.225\ m$, hence the range of values for which the slider will be stationary is $\boxed{-1.963 < y < 0.225\ (m)}$

Problem 10.133 The axial force on the thrust bearing is $F = 200\ lb$ and the dimension $b = 6\ in$. The uniform pressure exerted by the mating surface is $p = 7\ psi$ and the coefficient of kinetic friction is $\mu_k = 0.28$. If a moment $M = 360\ in\ lb$ is required to turn the shaft, what are the dimensions D_o and D_i?

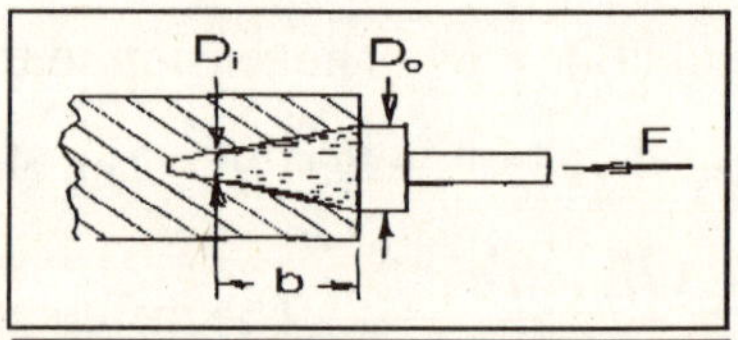

Solution The angle of the taper is $\alpha = 90 - \tan^{-1}\left(\frac{D_o - D_i}{2b}\right)$.

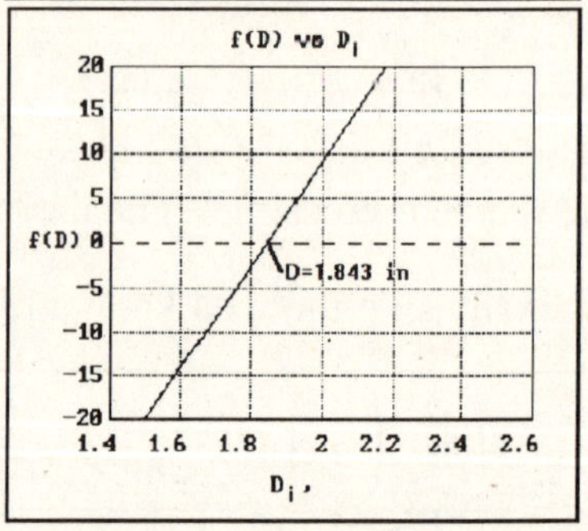

The active area of contact is $A = \frac{\pi(D_o^2 - D_i^2)}{4\cos\alpha}$, the area of a truncated cone. The normal force on the surface is $N = pA$, and the axial force is $F = N\cos\alpha$. The moment required to turn the shaft is $M = \frac{\mu_k F}{3\cos\alpha}\left(\frac{D_o^3 - D_i^3}{D_o^2 - D_i^2}\right)$. Substitute the axial force:

$M = \frac{\pi\mu_k p}{12\cos\alpha}\left(D_o^3 - D_i^3\right)$, from which $D_o^3 - D_i^3 = \frac{12M\cos\alpha}{\pi\mu_k p}$.

Eliminate D_o by the relation: $D_o^2 = \frac{4F}{\pi p} + D_i^2$, to obtain

$f(D_i) = \left(D_i^2 + \frac{4F}{\pi p}\right)^{\frac{3}{2}} - D_i^3 - \frac{12M\cos\alpha}{\pi\mu_k p} = 0$. Substitute values to obtain

$f(D_i) = \left(D_i^2 + 36.378\right)^{\frac{3}{2}} - D_i^3 - 701.58\cos\alpha = 0$, where $\alpha = 90 - \tan^{-1}\left(\frac{\sqrt{D_i^2 + 36.378} - D_i}{12}\right)$. The function $f(D_i)$ is graphed to determine the zero crossings. Tabulated values show a zero crossing at $\boxed{D_i = 1.834\ in}$. The outer diameter is obtained from $\boxed{D_o = \left(\frac{4F}{\pi p} + D_i^2\right)^{\frac{1}{2}} = 6.31\ in}$

Problem 10.134 The masses of crates A and B are 25 *kg* and 30 *kg* respectively. The coefficient of static friction between the contacting surfaces is $\mu_s = 0.34$. What is the largest value of α for which the crates will remain in equilibrium?

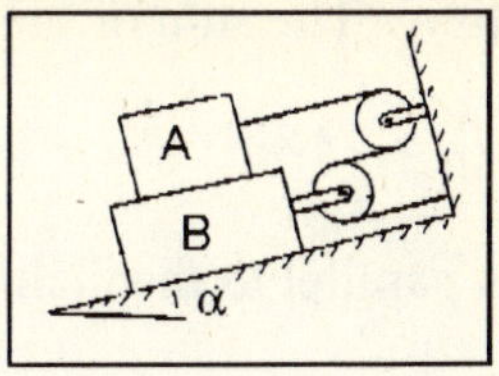

Solution Choose a coordinate system with the x-axis parallel to the inclined surface. Denote the tension in the cables by *T*. Suppose that at impending slip the lower box tends to move up the plane and the upper box tends to move down the plane Thus, for the lower box:

$\sum F_x = -W_B \sin\alpha - \mu_s(W_A + W_B)\cos\alpha - \mu_s W_A \cos\alpha + 2T = 0$. For the upper box, $\sum F_x = +\mu_s W_A \cos\alpha - W_A \sin\alpha + T = 0$. Eliminate *T* from the two equations, and reduce:

$(W_B - 2W_A)\sin\alpha + \mu_s(4W_A + W_B)\cos\alpha = 0$, from which

$$\alpha = \tan^{-1}\left(\frac{\mu_s(4W_A + W_B)}{(2W_A - W_B)}\right) = \tan^{-1}\left(\frac{0.34(1275.3)}{196.2}\right) = 65.65^o$$

Problem 10.135 The side of a soil embankment has a 45° slope. If the coefficient of static friction of soil on soil is $\mu_s = 0.6$, will the embankment be stable or will it collapse? If it will collapse, what is the smallest slope that can be stable?

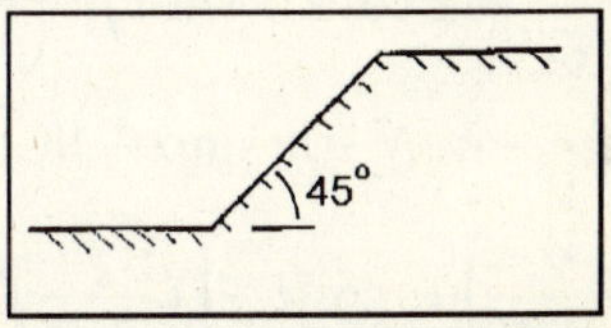

Solution The strategy is to analyze the free body diagram formed by isolating part of the embankment, as shown.

The sum of the force parallel to the slope are:

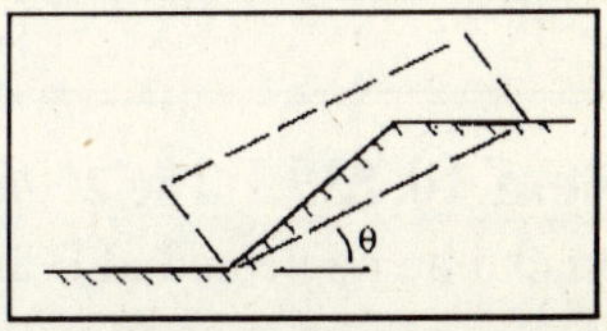

$\sum F_x = -W\sin\theta + \mu_s W\cos\theta = 0$, from which the required value of the coefficient of static friction is: $\mu_s = \tan\theta = \tan 45^o = 1$. Since the coefficient of static friction of soil on soil is less than the required value, the embankment will collapse. The smallest slope that will be stable is $\boxed{\alpha = \tan^{-1}(0.6) = 30.96 = 31^o}$ Note the similarity to the problem of the box on an incline (Problems 10.4 and 10.5).

Problem 10.136 The mass of the van is 2250 *kg* and the coefficient of static friction between its tires and the road is 0.6. If its front wheels are locked and its rear wheels can turn freely, what is the largest value of α for which it can remain in equilibrium?

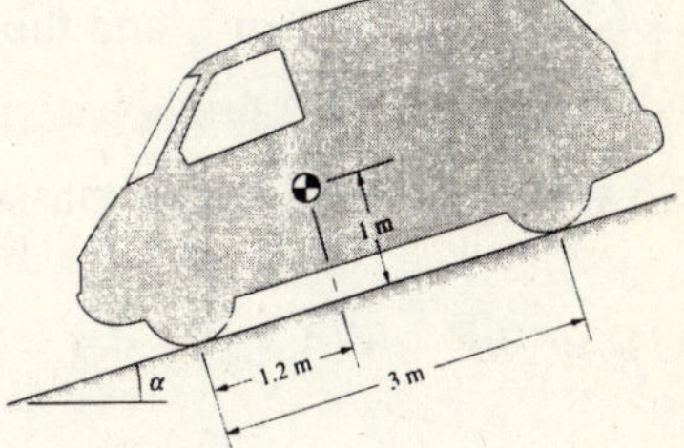

Solution Choose a coordinate system with the x-axis parallel to the incline. The weight of the van is $W = mg = 22072.5\ N$. The moment about the point of contact of the rear wheels is

$\sum M_R = (3 - 1.2)W\cos\alpha + 1W\sin\alpha - 3N = 0$, from which the normal force at the front wheels is $N = \dfrac{W(1.8\cos\alpha + \sin\alpha)}{3}$.

The sum of the forces parallel to the inclined surface is $\sum F_x = +\mu_s N - W\sin\alpha = 0$. Combine and reduce: $\mu_s\left(\frac{1.8}{3}\right)\cos\alpha + \left(\frac{\mu_s}{3} - 1\right)\sin\alpha = 0$, from which $\boxed{a = \tan^{-1}\left(\frac{1.8\mu_s}{3 - \mu_s}\right) = \tan^{-1}(0.45) = 24.2^o}$

Problem 10.137 In Problem 10.136, what is the largest value of α for which the van can remain in equilibrium if it points up the slope?

Solution The sum of the moments about the point of contact of the rear wheels is

$\sum M_R = -1.8W\cos\alpha + 1W\sin\alpha + 3N = 0$. The normal force is $N = \dfrac{W(1.8\cos\alpha - \sin\alpha)}{3}$. The sum of forces parallel to the incline is $\sum F_x = +\mu_s N - W\sin\alpha = 0$. Combine and reduce:

$$\frac{1.8\mu_s}{3}\cos\alpha - \left(\frac{\mu_s}{3} + 1\right)\sin\alpha = 0, \text{ from which } \boxed{\alpha = \tan^{-1}\left(\frac{1.8\mu_s}{\mu_s + 3}\right) = 16.7^o}$$

Problem 10.138 In Problem 10.136, suppose that a 300 *kg* crate of castings is loaded onto the van. The center of mass of the castings is 0.5 *m* in front of the rear axle and 1.5 *m* above the street. What is the largest value of α for which the van can remain in equilibrium?

Solution The weight of the crate is $W_C = 300g = 2943\ N$. The sum of moments about the rear wheels is $\sum M_R = +1.8W\cos\alpha + 1W\sin\alpha + 0.5W_C\cos\alpha + 1.5W_C\sin\alpha - 3N = 0$, from which

$N = \left(\dfrac{W}{3}\right)(1.8\cos\alpha + \sin\alpha) + \left(\dfrac{W_c}{3}\right)(0.5\cos\alpha + 1.5\sin\alpha)$. The sum of forces parallel to the road is

$\sum F_x = +\mu_s N - W\sin\alpha - W_C\sin\alpha = 0$. Combine and reduce

$$\left(0.6 + \frac{W_C}{6W}\right)\mu_s\cos\alpha + \left((\frac{\mu_s}{3} - 1) + \frac{W_C}{W}(\frac{\mu_s}{2} - 1)\right)\sin\alpha = 0, \text{ from which } \boxed{\alpha = \tan^{-1}\left(\frac{0.3733}{0.8933}\right) = 22.7^o}$$

Problem 10.139 The 20 *lb* homogenous object is supported at A and B. The distance $h = 4\ in$, friction can be neglected at B and the coefficient of static friction at A is 0.4. Determine the largest force F that can be exerted without causing the object to slip.

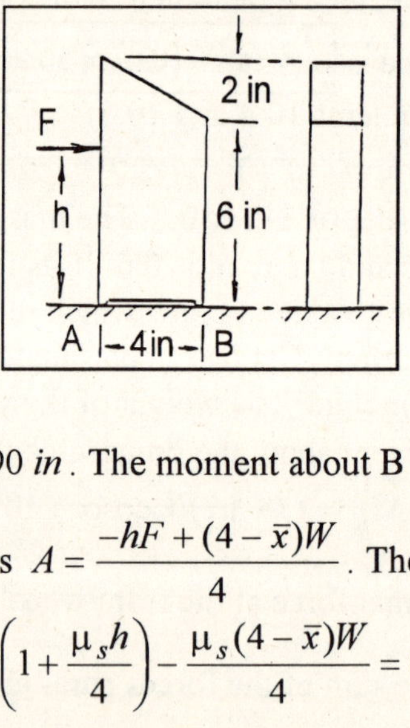

Solution Choose a coordinate system with origin at A and the x-axis parallel to the floor. Divide the object into a "rectangular" volume and a "triangular" (wedge) volume. The volume of the lower rectangular portion is $V_1 = 4(6)(1) = 24\ in^3$. The centroidal coordinates are $\bar{x}_1 = 2\ in, \bar{y}_1 = 3\ in$. The wedge has a volume $V_2 = \left(\dfrac{1}{2}\right)(4)(2)(1) = 4\ in^3$, and the centroid is at $\bar{x}_2 = (4/3)\ in$, $\bar{y}_2 = 6 + \dfrac{2}{3}\ in$. The center of mass is located at $\bar{x} = \dfrac{2(24) + (4/3)4}{28} = 1.90\ in$. The moment about B is $\sum M_B = -hF + W(4 - \bar{x}) - 4A = 0$, from which the normal force at A is $A = \dfrac{-hF + (4 - \bar{x})W}{4}$. The sum of the forces parallel to x is $\sum F_x = -\mu_s A + F = 0$, from which $F\left(1 + \dfrac{\mu_s h}{4}\right) - \dfrac{\mu_s(4 - \bar{x})W}{4} = 0$,

and $\boxed{F = \dfrac{\mu_s(4 - \bar{x})W}{(4 + \mu_s h)} = 2.993 = 2.99\ lb}$

Problem 10.140 In Problem 10.139, suppose that the coefficient of static friction at B is 0.36. What is the largest value of *h* for which the object will slip before it tips over?

Solution Use the solution to Problem 10.139, as applicable. Tipping is imminent when the normal force at A becomes zero. From the solution to Problem 10.139, $A = \dfrac{-hF + (4 - \bar{x})W}{4}$, from which $h_{tip} = \dfrac{(4 - \bar{x})W}{F}$. The sum of the forces parallel to the *y* axis is $\sum F_y = A + B - W = 0$, from which the normal force at $B = W - A$. The sum of the forces parallel to the x-axis $\sum F_x = -\mu_{sA}A - \mu_{sB}B + F = 0$, from which, for $A = 0$, $F = \mu_{sB}W$. Combine and reduce to obtain

$$\boxed{h_{tip} = \frac{(4 - \bar{x})}{\mu_{sB}} = 5.82\ in}$$

Problem 10.141 The 180 *lb* climber is supported in the "chimney" by the normal and friction forces exerted on his shoes and back. The static coefficients of friction between his shoes and the wall and between his back and the wall are 0.8 and 0.6, respectively. What is the minimum normal force his shoes must exert? Assume that slip impends at both surfaces.

Solution Choose a coordinate system with the x-axis horizontal and y-axis vertical. Let N_s be the normal force exerted by shoes, N_b the normal force exerted by his back. The sum of the forces:

$$\sum F_x = N_s \cos 4^o - N_b \cos 3^o - \mu_{ss} N_s \sin 4^o + \mu_{sb} N_b \sin 3^o = 0,$$

$$\sum F_y = N_s \sin 4^o + N_b \sin 3^o + \mu_{ss} N_s \cos 4^o + \mu_{sb} N_b \cos^o - W = 0.$$

Reduce to two simultaneous equations in two unknowns: $a_{11}N_s + a_{12}N_b = 0$, and $a_{21}N_s + a_{22}N_b = W$, where

$a_{11} = \cos 4^o - \mu_{ss} \sin 4^o = 0.9418$,

$a_{12} = -\cos 3^o + \mu_{sb} \sin 3^o = -0.9672$, $a_{21} = \sin 4^o + \mu_{ss} \cos 4^o = 0.8678$, and

$a_{22} = \sin 3^o + \mu_{sb} \cos 3^o = 0.6515$. The equations have the solutions $N_s = -\dfrac{a_{12}W}{\det}$, and $N_b = \dfrac{a_{11}W}{\det}$,

where $\det = a_{11}a_{22} - a_{12}a_{21} = 1.4529$ is the determinant of the coefficients. The result: $\boxed{N_s = 119.83\ lb}$

Problem 10.142 The sides of the 200 *lb* door fit loosely into grooves in the walls. Cables at A and B raise the door at a constant rate. The coefficient of kinetic friction between the door and the grooves is $\mu_k = 0.3$. What force must the cable A exert to continue raising the door at a constant rate if cable B breaks?

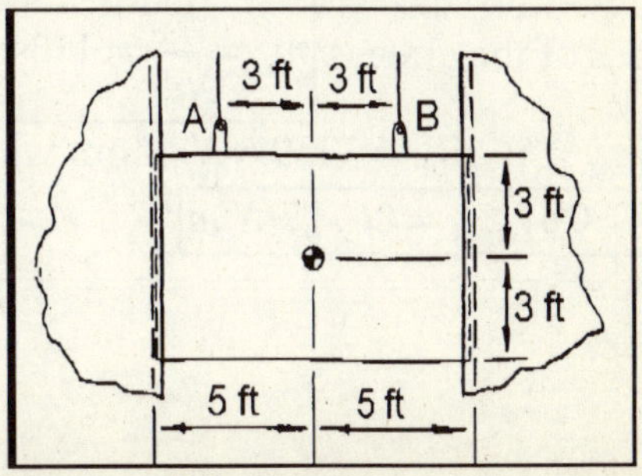

Solution Since the door fits loosely in the grooves, assume that the moment due to the unbalance when cable B breaks causes the door to contact the upper right and lower left corners. Thus the normal force on the sides occur at these corners. The friction forces at the corners oppose movement. The sum of the moments about the lower left corner is

$$\sum M_L = 2A - 5W + 6N_R - 10\mu_k N_R = 0.$$

Problem 10.142 continued on next page

continuation of Problem 10.142..........

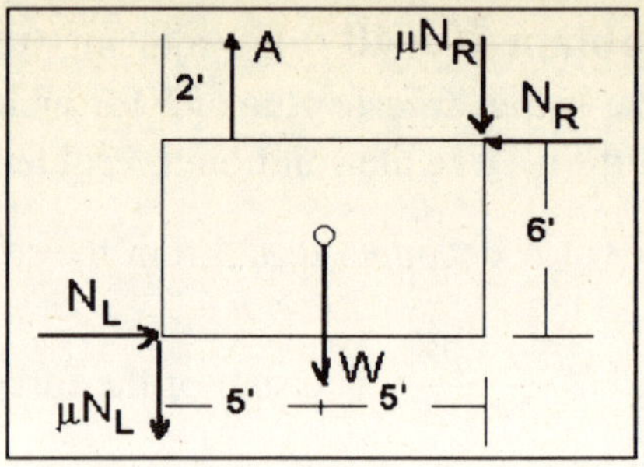

The sum of the forces: $\sum F_x = N_L - N_R = 0$, from which $N_L = N_R$, and $\sum F_y = A - W - \mu_k N_R - \mu_k N_L = 0$.

Combine to obtain the two simultaneous equations: $2A + (6 - 10\mu_k)N_R = 5W$, and $A - 2\mu_k N_R = W$. These have the solution: $A = -\frac{(2\mu_k)(5W)}{\det} - \frac{(6-10\mu_k)W}{\det}$, where $\det = 6(\mu_k - 1)$ is the determinant of the coefficients. Reducing:

$$\boxed{A = \frac{(-10\mu_k - (6 - 10\mu_k))W}{-6(1-\mu_k)} = \frac{W}{1-\mu_k} = 285.71\ lb}$$

Problem 10.143 The coefficient of static friction between the tires of the 1000 *kg* tractor and the ground and between the 450 *kg* crate and the ground are 0.8 and 0.3, respectively. Starting from rest, what torque must the tractor's engine exert on the rear wheels to cause the crate to move? (The front wheels can turn freely.)

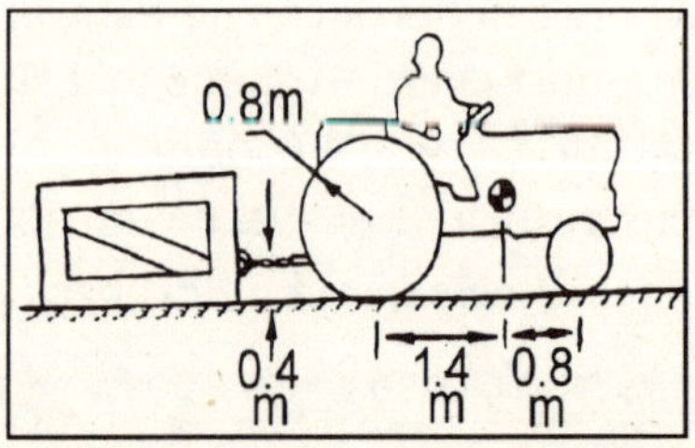

Solution The weight of the crate is $W = mg = 4414.5\ N$. The force required to produce imminent slip of the crate on level ground is $F_c = \mu_s W = 0.3(4414.5) = 1324.5\ N$. This is the friction force exerted by the ground on the tires, $\sum F_x = -F_c + f_{tires} = 0$.

The friction force is related to the torque about the axle (both wheels) by $\sum M_{axle} = T - 0.8 f_{tires} = 0$, from which $\boxed{T = 0.8F_c = 1059.5\ N\ m}$

Problem 10.144 In Problem 10.143, what is the most massive crate the tractor can cause to move from rest if its engine can exert sufficient torque? What torque is necessary?

Solution The weight of the tractor is $W_t = m_t g = 9810\ N$. The sum of the moments about the front wheels is $\sum M_F = +0.8W_t + 0.4\mu_{sc}W_c - 2.2N = 0$, where N is the normal force on the rear wheels. The sum of the forces at imminent tire slip is $\sum F_x = -\mu_{sc}W_c + \mu_{st}N = 0$, from which $N = \left(\frac{\mu_{sc}}{\mu_{st}}\right)W_c$.

Substitute into the first equation and reduce: $W_c = \frac{0.8\mu_{st}W_t}{(2.2\mu_{sc} - 0.4\mu_{sc}\mu_{st})} = 11131.9\ N$, from which the mass of the crate is $\boxed{m_c = \frac{W_c}{g} = 1134.75 = 1134.8\ kg}$. The friction force on the tires is $f_{tires} = \mu_{sc}W_c = 3339.6\ N$, from which the torque on the axle (both wheels) is $\boxed{T = 0.8 f_{tires} = 2671.7\ N\ m}$

==============================<>==============================

Problem 10.145 The mass of the vehicle is 900 *kg*, it has rear wheel drive, and the coefficient of static friction between its tires and the surface is 0.65. The coefficient of static friction between the crate and the surface is 0.4. If the vehicle attempts to pull the crate up the incline, what is the largest value of the mass of the crate for which it will slip up the incline before the vehicle's tires slip?

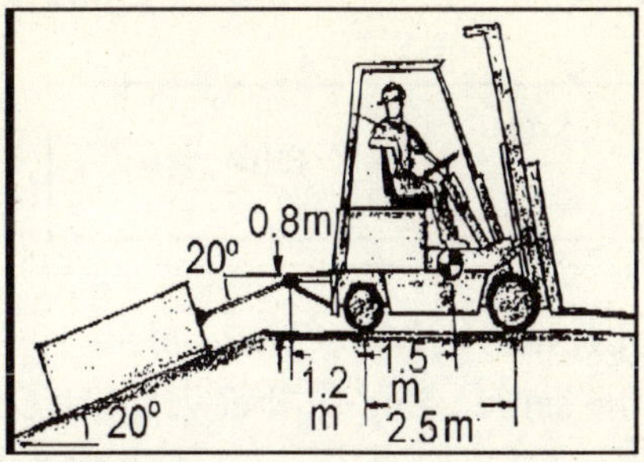

Solution The normal force between the crate and the incline is $N_c = W_c \cos\theta$, where $\theta = 20^o$. The drawbar force parallel to the incline is $F_d = -W_c \sin\theta - \mu_{sc} N_c$. For brevity write $\psi = \sin\theta + \mu_{sc}\cos\theta$. The horizontal component of the drawbar force at the tractor is $F_{dh} = F_d \cos\theta = -W_c\psi\cos\theta$. The vertical component of the drawbar force at the tractor is $F_{dv} = F_d \sin\theta = -W_c\psi\sin\theta$. The weight of the tractor is $W_t = 900g = 8829\ N$. The sum of the moments about the front wheels is $\sum M_F = 1W_t - 0.8F_{dh} - 3.7F_{dv} - 2.5N = 0$, from which $0 = W_t + (0.8\cos\theta + 3.7\sin\theta)W_c\psi - 2.5N = 0$.

The sum of the force parallel to the ground is $\sum F_x = F_{dh} + \mu_{st}N = 0$, from which $N = \dfrac{W_c\psi\cos\theta}{\mu_{st}}$.

Substitute and reduce: $W_c = \dfrac{\mu_{st}W_t}{((2.5 - 0.8\mu_{st})\cos\theta - 3.7\mu_{st}\sin\theta)(\sin\theta + \mu_{sc}\cos\theta)} = 7701.1\ N$. The mass of the crate is $\boxed{m_c = \dfrac{W_c}{g} = 785\ kg}$

==============================<>==============================

Problem 10.146 The collars A and B each have a mass of 2 *kg*. If friction between collar B and the bar is neglected, what minimum coefficient of static friction between collar A and the bars for the collars to remain in equilibrium in the postion shown?

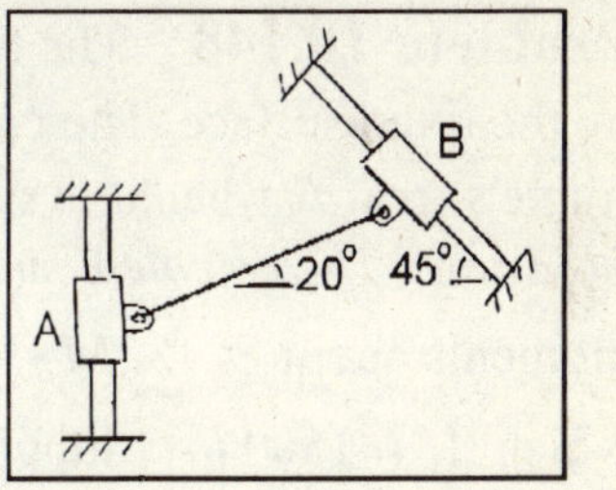

Solution The weight of each collar is $W = mg = 19.62\ N$. Denote $\theta = 45^o$, $\alpha = 20^o$. The unit vector parallel to the bar holding B is $\vec{e}_B = -\vec{i}\sin\theta + \vec{j}\cos\theta$. The weight of B is $\vec{W} = -\vec{j}|\vec{W}|$. The components of the weight of B parallel to the bar is $W_{PB} = (\vec{e}_B \cdot \vec{W})\vec{e}_B = -|\vec{W}|\cos\theta\ \vec{e}_B$. This force must be balanced by a component of the force in the connecting wire for B to remain stationary. The components of the tension in the wire are $\vec{T} = |\vec{T}|(\vec{i}\cos(180^o + \alpha) + \vec{j}\sin(180^o + \alpha)) = |\vec{T}|(-\vec{i}\cos\alpha - \vec{j}\sin\alpha)$, from which the component of $\vec{T}$ parallel to the bar supporting B is $\vec{T}_{PB} = (\vec{e}_B \cdot \vec{T})\vec{e}_B = |\vec{T}|\sin(\theta - \alpha)\ \vec{e}_B$. The sum of forces along $\vec{e}_B$: $|\vec{T}|\sin(\theta - \alpha)\sin\theta - |\vec{W}|\cos\theta\sin\theta = 0$, from which: $\dfrac{|\vec{W}|}{|\vec{T}|} = \dfrac{\sin(\theta - \alpha)}{\cos\theta}$. The unit vector parallel to the bar supporting A is $\vec{e}_A = \vec{j}$. The component of $\vec{T}$ parallel to the bar supporting A is $|\vec{T}_{PA}| = -\vec{j} \cdot \vec{T} = |\vec{T}|\sin\alpha$, and the force exerted by $\vec{T}$ on the slider A perpendicular to the bar is $|\vec{N}_A| = -\vec{i} \cdot |\vec{T}| = |\vec{T}|\cos\alpha$, where the negative sign is used because the tension at A is in opposition to the tension at B (tension is reversed).

Problem 10.146 continued on next page

continuation of Problem 10.146..........

The sum of forces parallel to the bar is $\sum F_A = +\mu_s|\vec{N}_A| - |\vec{W}| + |\vec{T}_{PA}| = 0$, from which

$\mu_s|\vec{T}|\cos\alpha - |\vec{W}| + |\vec{T}|\sin\alpha = 0$, and $\mu_s = \frac{|\vec{W}|}{|\vec{T}|}\left(\frac{1}{\cos\alpha}\right) - \tan\alpha$. Substitute and reduce:

$$\boxed{\mu_s = \left(\frac{\sin(\theta - \alpha)}{\cos\alpha\cos\theta}\right) - \tan\alpha = 0.272}$$

Problem 10.147 In Problem 10.146, if the coefficient of friction has the same value μ_s between the collars A and B and the bars, what minimum value of μ_s is necessary for the collars to remain in equilibrium in the position shown? Assume slip impends at A and B.

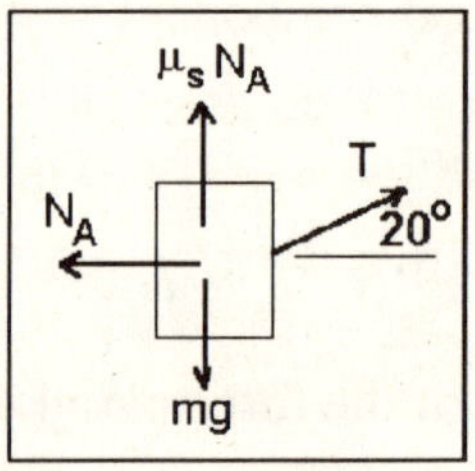

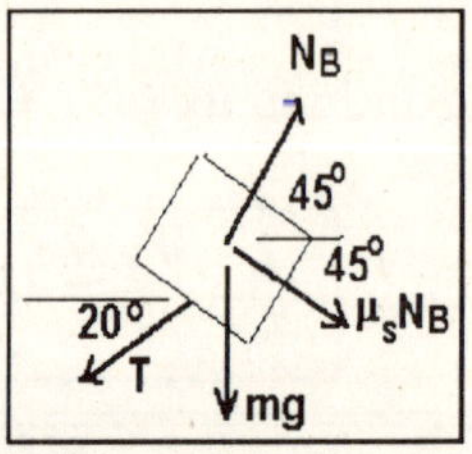

Solution The weight of each collar is $W = mg = 19.62\ N$. Denote $\theta = 45^o, \alpha = 20^o$. *Isolate Collar A.* The sum of forces:

$\sum F_x = T\cos\alpha - N_A = 0$, $\sum F_y = T\sin\alpha + \mu_s N_A - mg = 0$. *Isolate Collar B:* The sum of forces:

$\sum F_x = N_B\cos\beta + \mu_s N_B\cos\beta - T\cos\alpha = 0$.

$\sum F_y = N_B\sin\beta - \mu_s N_B\sin\beta - T\sin\alpha = 0$. These are four equations in four unknowns. Solve: $T = 45.4\ N$, $N_A = 42.6\ N$, $N_B = 55\ N$, and $\boxed{\mu_s = 0.0963}$. This is the minimum coefficient of friction required to maintain equlibrium.

Problem 10.148 The frame is supported by the normal and friction forces exerted on the plate at A by the fixed surface. The coefficient of static friction between the plate and wall is $\mu_s = 0.5$. Will the frame slip at A when it is subjected to the 600 *lb* load?

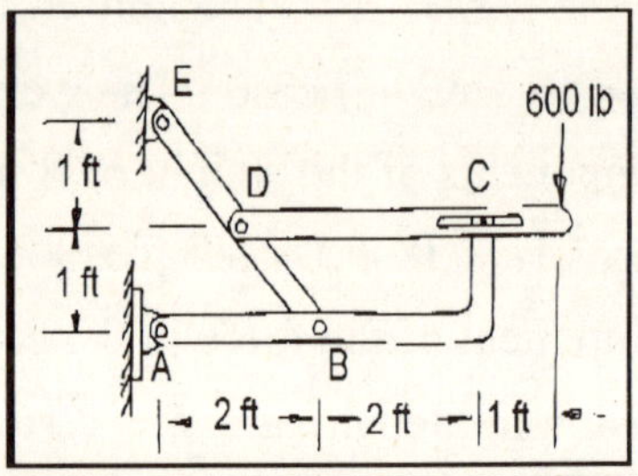

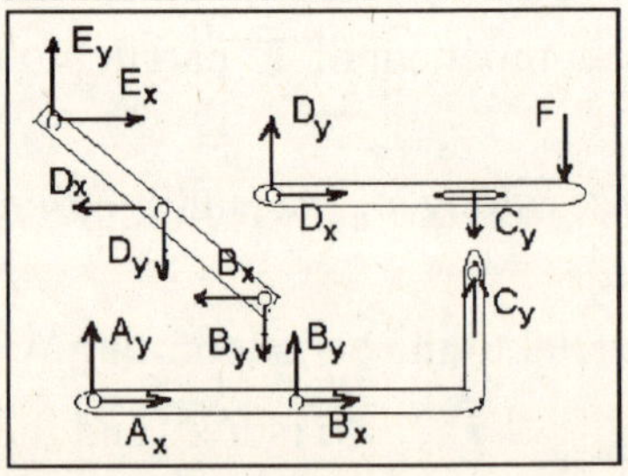

Solution *The complete structure as a free body:* The sum of the moments about E: $\sum M_E = -5(600) + 2A_x = 0$, from which $A_x = 1500\ lb$. Suppose that the structure does not slip at A. Then the friction force at A must be less than or equal to $f = \mu_s A_y = 0.5(1500) = 750\ lb$. The sum of the forces:

$\sum F_x = E_x + A_x = 0$, from which $E_x = -1500\ lb$. (1)

$\sum F_y = E_y + A_y - 600 = 0$.

Element DC: The sum of the moments about D:

$\sum M_D = -3C_y - 4(600) = 0$, from which $C_y = -800\ lb$.

The sum of the forces: $\sum F_y = -C_y - 600 + D_y = 0$, from which $D_y = -200\ lb$. $\sum F_x = D_x = 0$.

Problem 10.148 continued on next page

=========================<>=========================

continuation of Problem 10.148..........

Element EDB: The sum of the moments about E: (1) $\sum M_E = -1D_y - 2B_y - 2B_x = 0$, The sum of forces: (2) $\sum F_y = -B_y - D_y + E_y = 0$, , (3) $\sum F_x = -B_x + E_x = 0$, from which $B_x = -1500\ lb$, and equations (1), (2) and (3) above are now solvable, $B_y = 1600\ lb$, $A_y = -800\ lb$, and $E_y = 1400\ lb$. The vertical reaction at A exceeds the limit on the friction force reaction, <u>hence the pad at A will slip.</u>

The minimum coefficient of friction required maintain equilibrium is $\mu_s = \frac{A_y}{A_x} = \frac{800}{1500} = 0.533$

=========================<>=========================

Problem 10.149 In Problem 10.53 the suspended weight $W = 600\ lb$. The bars AB and AC have ball and socket supports at each end. Suppose that you want the ball and socket at C to be held in place by the normal and friction forces between the support and the wall. What minimum coefficient of friction is required?

Solution The strategy is to develop the unit vectors parallel to the bars and the rope, project the forces onto these unit vectors, and solve the associated equilibrium equations. The vector positions are $\vec{r}_A = 3\vec{i} + 2\vec{j} + 4\vec{k}$, $\vec{r}_B = 0\vec{i} + 0\vec{j} + 3\vec{k}$, $\vec{r}_C = 4\vec{i} + 0\vec{j} + 0\vec{k}$, and $\vec{r}_O = 0\vec{i} + 2\vec{j} + 0\vec{k}$. The unit vectors are defined: $\vec{e}_{AB} = \frac{\vec{r}_B - \vec{r}_A}{|\vec{r}_B - \vec{r}_A|}$. Apply this definition to obtain: $\vec{e}_{AB} = -0.8018\vec{i} - 0.5345\vec{j} - 0.2673\vec{k}$,

$\vec{e}_{AC} = 0.2182\vec{i} - 0.4364\vec{j} - 0.8729\vec{k}$, $\vec{e}_{AO} = -0.6\vec{i} + 0\vec{j} - 0.8\vec{k}$. The weight vector is $\vec{W} = -600\vec{j}$. The equilibrium conditions:

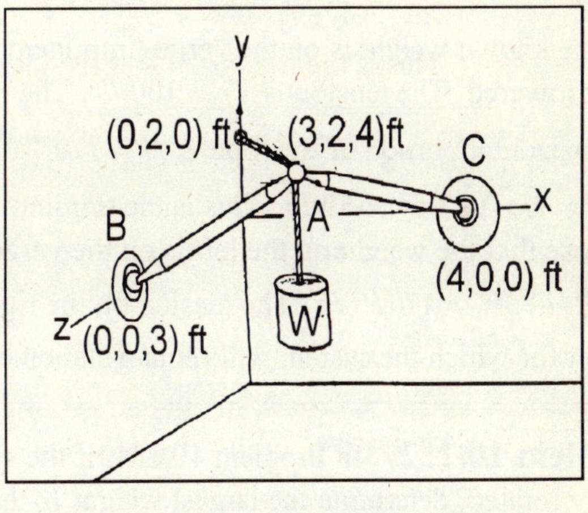

$\sum F_x = -0.2182|\vec{F}_{AC}| - 0.8018|\vec{F}_{AB}| - 0.6|\vec{F}_{AO}| = 0$,

$\sum F_y = -0.4364|\vec{F}_{AC}| - 0.5345|\vec{F}_{AB}| + 0|\vec{F}_{AO}| - |\vec{W}| = 0$

$\sum F_z = -0.8729|\vec{F}_{AC}| - 0.2673|\vec{F}_{AB}| - 0.8|\vec{F}_{AO}| = 0$

Solve: $|\vec{F}_{AC}| = -494.92\ lb$, $|\vec{F}_{AB}| = -718.4\ lb$, and $|\vec{F}_{AO}| = 780\ lb$. The projection of $|\vec{F}_{AC}|$ onto the unit vector is the force exerted on the wall at C: $\vec{F}_{AC} = -108.92\vec{i} + 216\vec{j} + 432\vec{k}$. The component normal to the wall is the normal force, $|\vec{N}_C| = 432\ lb$. The friction force is

$|\vec{f}_C| = \sqrt{108.92^2 + 216^2} = 242\ lb$. The coefficient of friction is $\mu_s = \frac{|\vec{f}_C|}{|\vec{N}_C|} = 0.559$ This is the minimum coefficient of static friction required to prevent slip.

=========================<>=========================

Problem 10.150 The axles of the tram are supported by journal bearings. The radius of the wheels is 75 *mm*, the radius of the axles is 15 *mm*, and the coefficient of kinetic friction between the axles and the bearings is $\mu_k = 0.14$. The mass of the tram and its load is 160 *kg*. If the weight of the tram and its load is evenly divided between the axles, what force *P* is necessary to push the tram at a constant speed?

Solution Assume that there are two bearings per axle. The weight of the tram is $W = mg = 1569.6\ N$. This load is divided between four bearings: $F = \frac{W}{4} = 392.4\ N$. The angle of kinetic friction is $\alpha = \tan^{-1}(\mu_k) = 7.97^o$. The moment required to turn each bearing at a constant rate is $M = Fr\sin\theta_k = 0.8161\ N\,m$, and the force per wheel is $P_w = \frac{M}{R} = \frac{0.8161}{0.075} = 10.88\ N$. The total force required to push the tram is $\boxed{P = 4P_w = 43.5\ N}$

Problem 10.151 If the two cylinders are held fixed, what is the range of *W* for which the weights will remain stationary?

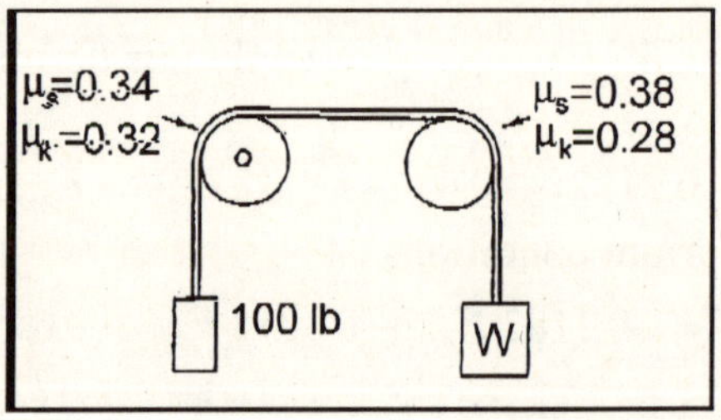

Solution Denote the tension in the horizontal part of the rope by H. The angle of contact with each cylinder is $\beta = \frac{\pi}{2}$. Begin on the left with the known weight. Suppose that the known weight is on the verge (imminent slip) of being lowered. The tension is $T_L = 100\ lb$. The tension in the hozizontal portion of the rope is $H = T_L e^{-0.34\beta} = 58.62\ lb$. The tension in the right part of the rope is $T_R = He^{-0.30\beta} = 36.59\ lb$. This is the minimum weight for which the system will remain stationary. Suppose that the weight on the left is on the verge of being raised. The horizontal tension is $H = T_L e^{0.34\beta} = 170.59\ lb$. The tension on the right is $T_R = He^{0.30\beta} = 273.3\ lb$. This is the maximum weight for which the system will remain stationary. Thus $\boxed{36.59 \le W \le 273.3\ \ (lb)}$

Problem 10.152 In Problem 10.151, if the system is initially stationary and the left cylinder is slowly rotated, determine the largest weight *W* that can be (a) raised; (b) lowered.

Solution Assume that the rope does not slip on the slowly rotating sylinder, but is always at the point of imminent slip. The tension in the horizontal part as the cylinder is rotated is $H = T_L e^{0.34\beta} = 170.6\ lb$, where the static coefficient of friction is used, since the rope does not slip. The tension in the right portion of the rope is $T_R = He^{-0.28\beta} = 109.88 = 110\ lb$, where the kinetic coefficient of friction is used, since the rope slips at a steady rate. Thus (a) $\boxed{W = 110\ lb}$. (b) If the weight W is slowly being lowered, the tension in the horizontal portion is also $H = T_L e^{0.34\beta} = 170.6\ lb$, and the tension in the right portion of the rope is $T_R = He^{0.28\beta} = 264.8\ lb$, from which $\boxed{W = 264.8\ lb}$

Problem 10.153 The chinning bar is secured on the door frame by turning the threaded collar C. The pitch of the threaded shaft is $p = 0.1\ in$ and the mean radius of the thread is $r = 0.5\ in$. The coefficients of friction between the thread and the mating groove are $\mu_s = 0.30$ and $\mu_k = 0.24$. The coefficient of static friction between the door frame and the ends of the bar is 0.8. What moment must be applied to the collar C for the bar to be able to support a 400 *lb* vertical load without slipping in the door frame?

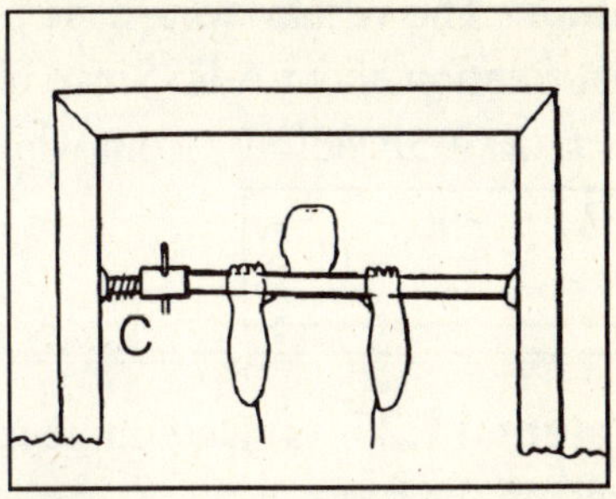

Solution The friction force required to support the bar on each side is 200 *lb*, from which $N = \frac{200}{\mu_s} = \frac{200}{0.80} = 250\ lb$. The angle of pitch is $\alpha = \tan^{-1}\left(\frac{0.1}{(2\pi)0.5}\right) = 1.823^o$. The static friction angle is $\theta_s = \tan^{-1}(0.3) = 16.7^o$. The moment required to turn the threaded shaft is $M = rN\tan(\theta_s + \alpha) = \left(\frac{0.5}{12}\right)(250)\tan 18.52^o = 3.49\ ft\ lb$

Problem 10.154 In Problem 10.153, what moment must be applied to the threaded collar to loosen it?

Solution Use the solution to Problem 10.153. The moment required to loosen the the collar is

$$M = rN\tan(\theta_s - \alpha) = \left(\frac{0.5}{12}\right)(250)\tan 14.88^o = 2.77\ ft\ lb$$

The following problems (Problem 11.1 through Problem 11.26) are to be solved using the principle of virtual work.

Problem 11.1 Determine the reaction at B.

Strategy: Subject the beam to a rotation about A.

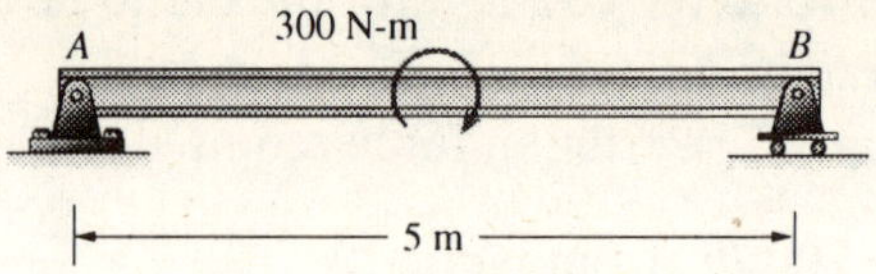

Solution The virtual work done by the moment in a virtual rotation about A is: Since $\delta x = 5\delta\theta$, then $\delta U = (-M + 5B)\delta\theta = 0$, from which

$$\boxed{B = \frac{M}{5} = \frac{300}{5} = 60\ N}$$

Problem 11.2 (a) Determine the moment exerted on the beam at A. (b) Determine the vertical force exerted on the beam at A.

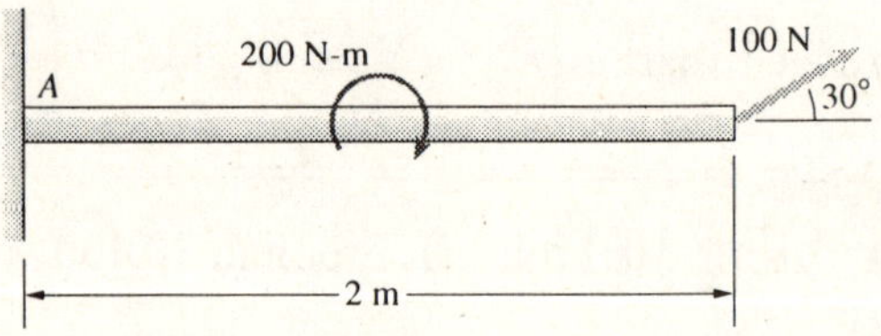

Solution (a) Perform a virtual rotation about A:

$\delta U = M_A\delta\theta - 200\delta\theta + 2(100)\sin 30^o\delta\theta = 0$, from which $(M_A - 200 + 2(100)\sin 30^o)\delta\theta = 0$, from which $\boxed{M_A = 200 - 2(50) = 100\ N\,m}$. (b) Perform a virtual translation of the bar in the y-direction:

$\delta U = A_y\delta y + 100(\sin 30^o)\delta y = 0$, from which $(A_y + 50)\delta y = 0$, or $\boxed{A_y = -50\ N}$

Problem 11.3 Determine the tension in the cable.

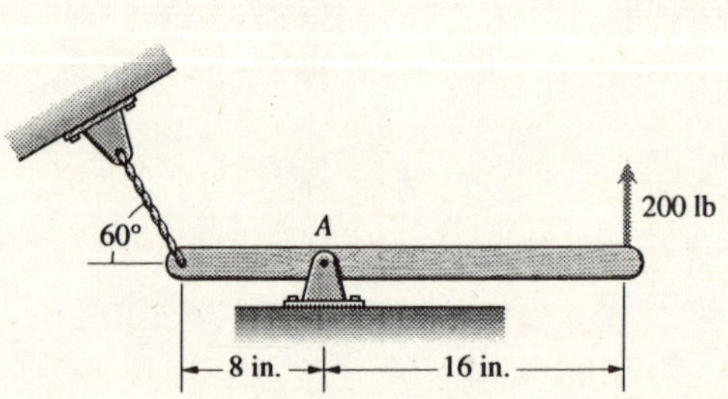

Solution Perform a virtual rotation about A.

$\delta U = -T\sin 60^o(8)\delta\theta + 16(200)\delta\theta = 0$, from which

$(-T(8)\sin 60^o + 16(200))\delta\theta = 0$,

$$\boxed{T = \frac{2(200)}{\sin 60^o} = 461.9\ lb}$$

Problem 11.4 The L-shaped bar is in equilibrium. Determine F.

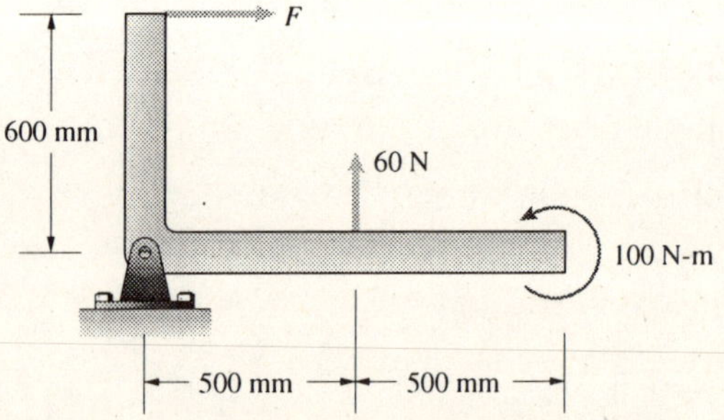

Solution Perform a virtual rotation about the pinned support: $\delta U = (100\delta\theta + 0.5(60)\delta\theta - 0.6F\delta\theta) = 0$, from which $(100 + 60(0.5) - 0.6F)\delta\theta = 0$, or

$$\boxed{F = \frac{100 + 0.5(60)}{0.6} = 216.7\ N}$$

Problem 11.5 If you were assigned to calculate the force exerted on the bolt by the pliers when the grips are subjected to forces F as shown in Fig. (a), you could carefully measure the dimensions, draw free body diagrams, and use the equilibrium equations. Another approach would be to measure the change in the distance between the jaws when the distance between the handles is changed by a small amount. If your measurements indicate that the distance d in. Fig. (b) decreases by 1 mm when D is decreased 8 mm, what is the approximate value of the force exerted on the bolt by each jaw when the forces F are applied? Why is the result approximate?

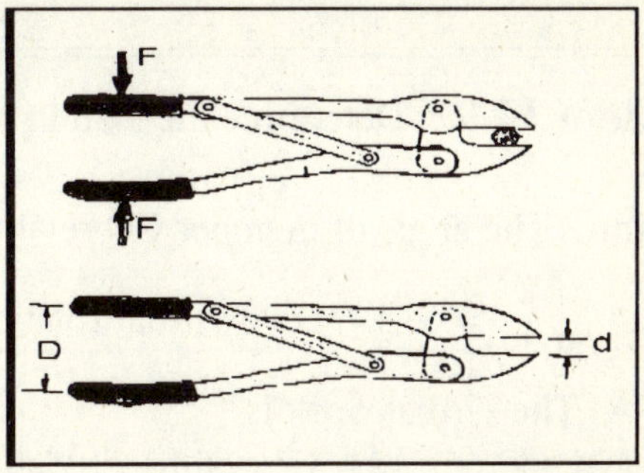

Solution Let L be the distance between the points of application of the forces F and the point of application of the gripping force at the jaw. The ratio of the motions indicates that the "effective" axis of rotation of a jaw is located $\frac{8}{9}L$ from the point of application of F. Perform a virtual rotation about this axis:

$\delta U = \left(\frac{8}{9}\right)LF\delta\theta - \left(\frac{1}{9}\right)Lf\delta\theta = 0$, from which

$(8F - f)\delta\theta = 0$, or $\boxed{f = 8F}$. This result is approximate because some work is done by the mechanism as the handle is closed. In addition, the closure of the handles produces a translation of one handle relative to the other in the direction required to close the jaws, and this translation does work since it is associated with moment about the effective axis; hence not all of the virtual work due to a virtual rotation is included in the above expression.

Problem 11.6 The mechanism is in equilibrium. Determine the force R in terms of F. All elements of of equal length L

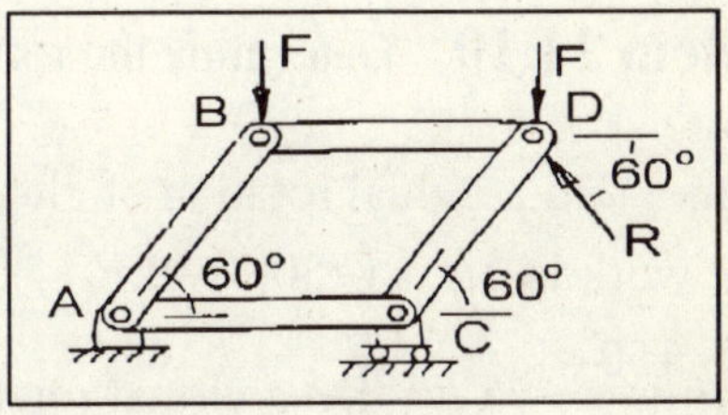

Solution For brevity, define $\alpha = 60^o$. Suppose that a virtual rotation of elements AB and CD occur: the virtual work is

$\delta U = -L\cos\alpha F\delta\theta - (L + L\cos\alpha)F\delta\theta +$

$R(L + L\cos\alpha)\sin\alpha\delta\theta + RL\sin\alpha\cos\alpha\delta\theta = 0$, from which $(-(1+2\cos\alpha)F + R\sin\alpha(1+2\cos\alpha))\delta\theta = 0$,

or $\boxed{R = \frac{F}{\sin\alpha} = 1.155F}$

Problem 11.7 Determine the axial load in member DE. All elements are of the same length L, except for $AE = 2L$.

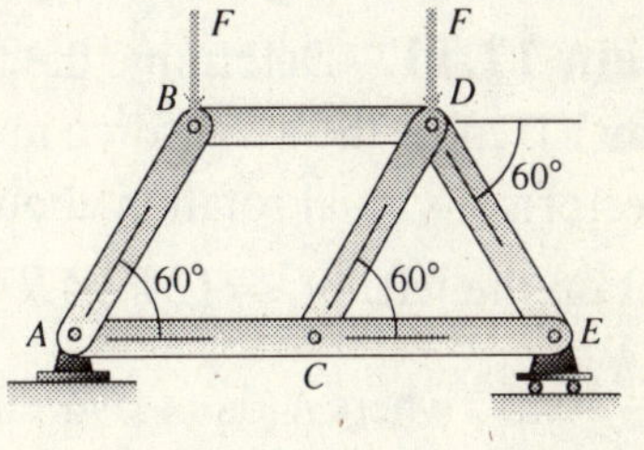

Solution For brevity, $\alpha = 60^o$. Denote the reaction in DE by R_{DE}. Perform a virtual rotation of elements AB and CD. The axial components in AB and CD do not contribute to the virtual work. The virtual work is

$\delta U = -L\cos\alpha F\delta\theta - (L + L\cos\alpha)F\delta\theta + R_{DE}(L + L\cos\alpha)\sin\alpha\delta\theta + R_{DE}L\sin\alpha\cos\alpha\delta\theta = 0$, from

which $\boxed{R_{DE} = \frac{F}{\sin\alpha} = 1.155F\ (C)}$

Problem 11.8 Determine the axial load in member BD of the structure in Problem 11.7

Solution Use the solution to Problem 11.7. All members are of equal length L, and for brevity denote $\alpha = 60^o$. Denote the reactions in BD by R_{BD}. Perform a virtual rotation of element AB about point A. The axial component of the reaction in AB does not contribute to the virtual work. The virtual work is

$\delta U = -LF\cos\alpha\delta\theta + LR_{BD}\sin\alpha\delta\theta = 0$, from which $\boxed{R_{BD} = \dfrac{F}{\tan\alpha} = 0.577F\ (C)}$

Problem 11.9 Determine the load in member CD of the frame.

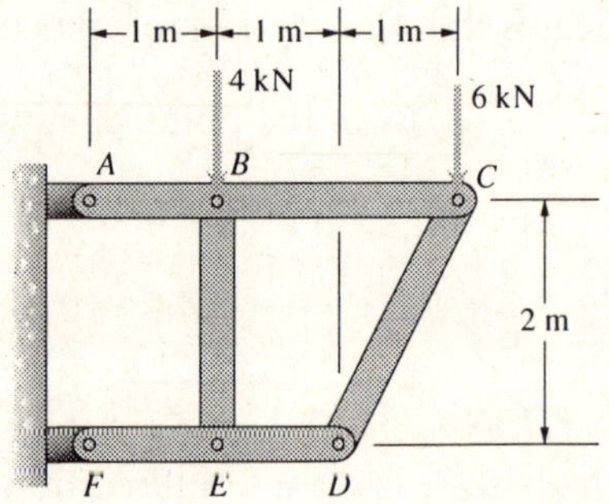

Solution The angle of member CD relative to the horizontal is $\alpha = \tan^{-1}\left(\dfrac{2}{1}\right) = 63.4^o$. Do a virtual rotation of element AC about A. The virtual work is

$\delta U = (-4 + R_{BE} - 18 + 3R_{CD}\sin\alpha)\delta\theta = 0$, from which

(1) $R_{BE} + 3R_{CD}\sin\alpha = 22$.

Do a virtual rotation of member FD about F. The virtual work is

$\delta U = (-R_{BE} - 2R_{CD}\sin\alpha)\delta\theta = 0$, from which

(2) $R_{BE} + 2R_{CD}\sin\alpha = 0$. These two simultaneous equations are solved to obtain $\boxed{R_{CD} = \dfrac{22}{\sin\alpha} = 24.6\ kN\ (C)}$

Problem 11.10 Determine the reaction at the roller support.

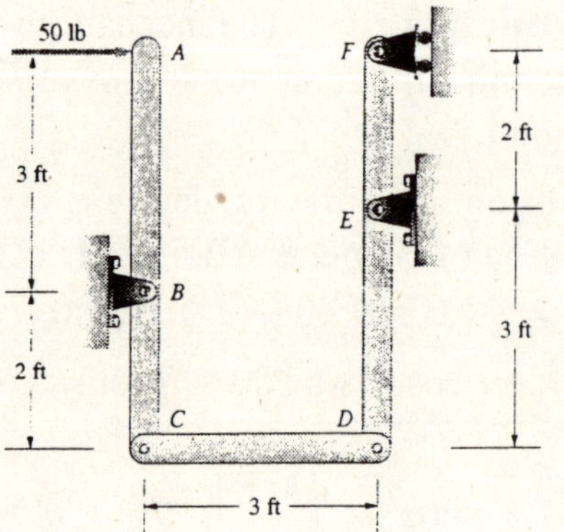

Solution Do a virtual rotation of element ABC about B. The virtual work is $\delta U = (-50(3) + 2R_{CD})\delta\theta = 0$, from which $R_{CD} = \dfrac{150}{2} = 75\ lb$. Do a virtual rotation of element DEF about point E. The virtual work is $\delta U = (2R_F - 3R_{CD})\delta\theta = 0$, from which $\boxed{R_F = \dfrac{3R_{CD}}{2} = 112.5\ lb}$

Problem 11.11 Determine the force P necessary for the mechanism to be in equilibrium.

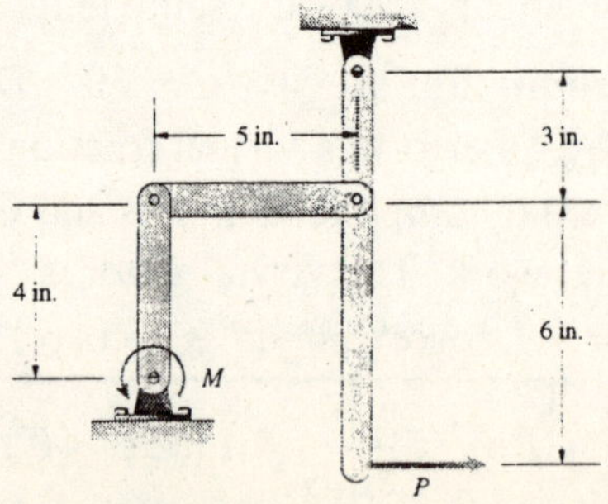

Solution Denote the axial force in the horizontal member by R_H. Perform a virtual rotation about the pin support of the element on the left: $\delta U = (M - 4R_H)\delta\theta = 0$, from which $R_H = \dfrac{M}{4}\ lb$, where M is measured in *in. lb*. Do a virtual rotation of the member on the right about the pin support.

$\delta U = (-3R_H + 9P)\delta\theta = 0$, from which $\boxed{P = \dfrac{3R_H}{9} = \dfrac{M}{12}\ lb}$

Problem 11.12

The system is in equilibrium. The total mass of the suspended load and assembly A is 120 *kg*. (a) By using equilibrium, determine the force *F*. (b) Using the result of Part (a) and the principle of virtual work, determine the distance the suspended load rises if the cable is pulled upward 300 *mm* at B.

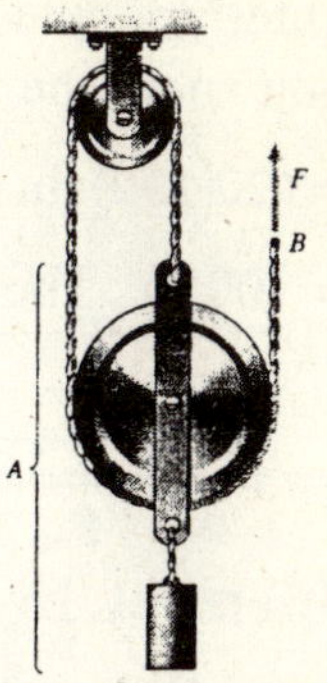

Solution The weight of assembly and load is $W = 1177.2\ N$. (a) The sum of the forces is $\sum F_Y = F - W + 2R = 0$, where *R* is the tension in the cable. The tension in the cable is *F* if the pulleys have no bearing losses. Thus $\boxed{F = \frac{W}{3} = 392.4\ N}$. (b) Do a virtual translation of the assembly A in the vertical direction. The work done is $\delta U = -W\delta y + F\delta x = 0$, from which the ratio $\frac{\delta x}{\delta y} = \frac{W}{F}$. The ratio of translations of the assembly A and the point B is $\frac{300}{y_A} = \frac{W}{F} = 3$, using the result of Part (a) for *F*. The assembly A rises $\boxed{y_A = \frac{300}{3} = 100\ mm}$ when point B is moved upward 300 *mm.*

Problem 11.13

Determine the force *P* necessary for the system to be in equilibrium.

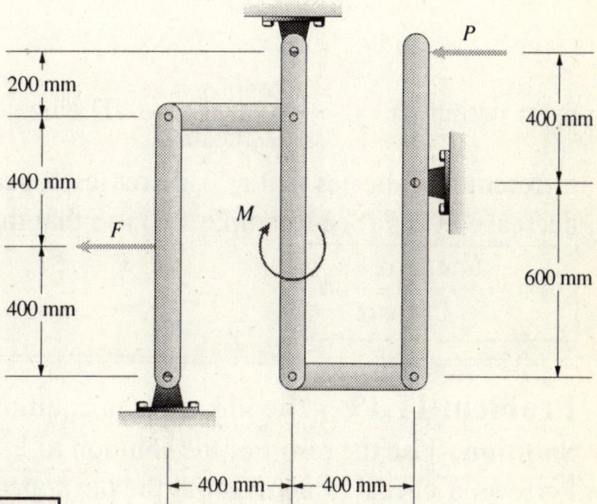

Solution Denote axial forces in the left horizontal member by R_1 and the right horizontal member by R_2. Do a virtual rotation of the leftmost vertical member about the pin support: $\delta U = (400F - 800R_1)\delta\theta = 0$, from which $R_1 = \frac{F}{2}$. Do a virtual rotation of the middle vertical member about the pin support: $\delta U = (-200R_1 - M + 1000R_2)\delta\theta = 0$, from which $R_2 = \frac{M + 200R_1}{1000} = \frac{M + 100F}{1000}$. Do a virtual rotation of the right vertical member about the pin support: $\delta U = (+400P - 600R_2)\delta\theta = 0$, from which

$$\boxed{P = \frac{3}{2}R_2 = \left(\frac{3}{2}\right)\frac{M + 100F}{1000} = \frac{3M}{2000} + \frac{3F}{20} = \left(\frac{3}{2}\right)M + \left(\frac{3}{20}\right)F}$$

Problem 11.14

The structure is subjected to a 20 *kN m* couple. Determine the horizontal reaction at C.

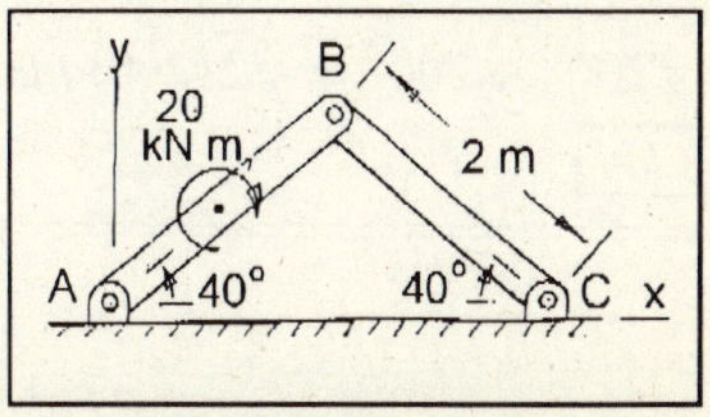

Solution The interior angle at B is 100°. Denote the axial force in BC by R_{BC}. Do a virtual rotation of member AB about A: $\delta U = \left(-20 + 2R_{BC}\sin 100^o\right)\delta\theta = 0$, from which $R_{BC} = \frac{20}{2\cos 10^o} = 10.15\ kN$. The horizontal component of the axial force is $\boxed{R_x = R_{BC}\cos 40^o = 7.778 = -7.8\ kN}$ (directed parallel to the negative x-axis.)

Problem 11.15 Determine the magnitude of the force exerted on the wall by the block at A.

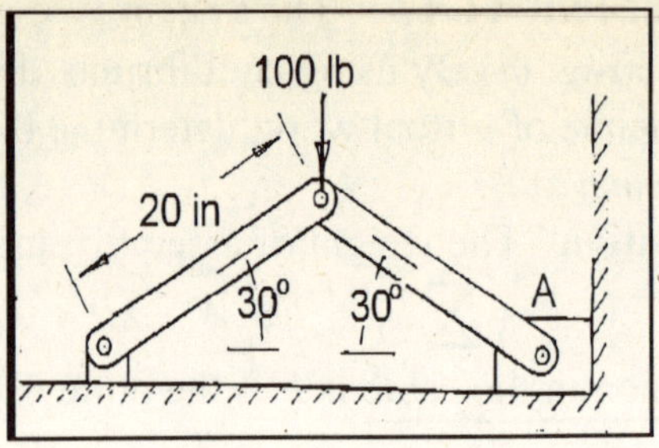

Solution Denote the axial force in the right element by R_A. The interior angle at the apex is 120°. Perform a virtual rotation about the left pin support:

$\delta U = \left(20(100)\sin 120^o - 20R_A \sin 120^o\right)\delta\theta = 0$, from which

$R_A = 100\ lb$. The horizontal component of the axial force on the wall is $\boxed{R_w = 100\cos 30^o = 86.6\ lb}$

Problem 11.16 Show that δx is related to $\delta\alpha$ by $\delta x = \dfrac{L_1 x \sin\alpha}{x - L_1\cos\alpha}\delta\alpha$. *Strategy:* Write the law of cosines in terms of α and take the derivative of the resulting equation with respect to α. (See Example 11.2)

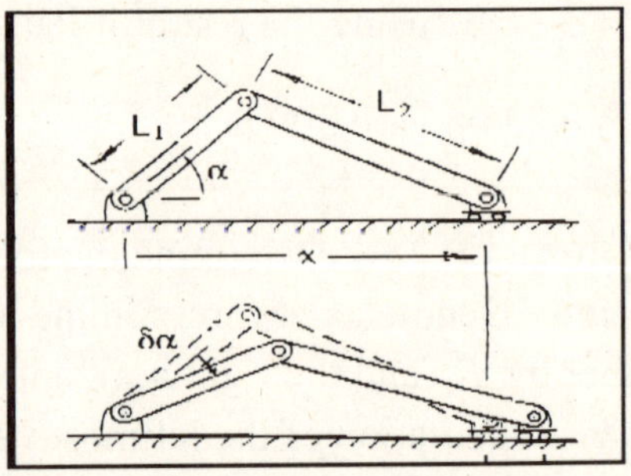

Solution Denote the horizontal distance from the pin support to the roller support by *x*. The law of cosines:

$L_2^2 = x^2 + L_1^2 - 2xL_1\cos\alpha$, from which

$x^2 - 2xL_1\cos\alpha + \left(L_1^2 - L_2^2\right) = 0$. Take the variation with respect to *x* and α, $dx(2x - 2L_1\cos\alpha) + 2xL_1\sin\alpha\, d\alpha = 0$,

from which $dx = -\dfrac{L_1 x \sin\alpha}{x - L_1\cos\alpha}d\alpha$ The negative sign for the differential indicates that as α increases, *x* decreases, as expected. However, the virtual rotation $\delta\alpha$ is a decrease in α, (See Example 11.1) so that the virtual rotation has the opposite sign to the variation,

$$\boxed{\delta x = \frac{L_1 x \sin\alpha}{x - L_1\cos\alpha}\delta\alpha}$$

Problem 11.17 The linkage is in equilibrium. What is *F*?

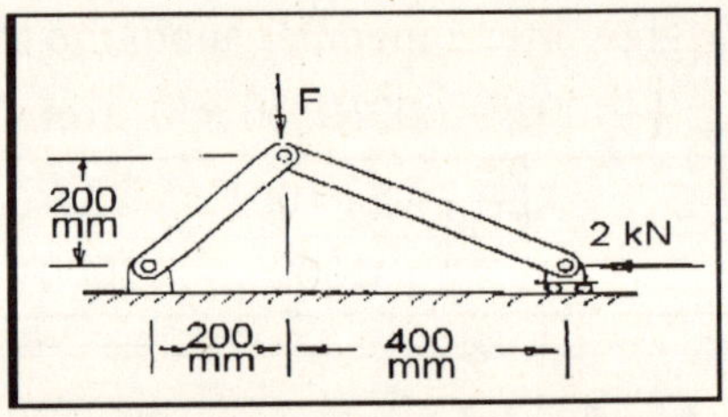

Solution Use the result of the solution to Problem 11.16. Perform a virtual rotation about the pin support of the left member. The virtual work is $\delta U = 200F\delta\alpha - 2\delta x = 0$, from which $F = \dfrac{1}{100}\dfrac{\delta x}{\delta\alpha}$. Use the result obtained in Problem 11.16,

$F = \dfrac{1}{100}\left(\dfrac{xL_1\sin\alpha}{x - L_1\cos\alpha}\right)$. From the dimensions given:

$L_1 = \sqrt{200^2 + 200^2} = \sqrt{2}(200)\ mm$, and $x = 600\ mm$ and the interior angle is $\alpha = 45^o$. From which $\boxed{F = 3\ kN}$

Problem 11.18

The linkage is in equilibrium. What is the force F?

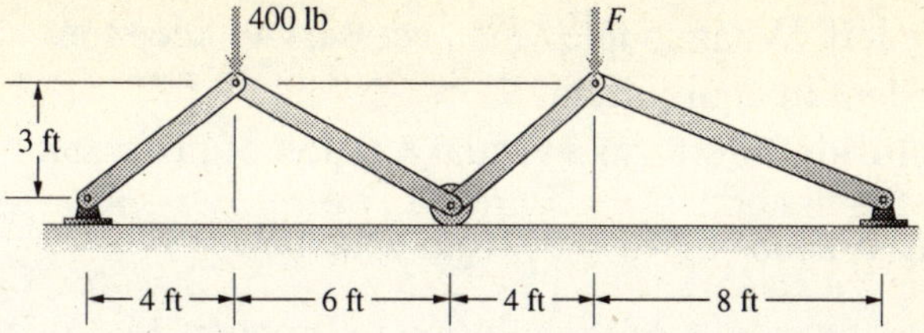

Solution Denote the horizontal reaction at the roller support by R_H. Perform a virtual rotation about the left pinned support:

$\delta U = 400(4)\delta\alpha - R_H\delta x$, from which

$R_H = 1600\dfrac{\delta\alpha}{\delta x}$. Use the result of the solution to Problem 11.16, $R_H = 1600\left(\dfrac{x - L_1\cos\alpha}{xL_1\sin\alpha}\right)$. From the dimensions given, $L_1 = \sqrt{3^2 + 4^2} = 5\ ft$, $x = 10\ ft$, and the interior angle is $\alpha = \tan^{-1}\left(\dfrac{3}{4}\right) = 36.9^o$, from which $R_H = 320\ lb$. Perform a virtual rotation about the right pin support: $\delta U = +8F\delta\theta - R_H\delta x = 0$, from which $F = \dfrac{R\,\delta x}{8\,\delta\theta}$. Use the result from Problem 11.16: $F = \dfrac{R_H}{8}\left(\dfrac{xL_1\sin\theta}{x - L_1\cos\theta}\right)$. From the dimensions given, $L_1 = \sqrt{8^2 + 3^2} = 8.54\ ft$, $x = 12\ ft$, and $\theta = \tan^{-1}\left(\dfrac{3}{8}\right) = 20.56^o$. From which the force is $\boxed{F = 360\ lb}$

Problem 11.19

The mechanism is in equilibrium. Neglect friction between the horizontal bar and the collar. Determine M in terms of F, α, and L.

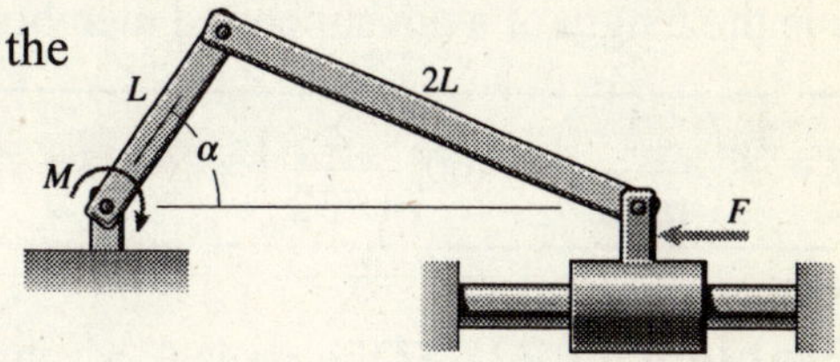

Solution Perform a virtual rotation about the left pin support.

$\delta U = M\delta\alpha - F\delta x = 0$, from which $M = F\dfrac{\delta x}{\delta\alpha}$. Use the results of the solution to Problem 11.16, $M = F\left(\dfrac{xL\sin\alpha}{x - L\cos\alpha}\right)$. From the dimensions given and the cosine law,

$4L^2 = x^2 + L^2 - 2Lx\cos\alpha$, from which

$x^2 - 2xL\cos\alpha - 3L^2 = 0$, which has the solution

$x = L\cos\alpha \pm \sqrt{L^2\cos^2\alpha + 3L^2} = L\left(\cos\alpha \pm \sqrt{\cos^2\alpha + 3}\right)$. Since a negative value of x has no meaning here, $x = L\left(\cos\alpha + \sqrt{\cos^2\alpha + 3}\right)$. Substitute into the expression for the moment, and reduce:

$M = FL\sin\alpha\left(1 + \dfrac{\cos\alpha}{\sqrt{\cos^2\alpha + 3}}\right)$. From the identity, $\cos^2\alpha = 1 - \sin^2\alpha$, an alternate form of the solution is $\boxed{M = FL\sin\alpha\left(1 + \dfrac{\cos\alpha}{\sqrt{4 - \sin^2\alpha}}\right)}$

Problem 11.20 The angle $\alpha = 20^o$ and the force exerted on the piston by pressure is 4 *kN* toward the left. What couple *M* is necessary to keep the system in equilibrium?

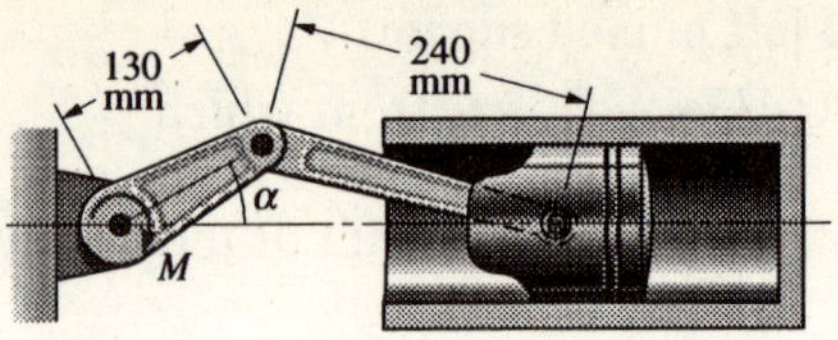

Solution Perform a virtual rotation of the crank:

$\delta U = M\delta\alpha + F\delta x = 0$, from which $M = F\dfrac{\delta x}{\delta\alpha}$.

Use the results of the solution to Problem 11.16,

$M = F\left(\dfrac{xL_1 \sin\alpha}{x - L_1 \cos\alpha}\right)$. From the dimensions

given, $L_1 = 130\ mm$, and from the cosine law, $240^2 = x^2 + 130^2 - 2(x)(130)\cos\alpha$, from which

$x^2 - 2(122.16)x - 40700 = 0$, which has the solutions $x = 358, \quad = -113.7\ mm$. Since a negative x has no meaning here, $x = 358\ mm$. Substituting and reducing: $\boxed{M = 269.97 = 270\ kN\ mm = 270\ N\ m}$

Problem 11.21 The structure is subjected to a 400 *N* load and is held in place by a horizontal cable. Determine the tension in the cable.

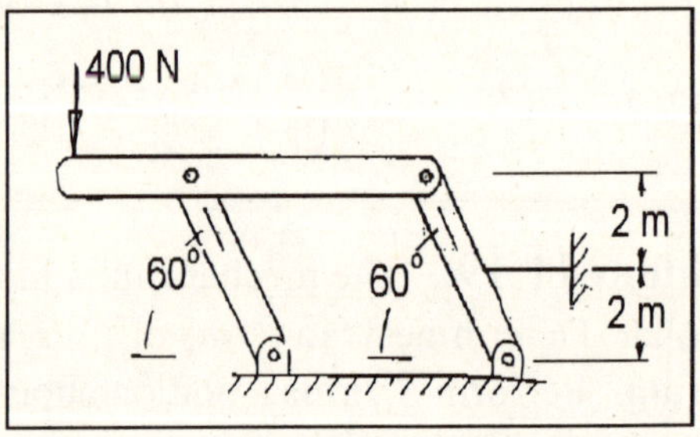

Solution Perform a virtual rotation about the pin supports. The load can only move vertically, since the structure is a parallelogram. In virtual motion, the point of attachment of the cable will move horizontally. The virtual work is

$\delta U = F\delta y - T\delta x = 0$, from which $T = F\dfrac{\delta y}{\delta x}$. Note that for a virtual rotation, $\dfrac{\delta y}{\delta\alpha} = L\cos\alpha$, and $\dfrac{\delta x}{\delta\alpha} = \dfrac{L}{2}\sin\alpha$, where

L is the length of a pin supported member. Thus $\dfrac{\delta y}{\delta x} = \dfrac{2}{\tan\alpha}$, and

$$\boxed{T = F\left(\frac{2}{\tan\alpha}\right) = 400\left(\frac{2}{1.732}\right) = 461.88 = 462\ N}$$

Problem 11.22 If the car jack is subjected to a force $F = 6.5\ kN$, what is the tension in the threaded shaft between B and D?

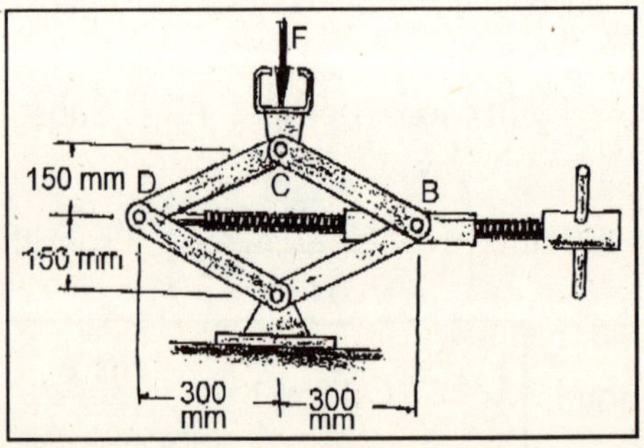

Solution Denote the tension in BD by *T*. Perform a virtual closing of the distance BD. The virtual work is

$\delta U = -F\delta y + T\delta x = 0$, from which $T = F\dfrac{\delta y}{\delta x}$. Half of the interior angle CDB is $\alpha = \tan^{-1}\left(\dfrac{150}{300}\right) = 26.56^o$. The vertical motion is $\delta y = L\cos\alpha\delta\alpha$, where L is the length of member DC. The horizontal motion is $\delta x = L\sin\alpha\delta\alpha$,

from which $\dfrac{\delta y}{\delta x} = \dfrac{1}{\tan\alpha}$. The tension is $\boxed{T = \dfrac{F}{\tan\alpha} = \dfrac{6.5}{0.5} = 13\ kN}$

Problem 11.23 What are the reactions at A and B? *Strategy:* Use the equilibrium equations to determine the horizontal components of the reactions and use the principle of virtual work to determine the vertical components.

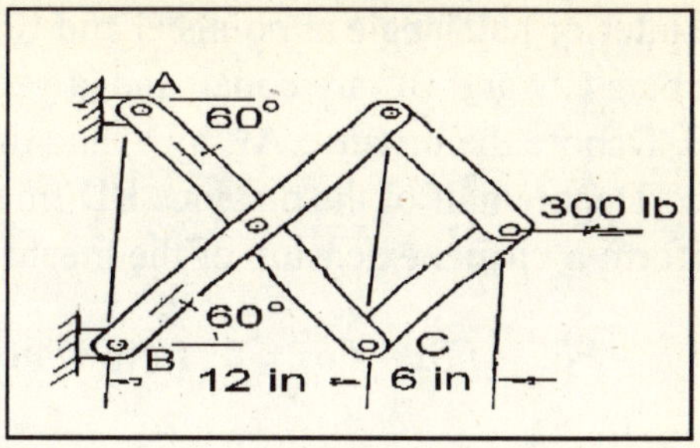

Solution Denote the angle $\alpha = 60^o$ and the distance AC by L. The sum of the moments about A is

$$\sum M_A = \left(\frac{12}{2\tan\alpha}\right)(300) + \left(\frac{12}{\tan\alpha}\right)B_x = 0,$$ from which

$\boxed{B_x = -150\ lb}$. The sum of the forces:

$\sum F_x = A_x + B_x + 300 = 0$, from which

$\boxed{A_x = -B_x - 300 = -150\ lb}$. The sum of the forces in the vertical direction is $\sum F_y = A_y + B_y = 0$, from which $A_y = -B_y$. Denote the distance between A and B by y and the distance from the wall to the point of application of the force as x. Then $x = \left(\frac{3}{2}\right)\sqrt{L^2 - y^2}$. Take the variation of both sides,

$\delta x = \left(\frac{3}{2}\right)\frac{-y\delta y}{\sqrt{L^2 - y^2}} = \left(\frac{1}{8}\right)(-y\delta y)$, $y = 12\tan\alpha$, from which $\frac{\delta x}{\delta y} = -\left(\frac{1}{8}\right)(12\tan\alpha)$. Perform a virtual elongation of the mechanism in the x-direction. The virtual work is

$\delta U = 300\delta x - A_y\frac{\delta y}{2} + B_y\frac{\delta y}{2} = 0$,from which $\left(\frac{1}{2}\right)(A_y - B_y) = 300\frac{\delta x}{\delta y} = A_y$.. The vertical reaction is

$\boxed{A_y = \left(\frac{3}{2}\right)300\tan\alpha = 779.4\ lb}$, and $\boxed{B_y = -A_y = -779.4\ lb}$

Problem 11.24 The device raises a load W by extending the hydraulic actuator DE. The bars AD and BC are each 2 m long and the distances $b = 1.4\ m$ and $h = 0.8\ m$. If $W = 4\ kN$, what force must the actuator exert to hold the load in equilibrium?

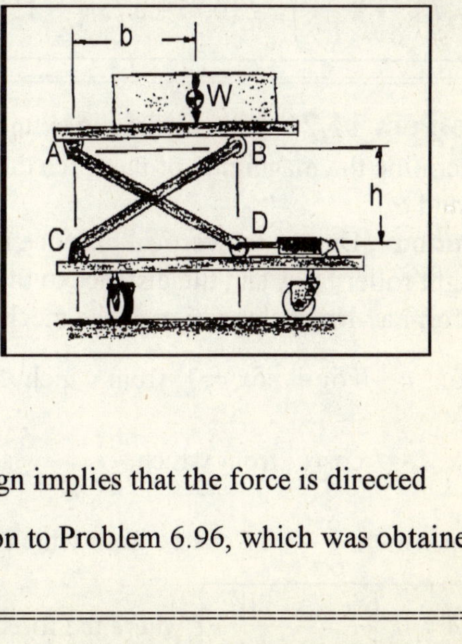

Solution Perform a virtual vertical displacement of the load. Denote the distance CD by x. The virtual work is

$\delta U = -W\delta h + D\delta x = 0$, from which $D = W\frac{\delta h}{\delta x}$. The distances are related: $h = \sqrt{L^2 - x^2}$, where L is the length of bar BC, from which $\frac{\delta h}{\delta x} = -\frac{x}{\sqrt{L^2 - x^2}} = -\frac{x}{h}$. Thus

$D = -W\frac{x}{h} = -W\frac{\sqrt{L^2 - h^2}}{h}$. Substitute numerical values:

$\boxed{D = -4\frac{\sqrt{4 - 0.8^2}}{0.8} = -9.165 = -9.17\ kN}$, where the negative sign implies that the force is directed parallel to the negative x-axis. (Note: Compare with the solution to Problem 6.96, which was obtained using an alternate approach.)

Problem 11.25 Determine the force P necessary for the mechanism to be in equilibrium.

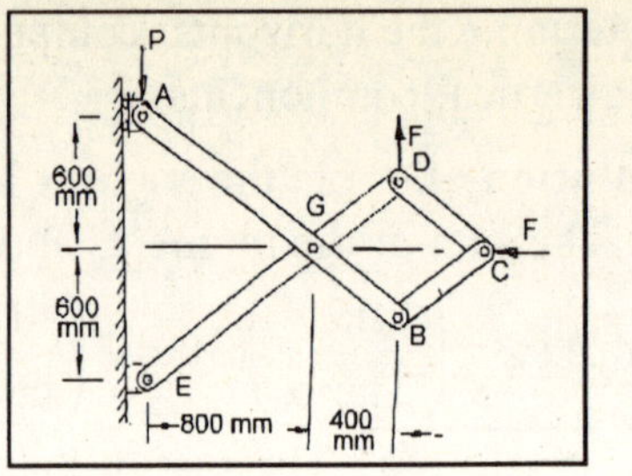

Solution The strategy is to relate the virtual displacements through the interior half-angle at points G and C. Assume that the lengths of the bars CD and BC are equal, and are equal to the segments BG and DG. Denote the distance AE by y, and the distance from the wall to C by x. Denote half of the distance BD (measured vertically) by z. Perform a virtual extension of the mechanism: the virtual work is $\delta U = -P\delta y - F\delta x + F\delta z = 0$, from which $P = F\left(\frac{\delta z}{\delta y} - \frac{\delta x}{\delta y}\right)$. The interior half angle AGE is $\alpha = \tan^{-1}\left(\frac{600}{800}\right) = 36.9^o$. The interior angles AGE, BGD, and DCB are equal, from geometry. The lengths of the bars AB and DE are $L_1 = \frac{1200}{\sin(90^o - \alpha)} = 1500\ mm$. The lengths of the segments DC and BC are $L_2 = 1500 - \frac{800}{\sin(90^o - \alpha)} = 500\ mm$. The distances AG and EG are $L_3 = \frac{800}{\sin(90^o - \alpha)} = 1000\ mm$. The distance x is related to the angle α by $x = (L_1 + L_2)\cos\alpha$, from which $\frac{\delta x}{\delta \alpha} = -(L_1 + L_2)\sin\alpha = -1200$. The distance z is related to the angle α by $z = L_1 \sin\alpha$, from which $\frac{\delta z}{\delta \alpha} = L_2 \cos\alpha = 1200$. The distance y is related to the angle α by $y = 2L_3 \sin\alpha$, from which $\frac{\delta y}{\delta \alpha} = 2L_3 \cos\alpha = 1600$. Combining these relations, $\frac{\delta x}{\delta y} = -0.75$, $\frac{\delta z}{\delta y} = 0.75$. The force P is given by

$$\boxed{P = F\left(\frac{\delta z}{\delta y} - \frac{\delta x}{\delta y}\right) = F(0.75 + 0.75) = -1.5F}$$ (directed parallel to the negative y-axis).

Problem 11.26 The hydraulic cylinder C exerts a horizontal force at A, raising the weight W. Determine the magnitude of the force the hydraulic cylinder must exert to support the weight in terms of W and α.

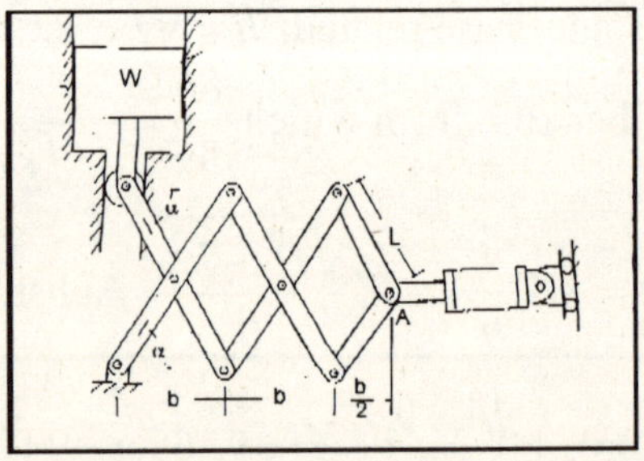

Solution Denote the vertical distance from the pin support to the weight roller by y, and the distance to the hydraulic cylinder by x. Perform a virtual elongation of the mechanism. The virtual work is $\delta U = -W\delta y + A\delta x = 0$, from which $A = W\frac{\delta y}{\delta x}$.

(1) $x = \frac{5}{2}L\cos\alpha$, from which $\frac{\delta x}{\delta x} = -\frac{5}{2}\sin\alpha$. (2) $y = 2L\sin\alpha$, from which $\frac{\delta y}{\delta \alpha} = 2L\cos\alpha$, from which $\frac{\delta y}{\delta x} = -\frac{2}{5\tan\alpha}$ The force that must be exerted by the cylinder is $\boxed{A = W\frac{\delta y}{\delta x} = -\frac{2W}{5\tan\alpha}}$, where the force is parallel with the negative x-axis.

================================<>================================

Problem 11.27 The potential energy of a conservative system is $V = 2x^3 + 3x^2 - 12x$. (a) For what values of x is the system in equilibrium? (b) Determine whether the equilibrium positions found in Part (a) are stable or unstable.

Solution (a) The system is in equilibrium when $\frac{dV}{dx} = 6x^2 + 6x - 12 = 0$. This is a quadratic equation in x: it is put into canonical form by dividing by 6, $x^2 + 2bx + c = 0$, from which $b = \frac{1}{2}$ and $c = -2$. The solutions are $\boxed{x = -b \pm \sqrt{b^2 - c} = 1, \quad = -2}$. (b) The stable and unstable positions are determined by the sign of the second derivative: $\frac{d^2V}{d^2x} = 12x + 6$. For $x = 1$, $\left[\frac{d^2V}{d^2x}\right]_{x=1} = 18 > 0$, and the equilibrium is stable at $x = 1$. For $x = -2$, $\left[\frac{d^2V}{d^2x}\right]_{x=-2} = -18 < 0$, and the equilibrium is unstable at $x = -2$.

================================<>================================

Problem 11.28 The potential energy of a conservative system is $V = 2q^3 - 21q^2 + 72q$. (a) For what values of q is the system in equilibrium? (b) Determine whether the equilibrium positions you found in Part(a) are stable or unstable.

Solution (a) The equilibrium positions are given by $\frac{dV}{dx} = 6q^2 - 42q + 72 = 0$. In canonical form: $q^2 + 2bq + c = 0$, where $b = -\frac{7}{2}$, and $c = 12$. The solutions: $\boxed{q = -b \pm \sqrt{b^2 - c} = 4, \quad = 3}$ (b) The second derivative is $\frac{d^2V}{dq^2} = 12q - 42$. For $q = 4$, $\left[\frac{d^2V}{dq^2}\right]_{q=4} = 6 > 0$, and the equilibrium is stable. For $q = 3$, $\left[\frac{d^2V}{dq^2}\right]_{q=3} = -6 < 0$, and the equilibrium is unstable.

================================<>================================

Problem 11.29 The weight $W = 100\ lb$ and the spring constant $k = 50\ lb/ft$. The spring is unstretched when $x = 0$. (a) Determine the value of x for which the weight is in equilibrium. (b) Determine whether the equilibrium position is stable or unstable.

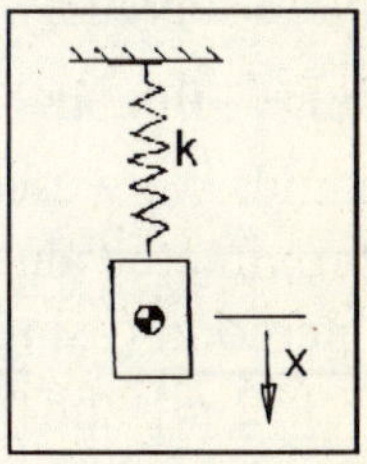

Solution Isolate the weight. The potential energy is $V_{weight}(x) = -\int W dx + C = -Wx + C$, where a positive displacement is downward. At $x = 0$, choose $V(x) = 0$, hence $C = 0$, and $V_{weight}(x) = -Wx$. Isolate the spring. The potential energy in the spring is $V_{spring}(x) = \int kx\,dx + C = \frac{1}{2}kx^2 + C$. Choose $x = 0$ as the datum point, $V_{spring}(0) = 0$, hence $C = 0$. (a) The total potential energy is $V_{total}(x) = -Wx + \frac{1}{2}kx^2$. The equilibrium point is given by $\frac{dV}{dx} = -W + kx = 0$, from which $\boxed{x = \frac{W}{k} = \frac{100}{50} = 2\ ft}$.

Problem 11.29 continued on next page

continuation of Problem 11.29..........

(b) The second derivative at $x = 2\ ft$ is $\left[\frac{d^2V}{dx^2}\right]_{x=2} = [k]_{x=2} = 50 > 0$, so the equilibrium is stable.

Problem 11.30 The nonlinear spring exerts a force $-kx + \varepsilon x^3$ on the mass, where k and ε are constants. Determine the potential energy V associated with the force exerted on the mass by the spring.

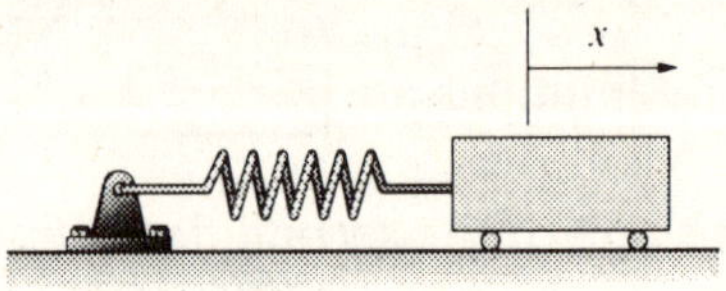

Solution The potential energy of the spring is

$V(x) = -\int (-kx + \varepsilon x^3)dx + C = \frac{1}{2}kx^2 - \frac{\varepsilon x^4}{4} + C$. Choose $x = 0$ as the datum point, $V(0) = 0$, hence $C = 0$, and the potential energy associated with the force is $\boxed{V(x) = \frac{1}{2}kx^2 - \frac{\varepsilon x^4}{4}}$

Problem 11.31 The 1 kg mass is suspended from the nonlinear spring described in Problem 11.30. The constants $k = 10$ and $\varepsilon = 1$, where x is in meters. (a) Show that the mass is in equilibrium when $x = 1.12\ m$ and when $x = 2.45\ m$. (b) Determine whether the equilibrium points are stable or unstable.

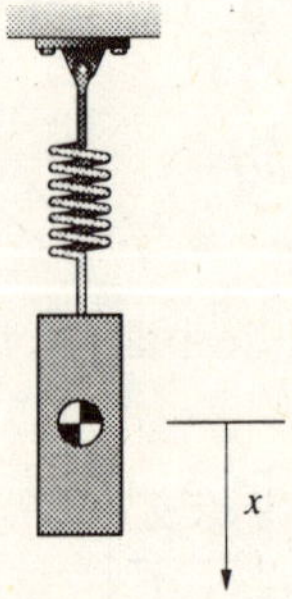

Solution In Problem 11.30 the potential energy of the spring is shown to be $V(x) = \frac{1}{2}kx^2 - \frac{\varepsilon x^4}{4}$. The potential energy of the mass is $V(x) = -\int W dx = -Wx$, where the datum is $x = 0$. The total potential energy is $V_{total}(x) = -Wx + \frac{1}{2}kx^2 - \frac{\varepsilon x^4}{4}$. The equilibrium points are determined from $\frac{dV}{dx} = -W + kx - \varepsilon x^3 = 0$. This cubic may be solved by iteration or by graphing the function $f(x) = -W + kx - \varepsilon x^3$ to find the zero crossings. Both methods were used here: the graph was used to get approximate values, and these values were then refined by iteration (using **TK Solver 2**). The results: $\boxed{x = 1.12\ m}$, and $\boxed{x = 2.45\ m}$. (b) The second derivative of the potential energy at $x = 1.12\ m$ is

$\left[\frac{d^2V}{dx^2}\right]_{x=1.12} = [10 - 3x^2]_{x=1.12} = 6.2 > 0$, and the system is stable. For $x = 2.45\ m$,

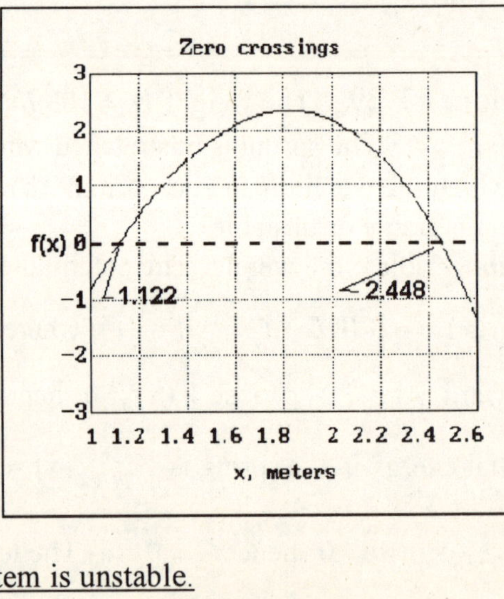

$\left[\frac{d^2V}{dx^2}\right]_{x=2.45} = [10 - 3x^2]_{x=2.45} = -7.98 < 0$, and the system is unstable.

Problem 11.32 Two straight segments of the bar are each of weight W and length L. Determine whether the equilibrium position shown is stable if (a) $0 < \alpha_o < 90^o$; (b) $\alpha_o = 0$; (c) $90^o < \alpha_o < 180^o$.

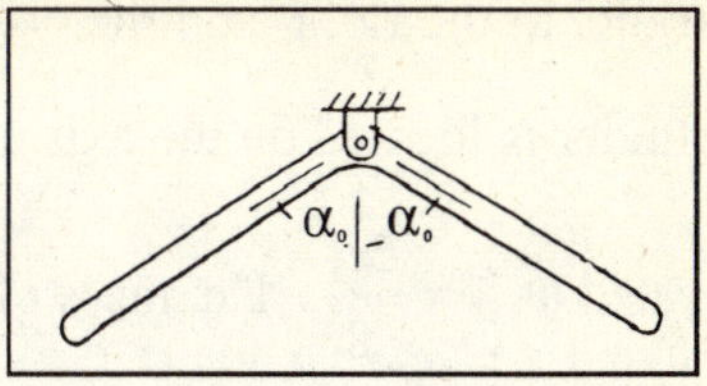

Solution From a heuristic argument, if the bars hang straight down ($\alpha_o = 0$) they are equivalent to one bar suspended at one end, and they should be stable in this position. If the bars are straight out ($\alpha_o = 90^o$) they are equivalent to a bar suspended at its midpoint, and as a balanced bar they should be neutrally stable. If the bars are positioned upward ($90^o < \alpha_o < 180^o$), the bars will be unbalanced everywhere, and the system will be unstable in every position. This heuristic reasoning suggests the strategy to treat the composite bar as if it were a mass point suspended (supported) at the pin support. The mass point will be located at the center of weight of the composite. It will behave like a pendulum under the action of gravity. Choose a coordinate system with origin at the pin support with the x-axis positive downward. Denote the right and left bars by the subscripts R and L. The center of weight of each bar has the coordinates: $\bar{x}_R = \frac{L}{2}\cos\alpha_o$, $\bar{y}_R = \frac{L}{2}\sin\alpha_o$; $\bar{x}_L = \frac{L}{2}\cos\alpha_o$, $\bar{y}_L = -\frac{L}{2}\sin\alpha_o$. The coordinates of the center of weight of the composite is $\bar{x} = \frac{W\bar{x}_R + W\bar{x}_L}{2W} = \frac{L}{2}\cos\alpha_o$, and $\bar{y} = \frac{W\bar{y}_R + W\bar{y}_L}{2W} = 0$. Suppose that a small angular displacement θ occurs. The potential energy of the equivalent system under this displacement is $V = -(2W)\left(\frac{L}{2}\cos\alpha_o\right)\cos\theta = -WL\cos\alpha_o\cos\theta$. The equilibrium points are determined by

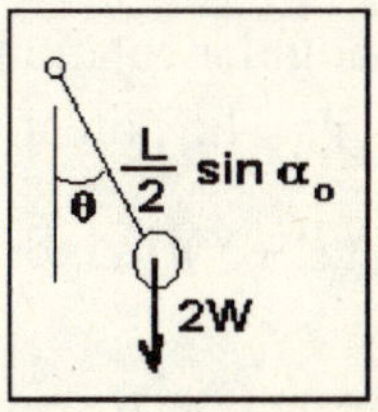

$\frac{dV}{d\theta} = WL\cos\alpha_o\sin\theta = 0$, from which $\theta = 0$, and $\theta = \pi$. The stability of the equilibrium points is determined from $\left[\frac{d^2V}{d\theta^2}\right]_{\theta=0} = \left[WL\cos\alpha_o\cos\theta\right]_{\theta=0} = WL\cos\alpha_o$. The stability depends upon the value of α_o: (a) in the interval $0 < \alpha_o < 90^o$, $WL\cos\alpha_o > 0$, and <u>the system is stable</u>.(b) At the point $\alpha_o = 0$, $WL > 0$, and <u>the system is stable</u>. (c) In the interval $90^o < \alpha_o < 180^o$, $WL\cos\alpha_o < 0$, and <u>the system is unstable.</u> If the <u>mechanical constraints permit the system to reach the equilibrium point at</u> $\theta = \pi$, the results above are reversed: (a) is unstable, (b) is unstable, and (c) is stable.

Problem 11.33 The homogenous composite object consists of a hemisphere and a cylinder. It is at rest on a plane surface. Show that the this equilibrium position is stable only if $L < \frac{R}{\sqrt{2}}$.

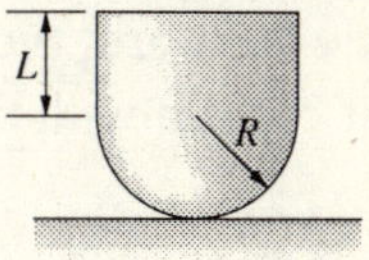

Solution An angular disturbance will cause the composite system to rock about the radial center of the hemisphere. The change in potential energy, if it occurs, must be a change in the height of the composite mass caused by motion about the radial center. This suggests the pendulum analogy: a point mass at the composite mass center, suspended (supported) from the radial center.

Problem 11.33 continued on next page..........

.............*continuation of solution to Problem 11.33.....*

Choose a coordinate system with the y-axis along the axis of the cylinder, positive upward, the x-axis parallel to the floor, and the origin at the point of contact with the floor. The center of mass of the cylinder is located on the axis of the cylinder at $y = \frac{L}{2} + R$. The center of mass of the hemisphere is located at $y = \frac{5R}{8}$. The mass of the cylinder is $m_{cyl} = \rho\pi R^2 L$, where ρ is the mass density in kg/m^3. The mass of the hemisphere is $\frac{2}{3}\pi\rho R^3$. The location of the mass centroid of the composite is

$$\bar{y} = \frac{\pi\rho R^2 L\left(\frac{L}{2} + R\right) + \frac{2}{3}\pi\rho R^3\left(\frac{5R}{8}\right)}{\pi\rho R^2 L + \left(\frac{2}{3}\right)\pi\rho R^3} = \frac{L\left(\frac{L}{2} + R\right) + \frac{5}{12}R^2}{L + \frac{2}{3}R}.$$

Suppose that the system is subjected to a small angular rotation about the point of contact with the floor. The length of the equivalent pendulum is $R - \bar{y}$. The potential energy due to this rotation is $V = -(R - \bar{y})W\cos\theta$. The point of equilibrium is $\frac{dV}{d\theta} = (R - \bar{y})W\sin\theta = 0$, from which $\theta = 0$ is a point of equilibrium. The stability is determined by $\left[\frac{d^2V}{d\theta^2}\right]_{\theta=0} = [(R - \bar{y})W\cos\theta]_{\theta=0} = (R - \bar{y})W$. Note that R and $\bar{y}$ are both positive numbers, from which, if $R > \bar{y}$, the system is stable; if $R < \bar{y}$, the system is unstable. Reduce algebraically:

$$R - \bar{y} = R - \frac{\left(L\left(\frac{L}{2} + R\right) + \frac{5}{12}R^2\right)}{L + \frac{2}{3}R} = \frac{RL + \frac{2}{3}R^2 - \frac{L^2}{2} - RL - \frac{5}{12}R^2}{L + \frac{2}{3}R} = \frac{\frac{3}{12}R^2 - \frac{L^2}{2}}{L + \frac{2}{3}R}.$$

Retain only the numerator, since the denominator must be positive always: The condition for stability is $\frac{3}{12}R^2 > \frac{L^2}{2}$.

Take the positive square root: $R > \sqrt{2}L$, or $\boxed{L < \frac{R}{\sqrt{2}}}$.

Problem 11.34

The homogenous object consists of a hemisphere and a cone. Show that this equilibrium position is stable only if $h < \sqrt{3}R$

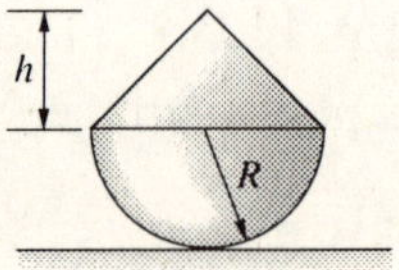

Solution .Use the same strategy used to solve Problem 11.33: a point mass at the composite mass center, suspended (supported) at the radial center. Choose a coordinate system with the y-axis along the axis of the cone, positive upward, the x-axis parallel to the floor, and the origin at the point of contact with the floor. The mass of the cone is $m_{cyl} = \frac{1}{3}\rho\pi R^2 h$, where ρ is the mass density. The center of mass of the cone is $y = \frac{h}{4} + R$. The mass of the hemisphere is $\frac{2}{3}\pi\rho R^3$. The center of mass of the hemisphere is located at $y = \frac{5R}{8}$.

Problem 11.34 continued on next page..........

............*continuation of solution to Problem 11.34.....*

The location of the mass centroid of the composite is

$$\bar{y} = \frac{\pi\rho R^2 \frac{h}{3}\left(\frac{h}{4}+R\right)+\frac{2}{3}\pi\rho R^3\left(\frac{5R}{8}\right)}{\pi\rho R^2\frac{h}{3}+\left(\frac{2}{3}\right)\pi\rho R^3} = \frac{\frac{h}{3}\left(\frac{h}{4}+R\right)+\frac{5}{12}R^2}{\frac{h}{3}+\frac{2}{3}R}.$$

For a small angular rotation θ the length of the equivalent pendulum is $R-\bar{y}$. The potential energy due to this rotation is $V = -(R-\bar{y})W\cos\theta$. The point of equilibrium is $\frac{dV}{d\theta} = (R-\bar{y})W\sin\theta = 0$, from which $\theta = 0$ is a point of equilibrium. The stability is determined by

$\left[\frac{d^2V}{d\theta^2}\right]_{\theta=0} = \left[(R-\bar{y})W\cos\theta\right]_{\theta=0} = (R-\bar{y})W$, from which, if $R > \bar{y}$, the system is stable; if $R < \bar{y}$, the system is unstable. Reduce: $R-\bar{y} = \frac{\frac{Rh}{3}+\frac{2}{3}R^2-\frac{h^2}{12}-\frac{Rh}{3}-\frac{5}{12}R^2}{\frac{h}{3}+\frac{2}{3}R} = \frac{\frac{R^2}{4}-\frac{h^2}{12}}{\frac{h}{3}+\frac{2R}{3}}$, from which, (since the denominator is always positive) if $\frac{R^2}{4}-\frac{h^2}{12} > 0$, the system is stable,. Thus $R^2 > \frac{h^2}{3}$, or, taking the positive square root: $\boxed{\sqrt{3}R > h}$

Problem 11.35

Problem 11.35 The homogenous bar has weight W and the spring is unstretched when the bar is vertical ($\alpha = 0$). (a) Use potential energy to show that the bar is in equilibrium when $\alpha = 0$. (b) Show that the equilibrium position $\alpha = 0$ is stable only if $2kL > W$.

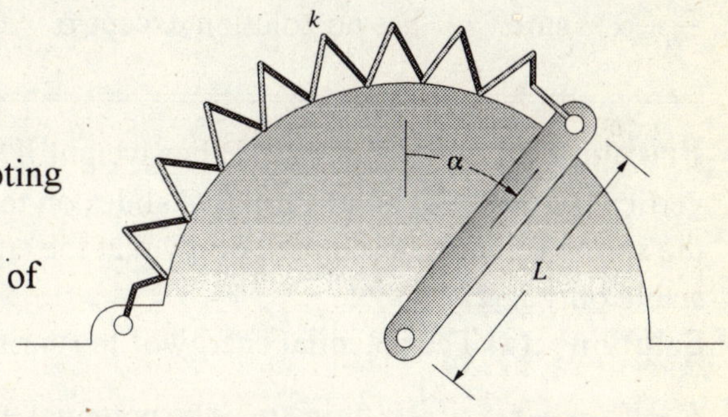

Solution (a) The potential energy of the spring is $V_{spring} = \int ksds = \frac{1}{2}ks^2$, where the datum is $\alpha = 0$. Noting $s = L\alpha$, then $V_{spring} = \frac{k}{2}L^2\alpha^2$ The height of the center of the bar relative to the pinned end is $\frac{L}{2}\cos\alpha$; using the pinned end as the datum: $V_{bar} = \frac{WL}{2}\cos\alpha$, from which $V_{tot} = \frac{k}{2}L^2\alpha^2 + \frac{WL}{2}\cos\alpha$. The equilibrium point is $\frac{dV_{tot}}{d\alpha} = kL^2\alpha - \frac{WL}{2}\sin\alpha = 0$, from which $\frac{2kL}{W}\alpha - \sin\alpha = 0$. This is a transcendental equation in α, with at least one solution $\boxed{\alpha = 0}$. (b) Stability is determined by

$\left[\frac{d^2V_{tot}}{d\alpha^2}\right]_{\alpha=0} = \left[\frac{2kL}{W} - \cos\alpha\right]_{\alpha=0} = \frac{2kL}{W} - 1$, from which if $\frac{2kL}{W} - 1 > 0$, the system is stable. Thus the condition for stability at $\alpha = 0$ is $\boxed{2kL > W}$

==================================<>==================================

Problem 11.36 Suppose that the bar in Problem 11.35 is in equilibrium when $\alpha = 20^o$. (a) Show that the constant $k = 0.490\frac{W}{L}$. (b) Determine whether or not the equilibrium position is stable.

Solution Use the results of the solution to Problem 11.35. The equilibrium condition is $\frac{2kL}{W}\alpha - \sin\alpha = 0$, from which, for $\alpha \neq 0$, divide by α: $\frac{2kL}{W} = \frac{\sin\alpha}{\alpha}$, or $k = \left(\frac{W}{L}\right)\frac{\sin\alpha}{2\alpha} = 0.490\left(\frac{W}{L}\right)$ (where α is in radians.) (b) The condition for stability is $\left[\frac{2kL}{W} - \cos\alpha\right]_{\alpha=20^o} > 0$, from which $2(0.4899) - 0.9397 > 0$, which is satisfied. The system is stable. *Check:* From a heuristic argument, two sets of equilibrium conditions apply: If the spring force is weaker than the gravity force in the neighborhood of $\alpha = 0$, the bar should rotate under the action of gravity until this tendency is balanced by the increased spring force, at which point by analogy with the mass-spring system (see Problem 11.29) the system should be in stable equilibrium. If, however, the spring force is greater that the gravity force in the neighborhood of $\alpha = 0$, the system should remain close to $\alpha = 0$, which will be a position of stable equilibrium. This heuristic argument is supported as follows: (1) The equilibrium condition $\frac{2kL}{W}\alpha - \sin\alpha = 0$, for $\alpha \neq 0$, is satisfied if $\frac{2kL}{W} = \frac{\sin\alpha}{\alpha}$, $\left(0 < \alpha < 90^o\right)$, which is the first set of equilibrium conditions. The condition for stability is $\frac{\sin\alpha}{\alpha} - \cos\alpha > 0$, which is satisfied for all $0 < \alpha < 90^o$, and the system is stable for $\alpha > 0$. Taking the limit as α goes to zero, the equilibrium position at $\alpha = 0$ is neutrally stable if $\frac{2kL}{W} = \frac{\sin\alpha}{\alpha}$. (2) If $2kL > W$, a second set of conditions apply: the $\alpha = 0$ position is stable (see Problem 11.35), and this is the only equilibrium position since $\frac{2kL}{W}\alpha - \sin\alpha = 0$ has no solution except $\alpha = 0$ for $\frac{2kL}{W} > 1$. *check.*

==================================<>==================================

Problem 11.37 The bar AB has weight W and length L. The spring is unstretched when the bar is vertical ($\alpha = 0$). The light collar C slides on the smooth vertical bar so that the spring remains horizontal. Show that the equilibrium position $\alpha = 0$ is stable only if $2kL > W$.

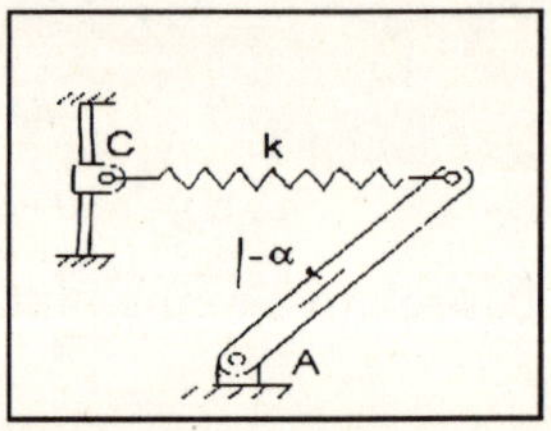

Solution (a) The potential energy of the spring is $V_{spring} = \frac{1}{2}ks^2 = \frac{1}{2}kL^2\sin^2\alpha$. The potential energy of the bar is $V_{bar} = \frac{WL}{2}\cos\alpha$, from which $V_{tot} = \frac{k}{2}L^2\sin^2\alpha + \frac{WL}{2}\cos\alpha$. The equilibrium point is $\frac{dV_{tot}}{d\alpha} = kL^2\sin\alpha\cos\alpha - \frac{WL}{2}\sin\alpha = 0$, from which $(\frac{2kL}{W}\cos\alpha - 1)\sin\alpha = 0$, which has two possible solutions: $\cos\alpha = \frac{W}{2kL}$, or $\alpha = 0$. (b) Stability is determined by

$$\left[\frac{d^2V}{d\alpha^2}\right]_{\alpha=0} = \left[-\frac{2kL}{W}\sin^2\alpha + \left(\frac{2kL}{W}\cos\alpha - 1\right)\cos\alpha\right]_{\alpha=0} = \left(\frac{2kL}{W} - 1\right).$$

If $\left(\frac{2kL}{W} - 1\right) > 0$, the system is stable. The condition for stability is $\boxed{2kL > W}$

==================================<>==================================

Problem 11.38 The bar AB in Problem 11.37 weighs 20 *lb*, its length is 4 *ft*, and the spring constant is $k = 5\ ft/lb$. (a) Determine the value of α in the range $0 < \alpha < 90^o$ for which the bar is in equilibrium. (b) Is the equilibrium position determined in Part (a) stable?

Solution Use the solution to Problem 11.37. The equilibrium condition is $(\frac{2kL}{W}\cos\alpha - 1)\sin\alpha = 0$, or

$$\boxed{\alpha = \cos^{-1}\left(\frac{W}{2kL}\right) = \cos^{-1}(0.5) = 60^o}$$

(b) The condition for stability is $\left[\frac{d^2V}{d\alpha^2}\right]_{\alpha=60} = \left[-\frac{2kL}{W}\sin^2\alpha + \left(\frac{2kL}{W}\cos\alpha - 1\right)\cos\alpha\right]_{\alpha=60} = -1.5 < 0$,

so the system is unstable.

Problem 11.39 The bar AB has weight W and length L. The spring is unstretched when the bar is vertical ($\alpha = 0$). The light collar C slides on the smooth vertical bar so that the spring remains vertical. Show that the equilibrium position $\alpha = 0$ is unstable.

Solution The potential energy of the spring is

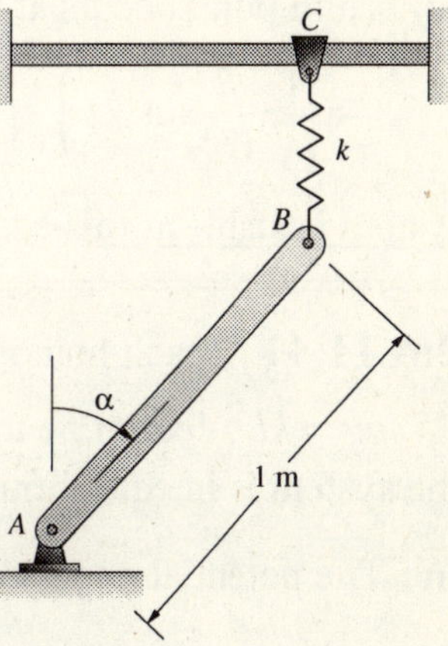

$V_{spring} = \int ksds = \frac{1}{2}ks^2$, where the datum is $\alpha = 0$. Noting $s = L(1-\cos\alpha)$, then $V_{spring} = \frac{k}{2}L^2(1-\cos\alpha)^2$. The height of the center of the bar above the pin joint is $\frac{L}{2}\cos\alpha$. With the pin joint as the datum, the potential energy of the bar is $V_{bar} = \frac{WL}{2}\cos\alpha$, from which

$V_{tot} = \frac{k}{2}L^2(1-\cos\alpha)^2 + \frac{WL}{2}\cos\alpha$. The equilibrium point is

$\frac{dV_{tot}}{d\alpha} = kL^2(1-\cos\alpha)\sin\alpha - \frac{WL}{2}\sin\alpha = 0$, from which $(\frac{2kL}{W}(1-\cos\alpha) - 1)\sin\alpha = 0$, which has at least one solution: $\alpha = 0$. Stability is determined by

$\left[\frac{d^2V}{d\alpha^2}\right]_{\alpha=0} = \left[\left(\frac{2kL}{W}\sin^2\alpha\right) + \left(\frac{2kL}{W}(1-\cos\alpha) - 1\right)\cos\alpha\right]_{\alpha=0} = -1$. Since $\left[\frac{d^2V}{d\alpha^2}\right]_{\alpha=0} = -1 < 0$, the system is unstable at the equilibrium position $\alpha = 0$.

Problem 11.40 The bar AB described in Problem 11.39 has a mass of 2 *kg* and the spring constant is $k = 80\ N/m$. (a) Determine the value in the range $0 < \alpha < 90^o$ for which the bar is in equilibrium. (b) Is the equilibrium position determined in Part (a) stable?

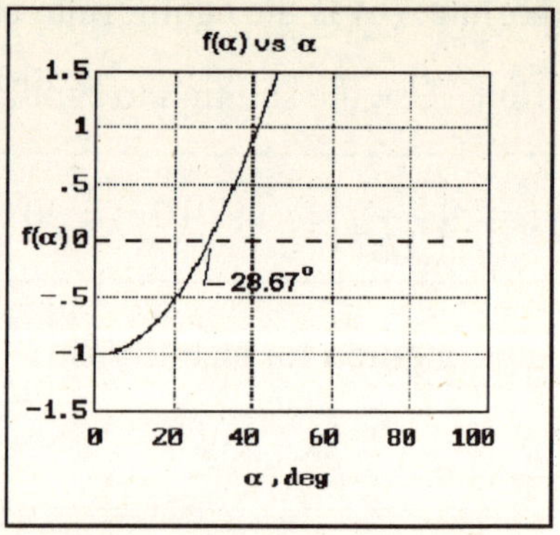

Solution Use the solution to Problem 11.39. The condition for equilibrium is $(\frac{2kL}{W}(1-\cos\alpha)-1)\sin\alpha = 0$, from which the non-zero position of equilibrium is determined by $(\frac{2kL}{W}(1-\cos\alpha)-1) = 0$ Substitute numerical values to obtain $\frac{2kL}{W} = 8.1549$. The zero crossing of a graph of $f(\alpha) = \left(\frac{2kL}{W}\right)(1-\cos\alpha) - 1$ was determined approximately over the interval $0 < \alpha < 90^o$, and this crossing value was then refined by iteration (using **TK Solver 2**). The equilibrium point occurs at : $\boxed{\alpha = 28.67 = 28.7^o}$. The condition for stability is

$$\left[\frac{d^2V}{d\alpha^2}\right]_{\alpha=\alpha_i} = \left[\left(\frac{2kL}{W}\sin^2\alpha\right)+\left(\frac{2kL}{W}(1-\cos\alpha)-1\right)\cos\alpha\right]_{\alpha=\alpha_i}$$

. For $\alpha = 28.7^o$, $\left[\frac{d^2V}{d\alpha^2}\right]_{\alpha=28.7} = 1.88 > 0$, so the system is stable at this equilibrium point

Problem 11.41 Each homogenous bar is of mass *m* and length L. The spring is unstretched when $\alpha = 0$. If $mg = kL$, determine the value of α in the range over $0 < \alpha < 90^o$ for which the system is in equilibrium.

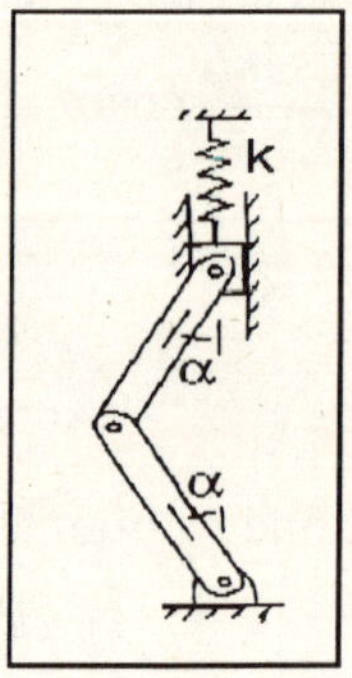

Solution The potential energy of the spring is $V = \frac{1}{2}ks^2$. Noting that $s = 2L(1-\cos\alpha)$, $V = 2kL^2(1-\cos\alpha)^2$. The potential energy of the bars is $V_{bars} = \frac{WL}{2}\cos\alpha + WL\cos\alpha + \frac{WL}{2}\cos\alpha = 2WL\cos\alpha$, where the datum point is the lower pin joint. The total energy is $V = 2kL^2(1-\cos\alpha)^2 + 2WL\cos\alpha$. The equilibrium condition is $\frac{dV}{d\alpha} = (\frac{2kL}{W}(1-\cos\alpha)-1)\sin\alpha = 0$. The non zero position of equilibrium is, when $W = kL$, $2(1-\cos\alpha)-1 = 0$. Reduce: $1-\cos\alpha = \frac{1}{2}$, from which $\cos\alpha = \frac{1}{2}$, $\boxed{\alpha = \cos^{-1}\left(\frac{1}{2}\right) = 60^o}$

Problem 11.42 Determine whether the equilibrium position found in Problem 11.41 is stable or unstable.

Solution Use the solution to Problem 11.41. The condition for an equilibrium point is $\frac{dV}{d\alpha} = (\frac{2kL}{W}(1-\cos\alpha)-1)\sin\alpha = 0$. The condition for stability is

$$\left[\frac{d^2V}{d\alpha^2}\right]_{\alpha=60^\circ} = \left[\frac{2kL}{W}\left(\sin^2\alpha - \cos^2\alpha\right)+\left(\frac{2kL}{W}-1\right)\cos\alpha\right]_{\alpha=60^\circ} = 1.5 > 0,$$ so the position is stable.

Problem 11.43 The mass of the bar is 10 *kg* and it is 1 *m* in length. The spring is unstretched when the bar is vertical ($\alpha = 0$) and the spring constant $k = 100\ N/m$. Determine the values of α at which the bar is in equilibrium.

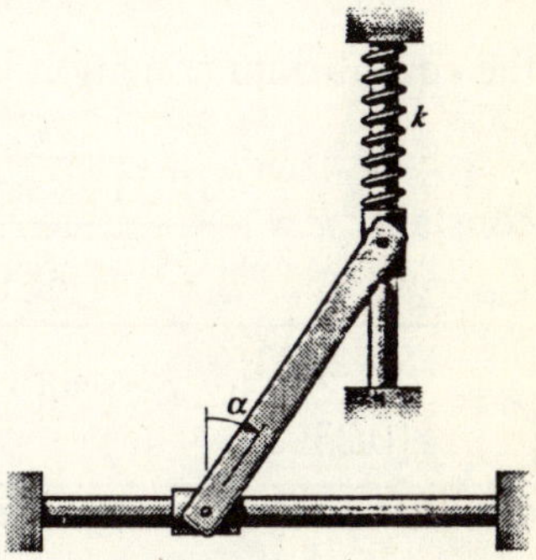

Solution The potential energy of the spring is $V = \frac{1}{2}ks^2$. Noting that $s = L(1-\cos\alpha)$, then $V = \frac{1}{2}kL^2(1-\cos\alpha)^2$. The potential energy of the bar is $V_{bar} = \frac{WL}{2}\cos\alpha$, where the datum point is the lower pin joint. From which $V_{tot} = \frac{kL^2}{2}(1-\cos\alpha)^2 + \frac{WL}{2}\cos\alpha$. The condition for equilibrium is $\frac{dV}{d\alpha} = \left(\frac{2kL}{W}(1-\cos\alpha)-1\right)\sin\alpha = 0$. The equilibrium points are $\boxed{\alpha = 0}$, and the value of α determined by $\frac{2kL}{W}(1-\cos\alpha)-1=0$, from which $\cos\alpha = 1-\frac{W}{2kL}$. Substitute numerical values: $\cos\alpha = 0.5095$, or $\boxed{\alpha = \cos^{-1}(0.5095) = 59.369 = 59.4^o}$

Problem 11.44 Determine whether the equilibrium positions of the bar in Problem 11.43 are stable or unstable.

Solution Use the solution to Problem 11.43. The equilibrium condition is $\frac{dV}{d\alpha} = \left(\frac{2kL}{W}(1-\cos\alpha)-1\right)\sin\alpha = 0$. The stability condition is determined by

$$\left[\frac{d^2V}{d\alpha^2}\right]_{\alpha=\alpha_i} = \left[\left(\frac{2kL}{W}\right)\left(\sin^2\alpha - \cos^2\alpha\right)+\left(\frac{2kL}{W}-1\right)\cos\alpha\right]_{\alpha=\alpha_i}.$$

For $\alpha = 0$, $\left[\frac{d^2V}{d\alpha^2}\right]_{\alpha=0} = -1 < 0$, so the equilibrium point is unstable. For $\alpha = 59.4^o$, $\left[\frac{d^2V}{d\alpha^2}\right]_{\alpha=59.4^o} = 1.51 > 0$, so the equilibrium point is stable.

==============================<>==============================

Problem 11.45 The bar weighs 15 *lb*. The spring is unstretched when $\alpha = 0$. The bar is in equilibrium when $\alpha = 30^o$. Determine the spring constant *k*.

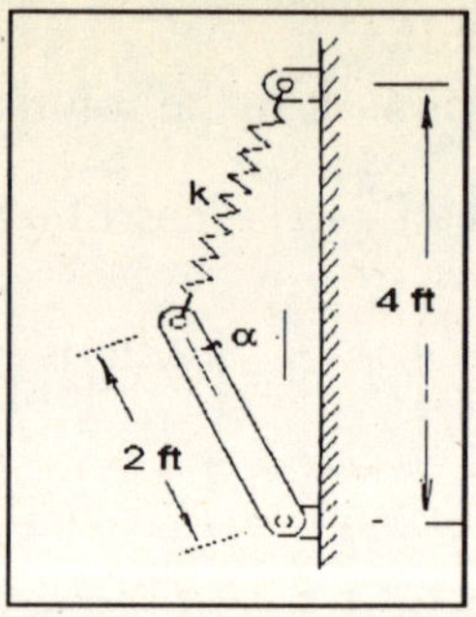

Solution From the cosine law, the length of the spring is $d^2 = 2^2 + 4^2 - 16\cos\alpha$, from which $d = 2\sqrt{5 - 4\cos\alpha}$. The spring extension is $\Delta = d - 2 = 2\left(\sqrt{5 - 4\cos\alpha} - 1\right)$. The potential energy of the spring is $V = \frac{k\Delta^2}{2}$. The potential energy of the bar is $V_{bar} = W\cos\alpha$. The total potential energy is $V_{tot} = \frac{k\Delta^2}{2} + W\cos\alpha$. Noting $\frac{d\Delta}{d\alpha} = \frac{4\sin\alpha}{\sqrt{5 - 4\cos\alpha}}$, the equilibrium condition is $\frac{dV}{d\alpha} = \frac{k(8\sin\alpha)\left(\sqrt{5 - 4\cos\alpha} - 1\right)}{\sqrt{5 - 4\cos\alpha}} - W\sin\alpha = 0$. Solve for the spring constant: $k = \frac{W\sqrt{5 - 4\cos\alpha}}{8\left(\sqrt{5 - 4\cos\alpha} - 1\right)}$. Substitute values: $\sqrt{5 - 4\cos\alpha} = 1.2393$,

$$\boxed{k = \frac{15(1.2393)}{8(0.2393)} = 9.71 \ lb/ft}$$

==============================<>==============================

Problem 11.46 Determine whether the equilibrium positions of the bar in Problem 11.45 are stable or unstable.

Solution Use the solution to Problem 11.45. The condition for equilibrium is $\frac{dV}{d\alpha} = \frac{k(8\sin\alpha)\left(\sqrt{5 - 4\cos\alpha} - 1\right)}{\sqrt{5 - 4\cos\alpha}} - W\sin\alpha = 0$. For brevity write $L = \sqrt{5 - 4\cos\alpha}$. The derivative is $\frac{dV}{d\alpha} = \left(\frac{8k(L-1)}{L} - W\right)\sin\alpha = \left(8k - W - \frac{8k}{L}\right)\sin\alpha$, from which the second derivative is $\frac{d^2V}{d\alpha^2} = \frac{16k}{L^3}\sin^2\alpha + (8k - W - \frac{8k}{L})\cos\alpha$. The condition for stability is determined from $\left[\frac{d^2V}{d\alpha^2}\right]_{\alpha=\alpha_i}$

For $\alpha = 0$, $\left[\frac{d^2V}{d\alpha^2}\right]_{\alpha=0} = -W = -15 < 0$, so that the equilibrium position is unstable. For $\alpha = 30^o$, $\left[\frac{d^2V}{d\alpha^2}\right]_{\alpha=30^o} = 20.4 > 0$, so that the equilibrium point is stable.

==============================<>==============================

Problem 11.47 Each bar is of weight W and the spring is unstretched when $\alpha = 90^o$. (a) Show that the system is in equilibrium when $\alpha = \sin^{-1}\left(\frac{W}{4kL}\right)$. (b) Is the equilibrium position described in Part (a) stable?

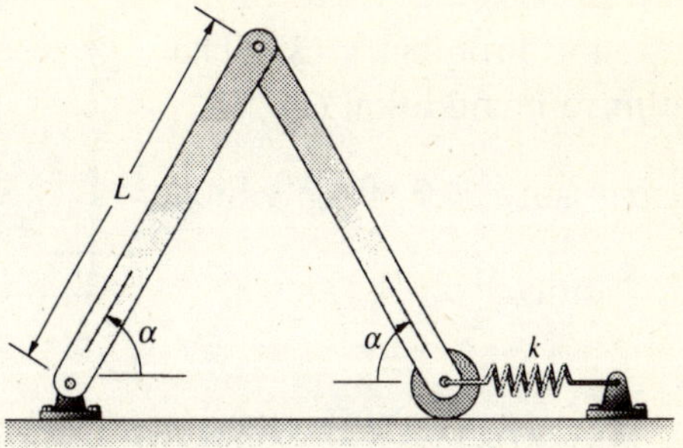

Solution The potential energy of the spring is $V = \frac{1}{2}ks^2$.

Noting $s = 2L\cos\alpha$, $V = 2kL^2\cos^2\alpha$. The potential energy of the bars is $V_{bars} = WL\sin\alpha$, where the datum is the lower pin joints. The total potential energy is $V_{tot} = 2kL^2\cos^2\alpha + WL\sin\alpha$. The condition for equilibrium is $\frac{dV}{d\alpha} = \left(1 - \frac{4kL}{W}\sin\alpha\right)\cos\alpha = 0$. This has the solutions $\alpha = \cos^{-1}(0) = 90^o$, and $\alpha = \sin^{-1}\left(\frac{W}{4kL}\right)$. (b) The second derivative is

$$\frac{d^2V}{d\alpha^2} = \left(\frac{4kL}{W}\right)\left(\sin^2\alpha - \cos^2\alpha\right) - \sin\alpha = \left(\frac{4kL}{W}\right)\left(2\sin^2\alpha - 1\right) - \sin\alpha$$

. The condition for stability:

$$\left[\frac{d^2V}{d\alpha^2}\right]_{\alpha=\sin^{-1}\left(\frac{W}{4kL}\right)} = \frac{1}{\sin\alpha}(2\sin^2\alpha - 1) - \sin\alpha = \frac{\sin^2\alpha - 1}{\sin\alpha} < 0,$$

since in the interval $0 < \alpha < 90^o, 0 < \sin\alpha < 1$. The equilibrium is unstable for this position.

Problem 11.48 The 1 kg mass is suspended from a nonlinear spring that exerts a force $-10x + x^3$, where x is in meters. (a) Draw a graph of the total potential energy of the system as a function of x from $x = 0$ to $x = 4\ m$. (b) Use your graph to estimate the equilibrium positions of the mass. (c) Determine whether the equilibrium positions you obtained in Part (b) are stable or unstable.

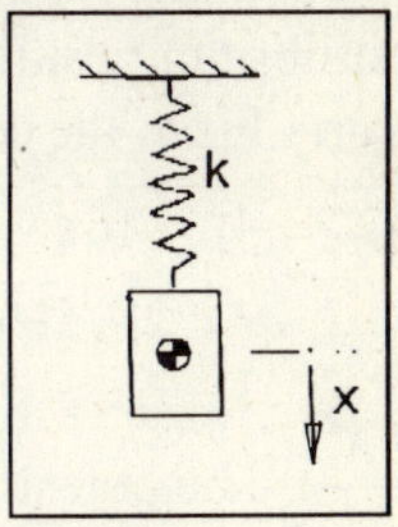

Solution As the mass moves downward, the potential energy of the weight decreases, and the potential energy in the spring increases. The potential energy of the spring is $V_{spring} = -\int(-10x + x^3)dx = 5x^2 - \frac{x^4}{4} + C$. Choose the datum point $x = 0$, hence $C = 0$. The potential energy of the mass is $V_{mass} = -mgx = -9.81x$. The total energy is $V_{tot} = 5x^2 - \frac{x^4}{4} - 9.81x$. The graph is shown. The equilibrium positions occur at local minima and maxima in the graph: at $\boxed{x = 1.12\ m}$ and $\boxed{x = 2.45\ m}$. The local minimum at $x = 1.12\ m$ means that this position is a stable equilibrium, and the local maximum at $x = 2.45\ m$ means that this position is an unstable equilibrium.

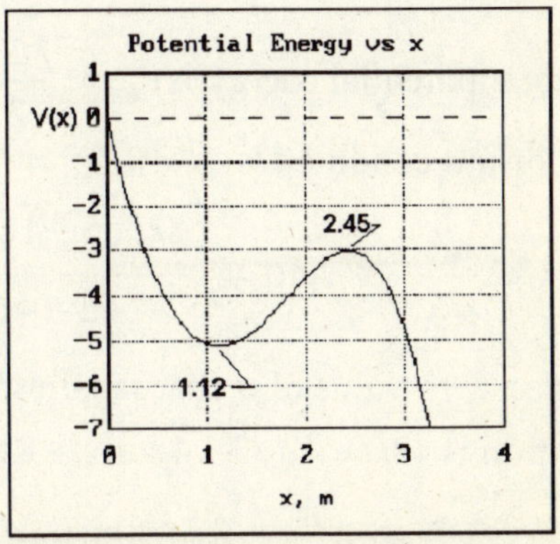

Problem 11.49 Suppose that the homogenous bar in Problem 11.35 weighs 20 *lb* and has length $L = 2\,ft$ and $k = 4\ lb/ft$. (a) Determine the values of α in the range $0 < \alpha < 90^o$ for which the bar is in equilibrium. (b) Is the equilibrium position found in Part(a) stable?

Solution Use the results of the solution to Problem 11.35. The equilibrium condition is

$$\frac{2kL}{W}\alpha - \sin\alpha = 0,$$ from which

$$\frac{2kL}{W} = \frac{\sin\alpha}{\alpha} = \frac{2(4)(2)}{20} = 0.8.$$

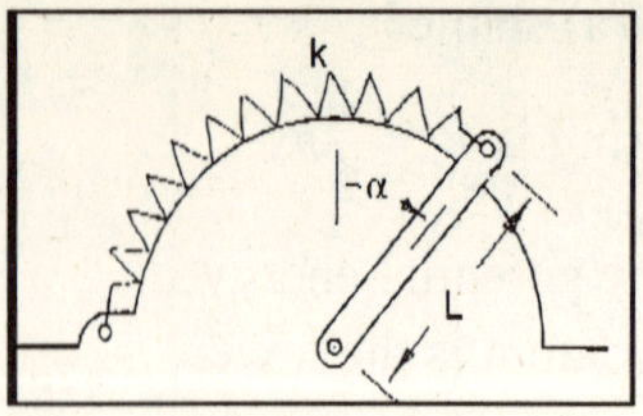

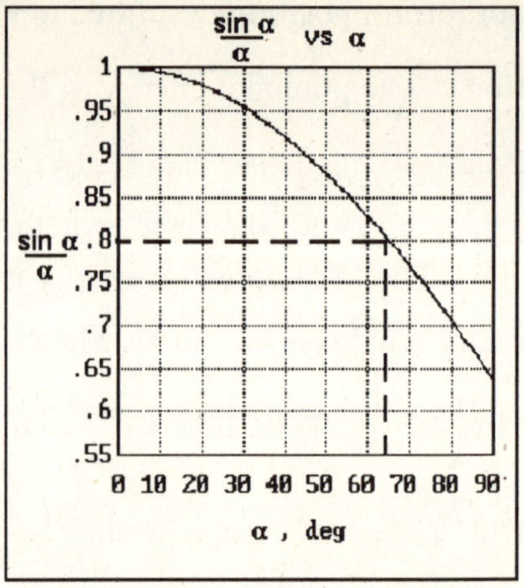

The ratio $f(\alpha) = \frac{\sin\alpha}{\alpha}$ (where α is in radians) is graphed against the α in degrees to get an approximate value of the angle which satisfies this condition, and then this value was refined by iteration (using **TK Solver 2**) to obtain $\boxed{\alpha = 64.8^o}$

(b) The condition for stability is $\frac{2kL}{W} - \cos\alpha > 0$, from which $0.8 - 0.4462 > 0$, which is satisfied. The system is stable. (See the solution to Problem 11.36, in which it is shown that this system is stable at every equilibrium point in the interval $0 < \alpha < 90^o$, and neutrally stable at $\alpha = 0$ when $\frac{2kL}{W} = \frac{\sin\alpha}{\alpha}$.)

Problem 11.50 The bar in Problem 11.45 weighs 15 *lb* and the spring is unstretched when $\alpha = 0$. The spring constant $k = 6\ lb/ft$. (a) Determine the value of α in the range $0 < \alpha < 90^o$ for which the bar is in equilibrium. (b) Is the equilibrium position found in Part (a) stable?

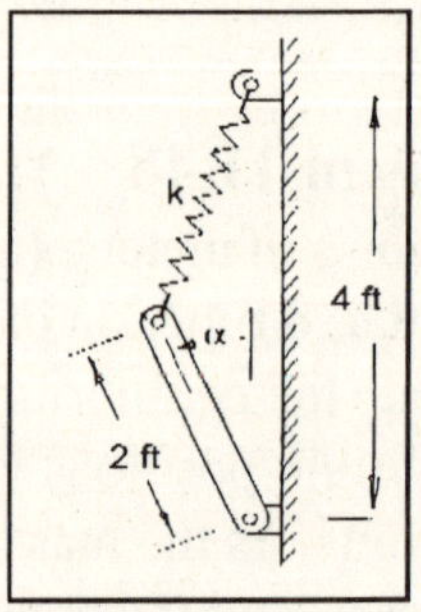

Solution From the cosine law, the length of the spring is $d^2 = 2^2 + 4^2 - 16\cos\alpha$, from which $d = 2\sqrt{5 - 4\cos\alpha}$. The spring extension is $\Delta = d - 2 = 2\left(\sqrt{5 - 4\cos\alpha} - 1\right)$. The potential energy of the spring is

$V = \frac{k\Delta^2}{2}$. The potential energy of the bar is $V_{bar} = W\cos\alpha$.

The total potential energy is $V_{tot} = \frac{k\Delta^2}{2} + W\cos\alpha$. The equilibrium condition is

$$\frac{dV}{d\alpha} = k\Delta\frac{d\Delta}{d\alpha} - W\sin\alpha = \left(\frac{8k\left(\sqrt{5 - 4\cos\alpha} - 1\right)}{\sqrt{5 - 4\cos\alpha}} - W\right)\sin\alpha = 0.$$

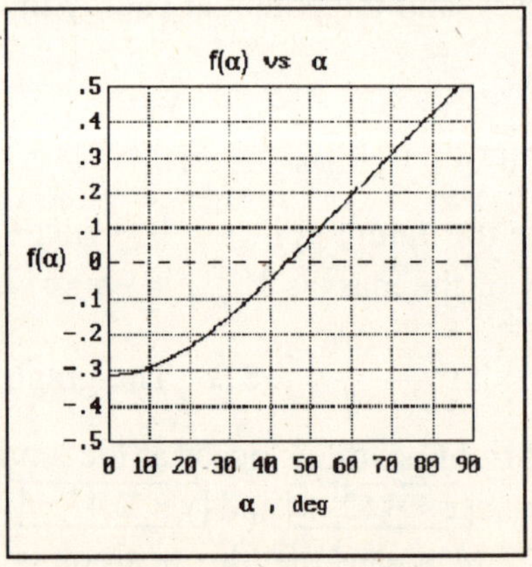

The non zero value of α that satisfies the equilibrium condition is $(8k - W)\sqrt{5 - 4\cos\alpha} - 8k = 0$.

Problem 11.50 continued on next page

continuation of Problem 11.50..........

A graph of the function $f(\alpha) = \left(1 - \frac{W}{8k}\right)\sqrt{5 - 4\cos\alpha} - 1$ was used to get an estimate of the zero crossing point value of α, and this estimate was then refined by iteration to obtain $\boxed{\alpha = 43.86 = 43.9^o}$.

(b) The stability of the equilibrium point may be determined from a graph of the potential energy in the neighborhood surrounding the equilibrium point. The graph of $V_{tot} = \frac{k\Delta^2}{2} + W\cos\alpha$, where $\Delta^2 = 4\left(\sqrt{5 - 4\cos\alpha} - 1\right)^2$, against the angle α is shown.

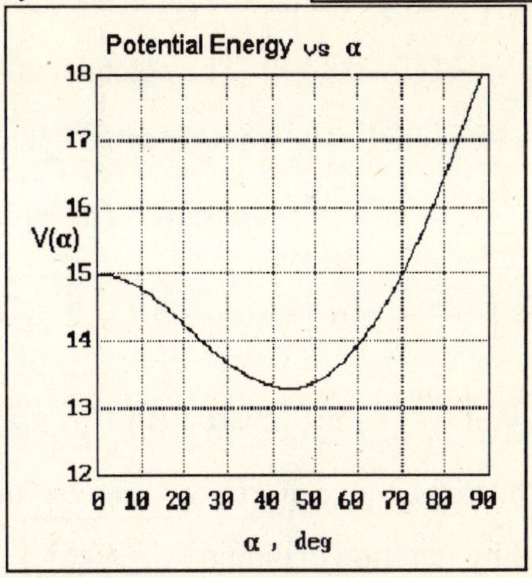

From the local minimum at $\alpha = 43.9^o$, the equilibrium is stable. *Check:* From the solution to Problem 11.46, the second derivative is

$$\frac{d^2V}{d\alpha^2} = \frac{16k}{L^3}\sin^2\alpha + (8k - W - \frac{8k}{L})\cos\alpha, \text{ where}$$

$L = \sqrt{5 - 4\cos\alpha}$. Stability of equilibrium is determined by $\left[\frac{d^2V}{d\alpha^2}\right]_{\alpha=43.9^o} = 14.99 > 0$, so the equilibrium is stable. *check.*

Problem 11.51 The homogenous bar has length L and mass $4m$.

(a) Determine the value of α in the range $0 < \alpha < 90^o$ for which the bar is in equilibrium. (b) Is the equilibrium position found in Part (a) stable?

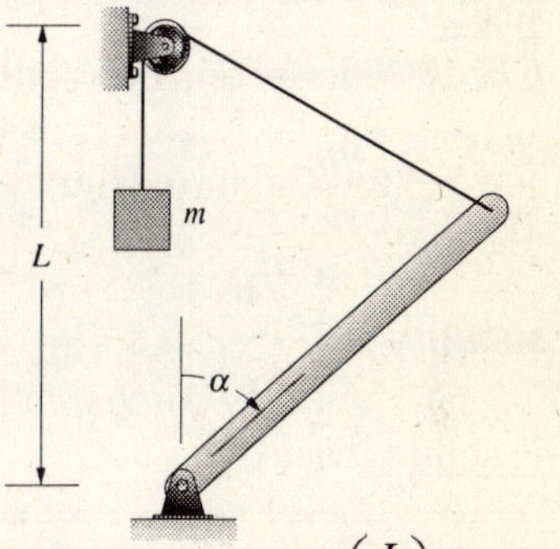

Solution The length of the cable between the bar and the pulley is $d = \sqrt{L^2 + L^2 - 2L^2\cos\alpha}. = \sqrt{2}L\sqrt{1 - \cos\alpha}$. Let the total length of the cable be L_{tot}. The potential energy of the mass m is $V_m = -mgx$ where $x = L_{tot} - \sqrt{2}L\sqrt{1 - \cos\alpha}$, where it is assumed that the weight is always "above the ground".at any allowed value of α. The potential energy of the bar is $V_{bar} = 4mg\left(\frac{L}{2}\right)\cos\alpha$.

The total energy is $V_{tot} = -mgL_{tot} + mg\sqrt{2}L\sqrt{1 - \cos\alpha} + 2mgL\cos\alpha$. The condition for equilibrium is $\frac{dV}{d\alpha} = \left(\frac{1}{\sqrt{2}\sqrt{1 - \cos\alpha}} - 2\right)\sin\alpha = 0$. The equilibrium points are determined by $\sin\alpha = 0$ and $\frac{1}{\sqrt{2}\sqrt{1 - \cos\alpha}} - 2 = 0$, from which $\boxed{\alpha = 0}$ and $\boxed{\alpha = \cos^{-1}\left(\frac{7}{8}\right) = 28.96 = 29^o}$. The stability of equilibrium is determined by $\frac{d^2V}{d\alpha^2} = -\frac{\sin^2\alpha}{2D^3} + \left(\frac{1}{D} - 2\sqrt{2}\right)\cos\alpha$, where for brevity $D = \sqrt{1 - \cos\alpha}$.

For $\alpha = 29^o$, $\left[\frac{d^2V}{d\alpha^2}\right]_{\alpha=29^o} = -2.65 < 0$, so the equilibrium point is unstable.

Problem 11.52 The 2 m long, 10 kg homogenous bar is pinned at A and at its midpoint B to light collars that slide on a smooth bar. The spring attached at A is unstretched when $\alpha = 0$ and its constant is $k = 1.2\ kN/m$. (a) Determine the value of α when the bar is in equilibrium. (b) Determine whether the equilibrium position found in Part (a) is stable.

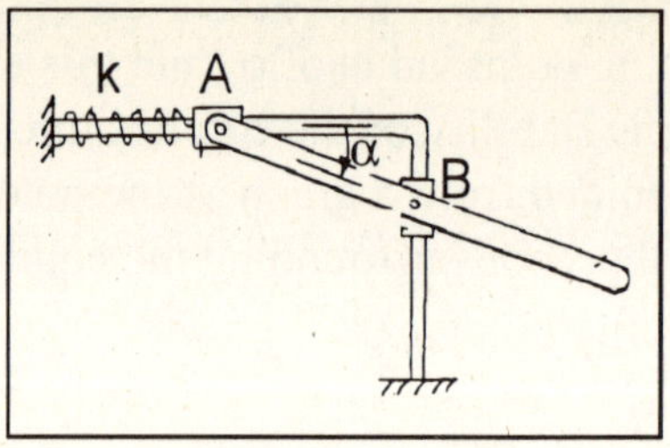

Solution The potential energy of the spring is

$V_{spring} = \frac{1}{2}k(1-\cos\alpha)^2$. The potential energy of the bar is

$V_{bar} = -mg\sin\alpha$. The total potential energy is

$V_{tot} = \frac{1}{2}k(1-\cos\alpha)^2 - mg\sin\alpha$. The equilibrium condition is

$\frac{dV}{d\alpha} = -\left(\frac{mg}{k} + \sin\alpha\right)\cos\alpha + \sin\alpha = 0$ The function

$f(\alpha) = -\left(\frac{mg}{k} + \sin\alpha\right)\cos\alpha + \sin\alpha$ is graphed over the interval

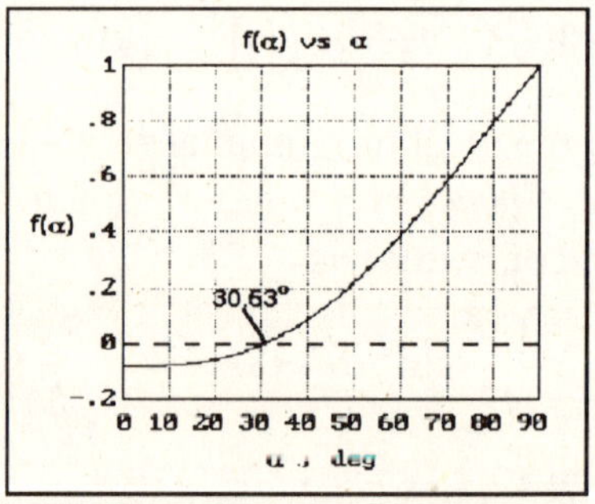

to estimate the value of α at the zero crossing, and this value is refined by iteration to obtain $\boxed{\alpha = 30.53^o}$.

(b) The graph of the potential energy

$V_{tot} = \frac{1}{2}k(1-\cos\alpha)^2 - mg\sin\alpha$ as a function of α shows a local minimum at about $\alpha = 30.5^o$, so <u>the equilibrium position is stable.</u>

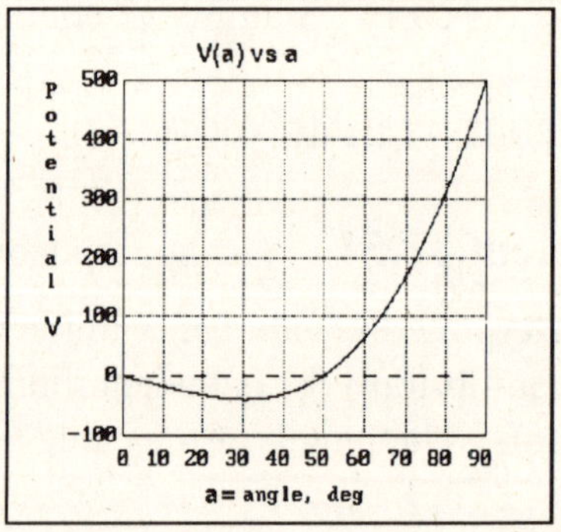

Check: The second derivative is

$\frac{d^2V}{d\alpha^2} = \left(\frac{mg}{k} + \sin\alpha\right)\sin\alpha - \cos^2\alpha + \cos\alpha$. The condition for stability is $\left[\frac{d^2V}{d\alpha^2}\right]_{a=30.5^o} = 0.42 > 0$, so <u>the equilibrium position is stable.</u> *check.*

Problem 11.53 The "rack and pinion" mechanism is used to exert a vertical force on a sample at A for a stamping operation. If a force $F = 30\ lb$ is exerted on the handle, use the principle of virtual work to determine the force exerted on the sample.

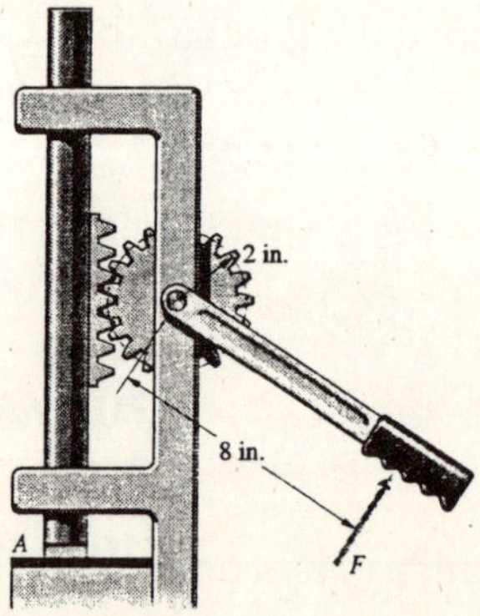

Solution Perform a virtual rotation of the handle. The virtual work is

$\delta U = 8F\delta\theta + A\delta x = 0$, from which $A = -8F\frac{\delta\theta}{\delta x}$. The angular rotation is related to the vertical translation by $2\delta\theta = -\delta x$, from which

$\frac{\delta\theta}{\delta x} = -\frac{1}{2}$, and $\boxed{A = 4F = 120\ lb}$

Problem 11.54

Problem 11.54 In an injection casting machine, a couple M applied to arm AB exerts a force on the injection piston at C. Given that the horizontal component of the force exerted at C is 4 kN, use the principle of virtual work to determine M.

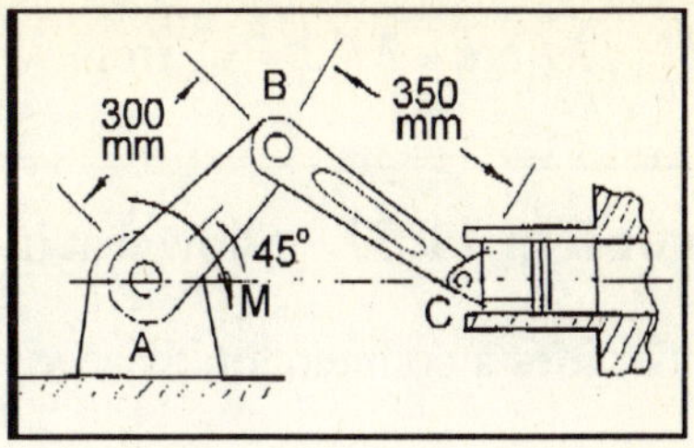

Solution Perform a virtual rotation of the crank. The virtual work is $\delta U = M\delta\theta + F\delta x = 0$, from which $M = -F\dfrac{\delta x}{\delta\theta}$. Denote the interior angle ACB by β, and the interior angle ABC by α.

From the sine law, $\dfrac{0.35}{\sin 45^o} = \dfrac{0.3}{\sin\beta}$, from which

$\beta = \sin^{-1}\left(\dfrac{0.3}{0.35}\sin 45^o\right) = 37.3^o$, and $\alpha = 180^o - 45^o - 37.3^o = 97.69^o$. The distance AC is

$\dfrac{x}{\sin\alpha} = \dfrac{0.35}{\sin 45^o}$, from which $x = 0.490\ m$. From the solution to Problem 11. 16,

$\dfrac{\delta x}{\delta\theta} = \dfrac{0.3(0.49)\sin 45^o}{0.49 - 0.3\cos 45^o} = 0.374$. The moment is $\boxed{M = -0.374F = -0.374(4) = -1.5\ kN\,m}$

Problem 11.55

Problem 11.55 The system is in equilibrium. The total weight of the suspended load and assembly A is 300 lb. (a) Using equilibrium, determine the force F. (b) Using the result of Part (a) and the principle of virtual work, determine the distance the suspended load rises if the cable is pulled downward 1 ft at B.

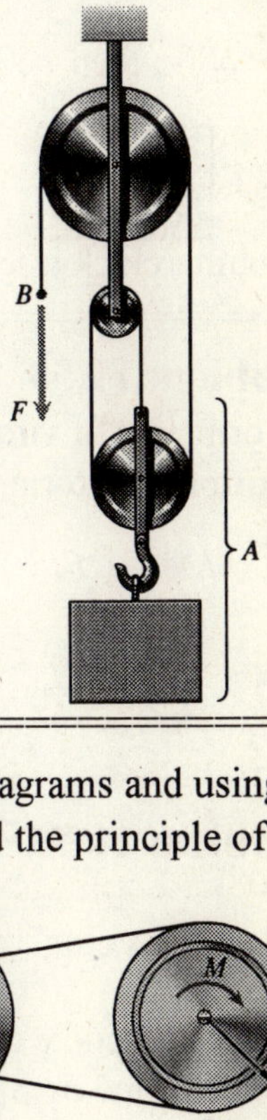

Solution (a) Isolate the assembly A. The sum of the forces:

$\sum F_y = -W - 3F = 0$, where F is the tension in the cable, from which

$\boxed{F = \dfrac{W}{3} = 100\ lb}$. (b) Perform a virtual translation of the assembly A in the vertical direction. The virtual work: $\delta U = -W\delta y + F\delta x = 0$, from which

$\dfrac{\delta x}{\delta y} = \dfrac{W}{F} = 3$. The ratio of translations of the assembly A and the point B is

$\dfrac{1}{y_A} = 3$, from which $\boxed{y_A = \dfrac{1}{3}\ ft}$

Problem 11.56

Problem 11.56 The system is in equilibrium. (a) By drawing the free body diagrams and using equilibrium equations, determine the couple M. (b) Using the result of Part (a) and the principle of virtual work, determine the angle though which the pulley B rotates if pulley A rotates through angle α.

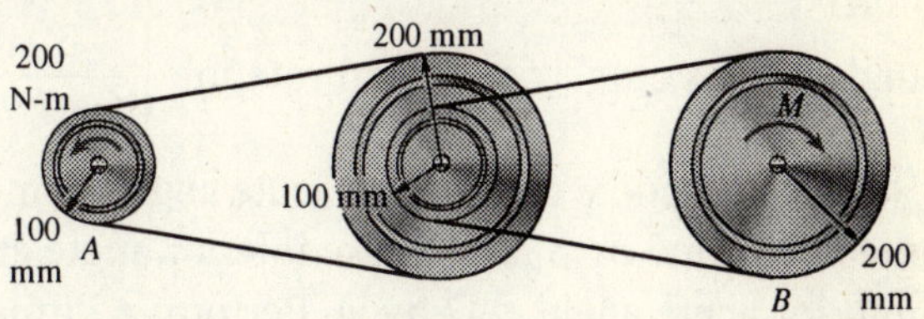

Solution The pulleys are frictionless and the belts do not slip. Denote the left pulley by A and the right pulley by B. Denote the upper and lower tensions in the belts at pulley A by T_3, T_4, at B by T_1, T_2. (a)

For pulley A: (1) $(T_3 - T_4)(0.1) = 200\ N\,m$, For pulley B (2) $M = (T_1 - T_2)(0.2)$. For the center pulley, (3) $(T_1 - T_2)(0.1) = (T_3 - T_4)(0.2)$. Combine and solve: $\boxed{M = (4)(200) = 800\ N\,m}$

Problem 11.56 continued on next page

continuation of Problem 11.56...........

(b) Perform a virtual rotation of the pulley A. The virtual work of the system is

$\delta U = M_1\delta\alpha - M\delta\theta = 0$, from which $\dfrac{\delta\theta}{\delta\alpha} = \dfrac{M_1}{M} = \dfrac{200}{800} = \dfrac{1}{4}$, from which $\boxed{\theta = \dfrac{\alpha}{4}}$

Problem 11.57 Show that if bar AB is subjected to a clockwise virtual rotation $\delta\alpha$, bar CD undergoes a counterclockwise virtual rotation $\left(\dfrac{b}{a}\right)\delta\alpha$.

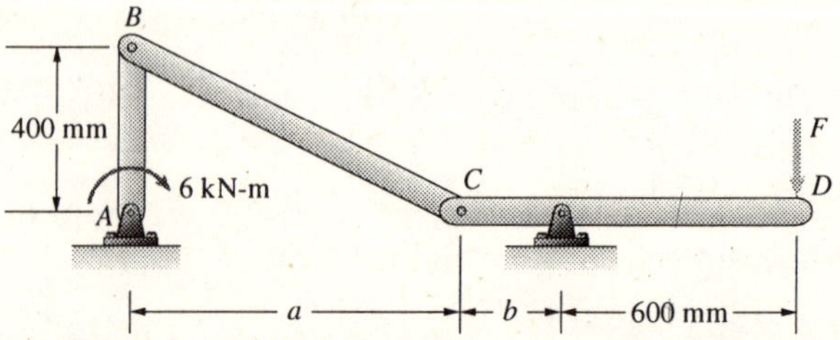

Solution Assume that the bar AB is vertical, so that the interior angle BAC is 90°. Denote the length of the bar AB as $R = 0.4\ m$ Denote the virtual vertical displacement of point C as δy ,the virtual rotation of the bar CD as $\delta\theta$, the interior angle BCA as β. Perform a virtual translation δX of the bar BC parallel to the bar. (Note: This parallel virtual motion is often a useful step where cranks- and connecting-rod-like mechanisms are involved.) The virtual displacements are related by $\delta X = -\delta y \sin\beta$, and $\delta X = R\delta\alpha\cos\beta$, from which $\delta y\sin\beta = -R\delta\alpha\cos\beta$. Noting that $\delta y = b\delta\theta$, then $b\delta\theta\sin\beta = -R\delta\alpha\cos\beta$, from which $\dfrac{\delta\theta}{\delta\alpha} = -\dfrac{R}{b\tan\beta}$. Substitute $\tan\beta = \dfrac{R}{a}$, from which $\boxed{\dfrac{\delta\theta}{\delta\alpha} = -\dfrac{a}{b}}$. The negative sign implies that the rotations are reversed: if one is clockwise the other is counterclockwise.

Problem 11.58 The system in Problem 11.57 is in equilibrium, $a = 800\ mm$ and $b = 400\ mm$. Use the principle of virtual work to determine the force F.

Solution Perform a virtual rotation of the crank at A. The virtual work done on the system is

$\delta U = M\delta\alpha + FL\delta\theta = 0$, where $L = 600\ mm$, from which $\dfrac{\delta\theta}{\delta\alpha} = -\dfrac{M}{FL}$. From the solution to Problem 11.57, $\dfrac{\delta\theta}{\delta\alpha} = -\dfrac{a}{b}$, from which $\boxed{F = \dfrac{Mb}{aL} = \dfrac{6(0.4)}{(0.8)(0.6)} = 5\ kN}$

Problem 11.59 Show that if bar AB is subjected to a clockwise virtual rotation $\delta\alpha$, bar CD undergoes a clockwise virtual rotation $\left(\dfrac{ad}{(ac + bc - bd)}\right)\delta\alpha$.

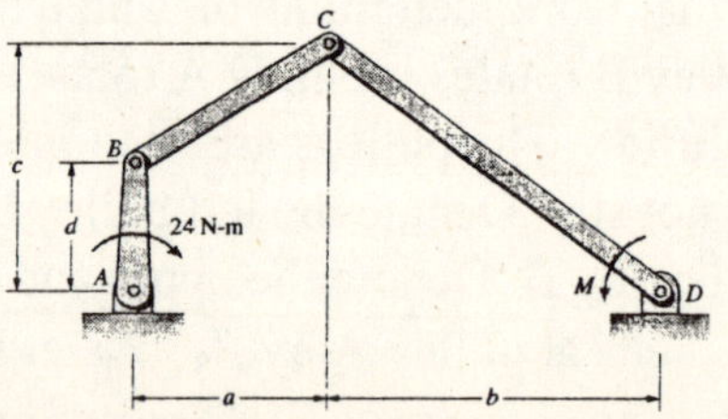

Solution Denote the interior acute angle formed by BC with the horizontal by β, the obtuse interior angle at C by γ, and the interior acute angle at D by ψ. Perform a virtual translation δX parallel to the bar BC. (Note: This is often a useful step where cranks- and connecting-rod-like mechanisms are involved.) Then $\delta X\cos\beta = d\delta\alpha$, and $\delta X\sin\gamma = \overline{CD}\delta\theta$, from which $\dfrac{\delta\theta}{\delta\alpha} = \dfrac{d\cos\beta}{\overline{CD}\sin\gamma}$.

Problem 11.59 continued on next page

continuation of Problem 11.59..........

Noting that $\sin\gamma = \sin(90^o - \beta + 90^o - \psi) = \sin(\beta + \psi) = \sin\beta\cos\psi + \sin\psi\cos\beta$, and $\cos\beta = \dfrac{a}{BC}$,

$\sin\psi = \dfrac{c}{CD}$, $\sin\beta = \dfrac{c-d}{BC}$, $\cos\psi = \dfrac{b}{CD}$. Substitute:

$$\boxed{\frac{\delta\theta}{\delta\alpha} = \frac{d\left(\dfrac{a}{BC}\right)}{CD\left(\dfrac{a}{BC}\times\dfrac{c}{CD} + \dfrac{b}{CD}\times\dfrac{(c-d)}{BC}\right)} = \frac{ad}{ac + bc - bd}}$$

Problem 11.60 The system in Problem 11.59 is in equilibrium, $a = 300\ mm$, $b = 350\ mm$, $c = 350\ mm$, and $d = 200\ mm$. Use the principle of virtual work to determine the couple M.

Solution Perform a virtual rotation of the crank at A. The virtual work is $\delta U = -M_1\delta a + M\delta\theta = 0$, from which $\dfrac{\delta\theta}{\delta\alpha} = \dfrac{M_1}{M}$. From the solution Problem 11.59, $\dfrac{\delta\theta}{\delta\alpha} = \dfrac{ad}{ac + bc - bd}$, from which

$$\boxed{M = \frac{M_1(ac + bc - bd)}{ad} = (24)2.625 = 63\ N\,m}$$

Problem 11.61 The string is unstretched when $\alpha = 90^o$. If $mg = \dfrac{bk}{2}$, determine the value in the range $0 < \alpha < 90^o$ for which the system is in equilibrium.

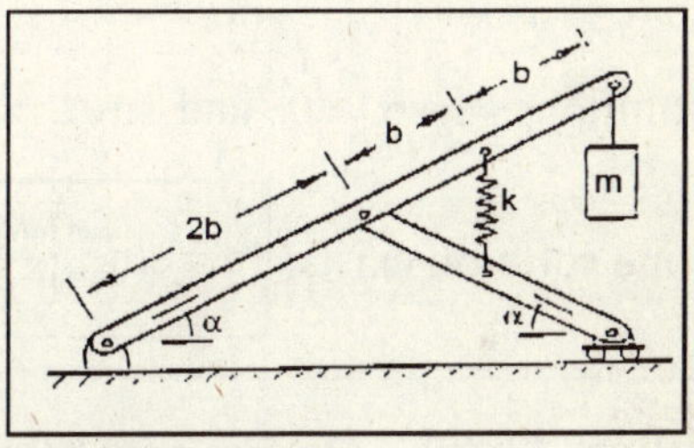

Solution Neglect the weight of the bars. The potential energy of the mass m is $V_{mass} = mg(4b\sin\alpha)$. The potential energy of the spring $V_{spring} = \dfrac{1}{2}k(\Delta s)^2$. The uncompressed length of the spring is $2b$. The compressed length is $(3b\sin\alpha - b\sin\alpha) = 2b\sin\alpha$, from which $\Delta s = 2b - (3b\sin\alpha - b\sin\alpha) = 2b(1 - \sin\alpha)$, from which $V_{spring} = 2kb^2(1 - \sin\alpha)^2$.

The total potential energy: $V_{tot} = \left(4mgb\sin\alpha + 2kb^2(1 - \sin\alpha)^2\right)$. The equilibrium condition is :

$\dfrac{dV}{d\alpha} = 4mgb\cos\alpha - 4kb^2\cos\alpha(1 - \sin\alpha) = 0$, from which $\left[\dfrac{dV}{d\alpha}\right]_{mg=\frac{bk}{2}} = \left(\sin\alpha - \dfrac{1}{2}\right)\cos\alpha = 0$.

Equilibrium occurs at $\cos\alpha = 0$, $\boxed{\alpha = 90^o}$ and $\sin\alpha = \dfrac{1}{2}$, $\boxed{\alpha = 30^o}$

Problem 11.62 Determine whether the equilibrium position found in Problem 11.61 is stable or unstable.

Solution From the solution to Problem 11.61 the derivative of the potential energy is

$$\frac{dV}{d\alpha} = 4mgb\cos\alpha - 4kb^2\cos\alpha(1-\sin\alpha)$$, from which

$$\left[\frac{d^2V}{d\alpha^2}\right] = -4mgb\sin\alpha + 4kb^2\sin\alpha(1-\sin\alpha) + 4kb^2\cos^2\alpha$$. The condition for stability is

$$\left[\frac{d^2V}{d\alpha^2}\right]_{\alpha=30^\circ} = (kb^2)(-1+3) > 0$$, so <u>the equilibrium position is stable.</u>

Problem 11.63 The spring is unstretched when $\alpha = 90^o$. Determine the value of α in the range $0 < \alpha < 90^o$ for which the system is in equilibrium.

Solution Choose a coordinate system such that the equilibrium position of the spring occurs at $x = 0$ and at $y = L$. The potential energy is $V = mgy + \frac{kx^2}{2}$. Noting that $y = L\sin\alpha$ and $x = L\cos\alpha$, then $V = mgL\sin\alpha + \frac{kL^2}{2}\cos^2\alpha$. The equilibrium condition is $\frac{dV}{d\alpha} = 0 = (mg - kL\sin\alpha)\cos\alpha = 0$. This has two solutions: $\cos\alpha = 0$, and $\sin\alpha = \frac{mg}{kL}$. In the interval $0 < \alpha < 90^o$, only one solution exists, $\boxed{\alpha = \sin^{-1}\left(\frac{mg}{kL}\right)}$

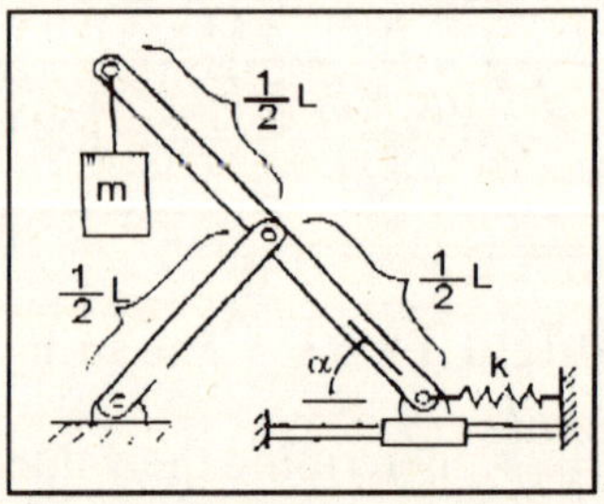

Problem 11.64 Determine whether the equilibrium position found in Problem 11.63 is stable or unstable.

Solution Use the solution to Problem 11.63. The stability condition is

$$\frac{d^2V}{d\alpha^2} = -mg\sin\alpha + kL(\sin^2\alpha - \cos^2\alpha) = kL(\sin^2\alpha - 1) < 0.$$ <u>The system is unstable.</u>